AF653137

Microbial Biotechnology and Ecology

Volume-1

Microbial Biotechnology and Ecology
Volume-1

Deepak Vyas
G.S. Paliwal
P.K. Khare
R.K. Gupta

2024
Daya Publishing House®
A Division of
Astral International Pvt. Ltd.
New Delhi – 110 002

© EDITORS

First Published, 2011

Reprinted, 2024

ISBN: 978-81-7035-694-3 (HB)

Disclaimer:

Every possible effort has been made to ensure that the information contained in this book is accurate at the time of going to press, and the publisher and author cannot accept responsibility for any errors or omissions, however caused. No responsibility for loss or damage occasioned to any person acting, or refraining from action, as a result of the material in this publication can be accepted by the editor, the publisher or the author. The Publisher is not associated with any product or vendor mentioned in the book. The contents of this work are intended to further general scientific research, understanding and discussion only. Readers should consult with a specialist where appropriate.

Every effort has been made to trace the owners of copyright material used in this book, if any. The author and the publisher will be grateful for any omission brought to their notice for acknowledgment in the future editions of the book.

All Rights reserved under International Copyright Conventions. No part of this publication may be reproduced, stored in a retrieval system, or transmitted in any form or by any means, electronic, mechanical, photocopying, recording or otherwise without the prior written consent of the publisher and the copyright owner.

Published by : **Daya Publishing House®**
A Division of
Astral International Pvt. Ltd.
– ISO 9001:2015 Certified Company –
4736/23, Ansari Road, Darya Ganj,
New Delhi-110 002
Ph. 011-43549197, 23278134
E-mail: info@astralint.com
Website: www.astralint.com

Printed at : Replika Press Pvt. Ltd.

Date of Birth: 19-09-1942 to 07-09-2006

– Biography –
Late Prof. K.M. Vyas

Born on 19th September, 1942 in village Dharoan then in district Hamirpur and presently in district Mahoba (UP) in the family of Smt Radhika Rani and Sri Rooplal Vyas. Krishna Murari had his primary education in village and for the secondary education he moved to a Tehsil place in Charkhari. He obtained BSc from Maharaja College, Chhatarpur and MSc (Botany) degree from Sagour University, Sagar in 1963. Young Krishna Murari joined as a lecturer in the Botany Department then Sagour University, Sagour now Dr. H.S.G. V.V., Sagar (M.P.). Simultaneously joined PhD programme under the supervision of Prof SB Saksena (FNA), a renowned mycologist of that time. He did his PhD on *"Studies on the Ecology and Physiology of Soil Microorganisms in Relation to Soil Borne Plant Diseases"*. For post doctoral work (1973-74), he went to Department of Biochemistry, University of Hull, England (UK). He worked on *"Production and Activity of Cell Wall Degrading Enzymes by Late Blight Fungus–Phytopthora infestens"* with Prof J Friend.

Foreign Visits

Visited and delivered lectures at various important Research Institutions and Universities of England, Scotland, Belgium and Holland, Imperial College of Science and Technology, London, Long Ashton Research Station, Bristol, University of Glasgow, International Agriculture Institute, Wageninghe, Phytopathologic Research Institute Baarn, Holland.

Outstanding Research Contributions

1. Successful control of *Phytophthora* diseases of pan by streptomycin and other antibiotics.
2. Antibiotics as a new group of plant growth regulators.
3. Control of mosquitoes by certain isolates of bacteria and fungi.
4. Fungal pathogenesis of plants with special reference of cell wall degrading enzymes and phenolics in fungal pathogenesis.

5. *Alectra parasitica*: a traditional drug for the treatment of leprosy and other skin diseases.
6. Fungal respiration with special reference to mycelial respiration of plant pathogenic fungi.
7. Rhizosphere, rhizoplane, phyllosphere and phylloplane microflora and their role in plant diseases.

Professor and Head

In August 1984 he became a professor and in the month of October he became Head of the Department. As soon as he assumed the duty of HOD he brought DRS programme from UGC to strengthen the teaching and research facilities. During the period of 17 years of headship he also brought DSA phase I and II, and laid a stone for Cosist (Assistant programme) under SAP programme of UGC.

Prof Vyas established a strong centre of research in Mycology and Microbiology at Sagar. Where he undertook and successfully completed various research projects funding by UGC, CSIR, ICMR etc. He has published his work in various journals of National and International repute. He has edited book "*Recent Advances in Microbiology*". More than 40 students have obtained PhD under his supervision. Many of his students are actively engaged in research work and are occupying prestigious position in country and abroad.

During his tenure department grow, and many learned people from country and abroad visited the department and deliver their talks. He organized many seminars/ symposia/ workshops and refresher courses. Beside his scientific tamper and aptitude he has got administrative skills and proved himself an administrator in various capacities.

Administrative Experiences

1. As Head of the Botany Department. Dr. H.S.G. V.V., Sagar: 17 years (November 1984 to October 2001).
2. As Dean Faculty of Life Science: 4 years (two terms: 1988-90 and 1996-98).
3. As University Proctor: 6 Years (1986-1982).
4. As Chairman Board of Studies in Botany: 6 years (two terms: 1984-87 and 1994-97).
5. As Chief Coordinator of University Examinations: (three terms: 1985-86, 1988-89, and 1994-1995).
6. As Coordinator UGC NET- Exam, December 1996, June 1997, December 1997 and June 1998.
7. As Member of Academic Council Dr. H.S. G. V.V., Sagar: 10 years.
8. As Member of Executive Council; As Dean Faculty of Life Science (1988-90) and as Senior Professor (1996-99).
9. As Member of Central Board of Studies in Botany, Bhopal (MP): 10 years.
10. As Vice Chancellor of Dr. H.S.G. V.V., Sagar on various occasions/ day (during short term leave vacancy).
11. As Programme Coordinator of DRS/ DSA phase I and II programme of UGC in Botany Dr. H.S.G. V.V., Sagar.
12. As Member, Board of Studies, RDC and Selection Committees of various Universities and Colleges: 14 years.
13. As Professor Incharge of Development, Dr. H.S.G. V.V., Sagar: 3 years.

14. As Chairman, Organizing Committee of various symposia and seminars.
15. As Coordinator of Refresher Course in Botany: (1995-1998).
16. As Director Development in Botany: 1 October 2001–30 April 2004.

He was retired on 30th April 2004. He was an amazing personality with strong will power. His fight against Hepatitis B was amazing. After obtaining superannuation from the university he joined as a Principal at Pt. Vasudev Prasad Tiwari Degree College, Jhansi. He worked for the welfare of humanity under the umbrella of Bundelkhand Janseva Samiti, Jhansi.

Prof. Vyas was a great source of inspiration for his students, family members. His untimely death created a vacuum not only in scientific community but also in the family. It is said that there is a woman behind every great man. In this case it is Mrs Urmila Vyas, since 1958 when she was married to Prof KM Vyas. She has sacrificed all her comforts and has taken over all the household responsibilities so that Prof KM Vyas could be free to devote all his time to academics. Any visitor to their home was deeply impressed by their modesty, simplicity and warmth of their hospitability.

The Vyas are a well knit family. They have well educated children. One son and three daughters. Son is Dr Deepak Vyas presently serving in the same Department (Botany), Dr. H.S.G. V.V., Sagar married with Mrs. Nisha Vyas who is lecturer in college and have two sons Siddhant and Vedant studying in school. Among the daughters, elder daughter is Dr Jyoti married with with Dr Sanjay Chaubey, having their own hospital at Jhansi and having two sons Rohan and Kartik, then Mrs Arti married with Sunil Sharma who is working as an area manager in private firm. Arti is lecturer in private college at Bhopal have two kids Shivani and Shreyansh. Younger daughter Mrs Bharti is married with Dr Anand Sharma. Who is working at Cambridge Hospital, London, England.

List of Publications

1. Vyas, A. D. Vyas and K.M. Vyas (2006). Microbial cellulases for Industrial application. Every man's Science XL (6) 411-416.
2. Sharma, C.D., A.N. Rai and K.M. Vyas (2006). An undescribed fungal genus of the forest flora of Madhya Pradesh J. Bot. Soc. Univ. Sagar, 41 : 94-98.
3. Vyas, A., A. Jain, D. Vyas and K. M. Vyas (2005). Production and partial characterization of cellulases by two *Aspergillus* spp. on pretreated low value ligno-cellulases. Proc. Natl. Acad. Sci. India, 75(B) p. 1-7.
4. Vyas, Ashish, D. Vyas and K.M. Vyas (2005). Biodiversity conservation through integrated participatory management J. Bot. Soc. Univ. Sagar, 40 : 178-189.
5. Dwivedi, O. P., R. K. Yadav, D. Vyas and K. M. Vyas (2004). Role of Potassium in the occurrence of VAM spores in the rhizosphre soils of Lantana species. In: Microbiology and Biotechnology for sustainable development. Ed. P.C. Jain, Pp. 248–253. CBS Publishers, New Delhi.
6. Yadav, R.K., O.P.Dwivedi, D. Vyas and K. M. Vyas (2004). Antifungal effect of *Microcystis* against Beetalvine *Phytophthora*. J. Basic and applied Myco. 3(1+2) : 105–107.
7. Vyas Ashish, Vyas Deepak and Vyas K.M. (2004). Production and optimization of cellulases on pretreated groundnut shell by *Aspergillus tenuis* AV49. Journal of Scientific and Industrial Research Vol. 64, Pp. 281-286.
8. Vyas, A., D. Vyas and K.M. Vyas (2004). Cellulolytic strains of *Aspergillu* and *Penicillium*. J. Bot.Soc.Univ. Sagar 39 66-71.

9. Yadav, R.K., O.P. Dwivedi, D. Vyas and K.M. Vyas (2003). Management of *Phytophthora* disease of betel vine by plant extracts J. Basic and Applied Mycol. 2 (1) : 79-82.
10. Agrawal, R., A.N. Rai and K.M. Vyas (2003). A new species of *Tretospora* from India. J. Basic and Applied Mycol. 2 (1) 48-49.
11. Soni, Anuradha, D. Vyas and K.M. Vyas (2003). Effect of VAM fungi on the growth and productivity of Soybean (*Glycine max.*) J. Basic and Applied Mycol. 2 (1) : 40-41.
12. Vyas, Ashish, D. Vyas and K.M. Vyas (2003). Screening of extra-cellular cellulase producing fungi from different lignocellulosic wastes. J. Basic and Applied Mycol. 2 (1) : 14-16.
13. Khare, N. and Vyas K.M. (2003). Effect of nitrogen sources on the production of pectolytic and cellulytic enzymes by *Rhizopus nodosus* and *Phytophthora nicotian*a J. Bot Soc. Univ. Sagar 36 : 87-94.
14. Dwivedi, O.P., R.K. Yadav, D. Vyas and K.M. Vyas (2002). Distribution of VAM fungi in the rhizosphere soils of Betel vine J. Phytopathol. 56 (2) : 228-229.
15. Dwivedi, O.P., Yadav, R.K., Vyas, D. and Vyas, K.M. (2002). Vesicular-Arbuscular Mycorrhizal Association with *Lantana* sp. In Sagar University Campus; in Different Disordered Systems INDIAS 2000. (eds) K. Fukawa, J.T. Watson, S.N. Saxena and S.K. Shrivastava, pp. 154-157. INDIAS Publication, Allahabad.
16. Vyas, D., O.P. Dwevedi, R.K. Yadav and K.M. Vyas (2002). Arbuscular Diversity of VAM Fungi. In frontiers of fungal diversity in India. (Eds.) G.P. Rao, C. Manoharachari, D.J. Bhat, R.C. Rajak and T.N. Lakhanpal pp. 873-889. International Book Distributing Company Lucknow, India.
17. Singh, N., Vyas, D., Gupta, U.S. and Vyas, K.M (1999). Physico-chemical properties of Sagar lake with special reference to *Microcystis* bloom in the proceeding of the New Millenium Conference on Retrospect of Indian Research on Environmental Pollution : Focus 21st Century, 38.
18. Sharma, C.D., Gadpandey, K.K., Firdousi, S.A., Rai, A.N. and Vyas, K.M. (1998). Three New species of *Cladosporium* from Madhya Pradesh, India. Indian Phytopath. 51 (2): 152-160.
19. Chaursia, J.P., Chaurasia, P. and Vyas, K.M. (1997). Effect of *Ocimum sanctum* leaf extract on betelvine *Phytophthora* disease; An ecofriendly new approach. Environment pollution in 21 Century (Ed). V.S. Bais. Narendra Publishing House, Delhi, p. 46-51.
20. Chaurasia, J.P. and Vyas, K.M (1997). *In vivo* evaluation of some homoeopathic drugs against betelvine *Phytophthora* disease, Indian Phytopath. 4:148-153.
21. Vyas, K.M., Chaurasia, J.P. and Mishra, A.S. (1996). Effect of hepar sulphate on betelvine *Phytopathora* disease. Madhya Bharti Journal, Dr HS Gour Vishwavidyalaya, Sagar, India 36A 40B: 65-71.
22. Vyas, K.M. and Firdousi, S.A. (1996). Leaf blight of *Terminalia tomentosa.* Indian Phytopathology (Accepted).
23. Vyas, K.M., Gadpandey, K.K., Sharma, C.D., Firdousi, S.A. and Rai, A.N. (1996). Two new Hyphomycetes from India. Mycological Research (Accepted).
24. Vyas, K.M., Sharma, C.D., Gadpandey, K.K., Rai, A.N. and Firdousi, S.A. (1996). Three new species of *Cladosporium* from Madhya Pradesh. Indian Phytopathological Society IARI, New Delhi (In press).

25. Vyas KM, Dubey MP and Sharma RS (1995). Low cost agronomic practice for soybean (*Glycine max*) cultivation in Madhya Pradesh. Indian Journal of Agriculture Sciences 65(10):743-745.
26. Vyas KM, Kumar R and Rai AN (1995). Two new *Alternaria* species from Uttar Pradesh. Acta Botanica Indica 23 (1): 151-153.
27. Vyas KM and Firdousi SA (1994). A new leaf spot disease of *Casearia tomentosa* caused by *Asteromella* species. Acta Botanica Indica 20: 120-121.
28. Vyas KM and Saxena AP (1993). Antimicrobial activity of *Alectra parasitica*, A rice var. *chitrakutensis* rau. Journal of Economic and Taxonomic Botany 17 (1) 55-59.
29. Vyas KM, Firdousi SA and Rai AN (1993). A new species of *Stenella* from India. Acta Botanica Indica 25: 131-132.
30. Vyas KM and Firdousi SA (1993). A new spot of *Miliusa tomentosa* Roxb. Acta Botanica Indica 21 (1): 124.
31. Vyas KM and Mehta A (1992). Degradation of plant tissues by fungal enzymes. In: Dubey NK and Sharma PD, Rastogi Publications, Meerut, India.
32. Vyas KM, Firdousi SA and Rai AN (1992). A new species of *Pseudocercospora* from India. Indian Phytopathology 45(4): 449-451.
33. Vyas KM, Firdousi SA and Rai AN (1992). A new species of *Pseudocercospora* from India. Acta Botanica Indica.
34. Vyas KM, Firdousi SA and Rai AN (1992). *Mycovellosiella adinae* Sp nov. from India. Indian Phytopathology 45(4): 451-452.
35. Vyas KM (1990). A new leaf spot disease of *Ficus begalensis* caused by *Coniothyrium olivaceum* from India. Acta Botanica Indica 1: 150.
36. Vyas KM, Firdousi SA and Rai NA (1990). *Pseudocercospora gymnosporinae* sp. nov. India. Mycological Research (UK in press).
37. Vyas KM, Firdousi SA and Rai NA (1990). A new host record of *Thielvia appendiculata* from India. Indian Phytopath 43(1): 124.
38. Vyas KM, Firdousi SA and Rai NA (1990). A new leaf spot disease of *Eleodendron glaucum* from India. Indian Phytopath 43(1): 122.
39. Vyas KM and Gupta DK (1989). Efficacy of *Bacillus subtilis* against Mosquito larvae (*Anophelis culicifacies*). Zeitschrift Fur Angewandte Zoologie (German Journal for Applied Zoology) 85-91.
40. Vyas KM and Chile SK (1989). Metabolism of ascorbic acid and *Piper betle* leaves under *Phytopathora* pathogenesis. Perspective in Mycological Research 1: 143-150.
41. Vyas KM and Tripathi HSS (1986). Production of toxin by *Xanthomaonas compestris* pv *betticola* causing leaf spot disease of *Piper betle* L. Acta Botanica Indica 14: 220-225.
42. Vyas KM and Saxena NK (1986). Microbial synthesis of 11-deoxy corticosteron from progesterone by a nutant of *Rhizophus nodosus* (Abst.) XIV International Congress of Microbiology: Menchester, England 7-13 September.
43. Vyas KM and Chile SK (1985). Electrolyte leakage from *Piper betle* leaves after infection with *Phytopathora parasitica* var *piperina*. Indian Phytopathology 38 (3): 529-530.

44. Vyas KM and Tripathi HSS (1985). Production of cellulolytic enzymes by *Xanthomonas compestris* pv *betticola*: *in vivo* and *in vitro*. Bull Bot. Soc. 33.

45. Vyas KM, Tripathi HSS, Saxena NK and Sadique Y (1985). Efficacy of hydroxyl stilbamidine isethionate USP against *Phytopathora parasitica* var *piperina*. Phytopathora Newsletters 13: 32-33.

46. Vyas KM, Saxena NK, Tripathi HSS and Sadique Y (1985). Complete control of *Phytopathora parasitica* var *piperina* by S-flurocytosine. Phytopathora News Letters 13: 30-31.

47. Vyas KM, Dubey GL and Dubey O (1985). Physico-chemical properties of linseed oil as affected by various seed borne fungi. Phytopath. z. 113: 66-70.

48. Vyas KM and Chile SK (1984). Efficacy of *Vinca rosea* extracts against protease from human pathogenic strain of *Trichophytom rubrum* sab. Hindustan Antibiotics Bull. 26: 114-116.

49. Vyas KM, Kher AK and Saxena SB (1984). Persistence of various fungicides in soil. Indian Phytopath 27(3): 574-576.

50. Vyas KM and Shrivastava A (1984). Effect of cholesterol on the toxicity of certain antibiotics against betelvine *Phytopathora*. Phytopathora News Letter 12 pp. 33.

51. Vyas KM, Soni, Saxena NK and Tripathi HSS (1984). Decline in the total reducing sugar content during the postharvest infection of *Lycopersicon lycopersicum* L. Karsgen fruits by *Bacillus fructodrestruens*. Bull Bot. Soc. 30-31: 69-71.

52. Vyas KM, Saxena NK and Tripathi HSS (1984). Cationic degradation of Streptomycine and Kanemycine by *Aspergillus niger* in broth cultures during fragmentation in shade flask. Hindustan Antibiotics Bulletin 26 (3 and 4) pp. 96-101.

53. Chile SK, Vyas KM and Singh R (1984). Induced resistance in *Piper betle* leaves in relation to *Phytopathora parasitica* var. *piperina*. Indian Phytopath. 37: 719-721.

54. Chile SK and Vyas KM (1984). Inhibitors of certain enzymes of *Botrydiplodia theobromae* in *Piper betle* leaf extract. Hindustan Antibiotics Bull. 26: 27-32.

55. Vyas KM and Chile SK (1984). Respiratory response of betlevine *Phytopathora* induced by various volatile oils. Jr Mycology Plant Path.

56. Vyas KM, Saxena NK and Tripathi HSS (1984). Accumulation of D.L. 3,4 dihydroxy-phenyalanine during the soft rot of *Lycopersicon esculentum* fruits caused by *Bacillus fructodestruens*. Abst. 36th Ann. Meeting Indian Phytopath (Hissar) 37: 401.

57. Vyas KM, Saxena NK and Tripathi HSS (1984). Postharvest accumulation of vitamin C induced by *Bacillus fructodestruens* in the *Lycopersicon esculentum* fruits. Abst. 36th Ann. Meeting Indian Phytopath, Hissar and paper published in Nat. Acad. Sci. Letters 7(4): 109-110.

58. Vyas KM and Shrivastava A (1984). Chemical control of betelvine *Phytopathora*. Abst. Symp. Chemical Control of Plant Disease. Indian Phytopath, Hissar.

59. Vyas KM and Singh R (1984). Nitrogen nutrition of *Phytopathora parasitica* var *piperina* (Dast) causing foot and leaf rot disease of *Piper betle* Linn. Proc. 71st Ind. Sci. Cong. Part III Abstract, 57: 26.

60. Vyas KM and Atri DC (1983). Pectolytic and cellulolytic activity during pathogenesis caused by *Botryodiplodia theobromae* Pat. In *Pyrus pyrifolia* Burm. fruits. Abst National Acad. Sci. (Biol.) Abst. No. 257 pp. 94.

61. Vyas KM and Chile SK (1983). Free amino acids in relation to *Phytopthora* leaf rot pathogenesis of Pan. Indian Phytopath 36(4): 626-628.

62. Vyas KM and Chile SK (1983). Dehydrogenase activity in *Piper betle* L. leaves infected with *Phytopathora parasitica* var. *piperina*. Indian Phytopath 36(4): 613-617.

63. Vyas KM and Sxena N (1983). Effect of antibiotics on certain cell wall degrading enzymes of *Rhizopus nodosus* and *Phytopathora nitotianae*. Hindustan Antibiotics Bulletin 25 (1 and 2) pp. 21-24.

64. Vyas KM and Shrivastava A (1983). In vitro efficacy of tetracycline against betelvine *Phytopathora*. Hindustan Antibiotics Bulletin 25 (1 and 2) pp. 15-17.

65. Vyas KM and Saxena AP (1983). Ethnobotany of Dhasan valley. Jr. of Econ. Tax. Bot., Jodhpur.

66. Vyas KM and Saxena AP (1983). Traditional treatment of leprosy and leucoderma by tribals from Bundelkhand UP. Bull. of Medico Ethn. Bot. Res. CCRAS, New Delhi.

67. Vyas KM and Thakur MS (1983). Production of plant growth regulators by some *Fusarium* species. Folia Microbial. 28: 124-129.

68. Vyas KM, Gupta DK and Shrivastava A (1983). In vitro antibacterial properties of some fixed oils. Proc 70th Ind. Sci Cong. Assoc 29.

69. Vyas, Chile SK and Shrivastava A (1982). Efficacy of some growth hormones on sporulation of betelvine *Phytopathora*. Bull. Bot. Soc. Univ. Saugar 29: 25-27.

70. Chile SK and Vyas KM (1982). Efficiency of carboxin in relation to *Phytopathora* leaf rot of *Piper betle*. Abstract 1st National Symposium of Synthetic Fungicides in Plant Disease Control in IARI, June 1982.

71. Chile SK, Vyas KM and Vyas RS (1982). Effect of culture filtrate of various fungi on sporulation of betelvine *Phytopathora*. Hind. Ant. Bull. 23: 25-26.

72. Chile SK and Vyas KM (1982). Effect of culture filtrate of various fungi on acidative enzymes of betelvine *Phytopathora*. Hind. Ant. Bull. 24: 26-28.

73. Vyas KM, Gupta DK and Chile SK (1982). A fruit rot disease of Momardiaca. Indian Phytopath 35(2): 351.

74. Vyas KM and Gupta DK (1982). Biological control of mosquito. IX Annual Conference of Indian Society of Human Genetics, Bhopal 58.

75. Vyas KM and Gupta DK (1982). Effect of *Aspergillus* mycotoxin on *Culex* larvae. Proc 69th Ind. Sci. Cong. Asso. Abstract 61.

76. Vyas KM and Thomas T (1982). On the abundance of *Trichoderma* species in the soil of Andaman and Nicobar Islands. Proc 69th Ind. Sci. Cong. Asso. Abstract 59.

77. Vyas KM and Saxena N (1982). Pectolytic enzymes of *Rhizopus nodosus* and *Phytopathora nictianae* in relation to electrolyte leakage and tissue maceration. Proc 69th Ind. Sci. Cong. Asso. Abstract 44.

78. Vyas KM and Saxena AP (1982). In vitro antimicrobial activity of some medicinal plants against human pathogens. Proc 69th Ind. Sci. Cong. Asso. Pt. III: 43-44.

79. Vyas KM, Singh R and Chile SK (1982). Nitrogen nutrition of *Phytopathora parasitica* var *piperina* (Dast) Causing foot rot and leaf rot diseases of *Piper betle* Linn. Jur. Biol. Res.

80. Vyas KM (1982). Peroxiodase and catalases in relation to betelvine. Phytopathora News Letters, USA 10: 8.

81. Vyas KM and Chile SK (1982). Breakdown of disease resistance in betelvine *Phytopathora* system by vitamin C. Phytopathora News Letters, USA 10: 6-7.

82. Vyas KM and Thakur MS (1982). Plant growth regulatory metabolites produced by fungi. Abstract Symp. Fermented Foods, Food Contaminants, Biofertlizers and Bioenergy pp. 54.

83. Vyas KM and Thoimas T (1982). *Gliocladium virens*- A new report from India. Geobios News reports 1: 45.

84. Vyas KM and Saxena AP (1981). *Martynia annua* Linn. A drug for Asthma, Itch and Eczema. Bull of Medico. Ethno Bot. Res. CCRAS, New Delhi II (3): 427- 429.

85. Vyas KM (1981). Problem and progress in *Phytopathora* diseases of Pan group discussion on improvement of betelvine cultivation. NBRI, Lucknow p 92-93.

86. Vyas KM, Thomas T and Gupta DK (1981). *Gilmaniella subornate*: A new report from India. Bull. Bot. Soc. Univ. of Saugar 28: 38.

87. Vyas KM, Thomas T and Gupta DK (1981). *Trichoderma pseudo-konningii* Rifai: A new report from India. Bull. Bot. Soc. Univ. of Saugar 28: 28.

88. Vyas KM (1981). Effect of culture filtrates of various fungi on sporulation of betelvine *Phytopathora*. Hind. Antibiot. Bot. 23: 25-26.

89. Vyas KM, Chourasia SC and Saxena SB (1981). Management of *Phytopathora* diseases of Pan (*Piper betle*). Abst. 3rd International Symposium on Plant Pathology (IPS) pp. 253.

90. Chile SK and Vyas KM (1981). Polyphenol oxidase in relation to betelvine *Phytopathora* pathogenesis of Pan. Phytopathora News Letters, USA 9: 40-41.

91. Vyas KM and Saxena AP (1981). Medicinal plants of Bundelkhand I.P preliminary ethnobotanical survey of Banda district UP. Proc. 68th Ind. Sci. Cong. Pt III (Abst): 147.

92. Vyas KM and Saxena AP (1981). Flora of Bundelkhand II. Hydrophytic tract of Ken-Baghain tract (Banda district UP). Ind. Soc. Weed. Sci. Abst: 40-41.

93. Vyas KM and Saxena AP (1981). Ethnobotanical records on infectious diseases from tribals of Banda district (UP). Ind. Soc. Weed. Sci. Abst: 40-41.

94. Vyas KM and Gupta DK (1981). Biological control of mosquitoes: I Effect of culture filtrate of certain *Aspergilli* on Anopheles larvae. Proc. 51 Session Nat. Acad. Sci.: 51.

95. Vyas KM and Chile SK (1981). Effect of the age of betel vine leaves on phenolics and *Phytopathora* rot. Ind. Sci. Cong.

96. Chile SK, Chourasia and Vyas KM (1981). Role of cuttings on the growth of the betel vine *Phytopathora*. Phytopathora News Letters, USA 8: 31.

97. Vyas KM and Chile SK (1980). Role of polyphenol oxidase in disease resistance with special reference to the betel vine *Phytopathora*. Symp. Plant Disease Problem at PAU, Ludhiana (Punjab) Abstract IIX.

98. Vyas KM, Chile SK and Saraf M (1980). Effect of some fungicides on sporulation of different isolates of betel vine *Phytopathora*. Bull. Bot. Univ. of Saugar, Sagar (MP) 27 pp. 16-18.

99. Vyas KM, Chourasia SC and Chile SK (1980). Role of infested soil and cuttings on the spread of betel vine *Phytopathora*. Phytopathora News Letters, USA 8: 28.

100. Vyas KM and Chile SK (1980). Role of infested soil and cuttings on the spread of betel vine *Phytopathora*. Symp. on Phytopathora Diseases of Tropical Plants (Calicut) pp. 22.

101. Vyas KM and Chile SK (1980). Varietal reaction of *Piper betle* leaves against different isolates of betel vine *Phytopathora*. Phytopathora News Letters, USA 8: 29.

102. Vyas KM, Chile SK and Vyas RS (1980). Two new diseases of cultivated plants. Indian Phytopath. 33(3): 493-495.

103. Vyas KM and Gupta M (1979). Studies on cell wall degrading enzymes in relation to banana fruit rot. Indian Phytopath. 32: 456- 457.

104. Vyas KM, Raghvan U and Saxena SB (1979). Studies on amino acid composition of mycelia of 14 of *Botryodiplodia theobromae* Pat. Bull. Bot. Soc. Univ. Saugar, 23-24: 21-25.

105. Vyas KM and Thakur MS (1979). Production of indole auxins by *Curvularia* spp. Abst. 66th Sess. Ind. Sci. Cong. p. 38.

106. Vyas KM and Dubey GL (1979). Physico-chemical properties of linseed oil as affected by various seed borne fungi. Abst. 66th Sess. Ind. Sci. Cong. p. 15.

107. Vyas KM and Saxena AP (1979). Flora of Bundelkhand I, A contribution to grasses of Banda and Hamirpur districts (UP). Bull. Bot. Soc. Univ. of Saugar, Sagar 25-26: 73-78.

108. Vyas KM and Chourasia SC (1979). Efficacy of certain antibiotics against betel vine *Phytopathora*. Phytopathora News Letters, USA 7: 28.

109. Vyas KM and Chourasia SC (1979). Leaf spot, a new disease of bakoli (*Mimusops elengi*) in Sagar, India 63: 806.

110. Vyas KM and Chourasia SC (1979). Oxidative metabolism under *Phytopathora* leaf rot pathogenesis of Pan. Phytopathora News Letters, USA 7: 27.

111. Vyas KM (1978). Effects of fungicides on the growth of *Phytopathora parasitica* var. *piperina* causing foot rot and leaf rot diseases of Pan. Proc. Nat. Acad. Sc. (Allahabad) 47 (B) III: 141-144.

112. Vyas KM and Chourasia SC (1978). Formation of cospores in vivo by *Phytopathora parasitica* var. *piperina*. Phytopathora News Letters USA, Feb. 1978, No. 6 pp. 42-43.

113. Vyas KM and Chourasia SC (1978). Leaf surface mycoflora in relation to *Phytopathora* leaf rot on Pan (*Piper betle*). Phytopathora News Letters USA, Feb. 1978, No. 6 pp. 40-41.

114. Vyas KM and Chourasia SC (1978). Host parasite interactions in betelvine *Phytopathora* varietal response to leaf rot pathogenesis. Phytopathora News Letters USA, Feb. 1978, No. 6 pp. 38-39.

115. Vyas KM (1978). Effect of fungal metabolites on the growth ot morphogenesis of crop plants. Abst. Ibid. pp. 43.

116. Vyas KM (1978). Studies on the molds and mycotoxins with special reference to mycoflora associated with the grains of jowar. Abst. All India Symposium on Physiology of Parasitism Jabalpur. pp. 20.

117. Vyas KM, Kazmi SM and Soni NK (1978). Studies on the antifungal properties of mamalian urine. Agriculture and Agro Industries Jur.

118. Chourasia SC, Vyas KM and Pathak RK (1977). In vitro inhibitory effects of certain fungicides in the growth of *Phytopathora parasitica* var. *piperina* causing foot rot and leaf rot diseases of Pan (*Piper betel* Linn.) Proc. Nat. Acad. Sci. Ind. 47(B) III: 141-144.

119. Vyas KM and Chourasia SC (1977). Effect of fungicides on the oxidative metabolism of *Phytopathora parasitica* var. *piperina* (Dast.) causing foot rot and leaf rot diseases of Pan (*Piper*

betel) in physiology of microorganism edited by KS Bilgrami Todays and Tomorrow Printers and Publishers, New Delhi 11-16.

120. Vyas KM and Jain SK (1977). Production of auxin like plant growth regulatory metabolites by soil fungi. In: physiology of microorganisms (Edited by KS Bilgrami), Todays and Tomorrow Printers and Publishers, New Delhi 331-340.
121. Vyas KM, Thind TS and Prakash V (1977). Effect of some antibiotics on the germination of corainder seeds. Ind Jr. of Exp. Biol. Vol. 15, No. 3, March 77 pp. 247-248.
122. Vyas KM and Chourasia SC (1977). In vitro effect of some volatile oils against *Phytopathora parasitica* var. *piperina*. Jr. Res. Ind. Med. Yoga and Homoeo. 12:3. 139-142.
123. Vyas KM and Chourasia SC (1977). The influence of age and starvation on endogenous respiration of mycelium of *Phytopathora parasitica* var. *piperina*. Acta Botanica Indica 5: 54-57, 77.
124. Vyas KM (1976). Respiratory characterstics of plant pathogenic fungi, *Ibid*.
125. Vyas KM (1976). Production of cell wall degrading enzymes by *Botryodiplodia theobromae*, *Ibid*.
126. Vyas KM (1976). Microbial metabolites and plant growth promoting activity of some antifungal antibiotic compounds, *Ibid*.
127. Vyas KM (1976). Microbial metabolites and plant growth production of indole auxins by soil fungi, *Ibid*.
128. Vyas KM (1976). Effect of fungicides on the oxidative metabolism of *Phytopathora parasitica* var. *piperina* (Dast) causing foot rot and leaf rot of Pan (*Piper betle*). In: Physiology of Microorganisms (Edited by KS Bilgrami), Today and Tomorrow Publications pp. 11-16.
129. Vyas KM (1976). Interactions of soil microorganisms with *Phytopathora parasitica* var. *piperina*. Abst. Symp. on Physiology of Microorganisms held at Bhagalpur Feb. 1976.
130. Vyas KM and Chourasia SC (1976). Activity of some volatile oils against *Phytopathora parasitica* var. *piperina*. Indian Drugs Jan 9-15.
131. Vyas KM and Chourasia SC (1976). Influence of culture media on fungal respiration. Hind. Antibiot. Bull. Vol. 18: 3 and 4.
132. Vyas KM and Mehta P (1975). Effect of plat growth regulators on production and activity of cellulase. Ind. J. Exptl. Biol. Vol 14, 206-208.
133. Vyas KM and Chourasia SC (1975). Studies on leaf rot of Pan (*Piper betle*) varietal factors in relation to disease resistance. Symp. on Plant Pathlogical problems held at Udaipur 5(1): 10-11.
134. Vyas KM, Mehta P and Saxena SB (1975). Pathological studies of fruit rot diseases of tamato caused by *Alternaria solani* and *A. tenuis*. Indian Phytopath. Vol. XXVIII No 2, 247-252.
135. Vyas KM, Mehta P and Saxena SB (1975). Metabolic changes during pathogenesis of tomato fruit rot diseases. Indian Phytopath. Vol. XXVIII No 2, 253-255.
136. Vyas KM and Soni NK (1975). Effect of antibiotics and fungicides on the mycelial respiration and growth of *Rhizoctonia capsici*. Ind. Jr. Exptl. Biol. Vol. 13, pp. 216-218.
137. Vyas KM, Mehta P and Saxena SB (1975). Production of pectolytic enzymes by *Alternaria* sp. of various culture media. J. Ind. Bot. Soc. 200-206.

138. Vyas KM and Soni NK (1975). Responses of *Rhizotonia solani* Kuhn and *Colletotrichum capsici* (Syd.) Butler and Bisby to various chemical substances. Madhya Bharti Univ. Sagar, Vol. 22-23 pp. 43-48.

139. Vyas KM, Mehta P and Saxena SB (1975). Effect of native carbon sources and pH on the cellulolytic enzymes of *Alternaria solani* and *A. tenuis*. Science and Culture Vol. 41 pp. 400-402.

140. Vyas KM and Chourasia SC (1975). Effect of vitamins on the mycelial growth of *Phytopathora parasitica* var. *piperina*. Madhya Bharti, vol. 22 and 23 pp. 63-65.

141. Vyas KM and Soni NK (1975). Respiratory response of *Gleosporium papayae* induced by various substances. Proc. Ind. Sci. Acad. Vol. 41, Part B. No. 5 pp. 458-461.

142. Vyas KM, Saxena RK and Saxena SB (1974). Physiology of *Tectona grandis* leaves infected with *Uncinula tectonae*. J. Indian Bot. Soc. 53: 256-270.

143. Vyas KM, Jain SK and Raghvan P (1974). Microbial metabolites and plant growth: Plant growth regulatory activity of culture filtrates of soil microorganisms and some antibiotic compounds. Abst. Symp. on Biology of Soil Microorganisms, Sagar pp. 35-36.

144. Vyas KM, Mehta P and Saxena SB (1974). Metabolic changes during pathogenesis of fruit rot diseases of tomato. MP Vigyan Acad. Bhopal Abst. pp. 7-8.

145. Vyas KM, Saxena SB and Soni NK (1974). Effect of antibiotics and fungicides on endogenous mycelial respiration and growth of *Rhizoctonia solani* Kuhn. and *Colletotrichum capsici* (Syd.) Butler and Bisby. MP Vigyan Acad. Bhopal Abst. pp. 5-6.

146. Vyas KM, Mehta P and Saxena SB (1974). Production of pectolytic and cellulolytic enzymes by *Alternaria* sp. during pathogenesis of tomato fruits. Hind. Antibiot. Bull. Vol. 16, No. 4 pp. 20-214.

147. Vyas KM, Mehta P and Saxena SB (1974). Effect of native carbon sources and pH on the pectolytic enzymes of *Alternaria* sp. during pathogenesis of tomato fruits. Hind. Antibiot. Bull. Vol. 40 Part B, No. 4 pp. 433-439.

148. Vyas KM, Pandey P and Soni NK (1974). Effect of antibiotics on endogenous mycelial respiration and growth of some plant pathogenic fungi. Hind. Antibiot. Bull. Vol. 16: 199-201.

149. Vyas KM and Soni NK (1973). Effect of the age of culture and starvation period on mycelial respiration. Bull. Bot. Soc., Univ. Saugar, Vol. 20 pp. 17-20.

150. Vyas KM and Saxena SB (1973). Studies on mycelial respiration of *Sclerotium rolfsii*. Proc. Ind. Nat. Acad. Vol. 39 Part B No. 5 569-575.

151. Vyas KM, Raghvan P and Jain SK (1973). Auxins like activity of the antibiotic aureofungin. Hind. Antibiot. Bull. Vol. 16 No 1 pp. 29-31.

152. Vyas KM and Jain SK (1973). Production of auxins by microorganisms. Hind. Antibiot. Bull. Vol. 11, pp 217-219.

153. Vyas KM and Soni NK (1973). Respiratory and growth responses of *Fusarium oxysporum* Sacc. induced by various substances. Ind J. Exptl. Biol. Vol. 11 pp. 217-219.

154. Vyas KM, Chourasia SC and Soni NK (1973). Efficiency of certain antibiotics against *Phytopathora parasitica* var. *piperina* causing leaf rot of Pan. Hind. Antibiot. Bull. Vol. 16 No. 1 pp. 4-8.

155. Vyas KM and Soni NK (1973). Effect of antifungal antibiotics on endogenous mycelial respiration of *Rhizoctonia solani*. Hind. Antibiot. Bull. Vol. 15, pp. 160-163.

156. Vyas KM, Raghavan P and Jain SK (1973). Synergistic interaction of cloremphenicol and IAA on the growth of *Avena* coleoptile sections. Hind. Antibiot. Bull. Vol. 15, pp. 81-83.

157. Vyas KM and Mishra AS (1972). Production of P-proline by soil microorganisms. Hind. Antibiot. Bull. Vol. 15 No 1 and 2 pp. 30-33.

158. Vyas KM, Soni NK and Saxena SB (1972). Effect of culture of various fungi on the mycelial respiration of *Fusarium oxysporum* Sacc. Hind. Antibiot. Bull. Vol. 14, pp. 179-180.

159. Vyas KM, Saxena SB and Jain BK (1972). Respiratory characteristics of some soil borne plant pathogenic fungi. J. Ind. Bot. Soc. 51: 92-96.

160. Vyas KM and Mehta P (1972). Control of fruit rot of tomatoes by phenolics and fungicides. Bull. Bot. Soc., Univ. Saugar, Vol. 19 pp. 35-37.

161. Vyas KM and Dubey GL (1972). Effect of plant growth regulators on the growth of *Sclerotium rolfsii*. Bull. Bot. Soc., Univ. Saugar, Vol. 19 pp. 35-37.

162. Vyas KM, Saxena SB and Soni SK (1972). Effect of plant growth regulators on oxygen uptake and growth of *Rhizoctonia solani* Kuhn and *Colletotrihum capsici* (Syd.) Butler and Bisby. Bull. Bot. Soc., Univ. Saugar, Vol. 19 pp. 24-27.

163. Vyas KM and Mehta P (1970-71). Production of cellulose and mycelial growth of *Alternaria solani* and *A. tenuis*. Bull. Bot. Soc., Univ. Saugar, Vol. 17 and 18 pp. 71-72.

164. Vyas KM and Soni NK (1970-71). Respiratory characteristics of mycelium of *Gleosporium papaya* P. Henn. Bull. Bot. Soc., Univ. Saugar, Vol. 17 and 18 pp. 48-50.

165. Vyas KM, Soni NK and Mishra AS (1970-71). Respiratory characteristics of *Gleosporium* infected papaya fruits in relation to disease development. Bull. Bot. Soc., Univ. Saugar, Vol. 17 and 18 pp. 48-50.

166. Vyas KM, Saxena SB and Atri DC (1970-71). Production of amino acids by soil fungi. Bull. Bot. Soc., Univ. Saugar, Vol. 17 and 18 pp. 31-34.

167. Vyas KM and Saxena SB (1962-64). The wood decaying fungi of Sagar. Madhya Bharti, Vol. 11-13 pp. 15-28.

List of Ph.D. Awarded Under the Supervision of Prof. K.M. Vyas

1. Mehta, Pradeep 1974. "Biochemical Investigations of Fruit-Rot Diseases of Tomatoes."
2. Jain, Satish Kumar 1975. "Studies on the Microbial Production of Auxins".
3. Chourasia, Shiv Charan 1976. "Studies on the Foot-Rot and Leaf-Rot Diseases of Pan (*Piper betle* L.) with Special Reference to Pathogenesis and Control Measures".
4. Raghavan (alias) Srinivasan, Parvati 1976. "Physiological Studies on the Effect of Antibiotics on Plants with Special Reference to their Growth Regulatory Effects"
5. Mishra, Jayendra Kumar 1979. Biochemical Studies on Some *Botryodiplodia* Rot with Special Reference to Cell-Wall Degrading Enzymes".
6. Arti, Dinesh Chandra 1980. "Study of Fruit-Rot Disease of *Pyrus pyrifolia* Caused by *Botryodiplodia theobromae*".
7. Dubey, Guljari Lal 1980."Fungal Succession in Relation to the Bio-deterioration of Linseed".

8. Thakur, Munna Singh 1981. "Studies on the Metabolities of Some Microorganisms and their Effects on Plant Growth".
9. Jain, Maya (nee) Singhai, Maya 1982. "Studies on Plant Growth Regulatory Activities of Certain Antibiotics".
10. Saxena, Neeraj 1982. "Studies on Cell Wall Degrading Enzymes in Relation to Fungal Pathogenesis".
11. Thomas, Thankamma 1982. "Studies on the Soil Fungi of the Andaman and Nicobar Islands".
12. Chiele, Satish Kumar 1983. "Studies on Host–Parasite Relationship with Special Reference to Betel–Vine *Phytopathora*".
13. Saxena, Anand Prakash 1983. "Studies on Some Medicinal Plants of Bundelkhand Region with Special Reference to their Antimicrobial Activity".
14. Gupta, Deepak Kumar 1984. "Studies on the Microbiological Control of Mosquito, Causing Malaria".
15. Shrivastava, Anjali 1985. "Studies on the Efficacy of Certain Antibiotics Against Betel-Vine *Phytopathora*".
16. Katare, Daya Shanker 1987. "Genetic Analysis of Metric and Quality Attributes in Tomato".
17. Saxena, Nitin Kumar 1987. "Microbial Transformation of Steroids".
18. Tripathi, Hari Har Sharran 1987. "Study on the Production and Activity of Toxic Metabolites by Certain Bacterial and Fungal Pathogens of *Piper betle* L".
19. Sadiq, Yasmin 1989. "Studies on the Role of Phenolic Compounds in Disease Resistance in Plants (with special reference to Betel-vine *Phytopathora*)".
20. Shrama, Pratap Bhanu 1991. "Study on the Effect of Cultural and Chemical Control Systems of Kharif Weeds in Tawa Command Area".
21. Firdousi, Shakeel Ahamad 1992. "Studies on the Fungal Disease of Some Important Forest Plant with Special Reference to the Foliage Diseases".
22. Sediq, Roohi 1992. "Studies on the Effect of Certain Factors on the Forest Vegetation of Sagar".
23. Tewari, Prem Narayan 1992. "Studies on Yield, Oil and Protein Content of Soybean as Affected by Fertility Levels, Different Stages of Sowing and Plant Density".
24. Malaiya, Ku. Seema 1992. "Antimicrobial Evaluation of Certain Medicinal Plants of Ethnobotanical-Importance".
25. Dubey, Mahadev Prasad 1993. "Effect of Soil Tilth Seed Inoculation and Sowing Methods on Growth and Yield of Soybean".
26. Goswami, Ravi Prakash 1994. "Studies on the Effect of Volatile Fungistatic Factors on Fungal Morphogenesis".
27. Dixit, Ashwini Kumar 1995. "Ethno-medicinal, Photochemical and Anti-microbial Studies of Some Medicinal Plants of Sagar District".
28. Chourasia, J.P. 1995. "Studies on the Management of Betalvine *Phytopathora* Diseases in Sagar".
29. Sharma, C.D. 1995. "Folicolus Fungal from the Forest Flora of M.P. with Special Reference to Deuteromycotina".

30. Goswami, Sita Ram 1995. "Response of Different Levels of Phosphetic Fertilizers and their Methods of Application on Growth Parameters Productivity and Quality of Soybean".
31. Malaiya, Manju 1998. "Studies on Antimicrobial Activity of Some Indian Traditional Remedies (Ethnomedicinal plants)".
32. Mishra, Avanish 1998. "Studies on the Biology of Some Ethno-medicinal Plants of Sagar Region".
33. Kaushal, A.K. 1998. "Role of Irrigation Schedule on Productivity of Sunflower".
34. Sthapak Smt. Madhu 2000. "Studies on the Fungal Diseases of Potato Under Storage Conditions at Sagar".
35. Nema Ku. Shushma 2000. "Studies on Some *Phytopathora* Species from Central India with Special Reference to their Reproduction".
36. Dwivedi, O.P 2003. 'Studies on Soil Microorganisms with Special Reference to Vesicular-Arbuscular Mycorrhizal (VAM) Fungal Association with Wheat Crop of Sagar Region".
37. Yadav, Rajesh 2003. "Studies on Integrated Management of Betelvine *Phytopathora*".
38. Agrawal, Roop Shikha 2004. "Studies on Fungal Diseases of Forest Plants of Mandla Region".
39. Jain, Ashish 2005. "Studies on Mushroom Cultivation with Special Reference to *Pleurotus* Species and their Marketing Potential in Sagar Region".
40. Vyas, Ashish 2005. "Studies on Microbial Biodiversity with Special Reference to their Cellulolytic Activity".
41. Soni Anuradha 2006. Studies on occurrence of VAM Fungi in leguminous crops of sagar.

Acknowledgements

A journey into unknown territory, this is how I felt at the onset of this task. The work undertaken has made me experience many ups and downs throughout the duration of compilation. Although I had been living with it days and nights but there are many people who have knowingly or unknowingly figured a major role in its completion. Although words are not sufficient to express my feelings but still I shall try to comprehend as far as I can and the feelings behind are really heartfelt.

This book is presented with obligation to the under mentioned with full acceptance of its limitations.

I thanks my Shredye Guru Ji Pt Omkar Dev Ji, whose wishes enabled me to reach this mile stone who showed me path to reach the destination and responsible for how to thank god.

I note down here with humility and pride that I am fortunate enough in having support of esteemed personalities like Prof. H.D. Kumar (FNA), BHU, Varanasi, Prof. D.P. Tiwari, Ex Head, Department of Botany, Govt. Science P.G. College, Jabalpur, Prof. C. Manoharachari (Emeritus Prof.), Department of Botany, Osmania University, Hyderabad, Prof. Anupam Dixit, Department of Botany, Allahabad University, Prof. S.P. Adhikari, Department of Biotechnology, Visva Bharti University, Shantiniketan (West Bengal), Prof K.L. Tiwari, Ex Head, Department of Biotechnology, Pt R.S. University, Raipur, Prof G.S. Paliwal (Emeritus Prof.) Senior Consultant National Afforestation and Eco-development Board, New Delhi, Prof H.K. Goswami, Ex Head, Department of Genetics, Barkatullah University, Bhopal and Prof. H.C. Lakshman, Head, Department of Botany, Karnataka University, Dharwad.

I am happy to remember my teachers who happened to be students of my father late Prof. K.M. Vyas, therefore I gratefully acknowledges Prof A.D. Adoni, Prof. T.R. Sahu, Prof. N.K. Soni, Prof. D.C. Atri, Prof. A.K. Kandya, Prof. P. Mehta, Prof. P.K. Khare, Prof. A.N. Rai, Prof. (Smt.) J. Dubey, Dr. A.S. Mishra, Dr. S.K. Yadav, Dr. N.P. Bhalla and my colleague Dr. A. Mehta, Dr. A. Biswas, Miss Poonam Dehariya and Dr. Ashok Shukla for their necessary help whenever I was in need.

I express sincere gratitude to all the authors who have contributed for this volume.

It gives me immense pleasure in acknowledging with sincere gratitude, the enthusiastic co-operation received from my students Dr. O.P. Dwivedi, Dr. Rajesh Yadav, Dr. Ashish Vyas, Dr. Anuradha Soni, Dr. Ashish Jain, Dr. Archana Dubey, Dr. Mahendra Mishra, Dr. Pradeep Singh, Dr. Prashant Soni, Dr. H.C. Gena, Pramod Richhariya, Meenakshi Mishra, Anjuli Chaubey, Javed Ahmad Wagay and Mrs. Anuradha Shukla.

I express a deep valuation to all the non-teaching staff especially Dr. Pradeep Tiwari, K.K. Dixit, J.P. Chaubey, P. Gujre and Shailesh Katare of the department for their time to time help and co-operations.

It is beyond the capacity of all the words to express my respect and profound affection for my mother Mrs. Urmila Vyas. Today whatever I am that is only by virtue of her love, care, affection and blessings. At this juncture I can not forget my wife Mrs. Nisha and children Sidhhant and Vedant and pain taken and sacrifices made by them can not be summed up. They remained as a source of constant inspiration; I bow down before their love and affection towards me. The love and affection extended by my sisters Dr. Jyoti, Arti and Bharti, and their family is memorable.

Last, but certainly not the least, I must thanks to Mr. Nafees of Perfect Computers for designing this cover page in a careful and elegant manner, which is miraculous. I also thank Mr. Anil Mittal of Daya Publishing House for publishing the volume a scientific tribute to my father.

Deepak Vyas

H.D. Kumar

Ph.D. (London). FASc. FNASc, FNA

Formerly Professor of Botany,
Dean, Faculty of Science, and Coordinator,
Biotechnology Programme
Banaras Hindu University

Phone (Res.): (0542) 2315180

214, Saketnagar, Naria,
VARANASi – 221 005

Foreword

Microbes, the real masters of life on Earth, are both a boon and bane for the planet. Our microbial friends live within our body, protecting us from disease, aiding digestion, and preserving immunity. On the other side, some of the deadliest diseases are caused by viruses and pathogenic bacteria. Whereas beneficial fungi are an asset to the industrialist and the environment, pathogenic fungi cause havoc and can devastate cereal crops and fruit trees.

The present volume is a Festschrift in honour of my long-term friend, late Prof. K.M. Vyas, a reputed mycologist with strong interest in microbiology. This volume includes some 50 chapters including 23 on microbiology, 13 on biotechnology and 14 on ecology. These chapters are authored by senior professors and younger scientists. The editors are practising researchers and I have pleasure in commending this Festschrift to all those interested in the exciting fields of mycology, microbiology and general biology.

H.D. Kumar

Preface

With the advent of new technologies, the knowledge of science is expanding in new dimensions every day. The frequent review of the state of our knowledge not only to keep ourselves up to date but also to provide over the various strategies ought to be adopted for further revolution of science in all its sphere.

As one know that presently existing life form are evolve from the microorganisms during the course of evolution. These microorganisms can adapt to virtually any physio-chemical environment and are found everywhere.

Despite their huge importance, the fact is that much of what we know about and microbial processes is based on the work done on only a harmful of species-weed like species which can be readily grown in laboratory culture. The vast majority of microorganisms found in nature (Perhaps over 95 per cent) have not so far been cultured but earnest attempts are under ways to grow as many of them as possible in culture.

One most striking property of many microorganisms is the highly promiscuous mode of gene transfer among unrelated or remotely related taxa. Such horizontal (lateral) gene transfers have been the subject of intense research activity in the past few years, in view their strong bearing on the applicability of the biological species concept and the Linnaean system of binominal nomenclature to prokaryotes.

Technological advancement has revolutionized and transform microbiology with biotechnology and created exciting opportunities for agriculture, medicine, industry and environment remediation the organism involved in these mechanisms mostly belongs to bacteria, fungi, cyanobacteria etc.

To keep pace with the growing knowledge in every field of science, our agenda for the coming few years should be to identify the key organizational and genetic principles underlying various morphogenetic processes in both prokaryotes and eukaryotes, while also pursuing the laborious elucidation of the molecular network that control the organizational of living matter.

Considering the importance of microorganism and their biotechnological processes are the major contents of the monograph. Beside microbiology and biotechnology few chapters on ecology are included.

Editors

Contents

— Volume 2 —

— BIOTECHNOLOGY —

List of Editors & Contributors

Editors

Dr. Deepak Vyas
Assistant Professor, Department of Botany, Dr. H.S.G. University, Sagar – 470 003, M.P.

Prof. G.S. Paliwal
Senior Consultant, Regional Centre of the National Afforestation and Eco-development Board, Ministry of Environment and Forests, Government of India, New Delhi

P.K. Khare
Professor, Department of Botany, Dr. H.S.G. University, Sagar – 470 003, M.P.

Dr. R. K Gupta
Assistant Professor, Department of Botany, Govt. P.G. College, Rhishikesh

Associate Editors

Prof. A.K. Pandey
Chairman, M.P. Govt. Private University Commission, Bhopal, M.P.

Dr. Jamaludin
Emeritus Prof. Department of Biological Sciences, R.D. University Jabalpur, M.P.

Dr. Neeraj Khare
Department of Botany, Dr. H.S.G. University, Sagar – 470 003, M.P.

Contributors

Adhikary, S.P.
Department of Biotechnology, Vishwa-Bharti University, Shantiniketan, West Bengal

Agrawal, Pooja
Department of Applied Microbiology and Biotechnology, Dr. H.S.G. University, Sagar – 470 003, M.P.

Akhtar, Nasim
Department of Biotechnology, GITAM University, Gandhi Nagar Campus, Rushikonda, Vishakapatnam – 530 045

Ashu
Department of Botany, Sahu Jain PG College, Najibabad – 246 763, U.P.

Atri, D.C.
Department of Botany, Dr. H.S.G. University, Sagar – 470 003, M.P.

Banerjee, Meenakshi
Laboratory of Algal Biotechnology, Department of Bioscience, Barkatullah University, Bhopal, M.P.

Behera, R.K.
Department of Botany, Berhampur University, Berhampur – 760 007, Orissa

Bharose, Ram
School of Forestry and Environment, SHIATS-Deemed University, Allahabad

Bhatt, R.K.
Indian Grassland and Fodder Research Institute, Jhansi – 284 003, U.P.

Bhattacharya, Abhishek
Bacteriology Lab, Department of Biological Sciences, Rani Durgavati University, Jabalpur, M.P.

Biradar, D.P.
Department of Agronomy, UAS, Dharwad – 580 005

Byatanal, Mahesh B.
Post Graduate Department and Research in Botany (Microbiology Lab), Karnataka University, Dharwad – 580 003

Chakravarty, Neha
National Research Centre for Agroforestry, Jhansi – 284 003, U.P.

Chaubey, Anjuli
Lab of Microbial Technology and Plant Pathology, Department of Botany, Dr. H.S.G. University, Sagar – 470 003, M.P.

Chauksay, P.
Department of Botany, Dr. H.S.G. University, Sagar – 470 003, M.P.

Chaurasia, Bhaskar
Department of Botany, Dr. H.S.G. University, Sagar – 470 003, M.P.

Chouhan, A.K.S.
Department of Botany, Dr. H.S.G. University, Sagar – 470 003, M.P.

David, Arun A.
School of Forestry and Environment, SHIATS-Deemed University, Allahabad

Dehariya, Poonam
Lab of Microbial Technology and Plant Pathology, Department of Botany, Dr. H.S.G. University, Sagar – 470 003, M.P.

Dikshit, Anupam
Biological Product Laboratory, Department of Botany, University of Allahabad, Allahabad –211 002

Diwedi, O.P.
Lab of Microbial Technology and Plant Pathology, Department of Botany, Dr. H.S.G. University, Sagar – 470 003, M.P.

Goswami, H.K.
24, Kaushalnagar, PO Misrod, Bhopal – 462 047, M.P.

Gupta, Pushpa
National JALMA Institute for Leprosy and Other Mycobacterial Diseases, Agra – 282 001

Gupta, Rajan Kumar
Department of Botany, Govt. P.G. College, Rishikesh, Uttarakhand

Gupta, U.S.
Department of Zoology, Dr. H.S.G. University, Sagar – 470 003, M.P.

Gupta, Umesh Dutta
National JALMA Institute for Leprosy and Other Mycobacterial Diseases, Agra – 282 001

Guru, S.D.
Department of Botany, Ranchi Women's College, Ranchi, Jharkhand

Hegde, Ganesh R.
Department of Botany, Sri Krishnadevaraya University, Anantapur – 515 0003

Jadhav, S.K.
School of Studies in Biotechnology, Pt. Ravishankar Shukla University, Raipur – 492 010,

Jain, Arun K.
Department of Anthropology and Tribal Development, GG University, Bilaspur, C.G.

Jain, Devendra
School of Life Sciences, Jaipur National University, Jaipur – 302 025, Rajasthan

Jamaluddin
Ex Group Coordinator and Director, Tropical Forest Research Institute, Jabalpur, M.P.

Jha, Anuradha
National Research Centre for Agroforestry, Jhansi – 284 003, U.P.

Kamalvanshi, Madhavi
National Research Centre for Agroforestry, Jhansi – 284 003, U.P.

Kamran, A.
Biological Product Laboratory, Department of Botany, University of Allahabad, Allahabad –211 002

Kango, Naveen
Department of Applied Microbiology and Biotechnology, Dr. H.S.G. University, Sagar – 470 003, M.P.

Karwa, Alka
Department of Biotechnology, S.G.B. Amravati University, Amravati – 444 602

Kaushik, B.D.
Division of Biotechnology, Netaji Subash Institute of Technology, Sector 3, Dwarka, New Delhi – 110 078

Khare, M.N.
Agricultural Research Station, Mandor, Jodhpur, Rajasthan (Ex. Dean and Professor Emeritus, Plant Pathology)

Khare, P.K.
Department of Botany, Dr. H.S.G. University, Sagar – 470 003, M.P.

Kumar, Anil
Principal Scientist (Plant Pathologist), National Research Centre for Agroforestry, Jhansi – 284 003, U.P.

Kumar, Mukesh
Department of Botany, Sahu Jain PG College, Najibabad – 246 763, U.P.

Kunjam, S.
School of Studies in Biotechnology, Pt. Ravishankar Shukla University, Raipur – 492 010,

Lakshman, H.C.
Department of Botany, Microbiology Laboratory, Karnataka University, Dharwad -- 580 003, Karnataka

Maya, C.
Department of Botany, Bangalore University, J.B. Campus, Bangalore – 560 056

Mehta, A.
Department of Botany, Dr. H.S.G. University, Sagar – 470 003, M.P.

Mehta, P.
Department of Botany, Dr. H.S.G. University, Sagar – 470 003, M.P.

Mishra, A.K.
Indian Grassland and Fodder Research Institute, Jhansi – 284 003, U.P.

Mishra, P.
Biological Product Laboratory, Department of Botany, University of Allahabad, Allahabad –211 002

Mishra, Arun Kumar
Laboratory of Microbial Genetics, Department of Botany, Banaras Hindu University, Varanasi – 221 005

Mishra, R.K.
Biological Product Laboratory, Department of Botany, University of Allahabad, Allahabad –211 002

Nema, Sushma
Senior Scientist, Plant Pathology, AICRP (S&N) J.N.K.V.V., Jabalpur

Pandey, A.K.
Chairman, M.P. Council Pvt University Ltd., Bhopal

Pandey, J.
Department of Botany, Banaras Hindu University, Varanasi – 221 005, U.P.

Pandey, R.K.
Biological Product Laboratory, Department of Botany, University of Allahabad, Allahabad – 211 002

Pandey, Usha
Faculty of Science and Technology, MG Kashividya Pith, Varanasi – 221 005, U.P.

Panigrahi, A.K.
Department of Botany, Berhampur University, Berhampur – 760 007, Orissa

Pataik, Anita R.
Department of Botany, Berhampur University, Berhampur – 760 007, Orissa

Pujari, Rajesh
Bacteriology Lab, Department of Biological Sciences, Rani Durgavati University, Jabalpur, M.P.

Rai, A.N.
Department of Botany, Dr. H.S.G. University, Sagar – 470 003, M.P.

Rai, Mahendra
Department of Biotechnology, S.G.B. Amravati University, Amravati – 444 602

Rajpurohit, T.S.
Pathologist, Oilseeds (Rajasthan Agricultural University)

Richhariya, Pramod K.
Lab of Microbial Technology and Plant Pathology, Department of Botany, Dr. H.S.G. University, Sagar – 470 003, M.P.

Rishi, Narayan
School of Virology, Amity University, Noida, U.P.

Rishi, Sushma
School of Virology, Amity University, Noida, U.P.

Roy, A.K.
Department of Botany, T.M. Bhagalpur University, Bhagalpur – 812 007

Sabannavar, Shweta J.
Post Graduate Department and Research in Botany (Microbiology Lab), Karnataka University, Dharwad – 580 003

Sahu, Alka
Department of Botany, Berhampur University, Berhampur – 760 007, Orissa

Satya
Lichenology Laboratory, National Botanical Research Institute (CSIR), Rana Pratap Marg, Lucknow – 226 001, U.P.

Sethi, S.K.
PG Department of Biotechnology, Utkal University, Bhubaneswar – 751 004, Orissa

Sharma, A.N.
Department of Anthropology, Dr. H.S.G. University, Sagar – 470 003, M.P.

Sharma, Anjana
Bacteriology Lab, Department of Biological Sciences, Rani Durgavati University, Jabalpur, M.P.

Shukla, Ashok
Lab of Microbial Technology and Plant Pathology, Department of Botany, Dr. H.S.G. University, Sagar – 470 003, M.P.

Shukla, C.S.
Department of Plant Pathology, Indira Gandhi Agricultural University, Raipur – 492 006, Chhattisgarh

Singh, A.
Laboratory of Microbial Genetics, Department of Botany, Banaras Hindu University, Varanasi – 221 005

Singh, A.N.
Department of Botany, T.M. Bhagalpur University, Bhagalpur – 812 007

Singh, Abhijeet
School of Life Sciences, Jaipur National University, Jaipur – 302 025, Rajasthan

Singh, P.
Laboratory of Microbial Genetics, Department of Botany, Banaras Hindu University, Varanasi – 221 005

Singh, P.K.
Department of Botany, Dr. H.S.G. University, Sagar – 470 003, M.P.

Singh, Siddhartha
Department of Botany, Dr. H.S.G. University, Sagar – 470 003, M.P.

Singh, Surendra
Centre of Advanced Study in Botany, Banaras Hindu University, Varanasi – 221 005, U.P.

Singh, Vijendra
Institute of Microbial Technology, Chandigrah – 160 036

Singh. S.S.
Laboratory of Microbial Genetics, Department of Botany, Banaras Hindu University, Varanasi – 221 005

Soni, Prashant
Department of Botany, Dr. H.S.G. University, Sagar – 470 003, M.P.

Srivastava, A.
Laboratory of Microbial Genetics, Department of Botany, Banaras Hindu University, Varanasi – 221 005

Suresh, B.
Department of Applied Genetics, Karnatak University, Dharwad – 580 003

Tamphasana, R.K.
Department of Applied Geology, Dr. H.S.G. University, Sagar – 470 003, M.P.

Thakur, A.S.
Department of Botany, Govt College, Khurai, Sagar – 470 117, M.P.

Thakur, M.P.
Department of Plant Pathology, Indira Gandhi Agricultural University, Raipur – 492 006, Chhattisgarh

Thangadurai, D.
Department of Botany, Karnatak University, Dharwad – 580 003

Thomas, H.
Department of Applied Geology, Dr. H.S.G. University, Sagar – 470 003, M.P.

Thomas, T.
School of Forestry and Environment, SHIATS-Deemed University, Allahabad

Tiwari, H.S.
Indian Grassland and Fodder Research Institute, Jhansi – 284 003, U.P.

Tiwari, K.L.
School of Studies in Biotechnology, Pt. Ravishankar Shukla University, Raipur – 492 010, C.G.

Tripathi, V.D.
National Research Centre for Agroforestry, Jhansi – 284 003, U.P.

Upadhyay, Mukesh K.
School of Life Sciences, Jaipur National University, Jaipur – 302 025, Rajasthan

Upadhyay, Saurabh
School of Forestry and Environment, SHIATS-Deemed University, Allahabad

Upreti, D.K.
Lichenology Laboratory, National Botanical Research Institute (CSIR), Rana Pratap Marg, Lucknow – 226 001, U.P.

Valho, MAA Pinheirho De Car
ISO Plexis Germplasm Bank, Department of Biology, University of Madeira, Funchal 9000-390, Portugal

Vasathaiah, Hemanth K.N.
Center for Viticulture and Small Fruit Research, Florida A and M University, 6505 Mahan Drive, Tallahassee, Florida 32317, USA

Verma, Naveen
Department of Botany, Dr. H.S.G. University, Sagar – 470 003, M.P.

Verma, Vidhi
Laboratory of Algal Biotechnology, Department of Bioscience, Barkatullah University, Bhopal, M.P.

Vyas, Ashish
Division of Biotechnology, Lovely Professional University, Ludhiana (Punjab)

Vyas, Deepak
Lab of Microbial Technology and Plant Pathology, Department of Botany, Dr. H.S.G. University, Sagar – 470 003, M.P.

Wagay, Javed Ahmad
Lab of Microbial Technology and Plant Pathology, Department of Botany, Dr. H.S.G. University, Sagar – 470 003, M.P.

Yadav, Rajesh
ABR PG College, Anpra, Sonbadhra, U.P.

Yadav, S.K.
Department of Botany, Dr. H.S.G. University, Sagar – 470 003, M.P.

Yadav, V.K.
Department of Plant Pathology, Indira Gandhi Agricultural University, Raipur – 492 006, Chhattisgarh

Microbiology

Microbial Biotechnology and Ecology (2011) *Pages* **1–28**
Editors: **Deepak Vyas, G.S. Paliwal, P.K.Khare and R.K Gupta**
Published by: **DAYA PUBLISHING HOUSE, NEW DELHI**

Chapter 1

Rhizobium Biofertilizer: Retrospect and Prospects

☆ *S.K. Sethi*[1] *and S.P. Adhikary*[2]

[1]*P.G. Department of Biotechnology, Utkal University, Bhubaneswar – 751 004*
[2]*Department of Biotechnology, Vishwa-Bharti University, Shantiniketan, West Bengal*

ABSTRACT

Basically nitrogen fixation is related to legume nodulation by the symbiotic association with rhizobia, which have the role to fix the atmospheric nitrogen and thus attain nitrogen in nitrogen deficient soils. Thus technology was developed for making rhizobial biofertilizers for nitrogen fixing plants, evaluate the growth promoting substances secreted by these bacteria and also educating about their unique benefits to the users, technique for mass multiplication and practicing the integrated nitrogen managements. The use of molecular genetic techniques could help in improvement the pathways and signal cascades involved in efficient and effective nodulation in legumes as well as non-leguminous plants. Details of the *Rhizobium* biofertilizer technology and its future prospects is reviewed in this chapter.

Introduction

India is an agricultural country. Nearly 70 per cent of the populations in rural areas are engaged in agriculture, cropping in low fertility soils, especially those poor in nitrogen contributing lower yield. Thus in modern agriculture use of chemical fertilizers has become indispensable. However, excessive use of chemical fertilizers and pesticides has generated several environmental problems including the green house effect, Ozone layer depletion and acidification of water. In some regions crop production has stagnated or even declined due to depletion in bioavailability of nitrogen. These problems can be tackled by use of biofertilizers and biopesticides, which are natural, ecologically sustainable and user friendly. Reports have shown that utilization of biological nitrogen fixation technology can decrease the use of urea-N and reduce the environmental problems to a considerable extent. About 80 per cent of stable biologically fixed nitrogen are a direct result of symbiotic interaction of diazotroph like rhizobia and some actinomycetes with the nitrogen as well as non-nitrogen fixing plants respectively (Bruijin *et al.*, 1995).

Rhizobia are gram negative, rod shaped, non-sporulating important soil microorganisms form symbiotic association with leguminous crops and contribute significant amount of fixed nitrogen. Bacteria of six genera namely *Rhizobium*, *Bradyrhizobium*, *Azorhizobium*, *Sinorhizobium* (recently changed to *Ensifer*), *Mesorhizobium* and *Allorhizobium* are of major agricultural importance because of their ability to form N_2- fixing nodules on the roots of leguminous plants.

Rhizobium and Nitrogen Fixation

Rhizobia-legume symbiosis has attracted world wide attention as it implies lesser dependence on expensive petroleum based nitrogenous fertilizers for legumes, thus reducing the chance of causing environmental pollution. The importance of this group of bacteria is further enhanced by the fact that the legumes nodulated by rhizobia include some of the most important crop plants of the world. Symbiotic nitrogen fixation is a complicated process involving reciprocal exchange of signals between rhizobia and the legumes leading to the coordinated gene-expression (Earl and Ausubel, 1993). Effective nodules formed by effective strains of rhizobia possess pink colour due to the presence of a pigment 'leghaemoglobin'. The nodule is merely a protective structure where bacteroids are the seat of nitrogen fixation. Nitrogenase is the enzyme which mediates the reduction of N_2 to NH_3. Nitrogenase can be obtained by the disintegration of bacteroids by mechanical process. This enzyme is made up of two components with both iron and molybdenum (Mo) having molecular weight of about 200,000 and second with iron and without Mo having molecular weight of about 65,000 (Quispel, 1974). The amount of leghaemoglobin and the extent of bacteroid tissue in nodules have a direct bearing on the amount of nitrogen fixed by legumes (Subba Rao, 1967; Verma and Bal, 1976).

Use of *Rhizobium* as Bioinoculant/Biofertilizer

Both inoculation success and failures at field levels have been well reported (Subba Rao and Tilak, 1977; Subba Rao, 1977). The failure to obtain the desired response may due to:

1. The presence of native ineffective strains which could not be displaced by introduced effective strains.
2. Presence of antagonists of rhizobia which minimize the number of rhizobia in the rhizosphere.
3. Availability of soil condition which limit symbiosis caused by acidity, alkalinity and other factors related to soil structure and mineral composition of soil.

Rhizobium is found in all types of soils in India. Reports have shown that utilization of *Rhizobium* as a biofertilizer for leguminous and non-leguminous crops in different soil conditions contribute significant amount of fixed nitrogen (Subba Rao, 1977). Increasing number of reports has also shown that Rhizobia can also act as PGPR in various crops (Hossain and Mårtensson, 2008). Summary of the work so far undertaken on this aspect is summarized in Table 1.1 and also are described as follows:

Survey of Different *Rhizobium* and their Host Specificity

It is possible to differentiate rhizobia on the basis of growth on defined substrates, as fast growers and slow growers. Studies have been done on morphological and physiological characters (colonial character, vitamin, carbohydrate and nitrogen nutrition, antibiotic sensitivities and infective attributes)

Table 1.1: Summary of Work on *Rhizobium* as Biofertilizer for Different Crops

Organism	*Strain*	*Isolated from*	*Growth Condition*	*Finding*	*Reference*
Rhizobium, *Brady rhizobium* sp.		Soybean *Glycine max* Arhar (*Cajanus cajan*), Chickpea (*Cicer arietinum*, *Lens culinaris*)	YEMA at 28°C for one week	The results of experiments done on all-India level showed that soybean responds spectacularly to *Rhizobium* application. Grain yield was increased up to 50 per cent over uninoculated control. Depending upon the agroclimatic conditions significant increase in yield over control of arhar, chickpea and masur was recorded. At certain sites, simple *Rhizobium* inoculation failed, however, pelleting inoculated seeds with lime or charcoal could significantly increase yield of arhar.	Subba Rao (1977)
Rhizobium, *Brady-rhizobium* sp.		*Cajanus cajan*, *Vigna mungo*, *Glycine max*	YEMA at 28°C for one week, purity tested by with Congored	Results of all India field experiments showed that all the leguminous crops responded to *Rhizobium* treatment. *Rhizobium* significantly increased grain yield of *Cajanus cajan* up to 16.4 per cent, *Vigna mungo* 2.4 per cent, *Glycine max* 65.9 per cent over uninoculated control.	Subba Rao (1984)
Rhizobium sp. and *Azospirillum* sp.				On field experiment of groundnut single or combined inoculation of *Azospirillum brasilense* and *Rhizobium* sp. with N_4(20 kg Nha^{-1}) and N_3(15 kg Nha^{-1}). height, nodulation in plant, dry matter, dry pod yield and N content of groundnut significantly increased in the highest level N treatment *i.e.* N (25 kg Nha^{-1}). For inoculation of groundnut with *A. brasilense* and *Rhizobium* sps. alone increased the growth and yield of crop and it was proved that these strains save up to 5 to 10 kg N ha^{-1}.	Wani *et al.* (1992)
Rhizobium sp.	Ie-108, SC-109, CO-57, CI-114	Wild and cultivated legumes Indigofera sp.	YEMA	Efficacy of Rhizobial strains was tested against groundnut. Maximum nodulation occurred due to inoculation of rhizobial isolates. Shoot and root length and total N content were significantly increased over uninoculated control showing they are effective in N-fixation in groundnut.	Salve and Ganawane (1992)
Rhizobium japonicum		*Glycine max*		Growth and yield of soybean forage in relation to inoculation at various nitrogen and phosphorous levels were reported. Seed treatment with *Rhizobium* culture resulted in increased production of dry matter as compared to control. Application of nitrogen @ 25kg/ha^{-1} with *Rhizobium* culture proved superior for growth and yield of soybean.	Dahiya *et al.* (1992)

Contd...

Table 1.1–Contd...

Organism	*Strain*	*Isolated from*	*Growth Condition*	*Finding*	*Reference*
Bradyrhizobium Japonicum		*Glycine max*		Seed inoculation with *B. japonicum* along with applying nitrogen and phosphorus each @ 30, 60kg and 120kg/ha respectively was carried out. Nitrogen application reduced nodulation by *Rhizobium* while phosphorous stimulated it on 30, 45 and 60 days of crop growth. N 30kg/ha and P 120 kg/ha with *B. japonicum* inoculation increased shoot dry wt. Maximum grain yield was obtained by application of 30 kg and 60 kg N/ha and P_2O_5 120 kg/ha.	Kurundkar *et al.* (1992)
Rhizobium sp.	NGR (L) VC-1163	Green gram (*Vigna radiata*)	YEMA	Eight strains of *Rhizobium* were isolated which were further tested for their survival under acid and alkaline condition. The strain INDE-2, a slow grower, and ALVG-2, a fast grower, isolated from coastal sand dune and mesophytic plain respectively. They showed compatively better survival under both acid and alkaline condition. Growth of VRNG-8 strain remained unaffected in alkaline range.	Biswal and Mishra (1992)
Bradyrhizobium japonicum		*Glycine max* Parvani isolate	YMA culture prepared with lignite based carrier	Seed inoculation was done with *B. japonicum* plus pelleting with carrier material and gypsum. Results were significant for increase in number of nodules and nodule dry wt/plant on 45·day of crop growth and dry wt. of plant at 35 and 45 days of crop growth. Seed treatment with ten times of the recommended dose recorded having higher number of nodules/plant than control. Seed inoculation plus pelleting with carrier gypsum, seed treatment with recommended dose of *Bradyrhizobium* and soil inoculation with *Rhizobium* enriched soil had given highest nodule dry wt/plant and grain yield.	Kawale *et al.* (1992)
B. japonicum	JS-2	Soybean (*Glycine max*) Parbhani isolate	YEM broth 3-4 days at 28°C	Seeds were inoculated with 10^2, 10^3, 10^4, 10^5 and 10^6 viable cells/seed rhizobial load significantly influenced all the growth and grain yield parameters of soybean except number of leaves and branches per plant.	Kurundkar *et al.* (1992)
Rhizobium sp.		Groundnut (*Arachis hypogea*)		Rhizobial culture with 40 per cent GAS (Gum arabic solution) was most effective. For groundnut lime pelleting with gum Arabic was however superior. Clay pelleting with plain water was at par. Clay pelleting with jaggery solution and gum Arabic solution yielded 13.19 q/ha and 7.66 q/ha respectively over control.	Chitriv (1992)

Contd...

Table 1.1–Contd...

Organism	Strain	Isolated from	Growth Condition	Finding	Reference
Rhizobium sp.	M-1, M-3, M-5, M-7 and S-8	Mung bean in the kharif season from Marathwada region	YEMA+ Congored	Combination M-5 strains of *Rhizobium* with 25kg N per hectare produced highest dry matter and nitrogen content in S-8 variety of Mung bean. The interaction between host varieties and *Rhizobium* strains revealed that M-1, M-3, M-5 and M-7 strains were superior over control.	Kore *et al.* (1992)
Bradyrhizobium japonicum	IARI-1 IARI-3	Soybean (*Glycine max*)	YEMA	A maximum nodule dry weight was recorded at 30 and 60 days of crop growth. Higher grain yield was recorded in *Rhizobium* treated plants than the control.	Kawale *et al.* (1992)
Rhizobium		*Cicer arietinum*		Maximum increase in the grain yield (16.33 per cent) was obtained with *Rhizobium* double coating.	Chopde *et al.* (1993)
Rhizobium sp.	C-235, Ca-2-Ca534 and CV-4	Chick pea (*Cicer arietinum*), collected from three districts of Haryana state	YEMA	378 native isolates were screened, of which 42 (11.1 per cent) showed antibiosis against homologous sensitive strain of CV-4 along with a standard strain Ca-181. 16 strains produced low molecular weight bacteriocin with mol.wt. above 12 KDa. The bacteriocin producing strains suppressed the growth of the sensitive strains under co-culture. Both the wild and the sensitive strains showed high nodulation ability. Co-inoculation of sensitive strain with a bacteriocin producing strains yielded 70-90 per cent nodules in the later. Bacteriocin producing strains failed to dominate the sensitive strain under co-growth condition as well as nodule competence. Bacteriogenic activity of a strain is beneficial in nodule competence among homologous strains.	Ahlawat and Dadarwal (1996)
Bradyrhizobium sp.	ARS-39	*Cajanus cajan*	YELAA (Yeast extract, L-arabinose agar)	A wild strain of *Bradyrhizobium* sp. showed high intrinsic resistance to sodium azide. Nitrogenase activity of its spontaneous mutants showed higher resistance to NaN_3. The percentage of mutants showing hyperactivity decreased with increase in azide resistance level.	Madhubala and Gaur (1996)

Contd...

Table 1.1–Contd...

Organism	*Strain*	*Isolated from*	*Growth Condition*	*Finding*	*Reference*
R. leguminosarum	PDR-14, HIM-1	Root exudates of *Phaseolus vulgaris*	YEMA plates at 28°C	Out of three nod regulators tested, a pigenin and naringenin enhanced the growth of *R. leguminosarum* while 7 hydroxy coumarin was found inhibitory in initial stages. Root exudates of two strains of host cultivars also stimulated the growth of rhizobia. Chemostatic response towards these flavonoids at 10^5 M and 10^{-6} M concentration ranged from 3.4 to 4.5. But this ratio towards root exudates was low.	Pathak and Khurana (1996)
Rhizobium sp.	Sb-3, SrS 35	*Sesbania bispinosa, S. rostrata*	YEMA at 28±1°C	Cross nodulation studies using all isolates of *S. bispinosa* and from *S. rostrata* showed no host species specificity as all isolates nodulated both the species. Some of the stem nodule isolates from *S. rostrata* was found to be more effective on both the host species resulting in higher gains in plant dry wt. Selected Hup^+ and Hup^- isolates were evaluated for symbiotic performance on both host species up to 50 days of plant growth. *S. bispinosa* inoculated with (Hup^-) isolate Sb-3 and SrS 35 resulted maximum gain in plant dry wt. 2 Hup^+ strains *viz.* SrR 83 and SVR8 were better in symbiotic effectivity but were not superior to Hup^-.	Saini *et al.* (1996)
Rhizobium sp.	Ca-181	Chick pea (*Cicer arietinum*)	YEMA plates at 28±1°C for 7 days	The nodule number and grain yield of *Cicer arietinum* was higher in *Rhizobium* treated plants in comparison to control indicating improvement in nodulation by native rhizobia. *Rhizobium* was effective in wilt control and promotes nodulation.	Khot *et al.* (1996)
Rhizobium sp.	VRF-10, VRF-57	*Vigna radiata*	YEMA	Two bacteriocin producing strains of *Rhizobium* sp. were studied for their nodule competitive ability on host legume green gram when used as co-inoculants in different proportions of 2 strains. VRF57 was the effective strain and formed higher proportion of nodules. The effective strain VRF57 besides producing bacteriocin was having other factors involved in nodule competence.	Goel *et al.* (1997)

Contd...

Table 1.1–Contd...

Organism	*Strain*	*Isolated from*	*Growth Condition*	*Finding*	*Reference*
Rhizobium sp., *Pseudomonas* sp., *Bacillus* sp.	*Rhizobium* sp. (Ca-181, Ca-534, Ca-313) *Pseudomonas* sp. (RS-13, RS-16, RS16B1 and RS-28) isolated fromgreen gram	Chick pea (*Cicer arietinum* L.)	YEMGA (Yeast extract mannitol glutamate agar)	These rhizobacteria when used as seed inoculant for chick pea some showed initial stunting effect on root seedling followed by root growth stimulation where as a few even prevented seed germination. The root growth stimulatory bacteria showed wide range of antifungal activity against the plant pathogenic fungi. The bacteria with antifungal activity exhibited enhanced production of flavonoids/flavones like compounds in the seedling roots, indicating their role in induction of systematic resistance in the plants as well as their possible role in rhizobial attachment for nodulation.	Parmar and Dadarwal (1997)
Bradyrhizobium sp.	S-24, AC-2, CC-1021, G-20, Br-9038, Br-9038U	*Vigna radiata*, *Acacia nilotica*, *Cajanus cajan*, *Cyamopsis tetragonoloba*	YEM medium	Strain Br-9038U was constructed by random Tn5 mutagenesis. For this plasmid PCAM 111 loaded with mini Tn5 containing gus-A gene, expressed from a constitutive PTAC promoter was used. The constructed strain formed blue colonies on YEMA plates containing X-GICA. The characteristic of the strain used is that it forms brown colour nodules on the host plants. This method was developed to monitor the nodule occupancy of the GUS marked strain of *Bradyrhizobium* sp. It was quicker, accurate and economical for nodule occupancy studies of the Gus marked *Bradyrhizobium* and is therefore better for ecological studies for rhizobia.	Sharma *et al.* (1997)
Bradyrhizobium sp.	Strain: USDA-110 112,31, 76,123,94 Strains used: NC-1005, SEMIA-566, CPAC-15, 7, S-204, S-335, CB-1809		7 days at 28°C in liquid medium	Parental strain SEMIA-566 produced little mucus. 59% of these isolates showed a colony diameter larger than parental strain. SEMIA-566 and S-340 showed resistance to the levels of antibiotics and only two (S-204 and S-335) were able to grow in presence of all antibiotics. Local strains showed a pattern closer to *B. elkanii*. In contrast to *B. elkanii* reference strain, *B. japonicum* was able to nodulate the Soybean. 32 strains fell into intermediate groups between the species *B. japonicum* and *B. elkanii*. CB-1809 and CPAC-7 were classified as *B. japonicum*.	Boddey and Hungria (1997)

Contd...

Table 1.1–Contd...

Organism	*Strain*	*Isolated from*	*Growth Condition*	*Finding*	*Reference*
Bradyrhizobium japonicum	SSF-4, 5, 6,7 and 8	Silt loam black in soil from Porur, near Chennai (13 N 8° E) with pH-7.2	The bacteria grow in Glutamate mdium (Bergerson, 1961) for 5 days	The strains were inoculated individually or with the type strain USDA 110 (collected from USA) at 1:1 ratio. Nodule occupancy determined by immunofluroscence and dot immunoblot assay revealed that *in vitro* condition SSF is more competitive than USDA 110 where as others were less competitive. In red soil both SSF-8 and USDA-110 were equally competitive where as in black soil SSF-8 competed better than USDA-110 and produced more nodules. In black soil field trial, when inoculated alone occupied the majority of the nodules and enhanced nodule dry weight and shoot biomass. SSF-8 was more competitive when the strains were co-inoculated.	Palaniappan *et al.* (1997
R. leguminosarum	DGC-R1, R4, Hv-1, Hv-7, DGC-L1, DL2, VBBL-7,-18, 19 and VBK1	*Trifoilum repens* L.	Strains were grown in sherwoods medium	Enzymatic evidences supported that arabinose and ribose were metabolized through pentose phosphate, 2-oxoglutamate and 1-2 keto-3 deoxyarabinose pathways and mediated repression of phospho-fructo-kinase in *Rhizobium leguminosarum.*	Ghosh and Mandal (1998)
Bradyrhizobium sp.	S1,S2,S3, S4,S5,S6, S7,S8 and S9- S16, GB-2	*Cyamopsis tetragonoloba*	YEM medium	Strain GB-2 tolerated up to 85mM NaCl, S1-S5 all tolerant to 0.5 per cent NaCl, S6-S9 tolerant to 1.5 per cent NaCl and S10-S16 tolerant to 2.5 per cent NaCl. GB-2 showed a slow growth rate arriving at stationary phase 144 h after incubation. Salt tolerant showed rapid growth arriving at stationary phase 36 h after incubation. S2, S6, S7 and S16 showed an acid metabolic pattern. Salt tolerant strains gained resistance to chloramphenicol and Tetracyclins and significancly enhance their capacity to oxidize C-sources by increasing growth rate, exopolysaccharide production and involved in adhesion resulting in a greater adapting capacity to colonize unfavourable saline environments.	Barboza *et al.* (2000)

Contd...

Table 1.1–Contd...

Organism	Strain	Isolated from	Growth Condition	Finding	Reference
Rhizobium sp.	Mo6 (SM400) and PP9038 (Tc20+Cm200)	Mung bean collected from Deptt. of Microb. CCS Haryana Agri. University, HIsar		Role of proline dehydrogenase (PrfDH) in nodulation competitiveness of rhizobia infecting mung bean was examined. Tn5 mutagenesis of four *Rhizobium* strains resulted in proline prototrophs of two strains Mo6 and PP9038 (PH 9022) which were deficient in ProDH activity. ProDH deficient mutants affected different N_2 fixing parameters such as nodule number, nodule fresh wt., root dry wt., shoot dry wt. and total shoot nitrogen under sterilized condition. ProDH mutants deficient and wild type showed comparable nodule occupancy when used individually as inoculant. However ProDH deficient mutants were found to be poor competitor when co-inoculated in ratio with the wild type.	Chaudhury *et al.* (2000)
Rhizobium sp.	S-1230	Root nodules of *Acacia tortilis* grown in a Senegalese soil sample	YMA (Vincent, 1970) and Minimal mineral MB+7 medium	Growth parameters of C- limited continuous cultures were grown in presence and absence of 342 mM NaCl. Culturability subjected to the double stress of starvation and salinity was reduced and a high percentage of cells entered the viable but non culturable state. All the starved cultures were capable of re-growth when nutrients become available, thus showing that this strain can withstand long periods of nutrients deprivation in soil while maintaining the capacity for an active metabolism and a potential infectiousness towards an appropriate host.	Lippi *et al.* (2000)
Rhizobium sp.	CM-6, CC-6, BC-1	*Brassica campestris* var. toria)	CCM slants (combined carbon medium) 28±2°C	Out of 50 isolates only 15 were found positive for *in vitro* nitrogenase activity Maximum activity was observed for isolates CM-6 (238.24 n moles). Only four isolates possessed the ability to excrete ammonia. Out of all strains only three benefited the crop higher than control. CM6 having high nitrogenase activity could not benefit the crop yield. In contrast to CC6, BC-1 was found to give maximum benefit to the crop. There was also significant increase in the oil content of seeds as well as total oil yield of the crop due to bacterization with diazotrophs. Most efficient strains were found to be BC-1 in terms of increase in oil content.	Paul and Shende (2000)

Contd...

Table 1.1–Contd...

Organism	*Strain*	*Isolated from*	*Growth Condition*	*Finding*	*Reference*
Azorhizobium sp.		*Pterocarpus santalinus* L.		The effect of (IBA) Indole-3-butyric acid treatment at 50, 250, 500 and 1000μg ml^{-1} and the rhizobial inoculation on red sanders increased nodulation and biomass production. The stumps treated with 250μg IBA ml^{-1} showed higher growth, nodulation, biomass and nitrogen content over the other treatments. The combination of IBA at 250 μgml^{-1} and the rhizobial inoculation at 30 and 40 d, growth of root stumps gave maximum nodulation, biomass and 'N' content in plants.	Rajsekhar and Reddy (2000)
Rhizobium and *Bradyrhizobium*		*Acacia auriculiformis*	YEMA slants grown at 28°C and stored at 4°C	Of the 17 strains from different sources tested 10 were able to produce visible nodules and their nodulation frequency was also different. Strains isolated from other species of *Acacia* were not compatible with *A. auriculiformis*. Nodule number appeared to be a poor indicator of the nitrogen fixing potential of strains.	Sharma and Ramamurthy (2000)
Rhizobium meliloti	M-15	*Cicer arietinum*	M-15 strain grown for 24 h. Culture diluted to 1: 50-1: 100 in RMM with out N-sources. RMM at 5-10mM, Naringenin (100mM) was the inducer.	Different sources of combined N were added at a concentration of 5-100mM to *Rhizobium* minimal medium (RMM). All the N sources except KNO_3 and KNO_2 inhibited the expression of *nod* genes. Sodium glutamate was the most potent inhibitor, followed by NH_4NO_3, Urea, NH_4Cl and $(NH_4)_2SO_4$. Sodium glutamate at 25 mM repressed the expression of *nod* ABC by 85 per cent as compared to 93, 29, 54 and 49 per cent at 100mM of NH_4No_3, NH_4Cl, $(NH_4)_2$ HPO_4 respectively. KNO_3 and KNO_2 did not affect the expression of *nod* genes even at higher concentration. These results indicate that combined N affects the nodulation process by inhibiting the expression of *nod* genes.	Sharma *et al.* (2000)
Rhizobia		*Acacia sp.*	YEMA all strains stored in 50% (v/v) glycerol at (–4°C)	The rhizobia were very diverse with respect to their cross-nodulation patterns as well as their physiological and biochemical properties. Phenotypic characteristics showed that isolates could fit into four clusters below boundary level of 0.85. Some strains grew at pH ranging from 4 to 9, tolerated a high salt concentration (3 per cent NaCl) and grew at a maximum temperature between 35 and 40°C.	Zerhari *et al.* (2000)

Contd...

Table 1.1–Contd...

Organism	*Strain*	*Isolated from*	*Growth Condition*	*Finding*	*Reference*
Sinorhizobium sp.		*Sesbania cannabina*	Nodules immersed in 95 per cent ethanol for 10 sec, sterilized in 0.1% (w/v) $HgCl_2$ for 5 min, crushed and streaked on YEMA with 0.0025 per cent Congored at 28°C for 3 days.	Eighteen rhizobial isolates nodulating *S. cannabina* were isolated from sugar cane rotation fields in southern Taiwan. The taxonomy of these isolates was investigated using polyphasic approach, including phenotypic characteristics, banding pattern of total proteins from SDS-PAGE, genomic fingerprint patterns from pulse field gel electrophoresis, amplified 16s rDNA restriction analysis (ARDRA), 16s rRNA gene sequencing and Nif H gene sequencing based on results of Protein-banding patterns, ARDRA and phenotypic characters. Seven isolates belonging to *Sinorhizobium* lineage based on phylogenetic analysis, and 11 isolates belong to *Agrobacterium-Rhizobium galegae* were more closely related to *R. huautlense.*	Chen and Lee (2001)
Bradyrhizobium sp.	SEMIA-566 SEMIA-5079, 5080, 587, 5019	Soybean (*Glycine max* L. *Merril*)	YEMA containing cycloheximide (50µg/ml) at 28°C	Fifty eight isolates showed slow growth rates and alkali reaction in medium containing mannitol as C source. Twenty had fast growth rates and an acid reaction in medium containing mannitol as C source. Most isolates did not tolerate pH-4.5 or high temperature (40°C). Very few isolates shared similar protein and lipopolysaccharides profiles. Therefore high level of diversity was detected. Some of isolates showed outstanding symbiotic performance.	Chen *et al.* (2002)
Sinorhizobium and *R. leguminosarum*		*Medicago sativa* L. and *Trifolium alexandrium*	YEMA for 3 days suspended in 10% sucrose solution and inoculated seed by submerging them for 30 min.	From 125 locations represented by 14 major soil types and agroclimates in country, *S. meliloti* occurred in almost all the soils while *R. leguminosarum* bv. *trifolii* occurred in very few soils. These nodule bacteria able to survive under extremes of the environment. Inoculation of these two legumes with their respective nodule bacteria improved the yield.	Gaur *et al.* (2002)

Contd...

Table 1.1–Contd...

Organism	*Strain*	*Isolated from*	*Growth Condition*	*Finding*	*Reference*
Rhizobium sp.	ICC4948 and ICC 5003	Chick pea, soils of CCS Haryana Agri. University, Hissar	YEMA	Out of 350 cultures, in 110 rhizobial isolates, both from HN (high nodulation) and LN (low nodulation) plant variants, proteins of more than 45kDa were present and slight variation in the presence of proteins of lower mol.wt. in different isolates was observed. In HN selection all the 13 major bands were present while in case of LN selection varieties two major bands (8th and 9th) were absent. All the isolates from the HN and LN selections were able to nodulate plants of their respective host. Isolates selected from HN or LN selections when cross infected to LN or HN plants also formed nodules.	Chaudhary *et al.* (2002)
Bradyrhizobium sp.	MR125S2-SMr8	*Vigna aconitifolia*	YEMA	Fifty strains isolated were tested both for swarm size and nodulation. Majority of the isolates (74 per cent) showed large stable swarm when compared to the parent 26 isolates with lost nodulation ability. The mutants whose swarm size was larger produced pinkish nodules. In culture medium the parent and mutants preferred KNO_3 as sole N source over yeast extract and did not grow on media containing glucose or maltose. When co-inoculated with the parent, the mutant had generally higher nodule occupancy.	Chakraborty *et al.* (2003)
Bradyrhizobium japonicum	R-25B, IRJ-2180 A	Soybean (*Glycine max*) from moist Savanna of Northern nigeria		R25B indigenous strain and mixture of R-25B+IrJ2180A were experimented. About 34 per cent of nodules were formed by mixture of introduced strains and only about 24 per cent nodule was observed in R25B which did not influence biomass and grain yield. The grain yield was not higher from the uninoculated soybean showing the of competitiveness among the introduced *Rhizobium* strains and native rhizobial population.	Okogus and Sanginga (2003)
Rhizobium sp.		Soil from *Pennisetum glaucum Sorghum bicolor, Vigna uniculata, Arachis hypogea* grown fields		Soils from these legume rotation crops were collected and analyzed for microbial diversity in rhizoplane communities. Total numbers of bacterial species were observed maximum at 14 days than at 7 days after sowing.	Alvey *et al.* (2003)

Contd...

Table 1.1–Contd...

Organism	*Strain*	*Isolated from*	*Growth Condition*	*Finding*	*Reference*
R. tropici	CIAT-899 PRF-81 H-20, H-12	Soil of Parana state, Brazil		Nodule occupancy increased from an average of 28% in the first experiment with inoculation of rhizobia, and 56 per cent after four inoculations. The strain H-12 and H-20 showed an increase of 437 and 465 kg ha^{-1}. Inoculations with superior strains also resulted in yield increases. DGGE analysis of soil extracts showed that the massive inoculation apparently did not affect the composition of bacterial community.	Hungria *et al.* (2003)
R. legumino-sarum	LX1-LX57 and 175 P4	*Pisum sativum* cv. tapper	YEM broth and growing on a gyratory shaker (150 rev/min.) at 28°C for 72h	The yield and N nutrition of pea inoculated with a superior *Rhizobium* strain was significantly enhanced by an apparently compatible AMF species. Compared to *Rhizobium* treatment, an apparently incompatible AMF sp. reduced the performance of an effective *Rhizobium* strain. In general co-inoculation treatments with effective *Rhizobium* strains and a compatible AMF species produced the best results. Total shoot dry matter, P-content and yield was enhanced.	Xavier and Germida (2003)
Rhizobium sp.	S1-S5 and P1-P5	Wild legumes of Shawara and Padubidri, Karnataka coast	YEMA	The strains isolated from four different cultivated legumes like cow pea, green gram, black gram and horse gram. Inoculation of S5 resulted in the highest increase of shoot biomass in cowpea while inoculation of P1-P5 induced the highest shoot biomass; P2 alone induced higher shoot biomass against uninoculated controls of horse gram, green gram and black gram.	Arun and Sridhar (2004)
Sinorhizobium fredii	USDA-205	*Glycine max* CV.	YEMA at 28°C	Twenty five soybean rhizobia were isolated from the soil above an altitude of 1500 m and all identified to be *S. fredii.* Their genetic diversity was characterized by 16S-23S rDNA, RFLP and RAPD analysis. All tested strains produced 2.1 kb 16S-23S rDNA ITS fragment. Tested strain could be differentiated into 11 ITS genotypes.	Chen *et al.* (2004)
Bradyrhizobium sp.	S6-1, 2, 3	Soybean	YEM broth medium	Effect of *Bradyrhizobium* strains on dry matter yield, nodulation and seed yield of soybean varieties was studied. A significant positive effect on growth, nodule number and yield of soybean was obtained. Strains S62 and S63 were more effective than strains S61. The protein content of seeds also increased after inoculation.	Egamber-diyeva *et al.* (2004)

Contd...

Table 1.1–Contd...

Organism	*Strain*	*Isolated from*	*Growth Condition*	*Finding*	*Reference*
Rhizobium sp.		From different soils	YEMA at 28°C for 4-5 days thermal shock: 45-55° C, pH-4.5-6.5, NaCl: 10000-100000 ppm	Out of 62 samples 9 showed thermal tolerance (55°C), 8 showed acidity tolerance (pH-4.5) and 11 showed tolerant to 40000 ppm NaCl. Some strains isolated from *Sesbania formosa*, *A. farnesiana* and also *D. sissoo* were adapted to grow on pH 12. Some strains tolerated salt concentration up to 5 (50,000 ppm).	Gupta *et al.* (2004)
Rhizobium, *Bradyrhizobium* and *Agrobacterium*	CA-18	Chick pea, mung bean, Pea and Sirato	YEMA Supplemented with Bromothymol blue and Congo red 28±2°C	Preliminary characterization of these isolates was done on the basis of plant infectivity test, acetylene reduction assay, C-source utilization, Phosphate solubilization, phytohormones and polysaccharide production. The plant infectivity test and acetylene reduction assay showed effective root nodule formation by all isolates on their respective hosts. All strains showed homology.	Hameed *et al.* (2004)
Rhizobium sp.		(*Trifolium ambiguum* M.B) perennial rhizomatous forage legumes	YM agar at 24°C for 4 days. Fresh YMA broth mixed with 80% glycerol solution (1:1) and preserved at –70°C for long time.	All 128 rhizobia strains were isolated from 4 sites from Kura clover. White colour clover rhizobia were finger printed using repetitive extragenic palindromic PCR. The nodulation specificity and phenotypic diversity of a subset of 13 isolates was determined. Percentage of similarity among 13 isolates ranged between 28 and 92 per cent. Three strains formed effective nodules on both Kura clover and White clover.	Beauregard *et al.* (2004)
B. japonicum and *B. elkani*	CPAC-7, 15	Soybean	YEMA, broth cultures applied to sterilized peat (pH raised previously with $CaCo_3$) to reach about 50% moisture, the mixture allowed to mature at room temperature for 30 days reaching $1.8x10^8$ to $2.0x10^9$ cells per gram	Maximum nodule number and dry weight of nodules observed in treatments CPAC-15, CPAC-7. Grain yield also increased in treated ones. All the strains inoculated in the first year presented a successful establishment in soybean nodules with occurrence of 100 per cent. In second year formed 90-97 per cent without inoculation, 45, 30, 19 and 2 per cent of the nodule occupancy observed in un-inoculated plots.	Mendes *et al.* (2004)

Contd...

Table 1.1–Contd...

Organism	*Strain*	*Isolated from*	*Growth Condition*	*Finding*	*Reference*
Rhizobium sp.		*Acacia* plant from 14 locations, 4 sites represent semi arid, sub-moist, moist and sub humid	YEMA plates 28°C for 2-14 days	Twenty four isolates were recovered from the root nodules of trap host species and from excavated nodules. Isolates were differentiated by growth rate and colony morphology. 68.5 per cent were fast growing acid producing rhizobia. 25.3 per cent were slow growing alkali producing rhizobia and fast growing, alkali producing (2.9 per cent) and slow growing acid producing strains (3.3 per cent). All isolates showed four types of colonies *e.g.* watery translucent, white translucent, dull glistening and milky (curdled) type.	Wolde-meskel *et al.* (2004)
Rhizobium sp.				During winter seeds (pea, green gram) treated with the liquid broth of *Rhizobium* containing 1.3x10^7 rhizobia/ml showed 71.4 per cent and 78.8 per cent increase in dry shoot wt. and grain yield of pea over control. In green gram grain yield was found to be increased to 60.8 per cent higher.	Barik and Talukdar (2004)
Rhizobium sp.	LR-1, BR-12, BR-8 and AR-10, PR-16, AR-1, GR-12, WR-10, GM-16	Green gram and black gram form six districts of Assam		Experimental result showed that AR-1 was most effective in terms of nitrogenase activity and PR-16 was the poorest nitrogen fixer. Among two crops green gram was more responsive to rhizobial inoculation than black gram and among seasons Kharif season was found to be most suitable for the cultivation of the pulses over Rabi season. Maximum yield was obtained with the strain AR-1, increased the grain yield by 35.16 per cent in black gram and 31.41 per cent in green gram in Kharif and 33.89 per cent and 32.43 per cent in black gram and green gram in Rabi season respectively.	Azad, (2004)
Rhizobium sp., PSM (*Bacillus polymyxa*)	TAL 1000	Groundnut from Manipur		*Rhizobium* and PSM inoculation increased the pod yield over control (992kg/ha) by 11.9 per cent individually and by 30 per cent in dual combination. When inoculated with lime @2.5 t/ha the percent increase in pod yield was less compared to inoculated plots. The results suggest that liming was not effective in increasing the efficiency of either the rhizobial strain TAL 1000 or the PSM strain. *Bacillus* inoculated alone increased pod yield 39 per cent by increase in pH of the soil which increased the available P content and exchangeable calcium content of the soil.	Raychaudhri *et al.* (2004)

Contd...

Table 1.1–Contd...

Organism	*Strain*	*Isolated from*	*Growth Condition*	*Finding*	*Reference*
Rhizobium sp.		Groundnut from Manipur		The yield attributes, number of pods, seed wt., number of nodules and nodule mass were found highest under *Rhizobium* + FYM treatment followed by FYM+15kg N. Yield attributes under *Rhizobium* treatment was also superior to control.	Singh *et al.* (2004)
Rhizobium sp.	NC-92	Groundnut from Manipur		Result revealed that the native rhizobial population was not much effective in comparison to the applied strain which enhanced the growth and pod yield. The pod yield showed 8.8 per cent increase over control. *Rhizobium* and PSM in combination can save 10 kg of nitrogen and and 10 Kg of P_2O_5 per hectare and increase the yield to maximum by 37.5 per cent over the recommended dose of N, P, K (20 kg N, 40 kg P_2O_5 and 30 kg K_2O).	Ngachan *et al.* (2004)
Rhizobium sp.	NC-92	Groundnut acid hill utisol of Manipur		NC-92 strain was found effective in increasing the pod yield of groundnut by a maximum 27.5 per cent in an acid hill soil of Manipur. The efficiency of the strain increased by liming in furrows @500 kg and 1000kg/ha and increase the pod yield by 88 and 126 per cent respectively over control.	Raychaudhri *et al.* (2004)
Sinorhizobium meliloti	L5-30 (Φ L54)	*Medicago sativa*		Using Tn5 mutagenesis of lysogenic *Sinorhizobium meliloti strain* L5-30 three neomycin resistatant transconjugants differing in phage resistant profiles were isolated. Two of them increased the dry wt. of plants and were capable of establishing root nodules. Where as the third one was effective. The bacterium phage interaction did not have observable consequences in *Medicago sativa* and *S. meliloti* symbiosis because it did not affect the number of nodules on *M. sativa* or plant dry wt.	Kowalski *et al.* (2004)
R. japonicum	SB-16	Soybean	YEM broth at 30±2°C	*Rhizobium* inoculant as seed treatment was found to be much more effective than seed inoculation. It increased the nodule number by 74.40 per cent, pod number by 110.18 per cent, dry matter by 67.48 per cent and the grain yield by 124.8 per cent over similar soil implant at 2.5 cm (along with the seed).	Chakradhar and Jauhri (2004)

Contd...

Table 1.1–Contd...

Organism	*Strain*	*Isolated from*	*Growth Condition*	*Finding*	*Reference*
Rhizobium sp.	CRM-8, CRM-11, CRM-15, CRM-9	Collected from greengram of different locations of Tamil nadu		Sixteen isolates were obtained from green gram. Growing the isolates on osmotic substrates PEG (polyethylene glycol) the osmotic pressure level was enhanced in the cells. Out of sixteen CRM-11 and CRM-15 tolerated higher OP levels and the ability of these isolates to grow under soil moisture levels of even 5 per cent was observed. Hence such isolates which are able to survive under high OP levels may tolerate less water potentials of soils.	Kumutha *et al.* (2004)
Rhizobium sp.	CRR-6, NIT, FACT, UAS-Bangalore	Redgram (Vamban-1)		Field expt. was conducted to study the influence of commercially available carrier based rhizobial inoculants and liquid rhizobial inoculant on nodulation and grain yield. Carrier based rhizobial culture (RR-6) had given maximum nodules (14 nodules/plant), plant dry wt. and grain yield (875 kg/ha) which was 28.48 per cent higher over uninoculated control (681kg/ha). The inoculant had given a grain yield of 763kg/ha and but other commercially available culture where as liquid inoculant had recorded a grain yield of 735kg/ha only.	Gunasekarn *et al.* (2004a)
Rhizobium+ PGPR+ PSB	COC-10	Blackgram		A field trial was conducted to study sites synersism between *Rhizobium*, PGPR (*Pseudomonas*) PSB (*Bacillus megaterium*) on nodulation and yield of black gram. Inoculation of *Rhizobium* alone had given a grain yielded of 655kg/ha whereas control (uninoculated) had recorded 472 kg only. Combination of all three organisms (*Rhizobium*+ PGPR+PSB) had recorded the maximum nodules, plant biomass and grain yield (760 kg/ha) which was 61% higher than individual inoculation of *Rhizobium* alone.	Gunasekarn *et al.* (2004b)
Rhizobium sp.		Soil samples from Bhiwani and Hissar in Haryana	YMA supplemented with 25μg/ml congored	Amplified and cloned PCR products from total DNA isolated from the soil samples, the total diversity within the resulting clone libraries was higher than that recovered from trap plants but differed depending on the PCR protocols and primer used. However some plant of selected genotype could not be obtained using the community approach, probably due to variable detection limits and limited clone library.	Sarita *et al.* (2005)

Contd...

Table 1.1–Contd...

Organism	*Strain*	*Isolated from*	*Growth Condition*	*Finding*	*Reference*
Mesorhzobium sp.	UPM-Ca 36+	Chickpea (*Cicer arietinum* L.) alkaline soil	Maintained on Tryptone yeast agar (TYA) slants 20-37°C and preserved 4°C	Tolerance to stress (temperature and pH) was evaluated at pH-5 to 9, heat shock-46°C and 60°C. SDS-PAGE analysis revealed qualitative and quantitative differences when isolates were submitted to temperature stress. A 60 kDa protein was over produced by all isolates were more tolerant to 20°C than to 37°C. The isolates more tolerant to temperature stress showed lower symbiotic efficiency.	Rodrigues *et al.* (2006)
Mesorhizobium sp.	Ca-181	*Cicer arietinum*	YEMA medium	The symbiotic effectiveness of *Mesorhizobium* sp. *cicer* strain was further improved at 80 d of plant growth on co-inoculation with 6 *Bacillus* strains. The shoot dry wt. ratio varied from 1.62 to 1.74 times those of *Mesorhizobium* inoculated treatments. Nitrogenase activity was increased in *Mesorhizobium* treated plant than that of control.	Sivaramaiah *et al.* (2007)
Rhizobium sp.	R-39, E-11 obtained from Elosmestari Ltd, Finland		YM agar medium	The potential use of legume bacteria in non-legumes by inoculating different rhizobial strains on to mixture of six botanically different non-nitrogen fixing plant species was analyzed. Inoculation with certain rhizobial strains increased biomass. A concentration of level 10^4 c.f.u (colony forming units) ml^{-1} proved to be the best for successful growth. *In vitro* study investigating some rhizobial strains have a capability to dissolve the fungal mycelium at the initial stage.	Hossain and Martensson (2008)
Bradyrhizobium sp.	C-145		Grown in YEM medium for 4-6 days at 28°C	In field experiments, only 9 per cent of total nodules were formed by bacteria inoculated by direct coating of seed, whereas 78 per cent of nodules were formed by bacteria inoculated in the furrow at seeding. In each case, the other nodules were formed by indigenous strains or by both strains indicating a positional advantage of native rhizobia or in-furrow inoculated rhizobia for nodulation in plant.	Bogino *et al.* (2008)

of rhizobia. Rhizobia are able to produce acid or alkali on YEMA plates. Based on this criterion, the fast growing are acid producers while the slow growing are grouped as non acid producers. The results have shown that fast growing pea and bean rhizobia come under a common species name *Rhizobium*. *Rhizobium* is host specific and their role varies for plant to plant. Induction of nodule formation by rhizobia involves interactions between two symbiotic partners including multiple regulatory signals between the bacterium and plant to coordinate expression of gene sets in both partners (Peters and Verma, 1990). These signals include plant factors that induce rhizobial nodulation (*nod*) genes, rhizobial nod factors that induce further reactions and are essential for development of nodules, hormones and other regulatory factors involved in symbiotic nitrogen fixation. For nodulation in plant the first step is secretion of flavonoid (a class of phenolic compounds), chalcone and conjugated isoflavonoid signals by host. Induction of the nod gene operon depends upon the structure of the phenolic compounds.

N_2 Fixation Mechanism and the *nif* Genes

The genes responsible for nitrogen fixation called *nif* genes include *nod* genes (for nodulation) responsible for nod factor synthesis, nodule development and also synthesis of nitrogen fixing apparatus called as *nif* and *fix* genes. The expression of nitrogen fixation genes is controlled by cascades of organized regulatory genes. Their role of action enables bacteria to sense optimal environmental conditions required for nitrogen fixation and to transmit this information to the level of gene expression.

Molecular Protocols for Increasing Nodulation and Productivity

Present era holds the crown of numerous discoveries regarding the molecular genetics of non-legumes and the extension of legume-*Rhizobium* symbiosis to leguminous as well as non-leguminous crops. Plazinski *et al.* (1985) successfully transferred gene associated with root hair curling (*hac* genes) from *R. trifolii* into a derivative of *R. trifolii* on plasmids pKT 230 and pRK 290. The root hair curling character could then be expressed on infection of rice seedlings. Okon, (1985) suggested that genetic modification of diazotrophs helps them to acquire the genes for the synthesis of pectin degrading enzyme that could facilitate the root invasion. Rolfe and Bender (1990) described further developments to evolve rhizobia for nodulation of non-legume. Kenedy *et al.* (1997) reported that an effective associative system of cereals could be assisted by the development of genetic tools based on the application of the *lacZ* and *gusA* fusion with the promoters of genes associated with the nitrogen fixation for leguminous and non-leguminous plants.

Mass Production Methods of *Rhizobium* Biofertilizer and Field Application Protocols

Thirty days old of selected legume plants were uprooted washed in distilled water and the well-formed, healthy pinkish nodule on the tap roots were carefully cut out. The nodules were immersed in 95 per cent ethanol for 10 sec, sterilized for 5 minutes in 0.1 per cent acidified mercuric chloride ($HgCl_2$ $^{-1}$g L^{-1}, Conc. HCl- 5ml.L^{-1}) and washed six times with sterile distilled water to get rid of the chemical (Chen and Lee, 2001). Each nodule was crushed using a sterile glass rod in a aliquot of sterile distilled water. Serial dilutions of the suspension were made and an aliquot of appropriate dilution was plated on Yeast-Extract Mannitol Agar medium (YEMA) and incubated at 28±2°C for 4-7 days (Bogino *et al.*, 2008). Distinct colonies were picked up and transferred to agar slants for further purification. Confirmation of the Rhizobia was ascertained by streaking on YEMA medium supplemented with congored 0.025 per cent (w/v) (Hameed *et al.*, 2004). The *Rhizobia* stand out as white, translucent colonies (Subba Rao, 1977). One week old Rhizobial colonies kept on YEM agar media (1.5 per cent agar) were used for preparation as inoculants. For this purpose loops of the respective colonies were inoculated in sterile YEMA medium containing K_2HPO_4-0.5g/l, $MgSO_4.7H_2O$-0.2g/l, NaCl-0.1g/l, Yeast Extract-0.4g/l, Mannitol-10.0g/l and pH 7.8, and the concentration of rhizobial suspension was 10^5 C.F.U/ml. Strains were routinely maintained on YEMA slants and kept at 4°C (Castro *et al.*, 1997).

For field experiments, healthy seeds of selected plants were surface sterilized with 95 per cent ethanol for 3-5 minutes followed by rinsing six times with sterile water. The seeds were then steeped in the respective rhizobial suspension. Seeds treated with distilled water were used as the control. All seeds were mixed gently in shade to bring the seeds and bacteria into close contact for 30 min.

Inoculation vs. Yield of Crops

In field demonstrations, the increases in yield of leguminous crops due to *Rhizobium* inoculation were as follows:

In India–20.36 per cent chickpea, 19.22 per cent pegion pea, 18.5 per cent mungbean, 14.29 per cent blackgram and 11.78 per cent lentil (Khurana *et al.*, 1997).

In Bangladesh–Lentil- 10-99 per cent, groundnut-8-41 per cent, soybean-18-71 per cent, chickpea-11-70 per cent, cowpea-21-26 per cent (Sattar *et al.*, 1997).

In Myanmar–chickpea (31-46 per cent), groundnut 29-44 per cent, blackgram-27-40 per cent, and butter bean-48 per cent (Thein and Hein, 1997).

In Nepal–lentil-10-80 per cent (Bhattarai *et al.*, 1997).

In Thailand–legume crops 43.5-12.6 per cent; In Vietnam- ground nut 3-47 per cent (Kongngoen *et al.*, 1997).

Future Prospective

Quantitative understanding of the ecological factors that control the future and performance of BNF systems in crop fields is essential for promotion and successful adoption of these technologies. More information about molecular genetic bases of rhizobial nodule symbiosis will improve understanding of root behavior towards microbes and provide information about the evolutionary relationships between plants process. Research in plant genes and molecules involved in *Rhizobium* is still not fully known. The interactions between host plants cells and micro-symbionts that lead to the expression of the various nodulin genes are a front line area of research today.

To induce most of the nodulins essential for functional symbiosis or signal exchanges between plant cells and bacteroids is required. Thus nodulation symbiosis and constantly evolving molecular biology techniques are now opening the way to identifying plant molecules that play an essential role in mechanisms involved in symbiotic interactions during root colonization by bacteria resources, such as rhizobial mutants are particularly valuable for such study, including search of signals and molecular approaches of the genes essential for the symbiosis. Research for the coming decades is to elucidate the various signals and signal transduction pathways operating during the nodule organogenesis and the regulation of expression of *nif* gene and *nod* gene cascades and their host specificity. Many aspects of the molecular signaling between rhizobia and leguminous plant have been already reported. Rhizobial lipochitin oligosaccharides (LCOs) that have been shown as the major determinant of the host specificity of the nodulation. It is also needed to understand the mechanisms of N gains and losses, and to identify and refine appropriate soil and crop management practices to maximize soil N and legume BNF in crop fields.

Conclusion

Biological nitrogen fixation is an important process in legume farming systems because it is an inexpensive source of nitrogen for increasing the productivity of crops. The nitrogen fixing biological systems such as *Rhizobium* were not able to adapt well in different ecological systems. So region specificity is important for development of rhizobia in addition to ecologically compatibility as well as fixing considerable amount of nitrogen. Besides nitrogen fixation, the *Rhizobium* biofertilizer also contribute the growth promoting substances. It is evidently clear now that the application of biological fertilizers were greatly involved in the accumulation of soil enzymes, which directly reflects on the soil fertility index. Thus use of biofertilizers in general, in consortia or even singly for specific crops, would reduce the cost of chemical fertilizers involved in crop production. The effective utilization of biological fertilizers for crops will not only provide economic benefits to the farmers but also improve and maintain the soil fertility and sustainability in the natural soil ecosystem.

References

Ahlawat, O. P. and Dadarwal, K. R. (1996). Bacteriocin production by *Rhizobium* sp. *cicer* and its role in nodule competence. *Indian Journal of Microbiology*, 36: 17-24.

Alvey, S., Yang, C. H., Buerkert, A. and Crowley, D. E. (2003). Cereal/legume rotation effects on Rhizosphere bacterial community structure in West African solis. *Biology and Fertility Soils*, 37: 73-82.

Arun, A. B. and Sridhar, K. R. (2004). Symbiotic performance of fast growing rhizobia isolated from the coastal sand dune legumes of west coast of India. *Biology and Fertility Soils*, 40: 435-439.

Azad, P. (2004). Screening of native *Rhizobium* strains for sustainable crop production in pulse-rice cropping system. In *Biotechnology in Sustainable and Organic Farming* (Eds. A. K. Yadav, S. R. Chaudhary and N. C. Talukdar) pp-244-248. Shree Publishers and Distributors, New Delhi.

Barboza, F., Correa, N. S. and Rosas, S. B. (2000). Metaboilc and physiological characteristics of salt-tolerant strains of *Bradyrhizobium* spp. *Biology and Fertility Soils*, 32: 368-373.

Barik, J. and Talukdar, N. C. (2004). Seasonal variation in response of pea (*Pisum sativum*) and Green gram (*Vigna radiata*) to *Rhizobium* inoculation in Assam. In *Biotechnology in Sustainable and Organic Farming* (Eds. A. K. Yadav, S. R. Chaudhary and N. C. Talukdar) pp-240-243. Shree Publishers and Distributors, New Delhi.

Bhattarai, S., Maskey, S. L., and Karna, R. L. (1997). On-farm experiments on rhizobial inoculants in Nepal. *In Extending Nitrogen Fixation Research to Farmers fields. Proceeding of an international workshop on Managing Nitrogen Fixation in the cropping systems of Asia.* 20-24 August 1996. ICRISAT Asia Centre, India. (Eds. O. P. Rupela, C. Johensen and D. F. Herridge. ICRISAT, Pantencheru, India). pp. 235-242.

Biswal, A. K. and Mishra, A. (1992). Selection of *Rhizobium* as biofertilizer for Greengram cultivation under acid and alkaline conditions. In *Biofertilizer Technology Transfer* (Ed. L. V. Gangawane) pp-121-126. Associated Publishing Co. New Delhi.

Beauregard, M.–S., Zheng, W. and Seguin, P. (2004). Diversity of *Trifolium ambiguum*-nodulating rhizobia from the lower Caucasus. *Biology and Fertility Soils*, 40: 128-135.

Boddey, L. H. and Hungria, M. (1997). Phenotypic grouping of Brazilian *Bradyrhizobum* Strains which nodulate Soybean. *Biology and Fertility Soils*, 25: 407-415.

Bogino, P. Banchio, E., Bonfiglio, C. and Giordano, W. (2008). Competitiveness of a *Bradyrhizobium* sp. strains in soils containing indigenous Rhizobia. *Current Microbiology* 56: 66-72.

Chakraborty, M., Khan, H. and Ramkrishna, K. (2003). Symbiotic properties of hypermotile mutants of *Bradyrhizobium* sp. (*Vigna aconitifolia*). *Indian Journal of Microbiology*, 43: 101-105.

Chakradhar, T. and Jauhri, K. S. (2004). Soil implant inoculants for increased crop production. In *Biotechnology in Sustainable and Organic Farming* (Eds. A. K. Yadav, S. R. Chaudhary and N. C. Talukdar) pp-214-217. Shree Publishers and Distributors, New Delhi.

Chaudhary, S., Dudeja, S. S., Sharma, H. R. and Khurana, A. L. (2000). Competitiveness of proline dehydrogenase of mungbean Rhizobia. *Indian Journal of Microbiology*, 40: 199-202.

Chaudhury, P., Khurana, A. L. and Dudeja, S. S. (2002). Heterogenecity of Rhizobia isolated from Chickpea nodulation variants. *Indian Journal of Microbiology*, 42: 195-200.

Chen, W., Huang, Q. and Xiong, X. (2004). Distribution and biodiversity of soybean rhizobia in the soils of Shennongjia forest reserve, China. *Biology and Fertility Soils*, 40: 306-312.

Chen, L. S., Figueredo, A., Villani, H., Michajluk, J. and Hungria, M. (2002). Diveristy and symbiotic effectiveness of rhizobia isolated from field-grown soybean nodules in paraguay. *Biology and Fertility Soils*, 35: 448-457.

Chitriv, A. J. and Chavan, O. G. (1992). Efficacy of different methods of inoculating Groundnut seed with *Rhizobium*. In *Biofertilizer Technology Transfer* (Ed. L. V. Gangawane) pp- 127-132. Associated Publishing Co. New Delhi.

Chen, W. M. and Lee, T. M. (2001). Genetic and phenotypic diversity of rhizobial isolates from sugarcane-*Sesbania cannabina*- rotation fields. *Biology and Fertility Soils*, 34: 14-20.

Chopde, S. T., Kolte, S. O. and Khan, Y. D. (1993). Effect of BGA and *Rhizobium* inoculation on wilt sensitive and resistant varieties of chickpea (*Cicer arietinum*). *Phykos*, 32:115-120.

Dahiya, D. R.., Hooda, R. S. and Hooda, I. S. (1992). Effect of inoculation, nitrogen and phosphorus levels on summer soybean forage. In *Biofertilizer Technology Transfer* (Ed. L. V. Gangawane) pp-101-105. Associated Publishing Co. New Delhi.

De Bruijin, F., Jing, Y. and Dazzo, F. (1995); Potential and pitfalls of trying to extend symbiotic interactions of nitrogen fixing organisms to presently non-nodulated plants such as rice. *Plant and Soil*, 174: 225-240.

Earl, C. D. and Ausubel, F. M. (1993). The genetic engineering of nitrogen fixation. *Nutrition Review*, 41: 1-6.

Egamberdiyeva, D., Quarshieva, D. and Davranov, K. (2004). Growth and yield of soybean varieties inoculated with *Bradyrhizobium* spp. in N-deficient calcareous soils. *Biology and Fertility Soils*, 40: 144-146.

Ghosh, A. C. and Mandal, N. C. (1998). Metabolism of pentose sugars in *Rhizobium leguminosarum bv. trifolii*. *Indian Journal of Experimental Biology*, 36: 1056-1057.

Gour, Y. D., Rewari, R. B., Bhatnagar, R. S., Narayan, K. P. and Sen, A. N. (2002). *Sinorhizobium meliloti* and *Rhizobium leguminosarum bv. trifoliii* in Indian soils and their inoculation response. *Indian Journal of Microbiology*, 42: 23-28.

Goel, A. K., Sindhu, S. S. and Dadarwal, K. R. (1997). Nodule competence between bacteriocin producing effective and in effective *Rhizobium* strains of *Vigna*. *Indian Journal of Microbiology*, 37, 51-52.

Gunasekaran, S., Balachandar, D. and Mohanasundaram, K. (2004) (a). Field evaluation of liquid and carrier based Rhizobial inoculants in Redgram. In *Biofertilizer Technology*. (Eds. S. Kannaiyan, K. Kumar and K. Gobindarajan) pp. 265-268. Scientific Publishers, Jodhpur, Rajasthan.

Gunasekaran, S., Balachandar, D. and Mohanasundaram, K. (2004) (b). Studies on synergism between *Rhizobium*, plant growth promoting rhizobacteria (PGPR) and phosphate solubilizing bacteria in blackgram. In *Biofertilizer Technology*. (Eds. S. Kannaiyan, K. Kumar and K. Gobindarajan) pp. 268-272. Scientific Publishers, Jodhpur, Rajasthan, India.

Gupta, S. B., Tedia, K., Singh, A., Anurag, Lakpale, N., Pal, A. and Chhonkar, P. K. (2004). Formulation of stress tolerant bioinoculants for rainfed cropping system. In *Biofertilizer Technology*. (Eds. S. Kannaiyan, K. Kumar and K. Gobindarajan) pp. 256-264. Scientific Publishers, Jodhpur, Rajasthan.

Hameed, S., Yasmin, S., Malik, K. A., Zafar, Y. and Hafeez, F. Y. (2004). *Rhizobium*, *Bradyrhizobium* and *Agrobacterium* strains isolated from cultivated legumes. *Biology and Fertility Soils*, 39: 179-185.

Hossain, M.S. and Mårtensson, A. (2008). Potential use of *Rhizobium* spp. to improve fitness of non-nitrogen-fixing plants. *Soil and Plant Science*, I: 1-7.

Hungria, M., Campo, R. J. and Mendes, I.C. (2003). Benefits of inoculation of the common bean (*Phaseolus vulgaris*) crop with efficient and competitive *Rhizobium tropici* strains. *Biology and Fertility Soils*, 39: 88-93.

Kawale, B. R., Thombre, P. A., Kurundkar, B. P. and Mayee, C. D. (1992). Efficacy of *Bradyrhizobium japonicum* inoculation methods in soybean under rainfed condition. In *Biofertilizer Technology* Transfer (Ed. L. V. Gangawane) pp. 107-110. Associated Publishing Co. New Delhi.

Kawale, B. R., Thombre, P. A., Kurundkar, B. P. and Mayee, C. D. (1992). Evaluation of the efficiency of the *Rhizobium* cultures for soybean varieties under rainfed condition. In *Biofertilizer Technology* Transfer (Ed. L. V. Gangawane) pp. 149-154. Associated Publishing Co. New Delhi.

Kennedy, I. R., Pereg, Gerk, L. L., Wood, C., Deaker, R., Gilchrest, K. and Katupitiya, S. (1997); Biological nitrogen fixation in non-leguminous field crops: facilitating the evolution of an effective association between *Azospirillum* and wheat. *Plant and Soil*, 194: 65-79.

Khurana, A. L., Namdeo, S. L., Patel, B. J. and Dudeja, S. S. (1997); On-farm experiments on Rhizobial inoculants–problems and possible solutions. In *Extending Nitrogen Fixation Research to Farmers Fields. Proceedings of an International workshop on Managing Nitrogen Fixation in the cropping Systems of Asia*. 20-24 August 1996. ICRISAT Asea Centre, India. (Eds. O. P. Rupela, C. Johensen and D. F. Herridge. ICRISAT, Pantencheru, India). pp. 217-226.

Khot, G. G., Tauro, P. and Dadarwal, K. R. (1996). Rhizobacteria from Chickpea (*Cicer arietinum* L.) Rhizosphere effective in wilt control and promote Nodulation. *Indian Journal of Microbiology*, 36, 217-222.

Kowalski, M., Malek, W., Czopska-Dolecka, J. and Szlachetka, M. (2004). The effect of rhizobiophases on *Sinorhizobum meliloti–Medicago sativa* symbiosis. *Biology and Fertility Soils*, 39: 292-294.

Kongngoen, S., Charoensaksiri, A., Kongngoen, R. and Chanaseni, C. (1997). On-farm experiments on Rhizobial inoculants in Thailand: Problems and likely solutions. In *Extending Nitrogen Fixation Research to Farmers Fields. Proceedings of an International workshop on Managing Nitrogen Fixation in the cropping Systems of Asia.* 20-24 August 1996. ICRISAT Asea Centre, India. (Eds. O. P. Rupela, C. Johensen and D. F. Herridge. ICRISAT, Pantencheru, India). pp. 243-248.

Kore, S. S., Deshmukh, G. V. and Jadav, V. T. (1992). Green house evaluation of *Rhizobium* strains of Mungbean. In *Biofertilizer Technology Transfer* (Ed. L. V. Gangawane) pp. 141-147. Associated Publishing Co. New Delhi.

Kurundkar, B. P., Kawale, B. R., Thombre, P. A., Mayee, C. D. and Chandrawanshi, R. D. (1992). Effect of nitrogen and phosphorus application on Rhizobial symbiosis in soybean. In *Biofertilizer Technology Transfer* (Ed. L. V. Gangawane) pp. 111-116. Associated Publishing Co. New Delhi.

Kurundkar, B. P., Thombre, P. A., Kawale, B. R., Mayee, C. D. and Chandrawanshi, R. D. (1992). Effect of Rhizobial load/seed on nodulation, vegetative growth and yield of soybean. In *Biofertilizer Technology Transfer* (Ed. L. V. Gangawane) pp. 117-120. Associated Publishing Co. New Delhi.

Kumutha, K., Sempavalan, J. and Vijila, K. (2004). Induction of Desiccation tolerance in greengram rhizobia. In *Biofertilizer Technology.* (Eds. S. Kannaiyan, K. Kumar and K. Gobindarajan) pp. 253-255. Scientific Publishers, Jodhpur, Rajasthan.

Lippi, D., Depaolis, M. R., DiMattia, E., Grego, S., Pietrosanti, T. and Cacciari, I. (2000). Effect of salinity on growth and starvation-survival of a tropical *Rhizobium* strain. *Biology and Fertility Soils*, 30: 276-283.

Madhuwala and Gaur, Y. D. (1996). Ex-planta Nitrogenase Activity of Bradyrhizobium sp. (Cajanus) and its Azide Resistant Mutants. *Indian Journal of Microbiology*, 36: 73-74.

Mendes, C. J., Hungria, M. and Vargas, M. A. T. (2004). Establishment of *Bradyrhizobium japonicum* and *B. elkanii* strains in a Brazilian Cerrado Oxisol. *Biology and Fertility Soils*, 40: 28-35.

Ngachan, S. V., Yadav, A. K., Raychaudhuri, S., Lokeswarsingh, K., Ingobisingh, L., Raghumanisingh, Th. and Sharma, B. K. (2004). *Rhizobium* application in groundnut in the farmers field. In *Biotechnology in Sustainable and Organic Farming* (Eds. A. K. Yadav, S. R. Chaudhary and N. C. Talukdar) pp. 229-231. Shree Publishers and Distributors, New Delhi.

Okon, Y. P. G. (1985); *Azospirillum* as a potential inoculant for agriculture. *TIB Technology*, 3: 223-228.

Okogun, J. A. and Sanginga, N. (2003). Can introduced and indigenous rhizobial strains compete for nodule formation by promiscuous soybean in the moist savanna agroecological zone of Nigeria?. *Biology and Fertility Soils*, 38: 26-31.

Pathak, D. V. and Khurana, A. L. (1996). Effect of Nod Gene Regulators on Growth and Chemotaxis of *Rhizobium leguminosarum bv. Phaseoli. Indian Journal of Microbiology*, 36: 89-92.

Peters, N. K. and Verma, D. P. S. (1990). Phenolics compounds as regulators of gene expression in plant-microbe interactions. *Molecular Plant Microbe Interactions*, 3: 4-8.

Palaniappan, S. P., Sudhisreedhar, P., Loganathan, P. and Thomas, J. (1997). Competitiveness of native *Bradyrhizobium japonicum* strains in two different soil types. *Biology and Fertility Soil*, 25: 279-284.

Paul, S., and Shende, R. (2000). Influence of ammonia excreting diazotrophs on growth and yield of rape seed. *Indian Journal of Microbiology*, 40: 255-258.

Parmar, N. and Dadarwal, K. R. (1997). Rhizobacteria from rhizosphere and rhizoplane of Chickpea (*Cicer arietinum* L.). *Indian Journal of Microbiology*, 37: 205-210.

Plazinski, J. Innes, R. and Rolfe, B. (1985). Expression of *Rhizobium trifolii* early nodulation genes on maize and rice plants. *Journal of Bacteriology*, 163: 812-815.

Raghuchaudhuri, M., Ngachan, C. V., Raychaudhuri, S. and Singh, A. L. (2004). Yield responses of groundnut to dual inoculation and liming in an acid hill ultisol of Manipur. In *Biotechnology in Sustainable and Organic Farming* (Eds. A. K. Yadav, S. R. Chaudhary and N. C. Talukdar) pp. 235-239. Shree Publishers and Distributors, New Delhi.

Rajsekhar, A. and Reddy, T. K. K. (2000). Effect of Indole-3-Butyric Acid (IBA) and *Rhizobium* on Red Sanders (*Pterocarpus santalinus* L.) *Indian Journal of Microbiology*, 40: 137-140.

Raychaudhuri, M., Kumar, K., Raychaudhuri, S. and Sanjeev Kumar. (2004). Efficiency of *Rhizobium* with liming in furrows in groundnut. In *Biotechnology in Sustainable and Organic Farming* (Eds. A. K. Yadav, S. R. Chaudhary and N. C. Talukdar) pp. 224-228. Shree Publishers and Distributors, New Delhi.

Rodrigues, C. S., Laranjo, M. and Oliveira, S. (2006). Effect of heat and pH stress in the growth of Chickpea Mesorhizobia. *Current Microbiology* 53: 1-7.

Rolfe, B. G. and Bender, G. L. (1990); Evolving a *Rhizobium* for non-legume nodulation. In: *Nitrogen Fixation: Achievements and Objectives* (Eds. P.M. Gresshoff, L. E. Roth, G. Stacey and W. E. Newton). Chapman and Hall, New York, pp. 829.

Sharma, P. K., Kambhoj, D. V., Rustagi, N. and Dogra, R. C. (1997). A Simple Method to study nodule occupancy of *Bradyrhizobium* sp. marked with *gus*A gene. *Indian Journal of Microbiology*, 37: 91-94.

Sharma, P.K., Srivastava, P., Upadhyay, K. K., Pathak, D. V., Dogra, R. C. and Kundu, B. S. (2000). Effect of Combined Nitrogen on the expression of nod genes in *Rhizobium* sp. (*Cicer*). *Indian Journal of Microbiology*, 40: 125-130.

Sattar, M. A., Khanam, D., Ahmad, S., Haider, M. R., Podder, A. K. and Bhuiyan, M. A. (1997). On-farm experiments on rhizobial inoculants in Bangladesh: Results, problems and possible solutions. In *Extending Nitrogen Fixation Research to Farmers Fields. Proceedings of an International workshop on Managing Nitrogen Fixation in the cropping Systems of Asia*. 20-24 August 1996. ICRISAT Asea Centre, India. (Eds. O. P. Rupela, C. Johensen and D. F. Herridge. ICRISAT, Pantencheru, India). pp. 201-216.

Sarita, S., Sharma, P. K., Priefer, U. B. and Prell, J. (2005). Direct amplification of rhizobial *nodC* sequences from soil total DNA and comparison to *nodC* diversity of root nodule isolates. *FEMS Mcrobiol Ecology*, 54: 1-11.

Saini, I., Sindhu, S. S. and Dadarwal, K. R. (1996). Uptake hydrogenase, nitrate respiration, *Explanta* Nitrogenase Expression and symbiotic effectivity of *Sesbania Rhizobia*. *Indian Journal of Microbiology*, 36: 93-98.

Sivaramaiah, N., Malik, D. K. and Sindhu, S. S. (2007). Improvement in symbiotic efficiency of chickpea (*Cicer arietinum*) by co-inoculation of *Bacillus* strains with *Mesorhizobium* sp. *Cicer*. *Indian Journal of Microbiology*, 47:51-56.

Subba Rao, N. (1977). Rhizobium and legume root nodulation. In *Soil Microbiology* (Eds. N. S. SubbaRao) pp. 166-228. Oxford and IBH Publishing Co. New Delhi.

Singh, K. P., Raychaudhuri, M. and Munda, G. C. (2004). Nitrogen economy in groundnut through *Rhizobium* in an acid hill soils of Manipur. In *Biotechnology in Sustainable and Organic Farming*

(Eds. A. K. Yadav, S. R. Chaudhary and N. C. Talukdar) pp. 232-234. Shree Publishers and Distributors, New Delhi.

Sharma, R. S. and Ramamurthy, V. (2000). Nodulation compatibility and growth promoting ability of *Rhizobium* and *Bradyrhizobium* strains for *Acacia auriculiformis*. *Indian Journal of Microbiology*, 40: 131-136.

Salve, P. B. and Gangawane, L. V. (1992). *Rhizobium* from wild legumes and nitrogen fixation in Groundnut. In *Biofertilizer Technology* Transfer (Ed. L. V. Gangawane) pp. 95-100. Associated Publishing Co. New Delhi.

Subba Rao, N. (1984). Rhizobium inoculant. In *Biofertilizer in Agriculture* (Eds. N. S. SubbaRao) pp-17-73. Oxford and IBH Publishing Co. New Delhi.

Thein, M. M. and Hein, M. (1997). Rhizobial inoculants production and their on-farm use in Myanmar. In *Extending Nitrogen Fixation Research to Farmers Fields. Proceedings of an International workshop on Managing Nitrogen Fixation in the cropping Systems of Asia.* 20-24 August 1996. ICRISAT Asea Centre, India. (Eds. O. P. Rupela, C. Johensen and D. F. Herridge. ICRISAT, Pantencheru, India). pp. 227-234.

Wani, P. V., Bambarse, L. B., Konde, B. K. and Indi, D. V. (1992). Effect of single and combined inoculation of *Azospirillum* and *Rhizobium* on the yield of Groundnut under different levels of nitrogen in field conditions. In *Biofertilizer Technology Transfer* (Ed. L. V. Gangawane) pp. 60-64. Associated Publishing Co. New Delhi.

Wolde-meskel, E., Berg, T., Peters, N. K. and Frostegård, Å. (2004). Nodulation status of native woody legumes and phenotypic characteristics of associated rhizobia in soils of southern Ethiopia. *Biology and Fertility Soils*, 40: 55-66.

Xavier, L. J. C. and Germida, J. J. (2003). Selective interactions between arbuscular mycorrhizal fungi and *Rhizobium leguminosarum bv. viceae* enhance pea yield and nutrition. *Biology and Fertility Soils*, 37: 261-267.

Zerhari, K., Aurag, J., Khabaya, B., Kharchaf, D. and Filali-Maltouf, A. (2000). Phenotypic characteristics of rhizobia isolates nodulating *Acacia* species in the arid and Saharan regions of Morocco. *Letter of Applied Microbiology*; 30: 351-357.

Microbial Biotechnology and Ecology (2011) *Pages* **29–33**
Editors: **Deepak Vyas, G.S. Paliwal, P.K.Khare and R.K Gupta**
Published by: **DAYA PUBLISHING HOUSE, NEW DELHI**

Chapter 2

Microbial Diversity: Its Role in Ecosystem Maintenance

☆ *Jamaluddin**

Emeritus Scientist,
Department Bioscience, R.D. University, Jabalpur, M.P.

ABSTRACT

Microorganisms are responsible for maintenance of nitrogen cycle. They have a unique role in the cycling of elements essential to plant life on the earth. They also serve as early warming system in biomonitoring global ecology, climate change, pollutants and habitat disturbance. They are the primary producers, and dominate the biogeochemical cycles in extreme environments and stresses and counteract them.

Microorganisms constituting the major part of biodiversity on earth are a major source of useful bioactive compounds and therefore emphasize the need for their *in situ* conservation cannot be overemphasized. Because of the uncertainty of long term security of this strategy and the isolation from nature of many microorganisms being problematic, *ex situ* collections alone can ensure conservation of microbial diversity for future investigations and use by man. Microbial diversity of the world is widely covered at an IUBS/IUMS Workshop in 1991.

Biodiversity has played a major role in the evolution and diversification of macroorganisms. As mutualists, mycorrhizae are either involved in nutrient supply or perform other biochemical processes on which they depend. Bacteria, fungi and protozoa in the guts of insects and herbivores performs crucial role in their digestive processes particularly in the breakdown of cellulose and lignin.

Formation of soil biomass through biodegradation of residue, breakdown of rock and nutrient cyling is performed mainly through fungi. In monitoring microorganisms with reference to

* Ex Group Coordinator and Director Tropical Forest Research Institute Jabalpur, M.P.....

conservation of biodiversity in macroorganisms, the maintenance of functional group rather than individual species can be presumed to be limiting except where a particular microorganisms if keystone species. Microorganisms also contribute to the maintenance of ecosystem structures through natural biocontrol. Plant pathogenic microorganisms can limit plant population. Similarly entomophagus fungi (microorganisms) can limit population of insects that otherwise become major pests.

Microorganisms play a major role in the maintenance of the global ecology through various biogeochemical cycles. Some of the bacteria produces oxygen and thereby reduce the CO_2. Supply other elements like carbon, nitrogen, sulphur and minerals are also maintained in the biosphere by microorganisms.

Microorganisms are beneficial in sustainable development particularly by increasing productivity of forest and agricultural lands through: (i) efficacious nitrogen-fixing strains into legume crops, (ii) enhancement of natural nitrogen fixation by application of cyanobacterial inocula, (iii) role of mutualists (mycorrhizas) in nutrient cycling in higher plants, (iv) the use of bacteria and fungi as biocontrol agents, (v) role of microbes in genetic manipulation, (vi) biotechnological aspects and (vii) CO_2 fixation, etc.

The forest ecosystems in Madhya Pradesh are highly rich in biological diversity. Though there are many variability of biological resources in M.P. yet the studies on microbial diversity are very few and peripheral. Variability in microbes occurring in sal, teak and miscellaneous forests are well as in mine disturbed ecosystem requires systematic study and documentation of effective beneficial microbial populations for biological rejuvenation of such disturbed areas. Fire and grazing in the forest not only disturb the forest and its regeneration but also affect the microbial population occurring in different habitat, which plays a key role in the maintenance of above ground biodiversity. The documentation and listing of such important functional groups is very essential for increasing the productivity of forests.

Introduction

Variability of microbes occurs in the global resources of microorganisms. As habitat destruction and species extinction rates accelerate, biotechnology is identifying new ways to use information and product from biota Management of microbial diversity is a serious task as there is little baseline information about the amount and distribution of microbes. In majority of case–microorganisms remain either on the fringes of biodiversity programmes or left out entirely. Microbial biodiversity provides the foundation for biotechnology, *i.e.* the basis for new product discovery, bioremediation and genetic manipulation of organisms for new commercial products and processes. Thus, retention and conservation of microbial genomic variation for future applications is not only an obligation of highest order but also of great importance for global economy.

Microorganisms constituting the major part of biodiversity on earth are a major source of useful biotic bioactive compounds and therefore, emphasize the need for their *in situ* conservation cannot be overemphasized. Because of the uncertainty of long term security of this strategy and the isolation from nature of many microorganisms being problematic, *ex situ* collections alone can ensure conservation of microbial diversity for future investigations and use by man. Microbial diversity of the world is widely covered at an IUBS/IUMS Workshop in 1991 (Hawksworth, 1991a,b; Hawksworth and Colwell, 1992).

Significance of Microbial Diversity

1. Microbial 'diversity must be treated as a global resource, to be indexed, used and preserved.
2. As habitat destruction and species extinction rates accelerate, biotechnology is identifying new ways to use information and products from the microbiota.
   - ✰ Basis for new product discovery
   - ✰ Bioremediation
   - ✰ Genetic manipulation organisms for new commercial products and processed.
3. Retention and conservation of microbial genomic variations for future applications.
4. Inventory of microbial species.

Species Diversity

Comparisons of species numbers between microorganisms reflected into several variations in species concept. This difficulty arises because sexual processes are either absent or *difficult* to detect in many microorganism groups. In practice micro biologists tend to be pragmatic and recognize species or strains on the basis of morphological, biochemical or molecular similarity. The process of assimilation, substrate utilization and cultural attributes are extensively used as the taxonomic criteria more in bacteria and yeasts than filamentous fungi and protozoa.

Genetic Diversity

The genetic diversity exhibited by microorganism group is very vast. The extent of genetic diversity now demonstrated between the higher ranks of microorganisms is also reflected at the species level.

Ecosystem Habitat Diversity

The variety of ecological niches exploited by the major groups of macro and microorganisms is directly related to their geological age. The greatest niche breadth is seen in the bacteria with declining sequence in the algae, protozoa and fungi. There is every reason to believe that regions and habitats with a maximum diversity of macroorganisms will also be rich in genetic base of microorganisms. A sequence of the larger number of host-specific parasites, mutualists and saprobes is considered for the conservation of macroorganisms diversity. Extreme environment also continues to be a rich source of previously unknown microorganisms, belonging to divers groups. Studies on the number of species of particular families and general of fungi in different geographic regions can particularly lead to the recognition of centres of diversity. Various edaphic and biotic factors also contribute towards change in species and genera.

Biodiversity Maintenance

Biodiversity has played a major role in the evolution and diversification of macroorganisms. As mutualists, mycorrhizae are either involved in nutrient supply or perform other biochemical processes on which they depend. Bacteria, fungi and protozoa in the guts of insects and herbivores performs crucial role in their digestive processes particularly in the breakdown of cellulose and lignin.

About 85 per cent of the earths vascular plants form mycorrhizas.:This association in life cycle of symbionts often obligate in nature, the mycorrhizas being crucial to the absorption of growth limiting *nutrients.* The very existence of many macroorganisms is consequently dependent on the continued availability of the requisite mutalistic microorganisms: In the absence of mutualistic microorganisms (crustess cerallinelgae), coral reef ecosystems simply cannot exist. Formation of soil biomass through biodegradation of residue, breakdown of rock and nutrient cycling is performed mainly through fungi. In monitoring microorganisms with reference to conservation of biodiversity in macroorganisms, the maintenance of functional group rather than individual species can be presumed to be limiting except where a particular microorganisms is keystone species. Microorganisms also contribute to the maintenance of ecosystem structures through natural biocontrol. Plant pathogenic microorganisms can limit plant population. Similarly entomophagus fungi (microorganisms) can limit population of insects that would otherwise become major pests.

Microorganisms in Biosphere Function

Microorganisms play a major role in the maintenance of the global ecology through various biogeochemical cycles. Some of the bacteria produces oxygen and thereby reduce the CO_2. Supply other elements like carbon, nitrogen, sulphur and minerals are also maintained in the biosphere by microorganisms.

Microorganisms are beneficial in sustainable development particularly by increasing productivity of forest and agricultural lands through: (i) efficacious nitrogen-fixing strains into legume crops, (ii) enhancement of natural nitrogen fixation by application of cyanobacterial inocula, (iii) role of mutualists (mycorrhizas) in nutrient cycling in higher plants, (iv) the use of bacteria and fungi as biocontrol agents, (v) role of microbes in genetic manipulation, (vi) biotechnological aspects and (vii) CO_2 fixation, etc.

Microbial Diversity in Madhya Pradesh

The biodiversity in Madhya Pradesh is very rich. The forest cover comprising sal forest, teak forest and miscellaneous forest, cover the vast area of the state. Due to rich diversity in the macro-organisms, microbial diversity is equally very high. The microbes belonging to different groups particularly fungi, bacteria, actinomycetes etc. live closely with the macro organisms in diverse habitat. Some of these organisms are causing diseases particularly fungal and bacterial diseases and they belong to different taxonomic groups. These microbes are capable of damaging leaf, stem roots, stored woods, flower and seeds etc.

The microbes are also capable of decomposing the forest litter and the major group of the microbes belongs to thermophilic or thermotolerant bacteria and fungi. The decomposition of and organic acid etc. (Satyanarayana *et al.*, 1992). The enzyme produced by this group of fungi are thermostable.

Scope of Microbial Diversity

The studies on microbial diversity are very scant, there is need to investigate and list out the fungi and bacteria and actinomycete from the soil and different habitat in forest ecosystem. There is need to identify the major functional groups of microorganisms responsible for biological diversity of macro-organisms particularly the mycorrhizae, N-fixers, phosphate solubilizing bacteria and many other biocontrol agents.

The microbes specifically involved in the biochemical process like bio-remediation, bio-indication as well as synthesis of important chemicals need identification and documentation. *Ex-situ* conservation of the microbes by making the life cultu Teherbaria is very essential to exploit their potential. *In situ* maintenance and marking of the potentially important microbes is also equally important.

Since there is no any specific account listing out the microbial diversity of M.P., therefore it is necessary to prepare a project, to document the relevant literature on microbial diversity and also to list out the significance microbe in different habitats.

References

Hawksworth, D.L. (1991a). The fungal dimension of biodiversity magnitude, significance and conservation. *Mycological Research,* 95 : 641-655.

Hawksworth, D.L. (1991b). *The biodiversity of microorganisms and Invertebrates: Its Role in sustainable Agriculture,* Wallingford: CAB International.

Hawksworth, D.L. and Colwell, RR (1992). Microbial diversity and biodiversity amongst microorganisms and its relevance. *Biodiversity and Conservation,* 1: 221–226.

Jamaluddin and Chandra, KK. (1999). Application of VAM technology in bamboo cultivation in mine soils. In: *Modern Approaches and Innovation in Soil Management*. Editors: D.I. Bagyaraj, Ajit Verma, K.K. Khanna and H.I. Kehri, Rastogi Publications, Meerut.

Microbial Biotechnology and Ecology (2011) *Pages* **34–40**
Editors: **Deepak Vyas, G.S. Paliwal, P.K.Khare and R.K Gupta**
Published by: **DAYA PUBLISHING HOUSE, NEW DELHI**

Chapter 3

Thermophilic Cyanobacteria: The Wonder Organisms

☆ *Vidhi Verma and Meenakshi Banerjee*

Laboratory of Algal Biotechnology,
Department of Bioscience, Barkatullah University, Bhopal, M.P

Introduction

After the evolution of life on Earth over 3 billion years ago several life forms have evolved since then. Living systems have to continuously interact with an environment that is not always kind to them. As a result they encounter various stresses at the organism and cellular level. Lazcano and Miller (1996); Ward and Castenholz (2000).

Such common niches that encompass a number of extreme physical stresses are found in the endolithic rock environment of hot and cold deserts with regard to temperature as well as marine environments and hot springs that are highly haline in nature. Bhaya *et al.* (2000); Norris and Castenholz (2006); Fleming and Castenholz (2007); Roeselers *et al.* (2007). In such environment microorganism must endure extremes of temperature, desiccation, low nutrient supply and low photon flux. Geothermal springs of temperature between 45-100°C are almost exclusively inhabited by thermophilic prokaryotic microorganisms, photosynthetic flexi bacteria and certain non-photosynthetic autotrophic and heterotrophic bacteria (Pentecost 2003, Banerjee and Castenholz 2007 (In Press), Norris and Castenholz 2007). In most extreme habitats the microbial communities which dominate are composed primarily of photosynthetic microorganism mainly the cyanobacteria.

Cyanobacteria, are a group of Gram-negative phototrophic prokaryotes which carryout oxygenic photosynthesis as well as nitrogen fixation. These blue green algae are ubiquitous in nature and have very simple nutrient requirement that allow these organisms to occupy highly diverse ecological niches. Maeda and Omata (2004); Banerjee and Verma (2007).

The hot water source contains most of the essential nutrients and permits a steady growth of cyanobacteria throughout the year. As the effluent moves away from the source of the spring the water cools, thus setting up a temperature gradient (Banerjee *et al.*, 2001; Banerjee and Sharma 2004). It is accompanied by concentration gradient of sulphides (decreasing) due to oxidation, oxygen (increasing) and pH (increasing) with concomitant gradual changes in the composition of the micro flora. Downstream as the temperature gradually decreases from 73°C to about 53°C, populations of unicellular and filamentous heterocystous cyanobacteria, forms deep green to reddish brown mats. These algae withstand 'or' tolerate a very high temperature. Castenholz (1969); (1977); Norris and Castenholz (2007) Banerjee and Sharma (2005). Thermophilic cyanobacteria have become an experimental system for the basic and applied research due to their unique properties. These include growth and temperature optima, habitat relationship, enzymatic abilities, and the study of pigments and the role in detecting evolution (Maheshwari *et al.*, 1987; Tomitani *et al.*, 1999; Whitton and Potts 2000).

Thermophilic Cyanobacteria

Temperature is an important environmental factor, which can create a series of challenges. A thermophile is an organism capable of living at temperature at 'or' near the maximum for the taxonomic group of which it is a part (Brock 1978; 1986; Ward and Castenholz 2000; Norris and Castenholz 2006). The blue-green algae are considered to be thermophilic when part or all of their optimal growth temperature range is above 45 °C (Castenholz 1969; Castenholz and Pierson 1995).

Thermophilic cyanobacteria have adopted themselves to wide range of ecological niches such as, thermal waters, geothermal hot spring, hot spring, volcano etc. (Madigan and Brock 1976). The thermal area is an area where warm water ascends through fissures, cervices and volcanic crust. Cold rainwater pours down into porous bedrock, is heated from the old magma chamber and fissures and ascends to the surface, as it gets hotter. Thermophilic cyanobacteria are predominant flora of thermal waters. The cyanobacteria are considered as thermophilic when their most favorable growth temperature range is above 45°C (Castenholz 1969). Setchell suggested the upper limit of temperature for cyanobacteria as 65° to 68°C and Lammerman described it as 69°C but we do have some cyanobacteria in the Yellowstone National Park USA grow at 95°C (Brock 1978; Castenholz 1996; Pentecost 2003; Norris and Castenholz 2006; Banerjee and Castenholz 2007) (In Press).

The morphological diversity of the cyanobacteria is considerable. Both unicellular and filamentous forms of thermophilic cyanobacteria are known and considerable variations with in these morphological types occur. Unicellular form like *Synechococcus elongates* found in cylindrical or ellipsoidal shape in thermal water where the temperature is above 75°C. *S. elongates* doesn't have heterocyst Desikachary (1959; Brock 1978; Castenholz 1969; Miller and Castenholz 2001; Bird and Wyman 2003). Filamentous type, like *Mastigocladus laminosus* forms single filament with one side branching and an intercalary heterocyst but hormogones are absent (Desikachary 1959; Stewart 1970; Wickstorm 1980; Jha *et al.*, 1986 Robinson *et al.*, 2007; Miller and Castenholz 2007). *Leptolyngbya*, a polyextremophile that can survive more than one extreme habitat like high temperature as well as high salt concentration, forms thin filaments with trichomes but without heterocyst (Rippka *et al.*, 1979; Anagnostidis and Komárek 1988; Albertano and Kovácik 1994; Castenholz 2001; Komárek and Anagnostidis 2005).

Distribution of Thermophilic Cyanobacteria

The water of hot springs may discharge at temperatures somewhat higher than the common boiling temperature, although superheated waters are uncommon except in volcanic regions (Ward and Castenholz 2000; Kiyoshi *et al.*, 2004). Generally thermal habitats are aquatic, and Earth is the only source of heat, for nearly all of these. In spite of that, solar radiation adequately increases the temperature in a few situations and the self-heating of organic material may bring localized temperature to the point of ignition Castenholz (1969). Hot springs and their drain ways provide the most adequate aquatic habitats for thermophilic blue-green algae. Cyanobacteria have been reported from thermal waters all over the world.

Natural geothermal areas are widely distributed around the Earth. Springs in which the water discharges at a temperature of over 45° C occur in every continent except Antarctica. The most studied biological sites are Yellowstone National Park Plateau of North America, Japan, Iceland, and North Island of New Zealand. The large numbers of hot springs are also widely scattered over most of the United States west of the Great Plains, the line of the Andes, Italy, Algeria-Tunisia, Greece, Turkey, portions of central Africa, central Asia, Indonesia, Philippines, Kamchatka Peninsula in Siberia, Naples area in Italy, Bakereshwar, West Bengal and Rajgir Bihar in India (Castenholz, 1969; Jana, 1970; 1973). Some prominent thermophilic cyanobacteria are *Chroococcales, Calothrix* sp., *C. thermalis, Aphanocapsa thermalis, Oscillatoria terebriformis, O. animalis, Spirulina* sp., *Phormidium laminosum, Mastigocladus laminosus, Synechococcus elongates, Synechococcus lividus, S. minervae, Phormidium tenue, Oscillatoria cf. terrbriformis* and *Leptolyngbya* that grow at very high temperatures.

Adaptation for Thermal Environment

These thermophilic cyanobacteria have very altered structural and metabolic differences from their normal counterparts that enable them to survive such adverse situations. Temperatures approaching 100°C normally denature proteins and nucleic acids, and increase the fluidity of membranes to lethal levels. Chlorophyll degrades above 75 °C, excluding photosynthesis (Rothschild *et al.*, 2001). In cyanobacteria lipids, nucleic acids and proteins are generally susceptible to heat and therefore there is no single factor that enables all thermophiles to grow at extreme temperature. High temperature increases the fluidity of membranes. To maintain optimal membrane fluidity the cell must adjust the composition of the membrane including the amount and type of lipids. The lipid membranes of thermophiles have saturated and straight chain of fatty acids to provide right degree of fluidity needed for membrane function. Some species contain paracrystalline surface layer that function as external protective barrier (Norris and Castenholz 2007; Rothschild *et al.*, 2001).

Temperature also effects the structure and function of proteins (Jaenicke, 1996). Ways that proteins have evolved to cope with high temperatures include increasing ion-pair content, forming higher-order oligomers and decreasing flexibility at room temperature. Decreasing the length of surface loops is also known; in particular those loops that connect elements of secondary structure, optimize electrostatic and hydrophobic interactions, and exchange amino acids to increase internal hydrophobicity and helix tendency of residues in α-helices. Heat shock proteins, chaperons are also present and are likely to play an important role in stabilizing and refolding of proteins as they begin to denature (Kallas and Castenholz, 1982a; Ferris *et al.*, 1997; Rothschild *et al.*, 2001).

DNA at high temperatures such as above 70°C is normally denatured. The DNA of hyperthermophiles is known to be more stable *in vivo* than that of a mesophile. Monovalent and divalent salts enhance the stability of nucleic acids because these salts screen the negative charges of the phosphate groups (Peak, *et al.*, 1995; Marguet, *et al.*, 1998; Rothschild, *et al.*, 2001). Thermophiles also have, histone like protein that protects DNA, and have reverse gyrase that is responsible for positive super coiling in DNA. The Guanine and Cytosine pair of nucleic acids is more thermostable than the Adenine and Thymine or Adenine and Urasil pairs because of the additional hydrogen bond.

But elevated G and C ratios are not found among thermophilic prokaryotes because of the stability of the chromosomal DNA, although thermo stability is correlated with G and C content of their ribosomal and transfer RNAs (Galtier *et al.*, 1997; 1999; Rothschild, *et al.*, 2001).

Importance

It is generally assumed that blue-green algae evolved in the early Precambrian period and were responsible for the first significant increase in atmospheric oxygen. Microfossils of cyanophyte like filaments have recently been found in early, middle, and late Precambrian strata. Fossils of the late Precambrian (about 1 billion years ago) resemble living members of the blue green algae (Swain, 1969; Schopf, 1968). The study of cyanobacteria from extreme thermophilic environments greatly expands our understanding of the survival of these organisms at very inhospitable environments. The cyanobacteria from extreme thermophilic environments have attracted great attention in recent years because of their ability to endure extreme environmental conditions and the regulatory mechanisms involved in the process of survival because, these microbes have several biotechnological applications. We can use thermophilic cyanobacteria as biofertilizers in region of very high temperatures because it is documented that thermophilic cyanobacteria like *Mastigocladus* can fix atmospheric nitrogen (Fogg, 1951; Stewart, 1970; Banerjee and Bhattacharya, 2001).

These microorganisms therefore produce unique biocatalyst that function under extreme condition comparable to those prevailing in various industrial processes. Some of the enzymes from extremophiles have already been purified and their genes successfully cloned in mesophilic host. These organisms therefore are very important biotechnologically with potential use in food, chemical and pharmaceutical industry as well as environmental biotechnology.

References

Albertano, P. and Kovácik, L. (1994). Is the genus *Leptolyngbya* (Cyanophyte) a homogeneous taxon? *Archiv für Hydrobiologie/Algological Studies* (75): 37–51.

Anagnostidis, K. and Komárek J. (1988). Modern approach to the classification system of cyanophytes 3–*Oscillatoriales*. *Archiv für Hydrobiologie/Algological Studies* (50–53) : 327–472.

Brock, T. D. (1976). Upper pH limit for the existence of blue-green algae: evolutionary and ecological implications. *Science*. (179): 480-483.

Brock, T. D. (1986). Introduction: An overview of the thermophiles. In *Thermophiles: General, Molecular and Applied Microbiology* (ed. Brock, T. D.), John Wiley and Sons, New York.; 1–16.

Brock, T. D. (1978) *Thermophilic Microorganisms and Life at High Temperatures*. Heidelberg, New York: Springer-Verlag.

Bargagli, R. Broady, F. A. and Walton, D. W. H. (1996). Preliminary investigation of the thermal biosystem of Mount Rittmon Fumaroles Northern Victoria Land Antarctica. *Antarctica Science*. (32): 239-250.

Banerjee, M. Verma, V. (2006). Nitrogen assimilation studies of four cyanobacteria from different ecological niches. *Indian Journal of Botanical Research*. 2 (2): 157-162.

Banerjee, M. Banik, J. and Bhattacharya, S. (2001). Influence of different physiological parameters on the N_2 fixation of *Mastigocladus*. *Journal of Nat. Con.* 14 (2): 8-12.

Banerjee, M. and Sharma, D. (2004). Regulation of phosphatase activity of *Chroococcidiopsis* isolates from two diverse habitats. Effect of light pH and temperature. *Applied Ecol. Environ. Res.* 2 (1): 71-82.

Banerjee M, Sharma D. (2005). Comparative studies on growth and phosphatase activity of endolithic cyanobacterial isolate *Chroococcidiopsis* from hot and cold deserts. *Journal of Microbiology and Biotechnology*. 15 (1): 125-130.

Banerjee, M. Castenholz, R.W. (2007). Characterization of a novel thermo-halophilic clade of cyanobacteria from saline thermal springs and endolithic habitats. *Environ. Microbiol* (In Press).

Bird C, Wyman M. Nitrate/Nitrite assimilation system of the marine picoplanktonic cyanobacterium *'Synechococcus'* sp. Strain WH 8103: (2003). Effect of nitrogen source and availability on gene expression. *Applied Environ. Microbiol*. 7009-7018.

Bhaya D, Schwarz R, Grossman A R. (2000). Molecular responses to environmental stress. In: Whitton BA. and Potts M. (eds) *Ecology of cyanobacteria, their diversity in time and space*. Kluwer Academic, Dordrecht.; 397- 442.

Castenholz, R. W. (1969). Thermophilic blue-green algae and the thermal environment. *Bacteriol. Rev.* 33:476–504.

Castenholz, R. W. (1977). The effect of sulfide on the blue-green algae of hot springs. II. Yellowstone National Park. *Microbiol Ecol*. (3) : 79-105.

Castenholz, R. W. and Pierson, B. K. (1995) *Ecology of Thermophilic Anoxygenic Phototrophs*. Kluwer, Dordrecht, The Netherlands.

Castenholz, R. W. (1996). Endemism and biodiversity of thermophilic cyanobacteria. *Nova Hedwigia* (112): 33–47.

Castenholz R.W. (2001). Oxygenic photosynthetic bacteria. In: D.R. Boone and R.W. Catenholz (eds) *Bergey's Manual of Systematic Bacteriology* (2nd edition), Springer?Verlag, New York: 473–600.

Desikachary T. V. (1959). *Cyanophyta*, Indian council of Agriculture Research. New Delhi.

Fogg, G. E. (1951). Studies on nitrogen fixation by blue-green algae. II. Nitrogen fixation by *Mastigocladus laminosus* Cohn. *J. Exp. Bot*. (2): 117.

Fleming DE, Castenholz RW. (2007). Effects of periodic desiccation on the synthesis of the UV-screening compound, scytonemin in cyanobacteria. *Environ. Microbiol*.; 9(6): 1448-1455.

Ferris, M. J. Nold, S. C. Revsbech, N. P., and Ward, D. M. (1997). Population structure and physiological changes within a hot spring microbial mat community following disturbance. *Appl Environ Microbiol*. (63): 1367-1374.

Galtier, N. and Lobry, J. R. (1997). Relationships between genomic G and C content, secondary structures and optimal growth temperature in prokaryotes. *J. Mol. Evol*. (44), 632.

Galtier, N., Tourasse, N. and Gouy, M. A (1999). Non-hyperthermophilic, common ancestor to extinct life forms. *Science* 283, 220–221.

Jana, B. B. (1970). On the study of limnology and bacteriology of the thermal springs of Bakreshwar *Ph.D. Thesis* Visva Bharti University Shantiniketan, India.

Jana, B. B. (1973). The thermal springs of Bakreshwar, India-physico-chemical conditions, Flora and Fauna. *Hydrobiologia*. (41): 291-307.

Jha M Ashok Kumar and H D Kumar (1986). Stimulation of *in vivo* nitrate reductase activity in the thermal cyanobacterium *Oscillatoria princeps* under microaerobic conditions. *Current Microbiology* (14): 145-147.

Jaenicke, R. (1996). Stability and folding of ultra stable proteins: eye lens crystallins and enzymes from thermophiles. *FASEB J.* (10): 84–92.

Kallas, T., Castenholz, R. W. (1982a). Internal pH and ATP-ADP pools in cyanobacterium *Synechococcus* sp. During exposure to growth–inhibiting low pH. *J. Bacteriol* (149): 229-236.

Komárek J. and Anagnostidis K. (2005). Cyanoprokaryota–2. Teil/2nd Part: Oscillatoriales. In: B. Büdel, L. Krienitz, G. Gärtner and M. Schagerl (ed.) *Süsswasserflora von Mitteleuropa* 19/2, Elsevier/ Spektrum, Heidelberg; 759 pp.

Kiyoshi O, Megumi M, Shino I, Kazuhisa O, and Masahiro I (2004). Circadian rhythms in the thermophilic cyanobacterium *Thermosynechococcus elongatus*: Compensation of Period Length over a Wide Temperature Range. *Journal of Bacteriology* (186): 4972-4977.

Lynn J. Rothschild and Rocco L Mancinelli (2001). Life in extreme environments. *Nature* (409): 1092-1101.

Madigan, M. T. Brock, T. D. (1977). Adaptation of hot spring phototrophs to reduced light intensity. *Arch. Microbiol.* (113): 111-120.

Marguet, E. and Forterre, P (1998). Protection of DNA by salts against thermo degradation at temperatures typical for hyperthermophiles. *Extremophiles* 2, 115–122.

Maeda, S. Omata, T. (2004). A novel gene (narM) required for expression of nitrate reductase activity in the cyanobacterium *Synechococcus elongates* strain PCC 7942. *J. Bacteriol.* (186): 2107-2114.

Norris, T. B. Castenholz, R. W. (2006). Endolithic photosynthetic communities with in ancient and recent travertine deposits in Yellowstone National Park. *FEMS Microbiol. Ecol.* 57 (3): 470–483.

Norris, T. B. Castenholz, R. W. (2007). Effect of environmental stressors on photosynthetic microorganisms in geothermal springs of Yellowstone National Park: Geothermal biology and Geochemistry in Yellowstone National Park. 221-234.

Peak, M. J. Robb, F. T. and Peak, J. G. (1995). Extreme resistance to thermally induced DNA backbone breaks in the hyperthermophilic archaeon *Pyrococcus furiosus*. *J. Bacteriol.* (177): 6316–6318.

Pentecost A. (2003). Cyanobacteria associated with hot spring travertines. *Can. J. Earth Sci.* (40): 1447-1457.

Rippka R., Deruelles J., Waterbury J.B., Herdman M. and Stanier R.Y. (1979). Generic assignments, strain histories and properties of pure cultures of cyanobacteria. *Journal of General Microbiology* (111): 1–61.

Robinson W. Brad Mealor, Augustus E., S Stevens, Jr. Edward and Mark Ospeck (2007). Measuring the Force Production of the Hormogonia of *Mastigocladus laminosus*. *Biophysical Journal* (93): 699-703.

Roeselers G, Norris TB, Castenholz RW, Rygaard S, Glud RN, Kühl M, Muyzer G. (2007). Diversity of photoautotrophic bacteria in microbial mats from Arctic hot springs (Greenland) *Environ. Microbiol.* 9(1): 26-38.

Schopf, J. W. (1968). Microflora of the Bitter Springs Formation, Late Precambrian, Central Australia. *J. Paleontol.* (42): 651-688.

Scott R. Miller Richard W. Castenholz, and Deana Pederse (2007). Phylogeography of the thermophilic cyanobacterium *Mastigocladus laminosus*. *Applied and Environmental Microbiology*. (73): 4751-4759.

Swain, F. M. (1969). Paleomicrobiology. *Annual Review Microbiol.* (23): 455-472.

Stewart W D P (1970). Nitrogen fixation by blue-green algae in Yellowstone thermal areas. *Phycologia.* (9): 261-268.

Tomitani A, Okada K, Miyashita H, Mathiys H C P, Ohno T and Tonak A. (1999). Chlorophyll *b* and phycobillins in the common ancestor of cyanobacteria and chloroplast. *Nature* (400): 159-162.

Wickstorm C E (1980). Distribution and physiological determinants of blue green algal nitrogen fixation along a thermogradient. *J. Phycol* (16): 436-443.

Ward, D. M. and Castenholz, R. W. (2000). In: *The Ecology of Cyanobacteria*, Eds. Whitton, B. A. and Potts, M. (Kluwer, Dordrecht, the Netherlands), pp. 37–59.

Microbial Biotechnology and Ecology (2011) *Pages* **41–59**
Editors: **Deepak Vyas, G.S. Paliwal, P.K.Khare and R.K Gupta**
Published by: **DAYA PUBLISHING HOUSE, NEW DELHI**

Chapter 4

Cyanobacteria: Phenotypic to the Genotypic Diversity

☆ *A. Srivastava, P. Singh, A. Singh, S.S. Singh and A.K. Mishra**

Laboratory of Microbial Genetics, Department of Botany, Banaras Hindu University, Varanasi – 221 005

ABSTRACT

Cyanobacteria, the pioneer model for evolution of photosynthetic machinery, have played an important role in the process of life sustenance on the planet. Their autotrophic nature in conjunction with the nitrogen fixation capacity of some of its members makes it an indispensable link in the carbon and nitrogen economy on the globe. They are the vital colonizers in all sorts of ecosystems varying from hot deserts to polar ice caps, rocky mountains to benthic locale, exposed soils to endosymbiotic milieu etc. Their caliber of such diverse habitation lies in the variety of morphological and physiological attributes that they can exhibit. For a long period, diversity assessment of cyanobacteria has been one of the prime objectives of microbiology researchers. Initially, morphological specifications in combination with cytochemical and physiological characteristics were the prime basis for cyanobacterial characterization. Any phenotypic character when considered alone is of less utility as it is generally shared by one or the other biological groups. Currently, increasing numeric and molecular taxonomic tools have widely aided, if not replaced the traditional morphological and physiological criteria. Protein, fatty acid and pigment analysis, isozyme profiling and nucleic acid studies are some of the presently used tools of taxonomy. Any particular trait when used in isolation, may lead to misleading results for identification, further classification and diversity assessment of the group. Therefore, a polyphasic top-to-bottom approach based on information reclaimed from techniques varying from microscopic observation to genome analysis provides better resolution at taxonomic level.

---

* E-mail: akmishraau@rediffmail.com.

Cyanobacterial Characteristics

Cyanobacteria (also called blue green algae) are gram negative photoautotrophic prokaryotes, stand at one of the major turnouts of evolutionary history. They are characterized by possession of machinery for synthesis of chlorophyll *a* and phycobilin. Apart from their flair for utilization of water (or H_2S under certain conditions) as an electron donor, some of its members also have a capacity for nitrogen fixation (Berman-Frank *et al.*, 2003; Bergman *et al.*, 1997). The evolutionary and ecological essence of cyanobacteria lies in the two above stated facets. Apart from chlorophyll *a* and *c*-phycocyanin production, it has capacity for synthesis of carotene and *c*-phycoerythrin (Prasanna *et al.*, 2007; Six *et al.*, 2004).

Cyanobacteria enjoy a very wide distribution. Most of them are freshwater forms except for a few like *Dermocarpa* and *Trichodesmium* which thrive well in marine locale. Terrestrial habitat is common for most of the species of *Nostoc* and *Oscillatoria*. Benthic habitation is also as common as planktonic one. Endophytic and symbiotic temperament of some distinguished species of this class such as *Nostoc, Anabaena, Chroococcus, Gloeocapsa, Scytonema,* and *Stigonema* forms a very fascinating aspect (Bergman *et al.*, 2007; Usher *et al.*, 2007; Carpenter and Foster, 2002; Perkins and Peters, 1993). The presence of cyanobacteria in freshwater blooms, marine ecosystems, soils and rice fields, within limestone, salt subjugated lands, deserts, polar environments and within other living forms highlight their propensity to survive in and moreover dominate the above niches (Whitton and Potts, 2000).

Members of cyanophyceae have a prokaryotic structure with an incipient type of nucleus, thylakoids and storage products called cyanophycean starch and cyanophycin granules (Fritsch, 1945; Kylin, 1943). Non existence of sex organs, gametes and flagellated zoospores grounds the way for vegetative and asexual reproduction. Vegetative reproduction occurs by cell division, fragmentation, hormogonium formation and asexual reproduction by means of akinetes, endospores, exospores, heterocysts and nannocytes.

As per the traditional botanical taxonomic conduct based on morphological and developmental attributes, cyanobacteria are currently alienated into five orders (Castenholz, 1989; Waterbury 1989; Waterbury and Rippka, 1989) that correspond closely to the five groups used by Stanier and his collaborators (Rippka *et al.*, 1979).

Chroococcales, the first order, comprises mainly of unicellular forms that reproduce by binary fission or budding but cell aggregates are equally prevalent. The members of order Pleurocapsales range from unicellular forms to cell aggregates that give rise to unicellular structures called 'baeocytes' (Pinevich *et al.*, 2008). They harbour an explicit feature of multiple fissions. All the undifferentiated filamentous cyanobacteria have been placed in the order Oscillatoriales. Trichome fragmentation or production of undifferentiated hormogonia (specialized reproductive trichomes) is the chief mode of reproduction. Members of the order Nostocales possess a unique property of cell differentiation. Normal cells can be specialized to form (a) heterocysts (specialized cells meant for nitrogen fixation), (b) akinetes and (c) hormogonia. Many members of this order are ecologically highly valued because of their role as converters of unusable dinitrogen to ammonia. The last order Stigonematales also comprises filamentous forms that can differentiate into a number of cellular subtypes. The only difference lies in the plane of binary fission as members of stigonematales can divide in multiple planes resulting in multiseriate trichomes (Whitton and Potts, 2000).

Characters such as prokaryotic cell construction, absence of division of labor, dearth of flagella and sexuality advocate their primitive nature. Paleontological, geological and isotopic geochemical substantiations have been instrumental in documenting the antiquity of cyanobacteria to early

Precambrian period. A considerably less rRNA sequence diversity has been reported among the members of this class as compared to other taxa (Giovannoni *et al.*, 1988). The above fact gives an obvious clue of a close knit phylogenetic relationship. Within the phylogenetic tree, members of Chroococcales and Oscillatoriales are omnipresent, members of Pleurocapsales show a single lineage and members of Nostocales and Stigonematales confirm a single coherent lineage. Cyanobacteria have been on the earth for over 3.5 billion years. During the course of the evolution, they have provided the major source of the oxygen and acquired several morphological, physiological, biochemical and molecular specialization, in order to survive best in diverse ecological conditions which ultimately leads to the evolution of diversified cyanobacteria.

Cyanobacterial Diversity

Cyanobacteria are a class that illustrates an assortment of diversities *viz.* morphological, habitat and physiological miscellany. The diversity it exhibits is visible in multitude of its metabolic tactics, motility, cell division, developmental biology and other structural and functional aspects of cell morphology. Thus, its diversity is seen at various levels *viz.* morphological, physiological and genetic. Morphological and physiological diversity can be a manifestation of either genetic or environmental factors. The correlations among all the factors taken into account reveal the overall similarity and contrast and this forms the basis of systematic groupings. The foundation for such groupings is the core fact that each and every organism is related through evolutionary decent (Luuc *et al.*, 1999). The investigation for International Code of Botanical Nomenclature (ICBN) included a wide range of eukaryotic and prokaryotic organisms and blue green algae (Geitler, 1932) were considered the largest prokaryotic group of organisms.

Morphological Diversity

Cyanobacteria are characterized by different types of cellular structures. Each order is characterized by some particular cell type. Unicellular, isopolar cyanobacteria occur in the order Chroococcales; pseudoparenchymatous in the order Pleurocapsales, and unicellular, heteropolar in the order Chamaesiphonales. Members of the order Oscillatoriales are multicellular, trichal and lack heterocyst while those of the order Nostocales have similar structure as that of oscillatoriales but with an added presence of heterocyst. Cyanobacteria of the order Stigonematales are also multicellular, trichal and heterocystous just like members of Nostocales, but their trichomes show proper branching (Komarek, 1994). Unicells may occur singly or grouped in colonies as reported in *Schizothrix*. Filamentous forms are also equally prevalent. The trichomes may be straight or coiled. Uniseriate and unbranched trichomes bearing absolutely identical cells are characteristic of the order Oscillatoriales. On the contrary, orders such as Nostocales and Stigonematales have trichomes bearing heterogeneous cellular composition. The branching pattern can be categorized as true or false. In true branching, cells in a filament can divide in more than one plane (*e.g. Haplosiphon*). In the case of false branching, division in a filament with a sheath occurs in such a way so as to give rise to two separate filaments that grow separately *e.g. Tolypothrix, Scytonema.*

In filamentous forms there are two modes of growth. In some filaments showing false branching patterns, growth is apical and the cells at the base die off simultaneous with growth. As a result of the above process, persistent sheaths present in some filamentous forms develop into a mass of old sheaths under the living material (*e.g. Scytonema*). In the other form of growth, cell divisions occur at the base of a sheathed trichome and consequently the apical cells become devoid of chlorophyll, become narrow, take the appearance of multicellular hair and their thylakoids rip apart in order to form liquid filled vacuoles (*e.g. Calothrix*).

Even cultivated strains of the same species (monospecies) do not always show exact similarity. They differ in their morphological features with respects to the varying growth conditions (Zapomelova, 2008; Gupta and Agarwal, 2006a, b; Lyra *et al.*, 2001; Nalewajko and Murphy, 2001; Gugger *et al.*, 2002a,b; Anand, 1988). Cyanophyceae exhibits lavish cellular and functional polymorphism in possession of different types of cells, with each one suited to a particular function.

This class enjoys cellular division of labour aided by a variety of cells. An akinete is a thick walled dormant cell derived from the enlargement of a vegetative cell (Moore *et al.*, 2005) and is much larger compared to latter. It depicts a resistant resting stage. The process of its derivation from vegetative cell is accompanied with immense thickening of cell wall so as to acquire entire protoplast within it. It is processed only under adverse conditions and circumstances leading to its formation vary among species and also among strains of monospecies (Whitton, 1992). Their occurrence is greatly constrained to heterocystous species, but may also be found in some non-filamentous species. Another specialized structure called gas vesicles/vacuoles are found in a number of aquatic species of many different genera such as *Anabaena* and are meant for buoyancy regulations. They are cylindrical in shape, filled with gas and are highly important for planktonic species and aid in adjustment of the cyanobacterium's vertical position in the water column (Walsby, 1987). Number of gas vesicles increases with reduction in light intensity and abating growth rate. Heterocysts are specialized nitrogen fixing cells that differentiate from vegetative cells in response to deprivation of fixed nitrogen *i.e.* ammonium or nitrate. In *Anabaena* spp., Nostoc spp. and many other strains, heterocysts differentiate at semiregular intervals along vegetative filaments, forming a spacing pattern. In *Gloeotrichia* spp., *Rivularia* spp. and other strains heterocyst occupy terminal position in the filaments while lateral heterocyst has been found in *Nostochopsis* sp. (Wolk 2000). The establishment and maintenance of the heterocyst spacing pattern depends upon two factors. Firstly, interaction between diffusible substances *i.e.* glutamine that originates from the heterocyst and move along the filament and secondly, other factors, presumably proteins that interact with diffusible substances and directly or indirectly control gene expression (Wolk 2000; Golden 2000; Mishra, 2003). All the cyanobacteria that form heterocyst are assorted in the orders Stigonematales and Nostocales which are characterized by branched and simple filaments respectively. Altogether they constitute a monophyletic group which shows very petite genetic variability (Anagnostids and Komarek, 1992). They may occur as solitary structures or in pairs as in *Anabaenopis*. They originate from vegetative cells that are slightly enlarged and possess pore at one or both poles. Such pores are usually present at single pole in case of terminal heterocysts. Differentiation of a vegetative cell to a heterocyst is unidirectional and the process cannot be reverted. Nitrogenase and other proteins involved in nitrogen fixation are synthesized within the heterocyst (Ernst *et al.*, 1992; Buikema *et al.*, 1991, 1993). They synthesize three additional cell walls, including one of glycolipid that forms a hydrophobic barrier to oxygen (Fay, 1992). Latter function is particularly important so as to protect the oxygen labile nitrogenase enzyme. For further enhancement of the above process they also degrade oxygen producing photosystem II within them. In conjugation with the above process, they up regulate the synthesis of glycolytic enzymes. Since the photosystem II is dismantled, heterocysts do not photosynthesize. They are provided with carbohydrates, mainly in the form of disaccharides, by vegetative cells. Heterocysts sustain photosystem I which permits ATP generation by cyclic photophosphorylation. In certain filamentous cyanobacteria, the function of nitrogen fixation is performed in typical cells called 'diazocytes' that may be joined together to form specialized portions of trichomes (Komárek and Anagnostidis, 2005). They posses the enzyme nitrogenase and are mainly found in non-heterocystous cyanobacteria such as *Trichodesmium, Oscillatoria* etc. (Ohki, 2008; El-Shehawy, 2003; Lugomela, 2002).

Hormogonia are short, motile chains of rather uniform cells (Desikachary, 1959), but this definition does not hold true for all hormogonia. More or less, they can be considered as motile filaments formed by some cyanobacteria in the family Nostocaceae which are responsible for conducting a number of indispensable physiological processes in cyanobacteria (Tandeau de Marsac, 1994). Being the smallest unit of infection, they are generally formed for the purpose of asexual reproduction in unicellular, filamentous forms and have affluent stores of nitrogen, phosphorus and other nutrients. Hormogonia can be dispersed either through their rapid motility on surface or even with the aid of gas vacuoles. Any unexpected environmental stress or transfer to a new medium can induce the differentiation of a cyanobacterium into hormogonium. In some genera such as *Rivularia*, a number of hormogonia are required for generation of a colony while in case of others such as *Nostoc*, colony originates from a single hormogonium. According to Herdman and Rippka (1988), a hormogonium possesses a lower number of genome copies as compared to vegetative cells. Nitrogen fixing plant cyanobacteria symbiosis (especially in case of genus *Nostoc*) is largely dependent on the formation of hormogonium. Plant host releases a 'hormogonium inducing factor' (HIF) that is recognized by the cyanobacterial symbiont. In response to this particular biochemical signal, the cyanobacterium differentiates into a hormogonium and further dedifferentiates back into vegetative cells after about 96 hours. By this time they manage to reach the desired site within their host.

Hormogonium is released from the end of a heterocystous trichome with the aid of a sacrificial cell called necridium. It facilitates division of some non-heterocystous forms and also formation of two daughter trichomes (Lamont, 1969).

Physiological and Habitat Diversity

Members of Cyanophyceae are ubiquitously present in numerous habitats ranging from hot springs to polar ice caps; dry soils to salt brines; plant roots to limestone etc. Their ability to inhabit all these extreme locales demonstrates the range of physiological modifications that they can undergo for survival. Each peculiar state of its existence reveals the amendments that it has to make at all possible levels of functional organization and hence is a mark of physiological diversity.

Cyanobacteria have been linked with freshwater blooms for times immemorial. A bloom generally comprises of only one or two species and the surface blooms are dominated by cyanobacteria capable of floatation with the help of gas filled cell inclusions called gas vacuoles. Members varying in structure from small filaments to large globular colonies can form blooms, for *e.g.* filamentous forms including *Anabaena, Nodularia, Gloeotricha, Oscillatoria, Spirulina* etc. and nonfilamentous forms such as *Microcystis, Coelosphaeria, Gomphosphaeria* etc. The filamentous forms often get twisted and entangled to form secondary aggregations (Reynolds and Walsby, 1975; Lewis, 1976) that leads to immense increase in the biomass. Shallow, well mixed, eutrophic lakes are dominated by *Oscillatoria*; clear water, temperate lakes are inhabited by *Aphanizomenon flos-aquae* and *Oscillatoria agardhii*; tropical waters experience a sequential growth of different cyanobacteria such as *Dactylococcopsis* (capable of growth in water at times of high nutrient availability and mixing and low light intensity), *Aphanotheceae* (growing under moderate to low nutrient availability, lesser turbulence and better light conditions), and some other genera such as *Microcystis aeruginosa* and *M. flos-aquae* are permanent inhabitants (Oliver and Ganf, 2000). Stratification in terms of nutrients, light intensity, turbidity etc. and seasonal periodicity determine the species dominating the habitat at a particular time. Appearance of bloom is not a sudden phenomenon, as it is normally thought. It is a physiological adaptation of already existing species in response to the changing environmental conditions. On calming down of water bodies carrying dispersed population of cyanobacteria, latter develop gas vacuoles and move upwards. There are several benefits associated with the buoyancy control such as, reduction in sedimentation losses (Reynolds, 1984), ability of better light harvest (Walsby *et al.*, 1997) and capacity to utilize the available nutrients in a better way (Ganf and Oliver, 1982). Phytoplanktonic forms also show a diversion from the normal nutrient cycle as an adaptation to the current requirement. Apart from carbohydrate accumulation they exhibit a lesser cell-specific quantum yield of photosynthesis (Turpin, 1991). The amount of limiting nutrient (nitrogen, phosphorus or others depending on the water body) within the cell gets highly reduced (Riegman and Mur, 1984) and an augmentation in the specific uptake rate of the limiting nutrient (Kromkamp, 1987; Riegman and Mur, 1984) is generally observed. Inclusion of many representatives from the family Nostocaceae indicate nitrogen fixation being performed under nitrogen limiting conditions during phytoplankton growth.

The oceanic environment is generally considered unproductive or oligotrophic and short of essential nutrients. Even then, former can sustain a variety of cyanobacteria ranging from picoplanktons (<5 mm), nanoplanktons (5–20 mm) to microplanktons (>5 mm). They act as major carbon fixers (*Synechococcus*, *Synechocystis* and *Prochlorococcus*) and also as dinitrogen fixers (*Trichodesmium*, *Nodularia* etc.) when present in nitrogen deplete waters. The localizers vary from unicellular (solitary–*Synechococcus*, *Aphanothece* and colonial–*Merismopedia*) to filamentous forms (heterocystous–*Anabaena*, *Richelia* and non-heterocystous–*Lyngbya*, *Spirulina*). Apart from showing adaptations to tolerate nutrient deficiency, most of them are also capable of tolerating saline conditions. Though nitrogen fixation gets highly hampered under increasing salt concentrations, other metabolic processes are stabilized by synthesis of compatible osmolytes (Reed and Stewart, 1985).

Cyanobacteria dwelling on and below soils withstand a variety of stresses such as desiccation, pH alterations, ultra violet radiation, low light intensity, mineral and nutrient deficiency and presence of herbicides and pesticides. Water and salt stress tolerance is prevalent in many species but they are not abundant in soils with pH less than 6. Contrary to the above fact, they can thrive well at low pH if the temperature is considerably high as in tropical soils. There have been prior documentations of use of cyanobacteria as a biofertilizer for reclamation of saline (usar) soils in India (Singh, 1961). *Microcoleus* is perhaps the most widespread cyanobacteria in saline soils (Whitton, 1990). According to Shibata (1969), some cyanobacteria accumulate colorless UV absorbing compounds which are members of a family of mycosporine-like amino acid derivatives (MAAs) (Garcia-Pichel and Castenholz, 1993; Sinha *et al.*, 1998). Cyanobacteria living on the soil surface generate dark coloration as an adaptive strategy against short wavelength insolation. The pigment 'scytonemin' is responsible for the dark colour and is capable of absorbing UV radiation (Garcia-Pichel and Castenholz, 1991). Those species which are present under soil are skilled for photosynthesizing at low photon flux densities (van Liere and Walsby, 1982). Studies have been performed on herbicide, fungicide and pesticide resistance in cyanobacteria (Peterson *et al.*, 1997; Das and Adhikary, 1996; Tiwari *et al.*, 1991). Results have shown that *Anabaena*, *Nostoc* and *Oscillatoria* are found to be tolerant to Fentin derivatives and sodium dithiocarbamate (Bisiach, 1970) and in general nitrogen fixing cyanobacteria are comparatively tolerant to 2,4 D (2,4-dichlorophenoxy acetic acid) (Leganes and Fernandez-Valiente, 1992).

Cyanobacterial occupants are diversely scattered from extremely hot to cold desert ecosystems. Desert dwellers can be divided into soil crust species, *e.g. Lyngbya*, *Phormidium*, *Plectonema* (Friedmann and Galun, 1974) etc. and *Nostoc commune*, *Anacystis montana*, *Oscillatoria lutea*, *Schizothrix calcicola* etc. particularly in cold deserts (Cameron, 1971); epilithic species such as *Gloeocapsa* spp., *Stigonema minutum*, *Calothrix* spp. etc. (Broady, 1989); and endolithic species such as *Hormathonema*, *Gloeocapsa* (Friedmann *et al.*, 1993). They have adopted several strategies to combat typical stresses found in such environments–EPS mucilage production for adhesion to substratum (Cameron and Devaney, 1970);

absorption of snow, water vapour and sheltered growth to overcome low humidity (Vestal, 1988; Palmer and Friedmann, 1990); production of trehalose and polyols to combat desiccation (Potts, 1994); and sunscreens such as caretenoids for tolerance of high light intensity (Downes *et al.*, 1993); reallocation of carbon and nitrogen, autotrophy and symbiosis under oligotrophic conditions (Friedmann and Ocampo, 1976; Vestal, 1988) and several other sunscreens to tolerate UV radiation (Garcia-Pichel and Castenholz, 1991).

Cyanobacteria are believed to not only survive but thrive optimally in harsh conditions like high latitude, glaciers, ice shelves, streams, ponds and lakes (Vincent, 2000). Some common species found in this extreme habitat are *Calothrix parietina* (Wharton *et al.*, 1983), *Phormidium frigidum, Nodularia harveyana, Lyngbya martensiana* etc. The importance of cyanobacteria is highlighted in areas like exposed moraines after the retreat of glaciers where they act as primary colonizers. They display a high degree of resilience as the optimum temperature for their growth ranges from 15 to 30°C and they still show tolerance for temperatures around 5°C. Apart from temperature regulatory adaptations, they share same desiccation, UV, light and salinity tolerance mechanisms as present in desert inhabiting cyanobacteria.

Some strains can tolerate free sulphide much better than other eukaryotic algae (Whitton and Potts, 2000; Padan and Cohen, 1982) and further many can even utilize H_2S in place of H_2O as a hydrogen donor (Cohen *et al.*, 1975). Under some circumstances, certain cyanobacteria show well elaborated circadian rhythms (Kondo and Ishiura, 1999) to accord with the fluctuating light and temperature during the day (Whitton and Potts, 2000).

Biochemical and Molecular Diversity

Taxonomy of cyanobacteria has been a highly debatable topic and has undergone several revisions (Turner *et al.*, 1999; Turner, 1997; Wilmotte and Golubic, 1991; Komarek, 1994). Bacteriological criteria have gained more importance than botanical criteria and major consideration for taxonomic determination depends heavily on morphological and physiological attributes. Latter are the consequences of various environmental and growth conditions. Sometimes these characters can even be lost during cultivation. Such contemplations among cyanobacteria lead to inaccurate identification and classification among taxa (Nelissen *et al.*, 1992). At present, cyanobacterial diversity assessment depends reliably on chemotaxonomic parameters including whole cell protein, isozyme analysis, fatty acid profiles, carotenoids or mycosporine like amino acids (Prasanna and Kaushik, 2005). These techniques are of particular interest in case of toxic cyanobacteria as toxicity is very insignificantly related to morphology. Therefore, differentiation of toxic and non-toxic cyanobacteria on morphological basis is barely possible (Golubic *et al.*, 2009). Difference in the protein profile has been utilized in number of instances. SDS-PAGE of whole cell protein has been used to obtain heterozygosity among planktonic filamentous cyanobacteria (Lyra *et al.*, 1997). MALDI-TOF MS analysis of oligopeptides within a natural population of *Microcystis* spp. revealed possession of a particular group of peptides by individual species (Fastner *et al.*, 2001). Cluster analysis among *Azolla* isolated cyanobionts (Sood *et al.*, 2007) and among *Anabaena* (from 6 *Azolla* species), *Notoc* (from *Anthoceros* sp., *Cycas* sp. and *Gunnera monoika*) and certain free living cyanobacteria (Sood *et al.*, 2008) disclosed taxonomic relationships based on whole cell protein profiles and molecular markers. Another important criterion is based on zymogram analysis and isozyme profile is considered a tool of utility for taxonomic studies at species and genus level (Holton, 1981). Heterogeneity in band pattern was observed between two unicellular strains, *Synechococcus* sp. strain PCC 6301 and *Synechococcus elongates* strain CCAP 1497/1 (Schenk *et al.*, 1973) where no common bands were found, in marine phytoplankton (Medlin *et al.*, 2000) and among axenic *Anabaena* sp. strains where five enzymes were studied (Stulp and Stam, 1984). α-esterase isozyme pattern was earlier studied in *Phormidium* strains (Klein *et al.*, 1973) and similarity obtained in zymograms of *Phormidium ectocarpi* strain PCC 7375 and *Phormidium persicinum* strain CCAP 1462/5 was used for their assignment to the same species (Wilmotte, 1994). Utility of phycoerythrin synthesis ability (Bryant, 1982) and fatty acid profile (Kenyon *et al.*, 1972) as a taxonomic marker has been well established. Fatty acid profiling has been used as a biochemical marker in various studies such as diversity assessment in 66 cyanobacetrial strains (Kenyon, 1972; Kenyon *et al.*, 1972), among free living strains of *Nostoc* and *Anabaena* (Caudales *et al.*, 1992) and marine picoplankton (Merritt *et al.*, 1992). A valuable unknown highly polar glycolipid was reported (Sallal *et al.*, 1990) which was found only in three heterocystous strains (namely *Anabaena Cylindrica*, *Nostoc canina* and *Nostoc muscorum*) studied (Wilmotte, 1994).

Still, analyses of photosynthetic pigment content, isoenzyme variation or differentiated cell culture may also be misleading because of the variable expression of cyanobacterial gene products in culture (Kato *et al.*, 1991; Rippka *et al.*, 1979). Morphological changes are based on factors such as cell dimensions and ecology which are not very reliable for distinct separation as per the species definition (Nubel *et al.*, 1997). This creates a vacuum in taxonomic studies which can be fulfilled by speculations based on molecular and genetic data. An appropriate blend of phylogenetic, phenotypic and genotypic analysis forms the basis of an unwavering form of taxonomy known as 'polyphasic taxonomy' (Luuc *et al.*, 1999; Vandamme *et al.*, 1996). A combination of genetic and phenotypic strategies aid in enhancement of methodology for complex natural community analysis (Rudi *et al.*, 2000; Nubel *et al.*, 1999). There are several examples of contradictions between phylogenetic results obtained on the basis of morphological and molecular data. For example, *Nostoc* PCC 7120 was formerly regarded as *Anabaena* species and *Nostoc* PCC 6720 was at one time described as *Anabaenopsis* (Henson *et al.*, 2002). On the basis of morphological characteristics and life cycle differences, *Nostoc* and *Anabaena* were traditionally placed in different groups (Turner *et al.*, 1999; Turner, 1997; Wilmotte, 1994). But now, 16S rRNA studies have shown that *Nostoc* and *Anabaena* are closely related. *Fischerella* spp. and *Chlorogloeopsis* spp. are placed in stigonematales order based on true branching; however, the two genera are clearly distinguished on the basis of 16S rRNA sequences (Wilmotte, 1994). From the above discussion it can made out that due to environmental influences on morphological characteristics, the use of morphological or physiological data or even biochemical data is insufficient. Under such circumstances, analysis of molecular data becomes the most effective and valued method of determining genetic variability and evolutionary relationships among cyanobacteria. Genotypic characters such as DNA-RNA hybridization and DNA base composition are a direct indication of relatedness of various organisms that can serve as taxonomic markers for phylogenetic purpose (Wood and Townsend, 1990; Lachance, 1981; Herdman *et al.*, 1979). Among molecular parameters, amplification and sequence analysis of hypervariable and conserved region of 16S rRNA, *nifH*, *nifD* and PC-IGS regions are most widely used to detect genetic relatedness and molecular phylogeny within the cyanobacteria. The 16S rRNA molecule contains variable regions (Woese, 1987); it is very highly conserved and can be instrumental in studying species identity (Fox *et al.*, 1992) or intra-species variation. *NifH* and *nifD* genes possess a highly conserved region as well as a great divergence in other regions. The molecular analysis of highly variable and conserved sequence of 16S rRNA and nitrogenase (*nifH* and *nifD*) genes (*i.e.* highly conserved throughout the evolution) will not only provide a stable genetic marker but will also serve to infer the phylogenetic relationship among the heterocystous cyanobacteria. DNA sequences of a fragment *nifH* and *nifD* from about diverse cyanobacterial strains were amplified, cloned and sequenced to determine the evolutionary relationship within cyanobacteria as a group

(Henson *et al.*, 2004; Zehr *et al.*, 1995). Studies examining the 16S rRNA have indicated that morphological characters may not coincide with molecular data in cyanobacteria.

PCR/RFLPs/RAPDs are among the most popular techniques employed by cyanobacteriologists to obtain more precise and genuine data about the cyanobacterial diversity and their phylogenetic relationship. PCR-amplified fragments can be used as substrate for RFLP or sequence analysis.

RAPD is a type of PCR reaction in which DNA segments are randomly amplified. Short arbitrary primers are used for binding to their specific sequence in a large template of genomic DNA (Bardakci, 2001). Its advantage lies in the fact that no prior knowledge of the entire genome is required for its operation. This technique allows random identification of polymorphisms at genetic level and aids in deducing phylogenetic relationships. This method has certain drawbacks which limit its application. From the band pattern obtained as a result of RAPD-PCR, it is difficult to make out whether a DNA segment is amplified from a heterozygous or a homozygous locus. The interpretation of RAPD results becomes difficult at times because of mismatching between the primer and the template resulting in total absence of PCR products. Apart from these technical downsides, it is cumbersome, non-reproducible at times and varies with handling due to involvement of various components as have been reported by number of workers (Hansen *et al.*, 1998; Newbury and Ford-Lloyd, 1993). A slightly modified and highly precise method of genomic fingerprinting involves the use of DNA primers corresponding to naturally occurring interspersed repetitive elements such as REP (Repetitive Extragenic Palindromic) and ERIC (Enterobacterial Repetitive Intergenic Consensus) sequences. PCR conducted by using both of these as primers is highly reproducible and comparatively simple for distinguishing closely related strains, inferring phylogenetic relationships and their diversity consideration in a variety of habitats (Lyra *et al.*, 2001; Rasmussen and Svenning, 1998).

RFLP analysis of DNA can be also used for genetic and phyletic analysis. This tool is of great help in identifying even minor variations in sequences if they are present in restriction sites. Another modification of this technique is 'Terminal Restriction Fragment Length Polymorphism (TRFLP). This technique was earlier used for characterization of bacterial communities in mixed-species samples, and is now used for other rhizoforms (Jin-Book *et al.*, 2004; Frias-Lopez *et al.*, 2003). Here, the primer pairs are labeled with fluorescent tags. After amplification with these specially tagged primers, PCR products are digested with restriction enzymes and the resulting sequences are visualized with a DNA sequencer.

Differentially, PCR products can also be analyzed through Denaturing Gradient Gel Electrophoresis (DGGE) (Boutte *et al.*, 2006). Here, non-descriptive similar sized PCR products can be differentiated as they move through increasingly higher concentrations of chemical denaturant. The weaker melting domains get fixed earlier as their threshold denaturant concentration is encountered earlier. This results in a pattern of bands in which each band represents a specific population.

RAPD has been used in identification of several organisms to strain level (Hadrys *et al.*, 1992). Neilan *et al.* (1995) performed RFLP of PCR products with a range of 4-base pair recognizing restriction endonucleases. 40 toxigenic strains of freshwater cyanobacteria belonging to genera *Anabaena, Microcystis, Aphanizomenon, Nodularia, Nostoc, Oscillatoria, Pseudomonas* and *Synechococcus* were analyzed; 28 equally parsimonious trees were obtained by using the DOLLOP programme from the PHYLIP package and the strains were clustered into two distinct groups. PCR fingerprinting using short tandemly repeated repetitive (STRR) sequence was performed by Zheng *et al.* (1999) to determine genetic diversity in cyanobacteria residing in different species of *Azolla*. The banding pattern depicted that cyanobacteria from *Euazolla* section showed similarity and this pattern was precisely different from that of *Rhizosperma* section. The occurrence of similar sized four clustered bands in cyanobacterial isolates from all *Azolla* species shows a common monophyletic origin. Costa *et al.* (2001) used RAPD with primers specific for the $tRNA^{leu}$ gene in *Nostoc* species present in symbiotic relationship with two species of thalloid bryophytes, *Anthoceros fusiformis* and *Blasia pusilla*. Results confirmed that several different strains of *Nostoc* are involved in symbiosis but single bryophyte thalli never contained colonies with a mixture of different intron sequences. RAPD analysis of 17 different cyanobacterial cultures derived from 6 different decamer primers (Jeberlin Prabina *et al.*, 2005) provide diagnostic fingerprints for each culture and their genetic distance (Perumal *et al.*, 2009). RAPD of genomic DNA using 7 different primers was used for identification and phylogenetic analysis of 12 cyanobacterial strains belonging to genera *Oscillatoria* and *Lyngbya* by Perumal *et al.* (2009). Some of the strains showed close genetic relationship while two *Lyngbya* strains were grouped separately.

16S rRNA/rDNA cloning, sequencing and further analytic reconstruction is one of the most popular bases for phylogenetic grouping and biodiversity assessment of cyanobacteria (Robertson, 2001; Turner *et al.*, 1999; Schmidt *et al.*, 1991; Giovannoni *et al.*, 1990; Giovannoni *et al.*, 1988). Comparative analysis with the aid of 16S rRNA gene sequence is a proper way to explore the discrepancy between strain collections and natural communities (Garcia-Pichel *et al.*, 1998, 1996). Application of DNA array hybridization using nonspecific labeling of the nucleic acid to 16S rDNA and 16S rRNA (Guschin *et al.*, 1997) is a powerful cyanobacterial appraisal tool. Rudi *et al.* (2000) investigated the presence and abundance of cyanobacteria in natural habitats using sequence specific labeled 16S rRNA gene probe. A good correlation was observed among the cyanobacterial groups *i.e.* *Nostoc*, *Microcystis* and *Planktothrix* spp. in the selected mesotrophic and eutrophic lakes. West and Adams (1997) did PCR amplification using either short arbitrary primers or primers specific for the regions flanking the 16S-23S rRNA internal transcriber spacer to evaluate the diversity of cyanobacterial symbionts from the hornwort *Phaeoceros laevis*. Even a single bryophyte thallus showed diverse populations of *Nostoc* strains but a single strain never occurred at more than one field site. *Anabaena* and *Aphanizomenon* were considered to be distinct on the basis of morphological characters (Gugger *et al.*, 2002a, b) but were shown to be highly similar on the basis of 16S rDNA profile (Gugger *et al.*, 2002; Lyra *et al.*, 2001). Thacker and Paul (2004) tried to use taxonomic relationships among several collections of *Lyngbya* spp. and *Symploca* spp. One of the various parameters taken into consideration was 16S rDNA sequence. A pair wise sequence divergence of 10 to 14 per cent was observed across 605 bp of 16S rDNA among the collected species (Thacker and Paul, 2004). Phylogenetic deduction on the basis of 16S rDNA sequences clearly voiced the reciprocal monophyly of *Lyngbya* and *Symploca*. Owing to the complexity of the 16S rDNA and 16S rRNA analysis, it is not suited for large-scale screening. Apart from 16S rRNA analysis, nitrogenase genes can also be employed for diversity estimation. Genes for nitrogenase enzyme complex are highly conserved and contain valuable taxonomic information (Ben-Porath and Zehr, 1994). Phylogenetic analysis of nucleotide sequence of *nifD* was used by Henson *et al.* (2004) to verify the distribution of heterocystous cyanobacteria belonging to subsections IV and V. Heterocystous cyanobacteria can be considered monophyletic but none of the subsections can be said to be monophyletic. Members of both subsections show intermixing in two sister clades and hence there is no meaning of two isolated subsections (Henson *et al.*, 2004). *NifH* analysis has been done to evaluate genetic diversity analysis of diazotrophs in the rice rhizosphere (Wartiainen *et al.*, 2008). Haande *et al.* (2008) characterized new isolates of *Cylinrospermopsis raciborskii* from Europe and Africa by sequencing *nifH* along with ITS1, PC-IGS and *rpoC1*. Within the African isolates and the European isolates, the amplified *nifH* gene showed 100 per cent sequence similarity while the similarity between the isolates from the two different continents was >93.3 per cent. All the parameters produced similar tree topology (Haande *et al.*, 2008). However 16S rRNA/rDNA and *nif* gene data show contamination with bacterial samples. To overcome this hindrance, study of phycocyanin gene has proved to be an influential tool. The PC operon consists of genes coding for 2 bilin subunits and 3 linker polypeptides (BelKnap and Hazelkorn, 1987). The intergeneric spacer (IGS) between the two bilin subunits–β (cpcB) and α (cpcA) is a highly variable region of DNA sequence and therefore serves prominently for cyanobacterial identification at strain level. Neilan *et al.* (1995) amplified the PC-IGS region of 44 cyanobacterial strains with specifically designed primers. The strains were distinctly grouped according to the results of RFLP data. In an attempt to validate the assumption that isolates able to grow in laboratory culture are representative of the entire diversity within the natural population, Hayes and Barker (1997) amplified the PC-IGS region of *Nodularia* strains taken from clonal cultures as well as natural populations. More variation was found among the culture isolates as compared to specimens from natural population. Bittencourt-Oliveira *et al.* (2001) used PC-IGS region amplification as a means to determine the genetic variability among 15 Brazilian strains of *Microcystis aeruginosa* and compare it with reference strains of *Microcystis aeruginosa*, *M. viridis*, and *M. wesenbergii*. Nucleotide sequences of the *cpc*BA IGS and flanking regions unambiguously stated that Brazilian strains classified as *M. aeruginosa* are phylogenetically diverse from the reference strains of *M. aeruginosa*. Phylogenetic analysis of *Anabaenopsis abijatae* and *Anabaenopsis elenkinii* from tropical inland water bodies of Kenya and Mexico showed low similarity value in PC-IGS sequences and thus the two clusters indicated two separate genera (Ballot *et al.*, 2008). Sequence analysis of ITS and *cpc*BA-IGS showed high level of differentiation between *Pseudoanabaena* isolates demonstrated considerable microdiversity (Acinas *et al.*, 2009).

Conclusion

Cyanobacteria are a diverse and cosmopolitan group possessing a number of unique properties. Their pliability for divergent atmospheric and vicinity situations is the major reason for their widespread existence. Members of this group are deemed as pioneers of life on earth. Their importance as primary producers, nitrogen fixers, and source of antibiotics, antifungal agents, restriction enzymes and pharmaceutically important products highlight the importance of their existence in biosphere. Their taxonomic classification has been based on morphological and physiological attributes that are prone to variations because of environmental and culture conditions. With this view, it has become important to analyze the characters and phylogeny of cyanobacteria on a larger canvas. Latter demands data analyses made on the basis of molecular parameters such DNA/RNA sequence of 16S ribosomal RNA, *nif* genes, phycocyanin genes etc. Advancements made in last two decades have enlightened the subject of cyanophycean description to a great extent. It is expected that an appropriate amalgamation of phenotypic, physiological and molecular surveillance will floor the way for cyanobacterial study to infinitesimal details.

Acknowledgement

Authors are thankful to the CSIR and DST, New Delhi for their financial support. Head, Department of Botany, BHU, Varanasi is gratefully acknowledged for their help and support.

References

Acinas, S.G., Haverkamp, H.A., Huisman, J. and Stal, L.J. (2009). Phenotypic and genetic diversification of *Pseudoanabaena* spp. (cyanobacteria). *The ISME J.* 3: 31-46.

Anagnostidis, K. and J. Komarek (1989). Modern approach to the classification system of Cyanophytes. 4–Nostocales. *Algological Studies* 56: 247–345.

Anand, N. (1988). Culture studies and taxonomy of blue-green algae–certain identification problems. *Arch. Hydrobiol. Suppl.* 80: 141–147.

Bardakci, F. (2001). Random amplified polymorphic DNA (RAPD) markers. *Turk J Biol.* 25: 185-196.

Bellknap, W.R. and Haselcorn, R. (1987). Cloning and light regulation of expression of the phycocyanin operon of the cyanobacterium *Anabaena. EMBO J.* 6: 871-884.

Ben-Porathj, Zehr J. (1994). Detection and characterization of cyanobacterial *nif*H genes. *Appl. Environ. Microbiol.* 60: 880-887.

Bergman, B. Gallon, J.R., Rai, A.N. and Stal, L.J. (1997). N_2 fixation by non-heterocystous cyanobacteria. *FEMS Microbiol Rev.* 19: 139-185.

Bergman, B., Rasmussen, U. and Rai, A.N. (2007). Cyanobacterial associations. In: Associative and Endophytic Nitrogen-Fixing Bacteria and Cyanobacterial Associations. Elmerich, C. and Newton, W.E. (ed) 5: 257–301. Dordrecht: Kluwer Acad.

Berman-Frank, I., Lundgren, P. and Falkowski, P. (2003). Nitrogen fixation and photosynthetic oxygen evolution in cyanobacteria. *Res Microbiol.* 154: 157-164.

Bisiach, M. (1970). Primi dati sulle infestazioni algali da cianoficee nelle risiae italiane. *Riso.* 19: 129-134.

Bittencourt-Oliveira, M.C., Oliveira, M. and Bolch, C.J.S. (2001). Genetic variability of brazilian strains of the *Microcystis aeruginosa* complex (Cyanobacteria/Cyanophyceae) using the phycocyanin intergenic spacer and flanking regions (*cpc*BA). *J. Phycol* 37: 810–818.

Broady, P.A. (1989). Survey of algae and other terrestrial biota at Edward VII peninsula, Marie Byrd Land. *Antarct Sci.* 1: 215-224.

Bryant, D.A. (1982). Phycoerythrocyanin and phycoerythrin: properties and occurrence in cyanobacteria. *J Gen Microbiol* 128: 835–844.

Boutte C., Grubisic, S. Balthasart, P. and Wilmotte, A. (2006). Testing of primers for the study of cyanobacterial molecular diversity by DGGE. *J. Microbiol Methods.* 65(3): 542-550.

Buikema, W.J and Haselkorn, R. (1991). Characterization of a gene controlling heterocyst differentiation in the cyanobacterium Anabaena 7120. *Genes Devel* 5: 321–330.

Cameron, R.E. and Devaney, J.R. (1970). Antarctic soil algal crusts. A scanning electron and microscope study. *Trans Am Microsc Soc.* 80: 264-273.

Carpenter, E.J and Foster, R.A. (2002). Marine Cyanobacterial Symbioses. In: Rai, A.N., Bergman, B. and Rasmussen, U. (eds) *Cyanobacteria in symbiosis.* pp. 11-17, Kluwer Academic Publishers, The Netherlands.

Castenholz, R.W. and Waterbury, J.B. (1989). Cyanobacteria (eds.) Staley, J.T., Bryant, M.P., Pfennig, N., and Holt, J.H. *Bergey's Manual of Systematic Bacteriology,* Vol. 3. Section 19: Oxygenic photosynthetic bacteria. Williams and Wilkins, Baltimore.

Caudales, R., Wells, J.M. andAntoine, A.D. (1992). Cellular fatty acid composition of symbiotic cyanobacteria isolated from the aquatic fern *Azolla. J Gen Microbiol* 138: 1489–1494.

Cohen, Y.J., Padan, E. and Shilo, M. (1975). Facultative anoxygenic photosynthesis in the cyanobacterium *Oscillatoria limnetica. J. Bacteriol.* 123: 855-863.

Costa, J.L., Paulsrud, P., Rikkinen, J. and Lindblad, P. (2001). Genetic Diversity of *Nostoc* Symbionts Endophytically Associated with Two Bryophyte Species. *Appl Env Microbiol.* 67(9): 4393-4396.

Das, M.K. and Adhikary, S.P. (1996). Toxicity of three pesticides to several rice-field cyanobacteria. *Tropical Agriculture* 73: 155-157.

Desikachary, T.V., (1959). Cyanophyta. Indian Council of Agricultural Research, New Delhi. 686 pp.

Downes, M.T., Hrstich, L. and Vincent, W.F. (1993). Extraction of chlorophyll and caretenoid pigments from Antarctic benthic mats for analysis by HPLC. *J. Appl Phycol.* 5;623-628.

El-Shehawy, R., Lugomela, C., Ernst, A. and Bergman, B. (2003). Diurnal expression of *hetR* and diazocyte development in the filamentous non-heterocystous cyanobacterium *Trichodesmium erythraeum*. *Microbiol* 149: 1139-1146.

Ernst, A., Black, T., Cai, Y., Panoff, J.M., Tiwari, D.N. and Wolk, C.P. (1992). Synthesis of nitrogenase in mutants of the cyanobacterium *Anabaena* sp. strain PCC 7120 affected in heterocyst development or metabolism. *J Bacteriol* 174: 6025–6032.

Fastern, J., Erhard, M. and Von Dohren, H. (2001). Determination of Oligopeptide Diversity within a Natural Population of *Microcystis* spp. (Cyanobacteria) by Typing Single Colonies by Matrix-Assisted Laser Desorption Ionization–Time of Flight Mass Spectrometry. *Appl Env Microbiol.* 67(11): 5069-5076.

Fay, P. (1992): Oxygen relations of nitrogen fixation in cyanobacteria. *Microbiol. Rev.* 56: 340-373.

Fox, G.E., Wisotzkey, J.D. and Jurtshuk. P. Jr. (1992). How close is close: 16S rRNA sequence identity may not be sufficient to guarantee species identity. *Int. J. System. Bacteriol.* 42: 166–170.

Frias-Lopez, J., Bonheyo, G.T., Jin, Q. and Fouke, B.W. (2003). Cyanobacteria associated with coral black band disease in Caribbean and Indo-Pacific reefs. *Appl Env Microbiol.* 69(4): 2409-2413.

Friedmann, E.I., Kappen, L., Meyer, M.A. and Nienow, J.A. (1993). Long-term productivity in the cryptoendolithic microbial community of the Ross desert, Antarctica. *Microb Ecol.* 25: 51-69.

Friedmann, E.I. and Ocampo, R. (1976). Endolithic blue-green algae in the dry valleys. Primary producers in the Antarctic desert ecosystem. *Sci.* 193: 1247-1249.

Fritsch, F.E. (1945). Structure and function of the algae, Vol II, Cambridge University Press.

Ganf, G.G. and Oliver, R.L. (1982). Vertical separation of light and available nutrients as a factor causing replacement of green algae by blue-green algae in the plankton of a stratified lake. *J. Ecol.* 70: 829-844.

Garcia-Pichel, F. and Castenholz, R.W. (1991). Characterization and biological impliations of scytonemin, a cyanobacterial sheath pigments. *J. Phycol.* 27: 395-409.

Garcia-Pichel, F. and Castenholz, R.W. (1993). Occurrence of UV-absorbing, mycosporine-like compounds among cyanobacterial isolates and an estimate of their screening capacity. *Appl Env Microbiol.* 59: 163-169.

Geitler, L. (1932). Cyanophyceae. In: Rabenhorst's Kryptogamenflora von Deutschland, Österreich und der Schweiz, Leipzig, Akad. Verslagsges. 1932. Reprinted 1971, New York, Johnson, pp. 1-1196.

Giovannoni, S.J., Turner, S., Olsen, G.J., Barns, S., Lane, D.J. and Pace, N.R. (1988). Evolutionary relationships among cyanobacteria and green chloroplasts. *J. Bacteriol* 170: 3584-3592.

Giovannoni, S.J., Britschgi, T.B., Moyer, C.L. and Field, K.G. (1990). Genetic diversity in Sargasso Sea bacterioplankton. *Nature.* 345: 60-63.

Golden, J.W., Whorff, L.L. and Wiest, D.R. (1991). Independent regulation of *nifHDK* operon transcription and DNA rearrangement during heterocyst differentiation in the cyanobacterium *Anabaena* sp. strain PCC 7120. *J. Bacteriol.* 173: 7098-7105.

Golubic, S., Abedb, R.M.M., Palinskac, K., Pauillacd, S., Chinaine, M. and Laurent, D. (2009). Marine toxic cyanobacteria: Diversity, environmental responses and hazards. *Toxicon* (in press).

Gugger, M., Lyra, C., Suominen, I., Tsitko, I., Humbert, J.F., Salkonoja-Salonen, M.S. and Sivonen, K. (2002a). Cellular fatty acids as chemotaxonomic markers of the genera *Anabaena, Aphanizomenon, Microcystis, Nostoc* and *Planktothrix* (cyanobacteria). *Int J Syst Evol Microbiol* 52: 1007–1015.

Gugger, M., Lyra, C., Henriksen, P., Coute, A., Humbert, J.F. and Sivonen, K., (2002b). Phylogenetic comparison of the cyanobacterial genera *Anabaena* nad *Aphanizomenon*. *Int J Syst Evol Microbiol.* 52: 1867-1880.

Gupta, S. and Agrawal, S.C. (2006a). Survival of blue-green and green algae under stress conditions. *Folia Microbiol.* 51: 121–128.

Gupta, S. and Agrawal, S.C. (2006b). Motility in Oscillatoria salina as affected by different factors. *Folia Microbiol.* 51 565–572.

Haande, S., Rhorlack, T., Ballot, K., Skulberg, R., Martin, B. and Wiedner, C. (2008). Genetic characterization of *Cylindrospermopsis raciborskii* (Nostocales, Cyanobacteria) isolates from Africa and Europe. *Harmful Algae.* 7(5): 692-701.

Hadrys, H., Balick, M. and Schierwater, B. (1992). Applications of random amplified polymorphic DNA (RAPD) in molecular ecology. *Mol. Ecol.* 1(1): 55-63.

Hayes, P.K. and Barker, G.L.A. (1997): Genetic diversity within Baltic Sea populations of *Nodularia* (Cyanobacteria). *J. Phycol.* 33: 919-923.

Henson, M., Hallden, C. and Sall, T. (1998): Error Rates and Polymorphism Frequencies for Three RAPD Protocols. Pl Mol Biol Reporter. 16(2): 139.

Henson, B.J., Watson, L.E. and Barnum, S.R. (2002). Molecular differentiation of the heterocystous cyanobacteria, *Nostoc* and *Anabaena*, based on complete *nifD* sequences. *Curr Microbiol* 45: 161–164.

Henson, B.J., Hesselbrock, S.M., Watson, L.E. and Barnum, S.R. (2004). Molecular phylogeny of the heterocystous cyanobacteria (subsections IV and V) based on *nifD Int. J. System. Evol. Microbio.*, 54: 493-497.

Herdman, M. and Rippka, R. (1988). Cellular differentiation: Hormogonia and baeocytes. *Methods Enzymol.* 167: 232-242.

Herdman, M., Janvier, M., Waterbuyr, J.B., Rippka, R. and Stanier, R.Y. (1979). DNA base composition of cyanobacteria. *J. Gen. Microbiol.* 111: 63–71.

Jeberlin Prabina, B., Kumar, K. and Kannaiyan, K. (2005). DNA amplification fingerprinting as a tool for checking genetic purity of strains in the cyanobacterial inoculum. World J. Microbiol. Biotechnol. 21: 629-634.

Jin-Book, K., Mi-Sook, M., Dong-Hun, L., Sung-Taik, L., Marco, B. and Chi-Kyung, K. (2004). Comparative analysis of cyanobacterial communities from polluted reservoirs in Korea. *J. Microbiol.* 42(3): 181-187.

Kato, T., Watanabe, M.F. and Watanabe, M. (1991). Allozyme divergence in microcystis and its taxonomic inference. *Arch. Hydrobiol*, 64: 129-140.

Kenyon, C.N. (1972). Fatty acid composition of unicellular strains of blue-green algae. *J Bacteriol* 109: 827–834.

Kenyon, C.N., Rippka, R. and Stanier, R.Y. (1972). Fatty acid composition and physiological properties of some filamentous blue-green algae. *Arch Mikrobiol* 83: 216–236.

Klein, S. Chapman, D.J. and Garber, E.D. (1973). Chemotaxonomy of the *Oscillatoria-Phormidium* complex. *Biochem Syst* 1: 173–177.

Komarek, J. (1994). Current trends and species delimitation in the cyanoprokaryote taxonomy. In: , Cyanobacteria, Morphology, Taxonomy, Ecology, Toxicology: Proceedings of the 12th Symposium of the International Association for Cyanophyte Research (IAC), (eds. Cronberg, G., O.M. Skulberg, and T. Willen), Sjoarp (Sweeden) 1992. *Algological Studies* 75: 11-29.

Komarek, J. and Anagnostidis, K. (2005). Cyanoprokaryota, 2. Teil: Oscillatoriales. Heidelberg: Elsevier/ Spektrum.

Kondo, T. and Ishiura, M. (1999). The circadian clocks of plants and cyanobacteria. *Trends in Pl Sci.* 4(5): 171-175.

Kromkamp, J. (1987). Formation and functional significance of storage products in cyanobacteria. *N Z J Mar Freshwater Res* 21: 457-465.

Kylin, H. (1943). Zur biochemie der rhodophyceen. Forh. Kgl. Fysiograf. Sallsk. I Lund 13b: 1-13.

Lachance, M.A. (1981). Genetic relatedness of heterocystous cyanobacteria by deoxyribonucleic acid-deoxyribonucleic acid reassociation. *Int. J. Syst. Bacteriol.* 31: 139–147.

Lamont, H.C. (1969): Sacrificial cell death and trichome breakage in an Oscillatorean Blue-Green Alga: the role of murein. *Arch Microbiol.* 69: 237-259.

Leganes, F.and Fernandez-Valiente, E. (1992). Effects of phenoxy acetic herbicides on growth, photosynthesis, and nitrogenase activity in cyanobacteria from rice fields. *Archives of Environmental Contamination and Toxicology*. 22: 130–134.

Lewis, W.M. (1976). Surface/Volume Ratio: implications for phytoplankton morphology. *Science*. 192: 885-887.

Lugomela, C., Lyimo, T., Bryceson, I., Semesi, A.K. and Bergman, B. (2002). *Trichodesmium* in coastal waters of Tanzania: diversity, seasonality, nitrogen and carbon fixation. *Hydrobiologia* 477: 1–13.

Luuc, Mur R., Skulberg, O.M. and Utkilen, H. (1999). Cyanobacteria in the environment. In: Chorus, I. and Bartram, J. (eds) Toxic cyanobacteria in water; A guide to their public health consequences, monitoring and management, WHO.

Lyra, C., Hantula, J., Vainio, E., Rapala, J., Rouhiainen, L. and Sivonen, K. (1997). Characterization of cyanobacteria by SDS-PAGE of whole-cell proteins and PCR/RFLP of the 16S rRNA gene. *Arch Microbiol.* 168: 176-184.

Lyra, C., Suomalainen, S., Gugger, M., Vezie, C., Sundman, P., Paulin, L. and Sivonen, K. (2001). Molecular characterization of planktic cyanobacteria of *Anabaena, Aphanizomenon, Microcystis*, and *Planktothrix* genera. *Int. J. Syst. Evol. Microbial.* 51: 513-526.

Medlin, L.K., Lange, M. and Nothig, E. (2000). Genetic diversity in the marine phytoplankton: a review and a consideration of Antarctic phytoplankton. *Antarctic Sci.* 12(3): 325-333.

Mishra, A.K. (2003): Fox- and Fix-mutants of Anabaena 7120 defective in heterocyst development and nitrogen fixation. *Algol Stud.* 108, pp. 75-85.

Moore, D., O'Donohue, M., Garnett, C., Critchley, C. and Shaw, G. (2005). Factors affecting akinete differentiation in *Cylindrospermopsis raciborskii* (Nostocales, Cyanobacteria), *Freshwater Biol.* 50(2), pp. 345–352.

Nalewajko, C. and Murphy, T.P. (2001). Effects of temperature, and availability of nitrogen and phosphorus on the abundance of Anabaena and Microcystis in Lake Biwa, Japan: an experimental approach. *Limnology* 2: 45–48.

Neilan, B.A., Jacobs, D. and Goodman, A.E. (1995). Genetic diversity and phylogeny of toxic cyanobacteria determined by DNA polymorphism within the phycocyanin locus. *Appl Environ Microbiol* 61: 3875-3883.

Nelissen, B., Wilmotte, R., De Baere, R., Haes, F., Van de Peer, Y., Neefs, J.M. and R. de Wachter (1992). Phylogenetic study of cyanobacteria on the basis of 16S ribosomal RNA sequences, *Belg. Journ. Bot.* 125: 210-213.

Newbury, J.H. and Ford-Lloyd, B.V. (1993). The use of RAPD for assessing variation in plants. *Pl Growth Reg.* 12(1-2): 43-51.

Nubel, U., Garcia-Pichel, F. and Muyzer, G. (1997). PCR primers to amplify 16S rRNA genes from cyanobacteria. *Applied and Environmental Microbiology.* 63: 3327-3332.

Nubel, U., Garcia-Pichel, F., Kuhl, M. and Muyzer, G. (1999). Quantifying microbial diversity: morphotypes, 16S rRNA genes, and carotenoids of oxygenic phototrophs in microbial mats. *Appl. Environ. Microbiol.* 65: 422-430.

Ohki, K. (2008). Intercellular localization of nitrogenase in a non-heterocystous cyanobacterium (cyanophyte) *Trichodesmium* sp. NIBB1067. *J Oceanography.* 64(2);211-216.

Oliver, R.L. and Ganf, G.G. (2000). Freshwater blooms. In: Whitton, B. A. and Potts, M. (eds) The Ecology of Cyanobacteria, pp. 149–194. Kluwer Academic Publishers.

Padan, E. and Cohne, Y. (1982). Anoxygenic photosynthesis. In: Carr, NG and Whitton, BA (eds) *The biology of cyanobacteria*.pp 215-235. Blackwell, Oxford and University of California Press, Berkeley.

Palmer, R.J.J. and Friedmann, E.I. (1990). Water relations and photosynthesis in the cryptoendolithic microbial habitat of hot and cold deserts. *Microb Ecol.* 19: 111-118.

Perumal, G.M., Ganesan, V. and Anand, N. (2009). Identification and phylogenetic analysis of filamentous Cyanobacteria using random amplified polymorphic DNA (RAPD) fingerprinting. *African J Biotech.* 8(6): 974-978.

Perkins, S.K. and Peters, G.A. (1993). The *Azolla-Anabaena* symbiosis: endophyte continuity in the *Azolla* life-cycle is facilitated by epidermal trichomes. I. Partitioning of the endophytic *Anabaena* into developing sporocarps. *New Phytol.* 123: 53-64.

Peterson, H.G., Boutin, C., Freemark, K.E. and Martin, P.A. (1997). Toxicity of hexazinone and diquat to green algae, diatoms, cyanobacteria and duckweed. *Aquatic Toxicol.* 39: 111-134.

Pinevich, A.V., Averina, S.G., Gavrilova, O.V. and Migunova, A.V. (2008). Baeocytes in the cyanobacterium Pleurocapsa sp.: Characterization of the differentiated cells produced by multiple fission. *Microbiol.* 77(1): 62-68.

Potts, M. (1994). Dessication resistance of prokaryotes. *Microbiol Revs.* 58: 755-805.

Prasanna, A. and Kaushik, B.D. (2005). Algal diversity- from morphology to molecules. In: Satyanarayana, T. and Johri, B.N. (eds) Microbial diversity: current perspectives and potential applications. pp. 323-344. International Publishing House Pvt. Ltd., New Delhi.

Prasanna, R., Sood, A., Suresh, A., Nayak, S. and Kaushik, B.D. (2007). Potentials and applications of algal pigments in biology and industry. *Act Biol Hungarica.* 49(1-2): 131-156.

Rasmussen, U. and Svenning, M.M. (1998). Fingerprinting of Cyanobacteria Based on PCR with Primers Derived from Short and Long Tandemly Repeated Repetitive Sequences.*Appl Env Microbiol.* 64(1): 265-272.

Reed, R.H. and Stewart, W.D.P. (1985). Osmotic adjustment and organic solute accumulation in unicellular cyanobacteria from planktonic and marine habitats. *Mar Biol.* 88: 1-9.

Reynolds, C.S. and Walsby, A.E. (1975). Water blooms. *Biol. Rev.,* 50: 437–481.

Riegman, R. and Mur, L.C. (1984). Regulation of phosphate uptake kinetics in *Oscillatoria agardhii. Arch Microbiol.* 139: 28-32.

Rippka, R., Deruelles, J., Waterbury, J.B., Herdman, M. and Stanier, R.Y. (1979). Genetic assignments, strain histories and properties of pure culture of cyanobacteria. *J.Gen. Microbiol.,* 111: 1-61.

Robertson, B.R., Tezuka, N.R. and Watanabe, M.M. (2001). Phylogenetic analyses of Synechococcus strains (cyanobacteria) using sequences of 16S rDNA and part of the phycocyanin operon reveal multiple evolutionary lines and reflect phycobilin content. *Int J Systm Evol Microbiol.* 51: 861-871.

Rudi, K., Skulberg, O.M., Skulberg, R. and Jacobsen, K.S. (2000). Application of Sequence-Specific Labeled 16S rRNA Gene Oligonucleotide Probes for Genetic Profiling of Cyanobacterial Abundance and Diversity by Array Hybridization. *Appl Env Microbiol.* 66(9): 4004-4011.

Sallal, A.K., Nimer, N.A. and Radwan, S.S. (1990). Lipid and fatty acid composition of freshwater cyanobacteria. *J Gen Microbiol* 136: 2043–2048.

Schenk, H.E.A., Hofer, I. and Metzner, H. (1973). Malat-Dehydrogenase Isoenzymbanden als potentielles chemotaxonomisches Kriterium für Cyanophyceen-Species. *Biochem System* 1: 179–184.

Schmidt, T.M., Delong, E.F. and Pace, N.R. (1991). Analysis of a marine picoplankton community by 16S rRNA gene cloning and sequencing. *J. Bacteriol.* 173(14): 4371-4378.

Shibata, K. (1969). Pigments and a UV-absorbing substance in corals and a blue green alga living in the Great Barrier Reef. *Plant Cell Physiol.* 10: 325-335.

Sinha, R.P., Klisch, M., Groniger, A. and Hader, D.P. (1998). Ultraviolet-absorbing/screening substances in cyanobacteria, phytoplankton and macroalgae. *J. Photochem. Photobiol. B: Biol.* 47: 83-94.

Six, C., Thomas, J.C., Brahmsha, B., Lemoine, Y. and Partensky, F. (2004). Photophysiology of the marine cyanobacterium *Synechococcus* sp. WH8102, a new model organism. *Aqua Microbiol Ecol.* 35: 17-29.

Sood, A., Prasanna, R. and Singh, P.K. (2007). Fingerprinting of freshly separated and cultured cyanobionts from different *Azolla* species using morphological and molecular markers. *Aqua Bot.* 2: 142-147.

Sood, A., Prasanna, R., Prasanna, B.M. and Singh, P.K. (2008). Genetic diversity among and within cultured cyanobionts of diverse species of *Azolla. Folia Microbiol.* 53(1): 35-43.

Stulp, B.K. and Stam, W.T. (1984). Genotypic relationships between strains of *Anabaena* (Cyanophyceae) and their correlation with morphological affinities. *Br Phycol J* 19: 287–301.

Tandeau de Marsac, N. (1994). Differentiation of hormogonia and relationships with other biological processes. In: Bryant, DA (ed) The molecular biology of cyanobacteria, pp. 825-842. Kluwer Academic Publishers, Dordrecht, The Netherlands.

Tiwari, D.N., Kumar, A. and Mishra, A.K. (1991). Use of cyanobacterial diazotrophic technology in rice agriculture. *Appl Biochem Biotechnol.* 28/29: 387-396.

Thacker, R.W. and Paul, V.J. (2004). Morphological, chemical and genetic diversity of tropical marine cyanobacteria *Lyngbya* spp. And *Symploca* spp. (Oscillatoriales). *Appl and Env Microbiol.* 70(6): 3305-3312.

Turner, S. (1997). Molecular systematics of oxygenic photosynthetic bacteria. *Plant syst Evol.* 11: 13-52.

Turner, S., Pryer, K., Miao, V. and Palmer, J. (1999). Investigating deep phylogenetic relationships among cyanobacteria and plastids by small subunit rRNA sequence analysis. *J. Eukaryote Microbiol.* 46: 327-338.

Turpin, D.H. (1991): Effect of inorganic N availability on algal photosynthesis and carbon metabolism. *J. Phycol* 27: 14-20.

Usher, K.M., Bergman, B. and Raven, J.A. (2007). Exploring cyanobacterial mutualisms. *Ann Rev Ecol Evol Systm.* 38: 255-273.

Vestal, J.R. (1988). Carbon metabolism in the cryptoendolithic microbiota in the Antarctic desert. *Appl Env Microbiol.* 54: 960-965.

Van Liere, L. and Walsby, A.E. (1982). Interactions of cyanobacteria with light. In: Carr, NG and Whitton BA (eds) The biology of Cyanobacteria, pp. 9-45. Blackwell Scientific Publications, Oxford.

Vandamme, P., Pot, B., Gillis, M., De Vos, P., Kersters, K. and Swings, J. 1996. Polyphasic taxonomy, a consensus approach to bacterial systematics. *Microbiol. Rev.* 60: 407-438.

Vincent, W.F. (2000). Cyanobacterial fominance in the polar regions. In: Whitton, BA and Potts, M (eds) The ecology of cyanobacteria. pp. 321-340. Kluwer Academic Publishers, The Netherlands.

Walsby, A.E., Hayes, P.K., Boje, R. and Stal, L.J. (1997). The selective advantage of buoyancy provided by gas vesicles for planktonic cyanobacteria in the Baltic Sea. *New Phytol.*, 136, 407–417.

Wartiainen, I., Eriksson, T., Zheng, W. and Rasmussen, U. (2008). Variation in the active diazotrophic community in rice paddy-*nif*H PCR-DGGE analysis of rhizosphere and bulk soil. *Appl Soil Ecol.* 39(1): 65-75.

Waterbury, J.B. (1989). Subsection II, order Pleurocapsales Geitler 1925, emend. Waterbury and Stanier 1978. In: Staley JT, Bryant MP, Pfennig N and Holt JG (eds) *Bergey's Manual of Systematic Bacteriology*, Vol 3, pp. 1746–1770. Williams and Wilkins Co., Baltimore.

Waterbury, J.B. and Rippka, R. (1989). Subsection I, order Chroococcales Wettstein 1924, emend. Rippka *et al.*, 1979. In: Staley JT, Bryant MP, Pfennig N and Holt JG (eds) Bergey's Manual of Systematic Bacteriology, Vol 3, pp. 1728–1746. Williams and Wilkins Co, Baltimore.

West, N.J. and Adams, D.G. (1997). Phenotypic and genotypic comparison of symbiotic and free-living cyanobacteria from a single field site. *Appl. Environ. Microbiol.* 63: 4479-4484.

Wharton, R.A., Parker, B.C. and Simmons, G.M. (1983). Distribution, species composition and morphological algal mats (stromatolites) in Antarctic dry valley lakes. *Phycologia.* 22: 355-365.

Whitton, B.A. (1990). *Microcoleus.* In: Kumar HD (ed) Phycotalk, pp. 173-182. Rastogi Publications, Meerut, India.

Whitton, B.A. (1992). Diversity, ecology, and taxonomy of the cyanobacteria. In: Mann NH and Carr NG (eds) Photosynthetic Prokaryots, pp. 1-51. Plenum, New York.

Whitton, B.A. and Potts, M. (2000). The ecology of cyanobacteria. Kluwer Academic Publishers, The Netherlands.

Wilmotte, A. (1994). Molecular evolution and taxonomy of cyanobacteria. In: Bryant D (ed) The molecular biolgy of cyanobacteria. Netherlands: Kluwer Academic Publishers, pp, 1-25.

Wilmotte, A. and Golubic, S. (1991). Morphological and genetic criteria in the taxonomy of cyanophyta/ cyanobacteria. *Arch. Hydribio.* Suppl. 64: 1-24.

Wilmotte, A. (1994). Molecular Evolution and Taxonomy of the Cyanobacteria. In: Bryant, A. (ed) The molecular biology of cyanobacteria, pp. 1-25. Kluwer Academic Publishers, The Netherlands.

Woese, C.R. (1987). Bacterial evolution. *Microbiol Rev.* 51(2): 221-271.

Wolk, C.P. (2000). Heterocyst formation in *Anabaena.* In: Brun, YV, Shimkets, LJ (eds) Prokaryotic development. American Society of Microbiology, pp 83-104, Washington, D.C.

Wood, A.M. and Townsend, D. (1990). DNA polymorphisms within the WH7803 serogroup of marine *Synechococcus* spp. (cyanobacteria). *J. Phycol.* 26: 576–585.

Zapomelova, E., Hrouzek, P., Rehakova, K., Sabacka, M., Stibal M., Caisova, L., Komarkova, J. and Lukesova, A. (2008). Morphological variability in selected heterocystous cyanobacterial strains as a response to varied temperature, light intensity and medium composition. *Folia Microbiol.* 53(4): 333-341.

Zehr, J. P., Mellon, M.T., Braun, S., Litaker, W., Steppe, T. and Paerl, H.W. (1995). Diversity of heterotrophic nitrogen Fixation genes in a marine cyanobacterial mat. *Appl Environ Microbiol* 61: 2527-2532.

Zehng, W.W., Nilsson, M., Bergman, B. and Rasmussen, U. (1999). Genetic diversity and classification of cyanobacteria in different Azolla species by the use of PCR fingerprinting. TAG Theor Appl Gen. 99(7-8): 1187-1193.

Microbial Biotechnology and Ecology (2011) *Pages* **60–93**
Editors: **Deepak Vyas, G.S. Paliwal, P.K.Khare and R.K Gupta**
Published by: **DAYA PUBLISHING HOUSE, NEW DELHI**

Chapter 5

Hydrogen Production by Cyanobacteria

☆ ***Deepak Vyas[1]* and Rajan Kumar Gupta[2]***

[1]Department of Botany, Dr. H.S. G. University, Sagar – 470 003, M.P.
[2]Government P.G. College, Rishikesh, Uttranchal

Introduction

First of all Jackson and Ellms (1896) reported that *Anabaena* species from a massachusetts reservoir immediately produced H_2 when placed in a sealed bottle. Whether H_2 production was by algae or by contaminationg bacteria is still unclear. There was a considerable gap in this area of research since the first report appeared. It was in 1942, that gaffron and Rubin demonstrated H_2 production by anaerobically incubated *Secnedesmus obliquus*, a green alga, and since then this alge has become a major focus of research for H_2 metabolism. Blue-green algae again came into focus when the occurrence of a reversible hydrogenase was demonstrated in a unicellular blue green alga *synechococcus elongatus* obviously after several hours of anaerobic preincubation (Frenkel *et al.*, 1950). Studies on H_2 metabolism by blue-green algae restarted only after the unequivocal report of N_2 fixation by this group of algae. The real H_2 production mediated by the nitrogenase enzyme complex was first of all reported in *Anabaena cylindrical* under in vitro conditions by Haystead *et al.* (1970). Subsequently H_2 production by intact filaments of *A. cylindrica* in the light under an atmosphere of argon and CO_2 was reported by Benemann and Weare (1974a,b). In the same year H_2 production by other cyanobacteria was independently reported by Russian worklers (Oshchepkov *et al.*, 1974). Since then extensive work has been done on various aspects of H_2 metabolism employing a variety of blue-green algal strains (Tamagnini *et al.*, 2002; Vyas and Kumar, 1995; Vyas and Gupta, 2003; Vyas, 2004) (Table 5.1).

* E-mail: dvyas64@yahoo.co.in

Enzymes Involved in H_2 Metabolism

Nitrogenase

Nitrogenase, a typical enzyme present in all the N_2–fixing microorganisms, forms H_2 both under in vivo and in vitro conditions with normal growth at the expense of dinitrogen, nitrogenase reduces N_2 to ammonia under consumption of electrons, protons, and ATP. Some of the reduced protons are evolved as H_2. H_2 formation by a blue-green algal nitrogenase was first of all demonstrated in vitro by Haystead and coworkers (1970). Nitrogenase has been isolated and purified by a number of workers (see Stewart, 1980; Lambert and Smith, 1981; fay and Van baalen, 1987).

Table 5.1: Important Reviews on H_2 Metabolism of Blue-green Algae (Cyanobacteria)

1.	Benemann 1977
2.	Benemann and Hallenbeck 1978
3.	Bothe *et al.*, 1978
4.	Schlegel and Schneider 1978
5.	Hallenbeck and Benemann 1979
6.	Benemann *et al.*, 1980
7.	Bothe *et al.*, 1980
8.	Lambert and Smith 1981
9.	Bothe 1982
10.	Kerfin and Boger 1982
11.	Kumazawa ans Mitsui 1982
12.	Houchins 1984
13.	Robinson *et al.*, 1986
14.	Calvin and Taylor 1989
15.	Tamagnini *et al.*, 2002
16.	Vyas and Kumar 1995
17.	Vyas and Gupta 2003
18.	Vyas, 2004

Hydrogenases

Nitrogenase/Hydrogenase

When nitrogenase specifically produces H_2 in an ATP-dependent reaction, such nitrogenase is preferably called as hydrogenase. This hydrogenase is basically a nitrogenase enzyme and shares all the properties of a typical nitrogenase.

Reversible Hydrogenase/Soluble Hydrogenase/Bidirectional Hydrogenase

This enzyme may catalyze either uptake or evolution of H_2 and is called reversible hydrogenase. Reversible hydrogenase couples to a low potential electron carrier (with Emnear- O.4.V). This is easily solubilized, evolves hydrogen from reduced methyl viologen, and catalyzes the reduction of phenazine methanosulfate, methylene blue, and dichloropheno lindophenol with PMS being the most effective electron acceptor.

Uptake Hydrogenase/Unidirectional Hydrogenase

This enzyme is membrane bound, saturated by low H_2 tension, and is capable of reducing methylene blue but not low-potential acceptors. In the heterocystous, nitrogen-fixing cyanobacteria, possession of a unidirectional hydrogenase enables them to catalyze the H_2 consumption reactions typical of anaerobically adapted algae, and at the same time, probably increases the efficiency of H_2 fixation in these organisms. This enzyme also takes part in the oxy-hydrogen reaction in a number of blue-green algae (Lambert and Smith, 1981b). Evidence has been presented that this hydrogenase is capable of recycling some of the energy lost as hydrogen by supplying nitrogenase with some form of reductant (Benemann and Weare, 1974a; Lambert and Smith, 1981b). This uptake "hydrogenase" enzyme is particularly active in heterocysts and is weakly or not at all expressed in NH^{+}_{2}–grown and heterocyst-free filaments of cyanobacteria (Lambert and Smith, 1981b).

The Capacity to Metabolize H_2 Species Distribution

Table 5.2 lists the names of blue-green algal species capable of metabolizing H_2. Earlier there was a belief that only heterocystous cyanobacteria fix N_2. But with the discovery that non-heterocystous forms fix N_2 under aerobic and/or anaerobic conditions, the number of N_2–fixer species has increased drastically. So far all the N_2-fixing species tested have shown the capacity of evolve H_2. Thus hydrogenase has now been reported from all sections of blue-green algae, *viz.*, unicellular, non-heterocystous filamentous and heterocystous-filamentous species. Nevertheless, the number of hydrogenases present in different species shows considerable differences. Thus it is possible for an organism to possess methylviologen- dependent hydrogen formation only under the specified growth conditions, nitrogenase and uptake hydrogenase only, nitrogenase only, uptake hydrogenase only or none of these activities under the conditions tested. Species like *Anabaena cylindrica* possess all three activities *viz.*, nitrogenase, uptake hydrogenase and reversible hydrogenase under N_2–fixing conditions, but methylviologen dependent H_2 formation, when grown in ammonium salts. Figure 5.1 outlines the general scheme showing the inter-relationship between vegetative cell and heterocyst.

Table 5.2: Occurrence of Hydrogenase in Blue-green Algae (Cyanobacteria)

	Species	*References*
A.	**Unicellular**	
	Anacystis nidulans	Eisbrenner *et al.*, 1981 Peschek 1979a
	Aphanothece holophytica	Belkin and Padan 1978
	Cyanothece sp. 7822	Van der Oost *et al.*, 1987
	Gloeocapsa species	Gallon *et al.*, 1974
	Gloeothece sps.6909	Van der Oost *et al.*, 1987
	Cyanophora paradoxa	Eisbrenner *et al.*, 1981
	Microcystis aeruginosa	Asada and Kawamura 1985
	Synechococcus elongatus	Frenkel *et al.*, 1950
	Synechococcus sp. strain Miami Bg043511	Arai and Mitsui 1987
	Synechococcus cedorum	Gerasimenko and Zavarzin 1981
	Synechococcus sp.7425	Van der Oost *et al.*, 1987
	Synechococcus sp.	Frenkel and Rieger 1951
B.	**Non heterocystous filamentous**	
	Lyngbya sp. 108	Kuwada and Ohta 1987
	Myxosarcina chroococcoides	Lambert and Smith 1980
	Oscillatoria limnetice	Belkin and Padan 1978
	Oscillatoria brevis	Lambert and Smith 1980
	Oscillatoria thiebautii	Scranton *et al.*, 1987
	Oscillatoria sp. Miami BG–7	Kumazawa and Mitsui 1981
	Plectonema boryanum	Padan and Schneider 1978
	Phormidium angustissimum	Gerasimenko and Zavarzin 1981
	Phormidium laminosum	Smith *et al.*, 1982
	Spirulina sp.	Liama *et al.*, 1979
	Schizothrix calciola	Lambert and Smith 1980

Contd...

Table 5.2–Contd...

	Species	*References*
C.	**Heterocystous-filamentous forms**	
	Anabaena cylindrica	Hattori 1963
	Anabaena sp. N-7363	Asada *et al.*, 1985
	Anabaena variabilis	Eisbrenner and Bothe 1979 Almon and Boger 1984
	Anabaena azollae	Peters *et al.*, 1977 Chen Ven *et al.*, 1983
	Anabaena sp.	Jones *et al.*, 1976
	Anabaena 7120	Peterson and Burris 1978
	Anabaena cycadeac	Perraju *et al.*, 1986
	Anabaena sp. strain CA	Xiankong *et al.*, 1984
	Anabaena sp. strain 1F	Xiankong *et al.*, 1984
	Anabaena oscillaroides	Paerl 1980
	Anabaena variabilis	Vyas 1992; Vyas and Kumar 1995; Vyas, 2004
	Anabaena sp.	Srivastava *et al.*, 1989
	Anabaena sp.	Chen *et al.*, 1989
	Aphanizomenon sp.	Paerl 1982
	Cylindrospermum licheniforme	Hirosawa and Wolk 1979
	Cylindrospermum sp.	Srivastava *et al.*, 1989
	Fischerella muscicola	Lambert and Smith 1980
	Fischerella sp.	Srivastava *et al.*, 1989
	Hapalosiphon sp.	Chen *et al.*, 1989
	Macrozamia communis	Daday and Smith 1987
	Mastigocladus laminosus	Ernst *et al.*, 1979 Singh and Kashyap 1988
	Nostoc spongifermae	Vyas 1992; Vyas and Kumar 1995; Vyas, 2004
	Nostoc sp.	Srivastava *et al.*, 1989
	Scytonema hofmanii	Srivastava *et al.*, 1989

Isolation, Purification and Characterization of Different Hydrogenases

Excepting nitrogenase, very little work has been done on the biochemical characterization of hydrogenase from cyanobacteria (Lambert and Smith, 1981b; Houchins, 1984; Rao and Hall, 1988). This has been mainly due to oxygen sensitivity of hydrogenase which ultimately creates problem during isolation (Houchins, 1984). However, now methods have been developed which yield good preparation of hydrogenase from a selected cyanobacteria (Rao and Hall, 1988). Due to limited investigations on hydrogenase, especially reversible hydrogenase, there still exists confusion about the exact number and localization of hydrogenase among cyanobacteria (Houchins, 1984; Rao and Hall, 1988). Extensive work is needed to obtain clearer picture about the hydrogenases in cyanobacteria. In this section a brief account is given for all types of hydrogenase including nitrogenase separately.

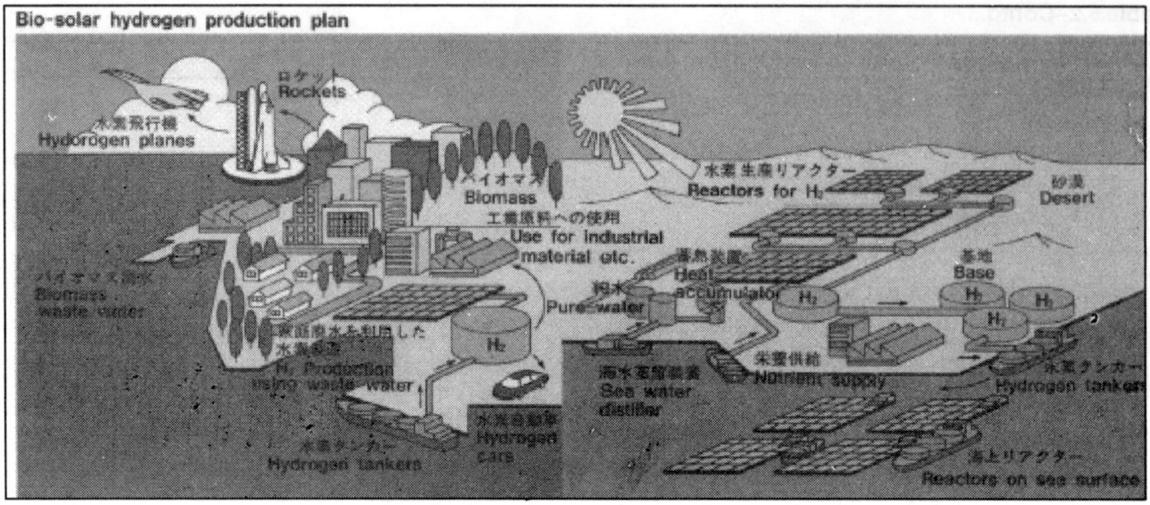

Figure 5.1: Use Cyanobacteria Combine Photosynthesis with Hydrogen Evolution Catalysed by Hydrogenases (Linblad, 2002)

Nitrogenase (ATP-Dependent Hydrogen Formation)

Nitrogenases from all organisms are very similar in their physical and chemical properties (Adams *et al.*, 1980; Bothe, 1982; Fay and Van Baalen, 1987; Mortenson, 1978; Mortenson and Chen, 1974; Mortenson and Thorneley, 1979; Stewart, 1980). The nitrogenase complex is comprised of two protein components. The larger component dinitrogenase (also called MoFe protein or component I) is responsible for reduction of substrate molecules. The MoFe component of *A. cylindrica* is an acidic tetramer (216,000 mol wt) with nonidentical subunits (52,600 and 55,000 mol wt); its isoelectric point lies between pH 4.72–4.99, and the Mo: Fe: s per mole ratio is 2.2:20.4:20 per cent in amino acid composition relatedness it resembles MoFe components from other N_2 fixers, particularly *Rhodospirillum rubrum* (Stewart, 1980; Fay and Van Baalen, 1987). The second coponent, dinitrogenase reductase, accepts electrons from donors such as ferredoxin, flavodoxin of dithionite and transfers these electrons to dinitrogenase with the concomitant hydrolysis of two molecules of ATP per electron transferred. The Fe protein (component II) is a highly labile dimer (60,000 mol wt; two identical subunits). The MoFe and Fe proteins of *A. cylindrica* and *P. boryanum* cross react with 65-90 per cent of the efficiency of homologous crosses (Stewart, 1980).

The six electron reduction of N_2 to 2 NH_3, therefore requires a minimum of 12 ATP molecules, making N_2 fixation an energetically expensive process. In addition to reducing N_2 to NH_3, dinitrogenase can reduce a number of other substrates including H^+ and acetylene.

The six electron reduction of N_2 to 2 NH_3, therefore requires a minimum of 12 ATP molecules, making N_2 fixation an energetically expensive process. In addition to reducing N_2 to NH_3, dinitrogenase can reduce a number of other substrates including H^+ and acetylene.

Both components of nitrogenase are repidly and irreversibly inactivated by O_2 (Mortenson, 1978; Mortenson and Thorneley, 1979). Organisms that fix N_2 in aerobic environments, therefore, must provide some means of protecting nitrogenase from oxygen. Cyanobacteria are the only organisms capable of simultaneous O_2 evolution and N_2 fixation, demonstrating that these processes can be compatible under suitable conditions.

Like the reduction of N_2, reduction of protons by nitrogenase requires ATP and a strong reductant. Reductant can be provided as reduced ferredoxin, flavodoxin or, as commonly employed in laboratory experiments, dithionite (Asada and Kawamura, 1985, 1986; Yu and Wolin, 1969). The evolution of H_2 by nitrogenase always occurs during N_2 fixation and a minimum of 25 per cent of total electron flux through nitrogenase is diverted to the reduction of H^+ even at very high partial pressures of N_2 and optimal levels of ATP and reductant (Houchins, 1984). If the rate of dinitrogenase turnover is slowed owing to limitation in the supply of ATP or reductant, the relative proportion of electrons diverted to H^+ reduction increases (Bothe, 1982). The evolution of H_2 by nitrogenase, therefore, consumes a substantial amount of the reductant and ATP utilized by nitrogenase.

Reversible Hydrogenase

Unlike "uptake" hydrogenase, reversible hydrogenase has been demonstrated in representatives of nearly every major cyanobacterial group (Houchins, 1984; Tables 5.3–5.8). Reactions catalyzed by reversible hydrogenase have been demonstrated in the unicellular organisms, *Synechococcus, Synechocystis* and *Aphanocapsa* (Asada and Kawamura, 1985; Asada *et al.*, 1987; Belkin and Padan, 1978; Mitsui *et al.*, 1979, 1985, 1987; Peschek, 1979a). in filamentous non-heterocystous strains such as *Spirulina, Lyngbya* and *Oscillatoria* (Belkin) and Padan, 1978; Kumazawa and Mitsui, 1982; Kuwada and Ohta, 1987, 1988, 1989, Llama *et al.*, 1979; Mitsui, 1978, 1979, 1981; Phlips and Mitsui, 1983a, b; Ramchandran and Mitsui, 1984; Scranton *et al.*, 1987; Smith *et al.*, 1982); and in heterocystous organisms including several strains of *Anabaena, Nostoc* and *Mastigocladus* (Almon and Boger, 1984; Asasda *et al.*, 1979; Benemann and Weare, 1974; Benemann *et al.*, 1982; Benemann and Weare, 1974; Benemann *et al.*, 1982; Chen *et al.*, 1989; Eisbrenner and Bothe,1979; Ernst *et al.*, 1979; Ewart and Smith, 1989; Houchins and Burris, 1981a, b,d; Laczko, 1980; Laczko and Brabas, 1981; Miura *et al.*, 1982; Miyamoto

Table 5.3: Profiles of Hydrogenase Activities in Different Cyanobacteria

Organism	*Growth Conditions*	*Nitrogenase Activity*	*H_2- Consumption*	*MV- Dependent H_2 Formation*	*References*
Anmacystis nidulans	Various	-	+	+[b]	Peschek (1979a,b,c)
Myxosarcina chrococcoides	NH^+_4-grown	-	-	-	Lambert and Smith (1980)
Oscillatoria brevis	NH^+_4-grown	-	-	+	Lambert and Smith (1980)
Plectonema boryanum	Air-grown	-	-	+	Weare and Benemann (1974)
	Anaerobic induction	+	-	-	Lambert and Smith (1980)
Schizothrix calcicola	NH^+_4-grown	-	-	-	Lambert and Smith (1980)
Anabaena cylindrica	N_2- grown	+	+	+	Bothe *et al.* (1977) Lambert and Smith (1980b)
	NH^+_4-grown	-	-	+	Lambert and Smith (1980)
Fischerella musciola	N_2–grown	+	+	+	Lambert and Smith (1980)

a: Modified from Lambert and Smith (1981b).

b: Activity was demonstrated to be membrane bound.

Table 5.4: Purification of Hydrogenase from *Spirulina maxima**

Step	Total Protein (mg)	Total Activity (units)[a]	Specific Activity (units/mg protein)	Purification (–fold)	Yield (Per cent)
Lysate	12,800	12,860	1.0	1.0	100.0
40,000 g supernatant	1,207	6,265	5.2	5.2	48.7
100,000 g supernatant	800	6,027	7.5	7.5	46.8
DE-52 eluate	107	3,750	35.0	35.0	29.0
Sephacryl S-200	6.6	1,453	112.4	112.4	11.3

* Modified from Llama *et al.* (1979).

[a]Activity expressed as mol H_2 evolved/hr.

Table 5.5: Partial Purification of *Oscillatoria limnetica* Hydrogenases*

Step	Total Protein (mg)	Total Activity (units)[a]	Specific Activity (units/mg protein)	Purification (–fold)	Recovery (Per cent)
Crude sonicate	5,190	4,418	0.85	1.0	100.0
35,000 g Supernatant	1,875	3,225	1.72	2.0	75.0
77,000 g supernatant	1,540	3,004	1.95	2.3	68.0
75 per cent $(NH_4)_2$ SO_4	311	928	2.98	3.5	21.0
First DEAT-Sephacel					
Form I	41	353	8.7	10.2	8.0
Form II	43	530	10.5	12.4	12.0
Second DEAT-Sephacel					
Form I	15	191	13.0	15.3	4.3
Form II	19	295	15.6	18.4	6.7
Sephacryl S-200					
Form I	6	106	17.6	20.7	2.3
Form II	6	158	25.8	30.3	3.6

* Modified from Belkin *et al.* (1981)

[a] Activity expressed as/mol H_2 evolved/hr.

et al., 1979, 1984; Spiller *et al.*, 1983; Tel-Or *et al.*, 1977, 1978; Weare *et al.*, 1980. In most cases, reversible hydrogenase was based on elution from a Sephacryl S-200 column for the enzyme from *Spirulina maxima*. The enzyme from *Anabaena* 7120 eluted in two peaks of molecular weight 113,000 and 165,000 (Houchins and Burris, 1981a,b). If this enzyme was first heated at 70°C for 1 h, an additional peak of 55,000 appeared suggesting that the larger forms might be multimers of a single polypeptide. Hallenbeck and Benemann (1978) have reported a molecular weight of 230,000 for the enzyme from *A. cylindrica*. Multiple forms of reversible hydrogenase which may or may not correspond to different for hydrogenase activity are invariably observed on polyacrylamide gels (Ward, 1970; Ewart and Smith, 1989, Kentemick *et al.*, 1989). Some of the characteristics of reversible hydrogenase are represented in Table 5.6.

Table 5.6: Properties of Reversible and Uptake Hydrogenases from *Anabaena* sp. Strain 7120*

	Uptake Hydrogenase	*Reversible Hydrogenase*
Cellular localization	heterocysts	both cell types
Subcellular localization	membrane-bound	cytoplasmic
Effect of NH_3 during growth	repressed	no effect
Response to anaerobic conditions during growth	increased activity	• large increase in activity
Effect of H_2	increased activity	no effect
Stability to atmospheric O_2 labels	irreversible inactivation	stable
Acceptor specificity	positive Em only	both positive and negative Em
Km for H_2	0.9/um	2.3/um
Molecular weight	56,000	55,000-230,000
Ki for CO	0.039 atm	0.0095 atm
Stability to 70°C heat treatment	$T_1/2$ = 12 min	stable
pH optimum for H_2 uptake	6.0, 8.5	9.0

*After Houchins (1984).

Table 5.7: Localization of the Reversible Hydrogenase in Aerobically and Microaerobically Grown *Anacystis nidulans* by the Immuno-Gold Labelling Technique*

	Cells Grown Aerobically	*Microaerobically*
Specific H_2- evolution activity[a]	269	408
Gold label associated with the cytoplasmic membrane	65±11	108±9
Gold label associated with the cytoplasm	26±11	28±8

* After Kenemich *et al.* (1989).

[a] The specific activity is given in nmol H_2 evolved/mg protein/h.

Partially purified preparations of reversible hydrogenase are usually quite resistant to irreversible inactivation by O_2. Samples of the enzyme from *Oscillatoria limnetica* (Belkin *et al.*, 1981) exposed to air for 50 days at 20 or –196°C retained about 80 per cent of the activity of an anaerobic control sample. Contrary to this result, the enzyme from *Mastigocladus* was considerably more O_2–sensitive and lost about 75 per cent of its activity relative to a control after 8 days under air (Riedar and Hall, 1981).

Two soluble hydrogenase activities were obtainable from cell extracts of the cyanobacterium *A. cylindrica*, one detectable by the tritium exchange assay, the other having a relatively low tritium exchange activity but catalyzing methylviologen dependent hydrogen formation. Their molecular weights by gel filtration chromatography, were 42,000 and 100,000 respectively (Ewart and Smith, 1989). Ewart and Smith (1989c) have futher raised antibodies in mice against the 42 Kd subunit of the soluble hydrogenase purified from *A. cylindrica*. The antibody did not cross-react with the 50 Kd protein, which appears to be necessary to confer methyl viologen-dependent reductive hydrogenase activity. Using western blotting procedures they found no evidence for the 42 Kd protein in heterocyst cells (Ewart and Smith, 1989a). Furthermore, although the protein was found only in the soluble fraction of vegetative cells from cultures grown in air/CO_2, antibody reactivity was also obtained with

Table 5.8: Purification of the Reversible Hydrogenase from *Anabaena nidulans**

Purification Step	*Specific Activity*[a]	*Enrichment*	*Total Activity (X 10³ U)*[b]	*Yield (Per cent)*
Crude extract	855	1.0	329	100
Supernatant after 20 per cent $(NH^-)_2$ SO_4 addition.	929	1.1	2.54	77
Linear KCL gradient on DEAE-cellulose fraction No. 10	2631	3.0	69	21
No. 11	3993	4.6	110	33
No. 12	2170	2.5	24	7
Octyl–sepharose CL-4B chromatography Fraction No. 4	16136	18.9	53	16
No. 5	214408	250.7	85	26

*After Kentemich *et al.* (1989).

a: The specific activity is given in nmol H_2/mg protein/h in the $Na_2S_2O_4$ and methyl viologen dependent evolution assay. The total activity refers to a 91 culture which was concentrated to 90 ml crude extract.

b. 1 U corresponds to 1 nmol H_2 evolved/h.

(Peschek, 1979a). A comparative account of reversible and uptake hydrogenase for Anabaena 7120 is presented in Table 5.6.

the particulate fraction when cultures were grown in nitrogen/4 per cent H_2/0.3 per cent CO_2. They have also developed an activity stain for the detection of cyanobacterial hydrogenase in polyacrylamide gels (Ewart and Smith, 1989b). In another study on *Anacystis nidulans*, Kentemich *et al.*, 1989 have reported isolation and purification of reversible hydrogenase (250-fold) by classical methods. Activity staining on gels obtained by native PAGE allowed to identify two bands. Antibodies were raised against the electrophoretically homogeneous protein. Crude extracts from the unicellular *Anacystis* and from heterocysts and vegetative cells of *Anabaena variabilis* showed precipitation bands of 56 and 17 Kd. From their study it appears that *Anacystis* has two different hydrogenases; the reversible or bidirectional hydrogenase which is located exclusively at the cytoplasmic membrane of the cells and thylakoid-bound enzyme which catalyzes only the uptake of H_2. The activity as per different purification steps is represented in Tables 5.6–5.8.

Inspite of many new reports appearing on reversible hydrogenase, the ambiguity still persists about the number and localization of this bidirectional enzyme. A systematic investigation is needed to reveal the exact nature and types of reversible hydrogenase in cyanobacteria.

Uptake Hydrogenase

Uptake hydrogenase has been demonstrated in all heterocystous cyanobacteria so far examined (Lambert and Smith, 1981b; Houchins, 1984; Rao and Hall, 1988). There exists great variations in activity with growth conditions and in some strains of *A. cylindrica*, activity is nearly undetectable under certain growth conditions (Bothe *et al.*, 1978; Jones and Bishop, 1976; Lambert and Smith, 1980). So far uptake hydrogenase has not been detected in any non-heterocystous N_2- fixers. The only non-heterocystous organism so far shown to possess uptake hydrogenase is the non–N_2–fixers unicell, *Anacystic nidulans* (Peschek, 1979a,b,c,).

The properties of cyanobacterial uptake hydrogenase in vitro are similar to those from other bacteria. Houchins and Burris (1981 a,b) have shown that when isolated heterocysts were broken with a French press, about 70 per cent of the recovered activity was membrane-bound and could be solubilized by detergent treatments or by prolonged sonication. The remaining 30 per cent of the activity that was released into the soluble fraction during the French press treatment resembled the membrane-bound uptake hydrogenase in physical and catalytic properties and is probably the same enzyme. The solubilized enzyme from Anabaena 7120 eluted from Sephadex G-100 with a molecular weight of 56,000. Co inhibition was competitive versus H_2 with a Ki of 0.039 atm. Exposure to O_2 results in both reversible inhibition and irreversible inactivation, the sensitivity to O_2 increasing with increasing disruption of the system (Houchins and Burris, 1981a,b; Lambert and Smith, 1981b). Irreversible inactivation of solubilized uptake hydrogenase displays biphasic kinetics with half-times of 2 and 14 min. Optimal O_2 levels for oxyhydrogen activity were 20 per cent for whole filaments, 10 per cent for isolated heterocysts and 2.5 per cent for heterocyst membranes (Houchins and Burris 1981d). In *Ancystis nidulans*, the oxyhydrogen reaction in isolated membranes was completely resistant to an O_2 concentration of 30 per cent.

Nitrogenase-Mediated H_2 Formation

Extensive studies have been conducted on H_2 production employing a number of N_2-fixing cyanobacteria including both heterocystous and non-heterocystous forms. A number of factors have been reported to affect H_2 formation (Lambert and Smith, 1981; Houchins, 1984). Excepting a few species, when cultures are grown in air, net H_2 formation is almost zero (Lambert and Smith, 1981; Xiankong *et al.*, 1983, 1984). This is mainly due to recycling of H_2 formed via "uptake" hydrogenase (Adams *et al.*, 1980; Bothe 1982; Bothe *et al.*, 1980; Peterson and Burris 1987a,b; Spiller *et al.*, 1978; Xiankong *et al.*, 1974; Vyas and Kumar, 1995). The basic strategy to obtain net H_2 formation is either to remove H_2 from the gas phase or to prevent H_2 incorporation, so that the assimilating power or to prevent N_2 incorporation, so that the assimilating power available to nitrogenase is all channelled into H_2 formation. A number of methods have been employed to optimize H_2 formation, and these allow the proper conditions for optimal gas production to be deduced.

Selection of Species and Strains

Studies of H_2 formation have focussed largely on the filamentous heterocystous cyanobacteria (Lambert and Smith, 1981b). *Anabaena cylindrica* has been consistently used as model organism for H_2 formation by a number of workers (Hallenbeck and Benemann, 1979; Lambert *et al.*, 1979). The rate of H_2 formation particularly in *A. cylindrica* obtained by various workers shows close similarity and reported differences if any, may be due to differences in culture and/or incubation conditions (Bothe *et al.*, 1978; Lambert and Smnith, 1981b; Tel-Or *et al.*, 1978). In general, the capacity to photoproduce H_2 varies widely depending on the levels of nitrogenase and hydrogenase activities of the cyanobacterium used. Although, any strain showing highest nitrogenase activity may or may not show the highest rate of H_2 formation. Anabaena CA, the fastest growing cyanobacterium, is reported to have the highest nitrogeanase activity amoongst all the cyanobacteria so for tested but it does not show the highest rate of H_2 formation (Xiankong *et al.*, 1983, 1984). On the other hand a few marine non-heterocystous cyanobacteria though having moderate level of nitrogenase activity show a very high rate of H_2 formation (Kumazawa and Mitsui, 1981, 1982, 1985; Mitsui *et al.*, 1977, 1979, 1985). Thus it is evident that rates of H_2 formation can vary widely with different species (Table 5.9). Therefore screening of many cyanobacteria isolated from diverse ecosystems may provide more suitable H_2 producers or strains for investigation. Non-heterocystous strains may be useful for biochemical studies on H_2

formation as well as H_2 formation may be enhanced as all the cells may contain nitrogenase unlike in heterocystous cyanobacteria.

Table 5.9: Rates of H_2 Production in Different Cyanobacteria[a]

Species	µmoles H_2 Produced per Hour		Method	References
	mg dry wt	mg Chl a		
Anabaena cylindrica	1.4	320	GLC[b]	Weissman and Benemann 1977
	0.17	0.7	GLC	Bothe *et al.*, 1977
	0.58	11.0	GLC	Daday *et al.*, 1977
	0.10	6.7	GLC	Jeffries *et al.*, 1978
				Lambert and Smith 1977
Anabaena Strain CA	40 u1			
Anabaena ambigua	-	8.5	GLC	Srivastava 1990
Anabaena cycadeae	-	1.56	GLC	Srivastava *et al.*, 1989
Anabaena (N–7363)	-	72	GLC	Asada *et al.*, 1979
Anabaena flosaquae	0.025/u1 pcv[d]		AMP[c]	Jones and Bishop 1976
Anabaena doliolum	-	1.90	GLC	Srivastava *et al.*, 1989
Calothrix membranacea	0.108		GLC	Lambert and Smith 1977
Calothrix scopulorum	0.128		GLC	Lambert and Smith 1977
Cylindrospermum majus	-	3.25	GLC	Srivastava 1990
Fischerella musciola	-	9.10	GLC	Srivastava 1990
Mastigocladus laminosus	8.5 ul	70 nmol	GLC	Benemann *et al.*, 1982
Nostoc muscorum	20mg	Chl a/h	GLC	Weisshaar and Boger 1983
Oscillatoria brevis	0.168	-	GLC	Lambert and Smith 1977
Oscillatoria sp.	0.26	260	GLC	Mitsui and Kumazaw 1977
Oscillatoria BG-7	0.183	-	GLC	Kumazawa and Mitsui 1985
Scytonema hofmanni	-	9.36	GLC	Srivastava *et al.*, 1989

a. Incubation conditions are not similar in all the cases.

b. GLC–Gas Liquid Chromatography

c. AMP–Amperometric

d. Pcv: Packed cell volume.

Factors Affecting Hydrogen Formation

As stated earlier a number of factors have been reported to affect nitrogenase-mediated H_2 formation. In this section a brief account is given for all those factors affecting the rate and duration of H_2 formation.

Growth Conditions Affecting H_2 Production

Gas Atmosphere for Growth

The most commonly used gas atmosphere for the growth of N_2 firming cultures is air supplemented with CO_2, growth is drastically slowed (Fay and Van Baalen, 1987; Stewart, 1980) and the characteristic blue-green coloration due to phycocyanin is reduced (Lambert *et al.*, 1979b). It has been demonstrated that cultures grown under limiting CO_2 conditions have H_2 photoproduction rates proportional to their growth rates (Jones and Bishop, 1976). The concentration of CO_2 also affects the rate of H_2 formation. A concentration below 5 per cent CO_2 is most suitable for growth as well as H_2 production (Benemann and Weare, 1974a; Lambert *et al.*, 1979b; Tel-Or *et al.*, 1978). It has been observed that both tested (Ernst *et al.*, 1979). It has been observed that both acetylene reduction and H_2 formation increased slighly. Azolla fronds when incubated in air + CO_2 showed higher H_2 formation than those growing on air alone (Banerjee *et al.*, 1989; Peters *et al.*, 1976). Tel-Or *et al.*, 1977 have also reported the induction of hydrogenases by $N_2/H_2/CO_2$ (75:20:5) enhanced nitrogenase and hydrogenase activities between days 4 and 8 growth of *A. cylindrica* and *Nostoc muscorum* (Tel–Or *et al.*, 1977). The maximum enhancement of nitrogenase activity was 3- to 5- fold and H_2 consumption and methyl viologen-dependent H_2 formation, 5 to 50-fold. This induction of hydrogenase has been observed in a number of other cyanobacteria (Eisbrenner *et al.*, 1979; 1981; Howarth and Codd, 1985) on the contrary increase in H_2 Production has also been observed following CO_2 deprivation in *A. cylindrica* (Jeffries *et al.*, 1978). Furthermore, obligated requirement of CO_2 at least in aged cultures for prolonged H_2 production has been demonstrated in a number of cyanobacteria (see Lambert and Smith, 1981b). In summary it seems that CO_2 is essential for continued H_2 production and this effect seems to be mediated via net CO_2 fixation.

Culture Medium

Culture medium used for growth of various cyanobacteria plays an important role in regulating the evolution of H_2 (Lambert and Smith, 1981; Stewart, 1980). The most commonly employed medium has been that of Allen and Arnon (1955). However, this medium does not constitute the minimal nutritional requirement for cyanobacteria and some of the trace elements are possibly unnecessary. However, *Azolla* fronds when incubated with different and Arnon's medium (Banerjee *et al.*, 1989; Kumar *et al.*, 1990). A number of other media have been used for various cyanobacteria; howerver BG–11 seems to be most suitable medium for growth and H_2 formation for a variety of cyanobacterial strains (Lambert and Smith, 1981b).

Many other constituents and added chemicals have been demonstrated to affect subsequent H_2 formation. Almost all the combined nitrogen sources *viz.*, NH_2^+, NO_3^-, NO_2^- and a few amino acids like glutamine, asparagine, arginine and glutamic acid inhibit H_2 formation. The inhibition of H_2 formation is solely due to loss of heterocyst and/or nitrogenase activity (Banerjee *et al.*, 1989; Bothe, 1982; Ernst *et al.*, 1979; Hounchins, 1984; Lambert *et al.*, 1979c; Srivastava, 1990; Stewart, 1980). Addition of exogenous NaH CO_3 to culture medium has been reported to stimulate H_2 formation in a number of cyanobacteria (Srivastava *et al.*, 1989). Similarly supplementation of a few metal ions and their concentration had been demonstrated to affect H_2 evolution Jeffries *et al.* (1978) showed that cultures grown with 5.0 mg of Fe^{3+}/litre produced H_2 at a rate about twice the rate of cultures with 0.5 mg Fe^{3+}/ litre. Nickel addition significantly affects up.

Light

Light quality and intensity both affect H_2 formation in a number of cyanobacteria. This effect is indirect and mainly through photosynthesis and nitrogenase activity (Stewart, 1980). A number of studies have demonstrated that pigment content of cyanobacteria can vary according to the vavelength characteristics of the light source used (Fay and Van Baalen, 1987; Frenkel *et al.*, 1950; Greenbaum, 1977; Kerfin and Boger, 1992; Lambert and Smith, 1980; Phlips and Mitsui, 1983b; Stewart, 1980). Chan Van *et al.* (1983) have demonstrated a close correlation between wavelength of light and photosynthesis together with H_2 production. Gogotov *et al.*, 1976 demonstrated that light of low intensity (2.5,10^{3}erg/cm^2/s) stimulates H_2 production by cell suspensions of *A. variabilis* in the presence of glucose, pyruvate and formate. The maximum rate of H_2 production in the presence of these substrates was observed at light intensities of 650, 1400 and 2250 erg cm^{-2} s^{-1} respectively. In another study it has been demonstrated that *A. cylindrica* sparged with argon gas produced H_2 continuously for 20 days under limited light conditions (6.0 W m^{-2}) and for 180 days under elevated light conditions (32 W m^{-2}) in the absence of exogenous nitrogen. Kumar and Kumar (1988) have observed that low light intensity favours H_2 production in *Anabaena* CA. In *Oscillatoria* sp. strain Miami BG-7, the rate of H_2 production saturates at low light intensity (*i.e.* 15–30 uE m^{-2} s^{-1}). Although so far no systematic studies on the effect of light intensity during growth on subsequent H_2 formation have been performed but studies so far conducted suggest that lower light intensity favours H_2 production.

Under dark, rate of H_2 production remains only about 10 per cent of the light incubated cultures. As yet, not a single species has been found to show sustained H_2 production shows almost similar trend in terms of duration as observed with nitrogenase activity.

Temperature

Excepting the thermophilic cyanobacterium *Mastigocladus laminosus*, in all the mesophilic cyanobacteria H_2 formation follows the pattern of growth and temperature (Lambert and Smith, 1981b). In marine *Oscillatoria* sp. strain Miami BG-7, the upper temperature limit for H_2 production has been found at 46°C.

H_2 evolution by the thermophilic cyanobacterium *Mastigocladus laminosus* has been studied by a few worker (Benemann *et al.*, 1982; Miura *et al.*, 1980, 1981, 1982; Miyamoto *et al.*, 197, 1984; Smith *et al.*, 1982). The optimal temperature for H_2 production was 44-49°C (Miura *et al.*, 1980).

Age of Culture

It has been realized lately that the age of cultures used for H_2 measurements is an important factor (Lambert and Smith, 1981). First of all Benemann and Weare (1974a) showed that H_2 formation in *A. cylindrica* is optimal when harvested at about 0.3 mg dry wt/ml of cyanobacterial concentration, after which a sharp decline is observed. This decline in H_2 formation is more pronounced than the observed decline in nitrogenase activity suggesting that hydrogen recycling capacity remains relatively constant. Similar finding was later on reported by Daday *et al.*, 1977. In *Nostoc muscorum* a similar optimum H_2 production in logarithmic growth was observed which correlated with changes in heterocyst frequency (Ernst *et al.*, 1979). The decline of H_2 evolution in aged cultures seems mainly due to gradual loss of nitrogenase activity and also the pigments of PS II (Lambert and Smith, 1981; Stewart, 1980).

Concentration of Culture

Concentration of culture has been implicated in regulating the net H_2 formation. H_2 formation has been shown to increase linearly with algal concentration to about 1.5 mg dry wt/ml suspension for *A. cylindrica* (Daday *et al.*, 1977). Subsequent decrease in H_2 formation is probably due to self shading of the culture.

Factors Employed to Enhance Nitrogenase-Catalyzed H_2 Production

Choice of Gas Atmosphere During Assay

Argon/CO_2

As most of the N_2-fixing cyanobacteria lack the capacity to produce H_2 under aerobic conditions, cultures are frequently incubated in argon in the absence of CO_2 to avoid generation of O_2 through photosynthesis during the experimental incubations (Lambert and Smith, 1981b). However, these conditions are not optimal for H_2 formation and generally result in lower rate of H_2 formation. It has been also demonstrated that the longer the incubation, the greater the requirement for CO_2 for prolonged H_2 formation (Chen *et al.*, 1989; Daday *et al.*, 1977; Ernst *et al.*, 1979; Kuwada and Ohta, 1989; Laczko, 1980; Lambert *et al.*, 1979; Lambert and Smith, 1980, 1981a,b). The reason for lower rate of H_2 formation under argon alone is primarily due to depletion of stored reductant (Lambert and Smith, 1981b). With argon alone there is no CO_2 fixation and hence stored reductant is gradually consumed in all other metabolic reactions including nitrogenase (Ernst *et al.*, 1979; Lambert *et al.*, 1979a).

To improve the efficiency of H_2 production, use of 99 per cent argon + 1 per cent CO_2 has been demonstrated to be the best alternative (Asada *et al.*, 1985; Banerjee *et al.*, 1989; Bothe *et al.*, 1978; Kerfin and Boger, 1982; Kosyak *et al.*, 1978; Kumazawa and Mitsui 1981; Scherer *et al.*, 1980; Tel-Or *et al.*, 1977, 1978). Preincubation under 99 per cent argon + 1 per cent CO_2 leads to multifold stimulation of H_2 production rate although depending upon the strains used (Lambert and Smith, 1981b). However, prolonged incubation under nitrogen starvation leading to the death and decay of cultures. Thus, this treatment has been found suitable for short term H_2 evolution study (Lambert and Smith, 1981b). However, investigators have shown that continued production of H_2 may be attained in the same culture if a very low concentration of NH^+_4 is added to the cultures so as to keep them under growing condition (Lambert *et al.*, 1979; Lambert and Smith, 1980; Mitsui, 1987; Mitsui *et al.*, 1985; Weaver *et al.*, 1980).

Dinitrogen (N_2)

To attain anaerobic condition, a number of workers have used N_2 gas in place of argon (see Lambert and Smith, 1981b), furthermore the effect of N_2 along with argon has also been tested. In either condition, N_2 inhibits H_2 formation due to competition of N_2 reduction with proton reduction (Houchins, 1984). N_2 at 4 per cent of the gas phase has been found to inhibit H_2 formation in argon by about 50 per cent, with complete inhibition occurring at 25 per cent N_2 for *A. cylindrica* (Jones and Bishop, 1979).

Oxygen

The effect of O_2 in argon or N_2 atmosphere has also been tested (Ernst *et al.*, 1979; Spiller *et al.*, 1978). Added oxygen inhibits H_2 formation in all the cyanobacteria (Asada and Kumazawa, 1986; Banerjee *et al.*, 1989; Daday *et al.*, 1977; Kumar and Kumar, 1990b,c; Lambert *et al.*, 1979c; Smith *et al.*, 1983, 1984). In an *Anabaena* sp., it was found that 5 per cent O_2 with argon stimulated H_2 production but this very concentration was inhibitory under N_2 (Asada *et al.*, 19790. The inhibition of H_2 formation by O_2 seems to be due to O_2 sensitivity of nitrogenase or by the facilitation of an oxyhydrogen reaction in heterocyst which diminishes net H_2 production (Ernst *et al.*, 1979; Spiller *et al.*, 1978; Xiankong *et al.*, 1983, 1984).

Carbon Monoxide/Acetylene

Carbon monoxide has been extensively employed in H_2 formation studies because it specifically inhibits N_2 fixation and C_2H_2 reduction but does not inhibit the nitrogenase mediated H_2 formation reaction (Bradbeer and Wilson, 1963). Similarly, C_2H_2, an inhibitor of hydrogenase in many bacteria, has been frequently used (Smith *et al.*, 1976). Stimulation of H_2 formation by the addition of CO and C_2H_2 in an argon gas phase was first of all shown in *A. cylindrica* by Bothe *et al.* (1977a). Subsequently, H_2 formation was obtained even in air with the use of CO + C_2H_2 for *A. cylindrica* (Daday *et al.*, 1977) and also in marine cyanobacteria (Lambert and Smith, 1981b). At optimal concentrations of CO and C_2H_2, H_2 formation occurred at the same rate in air after a lag phase of 2-3 has in argon gas phase (Lambert *et al.*, 1979c). A number of other workers have shown the same effect of CO + C_2H_2 on H_2 formation in various cyanobacteria (Miura *et al.*, 1981, 1982; Miyamoto *et al.*, 1984; Srivastava, 1990; Xiankong *et al.*, 1984; Figure 5.4). Increase of H_2 formation by CO + C_2H_2 seems to be mediated by number of ways. Bothe (1982) provided evidence that C_2H_2 inhibits by H_2 recycling and that this inhibition is increased somewhat by CO. There is also a report that pre-incubation with C_2H_2 alters the conformation of the nitrogenase complex. Still other workers have suggested that the effect of C_2H_2 is to alter the H_2 tension than C_2H_2 itself that causes a conformational activation of the enzyme (Scherer *et al.*, 1980).

Metabolic Inhibitors

DCMU (3-(3,4 dichlorophenyl) -1, 1-dimethylurea)

Out of various metabolic inhibitors, DCMU, an inhibitor of PS II, has been frequently used to optimize H_2 production (Lambert and Smith, 1981b). The possible effect of DCMU may be: (a) a reduction in H_2 formation by decreasing reductant supply to PS I from PS II, or (b) an increase in H_2 formation due to elimination of an oxyhydrogen reaction that uses photosynthetically generated O_2 and nitrogenase-mediated H_2. Both these effects have been observed in some cyanobacteria (Houchins, 1984). That a loss of reductant does take place by DCMU treatment was first of all proved by Benemann and Weare (1974b) was demonstrated H_2-dependent acetylene reduction in the presence of DCMU. Gogotov *et al.* (1976) have reported that H_2 evolution is not inhibited by DCMU in the presence of pyruvate or formate in *Anabaena variabilis*. Working with isolated heterocysts fo *Anabaena* sp. strain CA, it has been demonstrated that DCMU has no effect on H_2 formation (Kumar and Kumar, 1990b). Xiankong *et al.* (1983) have reported that DCMU addition immediately inhibits H_2 production in *Anabaena* CA. It seems that the effect of DCMU depends on the particular cyanobacterium used; those that produce little H_2 in the absence of DCMU and exhibit a vigorous oxyhydrogen reaction might be stimulated in H_2 formation by DCMU, whereas those in which H_2 and O_2 are not re-utilized until significant tensions accumulate in the environment, such as in the case of *A. cylindrica* will show no benefit from DCMU addition (Lambert and Smith, 1981b). As a whole if DCMU results in stimulation of H_2 formation, the effect is temporary. Furthermore, the effect of DCMU also depends on the level of endogenous reductant present in a particular alga.

Metronidazole (2-methyl-5-nitroimidazole-1-ethanol)

Tetley and Bishop (1979) first of all showed that metronidazole at 1-2 mM levels was a selective inhibitor of nitrogenase activity in *Anabaena* sp. Hydrogenases are insensitive to metronidazole. Although this chemical does not stimulate H_2 production but it offers a model system to distinguish between hydrogenase and nitrogenase activity.

MSO (L-Methionine-DL-Sulforimine)

MSO, an inhibitor of glutamine synthetase which depresses nitrogenase activity, has been found to increase H_2 production for shorter duration (Lambert and Smith, 1981b; Stewart, 1980; Srivastava, 1990). Lambert *et al.* (1979b) found stimulation of H_2 production by treatment of MSO (2µM) in *A. cylindrica* B 629. Use of MSO is important especially where cultures are grown at lower concentration of NH_4Cl for longer duration of H_2 production. MSO drepresses nitrogenase even in the presence of NH_4Cl and thus continued production of H_2 occurs.

Organic Carbon Sources

Cyanobacteria are obligate photoautotrophs and as such do not require any organic carbon sources for normal growth (Stewart, 1980). However, there are certain species which show stimulation of growth in light and a few show growth in dark in the presence of certain organic carbon sources. As most of the cyanobacteria show little or no H_2 formation in dark, the role of exogenous organic carbon sources seems important. Chan Van *et al.* (1983) have reported that pyruvate and glucose stimulate photohydrogen production in *Anabaena azollae*. Kumar and Kumar (1990a) did not observe stimulation of H_2 evolution by addition of fructose and erythrose in *Anabaena* CA. Srivastava *et al.* (1989) have reported significant stimulation of H_2 production in a number of N_2-fixing cyanobacteria. They have demonstrated stimulation of H_2 production by exogenously added carbon sources in the dark (Srivastava, 1990; Srivastava *et al.*, 1989). So far no attempt has been made to study the response of added organic carbon sources on H_2 production in any typical heterotrophic cyanobacteria. Differential response of organic carbon sources on H_2 formation seems to be related with transport of carbon sources by specific cyanobacterial species.

Nickel

Nickel has been demonstrated to be an essential metal for uptake hydrogenase activity (Almon and Boger, 1984; Daday and Smith, 1983; Daday *et al.*, 1985; Srivastava, 1990; Xiankong *et al.*, 1984; Vyas, 2004). In almost all the cyanobacteria tested, nickel addition stimulates uptake of H_2 and thus inhibition of H_2 formation takes place. Attempts are being made to isolate nickel-resistant mutant which might be uptake hydrogenase negative (Hup-) and therefore may prove better strain in terms of H_2 production.

Na_2S, Thiosulfate and Other Reducing Agents

A number of workers have tested the effect of a variety of reducing agents on H_2 production in a few cyanobacteria. It was found that Na_2S addition showed stimulation of H_2 production in *Nostoc muscorum* (Fry *et al.*, 1984). The results obtained by various workers are variable and as such no definite conclusion can be drawn on the role of these substances on H_2 production.

H_2 Production by Hydrogenase (Reversible Hydrogenase)

In comparison to nitrogenase-mediated H_2 formation, little work has been done on H_2- formation by hydrogenase (Houchins, 1984; Lambert and Smith, 1981b). Whereas nitrogenase catalyzed H_2 formation takes place solely in N_2 fixing forms, hydrogenase is present both in N_2-fixer and non-N_2 fixer forms. There is a lot of confusion about the exact found in vegetative cells even when grown in the absence of heterocysts it is unlikely that its function relates to nitrogen fixation. Furthermore there is still ambiguity about the exact number of reversible hydrogenases in cyanobacteria. However, now it is believed that two pools of hydrogenase exist, an oxygen-sensitive activity in vegetative cells and an oxygen-resistant activity in heterocysts.

Reversible hydrogenase activities are usually present in aerobically grown cyanobacteria prior to a dark, anaerobic adaptation period (Daday and Smith, 1979; Houchins and Burris, 1981a,b; Tel-Or *et al.*, 1978). Demonstration of activity requires reductive activation of the enzyme or removal of a tightly bound O_2 molecule. This activation can be achieved in intact cells by addition of dithionite and methyl viologen. The methyl viologen-dependent H_2 evolution activity in cyanobacteria usually increases 2-fold during a dark anaerobic adaptation period (Houchins and Burris, 1981a,b; Spiller *et al.*, 1983).

Little reversible hydrogenase activity develops in the presence of molecular oxygen. By far the most dramatic and sustained increase in activity can be obtained by removal of O_2 during continuous illumination. Increases in activity of one to three orders of magnitude occur over a period of hours to days when cultures are sparged with an anaerobic gas mixture during growth (Houchinc, 1984; Tel-Or *et al.*, 1977; Lambert and Smith, 1981b). The enzyme is not catalytically functional in vivo under these conditions but is maintained in an inactive state by photosynthetically produced O_2 (Houchins and Burris, 1981a,b; Houchins, 1984).

The presence of combined nitrogen in the growth medium has little effect on the level of reversible hydrogenase activity, but an indirect effect is observed in heterocystous organisms (Houchins, 1984). Though reversible hydrogenase occurs in both heterocysts and vegetative cells, the specific activity in heterocyst during aerobic growth is about 4-times that in vegetative cells (Houchins and Burris, 1981a, b). The microaerobic environment within the heterocyst probably leads to additional reversible hydrogenase synthesis.

Reversible hydrogenase also shows oxyhydrogen reaction in *Anabaena* 7120 when traces of O_2 are added although the reaction is repidly inactivated by O_2 levels as low as 0.1 per cent in contrast to the reaction catalyzed by uptake hydrogenase (Houchins and Burris, 1981a,b). CO_2-dependent H_2 uptake occurs at low light intensity, and higher light intensities leads to deadaptation and reversion to normal O_2-evolving photosynthesis. This reaction has been shown to be strictly light-dependent and is completely inhibited by the uncoupler CCCP, demonstrating a requirement for photophosphorylation.

Hydrogenase dependent H_2 evolution has been demonstrated in a number of species of cyanobacteria (Adams *et al.*, 1980; Asada and Kawamura, 1985; Asada *et al.*, 1987a,b,c; Belkin and Padan, 1978; Benemann *et al.*, 1982; Daday *et al.*, 1979; Daday and Smith, 1987; Ewart and Smith, 1989a,b,c; Houchins, 1984; Houchins, 1981d; Howarth and Codd, 1985; Laczko, 1980; Laczko and Barabas, 1981; Llama *et al.*, 1979; Mitsui, 1979b; Miyamoto *et al.*, 1982, 1984; Peschek, 1979a,b,c,d; Spiller *et al.*, 1983; Tel-Or *et al.*, 1977; Yagi, 1976). A few organic carbon sources have been reported to enhance reversible hydrogenase activity.

H_2 Metabolism in Symbiotic Cyanobacteria

Symbiotic cyanobacteria have not been widely employed for studies on H_2 production or uptake. However, a few workers have made attempts to screen the capacity of H_2 production/uptake in *Anabaena azollae*, *Anabaena cycadeae* and *Nostoc* sp. (from lichen) (Newton, 1976; Peters, *et al.*, many N_2 -fixing organisms, there was no short term effect of nitrate on H_2 production. H_2 production was observed with *Azolla* fronds grown for even 1 to 9 months on nitrate supplemented medium. H_2 production was found almost comparable in experiments where *Azolla* was grown under 1 per cent CO_2 in air or under 1 per cent CO_2, 20 per cent O_2 in argon.

Increase of light intensity showed parallel increase in the rate of H_2 production. Thus incubation of Azolla under 1200 ft-C showed the highest production of H_2. There was almost insignificant level of H_2 production under dark. Experiments done with O_2 under dark showed no production of H_2 suggesting the presence of an "uptake hydrogenase". *Anabaena azollae* freshly isolated from Azolla also showed active formation of H_2. A rate of 50.4 nmoles H_2/mg Chl. a/min was routinely obtained under Ar + CO_2. Addition of C_2H_2 and CO greatly stimulated H_2 formation. Overall studies conducted by Peters *et al.* (1976) are indicative of nitrogenase-mediated H_2 formation by *Azolla* plants or isolated *A. azollae*. Effects of a number of other factors on H_2 production have also been reported (Newton, 1976; Peters *et al.*, 1987).

Another water fern Azolla pinnata has been employed for studies on H_2 production by Banerjee *et al.*, 1989. With this water fern, highest H_2 evolution (5.40 nmol H_2/mg fresh wt at 24 h) was observed if the fronds were incubated in allen and Arnon's medium devoid of combined nitrogen sources. H_2 evolution was 4-5 times higher under light-anaerobic medium. Addition of NH_4Cl or KNO_3 (10mM) did not inhibit H_2 production. H_2 evolution was significantly stimulated by the exogenous addition of phosphate. 60 PPm phosphate addition elicited 6-8 times higher H_2 formation than the control. Furthermore, there were significant differences in the rate of H_2 production between pink and green fronds of *Azolla*. Pink frond always showed higher rate of H_2 formation than the green one. Although considerable activity (H_2 formation) was detected in the dark but addition of PS-II inhibitor completely suppressed H_2 formation. Possible presence of a soluble hydrogenase has been suggested though mainly based on the data of dark and NO^-_3- grown *Azolla* fronds. A number of workers have recently studied H_2 evolution by immobilized *A. azollae* (Brouers and Hall, 1986; Shi *et al.*, 1987; Srivastava, 1990). Immobilization elicited multifold stimulation of H_2 production.

Perraju *et al.*, 1986 have studied H_2 evolution and uptake in cycas *Anabaena* cycadeae association. They demonstrated that H_2 evolution was undetectable in free-living H_2 fixing cultures of *A. cycadeae* but intact coralloid roots showed a rate of 4umol H_2/mg Chl. a/h. Furthermore, it was found that H_2 uptake was absent in the coralloid root but free-living *A. cycadeae* showed an H_2 uptake rate of 28 umol formation (1.56 nmol H_2/ug Chl a/h) in *A. cycadeae* has this rate is almost 2-3 fold lower than the intact coralloid roots.

Other than Azolla and Cycasc, H_2 evolution and uptake studies have been performed with three N_2 fixing lichens *viz.*, *Peltigera membranacea*, *P. polydactyla* and *Lobaria pulmonaria* (Millbvank, 1981). It was found that H_2 evolved concomitant with N_2 fixation was recycled by means of an uptake hydrogenase, and in general the net evolution was zero or small at 5 and 15°C. At 25°C there was appreciable net evolution of H_2 in the *Peltigera* sp. but *Lobaria* showed no net evolution. Extensive study is essential to understand the mode of H_2 metabolism in symbiotic cyanobacteria in general.

H_2 Production by Immobilized Cyanobacteria

Immobilization is a method of conversion of enzyme or whole cells from a water soluble mobile state to a water insoluble immobile state, by trapping them in an insert, insoluble, solid support. In recent years immobilizing action of enzymes, microbial cells and other living systems is being used extensively for the large-scale production of drugs and other compounds (Brodelius, 1985). A number of matrixes are used for immobilization *viz.*, glass beads, agar-agar, alginate etc. although the use of particular system depends upon the nature of investigation. The most common procedure for immobilizing microbes is cell entrapment in natural polymers, in particular agar, shaped into blocks, layers or beads.

Immobilization of cyanobacteria for H_2 production is preferred over liquid culture mainly due to two reasons. First cyanobacteria require continuous agitation in liquid culture to keep cultures uniformaly suspended, since without it the organisms settle out and form clumps which exhibit no H_2 formation (Weissman and Benemann, 1977). Second, agitation causes filament breakage and structural degeneration of vegetative cells, both of which are associated with cessation of biophotolysis in liquid cultures.

H_2 production by immobilized cells of cyanobacteria has been studied by a few workers (Weissaman and Benemann, 1977; Lambert *et al.*, 1979; Kayanlo *et al.*, 1981; Ocjhiai *et al.*, 1983; Brouers and Hall, 1986; Robinson *et al.*, 1986; Kuwada and Ohta, 1987; Shi *et al.*, 1987). In almost all the studies so far conducted, immobilization of cultures showed significant stimulation of H_2 production (Table 5.10). The duration of H_2 evolution was also prolonged significantly. The immobilizing conditions have been found to vary from strains to strains. Similarly types of matrix used also show significant differences in rate as well as duration of H_2 production. For example, *Anabaena* N-7363 showed maximum hydrogen production 90.53 umole H_2 h^{-1} b^{-1} (wet gel) with 2 per cent agar gel at 30°C and 3000 lux light intensity (Kayano *et al.*, 1981). This rate was three times higher than that of free algae. On the other hand *Lyngbya* sp. strain 108 showed optimum H_2 production when immobilized on 4 per cent alginate; 0.5 M $CaCl^2$ and 0.11 mg dry microbial cells ml^{-1} gel (Kuwada and Ohta, 1987). The pH temperature, and light intensity were kept at 9.0, 30°C and 1000 lux. On the other hand *A. azollae* showed maximum yield of H_2 production when immobilized on polyvinyl (Brouers and Hall, 1986; Shi *et al.*, 1987). Higher yield with polyvinyl has been shown to be partly related to changes in membranes permeability induced by the immobilization process itself. A similar increase in yield of H_2 production was observed when membrane permeability of alginate immobilized *A. azollae* was increased by an acetone pretreatment (Brouers and Hall, 1986).

Table 5.10: H_2 Production by Immobilized Cyanobacteria

Test Organism	*Matrix Used*	*Increase % of Fold*	*Duration of H_2 Production*	*References*
Anabena cylindrica	Alginate	two-fold		Weissman and Benemann 1977
A. cylindrica B-629	Glass beads	5 per cent	upto 57 days	Lambert *et al.*, 1979a
Anabaena N-7363	Agar-agar	3-fold	upto 7 days	Kayano *et al.*, 1981
Phormidium sp.	SnO_2 OTE Calcium alginate	ND*	ND	Ochiai *et al.*, 1983
Anabaena azollae	Polyvinyl polyurethane and alginate	2-fold		Brouers and Hall 1986
Mastigocladus laminosus				
Lyngbya sp. strain 108	Calcium alginated	3.4 fold	5 days	Kuwada and Ohta 1987
Anabaena azollae	Polyurethane, polyvinyl and alginate	2-fold	ND	Shi *et al.*, 1987

* ND–not determined; –: Measurement of photocurrent.

The exact reasons pertaining to the enhancement of H_2 formation following immobilization are as yet not clearly known (Robinson *et al.*, 1986). However, unlike free-living cultures, immobilized cyanobacteria do not settle out and form clump. Moreover, cyanobacteria are protected from filament breakage and structural degeneration which occurs as a result of agitation. Furthermore, increase of heterocyst frequency in immobilized cyanobacteria may also account for the increased rate of H_2 production. Increase in heterocyst frequency has been reported in the cyanobacterium immobilized on calcium alginate beads (Kerby *et al.*, 1986).

The prolonged duration of H_2 formation following immobilization has been demonstrated to be a phenomenon related with growth (Srivastava, 1990). The non-growth period is prolonged especially in agar-agar and thus results in sustained H_2 production for a longer period. A potential limitation of immobilization for an organism is the problem of light penetration through the supporting matrix. However, as cyanobacteria themselves stick to the cubes or beads, the large surface area provided would thus ensure a uniform distribution of organisms and thus light penetration. Even if lesser light intensity favours H_2 production (Lambert and Smith, 1981b; Kuwada and Ohta, 1987). Stimulation of H_2 production following immobilization has also been claimed to be due to oxygen tolerance. Although immobilization does not prevent the inhibitory effects of O_2 but most probably it does not production by cyanobacterial system.

Immobilization and Development of Photochemical Fuel Cell

One of the first biological fuel cells designed to use H_2 was developed by Rohrback *et al.* (1962), who reported on the production of hydrogen from glucose by *Clostridium butyricum*. In this biological fuel cell, glucose was used as a nutrient; however, it is expensive as an energy resource. Solar energy is very attractive for energy production and Berk and Canfield (1964) reported one of the first photochemical fuel cells utilizing microorganisms. In the presence of light, *R. rubrum* produced hydrogen, which was oxidized on the surface of the anode. Later on, coupling of immobilized chloroplasts with *Clostridium butyricum* made possible water-splitting hydrogen evolution (Kayano *et al.*, 1981) and the hydrogen prduced was applied to a hydrogen oxygen fuel cell. A photo-current of 0.4–1.5 mA was obtained for 4 h from the photochemical fuel cell system using immobilized chloroplasts and *C. butyricum*.

Using the cyanobacterium *Anabaena* No. 7363, a photochemical fuel cell system was first developed by Kayano *et al.*, 1981. In this system the alga *Anabaena* No. 7363 was immobilized in 2 per cent agar gel and thus produced 3-fold higher H_2 in light (10,000 lux). The oxygen evolved was removed by reactor containing aerobic bacterium *Bacillus subtilis*. The H_2 evolved was passed through soda lime and the flow was regulated by a flow meter before reaching the anode chamber of the fuel cell. The hydrogen-oxygen fuel cell consisted of platnized platinum anode (10 x 50 cm), a porous active carbon cathode (7.5 x 8.0 x 2.5 cm) and the electrolyte (0.1 M phosphate buffer solution, pH 8.0). The anode and cathode were separated by an anion exchange membrane. The current, the anode potential and the cell voltage were measured by a millivolt-millimeter and displayed on a recorder. A photo-current of 15-30 mA was continuously produced for 7 days by the photochemical fuel cell consisting of the immobilized *Anabaena* reactor, the O_2-removing reactor and the hydrogen-oxygen fuel cell. The conversion ration of hydrogen to current was from 80 to 100 per cent.

Ochiai *et al.*, 1983 have developed semiconducter electrodes coated with living films of cyanobacteria. They have demonstrated that the intact *Phormidium* sp. cells immobilized on SnO_2 semiconductor electrode are capable of transferring electrons to SnO_2 in a light-dependent reaction. They have also demonstrated that the drying of a "wet" algal electrode at 50ºC for 60 min increases photocurrent output capacity by 100-fold.

Hydrogen Uptake

Uptake of H_2 occurs via uptake hydrogenase and/or reversible hydrogenase in all the cyanobacteria. Hydrogen functions as an electron donor for both respiratory and light dependent photosynthetic electron flow, the precise nature of which appears to vary from species to species. The characteristics of enzymes involved in uptake of H_2 are described elsewhere (Figures 5.1, 5.2 and Tables 5.6–5.8).

Like nitrogenase-mediated H_2 formation, uptake of H_2 is affected by gas atmospheres, culture medium, light intensity and quality, and age of the culture (Lambert and Smith, 1981b; Vyas and Kumar, 1995). Furthermore, the uptake is influenced by incubation conditions (Houchins, 1984). Combined nitrogen sources *viz.*, NO^-_3 NH^+_4 and amino acids completely inhibit uptake hydrogenase activity. Nickel has been demonstrated to be involved in regulating uptake hydrogenase activity (Smith *et al.*, 1985; Xinankong *et al.*, 1984; Vyas, 2004) Friedrich *et al.*, 1982 have reported that nickel is a constituent of soluble and particulate hydrogenase of *Alcaligenes eutrophus*. However, till now, nickel as a prosthetic group or co-factor of cyanobacterial hydrogenases has not been demonstrated by any workers (Houchins, 1984; Lambert and Smith, 1981b; Xiankong *et al.*, 1984). However, it has been suggested that nickel may be required for activation of an uptake hydrogenase, or for hydrogenase synthesis, or for synthesis of another protein which is involved in H_2 uptake (Almon and Boger, 1984; Daday and Smith, 1983; Daday *et al.*, 1985; Pederson *et al.*, 1986; Srivastava, 1990; Xiankong *et al.*, 1985; Pederson *et al.*, 1984; Srivastava, 1990; Xiankong *et al.*, 1984). For cyanobacteria uptake of H_2 has many possible functions, *viz.* (a) as a scavenger of O_2; (b) as a mechanism that prevents H_2 inhibition of nitrogenase; (c) in the production of ATP via H_2 oxidation (Knallgas reaction). Physiological function

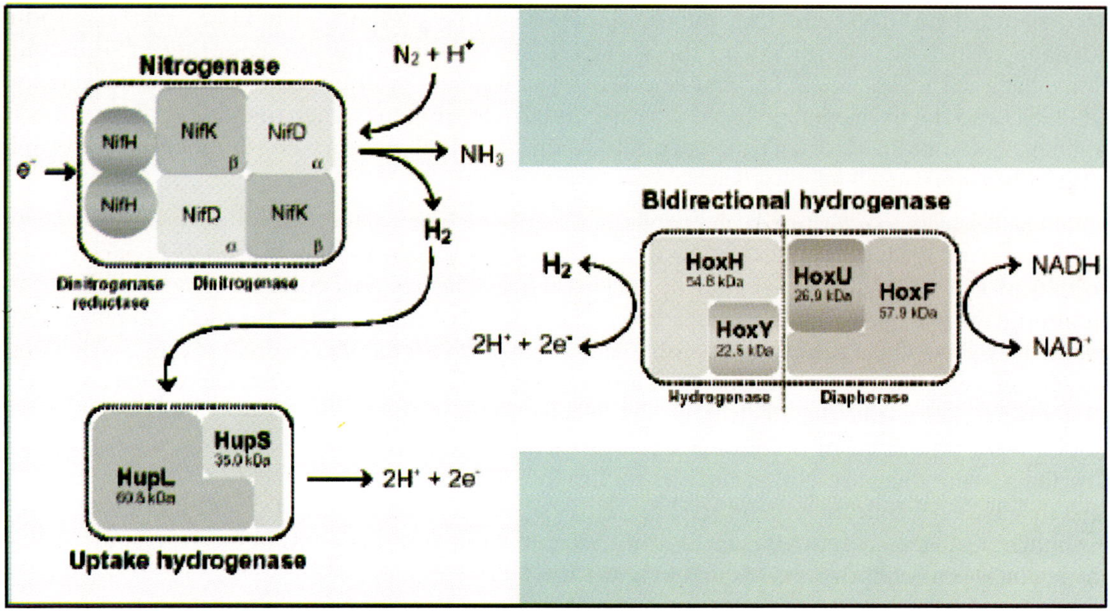

Figure 5.2: Cyanobacterial Enzymes Directly Involved in H_2 Metabolism

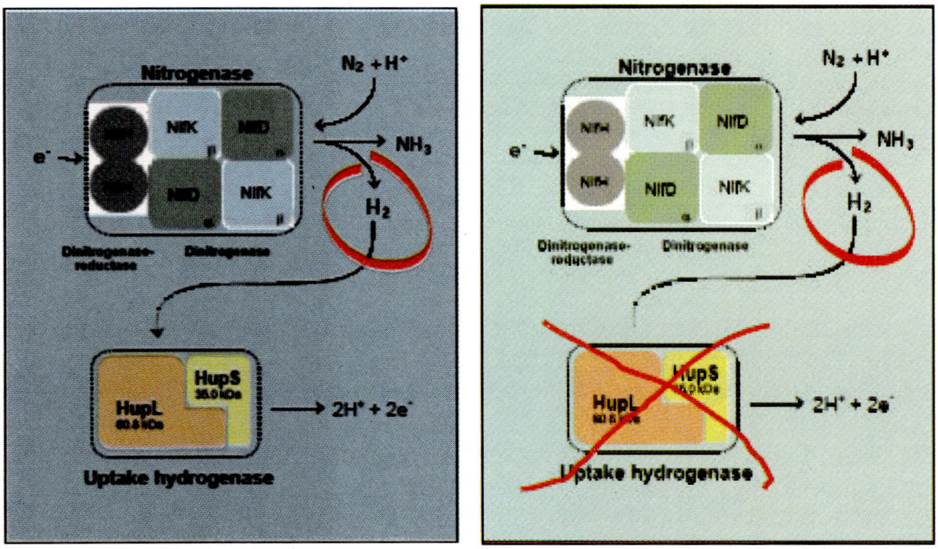

Knock out the uptake hydrogenase

Nitrogenase based **Nitrogenase based**

Figure 5.3: Two Different Approaches for Cyanobacterial Biohydrogen

of the uptake hydrogenase in cyanobacteria is to catalyse the consumption of hydrogen produced by nitrogenase (Happe *et al.*, 2000; Lindberg *et al.*, 2002). All these function have been demonstrated in one or another cyanobacteria (Houchins, 1984). Though uptake hydrogenase does increase the efficiency of nitrogenase, it adversely affects H_2 evolution activity.

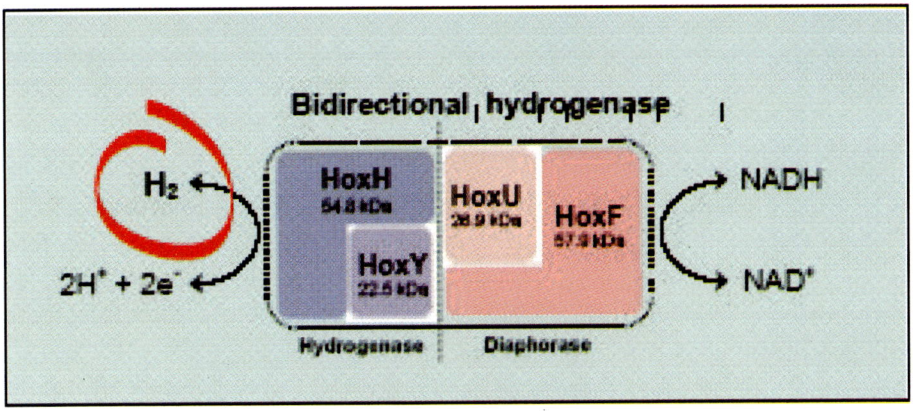

Figure 5.4: Hydrogenase Based + Fermentation (Linblad, 2002)

Cyanobacterial uptake hydrogenases consist of two subunits, encoded by hupS and hupL respectively the small subunit. HupS contains the iron-sulphur (Fe-S) cluster necessary for electron transfer to active site, which is located in the large subunit HupL. HupL contains two putatine binding sites (RXCGXC) necessary for the coordination as the nickel in the activity site (Happe *et al.*, 2000). Recently Vyas and Gupta (2007) reviewed transcriptional regulation of hydrogenases.

Bioengineering Aspects of H_2 Production

The proposal to produce hydrogen from water using solar energy and a biological catalyst has been the major impetus behind most recent research in the hydrogen metabolism of algae. On the basis of present knowledge of algal physiology two basic concepts have been considered: two stage systems and single-stage systems. In a two stage system photosynthetic CO_2 fixation would lead to O_2 evolution and bring about the accumulation of stable reduced carbon compounds which would then be transferred by pumping to an anaerobic chamber where a hydrogen fermentation would take place. In single stage systems hydrogen and oxygen would be released under a transparent covered collector either simultaneously or alternatively.

Two stage system has not yet been demonstrated in practice. One stage system has been employed for outdoor H_2 production (Benemann *et al.*, 1978). To date one oxygen stable biophotolysis catalyst has been demonstrated–the heterocystous nitrogen–fixing cyanobacteria–in which O_2 and H_2 production are separated spatially at the microscopic level. Use of these organisms has already demonstrated a catalytic, sustained, stoichiometric production of O_2 and H_2 using sunlight in an outdoor test (Benemann *et al.*, 1978). The experimnents involved sparging the cultures with a mixture of CO_2 (0.3 per cent), N_2 (1 per cent) and argon (balance) in vertical glass tubes containing 1-liter of culture. The low N_2 -concentrations maintained the algae in a healthy condition, substituting for the periodic additions of ammonia used in the laboratory experiments. H_2- production rates followed the sunlight intensities, decreasing to zero after sunset. A rate of 24 um/mg decreased but H_2 production continued upto 19 days.

A modified single-stage system has been suggested (Benemann *et al.*, 1978; Hallenbeck and Benemann, 1979). The vertical glass tubes fused in current experiments will have to be replaced by horizontal glass tubes, where gas flows over the surface of the algal cuture. This will minimize the energy required to pump the gas phase through the system. There is utmost need to make trials of such system at larger scale for large scale production of H_2.

Conclusion and Future Strategy for Photobiological H_2 Production

Studies so far made on H_2 metabolism on cyanobacteria reveall presence of H_2 metabolizing pathways in almost all the cyanobacteria. However, N_2-fixing cyanobacteria seem more suitable than non-N_2^+ fixers due to presence of the O_2 protection mechanism in them. Much more work is needed to understand exactly the biochemistry of hydrogenase.

There are a few aspects which need to be dealt with extensively in future studies. They are:

1. Isolation of efficient aerobic H_2 producers from various habitat,
2. Isolation of Hup-mutants,
3. Screening of depressed strains which can evolve H_2 even in the presence of combined nitrogen sources,
4. Isolation of some mutant which is defective in N_2 fixation but can produce H_2,
5. Thorough study on immobilization employing various matrices.

References

Asada, Y. and Kawamura, S. (1986). Aerobic hydrogen accumulation by a nitrogen-fixing cyanobacterium, *Anabaena* sp. *Appl. Environ. Microbiol.* 51: 1063-1066.

Belkin, S. and Padan, E. (1978). Hydrogen metabolism in the facultative anoxygenic cyanobacteria (blue-green algae) Oscillatoria limnetica and *Aphonothece halphytica*. *Arch. Microbiol.* 116: 109-111.

Benemann, J.R. and Weare, N.M. (1974a). Hydrogen evolution by nitrogen-fixing *Anabaena cylindrica* cultures. *Science* 184: 174-175.

Benemann, J. (1977). Hydrogen and methane production through microbial photosynthesis. In: *Living Systems as Energy Converters* (eds. Buvet, R., Allen, and Massue, J.P.) PP. 285. Elsevier/North Holland Biomedical Press, Amsterdam.

Benemann, J.R. and Hallenbeck, P.C. (1978). Basic and applied studies of hydrogenase in cyanobacteria. In: *Hydrogenases: Their Catalytic Activity, Structure and Functions* (eds.) Schlegel, H.G. and Schneider, K. P. 395. Goltze. Gotingen.

Benemann, J.R. and Weare, N.M. (1974b). Nitrogen fixation by *Anabaena cylindrica*. III. Hydrogen-supported nitrogenase activity. *Arch. Microbiol.* 101: 401-408.

Benemann, J.R. Miyamoto, K, Hallenbeck, P.C. and Murray, M.A. (1982). Hydrogenase activity in the thermophile *Mastigocladus laminosus*. *Biochem. Biophys. Res. Commun.*. 106: 1196–1202.

Benemann, J.R. Miyamoto, K. and Hallenbeck, P.C. (1980). Bioengineering aspects of biophotolysis. *Enzyme Microbiol. Technol.* 2: 103-111.

Benerjee, M., Kumar, A. and Kumar, H.D. (1989). Factors regulating nitrogenase activity and hydrogen evolution in *Azolla Anabaena* symbiosis. *Int. J. Hydrogen Energy* 14: 871-879.

Berk, R.S. and Canfield, J.H. (1964). Bioelectrochemical energy conversion. *Appl. Microbiol.* 12: 10-12.

Bothe, H., Neuer, G., Kalbe, I. and Eisbrenner, G. (1980). Electron donors and hydrogenase in nitrogen-fixing microorganisms. In: *Nitrogen Fixation* (Eds.) Stewart, W.D.P. and Gallon, J.R. pp. 83-112. Academic press, New york.

Bothe, H., Tennigkeit, J. and Eisbrenner, G. (1977b). The hydrogenase-nitrogenase relationship in the blue-green alga *Anabaena cylindrica. Planta* 133: 237-242.

Bothe, H, Distler, E. and Eisbrenner, G. (1978). Hydrogen metabolism in blue-green algae. *Biochimie* 60: 277-289.

Bothe, H. (1982). Nitrogen fixation. In: *The Biology of Cyanobacteria* (eds.) Carr, N.G. and whitton, B.A.) pp. 87-104. The University of California Press, Berkeley, Los Angeles.

Bothe, H., Tennigkeit, J. and Eisbrenner, G. (1977a). The utilization of molecular hydrogen by the blue-green alga *Anabaena cylindrica. Arch. Microbiol.* 114: 43-49.

Brodelius, P. (1985). The potential role of immobilization in plant cell biotechnology. *Trends Biotechnol.* 3: 280-285.

Brouers, M. and Hall, D.O. (1986). Ammonia and hydrogen production by immobilized cyanobacteria. *J. Biotechnol.* 3: 307-322.

Calvin, M. and Taylor, S.E. (1989). Fuels from algae. In: *Algal and Cyanobacterial Biotechnology* (eds.) Cresswell, R.C., Rees, T.A.V. and Shah, N. Longman Group, UK Ltd.

Chan, Van, N., Hai, C. and Gogotov, I.N. (1983). Photosynthesis and hydrogen metabolism in *Anabaena azollae. Mikrobiologiya* 52: 896-901.

Chen, C.P., Almon, H. and Boger, P. (1989). Physiological factors determining hydrogenase activity in nitrogen-fixing heterocystous cyanobacteria. *Plant Physiol.* 89: 1035-1038.

Daday, A. and Smith, G.D. (1983). The effect of nickel on the hydrogen metabolism of the cyanobacterium *Anabaena cylindrica. FEMS Microbiol. Lett.* 20: 327-330.

Daday, A. and Smith, G.D. (1987). The hydrogenase-nitrogenase relationship in a symbiotic cyanobacterium isolated from *Macrozamia communis* L. Johnson. *Aust. J. plant Physiol.* 14: 319-324.

Daday, A., Lambert, G.R. and Smith, G.D. (1979). Measurement in vivo of hydrogenase-catalysed hydrogen-evolution in the presence of nitrogenase enzyme in cyanobacteria. *Biochem. J.* 177: 139-144.

Daday, A., Mackerras, A.H. and Smith, G.D. (1985). The effect of nickel on hydrogen metabolism and nitrogen fixation in the cyanobacterium *Anabaena cylindrica. J. Gen. Microbiol.* 131: 231-238.

Daday, A., Platz, R.A. and Smith, G.D. (1977). Anaerobic and aerobic hydrogen gas formation by the blue-green alga *Anabaena cylindrica. Appl. Environ. Microbiol.* 34: 478-483.

Davis, D.D. and Stevenson, K.L. (1977). A recording gas microvolumeter. *J. Chem. Educ.* 54: 394-395.

Eisbrenner, G. and Bothe, H. (1979). Modes of electron tansfer from molecular hydrogen in *Anabaena cylindrica. Arch. Microbiol.* 123: 37-48.

Eisbrenner, G., Roos, P. and Bothe, H. (1981). The number of hydrogenases in cyanobacteria. *J. Gen. Microbiol.* 125: 385-390.

Ernst, A., Kerfin, W., Spiller, H. and Boger, P. (1979). External factors influencing light-induced hydrogen evolution by the blue-green alga *Nostoc muscorum*. *Z. Naturforsch.* 34: 820-825.

Ewart, G.D. and Smith, G.D. (1989a). Immunochemical analysis of the soluble hydrogenase from the cyanobacterium *Anabaena cylindrica*. *Biochim. Biophys. Acta* 997: 83-89.

Ewart, G.D. and Smith, G.D. (1989b). An activity stain for the detection of cyanobacterial hydrogenase in polyacrylamide gels. *Biochem. Intl.* (19: 89-97.

Ewart, G.D. and Smith, G.D. (1989c). Purification and properties of soluble hydrogenase from the cyanobacterium *Anabaena cylindrica*. *Arch Biochem. Biophys.* 268: 327-337.

Frenkel, A.H., Gaffron, H. and Battley, E.H. (1950). Photosynthesis and photoreduction in the blue-green alga, *Synechococcus elongatus* Nag. Biol. Bulletin. Marine Biological Laboratory, Woods Hole, Mass. 99: 157-162.

Fry, I., Robinson, A.E., Spath, S. and Packer, L. (1984). The role os sodium sulfide in anoxygenic photosynthesis and hydrogen production in the cyanobacterium *Nostoc muscorum*. *Biochem. Biophys. Res. Commun.* 123: 1138-1143.

Gerasimenko, L.M. and Zavarzin, G.A. (1981). Anabolic hydrogen uptake by cyanobacteria. *Mikrobiologiya* 50: 955-959.

Gogotov, I.N. and Kosyak, A.V. (1976). Hydrogen metabolism by *Anabaena variabilis* in the dark. *Mikrobiologiya* 45: 586-591.

Gogotov, I.V., Kosyak, A.V. and Krupenko, A.N. (1976). Hydrogen production by the cyanobacterium *Anabaena variabilis* in light. *Mikrobiologiya* 45: 941-945.

Greenbaum, E. (1977). The molecular mechanisms of photosynthetic hydrogen and oxygen production. In: *Biological Solar Energy Conversion* (Eds.) Mitsui, A., Miyachi, S., San Pietro, A. and Tamura, S. pp. 101-107. Academic Press, New York.

Hallenbeck, P.C. and Benemann, J.R. (1979). Hydrogen from algae. In: *Photosynthesis in Relation to Model Systems* (ed. Barber, J.). Elsevier/Noth-Holland Biomedical Press, Amsterdam.

Hallenbeck, P.C. Kochian, L. V. and Benemann, J.R. (1981). Hydrogen evolution catalyzed by hydrogenase in cultures of cyanobacteria. *Z. Naturforsch.* 36c: 97-92.

Happe, T., Schute, K. and Bohme, H. (2000). Transcriptional and mutational analysis of the uptake hydrogenase of the filamentous cyanobacterium *Anabaena variabilis* ATCC 29413. J. Bacteriol. 182: 1624–1631.

Hartman, H. and Krasna, A.I. (1963). Studies on the "adaptation" of hydrogenase in Scenedesmus. J. Biol. Chem. 238: 749-757.

Hattori, A. (1963). Microalgae and photosynthetic bacteria. Special Issue of plant and Cell physiology, University of Tokyo Press, pp. 485-492.

Haystead, A., Robinson, R. and Stewart, W.D.P. (1970). Nitrogenase activity in extracts of heterocystous blue-green algae. *Arch. Mikrobiol.* 82: 325-336.

Hillmer, P. and Gest, H. (1977). H_2 metabolism in the photosynthetic bacterium *Rhodopseudomonas capsulata*: H_2 production by growing cultures. *J. Bacteriol.* 129: 724-731.

Hirosawa, T. and wolk, C.P. (1979). Factors controlling the formation of akinetes adjacent to heterocyst in the cyanobacterium *Cylindrospermum licheniforme. J. Gen. Microbiol.* 114: 423-432.

Houchins, J. and Burris, R.H. (1981a). Comarative characterization of two distinct hydrogenases from *Anabaena* sp. strain 7120. *J. Bacteriol.* 146: 215-221.

Houchins, J.P. (1984). The physiology and biochemistry of hydrogen metabolism in cyanobacteria. *Biochim. Biophys. Acta* 768: 227-255.

Houchins, J.P. and Burris, R.H. (1981b). Occurrence and localization of two distinct hydrogenases in the heterocystous cyanobacterium *Anabaena* sp. strain 7120. *J. Bacteriol.* 146: 209-214.

Houchins, J.P. and Burris, R.H. (1981c). Light and dark reactions of the uptake hydrogenase from Anabaena 7120. *Plant Physiol.* 68: 712-716.

Houchins, J.P. and Burris, R.H. (1981d). Physiological reaction of the reversible hydrogenase from *Anabaena* 7120. *Plant Physiol.* 68: 717-721.

Howarth, D.D. and Codd G.A. (1985). The uptake and production of molecular hydrogen by unicellular cyanobacteria. *J. Gen. Microbiol.* 131: 1561-1570.

Jackson, D.D. and Ellms, J. W. (1886). On odors and tastes of surface waters, with special reference to *Anabaena*, a microscopical organism found in certain water supplies of Massachusetts. In the 1896

Jouanneau, Y., Kelley, B.C., Barlier, Y., Lespinat, P. A. and Vignais, P.M. (1980). Continuous monitoring by mass spectrometry of H_2 production and recycling in *Rhodopseudomonas capsulata. J. Bacteriol.* 143: 628-636.

Kayano, H., Karube, I., Matsunaga, T., Suzuki, S. and Nakayama, O. (1981). A photochemical fuel cell system using *Anabaena* N-7363. *Eur. J. Appl. Microbiol. Biotechnol.* 12: 1-5.

Kerby, N. W., Musgrave, S.C., Rowell, P., Shestakov, S.V. and Stewart, W.D.P. (1986). Photoproduction of ammonium by immobilized mutant strains of *Anabaena variabilis. Appl. Microbiol. Biotechnol.* 24: 42-46.

Kerfin, W. and Boger, P. (1982). Light-induced hydrogen evolution by blue-green algae (cyanobacteria) physiol. *Plant.* 54: 93-98.

Kerfin, W., Spiller, H., Ernst, A. and Boger, P. (1978). Properties of the hydrogen producing system in the blue-green alga Nostoc muscorum. In: Hydrogenases: Their Catalytic Activity, Structure and Function (eds.) Schlegel, H.G. and Schneider, K. pp. 381-386. E. Goltze KG. Gottingen.

Kosyak, A.V. Gogotov, I.N. and Kulakova, S.M. (1978). Photoproduction of hydrogen by the cyanobacterium *Anabaena cylindrica. Mikrobiologiya* 47: 605-610.

Krasna, A.I. (1978). Oxygen-stable hydrogenase and assay. *Methods Enzymol.* 53: 314-332.

Krasna, A.I. (1979). Hydrogenase: Properties and applications. *Enzyme Microbial Technol.* 1: 165-172.

Krasna, A.I. (1977). Catalytic and structural properties of the enzyme hydrogenase and its role in biophotolysis of water. In: *Biological Solar Energy Conversion* (eds.) Mitsui, A., Miyachi, S., San Pietro, A. and Tamura, S. P. 53. Academic Press, New York.

Kumar, D. (1986). Effects of amitrol on cyanobacterium *Nostoc linckia. J. Gen. Appl. Microbiol.* 32: 51-56.

Kumar, D. and Kumar, H.D. (1988). Light-dependent acetylene reduction and hydrogen production by the cyanobacterium, *Anabaena* sp. (strain CA). *Int. J. Hydrogen Energy* 13: 573–575.

Kumar, D. and Kumar, H.D. (1989). Effects of glucose and fructose on acetylene reduction, photosynthesis, respiration and hydrogen evolution by whole cells of the cyanobacterium *Anabaena* sp. strain CA. In *Proc. Intl. Phycotalk Symp.*, Dept. of Botany, Banaras hindu University, Varanasi.

Kumar, D. and Kumar, H.D. (1990b). Effects of some inhibitors and carbon sources on acetylene reduction and hydrogen production by isolated heterocysts of *Anabaena* sp. (strain CA). *British Phycol. J.* (Accepted).

Kumar, D. Kumar, H.D. (1990c). Protection of nitrogenase levels in dark-incubated cultures of *Anabaena* sp. strain CA by various carbon sources, and restoration of nitrogenase activity by oxygen. *Br. Phycol. J.* 25: (in press).

Kumar, D., Pandey, A.B. and Kumar, H.D. (1990a). Effect of some physiological factors on the nitrogenase and hydrogenase of the *Azolla-Anabeana azollae* association. *Int. J. Hydrogen Energy* 15: 313-317.

Kumazawa, S. and Mitsui, A. (1981). Characterization and optimization of hydrogen photoproduction by a saltwater bluegreen alga, *Oscillatoria* sp. miami BG7. I. Enhancement through limiting the supply of nitrogen nutrients. *Int. J. Hydrogen Energy* 6: 339-348.

Kumazawa, S. and Mitsui, A. (1985). Comparative reproduction activities between heterocystous cyanobacterium *Anabaena cylindrica* B 629 and non heterocystous cyanobacterium *Oscillatoria* sp. strain Miami BG 7. *Appl. Environ. Microbiol.* 50: 287-291.

Kumazawa, S. and Mitsui, A. (1982). Hydrogen metabolism of photosynthetic bacteria and algae. In: *Handbook of Biosolar Resources, Vol. I. Basic principles* (eds. Mitsui, A. and Black, C.C.O. pp. 299-316. CRC Press, Boca Raton, Florida.

Kuwada, Y. and Ohta, Y. (1987). Hydrogen production by an immobilized cyanobacterium, *Lyngbya* sp. *J. Ferment. Technol.* 65: 597-602.

Kuwada, Y. and Ohta, Y. (1989). Hydrogen production and carbohydrate consumption by *Lyngbya* sp. No. 108. *Agric. Biol. Chem.* 53: 2847-2851.

Kuwada, Y., Nakatsukasa, M. and Ohta, Y. (1988). Isolation and characterization of a non heterocystous cyanobacterium *Lyngbya* sp. isolate No. 108, for large quantity hydrogen production. *Agric. Biol. Chem.* 52: 1923-1928.

Laczko, I. (1980). Mechanism of hydrogen evolution by nitrogen starved *Anabaena cylindrica*. *Z. Pflanzenphysiol.* 100: 241-246.

Laczko, I. and Barabas, K. (1981). Hydrogen evolution by photobleached *Anabaena cylindrica*. *Planta* 153: 312-316.

Lambert, G.R. and Smith, G.D. (1977). Hydrogen formation by marine blue-green algae. *FEBS Letts.* 83: 159-162.

Lambert, G.R. and Smith, G.D. (1980). Hydrogen metabolism by filamentous cyanobacteria. *Arch. Biochem. Biophys.* 205: 36-50.

Lambert, G.R. and Smith, G.D. (1981a). Hydrogen uptake by the nitrogen-starved cyanobacterium *Anabaena cylindrica*. *Arch. Biochem. Biophys.* 211: 360-367.

Lambert, G.R. Daday, A. and Smith, G.D. (1979a). Hydrogen evolution from immobilized cultures of the cyanobacterium *Anabaena cylindrica. FEBS Letts.* 101: 125-128.

Lambert, G.R. Daday, A. and Smith, G.D. (1979b). Duration of hydrogen formation by *Anabaena cylindrica.* B 629 in atmospheres of argon, air and nitrogen. *Appl. Environ. Microbiol.* 38: 530-536.

Lambert, G.R. Smith, G.D. (1981b). The hydrogen metabolism of cyanobacteria (blue-green algae). *Biol. Rev.* 56: 589-660.

Lambert, G.R., Daday, A. and Smith, G.D. (1979c). Effects of ammonium ions, oxygen, carbon monoxide and acetylene on anaerobic and aerobic hydrogen formation by *Anabaena cylindrica.* B-629. *Appl. Environ. Microbiol.* 38: 521-529.

Liama, M.J., Serra, J.L., Rao, K.K. and Hall, D.O. (1979). Modes of reduction of nitrogenase in heterocysts isolated from *Anabaena* species. *Biochim. Biophys. Acta* 502: 298-308.

Lindberg, P., Schulz, K., Happe, T. and Lindblad, P. (2002). A hydrogen producing, hydrogenase free mutant-strain of Nostoc punetiforme ATEC 29133. *Int. J. Hydrogen Energy*, 27: 1291-1296.

Millbank, J.W. (1981). Nitrogenase and hydrogenase in cyanophilic lichens. *New Phytol.* 92: 221-228.

Mitsui, A. (1980a). Nitrogen fixation and hydrogen production by marine blue-green algae used as a source of food, feed and chemicals. In: *Proceedings of the International Symposium on Biological Application of Solar Energy* (eds.) Gnanam, A. and Kahn, J. pp. 109-115. Viking publ., New Delhi.

Mitsui, A. (1975a). Photoproduction of hydrogen via microbial and biochemical processes. In: *Symposium Proceedings of Hydrogen Energy Fundamental: Symposium Proceedings of Hydrogen Energy Fundamental* (ed.) Veziroglu, T.N. pp. 52 31-48. University of Miami Press, Miami.

Mitsui, A. (1975b). Hydrogen production via photosynthetic processes. In: Key Technologies for Hydrogen Energy Systems. (ed. Ohta, T.) pp. 75-92. Yokohama National University Press.

Mitsui, A. (1975c). Utilization of solar energy for hydrogen production by cell-free system of photosynthetic organisms. In: *Hydrogen Energy, Part A* (ed.) Veziroglu, T.N. pp. 309-316. Plenum publishing Corp., New York.

Mitsui, A. (1976). Bioconversion of solar energy in salt water photosynthetic hydrogen production system. In: *Proceedings of 1st World Hydrogen Energy Conference: Progress in World Hydrogen Energy projects and planning* (ed.) Veziroglu, T.N. pp. 4B: 77-99. University of Miami Press.

Mitsui, A. (1978a). Bio-solar hydrogen production. In: *Hydrogen Energy System*, Vol. 3 (eds.) Veziroglu, T.N. and Seifritz, W. pp. 1267-1291. Pergamon Press, Oxford, England.

Mitsui, A. (1978b). Solar energy conversion to hydrogen by bacteria and algae. In: *Bio-Energy* (ed.) Gottib Duttweiler-Institute. Zurich, pp. 205-236.

Mitsui, A. (1978c). Solar energy conversion in marine biological systems. In: An International Compendium (ed.) Veziroglu, T.N.. Hemisphere Publ. Co., Washington, 2: 1013-1017.

Mitsui, A. (1979a). Biosaline research: The use of photosynthetic marine organisms in food and feed production. In: *The Biosaline Concept* (eds.) Hollaender *et al.* pp. 177-215. Plenum Press, New York.

Mitsui, A. (1979b). Biological and biochemical hydrogen production. In: Solar Hydrogen Energy Systems (ed.) Ohta, T. pp. 171-191. Pergamon press, Oxford and New York.

Mitsui, A. (1980b). Salt water based biological solar energy conversion for fuel, chemicals, fertilizer, food and feed. In: *Proceedings of Bio-Energy* (80. World Congress and Exposition, (ed.) pp. 486-491. Bio-Energy Council, Washington, D.C.

Mitsui, A. (1981). *Hydrogen production by marine photosynthetic microorganisms*. Kitani, O. pp. 165-180. Japan Scientific Society Press, Tokyo.

Mitsui, A. (1987). Photobiological production of hydrogen from water and waste. In: *Proceedings of the Hydrogen Photoproduction workshop* (eds.) Seki, A., Morgan, B. and Takahashi, P.O. pp. 32-39.

Mitsui, A. and Black, C.C. (eds.) (1982). *CRC Handbook of Biosolar Resources. Vol. 1. Basic Principles*, Part 1, pp. 643. CRC Press, Boca Raton, Florida.

Mitsui, A. and Black, C.C. (eds.) (1982). *CRC Handbook of Biosolar Resources. Vol. 1. Basic Principles*, Part 2, CRC Press, Boca Raton, Florida, pp. 657.

Mitsui, A. and Kumazawa, S. (1978). Hydrogen production by marine photosynthetic organisms on a potential energy resource. In: *Biological Solar Energy Conversion* (eds.) Mitsui, A., Miyachi, S., San Pietro, A. and Tamura, S.. Academic Press, London and New York.

Mitsui, A. and Kumazawa, S. (1979). Solar energy biological conversion research. *J. Solar Energy* (in Japanese) 5: 55-67.

Mitsui, A. and Matsunaga, T. (1983). Hydrogen production by photosynthetic microorganisms. In: *Biomass Energy Conversion* (ed. Suzuki, S.) pp. 194-228. Kodansha Scientific Press, Tokyo.

Mitsui, A., Duerr, E. Kumazawa, S., Phlips, E. and Skjoldal, H. (1979). Biological solar energy conversion: hydrogen production and nitrogen fixation by marine blue-green algae. In: *Sun II. Proceedings of the International Solar Energy Society Silver Jubilee Congress, Vol. I* (eds.) Boer, K. and Glenn, B.H. pp. 31-35. Pergamon Press, New York and Oxford.

Mitsui, A., Kumazawa, S., Phlips, E.J., Reddy, K.J., Gill, K., Renuka, B.R. Kusumi, T., Reyes-Vasques, G., Miyazawa, K., Haynes, L., Ikemoto, h., Duerr, E., Leon, C.B. Rosner, D., Sesco, R. and Moffat, E. (1985). Mass cultivation of algae and photosynthetic bacteria: Concepts and Application. In: *Biotechnology and Bioprocess Engineering* (ed.) Ghosh, T.K. pp. 119-155. International Union of Pure and Applied Chemistry and Indian National Science Academy, United India Press, New Delhi.

Mitsui, A., Kumazawa, S., Takahashi, A., Ikemoto, H., Coa, S. and Arai, T. (1987). Relevance of the study of the strategy by which nitrogen-fixing unicellular cyanobacteria grow photoautotrophically to the hydrogen photoproduction research. *J. Hydrogen Energy Systems Society of Japan* 12: 39-49 (in Japanese).

Mitsui, A., Miyachi, S., San Pietro, A. and Tamura, S. (Eds.) (1977). *Biological Solar Energy Conversion*. Academic press, London and New York.

Mitsui, A., Philips, E.J., Kumaawa, S., Reddy, K.J., Ramachandran, S., Matsunaga, T., Haynes, L. and Ikemoto, H. (1983). Progress in research toward outdoor biological hydrogen production using solar energy, sea water and marine photosynthetic microorganisms. *Ann. NoY. Acad. Sci.* 413: 514-530.

Mitsui, A., Rosner, D., Kumazawa, S., Barciela, S., Phlips, E., Ramachandran S., Takahashi, A. and Richard, J. (1985). Hydrogen production from salt water by marine blue green algae and solar radiation. In: *Proceedings of Twenty-Second Space Congress* (ed.) Haise, F.W. pp. 11.7-11.14. Canaveral Council of Technical Societies, Revised form.

Miura, Y, Yokoyama, H., Kimura, M., Iwamoto, T. and Miyamoto, K. (1981). Effect of acetylene and carbon monoxide on hydrogen uptake and evolution by nitrogen-starved cultures of *Mastgocladus laminosus*. *Plant Cell Physiol.* 22:1375-1383.

Miura, Y., Yokoyama, H., Takahara, K. and Miyamoto, K. (1982). Effects of acetylene and carbon monoxide on long-term hydrogen production by *Mastgocladus laminosus*: A thermophilic blue-green alga. *J. Ferment Technol.* 60: 411-416.

Miura, Y.H., Kanaoka, K., Saito, S., Iwasa, K., Okazaki, M. and Komemushi, S. (1980). Hydrogen evolution by a thermophilic blue-green *Mastgocladus laminosus*. *Plant Cell Physiol.* 21: 149-156.

Miyamoto, K., Takahara- Ishikawa, K., Horie, H. and Miura, Y. (1984). Dark period nitrogen fixation by *Mastgocladus laminosus* and its application to a biophotolysis system. *J. Ferment. Technol.* 62: 13-18.

Miyamoto, K.,Hallenbeck, P.C. and Benemann, J.R. (1979). Hydrogen production by the thermophilic alga *Mastgocladus laminosus* : Effects of nitrogen, temperature and inhibitors of photosynthesis. *Appl. Environ. Microbiol.* 38: 440-446.

Mortensen, L.E. (1978). The role of dihydrogen and hydrogenase in nitrogen fixation. *Biochimie* 60: 219-223.

Mortensen, L.E. and Chen J.S. (1974). Hydrogenase. In: *Microbial Iron Metabolism, a Comprehensive Treatise* (ed.) Neilands, J.B. P. 231. Academic Press, London and New York.

Mortensen, L.E. and Thorneley, R.N.F. (1979). Structure and function of nitrogenase. *Ann. Rev. Biochem.* 48: 387-418.

Newton, J.W. (1976). Photoproduction of molecular hydrogen by a plant algal symbiotic system. *Science,* 191: 559-561.

Ochiai, H., Shibata, H., Sawa, Y., Shoga, M. and Ohta, S. (1983). Properties of semiconductor electrodes coated with living films of cyanobacteria. *Appl. Biochem. Biotechnol.* 8:289-303.

Ochial, H., Shibata, H., Sawa, Y. and Katoh, T. (1980). 'Living electrode' as a long-lived photoconverter for biophotolysis of water. *Proc. Natl. Acad. Sci.*, USA 77: 2442-2444.

Ormeland, R. (1983). Hydrogen metabolism by decomposing cyanobacterial aggregates in Bio-Soda Lake, Nevada (USA). *Appl. Environ. Microbiol.* 45: 1519-1525.

Oschchepkov, V.P., Nikitina, A. A., Gusev, M.V. and Krasnovskii A.A. (1974). Evolution of molecular hydrogen by cultures of blue-green algae. Doklady Akademii Nauk, *SSSR* 215: 557-560.

Oschchepkov, V.P., Nikitina, K.A., Gusev, MoV. and Krasnovskii A.A. (1973). Evolution of molecular hydrogen by blue-green algal cultures. Doklady Nauk *SSSR Biol.* 213: 739-742.

Padan, E. and Schuidiner, S. (1978). Energy transduction in the photosynthetic memberanes of the cyanobacterium blue-green alge *Plectonema boryanum*. *J. Biol. Chem.* 253: 3281-3286.

Paerl, H, W. and Bland, P.T. (1982). Localized tetrazolium reduction in relation to N_2 fixation, and CO_2 fixation, and H_2 uptake in aquatic filamentous cyanobacteria. Appl. Environ. *Microbiol.* 43: 218-226.

Paerl, H.W. (1980). Ecological rationale for hydrogen metabolism during aquatic blooms of the cyanobacterium *Anabaena*. *Oecologia* 47: 43-53.

Paerl, H.W. (1982). *In situ* hydrogen production and utilization by natural population of nitrogen-fixing blue-green algae. *Can. J. Bot.* 60: 2542-2546.

Peck, H.D. Jr. and Gest, H. (1956). A new procedure for assay of bacterial hydrogenase. *J. Bacteriol.* 71: 70-80.

Perderson, D.M. Daday, A. and Smith, G.D. (1986). The use of nickel to probe the role of hydrogen metabolism in cyanobacterial nitrogen fixation. *Biochimie* 68: 113-120.

Perraju, B.T. Rai, A.N., Kumar, A.P. and Singh, H.N. (1986). *Cycas circinalis-Anabaena cycadeae* symbiosis: Photosynthesis and the enzymes of nitrogen and hydrogen metabolism in symbiotic and cultued *Anabaena cycadeae*. *Symbiosis* 1: 239-250.

Peschek, G.A. (1978). Reduced sulfur and nitrogen compunds and molecular hydrogen as electron donors for anaerobic CO_2 photoreduction in *Anacystis nidulans*. *Arch. Microbiol.* 119: 203-215.

Peschek, G.A. (1979a). Anaerobic hydrogenase activity in *Anacystis nidulans*. Hydrogen dependent photoreduction and related reaction. *Biochim. Biophys. Acta* 548: 187-202.

Peschek, G.A. (1979b). Aerobic hydrogenase activity in *Anacystis nidulans*. The oxyhydrogen reaction. *Biochim. Biophys. Acta* 548: 203-215.

Peschek, G.A. (1979c). Evidence for functionally distinct hydrogenases in *Anacystis nidulans*. *Arch. Microbiol.* 123: 81-92.

Peschek, G.A. (1979d). Nitrate and nitrite reductase and hydrogenase in *Anacystis nidulans*. grown in Fe- and Mo-deficient media. *FEMS Microbiol. Letts.* 6: 371-374.

Peschek, G.A. (1980). Electron transport reactions in respiratory particles of hydrogenase- induced *Anacystis nidulans*. *Arch. Microbiol.* 125: 123-131.

Peters, G.A. Evans, W.R. and Toia, R.E. Jr. (1976). *Azolla anabaena* cycadeae relationship V. $^{15}N_2$ fixation, acetylene reduction and H_2 production. *Plant Physiol.* 59: 1021-1025.

Peterson, R.B. and Burris, R.H. (1978a). Hydrogen metabolism in isolated heterocysts of *Anabaena* 7120. *Arch. Microbiol.* 116: 125-132.

Peterson, R.B. and Wolk, C.P. (1978b). Localization of an uptake hydrogenase in *Anabaena. Plant Physiol.* 61: 688-691.

Phlips, E.J. and Mitsui, A. (1983a). Role of light intensity and temperature in the regulation of hydrogen photoproduction by the regulation of hydrogen photoproduction by the marine cyanobacterium Oscillatoria sp. strain Miami BG 7. *Appl. Environ. Microbiol.* 45: 1212-1220.

Phlips, E.J. and Mitsui, A. (1984). Development of hydrogen production activity in the marine blue-green alga *Oscillatoria* sp. Miami BG 7 under natural sunlight conditions. In: *Advances in Photosynthesis Research* (ed.) Sybesm, C. pp. 810-804 Vol. II, Martinus Nijhoff/Dr W. Junk Publishers, Hague,

Planchard, A., Mignot, L., Jouenne, T. and Junter, G. (1989). Photoproduction of molecular hydrogen by *Rhodospirillum rubrum*: immobilized in composite agar layer/microporous membrane structures. *Appl. Microbiol. Biotechnol.* 31: 49-54.

Ramachandran, S. and Mitsui, A. (1984). Effect of seawater quality on biomiass and hydrogen photoproduction by a marine blue-green alga *Oscillatoria* sp. Miami BG 7. In: *Advances in Photosynthesis Research* (ed.) Sybesma, C. Vol. II, pp. 805-808. Martinus Nijhoff/Dr W. Junk Publishers, Hague.

Ramachandran, S. and Mitsui, A. (1986). Biosystem catalyzed continuous hydrogen photoproduction from seawater using solar energy. In: *Proc. 6th International Symp. Hydrogen Energy* (ed.) Veziroglu, T.N. pp. 1371-1381.

Scherer, S., Kerfin, W. and Boger, P. (1980b). Increase of nitrogenase activity in the blue green alga *Nostoc muscorum* (cyanobacterium). *J. Bacteriol.* 144: 1017-1023.

Schlegel, H.G. and Schneider, K. (1978). Introductory report: distribution and physiological role of hydrogenases in microorganisms. In: *Hydrogenases: their Catalytic Activity Structure and Function* (eds.) Schlegel, H.G. and Schneider, K. p. 15. Goltze, Gottingen.

Scranton, M.I., Novelli, P.C. Michaels, A., Horrigan, S.G. and Carpenter, E.J. (1987). Hydrogen production and nitrogen fixation by *Oscillatoria thiebautii* during *in situ* incubations. *Limnol. Oceanogr.* 32: 998-1006.

Shi, D.J., Brouers, M., Hall, D.O. and Robins, R.J. (1987). The effects of immobilization on the biochemical, physiological and morphological features of *Anabaena azollae. Planta* 172: 298-308.

Singh, S. and Kashyap, A.K. (1988). Photoevolution of hydrogen during oxygenic photosynthesis of cyanobacterium *Nostoc muscorum. Biotech. Lett.* 10: 921-925.

Smith, R.L., Kumar, D., Xiankong, Z., Tabita, F.R. and Van Baalen, C. (1985). H_2, N_2 and O_2 metablism by isolated heterocysts from *Anabaena* sp. strain Ca. *J. Bacteriol.* 162: 565-570.

Smith, G.D., and Lambert, G.R. (1981). Outdoor biophotolytic system using the cyanobacterium, *Anabaena cylindrica* B 629. *Bioeng.* 23: 213-220.

Smith, G.D., Mualiem, A. and Hall, D.O. (1982). Hydrogenase-catalyzed photoproduction of hydrogen by photosystem I of the thermophilic blue-green algae *Mastigocladus laminosus* and *Phormidium laminosum. Photobiochem. Photobiophys.* 4: 307-320.

Spiller, H., Bookjans, G. and Shanmugam, K.T. (1983). Regulation of hydrogenase activity in vegetative cells of *Anabaena variabilis. J. Bacteriol.* 155: 129-137.

Spiller, H., Ernst, E., Kerfin, W. and Boger, P. (1978). Increase and stabilization of photoproduction of hydrogen in *Nostoc muscorum* by photosynthetic electron transport inhibitors. *Z. Naturforsch.* 33c: 541-547.

Srivastava, M. (1990). Regulation of nitrogen fixation and hydrogen production in *Anabaena* species. *Ph.D. Thesis*, Banaras Hindu University, Varanasi.

Srivastava, M., Sharma. A., Kumar, A. and Kumar, H.D. (1989). Hydrogen production by nitrogen-fixing cyanobacteria. In: *Proc. Intl. Phycotalk Symposium*, Department of Botany, Banaras Hindu University, Varanasi-221005, pp. 26-29.

Stewart, W.D.P. (1980). Some aspects of structure and function in N_2-fixing cyanobacteria. *Ann. Rev. Microbiol.* 34: 497-536.

Stuart, T.S. Herold, E.W. Jr. and Gaffron, H. (1972). A simple combination mass spectrometer inlet and oxygen electrode chamber for sampling gases dissolved in liquids. *Anal. Biochem.* 46: 91-100.

Tamagnini, P., Axelsson, R., Lindberg, P., Oxelfelt, F., Wunschiers, R. and Lindblad, P. (2002). Hydrogenases and Hydrogen Metabolism of Cyanobacteria. *Microbiology and Molecular Biology Reviews*. 66(1) : 1-20.

Tamiya, N., Kondo, Y., Kameyama, T. and Akabori, S. (1955). Determination of hydrogenase by the hydrogen evolution from reduced methyl viologen. *J. Biochem*. 42: 613-614.

Tel-Or, E., Luijk, L.W. and packer, L. (1977). An inducible hydrogenase in cyanobacteria enhances N_2 fixation. *FEBS Lett*. 78: 49- 52.

Tel-Or, E., Luijk, L.W. and Packer, L. (1978). Hydrogenase in N_2 fixing cyanobacteria. *Arch. Biochem. Biophys*. 185: 185-194.

Tetley, R.M. and Bishop, N.I. (1979). The differential action of metronidazole an nitrogen fixation, hydrogen metabolism, photosynthesis and respiration in *Anabaena* and *Scenedesmus*. *Biochim. Biophys. Acta* 546: 43-53.

Umbreit, W. W., Burris, R.H. and Stauffer, J.F. (1972). *Manometric and biochemical techniques*. 5th Edition. Burgess publ. C., Minneapolis, Minnesota, pp, 387.

Van Der Oost, J., Kanneworff, W.A. Krab, K. and Kraayenhof, R. (1987). Hydrogen metabolism of three unicellular nitrogen-fixing cyanobacteria. *FEMS Microbiol. Lett*. 48: 41-45.

Vyas, D. (2004). Cyanobacterial biodiversity and their hydrogen production potential. *J. Bot. Asoc. Univ. Sagar*. 39:55-65.

Vyas, D. and Gupta, R.K. (2003). Impact of ultraviolet B radiation on hydrogen production, and nitrogen fixation in *Cyanobacterium nostoc* sp. *J. Bot. Asoc. Univ. Sagar*. 36-38:89-97.

Vyas, D. and Kumar, H.D. (1995). Nitrogen fixation and hydrogen uptake in four cyanobacteria. *Int. J. Hydrogen Energy*. 20:163-168.

Vyas, D. and Gupta, R.K. (2007). Cynobacterial hydrogen metabolism with special reference to transcriptional regulation of hydrogenases. In: Advances in Applied Phycology. (eds.) Gupta, R.K. and Pandey, V.D. pp. 155–168. Daya Publishing House, New Delhi.

Wang, R., Healey, F.P. and Myers, J. (1971). Amperometric measurement of hydrogen evolution in *Chlamydomonas*. *Plant Physiol*. 48: 108-110.

Ward, M.A. (1970). Whole cell and cell free hydrogenases of algae. *Phytochemistry* 9: 108-266.

Weaver, P.F., Lien, S. and Seibert, M. (1980). Photobiological production of hydrogen. *Solar Energy* 24: 3-45.

Weisshaar, H. and Boger, P. (1983). Sulfide stimulation of light induced hydrogen evolution by the cyanobacterium Nostoc muscorum. *Z. Naturforsch*. 38: 237-242.

Weisshaar, H. and Boger, P. (1985). Pathways of hydrogen uptake in the cyanobacterium *Nostoc muscorum*. *Arch. Microbiol*.142: 349-353.

Weissman, J.C. and Benemann, J.R. (1977). Hydrogen production by nitrogen starved cultures of *Anabaena cylindrica*. *Appl. Environ. Microbiol*. 33: 123-131.

Xiankong, Z., Haskell, J.B., Tabita, F.R. and Van Baalen, C. (1983). Aerobic hydrogen production by the heterocystous cyanobacteria *Anabaena* spp. strains CA and 1F. *J. Bacteriol*. 156: 1118-1122.

Xiankong, Z., Tabita, F.R. and Van Baalen, C. (1984). Nickel control of hydrogen production and uptake in *Anabaena* spp. strains CA and 1F. *J. Gen. Microbiol.* 130: 1815-1818.

Yagi, T. (1976). Separation of hydrogenase-catalysed hydrogen evolution system from electron donating system by means of enzymic electric cell technique. *Proc. Natl. Acad. Sci.*, U.S.A.. 73: 2947-2949.

Yagi, T. (1977). Use of an enzymic electric cell and immobilized hydrogenase in the study of the biophotolysis of water to produce hydrogen. In: *Biological Solar Energy Conversion* (eds.) Mitsui, A., Miyachi, S., San Pietro, A. and Tamura, S. pp. 61-68. Academic Press, New York.

Yu, L. and Wolin, M.J. (1969). Hydrogenase measurement with photochemically reduced methyl viologen. *J. Bacteriol.* 98: 51-55.

Zajic, J.E., Kosaric, N. and Brosseau, J.D. (1978). Microbial production of hydrogen. *Adavances in Biochem. Engin.* 9: 57-109.

Microbial Biotechnology and Ecology (2011) *Pages* **94–149**
Editors: **Deepak Vyas, G.S. Paliwal, P.K.Khare and R.K Gupta**
Published by: **DAYA PUBLISHING HOUSE, NEW DELHI**

Chapter 6

Cyanobacterial Toxins

☆ ***U.S. Gupta*[1]*, Deepak Vyas*[2]* *and Rajan Kumar Gupta*[3]**

[1]*Department of Zoology,* [2]*Department of Botany,*
Dr. H.S. Gour University, Sagar – 470 003, M.P.
[3]*Government P.G. College, Rishikesh, Uttranchal*

Introduction

The cyanotoxins are a diverse group of natural toxins, both from the chemical and the toxicological points of view. In spite of their aquatic origin, most of the cyanotoxins that have been identified to date appear to be more hazardous to terrestrial mammals than to aquatic biota. Cyanobacteria produce a variety of unusual metabolites, the natural function of which is unclear, although some, perhaps only coincidentally, elicit effects upon other biota. Research has primarily focused on compounds that impact upon humans and livestock, either as toxins or as pharmaceutically useful substances. Further ranges of non-toxic products are also being found in cyanobacteria and the biochemical and pharmacological properties of these are totally unknown. An overview of the currently identified cyanotoxins is given and their toxicological properties are discussed. Studies on the occurrence, distribution and frequency of toxic cyanobacteria were conducted in a number of countries during the 1980s using mouse bioassay. Analytical methods suitable for quantitative toxin determination only became available in the late 1980s, but studies of specific cyanotoxins have been increasing since then. The results of both approaches indicate that neurotoxins are generally less common, except perhaps in some countries where they frequently cause lethal animal poisonings. In contrast, the cyclic peptide toxins (microcystins and nodularins) which primarily cause liver injury are more widespread and are very likely to occur if certain taxa of cyanobacteria are present. The data currently available on the occurrence of cyanotoxins is noteworthy, however, that current knowledge is clearly biased by the inconsistent distribution of research effort around the world, with studies from Asia, Africa and South

* E-mail: dvyas64@yahoo.co.in

America beginning to appear in the 1990s. Because the ecological role of the toxins is unclear, it is not possible to use a functional approach to study the factors that enhance toxicity. The relationships between environmental factors and toxin content and at the emerging understanding of genetic regulation of toxin production by cyanobacteria is increasing, and a better understanding of toxin function may provide a basis for predicting occurrence of toxicity in the future.

For assessing the health risk caused by cyanotoxins, an understanding of their persistence and degradation in aquatic environments is of crucial importance.

Classification

Mechanisms of cyanobacterial toxicity currently described and understood are very diverse and range from hepatotoxic, neurotoxic and dermatotoxic effects to general inhibition of protein synthesis. To assess the specific hazards of cyanobacterial toxins it is necessary to understand their chemical and physical properties, their occurrence in waters used by people, the regulation of their production, and their fate in the environment.

Cyanotoxins fall into three broad groups of chemical structure: cyclic peptides, alkaloids and lipopolysaccharides (LPS). An overview of the specific toxic substances within these broad groups that have been identified to date from different genera of cyanobacteria, together with their primary target organs in humans, is given in Table 6.1.

Table 6.1: General Features of the Cyanotoxins

Toxin Group	*Primary Target Organ in Mammals*	*Cyanobacterial Genera*
Cyclic peptides		
Microcystins	Liver	*Microcystis, Anabaena, Planktothrix (Oscillatoria), Nostoc, Hapalosiphon, Anabaenopsis*
Nodularin	Liver	*Nodularia*
Alkaloids		
Anatoxin-a	Nerve synapse	*Anabaena, Planktothrix (Oscillatoria), Aphanizomenon*
Anatoxin-a(S)	Nerve synapse	*Anabaena*
Aplysiatoxins	Skin	*Lyngbya, Schizothrix, Planktothrix (Oscillatoria)*
Cylindrospermopsins	Liver	*Cylindrospermopsis, Aphanizomenon, Umezakia*
Lyngbyatoxin-a	Skin, gastro-intestinal tract	*Lyngbya*
Saxitoxins	Nerve axons	*Anabaena, Aphanizomenon, Lyngbya, Cylindrospermopsis*
Lipopolysaccharides (LPS)	Potential irritant; affects any exposed tissue	All

Hepatotoxic Cyclic Peptides-Microcystins and Nodularins

Globally the most frequently found cyanobacterial toxins in blooms from fresh and brackish waters are the cyclic peptide toxins of the microcystin and nodularin family. They pose a major challenge for the production of safe drinking water from surface waters containing cyanobacteria with these toxins. In mouse bioassays, which traditionally have been used to screen toxicity of field and laboratory samples, cyanobacterial hepatotoxins (liver toxins) cause death by liver haemorrhage within a few hours of the acute doses. Microcystins have been characterised from planktonic *Anabaena, Microcystis, Oscillatoria (Planktothrix), Nostoc,* and *Anabaenopsis* species, and from terrestrial *Hapalosiphon* genera. Nodularin has been characterised only from *Nodularia spumigena.*

The cyclic peptides are comparatively large natural products, molecular weight (MW) >> 800-1,100, although small compared with many other cell oligopeptides and polypeptides (proteins) (MW > 10,000). They contain either five (nodularins) or seven (microcystins) amino acids, with the two terminal amino acids of the linear peptide being condensed (joined) to form a cyclic compound. They are water soluble and, except perhaps for a few somewhat more hydrophobic microcystins, are unable to penetrate directly the lipid membranes of animal, plant and bacterial cells. Therefore, to elicit their toxic effect, uptake into cells occurs through membrane transporters which otherwise carry essential biochemicals or nutrients, this restricts the target organ range in mammals largely to the liver. In aquatic environments, these toxins usually remain contained within the cyanobacterial cells and are only released in substantial amounts on cell lysis. Along with their high chemical stability and their water solubility, this containment has important implications for their environmental persistence and exposure to humans in surface water bodies.

The first chemical structures of cyanobacterial cyclic peptide toxins were identified in the early 1980s and the number of fully characterised toxin variants has greatly increased during the 1990s. The first such compounds found in freshwater cyanobacteria were cyclic heptapeptides. cyclo-(D-alanine^{1}-X^{2}-D-MeAsp^{3}-Z^{4}-Adda^{5}-D-glutamate^{6}-Mdha^{7}) in which X and Z are variable L amino acids, D-MeAsp^{3} is *D*-erythro-β-methylaspartic acid, and Mdha is *N*-methyldehydroalanine (Figure 6.1). The amino acid Adda, (2S,3S,8S,9S)-3-amino-9-methoxy-2,6,8-trimethyl-10-phenyldeca-4,6-dienoic acid, is the most unusual structure in this group of cyanobacterial cyclic peptide toxins.

These compounds were first isolated from the cyanobacterium *Microcystis aeruginosa* and therefore the toxins were named microcystins (Carmichael *et al.*, 1988). Structural variations have been reported in all seven amino acids, but most frequently with substitution of L-amino acids at positions 2 and 4, and demethylation of amino acids at positions 3 and/or 7 (Figure 6.1A). About 60 structural variants of microcystins have been characterised so far from bloom samples and isolated strains of cyanobacteria (Table 6.2).

In one species of brackish water cyanobacterium, an identically acting and structurally very similar, cyclic pentapeptide occurs. It has been named as nodularin after its producer, *Nodularia spumigena.* The chemical structure of nodularin is cyclo-(D-MeAsp^{1}-L-arginine^{2}-Adda^{3}-D-glutamate^{4}-Mdhb^{5}), in which Mdhb is 2-(methylamino)-2-dehydrobutyric acid (Figure 6.1B). A few naturally occurring variations of nodularins have been found: two demethylated variants, one with D-Asp^{1} instead of D-MeAsp^{1}, the other with DMAdda^{3} instead of Adda^{3}; and the non-toxic nodularin which has the 6Z-stereoisomer of Adda^{3} (Namikoshi *et al.*, 1994). The equivalent 6Z-Adda^{3} stereoisomer of microcystins is also non-toxic. In the marine sponge, *Theonella swinhoei,* a nodularin analogue called motuporin has been found. It differs from nodularin only by one amino acid, having hydrophobic L-Val in place of the polar L-Arg in nodularin (de Silva *et al.*, 1992). The toxin might be cyanobacterial in origin because the sponge is known to harbour cyanobacterial symbionts.

The mammalian toxicity of microcystins and nodularins is mediated through their strong binding to key cellular enzymes called protein phosphatases. In solution, microcystins and nodularins adopt a chemical "shape" that is similar, especially in the Adda-glutamate part of the cyanotoxin molecule (Rudolph-Böhner *et al.*, 1994; Annila *et al.*, 1996). Recent studies have shown that this region is crucial for interaction with the protein phosphatase protein molecule, and hence it is crucial for the toxicity of

Figure 3.1: The Structure of Cyclic Peptide Toxins and Cylindrospermopsin

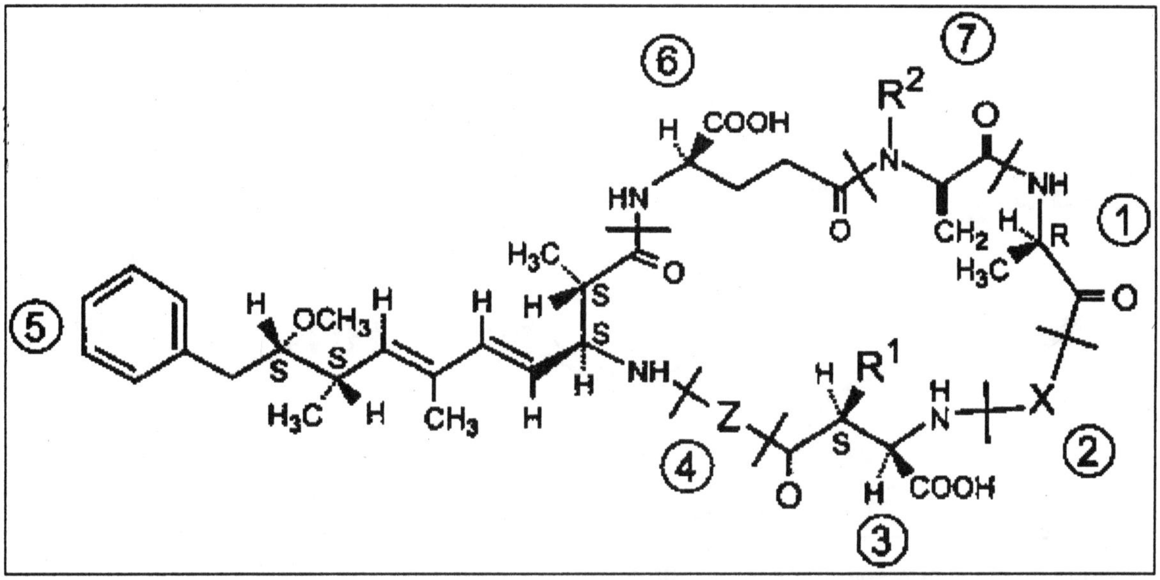

A: General structure of microcystins (MCYST), cyanobacterial heptapeptide hepatotoxins, showing the most frequently found variations. X and Z are variable L-amino acids (in MCYST-LR, X = L-Leusine (L) and Z = L-Arginine (R)); R^1 and R^2 are H (demethylmicrocystins) or CH_3; D-MeAsp is D-*erythro*-β-methylaspartic acid; Adda is (2*S*,3*S*,8*S*,9*S*)-3-amino-9-methoxy-2,6,8-trimethyl-10-phenyldeca-4,6-dienoic acid and Mdha is N-methyldehydroalanine (Dha = dehydroalanine) (see Table 3.2 for known microcystins); General structure of microcystins cyclo-(D-Ala^1-X^2-D-MeAsp^3-Z^4-Adda^5-D-Glu^6-Mdha^7)

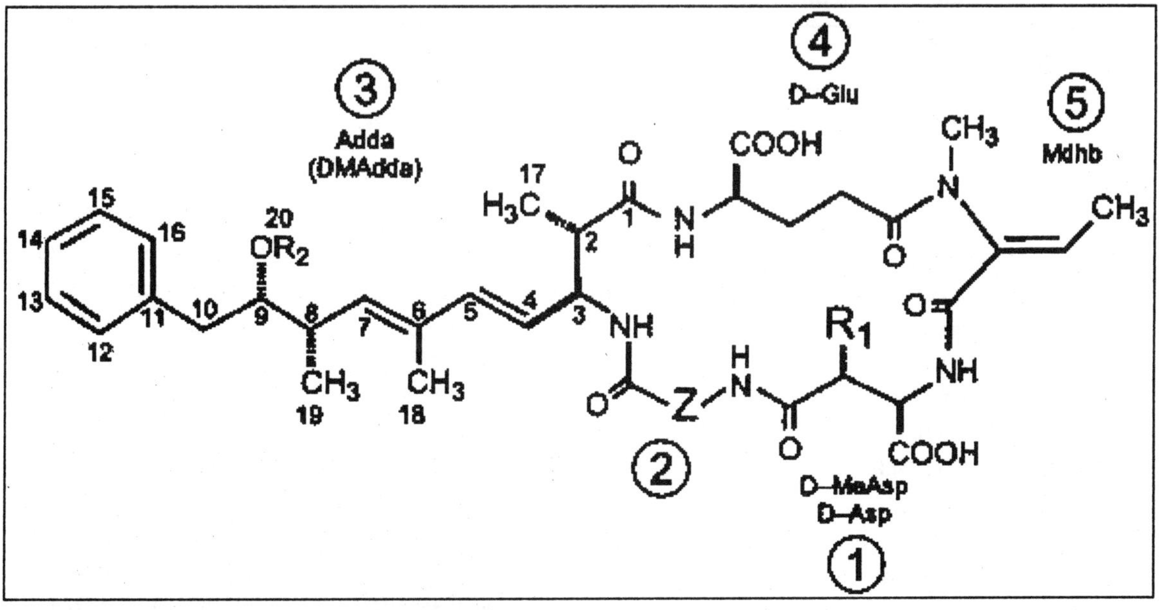

B: Structures of nodularins (Z = L-arginine) and motuporin (Z = L-Valine). Mdhb is N-methyldehydrobutyrin; Nodularin R_1,R_2 = CH_3; D-Asp^1 Nodularin R_1 = H, R_2 = CH_3; DMAdda^3 Nodularin R_1 = CH_3, R_2 = H; Structure of nodularins cyclo-(D-MeAsp^1-Z^2-Adda^3-D-Glu^4-Mdhb^5).

Contd...

Figure 3.2–Contd...

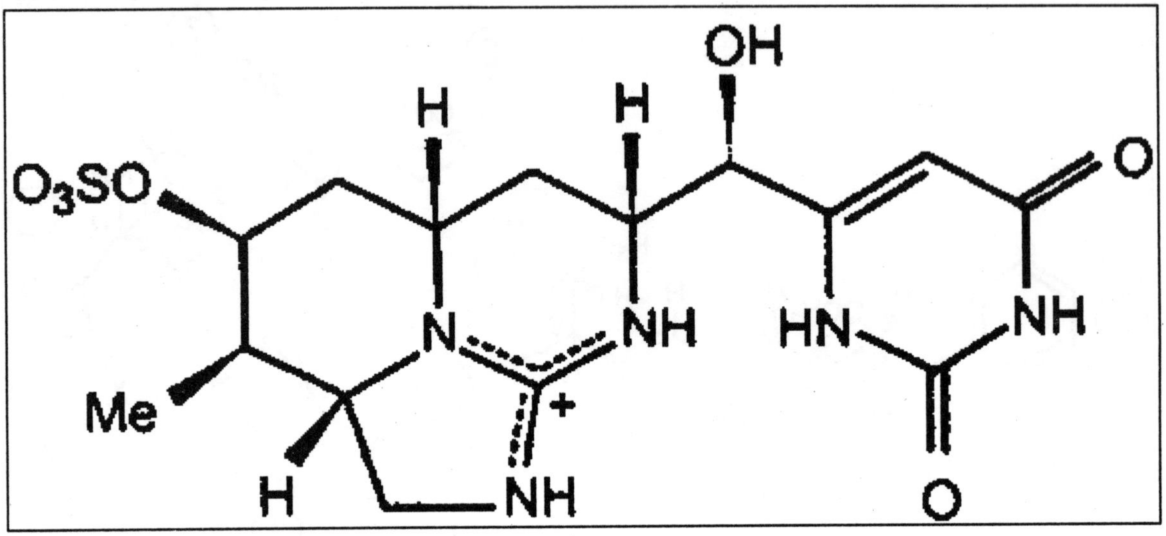

C: Structure of cylindrospermopsin; Cylindrospermopsin MW 415; $C_{15}H_{21}N_5O_7S$.

these cyanotoxins (Barford and Keller, 1994; Goldberg *et al.*, 1995). Microcystins show an additional characteristic of forming a covalent bond between the Mdha residue and the protein phosphatase molecule.

Table 6.2: The Microcystins (MCYST) Reported in the Scientific Literature

Microcystin	*Molecular Weight*	*Toxicity LD_{50}[2]*	*Organism[3]*	*Reference(s)*
MCYST-LA	909	50	*M. aeruginosa*[s], *M. viridis*[s]	Botes *et al.*, 1984; Kaya and Watanabe, 1990
MCYST-LAba	923	NR	*M. aeruginosa*[s]	Gathercole and Thiel, 1987
MCYST-LL	951	+	*M. aeruginosa*[b]	Craig *et al.*, 1993
MCYST-AR	952	250	*Microcystis* spp.[b]	Namikoshi *et al.*, 1992a
MCYST-YA	959	NR	*M. aeruginosa*[s]	Botes *et al.*, 1985
[D-Asp^{3},Dha^{7}]MCYST-LR	966	+	*M. aeruginosa*[s], [s] *Anabaena* sp.	Harada *et al.*, 1991b; Sivonen *et al.*, 1992a
[D-Asp^{3},Dha^{7}]MCYST-EE(OMe)	969	+	*Anabaena* sp.[s]	Namikoshi *et al.*, 1998
MCYST-VF	971	NR	*M. aeruginosa*[s]	Bateman *et al.*, 1995
(D-Asp^{3}]MCYST-LR	980	160-300	*A. flos-aquae*[s], *M. aeruginosa*[s], *M. viridis*[b], *O. agardhii*[s]	Krishnamurthy *et al.*, 1989; Cremer and Henning, 1991; Harada *et al.*, 1990b; 1991a; Luukkainen *et al.*, 1993
[Dha^{7}]MCYST-LR	980	250	*M. aeruginosa*[s], *Anabaena* sp.[s], *O. agardhii*[s]	Harada *et al.*, 1991b; Sivonen *et al.*, 1992a; Luukkainen *et al.*, 1993

Contd...

Table 6.2–Contd...

Microcystin	Molecular Weight	Toxicity LD_{50}^{2}	Organism[3]	Reference(s)
[DMAdda[5]]MCYST-LR	980	90-100	*Microcystis* spp.[b,s] *Nostoc* sp.	Namikoshi *et al.*, 1992a; Sivonen *et al.*, 1992b
[Dha[7]]MCYST-EE(OMe)	983	+	*Anabaena* sp.[s]	Namikoshi *et al.*, 1998
[D-Asp[3],Dha[7]]MCYST-E(OMe)E(OMe)	983	+	*Anabaena* sp.[s]	Namikoshi *et al.*, 1998
MCYST-LF	985	+	*M. aeruginosa*[s]	Azevedo *et al.*, 1994
MCYST-LR	994	50	*M. aeruginosa*[s], *A. flos-aquae*[s] *M. viridis*[s]	Botes *et al.*, 1985; Rinehart *et al.*, 1988; Krishnamyrthy *et al.*, 1989; Watanabe *et al.*, 1988
[D-Asp[3],D-Glu(OCH_3)[6]]MCYST-LR	994	NR	*A. flos-aquae*[s]	Sivonen *et al.*, 1992d
[(6Z)-Adda[5]]MCYST-LR	994	>1,200	*M. viridis*[b]	Harada *et al.*, 1990a,b
[Dha[7]]MCYST-E(OMe)E(OMe)	997	+	*Anabaena* sp.[s]	Namikoshi *et al.*, 1998
[L-Ser[7]]MCYST-LR	998	+	*Anabaena* sp.[s]	Namikoshi *et al.*, 1992c
MCYST-LY	1,001	90	*M. aeruginosa*[s]	Stoner *et al.*, 1989
[L-Ser[7]]MCYST-EE(OMe)	1,001	+	*Anabaena* sp.[s]	Namikoshi *et al.*, 1998
[D-Asp[3],Ser[7]]MCYST-E(OMe)E(OMe)	1,001	+	*Anabaena* sp.[s]	Namikoshi *et al.*, 1998
MCYST-HilR	1,008	100	*Microcystis* spp.[b]	Namikoshi *et al.*, 1995
[D-Asp[3],ADMAdda[5]]MCYST-LR	1,008	160	*Nostoc* sp.[s]	Sivonen *et al.*, 1990a; Namikoshi *et al.*, 1990
[D-Glu(OCH_3)[6]]MCYST-LR	1,008	>1,000	*A. flos-aguae*[s], *Microcystis sp.*[s]	Sivonen *et al.*, 1992d; Bateman *et al.*, 1995; Rinehart *et al.*, 1994
[D-Asp[3],Dha[7]]MCYST-RR	1,009	+	*O. agardhii*[b], *Anabaena* sp.[s], *M. aeruginosa*[s]	Krishnamyrthy *et al.*, 1989; Sivonen *et al.*, 1992a; Luukkainen *et al.*, 1994
[D-Asp[3], ADMAdda[5], Dhb[7]]MCYST-LR	1,009	+	*Nostoc sp.*[s]	Beattie *et al.*, 1998
[L-MeSer[7]]MCYST-LR	1,012	150	*Microcystis* spp.[b]	Namikoshi *et al., 1992a;* 1995
[Dha[7]]MCYST-FR	1,014	NR	*Microcystis sp.*[s]	Luukkainen *et al.*, 1994
[L-Ser[7]]MCYST-E(OMe)E(OMe)	1,015	+	*Anabaena* sp.[s]	Namikoshi *et al.*, 1998
[ADMAdda[5]]MCYST-LR	1,022	60	*Nostoc* sp.[s]	Sivonen *et al.*, 1990a; Namikoshi *et al.*, 1990
[D-Asp[3],ADMAdda[5]]MCYST-LHar	1,022	+	*Nostoc* sp.[s]	Sivonen *et al.*, 1992b
[D-Asp[3]]MCYST-RR	1,023	250	*O. agardhii*[s], *Anabaena* sp.[s], *M. aeruginosa*[s]	Meriluoto *et al.*, 1989; Sivonen *et al.*, 1992a; Luukkainen *et al.*, 1994
[Dha[7]]MCYST-RR	1,023	180	*M. aeruginosa*[s], *Anabaena* sp.[s], *O. agardhii*[s]	Kiviranta *et al.*, 1992; Sivonen *et al.*, 1992a; Luukkainen *et al.*, 1993
MCYST-LW	1,024	NR	*M. aeruginosa*[s]	Bateman *et al.*, 1995
MCYST-FR	1,028	250	*Microcystis* spp.[b]	Namikoshi *et al.*, 1992a

Contd...

Table 6.2–Contd...

Microcystin	*Molecular Weight*	*Toxicity* LD_{50}[2]	*Organism*[3]	*Reference(s)*
MCYST-M(O)R	1,028	700-800	*Microcystis* spp.[b]	Namikoshi *et al.*, 1992a
[Dha^{7}]MCYST-HphR	1,028	+	*Anabaena* sp.[s]	Namikoshi *et al.*, 1992b
[D-Asp^{3}, Dha^{7}]MCYST-HtyR	1,030	+	*Anabaena* sp.[s]	Namikoshi *et al.*, 1992b
[Dha^{7}]MCYST-YR	1,030	+	*M. aeruginosa*[s]	Sivonen *et al.*, 1992c
[D-Asp^{3}]MCYST-YR	1,030	+	*Microcystis* spp.[b]	Namikoshi *et al.*, 1992d
MCYST-YM(O)	1,035	56	*M. aeruginosa*[b]	Botes *et al.*, 1985; Elleman *et al.*, 1978
[ADMAdda^{5}]MCYST-LHar	1,036	60	*Nostoc* sp.[s]	Sivonen *et al.*, 1990a; Namikoshi *et al.*, 1990
MCYST-RR	1,037	600	*M. aeruginosa*[s], *M. viridis*[s], *Anabaena* sp.[s]	Kusumi *et al.*, 1987; Painuly *et al.*, 1988; Watanabe *et al.*, 1988; Sivonen *et al.*, 1992a
[(6Z)-Adda^{5}]MCYST-RR	1,037	>1,200	*M. viridis*[b]	Harada *et al.*, 1990a,b
[D-Ser^{1}, ADMAdda^{5}]MCYST-LR	1,038	+	*Nostoc* sp.[s]	Sivonen *et al.*, 1992b
[ADMAdda^{5},MeSer^{7}]MCYST-LR	1,040	+	*Nostoc* sp.[s]	Sivonen *et al.*, 1992b
[L-Ser^{7}]MCYST-RR	1,041	+	*Anabaena* sp.[s], *M. aeruginosa*[s/b]	Namikoshi *et al.*, 1992c; Luukkainen *et al.*, 1994
[D-Asp^{3},MeSer^{7}]MCYST-RR	1,041	+	*O. agardhii*[b]	Luukkainen *et al.*, 1993
MCYST-YR	1,044	70	*M. aeruginosa*[s], *M. viridis*[s]	Botes *et al.*, 1985; Watanabe *et al.*, 1988
[D-Asp^{3}]MCYST-HtyR	1,044	160-300	*A. flos-aquae*[s]	Harada *et al.*, 1991a
[Dha^{7}]MCYST-HtyR	1,044	+	*Anabaena* sp.[s]	Namikoshi *et al.*, 1992b
MCYST-(H_4)YR	1,048	NR	*Microcystis* spp.[b]	Namikoshi *et al.*, 1995
[D-Glu-$OC_2H_3(CH_3)OH^{6}$]MCYST-LR	1,052	>1,000	*Microcystis* spp.[b]	Namikoshi *et al.*, 1992a
[D-Asp^{3},ADMAdda^{5}, Dhb^{7}]MCYST-RR	1,052	+	*Nostoc* sp.[s]	Beattie *et al.*, 1998
MCYST-HtyR	1,058	80-100	*A. flos-aquae*[s]	Harada *et al.*, 1991a
[L-Ser^{7}]MCYST-HtyR	1,062	+	*Anabaena* sp.[s]	Namikoshi *et al.*, 1992b
MCYST-WR	1,067	150-200	*Microcystis* spp.[b]	Namikoshi *et al.*, 1992a
[D-Asp^{3},ADMAdda^{5},Dhb^{7}]MCYST-HtyR	1,073	+	*Nostoc* sp.[s]	Beattie *et al.*, 1998
[L-MeLan^{7}]MCYST-LR	1,115	1,000	*Microcystis* spp.[b]	Namikoshi *et al.*, 1995

Aba: Aminoisobutyric acid; ADMAdda: *O*-Acetyl-*O*-demethylAdda; Dha: Dehydroalanine; Dhb: Dehydrobutyrine; DMAdda: *O*-DemethylAdda; E(OMe): Glutamic acid methyl ester D; (H_4)Y: 1,2,3,4,-tetrahydrotyrosine; Har: Homoarginine; Hil: Homoisoleucine; Hph: Homophenylalanine; Hty: Homotyrosine; MeLan: *N*-Methyllanthionine; M(O): Methionine-*S*-oxide; MeSer: *N*-Methylserine; (6Z)-Adda: Stereoisomer of Adda at the D^{6} double bond.

1 Several partial structures of microcystins have been reported in addition to those shown in this table (see Boland *et al.*, 1993; Craig *et al.*, 1993; Jones *et al.*, 1995; Sivonen *et al.*, 1995)

2 Toxicity determined i.p. mouse (μg kg^{-1}); the LD_{50} value is the dose of toxin that kills 50 per cent of exposed animals; a '+' denotes a toxic result in a non-quantitative mouse bioassay or inhibition of protein phosphatase and 'NR' denotes 'Not reported'

3 An '*s*' denotes toxins isolated from culture samples and a '*b*' denotes toxins isolated from bloom samples

Most of the structural variants of microcystin and nodularin are highly toxic within a comparatively narrow range (intra-peritoneal (i.p.) mouse toxicities largely in the range 50-300 µg kg^{-1} body weight (bw); Table 6.2). Only a few non-toxic variants have been identified. In general, any structural modifications to the Adda-glutamate region of the toxin molecule, such as a change in isomerisation of the Adda-diene (6(E) to 6(Z)) or acylation of the glutamate, renders microcystins and nodularins non-toxic (Harada *et al.*, 1990 a,b; Rinehart *et al.*, 1994). Linear microcystins and nodularin are more than 100 times less toxic than the equivalent cyclic compounds. The linear microcystins are thought to be microcystin precursors and/or bacterial breakdown products (Choi *et al.*, 1993; Rinehart *et al.*, 1994; Bourne *et al.*, 1996).

Microcystins and nodularin have been characterised from axenic cyanobacterial strains (*i.e.* strains free of contaminating bacteria) and thus the cyanobacterial origin of these compounds is clear. At the present time, it is known that microcystins are produced by bloom forming species of *Microcystis, Anabaena, Oscillatoria* (*Planktothrix*), and *Nostoc* (Table 6.2), by a species of *Anabaenopsis* and by a soil isolate of *Haphalosiphon hibernicus.* Nodularins have been found, with the exception of the marine sponge *Theonella* already mentioned, only in *Nodularia spumigena* (see section 3.2 for more details). Further species may yet be demonstrated to produce microcystin.

Neurotoxic Alkaloids-Anatoxins and Saxitoxins

Mass occurrences of neurotoxic cyanobacteria have been reported from North America, Europe and Australia, where they have caused animal poisonings. In mouse bioassays death by respiratory arrest occurs rapidly (within 2-30 minutes). Three families of cyanobacterial neurotoxins are known:

- ✰ Anatoxin-a and homoanatoxin-a, which mimic the effect of acetyl choline,
- ✰ Anatoxin-a(S), which is an anticholinesterase, and
- ✰ Saxitoxins, also known as paralytic shellfish poisons (PSPs) in the marine literature, which block nerve cell sodium channels.

Anatoxin-a has been found in *Anabaena, Oscillatoria* and *Aphanizomenon,* homoanatoxin-a from *Oscillatoria,* anatoxin-a(S) from *Anabaena,* and saxitoxins from *Aphanizomenon, Anabaena, Lyngbya* and *Cylindrospermopsis.* Sixteen confirmed saxitoxins from cyanobacterial samples have been reported, some of which (*e.g.* the decarbamoyl-gonyautoxins) may be chemical breakdown products in some species.

The alkaloid toxins are diverse, both in their chemical structures and in their mammalian toxicities. Alkaloids, in general, are a broad group of heterocyclic nitrogenous compounds (*i.e.* they contain ring structures with at least one carbon-nitrogen bond) usually of low to moderate molecular weight (< 1,000). They are produced, in particular, by plants and by some bacteria, and are invariably bioactive and commonly toxic. The non-sulphated alkaloid toxins of freshwater cyanobacteria (anatoxins and saxitoxin) are all neurotoxins. The sulphated PSPs, C-toxins and gonyautoxins (sulphated derivatives of saxitoxin) are also neurotoxins, but the sulphated alkaloid cylindrospermopsin blocks protein synthesis with a major impact on liver cells. Some marine cyanobacteria also contain alkaloids (lyngbyatoxins, aplysiatoxins) which are dermatoxins (skin irritants), but have also been associated with gastro-enteritis and more general symptoms such as fever.

Alkaloids have varying chemical stabilities, often undergoing spontaneous transformations to by-products which may have higher or lower potencies than the parent toxin. Some are also susceptible to direct photolytic degradation.

Anatoxin-a

Anatoxin-a is a low molecular weight alkaloid (MW = 165), a secondary amine, 2-acetyl-9-azabicyclo(4-2-1)non-2-ene (Figure 6.2) (Devlin *et al.*, 1977). Anatoxin-a is produced by *Anabaena flos-aquae, Anabaena* spp. (*flos-aquae-lemmermannii* group), *Anabaena planktonica, Oscillatoria, Aphanizomenon* and *Cylindrospermum* (see section 3.2 for details). Homoanatoxin-a (MW = 179) is an anatoxin-a homologue isolated from an *Oscillatoria formosa* (*Phormidium formosum*) strain (Figure 6.2). It has a propionyl group at C-2 instead of the acetyl group in anatoxin-a (Skulberg *et al.*, 1992). The LD_{50} (lethal dose resulting in 50 per cent deaths) of anatoxin-a and homoanatoxin-a are 200–250 µg kg^{-1} bw (Devlin *et al.*, 1977; Carmichael *et al.*, 1990; Skulberg *et al.*, 1992).

Figure 6.2: The Chemical Structures of Cyanobacterial Neurotoxins, Anatoxin-a, Homoanatoxin-a, Anatoxin-a(S), and the General Structure of Saxitoxins. Sixteen different saxitoxins have been reported from cyanobacteria (for details see Table 6.3). MW = molecular weight

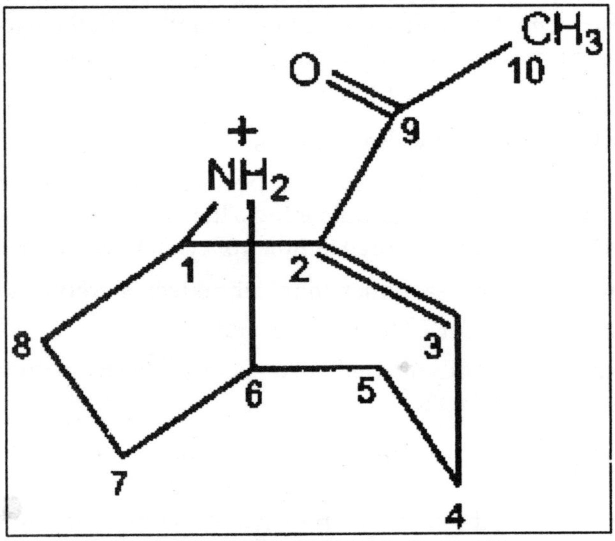

Anatoxin-a MW 165; $C_{10}H_{15}NO$

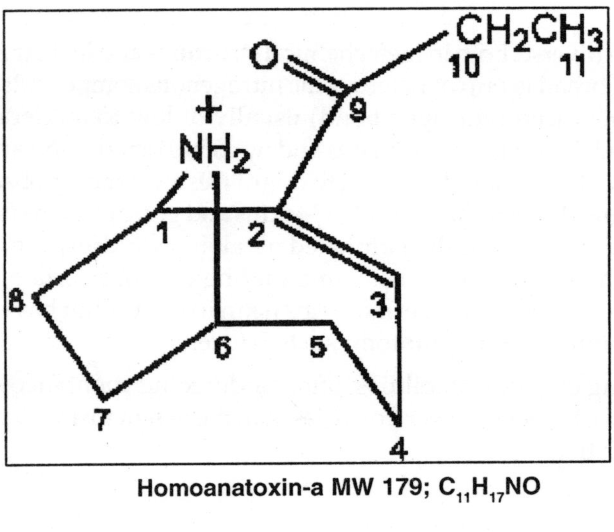

Homoanatoxin-a MW 179; $C_{11}H_{17}NO$

Contd...

Figure 6.2–Contd...

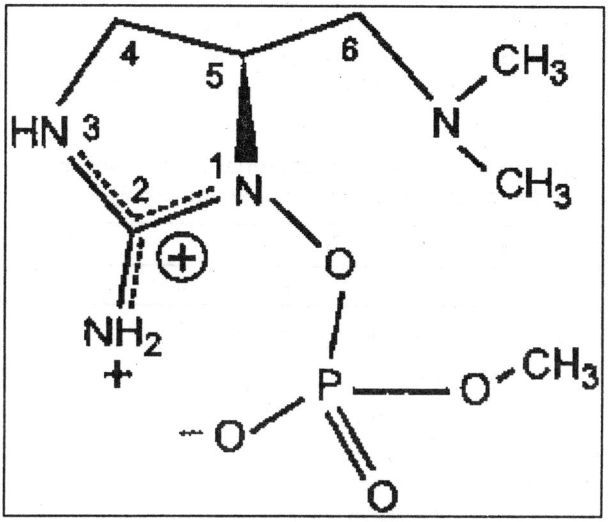

Anatoxin-a(S) MW 252; $C_7H_{17}N_4O_4P$

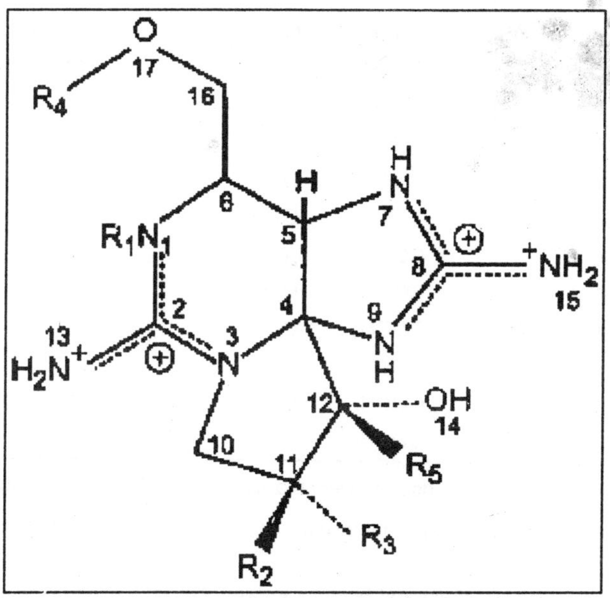

General structure of saxitoxins

Anatoxin-a(S)

Anatoxin-a(S) is a unique phosphate ester of a cyclic N-hydroxyguanine (MW = 252) (Figure 6.2) produced by *Anabaena flos-aquae* strain NRC 525-17 (Matsunaga *et al.*, 1989). It has more recently been identified in blooms and isolated strains of *Anabaena lemmermannii* (Henriksen *et al.*, 1997; Onodera *et al.*, 1997a). The LD_{50} of anatoxin-a(S) is 20 µg kg^{-1} bw (i.p. mouse) (Carmichael *et al.*, 1990). Structural variants of anatoxin-a(S) have not been detected.

Saxitoxins

Saxitoxins are a group of carbamate alkaloid neurotoxins which are either non-sulphated (saxitoxins–STX), singly sulphated (gonyautoxins–GTX) or doubly sulphated (C-toxins) (Figure 6.2 and Table 6.3). In addition, decarbamoyl variants and several new toxins have been identified in some species.

Table 6.3: Saxitoxins Reported from Cyanobacterial Strains and Bloom Samples (for the chemical structure see Figure 6.2)

Name of Toxin	*Variable Chemical Groups in Toxins*					*Cyanobacteria*			
	R_1	R_2	R_3	R_4	R_5	*Aph*[1]	*Ana*[2]	*Lyn*[3]	*Cyl*[4]
STX	H	H	H	$CONH_2$	OH	+	+		+
GTX2	H	H	OSO_3^-	$CONH_2$	OH		+		
GTX3	H	OSO_3^-	H	$CONH_2$	OH		+		
GTX5	H	H	H	$CONHSO_3^-$	OH		+		
C1	H	H	OSO_3^-	$CONHSO_3^-$	OH		+		
C2	H	OSO_3^-	H	$CONHSO_3^-$	OH		+		
NEO	OH	H	H	$CONH_2$	OH	+			+
GTX1	OH	H	OSO_3^-	$CONH_2$	OH		*		
GTX4	OH	OSO_3^-	H	$CONH_2$	OH		*		
GTX6	OH	H	H	$CONHSO_3^-$	OH		*		
dcSTX	H	H	H	H	OH		+	+	
dcGTX2	H	H	OSO_3^-	H	OH		+	+	
dcGTX3	H	OSO_3^-	H	H	OH		+	+	
LWTX1[3]	H	OSO_3^-	H	$COCH_3$	H			+	
LWTX2[3]	H	OSO_3^-	H	$COCH_3$	OH			+	
LWTX3[3]	H	H	OSO_3^-	$COCH_3$	OH			+	
LWTX4[3]	H	H	H	H	H			+	
LWTX5[3]	H	H	H	$COCH_3$	OH			+	
LWTX6[3]	H	H	H	$COCH_3$	H			+	

STX: Saxitoxin; GTX: Gonyautoxins; C: C-toxins; dcSTX: Decarbamoylsaxitoxin; LWTX: Lyngbya-wollei-toxins.

1 Toxins found in *Aphanizomenon flos-aquae,* New Hampshire, USA (Ikawa *et al.,* 1982; Mahmood and Carmichael, 1986).

2 Toxins reported in an *Anabaena circinalis* strain and bloom samples, Australia (Hump-age *et al.,* 1994; Negri *et al.,* 1995; Negri *et al.,* 1997). dcGTX2 and dcGTX3 are probably break down products of C1 and C2 in this species (Jones and Negri, 1997). An asterisk in this column denotes toxins reported by Humpage *et al.,* 1994 for *Anabaena circinalis* based on retention time data, but not confirmed by mass spectrometry, and not found in subsequent studies.

3 Toxins detected in *Lyngbya wollei,* USA (Onodera *et al.,* 1997b).

4 Toxins thus far found in *Cylindrospermopsis raciborskii,* Brazil (Lagos *et al.,* 1997).

Saxitoxins were originally isolated from shellfish where they are concentrated from marine dinoflagellates (so called "red tide" algae) and have caused deaths in humans (Anderson, 1994).

Saxitoxins have been found in the cyanobacteria *Aphanizomenon flos-aquae, Anabaena circinalis, Lyngbya wollei* and *Cylindrospermopsis raciborskii.* The North American *Aphanizomenon flos-aquae* strains NH-1 and NH-5 contain mostly neosaxitoxin and less saxitoxin (plus a few unidentified neurotoxins). *Anabaena circinalis* strains (from Australia) contain mostly C1 and C2 toxins, with lesser amounts of gonyautoxins 2 and 3. The freshwater cyanobacterium *Lyngbya wollei* produced three known and six new saxitoxin analogues. *Cylindrospermopsis raciborskii* in Brazil was found to contain mostly neosaxitoxin and a smaller amount of saxitoxin.

Other Neurotoxic Cyanobacteria

In marine *Trichodesmium* blooms from the Virgin Islands, a neurotoxic factor has been reported which was not anatoxin-a or anatoxin-a(S) but remains to be characterised (Hawser *et al.*, 1991).

Cytotoxic Alkaloids

In tropical and subtropical waters of Australia, the alkaloid hepatotoxin cylindrospermopsin with a completely different mechanism of toxicity has caused health problems in drinking water supplies. It is a cyclic guanidine alkaloid with a molecular weight of 415 (Figure 6.1C). It is produced by *Cylindrospermopsis raciborskii* (Hawkins *et al.*, 1985, 1997), *Umezakia natans* (Harada *et al.*, 1994) and *Aphanizomenon ovalisporum* (Banker *et al.*, 1997). In pure form, cylindrospermopsin mainly affects the liver, although crude extracts of *C. raciborskii* injected or given orally to mice also induce pathological symptoms in the kidneys, spleen, thymus and heart (see Chapter 4 for more details). Pure cylindrospermopsin has an LD_{50} in mice (i.p.) of 2.1 mg kg l^{-1} bw at 24 h and 0.2 mg kg l^{-1} bw at 5-6 days (Ohtani *et al.*, 1992). Recently, new structural variants of cylindrospermopsin have been isolated from an Australian strain of *C. raciborskii*, with one being identified as demethoxy-cylindrospermopsin (Chiswell *et al.*, 1999).

Dermatotoxic Alkaloids-Aplysiatoxins and Lyngbyatoxin

Benthic marine cyanobacteria such as *Lyngbya, Oscillatoria* and *Schizothrix* may produce toxins causing severe dermatitis among swimmers in contact with the filaments. The inflammatory activity of *Lyngbya* is caused by aplysiatoxins and debromoaplysiatoxin (Figure 6.3) which are potent tumour promoters and protein kinase C activators (Mynderse *et al.*, 1977; Fujiki *et al.*, 1990). Another strain of *Lyngbya majuscula* has caused dermatitis and severe oral and gastrointestinal inflammation. It was found to contain lyngbyatoxin-a (Figure 6.3) (Cardellina *et al.*, 1979). Debromoaplysiatoxin along with other toxic compounds has also been isolated from other Oscillatoriaceae, such as *Schizothrix calcicola* and *Oscillatoria nigroviridis.*

Irritant Toxins-Lipopolysaccharides

Weise *et al.* (1970) were the first to isolate LPS from the cyanobacterium *Anacystis nidulans* and numerous reports of endotoxins in cyanobacteria have followed. Lipopolysaccarides are generally found in the outer membrane of the cell wall of Gram negative bacteria, including cyanobacteria, where they form complexes with proteins and phospholipids. They are pyrogenic and toxic (Weckesser and Drews, 1979). Lipopolysaccarides, as the name implies, are condensed products of a sugar, usually a hexose, and a lipid, normally a hydroxy C_{14}-C_{18} fatty acid. The many structural variants of LPS are generally phylogenetically conserved, *i.e.* particular orders, genera and occasionally species, have identical or similar fatty acid and sugar components contained in their cell wall LPS. It is generally the fatty acid component of the LPS molecule that elicits an irritant of allergenic response in humans and mammals.

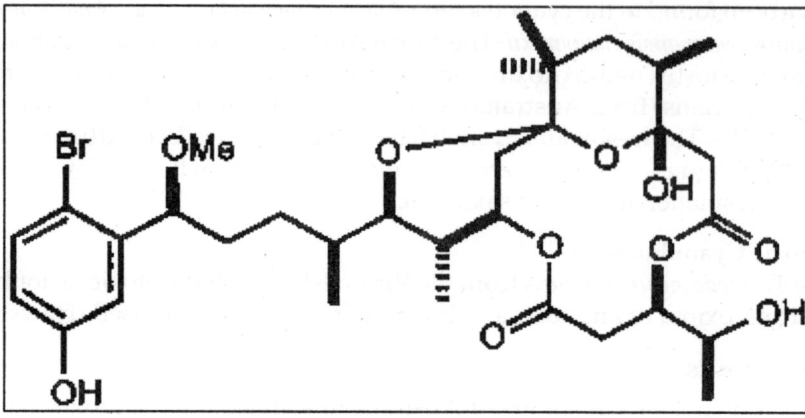

Debromoaplysiatoxin

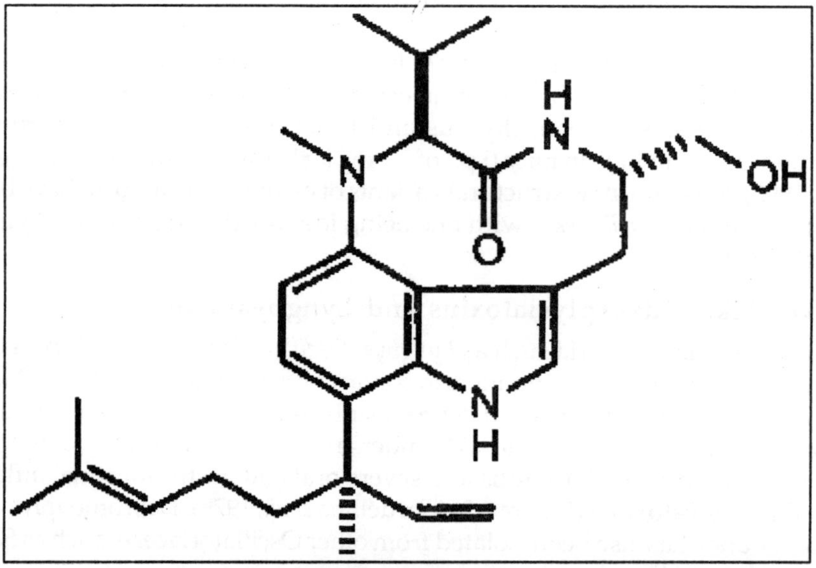

Lyngbyatoxin A

Figure 6.3: The Chemical Structures of Debromoaplysiatoxin and Lyngbiatoxin-a

Lipopolysaccharides are an integral component of the cell wall of all Gram negative bacteria, including cyanobacteria, and can elicit irritant and allergenic responses in human and animal tissues that come in contact with the compounds. There is considerable diversity of LPS composition among the cyanobacteria, but differences are largely related to phylogeny. Thus, different genera typically have distinct LPS compositions that are largely conserved within that genus (Kerr *et al.*, 1995). Cyanobacterial LPS are considerably less potent than LPS from pathogenic gram-negative bacteria such as, for example, *Salmonella*. The chemical stability of cyanobacterial LPS in surface waters is unknown.

Structurally, LPS is a complex polymer composed of four regions. Region I, the O-antigen region, consists of repeating oligosaccharide units that may vary in structure, with numerous combinations of different sugar residues and associated glycosidic linkages. As suggested by its name, the O-antigen also exhibits several antigenic determinants that constitute the receptor sites for a number of lysogenic bacteriophages. Regions II and III are the outer core and backbone of a core polysaccharide. There is generally only minor variation in core structure between species. The backbone of the polysaccharide is connected to a glycolipid, lipid A (Region IV), via a short link normally composed of 3-deoxy-D-mannoocmiosonic acid (KDO). Lipid A is a disaccharide of glucosamines highly substituted with phosphate, fatty acids and KDO, although the proportion of KDO is low or absent in cyanobacteria compared with other bacterial LPS. The lipid A component is also acetylated with amide and ester-linked hydroxy fatty acids.

Recent studies of the fatty acid composition of Australian species of cyanobacteria (Kerr *et al.*, 1995) show a range of β-OH fatty acids ranging in size from C_{10} to C_{22}. Normal, saturated and branched chain acids have been detected. There was a stark predominance of straight chain 14:0 and 18:0 β-OH acids in *Microcystis* strains that was quite distinct from *Anabaena* and *Nodularia* strains where 16:0 β-OH predominated the LPS fatty acid fraction.

Although comparatively poorly studied, cell wall components, particularly LPS endotoxins from cyanobacteria may contribute to human health problems associated with exposure to mass occurrences of cyanobacteria. The few results available indicate that cyanobacterial LPS is less toxic than the LPS of other bacteria, such as *Salmonella* (Keleti and Sykora, 1982; Raziuddin *et al.*, 1983). More studies are needed to evaluate the chemical structures and health risks of cyanobacterial LPS.

Other Bioactive Compounds

Cyanobacteria are known to produce several other bioactive compounds, some of which are of medical interest, as well as compounds toxic to other cyanobacteria, bacteria, algae and zooplankton. Severe intoxication of fish embryos by crude extracts of *Planktothrix agardhii* has been reported by Oberemm *et al.* (1997). Skulberg *et al.* (1994) reported the presence of an unidentified "protracted toxic effect" in cyanobacterial samples that caused death within 4-24 hours in mice. Whether this effect was due to a specific cyanotoxin is unclear.

Cyanobacteria have been found to be a rich source of biomedically interesting compounds and therefore screening programmes for new bioactivities are underway. Cyanobacteria are known to produce antitumour, antiviral, antibiotic and antifungal compounds. Of the cyanobacterial extracts screened by a Hawaiian research group, 0.8 per cent showed solid tumour selective cytotoxicity (Moore *et al.*, 1996). Depsipeptides (peptides with an ester linkage) called cryptophycins isolated from a cyanobacterium, *Nostoc* sp. strain GSV 224, are promising candidates for an anticancer drug (Trimurtulu *et al.*, 1995). Recently, several new cyclic or linear peptides and depsipeptides from cyanobacteria have been characterised. Some are protease inhibitors, but the biological activity of the others remains to be characterised (Namikoshi and Rinehart, 1996). Many of the cyanobacterial bioactive compounds possess structural similarities to natural products from marine invertebrates.

Occurrence of Cyanotoxins

Mass Occurrences of Toxic Cyanobacteria

The toxicity of cyanobacterial mass occurrences (blooms) was originally brought to the attention of scientists through reports of animal poisonings by farmers and veterinarians, with the first well documented case being reported from Australia in 1878 (Francis, 1878). In most, if not all, reported cases since that time, afflicted animals consumed water from water bodies where there was an obvious presence of a cyanobacterial scum on the water surface (see Ressom *et al.* (1994) and Yoo *et al.* (1995) for a list of reported animal poisonings). More recent measurements of cyanobacterial toxins using sensitive modern analytical methods have often revealed high frequencies of toxic blooms even when animal poisonings have not been reported (Table 6.4).

Table 6.4: Frequencies of Mass Occurrences of Toxic Cyanobacteria in Freshwaters

Country	*No. of Tested Samples*	*% of Toxic Samples*	*Type of Toxicity*	*Reference*
Australia	231	42	Hepatotoxic Neurotoxic	Baker and Humpage, 1994
Australia	31	84[1]	Neurotoxic	Negri *et al.*, 1997
Brazil	16	75	Hepatotoxic	Costa and Azevedo, 1994
Canada, Alberta	24	66	Hepatotoxic Neurotoxic	Gorham, 1962
Canada, Alberta	39	95	Hepatotoxic	Kotak *et al.*, 1993
Canada, Alberta (3 lakes)	226	74[1]	Hepatotoxic	Kotak *et al.*, 1995
Canada, Saskatchewan	50	10	Hepatotoxic Neurotoxic	Hammer, 1968
China	26	73	Hepatotoxic	Carmichael *et al.*, 1988b
Czech and Slovak Rep.	63	82	Hepatotoxic	Maršálek *et al.*, 1996
Denmark	296	82	Hepatotoxic SDF Neurotoxic	Henriksen *et al.*, 1996b
Former German Dem. Rep.	10	70	Hepatotoxic SDF	Henning and Kohl, 1981
Germany	533	72[1]	Hepatotoxic	Fastner, 1998
Germany	393	22	Neurotoxic	Bumke-Vogt, 1998
Greece	18	?	Hepatotoxic	Lanaras *et al.*, 1989
Finland	215	44	Hepatotoxic Neurotoxic	Sivonen, 1990a
France, Brittany	22	73[1]	Hepatotoxic	Vezie *et al.*, 1997
Hungary	50	66	Hepatotoxic	Törökné, 1991
Japan	23	39	Hepatotoxic	Watanabe and Oishi, 1980
Netherlands	10	90	Hepatotoxic	Leeuwangh *et al.*, 1983
Norway	64	92	Hepatotoxic Neurotoxic SDF	Skulberg *et al.*, 1994
Portugal	30	60	Hepatotoxic	Vasconcelos, 1994
Scandinavia	81	60	Hepatotoxic	Berg *et al.*, 1986
Sweden	331	47	Hepatotoxic Neurotoxic	Willén and Mattsson, 1997

Contd...

Table 6.4–Contd...

Country	No. of Tested Samples	% of Toxic Samples	Type of Toxicity	Reference
UK	50	48 28[1]	Hepatotoxic	Codd and Bell, 1996
USA, Minnesota	92	53	Unspecified Neurotoxic	Olson, 1960
USA, Wisconsin	102	25	Hepatotoxic Neurotoxic	Repavich *et al.*, 1990
Mean		59		
India			Hepatotoxic	Parkar *et al.*, 1997

1: HPLC was used to detect the toxin content of the samples.

Throughout the world, it appears that liver-toxic (hepatotoxic, microcystin-containing) freshwater blooms of cyanobacteria are more commonly found than neurotoxic blooms. Liver-toxic blooms have been reported from all continents and almost every part of the world where samples have been collected for analysis. Nevertheless, mass occurrences of neurotoxic cyanobacteria are common in some countries and these have been reported from North America, Europe and Australia, where they have caused several animal poisonings. Blooms which have caused both liver and kidney damage due to the toxin cylindrospermopsin (and possibly related cyanotoxins) have been reported in Australia, Japan, Israel and Hungary.

In recent years, surveys have been carried out in a number of countries in South America, Africa, Australia, Asia and Europe. The conclusion that can be drawn from these surveys is that toxic cyanobacteria are internationally ubiquitous, and that as further surveys are carried out more toxic cyanobacterial blooms and new toxic species will be discovered. This is particularly true of tropical and subtropical regions that are currently under-represented in the literature. It seems likely that every country in the world will have water bodies which support blooms of toxic cyanobacteria at some time or another. It is also important to note that mass occurrences of toxic cyanobacteria are not always associated with human activities causing pollution or "cultural eutrophication". For example, massive blooms of toxic cyanobacteria have been reported in Australian reservoirs with pristine or near-pristine catchments (watersheds), and toxic benthic cyanobacteria have killed cattle drinking from oligotrophic, high-alpine waters in Switzerland.

Species Composition and Variation Among Toxic Blooms

Cyanobacterial populations may be dominated by a single species or be composed of a variety of species, some of which may not be toxic. Even within a single-species bloom there may be a mixture of toxic and non-toxic strains. A strain is a specific genetic subgroup within a particular species, and each species may encompass tens or hundreds of strains, each with slightly different traits. Some strains are much more toxic than others, sometimes by more than three orders of magnitude. This can mean that one highly toxic strain, even when occurring in minor amounts amongst larger numbers of non-toxic strains, may render a bloom sample toxic (Sivonen *et al.*, 1989a,b; Bolch *et al.*, 1997; Vezie *et al.*, 1998).

Some of the studies shown in the table have been conducted over several years while others lasted only one season. The relative share of cyanobacteria in the samples varied; low frequency of cyanobacteria led to low percentages of toxic samples in some studies. In most of the studies the method used to detect toxicity is mouse bioassay, normally with a 4-hour time limit (or longer when slow death factors (SDF) have been included). SDF may indicate low hepatotoxicity of samples or other unknown toxicity.

Toxic and non-toxic strains from the same cyanobacterial species cannot be separated by microscopic identification. The use of molecular genetic methods, in particular the use of molecular probes and primers that target specific toxin production genes, will lead to the development of more precise identification methods for toxic cyanobacteria in the future. To confirm that a particular cyanobacterial strain is a toxin-producer, it is important to isolate a pure culture of that strain, preferably free of other bacteria; then to detect and quantify toxin concentrations in the pure culture (either by bioassay or chemical analysis); and, where possible, to purify and characterise fully the toxins (for such examples see Tables 6.2 and 6.3). It is likely that the list of confirmed toxic species will increase in the future due to the isolation of new species and strains, and because of the use of improved isolation, culturing and analytical methods.

Microcystis sp., commonly *Microcystis aeruginosa,* are linked most frequently to hepatotoxic blooms world-wide (see Tables 6.2 and 6.5 for details and references for all toxic species). *Microcystis viridis* and *Microcystis botrys* strains also have been shown to produce microcystins. As noted previously, *Microcystis* is a non-nitrogen-fixing genus which is often dominant under nutrient-rich conditions (especially where there is a significant supply of ammonia), although it also forms blooms in less polluted waters. Microcystin-producing *Anabaena* sp. have been reported from Canada, Denmark, Finland, France and Norway. A recent study from Egypt revealed that 25 per cent of 75 *Anabaena* and *Nostoc* strains isolated from soil, rice fields and water bodies contained microcystins. *Planktothrix agardhii* and *Planktothrix rubescens* (previously called *Oscillatoria agardhii* and *O. rubescens*) are common microcystin producers in the Northern Hemisphere; toxic strains of these have been isolated from blooms in Denmark, Finland and Norway. In addition, these species were frequently shown to be dominant in microcystin containing blooms in China, in Germany and in Sweden. In Swiss alpine lakes, *Oscillatoria limosa,* which is benthic (*i.e.* it grows attached to sediments and rocks), is a microcystin producer. In spite of the widespread occurrence of cyanobacterial blooms in Australia, *Planktothrix* blooms are rare there. This may be due to the higher temperature and tendency for elevated clay-derived turbidity in Australian water bodies.

The toxicity of the species listed in the table is in most cases verified by laboratory studies with isolated strains. A few bloom samples are also included from the new areas of occurrence where toxicity of the species is not verified by strain isolation but the toxins are determined in the bloom samples. The authors have suggested the listed species as the probable toxin producer (based on their dominance) but these reports should be treated as tentative until pure strains are studied.

Nostoc rivulare blooms in Texas, USA have caused poisoning of domestic and wild animals (Davidson, 1959) and, more recently, two unidentified *Nostoc* strains were shown to produce microcystins (Table 6.5).

The hepatotoxin, cylindrospermopsin, has been found in *Cylindrospermopsis raciborskii* in Australia and Hungary, in *Umezakia natans* in Japan, and in *Aphanizomenon ovalisporum* in Israel (Table 6.5). In spite of their occurrence in Europe, it appears that cylindrospermopsin-producing genera most commonly form toxic blooms in subtropical, tropical or arid zone water bodies. However, there have been reports of increasing occurrences of *Cylindrospermopsis raciborskii in* Europe and the USA (Padisák, 1997).

Table 6.5: Toxic Cyanobacteria Species and their Geographical Distribution

Toxic Species	*Cyanotoxin*	*Location*	*Reference(s)*
Anabaena flos-aquae	Microcystins	Canada	Krishnamurthy *et al.*, 1989; Harada *et al.*, 1991
Anabaena	Microcystins	Denmark	Henriksen *et al.*, 1996b
Anabaena spp.	Microcystins	Egypt	Yanni and Carmichael, 1997
Anabaena spp. (*flos-aquae, lemmermannii, circinalis*)	Microcystins	Finland	Sivonen *et al.*, 1990b; 1992a
Anabaena circinalis	Microcystins	France	Vezie *et al.*, 1998
Anabaena flos-aquae	Microcystins	Norway	Sivonen *et al.*, 1992a
Microcystis aeruginosa	Microcystins	Worldwide	Several; see Rinehart *et al.*, 1994 for a summary
M. viridis	Microcystins	Japan	Kusumi *et al.*, 1987; Watanabe *et al.*, 1986
M. botrys	Microcystins	Denmark	Henriksen *et al.*, 1996b
Planktothrix agardhii	Microcystins	China	Ueno *et al.*, 1996a
P. agardhii	Microcystins	Denmark	Henriksen *et al.*, 1996b
P. mougeotii	Microcystins	Denmark	Henriksen *et al.*, 1996b
P. agardhii	Microcystins	Finland	Sivonen, 1990b; Luukkainen *et al.*, 1993
P. agardhii	Microcystins	Norway	Krishnamurthy *et al.*, 1989; Meriluoto *et al.*, 1989
Oscillatoria limosa	Microcystins	Switzerland	Mez *et al.*, 1996
Nostoc sp.	Microcystins	Finland	Sivonen *et al.*, 1990a, 1992b
Nostoc sp.	Microcystins	England	Beattie *et al.*, 1998
Anabaenopsis millerii	Microcystins	Greece	Lanaras and Cook, 1994
Haphalosiphon hibernicus (soil isolate)	Microcystins	USA	Prinsep *et al.*, 1992
Nodularia spumigena	Nodularins	Australia	Baker and Humpage, 1994; Jones *et al.*, 1994
N. spumigena	Nodularins	Baltic Sea	Sivonen *et al.*, 1989b
N. spumigena	Nodularins	New Zealand	Carmichael *et al.*, 1988a; Rinehart *et al.*, 1988
Aphanizomenon ovalisporum	Cylindrospermopsin	Israel	Banker *et al.*, 1997
Cylindrospermopsis raciborskii	Cylindrospermopsin	Australia	Hawkins *et al.*, 1985; 1997
C. raciborskii	Cylindrospermopsin	Hungary	Törökné, 1997
Umezakia natans	Cylindrospermopsin	Japan	Harada *et al.*, 1994
Anabaena flos-aquae	Anatoxin-a	Canada	Carmichael *et al.*, 1975; Devlin *et al.*, 1977

Contd...

Table 6.5–Contd...

Toxic Species	Cyanotoxin	Location	Reference(s)
Anabaena spp.	Anatoxin-a	Finland	Sivonen *et al.*, 1989a
Anabaena blooms	Anatoxin-a	Germany	Bumke-Vogt, 1998
Anabaena sp.	Anatoxin-a	Ireland	James *et al.*, 1997
Anabaena sp.	Anatoxin-a (minor amounts)	Japan	Park *et al.*, 1993a
Anabaena planctonica bloom	Anatoxin-a	Italy	Bruno *et al.*, 1994
Aphanizomenon sp.	Anatoxin-a	Finland	Sivonen *et al.*, 1989a
Aphanizomenon blooms	Anatoxin-a	Germany	Bumke-Vogt, 1998
Cylindrospermum sp.	Anatoxin-a	Finland	Sivonen *et al.*, 1989a
Microcystis sp.	Anatoxin-a (minor amounts)	Japan	Park *et al.*, 1993a
Oscillatoria sp. benthic	Anatoxin-a	Scotland	Edwards *et al.*, 1992
Oscillatoria sp.	Anatoxin-a	Ireland	James *et al.*, 1997
Planktothrix sp.	Anatoxin-a	Finland	Sivonen *et al.*, 1989a
Planktothrix formosa	Homoanatoxin-a	Norway	Skulberg *et al.*, 1992
Anabaena flos-aquae	Anatoxin-a(S)	Canada	Matsunaga *et al.*, 1989; Mahmood and Carmichael, 1987
A. lemmermannii	Anatoxin-a(S)	Denmark	Henriksen *et al.*, 1997; Onodera *et al.*, 1997a
Anabaena circinalis	Saxitoxins	Australia	Humpage *et al.*, 1994; Negri *et al.*, 1995; 1997
Aphanizomenon flos-aquae	Saxitoxins	USA	Jackim and Gentile, 1968; Ikawa *et al.*, 1982; Mahmood and Carmichael, 1986
Cylindrospermopsis raciborskii	Saxitoxins	Brazil	Lagos *et al.*, 1997
Lyngbya wollei	Saxitoxins	USA	Carmichael *et al.*, 1997; Onodera *et al.*, 1997b
Microcystis		India	Parkar *et al.*, 1997

The neurotoxin, anatoxin-a, was first shown to be produced by *Anabaena flos-aquae* strains originating from Canada, and later by Finnish strains of unidentified *Anabaena* species, and in individual species of *Oscillatoria, Aphanizomenon*, and *Cylindrospermum*, by benthic *Oscillatoria* from Scotland, and by *Anabaena* and *Oscillatoria* in Ireland. It also was present in *Anabaena planctonica* blooms in Sardinia, Italy, in *Anabaena* and *Aphanizomenon* blooms in Germany, and in minor amounts in some Japanese bloom samples, as well as in *Anabaena* strains. Homoanatoxin-a has been characterised from an *Oscillatoria formosum* (*Phormidium formosum*) strain from Norway (Table 6.5). To date, anatoxin-a(S) has been found only from *Anabaena* species: *A. flos-aquae* in the USA and Scotland, and *A. lemmermannii* in Denmark.

Aphanizomenon flos-aquae blooms and strains were found to contain saxitoxins in the USA and this species was for a long time the only known saxitoxin producer amongst the cyanobacteria. More recently, saxitoxins have been shown to be common in Australian rivers and reservoirs and to be produced by *Anabaena circinalis*. In North America, a benthic freshwater *Lyngbya wollei* was found to produce saxitoxins, as was a strain of *Cylindrospermopsis raciborskii* in Brazil (Table 6.5).

Cyanotoxin Patterns in Strains and Species of Cyanobacteria

Cyanobacteria may produce several toxins simultaneously. In general, more than one microcystin has been characterised from the strains listed in Table 6.2. Among neurotoxic strains, several PSPs are found in the same strain, although there are considerable variations between species (Table 6.3). Furthermore, simultaneous neurotoxin and hepatotoxin production has been noted; the best example studied being the *Anabaena flos-aquae* strain NRC 525-17 which produces anatoxin-a(S) (Matsunaga *et al.*, 1989) and several microcystins (Harada *et al.*, 1991a).

Microcystin

Although many strains produce several microcystins simultaneously, usually only one or two of them are dominant in any single strain. Qualitative variation in the microcystins present is most frequently found among strains of *Anabaena* but also in *Microcystis* (Sivonen *et al.*, 1995). Some taxa have a number of microcystins in common, such as planktonic *Anabaena, Microcystis* and *Planktothrix* (*Oscillatoria*). However, there is evidence of microcystin variants that are typical for certain cyanobacterial taxa.

Planktothrix and some strains of *Anabaena* produce only demethylmicrocystins (Table 6.2). *Planktothrix* (*Oscillatoria*) isolates from Finland (13 strains studied and toxins fully characterised) seem to produce one of two types of microcystin (D-Asp^{3}-RR or Dha^{7}-RR) (Sivonen *et al.*, 1995).

German field samples dominated by *Planktothrix* have also shown these microcystins, with dominance of one major demethylated microcystin in populations of *P. rubescens*, and two or three of these variants in populations of *P. agardhii* (Fastner *et al.*, 1998).

Microcystis strains from Japan appear to contain chiefly microcystin-LR, -RR and -YR, with some cultures showing all three variants, and some strains being dominated by one of them. These three microcystins are the only variants reported in several studies on *M. aeruginosa* and *M. viridis* (*e.g.* Watanabe, 1996). Many of the microcystins listed in Table 6.2 have been found only in minor amounts or, to date, have been found to be produced only by individual isolates.

In natural samples which usually contain many strains, or more than one toxin-producing species, different combinations of microcystins can be found. For example, in a *Microcystis* bloom from Homer Lake, USA, 19 different microcystins were characterised (Namikoshi *et al.*, 1992a, 1995) and in one Australian bloom of *Microcystis aeruginosa*, 23 microcystins were detected by high pressure liquid chromatography (HPLC), none of which were microcystin-LR (Jones *et al.*, 1995).

Microcystin-LR is often mentioned as the most frequently occurring microcystin, although such observations may be biased by the fact that a chemical standard for the analysis of microcystin-LR was the earliest to be commercially available. It has been reported to be the major toxin in bloom and strain samples from Portugal (Vasconcelos *et al.*, 1995. 1996), France (Vezie *et al.*, 1997), Canada (Kotak *et al.*, 1993) and frequently co-occurring with microcystin-RR and -YR in Japan (Watanabe *et al.*, 1988, 1989).

Geographical patterns are indicated by some survey results. Wide variation among the L-amino acids of microcystins has been reported for South Africa (Scott, 1991), frequent presence of demethylmicrocystins (RR and LR) has been seen in Finnish strains (Sivonen *et al.*, 1995) and microcystins in Danish samples show wide variation (Henriksen, 1996a). In part, these patterns probably reflect regional differences in dominance of cyanobacterial species or strains. Water bodies with regular dominance of specific taxa are likely to exhibit characteristic patterns of microcystin variants.

Nodularin

The cyanotoxin, nodularin, is found in waters where *Nodularia spumigena* is present; the most prominent areas being the Baltic Sea and brackish water estuaries and coastal lakes of Australia and New Zealand. However, the best known *Nodularia spumigena* bloom location, Lake Alexandrina, Australia, has salinities which are only slightly elevated above normal river water and at levels still suitable for drinking water. The presence of variants of nodularin in environmental samples is usually rather insignificant. In the Baltic Sea, the collection of samples for several years has shown nodularin to be the major compound present. The same is true for the almost 90 hepatotoxic *Nodularia* strains isolated from the same source (Sivonen *et al.*, 1989b; Lehtimäki *et al.*, 1997). Analyses of several strains isolated from blooms across Australia have revealed similar results, with nodularin variants being found rarely, and then only at low relative abundance (Jones *et al.*, 1994; Blackburn *et al.*, 1997).

Saxitoxins

There is much diversity of saxitoxin distribution in cyanobacteria from around the world (Table 6.3). In addition, the lack of analytical standards for many saxitoxins has probably restricted findings in some countries. Saxitoxin-producing *Anabaena circinalis* blooms are widespread in Australian rivers and reservoirs, and the relative abundance of individual toxins is remarkably constant in toxin-producing strains, which is quite opposite to the microcystin-producing cyanobacteria. In all healthy *Anabaena circinalis* cultures examined in Australia, the relative composition of individual saxitoxins is very similar and dominated by C-toxins (Blackburn *et al.*, 1997). As blooms and cultures age, the proportion of decarbamoyl-gonyautoxins breakdown products increases at the expense of the C-toxins. In North American *Aphanizomenon flos-aquae* and in Brazilian *Cylindrospermopsis raciborskii* samples, mostly neosaxitoxin and smaller amounts of saxitoxin have been detected. In the case of the *Aphanizomenon flos-aquae* from North America, only bloom samples and strains from New Hampshire, USA have been found to contain saxitoxins. Mat-forming *Lyngbya wollei* from Alabama, USA was found to produce decarbamoyl saxitoxin (dcSTX), decarbamoylgonyautoxin-2 and -3 (dcGTX2 and dcGTX3) and six new saxitoxins.

Concentrations of Cyanotoxins in Surface Waters

Information on the concentrations of cyanotoxins in surface waters has been appearing only recently in the international literature. In early studies (pre-1990s), the toxicity of bloom samples was determined by mouse bioassays, but this method is unsuitable for measuring the low concentrations of cyanotoxins that usually prevail in cyanobacterial populations when they do not accumulate in scums. The development of better analytical methods, in the first instance HPLC and more recently enzyme linked immunosorbent assay (ELISA), (and for microcystins and nodularins also the protein phosphatase assay) has made the quantification of total and individual toxins possible.

The quantitative determination of toxin concentrations is mostly performed from lyophilised (freeze-dried) cultures, bloom samples or seston (particulate material suspended in water, which contains not only cyanobacterial cells but usually other algae, some zooplankton, and possibly inorganic material such as soil and sediment particles). Results are usually expressed as milligrams or micrograms of toxin per gram dry weight (dw). Whereas in cultures and bloom samples, the dry weight originates from cyanobacteria, it will encompass further particles (seston) in plankton samples taken outside of scum areas. The highest published concentrations of cyanotoxins from cyanobacterial bloom samples, measured by HPLC, are (Table 6.6):

- Microcystin–7,300 $\mu g\ g^{-1}$ dw from China and Portugal,
- Nodularin–18,000 $\mu g\ g^{-1}$ dw from the Baltic Sea,

- ☆ Cylindrospermopsin–5,500 μg g^{-1} dw from Australia,
- ☆ Anatoxin-a–4,400 μg g^{-1} dw from Finland,
- ☆ Saxitoxins–3,400 μg g^{-1} dw from Australia,
- ☆ Anatoxin-a(S)–3,300 μg g^{-1} dw from the USA.

Table 6.6: Toxin Concentrations Reported in Cyanobacterial Bloom or Water Samples

Location	*Period of Study*	*No. of Toxic Samples (Total no. of samples)*	*Toxins Identified*	*Range of Total Concentrations (μg g^{-1} dw, unless otherwise indicated)*	*Analysis Method*	*Reference*
			Microcystins			
Australia	1991	4	Microcystins, 24 unidentified	2,100-4,100[2]	HPLC	Jones *et al.*, 1995
Canada, Alberta	1990	37(50)	Microcystin-LR	4-610	HPLC	Kotak *et al.*, 1993
Canada, Alberta (3 lakes)	1990-93	168(226)	Microcystin-LR	1-1,550	HPLC	Kotak *et al.*, 1995
China	1988	5(10)	Microcystin-RR, -LR	200-7,300	HPLC	Zhang *et al.*, 1991
Czech and Slovak Rep.	1995-96	(63)	Microcystin-LR	4-6,835	HPLC	Maršálek *et al.*, 1996
Denmark	1992-94		Microcystin-RR, -LR	3-2,800	HPLC	Christoffersen, 1996
Denmark	1993-95	198(296)	Microcystins	5-1,900	HPLC	Henriksen *et al.*, 1996b
Finland	1994-95	17(20)	Microcystin-LR	> 10-800	HPLC	Lahti *et al.*, 1997
France	1994	16(22)	Microcystins	70-3,970	HPLC	Vezie *et al.*, 1997
France, L. Grand-Lieu	1994	19(30)	Microcystins	30-230	HPLC	Vezie *et al.*, 1998
Germany	1992	8(15)	Microcystin-LR	36-360	HPLC	Fastner, 1994
Germany	1993	17(18)	Microcystins	0.15-36[1,2]	ELISA	Ueno *et al.*, 1996b
Germany	1995-96	385(533)	Microcystins	1-5,000	HPLC	Fastner, 1998
Germany	1997	34	Microcystins, several	1-25,000[1]	HPLC	Chorus *et al.*, 1998
Japan	1990	12(14)	Microcystin-RR, -YR, -LR	160-950	HPLC	Watanabe *et al.*, 1992
Japan	1988-92	11(19)	Microcystin-RR, -YR, -LR, [Dha[7]] -LR	70-1,610	HPLC	Park *et al.*, 1993a
Japan, Lake Suwa	1980-91	13	Microcystin-RR, -YR, -LR	30-2,100	HPLC	Park *et al.*, 1993b
Japan	1986-88	4(4)	Microcystin-RR, -YR, -LR	100-860	HPLC	Shirai *et al.*, 1991
Japan	1992-95	18(22)	Microcystin-RR, -YR, -LR	0.04-480[1]	HPLC	Tsuji *et al.*, 1996
Japan	1993-95	46(57)	Microcystins	0.05-1,300[1,2]	ELISA	Ueno *et al.*, 1996b
Japan	1993-94	12(17)	Microcystins	0.06-94[1,2]	ELISA	Nagata *et al.*, 1997
Japan	1989-94	10(10)	Microcystins	300-15,600[1,2] 330-19,500[1,2]	ELISA HPLC	Nagata *et al.*, 1997

Contd...

Table 6.6–Contd...

Location	Period of Study	No. of Toxic Samples (Total no. of samples)	Toxins Identified	Range of Total Concentrations (μg g^{-1} dw, unless otherwise indicated)	Analysis Method	Reference
Portugal	1989-92	12(12)	Microcystin-LR plus six known and three unidentified microcystins	1,000-7,100	HPLC	Vasconcelos *et al.*, 1996
Portugal	1994-95	28(29)	Microcystins	0.1-37[1,2]	ELISA	Ueno *et al.*, 1996b
South Africa	1985-86		Microcystin-FR, -LR, -YR, -LA, -YA, -LAba	5-420	HPLC	Wicks and Thiel, 1990
South Africa	1988-89	9(9)	Microcystin-YR, -LR, -FR, -YA, -LA, -LAba	40-630	HPLC	Scott, 1991
UK	1992	3(3)	Microcystins, 3	17-131[1,2]	HPLC	Codd *et al.*, 1995
USA, Wisconsin	1993	9	Microcystins	1,900-12,800[2]	ELISA	McDermott *et al.*, 1995
			Nodularins			
Baltic Sea	1985-87	17(23)	Nodularin	< 100-2,400	HPLC	Sivonen *et al.*, 1989b
Baltic Sea	1990-91	6(16)	Nodularin	300-18,000	HPLC	Kononen *et al.*, 1993
Tasmania, Orielton Lagoon	1992-93	7(9)	Nodularin	2,000-3,500	HPLC	Jones *et al.*, 1994
			Anatoxins			
Denmark	1993-95	9(10)	Anatoxin-a(S)	4-3,300[3]	ChE inhibition assay	Henriksen *et al.*, 1997
Finland	1985-87	13(30)	Anatoxin-a	10-4,400	GC/MS	Sivonen *et al.*, 1989a
Finland and Japan		3(3)	Anatoxin-a	0.4-2,600	LC/MS	Harada *et al.*, 1993
Germany	1995-96	10(45)	Anatoxin-a	0.02-0.36[1]	GC-ECD	Bumke-Vogt, 1998
Ireland	1995	2(2)	Anatoxin-a	10-100	HPLC (fluorimetric)	James *et al.*, 1997
Japan	1988-92	9(14)	Anatoxin-a	0.3-16	TSP-LC/MS	Park *et al.*, 1993a
			Saxitoxins			
Australia	1990-92	11(11)	Saxitoxins	85-2,040	HPLC and FAB-MS	Humpage *et al.*, 1994
Australia	1992-94	24(31)	Saxitoxins	50-3,400	HPLC	Negri *et al.*, 1997
USA	1994	7(8)	Saxitoxins	5-60[4]	HPLC/ AOAC	Carmichael *et al.*, 1997

Dw: Dry weight; HPLC: High pressure liquid chromatography; ELISA: Enzyme linked immunosorbent assay; GC/MS: Gas chromatography/mass spectrometry; LC/MS: Liquid chromatography/mass spectrometry; GC-ECD: Gas chromatography-electron capture detection; TSP-LC/MS: Thermospray-liquid chromatography/mass spectrometry; FAB-MS: Fast atom bombardment-mass spectrometry; AOAC: Mouse bioassay done according to the Association of Official Analytical Chemists.

1: Given as μg l^{-1}; 2: Microcystin-LR used as standard; 3: Measured by enzyme inhibition;
4: Micrograms of STX equivalents.

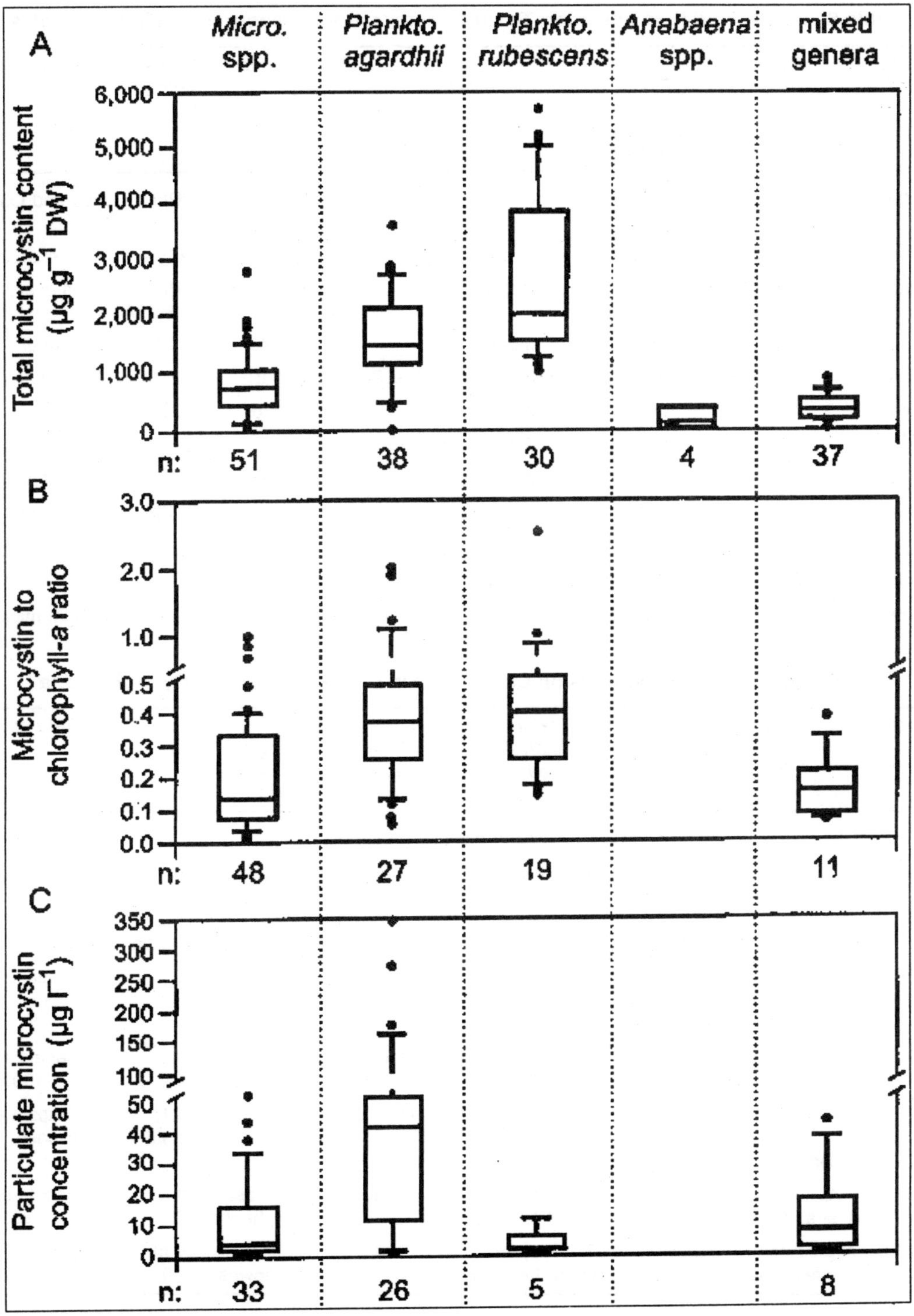
A
B
C
Micro. spp.
Plankto. agardhii
Plankto. rubescens
Anabaena spp.
mixed genera
Total microcystin content (µg g⁻¹ DW)
6,000
5,000
4,000
3,000
2,000
1,000
0
n: 51 38 30 4 37
Microcystin to chlorophyll-a ratio
3.0
2.0
1.0
0.5
0.4
0.3
0.2
0.1
0.0
n: 48 27 19 11
Particulate microcystin concentration (µg l⁻¹)
350
300
250
200
150
100
50
40
30
20
10
0
n: 33 26 5 8

Figure 6.4

Table 6.7: Dissolved (Extracellular) Toxin Concentrations Measured in Water Samples

Location	*Period of Study*	*No. of Toxic Samples (Total no. of samples)*	*Toxin Identified*	*Concentration (µg l⁻¹)*	*Analysis Method*	*Reference*
Australia		24	Microcystins	3-1.800[1]	HPLC	Jones and Orr, 1994
China	1993-94	130(835)	Microcystins	0.05-1.6	ELISA	Ueno *et al.*, 1996a
Finland	1993-94	38(38)	Microcystin-LR	0.06-0,21	HPLC and ELISA	Lahti *et al.*, 1997
Germany	1993-94	11(19)	Microcystins	0.07-0.76	ELISA	Ueno *et al.*, 1996b
Japan	1992-95	9(22)	Microcystin-RR, -YR, -LR	0.02-3.8	HPLC	Tsuji *et al.*, 1996
Japan	1993-95	26(38)	Microcystins	trace-5.6	ELISA	Ueno *et al.*, 1996b
Japan	1993-94	4(13)	Microcystins	0.08-0.8	ELISA	Nagata *et al.*, 1997
Thailand	1994	7(10)	Microcystins	0.08-0.35	ELISA	Ueno *et al.*, 1996b
USA, Wisconsin	1993	27(27)	Microcystins	0.07-200	ELISA	McDermott *et al.*, 1995

HPLC: High pressure liquid chromatography; ELISA: Enzyme linked immunosorbent assay.

1 High range concentrations following treatment of a large bloom with algicide, which released intracellular microcystins.

Table 6.8: Laboratory Studies on Cellular Toxin Concentrations in Cyanobacteria

Parameter	*Organism*	*Toxin(s)/ Analysis Method*	*Changes in Toxin Concentrations (dw)*	*Highest/ Lowest Toxin Production*	*Reference(s)*
		Temperature (°C)			
12.5-30	*Anabaena* spp. (2 strains), batch cultures	Microcystins HPLC	3.5-30 fold[1]	Highest at 25, lowest at 30; different toxins at different temperatures	Rapala *et al.*, 1997
10-28	*Anabaena* spp. (2 strains), continuous cultures	Microcystins HPLC	3-10 fold	Lowest at 10, highest at 25	Rapala and Sivonen, 1998
15-30	*Anabaena* spp. (2 strains), batch cultures	Anatoxin-a HPLC	3 fold	Lowest at 30, highest at 20	Rapala *et al.*, 1993
10-28	*Anabaena* spp. (2 strains), continuous cultures	Anatoxin-a HPLC	4-7 fold	Highest at 19-21, lowest at 10 and 28	Rapala and Sivonen, 1998
15-30	*Aphanizomenon* sp. (1 strain), batch cultures	Anatoxin-a HPLC	3 fold	Lowest at 30, highest at 20	Rapala *et al.*, 1993
10,25,34	*Microcystis aeruginosa* (1 strain), batch cultures	Microcystins mouse bioassay	5 fold	Highest toxicity at 25, lowest at 10	Codd and Poon, 1988

Contd...

Table 6.8–Contd...

Parameter	*Organism*	*Toxin(s)/ Analysis Method*	*Changes in Toxin Concentrations (dw)*	*Highest/ Lowest Toxin Production*	*Reference(s)*
15-35	*Microcystis aeruginosa* (1 strain) batch cultures	Microcystins HPLC mouse bioassay	4 fold	Highest toxicity at 20; different toxins at different temperatures	van der Westhuizen and Eloff, 1985; van der Westhuizen *et al.*, 1986
18,25,35	*Microcystis aeruginosa* (1 strain), batch cultures	Microcystins mouse bioassay	1.4 fold	Highest toxicity at 18, lowest at 32	Watanabe and Oishi, 1985
10-30	*Nodularia spumigena* (2 strains), batch cultures	Nodularin HPLC	3-4 fold	Highest at 20, lowest at 10 or 30	Lehtimäki *et al.*, 1994
7-28	*Nodularia spumigena* (1 strain), batch cultures	Nodularin HPLC	3 fold	Highest at 19	Lehtimäki *et al.*, 1997
15-30	*Oscillatoria agardhii* (2 strains), batch cultures	Microcystins HPLC	7 fold	Strain dependent; lowest at 30	Sivonen, 1990b
		Light ($\mu mol\ m^{-2}\ s^{-1}$)			
2-100 continuous	*Anabaena* spp. (2 strains), batch cultures	Microcystins HPLC	3 fold	Highest at 25	Rapala *et al.*, 1997
7, 19, 42 continuous	*Anabaena* spp. (2 strains), continuous cultures	Microcystins HPLC	2.5-15	Lowest at 10, highest at 25	Rapala and Sivonen, 1998
2-128 continuous	*Anabaena* spp. (2 strains), batch cultures	Anatoxin-a HPLC	3 fold	Highest at 26-44, lowest at 2	Rapala *et al.*, 1993
7, 19, 42 continuous	*Anabaena* spp. (2 strains), continuous cultures	Anatoxin-a HPLC	No effect	Highest at 19, lowest at 7	Rapala and Sivonen, 1998
2-128 continuous	*Aphanizomenon* sp. (1 strain), batch cultures	Anatoxin-a HPLC	4 fold	Highest at 128, lowest at 2	Rapala *et al.*, 1993
5-50 continuous	*Microcystis aeruginosa* (1 strain), batch cultures	Microcystins mouse bioassay	2.4 fold	Highest toxicity at 20	Codd and Poon, 1988
20-75 continuous	*Microcystis aeruginosa* (1 strain), continuous cultures	Microcystins HPLC	2.5 fold	Highest at 40	Utkilen and Gjølme, 1992
21-205 continuous	*Microcystis aeruginosa* (1 strain), batch cultures	Microcystins mouse bioassay	1.2 fold	Highest toxicity at 142, lowest at 21	van der Westhuizen and Eloff, 1985
7.5, 30, 75 continuous	*Microcystis aeruginosa* (1 strain), batch cultures	Microcystins mouse bioassay	3.8 fold	Highest toxicity at 30, lowest at 7.5	Watanabe and Oishi, 1985
25, 50, 80 continuous	*Nodularia spumigena* (2 strains), batch cultures	Nodularin HPLC	No difference		Lehtimäki *et al.*, 1994

Contd...

Table 6.8–Contd...

Parameter	Organism	Toxin(s)/ Analysis Method	Changes in Toxin Concentrations (dw)	Highest/ Lowest Toxin Production	Reference(s)
2-155 continuous	*Nodularia spumigena* (1 strain), batch cultures	Nodularin HPLC	50 fold[1]	Higher at high irradiances, minimal at 2	Lehtimäki *et al.*, 1997
12-95 continuous	*Oscillatoria agardhii* (2 strains), batch cultures	Microcystins HPLC	2.5 fold	Highest at 12-44	Sivonen, 1990b
		Phosphorus (mg P l⁻¹)			
0.05-5.5	*Anabaena* spp. (2 strains), batch cultures	Microcystins HPLC	5 fold	Highest at 5.5, lowest at 0.05	Rapala *et al.*, 1997
0.05-5.5	*Anabaena* spp. (2 strains), batch cultures	Anatoxin-a HPLC	No difference	No statistically significant differences	Rapala *et al.*, 1993
0.05-5.5	*Aphanizomenon* sp. (2 strains), batch cultures	Anatoxin-a HPLC	2 fold	Lowest at 0.05-0.1, highest at 0.5-5.5	Rapala *et al.*, 1993
BG-11 and medium without P	*Microcystis aeruginosa* (1 strain), batch cultures	Microcystins mouse bioassay	1.7 fold	Higher toxicity without P	Codd and Poon, 1988
0.0025, 0.025	*Microcystis aeruginosa* (1 strain), continuous cultures	Microcystins HPLC	2.3 fold	More toxin at 0.025	Utkilen and Gjølme, 1995
MA medium 1/1; dilutions 1/10, 1/20	*Microcystis aeruginosa* (1 strain), batch cultures	Microcystins mouse bioassay	Less than 1	Highest toxicity with the original medium	Watanabe and Oishi, 1985
0.1-5.5	*Oscillatoria agardhii* (2 strains), batch cultures	Microcystins HPLC	1.8-2.5 fold	Lowest toxin at 0.1	Sivonen, 1990b
0.3, 0.6, 1.0	*Nodularia spumigena* (2 strains), batch cultures	Nodularin HPLC	Less than 1 fold	Lowest at 0.3	Lehtimäki *et al.*, 1994
0-5.5	*Nodularia spumigena* (1 strain), batch cultures	Nodularin HPLC	4 fold	Lowest at 0-0.02, highest at 0.2-5.5	Lehtimäki *et al.*, 1997
		Nitrogen (mg N l⁻¹)			
BG-11 medium, medium without N	*Microcystis aeruginosa* (1 strain), batch cultures	Microcystins mouse bioassay	5 fold	Higher toxicity with the medium containing N	Codd and Poon, 1988
0.05-1	*Microcystis aeruginosa* (1 strain), continuous cultures	Microcystins HPLC	3 fold	Higher at high N	Utkilen and Gjølme, 1995
MA medium 1/1; dilutions 1/10, 1/20	*Microcystis aeruginosa* (1 strain), batch cultures	Microcystins mouse bioassay	2.5 fold	Highest toxicity with the original medium	Watanabe and Oishi, 1985
0.42-84	*Oscillatoria agardhii* (2 strains), batch cultures	Microcystins HPLC	5 fold	Higher at high N, lowest at low N	Sivonen, 1990b

Contd...

Table 6.8–Contd...

Parameter	*Organism*	*Toxin(s)/ Analysis Method*	*Changes in Toxin Concentrations (dw)*	*Highest/ Lowest Toxin Production*	*Reference(s)*
			Micronutrients		
Al, Cd, Cr, Cu, Fe, Mn, Ni, Sn, Zn; various concentrations	*Microcystis aeruginosa* (1 strain), batch cultures	Microcystins HPLC	1.7 fold	Less toxins at low Fe concentrations	Lukac and Aegerter, 1993
0.1-3.4 µg Fe l^{-1}	*Microcystis aeruginosa* (1 strain), continuous cultures	Microcystins HPLC	1.5 fold	More toxin at high Fe concentrations	Utkilen and Gjølme, 1995
0.03-1.2 µg Fe l^{-1}	*Microcystis aeruginosa* (1 strain), continuous cultures	Microcystins HPLC	0-3 fold	Less toxin at low Fe concentrations	Lyck *et al.*, 1966
			Salinity (‰)		
3, 5, 8, 11	*Nodularia spumigena* (*2* strains), batch cultures	Nodularin HPLC	No difference	No statistical difference	Lehtimäki *et al.*, 1994
0-30	*Nodularia spumigena* (1 strain), batch cultures	Nodularin HPLC	8 fold	Highest at 15, lowest at 0 and 30	Lehtimäki *et al.*, 1997
0-35	*Nodularia spumigena* (6 strains), batch cultures	Nodularin HPLC	5 fold	Highest at 12, lowest at 35	Blackburn *et al.*, 1996
			CO_2		
BG-11 medium, medium without CO_2	*Microcystis aeruginosa* (1 strain), batch culture	Microcystins mouse bioassay	6 fold	Higher toxicity with the medium containing CO_2	Codd and Poon, 1988
			pH		
1-14	*Microcystis aeruginosa* (1 strain), batch cultures	Microcystins mouse bioassay	1.8 fold	Toxicity highest at low and high pH	van der Westhuizen and Eloff, 1983

Dw: Dry weight; HPLC: High performance liquid chromatography.

1: When the growth of the strains was poor the amount of toxins was also very low (less than 0.1 mg g^{-1} dw of cells); when these cases were compared to maximal toxin production more than ten-fold differences could be seen

Biosynthesis

To understand how cyanotoxins are produced, it is necessary to study the biochemical and genetic basis of toxin production. Knowledge of the biosynthetic pathways of cyanotoxins is in its early stage and no complete biochemical pathways are known. Biosynthesis of several cyanotoxins has been studied by feeding labelled precursors to a cyanobacterial culture and following their incorporation into the carbon skeleton of the toxins. Shimizu *et al.* (1984) used an *Aphanizomenon flos-aquae* strain to study biosynthesis of saxitoxin analogues. They proposed a new pathway for neosaxitoxin biosynthesis, the key steps of which are the condensation of an acetate unit, or its derivative, to the amino group bearing an a-carbon of arginine or an equivalent, and a subsequent loss of the carboxyl carbon and imidazole ring formation on the adjacent carbonyl carbon. They established the origin of all the carbons in the toxin alkaloid ring system. The side-chain carbon was derived from methionine (Shimizu, 1986).

Anatoxin-a is related structurally to the tropane class of alkaloids found in higher plants. Based mainly on ^{14}C-labelled precursors and enzymatic studies, Gallon *et al.* (1990) and Gallon *et al.* (1994) suggested the biosynthesis of anatoxin-a to be analogous to that of tropanes. Anatoxin-a was proposed to be formed from ornithine/arginine via putrescine, which is oxidised to pyrroline, a precursor of anatoxin-a. Labelling experiments using ^{13}C NMR (nuclear magnetic resonance spectrometry) indicated that the carbon skeleton of anatoxin-a is derived from acetate and glutamate. The studies showed that C-1 of glutamic acid is retained during the transformation of anatoxin-a and not lost by decarboxylation, a finding incompatible with the tropane alkaloid theory (Hemscheidt *et al.*, 1995b). All of the carbons of anatoxin-a(S) are derived from amino acids. Three methyl carbons arise from L-methionine or other donors to the tetrahydrofolate C1 pool. L-arginine accounts for C-2, C-4, C-5 and C-6 carbons of the toxin (Moore *et al.*, 1992, 1993). The intermediate in the biosynthesis of anatoxin-a(S) from L arginine is (2*S*,4*S*)-4 hydroxyarginine (Hemscheidt *et al.*, 1995a). The structure of the cylindrospermopsin suggests a polyketide origin for the toxin (Moore *et al.*, 1993).

The origin of carbons in microcystin (Moore *et al.*, 1991) and in nodularin (Choi *et al.*, 1993; Rinehart *et al.*, 1994) have been studied by following the incorporation of labelled precursors into the toxins by NMR. Carbons C1-C8 of Adda in nodularin are acetate derived and the remaining carbons presumably originate from phenylalanine. Methyl groups in carbons 2, 4, 6, 8, and the O-methyl group in the Adda unit, originated from methionine. The D-Glu and L-Arg carbons C4-C5 were acetate derived, with C1-C2 being from glutamate. Methyldehydrobutyrine was possibly formed from threonine, its methyl group coming from methionine. The β-methylaspartic acid was found to originate from condensation of pyruvic acid (C3-C4) and acetyl-CoA (C1-C2) (Rinehart *et al.*, 1994). The studies on the carbon skeleton of nodularin, with some minor differences, agree with work on microcystin-LR by Moore *et al.* (1991). In their study, L-Leu and D-Ala units in microcystin had acetate incorporation. The dehydroamino acid in microcystin has been proposed to be formed from serine rather than from threonine (Rinehart *et al.*, 1994). Rinehart's group found linear nodularin, which was shown by culture experiments to be a precursor of cyclic nodularin. Three additional linear peptides were isolated from a bloom sample, one of them was possibly a precursor of cyclic microcystin-LR and the others possibly degradation products (Rinehart *et al.*, 1994).

Genetic Regulation of Cyanotoxin Production

The genes and enzymes involved in cyanotoxin production are still mostly unknown. The first molecular biological studies on toxic cyanobacteria investigated the possible involvement of plasmids in toxin production. Four toxic strains of *Microcystis aeruginosa* contained plasmids, and no plasmid could be shown in one toxic and in several non-toxic strains (Schwabe *et al.*, 1988). More recently, a similar study in Australia found no evidence for plasmid involvement in microcystin synthesis (Bolch *et al.*, 1997). Gallon *et al.* (1994) studied an anatoxin-a producing *Anabaena* strain NCR 44-1, which spontaneously became non-toxic. They found that the size of a plasmid was reduced in that non-toxic clone, but this work has not been repeated or confirmed.

Multi-enzyme complexes and peptide synthetase genes are involved in hepatotoxin production. Several cyclic and linear peptides, often with D-amino acids, are known to be produced, non-ribosomally, by multi-domain peptide synthetases via the so-called thiotemplate mechanism in bacteria and lower eukaryotes. The best characterised are the synthesis of gramicidin S and tyrocidin by *Bacillus*. Peptide synthetase genes have been detected and sequenced (partly) in *Microcystis aeruginosa*

(Meissner *et al.*, 1996) and in *Anabaena* (Rouhiainen *et al.*, 1994). Analogous polymerase chain reaction (PCR) products to the peptide synthetase genes have been shown by using DNA from *Microcystis* (Jacobs *et al.*, 1995; Arment and Carmichael, 1996) and *Nodularia* as a template. Dittman *et al.* (1997) showed, in knockout experiments, that peptide synthetase genes are responsible for microcystin production.

At least some strains which produce hepatotoxins also produce other small cyclic peptides (Namikoshi and Rinehart, 1996; Weckesser *et al.*, 1996) which are likely to be produced by nonribosomal peptide synthesis.

Fate in the Environment

Partitioning Between Cells and Water

It appears likely that cyanotoxins are produced and contained within the actively growing cyanobacterial cells (*i.e.* they are intracellular or particulate). Release to the surrounding water, to form dissolved toxin, appears to occur mostly, if not exclusively, during cell senescence, death and lysis, rather than by continuous excretion.

In laboratory studies, where both intracellular and dissolved toxins (microcystins/nodularin and saxitoxins) have been measured, it is generally the case that in healthy log phase cultures, less than 10-20 per cent of the total toxin pool is extracellular (Sivonen, 1990b; Lehtimäki *et al.*, 1997; Negri *et al.*, 1997; Rapala *et al.*, 1997). As cells enter stationary phase the increased rate of cell death may lead to an increase in the extracellular dissolved fraction. Even during log-phase cell growth in culture, a small percentage of cells in the population may be dying and lysing (and releasing intracellular toxins), even though there is an overall positive population growth. There are some indications that anatoxin-a may leak out of cells during growth especially in low light conditions. High concentrations of anatoxin-a, sometimes exceeding the intracellular pool of toxins, have been found in media in a batch culture study (Bumke-Vogt *et al.*, 1996).

In the field, healthy bloom populations produce little extracellular toxin. The range of measured concentrations for dissolved cyanotoxins, in all cases except those where a major bloom is obviously breaking down, is 0.1-10 $\mu g\ l^{-1}$ (Lindholm and Meriluoto, 1991; Jones and Orr, 1994; Tsuji *et al.*, 1996; Ueno *et al.*, 1996b; Lahti *et al.*, 1997b) Cell-bound concentrations are several orders of magnitude higher (Tables 6.6 and 6.7). In lakes or rivers, toxins liberated from cells are rapidly diluted by the large mass of water, especially if mixing of water by wind action or currents is vigorous (Jones and Orr, 1994). However, the concentration of dissolved toxins may be much higher in ageing or declining blooms. This is an important consideration for water treatment plant operators, because it means that removal of healthy cyanobacterial cells intact from the raw water supply may obviate or substantially reduce the need for additional adsorptive (activated carbon) or oxidative (ozone or chlorine) toxin removal processes.

The release of toxins from cells is enhanced by chemical treatments for the eradication of cyanobacteria, especially the use of algicides (either copper-based or organic herbicides). Treatment of a bloom with copper sulphate, for example, may lead to complete lysis of the bloom population within three days and release of all the toxins into the surrounding water (Berg *et al.*, 1987; Kenefick *et al.*, 1992; Jones and Orr, 1994). The efficacy of copper sulphate treatment is, however, very much dependent on water chemistry, especially alkalinity, pH and dissolved organic content.

Chemical Breakdown

The four main groups of cyanotoxins: microcystins, anatoxins, PSPs and cylindrospermopsins, exhibit quite different chemical stabilities and biological activities in water.

Microcystins

Microcystins, being cyclic peptides, are extremely stable and resistant to chemical hydrolysis or oxidation at near neutral pH. Microcystins and nodularin remain potent even after boiling. In natural waters and in the dark, microcystins may persist for months or years. At high temperatures (40 °C) and at elevated or low pH, slow hydrolysis has been observed, with the times to achieve greater than 90 per cent breakdown being approximately 10 weeks at pH 1 and greater than 12 weeks at pH 9 (Harada *et al.*, 1996). Rapid chemical hydrolysis occurs only under conditions that are unlikely to be attained outside the laboratory, *e.g.* 6M HCl at high temperature.

Microcystins can be oxidised by ozone and other strong oxidising agents, and degraded by intense ultra violet (UV) light. These processes have relevance for water treatment, although they are unlikely to contribute to degradation occurring in the natural environment.

In full sunlight, microcystins undergo slow photochemical breakdown and isomerisation, with the reaction rate being enhanced by the presence of water-soluble cell pigments, presumably phycobiliproteins (Tsuji *et al.*, 1993). In the presence of pigments the photochemical breakdown of microcystin in full sunlight can take as little as two weeks for greater than 90 per cent breakdown, or longer than six weeks, depending on the concentration of pigment (and presumably toxin, although this has not been tested). A more rapid breakdown under sunlight has been reported in the presence of humic substances (which can act as photosensitisers) in field concentrations ranging from 2-16 mg l^{-1} dissolved organic carbon (DOC). Approximately 40 per cent of the microcystins was degraded per day under summer conditions of insolation (Welker and Steinberg, 1998). In deeper or muddy waters, the rate of breakdown is likely to be considerably slower.

Anatoxins

Anatoxin-a is relatively stable in the dark, but in pure solution in the absence of pigments it undergoes rapid photochemical degradation in sunlight. Breakdown is further accelerated by alkaline conditions (Stevens and Krieger, 1991). The half-life for photochemical breakdown is 1-2 hours. Under normal day and night light conditions at pH 8 or pH 10, and at low initial concentrations (10 μg l^{-1}), the half-life for anatoxin-a breakdown was found to be approximately 14 days (Smith and Sutton, 1993). Anatoxin-a(S) decomposes rapidly in basic solutions but is relatively stable under neutral and acidic conditions (Matsunaga *et al.*, 1989).

Saxitoxins

In the dark at room temperature, saxitoxins undergo a series of slow chemical hydrolysis reactions. The C-toxins lose the N-sulphocarbamoyl group to form decarbamoyl gonyautoxins (dc-GTXs); while the dc-GTXs, GTXs and STXs slowly degrade to, as yet unidentified, non-toxic products. The half-lives for the breakdown reactions are in the order of 1-10 weeks, with more than three months often being required for greater than 90 per cent breakdown (Jones and Negri, 1997). Because dc-GTXs are much more toxic than C-toxins (by a factor of 10-100), a solution or water body containing a natural mixture of C-toxins and GTXs, for example from the lysis of an Australian bloom of *Anabaena circinalis*, will actually increase in toxicity over a period of up to three weeks, before toxicity begins to abate during the succeeding 2-3 months. Boiling an extract of *Anabaena* with predominant C-toxins may also substantially increase toxicity. Similar transformation reactions occur in living cells as they age in culture or in a natural bloom (Negri *et al.*, 1997). No detailed studies have been carried out on saxitoxin breakdown in sunlight, either with or without pigments.

Cylindrospermopsins

Cylindrospermopsin is relatively stable in the dark, with slow breakdown occurring at elevated temperature (50 °C) (Chiswell *et al.*, 1999). In sunlight and in the presence of cell pigments, breakdown occurs quite rapidly being more than 90 per cent complete within 2-3 days (Chiswell *et al.*, 1999). Pure cylindrospermopsin is relatively stable in sunlight.

Removal on Natural Sediments and Soils

Microcystins appear to be retained only weakly on natural suspended solids in rivers and reservoirs; usually no more than 20 per cent of the total microcystin concentration is adsorbed. In a laboratory experiment, some of the dissolved anatoxin-a and microcystins were reported by Rapala *et al.* (1993) to be adsorbed on lake sediments. Percolation through clay soils may provide some cyanotoxin removal, but this will depend greatly on the type of clay, surface charge, cation concentration of the water, etc. Cyanobacterial cells and microcystins were retained in soil columns, but less efficiently in sediment columns, in laboratory experiments simulating the fate of cyanobacterial toxins in artificial recharge of groundwater and bank filtration (Lahti *el al.*, 1996). No data are available for other cyanobacterial toxins, but some removal may be expected, again depending on the chemical conditions of soil and water.

Sedimentation of living cells without lysis, for example through grazing by zooplankton and sinking of faecal pellets, may lead to accumulation and persistence of toxin material in sediments, although this process has received little scientific attention. As discussed in more detail below, microcystins retained in intact cells may persist for several months. Cells deposited in sediments may be subject to fairly rapid breakdown by sediment bacteria and protozoa, with the resultant release of toxins.

Biodegradation

Microcystins

In spite of their chemical stability and resistance to eucaryotic and many bacterial peptidases, microcystins are susceptible to breakdown by aquatic bacteria found naturally in rivers and reservoirs. These bacteria appear to be reasonably common and widespread. Degradative bacteria have been found in sewage effluent (Lam *et al.*, 1995), lake water (Jones *et al.*, 1994; Cousins *et al.*, 1996; Lahti *et al.*, 1997a), lake sediment (Rapala *et al.*, 1994; Lahti *et al.*, 1997a) and river water (Jones *et al.*, 1994). Nonetheless, one Finnish study showed a complete lack of degradation of microcystin over a three-month period by an inoculum taken in winter from the Vantaanjoki river (Kiviranta *et al.*, 1991). There is usually an initial lag phase with little loss of microcystin and this period can be as short as two days or more than three weeks, depending on the water body, climatic conditions, the concentration of dissolved microcystin and in some cases, although not all, the previous bloom history of a lake (Jones *et al.*, 1994; Rapala *et al.*, 1994; Lahti *et al.*, 1997b). Once the biodegradation process commences, removal of microcystin can be more than 90 per cent complete within 2-10 days. This may vary depending on the water body, initial microcystin concentration and water temperature (Jones *et al.*, 1994; Lahti *et al.*, 1997b).

Jones *et al.* (1994) isolated a species of aquatic *Sphingomonas* that initiated ring-opening of microcystin-LR to produce linear (acyclo-)microcystin-LR as a transient intermediate (Bourne *et al.*, 1996). This compound was nearly 200 times less toxic than the parent toxin. The products of complete bacterial degradation were non-toxic to mice at doses up to 500 μg kg^{-1} (compared with an LD_{50} for microcystin-LR of about 60 μg kg^{-1}). The same bacterium, however, did not degrade the closely related cyclic pentapeptide nodularin. In a strain of *Pseudomonas aeruginosa* from a Japanese lake, microcystin degradation appeared to proceed by attack on the Adda side chain of microcystin (Takenaka and Watanabe, 1997). Several bacteria were isolated from lake water and sediment in Finland capable of degradation of microcystins and some strains also degraded nodularin. One strain was identified as a *Sphingomonas* sp. and two of the strains belonged to the beta-subgroup of *Proteobacteria*, although the genera remains to be determined (Lahti *et al.*, 1997a).

Other Cyanobacterial Toxins

Little work has been undertaken on the biodegradation of anatoxins, saxitoxins or cylindrospermopsin. Anatoxin-a may be readily degraded by bacteria associated with cyanobacterial filaments. Laboratory studies using non-axenic strains of cyanobacteria found low concentrations of dissolved anatoxin-a in the culture medium (Kiviranta *et al.*, 1991; Rapala *et al.*, 1993) whereas high concentrations of anatoxin-a were found in the medium of a continuous culture using an axenic strain (free of contaminating bacteria) of the same species (Rapala and Sivonen, 1998). A *Pseudomonas* sp. strain able to degrade anatoxin-a at a rate of 6-10 μg ml^{-1} per three days was isolated by Kiviranta *et al.* (1991). In the presence of lake sediment and natural bacteria, the half-life for breakdown of anatoxin-a in the laboratory was about five days (Smith and Sutton, 1993). In a recent study by Jones and Negri (1997) no bacterially-mediated degradation of saxitoxins from *Anabaena circinalis* was observed in a range of surface water samples.

Bioaccumulation

Microcystins bioaccumulate in common aquatic vertebrates and invertebrates, including fish (Carbis *et al.*, 1997; Beattie *et al.*, 1998), mussels (Eriksson *et al.*, 1989; Falconer *et al.*, 1992; Prepas *et al.*, 1997; Watanabe *et al.*, 1997) and zooplankton (Watanabe *et al.*, 1992). In mussels, the highest microcystin concentrations are found in the hepatopancreas, and in vertebrates they are found in the liver. Williams *et al.* (1997) have shown covalent binding and accumulation of microcystin-LR in salmon liver and crab larvae. Whether the levels of microcystin accumulation are sufficient to pose a risk to humans is uncertain, and will depend on levels of consumption and the severity of toxic blooms in the area where fish or shellfish are caught or collected. Common advice given by water authorities is that the viscera of the fish should not be eaten, but caution should be taken in all cases where major toxic blooms occur.

Saxitoxins from marine "red tide" dinoflagellates are well known for their propensity to bioaccumulate in marine vertebrates and invertebrates, often with disastrous consequences for animals and humans that consume them. Similarly, saxitoxins from the freshwater cyanobacterium *Anabaena circinalis* may bioaccumulate in an Australian species of freshwater mussel to concentrations exceeding international guidelines (Shumway *et al.*, 1995) during as little as seven days exposure to a cell density of 100,000 cells per ml of a toxigenic strain (Negri and Jones, 1995). This cell density is commonly encountered in natural blooms of this species.

Persistence and Stability in Cells

Culture studies indicate that microcystins and nodularin degrade only very slowly (time scale of weeks), if at all, whilst contained within living cells (Sivonen, 1990b; Lehtimäki *et al.*, 1994, 1997; Rapala *et al.*, 1997; Orr and Jones, 1998). Similarly, scums of *Microcystis aeruginosa* that dry on the shores of lakes may contain high concentrations of microcystin for several months (Jones *et al.*, 1995).

These toxins are released back into the water body when re-immersed. Thus there is the potential for significant localised concentrations of dissolved microcystin even in the absence of living cells or a recently collapsed bloom.

In a lake study carried out over two summer–autumn periods, Lahti *et al.* (1997b) found that dissolved microcystin was more persistent than particulate toxin, with 30 and 15 days respectively required for 90 per cent degradation to occur.

Impact on Aquatic Biota

Direct cyanobacterial poisoning of animals can occur by two routes: through consumption of cyanobacterial cells from the water, or indirectly through consumption of other animals that have themselves fed on cyanobacteria and accumulated cyanotoxins. Cyanotoxins are known to bioaccumulate in common aquatic vertebrates and invertebrates, including fish, mussels and zooplankton. Consequently, there is considerable potential for toxic effects to be magnified in aquatic food chains. Such toxicity biomagnification is well known for anthropogenic pollutants such as heavy metals and pesticides. There is no reason to suspect that the situation would be any different with natural cyanotoxins.

It is difficult to ascribe the deaths of natural populations of aquatic animals, especially fish, unequivocally to cyanotoxin poisoning. One of the main reasons for this is because the collapse of a large cyanobacterial bloom can lead to very low concentrations of oxygen in the water column as a consequence of bacterial metabolism; consequent fish deaths may be due to the anoxia. The best evidence for the potential for toxic effects on aquatic organisms comes from controlled laboratory trials with exposure of animals to toxic cyanobacteria or cell-free solutions of cyanotoxins.

Effects on Aquatic Bacteria

The influence of cyanobacterial toxins on bacteria is not fully understood and the scientific literature gives a number of contradictory statements. According to some authors neither an extract of *Microcystis aeruginosa* nor pure microcystin-LR have a biocidal effect on *Bacillus subtilis, Staphylococcus aureus, Escherichia coli* or *Pseudomonas hydrophila* (Foxall and Sasner, 1988). However, these limited tests should not be seen as general indicators of the potential impacts of cyanotoxins on aquatic bacteria. The majority of aquatic bacteria are yet to be cultured, and studies with common mammalian pathogens or "laboratory" bacteria should not be taken as all encompassing. It is quite possible that cyanotoxins impact on some species of aquatic bacteria and not others. Certainly, microcystins are not toxic to all bacteria because several species are known to degrade quite high concentrations of these toxins. It is even possible that the slow release of cyanotoxins from the cell surface or from senescent cells may stimulate associations of particular bacterial types which may even act as symbionts.

Attempts have been made to use bacterial toxicity tests (based on inhibition of bacterial phosphorescence) to screen for the presence of cyanotoxins, especially microcystins. However it appears that the inhibition of bacterial phosphorescence is not related to the commonly known cyanotoxins. It has been suggested that the negative effect may be related to the presence of unidentified LPS endotoxins in the cell wall of the cyanobacterial cells.

Effects on Zooplankton

Evidence of the potential effects of cyanotoxins on zooplankton from numerous studies, mostly in laboratory situations, is complex and inconsistent. The vast majority of published studies has been based on mouse bioassay data describing cyanobacterial toxicity, with only a few more recent studies having used analytical methods such as HPLC to quantify individual toxins. Overall, it appears that cyanobacteria may exhibit a deleterious effect on zooplankton, but the effect is highly variable between genera and species, and even between clones of individual zooplankton species. One of the main questions yet to be resolved is whether the observed inhibitory effects are due to the putative poor nutritional value of cyanobacteria, to the known cyanotoxins, or to other unidentified compounds. There is evidence in the literature to support all three effects as being significant, at least with particular species under experimental growth conditions. A major difference in study design is whether organisms are exposed to cyanotoxins dissolved in water, or fed with toxic cyanobacteria. The latter is likely to lead to a substantially higher dose. Furthermore, Jungmann and Benndorf (1994) reported that exposure of *Daphnia* to dissolved microcystins showed effects only at concentrations several orders of magnitude above those found in field samples. They did, however, observe toxicity to *Daphnia* by unidentified metabolites other than microcystins from *Microcystis.*

There is dramatic variation among zooplankton species in their response to toxic (and even non-toxic) cyanobacteria. For example, DeMott *et al.* (1991) showed that the four species of zooplankton differed in their sensitivity to hepatotoxins by almost two orders of magnitude, but toxicity was observed only at very high concentrations that are scarcely encountered in natural water bodies (48 h LC_{50} ranging from 450 to 21,400 μg of microcystin per litre). Snell (1980) found that there was a genotype-dependent response of the rotifer *Asplanchna girodi* to toxic *Anabaena flos-aquae* and *Lyngbya* sp. Hietala *et al.* (1997) observed a variation in susceptibility of more than three orders of magnitude in the acute toxicity of *Microcystis aeruginosa* to 10 clones of *Daphnia pulex.* Both DeMott *et al.* (1991) and Laurén-Määttä *et al.* (1997) suggested that clone and species differences between zooplankton susceptibilities to toxic cyanobacteria may lead to selection pressures in favour of resistant strains or species in water bodies where toxic cyanobacteria occur frequently.

Benndorf and Henning (1989) found that the toxicity of a field population of *Microcystis* was increased by the feeding activity of *Daphnia galeata* over a period of a few months. A possible explanation for this phenomenon is offered by DeMott *et al.* (1991) who demonstrated that a number of zooplankton species will avoid grazing on toxic cyanobacteria, but continue to graze on non-toxic species. Similar results have also been shown for grazing by the phytoplanktivorous fish *Tilapia* and silver carp. Thus, grazing pressure from zooplankton and some fish may lead to the selective enrichment of toxic cyanobacterial strains over time.

It is likely that under natural conditions in water bodies, certain species and strains of zooplankton may be affected by cyanotoxins, whereas others will be unaffected. As such, cyanotoxins may influence the zooplankton community structure, especially during times when cyanobacteria are dominant within the phytoplankton.

Effects on Fish

If fish are dosed with cyanotoxins by i.p. injections or by force-feeding, they develop similar symptoms of intoxication as laboratory mammals. The question relevant for field exposure is whether cyanotoxins enter healthy fish. For example, Tencalla *et al.* (1994) showed that gastrointestinal uptake by gavage (force-feeding) caused massive hepatic necrosis followed by fish deaths, whereas immersion of adults and juveniles in contaminated water did not cause toxic effects. Other reported evidence suggests that immersion in toxic cyanobacteria or cyanotoxins may be harmful to fish. Differences in sensitivity may be pronounced between species: goldfish were found to be nearly 30 times less susceptible to i.p. microcystin than mice (Sugaya *et al.*, 1990). Release of toxic compounds from mass developments of cyanobacteria was considered to be the cause of fish kills by Penaloza *et al.* (1990).

Histopathological investigations of fish deaths during cyanobacterial blooms in the UK, indicated that the cause of death was mostly due to damage of the gills, digestive tract and liver (Rodger *et al.*, 1994). The gill damage was probably caused by the high pH induced by cyanobacterial photosynthesis activity prior to the bloom collapse, together with the higher level of ammonia arising from the decomposition of the cyanobacteria. However, gill damage may have enhanced microcystin uptake and thus led to liver necrosis. Damage to gills by dissolved microcystin-LR has been shown experimentally in *Tilapia* and trout (Garcia, 1989; Gaete *et al.*, 1994; Bury *et al.*, 1996).

Other pathological symptoms ascribed to toxic cyanobacterial blooms include damage to the liver, heart, kidney, gills, skin and spleen (Garcia, 1989; Råbergh *et al.*, 1991). Garcia (1989) and Rodger *et al.* (1994) carried out experiments on trout, while Råbergh *et al.* (1991) experimented on carp. The latter study highlighted degenerative changes in kidney tubules and glomeruli. The effect of microcystins on European carp, *Cyprinus carpio*, under natural field conditions in Australia has been described by Carbis *et al.* (1997) as atrophy of hepatocytes, gills with pinpoint necrosis, epithelial ballooning, folded lamellar tips, exfoliation of the lamellar epithelium, elevated asparate aminotransferase activity and serum bilirubin concentrations. Laboratory studies indicate that dissolved microcystins may affect fish embryos (Oberemm *et al.*, 1997) and behaviour of fish (Baganz *et al.*, 1998). Recently Gupta and Gupta (2006) reported Microcystin toxicity in freshwater fish *Heteropneustes fossilis* (Bloch).

The most definitive effect of microcystin on fish concerns Atlantic Salmon reared in net pens in coastal waters of British Columbia and Washington State, USA. As yet unidentified microcystin-producing organisms produce a progressive degeneration of the liver in salmon smolts placed into open-water net pens (Anderson *et al.*, 1993). The disease, referred to as Net Pen Liver Disease (NPLD), has resulted in significant economic losses for the mariculture industry.

Effect on Fungi

The toxic cyanobacterium *Microcystis aeruginosa* was studied earlier for its antibacterial and antialgal properties. From the present study we deduce that though *Mycrocystis* extract is not only detrimental to fungal weeds but also kill *Pleurotus* mushrooms (Singh *et al.*, 2002), yet it should not be used either for irrigating of mushroom beds or controlling agents against fungal weeds occurred during the *Pleurotus* mushroom cultivation. Because, its toxin might spoil medicinal properties of the mushrooms, thus instead of getting good health people may suffer health hazards. Vyas (2007) reported deterimental effect of *Microcystis aeuroginosa* cell extract on *Phytopthora parasitica* and some species of AM fungi.

Table 6.9: Percent Growth Inhibition *in vitro* of Test Three *Pleurotus* spp. and Two Fungal Weeds by the Extract of Cyanobacterium *Microcystis eruginosa*

Extract Conc. (ppm)	*Percent Inhibition*				
	P.sajor-caju	*P.florida*	*P.ostreatus*	*A.niger*	*Penicillium sp.*
5	20	20.1	20.4	23.1	29.4
10	25.5	25.3	25.4	29.2	38.2
25	45.1	45.5	45.2	48.2	54.1
50	78.3	78.2	78.1	82.1	92.4
75	100	100	100	100	100

Table 6.10: Sporophore Initiation and Biological Efficiency of Three Test *Pleurotus* spp. Treated with Extract of Cyanobacterium *Microcystis aeruginosa* Under *in vivo* Condition

Extract Conc. (ppm)	*P.sajor-caju*		*P.florida*		*P.ostreatus*	
	Sporophore Initiation	*BE (%)*	*Sporophore Initiation*	*BE (%)*	*Sporophore Initiation*	*BE (%)*
Control	19	82.1	21	74.2	18	71.4
5	30	50.2	33	48.1	34	46.2
10	32	40.1	36	38.1	36	35.3
25	40	25.3	42	22.2	42	20.1
50	ND	ND	ND	ND	ND	ND
75	ND	ND	ND	ND	ND	ND

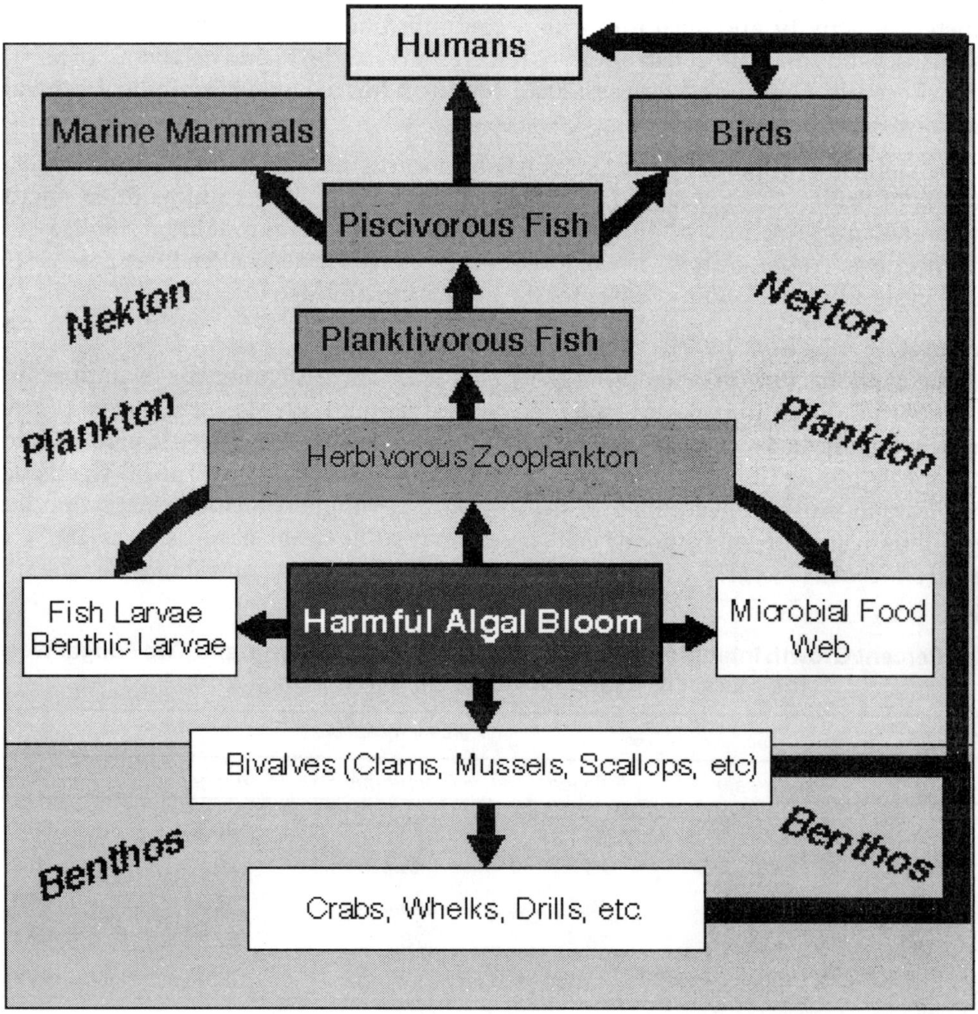

Figure 6.5

Table 6.11: Clinical Symptoms Produced by Cyanobacterial Toxins and their Median Lethal Doses (LD_{50}) (after Carmichael, 1994; Harada *et al.*, 1994)

Toxin Producing Cyanobacteria	*Types of Toxins*	*Symptoms*	*LD_{50}* (µg/Kg body wt.)*	*Test Organisms(s)*
		Neurotoxins		
Anabaena flosaquae	Anatoxin-a	Muscular fasciculation, decreased movement, collapse, cyanosis, convulsions, death	200	Mice, rat
Anabaena flosaquae	Anatoxin-a(s)	Hypersalvation, mucoid nasal, tremors, diarrhoea, cyanosis, death	20	Mice, pig, rat, duck
Aphanizomenon flosaquae, Anabaena circinalis	Saxitoxin and Neosaxitoxin	Irregular breathing, spasm, gasping, loss of coordination, tremors, death	10	Mice, rat
		Hepatotoxins		
Microcystis aenigiliosa M. viridis, M. wesenbergii, Nostoc sp., *Oscillatoria* sp., *Anabaena* sp.	Microcystins	Slow movement, increase in liver weight, hypovolemic shock, intrahepatic haemmorrage, death	50-100	Mice, rat
Nodularia spumigena	Nodularin	Slow movement, increase in liver weight, hypovolemic shock, intrahepatic haemmorrage, death	ND	Mice
Cylindrospermopsis raciborskii, Umezakia natans	Cylindros-permopsin	Liver swelling, hepatic necrosis, congestion in kidney and heart death	ND	Mice

Table 6.12: Species of Cyanophyceans with Recognized Toxin-producing Strains. The list presents species of cyanophyceans that have been confirmed lo have toxin-producing strains. The publication referred to represent selected key papers. The early publications cited cover primarily investigations of field cases of intoxications

Species	*References*
Unicellular	
Coeloshaerium kutzingianum Nag.	Fitch *et al.* (1934)
Gomphosphaeria lacustris Chod.	Gorham and Carmichael (1988)
Gumphosphaeria nageliana (Unger) Lemm.	Berg *et al.* (1986)
Microcystis areuginosa Kutz.	Hughes *et al.* (1958)
Microcystis cf. *botrys* Teil.	Berg *et al.* (1986)
Microcystis viridis (A. Br.) Lemm.	Watanabe *et al.* (1986)
Microcystis wesenbergii Kom.	Gorham and Carmichael (1988)
Synechococcus nageli sp. (strain Miami BCII 6S)	Mitsui *et al.* (1987)
Synechococcus nageli sp. (strain ATCC 18800)	Amann (1977)
Synechococcus sauvageau sp.	Lincoln and Carmichael (1981)

Contd...

Table 6.12–Contd...

Species	*References*
Multicellular	
Anabaena circinalis Rabcnh.	May and McBarron (1973)
Anabaena flos-aquae (Lyngb.) Breb.	Porter (1887)
Anabaena hassallii (Kutz.) Witttr.	Andrijuk *et al.* (1975)
Anabaena lemmermannii P. Richt.	Fitch *et al.* (1934)
Anabaena spiroides var. contracta Kleb.	Beasley *et al.* (1983)
Anabaena variabilis Kutz.	Andrijuk *et al.* (1975)
Anabaenopsis milleri Woron.	Lanaras *et al.* (1989)
Aphanizomenon flos-aquae (L.) Ralfs	Jackim and Gentile (1968)
Cylindrospermum Kutzing sp.	Sivonen *et al.* (1989)
Cylindrospermopsis raciborskii (Wolos.) Seenaya and Subba Raju	Hawkins *et al.* (1985)
Fischerella epiphytica Ghose	Ransom *et al.* (1978)
Gloeotrichia echinulata (J.E. Smith) P. Richter	Ingram and Prescott (1954)
Hapalosiphon fontinalis (Ag.) Born.	Moore *et al.* (1984)
Hormothamnion entromorphoides Grun.	Gerwick *et al.* (1986)
Lyngbya majuscula Harvey	Grauer and Arnold (1961)
Nodularia spumigena Mertens	Francis (1878)
Nostoc linckia (Roth) Born. et Flah.	Ransom *et al.* (1978)
Nostoc paludosum Kutz.	Andrijuk *et al.* (1975)
Nostoc rivulare Kutz.	Davidson (1959)
Nostoc zetterstedtii Areschoug	Mills and Wyatt (1974)
Oscillatoria acutissima Kuff.	Barchi *et al.* (1984)
Oscillatoria agardhii Gom.	Ostensvik *et al.* (1981)
Oscillatoria agardhii/rubescens group	Skulberg and Skulberg (1985)
Oscillatoria nigro-viridis Thwaites	Mynderse *et al.* (1977)
Oscillatoria vaucher sp.	Sivonen *et al.* (1989)
Oscillatoria formosa Bory (*Phormidium formosum* (Bory) Anagn. and Kom.)	Skulberg *et al.* (1992)
Pseudanabaena catenata Lauterb.	Gorham *et al.* (1982)
Schizothrix calcicola (Ag.) Gom.	Mynderse *et al.* (1977)
Scytonema pseudohofmnni Bharadw.	Moore *et al.* (1986)
Tolypothrix byssoidea (Hass.) Kirchn.	Barchi *et al.* (1983)
Trichodesmium erythraeum Ehrb.	Feldmann (1932)

Acknowledgements

D.V. gratefully acknowledge U.G.C. for awarding research award and financially support.

References

Anderson, D.M. (1994). Red tides. *Scientific American,* August, 52-58.

Anderson, R.J., Luu, H.A., Chen, D.Z.X., Holmes, C.F.B., Kent, M., LeBlanc, M., Taylor, F.J.R. and Williams, D.E. (1993). Chemical and biological evidence links microcystins to salmon "Netpen liver Disease". *Toxicon,* 31, 1315-1323.

Annila, A., Lehtimäki, J., Manila, K., Eriksson, J.E., Sivonen, K., Rantala, T.T. and Drakenberg, T. (1996). Solution structure of nodularin an inhibitor of serine/threonine specific protein phosphatases. *J. Bio. Chem.,* 271(28), 16695-16702.

Arment, A.R. and Carmichael, W.W. (1996). Evidence that microcystin is a thio-template product. *J. Phycol.,* 32, 591-597.

Azavedo, S.M.F.O., Evans, W.R., Carmichael, W.W. and Namikoshi, M. (1994). First report of microcystins from a Brazilian isolate of the cyanobacterium *Microcystis aeruginosa. J. App. Phycol.,* 6, 261-265.

Baganz, D., Staaks, G., Steinberg, C. (1998). Impact of the cyanobacteria toxin, microcystin-LR, on the behavior of zebrafish, *Danio rerio*: *Wat. Res.,* 32, 948-952.

Baker, P.D. and Humpage, A.R. (1994). Toxicity associated with commonly occurring cyanobacteria in surface waters of the Murray-Darling basin, Australia. *Aust. J. Mar. Freshwat. Res.,* 45, 773-786.

Banker, P.D., Carmeli, S., Hadas, O., Teltsch, B., Porat, R. and Sukenik, A. (1997). Identification of cylindrospermopsin in *Aphanizomenon ovalisporum* (*Cyanophyceae*) isolated from Lake Kinneret, Israel. *J. Phycol.,* 33, 613-616.

Barford, D. and Keller, J. (1994). Co-crystallization of the catalytic subunit of the serine/threonine specific protein phosphatase 1 from human in complex with microcystin-LR. *J Molecular Biol.,* 235, 763-766.

Bateman, K. P., Thibault, P., Douglas, D. J. and White, R. L. (1995). Mass spectral analyses of microcystins from toxic cyanobacteria using on-line chromatographic and electrophoretic separations. *J. Chromatog. A,* 712, 253-268.

Beattie, K.A, Kaya, K., Sano, T. and Codd, G.A. (1998). Three dehydrobutyrine (Dhb)-containing microcystins from the cyanobacterium *Nostoc* sp. *Phytochemistry,* 47(7), 1289-1292.

Benndorf, J. and Henning, M. (1989). *Daphnia* and toxic blooms of *Microcystis aeruginosa* in Bautzen Reservoir. *Int. Rev. ges. Hydrobiol.,* 74, 233-248.

Berg, K. Skulberg, O.M. and Skulberg, R. (1987). Effects of decaying toxic blue-green algae on water quality–a laboratory study. *Arch. Hydrobiol.,* 108, 549-563.

Berg, K., Skulberg, O.M., Skulberg, R., Underdal, B. and Willén, T. (1986). Observations of toxic blue-green algae (Cyanobacteria) in some Scandinavian lakes. *Acta. Vet. Scand.,* 27, 440-452.

Blackburn, S.I., Bolch, C.J., Jones, G.J., Negri, A.P. and Orr, P.T. (1997). Cyanobacterial blooms: why are they toxic? In: J.R.D. Davis (Ed.) *Managing Algal Blooms: Outcomes from the CSIRO Blue-Green Algal Research Program.* J.R.D. Davis (Ed.) CSIRO Land and Water, Canberra, 67-77.

Blackburn, S., McCausland, M.A., Bolch, C.J.S., Newman, S.J. and Jones, G.J. (1996). Effect of salinity on growth and toxin production in cultures of the bloom-forming cyanobacterium *Nodularia spumigena* from Australian waters. *Phycologia,* 35, 511-522.

Boland, M.P., Smillie, M.A., Chen, D.Z.X. and Holmes, C.F.B. (1993). A unified bioscreen for the detection of diarrhetic shellfish toxins and microcystins in marine and freshwater environments. *Toxicon*, 31, 1393-1405.

Bolch, C.J.S., Blackburn, S.I., Jones, G.J., Orr, P.T. and Grewe, P.M. (1997). Plasmid content and distribution in the toxic cyanobacterial genus *Microcystis* Kutzing ex Lemmermann (Cyanobacteria: Chroococcales). *Phycologia*, 36, 6-11.

Botes, D.P., Tuinman, A.A., Wessels, P.L., Viljoen, C.C. and Kruger, H. (1984). The structure of cyanoginosin-LA, a cyclic heptapeptide toxin from the cyanobacterium *Microcystis aeruginosa. J. Chem. Soc., Perkin Transactions*, I, 2311-2318.

Botes, D.P., Wessels, P.L., Kruger, H., Runnegar, M.T.C., Santikarn, S., Smith, R.J., Barna, J.C.J. and Williams, D.H. (1985). Structural studies on cyanoginosins-LR, -YR, -YA, and -YM, peptide toxins from *Microcystis aeruginosa. J. Chem. Soc., Perkin Transactions*, I, 2747-2748.

Bourne, D., Jones, G.J. Blakeley, R.L., Jones, A., Negri, A.P. and Riddles, P. (1996). Enzymatic pathway for the bacterial degradation of the cyanobacterial cyclic peptide toxin microcystin -LR. *Appl. Env. Microbiol.*, 62, 4086-4094.

Bruno, M., Barbini, D.A., Pierdominici, E., Serse, A.P. and Ioppolo, A. (1994). Anatoxin-a and previously unknown toxin in *Anabaena planctonica* from blooms found in Lake Mulargia (Italy). *Toxicon*, 32, 369-373.

Bumke-Vogt, C. (1998). Anatoxin-a and neurotoxic cyanobacteria in German lakes and reservoirs. *Env. Toxicol. Water Qual.* (In Press).

Bumke-Vogt, C., Mailahn, W., Retard, W. and Chorus, I. (1996). A highly sensitive analytical method for the neurotoxin anatoxin-a, using GC-ECD, and first application to laboratory cultures. *Phycologia*, 35(6, Supplement), 51-56.

Bury, N.R., Flik, G., Eddy, F.B. and Codd, G.A. (1996). The effects of cyanobacteria and the cyanobacterial toxin microcystin-LR on Ca^{2+} transport and Na^+/K^+–ATPase in Tilapia gills. *J. Exp. Biol.*, 199, 1319-1326.

Canter, H., Heaney, S.I. and Lund, J.W.G. (1990). The ecological significance of grazing on planktonic populations of cyanobacteria by the ciliate *Nassula. New Phytol.*, 114, 247-263.

Carbis, C.R., Rawlin, G.T., Grant, P., Mitchell, G.F., Anderson, J.W. and McCauley, I. (1997). A study of feral carp *Cyprinus carpio* L., exposed to *Microcystis aeruginosa* at Lake Mokoan, Australia, and possible implication on fish health. *J. Fish Diseases*, 20, 81-91.

Carbis, C.R. Simons, J.A., Mitchell, G.F., Anderson, J.W. and McCauley, I. (1994). A biochemical profile for predicting the chronic exposure of sheep to *Microcystis aeruginosa*, an hepatotoxic species of blue-green alga. *Res. Vet. Sci,*. 57(3), 310-316.

Cardellina II, J.H., Marner, F.J and Moore, R.E. (1979). Seaweed dermatitis, structure of lyngbyatoxin A. *Science*, 204, 193-195.

Carmichael, W.W. (1986). Algal toxins. In: J.A. Callow (Ed.) *Advances in Botanical Research, No. 12.* Academic Press, London, 47-101.

Carmichael, W.W. (1992). *A Status Report on Planktonic Cyanobacteria (Blue Green Algae) and their Toxins.* EPA/600/R-92/079, Environmental Monitoring Systems Laboratory, Office of Research and Development, US Environmental Protection Agency, Cincinnati, Ohio.

Carmichael, W.W. (1997). The Cyanotoxins. *Adv. Bot. Res.*, 27, 211-256.

Carmichael, W.W., Beasley, V.R., Bunner, D.L., Eloff, J.N., Falconer, I., Gorham, P., Harada, K.I., Krishnamurthy, T., Yu, M.J., Moore, R.E., Rinehart, K., Runnegar, M., Skulberg, O.M. and Watanabe, M. (1988a). Naming of cyclic heptapeptide toxins of cyanobacteria (blue-green algae). *Toxicon*, 26, 971-973.

Carmichael, W.W., Biggs, D.F. and Gorham, P.R. (1975). Toxicology and pharmacological action of *Anabaena flos-aquae* toxin. *Science*, 187, 542-544.

Carmichael, W.W., Eschedor, J.T, Patterson, G.M.L. and Moore R.E. (1988b). Toxicity and partial structure of a hepatotoxic peptide produced by the cyanobacterium *Nodularia spumigena* Mertens. *Appl. Env. Microbiol.*, 54, 2257-2263.

Carmichael, W.W., Evans, W.R., Yin, Q.Q., Bell, P. and Mocauklowski, E. (1997). Evidence for paralytic shellfish poisons in the freshwater cyanobacterium *Lyngbya wollei* (Farlow ex Gomont) comb. nov. *Appl. Environ. Microbiol.*, 63, 3104-3110.

Carmichael, W.W. and Gorham, P.R. (1981). The mosaic nature of toxic blooms of cyanobacteria. In: W.W. Carmichael (Ed.) *The Water Environment: Algal Toxins and Health*, Plenum Press, New York, 161-172.

Carmichael, W.W., Mahmood, N.A. and Hyde, E.G. (1990). Natural toxins from cyanobacteria (blue-green algae). In: S. Hall and G. Strichartz (Eds) *Marine Toxins, Origin, Structure and Molecular Pharmacology*, Vol. 418, American Chemical Society, Washington D.C., 87-106.

Carmichael, W.W., Yu, M. J., He, Z.R, He, J.W. and Yu, J-L. (1988c). Occurrence of the toxic cyanobacterium (blue-green alga) *Microcystis aeruginosa* in Central China. *Arch. Hydrobiol.*, 114, 21-30.

Chiswell, R.K., Shaw, G.R., Eaglesham, G.K., Smith, M.J., Norris, R.L., Seawright, A.A. and Moore, M.R. (1999). Stability of cylindrospermopsin, the toxin from the cyanobacterium *Cylindrospermopsis raciborskii*. Effects of pH, temperature and sunlight on decomposition. *Environ. Toxicol. Water Qual.* (In Press).

Choi, B.W., Namikoshi, M., Sun, F., Rinehart, K.L., Carmichael, W.W., Kaup, A.M., Evans, W.R. and Beasley, V.R. (1993). Isolation of linear peptides related to the hepatotoxins nodularin and microcystins. *Tetrahedron Letters*, 34, 7881-7884.

Chorus, I. (1998). Die Blaualgen ("Cyanobakterien")–eine Berliner Besonderheit welcher Zusammenhang besteht zwischen dem Massenhaften Auftreten von Blaualgen und der Nutzung von Badegewässern? In: *Nachhaltigkeit im Wasserwesen in der Mitte Europas*. Symposium 17-19 June, 1998, Berlin (In Press).

Christoffersen, K. (1996). Ecological implications of cyanobacterial toxins in aquatic food webs. *Phycologia*, 35(6 Supplement), 42-50.

Codd, G.A. and Bell, S.G. (1996). *The Occurrence and Fate of Blue-green Algae in Fresh-waters*. National Rivers Authority Research and Development Report 29, Her Majesty's Stationery Office, London, 30 pp.

Codd, G.A. and Poon, G.K. (1988). Cyanobacterial toxins. In: J. G. Gallon, and L. J. Rogers (Eds) *Proc. Phytochem. Soc. Europe*, 28, Oxford University Press, Oxford, 283-296.

Codd, G.A., Edwards, C., Beattie, K.A., Lawton, L.A., Campell, D.L. and Bell, S.G. (1995). Toxins from blue-green algae. The Pringsheim Lecture. In: W. Wiessner, E. Schnepf and R. C. Starr (Eds) *Algae, Environment and Human Affairs*. Biopress, Bristol, 1-17.

Costa, M.S. and Azevedo, S.M.F.O. (1994). Implantacao de um banco de culturas de cianoficeas toxicas. *Iheringia*, 45, 69-74 (in Portuguese).

Cousins, I.T., Healing, D.J., James, H.A., Sutton, A. (1996). Biodegradation of microcystin-LR by indigenous mixed bacterial populations. *Wat. Res.*, 30, 481-485.

Craig, M., McCready, L., Luu, H.A., Smillie, M.A., Dubord, P. and Holmes, C.F.B. (1993). Identification and characterization of hydrophobic microcystins in Canadian freshwater cyanobacteria. *Toxicon*, 31, 1541-1549.

Cremer, J. and Henning, K. (1991). Application of reversed-phase medium-pressure liquid chromatography to the isolation, separation and amino acid analysis of two closely related peptide toxins of the cyanobacterium *Microcystis aeruginosa* strain PCC 7806. *J. Chromatog.*, 587, 71-80.

Davidson, F.F. (1959). Poisoning of wild and domestic animals by a toxic waterbloom of *Nostoc rivulare* Kuetz. *J. Am. Water Works Ass.*, 51, 1277-1287.

DeMott, W.R. Zhang, Q. Carmichael, W.W. (1991). Effects of toxic cyanobacteria and purified toxins on the survival and feeding of a copepod and three species of *Daphnia. Limnol. Oceanog.*, 36(7), 1346-1357.

de Silva, E.D., Williams, D.E., Andersen, R.J., Klix, H., Holmes, C.F.B. and Alien, T.M. (1992). Motuporin, a potent protein phosphatase inhibitor isolated from the Papua New Guinea sponge *Theonella swinhoei* Gray. *Tetrahedron Letters*, 33, 1561-1564.

Devlin, J.P., Edwards, O.E., Gorham, P.R., Hunter, M.R., Pike, R.K. and Stavric, B. (1977). Anatoxin-a, a toxic alkaloid from *Anabaena flos-aquae* NCR-44h. *Cana. J. of Chem.*, 55, 1367-1371.

Dittman, E., Neilan, B.A., Erhard, M., v. Döhren, H. and Börner, T. (1997). Insertional mutagenesis of a peptide synthetase gene which is responsible for hepatotoxin production in the cyanobacterium *Microcystis aeruginosa* PCC 7806. *Mol. Microbiol.*, 26, 779-787.

Edwards, C., Beattie, K.A., Scrimgeour, C.M. and Codd, G.A. (1992). Identification of anatoxin-a in benthic cyanobacteria (blue-green algae) and in associated dog poisonings at Loch Insh, Scotland. *Toxicon*, 30, 1165-1175.

Ekman-Ekebom, M., Kauppi, M., Sivonen, K., Niemi, M. and Lepistö, L. (1992). Toxic cyanobacteria in some Finnish lakes. *Environ. Toxicol. Water Qual.*, 7, 201-213.

Elleman, T.C., Falconer, I.R., Jackson, A.R.B., and Runnegar, M.T. 1978). Isolation, characterization and pathology of the toxin from a *Microcystis aeruginosa* (=*Anacystis cyanea*) bloom. *Aust. J. Biol. Sci.*, 31, 209-218.

Endean, R., Monks, S.A., Griffith, J.K. and Llewellyn, L.E. (1993). Apparent relationship between toxins elaborated by the cyanobacterium *Trichodesmium erythraeum* and those present in the flesh of the narrow-barred Spanish mackerel *Scomberomorus commersoni. Toxicon*, 31, 1155-1165.

Entzeroth, M., Blackman, A.J., Mynderse, J.S. and Moore, R.E. (1985). Structures and stereochemistries of oscillatoxin B, 31-noroscillatoxin B, oscillatoxin D, and 30-methyloscillatoxin D. *J. Organic Chem.*, 50, 1255-1259.

Eriksson, J.E., Meriluoto, J.A., Lindholm, T. (1989). Accumulation of peptide toxin from the cyanobacterium *Oscillatoria agardhii* in the freshwater mussel *Anadonta cygnea. Hydrobiologia*, 183, 211-216.

Falconer, I.R., Choice, A. and Hosja, W. (1992). Toxicity of the edible mussel (*Mytilus edulis*) growing naturally in an estuary during a water-bloom of the blue-green alga *Nodularia spumigena J. Environ. Toxicol. Water Qual.*, 7,119-123.

Fastner, J. (1994). First evidence on the occurrence of microcystin-LR in Berlin and Brandenburg lakes. In: G. A. Codd, T. M. Jefferies, C. W. Keevil and E. Potter (Eds) *Detection Methods for Cyanobacterial Toxins.* The Royal Society of Chemistry, Cambridge, 149-151.

Fastner, J. (1998). Microcystins in German fresh-water bodies. *Environ. Toxicol. Water Qual.* (In Press).

Fastner, J., Neumann, U. and Erhard, M. (1998). Patterns of different microcystins in field samples dominated by different species of cyanobacteria. In: I. Chorus, (Ed.) *Cyanotoxins–Occurrence, Effects, Controlling Factors.* Springer Verlag, (In Press).

Foxall, T.L. and Sasner J.J. (1988). Effect of a hepatic toxin from the cyanophyte *Microcystis aeruginosa.* In: W.W. Carmichael (Ed.) *The Water Environment. Algal Toxins and Health.* Plenum Press, New York.

Francis, G. (1878). Poisonous Australian lake. *Nature*, 18, 11-12.

Fujiki, H., Suganuma, M., Suguri, H., Yoshizawa, S., Takagi, K., Nakayasu, M., Ojika, M., Yamada, K., Yasumoto, T., Moore, R.E. and Sugimura, T. (1990). New tumor promoters from marine natural products. In: S. Hall and G. Strichartz (Eds) *Marine Toxins, Origin, Structure and Molecular Pharmacology*, Vol. 418, American Chemical Society, Washington D.C., 232-240.

Fulton, R.S. (1988). Resistence to blue-green algal toxin by *Bosmina longirostris. J. Plankton Res.*, 10(4), 771-778.

Gaete, V., Canelo, E., Lagos, N., and Zambrano, F. (1994). Inhibitory effects of *Microcystis aeruginosa* toxin on ion pumps of the gill of freshwater fish. *Toxicon.*, 82, 121-127.

Gallon, J.R., Chit, K.N. and Brown, E.G. (1990). Biosynthesis of the tropane-related cyanobacterial toxin anatoxin-a: role of ornithine decarboxylase. *Phytochemistry*, 29, 1107-1 111.

Gallon J.R., Kittakoop, O. and Brown, E.G. (1994). Biosynthesis of anatoxin-a by *Anabaena flos-aquae:* examination of primary enzymic steps. *Phytochemistry*, 35, 1195-1203.

Garcia, G.O. (1989). Toxicity of the cyanobacterium *Microcystis aeruginosa* strain 7820 to trout and tilapia: a clinical and histopathological study. MSc Thesis, University of Stirling, UK.

Gathercole, P.S. and Thiel, P.G. (1987). Liquid chromatographic determination of the cyanoginosins, toxins produced by the cyanobacterium *Microcystis aeruginosa. J. Chromatog.*, 408, 435-440.

Gentile, J.H. and Maloney, T.E. (1969). Toxicity and environmental requirements of a strain of *Aphanizomenon flos-aquae* (L.) Ralfs. *Can. J. Microbiol.*, 15,165-173.

Goldberg, J., Huang, H.B., Kwon, Y.G., Greengard, P., Nairn, A.C. and Kuriyan, J. (1995). Three-dimensional structure of the catalytic subunit of protein serine/threonine phosphatase-1. *Nature*, 376, 745-753.

Gorham, P.R. (1962). Laboratory studies on the toxins produced by waterblooms of blue-green algae. *Am. J. Public Health*, 52, 2100-2105.

Gupta, U.S. and Guha, S. (2006). Microcystin toxicitiy in a freshwater fish, *Heteropneustes fossilis* (Bloch). *Current Science*, 19: 1261–1271.

Hahn, S.T. and Capra, M.F. (1992). The cyanobacterium *Oscillatoria erythraea* a potential source of toxin in the ciguatera food-chain. *Food Additives and Contaminants*, 9, 351-355.

Hammer, U.T. (1968). Toxic blue-green algae in Saskatchewan. *Can. Vet. J.*, 9, 221-229.

Harada, K-I., Matsuura, K., Suzuki, M., Watanabe, M.F., Oishi, S., Dahlem, A.M., Beasley, V.R. and Carmichael, W.W. (1990b). Isolation and characterization of the minor components associated with microcystins-LR and -RR in the cyanobacterium (blue-green algae). *Toxicon*, 28, 55-64.

Harada, K.I., Nagai, H., Kimura, Y., Suzuki, M., Park, H.D., Watanabe, M.F., Luukkainen, R., Sivonen, K. and Carmichael, W.W. (1993). Liquid chromatography/mass spectrometric detection of anatoxin-a, a neurotoxin from cyanobacteria. *Tetrahedron*, 49, 9251-9260.

Harada, K.I., Ogawa, K., Kimura, Y., Murata, H., Suzuki, M., Thorn, P.M., Evans, W.R. and Carmichael, W.W. (199 la). Microcystins from *Anabaena flos-aquae* NRC 525-17. *Chem. Res. Toxicol.*, 4, 535-540.

Harada, K.I., Ogawa, K., Matsuura, K., Murata, H., Suzuki, M., Watanabe, M.F., Itezono, Y. and Nakayama, N. (1990a). Structural determination of geometrical isomers of microcystins-LR and -RR from cyanobacteria by two-dimensional NMR spectroscopic techniques. *Chemical Research in Toxicology*. 3, 473-481.

Harada, K.-I., Ogawa, K., Matsuura, K., Nagai, H., Murata, H., Suzuki, M., Itezono, Y., Nakayma, N., Shirai, M. and Nakano, M. (1991b). Isolation of two toxic heptapeptide microcystins from an axenic strain of *Microcystis aeruginosa*, K-139. *Toxicon*, 29, 479-489.

Harada, K.I., Ohtani, I., Iwamoto, K., Suzuki, M., Watanabe, M.F., Watanabe, M. and Terao, K. (1994). Isolation of cylindrospermopsin from a cyanobacterium *Umezakia natans* and its screening method. *Toxicon*, 32, 73-84.

Harada, K-I., Tsuji, K, Watanabe, M.F. (1996). Stability of microcystins from cyanobacteria–III. Effect of pH and temperature. *Phycologia*, 35(6 Supplement), 83-88.

Hashimoto, Y., Kamiya, H., Yamazato, K. and Nozawa, K. (1976). Occurrence of a toxic blue-green alga inducing skin dermatitis in Okinawa. In: A. Ohsaka, K. Hayashi and Y. Sawai (Eds) *Animal, Plant, and Microbial Toxins*. Vol. 1, Plenum Publishing, New York, 333-338.

Hawkins, P. and Lampert, W. (1989). The effect of *Daphnia* body size on filtering rate inhibition in the presence of a filamentous cyanobacterium. *Limnol. Oceanog.*, 34(6), 1084-1089.

Hawkins, P.R., Chandrasena, N.R., Jones, G.J., Humpage, A.R. and Falconer, I.R. (1997). Isolation and toxicity of *Cylindrospermopsis raciborskii* from an ornamental lake. *Toxicon*, 35(3), 341-346.

Hawkins, P.R., Runnegar, M.T.C., Jackson, A.R.B. and Falconer, I.R. (1985). Severe hepatotoxicity caused by the tropical cyanobacterium (blue-green alga) *Cylindrospermopsis raciborskii* (Woloszynska) Seenaya and Subba Raju isolated from a domestic water supply reservoir. *Appl. Environ. Microbiol.*, 50, 1292-1295.

Hawser, S.P., Codd, G.A., Carpenter, E.J. and Capone, D.G. (1991). A neurotoxic factor associated with the bloom-forming cyanobacterium *Trichodesmium*. *Toxicon*, 29, 277-278.

Hemscheidt, T., Burgoyne, D.L. and Moore, R.E. (1995a). Biosynthesis of anatoxin-. (2*S*,4*S*)-4-hydroxyarginine as an intermediate. *J. Chem. Soc., Chemical Communications*, 205-206.

Hemscheidt T., Rapala, J., Sivonen, K. and Skulberg, O.M. (1995b). Biosynthesis of anatoxin-a in *Anabaena flos-aquae* and homoanatoxin-a in *Oscillatoria formosa*. *J. Chem. Soc., Chemical Communications*, 1361-1362.

Henning, M. and Kohl, J.G. (1981). Toxic blue-green algae water blooms found in some lakes in the German Democratic Republic. *Int. Rev. ges. Hydrobiol.*, 66, 553-561.

Henriksen, P. (1996a). Microcystin profiles and contents in Danish populations of cyanobacteria/ blue-green algae as determined by HPLC, *Phycologia, (Supplement)*, 35(6), 102-110.

Henriksen, P. (1996b). Toxic cyanobacteria/blue-green algae in Danish freshwaters. Thesis, Dept. of Phycology, University of Copenhagen.

Henriksen, P., Carmichael, W.W., An, J. and Moestrup, Ø. (1997). Detection of an anatoxin-a(s)-like anticholinesterase in natural blooms and cultures of cyanobacteria/blue-green algae from Danish lakes and in the stomach contents of poisoned birds. *Toxicon*, 35, 901-913.

Henriksen, P., Kaas, H., Damsø, L. and Moestrup, Ø. (1995). Toxic cyanobacteria (blue-green algae) in Danish freshwaters. Poster presented in 1st International Congress on Toxic Cyanobacteria (Blue-green Algae), August, Rnne, Denmark, 20-24.

Henriksen, P., Sivonen, K., Kaas, H. and Moestrup, Ø. (1998). Toxic cyanobacteria/blue-green algae in Danish freshwaters. *J. Phycol.* (In Press).

Hietala, J., Laurén-Määttä, C. and Walls, M. (1997). Sensitivity of *Daphnia* to toxic cyanobacteria: effects of genotype and temperature. *Freshwat. Biol.*, 37, 299-306.

Humpage, A.R., Rositano, J., Bretag, A.H., Brown, R., Baler, P.D., Nicholson, B.C. and Steffensen, D.A. (1994). Paralytic shellfish poisons from Australian cyanobacterial blooms. *Aust. J. Mar. Freshwater Res.*, 45(5), 761-771.

Ikawa, M., Wegener, K., Foxall, T.L. and Sasner, J.J. (1982). Comparison of the toxins of the blue-green alga *Aphanizomenon flos-aquae* with the *Gonyaulax* toxins. *Toxicon*. 20, 747-752.

Jackim, E. and Gentile, J. (1968). Toxins of a blue-green alga, similarity to saxitoxin. *Science*, 162, 915-916.

Jacobs, D., Bass, R.A. and Neilan, B.A. (1995). Isolation and characterization of candidate genes responsible for toxins biosynthesis in cyanobacteria. Abstract, Presented at the Vth Cyanobacterial Molecular Biology Workshop, July, 1995, Asilomar, CA, USA, 98, 21-25.

James, K.J., Sherlock, I.R. and Stack, M.A. (1997). Anatoxin-a in Irish freshwater and cyanobacteria, determined using a new fluorimetric liquid chromatographic method. *Toxicon*, 35(6), 963-971.

Jones, G.J. and Falconer, I.R. (1994). Factors affecting the production of toxins by cyanobacteria. Final Grant Report to the Land and Water Resources Research and Development Corporation, Canberra, Australia.

Jones, G.J. and Negri, A.P. (1997). Persistence and degradation of cyanobacterial paralytic shellfish poisons (PSPs) in freshwaters. *Wat. Res.*, 31, 525-533.

Jones, G.J. and Orr, P.T. (1994). Release and degradation of microcystin following algicide treatment of a *Microcystis aeruginosa* bloom in a recreational lake, as determined by HPLC and protein phosphatase inhibition assay. *Wat. Res.*, 28, 871-876.

Jones, G.J., Blackburn, S.I. and Parker, N.S. (1994). A toxic bloom of *Nodularia spumigena* Mertens in Orielton Lagoon, Tasmania. *Australian J. Mar. Freshwat. Res.*, 45, 787-800. 1783

Jones, G.J., Falconer, I.F. and Wilkins, R.M. (1995). Persistence of cyclic peptide toxins in dried cyanobacterial crusts from Lake Mokoan, Australia. *Environ. Toxicol. Water Qual.*, 10, 19-24.

Jungmann, D. (1992). Toxic compounds isolated from *Microcystis* PCC 7806 that are more active against *Daphnia* than two microcystins. *Limnol. Oceanog.*, 37, 1777-1783.

Jungmann, D. and Benndorf, J. (1994). Toxicity to *Daphnia* of a compound extracted from laboratory and natural *Microcystis* spp., and the role of microcystins. *Freshwat. Biol.*, 32, 13-20.

Kaya, K. and Watanabe, M.M. (1990). Microcystin composition of an axenic clonal strain of *Microcystis viridis* and *Microcystis viridis*–containing waterblooms in Japanese freshwaters. *J. App. Phycol.*, 2, 173-178.

Keleti, G. and Sykora, J.L. (1982). Production and properties of cyanobacterial endotoxins. *Appl. Environ. Microbiol.*, 43, 104–109.

Kenefick, S.L., Hrudey, S.E., Peterson, H.G. and Prepas, E.E. (1992). Toxin release from *Microcystis aeruginosa* after chemical treatment. *Wat. Sci. Tech.*, 27, 433-440.

Kerr, T., Jones, G.J. and Negri, A. (1995). Hydroxy fatty acid analysis of lipopolysaccharides in cyanobacteria. CSIRO Division of Water Resources, Australia, Summer Studentship report, 73 pp.

Kiviranta, J., Sivonen, K., Luukkainen, R., Lahti K. and Niemelä, S. I. (1991). Production and biodegradation of cyanobacterial toxins; a laboratory study, *Arch. Hydrobiol.*, 121, 281-294.

Kononen, K., Sivonen, K. and Lehtimäki, J. (1993). Toxicity of phytoplankton blooms in the Gulf of Finland and Gulf of Bothnia, Baltic Sea. In: T. J. Smayda and Y. Shimizu (Eds) *Toxic Phytoplankton Blooms in the Sea*, Elsevier Scientific Publications, Amsterdam, 269-274.

Kotak, B.G., Kenefick, S.L., Fritz, D.L., Rousseaux, C.G., Prepas, E.E. and Hrudey, S.E. (1993). Occurrence and toxicological evaluation of cyanobacterial toxins in Alberta lakes and farm dugouts. *Wat. Res.*, 27, 5-506.

Kotak, B.G., Lam, A.K.Y., Prepas, E.E., Kenefick, S.L. and Hrudey, S.E. (1995). Variability of the hepatotoxin microcystin-LR in hypereutrophic drinking water lakes. *J. Phycol.*, 31, 248-263.

Krishnamyrthy, T., Szafraniec, L., Hunt, D.F., Shabanowitz, J., Yates, J.R., Hauer, C.R., Carmichael, W.W., Skulberg, O., Codd, G.A. and Missler, S. (1989). Structural characterization of toxic cyclic peptides from blue-green algae by tandem mass spectrometry. *Proceedings of National Academy of Sciences USA.*, 86, 770-774.

Kusumi, T., Ooi, T., Watanabe, M.M., Takahashi, H. and Kakisawa, H. (1987). Cyanoviridin RR, a toxin from the cyanobacterium (blue-green alga) *Microcystis viridis. Tetrahedron Letters*, 28, 4695-4698.

Lagos, N., Liberona, J.L., Andrinolo, D., Zagatto, P.A., Soares, R.M. and Azevedo, S.M.F.Q. (1997). First evidence of paralytic shellfish toxins in the freshwater cyanobacterium *Cylindrospermopsis raciborskii* isolated from Brazil. Abstract, VIII International Conference on Harmful Algae, June 25-29, 1997, Vigo, Spain, 115.

Lahti, K., Kilponen, J., Kivimäki, A.L., Erkomaa, K. and Sivonen, K. (1996). Removal of cyanobacteria and their hepatotoxins from raw water in soil and sediment columns. In: A.L. Kivimäki and T. Suokko (Eds) *Artificial Recharge of Groundwater*. NHP Report No. 38, Helsinki, 187-195.

Lahti, K., Niemi, M. R., Rapala J. and Sivonen, K. (1997a). Biodegradation of cyanobacterial hepatotoxins–characterization of toxin degrading bacteria. Proceedings of the VII International Conference on Harmful Algae.

Lahti, K., Rapala, J., Färdig, M., Niemelä, M. and Sivonen, K. (1997b). Persistence of cyanobacterial hepatotoxin, microcystin-LR, in particulate material and dissolved in lake water. *Wat. Res.*, 31(5), 1005-1012.

Lam, A.K.-Y., Fedorak, P.M. and Prepas, E.E. (1995). Biotransformation of the cyanobacterial hepatotoxin microcystin-LR, as determined by HPLC and protein phosphatase bioassay. *Environ. Sci. Technol.*, 29(1), 242-246.

Lampert, W. (1981). Inhibitory and toxic effect of blue-green algae on *Daphnia. Rev. Ges. Hydrobiol.*, 66, 285-298.

Lanaras, T. and Cook, C.M. (1994). Toxin extraction from an *Anabaenopsis milleri*–dominated bloom. *Sci. Total Environ.*, 142, 163-169.

Lanaras, T., Tsitsamis, S., Chlichlia, C. and Cook, C.M. (1989). Toxic cyanobacteria in Greek freshwaters. *J. Applied Phycol.*, 1, 67-73.

Laurén-Määttä, C., Hietala, J. and Walls, M. (1997). Responses of *Daphnia pulex* populations to toxic cyanobacteria. *Freshwat. Biol.*, 37, 635-647.

Lawton, L.A., Beattie, K.A., Hawser, S.P., Campbell, D.L. and Codd, G.A. (1994). Evaluation of assay methods for the determination of cyanobacterial hepatotoxicity. In: G.A Codd, T.M. Jeffries, C.W. Keevil, E. Potter (Eds) *Detection Methods for Cyanobacterial Toxins.* Proceedings of the First International Symposium on Detection Methods for Cyanobacterial Toxins, September 1993 The Royal Society of Chemistry, Cambridge, 111-116.

Leeuwangh, P., Kappers, F.I., Dekker, M. and Koerselman, W. (1983). Toxicity of cyanobacteria in Dutch lakes and reservoirs. *Aquat. Toxicol.*, 4, 63-72.

Lehtimäki, J., Moisander, P., Sivonen, K. and Kononen, K. (1997). Growth, nitrogen fixation, and nodularin production by two Baltic Sea cyanobacteria. *Appl. Environ. Microbiol.*, 63, 1647-1656.

Lehtimäki, J., Sivonen, K., Luukkainen, R. and Niemelä, S.I. (1994). The effects of incubation time, temperature, light, salinity, and phosphorus on growth and hepatotoxin production by *Nodularia* strains. *Arch Hydrobiol.*, 130, 269-282.

Lindholm, T. and Meriluoto, J.A.O. (1991). Recurrent depth maxima of the hepatotoxic cyanobacterium *Oscillatoria agardhii. Can. J. Fish. Aquat. Sci.*, 48, 1629-1634.

Lukac, M. and Aegerter, R. (1993). Influence of trace metals on growth and toxin production of *Microcystis aeruginosa. Toxicon*, 31, 293-305.

Luukkainen, R., Namikoshi, M., Sivonen, K., Rinehart, K.L. and Niemelä, S.I. (1994). Isolation and identification of 12 microcystins from four strains and two bloom samples of *Microcystis* spp.: structure of a new hepatotoxin. *Toxicon.*, 32, 133-139.

Luukkainen R., Sivonen, K., Namikoshi, M., Färdig, M., Rinehart, K.L. and Niemelä, S.I. (1993). Isolation and identification of eight microcystins from 13 *Oscillatoria agardhii* strains: structure of a new microcystin. *Appl. Environ. Microbiol.*, 59, 2204-2209.

Lyck, S., Gjølme, N. and Utkilen, H. (1996). Irom-starvation increases toxicity of *Microcystis aeruginosa* CYA 22/1 (Chroococcales, Cyanophyceae). *Phycologia*, 35(6 Supplement), 120-124.

Mahmood, N.A. and Carmichael, W.W. (1986). Paralytic shellfish poisons produced by the freshwater cyanobacterium *Aphanizomenon flos-aquae* NH-5.*Toxicon*, 24, 175-186.

Mahmood, N.A. and Carmichael, W.W. (1987). Anatoxin-a(S), an anticholinesterase from the cyanobacterium *Anabaena flos-aquae* NCR-525-17.*Toxicon*, 25, 1221-1227.

Marš álek, B., LeDu, A., Turánek, J. and Machala, M. (1995). The selection of bioassays for cyanotoxin detection. In: Proceedings 1st International Congress on Toxic Cyanobacteria, 20-24 August 1995, Rønne, Denmark, 45.

Marš álek, B., Turánek J. and Hindák F. (1996). Toxic cyanobacteria in Czech and Slovak republic. In: Proceedings 1st. European Phycological Congress, Cologne, August 11-18, 1996, *Eur. J. Phycol. Suppl.*, 42.

Matsunaga, S., Moore, R.E., Niemczura, W.P. and Carmichael, W.W. (1989). Anatoxin-a(s), a potent anticholinesterase from *Anabaena flos-aquae*. *J. Amer. Chem. Soc.*, 111, 8021-8023.

Mattson, R. and Willén, T. (1986). *Toxinbildande blågröna alger i svenska insjöar 1985* (Toxin producing blue-green algae in Swedish lakes in 1985). Naturvårdsverket rapport 3096 (2)/1986. Laboratoriet för miljökontroll, Uppsala, ISSN 0282-7298.26. pp. (In Swedish, English summary).

Matveev, V., Matveeva, L., and Jones, G.J. (1994a). Study of the ability of *Daphnia carinata* King to control phytoplankton and resist cyanobacterial toxicity: implications for biomanipulation in Australia. *Aust. J. Mar. Freshwat. Res.*, 45, 889-904.

Matveev, V., Matveeva, L. and Jones, G.J. (1994b). Phytoplankton stimulation by mosquito fish in the presence of large *Daphnia*. *Ver. Verein. fur Theor. Ang. Limnol.*, 25, 2193-2197.

McDermott, C.M., Feola, R. and Plude, J. (1995). Detection of cyanobacterial toxins (microcystins) in waters of northeastern Wisconsin by a new immunoassay technique. *Toxicon.*, 33, 1433-1442.

Meissner, K., Dittman, E. and Börner, T. (1996). Toxic and non-toxic strains of the cyanobacterium *Microcystis aeruginosa* contain sequences homologous to peptide synthetase genes. *FEMS Microbiology Letters*, 135, 295-303.

Meriluoto, J.A.O., Sandström, A., Eriksson, J.E., Remaud, G., Grey Craig, A. and Chattopadhyaya, J. (1989). Structure and toxicity of a peptide hepatotoxin from the cyanobacterium *Oscillatoria agardhii*. *Toxicon*, 27, 1021-1034.

Mez, K., Hanselmann, K., Naegeli, H. and Preisig, H.R. (1996). Protein phosphatase-inhibiting activity in cyanobacteria from alpine lakes in Switzerland. *Phycologia*, 35(6 Supplement), 133-139.

Moikeha, S.N. and Chu, G.W. (1971). Dermatitis-producing alga *Lyngbya majuscula* Gomont in Hawaii. II. Biological properties of the toxic factor. *J. Phycol.*, 7, 8-13.

Moore B.S., Ohtani, I., de Koning, C.B., Moore, R.E. and Carmichael, W.W. (1992). Biosynthesis of Anatoxin-a(s). Origin of the carbons. *Tetrahedron Letters*. 33, 6595-6598.

Moore, R.E., Corbett, T., Patterson, G.M.L. and Valeriote, F.A. (1996). The search for new antitumor drugs from blue-green algae. *Current Pharm. Design*, 2, 317-330.

Moore, R.E., Lu Chen, J., Moore, B.S. and Patterson, G.M.L. (1991). Biosynthesis of microcystin-LR. Origin of the carbons in the Adda and Masp units. *J. Amer. Chem. Soc.*, 113, 5083-5084.

Moore, R.E., Ohtani, I., Moore, B.S., de Koning, C.B., Yoshida, W.Y., Runnegar, M.T. and Carmichael, W.W. (1993). Cyanobacterial toxins. *Gazzetta Chimica Italiana*, 123, 329-336.

Mynderse, J.S., Moore, R.E., Kashiwagi, M. and Norton, T.R. (1977). Antileukemia activity in the Oscillatoriaceae, isolation of debromoaplysiatoxin from *Lyngbya*. *Science*, 196, 538-540.

Nagata, S., Tsutsumi, T., Hasegawa, A., Yoshida, F., Ueno, Y. and Watanabe, M.F. (1997). Enzyme immunoassay for direct determination of microcystins in environmental water. *J. AOAC Int.*, 80, 408-417.

Namikoshi, M., and Rinehart, K.L. (1996). Bioactive compounds produced by cyanobacteria. *J. Ind. Microbiol.*, 17, 373-384.

Namikoshi, M., Choi, B.W., Sakai, R., Sun, F., Rinehart, K.L., Carmichael, W.W., Evans, W.R., Cruz, P., Munro, M.H.G. and Blunt, J.W. (1994). New nodularins, A general method for structure assignment. *J. Org. Chem.*, 59, 2349-2357.

Namikoshi, M., Rinehart, K.L., Sakai, R., Sivonen, K. and Carmichael, W.W. (1990). Structures of three new cyclic hepatotoxins produced by the cyanobacterium (blue-green alga) *Nostoc* sp. strain 152. *J. Org. Chem.*, 55, 6135-6139.

Namikoshi, M., Rinehart, K.L., Sakai, R., Stotts, R.R., Dahlem, A.M., Beasley, V.R., Carmichael, W.W. and Evans, W.R. (1992). Identification of 12 hepatotoxins from a Homer lake bloom of the cyanobacteria *Microcystis aeruginosa, Microcystis viridis, Microcystis wesenbergii;* nine new microcystins. *J. Org. Chem.*, 57, 866-872.

Namikoshi, M., Sivonen, K., Evans, W.R., Carmichael, W.W., Rouhiainen, L., Luukkainen, R. and Rinehart, K.L. (1992b). Structures of three new homotyrosine-containing microcystins and a new homophenylalanine variant from *Anabaena* sp. strain 66. *Chem. Res. Toxicol.*, 5, 661-666.

Namikoshi, M., Sivonen, K., Evans, W.R., Carmichael, W.W., Sun, F., Rouhiainen, L., Luukkainen, R. and Rinehart, K.L. (1992c). Two new L-serine variants of microcystins-LR and -RR from *Anabaena* sp. strain 202 A1 and 202 A2. *Toxicon*, 30, 1457-1464.

Namikoshi, M., Sivonen, K., Evans, W.R., Sun, F., Carmichael, W.W. and Rinehart, K.L. (1992d). Isolation and structures of microcystins from a cyanobacterial water bloom (Finland). *Toxicon*, 30, 1473-1479.

Namikoshi, M., Sun, F., Choi, B.W., Rinehart, K.L., Carmichael, W.W., Evans, W.R. and Beasley, V.R. (1995). Seven more microcystins from Homer lake cells: application of the general method for structure assignment of peptides containing,-dehydroamino acid unit(s). *J. Org. Chem.*, 60, 3671-3679.

Namikoshi, M., Yuan, M., Sivonen, K., Carmichael, W.W., Rinehart, K.L., Rouhiainen, L., Sun, F., Brittain, S. and Otsuki, A. (1998). Seven new microcystins possessing two L-glutamic acit units, isolated from *Anabaena* sp. strain 186. *Chem. Res. Toxicol.*, 11, 143-149.

Negri, A.P. and Jones, G.J. (1995). Bioaccumulation of paralytic shellfish poisoning (PSP) toxins from the cyanobacterium *Anabaena circinalis* by the freshwater mussel *Alathyria condola. Toxicon*, 33(5), 667-678.

Negri, A.P., Jones, G.J., Blackburn, S.I., Oshima, Y. and Onodera, H. (1997). Effect of culture and bloom development and of sample storage on paralytic shellfish poisons in the cyanobacterium *Anabaena circinalis. J. Phycol.*, 33, 26-35.

Negri, A.P., Jones, G.J. and Hindmarsh, M. (1995). Sheep mortality associated with Paralytic Shellfish poisoning toxins from the cyanobacterium *Anabaena circinalis. Toxicon*, 33, 1321-1329.

Oberemm, A., Fastner, J. and Steinberg, C. (1997). Effects of microcystin-LR and cyanobacterial crude extracts on embryo-larval development of zebrafish (*Danio rerio*). *Wat. Res.*, 31(11), 2918-2921.

Ohtani, I., Moore, R.E. and Runnegar, M.T.C. (1992). Cylindrospermopsin, a potent hepatotoxin from the blue-green alga *Cylindrospermopsis raciborskii. J. Amer. Chem. Soc.*, 114, 7941-7942.

Olson, T.A. (1960). Water poisoning–a study of poisonous algae blooms in Minnesota. *Amer. J. Public Health*, 50, 883-884.

Onodera, H., Oshima, Y., Henriksen, P. and Yasumoto, T. (1997a). Confirmation of anatoxin-a(s) in the cyanobacterium *Anabaena lemmermannii* as the cause of bird kills in Danish lakes. *Toxicon*, 35, 1645-1648.

Onodera, H., Satake, M., Oshima, Y., Yasumoto, T. and Carmichael, W.W. (1997b). New saxitoxin analogues from the freshwater filamentous cyanobacterium *Lyngbya wollei*. *Natural Toxins* (In Press).

Orr, P.T. and Jones, G.J. (1998). Relationship between microcystin production and cell division rates in nitrogen-limited *Microcystis aeruginosa* cultures. *Limnol. Oceanogr.*, 43(8), (In Press).

Padisák, J. (1997). *Cylindrospermopsis raciborskii* (Woloszynska) Seenayya et Subba Raju, an expanding, highly adaptive cyanobacterium: worldwide distribution and review of its ecology. *Arch. Hydrobiol., Suppl.*, 107, 563-593.

Painuly, P., Perez, R., Fukai, T. and Shimizu, Y. (1988). The structure of a cyclic peptide toxin, cyanogenosin-RR from *Microcystis aeruginosa*. *Tetrahedron Letters*, 29, 11-14.

Park, H.-D., Watanabe, M.F., Harada, K.-I., Nagai, H., Suzuki, M., Watanabe, M. and Hayashi, H. (1993a). Hepatotoxin (microcystin) and neurotoxin (anatoxin-a) contained in natural blooms and strains of cyanobacteria from Japanese freshwaters. *Natural Toxins*, 1, 353-360.

Park, H.-D., Watanabe, M.F., Harada, K.-I., Suzuki, M., Hayashi, H. and Okino, T. (1993b). Seasonal variations of *Microcystis* species and toxic heptapeptide microcystins in Lake Suwa. *Environ. Toxicol. Water Qual.*, 8, 425-435.

Parkar, D.L., Kumar, H.D., Rai, L.C. and Singh, J.B. (1997). Potassium salts inhibits growth of the cyanobacteria *Microcystis* spp. in pond water and defined media. Implication for control of Microcystin producing aquatic bloom. *Applied and Environmental Microbiology*, 2324-2329.

Penaloza, R., Rojas, M., Vila, I. and Zambrano, F. (1990). Toxicity of a soluble peptide from *Microcystis* sp. to zooplankton and fish. *Freshwat. Biol.*, 24(2), 223-240.

Prepas, E.E., Kotak, E.G., Campbell, L.M., Evans, J.C., Hrudey, S.E. and Holmes, C.F.B. (1997). Accumulation and elimination of cyanobacterial hepatotoxins by the freshwater clam *Anodonta grandis simpsoniana*. *Can. J. Fish. Aquat. Sci.*, 54, 41-46.

Prinsep, M.R., Caplan, F.R., Moore, R.E., Patterson, G.M.L., Honkanen, R.E. and Boynton, A.L. (1992). Microcystin-LR from a blue-green alga belonging to the Stignonematales. *Phytochemistry*, 31, 1247-1248.

Råbergh, C.M.I., Bylund, G., Eriksson, J.E. (1991). Histopathological effect of microcystin LR a cyclic polypeptide from the cyanobacterium *Microcystis aeruginosa* on common carp (*Cyprinus carpio L.*). *Aquat. Tox.*, 20, 131-146.

Rapala, J. and Sivonen, K. (1998). Assessment of environmental conditions that favour hepatotoxic and neurotoxic *Anabaena* spp. strains in cultured under light-limitation at different temperatures. *Microbial. Ecol.*, 36, 181-192.

Rapala, J., Lahti, K., Sivonen, K. and Niemelä, S. I. (1994). Biodegradability and adsorption on lake sediments of cyanobacterial hepatotoxins and anatoxin-a. *Letters in Applied Microbiol.*, 19, 423-428.

Rapala, J., Sivonen, K., Luukkainen, R. and Niemelä, S.I. (1993). Anatoxin-a concentration in *Anabaena* and *Aphanizomenon* at different environmental conditions and comparison of growth by toxic and non-toxic *Anabaena* strains, a laboratory study, *J. App. Phycol.*, 5, 581-591.

Rapala, J., Sivonen, K., Lyra, C. and Niemelä, S.I. (1997). Variation of microcystins, cyanobacterial hepatotoxins, in *Anabaena* spp. as a function of growth stimuli. *App. Environ. Microbiol.*, 64, 2206-2212.

Raziuddin, S., Siegelman, H.W. and Tornabene, T.G. (1983). Lipopolysaccharides of the cyanobacterium *Microcystis aeruginosa*. *Eur. J. Biochem.*, 137, 333-336.

Reinikainen, M., Kiviranta, J., Ulvi, V. and Niku-Paavola, M.L. (1995). Acute toxic effects of a novel cyanobacterial toxin on the crustaceans *Artemia salina* and *Daphnia pulex*. *Arch. Hydrobiol.*, 133, 61-69.

Repavich, W.M., Sonzogni, W.C., Standridge, J.H., Wedepohl, R.E. and Meisner L.F. (1990). Cyanobacteria (blue-green algae) in Wisconsin waters: acute and chronic toxicity. *Wat. Res.*, 24, 225-231.

Ressom, R., Soong, F.S., Fitzgerald, J., Turczynowicz, L., El Saadi, O., Roder, D., Maynard, T. and Falconer, I. (1994). *Health Effects of Toxic Cyanobacteria (Blue-green Algae)*. National Health and Medical Council, The Australian Government Publishing Service, Canberra, Australia, 108 pp.

Rinehart, K.L., Harada, K.-I., Namikoshi, M., Chen, C., Harvis, C.A., Munro, M.H.G., Blunt, J.W., Mulligan, P.E., Beasley, V.R., Dahlem, A.M. and Carmichael, W.W. (1988). Nodularin, microcystin, and the configuration of Adda. *J. Amer. Chem. Soc.*, 110, 8557-8558.

Rinehart, K.L., Namikoshi, M. and Choi, B.W. (1994). Structure and biosynthesis of toxins from blue-green algae (Cyanobacteria). *J. App. Phycol.*, 6, 159-176.

Rodger, H.D., Turnbull, T., Edwards, C. and Codd, G.A. (1994). Cyanobacterial bloom associated pathology in brown trout *Salmo trutta L.* in Loch Leven, Scotland. *J. Fish Dis.*, 17, 177-181.

Rouhiainen L., Buikema, W., Paulin, L., Fonstein, M., Sivonen, K. and Haselkorn, R. (1994). Peptide synthetase genes from hepatotoxin producing Cyanobacteria. Abstract In: VIII International Symposium on Phototrophic Prokaryotes, 10-15, September, Urbino, Italy, 26.

Rudolph-Böhner, S., Mierke, D.F. and Moroder, L. (1994). Molecular structure of cyanobacterial tumour-promoting microcystins. *FEBS Letters*, 349, 319-323.

Schwabe W., Weihe, A., Börner, T., Henning, M. and Kohl, J.G. (1988). Plasmids in toxic and nontoxic strains of the cyanobacterium *Microcystis aeruginosa*. *Current Microbiology*, 17, 133-137.

Scott, W.E. (1991). Occurrence and significance of toxic Cyanobacteria in Southern Africa. *Wat. Sci. Techno.*, 23, 175-180.

Shimizu, Y. (1986). Toxigenesis and biosynthesis of saxitoxin analogues. *Pure Appl. Chem.*, 58, 257-262.

Shimizu, Y., Norte, M., Hori, A., Genenah, A. and Kobayashi, M. (1984). Biosynthesis of saxitoxin analogues: the unexpected pathway. *J. Amer. Chem. Soc.*, 106, 6433-6434.

Shirai, M., Ohtake, A., Sano, T., Matsumoto, S., Sakamoto, T., Sato, A., Aida, T., Harada, K.I., Shimida, T., Suzuki, M. and Nakano, M. (1991). Toxicity and toxins of natural blooms and isolated strains of *Microcystis* spp. (Cyanobacteria) and improved procedure for purification of cultures. *App. Environ. Microbiol.*, 57, 1241-1245.

Shumway, S.E., van Egmond, H.P., Hurst, J.W. and Bean, L.L. (1995). Management of shellfish resources. In: G.M. Hallegraeff, D.M. Anderson, A.D. Cembella and H.O Enevoldsen [Eds] *Manual on Harmful Marine Microalgae*. UNESCO Intergovernmental Oceanographic Commission, Manuals and Guides 33, United Nations Educational, Scientific and Cultural Organization, 433-457.

Singh, N., Jain, A.K. and Vyas, D. (2002). Effect of cyanobacterial extract on oyster mushroom *Pleurotus* Sojor Caju. *J. Mushroom Res.*, 11: 85–87.

Sivonen, K. (1990a). *Toxic Cyanobacteria in Finnish Freshwaters and the Baltic Sea.* Reports from Department of Microbiology, 39, University of Helsinki, Hakapaino, Helsinki, 87 pp.

Sivonen, K. (1990b). Effects of light, temperature, nitrate, orthophosphate, and bacteria on growth of and hepatotoxin production by *Oscillatoria agardhii* strains. *App. Environ. Microbiol.*, 56, 2658-2666.

Sivonen, K. (1996). Cyanobacterial toxins and toxin production. *Phycologia,* 35(6 Supplement), 12-24.

Sivonen, K., Carmichael, W.W., Namikoshi, M., Rinehart, K.L., Dahlem, A.M. and Niemelä, S.I. (1990a). Isolation and characterization of hepatotoxic microcystin homologues from the filamentous freshwater cyanobacterium *Nostoc* sp. strain 152. *App. Env. Microbiol.*, 56, 2650-2657.

Sivonen, K., Himberg, K., Luukkainen, R., Niemelä, S.I., Poon, G.K. and Codd, G.A. (1989a). Preliminary characterization of neurotoxic blooms and strains from Finland. *Tax. Assess.*, 4, 339-352.

Sivonen, K., Kononen, K., Carmichael, W.W., Dahlem, A.M., Rinehart, K.L., Kiviranta, J. and Niemelä, S.I. (1989b). Occurrence of the hepatotoxic cyanobacterium *Nodularia spumigena* in the Baltic Sea and the structure of the toxin. *App. Env. Microbiol.*, 55, 1990-1995.

Sivonen, K., Namikoshi, M., Evans, W.R., Carmichael, W.W., Sun, F., Rouhiainen, L., Luukkainen, R. and K.L. Rinehart. (1992a). Isolation and characterization of a variety of microcystins from seven strains of the cyanobacterial genus *Anabaena. App. Env. Microbiol.*, 58, 2495-2500.

Sivonen, K., Namikoshi, M., Evans, W.R., Färdig, M., Carmichael, W.W. and Rinehart, K.L. (1992b). Three new microcystins, cyclic heptapeptide hepatotoxins, from *Nostoc* sp. strain 152. *Chem. Res. Toxicol.*, 5, 464-469.

Sivonen, K., Namikoshi, M., Evans, W.R., Gromov, B.V., Carmichael, W.W. and Rinehart, K.L. (1992c). Isolation and structures of five microcystins from a Russian *Microcystis aeruginosa* strain CALU 972. *Toxicon.*, 30,1481-1485.

Sivonen, K., Namikoshi, M., Luukkainen, R., Färdig, M., Rouhiainen, L., Evans, W.R., Carmichael, W.W., Rinehart, K.L. and Niemelä, S.I. (1995). Variation of cyanobacterial hepatotoxins in Finland. In: M. Munawar and M. Luotola (Eds) *The Contaminants in the Nordic Ecosystem, Dynamics, Processes and Fate.* Ecovision World Monograph Series, SPB Academic Publishing, Amsterdam, 163-169.

Sivonen, K., Niemelä, S.I., Niemi, R.M., Lepistö, L., Luoma, T.H. and Räsänen, L.A. (1990b). Toxic cyanobacteria (blue-green algae) in Finnish fresh and coastal waters. *Hydrobiologia,* 190, 267-275.

Sivonen, K., Skulberg, O.M., Namikoshi, M., Evans, W.R., Carmichael, W.W. and Rinehart, K.L. (1992d). Two methyl ester derivates of microcystins, cyclic heptapeptide hepatotoxins, isolated from *Anabaena flos-aquae* strain CYA 83/1. *Toxicon.*, 30, 1465-1471.

Skulberg, O.M., Carmichael, W.W., Anderson, R.A., Matsunaga, S., Moore, R.E. and Skulberg, R. (1992). Investigations of a neurotoxic Oscillatorialean strain (cyanophyceae) and its toxin. Isolation and characterization of homoanatoxin-a. *Env. Toxicol. Chem.*, 11, 321-329.

Skulberg, O.M., Underdal, B. and Utkilen, H. (1994). Toxic waterblooms with cyanophytes in Norway–current knowledge. *Algol. Stud.*, 75, 279-289.

Smith, C. and Sutton, A. (1993). The persistence of anatoxin-a in reservoir water. Foundation for Water Research, UK Report No. FR0427.

Snell, T.W. (1980). Blue-green algae and selection in rotifer populations. *Oecologia Berlin*, 46, 343-346.

Stevens, D.K. and Krieger, R.I. (1991). Stability studies on the cyanobacterial nicotinic alkaloid anatoxin-a. *Toxicon,* 29, 167-179.

Stoner, R.D., Adams, W.H., Slatkin, D.N. and Siegelman, H.W. (1989). The effects of single L-amino acid substitutions on the lethal potencies of the microcystins. *Toxicon.,* 27, 825-828.

Sugaya, Y., Yasuno, M. and Yani, T. (1990). Effects of toxic *Microcystis viridis* and isolated toxins on goldfish. *Jpn J. Limnol.,* 51(3), 149-153.

Sugiura, N. Nishimura, O. Inamori, Y. Ouchiyama, T. Sudo, R. (1997). Grazing characteristics of musty-odour-compound-producing *Phormidium tenue* by a microflagellate, *Monasguttula. Wat. Res.,* 31(11), 2792-2796.

Swoboda, U.K., Dow, C.S., Chaivimol, J., Smith, N. and Pound, B.P. (1994). Alternatives to the Mouse Bioassay for Cyanobacterial Toxicity Assessment. In: G.A. Codd, T.M. Jeffries, C.W. Keevil and E. Potter (Eds) *Detection Methods for Cyanobacterial Toxins.* Proceedings of the First International Symposium on Detection Methods for Cyanobacterial Toxins, September 1993, The Royal Society of Chemistry, 106-110.

Takenaka, S. and Watanabe, M.F. (1997). Microcystin-LR degradation by *Pseudomonas aeruginosa* alkaline phosphatase. *Chemosphere,* 34, 749-757.

Tencalla, F.G., Dietrich, D.R. and Schlatter, C. (1994). Toxicity of *Microcystis aeruginosa* peptide toxin to yearling rainbow trout (*Oncorhynchus mykiss*). *Aqua. Toxicol.,* 30(3), 215-224.

Todorova, A. and Jttner, F. (1996). Ecotoxicological analysis of nostocyclamide, a modified cyclic hexapeptide from *Nostoc. Phycologia,* 35(6 Supplement), 183-188.

Törökné, A.K. (1991). Toxin producing cyanobacteria in Hungarian freshwaters. Doctoral Dissertation, Budapest, Hungary.

Törökné, A.K. (1997). Interlaboratory trial using Thamnotox kit for detecting cyanobacterial toxins. In: *Proceedings of VIII International Conference on Harmful Algae,* Vigo, Spain, 114.

Trimurtulu, G., Ogino, J., Helzel, C.E., Husebo, T.L., Jensen, C.M., Larsen, L.K., Patterson, G.M.L., Moore, R.E., Mooberry, S.L., Corbett, T.H. and Valeriote, F.A. (1995). Structure determination, conformational analysis, chemical stability studies, and antitumor evaluation of the cryptophycins. Isolation of 18 new analogs from *Nostoc* sp. strain GSV 224. *J. Amer. Chem. Soc.,* 117, 12030-12049.

Tsuji, K., Naito, S., Kondo, F., Ishikawa, N., Watanabe, M.F., Suzuki, M., Harada, K-I. (1993). Stability of microcystins from cyanobacteria: Effect of light on decomposition and isomerization. *Environ. Sci. Technol.* 28, 173-177.

Tsuji, K., Setsuda, S., Watanuki, T., Kondo, F., Nakazawa, H., Suzuki, M. and Harada, K.I. (1996). Microcystin levels during 1992-95 for lakes Sagami and Tsukui Japan. *Natural Toxins,* 4, 189-194.

Utkilen, H. and Gjølme, N. (1992). Toxin production by *Microcystis aeruginosa* as a function of light in continuous cultures and its ecological significance. *App. Environ. Microbiol.,* 58, 1321-1325.

Utkilen, H. and Gjølme, N. (1995). Iron-stimulated toxin production in *Microcystis aeruginosa. App. Environ. Microbiol.,* 61, 797-800.

Ueno, Y., Nagata, S., Tsutsumi, T., Hasegawa, A., Watanabe, M.F., Park, H.D., Chen, G.C., Chen, G. and Yu, S.Z. (1996a). Detection of microcystins, a blue-green algal hepatotoxin, in drinking water sampled in Haimen and Fusui, endemic areas of primary liver cancer in China, by highly sensitive immunoassay. *Carcinogenesis,* 17, 1317-1321.

Ueno, Y., Nagata, S., Tsutsumi, T., Hasegawa, A., Yoshida, F., Suttajit, M., Putsch M. and Vasconcelos, V. (1996b). Survey of microcystins in environmental water by a highly sensitive immunoassay based on monoclonal antibody. *Natural Toxins*, 4, 271-276.

Van der Westhuizen, A.J. and Eloff, J.N. (1983). Effect of culture age and pH of culture medium on the growth and toxicity of the blue green alga *Microcystis aeruginosa. Zeit. Planzenphysiol.*, 110, 157-163.

Van der Westhuizen, A.J. and Eloff, J.N. (1985). Effect of temperature and light on the toxicity and growth of the blue-green alga *Microcystis aeruginosa* (UV-006). *Planta*, 163, 55-59.

Van der Westhuizen, A.J., Eloff, J.N. and Krüger, G.H.J. (1986). Effect of temperature and light (fluence rate) on the composition of the toxin of the cyanobacterium *Microcystis aeruginosa* (UV-006). *Arch. Hydrobiol.*, 108, 145-154.

Vasconcelos, V.M. (1994). Toxic cyanobacteria (blue-green algae) in Portuguese freshwaters. *Arch Hydrobiol.*, 130, 439-451.

Vasconcelos, V.M., Sivonen, K., Evans, W.R., Carmichael, W.W. and Namikoshi, M. (1995). Isolation and characterization of microcystins (heptapetide hepatotoxins) from Portuguese strains of *Microcystis aeruginosa* Kutz emend Elekin. *Arch. Hydrobiol.*, 134, 295-305.

Vasconcelos, V.M., Sivonen, K., Evans, W.R., Carmichael, W.W. and Namikoshi, M. (1996). Microcystin (heptapeptide hepatotoxins) diversity in cyanobacterial blooms collected in Portuguese freshwaters. *Wat. Res.*, 30, 2377-2384.

Vezie, C., Brient, L., Sivonen, K., Betru, G., Lefeuvre, J.C. and Salkinoja-Salonen, M. (1997). Occurrence of microcystins containing cyanobacterial blooms in freshwaters of Brittany (France). *Arch. Hydrobiol.*, 139, 401-413.

Vezie, C., Brient, L., Sivonen, K., Bertru, G., Lefeuvre, J.C. and Salkinoja-Salonen, M. (1998). Variation of microcystin content of cyanobacterial blooms and isolated strains in Grand-Lieu lake (France). *Microbiol. Ecol.*, 35, 126-135.

Von Elert, E. and Juttner, F. (1996). Factors influencing the allelopathic activity of the planktonic cyanobacterium *Trichormus doliolum. Phycologia*, 35(6 Supplement), 68-73.

Vyas, D. (2007). Fungicidal activity of cyanobacterium microcystis aeruginosa. Abstract in National Symposium on Plant Pathogens: Exploitation and Management and 59th Annual Meeting of Indian Phytopathology, pp. 66.

Watanabe, M. F. (1996). Production of Microcystins. In: M.F. Watanabe, H. Harada, W. W. Carmichael and H. Fujiki, (Eds) *Toxic Microcystis*. CRC Press, London, 262 pp.

Watanabe, M.F. and Oishi, S. (1980). Toxicities of *Microcystis aeruginosa* collected from some lakes, reservoirs, ponds and moat in Tokyo and adjacent regions. *J. Limnol.*, 41, 5-9.

Watanabe, M.F. and Oishi, S. (1985). Effects of environmental factors on toxicity of a cyanobacterium (*Microcystis aeruginosa*) under culture conditions. *App. Environ. Microbiol.*, 49, 1342-1344.

Watanabe, M.F., Harada, K.I., Matsuura, K., Oishi, S., Watanabe, Y. and Suzuki, M. (1989). Heptapeptide toxins contained in natural samples of *Microcystis* species. *Tox. Assess.*, 4, 487-497.

Watanabe M.F., Oishi, S., Harada, K-.I., Matsuura, K., Kawai, H. and Suzuki, M. (1988). Toxins contained in *Microcystis* species of cyanobacteria (blue-green algae). *Toxicon*, 26, 1017-1025.

Watanabe, M.F., Oishi, S., Watanabe, Y. and Watanabe, M. (1986). Strong probability of lethal toxicity in the blue-green alga *Microcystis viridis* Lemmermann. *J. Phycol.*, 22, 552-556.

Watanabe, M.F., Park, H-D., Kondo, F., Harada, K-I., Hayashi, H. and Okino, T. (1997). Identification and estimation of microcystins in freshwater mussels. *Nat. Toxins*, 5, 31-35.

Watanabe, M.M., Kaya, K. and Takamura, N. (1992). Fate of the toxic cyclic heptapeptides, the microcystins, from blooms of *Microcystis* (cyanobacteria) in a hypertrophic lake. *J. Phycol.*, 28, 761-767.

Watanabe, M.M., Zhang, X and Kaya, K. (1996). Fate of toxic cyclic heptapeptides, microcystins, in toxic cyanobacteria upon grazing by the mixotrophic flagellate *Poterioochromonas malhamensis* (Ochromonadales, Chrysophyceae). *Phycologia*, 35(6 Supplement), 203-206.

Weckesser, J. and Drews, G. (1979). Lipopolysaccharides of photosynthetic prokaryotes. *Ann. Rev. Microbiol.*, 33, 215-239.

Weckesser, J., Martin, C. and Jakobi, C. (1996). Cyanopeptolins, depsipeptides from cyanobacteria. *System. App. Microbiol.*, 19, 133-138.

Weise, G., Drews, G., Jann, B. and Jann, K. (1970). Identification and analysis of lipopolysaccharide in cell walls of the blue-green algae *Anacystis nidulans*. *Arch. Microbiol.*, 71, 89-98.

Welker, M. and Steinberg, C.E.W. (1998). Indirect photolysis of cyanotoxins: one possible mechanism of their low persistence. *Wat. Res.* (In Press).

Wicks, R.J. and Thiel, P.G. (1990). Environmental factors affecting the production of peptide toxins in floating scums of cyanobacterium *Microcystis aeruginosa* in a hypertrophic African reservoir. *Environ. Sci. Technol.*, 24, 1413-1418.

Wiedner, C. and Nixdorf, B. (1997). *Verbreitung und Steuerung der Entwicklung von toxischen und nicht-toxischen Cyanobakterien in ostbrandenburgischen Gewässern unterschiedlicher Trophie und Hydrographie im Zuge der reduzierten Belastung*. WaBoLu-Hefte 4/97, Berlin ISSN 0175-4211.

Willn, T. and Mattsson, R. (1997). Water-blooming and toxin-producing cyanobacteria in Swedish fresh and brackish waters, 1981-1995. *Hydrobiologia*, 353, 181-192.

Williams, D.E., Craig, M., Dawe, S.C., Kent, M.C., Holmes, C.F.B. and Anderson, R.J. (1997). Evidence for a covalently bound form of microcystin-LR in salmon larvae and dungeness crab larvae. *Chem. Res. Toxicol.*, 10, 463-469.

Yanni, Y.G. and Carmichael, W.W. (1997). Screening of cyanobacteria isolated from soil, rice fields and water resources of the Nile Delta for production of cyanotoxins. Abstract, In: VIII International Conference on Harmful Algae, 25-29 June, 1997 Vigo, Spain, 216.

Yoo, S., Carmichael, W.W., Hoehn, R.C. and Hrudey, S.E. (1995). *Cyanobacterial (Blue-green Algal) Toxins: A Resource Guide.* American Water Works Association Research Foundation, Denver, Colorado.

Zhang, Q.X., Carmichael, W.W., Yu, M.J. and Li, S.H. (1991). Cyclic peptide hepatotoxins from freshwater cyanobacterial (blue-green algae) waterblooms collected in central China. *Env. Toxicol. Chem.*, 10, 313-321.

Zhang, X. Watanabe, M.M., Inouye, I. (1996). Light and electron microscopy of grazing by flagellate *Poterioochromonas malhamensis* (Chrysophyceae) on a range of phytoplankton taxa. *J. Phycol.*, 32, 37-46.

Microbial Biotechnology and Ecology (2011) *Pages* **150–161**
Editors: **Deepak Vyas, G.S. Paliwal, P.K.Khare and R.K Gupta**
Published by: **DAYA PUBLISHING HOUSE, NEW DELHI**

Chapter 7

The Toxins of Cyanobacteria

☆ ***Mukesh Kumar***

Department of Botany, Sahu Jain P.G. College, Najibabad – 246 763, U.P.

ABSTRACT

Cyanobacteria is a group of extraordinarily diverse Gram-negative prokaryotes that originated 3.5 billion years ago on the earth. They are prokaryotes with oxygen evolving photosynthetic system and are found in a wide range of environmental conditions including extreme ones. They constitute a larger and fascinating group of photosynthetic and nitrogen fixing prokaryotes. They colonize most aquatic and terrestrial ecosystems and are considered to be one of the potential organisms which can be useful to mankind in various ways. The morphology of Cyanobacteria ranges from unicellular to branch filamentous with a maximum of three cell types: vegetative cells, akinetes and heterocysts. Massive growth of Cyanobacteria in ponds, lakes, reservoirs or other freshwater systems has become serious water quality problem which also threaten human and animal health. Blooms can discolour water, form surface scums, produce unpleasant taste and odor to water, and create problems to aquatic life. They may vary in colour from green to blue, red, brown, dark green or black. Some of the cyanobacterial species are harmless while other produces toxins affecting nervous system and the liver of the animals. Many cyanobacterial species produce toxins that affect animals and humans. Cyanobacterial toxins are the group of organic compounds with very diverse chemical structure. They are divided into two groups *i.e.* cyanotoxins and biotoxins. Cyanotoxins are not lethal for people and animals. They are enzymes, antibiotics and anticarcinogenic factors with very complicated chemical structure. Biotoxins are very toxic for people and can even cause lethal effect. Cyanobacterial toxins can be controlled by the use of chemicals, biological means or by the use of microorganisms.

Introduction

The Cyanobacteria

Cyanobacteria were among the pioneer organisms of the early earth. They probably evolved as a group of organisms about 2,000 million years before the eukaryotic micro algae (Bisby 1995). They have the distinction of being the oldest known fossils, more than 3.5 billion years old. It is a group of extraordinarily diverse gram-negative prokaryotes that originated 3.5 billion years ago. Their diversity ranges from unicellular to multicellular, coccoid to branched filaments, nearly colourless to intensely pigmented, autotrophic to heterotrophic, psychrophilic to thermophilic, acidophilic to alkylophilic, planktonic to barophilic, freshwater to marine including hypersaline (salt pans). They are found both free living and as endosymbionts. Adapted to a wide range of environmental factors including extreme ones, these colonize most aquatic and terrestrial ecosystems. They are considered to be one of the potential organisms which can be useful to mankind in various ways. Based on the International Code of Botanical Nomenclature the class cyanophyceae contains about 150 genera and 2000 species. The micro algae investigated by Phycologists under ICBN (Greuter *et al.*, 1994) included organisms of both eukaryotic and prokaryotic cell types. The blue-green algae, constituted the largest group of the later category. The prokaryotic nature of these organisms and their fairly close relationship with eubacteria made work under provisions of the International Code of Botanical Nomenclature (Sneath 1992, Rippka *et al.*, 1999, Waterbury 1992).

Cyanobacteria are prokaryotes with oxygen evolving photosynthetic system and are found in a wide range of environmental conditions. Cyanobacteria inhabit a wide range of environments and ecological habitats, constitute a larger and fascinating group of photosynthetic and nitrogen fixing prokaryotes. Their morphology ranges from unicellular to branch filamentous with a maximum of three cell types: vegetative cells, akinetes and heterocysts. They occur naturally in ponds and other water reservoirs. The most important factor, which favors their distribution widely through out the world is their ability to survive in the extreme conditions of the environment. Along with this some environmental factors also play an important role in the growth of cyanobacteria as light, temperature, pH etc. The recent trend among microbiologist is to call these organisms cyanobacteria in order to emphasize their close relationship with bacteria. In fact, they are best defined as microorganisms that contain a typically prokaryotic cell, a photosynthetic apparatus structurally and functionally similar to that located within chloroplast of the eukaryotic red algae. One general attribute of most cyanobacteria is that they are morphologically conservative but metabolically versatile. Conceptual advances in microbiology during the 20th century included the realization that a discontinuity exits between prokaryotic and eukaryotic cellular organisms within the organization of their cells.

Cyanobacterial Toxins

A toxin is a poisonous substance produced by living cells or organisms. Although technically man is a living organism, man-made substances created by artificial processes usually aren't considered toxins by this definition. A toxin is simply a poison that can affect the body by internal or external means. A toxin can be a chemical which occurs naturally or in synthetic form. They can be small molecules, peptides, or proteins that are capable of causing disease on contact with or absorption by body tissues interacting with biological macromolecules such as enzymes or cellular receptors. Toxins vary greatly in their severity, ranging from usually minor and acute (as in a bee sting) to almost immediately deadly (as in botulinum toxin). External toxins, either chemical or microbial, enter the body through food, water, air, or physical contact with the skin or mucous membranes. Internal toxins, as the free radicals, are produced inside the body through normal metabolic processes or through the decomposition of foods in the small and large intestines. Bacterial toxins and yeast overgrowth can also form in cases of chronic constipation.

More than 120,000 human-made chemicals have been introduced into the environment, in one form or another, and this number continues to grow each year at a phenomenal rate. At the same time, microbial toxins, being influenced by the vast numbers of chemicals, are mutating beyond belief. Each category of microbes produces species that generates toxins in host cells. Evidence is proving a definitive link between the accumulation of toxins in body tissues and the development of chronic diseases. Massive growth of cyanobacteria in ponds, lakes, reservoirs or other freshwater systems has become serious water quality problem which also threaten human and animal health. Blooms of *Microcystis* species are known as one of the most common worldwide problems of freshwater bodies. Blooms can discolour water, form surface scums, produce unpleasant taste and odor to water, and create problems to aquatic life. They may vary in colour from green to blu, , red, brown, dark green or black. Species of cyanobacteria or blue-green algae may dominate and increase excessively in water bloom conditions.

Some of the cyanobacterial species are harmless while other produces toxins affecting nervous system and the liver of the animals. Many cyanobacterial species produce toxins that affect animals and humans. Consumption of contaminated water by people in any form may cause cyanobacterial toxicity. The most frequently adverse health affects are caused by drinking water containing the toxins or by recreational water contact. Cyanobacteria are particularly troublesome organisms because apart from producing large amount of biomass, they secrete very toxic substances to water that are known as cyanobacterial toxins. Any release of toxin into surrounding water can present a significant hazard to humans drinking the same water.

Cyanobacterial toxins are the group of organic compounds with very diverse chemical structure. They are divided into two groups *i.e.* cyanotoxins and biotoxins. Cyanotoxins are not lethal for people and animals. They are enzymes, antibiotics and anticarcinogenic factors with very complicated chemical structure. Biotoxins are very toxic for people and can even cause lethal effect. They are divided into neurotoxins (affecting nervous system), hepatotoxins (affecting liver) and dermotoxins (affecting skin).

High pH of water, scarcity of CO_2 dissolved in water and high amount of phosphorus is conductive to the production of the biomass of cyanobacteria (Carmichael 1994, Gajdek 2000). Cyanobacteria survive in the periods of unfavorable conditions in the form of spores and then, in favorable conditions quickly achieve quantitative mass development. In this time they give water distinct colour, referred to as bloom.

Production and Structure of Cyanobacterial

One of the groups of toxins produced and released by cyanobacteria is called microcystin named from the species *Microcystis aeruginosa*. The microcystins are a group of cyclic heptapeptide hepatotoxins produced by a number of cyanobacterial genera. The peptide ring is made up of two protein amino acids and five non-protein amino acids. It is the two protein amino acids that distinguish microcystin types from each other, while the other amino acids are relatively constant.

By using amino acid single letter code classification, each microcystin is designated a name depending on the variable amino acids which complete their structure. For instance, one of the most common toxins found in water supplies around the world, microcystin-LR contains the amino acids Leucine (L) and Arginine (R) in these variable positions.

The most common of the toxins, microcystins are also the ones most often responsible for poisoning animals and humans who come into contact with toxic blooms. They are extremely stable in water because of their chemical structure, surviving in both warm and cold water and can tolerate radical changes in water chemistry, including pH. To date, approximately 75 different kinds of microcystins have been discovered.

These hepatotoxins inhibit the protein phosphatases inside hepatocytes. They damage the liver by affecting the maintenance of the cytoskeleton by disrupting the balance of phosphate groups on cytoskeletal proteins. The hepatotoxins cause the cytoskeleton to collapse, causing the hepatocytes to collapse inwards. The hepatocytes and also the capillary cells then pull apart, spilling blood into the liver. The blood pools in the liver, causing death.

Another cyanobacterial toxin is *Nodularia spumigena*, characterized as Nodularin. In 1878, George Francis issued the first scientific report of the potent toxicity of Nodularin cyanobacterial blooms. The structure of nodularin is closely related to that of the potent cyclic heptapeptide hepatotoxins, the microcystins. The difference lies in that nodularin is composed of only five amino acids in the peptide ring. Not only are the structures of these two cyanobacterial toxins similar, but they also show the same hepatotoxic effects through the potent inhibition of protein phosphatases.

It is speculated that low-level exposure to these toxins may promote the development of cancer in the liver and other chronic disorders of the gastrointestinal tract. This is likely to occur because the protein phosphatases that are inhibited by hepatotoxins play an important role in regulating cell division. While not initiating the cancers, these toxins have been shown to hasten the development of cancer in animals. Carmichael suggests that "the extraordinarily high rates of liver cancer in parts of China may be tied to the cyanobacterial toxins in water."

Analytical testing confirms the amount of cyanobacteria toxin in water samples. In recent years, HPLC, ELISA, and the protein phosphatase assay have made the quantification of total and individual toxins possible. Lyophilized cultures and bloom samples are used to determine the quantitative concentration of toxin. Results are expressed as milligrams or micrograms of toxin per gram of dry weight.

Microcystin LR

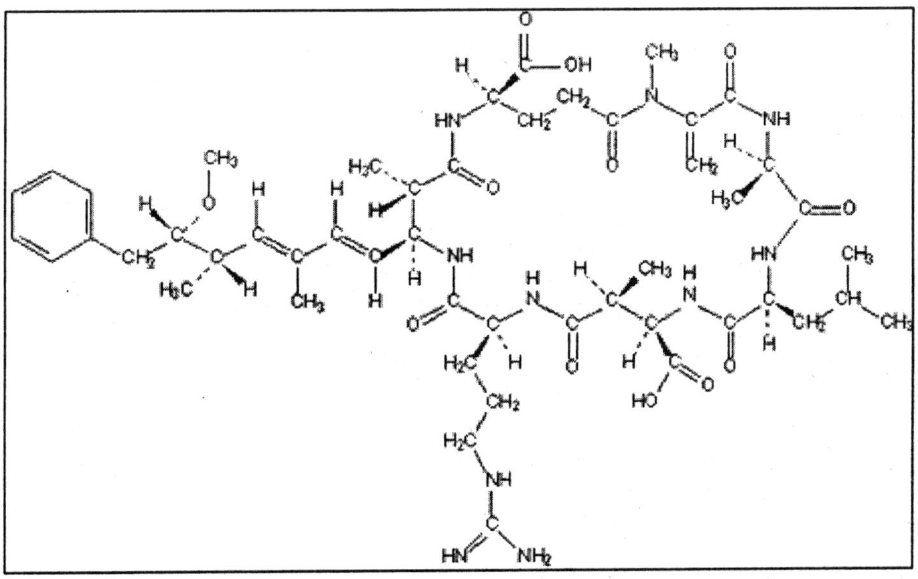

Microcystin LR

Potent inhibitor of protein phosphatase types 1 and 2A; has no effect on protein kinase.

Microcystin RR

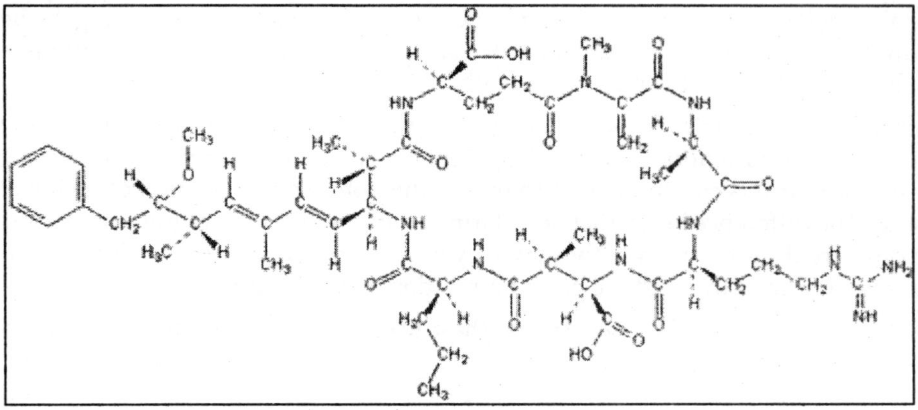

Microcystin RR
Inhibitor of protein phosphatase type 2A with lower toxicity than microcystin LR.

Microcystin LA

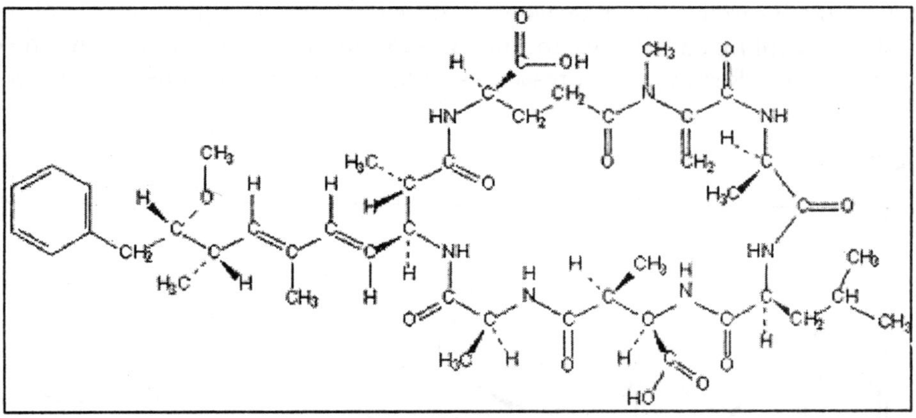

Microcystin LA
Inhibitor protein phosphatase type 2A and protein phosphatase type 3 more potently than protein phosphatase type 1 (order of potency PP2A > PP3 > PP1)

Cyanobacteria are capable of producing two kinds of toxin, the cyclic peptide hepatotoxin and the alkaloid neurotoxin. Serious illness such as hepatoenteritis, a symptomatic pneumonia and dermatitis may result from consumption of, or contact with water contaminated with toxin producing cyanobacteria (Hawkins *et al.*, 1985, Turner *et al.*, 1990). The neurotoxins include anatoxin-a, a depolarizing neuromuscular blocking agent; anatoxin-a [s], an anti–cholinesterase; and saxitoxin and neosaxitoxin that inhibit nerve conduction by blocking sodium channels (Carmicheal 1994).

Toxicity due to Cyanobacterial Toxins

Cyanobacterial toxins in drinking or recreational waters are also hazardous to human health. In India, the warm water temperature promotes dense *Microcystis* growth almost throughout the year.

Microcystin YR

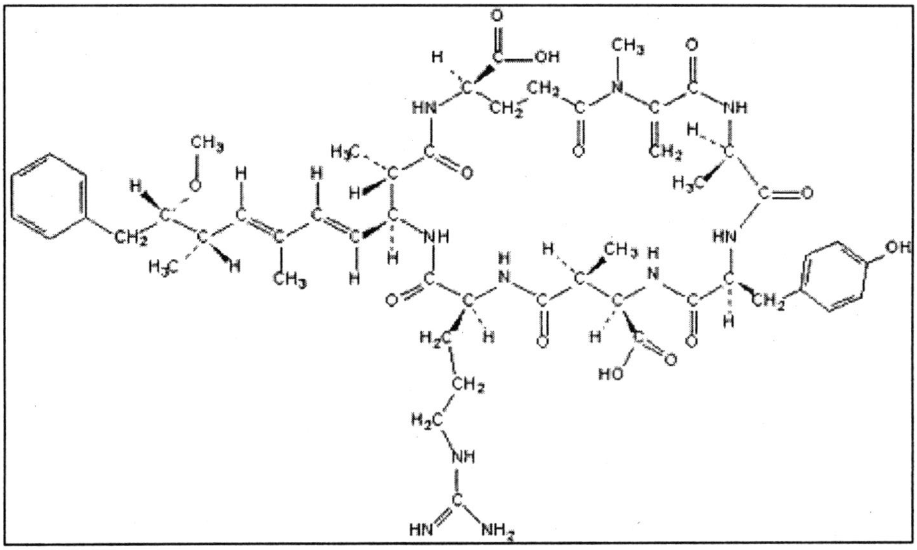

Microcystin YR
Inhibitor protein phosphatase type 2A and protein phosphatase type 1.

Cylindrospermopsin

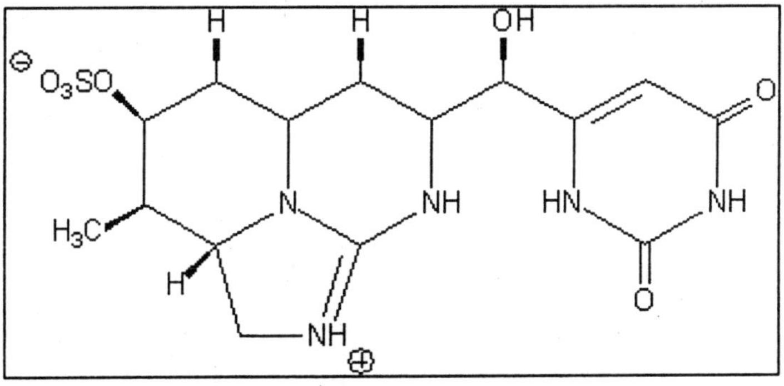

Cylindrospermopsin
Inhibition of protein synthesis. Target organ = liver

Although there is no evidence of human or animal lethality associated with these blooms, the occurrence of toxic blooms has been described in a few cases. In an epidemiologic study, it was found that patients with a history of bathing in *Microcystis*-infested ponds developed acute rhinosporidiosis, a disease caused by "pathogenic" strains of *Microcystis*. Several *Microcystis*-dominant phytoplankton materials collected from Central India have also been found to be toxic to the crustacean zooplankton *Moina macrocopa*. Accordingly, there have been many reports of the intoxication of birds, fish and other animals by cyanobacterial toxins (Vasconceles *et al.*, 2001, Alonso-Andicoberry *et al.*, 2002, Best *et al.*, 2002, Romanowska-Duda *et al.*, 2002, Krienitz *et al.*, 2003). As stated before, blooms of cyanobacteria

Nodularin

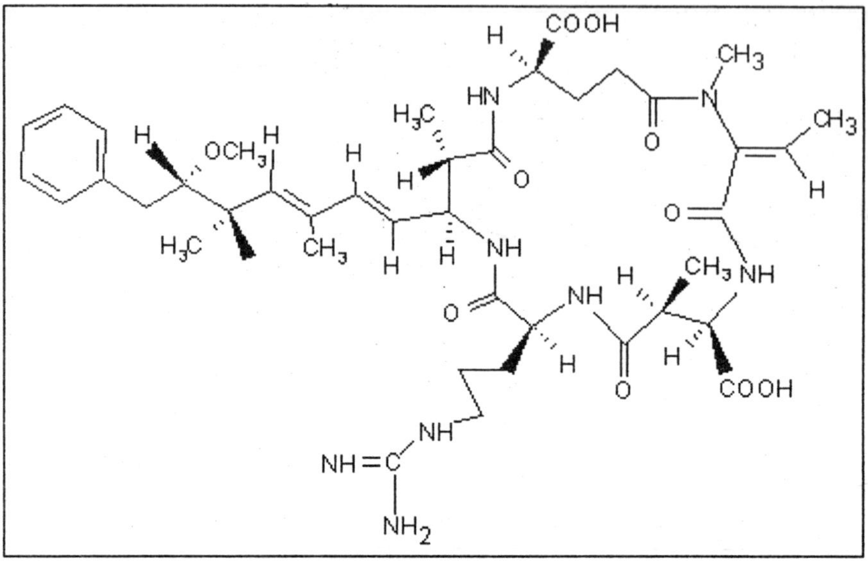

Nodularin
Potent inhibitor of protein phosphatases types 1 and 2A.

usually follow enrichment by nutrients such as phosphates and nitrates in the water. Most of these nutrients are derived from human wastes such as sewage and detergents, industrial pollution, run-off of fertilizers from agricultural land, and the input of animal or bird wastes from intensive farming (Bell and Codd 1994). Illnesses caused by cyanobacterial toxins to humans fall into three categories; gastroenteritis and related diseases, allergic and irritation reaction, and liver diseases. Microcystins have also been implicated as tumour-promoting substances (An and Carmichael 1994, Trogen *et al.*, 1996).

Impact of Toxins on Health of Man and his Pets

Microcystins are hepatotoxins (toxins that acts upon the liver) and known tumor promoters. If people drink water contaminated by microcystins, symptoms of exposure include nausea, vomiting and, in very rare but severe cases, acute liver failure. Reported health effects from cyanobacteria in humans are highly uncommon in the United States. Although the likelihood of people being affected by a *Microcystis* bloom is low, minor skin irritation can occur with contact, and gastrointestinal discomfort can also occur if water from a bloom is ingested.

Through the recreational use of contaminated water, cyanobacterial blooms of *Microcystis*, *Anabaena*, and others have been linked to incidence of human illness in many countries, but no fatalities have been reported (Lambert *et al.*, 1994b). In Canada, human illnesses have been reported in Saskatchewan, with symptoms including stomach cramps, vomiting, diarrhoea, fever, headache, pains in muscles and joints, and weakness (Dillenberg and Dehnel 1960). Similar symptoms as well as skin, eye, and throat irritation and allergic responses to cyanobacterial toxins in water have also been reported in other countries. The reported instances of illnesses are few, but, because they are difficult to diagnose, such illnesses may in fact be more common than has been reported (Carmichael and Falconer 1993).

Factors Affecting the Cyanobacterial Blooms

Physical Factors

Temperature

In general, cyanobacteria prefer warm conditions, and low temperatures are one of the major factors that end cyanobacterial blooms. Temperature alone may only partly determine bloom formation and it is accepted that a combination of factors are responsible for a bloom to develop. These are increasing temperatures, decreasing nutrients and increased water column stability. Van der Westhuizen and Eloff (1985) determined that temperature has the most pronounced effect on toxicity. At temperatures of 32°C and 36°C cell culture's toxicity was 1.6 and 4 times less than cells cultured at 28°C, suggesting that highest growth rate is not correlated with highest toxicity.

Light and Buoyancy

The effect of light intensity on the fine structure of *Microcystis aeruginosa* under laboratory conditions has been investigated. The optimal growth rate for *M. aeruginosa* cells was at 3,600-18,000 lux. At light levels in the excess of 18,000 lux the growth rate declined rapidly. Buoyancy is regulated by a number of mechanisms, such as the form of stored carbohydrates and turgor pressure regulation. Compositional changes in the protein: carbohydrate ratios during buoyancy reversals suggest a complex relationship between light and nutrients (N: P) (Villareal and Carpenter 2003). It however, seems that the regulation of gas vacuole synthesis is the most important. This almost unique feature of cyanobacteria gives these organisms a significant advantage over other phytoplanktons. In turbulent waters cyanobacteria loose this advantage and often this characteristic is used to control their blooms (Grobbelaar 2002).

Chemical Factors

Nitrogen and Phosphorus Ratios

It is becoming increasingly apparent that, notwithstanding the prevailing nitrogen and phosphorus ratio, the phytoplankton assemblage may be significantly altered through biomanipulation, and without any change whatsoever to the ambient availability of nitrogen and phosphorus (Harding and Wright 1999). Carmichael (1986) demonstrated that the omission of nitrogen causes approximately tenfold decrease in toxicity.

Iron and Zinc

Certain metal ions such as Zn^{2+} and Fe^{2+} significantly influence toxin yield. Zn^{2+} is involved in the hydrolysis of phosphate esters, the replication and transcription of nucleic acids, and the hydration and dehydration of CO_2 (Sunda 1991). All cyanobacteria require Fe^{2+} for important physiological functions such as photosynthesis, nitrogen assimilation, respiration and chlorophyll synthesis (Boyer *et al.*, 1987). It is not yet clear how Fe^{2+} deficiency modulates microcystin production, but it has been noted that as Cyanobacteria experience iron stress, they appear to compensate for some of the effects of iron loss by synthesizing new polypeptides (Lukac and Aegerter 1993).

Prevention and Control of Cyanobacterial Toxins

Chemical Control

Traditional methods for controlling algal blooms include use of copper sulfate and urea which have not only been found to be extremely expensive but cause adverse effects on aquatic biota when employed for the control of blooms in recreational water. Such treated water may also be unhealthy for further use. Few researchers were able to provide some basic information about the role of K^+ in controlling *Microcystis* growth. They conducted an extensive survey of over 21 Indian ponds and assayed the concentrations of various heavy metals (Cu, Cr, Ni, Zn, Pb, Mn), cations (Na^+, K^+, Ca^+) and nutrients (ammonium, phosphate, nitrate) commonly present in pond water of Varanasi, India. They came across very interesting results that Na^+ concentrations remain constant in almost all ponds surveyed but the K^+ concentration varied. Keeping in mind the above finding, these workers provided evidence that NaCl stimulates while KCl strongly inhibits the growth of *Microcystis* in both pond water and in defined media.

Biological Control

One alternative approach to the control of algal blooms involves the use of biological control (biocontrol) agents. Predatory bacteria that are indigenous to the lake environment have been isolated from algal blooms. Daft *et al.* (1985) proposed the following seven attributes that defined a good predatory bacterial agent:

1. Adaptability to variations in physical conditions
2. Ability to search for or trap prey
3. Capacity and ability to multiply
4. Prey consumption
5. Ability to survive low prey densities (switch or adapt to other food sources)
6. Wide host range, and
7. Ability to respond to changes in the host.

The practice of introduction of foreign microbial agents has raised some concern with regards to environmental safety due to the so-called host specificity paradigm involving host switching (HS) and host range expansion (HRE) (Secord 2003). Biocontrol of cyanobacteria, like other control measures for nuisance organisms, is often viewed with caution. This may be attributed to the experiences of plant pathologists who observed the destruction of important crops such as chestnut blight in the United States and potato blight in Ireland after the accidental release of pathogens (Atlas and Bartha 1998).

Viral pathogens would be ideal as biocontrol agents as they are target selective and specific for nuisance cyanobacteria. However, bacterial agents are considered more suitable than viruses as biocontrol agents because bacteria can survive on alternate food sources during non-bloom periods and the possibility of mutation within the host is not problematic, as bacterial predation is not reliant on unique attachment receptors (Rashidan and Bird 2001).

The Use of Microorganisms to Control Cyanobacterial Blooms

In the natural environment, there are predatory microorganisms that are antagonistic towards particular nuisance organisms (weeds, cyanobacteria) thus providing a natural means of controlling these nuisance organisms. Microbial agents antagonistic to algae (bacteria, fungi, virus and protozoa) have been isolated from harmful algal blooms (Burnham *et al.*, 1981, Ashton and Robarts, 1987, Yamamoto *et al.*, 1998, Sigee *et al.*, 1999, Walker and Higginbotham 2000, Rashidan and Bird 2001, Nakamura *et al.*, 2003a, Choi *et al.*, 2005). These microbial agents may play a major role in the prevention, regulation and termination of harmful algal blooms. Such microbial populations are called microbial herbicides (Atlas and Bartha 1998). The biological control of cyanobacteria provides a potential control measure to reduce the population of nuisance algal blooms to manageable levels. In many cases these bacterial agents are species- or genus-specific (Rashidan and Bird 2001), while others attack a variety of cyanobacteria (Daft *et al.*, 1985). The predatory bacteria are classified as members of the *Bacteroides-Cytophaga- Flavobacterium*, ranging from *Bacillus* spp. to *Flexibacter* spp., *Cytophaga* and *Myxobacteria*.

Choi *et al.* (2005) isolated the bacterium, *Streptomyces neyagawaensis*, which had a *Microcystis*-killing ability, from the sediment of a eutrophic lake in Korea. Under natural conditions, *Cytophaga* spp. was implicated in the demise of marine red tides caused by the flagellate *Chatonella* spp. in the Seto Inland Sea of Japan. *Bacillus cereus* N14 was isolated by Nakamura *et al.* (2003a) from a eutrophic lake in Japan and caused lysis of the *viridis*. Caiola and Pellegrini (1984) showed cells of *Microcystis aeruginosa* that were infected and lysed by *Bdellovibrio*-like bacteria in bloom containing water samples from Lake Varse Italy. Blakeman and Fokkerna (1982) observed that naturally occurring, resident microorganisms adapt to survive and grow in a specific habitat. If these organisms were effective antagonists against a pathogen, they would be preferred for biocontrol purposes. Organisms from other habitats, which may be equally antagonistic to the pathogen, would be less likely to survive, and consequently would have to be reapplied more frequently. The same would be true in other habitats, such as where antagonists are used to control cyanobacterial blooms.

However, there are disadvantages, which include the limited destruction of the target organism, limited survival of the microbial agent or its removal by other organisms, problems of large scale production, storage and application, as well as reluctance to apply microbial agents in a field environment.

References

Alonso-Andicoberry, C., Garcia-Villada, L., Lopez-Rodas, V. and Costas, E. (2002). Catastrophic mortality of flamingos in a Spanish National Park caused by cyanobacteria. *Veterinary Record* 151: 706-707.

An, J. and Carmichael, W.W. (1994). Use of a colorimetric protein phosphatase inhibition assay and enzyme linked immunosorbentassay for the study of microcystins and nodularins. *Toxicon* 32: 1495-1507.

Ashton, P.J. and Robarts, R.D. (1987). Apparent predation of *Microcystis aeruginosa* kutz emend elenkin by a Saprospira-like bacterium in a hypertrophic lake (Hartbeespoort dam, South Africa). *J Limnol Soc* S. Afr 13: 44-47.

Atlas, R.M. and Bartha, R. (1998). Microbial Ecology: Fundamentals and Applications. 4th edition. Benjamin/Cummings Science Publishers, 2725 Sand/Hill Road, Menlo Park, California 94025. p. 698.

Bell, S.G. and Codd, G.A. (1994). Cyanobacterial toxins and health hazards. Reviews in Medical Microbiology 5: 256-264.

Best, J.H., Pflugmacher, S., Wiegand, C., Eddy, F.B., Metcalf, J.S. and Codd, G.A. (2002). Effects on enteric bacterial and cyanobacterial lipopolysaccharides, and of microcystin-LR, on glutathione transferase activities in zebra fish (*Danio rerio*). *Aquatic Toxicol* 60: 223-231.

Bisby, F.A. (1995). Characterization of Biodiversity pp. 21-106 *In:* Global Biodiversity Assessment. (Eds.) Heywood V H and Waston R T. Cambridge University Press, Cambridge.

Blakeman, J.P. and Fokkerna, N.J. (1982). Potential for biological control of plant diseases on the phylloplane. *Ann Rev Phyopathol* 20: 167-192.

Boyer, G.L., Gillam, A.H. and Trick, C.G. (1987). Iron chelation and uptake. *In*: The Cyanobacteria. (Eds.) Van Baalen C and Fay P. Elsevier Science: New York, pp. 415-436.

Burnham, J.C., Collart, S.A. and Highison, B.W. (1981). Entrapment and lysis of the cyanobacterium *Phormidum luridum* by aqueous colonies of *Myxococcus xanthus* PCO2. *Arch Microbiol* 129: 285-294.

Caiola, M.G. and Pellegrini, S. (1984). Lysis of *Microcystis aeruginosa* (Kutz) by Bdellovibrio-like bacteria. *J Phycol* 20: 471-475

Choi, H.J., Kim, B.H., Kim, J.D. and Han, M.S. (2005). *Streptomyces neyagawaensis* as a control for the hazardous biomass of (cyanobacteria) in eutrophic freshwaters. *Biol Control* 33: 335-343.

Carmichael, W.W. (1994). The toxins of Cyanobacteria. Scientific American 270: 78-86.

Carmichael, W.W. and Falconer, I.R. (1993). Diseases related to freshwater blue-green algal toxins, and control measures. *In*: Falconer IR, (ed.) Algal Toxins in Seafood and Drinking Water. London, Academic Press, pp. 187-209.

Daft, M.J., Burnham, J.C. and Yamamato, Y. (1985). Algal blooms: conesquences and potential cures. J. Appl. Bacteriol Symp Suppl 175S-186S.

Daft, M.J., Mc Cord, S.B. and Stewart, W.D.P. (1975). Ecological studies on algal lysing bacteria in freshwaters. *Freshwater Biol* 5: 577-596.

Dillenberg, H.O. and Dehnel, M.K. (1960). Toxic water bloom in Saskatchewan. *Canadian Med Asso Journal* 83:1151-1154.

Gajdek, P. (2000). Mikrocystyny sinic W zbiornikach wodnych. Wiadomollci Chemiczne 54: 637-650.

Greuter, W., Barrie, F., Burdet, H.M., Chaloner, W.G., Demoulin, V., Hawksworth, D.L., Jorgensen, P.M., Nicholson, D.H., Silva, P.C., Trehane, P. and Mcneill, J. (1994). International Code of Botanical Nomenclature (Tokyo code), Koletz Scientific Books, Konigstein.

Grobbelaar, J.U. (2002). Dynamics of toxin production by Cyanobacteria. WRC Report: K5/1029/0/1.

Harding, W.R. and Wright, S. (1999). Initial findings regarding changes in phyto and zooplankton composition and abundance following the temporary drawdown and refilling of a shallow, hypertrophic South African costal lake. *J Lake Reservoir Manage* 15: 47-53.

Hawkins, P.R., Runnegar, M.T.C., Jackson, A.R.B. and Falconer, I.R. (1985). Severe hepatoxicity caused by the tropical cyanobacterium (Blue-Green alga) *Cylindrospermopsis reciborskii* (Wolonszynzka) Seenaya and Subba Raju isolated from a domestic water supply reservoir *Appl Environ Microbiol* 50: 1292-1295.

Krienitz, L., Ballot, A., Kotut, K., Wiegand, C., Putz, S., Metcalf, J.S., Codd, G.A. and Pflugmacher, S. (2003). Contribution of hot spring cyanobacteria to the mysterious deaths of Lesser flamingos at Lake Bogoria, Kenya. FEMS *Microbiol. Ecol* 43: 141-148.

Lambert, T.W., Holmes, C.F.B. and Hrudey, S.E. (1994). Microcystin class of toxins: health effects and safety of drinking water supplies. *Environmental review* 2:167-186.

Lukac, M. and Aegerter, R. (1993). Influence of trace metals on growth and toxin production of *Microcystis aeruginosa*. *Toxicon* 31: 293-305.

Nakamura, N., Nakano, K., Sungira, N. and Matsumura, M. (2003). A novel control process of cyanobacterial bloom using cyanobacteriolytic bacteria immobilized in floating biodegradable plastic carriers. *Environ. Technol* 24: 1569-1576.

Rashidan, K.K. and Bird, D.F. (2001). Role of predatory bacteria in the termination of a cyanobacterial bloom. *Microbial Ecol* 41: 97-105.

Rinehart, K.L. *et al.* (1988). Nodularin, microcystin and the configuration of Adda. *J. Am. Chem. Soc.,* 110(25): 8557-8558.

Rippka, R., Deruelles, J., Waterbury, J.B., Herdman, M. and Stanier, R.Y. (1979). Generic assignments strain histories and properties of pure cultures of cyanobacteria. *J Gen Microbiol* 111: 1–61.

Romanowska-Duda, Z., Mankiewicz, J., Tarczynska, M., Walter, Z. and Zalewski, M. (2002). The effect of toxic cyanobacteria (blue-green algae) on water plants and animal cells. *Polish J. Environ Studies* 11: 561- 566.

Secord, D. (2003). Biological control of marine invasive species: cautionary tales and land-based lesions. *Biol. Invasions* 5: 117-131.

Sigee, D.C., Glenn, R., Andrews, M.J., Bellinger, E.G., Butler, R.D., Epton, H.A.S. and Hendry, R.D. (1999). Biological control of cyanobacteria: principles and possibilities. *In*: The Ecological Bases for Lake and Reservoir Management, (Eds.) Harper DM, Brierley, Ferguson AJD, Philips G. Kluwer Academic Publishers, Netherlands 395 (396): 161- 172.

Sneath, P.H.A. (1992). International Code of Nomenclature of Bacteria, 1990 Revision. Amer Soc Microbiol Washington, D C.

Sunda, W.G. (1991). Trace metal interactions with marine phytoplankton. *Biol. Oceanogr* 6: 442-511.

Turner, P.C., Gammie, A.J., Hollinrake, K. and Codd, G.A. (1990). Pneumonia associated with contact with Cyanobacteria. *Brit Med J* 300: 1440- 1441.

Trogen, G.B., Annila, A., Eriksson, J., Kontteli, M., Meriluoto, J., Sethson, I., Zdunek, J. and Edlung, U. (1996). Conformational studies of microcystin-LR using NMR spectroscopy and molecular dynamics calculations. *Biochemistry* 35: 3197-3205.

Villareal, T.A. and Carpenter, E.J. (2003). Bouyancy regulation and the potential for vertical migration in the oceanic cyanobacterium *Trichodesmium*. *Microbial Ecol* 45: 1-10.

Van der, Westhuizen, A.J. and Eloff, J.N. (1985). Effects of temperature and light on toxicity and growth of the blue-green alga *Microcystis aeruginosa*[UV-006]. *Planta* 163: 55-59.

Vasconceles, V., Olivera, S. and Teles, F.O. (2001). Impact of a toxic and a non-toxic strain of *Microcystis aeruginosa* on the crayfish *Procambarus clarkii*. *Toxicon* 39: 1461-1470.

Waterbury, J.B. (1992). The Cyanobacteria isolation, purification and identification pp. 2058–2078. *In:* Balows H G, truper M, Dworkin M, Harder W and Schleifer K H (eds) *The Prokaryotes.* Second edition, Volume II. Springer Verlag, New York.

Yamamoto, Y., Kouchiwa, T., Hodoki, Y., Hotta, K., Uchida, H. and Harada, K. (1998). Distribution and identification of actinomycetes lysing cyanobacteria in a eutrophic lake. *J Appl Phycol* 10: 391-397.

Microbial Biotechnology and Ecology (2011) *Pages* **162–167**
Editors: Deepak Vyas, G.S. Paliwal, P.K.Khare and R.K Gupta
Published by: DAYA PUBLISHING HOUSE, NEW DELHI

Chapter 8

Antibacterial Potential of Diazotrophic Cyanobacteria: A Natural Strategy to Maintain Nitrogen Budget in Tropical Lowlands

☆ *Usha Pandey*

Faculty of Science and Technology, M.G. Kashividya Pith, Varanasi – 221 005, U.P.

ABSTRACT

Cyanobacteria are now considered as one of the most potential groups of organisms which can be useful to mankind in numerous ways. They are important component of soil microflora and contribute to nutrient enrichment in a variety of habitats including agricultural lands. For this reason, many species of cyanobacteria are used in the management of soil fertility. This study examined the antibacterial potential of four diazotrophic cyanobacteria namely *Oscillatoria limnetica, O. limosa, Nostoc calcicola,* and *N. piscinale* against the denitrifying bacterium *Pseudomonas denitrificans.* Addition of cell extracts (500 mg L^{-1}) of cyanobacteria in the growth medium resulted significantly lower growth and denitrification by *P. denitrificans. O. limnetica* showed superiority over rest of the species in this respect. Inhibition of growth of *P denitrificans* and its denitrification potential reduce N–losses in the ecosystem. This antagonistic potential of diazotrophic cyanobacteria can be exploited for management of soil fertility in tropical water–logged soils. On global scale, this natural strategy of cyanobacteria could help regulating N–cycle.

Introduction

Cyanobacteria, which constitute a versetille group of microorganisms, occur in divers habitats ranging from alkaline hot springs to permanent snow fields in poles. In addition to being primary producer, many cyanobacteria are capable of fixing molecular nitrogen. In tropics, where nitrogen often becomes a limiting factor for productivity in agriculture, diazotrophic cyanobacteria contribute significantly in maintaining nitrogen budget and fertility of agricultural lands. However, the process of denitrification, which converts nitrate into molecular nitrogen returning it back to the atmosphere, create an imbalance in soil nitrogen economy. The denitrification is especially rapid in tropical lowlands where elevated temperature coupled with low oxygen tension due to frequent water logging promotes the growth of denitrifying anaerobes. Spatial and temporal variations in field conditions and thus denitrifying activity are also caused by cycles of wetting and drying, which promote nitrification followed by denitrification (Knowles, 1982).

Denitrification is a process of dissimilatory reduction of NO_3^- and NO_2^- in which the major product is nitrous oxide (N_2O), which may further reduced to dinitrogen (N_2). In fact, N_2O is a free intermediate in the reduction of other nitrogen oxides to dinitrogen. When denitrification occurs under conditions favourable for this process, N_2O is produced only transiently. Denitrification is a major mechanism of loss of nitrogen in fertilizer use and release of N_2O to the atmosphere, where it involves in stratospheric reactions resulting in the depletion of ozone. Interestingly, however, denitrification is an essential mechanism to maintain global nitrogen budget (N–cycle). Nitrous oxide reductase (N_2O R) enzyme plays a significant role in reduction of nitrogen oxides. This is a membrane bound enzyme reported in *Pseudomonas perfectomarinus* (Payne *et al.*, 1971). The enzyme was present in the soluble fraction and, although extremely labile, has a half life of about 1h at room temperature. McCarthy *et al.* (1981) reported that electron transport to N_2O R generates a membrane potential only slightly smaller than that associated with O_2 utilization. It involves cytochromes of type b and c, which are membrane associated cytochromes in *Pseudomonas denitrificans*. Furthermore, continuous culture studies of energy yields in this organism growing on various nitrogen oxides suggest that energy conservation is associated with the use of N_2O as electron acceptor (Matsubara, 1975).

Certain cyanobacteria are known for their toxin production properties (McGregor and Fabbro, 2000; Pandey and Pandey, 2002). The capacity of some strains to release extracellular compounds that inhibit the growth of other microorganisms has now been considered as an important factor contributing to the formation and maintenance of cyanobacterial blooms (Bormans, 1999). Such mechanisms could be major determinants regulating nutrient balance, especially N–budgeting of a particular habitat. The antibacterial potential of cyanobacteria can affect the co–existing populations of microorganisms including denitrifying bacteria to minimize niche overlap and resource sharing and conservation. Such aspects of cyanobacteria are relatively less explored. In particular, no data so far is available on antagonistic potential of cyanobacteria against denitrifying anaerobes in Indian tropics. The present study is an effort in this direction with its possible relevance in maintaining nitrogen budget of tropical lowlands. In the present study, four strains of cyanobacteria namely, *Oscillatoria limnetica*, *Oscillatoria limosa*, *Nostoc calcicola*, and *Nostoc piscinale* were chosen for their antibacterial potential towards *Pseudomonas denitrificans*, an important denitrifying bacterium.

Materials and Methods

Cyanobacterial samples were collected from natural habitats and standard microbial techniques were employed for selection, isolation and clonal propagation of cyanobacteria. In brief, cyanobacterial mats collected from natural habitats were crushed with sterilized glass beads, diluted in sterilized double distilled water and spread on agar plates containing BG–11 medium and incubated until visible colonies appeared (Santra, 1993). Identification was made following Desikachary (1959). Colonies were picked up using sterilized micro-capillaries and transferred to culture tubes containing

10 ml liquid BG–11 medium. Growth was visible in tubes after 10 days of incubation. Such clonal cutures were washed thrice with sterilized double distilled water through repeated centrifugation and subsequently transferred to fresh BG–11 medium to obtain desirable population size. Cultures were incubated at 24 ± 1°C and illuminated for 10 days with cool day light fluorescent tubes (intensity 14.4 Wm^{-2}) on surface of the culture vessel with a 18/6 h night/dark cycle.

For the preparation of cyanobacterial cell extract, 10 g of 8 days old laboratory grown filaments of different strains were taken separately. Aqueous extracts of cyanobacteria, prepared as described in Pandey and Pandey (2002), were filtered separately through Whatman filter paper and filtrates were used for experiment.

The inhibitory effect of cell extracts of cyanobacteria on *Pseudomonas denitrificans* was studied following plate diffusion assay (Deora *et al.*, 2007). Actively growing bacterial culture was incorporated into presterilized cooled molten agar medium mixed thoroughly and dispersed into properly sterilized petri plates. After solidification, wells of 0.5 mm diameter were made using a sterilized cork borer and 10 ml of cyanobacterial extracts were carefully poured into wells. These plates were kept for 24h in bacteriological incubator at 28 °C. The inhibition zones formed due to allelopathic effect of cell extracts were measured. Bacterial growth was measured by Nephelo turbidity meter in the presence and absence of cyanobacterial cell extracts. Enzyme nitrous oxide reductase (N_2O R) was measured following the method of Kristjansson and Hollocher (1980). Percent reduction in denitrification was expressed in terms of loss of activity of nitrous oxide reductase in the presence of cyanobacterial cell extracts. Nitrous oxide reductase is the terminal enzyme that converts nitrous oxide to dinitrogen.

Data presented here represent the mean of three independent experiments. Variability is expressed in terms of standard error and paired t–test was used to represent significant differences from respective control.

Results and Discussion

Mean diameters of inhibition zone formed due to the antagonistic effects of cyanobacterial extracts are presented in Table 8.1. All the tested cyanobacteria inhibited the growth of *P. denitrificans* and the effect was maximum in case of *O. limnetica* followed by *O. limosa, N. piscinale* and minimum in *N. calcicola*. Suikkanen *et al.* (2006) have investigated the allelopathic effects of three cyanobacteria on phytoplankton and suggested it as a mechanism for the formation and maintenance of blooms by removing potential competitors. All the strains of cyanobacteria considered in this study are diazotrophs and commonly found in agricultural fields.

Cell extracts of all the four cyanobacteria were inhibitory to growth of *Pseudomonas denitrificans. Oscillatoria limnetica* cell extract was found to be the most effective in this respect (Table 8.2). At 500 ml L^{-1} concentration of *O. limnetica* cell extract the growth of the bacterium was almost negligible. For *O. limosa, N. calcicola* and *N. piscinale,* the effectiveness remained over 95 per cent at same concentration of cell extracts. These trends were consistent with those recorded for enzyme nitrous oxide reductase (N_2O R) activity (Figure 8.1). Denitrification measured in terms of

Table 8.1: Size of Inhibition Zones Formed in Cultures of *P. denitrificans* Resulting from Cyanobacterial Extracts. Values are mean ± 1SE.

Cyanobacteria	*Diameter of Inhibition Zone (mm)*
Oscillatoria limnetica	19.5 ± 1.35
Oscillatoria limosa	15.0 ± 1.26
Nostoc calcicola	9.5 ± 1.02
Nostoc piscinale	11.0 ± 1.04

nitrous oxide reductase activity was reduced by 78 per cent at 500 mg L^{-1} cell extract concentration of *O. limnetica*. For *O. limosa*, *N. calcicola* and *N. piscinale*, percent reductions in denitrification, in terms of N_2O R activity were 70, 61 and 68 respectively (Figure 8.1).

Table 8.2: Effect of Cyanobacterial Cell Extracts (500 mg L^{-1}) on Growth of *Pseudomonas denitrificans*

Time (h)	*Control*	*Growth of P. denitrificans (NTU)**			
		O. limnetica Cell Extract	*O. limosa Cell Extract*	*N. calcicola Cell Extract*	*N. piscinale Cell Extract*
0	0.03	0.03	0.03	0.03	0.03
24	3.5	0.04[a]	0.08[a]	0.09[a]	0.08[a]
48	7.5	0.04[a]	0.18[a]	0.23[a]	0.21[a]
72	7.9	0.04[a]	0.18[a]	0.23[a]	0.21[a]

*NTU: Nephelo Turbidity Unit.

a: Differences from respective controls are significant at $p < 0.01$ (analysis performed using paired t–test).

Denitrification is an essential process as for as N–cycle is concerned. However, from standpoint of agriculrure, denitrification is an undesirable process as it results in loss of nitrogen from the soil and hence a decline in nutrient available for plant growth. Because of the changing conditions and pulsed input associated with both natural and agricultural soils, denitrifying bacteria undergo flushes

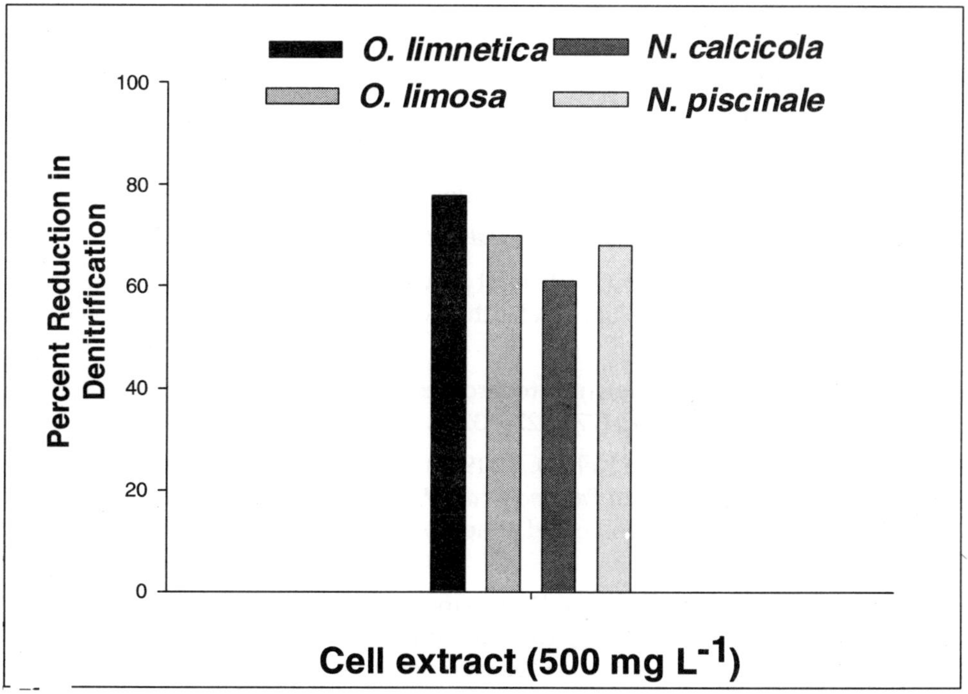

Figure 8.1: Percent Reduction in Denitrification (Measured in terms of N_2O R activity) in the Presence of Cyanobacterial Cell Extracts

of activity of short or long duration. Tropical lowland soils generally act as source of N_2O due to conditions favorable for denitrification. They can however act as N_2O sinks under complete anaerobic condition, readly availability of organic carbon and absence of NO_3^- (Knowles, 1982). The problem of denitrification is specially acute in tropical lowlands where elevated temperature, abundance of organic matter, neutral to alkaline pH and limited oxygen supply due to frequent water–logging, all together accelerate the process of denitrification. Data generated in this study suggest that continuous association of such cyanobacteria as *O. limnetica*, *O. limosa*, *N. calcicola* and *N. piscinale* with *P. denitrificans* may help maintaining nitrogen economy of lowland crop fields. As indicated in terms of growth and N_2O R activity, suppression of growth of *P. denitrificans* due to antagonistic activity of these diazotrophs suggests that these cyanobacteria have evolved strategy to maintain high nitrogen status of soil not only by fixing atmospheric N but also by conserving the fixed N in lowlands of tropics by reducing the growth of denitrifying anaerobes. In paddy fields, massive cyanobacterial mats could play significant role in inhibiting the growth of denitrifying anaerobes besides fixing atmospheric N_2.

Potential management of denitrification is essential for maintaining soil fertility stability especially in tropical water–logged soils. So far, the manipulation of soil conditions to inhibit denitrification is being used as the most general practice for management of denitrification in agriculture. Although more data may be required for field–scale verification, the present study clearly indicates that the antagonistic potential of diazotrophic cyanobacteria can be exploited as a management tool for N–addition as well as for minimizing denitrification–linked N–losses in agricultural fields. The study forms the first report of alleopathic effects of cyanobacteria towards denitrifying bacterium. This has relevance in budgeting the global nitrogen cycle as well as in generating genetically important strains of cyanobacteria to reduce N loss due to denitrification along with fixing dinitrogen.

References

Bormans, M. (1999). Controlling algal blooms in the Fitzroy River. *River for the Future* (10): 26–29.

Deora, G. S., Bhati, D. and Jain, N. S. (2007). Antibiotic effects of certain bryophytes on *Agrobacterium tumifacians*. *Journal of Pure and Applied Microbiology* (1): 215–219.

Desikachary, T. V. (1959). *Cyanobacteria*. Indian council of Agricultural Research, New Delhi. 1–686.

Knowles, R. (1982). Denitrification. *Microbiological Reviews* (46): 43–70.

Kristjansson, J. K. and T. C. Hollocher (1980). First practical assay for soluble nitrous oxide reductase of denitrifying bacteria and a partial kinetic characterization. *Journal of Biological Chemistry* (255): 704–707.

Matsubara, T. (1975). The participation of cytochromes in the reduction of N_2O to N_2 by denitrifying bacterium. *Journal of Biochemistry* (77): 627–632.

McCarthy, J. E. G., Ferguson, S. J. and Kell, D. B. (1981). Estimation with an iron selective electrode of the membrane potential in cells of *Paracoccus denitrificans* from the uptake of the butyltriphenyl–phosphorium cation during aerobic and anaerobic respiration. *Biochemistry Journal* (196): 311–321.

McGregor, G. B. and Fabbro, L. D. (2000). Dominance of *cylindrospermopsis raciborskii* (Nostocales, Cyanoprokaryota) in Queensland tropical and subtropical reservoirs : Implications for monitoring and management. *Lakes and Reservoirs : Research and Management* (5): 195–205.

Pandey, U. and Pandey, J. (2002). Antibacterial properties of cyanobacteria : A cost effective and ecofriendly approach to control bacterial leaf spot disease of chilli. *Current Science* (82): 262–264.

Payne, W. J. Riley, P. S. and Cox, C. D. (1971). Separate nitrite, nitric oxide and nitrogen oxide reducing fractions from *Pseudomonas perfectomarinus*. *Journal of Bacteriology* (106): 356–361.

Santra, S. C. (1993). *Biology of Rice Field Blue Green Algae*. Daya Publishing House. Delhi. 1- 84.

Suikkanen, S., Engström–Öst, J., Jokela, J., Sivonen, K. and Viitasalo, M. (2006). Allelopathy of Baltic Sea cyanobacteria: no evidence for the role of nodularin. *Journal of Plankton Research* (28): 543–550.

Microbial Biotechnology and Ecology (2011) Pages 168–188
Editors: **Deepak Vyas, G.S. Paliwal, P.K.Khare and R.K Gupta**
Published by: **DAYA PUBLISHING HOUSE, NEW DELHI**

Chapter 9

Histochemical Studies on Some Petro-Plants Infected with AM Fungus (*Glomus fasciculatum*)

☆ ***H.C. Lakshman***

Department of Botany, Microbiology Laboratory, Karnataka University, Dharwad – 580 003, Karnataka

Introduction

Imagine a day when you get up in the morning and read the newspaper headlines. 'The petroleum company has gone on strike' with this thought in your mind you start for your work. You are in your car driving towards your workplace and your car gives it up, you have to walk and further the entire day filled with hardship. Now multiply this intense feeling with the possibility of it lasting forever. In such a nightmare our so-called valuation lifestyle, rather our very existence shall be in a dilemma. We do not intent to exaggerate this worst-case scenario but to prove the grievousness of energy crisis.

The probability of such an extremity is evident in the future. This is because ever since the beginning of time Man has been depending extensively on fossil fuel for his every requirement and as time passed this requirement increased many folds and so did his dependency on this limited source of energy. As a result by the year 2020 the total fossil fuel supply will fall to a meager 10 per cent.

To meet this acute shortage, the technology aimed at the utilization of renewable energy that can serve to replace fossil fuel hydrocarbons needs to be expanded. The basic composition of these fossil fuels is hydrocarbons. In 1979 Dr. Calvin for the first time discovered hydrocarbons in latex of some plants, these hydrocarbons from plants have a great potential of replacing the fossil fuel hydrocarbon. Thus started the pioneering work towards the very obvious renewable source of petroleum, in the future–"The petrocrops"/petroplants. There are about 400 species of petrocrop plants in India itself, which are hoped to yield 40-45 barrels of fuel per acre.

A histochemical process is essential to study the structural organization of the vesicular-arbuscular mycorrhiza in cells and tissues. This technique is one of the useful tools in understanding basic biology of AM fungi and their interactions at the anatomical and physiological levels with host cells/tissues. Histochemical protocols have been integral components of investigation on AM fungi and their host plants. Gallaud (1905) for the first time explained, how fungal components could be stained by using cotton blue. AM inoculated strawberries gained the root sections by using Lactic acid and acid fuchsine with pioric acid O'Brien and McNaughton (1928). Daft and Nicolson (1966) used methylene blue. Such staining techniques were also proposed by other workers (Furlan *et al.*, 1973; Mossoe and Hopper, 1975; Davis and Menge, 1981). Philips and Hayman (1970) had given modified and improved staining procedure for rapid assessment of infection of root with vesicular arbuscular mycorrhizae. Kessler and Blank (1972) demonstrated the localization of phenolic substances and polyphosphates in *Eucalyptus* roots colonized by ecto and endo and ecto-endo mycorrhiza. Many workers are of the opinion that the localization of different chemical substances formed during physiological process are mainly due to AMF symbiosis (Mosse, 1973; Cox and Sanders, 1974; Kinden and Brown, 1975; Bonfante, Fosolo and Semerini, 1977). McDonald and Lewis (1978) reported the occurrence of some acid phosphates and sodium dehydrogenase activity in onion roots colonized with *Glomus mosseae*. Gianinazzi *et al.* (1979) have reported the acid phosphatase and alkaline phosphatase activity in onion roots colonized with *Glomus* species. Lakshman (1996); Senthlikumar and Krishnamurthy, (1999) have studied the peroxidase and acid phosphatase activities in some important timber trees species and ground orchid (*Spathoglottis*) with special reference to plant colonized with mycorrhizae. Solaiman *et al.* (1999) detected polyphosphates in intraradical and extraradical hyphae of *Gigaspora margarita*. Bago *et al.* (2002) have investigated the translocation of fungal storage lipids in mycorrhiza colonized roots of some important plants. Understanding mycorrhizal symbosis is of great importance, as the fungal symbionts may play a key role in determining petroplants distribution and diversity. The relationship of petroplants with fungi is relatively unique in the plant kingdom. Petroplants such as *Jatropha curcas* L., *Jatropha gossypifolia* L., *Ricinus communis* L. and *Madhuca indica Gmel.* vary in their fungal specificity, both among the genera and over the course of their life cycle. AM mycorrhizal fungi are found intracellularly in cells of the cortex, and are confined to the roots (Hardley, 1992).

Histochemical preparations enable the identification and localization of certain classes of chemical substances. Morphologically the fungus penetrate the plant cell wall and develops an intimate contact with the plant protoplast in response to the infection, the fungus modifies its behavior to control and exploit its own advantage, Visualization of find structural details in cells interacting in mycorrhizal system is of major importance, structural-functional relationships can be understood when the structural organization is known. Conventional techniques could be used to complement other findings. Knowledge of the special distribution of macromolecular components in tissue and cells is another requisite for understanding of arbuscular mycorrhiza fungal interactions.

Materials and Methods

Seeds of four petro plants *viz., Jatropha curcas* L., *Jatropha gossypifolia* L., *Ricinus communis* L. and *Madhuca indica Gmel.* were sown in earthen pots measuring 30×30 diameter. Each pot was filled with 4kg of sterilized garden soil containing 3 parts of soil and one part of pure sand in the ratio (3:1). When the seedlings were one week old, a 15g mixed inoculum of *Glomus fasciculatum* was inoculated to each pot in triplicate and were maintained in green house. Experiments were conducted after determining optimum VA-mycorrhizal colonization after 60, 120 and 180 days. Root bits were washed thoroughly in distilled water and were fixed in Corony's fluid (6 parts of ethyl alcohol +3 parts of chloroform +1 part of acetic acid) for fourty eight hrs and then washed with water before storing in 70 per cent alcohol. The fixed root bits were then dehydrated in a series of alcohol grades for two-hour interval between each grade (50 per cent alcohol, 70 per cent alcohol, 80 per cent alcohol). The fixed root bits were then dehydrated a series of alcohol grades for two-hour interval between each grade (50 per cent alcohol + butanol), 100 per cent alcohol + butanol (1:1), 100 per cent alcohol, 3:1 (alcohol + butanol), 100 per cent alcohol + butanol (1:3), pure butanol-2 changes. The dehydrated root bits were subjected to infiltration, and embedded in paraffin wax at 56°C to 58°C according to Johanson, (1940).

Infiltration, Embedding and Sectioning

The dehydrated root pieces were transferred to small vials containing pure butanol, then pieces of paraffin were added until the solvent (butanol) become saturated at room temperature then incubated by placing under table lamp (45 w) for 12 hrs. Then the vials containing material were transferred to oven and its temperature was at 58°C climate the traces of butanol. Subsequently, six to eight changes were given for all samples with an interval of 3 hrs was maintained for each molten paraffin change. Root bits were then embedded in paraffin. The paraffin blocks containing root bits were fixed to the wooden blocks. Serial microtome sections of 6 mm thickness were taken by using Lipstaw rotary microtome. The paraffin ribbons were affixed on to the slides flooded with gelatin adhesive. (Adhesive–1g of granular gelatin + 100ml of warm distilled water + pinch of potassium dichromate). The slides along with paraffin ribbons were kept on hot plate so that the sections were stretched to their original size. All the slides were labeled properly and kept in dust free boxes for drying for about 7 days. The slides were then subjected for staining. The following histochemical tests were conducted to understand the nature of cell wall and cellular components of the infected mycorrhizal roots.

Confirmation Test for Starch (Johanson, 1940)

In the present investigation this test is used to confirm the nature of Periodic Acid Schiff's (PAS) positive grains. It do not show that positive grains in the fungal hyphae and vesicles only a few small drops of Iodine is seen in arbuscules.

Test for Chitin

1000 ml of saturated potassium hydroxide solution was boiled. Hydrated sections were treated with warm KOH for 5-10 minutes and washed in 90 per cent alcohol. Chitin has been transformed into chitosan. It is tested later on with weak iodine solution. Red violet colouration indicates the presence of chitin.

Test for Total Proteins

This technique is excellent for the qualitative determination of proteins present in the cells and tissues.

Procedure

Deparaffinized sections were brought to water level and then immersed in mercuric bromophenol blue solution (prepared by adding 10g mercuric chloride and 100 mg of bromophenol blue in 100 ml of absolute alcohol) at room temperature for about 30 minutes. Stained sections were treated with 0.5 per cent acetic acid for about 5-10 min. which removes excess dye. On immersion in tap water, the stain in the sections became blue alkaline form. The sections were air dried, dehydrated in n-butanol, cleared in xylene and mounted with D.P.X.

Colour Indication

AMF fungus colonized root thin section sites of proteins stained blue.

Results

Blue colouration indicates the presence of total proteins in mycorrhizal roots. Blue colouration is not detected in non-mycorrhizal roots.

Test for Polysaccharides

PAS Method (Feder and O'Brien, 1968): Total inosoluble polysaccharides are localized by employing Periodic Acid Schiff's (PAS) method. In this method periodic acid ($IIIO_4$) will oxidize 1,2-glycol groups to produce aldehyde groups. Aldehyde groups thus formed react with leucobasic-fuchsin of Schiff's reagent to produce a visible purplish red (magenta) colour. Periodic Acid Schiff's reaction is highly recommended for the localization of total insoluble polysaccharides because of following characteristics.

1. Does not cause the breakage of polysaccharide chains.
2. Specific to polysaccharides.
3. Offers least interface and gives no false localization.
4. Forms intense and stable colour complex.

Preparations of Reagents

Periodic Acid 0.5 per cent

500 mg of periodic acid powder was dissolved in 100 ml of distilled water.

Schiff's Reagent (for Polysaccharides)

1g of basic fuchsin was dissolved in 100 ml of 0.15N HCl, agitated for three hours and then added with 1gm sodium metabisulphate and kept overnight. Treatment of this solution with activated charcoal produces straw colour Schiff's reagent.

Bleach (for Polysaccharides)

0.5 mg of sodium metabisulphate is dissolved in 95 ml of distilled water to which 5ml of 1N HCl is added.

Staining Procedure

1. Hydrated root sections were incubated in 0.5 per cent periodic acid for 15 minutes at room temperature.
2. Sections were washed in running water for about 5 minutes and incubated in Schiff's reagent for 30 minutes.
3. After incubation sections were rinsed with distilled water and were bleached with 2 per cent sodium meta bisulphate for a minute, dehydrated in alcohol series, cleared in xylene and mounted with DPX.

Results

The magenta red colouration of mycorrhizal roots indicates the presence of insoluble polysaccharides. Magenta red coloration was not detected in non-mycorrhizal roots.

Detection of DNA

Feulgen Method (Gomorol 1952): This method is based on Schiff's reaction for staining aldehydes, released from deoxyribose sugar, after the removal of purine at the level of purine-deoxyribose glucoside bonds of DNA by hydrolyzing in 1N HCl.

Staining Procedure

1. Hydrated sections were hydrolysed in 1N HCl at 60°C for 13 minutes.
2. Sections are rinsed in water and incubated with Schiff's reagent at room temperature for 30 minutes.
3. After rinsing in water sections are treated with bleach and again rinsed in water dehydrated, cleared in xylene and mounted in DPX colour indication.

Results

AMF colonized roots DNA appeared magenta–red and non-mycorrhizal roots do not colored with magenta–red.

Test for RNA: Toluidine Blue Method (O'Brien and McCully, 1981)

Preparation of Reagent

50 mg toluidine blue is dissolved in 100 ml of 0.05 M citrate phosphate buffer at pH 4.4.

Stock Solutions

1. 0.1 M solution of citric acid (19.21 gms in 1000 ml)
2. 0.2 M solution of disodium phosphate (53.65 gms of $Na_2HPO_{4,}7H_2O$ or $Na_2HPO_4 12H_2O$ in 1000 ml).
3. Citrate phosphate buffer 27.8 ml of B, diluted to a total of 100 ml with distilled water.

Staining Procedure

1. Deparaffinized and hydrated sections were immersed in 0.05 per cent toluidine blue for 5 minutes.
2. Sections were rinsed in distilled water, air dried, cleared in xylene and mounted in DPX.

Colour Indication

RNA appears purple and non-mycorrhizal root donot showed purple colour.

Test for Total Lipids

Free hand sections of fresh material were used. Root sections were treated with ethylene glycol for five minutes. Ethylene glycol was blotted out and treated with Sudan black B for ten minutes. Slides were washed 3-4 times in distilled water and sections were mounted on cleaned slides with glycerine jelly.

Control Test

Fresh freehand sections have been treated with heat cold alcohol in the reaction.

Preparation of Reagents

Sudan Black–B

0.7 g of Sudan black-B powder was dissolved in 100 ml of warm Ethylene glycol. Then the hot solution was filtered, cooled and stored in freezer. Free hand out sections of VAM colonized roots were used to prepare the histochemical slides for the localization of lipids and peroxidase.

Glycerine Jelly

10 g of Gelatin powder was dissolved in 60 ml of distilled water and followed by the addition of 70 ml of glycerol and 250mg phenol crystals.

Staining Procedure

VAM colonized fresh roots were selected and thoroughly washed with tap water and then with distilled water. The roots were macerated with 10 per cent KOH and washed with distilled water. The roots were treated with ethylene glycol for five minutes, excess ethylene glycol was blotted out and treated with Sudan black-B for 10 minutes, slides were washed 3-4 times in distilled water and sections were mounted on clean slides with glycerine jelly.

Results

The lipids appear blue to blue black colour in mycorrhizal roots. In non mycorrhizal roots blue to blue black colouration was not detected.

Test for Neutral Lipids

For this test freshly VAM colonized roots were used. Root sections were treated with ethylene glycol for five minutes. Ethylene glycol was blotted and later treated with III Sudan bland and IV Sudan black separately. The lipid bodies in hyphae, arbuscule and vesicles stained blue-black colouration indicates the presence of lipids.

Control Test

Sections are treated with cold pyridine for 30 minutes later they are transferred to hot pyridine for 30 minutes later they are transferred to hot pyridine at 60°C for 24 hrs then rinsed in water and stained in sudan black-B. The areas of lipids do not stain any colour *i.e.*, non-mycorrhizal root samples.

Toluidene Blue (for Polyphosphates)

0.05 per cent toluidene blue in citrate phosphate buffer of pH 4.4.

Results

Red coloured granular depositions indicate the presence of polyphosphates.

Test for Peroxidase Enzyme

Diaminobenzidinete trichloride (DAB) method (Graham and Karnovasky 1966): Root sections were incubated at room temperature for about 10-20 minutes in reactant solution and later they were washed thoroughly with distilled water and mounted in glycerine jelly.

Reactant solution was prepared by mixing 5 mg of 3.3 diaminobenzidinete trichloride (DAB) in 10 ml of 0.05M warm Tris-HCl buffer of 7.6pH and to this added 0.2ml of freshly diluted 1 per cent H_2O_2 then cooled, filtered and stored in brown bottle.

Results

Brown-black colour in mycorrhizal roots. Brown-black color was not detected in non-mycorrhizal roots.

Test for Polyphosphates

Fresh mycorrhizal infected root sections were used to localize polyphosphates. Sections were passed in to 0.1 per cent Toludine blue at pH-1-2 at room temperature. Sections rinsed in distilled water and mounted on clean glass slides with glycerine jelly. Red coloured granular deposition in hyphae arbuscules and vesicles gives the presence of polyphosphates.

Control Test

Fresh free hand sections have been treated with heat cold alcohol in the reaction.

Test for Cytochrome Oxidase Enzyme

Free hand sections of Fresh material were used. Sections were treated with stock solution A and B for one hour washed in distilled water and mounted on slides diluted glycerine jelly.

Stock solutions consists of (*a*) 315gms of Sodium phosphate dibasic dissolved and made it one litre and (b) 3.026 gms of Potassium phosphate monobasic dissolved and made it 1 liter. Mixed as shown below to get a particular pH?

Control Test

Fresh free hand sections have been treated with heat cold alcohol in the reaction.

Enzyme Esterase (Gomori 1950)

Esterases are enzymes which are capable of hydrolyzing esters, with in this group there are many different types of esters, acting upon a number of difficult substances. The method can be carried and employing the diazonium salt fast blue B as the coupling agent.

Reagent Required

1. Napthyl acetate, 2. Acetone, 3. 0.2 N phosphate buffer pH 7.4, 4. Diazonium salt fast blue B. Napthol acetate is dissolved in the acetone and the phosphate buffer was added, and the solution filtered and used immediately.

Procedure

Root sections were brought down to water placed incubating medium for 15-30 minutes at room temperature. Washed the sections on tap water for three minutes and counted in glycerine jelly. Reddish brown colour stained sections shows presence of enzyme esterase.

Test for Succinic Dehydrogenase

Freshly colonized mycorrhizal sections were flooded with incubating media 'a' and 'b' for 30 minutes each. After draining the media sections were mounted on clean slides with glycerin jelly.

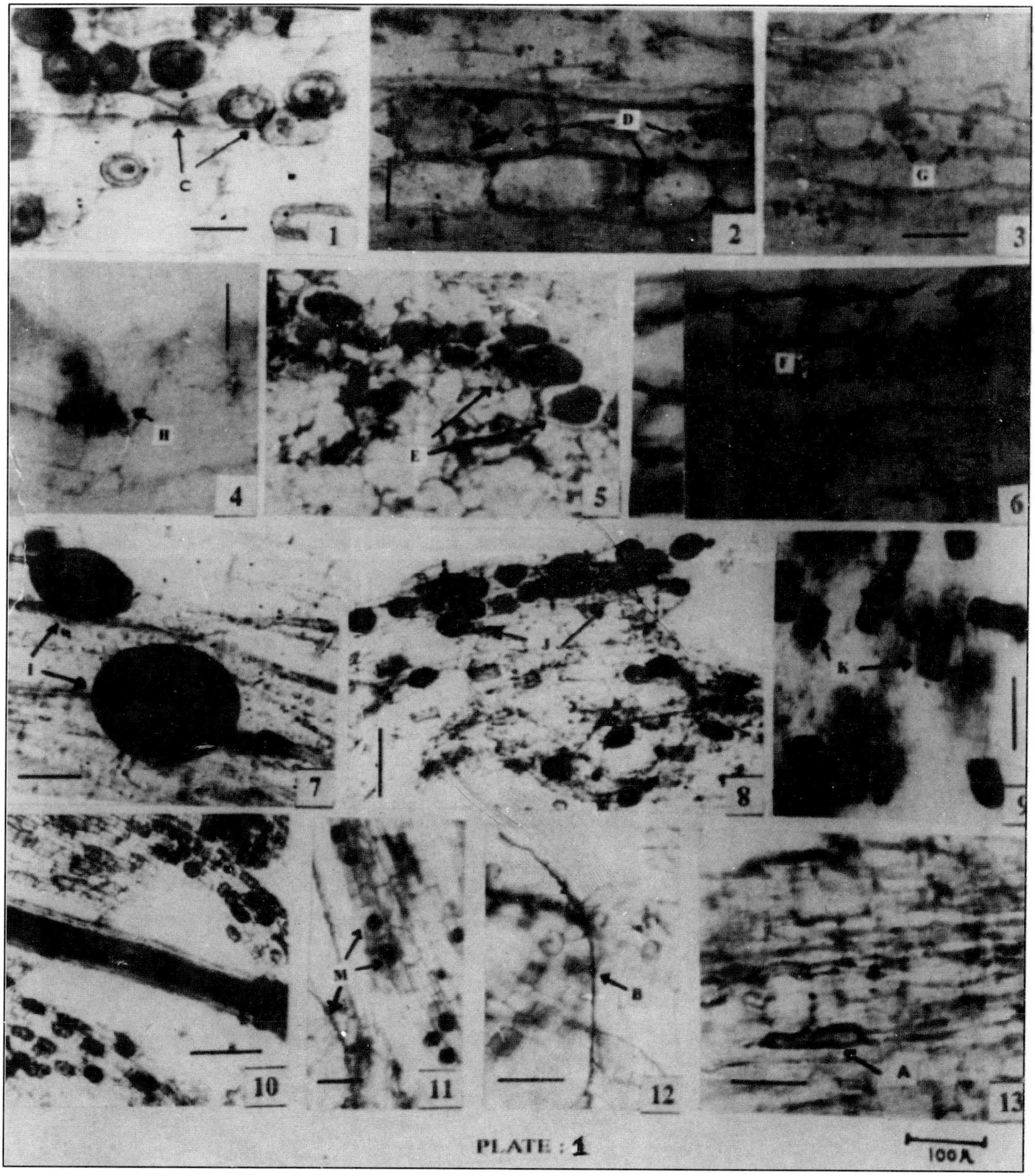

Plate 9.1

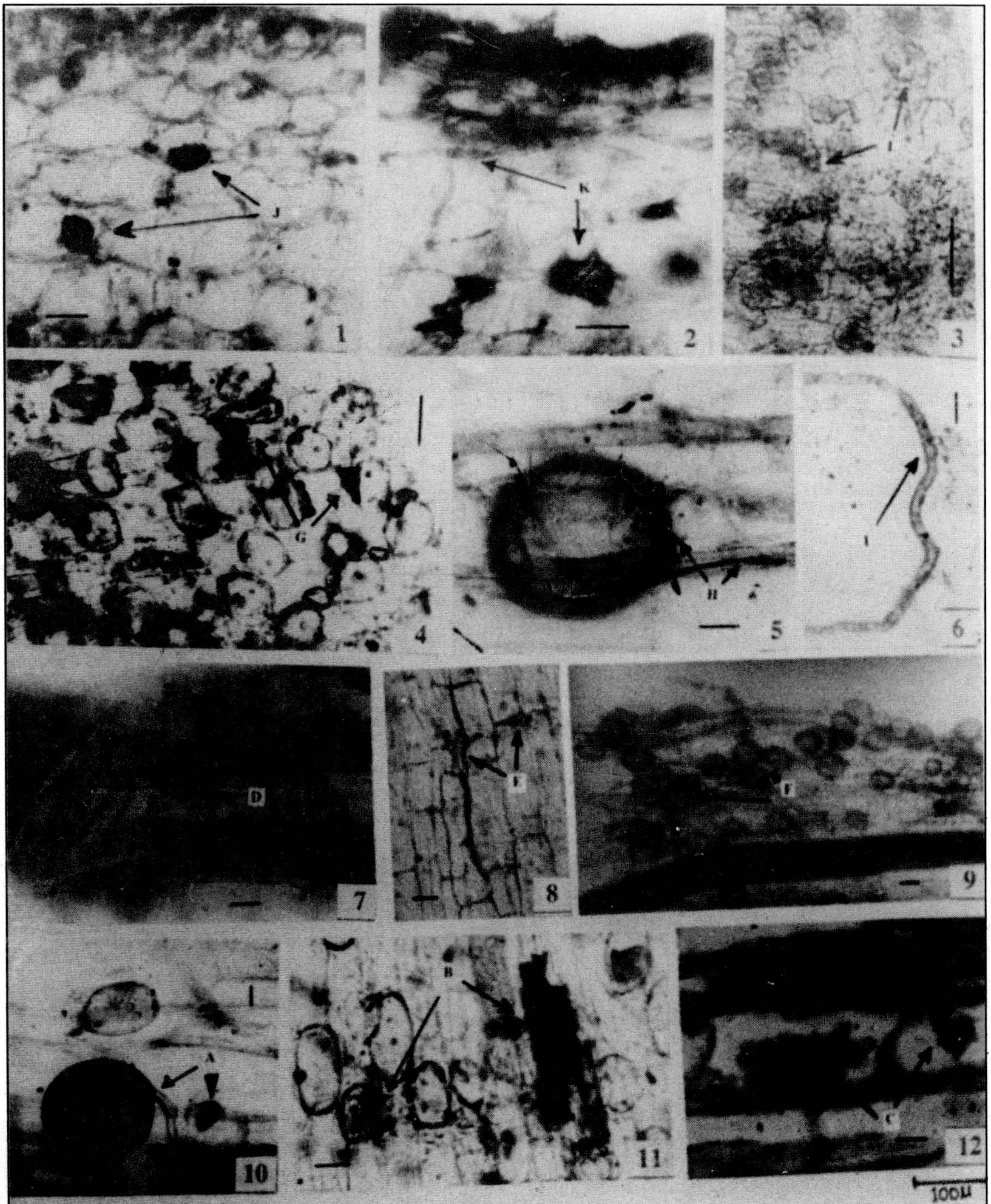

Plate 9.2

Plate 9.3

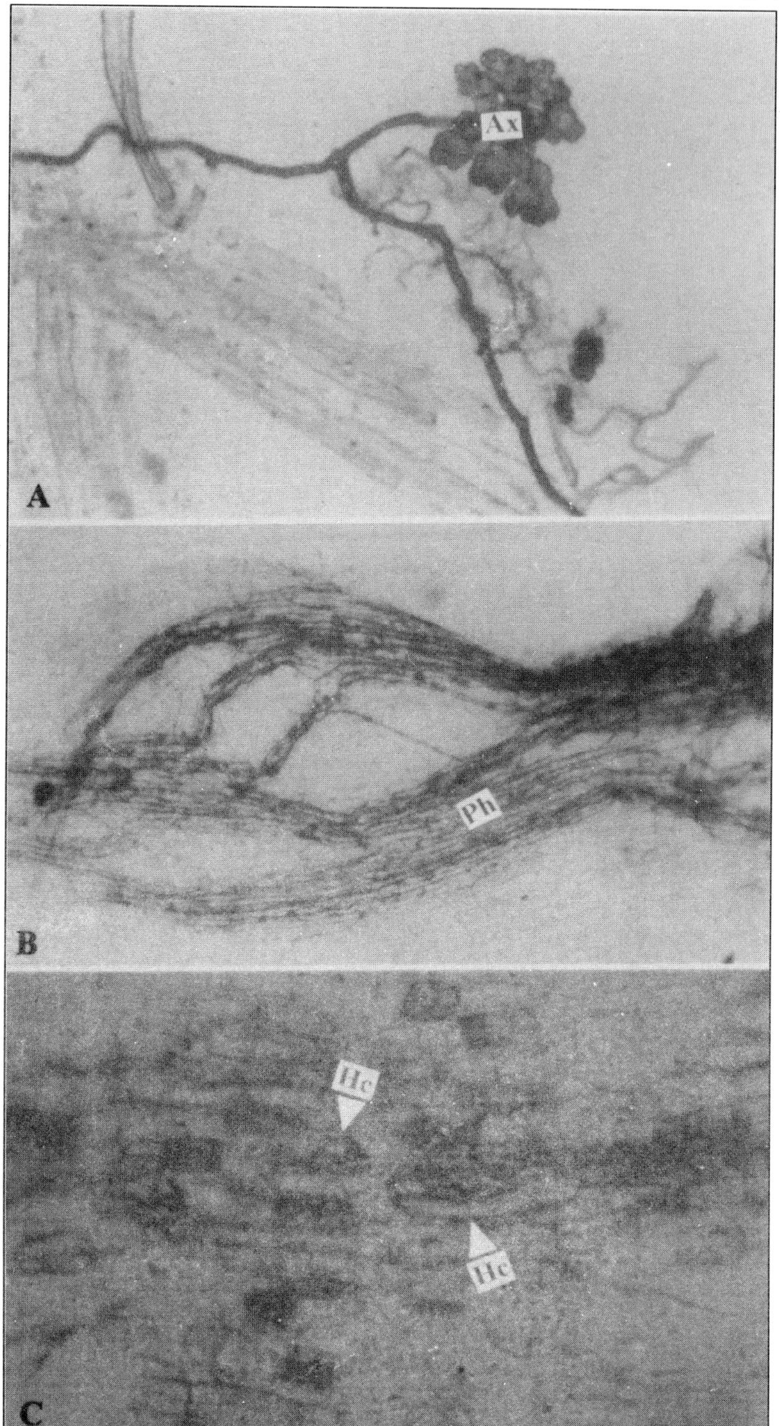

Plate 9.4

Plate 9.5

Plate 9.6

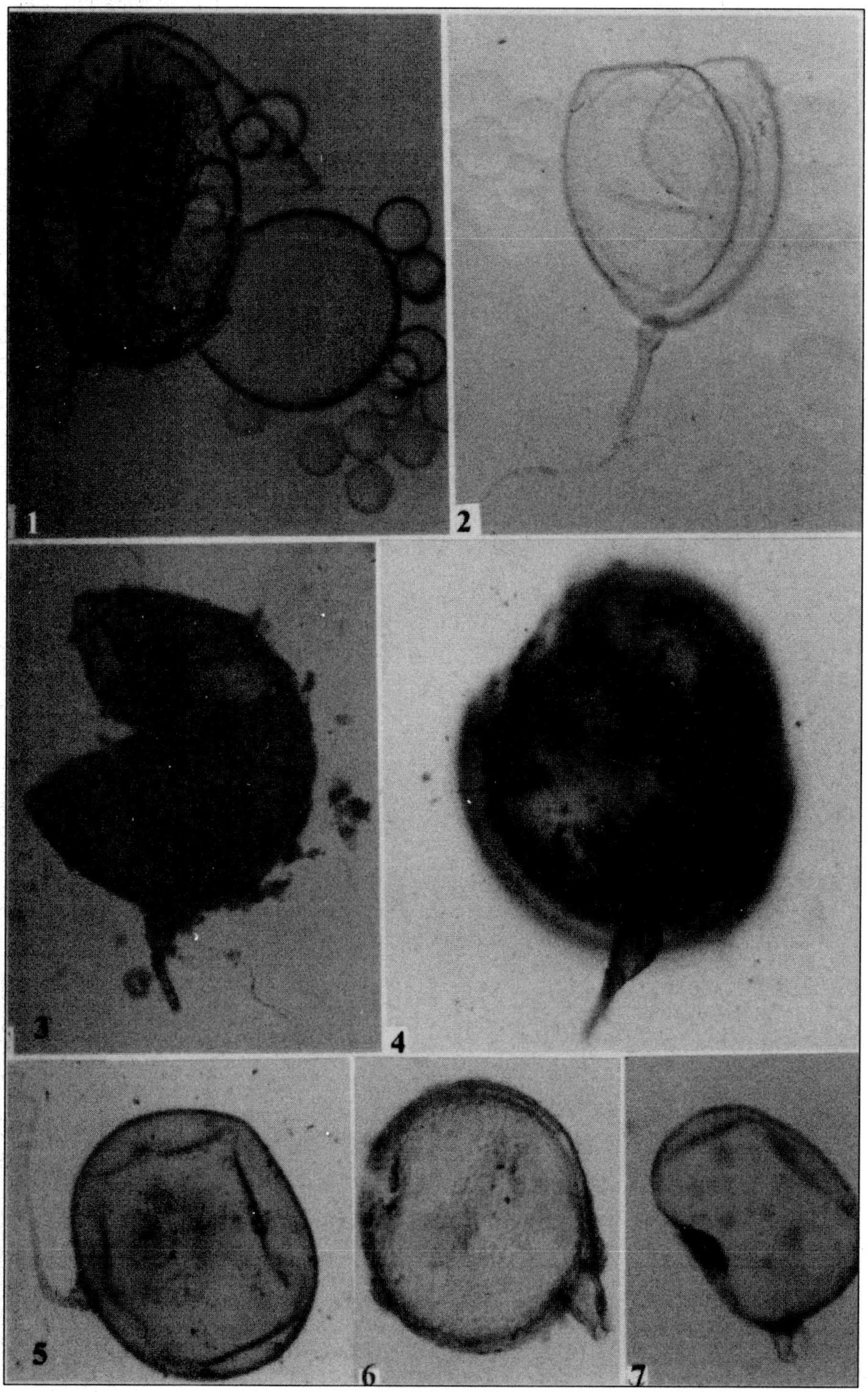

Plate 9.7

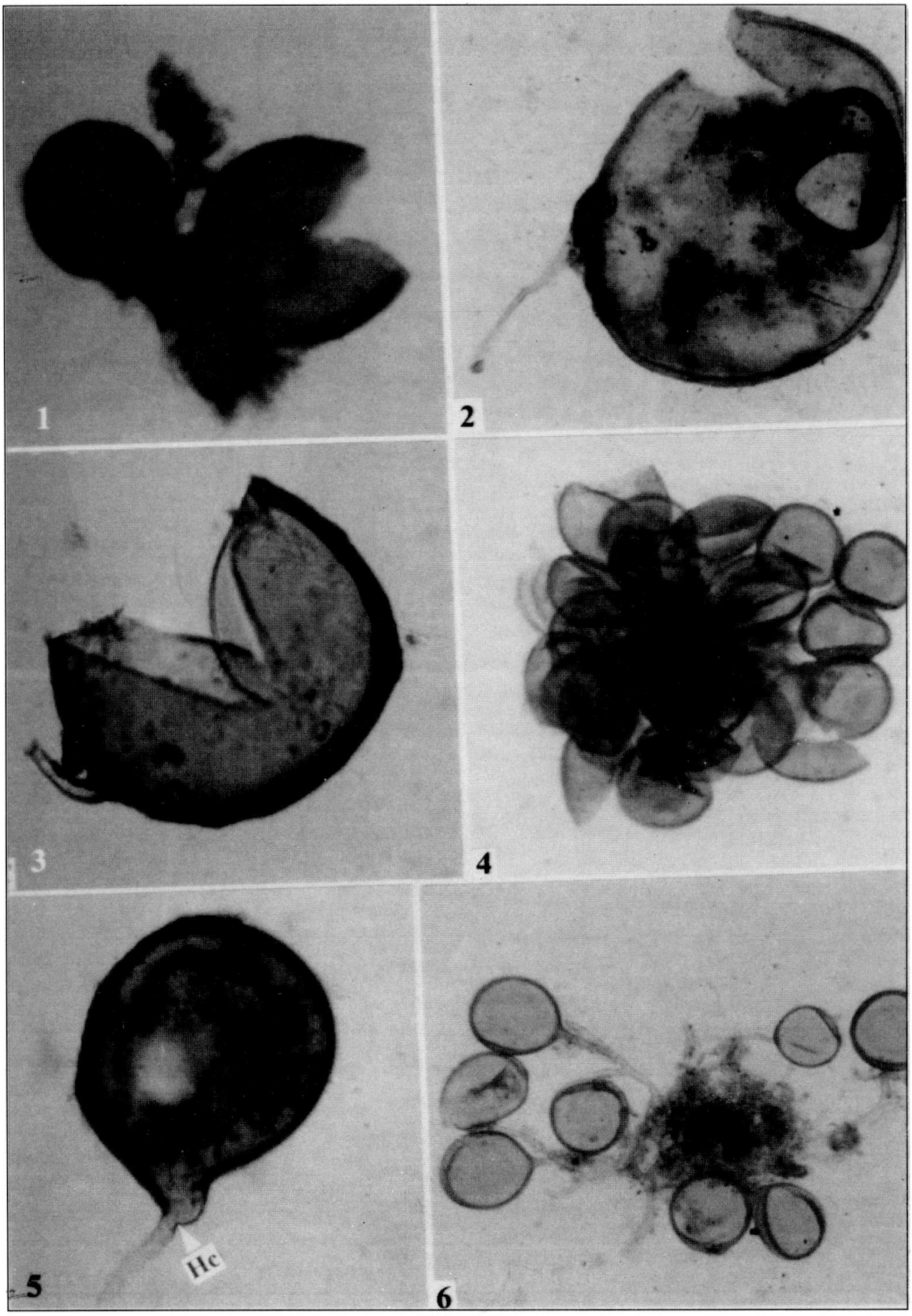

Plate 9.8

Media 'a'

6.75 g of sodium succinate was dissolved in 8 ml of water to which 0.05ml of 1N HCl was added. The pH was adjusted to 7.1 with 1N HCl.

Media 'b'

10 mg (powder) of NBT (4 mg/ml) was dissolved in 2.5 ml of distilled water with 0.2 M tris-buffer at pH 7.4 to this 1 ml of 0.05 ml of $MgCl_2$ was added followed by the addition of 3 ml distilled water Dark purple colouration of fungal components indicates the localization of succinic dehydragenase.

Test for Acid Phosphatase

Sections were incubated for 4-5 hrs in a substrate consisting of 0.1 M acetate buffer, plus 1mg ml^{-1} sodium 2 naphthyl acid phosphate, 1 mg ml^{-1} fast Garnet, 0.05 per cent $MgCl_2$ at pH 4 to 5 rinsed in distilled water. Extra pigmented root sections were bleached and washed in distilled water and mounted on clean slides with diluted glycerin jelly.

Test for Alkaline Phosphatase

Freshly colonized mycorrhizal sections were incubated overnight in a substrate. It consists of 0.05 tris-citric acid buffer at pH 8.5-9.2, plus 1mg ml^{-1} sodium 2-naphthyl acid (alkaline). 1 mg ml^{-1} fast blue, 0.05 per cent $MgCl_2$. Extra pigmented root sections were bleached and washed in distilled water and mounted on clean slide with diluted glycerin jelly.

Results and Discussion

Recent upsurge of interest and research on mycorrhiza associated plants exhibited various biochemical substances in their roots. It was noticed that translocation of materials takes place from host to fungus and vice versa. Specific observations such as localization of different chemical constituents in different components of VAM were made in the present study. Histochemical slides of freshly colonized roots clearly revealed the localization of RNA, polysaccharides (Plate 9.1: 1, 4, 6), polyphosphates (Plate 9.2: 5, 6 and 10) and protein, whereas lipids and peroxidase were localized by employing hand out section of fresh roots (Plate 9.1: 7 to 9). In the present study, mature arbuscules stained darkly for proteins test and remained negative in vesicles (Plate 9.1: 3) the inter/intracellular hyphae and mature arbuscules showed positive test for PAS, indicating high concentration of insoluble polysaccharide (Plate 9.1: 2 and 6 and Plate 9.2: 12). RNA was localized only in the arbuscules (Plate 9.1: 13). Accumulation of lipids in VAM localized roots imparts black colour, which had strong affinity for the general lipid stain called Sudan black-B. Accumulation of lipids in the form of very thick dark droplets was observed in hyphae and vesicles of mycorrhizal roots (Plate 9.1: 7, 10-11). Plate (2) shows the discontinuous localization of lipid droplets in hyphae which passes to the vesicles. Arbuscules shows the localization of lipid droplets in hyphae which passes to the vesicles. Arbuscules shows the localization of peroxidase (Plate 9.2: 1-3) and remain absent in vesicles. Cytochrome oxydase (Plate 2:4-5) Succinic dehydrogenase indiscontinuous arbuscule (Plate 9.2.2 and 4). Alkaline and acid phosphate are in (Plate 9.2: 7-8, 10-10). Plate 9.3 showing important petro plants selected by Lakshman (1996) these were 1) *Jatropha gossifolia* 2) *Jatropha curcas*, 3) enlarged *J. curcas*, 4) *Ricinis communis* var. 5) *Madhuca indica* 6) *Ricinis communis* var (Rosa). Plate 9.4 showing the macerated root sections and stained : A. axially cell with hyphae from *madhuca indica* soil sample B: numerous small vesicles hyphae from *Jatropha curcas* soil sample. C: Large vesicles with hyphae from *Ricinus communis*. (Plate 9.5) Inoculation of VAM with rock phosphate treatments 1:(A, B, C, D) showing the effect of *Glomus mosseae* with different levels of rock phosphate on *Jatropha gossifolia* 2: (A,B,C) showing the effect of

Glomus mosseae with different levels of rock phosphate on *Jatropha curcas*. 3: (A,B,C) showing the effect of *Glomus mosseae* with different levels of rock phosphate on *Madhuca indica*. (Plate 9.6) 1: (A,B) showing the growth response of *Ricinus communis* inoculated and non-inoculated plant. 2: (A,B,C,D) showing the effect of *Glomus mosseae* with different levels of rock phosphate on *Ricinus communis*. Some of the important AMF spores recovered from the rhizosphere soils of experimental plants. These were (Plate 9.7) 1: *Gigaspora margarita*, 2: *Gigaspora rosea*, 3: *Gigaspora decipiens*, 4: *Gigaspora gigentea*, 4: *Glomus fasiculatum*, 5: *Glomus mosseae*, 6: *Glomus lactum*. (Plate 9.8) 1: *Glomus citricola*, 2: *Glomus macrocarpum*, 3: *Scutellospora calospora*, 4: *Glomus aggregatum*, 5: *Glomus fulvum*, 6: *Glomus microcarpum*. showing the effect of Similar cytological changes were reported earlier. (Werner 1992; Senthilkumar and Krishnamurthy, 1999). The cytochemical study using coomassie brilliant blue (Fisher, 1968) showed positive staining of fungal hyphae, indicating high protein content (Plate No.1).

Striking increase of biochemicals, due to metabolic activities between host and fungus lead to biochemical reactions. This was detected and determined by staining procedure. The outcome of present work enabled us to understand the transformation of series of chemical substances due to host and symbiont interaction (Table 9.1), Lakshman (1996) who have demonstrated clearly that the various parameters in four Petrochemical plants. The results clearly demonstrated the translocation of materials from host to fungus and vice-versa. Basic proteins appeared to be most common type detected in immature arbuscules, vesicles and hyphal cytoplasm; these results are in conformity with that of Neneo, (1981). The periodic acid schiffs (PAS) reaction of polysaccharide with the arbuscular walls shows osmophilic and acidic properties and suggests that the wall is primarily made up of glycolipid in composition. The absence of starch grains in arbuscules and presence of starch grains in host cells lead to the conclusion that the carbohydrate materials has been taken by arbuscules, but not in the form of starch. Similar findings were also reported by the earlier workers in *Eucalyptsu* tree (Ling Lee *et al.*, 1977; Weete, 1974). The present study clearly demonstrates that the mycorrhizal roots were considerably large and cells were densely stained with PAS than non-mycorrhizal roots, this indicates that they are rich in polysaccharides (Junifer and Roberts, 1961; Harris and Northcote, 1971; Scannerini and Bonafonte-fasalo, 1979; Lakshman 1999, Inchal, 2002, Firoz, 2002). It has been pointed out that, VAM fungal hyphae and vesicles are rich in osmophillic substances (Bevente *et al.*, 1975; Cox *et al.*, 1975; Harley, 1975; Kinden and Brown, 1975; Holley Peterson, 1979). The accumulation of lipids in hyphae, vesicles and arbuscules has been observed in apple and Ericaceae members (Mosse, 1957). The present investigation supports the work of earlier workers that the vesicles and hyphae contain rich lipids probably of triglyceride. The vesicles acts as storage sink of lipids for the host plant (Cox *et al.*, 1975, Brannan and Losel, 1978). More accumulation of lipid in vesicles resulted in the transformation into triglycerides, monoglycerides, fats, sterols, etc. the potential role of polyphosphate uptake in hyphae is well documented by Callow *et al.* (1978). The VAM fungi synthesizes polyphosphates in granular form and transfer the same to arbusculus and ensure its availability to the host cell. Several workers demonstrated that the uptake and storage of phosphate might be a common phenomenon of all mutualistic mycorrhizal associations. Hence mycorrhizal plant utilizes more phosphate than non-mycorrhizal plants. Polyphosphate is a major 'P' reserve in VA-mycorrhizae. Root cortical cells of mycorrhizal plants contained arbuscules with high 'P' concentration. Whereas polyphosphate granules are commonly found in young proliferating arbuscules (Grey and Gerdmann, 1969; Bowen, 1968; Mosse 1973a). the peroxidiase activity was detected only in senescing arbuscules of all the experimental plants. It seems that mycorrhizal fungi colonize the host tissue by combination of mechanical interaction and enzymatic mechanism (Gianinazzi-Person *et al.*, 1996). Thus, VAM fungi can play an important role in developing the pathogen resistance host plant. The peroxidase

Table 9.1: Histochemistry Tests in Infected Roots of *Jatropha curcas* Plant Inoculated with *Glomus fasciculatum*

Sl.No.	Metabolic Localized	Test Reagent	Colour Indication	Localization in Fungal Structure			Reference
				Arbuscules	Hyphae	Vesicles	
1.	Polysaccharides	PAS	Magenta Red	+	–	–	Johnsen, 1992
2.	Starch	IKI solution	dissolved				Johnsen, 1940
3.	Chitin	KOH-IKI	Red Violet				Johnsen, 1940
4.	Nucleic acids (DNA)	Feulgen	Purple blue	+	+	+	
5.	RNA	Toluidine blue	Purple	+	–	–	O, Brein and McCully; 1981
6.	Protein (Total)	Mercuric bromophenol blue	Blue Blue blak	+	+	+	Pearse, 1960
7.	Lipids (Total)	Sudan black-B		+	+	+	Johnson, 1962
8.	Neutral Lipids	Sudan III or IV		+	+	+	Laxman, 1996
9.	Polyphosphates	Toluidine blue	Blue purple	+	+	+	Ashford *et al.*, 1975
10.	Cellulose	Calcoflour white	Fluorescence	–	–	–	Hughes and McCully, 1975
			Type of enzymes				
11.	Cytochrome oxidase	Nadireagent	Room temp. Bluish Purple	–	+	+	Chayen *etal.*, 1969.
12.	Peroxidase	Bensidine	blue turns brown	+	–	–	Jensen, 1962
13.	Succinic dehydrogense	Sodium succinate	Room temp. Bluish Purple	+	–	–	Bancroft, 1975
14.	Acid Phosphatase	Sodium B glycero phosphate	violet black	+	+	–	Pears, 1972
15.	Alkaline Phosphatase	Sodium B glycero phosphate	violet black	+	+	–	Pears, 1972

activity was detected only in senescing arbuscules in the present study and similar to the reports of Nemec (1981); Lakshman (1996) and Shanthikumar and Krishnamurthy (1999). It is concluded that the different chemical substances were accumulated in VAM fungi and and transferred to the host tissue. In finger millet, VAM colonized roots with arbuscules lead to be the path for the exchange of materials from fungus to host and vice versa.

References

Ashford, A. E, Ling lee, M and Chilvers G. A. (1975). Polyphosphate in mycorrhizas; A cytochemical demonstration. *New Phytol* 74: 447-453.

Bevege, D. I. (1968). A rapid technique for clearing tannins and staining intact roots for detection of mycorrhizas caused by *Endogone* spp., and some records of infection in Australian plants. *Trans. Bt. Mycol. Soc.* 51: 808-811.

Bevege, D. I. and Bowen, G. D. (1975). Endogone strain and host plant differences in development of vesicular-arbuscular mycorrhizas; 77-86 In. (Eds.) Sanders, F. E mosse, B and Tinker, P. B. *Endomycorrhizas.* Academic press, London. pp. 626.

Bonfanate-Fasolo, P. and Scannerient (1977). Cytological study of the VA-mycorrhiza in *orinthogallum umbellatum* L. *Allionia.* 22: 15-21.

Bonfante-Fasolo, P. and Scannerini, S. (1977). A cytochemical study of the vesicular-arbascular mycorrhiza clover. *Aust. J. Bot.*, 30: 485.

Callow, J. A. , Capaccio, L. C. M. , Parish, G. , and Tinker, P. B. (1978). Detection and estimation of polyphosphate in vesicular-arbuscular mycorrhizas. *New Phytol.* 80: 125-134.

Carling, D. E. , and white, J. A, and Brown, M. F. (1977). The influence of fixation procedure on the ultrastructure of the host endophyto interace of vascular-arbuscular mycorrhizae *Can. J. Bot.* 71: 48-51.

Chayen, J. L. Bitensky. R. G Butcher, and L. poulte. (1969). *A guide to practical histochemical.* Oliver and Boyd, Edinburgh, pp. 35.

Cox, G. , and Sanders, F. 1974. Ultrastructure of the host-fungus interface in vesicular-arbascular mycorrhiza. *New Phytol.* 73: 901-912.

Daft. M. J. , and Nicolson, T. H. (1966). Effect of Endogone mycorrhiza on plant growth. *New Phytol.* 65: 343-350.

Davis, R. M. , and Menge, J. A. (1981). Phytophthora parasitica inoculation intensity of vesicular-arbuscular mycorrhize in citrus. *New Phytol.* 87: 705-715.

Feder. N and O'Brien, T. (1968). Plant microtechnique some principals of new methods. *Am. J. Bot.* 55: 123.

Fisher, D. B. (1968). Protein staining of ribboned epon sections for light microscropy. *Histomchemie.* 16: 82-96.

Furlan, V. , and Fortin, J. Andre (1973). Formation of endomycorrhizae by Endogone calospora in Allium cepa under three temperature regimes. *Naturaliste Can.* 100: 467-477.

Gallaud, Etude. (1905). sur les mycorhizes endotrophes, *Rev. Gen. Bot.* , 17: 5-15.

Gianinazzi, S. , Gianiazzi-Pearson. , and Dexheimer, J. (1979). Enzymatic studies on the metabolism of vesicular-arbuscular mycorrhiza. III. Ultrastructural localization of acid and alkaline phosphatase in Onion roots fected by Glomus mosseae (Nicol. and Gerd). *New Phytol.* 82: 127-132.

Hardley, G. (1992). Orchid Mycorrhiza. In *Orchid Biology: Reviews and prospectives II*, pp. 83-113, edited by A. J. Corness, New York : University Press.

Harold, F. M. (1966). Inorganic polyphosphates in biology; structure, metabolism and function. *Bact. Rev.* 30: 772-976.

Holley, J. D., and Peterson, R. L. (1979). Development of a vesicular-arbascular mycorrhiza in bean roots. *Can. J. Bot.* 57: 1960-1978.

Jensen, W. A. (1962). *Botanical Histochemistry*. W. H. Freeman and Francisco. pp. 352.

Kessler, K. J, . and Blank, R. W. (1972). Endogone. Sporocarps associated with sugar maple. *Mycologia*. 64: 634-638.

Kinden, D. A. , and Brown, M. F. (1975). Electron microscope of vesicular-arbuscular mycorrhizae of yellow popular. III. Host-endophyte interaction during arbuscular development. *Can. J. Microbiol.* 21: 1930-1939.

Lakshman. H. C. (1996). Studies and histochemical on some tree species infected with Am-fungi in; *Ph.D. Thesis* Karnatak university Dharwad, India. pp. 249.

Lakshman. H. C. and Inchal. R. F. and Mulla. F. I. (2002). Histochemical studies on some medicinal plant infected with Am-fungi. *Eco. plan-Environment*. 91(4): 71-75.

Lev, R and Spicer S. S. (1964). Specific staining of sulphate group with alcian blue at low PH. *Histochem. Cytochem*. 12-309-312.

Lillie, R. D. (1965). *Histopathologic Technic and Practical Histochemistry*. Blakiston N. Y. 3rd edition.

Ling-Lee, Chilvers, G. A and Ashford. A. F. (1975a). A histochemical study of phenolic materials in mycorrhizal and infeeted roots of Eucalyptus fasitigiata deane and maden. *New Phytol*. 78-313-318.

Ling-lee, M. A. Ashford, and G. A. Chilvers. (1977). A histochemical study of saccharide distribution in eucalypt mycorrhizas. *New Phytol* 78-329-335.

Mosse, B. , and Hopper, C. (1975). Vesicular-arbuscular mycorrhizal infection in root organ cultures. *Physiol. Plant Pathol*. 5: 215-223.

Nemec, S. (1981). Histochemical characteristics of Glomus etunicatus infection of citrus *limon fibrous* root; *Can. J. Bot*. p. 609-617.

Pearse, A. G. E. (1968). *Histochemistry: Theoretical and Applied*, Vol. 1 Churchill. London. PP. 412.

Pearse, A. G. E. (1972). *Histochemistry: Theortical and Applied*, Vol. 2 Churcchill. Living stone Edinburgh, PP. 276.

Phillips, J. M. , and Hayman, D. S. (1970). Improved procedures for clearing roots and staining parasitic and vesicular-arbuscular mycorrhizal fungi for rapid assessment of infection. *Trans Bt. Mycol. Soc*. 55: 158-161.

Rao, J. (1973). Cytochemical localization of peroxidase in plant cell. *Physiol Plant*. 28: 132-133.

Scannerini, S. and Bonfante-Fasolo, P. (1979). Ultrastructural cytochemical demonstration. Of polysaccharides and proteins with in the host-arbuscul interfacical matrirx in an endomycorrhizae. *New Phytol.* 83: 87-94.

Scannerini, S. and P. Bonfante-Fasolo. (1977). Ultrastructural and cytochemical features of the matrix material in a vesicular-arbuscular. *Mycorrhiza caryologia*, 30(4): 499-500.

Senthlikumar and Krishnamurty. (1999). Peroxidase and acid Phosphatase activity in ground orchid Spathoglottis with reference to mycorrhiza. *Mycorrhiza News*, 10 (4): 11-13.

Werner, D. (1992). Orchid mycorrhiza. In: *Symbosis of Plants and Microbes*, pp. 367-380. Chapman and Hall, London.

Microbial Biotechnology and Ecology (2011) *Pages* **189–200**
Editors: **Deepak Vyas, G.S. Paliwal, P.K.Khare and R.K Gupta**
Published by: **DAYA PUBLISHING HOUSE, NEW DELHI**

Chapter 10

Biodiversity of VAM Fungi Associated with Some Common Medicinal Plants of Bihar

☆ ***A.K. Roy* and A.N. Singh***

Department of Botany, T.M. Bhagalpur University, Bhagalpur – 812 007

Introduction

Arbuscular mycorrhizal fungi are well distributed in both cultivated as well as barren areas. However, different types of soil show variation in the number and type of spores (Hayman, 1978). Distribution and diversity of AM fungi have been reported in diverse ecological conditions such as arctic (Kothari *et al.*, 1997), temperate (Smith and Bowen, 1979) and tropical regions (Singh *et al.*, 1994). Further these have been reported to occur in non-cultivated soil, cultivated soil, mining areas (Roy *et al.*, 2004), desert soils (Shanker *et al.*, 1990), scrub and grassland (Mukerji *et al.*, 1984). Moreover, these endophytes may also survive for years inside the living roots but the species composition of VAM population is found to vary with soil nature.

Thus, diversity is influenced by the several abiotic and biotic components including edaphic factors in an ecosystem which includes soil pH, soil moisture, soil temperature, soil depth, soil nutrients, soil disturbance, soil porosity, soil fertility and soil organic matter and climatic condition such as rainfall, atmospheric temperature, seasonal variation, humidity, light intensity, physical movement, and plant susceptibility. In general diversity of mycorrhizal fungi has also been observed in the same plant and also within the same plant species at different places (Allen and Booalis, 1983). Beside these, mycorrhizal activity like germination rate, growth of hyphal branching, recognition of the host,

---

* E-mail: botanyakr@yahoo.co.in.

penetration of the roots, growth of hyphae into the soil and sporulation have also been regulated by plant root exudates (Koske and Gemma, 1992).

The species of AM fungi are influenced by the soil temperature as *Gigaspora decipiens* produces highest number of spores at 21°C day temperature (Hari Kumar and Bagyaraj, 1988), where as, *Scutellospora calospora* prefers cooler northern zone (Koske, 1987). *Glomus* and *Gigaspora* were among the most commonly occurring genera in both tropical and temperate climate (Nicolson and Schenck, 1979). Trinick (1977) and Theodorou (1978) studied that the root colonization and spore number were less with higher level of moisture content. Similarly, more germination of spores in *Glomus epigeum* was observed at moisture levels above field capacity (Daniel and Trappe, 1980). The soil pH influences the VAM population in the forest soil (Abbott and Robson, 1991). Kruckelmann (1975) recorded that VAM spores were more influenced by soil pH than by any other factor.

The forest regions under study are found to have continuous pressure of human activities like implementation of developing projects, forest fire, over exploitation of medicinal plants and illegal felling of trees. The attempt to destroy forests and conversion of it in cultivable land results not only in the depletion of certain valuable taxa but also leads to substantial reduction in the density and extent of forest cover. During the course of unabated depletion of forest, there is possibility of drastic declination in the number of some valuable plants leading towards the category of rare or vulnerable. Medicinal plants growing under natural habitat also get victim of such natural and man made disasters which require regular surveillance along with strategy for their conservation and propagations. Growth promoting rhizosphere microbes or root symbionts associated with them may be exploited for the said purposes. Keeping this in view an attempt has been made here to study the present status of medicinal plants and diversity of root associated fungi *i.e.*, VAMF in the forest regions of eastern Bihar.

Methods Adopted

Extensive survey of forest areas of eastern Bihar (Bhagalpur, Banka and Jamui districts) was made at a regular interval of 3 months to get clear picture of diversification of medicinal plants in different seasons continuously for complete one year. The purpose of the survey was to collect detailed information *i.e.* distribution and population density, frequency, local name, useful parts of medicinal plants and their uses for health care. Collection of rhizosphere soil and root sample was also made during the survey for screening of AMF spore population and per cent root colonization. The frequency percentage and density of medicinal plants were also calculated with the help of 30 × 30m^2 quadrate by the following formula:

Frequency Percentage

$$\frac{\text{Total number of quadrates in which species occurred}}{\text{Total number of quadrates studied}} \times 100$$

Density

$$\frac{\text{Total number of quadrates studied}}{\text{Total number of individuals of the species}} \times 100$$

Documentation of Rare/Vulnerable Medicinal Plants

On the basis of frequency and density of individual plant species in the surveyed area, list of rare (taxon less in number or risk) and vulnerable (taxa of endangered category for a particular area) medicinal plant species was prepared.

Rhizosphere soil and feeder roots were collected separately in polythene bags at about 15 cm depth around the plant. Soil and root test samples were brought to the laboratory for assessment of mycorrhizal association and physico-chemical analysis of soil. Screening of soil was done by Wet Sieving and Decantation Method of Gerdemann and Nicolson (1963). 10g of soil samples were mixed in 50 ml luke warm tap water in 200 ml beaker, stirred the solution 1-2 min. properly and left it for 3 min. for setting of heavier particles. Suspension was allowed to pass through the sieves of various measurements (710 mm–45 mm) with course sieve on the top and fine at the bottom. This process was repeated twice or thrice, so that whole suspension decanted through mesh and majority of spores isolated from soil samples. Debris retained by the sieve was washed under a stream of water. The suspension was transferred into petriplates separately. The content was examined for the presence of spores or sporocarp under the dissecting or compound microscope. The spore population was determined either number of spores/gm. dried soil or/10 gm. dried soil.

Spores were isolated and identified on the basis of morphological features *i.e.*, spore colour, spore size, shape, thickness of the wall, lamination, number of the wall, attachment pattern of subtending hyphae and content of the spores as ascribed in the manual of Schenck and Perez (1988). Phillips and Hayman (1970) method was followed for determining per cent of AMF colonization. The root sample were washed carefully and cut into small pieces (1 cm), which were boiled in 10 per cent KOH solution for one hour, washed for 3 times and bleached in alkaline hydrogen peroxide (5 per cent NH_4OH, 0.5 per cent H_2O_2, H_2O v/v) for 12 hours. The root bits were again washed 2-3 times in tap water. The trypan blue (0.05 per cent) was used for staining and lactophenol for mounting. The treated root bits were tapped softly and examined AMF infection in terms of percentage root colonization (per cent RC) which was determined by the following formula.

$$\% \text{ Root colonization} = \frac{\text{Number of segments observed}}{\text{Number of positive segments}} \times 100$$

The garden soil was dried on room temperature, sieved and autoclaved at 15 lb./inch^2 pressure for 3 hours for 3 consecutive days. The soil was placed in oven for 3 days at 90°C. After proper sieving of sand treated with 10 per cent HCl left for 3 days and was then washed till whole HCl was removed.

Results and Discussion

Survey of Forest Regions and their Soil Profile

During the course of study, survey of forest areas of eastern Bihar (Bhagalpur, Banka and Jamui districts) was made at a regular interval of three months for a year. Field trips were arranged so as to cover most of the forest and hilly tracts of all the study sites.

The land surface as a whole is apparently a level tract sloping gently from south to north in Bhagalpur and Banka districts, where as Jamui district is full of hilly tracts covering major forest areas. On the basis of soil texture of the area understudy was classified into Clay, Loam, Sand or Sandy Loam. The area enjoys the monsoon climate with a hot summer and a pleasant winter season. The average annual rainfall of the area is 1131.26 mm with 82 per cent of the precipitation falling during the monsoon months of June to September.

Since the soil of this region is complex in origin, it possesses a very heterogeneous flora, which is essentially tropophilous but a tendency towards xerophilous structure in many of its species. The types of forest of the surveyed areas are categorized as follows: Bhagalpur and Banka Districts: Northern dry mixed deciduous forest, Jamui District: Moist deciduous Sal, *Boswellia* and Aegle forests. A large number of forest produce *viz.* Timber, Gum, Kendu leaf, Bamboo, Medicinal plants, fruits etc. has been reported from this region as the main source of income for local inhabitants but the present investigation is mainly concerned with the present status of medicinal plants.

A total number of 113 medicinal plants were recorded out of which 10 plants *viz.Achyranthes aspera, Andrographis paniculata, Asparagus racemosus, Catharanthus roseus, Costus specious, Euphorbia thymifolia, Rauvolfia serpentina, Rauvolfia tetraphylla, Woodfordia fruticosa* and *Withania somnifera* were found growing invariably in all the forest regions, which were selected for screening of their root symbionts *i.e.*,VAM fungi in different soil profile.

The soil types are almost same in all the surveyed sites *i.e.* alluvial, sandy loam and sandy silt. Macro- and micronutrients *viz.*, N, K, P, Na, Ca, organic carbon and organic matter of soil samples and their pH were determined and results are depicted in Table 10.1. The soil pH was found to range in alkaline side except the soil of Jamui district where it was 6.5 tending towards acidic side. The organic carbon of the soil was noticed to range between 3.0–4.1 per cent whereas organic matter ranged between 6.5–7.05. A comparative study of the data particularly related with N, K, P, Ca and Na contents of the soil samples reflects that the soil of Jamui and Bhagalpur have almost similar status whereas Banka soil is slightly poorer in nutrients. On the basis of quantity of macro and microelements *i.e.*, OC, OM, N, K, P, Ca and Na, soil samples of different sites may be categorised as good–best Banka > Bhagalpur > Jamui.

Table 10.1: Showing the Chemical Profile of Soil Samples Collected from Sites Under Study

Soil samples	*pH*	*OC (%)*	*OM (%)*	*P (%)*	*N (%)*	*K (%)*	*Ca (%)*	*Na (%)*
Bhagalpur	7.88	4.05	7.02	0.260	0.056	0.237	0.531	0.443
Banka	7.2	3.09	6.5	0.230	0.043	0.226	0.482	0.422
Jamui	6.5	4.10	7.05	0.265	0.055	0.250	0.602	0.472

OC: Organic carbon; OM: Organic matter; P: Phosphorus; N: Nitrogen; K: Potassium; Ca: Calcium; Na: Sodium.

Screening of Rhizosphere Soil Samples of Medicinal Plants Understudy

The rhizosphere soil sample and feeder roots of 10 commonly growing medicinal plants were collected to determine the root colonization, spore population and diversity of AM fungi. Results are presented in Tables 10.2–10.4. The study of root colonization is necessary to observe the degree of infectivity and to describe the morphology of specific mycorrhizal structure formed within the roots. It is also essential to evaluate the responses of various treatments such as the effectiveness of different inoculants on plant growth and the extent of host specificity in different host fungus combination. Furthermore, the anatomical features of some of the fungal structure inside the roots are diagnostic for certain species. It may be possible to differentiate AM fungi responsible for root colonization.

Root Colonization (RC)

The root samples of 10 medicinal plants were examined for percentage colonization. The results (Tables 10.2–10.4) show that in Bhagalpur region, the maximum RC was recorded in *A. racemosus i.e.*, 75 per cent, 83 per cent, 78 per cent and 70 per cent during 1st, 2nd, 3rd and 4th quarters respectively with yearly average of 76.5 per cent whereas it was found minimum in *E. thymifolia i.e.* 9 per cent, 15 per cent, 12 per cent and 10 per cent during 1st, 2nd, 3rd and 4th quarters respectively with yearly average of 11.5 per cent. In Banka region the maximum RC was recorded in *A. racemosus i.e.* yearly average of 75.25 per cent whereas *E. thymifolia* showed only 10 per cent root colonization. In Jamui forest region again *A. racemosus* plant showed maximum RC *i.e.*, 82 per cent while it was minimum *i.e.*, 14.5 per cent in case of *E.thymifolia*. The results also make it clear that the medicinal plants of Jamui region were found to have maximum fascination towards root infection followed by Bhagalpur and Banka sites. The reason behind this might be due to variation in soil texture.

Table 10.2: Showing Per cent Root Colonization and Spore Population of Medicinal Plants Understudy in Bhagalpur Region

Plant Species	*Jan-Mar* (Q_1)		*Apr-Jun* (Q_2)		*Jul-Sep* (Q_3)		*Oct-Dec* (Q_4)		*Yearly Average*	
	RC	*SP*	*RC*	*SP*	*RC*	*SP*	*RC*	*SP*	*RC*	*SP*
Achyranthes aspera	12	21	20	28	17	31	14	23	15.75	25.75
Andrographis paniculata	40	27	55	35	48	34	42	29	46.25	31.25
Asparagus racemosus	75	20	83	24	78	27	70	19	76.5	22.50
Catharanthus roseus	30	22	48	28	46	30	39	23	40.75	25.75
Costus specious	38	31	47	23	43	21	39	16	41.75	22.75
Euphorbia thymifolia	9	10	15	15	12	17	10	12	11.50	13.50
Rauvolfia serpentine	40	23	53	32	50	27	47	25	47.50	26.75
Rauvolfia tetraphylla	43	21	57	28	53	30	45	22	49.50	25.25
Woodfordia fruticosa	62	28	69	32	64	33	60	30	63.75	30.75
Withania somnifera	70	32	78	43	73	37	69	31	72.50	35.75

RC: Root Colonization; SP: Spore Population; Q: Quarter.

Table 10.3: Showing Per cent Root Colonization and Spore Population of Medicinal Plants Understudy in Banka Region

Plant Species	*Jan-Mar* (Q_1)		*Apr-Jun* (Q_2)		*Jul-Sep* (Q_3)		*Oct-Dec* (Q_4)		*Yearly Average*	
	RC	*SP*	*RC*	*SP*	*RC*	*SP*	*RC*	*SP*	*RC*	*SP*
Achyranthes aspera	10	19	17	28	15	26	12	21	13.50	23.50
Andrographis paniculata	37	26	49	35	40	33	38	28	41.00	30.50
Asparagus racemosus	70	18	82	24	77	20	72	20	75.25	20.50
Catharanthus roseus	29	20	37	28	33	24	30	19	32.25	22.75
Euphorbia thymifolia	7	8	13	15	11	14	9	12	10.00	12.25
Costus speciosus.	36	15	44	23	40	21	38	17	39.50	19.00
Rauvolfia serpentine	37	25	48	32	41	29	38	27	41.00	28.25
Rauvolfia tetraphylla	41	17	54	28	48	23	42	19	46.25	21.75
Woodfordia fruticosa	60	20	72	32	65	30	61	22	64.50	26.00
Withania somnifera	66	29	75	37	71	32	68	30	70.00	32.00

RC: Root Colonization; SP: Spore Population; Q: Quarter.

Table 10.4: Showing Per cent Root Colonization and Spore Population of Medicinal Plants Understudy in Jamui Region

Plant Species	Jan-Mar (Q_1)		Apr-Jun (Q_2)		Jul-Sep (Q_3)		Oct-Dec (Q_4)		Yearly Average	
	RC	SP	RC	SP	RC	SP	RC	SP	RC	SP
Achyranthes aspera	14	24	20	32	18	30	15	24	16.75	27.50
Andrographis paniculata	47	30	58	43	50	40	48	32	50.75	36.25
Asparagus racemosus	80	18	87	29	82	24	79	20	82.00	22.75
Catharanthus roseus	33	23	40	34	38	31	34	27	36.25	28.75
Euphorbia thymifolia	11	8	19	12	16	14	12	9	14.50	10.75
Costus speciosus	42	15	49	25	43	24	40	19	43.50	20.75
Rauvolfia serpentina	42	24	51	37	47	32	43	28	45.75	30.25
Rauvolfia tetraphylla	45	23	53	34	51	30	46	27	48.75	28.50
Woodfordia fruticosa	60	27	68	38	63	32	61	25	63.00	30.50
Withania somnifera	68	34	74	42	72	17	69	32	70.75	31.25

RC: Root Colonization; SP: Spore Population; Q: Quarter.

Spore Population (SP)

The rhizosphere soil samples of all the 10 medicinal plants were collected from survey sites at the interval of 3 months *i.e.*, January–March, April–June, July–September and October–December and were analyzed for AMF spore population for getting annual status of mycorrhizal population under the influence of variable survey site factors. Results are depicted in Tables 10.2–10.4. Spore population was noticed to vary with sites, plants, and period of study. In Bhagalpur region, *W. somnifera* showed presence of maximum number of spores/10 gm dry soil *i.e.* 32, 43, 37 and 31 during 1st, 2nd, 3rd and 4th quarters respectively with yearly average of 36 whereas the minimum SP was recorded in *E. thymifolia* with yearly average of 14. In Banka the maximum SP was recorded in *W. somnifera* with yearly average of 32, whereas it was found minimum in *E. thymifolia* with yearly average of 13. Similarly in Jamui samples it was highest (43) in case of *A. paniculata* and recorded minimum (8) in case of *E.thymifolia*. However, definite correlation could not be established between root colonization and spore population as *A. racemosus* showed maximum per cent of root colonization in all sites whereas spore population was highest in case of *W. somnifera* both in Bhagalpur and Banka soil samples and *A. paniculata* in Jamui.

Seasonal Variations

Investigation on seasonal variation of AM fungi has also been carried out by several earlier workers including Srinivas *et al.* (1988); Rachel *et al.* (1990); Mago and Mukerji (1994) and they confirmed variation in the incidence of AM fungi during different time periods of the year. Such seasonal variation in AMF population also affects the establishment of plants under field conditions (Boerner, 1986). There are reports on seasonal diversity in AM fungi associated with crops and other plant species (McGee, 1989; Lopez and Honrubia, 1992; Johnson–Green *et al.*, 1995 and Siguenza *et al.*, 1996); however, fragmentary report is available so far with regard to medicinal plants under natural habitat. In order to record it medicinal plants were screened from all the sites under study at a regular interval of 3 months representing different seasons such as Autumn (Jan-March), Summer (Apr-June), Rainy

(July- Sept) and Winter (Oct-Dec). The results are shown in Tables 10.2–10.4. The observation reveals that the degree of infectivity in terms of root colonization and spore population of AMF varied with seasons and also to some extent to the type of plants. The comparative study of results reflects that the summer season favours both the number of spores and root colonization in case of almost all plants. Of all *A. racemosus* showed maximum RC (83 per cent) during this period but spore population in rainy season. In general the spore population was found not always highest in the summer but in some cases it was also noticed maximum in rainy period. The winter season did not support comparatively better RC and SP than any other seasons in case of all plants. However, *Achyranthus aspera, Andrographis paniculata,* and *E. thymifolia* showed presence of minimum number of spores during autumn. The results clearly point out in general that the maximum RC and SP were noticed during summer followed by rainy, autumn and winter season. Mohan Kumar and Mahadevan (1984) have also recorded similar result that more number of spores were present in soil in summer whereas least number in rainy season. Maximum number of spores is encountered during July and October and further decreased in the subsequent months. To the contrary Verma (1999) observed that low temperature of winter season favoured the colonization of fine endophyte (*Glomus tenue*) in root cortex of *Eupatorium* species.

Taxonomy and Distribution of AMF

Morton and Benny (1990) proposed a revised classification of arbuscular mycorrhizal fungi with a new order Glomales, and two sub orders Glomineae and Gigasporineae and two new families Acaulosporaceae and Gigasporaceae. Gigasporaceae includes *Gigaspora* and *Scutellospora,* Glomaceae includes *Sclerocystis* and *Glomus* and Acaulosporaceae includes *Acaulospora* and *Entrophospora.* Morton (1998) has listed 126 species of AM fungi which includes 22 species of *Acaulospora,* 3 species of *Entrophospora,* 6 species of *Gigaspora,* 67 species of *Glomus,* 9 species of *Sclerocystis* and 19 species of *Scutellospora.* In India, important contributions on the taxonomy of AM fungi were made by Bakshi (1974) and Mukerji and Kapoor (1990). Since vesicles and arbuscules formation is not uniform in the genera of AM fungi. Berch (1986) proposed the use of term vesicular arbuscular mycorrhizae (VAM) for the genera *Glomus, Sclerocystis, Acaulospora* and *Entrophospora,* which produce both vesicles and arbuscules, whereas arbuscular mycorrhizae (AM) for the genera *Gigaspora* and *Scutellospora* which form arbuscules only.

A large number of species belonging to five genera of AMF *viz. Glomus aggregatum, Glomus australe, Glomus deserticola, Glomus fasciculatum, Glomus geosporum, Entrophosphora colombiana, Scutellospora calospora, Sclerocystis pakistanika, Gigaspora margarita, Glomus multicauli,* and *Glomus maculosum* were isolated from selected medicinal plants of forest areas of eastern Bihar.

The fungi were identified on the basis of morphology of Chlamydospore, Azygospore, Zygospore, subcellular structures, subtending hyphae and lamination/ornamentation of spore. One species each of *Sclerocystis, Entrophosphora, Gigaspora* and *Scutellospora* and seven species of *Glomus* were found to constitute all together 11 species of VAM fungal flora. Out of five genera, *Glomus* was found to present dominantly in all the surveyed forest areas. Several earlier workers (Kumaran and Santha Krishnan, 1995; Lekha *et al.,* 1995; Mehrotra, 1998; Trimurthulu and Johri, 1998) have also reported the dominance of *Glomus* in different plants. The identification of AM fungi is mainly based on the morphology of the resting spores. However, many investigators realized that they are insufficient to identify the genus or species. Spore wall structure and ornamentation characters are very frequently used to identify the species (Tiwari *et al.,* 1982). Some efforts have also been made to identify the AMF spore through serological tests (Aldwell *et al.,* 1983). Mosse (1990) stated that sequencing of DNA and RNA helps in determining the taxonomy of AM fungi. Allen (1991) investigated Restriction Fragment Length

Polymorphism (RFLP) technique is found highly specific and useful to differentiate AM fungi. Recently, Bonito *et al.* (1995) employed the PCR technology for the detection of an AM fungus in roots of different plant species.

Table 10.5: Distribution of AMF Species in Selected Medicinal Plants in Survey Sites

Plant Species	Bhagalpur	Banka	Jamui
Achyranthes aspera	*Glomus aggregatum, Glomus fasciculatum*	*Glomu smaculosum, Gigaspora margarita*	*Scutellospora calospora, Glomus deserticola*
Andrographis paniculata	*Glomus deserticola, Sclerocystis pakistanica*	*Glomus deserticola, G.multicauli*	*Glomus aggregatum, Glomus fasciculatum*
Asparagus racemosus	*Glomus maculosum, Gigaspora margarita*	*Scutellospora calospora, Glomus deserticola*	*Glomus deserticola, G.multicauli,*
Catharanthus roseus	*Glomus multicauli, Sclerocystis pakistanica*	*Glomus australe, Gigaspora margarita*	*Glomus multicauli, Sclerocystis pakistanica*
Euphobia thymifolia	*Scutellospora calospora, Glomus deserticola*	*Scutellospora calospora, Glomus deserticola*	*Glomus aggregatum, Glomus fasciculatum*
Costus specious	*Glomus australe, Gigaspora margarita*	*Entrophosphora colombiana, Glomus aggregatum*	*Glomus maculosum, Gigaspora margarita*
Rauvolfia serpentina	*Glomus multicauli, Sclerocystis pakistanica*	*Entrophosphora colombiana, Glomus aggregatum*	*Glomus australe Gigaspora margarita*
Rauvolfia tetraphylla.	*Glomus fasciculatum, Gigaspora margarita*	*Glomus multicauli, Sclerocystis pakistanica*	*Glomus geosporum, G.maculosum*
Woodfordia fruticosa	*Entrophosphora colombiana Glomus aggregatum*	*Glomus geosporum, G.maculosum*	*Entrophosphora colombiana, Glomus aggregatum*
Withania somnifera	*Glomus geosporum, G.maculosum*	*Glomus geosporum, G.maculosum*	*Entrophosphora colombiana, Glomus aggregatum*

Earlier scanty reports available so far confirms the AMF role in increasing the level of active components of the plants but in the last decade some workers reported the role of AMF on medicinal plants (Harley and Harley, 1987; Campurbi *et al.*, 1990; Sharma and Roy, 1991; Chaterjee, 1992; Harikumar, 2001; Babu and Manoharcharya, 2003;). Chaterjee (1992) studied the possibility to improve the quality of some important medicinal plants such as *Catharanthus roseus*, *Papaver* sp. and *Ageratum* sp. by the association of AMF. Babu and Manoharacharya (2003) also reported the presence of AM fungi in the roots of alkaloid rich medicinal plants (about 15 medicinal plant species of different families). Reports indicated the association of many AMF species with medicinal plants. In *Euphorbia rosea,* two *Glomus* species were reported *i.e. G. pubescence* and *G. monosporus*, *Croton sparsiflorus* showed the mycorrhizal association of *Glomus darus*, *Acalypha indica* showed the mycorrhizal association of *Glomus pubescence. Euphorbia heterophylla* was mycorrhizal for *Glomus citricolum* (Raghupathy, Mohan Kumar and Mahadevan, 1984). In *Phyllanthus fraturnus* invariable mycorrhizal colonization (75 per cent–80 per cent) with secondary root was noticed (Mulani *et al.*, 2002). *Acaulospora* sp., *Entrophospora* sp., *Glomus* sp., and *Sclerocystis* sp. were found to be colonizing in the root and root soil sample of *Abrus precatorious* and *Asparagus racemosus. Gigaspora* sp., *Acaulospora* sp., *Glomus* sp. and *Sclerocystis* sp. were isolated from the roots of *Withania somnifera* (Babu and Manoharacharya, 2003). Six species of AM fungi were found to colonize the roots of alkaloid bearing medicinal plants (Ratti and Janardhanan, 1995). *Convolvulus arvensis* (100 per cent) was found associated with VAM fungi (Rani and Bhaduria, 2001). During the present investigation different AMF spores were isolated from 10 selected medicinal plants. The results show that seven species of *Glomus* was found to be associated with most of the plants along with some other species. The results are depicted in the Table 10.5.

During the course of study it was noticed that most of the plants having association of minimum two AMF species, which clearly indicates the presence of mixed AMF in the rhizosphere soil of medicinal plants of survey sites. However, *Glomus* was recorded as the dominant AMF in all the three survey sites. Several earlier workers (Lekha *et al.*, 1995; Mehrotra, 1998; Trimurthulu and Johri. 1998) have also reported the dominance of *Glomus* in different plants.

Acknowledgements

Authors are thankful to Head, University Department of Botany, T.M. Bhagalpur University, Bhagalpur for providing laboratory facilities.

References

Abbott, L.K. and Robson, A.D. (1991). Field management of VAM fungi. In Keister D.L. and Cregan P.B. (eds.). *The Rhizosphere and Plant Growth*, Kluwer Academic publishers, Dordrecht, London, pp.355-362.

Abbott, L.K. and Robson, A.D. (1991). Field management of VAM fungi. In Keister D.L. and Cregan P.B. (eds.). *The Rhizosphere and Plant Growth*, Kluwer Academic publishers, Dordrecht, London, pp. 355-362.

Aldwell, F.E.B., Hall, I.R. and Smith (1983). The identification of VAM fungi using immuno-fluorescence. *Soil Biol. Biochem.*, 15:439-445.

Allen, M.F. (1991). *The Ecology of Mycorrhizae*. Cambridge University press, pp.184.

Babu, K.S. and Manoharachary, C. (2003). Occurrence of AM fungi in rhizosphere soils of some medicinal plants. *Indian Phytopath.*, 56(2): 223-227.

Bakshi, B.K. (1974). *Mycorrhiza and its rôle in forestry*, PL 480 Project report. Forest Research Institute, Dehradun, pp. 89.

Berch, S.M. (1986). Endogonaceae taxonomy, specificity, fossil record, phylogeny. *Front. Appl. Microbiol.*, 2: 161-188.

Boerner, R.E.J. (1986). Seasonal nutrient dynamics, nutrient re-absorption and mycorrhizal infection intensity of two perennial forest herbs. *Amer.J. Bot.*, 73: 1249-1257.

Bonito, R.D., Elliot, M.L. and Des Jardin, E.A. (1995). Detection of an AM fungus in roots of different plant species with the PCR. *Appl. Environ. Microbiol.* 61(7): 2809-2810.

Camprubi, A.; Estann, V.; Calvet.; Pera, J. (1990). Infectivity and affectivity of *Glomns mosseae* Mycorrhizal in four different species of medicinal plants, *Symbiosis (Rehorot)*, 9 (1-3): 305–307.

Chatterjee, T. (1992). In Abstract vol. of *Nat. sym. Management of Microbes*, Allahabad University.

Daniels, B.A. and Trappe, J.M. (1980). *Mycologia*, 72: 457-471.

Gerdemann, J.W. and Nicolsson, T. H. (1963). Spores of mycorrhizal *Endogone* species extracted from soil by wet sieving and decanting technique. *Transactions of British Mycological Society*, 46: 235-243.

Hari Kumar, K. M. and Bagyaraj, D. J. (1988). Effect of intercropping and organic Soil amendments on native VAM. Mycorrhizae for Green Asia. *Proceedings of the first Asian Conference on Mycorrhizae*, pp. 207-208.

Hari kumar, V.S. (2001). Arbuscular Mycorrhizal synthesis in some wetland plants in Kerala *Mycorrhizal News*. 12(4): 14–15.

Harly, J.L. and Harley, E.L. (1987). A checklist of mycorrhiza in the British flora. The phytologist. *Can. J. Bot.*, 73(9): 1318–1327.

Hayman, D. S. (1978). *Endomycorrhizal* pp. 401–442 in interaction between non-pathogenic sol microorganizeam and plants Edited by X. R. Dommergues and S. K. Kyup, Elsevier, Amsterdam (1978).

Johnson-Green, P.C., Kenkel, N.C. and Booth, T. (1995). The distribution and phenology of arbuscular mycorrhizae along an inland salinity gradient.*Can.J. Bot.* 73 (9): 1318-1327.

Koske R. E. and Gemma, J. N. (1992). In mycorrhizal functioning: A Integrative plant fungal process (eds., Allen M. F.), Chapman and Hall, New York, USA, pp. 3–36.

Koske, R. E. (1987). Distribution of VAM fungi along a latitudinal temperature gradient. *Mycologia*, 79: 55–68.

Kothari, I. L. Udhyaya Joseph and Patel, C. R. (1997). Mycorrhizal biodiversity of petro effluents irrigated fields. *J. Mycol. Plant Pathol.*, 27(3): 275–278.

Kruckalmann, H.W. (1975). Effect of fertilizer, soils tillage and plant species on the frequency of *Endogone* chlamydospores and mycorrhizal infection in arable soils, In: *Endomycorrhizas* (Eds., Samders F. E., Mosse, B. and Tinker P. B.) Academic Press, London, pp. 511–525.

Kumaran, K. and Santha Krishnan, P. (1995). Vesicular arbuscular mycorrhizal fungi in tea (*Camellia sinensis* L.O. Kuntz.) soil. In: *Mycorrhizae: Biofertilizer for the Future* (eds.) Alok Adholeya, Sujan Singh. *Proceedings of the Third National Conference on Mycorrhiza*, pp. 33-37.

Lekha K. S., prasad, Joseph P. and Vejayan M. (1995). *Glomus fasciculatum* a predominant VAM fungus associated with Black pepper in forest soil of Kerala. In: *Mycorrhizae: Biofertilizer for the Future. The Proceedings of Third National Congress on Mycorrhizal* (Eds. Adholeya, Alok and Singh, Sujan) TERI, pp. 81-85.

Lopez–Sanchez E. and Honrubia M. (1992). Seasonal variation of VAM in eroded soils from southern Spain, *Mycorrhiza*, 2: 33–39.

Mago, P. and Mukerji, K. G. (1994). VAM in Lamiaceae: I. Seasonal variation in some member. *Phytomorphology*, 44(102): 83–88.

Mc Gee, P. A. (1989). Variation in propagule members of VAM fungi in a semi–arid soil. *Mucol. Res.* 92: 28–33.

Mehrotra, V. S. (1998). Arbuscular mycorrhizal associations of plants colonizing coalmine spoil in India. *Science*, Cambridge, 130: 124.

Mohan Kumar, V. and Mahadevan, A. (1984). Do secondary substances inhibit mycorrhizal association. *Current Science*, 53: 377-378.

Morton, J. B. and Benny, G. L. (1990). Revised classification of AMF (Zygomycetes). A new order Glomales two new suborders Glomineae and Gigasporineae and two new families Acaulosporaceae and Gigasporaceae with identification of Glomaceae. *Mycotaxon*, 37: 471–494.

Morton, J.B. (1998). Taxonomy of VAM fungi: Classification, nomenclature and identification. *Mycotaxon*, 32(Jul-Sept). 267-324.

Mosse, B. (1990). Mycorrhizal symbiosis and plant growth. *Procedings of the 2nd National Congress on Mycorrhizal*. (Eds. D.J. Bagyaraj and A.Manjunath). Bangalore (Abstract).

Mukerji, K. G. and Kapoor, A. (1990). Taxonomy of VAM fungi with special reference to Indian taxa. In G. P. Agrawal (Ed) *Perspectives in Mycological Research*. Today and Tomorrow's Printers and Publishers, New Delhi, PP. 7-16.

Mukherjee, K.G., Sabharwal, A., Kochar, B. and Ardey, J. (1984). VAM. Concepts and Advances. In K.G.V.P. Agnihotri and R.P.Singh (Eds.) *Progress in Microbial Ecology*, pp. 489-525.

Mulani, R.M., Prabhu, R. Rajendra and Manjusha Dinkaran (2002). Occurrence of VAM in the roots of *Phyllanthus fraternus* Webster. *Mycorrhizal News*, 14(2): 11-14.

Nicolson, T.H. and Schenck, N. C. (1979). *Mycologia*, 71: 178-198.

Phillips, J. M., Hayman, D.S. (1970). Improved procedures for clearing roots and staining parasites and vesicular-arbuscular mycorrhizal fungi for rapid assessment of infection. *Transactions of the British Mycological Society* 55: 15-161.

Raghupathy, S.; Mohankumar, V. and Mahadevan, A. (1988). Distribution of VAM in Thanjavur district flora pp. 96–98. In: *Mycorrhizae for Green Asia*, edited by A. Mahadevan; N. Raman and K. Natarajan Madras; University of Madras (351pp.)

Rachel, E. K., Reddy, S. R. and Reddy.S. M. (1990). Seasonal variation in the colonization and spore population of VAM fungi in semi arid of A. P., *Nat. Acad. Sci. Letters.*, pp. 13–16.

Rani, V. and Bhaduria, S. (2001). VAM association with some medicinal plants growing on alkaline soils of Mainpuri district, Uttar Pradesh, *Mycorrhiza News*, 13(2): 12–14.

Ratti, Neelima and Janardhanan, K. K. (1995). VAM association in some alkaloid bearing plants. Mycorrhizae : biofertilizers for the future, *Third National Conference on mycorrhiza* (eds. Adholeya, Alok and Singh, Sujan) pp. 407–409.

Roy, A. K., Singh, A. N. and A. Arshi (2004). Dynamics of rhizosphere microflora in nutrient deficient OBD soil of ECL. *Proceedings of National Symposium on Current Perspective in Stress Biology*. 6–8 Feb. 2004, NBU, Siliguri.

Schenck, N. C. and Perez, Y. (1988). *Manual of the Identification of VA Mycorrhizal Fungi*, University of Florida, Gainesville, GL, 245pp.

Shanker, A., Mathew J. Neeraj Kaur, Mehrotra, R. S. and Verma, A. (1990). Mycorrhizal status of some desert plants and their physiological significance. In: *Trends in Mycorrhizal Research. Proceeding of the National Conference on Mycorrhizae*, HAU, Hissar, pp. 160–161.

Sharma, M. and Roy, A. K. (1991). VAM in hypoglycaemic plants. *Nat. Acad. Sci. Letters*, 14(12): 467–468.

Siguenza, C., Espejel, I. And Allen, E. B. (1996). Seasonality of mycorrhizae in coastal sand dunes of Baja California. *Mycorrhiza*, 6(2): 151–157.

Singh, C. S. Deepa Jha and Jha D. (1994). Mass inoculum production of VAM effect of various bacteriological media and fertilizer solutions. *Microbiol. Res.*, 149: 27–29.

Smith, S. E. and Bowen, G. D. (1979). Soil temperature, mycorrhizal infection and nodulation of *Medicago trumcalila* and *Trifolium subterraneum*. *Soil Biol and Biochemistry*. 11: 469–473.

Srinivas, K. Shanmugam, N. and Ramraj, B. (1988). Effect of VAM fungi on growth and nutrient uptake of forest tree seedlings. *Proc. 1st Asian conf. Mycorrhizae*, Madras. India. pp. 294–297.

Tiwari, J. P., Skorpod, W. P., Mukherji, K. G. and Mishra, S. (1982). *Trans. Br. Mycol. Soc.* 76: 207–304.

Trinick, M. J. (1977). *New Phytologist*, 78: 207–304.

Trimurthulu, N. and Johri, B. N. (1998). Prevalence and distribution of VAM spore population in different tarai soils of Uttar Pradesh. *J. Mycol. Pl. Pathol.* Vol. 28 No. 3, 236–239.

Verma, A. (1999). In Verma, A., Hock, B. (eds.) *Mycorrhiza*. Springer-Verlag, pp. 521–556.

Microbial Biotechnology and Ecology (2011) *Pages* **201–208**
Editors: **Deepak Vyas, G.S. Paliwal, P.K.Khare and R.K Gupta**
Published by: **DAYA PUBLISHING HOUSE, NEW DELHI**

Chapter 11

Association of Vesicular Arbuscular Mycorrhizal Fungi with Ornamental Plants

☆ ***Bhaskar Chaurasia and P.K. Khare***

Department of Botany, Dr. H.S. Gour Vishwavidyalaya,
Sagar – 470 003, M.P.

ABSTRACT

The present chapter deals with the analysis of root and rhizosphere soil samples for composition of VAM population in ten selected ornamental plants. The VAM colonization in the root samples ranged from 63 to 100 per cent with maximum colonization in *Narcissus poeticus*. The maximum and minimum spore population in the rhizosphere soil were found in *Narcissus poeticus* (126 spores 25^{-1} g soil) and *Rosa* sp. (77 spore 25^{-1} g soil), respectively. Further, maximum number of vesicles (47 vesicles cm^{-1} root bit) were also observed in *Narcissus poeticus* and the lowest (21 vesicles cm^{-1} root bit) in *Rosa* sp.

A total number of 23 VAM species were identified belonging to five genera *i.e. Acaulospora, Gigaspora, Glomus, Sclerocystis* and *Scutellispora*. Genus *Glomus* was found dominant in the rhizosphere soil samples of all the ornamental plant species. *Glomus* species constituted 66 per cent of total VAM species composition followed by *Sclerocystis* (13 per cent), *Acaulospora* (13 per cent), *Gigaspora* (4 per cent) and *Scutellispora* (4 per cent). Alongwith three species of *Glomus i.e. G. aggregatum, G. ambisporum, G. fasciculatum, Acaulospora scrobiculata* and *Sclerocystis pakistanika* were found most frequent and abundant VAM species in the rhizosphere soil of all ornamental plants studied.

Observation of the present investigation indicates that VAM population attributes *i.e.* no. of spores and vesicles, species diversity and colonization, were more in herbaceous plants as compared to woody species.

Introduction

Vesicular Arbuscular Mycorrhizal (VAM) fungi have extremely wide host range due to their non specificity. Vascular plants of all life forms association with VAM fungi and its importance for plant growth have been fully appreciated. The VAM fungi exhibit a perfect symbiosis and this interaction in turn increase plant growth, enhance accumulation of plant nutrients through greater soil exploration (Abbot and Robson, 1989) and improve soil health (Hodge *et al.*, 2001). The primary benefit for the plant is the increase in phosphorus and nutrient uptake from soil, although protection against certain pathogens, production of plant growth hormones and the increased solubility of soil minerals are also regarded as important benefits (Ahmadjian, 1986). Further, due to their wide ecological amplitude, they are found in diverse habitats ranging from arctic to tropics (Dalpe and Aiken, 1998; Chaurasia, 2001), arid to aquatic (Ba *et al.*, 1996) and climax to highly disturbed communities (Chaurasia *et al.*, 2007; Sastry and Johri, 1999). In view of the above importance and needs of VAM associations and also the paucity of work done there is a need to screen each and every habitat for VAM population and their distribution. The present work was carried out to study the isolation, and identification of indigenous VAM species associated with the rhizosphere soil of ornamental plants.

Materials and Methods

Study Site

The Botanic Garden of Dr. H.S. Gour Vishwavidyalaya is recognized by Botanic Garden Conservation International. This site was selected with a view that it consists of plants from different climatic region *i.e.*, Himalayan, Southern, Sea shore, desert and vegetation of Central India. Since it is actively managed for soil and fertility, the soil is clay loam. A number of forest, ornamental and rare plants are maintained and propagated here. Plants from all facets of groups *i.e.*, angiosperms, gymnosperms, pteridophytes, bryophytes are growing here luxuriantly. Further, plants from different habitats *i.e.*, forest, grassland, desert, aquatic, high altitude regions, are maintained.

Collection of Soil and Root

The soil and root samples were randomly collected from rhizosphere of selected plants. Soil samples were taken by digging out a small amount of soil close to a plant to a depth of 10-15 cm after removing the top soil. The samples were taken in three replicates. All the samples were brought to the laboratory in labelled polyethylene bags.

Roots were excavated carefully with the help of hoe and care was taken to prevent damage to the fine roots. Fine feeder roots were separated and lightly shaked. Root samples were collected in three replicates from each plant species and kept in labelled polyethylene bags.

Processing and Assessment of Colonization of Root Samples

Freshly collected root samples were rinsed with tap water several times to wash the roots thoroughly. Roots were cut into small bits of about 1 cm pieces and modified procedure (Kormanik *et al.*, 1980) was used for clearing and staining. Alternatively presence or absence of colonization was recorded in each of the 10 pieces. Minimum 100 root segments were used for this method.

Extraction, Estimation and Identification of VAM Spores

Gerdemann and Nicolson's (1963) wet- sieving and decanting procedure was adopted for isolation of VAM spores. For estimation of VAM spores, a modified method of Smith and Skipper (1979) by Gour and Adholeya (1994) was followed. VAM fungal identification was done following different identification keys (Hall, 1984; Morton, 1988; Schenck and Perez, 1990; Mehrotra and Baijal, 1994).

Analysis of Data

VAM colonization (per cent): The per cent VAM colonization was calculated by using following formula:

$$\text{Per cent VAM colonization} = \frac{\text{Total no. of infected root pieces}}{\text{Total no. of root pieces studied}} \times 100$$

Relative density: Relative density of VAM species with each host species was calculated following the under mentioned formula.

$$\text{Relative Density} = \frac{\text{No. of VAM species associated with the host species}}{\text{Total no. of VAM species associated with the entire host considered}} \times 100$$

Results and Discussion

As evident from the results that there existed consistency in almost all ornamental plant species in having higher VAM colonization, number of vesicles and spore population (Table 11.1).

Narcissus poeticus showed 100 per cent VAM colonization and maximum number of vesicles (47 cm^{-1} root bit) in their roots and highest VAM spore population (126 spores $25g^{-1}$ soil) occurred in its rhizosphere soil. *Rosa* sp. comes out to be the poorest host for VAM fungi as the colonization, number of spores, vesicles (Figure 11.1) and number of VAM species associated with it were minimum (Table 11.1). All ornamental plants selected for study are herbaceous in nature except *Rosa* species. Several workers reported that herbaceous plants are extensively colonized by VAM fungi (Jain *et al.*, 1997, Chaurasia and Khare, 1999).

Table 11.1: Different Attributes of VAM Fungi Associated with Some Ornamental Plants

Sl.No.	*Name of Species*	*Colonization (per cent)*	*Average Number of Vesicles (cm^{-1} root bit.)*	*Spore Population ($25g^{-1}$ soil)*	*Total Number VAM Fungal Species*	*Dominant VAM Species*
1.	*Calendula officinalis*	75	37	105	12	ASCB, LAGR, LMCC
2.	*Chrysanthemum morifolium*	85	39	80	16	LFSC, ASCB, LAGR, SPKS
3.	*Dahelia pinnata*	95	24	105	10	ASCB, LABS, LFSC, SPKS
4.	*Dianthus caryophyllatus*	84	42	91	9	LAGR, LFSC, LABS, SPKS
5.	*Gerbera gamsoni*	87	41	110	15	LAGR, LFSC, ASCB, LMSS, SSNS
6.	*Narcissus poeticus*	100	47	126	20	ASCB, LMCC, LAGR, LFSC, SSNS
7.	*Phlox* sp.	94	40	95	9	ASCB, LAGR, LFSC, SPKS
8.	*Polianthes tuberosa*	66	37	87	7	LAGR, LABS, LFSC, SSNS
9.	*Rosa* sp.	63	21	77	6	LFSC, SPKS, LAGR, LABS
10.	*Tagates erectus*	74	28	102	12	ASCB, LHTS, LFSC, SPKS

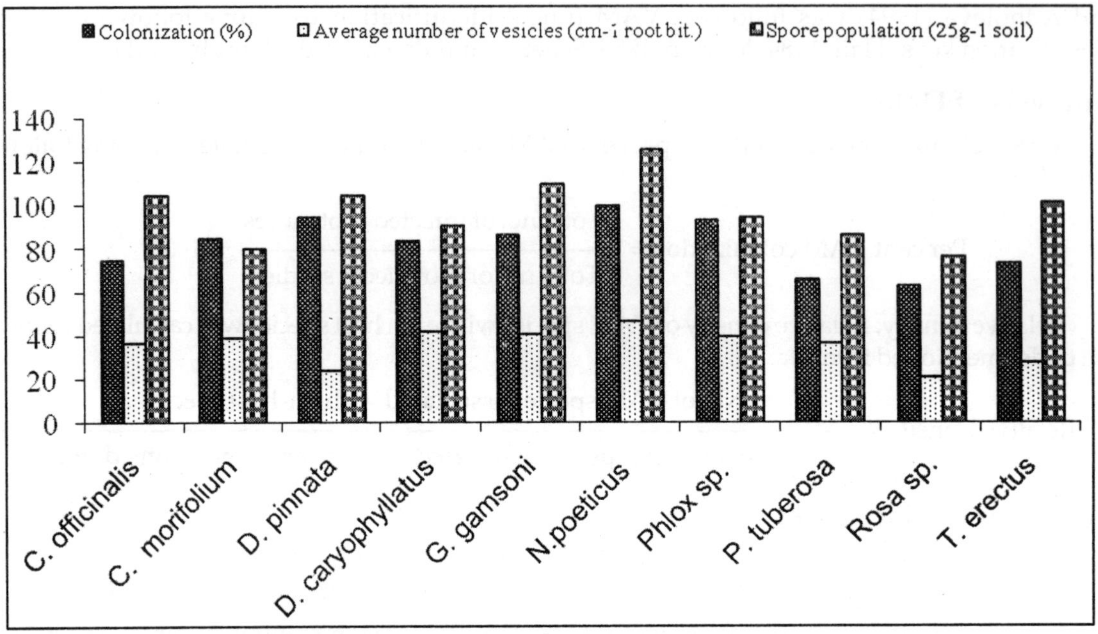

Figure 11.1: Different Attributes of VAM Fungi in the Rhizosphere Soil of Different Ornamental Plants at Botanical Garden

23 VAM species were recovered from rhizosphere soil of different ornamental plants at Botanical Garden. About 66 per cent of total VAM isolates of VAM fungi included *Glomus* species followed by *Acaulospora* (3 species), *Sclerocystis* (3 species), *Gigaspora* and *Scutellispora* (each with one species) (Table 11.2 and Figure 11.2).

Several workers reported predominance of *Glomus* species in different habitats (Bhadauria and Yadav, 1999; Lingon *et al.*, 1999, Chaurasia 2001, Chaurasia *et al.*, 2005 a and b; 2007). Predominance of *Glomus* at the present study sites as well as the reports given by a number of workers cannot be assertively interpreted due to paucity of data pertaining to the physiology and adaptiveness of this species.

Five other species *e.g. Calendula officinalis, Dahelia pinnata, Gerbera gamsoni, Narsissus poeticus* and *Tagatus erectus* possessed higher VAM spore population (>100 spore $25g^{-1}$ soil). It is evident from results that *A. scrobiculata, G. aggregatum, G. ambisporum, G. fasciculatum* and *S. pakistanica* were the dominant VAM species (Tables 11.2 and 11.3). Except for *Glomus* species other VAM species showed a poor frequency distribution. Similar to present findings, several workers reported predominance of *Glomus* species *viz. G. aggregatum* and *G. fasciculatum* species in different rhizosphere soil (Kunwar *et al.*, 1999; Lingon *et al.*, 1999, Chaurasia *et al.*, 2005 a and b; 2007). Besides these two *Glomus* species some other *Glomus* species were frequently distributed in the rhizosphere soils. Wide occurrence and distrubution of *Glomus* species in the rhizosphere soil may be attributed to its early evolution and can be considered one of the few plant fungus associations with a fossil record (Pirozynski and Dalpe, 1989; Taylor *et al.*, 1995; Phipps and Taylor, 1996; Helgason *et al.*, 1998). Allen (1991) reported that uneven distribution of VAM fungi in rhizosphere of different species may be dependent on the nature of host, host growth pattern and age of plant.

Table 11.2: Distribution of Different VAM Fungi in Rhizosphere Soil of some Important Ornamental Plant Species at Botany Garden

Sl.No.	VAM Sp. Code	Plant Species										Frequency %
		Calendula officinalis	*Chrysan-themum morifolium*	*Dahelia pinnata*	*Dianthus caryo-phyllatus*	*Gerbera gamsoni*	*Naricissus poeticus*	*Phlox* sp.	*Polianthes tuberosa*	*Rosa* sp.	*Tagates erectus*	
1.	ASCB	+	+	+	+	+	+	+	+	+	+	100.00
2.	ASPN	–	–	–	–	–	–	+	–	–	–	10.00
3.	AUDL	+	+	–	–	–	+	–	–	–	–	30.00
4.	GRSA	+	+	–	–	–	+	–	–	–	+	40.00
5.	LAGR	+	+	+	+	+	+	+	+	+	+	100.00
6.	LABS	+	+	+	+	+	+	+	+	+	+	100.00
7.	LCST	+	+	–	+	+	+	–	–	–	–	50.00
8.	LETC	–	+	–	–	+	+	+	–	–	–	40.00
9.	LFSC	+	+	+	+	+	+	+	+	+	+	100.00
10.	LGSP.	+	+	–	–	–	+	+	–	–	+	40.00
11.	LHTS	–	+	+	+	–	+	–	–	–	+	50.00
12.	LINR	–	+	+	–	+	+	–	–	–	+	50.00
13.	LMCC	+	+	–	–	+	+	–	–	–	–	40.00
14.	LMAG	–	–	–	–	+	+	+	–	–	–	30.00
15.	LMSS	–	+	+	–	+	+	–	–	–	+	50.00
16.	LOCT	–	–	+	+	–	+	–	–	–	–	30.00
17.	LTNB	–	+	–	–	–	+	–	–	–	+	30.00
18.	G.sp.B.	+	–	–	+	+	–	–	–	–	–	30.00
19.	G.sp.C.	–	–	–	–	–	–	–	+	+	–	20.00
20.	SMRC	–	–	–	–	+	+	–	–	–	+	30.00
21.	SPKS	+	+	+	+	+	+	+	+	+	+	100.00
22.	SSNS	+	+	–	–	+	+	–	+	–	–	50.00
23.	CARN	–	–	+	–	+	+	–	–	–	–	30.00

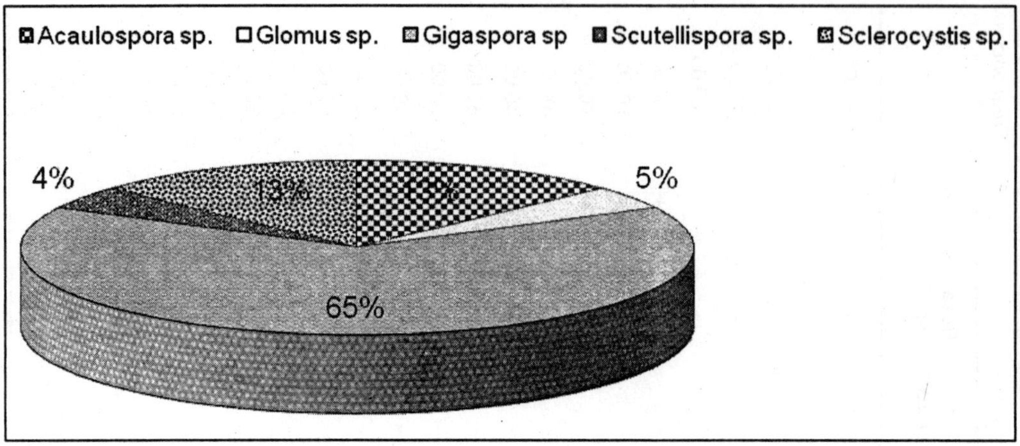

Figure 11.2: VAM Species Composition in the Rhizosphere Soil in the Ornamental Plants

Table 11.3: Association of VAM Species and their Relative Densities with Different Ornamental Plants

Sl.No.	Name of Plants Species	No. of VAM Species Associated	Relative Density
1.	*Calendula officinalis*	12	10.34
2.	*Chrysanthemum morifolium*	16	13.79
3.	*Dahelia pinnata*	10	8.62
4.	*Dianthus caryophyllatus*	9	7.75
5.	*Gerbera gamsoni*	15	12.93
6.	*Narcissus poeticus*	20	17.24
7.	*Phlox* sp.	9	7.75
8.	*Polianthes tuberosa*	7	6.03
9.	*Rosa* sp.	6	5.17
10.	*Tagates erectus*	12	10.34

In conclusion the important observation of the present investigations are (1) there is no correlation among per cent colonization, spore population, species diversity and (2) herbaceous species having higher VAM attributes in comparison to woody ornamental plants. This report gives a preliminary account of VAM fungi associated with ornamental plant species. Ornamental plants are economically important and role of VAM fungi in their productivity still needs considerable attention.

References

Abbott, L.K. and Robson, A.D. (1984). The effect of mycorrhizae on plant growth, 113-130. In: *VA mycorrhiza,* (eds) CL, Powell, D. J. Bagyaraj. CRC Press, Boca Raton, Florida USA..

Ahmadjian, V. (1986). *Symbiosis*. University Press of New England, Hanover, New Hampshire, pp. 212.

Allen, M. F. (1991). *The ecology of mycorrhizae*. Cambridge University Press England.

Ba A.M., Dalpe Y., Guissou T. (1996). Glomales of Acacia holosericea and Acacia mangium. *Bios For Trop*. 25: 5-18.

Bhadauria, S. and Yadav, R. (1999). Vesicular-arbuscular mycorrhizal associations in fuel wood trees growing in alkaline soil, *Mycorrhiza News*, 10 (4): 14-15.

Chaurasia B. (2001). Ecological study of tropical forests with special reference to vesicular Arbuscular mycorrhizal (VAM) association. *Ph. D. Theis*, Dr. H.S. Gour V.V. Sagar, M.P.

Chaurasia B. and Khare P. K. (1999). VAM fungi associations in a protected grassland. *Mycorrhiza News*, 11 (1): 13-14.

Chaurasia B, Pandey A. and Palni L.M.S. (2007). Arbuscular mycorrhizal species diversity in Kumaun region of Indian Central Hiamlaya. In: *The Mycorrhizae: Diversity, Ecology and Application* (eds) Tiwari M. and Sati S. C., Daya Publishing House, New Delhi-110 035. 286-294.

Chaurasia B, Pandey A. and Palni L.M.S. (2005a). Occurrence of arbuscular mycorrhizae in the rhizosphere of Himalalayan Yew (*Taxus baccata* L. subsp. Wallichiana (Zucc.) Pilger)- a case study. In: *Basic Research and Applications of Mycorrhizae*. (eds) Podila G.K. and Verma A.K., IK International Publication, New Delhi, pp. 26-35.

Chaurasia B, Pandey A. and Palni L.M.S. (2005b). Distribution, colonization and diversity of arbuscular mycorrhizal fungi in Rododendrons of Central Himalayan region of India. *Forest Ecology and Management* 207: 315-324.

Dalpe Y. and Aiken S.G. (1998). Arbuscualr mycorrhizal fungi associated with *Festuca* species in the Canadian High Arctic. *Canadian Journal of Botany* 76: 1930-1938.

Gerdemann, J.W. and Nicolson, T.H. (1963). Spores of mycorrhizal Endogone species extracted from soil by wet sieving and decanting. *Trans Br. Mycol. Soc.* 46: 235-244.

Gour, A., Adholeya, A., Mukerji, K. G. (1998). A comparison of AM fungi inoculants using *Capsicum* and *Polianthes* in marginal soil amended with organic matter. *Mycorrhiza*, 7 (6): 307-312.

Hall, I.R. (1984). Taxonomy of VA mycorrhizal fungi, pp.57-94. In: *VA mycorrhiza*, ed. by C.L.Powell and D.J. Bagyaraj. CRC Press, Boca Raton, FL.

Helgason, T., Daniell, T. J., Husband, R., Fitter, A. H. and Young, J. P. W. (1998). Ploughing up the wood wide web ?. *Nature*. 394 (30): 431.

Hodge A., Campbell C.D., and Fitter A.H. (2001). An arbuscular mycorrhizal funbgus accelerates decomposition and acquires nitrogen directly from organic material. *Nature* 413: 297-299.

Jain, R.K., Hasan, N., Singh,R.K., and Pandey, S.N. (1997). Vesicular-arbuscular mycorrhizal (VAM) associations in some weeds of forage legumes, *Mycorrhiza News*, 9(1): 10-12.

Kormanik, P.P., Bryan, W.C. and Schultz, R.C. (1980). Procedure and equipment for staining large number of plant roots for endomycorrhizal assay. *Can. J. Microbiol.* 26: 536-538.

Kunwar, J.K., Reddy, J.P.M. and Manoharachary, C. (1999). Occurrance and distribution of AMF associated with garlic rhizosphere soil. *Mycorrhiza New.* 11(2): 4-6.

Lingan, V.K., Tholkappian, D., and Sundaram, M.D. (1999). VA mycorrhizal fungi occurring in the mangrove vegetation of Pichavaram forest. *Mycorrhiza News*. Vol.1 (2): 6-7.

Mehrotra, V.S. and Baijal, U. (1994). Advances in the taxonomy of vesicular-arbuscular mycorrhizal fungi. *Biotechnology in India*. pp. 227-286.

Morton, J.B. (1988). Taxonomy of VA mycorrhizal fungi: Classification nomenclature, and indetification. *Mycotaxon.* 32: 267-324.

Phipps, C.J. and Taylor, T.N. (1996). Mixed arbuscular mycorrhizae from the triassic of Antartica. *Mycologia.* 88(5): 707-714.

Pirozynski, K.A. and Dalpe, Y. (1989). Geological history of the Glomaceae with particular reference to mycorrizal symbiosis. *Symbiosis,* 7: 1-36.

Sastry M.S.R. and Johri B.N. (1999) Arbuscular mycorrhizal fungal diversity of stressed soils of Bailadila iron ore sites in Bastar region of Madhya Pradesh. Current Science 77: 1095-1100.

Schenck, N.C. and Perez, Y. (1990). A manual for identification of vesicular-arbuscular mycorrhizal fungi. INVAM University of Florida. Gainesville, FL. 286 pp.

Smith, G.W. and Skipper, H.D. (1979). Comparison of methods to extract spores of vesicular-arbuscular-mycorrhizal fungi. *Soil. Sci. Soc. J.* 43: 722-725.

Taylor, T.N., Remy, W., Hass, H. and Kerp., H. (1995). Fossil arbuscular mycorrhizae from the early Devonian. *Mycologia.* 87: 560-573.

Microbial Biotechnology and Ecology (2011) *Pages* **209–224**
Editors: **Deepak Vyas, G.S. Paliwal, P.K.Khare and R.K Gupta**
Published by: **DAYA PUBLISHING HOUSE, NEW DELHI**

Chapter 12

Perspective of AM Fungi in Agroforestry Systems

☆ *Ashok Shukla*[1]*, *A. Kumar*[1], *D. Vyas*[2], *A. Jha*[1], *M. Kamalvanshi*[1] and *N. Chakravarty*[1]

[1]*National Research Centre for Agroforestry, Jhansi – 284 003, U.P.*
[2]*Dr. H.S. Gour University, Sagar – 470 003, M.P.*

ABSTRACT

Arbuscular mycorrhizal (AM) fungi form symbiotic association with most economically important plants. These fungi improve plant growth under low fertility conditions, increase tolerance against plant pathogens, improve water balance of the plants, contribute to the formation of soil structure and help plants to establish in new areas. In agroforestry systems trees are deliberately planted on the same unit of land with agricultural crops for rehabilitation of the degraded areas. The potential benefit of AM fungi in rehabilitation of degraded lands by use of agroforestry system is more apparent than ever before. The need to increase food, fiber and fuel wood production to keep pace with the fast growing population is crucial. The low biomass production of agroforestry tree species in degraded areas can, therefore, be circumvented by the use of AM fungi. The role of AM fungi in enhancing plant growth and yield, resistance to drought and salinity and tolerance to pathogens is well documented, but little is known about their interactions with agroforestry systems. The present chapter enlighten the potential benefit of AM fungi in rehabilitation of degraded lands by use of agroforestry systems.

Introduction

In 1885 A. B. Frank was the first to describe the symbiosis between a fungus and the roots of trees and he coined the term *Mycorrhiza.* The term originates from the Greek *mycos*, meaning 'fungus' and

* E-mail: ashokshukla06@gmail.com.

rhiza, meaning 'root'. Mycorrhiza is a symbiotic mutualistic relationship between special soil fungi and fine plant roots; it is neither the fungus nor the root, but rather the structure formed from these two partners. Since the association is mutualistic, both organisms benefit from the association. The fungus receives carbohydrates (sugars) and growth factors from the plant, which in turn receives many benefits, including increased nutrient absorption. In this association, the fungus takes over the role of the plant's root hairs and acts as an extension of the root system.

Mycorrhizal fungi are an important group of soil-borne microorganisms that contribute substantially to the establishment, productivity, and longevity of natural or man-made ecosystems (Harley and Smith, 1983). These fungi form symbiotic association with most terrestrial plant families (Trappe, 1977). In natural ecosystems mycorrhizal fungi can colonize much of the root system. Colonization is restricted to the root cortex and does not enter the vascular cylinder. The symbiosis is so well balanced that, although many of the host cells are invaded by the fungal endophyte, there is no visible tissue damage and under certain conditions it enhances the growth and vigour of the host plant. Because most economically important plants form mycorrhizae, the subject is currently attracting much attention in agricultural, horticultural, and forestry research. There are different types of mycorrhizae. The present chapter is concerned with AM fungi, information available in literature on its benefits, occurrence, taxonomy, ecology, culturing techniques and their role in agroforestry are being reviewed and being presented here, in short.

Arbuscular mycorrhizal fungi (AMF) are by far the most abundant of all mycorrhizas. They are characterized by the formation of an extraradical mycelium and branched haustorial structures within the cortical cells termed arbuscules. The latter are the main sites of exchange between the two symbiotic partners (Hock and Varma, 1995). In contrast to ectomycorrhizal fungi, AMF are considered to be obligate biotrophic microorganisms and cannot be cultured axenically. The general differentiation between symbiosis and pathogenicity is not easy and this is particularly true for mycorrhizas. Depending on environmental conditions, root colonization with AM fungi can either stimulate or inhibit plant growth (Sanders, 1993) and during the formation of the AM symbiosis several defence responses have been observed in the plant (Gianinazzi-Pearson *et al.*, 1996; García-Garrido and Ocampo, 2002).

Beneficial Effects of Mycorrhizae

The most prominent effect of AM fungi is that it plays an important role in enhancing the plant growth and yield due to an increase supply of phosphorus to the host plant. Mycorrhizal plants can absorb and accumulate several times more phosphate from the soil or solution than non–mycorrhizal plants. However uptake of nitrogen, zinc, copper and minor nutrients are enhanced as well. Improved plant nutrition is due to (i) increased root surface through extraradical hyphae, (ii) degradation of organic material and (iii) alteration of the microbial composition in the rhizosphere (Marschner, 1998; Hodge and Campbell, 2001). In exchange for the nutrients, the fungus is supplied with photosynthates (about 20 per cent of the plant assimilation capacity) and these assimilates are either respired or released into the soil. As a result of higher carbon drain, microbial rhizosphere composition is altered and therefore the term "mycorrhizosphere" was coined (Rambelli, 1973). As the majority of all plants are mycorrhizal, mycorrhizosphere versus rhizosphere might be the rule rather than the exception (Linderman, 1988). In addition, mycorrhizas increase drought resistance (Augé and Stodola, 1990) and heavy metal tolerance (Brundrett, 1991), improve soil aggregation (Andrade *et al.*, 1998) and promote overall plant health (Linderman, 1992).

Mycorrhiza increase root surface area for water and nutrients uptake. The use of mycorrhizal bio-fertilizer helps to improve higher branching of plant roots, and the mycorrhizal hyphae grow from the root to soil enabling the plant roots to contact with wider area of soil surface, hence, increasing the absorbing area for water and nutrients absorption of the plant root system. Therefore, plants with mycorrhizal association will have higher efficiency for nutrients absorption, such as nitrogen, phosphorus, potassium, calcium, magnesium, zinc and copper and also increase plant resistance to drought. Benefits of mycorrhizal bio-fertilizer are as follows:

- ✰ Allow plants to take up nutrients in unavailable forms or nutrients that are fixed to the soil, especially phosphorus which are immobile in the soil. In addition, mycorrhiza help to absorb other organic substances that are not fully soluble for plants to use, and also help to absorb and dissolve other nutrients for plants.
- ✰ Enhance plant growth, improve crop yield and increase income for the farmers.
- ✰ Reduce the use of chemical fertilizer.
- ✰ Improve plant resistance to root rot and collar rot diseases. Mycorrhizal association in plant roots will help plant to resist root rot and collar rot diseases caused by other fungi.
- ✰ Enhance water transport in plants, decrease transplant injury and promote establishment of plants in wasteland. Mycorrhizal plants recover better following moisture stress, more efficient in water use and frequently have higher root-shoot ratio than non-mycorrhizal plants.
- ✰ Increase plant growth by 20-30 per cent in terms of bio-mass.
- ✰ Enable the plants to grow and survive better under various stress conditions, like those in coal spoils, sand dunes, saline-alkaline soils, eroded- degraded sites and lands polluted by industrial wastes.

Table 12.1: Different Types of Mycorrhizae and their Hosts

Types of Mycorrhizae	*Host Plant/Families*
Ecto	Pinaceae, Fagaceae, Betulaceae, Leguminaceae, mostly Caesalpinoid legumes, Salicaceae, Tiliaceae, Rosaceae, Juglandaceae, etc.
Arbuscular	Majority of the plants including those important in agriculture, horticulture, pasture and tropical forests
Ericoid	Ericaceae and Epacridaceae
Orchidaceous	Orchidaceae
Arbutoid	Arbutus and Monotropha

Occurrence

VAM fungi have the widest host range and distribution of all the mycorrhizal associations. It is estimated that about 90 per cent of vascular plants normally establish mutualistic relationship with VAM fungi. VAM fungi have been observed in 1000 genera of plants representing some 200 families. There are at least 300,000 receptive hosts in the world flora (Kendrick and Berch, 1985) and there are about 120 species of VAM fungi (Schenck and Perez, 1987). If the hosts are divided up evenly among the fungi with no overlap in host range, each fungus would have more than 2500 potential partners. We know that host ranges overlap extensively, suggesting that some individual VAM fungi may well have access to thousands of hosts (Kendrick and Berch, 1985).

According to Gerdemann (1975), it is easier to list most plant families that do not form VAM than to list those that do. Families not forming vesicular arbuscular mycorrhizal fungi include Pinaceae, Betulaceae, Orchidaceae, Fumariaceae, Commelinaceae, Urticaceae, and Ericaceae (Table 12.2). Families that rarely form vesicular-arbuscular mycorrhiza include the Brassicaceae, Chenopodiaceae, Polygonaceae, and Cyperaceae. Families that form both ectomycorrhizae and vesicular arbuscular mycorrhizal fungi include Juglandiaceae, Tilliaceae, Myrtaceae, Salicaceae, Fagaceae, and Caesalpiniaceae (Gerdemann, 1975). Important crops with VAM fungi include wheat, maize, all millets, potatoes, beans, soybeans, tomatoes, apples, oranges, grapes, banana, castor, tobacco, tea, coffee, cocoa, sugarcane, mango, asparagus, rubber, cardamom, pepper, etc. Most of the tropical rain forest trees are arbuscular mycorrhizal (Janos, 1983). Harley (1969) has listed the gymnosperms in which VAM fungi have been observed. They are found in Pteridophytes (Cooper, 1976) and Bryophytes (Parke and Linderman, 1980). Recently, arbuscular mycorrhizal colonization has been reported in floating (Bagyaraj *et al.*, 1979) and submerged aquatic plants (Clayton and Bagyaraj, 1984). Usually arbuscular mycorrhizae are confined to the roots; they have been reported in diverse structures such as the modified leaves of water fern *Salvinia cucullata* (Bagyaraj *et al.*, 1979), fruiting peg of peanut (Graw and Rehm, 1977) and modified scale like leaves and rhizomes of ginger and canna (Selvaraj *et al.*, 1986).

Table 12.2: Important Plant Families and Tree Genera Known to Form Mycorrhizal Associations

Family	*Genus*	*ECM*	*VAM*
Betulaceae	*Alnus* (alder)	√	√
	Betula (birch)	√	√
Caesalpinoideae	*Cassia*	√	
	Non N-fixing trees	√	√
Casuarinaceae	*Casuarina*	√	√
Dipterocarpaceae	*Shorea*	√	√
Ebenaceae	*Diospyros*		√
Fagaceae	*Fagus* (beech)	√	
	Quercus (oak)	√	
Juglandaceae	*Carya* (hickory, pecan)	√	
Meliaceae	*Cedrela*		√
	Khaya (African mahogany)		√
	Swietenia (mahogany)		√
Mimosoldeae	*Acacia*		√
	Albizia		√
Myrtaccae	*Eucalyptus*	√	√
Paplionoideae	*Dalbergia* (rosewood)		√
	Pterocarpus		
Rosaceae	*Malus* (apple)		√
Rutaceae	*Citrus* (orange, lemon)		√
Sallcaceae	*Salix* (willow)	√	√
Commercial crops	*Caria* (papaya)		√
	Coffee (coffee)		√
	Hevea (rubber)		√
	Most palms		√

VAM fungi, in addition to their widespread distribution throughout the plant kingdom, are also geographically ubiquitous and occur in plants growing in arctic, temperate, and tropical regions (Mosse *et al.*, 1981). In general, VAM population is more in cultivated soil compared to virgin soil (Mosse and Bowen, 1968). They are mostly seen in the top 15-30 cm of soil, and their numbers decrease markedly below the top 15 cm (Redhead, 1977). They are normally not found in depths beyond the normal root range of plants (Mosse *et al.*, 1981). The distribution of species of VAM fungi varies with climatic and edaphic environment as well as with land use. For example, *Acaulospora laevis* is common in Western Australia (Abbott and Robson, 1977) and New Zealand (Mosse and Bowen, 1968) but occurs less frequently in soils of Eastern Australia (Mosse and Bowen, 1968). *Glomus* species appear to have widest distribution. *Gigaspora* and *Sclerocystis* species are more common in tropical soils. *Acaulospora* seems to be better adapted to soils with pH <5.0. In fact, certain VAM fungi have been linked to particular kinds of soil: *Glomus mosseae* with fine textured, fertile, high pH soils; *Acaulospora leavis* with coarse-textured, acid soils, and *Gigaspora* species with sand dune soils (Kendrick and Berch, 1985).

Classification of VA Mycorrhizal Fungi

VAM fungi, being obligate biotrophs, do not grow on synthetic media and hence are classified according to morphological characteristics of the spores formed in the soil. One hundred and twenty species have been described by Schenck and Perez (1987), and they are all grouped in a single family, the Endogonaceae. The family is considered to be in an order of its own, the Endogonales tentatively placed in the class Zygomycetes in the subdivision Zygomycotina (Benjamin, 1979). The seven genera, *Endogone, Glomus, Sclerocystis, Entrophospora, Acaulospora, Scutellospora and Gigaspora,* presently constitute the Endogonaceae. Of these, only *Endogone* forms zygospores characteristic of Zygomycetes, while others probably lack sexual reproduction, species of *Endogone* do not form VAM fungal association. The other six genera form vesicular-arbuscular mycorrhizal associations.

The main features of VAM fungi are summarized by Trappe and Schenck (1982) and Hall (1984). Hall and Abbott (1983) compiled photographic transparencies illustrating diagnostic features of the Endogonaceae. Walker (1983) has suggested a standardized terminology for use in description of endogonaceous spore walls involving a stylized graphic representation of the wall structure ("murograph"). This concept is recommended for future descriptions of new species. Recently, Schenck and Perez (1987) published a manual for the identification of VAM fungi.

Glomus, the most common genus of VAM fungi, has over 50 species, which form globose, ellipsoid or rather irregularly shaped spores that range from 20-400 µm. These spores are thick-walled (up to 30 µm) and hyaline, yellow, red-brown, brown or black. Spores are attached to a single subtending hypha (or to two, three, or more in exceptional species) and produced in the soil near plant roots, at the soil surface, or occasionally in roots, and they may form singly, in groups of a few to many, or in large aggregates called sporocarps. Sporocarps are most commonly found in undisturbed forest communities with perennial plants and accumulated organic material. Sporocarpic species, when grown on plants in pot culture, usually produce spores singly or in small aggregates. Some *Glomus* species (*e.g. Glomus lacteum*) are not known to form spores in sporocarps, while others form spores outside the plant root only infrequently (*e.g. Glomus intraradices*). The spores of most species germinate by emergence of a germ tube directly through the subtending hypha.

Sclerocystis species produce chlamydospores that are very similar to those of *Glomus*, except that they tend to be clavate rather than globose. All *sclerocystis* species form spores only in sporocarps up to 700 µm in diameter. The spores, which can be up to 125 µm long, develop in a single layer, radially arranged around a central plexus of sterile and sporogenous hyphae. The sporocarps may group together in masses up to several cm in diameter on the soil surface, on leaves, twigs, and mosses. There is no published information on the germination of *Sclerocystis* spores.

Spores of the genus *Acaulospora* appear to form laterally on the neck of a small thin-walled saccule, called variously a mother spore, a vesicle, a hyphal terminus, a soporiferous saccule, or a sporogenous saccule (Walker, 1987). When mature, the spore does not display a subtending hypha because the vesicular structure that gives rise to it loses its cytoplasm and collapses during development of the spore. The spores are globose or ellipsoid, from <100 to >400 µm in diameter, and hyaline, yellow, or reddish-brown. The surface of the sporewall may be ornamented with pits, projections of various shapes, folds, spines, or reticulations, and the wall is up to 12 µm thick. One species of *Acaulospora* is now known to produce sporocarps, but in other species, spores are normally found in the soil of the rhizosphere, *Entrophospora* is more or less similar to *Acaulospora* except that the spore is formed inside the parent hypha just below the vesicle.

Scutellospora spores can be greater than 600 µm in diameter, and the smallest ones are just under 200 µm in diameter. Spores develop terminally on a bulbous hyphal suspensor, which remains attached at maturity and may bear short lateral projections. Spores are usually globose but can be ovoid, obovoid, pyriform or irregular. Spores are hyaline, white, yellow, pinkish, grey-green, or brown. Spore wall structure is composed of at least two wall groups, with one or more flexible membranous or coriaceous walls in the inner group or groups. The genus is characterized in particular by its mode of spore germination. Germination is by means of one or more germ tubes produced near the spore base from a germination shield formed within or upon a flexible inner wall. Thin walled, knobby, or broadly papillate auxiliary cells (vesicles) are borne in soil on straight or coiled hyphae, formed singly or in clusters. *Gigaspora* produces spores which are superficially similar to those of *Scutellospora* spores. The striking difference is the germination of spores, which is by direct growth of one or more germ tubes through the spore wall. It is possible that this genus evolved from *Scutellospora* (Walker, 1987).

Another fungus, or group of fungi, that forms endomycorrhizae has been referred to in the literature as "the fine endophyte," since its hyphae are <4 µm wide, compared to 10-20 µm for typical VAM fungi. This anomalous endophyte has been described as *Glomus tenue*, though it does not resemble the other species of *Glomus*. Its spores are 10-15 µm in diameter, and when young are hyaline and indistinguishable from the vesicles that form in the roots; when mature, however, they are pigmented. Because of their small size as compared to those of other VAM fungi, the spores of *Glomus tenue* are not often recorded in nature, though mycorrhizae of this type are common, particularly in dry regions.

Morphology and Histology

Arbuscular mycorrhizal roots are surrounded by an extensive loose hyphal network that may extend into the soil for as much as 1 cm. Hyphae also grow along root surface tightly appressed to the epidermis. However, they are never abundant enough to form a fungal mantle comparable to that of ectotrophic mycorrhizae. Vesicles and large thick walled spores are borne on hyphae in the soil. During mycorrhiza formation, the AM fungus undergoes several developmental stages. In the asymbiotic stage, spores germinate and AM fungi show limited hyphal development in the absence of a host plant. However, in the presence of root exudates, they switch to the presymbiotic stage which is characterized by extensive hyphal branching (Figure 1; Buee *et al.*, 2000). Subsequent to hyphal branching, the fungus contacts the root surface via appressorium formation before hyphal penetration of the root epidermis. This is followed by symbiotic colonization of the root cortex tissue, which involves formation of intracellular arbuscules (tree-like heavily branched structures) or hyphal coils and concomitantly, the production of a sporulative extraradical mycelium (Smith and Read, 1997).

Histology of mycorrhizal roots can easily be observed by examining whole mounts under relatively low power of a compound microscope. The hyphae penetrate the epidermal cells of young roots behind the meristematic region. The fungus grows throughout the cortex, but it does not invade the endodermis, stele or root meristem. Cells containing chloroplast are not infected. Shortly after infection, the fungus forms arbuscules within cortical cells. Vesicles are inter or intracellular, depending on the host species. Some species of the endophyte produce vesicles that remain relatively thin walled. In others the vesicles become thick walled, and do not differ in any important respect from chlamydospores produced in the soil. Vesicles function as food-storage organs, or if they develop into thick-walled chlamydospores, they may function as reproductive structures.

Production of VAM Inoculum

Identification of efficient indigenous strains of VAM fungi along with suitable host and substrate is a pre-requisite for mass production of VAM inocula. Plant growth period of 2-3 months gives a large crop of mycorrhizal spores to produce sizable amount of inoculum. VAM inoculum can also be produced under sterile environment by using hydroponic cultures and transferred organ culture techniques. Although, these techniques have not reached commercialization due to certain limitations.

Inoculation Methods

For successful mycorrhizal infection, the inoculum is placed in the root zone of actively growing plants, which are not heavily fertilized. The quantity of infection by the introduced endophytes depends on placement, inoculum potential, inoculum density, type of inoculum and environmental factors. Various methods proposed for placement of the inoculum are as follows:

Layering or Pads

Inoculum is placed in layers or pads beneath seeds such that the roots penetrating the inoculum will get infected. Presently, it is one of the most effective methods of inoculating plants with VAM fungi. Layering on small and large scale can be done by using manual planters and tractor drawn seeders, respectively.

Banding or Side Dressing

In this method, VAM inoculum is placed to sides of seedling or seeds in the zone of root proliferation. Banding mycorrhizal inoculum 5 cm deep and 5 cm away from 2 month old citrus seedling was found to be as effective as placing the inoculum directly on the roots at transplanting and more effective than seed pelleting.

Mixing Soil with Inoculum

Mixing mycorrhizal inoculum with soil is the most natural method for inoculating plants. However, it requires large amount of inoculum to obtain rapid infection. Generally, this method is used for inoculating plants in nursery.

Seed Pelleting

In this method, inoculum is extracted from a pot culture by wet sieving and decantation, root retained on 1 mm sieve are fragmented and added to the inoculum. Five ml aqueous solutions (1 per cent w/v) of methyl cellulose and calcium carbonate are mixed with 35 ml of mycorrhizal sievings. This material is poured over seeds (amount depending upon the crop species), mixed thoroughly and seeds are dried.

Pre-inoculation of Transplanted Seedlings

In this method, inoculum is placed in the seedbeds and seedlings are transplanted when these get adequately infected. The method is useful with crops like onion and tomato.

Ecology

Factors Affecting VAM Symbiosis in Field

Initial studies have shown that temperature, moisture, pH, amendments and some management practices like fallowing, cropping pattern, fertilizers and pesticides may all affect the conduciveness of soil to mycorrhizal symbiosis. In general, long fallowing, excessive secondary tillage, high inputs of fertilizers and pesticides especially broad-spectrum fumigants, substituted aromatic hydrocarbons and benzimidazoles leads to a decline in the inoculum and VAM colonization. Further, maintaining a plant cover on soil help in building up the inoculum of VAM in soil. Since, crop species differ in association with VAM fungi, selection of appropriate crop/cultures in the cropping sequences can be expected to help raise the population of VAM in soil.

Biological Interactions with VAM Fungi

In 1904, Hiltner coined the term *rhizosphere* to denote the region of the soil subjected to the influence of plant roots and characterized by intense microbial activity. Many workers have subsequently shown that the microflora of the rhizosphere differs both quantitatively and qualitatively from that in the soil beyond the influence of the root (Parkinson, 1967; Bowen, 1976; Bagyaraj *et al.*, 1972). The rhizosphere effect was greatest with bacteria followed by actinomycetes and fungi. The increased microbial activity in the rhizosphere has been attributed to the extra nutrients available in the region. The extra nutrient supply comes from the root exudates, sloughed-off parts of the root cap, mucigel, root hair residues and abraded epidermal cells (Rovira, 1979 and Tinker, 1980). The microflora of the rhizosphere could influence plant growth in many ways as all plant nutrients pass through this region (Rovira, 1967 and Nye, 1977).

Of the various microorganisms colonizing the rhizosphere, VAM fungi occupy a unique ecological position as they are partly inside and partly outside the host. The internal phase of a mycorrhizal fungus does not encounter competition and antagonism from other soil microorganisms and has an ensured source of nutrients from the host. This advantage enables them to achieve a large and more functional biomass in more intimate contact with the root and thus increases their chances of exerting a greater effect on plants than other microbial species restricted only to the rhizosphere.

Certain generalizations and conclusions about the interactions of VAM fungi with other soil organisms can be drawn. VAM markedly improve nodulation and nitrogen fixation by legume bacteria mainly by providing high phosphorus required for the fixation process. Mycorrhizal plants also allow introduced populations of beneficial soil organisms like *Azotobacter* and phosphate solubilising bacteria to maintain higher numbers than around non-mycorrhizal plants and to exert synergistic effects on plant growth. In general, VAM fungi decrease the severity of root diseases. Summary of some important interactions between VAM fungi and beneficial soil organisms are presented in Table 12.3.

Table 12.3: Summary of Important Interactions Between VAM Fungi and Beneficial Soil Microorganisms

Groups	Species	Interactions
N-fixation bacteria	*Rhizobium* sp.	Increased VAM infection level, nodulation, N-fixation and plant growth
Free living N-fixation bacteria	*Azotobacter* sp. *Beijerinkia* sp. *Clostridium* sp. *Azospirillum* sp.	Increased VAM infection level, AM spore production and bacterial population
Phosphate solubilizing bacteria	*Pseudomonas* sp. *Agrobacterium* sp.	Good bacterial population in rhizosphere of mycorrhizal plants
Phosphate solubilizing fungi	*Aspergillus niger* *Penicillum funiculosum*	Availability of P increased by P-solubilizing fungi and uptake increased by VAM

Research Work Carried Out by Authors at NRCAF, Jhansi

The role of VAM fungi in enhancing plant growth and yield, resistance to drought and salinity and tolerance to pathogens is well documented, but little is known about their interactions with agroforestry systems. Studies are being carried out in this regard at National Research Centre for Agroforestry, Jhansi. The results obtained so far showed the presence of VAM fungi in roots of 19 forestry trees, 7 horticultural trees, 10 field crops and 8 fodder crops. The colonization index varied from 20.4 to 95.9 per cent (Kumar, 1992). Local soils were found to be rich in VAM spore populations. *Glomus* was predominant genus, apart from this *Acaulospora*, *Gigaspora* and *Sclerocystis* were also present (Kumar *et al.*, 2006; Jha *et al.*, 2008 and Kamalvanshi *et al.*, 2008). Cultures of 11 *Glomus* and 2 *Acaulospora* species were purified from tree rhizosphere. Inoculation of tree seedlings (*Acacia nilotica*, *Pongamia pinnata*, *Jatropha curcas*, *Eucalyptus tererticornis*, *Albizia procera*, *Dalbergia sissoo*, *Leucaena leucocephala*, *Emblica officinalis*, *Zizyphus mauritiana*, *Buchanania lanzan* and *Cordia dichotoma*) and intercrops (*Phaseolus mungo*, *P. vigna*, *Zea mays*, *Triticum aestivum*, *Cicer arietinum* and *Pisum sativum*) with suitable VAM species recorded substantial increase in plant growth and phosphorus (P) uptake (Kumar *et al.*, 2007). This resulted in a benefit comparable to high input levels of P in *P. mungo*, *T. aestivum*, *E. tererticornis* and *A. procera* at much lower P concentrations (Shukla *et al.*, 2008).

In agroforestry systems, trees acted as VAM inoculum reservoir for intercrops (Kumar *et al.*, 2007). Cross inoculation test performed by using VAM fungi isolated from trees showed that these could colonize the roots of important rainy season (*P. mungo*, *P. vigna* and *Z. mays*) and winter (*C. arietinum*, *P. sativum* and *T. aestivum*) intercrops (Kumar *et al.*, 2007). VAM fungi also cross infected the plant species under agroforestry situations. Intercropping increased VAM activity in rhizosphere of trees (Kumar *et al.*, 2007). Tree shade reduced the rate of mycorrhization of intercrops, especially during winter months. VAM inoculations enhanced the growth and phosphorus uptake of intercrops in shade and the tree canopy management increased this advantage. Use of excessive shading (25 per cent of full sun or more) in nurseries reduced mycorrhization of tree seedlings (Shukla *et al.*, 2008). VAM activity in terms of spore population and mycorrhizal diversity was more in light soils as compared to heavy soils. Pruning/canopy management of trees reduced their VAM colonization (Shukla, 2009). Excessive stagnation of water during rainy season in plantations (*E. officinalis* and

Table 12.4: List of Efficient Vesicular Arbuscular Mycorrhizal Fungi for Different Plant Species

Plant Species	Efficient Mycorrhizal Fungus/Fungi	Mycorrhizal Fungi Tested	Parameters Evaluated	References
		Agriculture crops		
Allium cepa (Onion)	*Gigaspora margarita*	*G. margarita*, *G. calospora*, and a known VAM fungus	Leaf number, dry matter, P-uptake, bulb yield	Ramana and Babu (1999)
	G. calospora	*G. margarita*, *G. calospora*, and a known VAM fungus	Root colonization	Ramana and Babu (1999)
Amorphophallus paeonufolius	*Glomus mosseae* + *G. aggregatum*	*G. mosseae*, *G. aggregatum*, *Gigaspora albida*, *Pisolithus tinctorius*	Tuber yield/plant	Ganesan and Mahadevan (1994)
Ananas comosus	*Glomus fasciculatum*	*G. fasciculatum*, *Glomus* sp., *Acaulospora* sp., *Scutellospora* sp.	Plant growth, nutrition	Jaizme-Vega and Azcon (1995)
Acacia nilotica	*Glomous mosseae* + *Gigaspora gilmorei*	*G. mosseae*, *G. constrictum*, *Gigaspora gilmorei*, and their combinations	Collar diameter, root–shoot dry weight and length	Mandal and Kaushik (1994)
Acacia auriculaeformis	*Glomus etunicatum*, *G. macrocarpum*	*G. etunicatum*, *G. macrocarpum*, *G. fasciculatum*, *G. mosseae*	Height, diameter, biomass, P-uptake	Aggangan, Lorilla, and Cruz (1990)
Acer saccharum (Sugar maple)	*Glomus macrocarpum*	*G. macrocarpum*, *G. clarum*, *G. epigaeum*, *G. mosseae*	Leaf number, mean lateral root number	Reid, Parker, Mitchell *et al.* (1988)
Azadirachta indica (neem)	*Glomus mosseae*	*Glomus mosseae*, *G. fasciculatum*, and seven other VAM fungi	Plant height, stem girth, biomass, P-content, zinc concentration, root colonizations	Sumana and Bagyaraj (1999)
Cajanus cajan	*Glomus clarum*	*G. fasciculatum* and seven other VAM fungi	Plant growth	Diederichs (1992)
	Glomus etunicatum	*G. etunicatum*, *Gigaspora margarita*	Shoot dry matter, nutrient uptake	Ahiabor and Hirata (1994)
Capsicum	*Glomus intraradices*	*G. intraradices*, indigenous mixed culture, commercial inoculum (Mycorise and C)	Fruit yield	Gaur, Adholeya, and Mukerji (1998)
Carica papaya	*Glomus fasciculatum*	*G. fasciculatum*, *Glomus* sp., *Acaulospora* sp., *Scutellospora* sp.	Plant growth, nutrition	Jaizme-vega and Azcon (1995)
Cacocasia esculenta	*Glomus mosseae* + *G. aggregatum*	*G. mosseae*, *G. aggregatum*, *Gigaspora albida*, Pisolithus tinctorius	Tuber yield/plant	Ganesan and Mahadevan (1994)
Cucumis sativus	*Glomus caledonium*	*G. caledonium*, *Glomus* sp.	P-uptake	Joner and Jokobsen (1994)

Contd...

Table 12.4–Contd...

Plant Species	*Efficient Mycorrhizal Fungus/Fungi*	*Mycorrhizal Fungi Tested*	*Parameters Evaluated*	*References*
Eleusine coracana	*Glomus caledonicum*	*G. caledonicum, G. mosseae, G. fasciculatum, G. epigaeum* (*G. versiforme*), *Gigaspora calospora, G. margarita*	Mycorrhizal efficacy, root colonization	Tewari, Johri, and Tandon (1993)
Fragaria xananassa (micropropagated)	*Glomus versiforme* 36-366	*G. versiforme* (2 isolates), *G. mosseae, G. etunicatum*	Yield, production of runners	Taube-Baab and Baltruschat (1996)
Glycine max	*Acaulsopora scrobiculata*	*A. scrobiculata, Glomus intraradices,* and four other AM fungi	Plant growth, crop yield	Vasuvat, Nopamornbodi, and Thamsurakul (1987)
Lycopersicum esculentum	*Glomus etunicatum*	*Glomus etunicatum, G. mosseae*	Shoot dry weight, plant height	McGraw and Schenck (1981)
cv. Manapal	*G. mosseae*	*Glomus etunicatum, G. mosseae*	Root colonization	McGraw and Schenck (1981)
cv. Walter	*G. mosseae*	*Glomus etunicatum, G. mosseae*	Shoot weight, plant height	McGraw and Schenck (1981)
Manihot esculenta (cassava)	*G. fasciculatum*	*G. fasciculatum,* G. *mosseae, G. constrictum, Glomus etunicatum, Acaulsopora morrowea*	Root colonization, plant weight, shoot and root dry weight	Sivaprasad, Sulochana and Nair (1990)
Manihot esculenta (cassava)	*G. mosseae + G.aggregatum*	*G. mosseae, G.aggregatum, Gigaspora albida, Pisolithus tinctorius*	Tuber yield/plant	Ganesan and Mahadevan (1994)
Medicago sativa	*Glomus intraradices*	*G. intraradices, G. mosseae, Gigaspora margarita*	Root colonization, fungal sporulation	Douds, Galvez, Becard *et al.* (1998)
Musa acuminata	*Glomus fasciculatum*	*G. fasciculatum, Glomus* sp., *Acaulospora* sp., *Scutellospora* sp.	Plant growth, nutrition	Jaisme-Vega and Azcon (1995)
Oryza sativa cv. Prakash	*Glomus intraradices*	*G. intraradices, G. fasciculatum*	Grain yield	Secilia and Bagyaraj (1994)
Orzya sativa (upland rice)	*Acaulospora spinosa*	*Acaulospora spinosa, A. scrobiculata*	Plant biomass, grain yield, root colonization	Ammani and Rao (1996)
Persea Americana	*Glomus fasciculatum*	*G. fasciculatum, Glomus* sp., *Acaulospora* sp., *Scutellospora* sp.	Plant growth, nutrition	Jaizme-vega and Azcon (1995)
Polianthes tuberosa (rajnigandha)	*Glomus intraradices*	*G. intraradices*, indigenous mixed cultures, commercial inoculum (Mycorise)	Fruit yield, spike length	Gaur, Adholeya, and Mukerji (1998)

Contd...

Table 12.4–Contd...

Plant Species	*Efficient Mycorrhizal Fungus/Fungi*	*Mycorrhizal Fungi Tested*	*Parameters Evaluated*	*References*
Trifolium (Clover)	*Acaulospora trappei*	*A. trappei, A. laevis, Acaulospora* sp.,	Root colonization	Gazey, Abbott, and Robson (1992)
Triticum aestivum wheat (var. Swift)	*Glomus intraradices*	*G. intraradices, Gigaspora margarita*	Plant yield, number of grains/spike	Asif Khan, Khan *et al.* (1995)
Vigna mungo (black gram)	*Glomus epigaeum* (*G. versiforme*)	*G. epigaeum, Acaulospora spinosa, A. morrowae*	Root colonization, shoot dry weight, N and P concentrations	Rao and Rao (1996)
Vigna radiata (green gram)	*Glomus epigaeum* (*G. versiforme*)	*G. epigaeum, Acaulospora spinosa, A. morrowae*	Root colonization, shoot dry weight, N and P concentrations	Rao and Rao (1996)
Vigna unguiculata	*G. etunicatum*	*G. etunicatum, Gigaspora margarita*	Shoot dry matter, nutrient uptake	Ahiabor and Hirata (1994)
Vigna unguiculata cv. Katumanu K80	*G. etunicatum*	*Glomus etunicatum, G. claroideum, Glomus* sp.	Top dry weight	Obura, Skipper, and Wagner (1987)
Vigna unguiculata cv. California No. 5 black eye	*G. claroideum*	*Glomus etunicatum, G. claroideum, Glomus* sp.	Top dry weight	Obura, Skipper, and Wanger (1987)
Zea mays (pioneer 3905)	*Glomus etunicatum*	*G. etunicatum, G. mosseae, G. aggregatum, G. versiforme*	Leaf mass, protein concentrations	Boucher, Dalpe and Charest (1999)
Nephrolepis exaltata var. whitmanii (fern) micropropagated plants	*Glomus intraradices, G. clarum*	*G. intraradices, G. clarum, G. versiculiferum, G. versiforme*	Root infection	Ponton, Piche, Parent *et al.* (1990)
		Tree Crops		
Malus domestica (apple)	*Glomus mosseae*	*G. mosseae, G. macrocarpum*	Root colonization, plant height, plant growth	Miller, Bodmer, and Schuepp (1989)
Hevea brasiliensis	*Glomus clarum, Glomus* sp.	*G. clarum, Glomus* sp., *A. spinosa, Gigaspora margarita, Entrophospora colombiana, Glomus* sp., and three other species	Dry weight	Ikram, Mahmud, and Othman (1993)
Tectona grandis	*Glomus leptotichus*	Not mentioned	Seedling growth parameters	Rajan, Reddy, and Bagyaraj (2000)
Zizyphus mauritiana	*Glomus fasciculatum*	*G. fasciculatum, G. mosseae, Scutellospora calospora*	Growth, root colonization, nutrient uptake	Mathur and Vyas (1996)
Citrus jambhiri (rough lemon)	*Glomus fasciculatum*	*G. fasciculatum, G. mosseae, G. etunicatum*	Plant height, top weight, root weight, chlamydospores in soil	Nemec (1979)
Persea American (avocado)	*Glomus fasciculatum*	*Glomus fasciculatum, Acaulospora* sp., *Scutellospora* sp.	Plantlet growth and nutrition	Jaizme-vega and Azcon (1995)

Hardwikia binata) was found to be harmful for colonization of plants. *Glomus* was more adapted to such conditions in *E. officinalis* plantations (Kumar *et al.*, 2009). Inoculation of intercrops (*P. mungo* and *T. aestivum*) and tree seedlings (*E. tererticornis* and *A. procera*) with a suitable VAM consortium was found more beneficial than individual VAM fungi in *P. mungo, T. aestivum, E. tererticornis* and *A. procera*.

Conclusions

Mycorrhizal symbiosis has lately created a keen interest among biological scientists of the country involving mycologists, soil scientists, agronomist, plant physiologists, plant geneticists, environmentalist, foresters and microbiologists alike. Major efforts are now being concentrated towards the production of bulk inoculum of mycorrhizal bio-fertilizers for use in forestry, horticulture, floriculture, agriculture, etc. While many ectomycorrhizal fungi have been brought in pure cultures, VAM fungi till now, defied all attempts to bring them in pure culture on synthetic media. Many efforts are being made all over the world to decode the genetic barrier to grow these fungi on synthetic media. In India, though fundamental studies on molecular and biochemical level are scarce, much headway has been achieved in selecting suitable VAM fungi for different crops and for different site conditions (Table 12.4). Special attempts are being made to determine the role of VAM fungi in ameliorating soils polluted with industrial wastes. Subjects on which more information is needed are as follows:

- ☆ Distribution of mycorrhizal fungi under varying edapho-climatic conditions.
- ☆ Plant growth responses and mycorrhizal dependency.
- ☆ Role of mycorrhiza on difficult/disturbed sites.
- ☆ Tripartite biological interactions involving rhizosphere flora, mycorrhizal fungi and plant roots.
- ☆ Physiology and biochemistry.
- ☆ Management of pests, mycorrhizae and biocides for increased plant productivity.
- ☆ Soil/water pollution and mycorrhizae.

References

Abbott, L.K. and Robson, A.D. (1977). Growth stimulation of subterranean clover with vesicular arbuscular mycorrhizas. *Ausraliant Journal of Agricultural Research* 28: 639-649.

Bagyaraj, D.J. and Rangaswami, G. (1972). Studies on the rhizosphere microflora of *Eleusine coracana* under the influence of foliar chemicals sprays. I. Quantitative incidence of microorganisms. *Madras Agricultural Journal* 59: 517.

Bagyaraj, D.J., Manjunath, A. and Patil, R.B. (1979). Occurrence of vesicular-arbuscular mycorrhizas in some tropical aquatic plants. *Transactions of British Mycological Society* 72: 164-167.

Benjamin, R.K. (1979). Zygomycetes and their spores. in B Kendricks (ed). The Whole Fungus. National Museums of Canada, Ottawa. pp. 573-616.

Bowen, G.D. and Rovira, A.D. (1976. Microbial colonization of plant roots. *Annual Review of Phytopathology* 14: 121-144.

Clayton, J.S. and Bagyaraj, D.J. (1984). Vesicular-arbuscular mycorrhizas in submerged aquatic plants of New Zealand. *Aquatic Botany* 19: 251-262.

Cooper, K.M. (1976). A field survey of mycorrhizas in New Zealand ferns. *New Zealand Journal of Botany* 14: 169-181.

Frank, A.B. (1985). Uber die auf Werzelsymbiose beruhende Ernahrung gewisser Baume durch unterirdische. Pilze. *Ber. Dtsch. Bot. Ges.* 3: 128-145.

Gerdemann, JW. (1975). Vesciular arbuscular mycorrhizae. in JG Torrey and Glarkson DT (eds) *The Development and Function of Roots*. Academic Press, London. pp. 576-591.

Graw, D. and Rehm, S. (1977). Vesicular-arbuscular Mycorrhiza in den Fruhttragern von *Arachis hypogea. L.Z. Acker Und Pflanzenbau* 144: 75-78.

Hall, I.R. and Abbott, L.K. (1983). Photographic slide collection illustrating features of Endogonaceae. *Bulletin of British Mycological Society* 17: 54-55.

Hall, I.R. (1984). Taxonomy of VA mycorrhizal fungi. in CL Powell and Bagyaraj DJ (eds) VA mycorrhiza. CRC Press, Boca Raton, FL. pp. 57-94.

Harley, J.L. (1969). *The Biology of Mycorrhiza*, 2nd ed. Leonard Hill, London.

Harley, J.L. and Smith, S.E. (1983). *Mycorrhizal Symbiosis*, 1st edn. Academic, London.

Hiltner, L. (1904). Uber neure Erfahrungen und Probleme auf dem Gebiet der Bodenbakteriologie und unter besonderer Beruksichtigung der Grundingung und Brache. *Arb Dtsch Landwirtsch Gesellschaftswiss* 98: 59-78.

Janos, DP. (1983). Tropical Mycorrhizas. In SL Sutton, TC Whitemore and AC Chadwick (eds) *Tropical Rain Forest: Ecology and Management*. Blackwell Scientific Publications, Oxford. pp. 327-345.

Jha, A, Kumar, A., Saxena, R. K., Kamalvanshi, M. and Shukla, A. (2008). Occurrence of Arbuscular Mycorrhizal Fungi in Selected Agroforestry Tree Species of Bundelkhand Region. National Symposium on Agroforestry Knowledge for Sustainability, Climate Moderation and Challenges Ahead. Dec.15-17. NRCAF, Jhansi. 60.

Kamalvanshi, M., Kumar, A., Jha, A. and Shukla, A. (2008). Occurrence of arbuscular mycorrhizal (AM) fungal species in *Pongamia pinnata, Jatropha curcas* and *J. gossypifolia.* National Symposium on Agroforestry Knowledge for Sustainability, Climate Moderation and Challenges Ahead. Dec.15-17. NRCAF, Jhansi 167.

Kendrick, B. and Berch, S. (1985). Mycorrhizae: Applications in Agriculture and Forestry. in CW Robinson (ed) *Comprehensive Biotechnology*, Vol. 4. Pergamon Press, Oxford. pp.109-150.

Kumar, A. (1992). Occurrence of VAM among some plant important in agroforestry under semiarid conditions. *Agroforestry Newsletter* 4(1): 7.

Kumar, A., Hashmi, S., Shukla, A. and Jha, A. (2006). Colonization of some important intercrops by Vesicular Arbuscular Mycorrhizae isolated from minor fruit crops. National Symposium on Agroforestry for Livelihood Security, Environment Protection and Bio fuel Production. Dec.16 - 18. NRCAF, Jhansi. 20.

Kumar, A., Hashmi, S. and Shukla, A. (2006). Occurrence of vesicular arbuscular mycorrhizae in *Buchanania lanzan* and *Cordia dichotoma. Range Management and Agroforestry* 27(1): 84-86.

Kumar, A., Shukla, A. Hashmi, S., Jha, A. and Kamalvanshi, M. (2007). Studies on arbuscular mycorrhizal fungi in agroforestry. *Agroforestry Newsletter* January- March, 2008, Vol. 20(1) pp. 5.

Kumar, A., Hashmi, S., Shukla, A. and Chaturvedi, O.P. (2007). Effect of intercropping on inoculum potential of arbuscular mycorrhizal fungi in agroforestry systems. *Annals of Arid Zone* 46(1): 31-36.

Kumar, A., Shukla, A., Hashmi, S. and Tewari, R.K. (2007). Effect of trees on colonization of intercrops by vesicular arbuscular mycorrhizae in agroforestry systems. *Indian Journal of Agricultural Sciences* 77(5): 291-298.

Mosse, B. and Bowen, G.D. (1968). The distribution of Endogone spores in some Australian and New Zeland soils and in the experimental field soil at Rothamsted. *Transactions of British Mycological Society* 51: 485-492.

Mosse, B., Stribley, D.P. and Le Tacon, F. (1981). Ecology of mycorrhizae and mycorrhizal fungi In M Alexander M (ed) *Advances in Microbial Ecology*. Plenum Press, New York. pp. 5: 137-209.

Nye, P.H. (1977). The rate limiting step in plant nutrient absorption from soil. *Soil Science*. 123: 292.

Parke, J.L. and Linderman, R.G. (1980). Association of vesicular-arbuscular mycorrhizal fungi with the moss *Funaria hygrometrica. Canadian Journal of Botany* 58: 1898-1904.

Parkinson, D. (1967). Soil microorganisms and plant roots in A Urges, F Raw (eds) *Soil Biology*. Academic Press, London, 449.

Redhead, J.F. (1977). Endotrophic mycorrhizas in Nigeria. Species of the Endogonaceae and Their distribution. *Transactions of British Mycological Society* 69: 275-280.

Rovira, A.D. and McDougall, B.M. (1967). Microbiological and biochemical aspects of the rhizosphere in: AD McLaren, GH Peterson (eds) *Soil Biochemistry*. Marcel Dekker, New York. pp.1: 417-493.

Rovira, A.D. (1979). Biology of soil-root interface in JL Warley, SR Russel (eds) *The Soil-Root Interface*. Academic Press, London 145-160.

Schenck, NC and Perez, Y. (1987). *Manual of the identification of VA Mycorrhizal Fungi*. University of Florida, Gainesville, Florida, USA. pp. 1-245.

Selvaraj, T., Kannan, K. and Lakshminarasimhan, C. (1986). Vesicular-arbuscular mycorrhizal fungi in roots and scale-like leaves of *Canna indica* L. (Cannaceae). *Current Science* 55: 728-730.

Singh, S. (2001). Mycorrhizal dependency, Part 1: Selection of efficient mycorrhizal fungi. *Mycorrhiza News* 13(1): 2-16.

Shukla, A., Kumar, A., Jha, A., Chaturvedi, O.P., Prasad, R. and Ajit (2008). Effect of shade on arbuscular mycorrhizal colonization and growth of crops and tree seedlings in Central India. *Agroforestry Systems* doi: 10.1007/s10457-008-9182-x.

Shukla, A., Kumar, A., Jha, A., Kamalvanshi, M. and Gupta, G. (2008). The effects of phosphorus application on growth and arbuscular mycorrhizal colonization of crops and tree seedlings. National Symposium on Agroforestry Knowledge for Sustainability, Climate Moderation and Challenges Ahead. Dec.15-17. NRCAF, Jhansi 41.

Shukla, A. (2009). Effect of tree introduction in agricultural fields on vesicular-arbuscular mycorrhizal status of component crops of agroforestry systems. *Ph.D Theis*, Bundelkhand University, Jhansi.

Tinker, P.B. (1980). Role of rhizosphere microorganism in phosphorus uptake by plants in FE Khasawneh, EC Sample and EJ Kamprath (eds) *The Role of Phosphorus in Agriculture*. American Society of Agronomy, Madison, wisc. pp. 617-647.

Trappe, J.M. (1977). "Biogeography of hypogeous fungi: Trees, mammals and continental drift" In HE Bigelow and EG Simmons (eds) *Abstr. 2nd Int. Mycol. Congr. Inc.* University of South Florida, Tampa. pp. 675.

Trappe, J.M. and Schenck, N.C. (1982). Taxonomy of the fungi forming endomycorrhizae. A. Vesicular-arbuscular mycorrhizal fungi (Endogonales) in NC Schenck (ed) Methods and Principles of Mycorrhizal Research. *American Phytopathological Society Publication*, St. Paul, MN. pp.1-10.

Walker, C. (1983). Taxonomic concepts in the Endogonaceae: Sproe wall characteristics in species description. *Mycotaxon* 18: 443-455.

Microbial Biotechnology and Ecology (2011) *Pages* 225–255
Editors: Deepak Vyas, G.S. Paliwal, P.K.Khare and R.K Gupta
Published by: DAYA PUBLISHING HOUSE, NEW DELHI

Chapter 13

Arbuscular Mycorrhizal Fungi: The Hidden Heroes of the Soil

☆ *Deepak Vyas[1]*, Meenakshi Mishra[1], Pradeep Singh[1], Ashok Shukla[1] and R.K. Gupta[2]*

[1]*Lab of Microbial Technology and Plant Pathology, Department of Botany Dr. H.S. Gour University Sagar – 470 003, M.P.*
[2]*Govt. P.G. College, Rishikesh, Uttarakhand*

ABSTRACT

Soil is not a pile of dirt but is a treasure house of earthy materials and home of many microorganisms as it was rightly said that "we think that we are standing on the earth but reality is that we are standing on the roof of another world". Rhizospheric region of the soil is one of the important room where plant, soil and microbes interact. However, these interactions are of many types but among these, mutualistic interactions are of great importance. Arbuscular mycorrhizal symbiosis is one of the mutualistic relationship among the plant, soil and other microbes.

In this review we have tried to summerize some of the multifacet roles performed by the AMF fungi in the soil for the benefit and better plant growth, soil remediation, disease control, dialogue among the co-organisms and their importance in the modern biology such as molecular biology, genetics, genomics, proteomics and fluxomics.

Introduction

AM symbioses are the most widespread mycorrhizal association, with a very long evolutionary history. Fossil survey and molecular data have demonstrated that the evolutionary history of AM fungi goes back at least to the Ordovician, about 460 million years ago (Redecker, 2002). The earliest

* E-mail: dvyas64@yahoo.co.in.

land plants show an association with fungi that formed vesicles and arbuscules that was very similar to today's AM fungi (Nicolson, 1975; Remy *et al.*, 1994). Recently Vyas *et al.* (2007-2008) reported occurrence of arbuscules and vesicles in bryophytes. This type of plant-fungal association is formed with 80 per cent of angiosperms (Harley and Harley, 1987). The fungi are obligate symbionts and they need a living plant root (or equivalent structure) in order to grow and reproduce. AM fungal development in roots of host plants starts when fungal hyphae grow from spores or from colonized roots toward the uncolonized roots. After contact of the hyphae with the root surface, the fungus is stimulated to change in morphology from an original simple branching pattern to irregularly septate pattern with reduced interhyphal spacing (Giovannetti *et al.*, 1993; Harrison, 1998). The fungus produces swollen appressoria on the root surface and spreads between and into the root cortical cells. There are two major morphological types of AM, *Arum* and *Paris* type, they are characterised by differences in fungal development within the roots (Gallaud, 1905; Smith and Smith, 1997). In general, internal hyphae branch and make four recognizable structures, intracellular hyphae that may be coiled, intercellular hyphae, arbuscules and spherical or ovoid vesicles (Smith and Smith, 1997). In *Arum-type* associations, intercellular hyphae penetrate between the root cortical cells and spread rapidly, then lateral branch-hyphae penetrate the root cortical cells and branch dichotomously and produce arbuscules. In *Paris-type* associations intracellular hyphae spread from cell to cell within the root cortex and form extensive intracellular hyphal coils and arbusculate coils (Gallaud, 1905). The types of internal structures that develop depend on the plant/fungal combination (Smith and Smith, 1997; Cavagnaro *et al.*, 2001).

About 83 per cent of dicotyledonous and 79 per cent of monocotyledonous plants is associated with mycorrhizal fungi (Wilcox, 1991). Mycorrhizal associations are found in a very wide range of habitats, including aquatic ecosystems, deserts, lowland tropical rain forests, high altitudes, high latitudes and in canopy epiphytes (Allen, 1991). Bi-directional movement of nutrients characterizes the fungus-plant symbioses, where carbon flows to the fungus and inorganic nutrients move to the plant, thereby providing a critical linkage between the plant root and soil. This association is a survival mechanism for both the fungi and plants, allowing each to survive in different environments (Gupta *et al.*, 2000). Mycorrhizal plants, in comparison with non mycorrhizal plants, have greater nutrient uptake because they possess a network of external hyphae (Sanders and Sheikh, 1983). The hyphae are the interface between soil and plant and have a large surface area that acts as an extension of the root absorbing area (Rhodes and Gerdemann, 1975; Owusu-Bennoah and Wild, 1979; Li *et al.*, 1991). This not only increases the volume of soil from which nutrients are absorbed, but also overcomes problem of depletion of nutrients (Nurlaeny *et al.*, 1996; Smith and Read, 1997) and water (Marulanda *et al.*, 2003) depletion close to actively absorbing roots, and plays a significant role in stabilizing soil structure. Mycorrhizal fungal associations have several advantages for their hosts, including increased growth and yield and reproductive success due to enhanced nutrient acquisition (Mosse *et al.*, 1973; Diederichs, 1990; Lewis and Koide, 1990; Stanley *et al.*, 1993). They may also increase disease and pest resistance, improve water relations (Allen and Allen, 1986; Davies *et al.*, 1993; Subramanian *et al.*, 1997), soil structure (Tisdall and Oades, 1979; Thomas *et al.*, 1986; Degens *et al.*, 1994; Beaden and Petersen, 2000) and tolerance of extreme pH (Sidhu and Behl, 1997; Douds *et al.*, 2000). This literature review will concentrate on structure and function of AM and their potential role in revegetating saline soils.

AMF in Carbon Sequestration

Arbuscular mycorrhizal fungi (AMF) comprising fungi in order Glomales (Zygomycota), ubiquitous symbionts of the majority of higher plants. AMF have been shown to have numerous effect on plant physiology and plant communities (Allen 1991; Smith and Read, 1997), which can lead to indirect effect on soil carbon storage. AMF are also very important in the process of soil aggregate stabilization (Tisdall and Oades, 1982). Relatively labile carbon can be protected inside soil aggregates which means AMF have yet another indirect influence on soil carbon storage (Miller and Jastrow, 1992).

Recently, a glycoprotein named glomalin was discovered using specialized protocols for soil that revealed amount up to several mg protein per gram of soil (Wright and Upadhyaya, 1996). Glomalin is produced by hyphae of all members of AM genera, but not by other groups of soil fungi so far tested. Although concentration of glomalin in soil seem to be responsive to global change factors such as elevated atmosphere CO_2 (Rillig *et al.*, 1999).

ERH (Extra Radical Hyphae) and Soil Aggregation

Indeed the influences of ERH on soil aggregation might be even more important to the carbon stock than the influence of the hyphal standing crop along (Miller *et al.*, 2000). Through their role in soil microaggregate stabilization, the ERH appear to contribute to the certain of an aggregate hierarchy. In doing so they help to create mechanism for increasing the residence time of organic debris within soil macroaggregates.

Mechanistically, the ERH contribution to soil aggregation can be viewed as 'Stioky string bag' mechanism, in which the hyphae help to etangle and enmesh soil particles to form macroaggregate structures (Miller *et al.*, 2000; Oades *et al.*, 1991). The physical dimensions of ERH allow them to grow and to remify through soil pores the size of those between macroaggregates. The contribution of AMF hyphae to carbon cycling lies not only with ERH but also, with exudates from hyphae. Arbuscular mycorrhizal fungi (AMF) take up photosynthetically fixed carbon from plant roots and translocate it to their external myceliumAMF are also responsible for directing the movement of huge quantities of photosynthate to the soil (for review, Douds *et al.*, 2000; Graham, 2000). Carbon in the root flows from plant to fungus in the form of sugars (Shachar-Hill *et al.*, 1995; Solaiman and Saito, 1997), and together with the transfer of mineral nutrients from fungus to root (Koide and Schreiner, 1992; George *et al.*, 1995; Jakobsen, 1995), this is the nutritional mainstay of what is arguably the world's most important mutualistic symbiosis. AM fungi obtain most or all of their carbon within the host root. Here, they acquire hexose and transform it into trehalose and glycogen, typical fungal carbohydrates (Shachar-Hill *et al.*, 1995). Triacylglyceride (TAG) is the main form of stored carbon in AM fungi (Beilby and Kidby, 1980; Jabaji-Hare, 1998), and this is mostly or exclusively made in the intraradical mycelium (IRM; Pfeffer *et al.*, 1999). Some of this stored lipid flows from the IRM to the extraradical mycelium (ERM; Pfeffer *et al.*, 1999), and in vivo microscopic observations indicate that the rate of export is sufficient to account for the high levels of *stored* lipid in the ERM (Bago *et al.*, 2002). The glyoxylate cycle is active in the ERM (Lammers *et al.*, 2001), and this pathway appears to be important in using exported TAG to make carbohydrate in the ERM.

The structure of glomalin has not been completely defined. The molecule appears to be a complex of a repeated monomeric structures bound together by hydrophobic interactions (Nichols, 2003) that, attaches to soil to help stabilize aggregates. The molecule contains tightly bound iron (0.04-8.8 per cent) (Rillig *et al.*, 2001; Nichols, 2003; Wright and Upadhyaya, 1998), and does not contain phenolic compounds such as tannins (Rillig *et al.*, 2001). Preliminary evidence (Wright and Upadhyaya, 1998; Nichols, 2003) suggests that cations are bound to glomalin in amounts that vary *for* different soils.

Rhizodeposition

The release of carbon compounds from living plant roots rhizodeposition into the surrounding soil is a ubiquitous phenomenon (Curl and Trueglove, 1986). The loss of C from root epidermal and cortical cells leads to a proliferation of microorganisms within endorhizosphere on the surface (rhizoplane) and outside the root (ectorhizosphere). C release also result in the rhizosphere having different chemical, physical and biological characteristics to the bulk soil (Barber, 1995).Theoretically, almost any soluble component present inside the root can be lost to the rhizosphere; however current evidence suggests that exudation is dominated by low molecular weight solutes such as sugars, amino acids and organic acids that are present in the cytoplasm at high concentrations (Farrar *et al.*, 2003). Influence of AM formation upon roots exudates following AM colonization root exudation patterns may be expected to alter because the AM fungus is a considerable C sink (Douds *et al.*, 2000; Graham, 2000). AM colonization alters the carbohydrate metabolism of the roots (Shachar-Hill *et al.*, 1995; Douds *et al.*, 2000; Bago *et al.*, 2003) and increases root respiration (Douds *et al.*, 2000; Shachar-Hill *et al.*, 1995). Hyphal exudation from the AMF will also occur.

AMF in Nutrient Sequestration

Enhanced uptake of P is generally regarded as the most important benefit that AMF provide to their host plant and plant P is often the main in plant fungal relationship (Thompson, 1987; Smith and Read, 1997; Graham, 2000). AMF can play a significant role in crop P nutrition, increasing total uptake and in some cases P use efficiency (Koide *et al.*, 2000). This may be associated with increased growth and yield (Vosatka, 1995, Ibibijen *et al.*, 1996; Koide *et al.*, 2000).

Where colonization by AMF is disrupted, uptake of P, growth and in some cases yield can be significantly reduced (Thompson, 1987, 1991 and 1994).

In many cases, this is due to a high concentration of (phyto) available soil P (Bethlenfalvay and Barea, 1994; Hetrick *et al.*, 1996; Thingstrup *et al.*, 1998; Soren Sen *et al.*, 2005). Under such conditions, the colonization of roots by AMF is often suppressed (Jensen and Jakobsen, 1980; Al-Karaki and Clark, 1999; Kahiluoto *et al.*, 2001). Where strong AMF colonization still occurs under condition of high soil P concentration it may reduce crop growth (Gavito and Varela, 1995; Kahiluoto *et al.*, 2001).

AMF play significant role in the uptake of other nutrients by the host plant, zinc (Zn) nutrition is most commonly reported as being influenced by the AM association, though uptake of copper (Cu), iron, N, K, Ca and Mg have been reported being as enhanced (Smith and Read, 1997; Clark and Zeto, 2000). AMF may also enhance plant uptake of N from organic sources (Hodge *et al.*, 2001). In many cases AMF cause a change in the absorption of soil nutrients by host simultaneously, though the effect on different nutrients is rarely the some (Lambert *et al.*, 1979, Kothari *et al.*, 1990; Wellings *et al.*, 1991; Thompson, 1987, 1991, 1994; Azaizeh *et al.*, 1995; Smith and Read, 1997; Srivastava *et al.*, 2002; Mohammad *et al.*, 2003).

The apparently contradictory evidence regarding the effect of AMF on plant nutrient absorption may be connected to the increasing realization that there is a degree of selectivity between host and the fungi and that different AMF have varying effects on different plant species, from strongly positive increases in nutrient uptake and or growth to strongly negative. (Monzon and Azcon, 1996; Bever *et al.*, 2001; Vander Heijden, 2002; O'Connor *et al.*, 2002.; Munkvold *et al.*, 2004). There is a good evidence for a substantial capacity of external hyphae to absorb NH_4-N and deliver it to the host plant. In the hyphosphere as in the rhizosphere, uptake of NH_4-N is associated with substrate acidification.

Increase of K/Na Ratio

Some previous studies on the effects of salinity on mycorrhizal plants have shown that AM roots had higher Na concentrations but also higher K concentrations and thus maintained a high K/Na compared to non-mycorrhizal plants. Scientists have found that mycorrhizal roots of salt grass plant

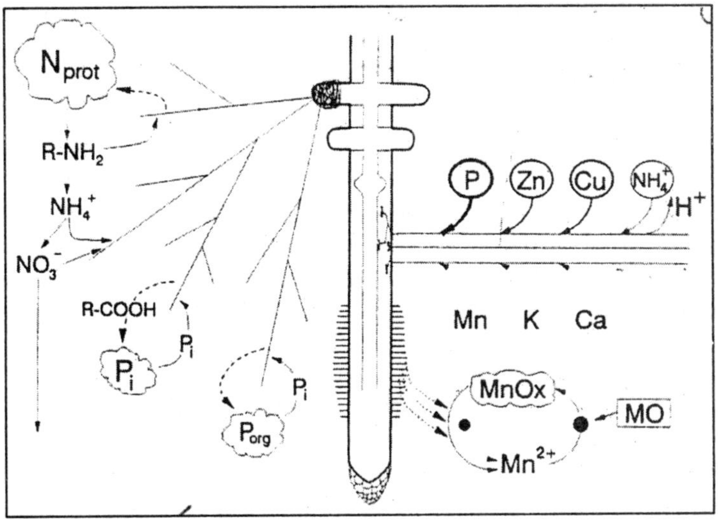

Figure 13.1: Schematic Presentation of some Established Components (Processes) of the Nutrient Dynamics in the 'Hyphosphere' of Ecto- and VA-mycorrhizal Roots. N_{prot}: Protein Nitrogen; MO: Microorganism. (H. Narchner, 1991)

(*Distichlis spicata*) had higher sodium, potassium and phosphorus concentrations than non-mycorrhizal roots. In contrast, the results of other studies have shown that Na uptake decreased in AM plants compared to controls. Sodium content of shoots of mycorrhizal halophytic *Aster tripolium* plants was lower than non-mycorrhizal plants under salinity stress (Rozema *et al.*, 1986). K/Na ratio increased in mycorrhizal barley (moderate salinity tolerant plant) at high levels of salinity by decreasing Na concentration, rather than by increasing K concentration (Mohammad *et al.*, 2003). Dwivedi 2003 and Soni 2006, reported potassium play an important role in mycorrhization in wheat, legumes and rice plants.

Increased Uptake of Other Mineral Nutrients

Concentrations of Ca, Mg and Zn in onion plants inoculated with *Glomus fasciculatum* increased in saline conditions and improved the nutritional status, which was at least partially responsible for increased plant growth (Ojala *et al.*, 1983). The improved growth and nutrient acquisition (P, K, Zn, Cu and Fe) in tomato demonstrate the potential of AM fungi for protecting plants against salt stress in arid and semiarid areas (Al-Karaki, 2000; Al-Karaki and Hammad, 2001). Although effects of AM fungi in increasing some toxic elements (*e.g.*, Na, CI and Mn) have been reported (Pfeiffer and Bloss, 1988; Cantrell and Linderman, 2001), a direct effect of AM fungi in rednenon of others has also been reported. Mycorrhizal fungi decreased sodium concentration in barley (*Hordeum vulgare*) when grow in saline conditions (Mohammad *et al.*, 2003), and Mn can be reduced in mycorrhizal plants when it occurs at toxic levels compared to non-mycorrhizal plants (Sanders and Fitter, 1992; Cardoso *et al.*, 2003).

AMF in Metal Sequestration

The use of living organism for the remediation of soils contaminated with heavy metals, radionuclide or polycyclic hydrocarbon is known as bioremediation (Kumar *et al.*, 1995; Brooks and

Robinson, 1998; Salt *et al.*, 1998, Baker *et al.*, 2000). AM fungi are involved in bioremediation through phytoremediation, the technique based on the use of plants for soil remediation (Leyval *et al.*, 1997; Turnau *et al.*, 2005). Depending on the type of pollutant different strategy for phytoremediation, such as phytostabilization, phytodegradation and phytoextraction has been used. For phytoremediation of soil polluted with heavy metals, the phytostabilization strategy involves the immobilization of heavy metals in the soil by establishing plants.

AMF can help phytoremediation activities, particularly in phytostabilization (Goncalves *et al.*, 1997; Leyval *et al.*, 1997, 2002; Orlowska *et al.*, 2002; Regvar *et al.*, 2003; Turnau *et al.*, 2005). Among possible mechanism by which AM fungi improve the resistance of plants to heavy metal is the ability of AM fungi to sequester heavy metals through the production of chelates or by absorption AM plants typically less translocate heavy metal to their shoots than the corresponding non AM control. Among the diverse type of mycorrhizosphere interactions known to benefit plant growth and health, those related to phytoremediation process. As rhizobacteria and AM fungi interact synergistically to the benefit of phytoremediation. A key point in phytoremediation is the use of heavy metal adapted microbes, soil microbial diversity and activity both negative affected by excessive concentratic of heavy metals. Indigenous bacterial population (Giller *et al.*, 1998) and AM fungi (Del Val *et al.*, 1999) must be adapted to metal toxicity and have evolved abilities to enable them to survive in polluted soil.

Various authors have reported isolating spores of arbuscular mycorrhizal fungal taxa such as *Glomus* and *Gigaspora* associated with most of the plants growing in heavy metal polluted habitats (Raman *et al.*, 1993; Chaudhry *et al.*, 1999). Identified *Glomus* and *Gigaspora* spp. in the mycorrhizospheres of fourteen plant species colonising a magnesite mine spoil in India. Whereas *Glomus mosseae* was isolated only and Duek *et al.* (1986) isolated *Glomus fasciculatum* alone from the heavy metal polluted soils. Pawlowska *et al.* (1996) surveyed a calamine spoil mound rich in Cd, Pb and Zn in Poland and recovered spores of *Glomus aggregatum*, *G. fasciculatum* and *Entrophospora* spp. from the mycorrhizospheres of the plants growing on spoil. Joner and Leyval (1997) reported that extra-radical hyphae of AM fungus G. *mosseae* can transport Cd from soil to subterranean clover plants growing in compartmented pots, but that transfer from fungus to plant is restricted due to fungal immobilization. It was also reported also reported no restriction of fungal hyphal growth into soil with high extractable Cd levels. It have been also showed very little, if any, translocation of Zn absorbed by mycorrhizal maize seedlings grown in contaminated soil, to the shoots. It was studied that localization of heavy metals within the fungal mycelium and mycorrhizal roots of *Euphorbia cyparissias* from Zn contaminated wastes and found higher concentrations of Zn as crystaloids deposited within the fungal mycelium and cortical cells of mycorrhizal roots. Studies by various researchers (Galli *et al.*, 1994; Hetrick *et al.*, 1994) have shown that mycorrhizal fungal ecotypes from heavy metal contaminated sites seem to be more tolerant to heavy metals (and have developed resistance) than reference strains from uncontaminated soils Galli *et al.* (1995) reported that although there was an increase in the contents of cystein, gamma EC and GSH in the mycorrhizal maize roots grown in quarto sand with added Cu, no differences in Cu uptake were detected between non-mycorrhizal and mycorrhizal plants. These results do not support the idea that AM fungi protects maize from Cu-toxicity. Mycorrhizae are also known to produce growth stimulating substances for plants, thus encouraging mineral nutrition and increased growth and biomass necessary for phytoremediation to become commercially viable strategy for decontamination of polluted soils. In addition to the damaging effects on plants, the effect of heavy metals on the soil microorganisms and soil microbial activity also need to be considered. Various soil factors such as the clay contents and mobility of heavy metals effect plants and soil biota. As metal uptake by plant roots depends on soil and their associated symbionts, it is important to monitor metal mobility and availability to plant and its symbionts when assessing the effect of soil contamination on plant uptake and related phytotoxic effects The prospect of symbionts existing in heavy metal contaminated soils has important implications for phytoremediation.. The potentials of phytoremediation of contaminated soil can be enhanced by inoculating hyper-accumulator plants with mycorrhizal fungi most appropriate for contaminated site. It is further suggested that the potential of phytoremediation of contaminated soil can be enhanced by inoculating hyper-accumulator plants with mycorrhiza.

AMF as Phytoprotector

The establishment of AM fungi in plant roots has been shown to reduce damage caused by soil borne plant pathogens with an enhancement of plant resistance/tolerance in mycorrhizal plants. The effectiveness of AM in biocontrol is dependent on the AM fungus involved as well as the substrate and the host plant (Azcon–Aguliar and Barea, 1996; Linderman, 1994, 2000, Elmer 2002).

Several AM fungal species have been found to control soil borne pathogens, for example under green house conditions.

Glomus faciculatum and *Gigaspora margarita* are shown to reduce root rot disease caused by *Fusarium oxysporum f.* sp. *asparagi* and *Helicobasidium mompa* in asparagus (*Asparagus officinalis L.*) (Matsubsara *et al.*, 2000; Matsubara *et al.*, 2001, Singh 2008).

Glomus clarum is shown to reduce root necrosis due to *Rhizcotonia solani* in cow pea (*Vigna unguiculata L.*) (Abdel-Fattah and Shabana, 2002).

In pasteurised soil AM fungi have been shown to decrease the root damage caused by root rot fungus *Cylindrocladium spathiphylli* in bananas, although the pathogen decreased the intensity of AM fungal root colonization (Declerck *et al.*, 2002). Since AM are formed in the roots of plants, maximum attention has been paid towards mycorrhizae disease incidence interactions in relation to soil and root borne pathogens (Table 13.1). AM fungi impart resistance to host plant against root pathogens. Root pathogens are major limiting factor for plant production. Most widespread root borne pathogens belong to genera, *Phytophthora, Fusarium, Pythium* and *Rhizoctonia*. They kill roots or reduce their ability to absorb water and nutrients by penetrating root tissues and producing toxins.

Different AM fungi are known to confer variable tolerance to *Phytophthora parasitica*. Davis and Menge (1980, 1981) used five different AM fungal species on three different plants. They observed that maximum resistance to the pathogens was due to two AM fungi namely *G. fasciculatum* and *G. constrictum* in *Citrus* sp. Arbuscular mycorrhizal peanut plants become resistant to root rot pathogen *Sclerotium rolfsii* (Krishna and Bagyaraj, 1983). Number of sclerotia produced by the pathogen were reduced on the mycorrhizal roots. Biomass production and phosphorus content were maximum in mycorrhizal plants and minimum in pathogen inoculated plants.

Take all disease of wheat, caused by *Guamannomyces graminis* var. tritici was reduced when mycorrhizal fungus was inoculated in plants. The disease is favoured by inadequate plant nutrition leading to mineral deficiency especially phosphorus (Graham and Menge, 1981).

AM fungi are able to increase nutritional status of the host plant thereby improving resistance against the pathogens. Recently Singh *et al.* (2008, 2009, 2010) reported inoculation of AM fungi before *Fusarium oxysporum* causing wilt disease protect the tomato plants. AM fungi and *Trichoderma* sp. produced more significant biocontrol effect (Singh, 2008). Vyas *et al.* (2010) reviewed mechanism phytoprotection by AM fungi in various plant diseases.

Table 13.1: Inhibitory Influence of VAM Fungi on Soil-Borne Fungal Pathogens

Host	Disease	Pathogen	Reference
Tobacco	Root rot	*Phytophthora cinnamomae*	Bartschi *et al.*, 1981
Onion	Wilt	*Pyrenochaeta terrestris*	Becker, 1976
Onion	Wilt	*Fusarium oxysporum f. sp. cepi*	Dehne, 1982
Tomato	Wilt	*Fusarium oxysporum f.sp. lycopersici*	Dehne and schonbeck, 1979
Wheat	Take-all disease	*Gaeumannomyces graminis var tritici*	Graham and Menge, 1982
Soybean	Root rot	*Macrophomia phaseolina, Rhizoctonia solani, Fusarium solani*	Zambolim and Schenck, 1983
Brassica	Root rot	*Olpidium brassicae*	Schonbeck, 1979
Casia tora	Wilt	*Fusarium oxysporum*	Chakrabarty and Mishra, 1986
Tomato	Root rot	*Fusarium oxysporum f.sp. redicis lycopersici*	Caron *et al.*, 1986
Brinjal	Wilt	*Verticillium albo-atrum*	Melo *et al.*, 1985

P Gossling, A Hodge, G Goodlaur and GD Bending, 2005.

Table 13.2: Inhibitory Effect of VAM Fungi on Nematodes

Host	Disease	Nematode	Reference
Phaseolus vulgaris	Root knot	*Meloidogyne javanica*	Oliveira and Zambolim, 1987
Barley	Root decline	*Aphalenchus avenae*	Hussey and Roncadori, 1981
Soybean	Root knot	*Meloidogyne spp.*	Kellam and Schenck, 1980
Tobacco	Root knot	*M. hapla*	Sikora
Tomato, Oat	Root knot	*M. incognita*	Schonbeck, 1975
Cotton	Root knot	*M. incognita*	Roncadori and Hussey, 1977
Tomato	Root knot	*M.incognita, M. Javanica*	Bagyaraj *et al.*, 1979
Soybean	Root knot	*M. incognita*	Kellam and Schenck, 1980
Cotton	Root knot	*M. incognita*	Salch and Sikora, 1984
Tomato	Root knot	*M. incognita*	Suresh and Bagyaraj, 1984
Cotton	Root knot	*M. incognita*	Smith *et al.*, 1986
Beans	Root knot	*M. javanica*	Olivera and Zambolim, 1987

KG Mukerji, BP Chamola and Mamta Sharma, 1997

Mycorrhiza-induced Resistance

Different mechanisms have been shown to play a role in plant protection by AMF, namely, improved plant nutrition, damage compensation, competition for colonization sites or photosynthates, changes in the root system, changes in rhizosphere microbial populations, and activation of plant defense mechanisms. Several mechanisms can be operative simultaneously, with contributions depending on environmental conditions, timing of the interaction and partners involved (Azcon Aguilar and Barea, 1996).

There is evidence for the accumulation of defensive plant compounds related to mycorrhization, although to a much lower extent than in plant-pathogen interactions. Accumulation of reactive oxygen species, activation of phenylpropanoid metabolism, and accumulation of specific isoforms of hydrolytic enzymes such as chitinases and glucanases has been reported in mycorrhizal roots. There reactions, however, are generally localized, suggesting. Concerning aboveground effects, accumulation of insect antifeedant compounds and transcriptional regulation of defense-related genes (Pozo *et al.*, unpublished data) have been described in the shoots of mycorrhizal plants. Furthermore, the volatile blends released by AM plants can be more attractive to aphid parasitoids than those from non-mycorrhizas. Nevertheless, accumulation of PR proteins, salicyclic acid, or expression of marker genes associated with systemic acquired resistance has not been reported in systemic tissues.

AMF as Communicators

The most important AM fungal structure for plant nutrition is represented by the extra radical mycelium spreading from mycorrhizal roots into the surrounding soil which is able to uptake nutrients N, P, S, Ca, K, Fe, Cu, Zn and to transfer them to root cells (Smith and Read, 1997; Cox *et al.*, 1980). Mycorrhizal mycelium has been investigated in different experimental studies, based on either destructive extraction from soil or root observation chambers or in vitro system which yielded only qualitative data on its structure and growth (Jakobson *et al.*, 1990; Jones *et al.*, 1998).

The first visualization of intact AM mycelium extending from mycorrhizal roots into the extraradical environment was obtained by means of a bidimensional model system which utilized two cellulose esters members "sandwiched" around the roots of individual plantlets. After only seven days growth, a fine network of extramatrical hyphae growing on the membrane was visible to the naked eye, and its length extended from 5169 to 7471 mm (hyphal length in *Thymus vulgaris* and *Allium porum*, respectively).

The experimental system divided to visualize the mycorrhizal mycelium also evidenced that the mechanism allowing the formation of the network was self recognition and hyphal anastomosis.

It is important to stress that the viability of mycorrhizal networks was 100 per cent and that all the anastomoses showed protoplasmic continuity and nuclear occurrence in hyphal bridges, confirming the occurrence of nuclear exchange also during fusion between extraradical symbiotic hyphae. AM fungi are known to infect a wide range of host species. They have a large geographical distribution (Malloch *et al.*, 1980), being found even in the Arctic tundras and the Antarctic region (DeMars and Boerner 1995b; Gardes and Dahlberg 1996). Unlike most ectomycorrhizal species, AM are not host specific. This enables them to form associations with a large number of plant species.

AM fungi regulate plant communities by affecting competition, composition and succession (Allen and Allen 1984; Kumar *et al.*, 1999). Limited resources and the struggle of the plants for a share of these is the primary selection pressure operating on plant species (St. John and Coleman, 1983). In competition between plants, mycorrhizae in the soil favour the growth of one species and are detrimental to other competing species. Fitter (1977) demonstrated this in competition between two grasses *Lolium perenne* and *Holcus lanatus*. Inoculation with mycorrhizae favoured the growth of *H. lanatus*. This was an indirect effect as infection with mycorrhizae reduced the root length of *L. perenne* by 40 per cent. AM may regulate competition between plants by making available to mycorrhizal plants, resources that are not available to non-mycorrhizal neighbours (Allen and Allen, 1984). AM symbiosis increases intraspecific competition (Facelli *et al.*, 1999). As a result, density of individuals of a single species would be reduced thereby allowing the co-existence of individuals of different species. This would lead to an increase in species diversity.

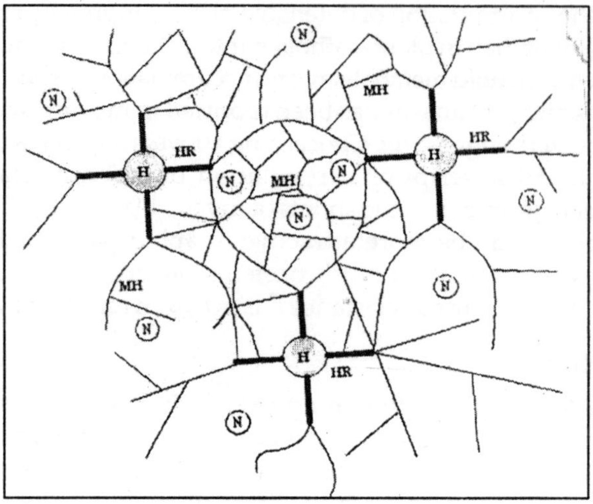

Figure 13.2: AM Hyphae Form a Network the Wood-Wide Web

Table 13.3: Extension and Interconnectedness of Extraradical Mycelial Networks Produced by AM Fungi Living

Plant Species/Fungal Species	*Hyphal Density (mm mm⁻²)*	*No. of Anastomoses per Hyphal Length*	*Anastomosis Frequence (per cent)*	*Reference*
Allium porum/Glomus mosseae	2.7	4.6	75.0	Giovannetti M *et al.*, 1999
Allium porum/Glomus mosseae	3.5	3.8	59.3	Giovannetti M *et al.*, 2004
*Daucus carota/Gigaspora margarita**	-	0.0075	9.8	De la Providencia IE *et al.*, 2005
*Daucus carota/Gigaspora rosea**	-	0.012	4.2	De la Providencia IE *et al.*, 2005
*Daucus carota/Glomus hoi**	-	0.057	100	De la Providencia IE *et al.*, 2005
*Daucus carota/intraradices**	-	0.076	100	Giovannetti M *et al.*, 2004
Daucus carota/Glomus mosseae	3.9	2.5	45.5	Giovannetti M *et al.*, 2004
*Daucus carota/Glomus proliferum**	-	0.066	100	De la Providencia IE *et al.*, 2005
*Daucus carota/Scutellospora reticulate**	-	0.0079	5.2	De la Providencia IE *et al.*, 2005
Gossypium hirsutum/Glomus mosseae	6.8	6.2	53.1	Giovannetti M *et al.*, 2004
Lactuca sativa/Glomus mosseae	2.9	3.0	63.8	Giovannetti M *et al.*, 2004
Petrosslinum cripsum/Glomus caledonium	3.8	1.5	18.6	Unpublished data
Petroselinum cripsum/Glomus intradices	2.3	5.5	56.9	Unpublished data
Petroselinum cripsum/Glomus mosseae	3.5	4.7	62.3	Unpublished data
Prunus cerasifera/Glomus mosseae	2.4	5.1	64.0	Giovannetti M *et al.*, 2001
Solanum melongena/Glomus mosseae	4.1	2.1	47.0	Giovannetti M *et al.*, 2004
Thymusvulgaris/Glomus mosseae	2.1	5.1	78.0	Giovannetti M *et al.*, 2001

*: Ri-TDNA transformed carrot root.

M Giovanetti, L Avio, P Fortuna, EP Sbrana, P Sterani, 2006.

Mycorrhizae govern species composition in communities by influencing plant fitness at the establishment phase. AM prevent non-mycorrhizal plants from growing in soils colonized by them. This has a selective advantage for the fungus. Maintaining a high proportion of compatible host species at the expense of non-compatible species provides the fungus with an undisturbed carbon supply (Francis and Read, 1994).

Succession is a chain of predictable processes whose course is influenced by nutrient availability. Mycorrhizae, owing to their role in nutrient uptake, may play an important part in determining the rate and direction of the process (Smith and Read, 1997a). They influence the outcome of succession by amending the composition of species or by affecting species diversity (Gange *et al.*, 1990).

AM fungi have been reported to be active in mediating nutrient transfer among plants (Chiariello *et al.*, 1982; Francis and Read, 1984; Grime *et al.*, 1987; Watkins *et al.*, 1996; Graves *et al.*, 1997; Lerat *et al.*, 2002) mainly through the extensive mycelial networks, which due to the lack host specificity may link the roots of contiguous plant species (Graves, *et al.*, 1997; Van der Heijden 1998). Recent studies showed a novel mechanism by which plants may become interconnected, that is hyphal fusion between extraradical hyphae originating from different individual plant root systems of different species, genera and families (Giovannetti, 2004).

AMF in Biotechnology

Introduction of AMF on micropropagate plantlets improves its establishment and growth in the field. Unfortunately not much work has been conduced in this line. But the few convincing reports decisively prove that successful hardening and *ex vitro* establishments of plantlets could be achieved by inoculation with AM fungi at the time of plants out.

One of the earliest report on the effect of growth and leaf mineral content of two apple clones propagated *in vitro* were increased substantially. Similarly enhanced effect on the growth of *in vitro* cultured strawberry plantlets could be achieved due to the association of AMF. The rooting of plantlets of garlic regenerated from called was significantly enhanced due to inoculation with *Glomus mosseae.* The transplant success and growth of *Robus idaeus* and *Pistacia integerrima* were achieved with mycorrhiza inoculation.

Effect of AMF on these culture flowers is also encouraging. In a different study at Vellayani, evaluated the effect of AMF in enhancing of survival, growth characteristic and uptake of nutrients by micropropagated Anthurium. AMF was incorporated into the rooting medium at the time of planting out through infected root bits. The study was conducted with *Glomus constrictum* and *G. etunicatum.* Micropropagated avocado plants generally exhibit low survival and very slow rate of growth during acclimatization. Inoculation of *G. fasciculatum* on micropropagated avocado planlets improved the formation of well developed root system, shoot growth, shoot root ratio and NPK content in plant tissue which helped plants to tolerate environmental stress transplanting. It was concluded that inoculation of AMF seems to be the key factor for subsequent growth and development of micropropagated plants of avocado.

AMF in Genetics

"Several constitutively expressed genes were shown to be expressed at the rhizodermis level, including root hairs and around the root tip. Moreover, a loss-of-function mutant defective in two *Phtl* transporters, normally expressed at the root periphery, in *Arabidopsis* exhibited a strong reduction of phosphate uptake by 75 per cent (Shin *et al.*, 2008), strongly indicating a role for *Phtl* transporters in the 'direct' uptake pathway. In the AM symbiosis, phosphate uptake by the fungus from soil is the first step in the process of phosphate transport to the root. Two AMF phosphate transporters (*GvPT* and *GiPT* from *Glomus versiforme* and *Glomus intraradices*, respectively (Harrison *et al.*, 1995; Maldonado-Mendoja *et al.*, 2001) form a common clade with transporters identified in fungal species from the Basidiomycota and Ascomycota. At least two members of this clade (pH084 and *GvPT*) exhibit high affinity towards phosphate, with *Km* values of 8 µM and 18 µM, respectively (Harrison *et al.*, 1995; Bun-Ya *et al.*, 1991). The *GvPT* and. *GiPT* genes are predominantly expressed in the extraradical fungal mycelium exposed to micromolar phosphate concentrations (Maldonado-Mendoja *et al.*, 2001). And so their encoded proteins are likely to participate in phosphate uptake at the fungus-soil interface.

AMF in Genomics and Proteomics

In recent years, outstanding molecular approaches have been used for the identification of genes and functions involved in plant-microbe endosymbioses. Following the first completion of genome sequencing projects, biological research has developed high throughout genetic programs with multiparallel analyses of gene transcripts, proteins and metabolites, that are now tentatively transposed to the world of mycorrhizal and rhizobial symbioses (Franken and Requena 2001; Colebatch *et al.*, 2002a, 2002b).

Table 13.4: Proteomic Studies in the Field of Plant-Microbe Symbioses

Biological Models	*Proteins Characterized by 2 DE*	*Identified Proteins*	*References*
	Endomycorrhizal symbioses		
Allium cepa/Glomus mosseae	15	0	Garcia-Garrido *et al.* (1993)
Nicotiana tabacum/G. mosseae/ Glomus instraradices	34	0	Dumans-Gaudot *et al.* (1994)
Lycopersicon esculentum/G. intraradices	5	0	Simoneau *et al.* (1994)
Pisum sativum/G. mosseae	42	0	Samra *et al.* (1997)
L. esculentum/G. mosseae	44	0	Benabdellah *et al.* (1998)
L. esculentum/G. mosseae	14	0	Dassi *et al.* (1999)
L. esculentum/G. mosseae	26	1	Benabdellah *et al.* (2000)
Triticum aestivum/G. mosseae	1	1	Fester *et al.* (2003)
Medicago truncatula/G. mosseae	55	8	Bestel-Corre *et al.* (2002)
P. sativum/G. mosseae	7	6	Repetto *et al.* (2003)
M. truncatula/G. mosseae	34	6	Bestel-Corre (2002)
Ri T-DNA Daucus carota/G. Intraradices	8	2	Bestel-Corre (2003)
G. intraradices	450	6	Dumas-Gaudot *et al.* (2003b)
Acaulospora laevis/Gigaspora rosea/ Scutellospora castanea/G. mosseae	12	0	Samra *et al.* (1996)

Franken and Requena, 2001

AM fungi bear large and highly repeated genomes, a feature that completely hinders the development of genomic analyses. Due to their obligate symbiotic status, the amount of accessible mRNA material is very limited; however, many limitations have been overcome with the help of polymerase chain reaction (PCR)-based methods (Franken and Requena, 2001) and *in vitro* monoxenic cultures (St-Arnaud *et al.*, 1996). Several EST libraries have thus been constructed using activated spores of *Gigaspora rosea* (Stommel *et al.*, 2001; Tamasloukht *et al.*, 2003), *Glomus mosseae* (Requena *et al.*, 2002), presymbiotic mycelium of *Gigaspora margarita* (Lanfranco *et al.*, 2000) and extra radical hyphae of *Glomus intradices* (Sawaki and Saito, 2001).

Concerning mycorrhizas, pioneering studies were achieved with ectomycorrhizas in the early 1990s (Hilbert *et al.*, 1991; Simoneau *et al.*, 1993; Burgess *et al.*, 1995; Burgess and Dell, 1996) allowing researchers to detect symbiosis-related (SR) polypeptides, up-regulated or newly induced in mycorrhizal roots, as well as down-regulated polypeptides, by comparison to control roots and mycelium extracts. Recently some SR proteins were identified by mass spectrometry and N-terminal sequencing (Tarka *et al.*, 2000), as well as mycelial proteins (Vallorani *et al.*, 2000). Arbuscular mycorrhizas, a similar progression was followed, from descriptive studies (Garcia-Garrido *et al.*, 1993; Dumas-Gaudot *et al.*, 1994; Simoneau *et al.*, 1994; Samra *et al.*, 1997; Benabdellah *et al.*, 1998; Dassi *et al.*, 1999) to the first identifications of a vacuolar H^+-ATPase and a Myk15 protein with an unknown function (Benabdellah *et al.*, 2000; Fester *et al.*, 2003). Additionally, protein profiles of both dormant and germinated spores of several fungi were compared (Samra *et al.*, 1996). Now, proteome analysis is used as a powerful tool to reveal more and more proteins involved in mycorrhiza development and functioning including proteins involved in defence response, root physiology and the respiratory pathway (Dumas-Gaudot *et al.*, 2001; Bestel-Corre *et al.*, 2002).

Fluxomics

The goal of fluxomics is to quantify all the metabolic fluxes in a cell, tissue or organism (Sauer *et al.*, 1999; Sauer, 2004). Investigation of the flow of matter through biochemical systems has always been central to the study of metabolism, and analysis of the rates of metabolic transformations–the study of enzyme kinetics–is likewise a long-established aspect of understanding any biological system in detail. However, the conceptual, experimental and computational tools for quantifying the integrated functioning of metabolic networks began to become available only in recent decades, and are still very much under development. Like many other omic approaches, fluxomics has yet to attain the goal of generating comprehensive system-wide data sets. However, progress in the last 10 year has been rapid and fluxomics has grown beyond is origins in bacterial systems and has begun to make significant contributions to the study of plant systems (Kruger *et al.*, 2003; Schwender *et al.*, 2004a; Ratcliffe and Shachar-Hill, 2006).

The analysis of multiple flows through a network involves both direct and indirect determination of metabolic and transport fluxes. Direct determination fluxes involves individually measuring the rates of substrate uptake, product secretion, and the accumulation of storage or structural compounds (lipids, carbohydrates and proteins) of known composition. Indirect determination of fluxes is performed in two ways. In the first method, fluxes measured directly are used to deduce other net fluxes; this is done by balancing the influxes and effluxes from individual metabolite pools using the known stoichiometries of biochemical reactions (flux balancing). The second method is based on interpreting the results of labeling experiments. Labeling measurements using radioactive isotypes are made by fractionation or chromatographic separation methods followed by scintillation counting of different intermediate and product metabolites. Measurements of stable isotopic labeling usually involve ^{13}C (or less commonly ^{15}N or ^{2}H) and are made by nuclear magnetic resonance (NMR) spectroscopy and mass spectrometry. The interpretation of labeling and flux measurement data in terms of multiple fluxes is nontrivial and almost always involves computer-aided modeling. Models are used to estimate fluxes by finding the values that result in a best fit of computed (simulated) to experimental results.

This reliance on fitting to a metabolic model means that a fluxomic investigation requires some prior knowledge of and assumptions about the metabolic architecture of the system. This knowledge is certainly much less complete in mycorrhizas than in many bacteria or model fungi and plants. However, progress in delineating metabolic and transport networks in mycorrhizal systems has been steady in recent years and has reached a point where models can be constructed for the quantitative interpretation of labeling data in the best-studied cases (Chalot and Brun, 1998; Bago *et al.*, 2000; Bucking and Shachar-Hill, 2005; Govindarajulu *et al.*, 2005). The outline of a working model of central metabolism and transport in the arbuscular mycorrhizal symbiosis that could serve as a starting point for flux analysis is shown in Figure 13.3. The advent of mycorrhizal plant and fungal genome sequences will be of enormous help in building and in filling in the molecular mechanisms of such models. When transcript and proteomic data sets for mycorrhizas become much more complete than they are at present–which can be expected to happen sooner rather than later–they too will be important in defining and validating model networks for use in flux analysis.

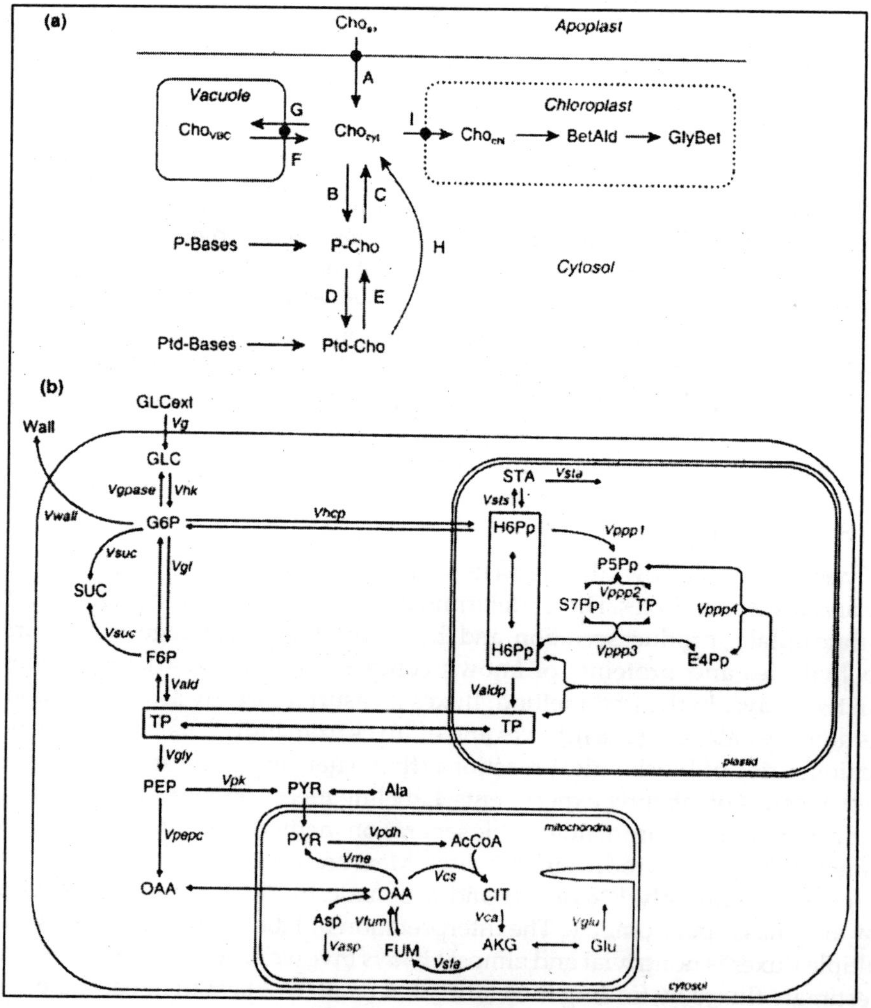

Figure 13.3: A Model Used in a Dynamic Laboring of the Glycine Betaine Synthesis in Transgenic Tobacco (*Nocotiana tabacum*) (Hill, 2000)

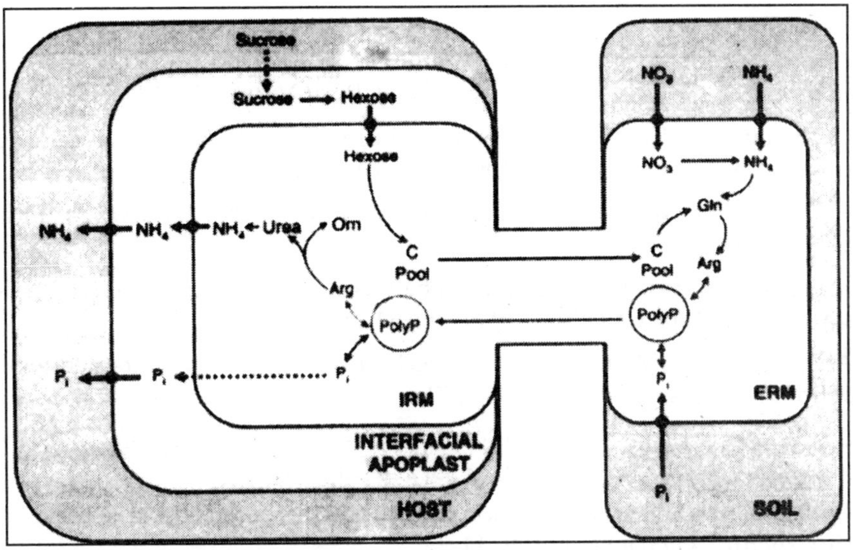

Figure 13.6: A Model of Nitrogen Flow in the Arbuscular Mycorrhizal Symbiosis, Suitable for Initiating a Fluxomics Analysis (Hill, 2000)

Conclusions

1. Soil is the major sink of carbon and AM fungi have shown that, they have also potential to sequester the carbon from the soil.
2. Deposition of heavy metal in the soil is troublesome affair which resulted into the toxicity, which is not only harmful for the plants but also unsafe for organism dependent on their food from plants. Here also AM fungi provide release from the heavy metal toxicity by sequestring heavy metals from the soil.
3. Control of plant disease was remain challenge for scientist and farmer both. Use of pesticide no longer remain the healthy affair, therefore, biological control was searched as an alternative, among the biocontrol agents, role of AMF have been recognized a fruitful one.
4. As it has been established that microbes and plants have very strong communication in the from of signal transduction, here also AMF provided evidences that they are best communicators in microbial world and have developed wood wide-web for their communication.
5. Biotechnology is the talk of hour and therefore role of AMF fungi have also been explored and recognised in the field of morden biology and have been tried to elucidate their relevance in biotechnology, molecular biology, genomics, proteomics and fluxomics.

Acknowledgements

Authors are thankful to Head, Department of Botany, Dr. H.S. Gour University, Sagar, RKG and AKS thankfully acknowledge UGC and DS Kothari for financial assistance respectively.

References

Al-Karaki GN and Clark RB. (1999). Varied rates of mycorrhizal inoculum on growth and nutrient acquisition by barley grown with drought stress. *J. Plant Nutrion* 22: 1775-1784.

Al-Karaki GN and Hammad R. (2001). Mycorrhizal influence on fruit yield and mineral content of tomato grown under salt stress. *Journal of Plant Nutrition* 24: 1311-1323.

Al-Karaki GN. (2000). Growth of mycorrhizal tomato and mineral acquisition under salt stress, *Mycorrhiza* 10: 51-54.

Allen EB and Allen MF. (1984). The competition between plants of different succession all stages: mycorrhizal as regulators. *Canadian J. of Botany* 62: 2625-2692.

Allen EB and Allen MF. (1986). Water relations of xeric grasses in the field: interaction of mycorrhizas and competition. *New Phytologist* 104: 559-571.

Allen EB and Cunningham GL. (1983). Effects of vesicular-arbuscular mycorrhizae on *Distichlis spicata* under three salinity levels. *New phytologist* 93: 227-236.

Allen R and Read DJ. (1994). The contribution of mycorrhizal fungi to the determination of plant community structure. *Plant and soil* 159 : 11-25.

Allen MF. (1991). *The Ecology of Mycorrhizae*. Cambridge: Cambridge University Press.

Azaizeh HA, Marschner H, Romheld V, Wittenmayer L. (1995). Effects of vesicular-arbuscular mycorrhizal fungus and other soil microorganisms on growth, mineral nutrient acquisition and root exudation of soil grown maize plants. *Mycorrhiza* 5: 321-327.

Azcon-Aguilar C and Barea JM. (1996). Arbuscular mycorrhizas and biological control of soil borne plant pathogens- an overview of the mechanisms involved. *Mycorrhiza* 6: 457-458.

Bago B, Azcon-Aguliar C, Goulet A, Piche Y. (1998). Branched absorbing structure (BAS): a feature of the extradical mycelium of symbiotic arbusculsar mycorrhizal fungi. *New Phytol* 139: 375-388.

Bago B, Pfeffer PE, Douds DD, Brouillette J, Becard G, Shachar-Hill Y. (1999). Carbon metabolism in spores of the arbuscular mycorrhizal fungus *Glomus intraradices* as revealed by nuclear magnetic resonance spectroscopy. *Plant Physiolz*121: 263-271.

Bago B, Pfeffer PE, Shachar-Hill Y. (2000). Carbon metabolism and transport in arbuscular mycorrhizas. *Plant Physiol* 128: 949-957.

Bago B, Zipfel W, Williams R, Jun J. Arreola R, Lammer P, Pfeffer PE, Shachar-Hill Y. (2002). Translocation and utilization of fungal lipid in the arbuscular mycorrhizal symbiosis. *Plant Physiol* 128: 108-124.

Bago B, Pfeffer PE, Abubaker J, Jun j, Allen JW, Brouillette, Douds DD, Lammers PJ and Sachar Hill. (1995). Carbon export from arbuscular mycorrhizal roots involves the translocation of carbohydrate as the translocation of carbohydrate as well as lipid. *Plant physiology*131: 1469-1507.

Bagyaraj DJ. (1984). Biological interactions with mycorrhizal fungi: I*n VA mycorrhiza* (eds). CLO Powell and DJ Bagyaraj (New York: CRC Press), 131-153.

Bagyaraj DJ, Manjunath A and Reddy DDR (1979). Interaction of vesicular arbuscular mycorrhizae with root knot nematode in tomato. *Plant soil.* 51: 397-403.

Baker AJM, MC Grath SP, Reeves RD, Smith JAC. (2000). Metal hyperaccumulator plants: a review of the ecology and physiology of a biological resource for phytoremediation of metal polluted soils. In: *Phytoremediation of contaminated soil and water*. (eds., Terry N, Banuelos G, Vangronsveld J,). Boca Ratton, Fl, USA: CRC Press, 85-107.

Barber S (1975). Soil nutrient bioavailability: A mechanistic approach. New York USA, John Wiley and Sons.

Bartschi H, Gianinazzi Pearson V and Vegh I. (1981). Vesicular arbuscular mycorrhizae formation and root disease (*Phytophthora cinnamomi*). development in *Chamaecyparis lawso. Phytopatho. Z.* 102: 213-218.

Beaden BN and Peterson L. (2000). Influence of arbuscular mycorrhizal fungi on soils structure and aggregate stability of vertisol. *Plant and Soil*. 218: 173-188.

Becker WN (1976). Quantification of onion vesicular-arbuscular mycorrhiza and their resistance to *Pyrenochaeta- terretris*. Ph.d. dissertation. Duke Uni, Durham.

Beilby JP, Kidby DK. (1980). Biochemistry of ungerminated and germinated spores of the vesicular-arbuscular mycorrhizal fungus, *Glomus caledonium*: changes in neutral and polar lipids. *J. Lipid Res*, 21: 739-750.

Benabdellah K, Azcon-Aguilar C, Ferrol N. (1998). Soluble and membrane symbiosis related polypeptides associated with the development of arbuscular-mycorrhizas in tomato (*Lycopersicon-esculentum*). *New phytol.*140: 15-143.

Benabdellah K, Azcon-Aguilar C, Ferrol N. (2000). Alterations in the plasma membrane polypeptide pattern of tomato roots (*Lycopersicon esculentum*). during the development of arbuscular mycorrhiza. *J Ex. Bot*. 51: 747-754.

Bestel Corre G, Dumas-Gaudot E, Gianinazzi S. (2004). Proteomics as at pool to monitor plant–microbe endosymbioses in the rhizosphere, 14: 1-10.

Bestel–corre G. (2002). La proteomique: un outil d' etude des interactions dans la rhizosphere. Identification de proteines reliees a la mycorrhization et a nodulation et evaluation de I' impact boues de I' impact de boues d' epuration sur les symbiosis. PhD Thesis. Universite de Bourgogne, Dijon.

Bethlenfalavy, GJ and Barea JM. (1994). Mycorrhizae in sustainable agricultural. J. Effects on seed yield and soil aggregation. *Am. J. Agric.*9: 157-161.

Bever JD, Schulatz PA, Pringle A, Morton JB. (2001). Arbuscular mycorrhizal fungi: more diverse than meets the eye, and the ecological tale of why. *Bioscience*, 51: 923-931.

Blilou I, Ocampo JA, Garcia-Garrido JM. (2000). Induction of Ltp (lipid transfer protein). and Pal (phenylalanine ammonia-lyase). gene expression in rise roots colonized by the arbuscular mycorrhizal fungus *Glomus mossease*. *J. Exp. Bot.*, 51: 1969-77.

Brooks RR and Robinson BH. (1998). The potential use of hyperaccumulators and other plants for phytomining. In: Brooks RR, ed., *Plants that hyperaccumulate heavy metals their role in phytoremediation, microbiology, archeology, mineral exploration, and phytomining*. Cambridge: CAB International, 327-356.

Bucking H, Sachar Hill Y. (2005). Phosphate uptake, transport and transfer by the arbuscular mycorrhizal fungus *Glomus* Abdel- Fattah GM and Shabana YM. (2002). Efficacy of the arbuscular mycorrhizal fungus *Glomus* in protection of cowpea plants against root rot pathogen *Rhictonia solani. Journal of Plant Disease Protection* 109: 207-215.

Bun-Ya M. (1991). The pho84 gene of *Saccharomyces cerevisiae* encodes an inorganic phosphate transporter.*Mol.Cell. Biol.*11: 3229-3238.

Burgess T, Laurent P, Dell B, Malajczuk N, Martin F. (1995). Effect of fungal isolate aggressivity on the biosynthesis of symbiosis related polypeptides in differentiating eucalypt mycorrhizas.*Can.J. Bot.*195: 408-417.

Burgess T and Dell B. (1996). Changes in protein biosynthesis during the differentiation of *Pisolithus-Eucalyptus grandios* ectomycorrhiza. *Planta* 74: 553-560.

Cantrell IC and Linderman RG. (2001). Preinoculation of lettuce and onion with VA mycorrhizal fungi reduces deleterious effects of soil salinity. *Plant and Soil* 223: 269-281.

Cantrell IC and Linderman RG. (2001). Preinoculation of lettuce and onion with VA mycorrhizal fungi reduces deleterious effects of soil salinity. *Journal of Environmental Quality* 30: 1822-1828.

Canvagnaro TR, Gao LL, Smith FA, Smith SE. (2001). Morphology of arbuscular mycorrhizas is influenced by fungal identity. *New Phytologist* 137: 373-388.

Cardoso E, Navarro RB, Nogueira MA. (2003). Charges in manganese uptake and translocation by mycorrhizal soyabean under increasing Mn doses. *Revista Brasileira De Ciencia Do Solo*, 27: 415-423.

Caron M, Fortin JA and Richard C. (1986). Effect of phosphorous concentration and *Glomus intraradices* on *Fusarium* crown and root rot of tomatoes. *Phytopathology* 76: 942-946.

Chakravarty P and Mishra RR. (1986). Influence of enditrophic mycorrhizae on the fusarial wilt of *Cassia tora L. J.phytopatho.*115: 130-133.

Chalot M, Blaudez D, Brun A. (2006). Ammonia: a candidate for nitrogen transfer at the mycorrhizal interface. *Trends in Plant Science*, 11: 263-266.

Chalot M and Brun A. (1998). Physiology of organic nitrogen acquisition by ectomycorrhizal fungi and ectomycorrhizas. *FEMS Microbiology Reviews* 22: 21-44.

Chaudhary TM, Hill, Khan AG, Duek C. (1999). Colonization of iron and zinc-contaminated dumped filter-cake waste by microbes, plants and association mycorrhizae. In: Remediation and management of Degraded Land. (eds Wong MH, Wong JWC Baker, AJM). CRC press, Boca, Chap, 27: 275-283.

Chiariello N, Hickman JC, Mooney HA. (1982). Endomycorrhizal role for interspecific transfer of phosphorous in a community of annual plants, *Science* 217: 941-943.

Clark RB and Zeto SK. (2000). Mineral acquisition by arbuscular mycorrhizal plants. *J. Plant Nutron.* 23: 867-902.

Colebatch G, Trevaskis B, Udvardi M. (2002a). Functional genomics: tool of the trade. *New phytol.* 153: 27-36.

Cox G, Moran KJ, Sanders F, Nockolads C, Tinker PB. (1980). Translocation and transfer of nutrients in vesicular-arbuscular nutrients in vesicular arbuscular mycorrhizas III Polyphosphate granules and phosphorous translocation. *New Phytologist*. 84: 649-659.

Curl EA and Trueglove B. (1986). The rhizosphere advanced series agricultural series in agriculture science 15. Berlin. Germany: Springer–Verlag.

Dassi B, Samara A, Dumas-Gaudat E, Gianinazzi S. (1999). Different polypepetide profiles from tomato polypeptide profiles from tomato roots following interaction with arbuscular mycorrhizal (*Glomus mosseae*). or pathogenic (*Phytophthora parasitica*). fungi.*Symbiosis*.26: 65-77.

Davies FT, Jr.Potter JR, Linderman RG. (1993). Drought resistance of mycorrhizal pepper plants independent of leaf p concentration–response in gas exchange and water relations. *Physiologia-Plantarum* 87: 45-53.

Davis RM and Menge JA. (1981). *Phytophthora parasitica* inoculation and intensity of vesicular–arbuscular mycorrhizae in citrus .*New Phytol*. 87: 705-715.

De la Providenica IE, de Souza FA, Fernandez F, Sejalon Delmaş N, Declerck F, Sejalon Delmas N, Declerck S. (2005). Arbuscular mycorrhizal fungi reveal distnict patterns of anastomosis formation and hyphal healing mechanism between different pathogenic groups. *New phytol*. 165: 261-271.

De mars BG and Boerner REJ. (1995). Arbuscular mycorrhizal development in three crucifers. *Mycorrhiza*. 5: 405-408.

Declerck S, Risede JM and Rufyikiri G and Delvaux B. (2002). Effect of arbuscular mycorrhizal fungi on severity of root rot of bananas caused by *Cylindrocladium Spathiphylli*. *Plant Pathology* 51: 109-115.

Degans B, Sparling G, Abott LK. (1994). The contribution from hyphae,roots and organic carbon constituients to the aggregation of a sandy loam under long term clover based and grass pastures., *European Journal of Soil Science* 45: 459-490.

Dehne HW and Schonbeck F (1979). Untersuchungen zun Einfluss der Endotrophen Mycorrhiza auf pflanzen krankheiten.II phenolstaff-wechsel und Linifizierung (The influence of endtrophic mycorrhiza on plant diseases.II.phenol metabolism and lignifications. *Phytopathol.Z*.95: 210-216.

Dehne HW. (1982). Interaction between vesicular–arbuscular mycorrhizal fungi and plant pathogens. *Phytopathology*.72: 1115-1119.

Del Val C, Barea JM and Azcon-Aguilar C. (1999). Diversity of arbuscular mycorohizal fungus population in heavy metal contaminated soil. *Applied and Environmental Microbiology* 65: 718-723.

Dewivedi OP and Vyas D. (2002). Effect of potassium on the occurrence of AM fungi. *Basic and Appli. Mycol*. 1: 233-235.

Diedrich C. (1990). Improved growth of *Cajanus cajan* (L). Millsp. In an unsterile tropical soil by three mycorrhizal fungi.*Plant and soil* 123: 261-266.

Douds DD, Pfeffer PE, Shachar-Hill Y. (2000). Carbon partitioning, cost eds. Arbuscular mycorrhizae. In: Douds DD, Kapulnik Y, eds. *Arbuscular mycorrhizas physiology and function*. Dordrecht, The Netherlands: Kluwer Academic Publishers, 197-130.

Douds DD, Gadkar V, Adholeya A. (2000). Production of VAM fungus biofertilizer. In: Mukerji KG, Chamola BP, Singh J, Mycorrhizal Biology. New York Academic Publisher.

Duek TA, Viser P, Ernst WHO, Schat H. (1986). Vesicular-arbuscular-mycorrhizae decrease zinc toxicity to grasses growing in zinc-polluted soil. *Soil Biol. Biochem.* 18: 331-333.

Dumas–Gaudot E, Bestel Corre G, Valot B, Lenogue S, Amiour N, St-Arnaud M, Fontain B, Dieu M, Raes M, Gianiazzi S. (2003b). Use of in vitro grown *Glomus intraradices* to sensor the effect of agricultural amendment. In proceeding of the IVth international congress on mycorrhiza, Montreal.

Dumas Gaudot E, Guillaume P, Tahiri-Alaoui A, Gianinazzi–pearson V, Gianinazzi S. (1994). Changes in polypeptide patterns in tobacco root colonized by two *Glomus* species.*Mycorrhiza* 4: 215-221.

Dumas-Gaudat E, Bestel-Corre G, Gianinazzi S. (2001). Proteomics, a powerful approach towards under standing functional plant root interactions with arbuscular mycorrhizal fungi: In: Pandilai SG (ed). Recent research developments in plant biology. Research Signopost, Trivandrum, India, pp. 95-104.

Dwivedi OP. (2003). Studies in soil microorganisms with special reference to vesicular-arbuscular mycorrhizal (VAM) fungal association with wheat crop of Sagar region. Ph.D. Thesis. Dr. H.S. Gour University, Sagar (M.P.). 176.

Elmer WH. (2002). Influence of formononetin and NaCl on mycorrhizal colonization and *Fusarium* crown and root rot of asparagus. *Plant Disease* 86: 1318-1324.

Facelli E, Facelli JM, Smith SE, Mclaughlin MJ. (1999). Interaotive effects of arbuscular mycorrhizal symbiosis intraspecific competition and resource availability on Trifolium subterranean cv. Mt. Barker. *New Phytol.* 141(3): 535-547.

Farrar SC, Hawes M, Jones D, Lindo S. (2003). How roots control the flux of carbon to the rhizosphere. *Ecology* 84: 827-833.

Fester MK and Strack D. (2003). Amycorrhizae–responsive protein in wheat roots. *Mycorrhiza* 12: 219-222.

Francis R and Read DJ. (1994). The contributions of mycorrhizal fungi to the determination of plant community structure. *Plant and soil* 159 : 11-25.

Francis Rand Read DJ. (1984). Direct transfer of carbon between plants connected by vesicular-arbuscular mycorrhizal mycelium *Nature*; 307: 53-56.

Franken P and Requena N. (2001). Analysis of gene expression in arbuscular mycorrhizas: New approaches and challenges. New *phytol.* 150: 517-523.

Gallaud I (1905). Etudes surles mycorrhizes endotrophes.Revue Generale de Botanique.17: 5-48,66-83,123-135;223-239-313-325,425-433,479-500.

Galli U, Schuepp H, Brunold C. (1994). Heavy metal binding by mycorrhizal fungi. *Physiol. Plantarum* 92: 364-368.

Garcia- Garrido JM, Toro N, Ocampo JA. (1993). Presence of specific polypeptides in onion roots colonized by *Glomus mosseae*. *Mycorrhizae.* 2: 175-177.

Garcia-Garrido JM, Ocampo JA. (2002). Regulation of the plant defence response in arbuscular mycorrhizal symbiosis. *J. Exp. Bot.* 53: 1377-1386.

Gavito ME and Varela L. (1995). Response of criolla maize to single and mixed species inoculation of arbuscular mycorrhizal fungi. *Plant soil.* 176: 101-105.

George E, Marschner H, Jakobsen I. (1995). Role of arbuscular mycorrhizae fungi in uptake of phosphorus and nitrogen from soil. *Crit Rev. Biotechnol* 15: 257-270.

Gianinazzi-Pearson V. (1996). Plant cell to arbuscular mycorrhizal fungi: getting to the roots of the symbiosis. *Plant Cell.* 8: 1871-83.

Giller K. (1998). Toxicity of heavy metals to microorganisms and microbial processes in agricultural soils: reviews: *Soil Biology and Biochemistry* 30: 1389-1414.

Giovannetti, Sbrana C, Avio L, Citernesi AS, Logi C. (1993). Differential hyphal morphogenesis in arbuscular mycorrhizal fungi during pre–infection stages. *New Phytologist* 125: 587-593.

Giovannetti M, Sbrana C, Avio L. (2004). Patterns of below ground plant interconnection established by means of arbuscular mycorrhizal networks. *New Phytologist.* 164: 175-181.

Giovannetti M, Azolini D, Citer nesi AS. (1999). Anastomosis formation and nuclear and protoplasmic exchange in arbuscular mycorrhizal fungi. *Applied Eviron Microbiol.* 69: 616-624.

Giovannetti M, Fortuna P, Citernesi AS, Morini S, Nuti MP. (2001). The ocurrence of anastomosis formation and nuclear exchange in arbuscular mycorrhizal network. *New phytol.*151: 171-124.

Goncalves SC, Goncalves MT, Freitas H, Martin-S, Loucao MA. (1997). Mycorrhizae in a Portuguese serpentine community. In: *The ecology of ultarmafic and metalliferous areas,* (eds. Jaffre T, Reeves RD, Becquer T), Proceedings of the second International Conference on Serpentine Ecology in Noumea, 87-89.

Gonzalez-Chavez C, D'Haen J, Vangronsveld J, Dodd JC. (2001). Copper sorption and accumulation by the extraradical mycelium of different *Glomus* spp. (arbuscular mycorrhizal fungi) isolated from the same polluted soil. *Plant and soil,* 240: 287-297.

Govindarajulu M, Pfeffer P, Jin HR, Abubaker J, Douds DD, Allen JW, Bucking H, Lammers PJ, Shachar-Hill Y. (2005). Nitrogen transfer in the arbuscular mycorrhizal symbiosis. *Nature* 435: 819-823.

Graham JH and Menge JA. (1982). Influence of vesicular-arbuscular mycorrhizae and soil phosphorous on the take all disease of wheat. Phytopathplogy 72: 95-98.

Graham JH, Leonard RT and Menge JA. (1981). Membrance mediated decrease in root exudation responsible for phosphorous inhibition of vesicular-arbuscular mycorrhiza formation. *Plant Physiology.* 68: 548-552.

Graham JH. (1982). Effect of citrus root exudates on germination of chlamydospores of the vesicular-arbuscular mycorrhizal fungus *Glomus epigacum. Mycologia* 74: 831-835.

Graham JH.. (2000). Assessing costs of arbuscular mycorrhizal symbiosis in agroecosystems. In: Podila GK, Douds DD, eds. *Current advances in mycorrhizae research.* St. Paul, MN, USA: American Phytopathological Society Press, pp. 127-140.

Graves JD, Walkins NK, Fitter AH, Robinson D, Scrimgeour C. (1997). Interspecific transfer of carbon between plants linked by a common mycorrhizal network. *Plant Soil* 192: 153-159.

Grime JP, Mannkey JML, Hillersh, Read DJ. (1987). Floristic diversity in a model system using experimental microcosms, *Nature* 328: 420-422.

Gupta V, Satyanarayana T, Sandeep G. (2000). General aspects of mycorrhiza. In: Mukerji KG, Chamola BP, Singh J, Mycorrhizal *Biology*. New York: Kluwer.

Harley JL and Harley EL. (1987). A check list of mycorrhizae in British flora. *New Phytologist* 105: 1-102.

Harrison MJ. (1999). Molecular and cellular aspects of the arbuscular mycorrhizal symbiosis. *Ann. Rev. Plant Physiol. Plant Mol. Biol.* 50: 361-89.

Harrison MJ andVan Burren ML. (1995). A phosphate transporter from the mycorrhizal fungus *Glomus vesiforme*. *Nature* 378: 626-629.

Harrison MJ. (2005). Signaling in the arbuscular mycorrhizal symbiosis. *Annu. Rev. Microbiol.* 59: 19-42.

Harrison MJ. (1998). Development of the arbuscular mycorrhizal symbiosis. *Current opinion in plant Biology* 1: 360-365.

Hetrick BAD, Wilson GWT and Todd TC. (1996). Mycorrhiza response in wheat cultivars, relationship to phosphorous, *Can. J. Bot.* 74: 19-25.

Hetrick BAD, Wilson GWT, Figge DH. (1994). The influence of mycorrhizal symbiosis and fertilizer amendment on establishment of vegetation in heavy metal mine spoil. *Environ. Pollut.* 86: 171-179.

Hilbert JL, Cista G, Martin F. (1991). Ectomycorrhizin synthesis and polypeptide changes during the early stage of *Eucalyptus* mycorrhiza development. *Plant Physiology* 97: 977-984.

Hill. (2007). Quantifying flows through metabolic networks and the propects for fluxomic studies of mycorrhizas, *New phytologist* 175: 235-240.

Hodge A, Campbell CD, Fitter AH. (2000). An arbuscular mycorrhizal fungus accelerates decomposition and acquires nitrogen directly from organic material. *Nature.* 413: 297-299.

Hussey RS and Roncodori RW. (1981). Influence of Aphelenchus avenae on vesicular-arbuscular endomycorrhizal growth response in cotton. *J.Nematol.*13: 48-52.

Ibijbijen J, Urguiaga S, Ismaili M, Alves BJR and Boddey MR. (1996). Effect of arbuscular mycorrhizas an uptake of nitrogen by *Brachiaria arrecta* and Sorghum vulgaris from soils labeled for several years with ^{15}N.*New phytologist* 134.

Jabaji- Hare S. (1998). Lipid and profiles of some vesicular arbuscular mycorrhizal fungi: Contribution to taxonomy. *Mycologia* 80: 622-629.

Jackson I, Rosendahl L. (1990). Carbon flow into soil and external hyphae from roots of mycorrhizal cucumber plants. *New phytol.*115: 77-83.

Jacobsen DJ, Beurkensk, Klomparens KL. (1998). Microscopic and Ultrastructural examination of vegetative incompatibility in partial diploids heterozygous at heterozygous loci in *Neurospora crasa, Fung Genet Biol.*23: 45-46.

Jakobsen I. (1995). Transport of phosphorus and carbon in VA mycorrhizas. In Mycorrhiza: Structure, Function, Molecular Biology and Biotechnology. (eds A Varma, B Hock), Springer-Verlag, Berlin, pp. 297-323.

Jensen A and Jakobsen I. (1980). The occurance of vesicular-arbuscular mycorrhiza in barley and wheat grow in some Danish soils with different fertilizer treatments, *Plant soil.* 55: 403-414.

Joner EL and Levyal C. (1997). Uptake of ^{109}cd by roots and hyphae of a Glomus mosseae/Trifolium subterranean mycorrhiza frm soil amended with high and low concentrations of cadmium. *Phytol.* 135: 353-360.

Jones MD, Durall DM, Tinker PB. (1998). Comparison of arbuscular and ectomycorrhizal *Eucalyptus coccifera*: Growth response, phosphorous uptake efficiency and external hyphal production. *New Phytol* 140: 125-134.

Kahiluoto H, Ketoja E and Vestberg M, Saarela I. (2001). Promotion of AM utilization through reduced P fertilization Z. Field studies. *Plant Soil* 231: 65-79.

Kahiluoto H, Vestberg M, Olesen JE, Eltun R, Gording MS, Jensen ES and Kopke U. (1999). Impact of cropping system of mycorrhizae. In: *Designing and Testing Crop Rotations for Organic Farming.* Proceedings from an International Workshop Danish Centre for Organic Farming (DAECOF). 305-309 DAECOF.

Kellam MK and Schenck NC. (1980). Interaction between a vesiculsr arbuscular mycorrhizal fungus and root knot nematode on soyabean. *Phytopathology* 70: 293-296.

Khaossad T, Garcia-Garrido JM, Steinkelner S, Vierheilig. (2007). Take all disease is systemically reduced in roots of mycorrhizal barley plants. *Soil Biol. Biochem.* 39: 727-734.

Koide RT, Goff MD, Dickie IA. (2000). Component growth efficiencies of mycorrhizal and non-mycorrhizal plant. *New Phyto.* 148: 163-168.

Koide RT, Schreiner RP. (1992). Regulation of the vesicular-arbuscular mycorrhizal symbiosis. *Annu Rev. Plant Physiol Plant Mol Biol* 43: 557-581.

Kothari SK, Marschner H, Romheld V. (1990). Direct and Indirect effects of VAMF and rhizospheric-microorganism on acquisition of mineral nutrients by maize (*Zea mays L.*). in calcareous soil. *New Phyto*l. 116: 637-645.

Krishna KR and Bagyaraj DJ. (1983). Interaction between Glomus -fasisculatum and Sclerotium-rolfsii in peanut. *Can.J.Bot.*61: 2349-2351.

Kruger NJ, Ratcliffe RG, Roscher A. (2003). Quantitative approaches for analyzing fluxes through plant metabolic networks using NMR and stable isotope labeling. *Phytochemistry Reviews* 2: 17-30.

Kumar P, Duschenkov V, Motto H and Raskin I. (1995). Phytoextraction: the use of plants to remove heavy metals from soils. *Environmental Science and Technology* 29: 1232-1238.

Lambert DH, Baker DJ, Code H (1979). The role of mycorrhize in the interactions of phosphorous with Zn, Cu and Other elements. *Soil Sci. Am. J.* 43: 976-980.

Lammers PJ, Jun J, Abubaker J, Arreola R, Gopalan A, Bago B, Hernandez-Sebastia C, Allen JW, Douds DD, Pfeffer PE. (2001). The glyoxylate cycle in an arbuscular mycorrhizal fungus: gene expression and carbon flow. *Plant Physiol* 127: 1287-1298.

Lanfranco L, Gabella S, Bonfante P. (2000). EST as a useful tool for studying gene expression in arbuscular mycorrhizal fungi.In: Procedings the third International Congress on Symbiosis. Marburg.

Lanfranco L, Novero M, Bonfante P. (2005). The mycorrhizal fungus *Gigaspora margarita* possesses a CuZn superoxide dismutase that is up-regulated during symbiosis with legume hosts. *Plant Physiol.* 137: 1319-1330.

Lerat S, Gauci R, Catford JG, Vierheilig H, Pichey Y, Lapointe L. (2002). C-14 transfer between the spring epehemeral Erythronium americanum and sugar maple saplings via arbuscular mycorrhizal fungi in natural strands. *Oceologia.* 132: 181-187.

Lerat S, Gauci R, Catford JG, Vierheilig H, Pichey, Lapointe L. (2002). C-14 transfer between the spring ephemeral Erythronium americanum and sugar mapl Saplings via arbuscular mycorrhizal fungi in natural sands. *Oecologia;* 132: 181-187.

Lewis JD and Koide RT. (1990). Phosphorous supply mycorrhizal infection and plant offspring vigour. *Functional Ecology* 4: 695-702.

Leyval C, Turnau K and Haselwandter K. (1997). Effect of heavy metal pollution on mycorrhizal colonization and function: physiological, ecological and applied aspects, *Mycorrhizae* 7: 139-153.

Leyval C, Turnau K and Haselwandter K. (2002). Potential of arbuscular mycorrhizal fungi for bioremediation. In: *Mycorrhiza technology in agriculture from genes to bioproducts* (eds. Gianinazzi S, Schuepp H, Barea JM, Haselwandtark). Basal, Switzerland: Birkhauser Verlag, pp. 175-186.

Li X-L, George E, Marschner H. (1991). Extension of the phosphorous, depletion zone in VA-mycorrhizal white clover in calcareous soil. *Plant and Soil* 136: 41-48.

Linderman RG. (1994). Role of VAM fungi in biocontrol. In: *Mycorrhizae and plant health.* (Eds. Pfleger FL, Linderman RG.). St Paul: APS Press, pp. 1-25.

Liu J, Blaylock LA, Van den, Bosch KA, Endre G, Cho J, Harrison MJ. (2003). Transcript profiling coupled with spatial expression analyses reveals genes involved in district developmental stages of an arbuscular mycorrhizal symbiosis. *Plant cell.* 15: 21016-21023.

Macaskie LE, Dean ACR. (1990). Metal-sequestering biochemicals. In: *Biosorption of Heavy Metals* (Ed., Volesky, B.), CRC Press, Boca Raton, Florida, pp. 200-248.

Mada RJ, Bagyaraj DJ. (1993). Root exudation from *Leucaena leucocephala* in relation to mycorrhizal colonization. World *Journal of Microbiology and Biotechnology.* 9: 342-344.

Maldnado-Mendoja IE. (2001). A phosphate transporter gene from the extraradical mycelium of an arbuscular–mycorrhizal fungus Glomus intraradices is regulated in response to phosphate in the environment. Mol. *Plant* Microbe *Interact.* 14: 1140-1148.

Marschner H. (1991). Mechanism of adaptation of plants toacid soils. *Plant and soil* 134: 1-20.

Marulanda A, Azcon R, Ruiz-Loza JM. (2003). Contribution of six arbuscular mycorrhizal fungal isolates to water uptake by *Lactuca sativa* plants under drought stress. *Physiologia- Plantarum*119: 1-8.

Mastubar AY, Ohba N and Fukui H. (2001). Effect of arbuscular mycorrhizal fungus infection on the incidence of *Fusarium* root rot in asparagus seedlings. *Journal of Japanese Society of Horticultural Sciences* 70: 202-206.

Mastubara Y, Kayukawa Y, Yano M, Fukui H. (2000). Tolerance of asparagus seedlings infected with arbuscular mycorrhizal fungus to violet root rot caused by *Helicobasidium mompa. J. Japn. Hortic. Sci.* 69: 552-556.

Melo IS, Costa CP and Silviera APD. (1985). Effect of vesicular arbuscular mycorrhizae on aubergine wilt caused by verticillium albo-atrum Reinke and Berth. *Summa Phytopathologia*. 11: 173-179.

Miller RM and Jastraw JD. (2004). Mycorrhizal fungi influence soil structure. In: kapulink Y, Douds DD, Arbuscular mycorrhizas-*Physiology and function*. Dordrecht: Kluwer Academic Publishers.3-19.

Miller RM and Jastrow JD. (1992). The role of mycorrhizal fungi in soil conservation. In: GJ Bethlenfalvay, RG Linderman (eds.), Mycorrhizae in sustainable agriculture. ASAspec. Publ. No. 54 American society of agronomy, Madison WI.pp. 29-44.

Miller RM and Jastrouw JD. (2000). Mycorrhizal fungi influence soil structure. In *Arbuscular Mycorrhiza: Physiology and Function*, 4-8, Kluwer Academic Publications.

Millner PD and Kitt DG. (1992). The Beltsville methord for soilless production of vesicular arbuscular-mycorrhizal fungi. *Mycorrhiza* 2: 9-15.

Mishra M, Dubey A, Singh P and Vyas D. (2006). VAM association in *Catharanthus roseu.J.Basic.Appl.Mycol.* 5(I and II): 57-59.

Mohammad MJ, Malkwai HI, Shibli. (2003). Effects of mycorrhizal fungi and phosphorous fertilization on growth and nutrient uptake of barley grown on soils with different levels of salts. *J. Plant Nuter.* 26: 125-137.

Monzon A and Azcon R. (1996). Relevance of mycorrhizal fungal origin and host plant genotype to inducing growth and nutrient uptake in *Medicago Species. Agric. Ecosyst. Envion.* 60: 9-15.

Mosse B (1973). Advances in the study of vesicular arbuscular mycorrhizal fungi; *Annu. Rev. Phytopathol*. 11: 171-196.

Munkvold L, Kjoller R, Vestberg M, Rosendahl S, Jakobsen I. (2004). High functional diversity with in species of arbuscular-mycorrhizal fungi, *New phyto.* 164: 357-364.

Nichols K. (2003). Characterization of Glomalin-A Glycoprotein Produced by Arbuscular- mycorrhizal fungi. PhD Dissertation, University of Maryland, College Park, Maryland.

Nichols K, Wright S, Schmidt W, Cavigelli M, Dzantor E. (2002). Carbron contribution and Characteristics opf humid acid, fulvic acid, particulate organic matter, and glomalin in ecosystems. In: *Proceedings of Humic Substances: Nature's most versatile materials,* International Humic Substances Society, Boston Massachusetts, pp. 365-367.

Nicoloson TH (1975). Evolution of vesicular-arbuscular mycorrhizas In: SandersE, MosseB, Tinker PB, *Endomycorrhizas*. landon: Academic press. p. 25-34.

Nurlaeny N, Marschner H, George E. (1996). Effects of liming and mycorrhizal colonization on soil phosphate depletion and phosphate uptake by maize (*Zea mays* L.). and soyabean (*Glycine max* L.) grown in two tropical soils. *Plant and Soil* 181: 275-285.

Oades JM and Waters AG. (1991). Aggregate hierarchy in soils. *Aust. J. Soil. Res.*, 29: 815-828.

Ocampo JA and Azcon R. (1985). Relationship between the concentration of sugars in the roots and VA mycorrhizal infection. *Plant and Soil* 86: 95-100.

Ojala JC and Linderman RG. (2001). Preinoculation of Lettuce and onio with VA mycorrhizal fungi reduces deleterious effect of soil salinity. *Plant and Soil*.223: 269-281.

Olivera AAR and Zambolim L. (1987). Interacao entro o fungo endomicorrizicio *Glomus facicculatum* e o nematoide de galhas Meloidogyne javanica em feijoeiro com raiz partida. *Fitopathol. Bras.*12: 222-225.

Orlowska E, Zubek SZ, Jurkiewic ZA, Szarek- Lukaszewska G. (2002). Influence of restoration on arbuscular mycorrhiza of *Biscutella laevigata L.* (Brassicaceae). and *plantago lanceolata L.* (Plantaginaceae) from calamine spoil mounds. *Mycorrhiza* 12: 153-160.

Owusu-Bennoah E, Wild A (1979). Autoradiography of the depletion zone of phosphate around onion roots in the presence of vesicular-arbuscular mycorrhiza. *New Phytologist* 82: 133-140.

Pawloska TEJ, Blaszkowski and Riihling A. (1996). The mycorrhizal status of plants colonizing a calamine spoil mound in southern Poland. *Mycorrhiza* 6: 499-505.

Pfeffer PE, Douds DD, Becard G, Shachar-Hill Y. (1999). Carbon uptake and the metabolism and transport of lipids in and arbuscular mycorrhiza. *Plant Physiol* 120: 587-598.

Pfeiffer CM and Bloss HE. (1988). Growth and nutrition of guayule (*Parthenium argenatum*) in a saline soil as influenced by esicular-arbuscular mycorrhiza and phosphorous fertilizer. *New phytologist* 108: 315-321.

Rancadori RW and Hussey RS (1977). Interaction of the endomycorrhizal fungus *Gigaspora-margarita* and root knot nem atode on cotton. *Phytopathology* 67: 1507-1511.

Raticliffe RG and Shachar Hill. (2006). Measuring multiple fluxes through plant metabolic networks. *Plant journal* 45: 495-511.

Ratnayake M, Leonard RT, Menge JA (1978). Root exudation in relation to supply of phosphorus and its possible relevance to mycorrhizal formulation. *New Phytologist* 81: 543-552.

Redecker D, Kodner R, and Graham LE. (2000). Glomalean fungi from Ordovician. *Science* 289: 1920-1921.

Regvar M, Vogel K, Irgel N, Wraber T, Hildebrandt U, Wilde P, Bothe H. (2003). Colonization of pennycresses (Th laspi spp.) of the Brassicaceae by arbuscular mycorrhizal fungi. *Journal of Plant Physiology* 160: 615-626.

Remy W, Taylor TN, Hass H, Kerp H. (1994). Four hundred–million year–old vesicular arbuscular mycorrhizae. *Proceedings of the National Academy of Sciences*, USA 91: 11841-11843.

Repett O, Bestel-Corre G, Dumas Gaudot E, Berta G, Gianinazzi-PearsonV, Gianinazzi S. (2003). Targeted proteomics to identify cadmium induced protein modification in *Glomus–mosseae* pea roots.*New phytol*. 157: 555-567.

Requena N, Mann P, Hampp R, Franken P. (2002). Early development regulated genes in the arbuscular mycorrhizal fungus *Glomus mosseae*: Identification of GmGINI, aC-terminus of metazoan hedgehog proteins. *Plant soil* 244: 129-139.

Rhodes H and Gerdemann JW (1975). Phosphate uptake zones of mycorrhizal and non-mycorrhizal onions. New Phytologist 75: 555-561.

Rilling MC, Field CB, Allen MF. (1999). Soil biota responses to long term atmospheric CO_2 enrichment in two Califorina annual grasslands. *Oceologia* (Berlin). 119: 572-577.

Rilling MC, Wright SF, Nichols KA, Schmidt WF, Torns MS. (2001). Large contribution of arbuscular mycorrhizal fungi to soil carbon pools in tropical forest soils. *Plant soil* 233: 167-177.

Rilling, MC, Ramsey PW, Morriis S, Paul EA. (2003). Glomalin, an arbuscular-mycorrhizal fungi soil protein, respond to land use change. *Plant and Soil,* 253: 293-299.

Rozema J, Arp W, Vandiggelen J,Vanesbroek M,Broekman R, Punte H. (1986). Occurrence and ecological significance of vesicular arbuscular mycorrhizae in the salt Marsh environment. *Acta Botanica Neerlandica* 35: 457-467.

Saleh H and Sikora RA. (1984). Relationship between *lomus facicculatum* root colonization of cotton and its effect on *Meloidogyne incognita. Nematologica* 30: 230-237.

Salt DE, Smith RD and Raskin I. (1998). Phytoremediation. Annual Review of Plant *Physiology and Plant Molecular Biology* 49: 643-668.

Samara A, Dumas- Gaudot E, Gianinazzi S. (1997). Detection of symbiosis-related polypeptides during the early stages of the establishment of arbuscular mycorrhiza between *Glomus-mosseae* and *Pisum-sativum* roots. New phytol.135: 711-722.

Samara A, Dumas-Gaudot E, Gianinazzi-Pearson V, Gianinazzi S. (1996). Soluble proteins and polypeptide profiles of spores of arbuscular mycorrhizal fungi. Interspecific variability and effect of host (myc+). and non-host (myc-). *Pisum sativum* root exudates. *Agronomie* 16: 709-719.

Sanders FE and Sheikh NA. (1983). The development of vesicular-arbuscular mycorrhizal in plant root systems. *Plant and Soil* 71: 223-246.

Sanders IR and Fitter AH. (1992). The ecology and functioning of vesicular arbuscular mycorrhizas in coexisting grassland species I. Seasonal patterns of mycorrhizal occurrence and morphology. *New Phytologist* 120: 517-524.

Sauer U. (2004). High-throughput phenomics: experimental methods for mapping fluxomes. *Current Opinion in Biotechnology* 15: 58-63.

Sauer U, Lasko DR, Flaux J, Hochuli M, Glaser R, Szyperski T, Wuthrich K, Bailey JE. (1999). Metabolic flux ratio analysis of genetic and environmental modulations of *Escherichia coli* central carbon metabolism. *Journal of Bacteriology* 181: 6679-6688.

Schenck NC. (1981). Can mycorrhiza control root disease? *Plant Dis. Rep.* 65: 230-234.

Schenck NC, Ridings WH and Cornell JA (1977). Interaction of endomycorrhizal fungi and *Phytophthora parasitica* of two citrus root stocks, in *Proceedings of the third North American Conference on Mycorrhizae* (Oregon: Corvallis). 9.

Schoenbeck F (1979). Endmycorrhizas in relation to plant diseases.In soil-borne plant pathogens (B. Schippers and W. Gam, eds).Academic press, New York pp. 271-280.

Schoenbeck F and Dehne HW. (1981). Mycorrhiza and Plant Health, *Gesunde Pflanzen.* 33: 186-190.

Schonbeck F. (1980). Endomycorrhiza: Ecology, function and phytopathological aspects: *Forum Microbial* 3: 90-96.

Schwab SM, Leonard RT, Menge JA. (1984). Quantitative and qualitative comparison of root exudates of mycorrhizal and nonmycorrhizal plant species. C*anadian Journal of Botany* 62: 1227-1231.

Schwab SM, Menge JA, Leonard RT. (1983). Quantitative and qualitative effects on phosphorus on extracts and exudates of sudangrass roots in relation to vesicular-arbuscular mycorrhiza formation. *Plant Physiology* 73: 761-765.

Schwender J, Goffman F, Ohlrogge JB, Shachar-Hill Y. (2004a). Rubisco without the Calvin cycle improves the carbon efficiency of developing green seeds. *Nature* 432: 779-782.

Shachar-Hill Y, Pfeffer PE, Douds DD, Osman SF, Doner LW, Ratcliffe RG. (1995). Partitioning of intermediary carbon metabolism in vesicular-arbuscular mycorrhizal leck. *Plant Physiology* 108: 7-15.

Sidhu OP and Behl HM. (1997). Response of three *Glomus* species on the growth of *Prosopis Juliflora* Qwartz at high pH levels. *Symbiosis* 23: 23-24.

Sikora RA and Schonbeck F (1975). Effect of vesicular–arbuscular mycorrhiza (*Endogone mosseae*) on the population dynamics of the root knot nematode (Meloidogyne–incognita and Meloidogyne hapla). Proceedings of VIII International congress protection. 5: 158-166.

Simoneau P, Louisy-Lois N, Plenchettec, Strullu DG. (1994). Accumulation of new polypeptides in Ri-tdna-transformed roots of tomato (*Lycopersicon-esculentum*) during the development of vesicular-arbuscular mycorrhizae. Appl. *Environ Microbiol* 60: 1810-1813.

Simoneau P, Viemont JD, Moreau JC, Strulu DJ. (1993). Symbiosis is related polypeptides associated with the early stages of ectomycorrhiza organogenesis in birch (*Betula pendula* Roth). *New Phytol* 124: 495-504.

Sin H. (2004). Phosphate transport in Arabidopsis: pht1: and pht1;4 play amajor role in phosphate acquisition from both low and high phosphate environments. *Plant J.* 39: 629-642.

Singh P and Vyas D. (2008). To study microbial interaction of below ground organism with special reference to *Fusarium* wilt of Chickpea. Ph.D. Thesis. Dr. H.S.G.University, Sagar.

Smith FA and Smith SE. (1997). Structural diversity in (vesicular)-arbuscular mycorrhizal symbiosis. *New Phytologist.* 151: 469-475.

Smith GS, Hussey RS and Roncadori RW. (1986). Penetration and postinfection development of *Meloidogyne- incognita* on cotton as affectd by *Glomus intraradices* and phosphorous.*J.Nematology.* 18: 429-435.

Smith SE and Read DJ. (1997). Mycorrhizal symbiosis. Academic Press, London.

Solaimain MD, Satio M. (1997). Use of sugars by intraradical hyphae of arbuscular mycorrhizal fungi revealed by radiorespirometry. *New phytol.* 136: 533-538.

Soni P and Vyas D. (2008). Studies on some medicinal herbs with special reference to nutrient uptake by VAM fungi. Ph.D. Thesis. Dr. H.S.G. University, Sagar.

Sorensen N, Larsen J, Jakobsen I. (2005). Mycorrhiza formation and nutrient concentration in looks (*Allium porrum*) in relation to previous crop and cover crop management on high P soils. *Plant soil* 273: 101-114.

Srivastava AK, Singh S, Marathe RA. (2002). Organic citrus, soil fertility and plant nutrition. *Sustai.agric*19: 5-29.

Stanley MR, Koide RT, Shumway DL. (1993). Mycorrhizal symbiosis increases growth, reproduction and recruitment of Abutilon theophrasti Medicin the field. *Oecologia* 94: 30-35.

St-Arnaud M, Hamel C, Vimard B,Caron M, Fortin A. (1996). Enhanced hyphal growth and spore production of the arbuscular mycorrhizal fungus Glomus intraradices in an vitro system in the absence of host roots. *Mycol. Res.* 100: 328-332.

Stommel M, Mann P, Franken P. (2001). Construction and analysis of an EST library using RNA from dormant spores of the arbuscular mycorrhizal fungus *Gigaspora rosea. Mycorrhiza* 10: 281-285.

Subramanian KS, Charest C, Dwyer LM, Hamilton RI. (1997). Effects of arbuscular mycorrhizae on leaf water potential, sugar content and phosphorous content during drought and recovery of maize. *Canadian journal of Botany*.75: 1582-1591.

Suresh CK and Bagyaraj DJ. (1984). Interaction between a vesicular-arbuscular and a root-knot nematode and its effecton growth and chemical composition of tomato. *Nematol.* Medit. 12: 31-39.

Swaki H and Satio M. (2001). Expressed genes in the extraradical hyphae of an arbuscular mycorrhizal fungus, *Glomus intraradices* in the symbiotic phase. FEMS. *Microbiol Lett*. 195: 109-113.

Tamasloukht M, Sejalon-Delmas N, Kluever A, Jauneau A, Roux C, Becard G, Franken P. (2003). Root factors induce mitochondrial related gene expression and fungal respiration during the devlopment switch from asymbiosis to presymbiosis in the arbuscular mycorrhizal fungus *Gigaspora-rosea. Plant physiology*, 131: 1468-1478.

Tarkka MT, Nyman TA, Kallkinen N, Raudaskoski M. (2000). Scots pine expresses short-root specific peroxidases during development. *Eur.J.Biochem*. 267: 1-8

Thingstrup I, Rubaek G, Sibbesen E, Jakobsen I. (1998). Flax (*Linum uritatiusim* L.). depends on arbuscular mycorrhizal fungi for growth and P uptake at intermediate but not high soil P levels in the field. *Plant soil*. 20: 37-46.

Thomas RS, Dakessian S, Ames RN, Brown MS, Bethlenfalvay GJ. (1986). Aggregation of a silty clay loam soil by mycorrhizal onion roots. Society of America, Journal 50: 1494-1499.

Thompson JP. (1987). Decline of vesicular arbuscular mycorrhizae in long fallow disorder of field crops and its expression in phosphorus deficiency of sunflower, *Aust. J. Agric. Res.* 38: 847-86.

Thompson JP. (1994). Inoculation with vesicular-arbuscular mycorrhizal fungi from cropped soil overcomes long-fallow disorder of linseed (*Linum usitatissimum L.*). by improving P and Zn uptake, *Soil Biol. Biochem*. 26: 1133-1143.

Thompson JP. (1991). Improving the mycorrhizal condition on the soil through cultural practices and effects on growth and P uptake in plants. In: *Phosphorous Nutrition of Grain Legumes in the Semi Arid Tropics*, (eds C. Johansen, Lee, K.K and Sahrawat, K.). International crops Research institute for semi arid Tropics, Andhra Patancheru, India pp. 117-137.

Tisdall JM and Behl HM. (1997). Organic matter and water stable aggregates in soils. *Journal of soil science*. 33: 141-163.

Tisdall JM and Oades JM (1979). Stabilization of soil aggregates by the root system rye grass. *Australian journal of soil Research* 17: 429-441.

Turman W, Bennett MH, Kubigsetelting I, Turnbull C, Grant M. (2007). Arabidopsis systemic immunity uses conserved defense signaling pathways and is mediated by Jasmonates. *Proc. National Acad. Sci. USA*. 104: 1075-1080.

Turnau K, Jurkiewicz A, Lingua G, Barea JM, Gianinazzi-Pearson V(2005). Role of arbuscular mycorrhiza and associated microorganisms in phytoremediation of heavy metal polluted sites. *Trace element in the environment biochemistry, biotechnology and bioremediation*. CRC Press/Lawis Publishers (in press).

Vallorani L, Bernardini F, Sacconic C, Pierleoni R, Pieretti B, Piccoli G, Buffalini M, Stocchi V. (2000). Identification of Tuber *borchii* vittad. Mycelium proteins separated by two dimensional polyacrylamide gel electrophoresis using amino acid analysis and sequence tagging. *Electrophoresis* 21: 3710-3716.

Van der Heijden. (2002). Arbuscular mycorrhizal fungi as determinant of plant diversity, in search for underlying mechanisms and general principles. In: *Mycorrhizal Ecology*. (eds. MGA Van der Heijdenand IR Sanders), *Ecological studies*. 157. Springer Verlag, Berlin, Germany, 243-265.

Van der Heijden MGA, Moutoglis P, Streitwolf, Klironomos JN, Ursic M, Engel R, Boller T, Wiemken A. (1998). Mycorrhizal fungal diversity determines plant biodiversity, ecosystem variability and productivity. *Nature*. 396: 69-72.

Van der Heijden, MGA, Wiemken A, Sanders IR. (2003). Different arbuscular mycorrhizal fungi alter coexistence and resource distribution between co-occurring structure. *Ecology*.79: 2082-2091.

Vyas D, Mishra M, Singh P, Soni P. (2006). Studies on mycorrhizal association in wheat.*Indian phytopath*. 59(2): 174-179.

Vyas D and Dwiedi OP. (2002). Effect of potassium on the occurrence of VAM fungi.*J. Basic.Appl.Mycol*1(2): 233-235.

Vyas D and Soni P. (2007). Arbuscular mycorrhizal fungi associated with important medicinal plants of Sagar. *Indian phytopathology* 60(1): 52-57.

Vyas D, Dubey A, Singh P. (2006). VA Mycorrhizal Fungi in tropical monsoonic grassland. *J.Basic. Mycol*. 5(I and II): 78-81.

Vyas D, Dubey A, Soni A, Mishra M and Singh P. (2007). Arbuscular mycorrhizal fungi in early land plants. *Mycorrhiza News* 19(2).

Vyas D, Soni A, Singh P. (2007). Differential effects of pesticides on occurrence of Vam fungi associated with some leguminous plants. *J.Basic Appl. Mycol*. 6(I and II): 143-150.

Watkins NK, Fitter AH, Graves JD, Robinson D. (1996). Carbon transfer between C_3 and C_4 plants linked by a common mycorrhizal network, quantified using stable carbon isotopes. *Soil Biol Biochem*. 28: 471-477.

Wellings NP, Wearing AH, Thompson JP. (1991). Vesicular-arbuscułr mycorrhizal (VAM) improve phosphorous and zinc nutrition and growth of pigeonpea in vertisol. *Austt J. Agric Res*. 42: 835-845.

Whipps JM. (2004). Prospects and limitation for mycorrhiza in biocontrol of root pathogen. *Can J. Bot*. 82: 1198-1227.

Wilcox HE. (1991). Mycorrhizae. In: Waisel Y, Eshel A, Kafkafi U. *The Plant Root,* The Hidden Half. New York: Marcel Dekker. 731-765.

Wright SF and Upadhayaya A. (1996). Extraction of an abundant and unusual protection from soil and comparison with hyphal protection of arbuscular mycorrhizal fungi. *Soil Sci.* 161: 575-586

Wright SF, Upadhyaya A. (1998). A survey of soil for aggregate stability and glomlin, a glycoprotein produced by hyphae of arbuscular-mycorrhizal fungi. *Plant and Soil* 1: 97-107.

Yao MK, Desilets, Charales MT, Boulanger R and Tweddell RJ. (2003). Effect of mycorrhization on the accumulation of rishitin and *solavetivone* in potato plantlets challenged with *Rhizoctonia solani. Mycorrhiza* 13: 333-336.

Zambolim L and Schenck NC. (1983). Reduction of the effects of pathogenics root-infecting fungi on soyabean by the mycorrhizal fungus. *Glomus mosseae. Phytopathology* 73: 1402-1405.

Microbial Biotechnology and Ecology (2011) *Pages* 256–267
Editors: Deepak Vyas, G.S. Paliwal, P.K.Khare and R.K Gupta
Published by: DAYA PUBLISHING HOUSE, NEW DELHI

Chapter 14

Cumulative Effect of VAM Fungi

☆ *O.P. Dwivedi, Meenakshi Mishra, Pradeep K. Singh, Pramod Kumar Richhariya, Prashant Soni and Deepak Vyas*

Lab of Microbial Technology and Plant Pathology, Department of Botany, Dr. H.S. Gour University, Sagar – 470 003, M.P.

ABSTRACT

The studies on effect of chemical fertilizer, VAM and other biofertilizer (*viz*. Farm Yard Manure (FYM), Phosphate Solubilizing Bacteria (PSB), *Pseudomonas* and *Azotobacter*, etc.), individually and in combination on the growth responses in wheat Cv. C-306 were carried out. For the present study, two experiments were conducted; (1) field experiments with VAM and chemical fertilizer; (2) field application of VAM and other biofertilizers. Results revealed that chemical fertilizer, VAM and other biofertilizers inoculation greatly influence the growth parameter i.e. germination, heading, maturity, plant height and yield as compared to controlled wheat (Cv. C-306). As far as the growth parameter is concerned in both experiments, VAM inoculation was found to be more effective in comparison to other inoculants and chemical fertilizers. Thus it may be concluded that VAM has greater impact on the growth and yield of wheat Cv. C-306 in the soils, which are already saturated with chemical fertilizer. So, there is no need to add any more chemical fertilizer in these soils. Use of VAM and phosphobacterin should be increased, because these organisms are not only responsible for the soil health but also are responsible for the good growth and better grain yield of wheat crop. Out of all the selected biofertilizers VAM was found best for good growth and higher grain yield of wheat.

Introduction

Phosphorus plays a key role in the balanced nutrition of plants; however, the native soil phosphorus is mostly unavailable because of its low solubility. It's deficiency in soil is met by the addition of phosphatic fertilizers but the efficiency of utilization of the added phosphorus by plants is very low. Microorganisms play an important role in mineralization and solubilization of phosphorus in different environments.

In recent past, the use of phosphatic fertilizers has increased to a very great extent. However, due to their heavy cost and pollution problems associated with them, their use especially in developing countries is loosing its relevance. Under the circumstances, availability of a system which can improve the fertilizer efficiency will be a great asset to the agriculture in these countries. According to Mosse (1973) 75 per cent of the phosphorus applied to the crops is not utilized by them but gets converted in to forms unavailable to plants. It has been estimated that in soils of high pH, a major portion of the fertilizers applied becomes non-usable via chemical reactions. VAM fungi have been found to be very effective under such conditions, because of their ability to make use of extremely small quantities of fertilizer.

That is the reason why VAM fungi are being regarded as biofertilizers and their use in agriculture is being encouraged. These fungi have been reported from almost all the important agricultural crops of the tropical countries. Mycorrhizal symbiosis may play an important role in the tropical agriculture because in the tropics, the soils are phosphorus deficient as well as phosphorus fixing (Dobin, 1980).

During last twenty years, much attention has been paid to VAM fungi as a tool for improving the growth of agricultural crops and horticultural plants (Tinker, 1978; Gianinazzi *et al.*, 1989, 1990a, b). The symbiotic system of these fungi has also been exploited to save the costly phosphatic fertilizers (Abbott and Robson, 1982 a, b). Their importance in natural ecosystem (Koske, 1981) and in revegitation of the disturbed lands has also been recognized (Reeves *et al.*, 1979). There is growing evidence that VAM can increase plant growth, especially in infertile soils (Bowen, 1978; Azcon-Aguilar *et al.*, 1982; Santhi and Kothandaraman, 1995) and that such growth increases are the result of an enhanced ability of the infected roots to absorb more nutrients. The uptake of phosphorus from rock phosphate and phosphatic fertilizers has been increased in plants by the inoculation of soils with mycorrhizal fungi, if the number of mycorrhizae in the soil is initially very low (Kucey, 1987).

Researches in the last few decades have established that VAM and phosphate solubilising bacteria (PSB) are the two groups of soil microorganisms associated with phosphorus (P) nutrition of plants. VAM fungi or obligate symbiotic association with plant roots and increase P availability by mobilizing the P with the help of their extramatrical hyphae, particularly in soil with less available P (Mosse, 1973). On the other hand, PSB live freely in soil and solubilise insoluble form of P with the help of different organic acids, secreted by them and thus increase the quantity of available P in soil for absorption by plant roots (Banik and Dey, 1962). Growth of wheat has been increased by inoculating VAM in sterilised soils (Thompson, 1990). Similarly, inoculation with PSB increased growth of different crops, particularly when rock phosphate and PSB were applied together (Gaur, 1987). However, the magnitude of plant growth may vary depending on the type of VAM, PSB strain, crop, amount of P present in soil and presence of coinoculant. Improved effects of *Rhizobium* as a nitrogen fixing agent have been achieved when it was used in certain cases along with *Azotobacter* (Rawat and Sanoria, 1976, 1977; Tosh and Sanoria, 1978). The role of *Azotobacter* in stimulating the crop growth appears to be a controversial issue. *Azotobacter* spp. produces auxins or hormones which accelerate the plant development and are also antagonistic to soil borne pathogens. Since the favorable response has been obtained only under the condition of high fertility (Mehrotra and Lehri, 1971). Allison (1947) referred to the positive response observed by several Russian and other workers in the crop yield due to *Azotobacter* inoculation. The subject of bacterial fertilizers should not be neglected and their value should not be dismissed as negligible. According to Alexander (1961), it is conceivable that nitrogen fixation takes place in the root zone, a region of intense microbiological activity and an area in which carbohydrate from root excretions and decomposition are abundant.

Strain variation, soil condition such as moisture, organic matter, phosphate content, pH and soil type also influence the result of bacterial fertilizers (Sundara Rao *et al.*, 1963). The combined inoculation of PSB and VAM has given encouraging results in many crop plants (Raj *et al.*, 1981).

Wheat is vital to the health of people and animals in a predominantly vegetarian population. Therefore the experiment were carried out to ascertain the effects of inoculated bioagents (*viz.* VAM, PSB, *Pseudomonas, Azotobacter*), farm yard manure (FYM) and chemical fertilizer (N, P, K) on growth and yield of wheat cultivar C-306.

Materials and Methods

The efficacy of selected combination of bioagents including VAM fungi and chemical fertilizer in improving the growth and yield of wheat (Cultivar C-306) was evaluated under field conditions. The field experiments were carried out at departmental Botanical Garden, Sagar University, Sagar on $1m^2$ micro-plots laid out in a randomized block design during crop seasons.

In the present study two experiments were conducted; (1) field experiment with VAM and chemical fertilizer, (2) field application of Bioagents including VAM fungi.

Mineral Manuring

For the chemical treatment, doses of nitrogen was applied in the form of urea (NH_2CONH_2), phosphorus in the form of rock phosphate (P_2O_5) and potash applied in the form of Potassium oxide (K_2O) as chemical fertilizer in test micro-plots. Table 14.1 gives a summary of the amount of the mineral fertilization.

Table 14.1: Amount of Chemical Fertilizers Used in Test Microplot for Wheat Cultivar C-306

Fertilizer	*Amount (Kgha^{-1})*
N (NH_2CONH_2)	60
P (P_2O_5)	60
K (K_2O)	100

Inoculation Procedure of Bioagents for Field Experiment

For the experiment, inoculation procedure of bioagents is given below:

VAM Fungal Inoculation

The VAM inoculum containing *Glomus mosseae* (20 spore/gm soil), which was consisted of infected roots and soil from pot culture on wheat, was uniformly spread over the soil @ 500 gm/micro-plot.

FYM Treatment

Sufficient steam sterilized form yard manure (FYM) (500 gm/micro-plot) was applied uniformly to all the micro-plots before mycorrhizal inoculation.

PSB Inoculation

The carrier based inoculant culture containing *Bacillus subtilis* was used as a source of PSB.

***Pseudomonas* Inoculation**

The charcoal based inoculum of fluorescent *Pseudomonas* (GRP_3) was used as a source of *Pseudomonas*.

***Azotobacter* Inoculation**

The peat based effective mutant of *Azotobacter chroococcum* was used as a source of *Azotobacter*. For all the micro-plots 15gm. of wheat (Cv. C-306) seeds were used for the sowing. In the case of bacterial treatment, seeds were treated with 2.5gm inoculant of all three bacterial cultures (*viz.* PSB,

Pseudomonas, Azotobacter) by preparing a slurry of the culture. Seeds of wheat were uniformly sprayed with a suspension of bacterial culture.

Treatments were done just before sowing, and sowing was done in row. Colonization of bioagents were studied in non-sterile soil. Along with each treatment series, control was maintained. All these bioagtents except VAM fungi were obtained from J.N.K.V. Field Station Ratouna, Sagar for present study. For each treatment there were three replications. The micro-plots were irrigated with equal amount of water as and when required. Observations on germination, heading, maturity, plant height, yield and physico-chemical properties of rhizosphere soil were recorded.

Physico-chemical Analysis of Soil

The soils were analysed for their nitrogen (A.O.A.C., 1965), potassium (Jackson, 1973) and phosphorus by the method of Jackson (1967).

Statistical Analysis

The data were statistically analysed and the minimum difference required for significance (C.D.) at 5 per cent level of probability was calculated and recorded (Gupta and Kapoor, 1997).

Results

Physico-chemical Properties of Soil

Nitrogen (N)

In experiment 1, the available soil N was maximum in chemical fertilizer treated micro-plot (536 kgha^{-1}) followed by control micro-plot (426.75 kgha^{-1}) and minimum (Table 14.2) in VAM inoculated micro-plot (380 kgha^{-1}) whereas, in experiment 2, the maximum N was recorded in FYM + PSB + *Pseudomonas* treated micro-plot (610.50 kgha^{-1}) followed by FYM + PSB and FYM + *Azotobacter* + PSB treated micro-plot (594 kgha^{-1}) then FYM and FYM + *Pseudomonos* treated micro-plot (585.75 kgha^{-1}), FYM + *Azotobacter* treated micro-plot (528 kgha^{-1}), FYM + VAM inoculated micro-plot (426.75 kgha^{-1}) and minimum was recorded (Table 14.4) in control micro-plot (420.51 kgha^{-1}).

Table 14.2: Effect of Chemical Fertilizer and VAM on Nutrient (N-P-K) in the Rhizosphere Soil of Wheat Cultivar C-306 Cultivated at Departmental Botanical Garden

Sl.No.	*Treatment*	*Nitrogen (kgha^{-1})*	*Phosphorus (kgha^{-1})*	*Potassium (kgha^{-1})*
1.	Control (Soil alone)	426.75	14.8	260.0
2	Chemical fertilizer	536.00	12.8	472.0
3	VAM inoculum	380.00	6.4	172.0

Phosphorus (P)

In the experiment 1, maximum soil phosphorus was recorded with control micro-plot (14.8 kgha^{-1}) followed by chemical fertilizer treated micro-plot (12.8 kgha^{-1}) and minimum was recorded (Table 14.2) with VAM inoculated micro-plot (6.4 kgha^{-1}) whereas, in experiment 2, Table 14.4 shows the maximum soil P with FYM + *Azotobacter* treated micro-plot (16 kgha^{-1}) followed by control micro-plot (13.7 kgha^{-1}) then FYM + PSB + *Pseudomonas* treated micro-plot (12 kgha^{-1}), FYM + *Pseudomonas* treated micro-plot (10 kgha^{-1}); FYM + PSB treated, FYM alone and FYM + *Azotobacter* + PSB treated micro-plot (9.6 kgha^{-1}) and minimum was recorded with FYM + VAM inoculated micro-plot (7.4 kgha^{-1}).

Potassium (K)

In experiment 1, the maximum soil potassium was recorded with chemical fertilizer treated micro-plot (472 $kgha^{-1}$), followed by control micro-plot (260 $kgha^{-1}$) and minimum (Table 14.2) was recorded with VAM inoculated micro-plot (172 $kgha^{-1}$).

In experiment 2, the maximum soil K was recorded with FYM + *Pseudomonas* treated micro-plot (312.6 $kgha^{-1}$) followed by control micro-plot (312 $kgha^{-1}$) then FYM treated micro-plot (308 $kgha^{-1}$), FYM + PSB + *Pseudomonas* treated micro-plot (291.2 $kgha^{-1}$), FYM + PSB treated micro-plot (240.8 $kgha^{-1}$), FYM + *Azotobacter* +PSB treated micro-plot (205 $kgha^{-1}$), FYM + *Azotobacter* treated micro-plot (145.6 $kgha^{-1}$) and minimum was recorded (Table 14.4) with FYM+VAM inoculated micro-plot (121 $kgha^{-1}$).

Growth Parameters of Wheat (Cv. C-306)

Germination

In experiment 1, the highest per cent germination was recorded with VAM inoculated plants (79 per cent) followed by chemical fertilizer treated plants (64 per cent) and lowest per cent germination was recorded (Table 14.3) with control plants (55 per cent).

Table 14.3: Effect of Chemical Fertilizer and VAM Inoculum on the Growth Parameter and Yield of Wheat Cultivar C-306 Cultivated at Departmental Botanical Garden

Sl.No.	Treatment	Germination (per cent)	Heading (Days)	Maturity (Days)	Plant Height (cm)	Yield (kg ha^{-1})
1.	Control (Soil alone)	55.0	74	128	37.74	232.1
2	Chemical fertilizer	64.0	77	132	60.49	625.0
3	VAM inoculum	79.0	81	137	86.13	725.0
	CD ($P<0.05$)	3.84	ND	ND	5.16	48.67

In experiment 2, the highest germination was recorded with FYM + VAM inoculated plant (81 per cent), followed by FYM + *Azotobacter* (75 per cent), FYM + *Azotobacter* + PSB treated plant (72 per cent), FYM + PSB + *Pseudomonas* treated plants (69 per cent), FYM + *Pseudomonas* and FYM alone treated plant (65 per cent) and lowest was recorded (Table 14.5) with control plant (52 per cent).

Heading

In experiment 1, the heading appeared after 81 days with VAM inoculated plants which was higher time period followed by chemical fertilizer treated plants (77d), whereas, control plants required lesser time period *i.e.* 74 days for the appearance of headings (Table 14.3).

The experiment 2, revealed that headings appeared after higher time period *i.e.* 84 days with FYM + *Azotobacter* and FYM + *Azotobacter* + PSB treated plants followed by FYM + VAM inoculated plant (83d) then FYM + PSB treated plant (80d), FMY + *Pseudomonas* treated plants (78d), FYM + PSB + *Pseudomonas* treated plants (76d), FYM treated plant (74d); whereas, control plant require lesser time period *i.e.* 71 days for the appearance of headings (Table 14.5).

Maturity

In experiment 1, the rate of maturity was achieved with VAM inoculated plant after 137 days in maximum time period followed by chemical fertilizer treated plant (132d) and with control plants rate of maturity was achieved after 128 days which was minimum time period in comparison to other *i.e.* VAM inoculated and chemical fertilizer treated plants respectively (Table 14.3).

In experiment 2, the rate of maturity was achieved with FYM + VAM inoculated plant, FYM + *Azotobacter* treated plants, FYM + *Azotobacter* + PSB inoculated plants after 138 days, which was maximum time period followed by FYM + PSB treated plant (137d) then FYM + PSB + *Pseudomonas* treated plant (136d), FYM alone treated plant (134d), control plant (132d) and minimum time period of 130 days was required to achieve maturity with FYM + *Pseudomonas* treated plants (Table 14.5).

Table 14.4: Effect of Bioagent on Nutrient (N-P-K) in the Rhizosphere Soil of Wheat Cultivar C-306 Cultivated at Departmental Botanical Garden

Sl.No.	*Treatment*	*Nitrogen (Kg ha^{-1})*	*Phosphorus (Kg ha^{-1})*	*Potassium (Kg ha^{-1})*
1.	FYM	585.75	9.6	308.0
2.	FYM + VAM	426.75	7.4	121.0
3.	FYM + PSB + *Pseudomonas*	610.50	12.0	291.2
4.	FYM + *Pseudomonas*	585.75	10.0	312.6
5.	FYM + PSB	594.00	9.6	240.8
6.	FYM + *Azotobacter*	528.00	16.0	145.6
7.	FYM + PSB + *Azotobacter*	594.00	9.6	205.0
8.	Control (Soil alone)	420.51	13.7	312.0

Table 14.5: Effect of Bioagent on Growth Parameter and Yield of Wheat Cultivar C-306 Cultivated at Departmental Botanical Garden

S.No.	*Treatment*	*Germination (per cent)*	*Heading (Days)*	*Maturity (Days)*	*Plant Height (cm)*	*Yield (Kgha^{-1})*
1.	FYM	65.0	74	134	61.2	717
2.	FYM + VAM	81.0	83	138	101.4	1350
3.	FYM + PSB + *Pseudomonas*	69.0	76	136	70.4	1200
4.	FYM + *Pseudomonas*	65.0	78	130	87.3	1283
5.	FYM + PSB	70.0	80	137	84.2	1250
6.	FYM + *Azotobacter*	75.0	84	138	98.7	1317
7.	FYM + PSB + *Azotobacter*	72.0	84	138	94.1	1125
8.	Control (Soil alone)	52.0	71	132	40.7	472
	CD ($P < 0.05$)	3.86	ND	ND	4.55	57.33

Plant Height

In experiment 1, the tallest plant was recorded in VAM inoculated pot (86.13cm.) followed by chemical fertilizer treated plant (60.49 cm.) and shortest plant was recorded (Table 14.3) with control (37.74 cm.).

However, in experiment 2, plant showed tallest height with FYM + VAM inoculated plant (101.4 cm.), followed by FYM + *Azotobacter* treated plant (98.7 cm.), FYM + *Azotobacter* + PSB treated plant (94.1 cm.), FYM + *Pseudomonas* treated plant (87.3 cm.), FYM + PSB treated plant (84.2 cm.), FYM + PSB

+ *Pseudomonas* treated plant (70.4 cm.), FYM treated plant (61.2 cm.) and shortest height was recorded with control plant (40.7 cm.) (Table 14.5).

Grain Yield

For the experiment 1, the maximum grain yield was recorded with VAM inoculated plant (725 kgha^{-1}), followed by chemical fertilization plant (625 kgha^{-1}) and minimum was obtained (Table 14.3) with control plants (232.1 kgha^{-1}).

In experiment 2, the maximum yield was recorded with FYM + VAM inoculated plant (1350 kgha^{-1}), followed by FYM + *Azotobacter* treated plant (1317 kgha^{-1}), FYM + *Pseudomonas* treated plant (1283 kgha^{-1}), FYM + PSB treated plant (1250 kgha^{-1}), FYM + PSB + *Pseudomonas* treated plant (1200 kgha^{-1}), FYM + *Azotobacter* + PSB (1125 kgha^{-1}), FYM treated plant (717 kgha^{-1}) and minimum grain yield was obtained (Table 14.5) with control plant (472 kgha^{-1}).

Discussion

Decreasing yields in continuous wheat mono-culture can be compensated only to a limited extent by agronomic practices such as regular supply of straw and green manure (Kurten and Range, 1980). This problem of continuous cropping of winter wheat is documented in diverse investigations (Debruck, 1974; Diereks *et al.*, 1980; Gutser *et al.*, 1981; Heyland, 1975; Pommer *et al.*, 1979). The activity of soil microorganism in continuous wheat mono-culture is limited. According to Beck (1974) the reduced microbial soil activity in continuous wheat mono-culture might result in a smaller N-supply by mineralization and thus in an insufficient N-supply for the plants.

Despite the fact that the favourable influence of VAM on the nutrient uptake and drought resistance of plants is generally accepted, relatively little is known about the occurrence and importance of these fungi in intensive agriculture, although several investigation have already been performed (Baltruschat and Dehne, 1982; Black and Tinker, 1979; Hayman, 1975; Jensen and Jacobsen, 1980; Ruissen, 1982; Schenck and Kinloch, 1980).

The results of present study indicate that, the rhizosphere soils of micro-plots are saturated with chemical substances particularly N-P-K, therefore addition of more N-P-K was not found as effective as it was expected. However, the rhizosphere soils were supplemented with VAM fungi (*Glomus mosseae*) inspite of that presence of indigenous VAM population, identified as species of *Glomus*. The addition of *G. mosseae* in the rhizosphere soils of wheat plant did not only develop intensive colonization but also helped in the nutrient uptake, resulting in to better growth of the plant and higher grain yield.

The role of VAM fungi in mineral nutrition has been discussed in depth in recent past (Rhodes and Gerdemann, 1980; Smith, 1980; Bowen and Smith, 1981). It is now well established that mycorrhizae can improve P nutrition of a host and it is likely that they can also assist in the uptake of other ions.

However, evidences to suggest that VAM affect the nitrogen nutrition of plants directly and cause an increased uptake resulting in higher N content in them are also available (Bowen and Smith, 1981; Ames *et al.*, 1983; Jensen, 1982).

The yield and uptake of nitrogen and phosphorus was substantially increased due to use of inoculants (Kundu and Gaur, 1980c). In non-sterilized soil, the highest increase in grain yield was recorded with FYM + VAM inoculated plants followed by FYM + *Azotobacter* treated plants. Mixed inoculation yielded better results than single inoculant. The becterization effect was well marked when the soil was amended with FYM. The better response in sterilized soil may be due to lack of competition and availability of more nutrients to the introduced PSB. It has been established that application of organic matter improves the physical, chemical and biological properties of soil (Gaur, 1982; Gaur *et al.*, 1984). However, other combination fails to gave high yield which may be attributed to the competition for the growth among the microbes or antagonistic activity between/among them. Pavlovich (1959) reported an adverse effect of *Bacillus subtilis* and *B. mesentericus* and a beneficial effect of *B. megaterium* on *Azotobacter* growth. Chan *et al.* (1970) observed inhibition of *Azotobacter* by *Pseudomonas putida* and the inhibition of *Azotobacter* was attributed to their sensitivity to the acidity produced. The antagonism of *Pseudomonas* sp. to *Azotobacter* (Chan *et al.*, 1963) and that of *Bacillus circulans* to fungi such as *Alternaria solani, Helminthosporium oryzae, Rhizoctonia solani, Fusarium udam* (Vasantharajan, 1964) were reported.

The phosphorus solubilising bacteria have the capacity to dissolve the insoluble phosphatic compounds present in soil in the form of tricalcium, aluminium and iron phosphates and also solubilize rock phosphate, bone meal and basic slag. The increases in phosphorus uptake by plants have been made possible by combining the abilities of the phosphorus solubilising microorganisms and the efficient P uptake mechanisms of the VAM fungi (Azcon *et al.*, 1976; Raj *et al.*, 1981; Kucey, 1987).

Increase in yield and phosphate uptake by wheat crops inoculated with phosphobacterin were reported by several workers (Arao *et al.*, 1956; Panova, 1956; Andronova, 1958; Kudzin and Yaroshevich, 1958; Afanaseva and Andronova, 1959; Timoriu *et al.*, 1964). Besides increase in yield, the increase in protein content has also been reported (Kudashev, 1956; Ivchenko, 1961). The increase in gliadin/gluten ratio (Pinevich, 1956) improved quality of grains, and increased gluten content (Ivchenko, 1961), better growth and germination (Mishustin and Naumova, 1956) have been observed. Smallii (1958) reported increased P_2O_5 uptake by wheat plants and grain yield by inoculation with *Pseudomonas. Bacterium glutinsus* and *Bacillus magaterium*. On the other hand, some workers did not get any response of wheat crop to phosphobacterin inoculation (Bystrova, 1958; Lauzne, 1960; Smith *et al.*, 1961).

The combined impact of inoculation of all bioagents tested in the present study on growth and yield response of wheat Cv. C-306 has not so far been found in the literature.

Conclusion

It is concluded here that VAM has greater impact on the growth and yield of wheat Cv. C-306 in the soils, which are already saturated with chemical nutrients. This is suggested here that there is no need to go on pouring chemical fertilizer in the soils, which instead of making the soil more fertile forced it to become barren.

Use of VAM and phosphobacterin should be increased, because these organisms are not only responsible for the soil health but also responsible for the good growth and better yield of wheat.

References

A.O.A.C. (1965). *Official Methods of Analysis of the Association of Official Agricultural Chemists.* 8th (ed.) Vol. I. Minnesota. U.S.A.: 523.

Abbott, L.K. and Robson, A.D. (1978). Growth of subterranean clover in relation to the formation of endomycorrhiza by introduced and indigenous fungi in field soil. *New Phytol.* 81: 575-585.

Afanaseva, A.L. and Andronowa, T.M. (1959). Experiment on use of phosphobacterin on the soils of the Omek region. *Udobr. Urozh.* 5: 44-47.

Alexander, M. (1961). *Introduction to Soil Microbiology*. John Wiley and Sons, New York and London.

Allison, F.E. (1947). *Azotobacter* inoculation of crops (1) *Historical. Soil Sci.* 64: 413-429.

Ames, R., Reid, C., Porter, L. and Cambardella, C. (1983). Hyphal uptake and transport of nitrogen from two 15 N-labelled sources by *Glomus mosseae*, a vesicular-arbuscular mycorrhizal fungus. *New Phytol.* 95: 381.

Andronova, T.M. (1958). Effectiveness of Phosphobacterin. Bull. nauchtech, Inform. Sibir.Nauchissled. *Inst. Sel. Khoz.* 2: 14-19.

Arao, I., Kan, Y.S. and Sihe, T.L. (1956). Bacterial transformation of insoluble organic and inorganic phosphorus compounds in to soluble phosphate in soils. Part II. Results obtained in pot and field experiments to test the practical value of inoculating wheat with *Bacillus megatherium* var. *phosphaticum. Kungchuling Acta Pedol. Sir.* 4: 179-184.

Azcon, R., Barea, J.M. and Hayman, D.S. (1976). Utilization of rock phosphate in alkaline soils by plants inoculated with mycorrhizal fungi and phosphate solubilizing bacteria. *Soil Biol. Biochem.* 8: 135-138.

Azcon-Aguilar, C., Barea, J. M., Azcon, R. and Olivares, J. (1982). Effectiveness of *Rhizobium* and VA-mycorrhiza in the introduction of *Hedysarum coronarium* in a new habitat. *Agric. Environ.* 7: 199-206.

Baltruschat, H. and Dehne, H.W. (1982). Zum Auftreten der endotrophen mycorrhizen in verschiedenen Getreidefruchtfolgen und Dungung- ssystemen. *Med. Fac. Landbouww. Rijksuniv. Gent*, 47:831-839.

Banik, S. and Dey, B.K. (1962). Plant soil. 69: 353.

Beck, T. (1974). Der Einfluβ langjahriger Monokultur auf die Bodenbelebung im Vergleich zur Fruchtfolge. Landw. Forschg. 31. *Sonderheft.* II: 268-276.

Black, R.L.B. and Tinker, P.B. (1979). The development of endomycorrhizal root systems. II. effect of agronomic factors and soil conditions on the development of vesicular–arbuscular mycorrhizal infection in barley and on spore density. *New Phytol.* 83: 401-413.

Bowen, G.D. and Smith, S.E. (1981). The effects of mycorrhizas on nitrogen uptake by plants. *Ecol. Bull. (Stockholm).* 33: 237.

Bystrova, N.A. (1958). The effect of phosphobacterin on the development of brown rust and increase in grain yield of winter wheat. *Trudv. vesesovuz Inst. Zashch. Rest.* 13: 68-70.

Chan, E.C.S., Basavanand, P. and Liivak, T. (1970). The growth inhibition of *Azotobacter chroococcum* by *Pseudomonas* sp. *Can. J. Microbiol.* 16: 9-16.

Chan, E.C.S., Rouatt, J.W. and Ketznelson, H. (1963). The influence of soil and root extracts on the associative growth of selected soil bacteria. *Can. J. Microbiol.* 9: 187.

Debruck, J. (1974). Intensiver Weizenbau-Zu Fragen der Fruchtfolge und Pflanzenges-undheit. Kali-Briefe. 6. *Folge.* 114.

Diercks, R., Bachthaler, G. and Pommer, G. (1980). Langjahrige Auswirkungen unterschiedlicher Fruchtfolge und Anbausysteme auf Ertrag und schaderregerbefall von winterweizen und sommergerste. *Z. Acker und Pflanzenbau.* 149: 454-471.

Dobin, R. (1980). Phosphorus deficiency in tropical soil as a constraint on agricultural output. In: Priotities for Alleviatting soil related constraints to food production in the Tropics. *IRRI:* 217-232.

Gaur, A.C. (1982). *A Practical Manual of Rural Composting*. Food and Agriculture Organization of United Nations: 102.

Gaur, A.C. (1987). In: *Mycorrhiza Round Table*. (eds.) A.K.Verma, K.J. Mukerji, K.V.B.R. Tilak and J. Raj. Proceedings of a National Workshop, J.N.U., New Delhi: 505.

Gaur, A.C., Neelakantan, S. and Dargan, K.S. (1984). *Organic Manures*. ICAR. New Delhi: 159.

Gianinazzi, S., Gianinazzi-Pearson, V. and Trouvelot, A. (1990a). Potentialities and procedures for use of endomycorrhizas with special emphasis on high value crops. In: *Biotechnology for fungi for improving plant growth*. (eds.) J.M. Whipps and B. Lumsden. Cambridge University Press. Cambridge U.K.: 41-54.

Gianinazzi, S., Trouvelot, A. and Gianinazzi-Pearson, V. (1989). Conceptual approaches in agriculture for the rational use of VA-endomycorrhizae in agriculture: Possibilities and limitations. *Agric. Ecosys. and Environments*. 29: 153-161.

Gianinazzi, S., Trouvelot, A. and Gianinazzi-Pearson, V. (1990b). Role and use of mycorrhizas in horticultural crop production. *Plenary lecture at the XXIII International Horticultural Congress*, I.S.H.S. Italy. (1990: 25-30.

Gupta, S.C. and Kapoor, V.K. (1997). *Fundamentals of applied statistics* (III ed.) Sultan Chand and Sons. New Delhi: 6.1-6.113.

Gutser, R., Amberger, A. and Hoffmann, G.M. (1981). Gesundung einer Winter-weizen- Monokultur durch dungung und pflanzenschutz. Landw. Forsch, Sonderheft. 38: 693-708.

Hayman, D.S. (1975). The Occurrence of Mycorrhiza in crops as affected by soil fertility. In *Endomycorrhizas*. (eds.) F.E. Sanders, B. Mosse and P.B. Tinker. Academic Press. London: 495-509.

Heyland, K.U. (1975). Uber den Einfluβ der zunehmenden sommerweizen-Monokultur auf den kornertrag und die kornertragsstruktur. *Z. Acker und Pflanzenbau*. 140: 117-129.

Ivchenko, N.P. (1961). Effect of fertilizers on the quality of the grain of spring wheat in the ural zone. *Agrobiologiya*. 2: 293-295.

Jackson, M.L. (1967). *Soil Chemical Analysis*. Prentice Hall of India (Ltd.) New Delhi: 428.

Jackson, M.L. (1973). *Soil Chemical Analysis*. Prentice Hall of India (Pvt. Ltd.) New Delhi.

Jensen, A. (1982). Influence of four vesicular-arbuscular mycorrhizal fungi on nutrient uptake and growth in barley (*Hordeum vulgare*). *New Phytol*. 90: 45-50.

Jensen, A. and Jacobsen, I. (1980). The occurrence of vesicular-arbuscular mycorrhiza in barley and wheat grown in some Danish soils with different fertilizer treatments. *Plant and Soil*. 55: 403-414.

Koske, R.E. (1981). A preliminary study of interaction between species of vesicular-arbuscular fungi in a sand-dune. *Trans. Br. Mycol. Soc*. 76: 411-418.

Kucey, R.M.N. (1987). Increased phosphorus uptake by wheat and field beans inoculated with a phosphorus solubilizing *Penicillium bilaji* strain and with vesicular-arbuscular mycorrhizal fungi. *Appl. Env. Microbiol*. 53: 2699-2703.

Kudashev, I.S. (1956). The effect of phosphobacterin on the yields and protein content in grains of autumn wheat, maiz and soyabean. *Dokl. Akad. Skh. Nauk*. 8: 20-23.

Kudzin, Y.K. and Yaroshevich, T.V. (1958). The use of phosphobacterin for winter wheat on chernozems. *Udobr. Urozh*. 9: 19-20.

Kundu, B.S. and Gaur, A.C. (1980c). Establishment of nitrogen fixing and phosphate dissolving bacteria in rhizosphere and their effect on yield and nutrient uptake of wheat crop. *Plant and Soil*. 57: 223-230.

Kurten, P.W. and Range, W. (1980). Results of an 8 years fertilization comprising winter wheat and winter barley in cereal rotation and one-crop system. *Landw. Forsch*. 33: 385-407.

Lauzne, E. (1960). Inter-relation of phosphorus bacteria and some cultivated plants. *Trudv. Inst. Microbiol. Akad. Nauk. Latv. SSR*. 11: 153-167.

Mehrotra, C.L. and Lehri, L.K. (1971). Effect of *Azotobacter* Inoculation on crop yields. *J. Indian Soc. Soil. Sci*. (19): 243-248.

Mishustin, E.N. and Naumova, A.N. (1956). The use of bacterial fertilizers in sowing vegetable seeds in to peat manure nutrient cubes. *Mikrobiologiya*. 25: 41-48.

Mosse, B. (1973). Advances in the study of vesicular-arbuscular mycorrhiza. *Annu. Rev. Phytopathol*. 11: 171-194.

Panova, A.V. (1956). Trials with bacterial fertilizers. *Semledelie*. 12: 85-86.

Pavlovich, D.Y. (1959). Interrelations of *Azotobacter* with other groups of soil microorganisms. Trudy. Inst. Mikrobiol. *Akad. Nauk. Latv. SSR*. 8: 127-143.

Pinevich, V.V. (1956). The effect of granular phosphorus fertilizers and phosphobacterin on the composition of wheat grain. *Dokl. Akad. Nauk*. 108: 157-159.

Pommer, G., Oppitz, K. and Fink, K. (1979). Winterweizensorten in Fruchtwechsel und Daueranbau-Ertagsrelationen und Einfluβ produktionstechnischer MaBnahmen. *Z. Acker-und Pflanzenbau*. 148: 214-229.

Raj, J., Bagyaraj, D.J. and Manjunath, A. (1981). Influence of soil inoculation with vesicular-arbuscular mycorrhiza and a phosphate dissolving bacterium on plant growth and ^{32}P uptake. *Soil Biol. Biochem*. 13: 105-108.

Rawat, A.K. and Sanoria, C.L. (1976). Effect of *Rhizobium, Azotobacter* and *Bejerinckia* inoculation on *Cicer arietinum* var. type I. *Curr. Sci*. 45: 665-666.

Rawat, A.K. and Sanoria, C.L. (1977). Influence of seed bacterization on nodulation, yield and amino acid composition of used protein of *Cicer arietinum* var. type I. *Curr. Sci*. 46: 373-375.

Reeves, F.B., Wagner, D., Moorman, T. and Kiel, J. (1979). The role of endomycorrhizas in revegetation practices in the semi-arid west. I. A comparison of incidence of mycorrhizae in severely disturbed and natural environments. *Am. J. Bot*. 66: 6-13.

Rhodes, L.H. and Gerdemann, J.W. (1980). Nutrient translocation in vesicular–arbuscular mycorrhizae, in cellular interactions in symbiosis and parasitism. (eds.) C.B. Cook, P.W. Pappas and E.D. Rudolph. The Ohio State University Press, *Columbus:* 173.

Ruissen, M.A. (1982). The development and significance of vesicular-arbuscular mycorrhizas as influenced by agricultural practices. *Thesis*, Agricultural University, Waageningen. The Netherlands: 111.

Santhi, R. and Kothandaraman, G.V. (1995). Effect of endomycorrhizal colonization and phosphorus sources on the phosphorus recovery, nutrient uptake and growth of green gram. *Indian J. Agric. Res*. 29: 209-214.

Schenck, N.C. and Kinloch, R.A. (1980). Incidence of mycorrhizal fungi on six field crops in monoculture on newly cleared woodland site. *Mycologia*. 72: 445-456.

Smallii, V.T. (1958). The effect of rhizosphere bacteria on the growth and productivity of wheat. Akad. Nauk. Ukrain, SSR. *Kiev:* 153-159.

Smith, J.H., Allison, F.E. and Soulides, D.A. (1961). Evaluation of Phosphobacterin as a soil inoculant. *Soil Sci. Soc. Am. Proc.* 25: 109-111.

Smith, T.F. (1980). The effect of season and crop rotation on the abundance of spores of vesicular-arbuscular (VA) mycorrhizal endophytes. *Plant and Soil*. 57: 475-479.

Sundara Rao, W.V.B., Bajpai, P.D., Sharma, J.P. and Subbaiah, B.V. (1963). Solubilization of phosphates by phosphorus solubilizing organism using P^{32} as tracer and the influence of seed bacterization on the uptake by the crop. *J. Indian Soc. Soil. Sci.* 11: 209-219.

Thompson, J.P. (1990). *Soil Biol. Biochem*. 22: 229.

Timoriu, G., Rosu, G. and Ciobana, V. (1964). Contribution to the study of the influence of bacterial fertilizers on the yield of certain varieties of autumn wheat. Inst. Agron. "I. Ionescu de la Brad". Iasi Luc Stiint. 2: 67-72.

Tinker, P.B. (1978). Effect of vesicular–arbuscular mycorrhizas on plant nutrition and plant growth. *Physiol. Veg*. 16: 743-749.

Tosh, B.N. and Sanoria, C.L. (1978). Combined effect of *Rhizobium* and *Azotobacter* on soil properties and yield of gram (*Cicer arietinum* L.) Varieties. *Fert. Technol*. 15: 35-37.

Vasantharajan, V.N. (1964). Inhibition of plant pathogenic fungi by *B. circulance* solubilizing phosphorus. *Indian J. Microbiol*. 4: 27.

Microbial Biotechnology and Ecology (2011) *Pages* 268–274
Editors: Deepak Vyas, G.S. Paliwal, P.K.Khare and R.K Gupta
Published by: DAYA PUBLISHING HOUSE, NEW DELHI

Chapter 15

Some Additions to the Foliicolous Fungi from the Forest Flora of Madhya Pradesh, India

☆ *P. Chauksay, A.N. Rai* and Naveen Verma*

Department of Botany, Dr. H.S. Gaur University, Sagar – 470 003, M.P.

ABSTRACT

During frequent surveys for phytoparasitic foliicolous microfungi, two interesting specimens were collected from Sagar Forest Circle of Madhya Pradesh, which upon detailed examination proved to be undescribed taxa of *Cytoplea palmacearum* sp.nov., and *Diplodia zizyphiae* sp.nov. infecting *Phoenix sylvestris* Roxb. (Palmaceae) and *Zizyphus oenoplia* Mill. (Rhamnaceae). These are illustrated, described and compared with closely related species in the genera.

Cytoplea palmacearum Chauksay *et al.*, sp. nov. (Figure 15.1)

Maculae amphigenosae, irregulares, disperseae per totam superficiem folii, atro brunneae. Coloniae, pleraeque epiphyllosae, raro hypogenae, atro. Mycelium hypharum immersum, ramosum, septatum, brunneae. Conidiomata immersum et demum erumpentia, subepidermalis, subglobosa vel irregularis, pallide brunneae, usque 550 µm diam. Ostiolum et conidiophora absentia. Conidiogenosae, ampulliformes, hyallina vel subhyllina, laevia, 2.5-8 × 3.5-6 µm. Conidia simplicia, solitaria, aseptata, ovalis vel fusiformia vel ellipsoiformia, laevia, tenuibus parietibus eguttulate, pallide olivacea, 2.5-13 × 1.5-5.

* E-mail: anrai_mycology@yahoo.com.

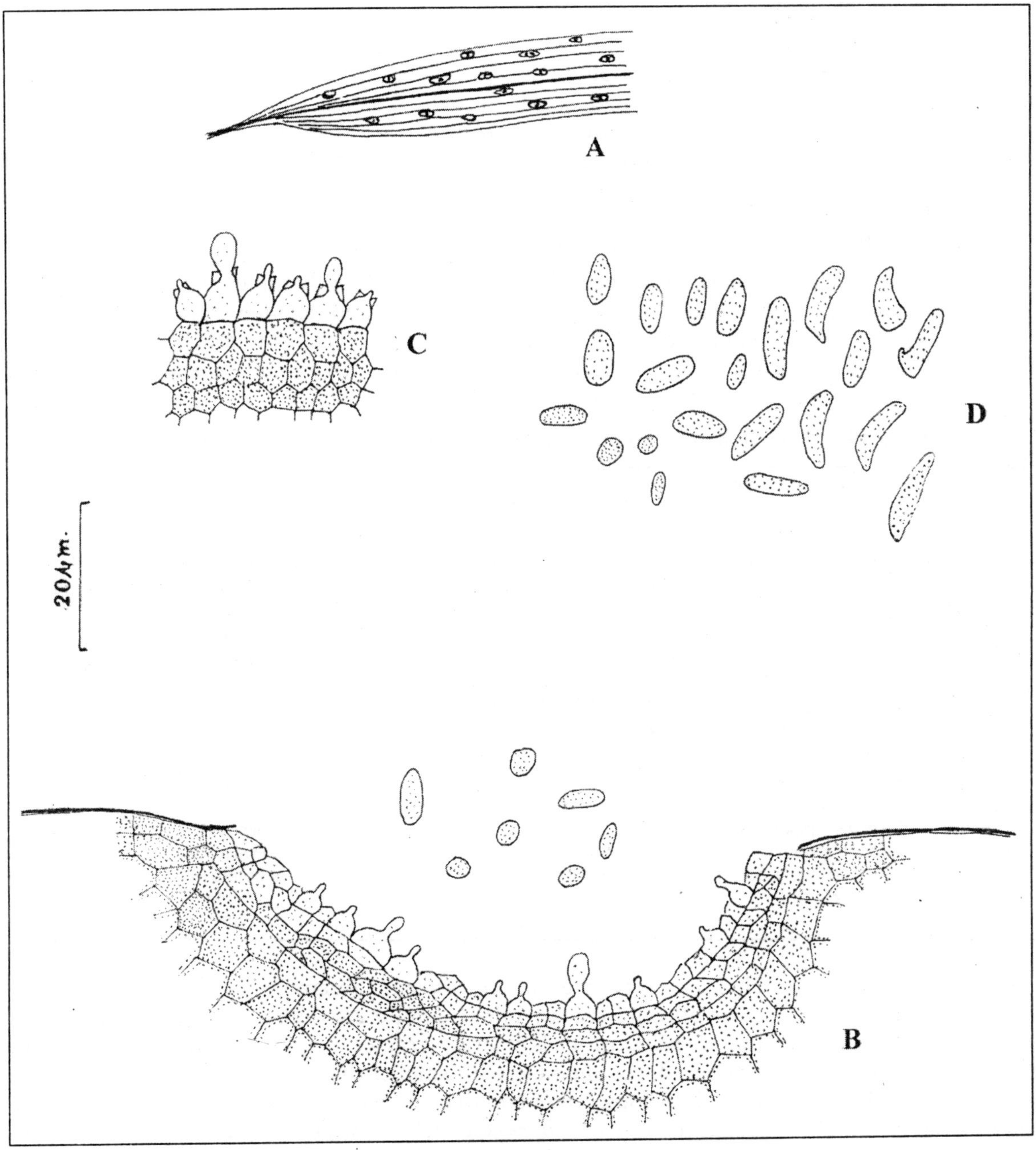

Figure 15.1: ***Cytoplea palmacearum*** **sp.nov.**

A: Symptom; B: Conidiomata; C: Conidiogenous cells; D: Conidia

Hab. In foliis vivis *Phoenix sylvestris* Roxb. (Palmaceae), October 1988 Gopalpura (South Sagar Forest Division), leg. M. Gupta. RG 52, typus, IMI 333066.

Lesions amphigenous, irregular, distributed all over the leaf surface, blackish brown. Colonies predominantly epiphyllous, rarely hypophyllous represented by very hard and warty dots, black. Mycelium of hyphae immersed, branched, septate, brown. Conidiomata immersed and strongly erumpent, subepidermal, subspherical to irregular light brown, up to 550 μm in diam. Ostioles absent, dehiscence by irregular fissured, rupturing in the central region of the upper wall. Conidiophores absent. Conidiogenous cells enteroblastic discrete, determinate, ampulliform, hyaline to subhyaline, smooth, 2.5-8 × 3.5-6 μm. Conidia simple, solitary, aseptate, oval to fusiform to ellipsoidal, smooth, thin walled, eguttulate, light olivaceous, 2.5-13 × 1.5-5 μm (Figure 15.1).

Only a few species of *Cytoplea* have been described so far, the only full treatment of *Cytoplea* is given by Petrak and Sydow (1927), and the most authoritative and detailed work is presented by Sutton (1980). Dr. E. Punithalingam of C.A.B. International Mycological Institute, in his expert opinion reported it *Cytoplea* sp. near *C. orientalis* Sutton (Sutton, 1980). Thus for convenience the proposed species is compared with *C. orientalis* Sutton (Table 15.1).

Table 15.1: Comparative Account *C. palmacearum* sp. nov. with Related Species

Species	*Spot and Colonies*	*Conidiophores*			*Conidia*		
		Structure	*Colour*	*Size (μm diam)*	*Structure*	*Colour*	*Size (μm)*
C. orientalis	Amphigenous, lenticular to ellipsoid, seprate, grayish brown with a paler brown centre.	Eustromatic, immersed to strongly erumpent, epidermal, lenticular to pulvinate, multilocular, wall thin and dark, ostiole absent, dehiscence by irregular breakdown of upper wall.	Dark brown	750	Fusiform, aseptate, verruculose, thin walled.	Pale brown	9-11 x 3-4.
C. palmacearum sp. nov.	Amphigenous, irregular blackish brown. Colonies epiphyllous, rarely hypophyllous, hard and warty dots, black	Immersed to strongly erumpent, sub epidermal, sub spherical to irregular ostiole absent, dehiscence by irregular fissure in the central region of the upper wall.	Light brown	Upto 550	Oval to fusiform to ellipsoid, thin walled, aseptate, eguttulate, smooth.	Light olivaceous	2.5-13 x 1.5-5.

A critical look to the table 1 shows that the authors' collection resembles a little bit to *C. orientalis* in infection spots and structure of conidomata while it is quite dissimilar in other morphological characters of taxonomic values. *C. palmacearum* is at large variance in having epiphyllous, hard and warty black colonies, light coloured, conidiomata with smaller diameter and oval to fusiform to ellipsoid, smooth walled, light olivaceous, larger and thicker conidia as against *C. orientalis* having amphyllous, rarely hypophyllous colonies with dark coloured and large diameter of conidiomata and fusiform, verruculose, pale brown comparatively smaller and thinner conidia. The forgoing discussion, therefore, concludes that authors' taxon has its own identity and merits description and illustration as a new species.

Diplodia zizyphiae Chauksay, *et al.*, sp. nov. (Figure 15.2)

Maculae, epigenosae, parvae circulares vel irregulares, interdem carbonaceous, dispersae per totam superficiem folii, brunneae atro. Mycelium hypharum immersum vel partim superficiale, globosa, unilocularis, fusce parietibus, extra laminae fusce brunneae, fusce parietibus sclerotiodeis formatis, medio laminae tenuibus parietibus hyaline cellulae et interius laminae tenuibus parietibus hyaline cellulae, 800 µm dimetro, ostiolata singularis, circularis, simplicia, centralis. Conidiophori cylindreta, formatus interius cellulae pycnidial parietibus, eramosa, septati, usque. 1-3 transverse septati, leavibus parietibus, erecta vel suberecta, hyalinae vel pallide olivacea, 26-75 × 5-8.5 µm. Conidiogenosae cellulae holoblastae, integratae, cylindratae, laevia, formatus singula apicale conidium, hyalinae vel pallide olivacea. Conidia simplicia, solitaria, aseptata, laevia, intio hyaline quam fusco brunneae et modreta 1 euseptata, fusce parietibus, subglobosa vel ovalis vel oblonga vel ellipsoiformia, apices et bases obtuse, 5-34.5 × 4.5-26 µm

Hab. In foliis vivis *Zizyphus oenoplia* Mill. (Rhamnaceae), January, 1989, Gopalpura (South Sagar Forest Division), leg. M. Gupta, RC 57, typus, IMI 333075.

Lesions epigenous, small circular to irregular, coal-tar like, distributed all over the leaf surface, brownish black, colonies epiphyllous, conferred to the center of the spot, represented by very fine dots, black. Mycelium of hyphae immersed branched, septate, brown. Conidomata pycnidial, immersed to partly superficial, globose to sub-globose, unilocular, thick walled, outer layer dark brown thick walled textura angularis, median layer light brown of thin walled cells and inner layer thin walled, hyaline cells, 800 µm in diam., ostiole single, circular, simple, central. Conidiophores cylindrical, formed from the inner cells of pycnidial wall, unbranched, septate, upto 1-3 transversely septate, smooth walled, erect to sub-erect, hayaline to light brown. Conidia simple, solitary, aseptate, smooth, at first hyaline then dark brown and medially 1 equiseptate, thick walled, subglobose to oval to oblong to ellipsoidal, apex and base obtuse, 5-34.5 × 4.5-26 µm (Figure 15.2)

Hab. on living leaves of *Zizyphus oenoplia* Mill. (Rhamnaceae), January, 1989, Gopalpura (South Sagar Forest Division), leg. M. Gupta, RC 57, type, IMI 333075.

Petrak and Sydow (1927) placed *Shaeria malorum* Berk. synonymy of *Botryodiploadia malorum* (Berk.) Petrak and Sydow but these names are regarded as synonyms of *Diplodia mutila* by Stevens (1933) who quoted an extensive synonymy. Shoemaker (1964) has clarified the identity of certain *Botryodiploadia* spp. in which *S. stenvensii* Shoem., was regarded as the ascomycetous state of *D. mutila*. Zambettakis (1954) in a reassessment of generic concepts in the didymosporous sphaeropsidales erroneously adopted *D. conigena* Desm. as the type for *Diplodia* and referred *D. mutila* to the genus *Metadiplodia* Syd., the genus is need of further revision. A number of *Diplodia* species are reported from different host families but for convenience the authors' collection is compared with *D. mutila* Fr. apud Mont. (Sutton, 1980) in Table 15.2.

A critical analysis of the Table 15.2 shows that the proposed species resembles *D. mutila* in a few characters such as structure of conidiomata, conidiophores and conidia while it is drastically dissimilar from *D. mutila* (*D. zizyphiae* has larger diam. of conidiomata, unbranched, longer, thicker and comparatively dark coloured conidiophores, conidia subglobose to oval to oblong to ellipsoid with obtuse base and apex having comparatively longer but considerably wider (almost just double) conidial dimensions as against *D. mutila*) so much so that it needs its description and illustration as a new species.

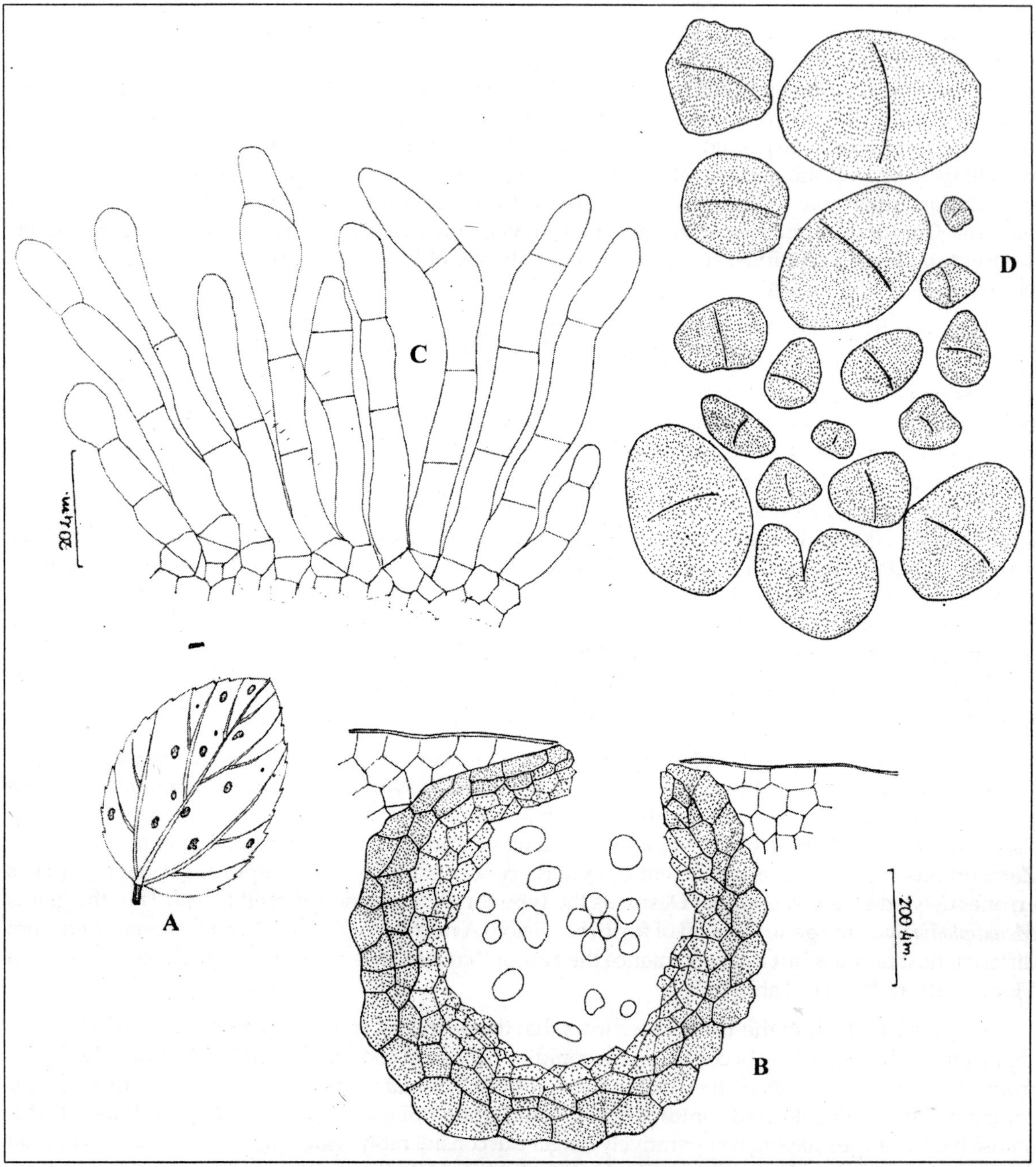

Figure 15.2: ***Diplodia zizyphiae*** **sp.nov.**

A: Symptom; B: Conidiomata; C: Conidiophores; D: Conidia

Table 15.2: Comparative Account *D. zizyphiae* sp. nov. with Related Species

Species	Spot and Colonies	Conidiomata			Conidiophores			Conidia		
		Structure	Colour	Size (μm diam.)	Structure	Colour	Size (μm diam.)	Structure	Colour	Size (μm)
D. mutila	–	Pycnidial, immersed, separated to aggregate, globose, unilocular, thick walled, single ostiole, papillate.	Dark brown to black	Upto 600	Cylindrical, branched, septate, smooth	Hyaline	45 x 3-5	Aseptate, centrally guttulate, thick walled, smooth, medianly 1 euseptate, apex obtuse and bases tr.	First hyaline then dark brown	27-31 x 12-13.5.
D. zizyphiae sp. nov.	Epigenous, circular to irregular, coaltar like, brownish black. Colonies epiphyllous black.	Pycnidial immersed to partly superficial, globose to subglobose, unilocular, thick walled, single ostiole	Dark brown	Upto 800	Cylindrical, unbranched, erect to sub erect, 1-3 trans. Septate, smooth.	Hyaline to light olivaceous	26-75 x 5-8.5.	Sub globose to oval to oblong to ellipsoid, aseptate but medianly 1 euseptate, thick walled, smooth, apex and base obtuse.	First hyaline then dark brown.	5-34.5 x 4.5-26.

Acknowledgements

The authors are grateful to the Director, International Mycological Institute, Kew, (U.K.) for the accession of the deposited fungal specimens and the Head, Department of Botany, Dr. H.S. Gour University, Sagar for providing facilities. The financial assistance from UGC Research Award Scheme to ANR is thankfully acknowledged.

References

Petrak, F. and Sydow, H. (1927). Die Gattungen der Pyrenomyzeten, Shaerospideen und Melanconieem. *Beih. Rep. Spec. nov. Regni Veg.*, 42: 1-551.

Sutton, B.C. (1980). *The Coelomycete.* Fungi Imperfecti with Pycnidia, Acervuli and Stromata. C.M.I., Kew, England.

Stevens, N.E. (1933). Two apple black rot fungi in the United States. *Mycologia*, 25: 536-548.

Shoemaker, R.A. (1964). Conidial state of some *Btryosphaeria* species on *Vitis and Quercus*. *Can. J. Bot.*, 42: 1297-1301.

Zambettakis, E.C. (1954). Recherches sur ia systematique des "Sphaeropsidales-Phaeodidymae". *Bull. Soc. Mycol. Fr.*, 70: 219-350.

Microbial Biotechnology and Ecology (2011) *Pages* **275–292**
Editors: **Deepak Vyas, G.S. Paliwal, P.K.Khare and R.K Gupta**
Published by: **DAYA PUBLISHING HOUSE, NEW DELHI**

Chapter 16

Current Status of Sesame and Niger Diseases and Their Management

☆ ***Sushma Nema[1], T.S. Rajpurohit[2] and M.N. Khare[3]***

[1]Senior Scientist, Plant Pathology, AICRP (S&N) JNKVV, Jabalpur
[2]Pathologist, Oilseeds (Rajasthan Agricultural University), Agricultural Research Station, Mandor, Jodhpur
[3]Ex. Dean and Professor Emeritus, Plant Pathology

Both Sesame and Niger suffer due to diseases caused by fungi, bacteria, viruses, nematodes and phytoplasma making the host weak resulting in losses both in quantity and quality. The pathogens are both seed borne and soil borne. The important diseases of sesame are damping off (*Pythium aphanidermatum, Thielavia terricola var. minor*); foot rot (*Sclerotium rolfsii*); wilt (*Fusarium oxysporum* f. sp. *sesami*); root rot, stem rot, charcoal rot, blight, seedling rot [(*Rhizoctonia bataticola*) (*Macrophomina phaseolina*)]; Phytophthora blight (*Phytophthora nicotianae var. parasitica*); Alternaria blight (*Alternaria sesami*); Cercospora leaf spot (*Cercospora sesami* and *C. sesamicola*); Corynespora blight (*Corynespora cassicola*); Botryosphaeria blight (*Botryosphaeria ribis*); *Cylinderosporium* leaf spot (*Cylinderosporium sesami*); anthracnose (*Colletotrichum* sp.); powdery mildew (*Leveillula taurica, Oidium* sp., *Sphaerotheca fuliginea*); Phoma leaf blight, blight, stem black (*Phoma sesamina, P. variosporeae*); bacterial leaf spot (*Pseudomonas syringae pv. sesami*); bacterial blight (*Xanthomonas campestris pv. sesami*) Leaf curl (*Nicotiana virus* 10); Phyllody (Phytoplasma) etc. Several other microorganisms have also been found associated with sesame seeds.

The incidence of important diseases vary from state to state based on agroclimatic situations. The diseases of sesame and Niger and their management are summarised below:

Diseases of Sesame

Fungal Diseases

About 72 fungi are responsible for causing diseases in sesame. Several such fungi have been observed associated with seed and soil which play a vital role in disease cycle.

Damping Off [*Pythium aphanidermatum* (Eds.) *Fitz.; Thielavia terricola* var. *minor* Rayss and Borut.]

Damping off of seedlings is important under high soil moisture both at pre and post emergence. The radicle and plumule get infected before emergence and no seedlings are formed. At post-emergence stage, the stem close to the soil becomes water soaked, shrinks, rots and the seedling topples down. The pathogens are soil borne *T. terricola* also causes root rot.

Foot Rot (*Sclerotium rolfsii* Sacc.)

Mishra and Khare (1970) first reported the disease from Jabalpur. The disease has been observed at different sesame growing areas. The basal part of the stem develops brown to black lesion due to necrosis of tissues. The affected stem shrivels and the plant dies. White mycelium is seen around the diseased part which develops mustard like sclerotia. The disease is severe under high humidity and high temperature. Leaf infection has also been observed.

Wilt [*Fusarium oxysporum f. sp. sesami* (Zapro.) *Castellani*]

The disease was first reported by Butler (1918). The disease is common in all the sesame growing areas. Loss upto 20 per cent has been observed due to wilt. As the diseased plants die before maturity heavy losses are possible. The leaves of diseased plants turn yellowish, droop down and dry. The vascular bundles turn brown and this brown discolouration starts from roots and gets extended upto the tip.

The fungus is soil borne as well as seed borne. The seed borne inoculum leads to infection of seedling at very early stage, resulting in mortality, depending upon the inoculum potential. The soil borne inoculum attacks and enters through finer roots.

Root Rot, Stem Rot, Charcoal Rot, Blight, Seedling Rot (*Macrophomina phaseolina* (Maubl.) Ashby (Pycnidial stage); *Rhizoctonia bataticola* Taub. Butler) (Mycelial and Sclerotial Stage)

These diseases are very important as infection occurs from seed germination and emergence to adult plant stage. The losses incurred are even upto 100 per cent. Murugesan *et al.* (1978) observed 1.8 kg yield loss per hectare at every one per cent increase in disease intensity. *M. phaseolina* causes seedling rot, stalk rot, root rot, charcoal rot and blight in sesame (Ali Ahmad and Saidawi, 1998). It is of major importance in West Bengal (Anonymous, 2001). The infected seeds are discoloured, on germination, seed rot occurs or the seedlings get damaged and dry. Brown dark spots develop on the stem which turn into lesions. The stem breaks. A number of pycnidia are formed in the lesion on the stem. Sclerotia develop in the tissues of diseased root and stem. On splitting it appears as if charcoal powder has bean sprinkled on the tissues, hence the disease is called charcoal rot. Black spots of various sizes are formed on the leaves which coalesce. The infected leaves are dropped off. Similar spots are also formed on capsules and sclerotia as well as pycnidia develop on them. Blackish hyphae are seen on the seed surface, associated with seed coat.

The pathogen is both seed and soil borne. The sclerotia survive for many years in soil. The seed borne infection is both externally and internally associated. Soil and seed borne inoculum is responsible for recurrence of the disease and seed borne inoculum helps in long distance transmission. The spread of the disease is through movement of workers, animals, water and equipments used in cultural management in the field. Seed to plant infection is systemic as well as local through seed borne inoculum and infected inert matter. Plant to seed transmission is through capsule. The pathogen has a very wide host range. High temperature and humidity favour the disease development. Phosphorus in soil checks the disease severity. Appressoria formation has bean observed on the epidermis of infected seedlings which helped the mechanical and chemical penetration. The hyphae extend inter and intracellularly from roots to stem, fruits and seeds. Gupta (1992) negatively correlated the production of volatile aldehyde compounds with the extent of seed infection. The infection of sesame plants was positively correlated with percentage of infected seeds. Singh and Singh (1990) examined sesame seed samples from 18 districts of Rajasthan and found 108 samples to be infected with 1-67 per cent incidence. Microsclerotia of the pathogen were present in the seed. Gupta and Cheema (1990) positively correlated the number of microsclerotia on sesame seeds with plant infection and negatively with seed germination, dry matter production and root and shoot length of seedlings. Kushi and Khare (1979) examined several seed samples of 21 sesame varieties and found 17 fungi associated of which *M. phaseolina* was the most important along with *Corynespora cossicola* and *A. sesami*. Karunanithi *et al.* (1999) established three groups of *M. phaseolina* isolates based on colony characters and virulence.

Phytophthora Blight (*Phytophthora nicotianae var. parasitica* Dastur)

The disease was first observed by Dastur in India. It is the most serious disease in northen India described by Butler (1918). High humidity and heavy rains increase the disease severity resulting in even 80 per cent seedling mortality (Singh *et al.*, 1976). The plants get diseased at all stages of crop growth. The disease starts from the base of the plants and gradually moves up covering the stem which turns black; Infloresence, flowers and fruits are also attacked by the fungus. The first symptom of the disease appears as brown, water soaked spots on leaves and seedlings in very early stage. In old plant, leaves develop water soaked spots which coalesce and give blight appearance. Many times the disease is so severe that nothing is left in the field if attack is at seedling stage and continuous rains occur for about a week. The partially affected plants produce shrivelled smaller seeds with brown discolourations, severely infected plants dry up.

The pathogen survives in soil during unfavourable period in the form of dormant mycelium and or in the form of chlamydospores/oospores. It is seed borne, the mycelium remains in embryo. The recurrence of the disease is through these two sources. Once the plants are attacked, sporangia are produced by the mycelium which spread the disease. Zoospores are formed in the sporangium, which are oozed out and infect the host plants. High humidity and 28-30°C temperatures favour the disease spread. Heavy soil and severe rains are responsible for high disease intensity low lying fields with poor or no drainage get the disease with very high severity.

The disease is both seed and soil borne, hence effective sanitation, rotation with other suitable crops and seed treatment are essential. Resistant sources are also available. In Sri Lanka, Pathirana (1992) obtained a high yielding mutant tolerant to Phytophthora blight by Gamma irradiation of 450 and 600 Gy.

Alternaria Blight (*Alternaria sesami* Zimm.)

The disease was first reported by Mohanti and Behera (1958) in Orissa, now the disease is prevalent in all sesame growing states in the country. Both typical leaf spot and blight symptoms are produced by the fungus. Losses upto 50 per cent have been reported due to this disease. The disease occurs at all the stages of crop growth however, Ojiambo *et al.* (1999) observed the plants to be least susceptible at 4 weeks and most susceptible at 8-10 weeks age. Dark brown spots develop on cotyledons. Water soaked, circular or irregular brown spots develop on leaf. Brown stripes are formed on stem. Capsules are also infected resulting in seed infection.

The pathogen is seed borne and the fungus survives in seed coat as well as in embryo. The recurrence of the disease is due to this inoculum. The cotyledons are infected on which conidia are produced which attack the foliage of other plants through air and the disease gets spread. The pathogen also survives in crop residues in soil and helps in disease recurrence. Temperature 20-30°C and pH 5-7 favour the disease development. *A. sesami* produced PGTE, PTE, PG enzymes in vitro (Rajpurohit and Prasad, 1982). Cultural studies *viz:* effect of different media, temperature, level of pH, Effect of Carbon and Nitrogen on growth and sporulation of *A. sesami* were carried out in detail (Rajpurohit and Prasad, 1982; Rajpurohit *et al.*, 1983). The possibility of races has been expected by some workers. A description of morphology and taxonomy of Sesamum Alternaria given by Rajpurohit (1981).

For the detection of *A. sesami* associated with seed, oat meal agar medium is the best. The seed infection has been observed from 9 to 25 per cent. Seed to plant transmission of the pathogen is important as 8 per cent infected seeds resulting in high incidence of the disease in field. According to Ojiambo *et al.* (2000) the tolerance level should be less than 2 per cent. In Karnataka it causes yield loss of 0.1 to 5.7 g seed per 100 capsules (Siddaramaiah, 1981). Seed infection reduces viability of seeds.

Rao and Vijayalakshmi (2000) tested nutritional requirements of *A . sesami* and its ability to produce phytotoxic and antibacterial metabolites. Chloroform extracts of the fungus exhibited antibacterial activity. The culture filtrate contained tenuazonic acid.

The pathogen survives in seeds as well as in soil on residues for more than one year as mycelium or chlamydospores. Excessive rainfall favours occurrence and spread of the disease (Maiti *et al.*, 1988) Cumulative rains, rainy days and relative humidity for many days positively correlated whereas maximum and minimum temperature had negative co-relation with Alternaria leaf spot (Naik *et al.*, 2003). The time of plant to seed infection has been determined by Ojiambo *et al.* (1999) in Kenya, which was maximum in plants inoculated between 8–10 weeks age and least when inoculated at 4, 6 and 12 weeks age.

Cercospora Leaf Spot (*Cercospora sesami* Zimm. and *C. sesamicola*)

The disease is prevalent in all sesame-growing areas but more severe and important in Madhya Pradesh, Uttar Pradesh, Orissa, Tamil Nadu and Karnataka. Losses upto 60 per cent have been reported due to the disease. C. *sesami* produces conidia which attack both the sides of leaves by forming water soaked spots which later become circular or irregular. Initially the leaf spots are whitish to light brown, later turning dark brown. The leaf spots coalesce and occupy a large area of lamina giving blight appearance. The infected leaves dry and drop off. The disease spread to petioles, stem and pods on which linear, dark coloured, deep-seated lesions are formed. *C. sesamicola* produces brown angular spots (Mohanti, 1958).

The pathogen is both seed and soil borne survives in the plant debris in soil as well as in and on seed for more than one year and acts as primary inoculum in the recurrence of the disease. The secondary invasion or spread of the pathogen is through conidia from plant to plant and field to field by wind. Plant to seed infection is through capsule.

Corynespora Blight [*Corynespora cassiicola* (Berk. and Curt.) Wei]

The disease was first observed at Bhubaneshwar. It has been reported from different sesame growing areas in this country. All the plant parts except root are attacked but stem and leaves are badly affected. On leaves brown specks are formed which later turn into large spots. At advanced stage of infection the leaves curl and drop off. On stem, purple brown specks develop which turn into elongated, scattered, irregular lesions, which enlarge at both the ends. Lesions turn into canker with straw coloured central area surrounded by dark brown margin. The infected stem bends and cracks. When both stem and leaves get infected plant dies prematurely. On capsules circular to elongate, minute, sunken, black spots develop which cover the whole surface. The infected seeds turn brown and become light. The disease reduces the yield. Low temperature and high humidity favour the disease development.

The pathogen is seed and soil borne. It is internally and externally seed borne. The fungus survives in soil in infected crop residues. Both the sources are responsible for recurrence of the disease. The spread occurs through conidia formed after the primary inoculum infects plants, gets established and produces conidia. The pathogen has several other hosts like pulses, tomato etc. also on which it causes diseases, survives and attacks sesame. *C. cassicola* is highly seed borne. In seed stored at 26± 2°C and 50 per cent RH, the infection decreased from 89.8 to 8 per cent in seven months and in ten months the fungus was fully inactive (Navas and Subero, 1995).

Botryosphaeria Blight [*Botryosphaeria ribis* (Gross) Duggar]

The disease was first observed by Shukla and Vyas (1977) at Jabalpur. The disease appears towards crop maturity and causes heavy losses in yield. All above ground plant parts are infected and brownish specks are formed which later become irregular lesions. They coalesce and enlarge 2-25 mm in length. On leaves the infection starts from the margin, the spots are brown but later the central part turns black. The spots even cover the whole leaf. On capsules small, black sunken spots appear which increase under high humidity and temperature. The seeds inside the capsule get infected and turn brown.

The disease is both soil borne and seed borne. The pathogen survives on diseased plant refuge in soil and is associated with seed externally and internally. These sources act as primary inoculum for recurrence of the disease. Ascospores and conidia formed on diseased parts spread the disease through air.

Cylindrosporium Leaf Spot (*Cylindrosporium sesami* Hansford)

The disease has been reported in 1965 from Gujarat. Water soaked spots are formed on leaves which later become darker and angular limited by veins. Necrosis occurs. Black acervuli of the fungus are formed on the upper surface in these spots and conidia are formed in large number. For the growth of the fungus 27°C temperature is the best. Sclerotia are formed which get mixed with the seed as inert matter and are responsible for recurrence of the disease alongwith those present in the infected crop residues in soil. The spread of the disease is through conidia formed on infected plant parts of the standing crop.

Anthracnose (*Colletotrichum* sp.)

Anthracnose has been reported from several sesame growing countries. In India, it was first reported from Uttar Pradesh in 1953. Infection occurs at all stages of crop growth and all the above ground parts attacked. Water soaked spots appear which turn brown and enlarge irregularly. The margins are dark brown. Black acervuli develop on the diseased portions. Stem also develops symptoms. The cortex is separated and woody part gets exposed. The infected plants dry. Capsules are either not formed on diseased plants or they are very few. The pathogen is seed borne and also survives in infected plant parts in soil and they help in recurrence of the disease. The spread is through conidia.

Powdery Mildew (*Leveillula taurica, Oidium* sp., *Sphaerotheca fuliginia*)

Powdery mildew is an important in all the sesame growing areas. The disease is severe under dry conditions. Small circular areas with whitish powder appear on leaves, later stem, leaves and capsules get fully covered with whitish powdery mass. The leaves start drying and fall off. The seeds produced from diseased plants are small and wrinkled. The disease is not seed borne. The pathogen has a wide host range and is always present in nature.

Rajpurohit (1993) has reported powdery mildew caused by *Erysiphe orontii* from Rajasthan. Ganesh *et al.* (1992) have found *Erysiphe cichoracearum* D.C. to be the casual fungus of powdery mildew in Tamil Nadu.

Phoma Leaf Blight and Black Stem (*Phoma sesamina* Sacc., *Phoma variosporeae* Shreemali)

Shreemali (1978) has reported *P. variosporeae* and *P. sesamina* from various places in India. The ashy spots surrounded by brown margin are formed on the leaf margins, which later cover entire leaf. The stem also develop similar symptoms. Black pycnidia are formed on the infected portions.

Bacterial

Bacterial Leaf Spot (*Pseudomonas syringae* pv. *sesami* Malkoff.)

The disease occurs in all the sesame growing areas of the world. In India the disease was first reported by Durgapal and Rao (1967) and is prevalent in Rajasthan, Uttar Pradesh, Madhya Pradesh, Tamil Nadu, Andhra Pradesh etc. The disease is also called black rot. The pathogen is seed borne. The pathogen has two physiological races. The most suited temperature for the growth is 30°C. Seed to plant transmission of the pathogen is systemic as well as local. Stomata and wounds are the entry points for the bacterium in disease spread. High temperature and frequent rain showers favour the disease development.

The crop is attacked at all stages of crop growth but it is more devastating at seedling stage. The disease starts from petiole, moves to leaf blade and small angular brown to black spots with dark margins are formed. They are confined by the veins. Bacterial exudates can be observed on the lower surface of the leaf. Under high humidity and temperature, the spots increase and coalesce. The diseased leaves drop off. On stem, black lesions are formed which girdle the whole stem and the plant dies. The capsules become black and seeds are not formed in them, but seeds, are formed in partially infected capsules which are discarded. The pathogen is internally and externally seed borne. Pandya *et al.* (2000) observed a decrease in N,P,K, Ca and increase in Mg content in infected leaves as disease intensity increased.

Bacterial Blight (*Xanthomonas campestris* pv. *sesami* Sabet and Dowson)

Sabet and Dowson (1960) reported the disease from Sudan first. At Jabalpur 10-25 per cent plants were found diseased in local and exotic varieties. The pathogen is highly seed borne and the recurrence of the disease is through infected seeds. The seed to plant transmission is both systemic and local. The spread of the disease is through air borne inoculum and rain splashes.

When infected seeds are sown, the colyledonary leaves exhibit small water soaked spots on the lower surface. These spots turn black. Under severe infection, the cotyledons dry and drop off. Typical dark brown to black angular spots appear on leaves which coalesce and give blight appearance. Bacterial ooze comes out from the affected parts under favourable conditions. Stem is also infected forming chlorotic spots which later turn dark brown and exude bacterial ooze. Capsules also develop redish brown raised spots (Wasnikar *et al.*, 1991). The bacterium also attacks certain weeds on which it survives till next season. Younger leaves are more susceptible. High humidity and 20-26°C temperature favour the disease. Higher dose of N increases the disease severity. Poor seed germination and damping off have been observed in Assam due to the bacterium by Phookan and Hazarika (1994).

Viral

Leaf Curl (Nicotiana Virus 10)

Leaf curl is a very harmful disease of sesame occurring in many sesame growing countries and sesame growing regions in India. As the diseased plant dies before flowering, 60-100 per cent losses have been observed. The virus has several hosts like tomato, tobacco, papaya etc. and is transmitted by white fly (*Bemisia tabaci* Gen.) hence the recurrence and spread of the disease take place through white fly obtaining the virus from such sources.

MLO

Phyllody (*Mycoplasma Like Organism- Phytoplasma*)

Phyllody is one of the most serious and wide spread disease of sesame and occurs in sesame growing states, Rajasthan, Madhya Pradesh, Tamil Nadu, Orissa, Maharashtra, Uttar Pradesh, Gujarat, Andhra Pradesh, Haryana, Punjab every year. The disease incidence ranges 1 to 100 per cent. Considerable losses are caused due to the disease. The disease was earlier considered to be due to physiological and genetic disorder, then due to virus and finally MLO. Murugesan *et al.* (1978) reported 8.36 kg/ha loss in yield on each one per cent increase in disease incidence. No capsules are formed on severely infected plants.

The disease symptoms of phyllody get manifested at flowering which get changed in leafy structures with abundant vegetative growth. Petals enlarge and turn green. The filaments of stamens are flattened and become green. Anthers also turn green and have no functional pollen. Ovules in the ovary get small petiole like growth. The diseased plants develop more branches with smaller leaves with clear veins.

The disease is transmitted by leaf hopper *Orosius albicinctus*, it has 15-23 day incubation period. The MLO survives in the leaf hopper throughout its life. It attacks various crops as well as weeds (82 plant species) throughout the year and is transmitted to sesame plant as the leaf hopper is also present. The MLO is transmitted fast by the leaf hopper at 25-40°C. Increased Indole Acetic Acid (IAA) due to infection is responsible for proliferation of vegetative and reproductive parts.

Application of N (15-45 kg/ha) imparted resistance to phyllody when coupled with medium dosage of P (15-30 kg/ha) (Borkar and Krishna, 2000).

Other Diseases

Besides above diseases, there are many other diseases which are not of great importance at present but close eye is needed so that they do not become limiting factor in sesame production.

Choanophora Wet Rot (*Choanephora cucurbitarum* Thax.)

The disease becomes severe in heavy rains. Small water soaked spots are formed on leaves which coalesce. Under high humidity whitish hyphae cover the diseased parts.

Drechslera Blight [*Drechslera sesami* (Miyake) Richa and Frasier]

The disease is prevalent in all the sesame growing areas. Small brownish elongated spots (0.2 × 2.0 cm) are formed on leaves and (1x4 cm) on stem. Capsules are also infected in which the seeds get spoiled. The spots are of V shape and later give blight appearance. *D. sesami* is seed borne.

Myrothecium Leaf Spot (*Myrothecium roridum* Tode ex Fries)

Small light brown circular spots are formed with darker margin and ashy central portion. Black spordochia are formed in the centre of the spot. The diseased portion is blown off by wind and shot hole symptoms develop. The pathogen is seed borne.

Synchytrium Gall Disease (*Synchytrium sesamicola*)

The disease has been reported from Bihar and Uttar Pradesh. Galls are formed due to hypertrophied host tissues.

Cyst Nematode Disease (*Heterodera* sp.)

The diseased plants remain stunted and roots do not develop fully. Capsules are either not formed or if formed, they remain small with small shrivelled seeds. The disease was first reported from Rajasthan.

Seed-borne Microorganisms

Several scientists (Kushi and Khare, 1979; Singh and Singh, 1990; Mahatma *et al.*, 2001) have examined the sesame seeds of different varieties and locations and observed the following fungi and bacteria associated with seeds. They are internally seed borne, externally seed borne, extraembryal, intraembryal, concomitant contaminant, mixed with the seed as inert matter or associated with inert matter *viz.*

Alternaria alternata, A. longissima, A. sesami, A. sesamicola, Aspergillus flavus, A. niger, A. ochraceus, A. tamarii, A. terreus, Beltrama rhombica, Botrytis cinerea, Cercospora sesami, C. sesamicola, Chaetomium globosum, Cladosporium herbarum, Colletotrichum dematium, Corynespora cassicola, Curvularia lunata, Cylindrosporium sesami, Drechslera hawaiensis, D.sesami, Epicocum nigram, Fusarium moniliforme, F. oxysporum, F.semitectum, F.solani, Macrophomina phaseolina, Mucar haematis, Myrothecium roridum, Penicillium chrysogenum, P. citrinum, P. purpurogenum, Phoma sp., Phytophthora parasitica, Pleospora infectoria, Pseudomonas syringae, Rhizopus nigricans, R. oryzae, R. stolonifer, Setophaeria sp., *Stemphylium* sp., *Syncephalasturm racemosum, Trichoderma hamatum, T. viride, Ulocladium* sp., *Xanthomonas campestris* etc.

Management

Cultural Practices

It is necessary to follow suitable phytosanitary measures, crop rotations in endemic areas, proper date of planting, integrated fertilizer management, application of required micro nutrients, desired plant population, removal of weeds etc. Sesame grown as mixed or inter cropped with moth bear or green gram 1:1 gave less incidence of Macrophomina stem and root rot of sesame (Rajpurohit, 2003).

Chemical Control

Seed treatment with suitable chemicals like Thiram (3g/kg), Bavistin (2g/kg), Captan (3g/kg), Thiram + Bavistin (2g + 1g) should be done before planting (Rajpurohit, 2005). Foliar fungal diseases can be checked by the spray of Mancozeb (0.2 per cent), Zineb (0.2 per cent), Copper oxychloride (0.2 per cent). Soil treatment with Benlate or Rizolax-T (0.04-0.1 per cent) has been recommended (Gabr *et al.*, 1998). Powdery mildew can be checked by spraying sulfex (0.3 per cent) (Rajpurohit, 1993). Efficacy of plant product and new chemicals against *A. sesami* reported by Rajpurohit (2004).

To control bacterial diseases seed, treatment with Streptocycline (250-1000 ppm) is recommended. For hot water treatment, seeds are dipped in water at 51°C for 10 minutes then dipped for 6-9 hours in water having 0.025 per cent Agrimycin 100. The foliar infection can be checked by spraying streptocycline.

For the control of leaf curl white fly needs to be checked by the application of 0.1 per cent metasystox as spray or any other suitable insecticide. Foliar spray of Mancozeb + Methyl demeton reduced Alternaria blight, Phyllody and leaf curl diseases of sesame (Rajpurohit, 2004). Spray of 0.05 per cent. Dimethoate, Fenthion and Monocrotophos has been recommended to control *Orosius albicinctus* the transmitting insect of phyllody MLO. Soil application of Phorate granules 11kg/ha combined with 0.03 per cent Dimethoate spray checks the disease. Management of Macrophomina stem and root rot through fungicidal seed treatment and varietal resistance reported by Rajpurohit (1997).

Integrated Disease Management

Integrated disease management module comprised seed treatment with Thiram 0.2 per cent + Carbendazim 50 WP 0.1 per cent + foliar spray of Mancozeb 0.2 per cent + Endosulfan 0.07 per cent at 30-45 DAS and 45-55 DAS reduced Macrophomina stem and root rot from 24.54 to 11.93 per cent, Alternaria leaf spot 41.73 to 10.76 per cent, Phyllody from 3.20 to 0.64 per cent, Leaf curl 3.3 to 0.47 per cent. This module recorded seed yield of 693 kg/ha with 2.92 BC ratio (Rajpurohit and Bishnoi, 2004).

Biological Control

Three isolates of *Bacillus subtilis* suppressed wilt and root rot diseases when applied in artificially infested soil (Gabr *et al.*, 1998). Seed treatment with *Trichoderma viride* (0.4 per cent) was effective in checking *M. phaseolina* (Rajpurohit, 1995). Application of phosphorus solublizing bacteria (PSB) alongwith neem cake or 50 per cent NPK + FYM or NPK + PSB reduced phytopthora blight (Verma and Bajpai, 2001).

From the experiments conducted under All India Coordinated Project on Sesame during the last three years, it was inferred that seed treatment with *T. viride* or *T. harzianum* (0.4 per cent) and Carbendazim 25 SD (0.25 per cent) + Thiram (0.15 per cent) minimized the incidence of Macrophomina stem and root rot; Seed treatment with *Pseudmonas fluorescens* (0.4 per cent) and Contaf (0.05 per cent) effectively managed Cercospora leaf spot incidence (Anonymous, 2001). Soil amendment with neem cake @ 500 kg/ha+ seed treatment with *I. viride* (0.47 per cent) + soil application of *T. viride* @ 2.5 kg/ha was found highly effective for the management of soil borne diseases.

Resistance

Resistance source is the most important to control the diseases. Selim *et al.* (1976a) have reported susceptibility in mature plants to *M. phaseolina* to be dominant over tolerance, controlled by 1, 2 or 3 pairs of genes. Selim *et al.* (1976 b) have reported that in mature plants the tolerance to wilt (*F. oxysporium* f. sp. *sesami*) is governed by 1 or 2 dominant pairs of genes and in one cross it was governed by 3 pairs of genes. Tolerance was recessive. The resistance to powdery mildew (*Erysiphe cichoracearum*) is controlled by two major complimentary genes (Krishnaswami *et al.*, 1983). Mishra *et al.* (1994) while studying the inheritance of leaf curl disease observed both additive × additive and/or additive x dominance types of gene interaction. The resistance to phyllody disease in *Sesame alatum* is due to insect resistance (Parani *et al.*, 1996). The available genotypes resistant to single disease is given in Table 16.1 and to multiple diseases in Table 16.2.

Abd El, Moneem *et al.* (1997) evaluated 112 somaclones obtained from epicotyl segments of variety Giza 25 grown in tissue culture. S_2 clones had higher resistance to root rot, charcoal rot and wilt diseases. The resistance is attributed to increased contents of either peroxidase or phenolic compounds in roots.

Table 16.1: Varieties/Lines of Sesame Resistant to Different Diseases

Disease	Resistant Varieties/Lines	Reference
Wilt	JT-7, N-32, MB-12	Vyas *et al.* (1981)
Macrophomina diseases	AT-9, HT-16, RT-46, ST-58, Gwalior-5, ES-25, ES-105, ES-124, IS-156, IS-190	Gupta (1995)
	Line-144, 885,	Saharan and Chand (1988)
	BAG-59, 21, 65, 9, 5, 11	Li *et al.* (1991)
	RT-46, JTS-8, SI-249, SI-254-1	Arriel *et al.* (1998)
	EC-370932, EC-370929, EC-351832, ES379-3-84, IS-97, IS-100 B, IS-646-1	Anonymous (2001) Rajpurohit *et al.* (2003)
Phytophthora blight	75A/1-1/2-1	Gemawat and Prasad (1964)
	M-3-1, M-3-2, No.2-39, No.66-193	Saharan and Chand (1988)
	Spring borok	Sasikumar and Sardana (1988)
Alternaria blight	No.-2, No.-4, J-77, E-8	Saharan and Chand (1988)
	EC-370929, EC-370932	Rajpurohit and Solanki (2002)
Cercospora leaf spot	IS-4, 14, 15, 21, 29, No.-41, No.-41A, No.-41B, No-128, No.128A	Kushwaha and Kaushal (1970), Vyas *et al.* (1981)
	L-36, L-45, RSB-1, HT-1, SI-2373, SI-1501, SI-1032, X-79-6, TMV-6, S-1683,	Saharan and Chand (1988)
	BIC-7-2, Sidhi-54, Rewa-114, Seoni Malwa	Tripathi *et al.* (1994)
	65B58, 60/2/3-1-8B, 69B-392, 73A-96B	Poswal and Misari (1994)
Powdery mildew	CO-1, BS-129, BS-518-6G, X-74-24, SI-3170, 59-1-1, IS-401	Ganesh *et al.* (1992) Shadakshari *et al.* (1981)
	OMT 30 and DORS 101	Rajpurohit (1993)
Synchytrium gall	JT-66-276, SP-70-23	Variar *et al.* (1981)
Bacterial leaf spot	Maporal, Ajimo Atar S-5, Morada, Almora white, C-24, C-87, C-96, M-31, NP-8, NP-23, Punjab-1, SSM-34, SSM-44, SSM-49, EC-4094, EC-13536, EC-14536, EC-20783, EC-20785, EC-20787,	Urdaneta and Mazzani (1976)
	N-63-177-74-145, TC-325, T-58	Duragapal *et al.* (1969)
Bacterial blight	T-58	Saharan and Chand (1988)
	Tozi-3, S-76F2-22, K-112	Abdal Rahim and Adam (1990)
Phyllody	RT-46, RT-54, RT-103, RT-125, VRI-1	Anonymous (2001)

Diseases of Niger and their Management

Niger is a crop of dry areas grown mostly by tribals in interior places due to which desired attention has not been given on the biotic and abiotic stresses. Now the crop is gaining importance and studies are being made on disease aspects. The diseases observed are given herewith. The important diseases of niger are ozonium wilt (*Ozonium texanum* var. *parasiticum*); collar rot (*Sclerotium rolfsii*); Macrophomina rot and blight (*Macrophomina phaseolina*); Cercospora leaf spot (*Cercospora guizoticola*), powdery mildew (*Sphaerotheca* sp.); rust (*Puccinia guizotiae*); Bacterial leaf spot (*X. campestris pv guizotiae*) and Cuscuta as *Phanerogamic* parasite.

Table 16.2: Sesame Varieties/Lines Resistant to Multiple Diseases

Disease	*Resistant Varieties/Lines*	*Reference*
Phyllody + wilt	Mulayanum	Mehetre *et al.* (1994)
Powdery mildew + Bacterial leaf spot	ES-277	Shadakshari *et al.* (1989)
Phytophthora blight + Alternaria blight	Suwonkkae	Lee *et al.* (1992)
Phytophthora blight + Alternaria blight + Cercospora blight	Sesame radiatum	Malaguti (1973)
Fusarium wilt + Macrophomina rot + Rhizoctonia solani root rot	87-AN, Golmarmara, Muganli, Ozberk-82, 9/10-1-2, 1/10-2-1	Alac *et al.* (1994)
Powdery mildew + Bacterial wilt + Cercospora blight + Fusarium stem rot	RAU-8817-4, CO-1, AT-6	Ram *et al.* (1995)
Macrophomina rot + Bacterial spot + Phyllody + Powdery mildew	Rajeshwari	Gangakishan *et al.* (1989)

Ozonium Wilt (*Ozonium texanum* var. *parasiticum* Thirum.)

The disease was reported by Rao and Pavgi (1975) around Varanasi where it appeared in great severity. Heavy losses are incurred as the diseased plants dry. Necrotic lesions develop on stem of well grown plants near the soil level, which gradually extend upwards. Whitish fungal mycelium grows on these necrotic areas under high humidity. The branches, leaves, inflorescence etc. become soft and start rotting. The diseased stem breaks. The mycelium grows on dead plants if enough moisture is present. Root rot is also commonly observed.The fungus forms rhizomorphs and sclerotia. The sclerotia are oval, grey to dark brown. The recurrence takes place through soil borne sclerotia and mycelium.

Collar Rot (*Sclerotium rolfsii* Sacc.)

The disease was first reported from Dharwad (Karnataka) by Siddarmaiah *et al.* (1979). The tissues of collar region become soft and depressed. White fungus grows on the diseased part and forms mustard seed like sclerotia. The diseased plants turn yellow and dry. The perfect stage of the fungus is *Corticium rolfsii*. The disease is soil borne. As the plant number is reduced, the disease causes yield losses.

Macrophomina Rot and Blight [*Macrophomina phaseolina* (Maubl.) Ashby]

The sclerotial stage of the fungus is *Rhizoctonia bataticola*. It is seed and soil borne. Typical root rot, stem rot, charcoal rot, leaf blight symptoms are produced. The recurrence of the disease takes place through seed and soil borne inoculum, later spread is through workers, tools and insects like *Melanoagromyza* stem borer.

Damping Off, Root Rot (*Rhizoctonia solani*)

The fungus attacks stem of the seedling at ground level and makes it water soaked soft and incapable of supporting the seedling, which falls over and dies. On older seedling elongated brownish black lesions appear which increase in length and width girdling the stem often the roots are blackened due to the fungal attack resulting in root rot. The disease is favoured by cold and wet weather.

Cercospora Leaf Spot (*Cercospora guizoticola*)

The disease was reported by Rao (1962). This disease is prevalent in all the niger growing areas and caused yield reduction. The disease is severe under warm and humid weather. Small straw coloured to brownish spots are formed on both the leaf surfaces. Later the spots increase in number and size and cover the entire lamina and the leaves start dropping off. Elongated dark brown spots are produced on the stem. The capsules are also affected.

The pathogen is seed and soil borne. The primary infection occurs through seed borne inoculum as well as the inoculum present on diseased plant left over in the soil. The spread of the disease is through conidia formed on infected plant parts. The fungus is active throughout the year where niger is grown in *rabi* and *kharif* seasons. The pathogen remains active on other collateral hosts also.

Alternaria Leaf Spot (*Alternaria* sp.)

Spots are brown to blackish, with grey centre circular to irregular with concentric rings on leaf. If floral buds are affected, they become black and fails to open. The pathogen is seed and soil borne.

Curvularia Leaf Spot (*Curvularia lunata*)

Small circular to irregular brown to reddish brown spots are formed on leaves, which later coalesce to form larger spots. The leaf turn yellow, dries and defoliates. The pathogen is seed borne and survives in plant debris.

Powdery Mildew (*Sphaerotheca* sp.)

The disease occurs in *rabi* as well as *kharif* season but it is more severe in *rabi*. The yield loses are due to early defoliation as a result of the disease. All the aerial parts may be affected. Small cottony spot develops on the leaves, which gradually covers the whole lamina. The diseased leaves turn yellow and drop off. The seed formed on diseased plants are small and shrivelled.

Rust (*Puccinia guizotiae*)

Rust was first reported from Ethiopia (Yirgu, 1964 a). Brown coloured pustules are formed on the lower leaf surface and the corresponding upper surface becomes chlorotic. Later teletospores are formed.

Bacterial Leaf Spot (*Xanthomonas campestris* pv. *guizotiae*)

This disease was first reported from Ethiopia by Yirgu (1946 b). Small brownish spots are formed on leaves which are surrounded by yellow halo. Most of the spots are on the leaf margin. The spots increase, coalesce and give appearance of blight. Under severe condition all the leaves get destroyed.

Cuscuta (*Cuscuta hyalina* and *C. chinensis*)

Cuscuta, a phaneroganic plant parasite has been observed to attack Niger badly. The plants remain stunted, pale yellow and bear a very small number of flowers and capitula due to the association of Cuscuta. It is a total stem parasite depending for shelter and food totally on the host. The food is obtained through haustoria from the host. Usually the Cuscuta seed gets mixed with Niger seed and such mixtures are planted. The Cuscuta seeds can be removed by sieving.

Seed Borne Microorganisms

Several fungi have been reported seed borne like *Alternaria* sp., *Aspergillus flavus*, *A. niger*, *Rhizopus nigracans*, *Cercospora guizoticola*, *M. phaseolina*, *R. batalicola* etc. (Siddarmaiah *et al.*, 1980 a, b). Besides fungi, bacteria like *Xanthomonas campestris* are also seed borne.

Management

(i) Cultural Practices

Sl.No.	*Practice*	*Diseases Managed*
1.	Pathogen free site	All diseases
2.	Deep ploughing in summer	Collar rot, root rot, wilt
3.	Pathogen free seed	Seed borne diseases
4.	Manipulation in planting	
	(a) Shallow planting (1")	Diseases caused by *Rhizoctonia solani* and *R.bataticola*
	(b) Space Planting 30 x 10 cm (3 lakh plants/ha)	– do –
5.	Irrigation whenever necessary to avoid stress condition	*Rhizoctonia bataticola*
6.	Mixed cropping with Kodo, Kutki, Mung	*Cercospora, Curvularia, Alternaria* leaf spot, powdery mildew and rust
7.	Hot water treatment at 52° for 10 minutes	Bacterial leaf spot
8.	Proper and timely rouging, uprooting and destruction of diseased plant	Root rot, collar rot, phyllody
9.	Clean cultivation, destruction of collateral host plant debris	All diseases
10.	Sieving of seed before sowing	Cuscuta
11.	Removal of Cuscuta infested niger seedlings	Cuscuta

(ii) Use of Resistant Varieties

Variety	*Resistant to*	
	Alternaria Leaf Spot	*Cuscuta*
IGP-76	+	-
Deomali	+	-
GA-11	+	-
ONS-8	+	+ (Moderately susceptible)

(iii) Chemical Protection from Pathogen

Disease	*Chemical*	*Mode of Application*
Root rot and wilt	Thiram (0.15 per cent) + Captan (0.15 per cent) in 1:1 + Urea (2 per cent)	2-3 times soil drenching, in infected area and one meter around, along with the plant using sprayer without nozzle from the initiation of disease at 7 days interval
Leaf spots caused by *Alternaria Cercospora* and *Curvularia*	Zineb (0.25 per cent) or Mancozeb (0.25 per cent) or Topsin-M (0.1 per cent) or Bavistin (0.1 per cent)	3 sprays as and when disease appear at 15 days interval

Contd...

(iii)–Contd...

Disease	Chemical	Mode of Application
Bacterial leaf spot	Agrimycin 100 (250 ppm) or Streptocycline (500ppm) or plantomycin (100 ppm)	– do –
Powdery mildew	Sulfex (0.3 per cent)	2 sprays starting from the initiation of disease at 10 days interval
Phyllody	Phorate 10 G (10 kg/ha)	Soil application
	Dimethoate (0.3 per cent) or Monocrotophos (0.3 per cent)	2-3 foliar sprays at 45, 60, 70 days after sowing
Cuscuta	Pendemethlene	Pre-emergence soil application

References

Abd El Moneem, K.M.H., El Farash, E.M. and Fahmy, F.G. (1997). In vitro selection for higher yielding somaclones resistant to charcoal rot and wilt disease complex in sesame. *Assiut Journal of Agricultural Sciences* 28: 201-224.

Abdel Rahim, A.M. and Adam, F.S. (1990). Some etiological characteristics of *Xanthomonas campestris pv. sesami* and disease reaction of different sesame cultivars. *Journal of Phytopathology* 129: 165-169.

Al Ahmad, M. and Saidawi, A. (1998). Macrophomina (charcoal) root rot of sesame in Syria. *Arali Journal of Plant Protection* 6 (2): 88-93.

Anonymous. (2001). Annual Progress Report Sesame and Niger 2000-2001. All India Coordinated Research Project (Sesame and Niger), J.N. Agricultural University Campus, Jabalpur, M.P. pp. 152.

Arriel, N.H.C., Moreira, J. de, A.N., Guedes, A.R., Costa, I.T., Da, Tarso, F.P. de, da Costa, I.T. and da, Tarso, F.P. (1998). Agronomic and botanical evaluation of the live collection of sesame germplasm at Embrapa Algodao. Pesquisa em Andamento, Embraba Algodao. No. 91 pp. 9.

Atac, A., Cetin, V. And Ipkin, B. (1994). Studies on root and crown rot disease of sesame. Bitki Koruma Bulletin 34: 43-54.

Borkar, S.G. and Krishna, A. (2000). Effect of fertilizers on development of phyllody in Sesame. *Indian Journal of Mycology and Plant Pathology*. 30: 422-423.

Butler, E.J. (1918). Fungi and Disease in Plants. Thaker Spink and Co. Calcutta. pp. 547.

Durgapal, J.C. and Rao, Y.P. (1967). Bacterial leaf spot of *Sesamum orientale in India. Indian Phytopath.* 20: 178-179.

Durgapal, J.C., Patel, P.N. and Rao, Y.P. (1969). Resistance in crops to bacterial diseases in India. Evaluation of Sesame for resistance to bacterial leaf spot incited by *Pseudomonas sesami. Indian Phytopathology*. 22: 292-294.

Gabr, M.R., Hussein, N.A., Saleh, O.I. and Khalil, M.A. (1998). Susceptibility of certain varieties and genotypes and control of wilt and root rot diseases of sesame attributed to *Fusarium oxysporum* f. sp. *sesami* and *Macrophomina phaseolina*. *Egyptian Journal of Microbiology* 33: 403-428.

Ganesh, S.K., Kandasamy, G., Kousalya, G., Sivarman, M.R. and Gangadharan, K. (1992). Field evaluation of (*Erysiphe cichoracearum* D.C.). *Madras Agric.* 79: 115-117.

Gangakishan, A., Jagadishwar, K., Raju, C.S. and Sudarshanum, A. (1989). Rajeshwari- a high yielding white seeded variety of sesame for Andhra Pradesh. *Indian Farming* 39(3): 7-8.

Gemavat, P.D. and Prasad, N. (1964). Further studies on Phytophthora blight of Sesame. *Indian Phytopathology*. 17: 273-283.

Gupta, I.J. and Cheema, H.S. (1990). Effect of microsclerotia of *Macrophomina phaseolina* and seed dressers on germination and vigour of sesame seed. *Seed Research* 18: 169-172.

Gupta, L.J. (1992). Change in the production of volatile aldehyde compounds in Macrophomina infected sesame (*Sesamum indicum*) seed. *Seed Research* 20: 27-30.

Gupta, R.B.L. (1995). Evaluation of sesame genotypes for resistance to root rot and oozing. *Indian Phytopathology* 48; 194-195.

Karunanithi, K. Muthusamy, M. and Seetharaman, K. (1999). The cultural and pathogenic variability among the isolates of *Macrophomina phaseolina* causing root rot of sesame. *Plant Disease Research* 14: 113-117.

Krishinaswami, S., Rao, A.V., Appadurai, R. and Venkatarao, A. (1983). A preliminary study on the inheritance of powdery mildew disease in Sesame. Proc. National Seminar on breeding crop plants for resistance to pests and diseases. TNAU, Coimbatore.

Kushi, K.K. and Khare, M.N. (1979). Seed borne fungi of sesame and their significance. *Seed Research* 7: 48-53.

Kushwaha, U.S. and Kaushal, P.K. (1970). Reaction of Sesame varieties to Cercospora leaf spot in M.P. *Mysore Journal of Agricultural Sciences*. 4: 228-230.

Lee, J.I., Kang, C.W., Lee, B.H. and Park, R.K. (1992). A nonbranching good quality and high yielding sesame variety 'Suwonkkae'. Research Report R.D.A.U. and I.C. 34: 91-96.

Li, L.L., Wang. S.Y., Fang, X.P., Hyang, Z.H., Wang, S.T., Li, M.L. and Cui, M.Q. (1991). Identification of *Macrophomina phaseolina* resistant germplasm of sesame in China. *Oil crops of China*. No. 1: 3-6.

Mahatma, L., Singh, S.D. and Lodha, P.C. (2001). Pathogenic seed mycoflora of sesame (*Sesamum indicum L.*) *Journal of Mycol. Pl. Pathol.* 31: 377-379.

Malaguti, G. (1973). Leaf diseases of sesame (*Sesamum indicum L.*) in Venezuela. Rev. de la Fac. de. *Agronomia* 7: 109-125.

Mahetre. S.S., Ghatge, R.D. and Lad, S.K. (1994). Wild *Sesame mulayanum*: a source of multiple disease resistance. *Annals of Agricultural Research* 15: 243-244.

Mishra, A.K., Tiwari, R.C., Pathak, K.N., Ali, S.A. and Ghuraya, R.S. (1994). Inheritance of leaf curl disease in *Sesamum indicum* L. *Crop Research* Hisar 8: 131-136.

Mishra, R.P. and Khare., M.N. (1970). A sclerotial leaf and stem rot of mesta. *Indian Phytopath.* 23: 706-708.

Mohanti, N.N. (1958). Cercospora leaf spot of sesame. *Indian Phytopathology*. 11: 186-187.

Mohanti, N.N. and Behera, B.C. (1958). Blight of sesame (*Sesame orientale* L.) caused by *Alternaria sesami* (Kawamara) Comb. *Current Science* 27: 492-493.

Murugesan, M., Shanmugam, N., Menon, P.P.V., Arokiaraj, A., Dhamo, K.P. and Kochubabu, M. (1978). Statistical arrangement of yield loss of Sesame due to insect pests and diseases. *Madras Agric.* J. 65: 290-295.

Navas, M. and Subero, L.J. (1995). Survival of *Corynespora cassicola* in sesame seeds under two storage temperatures. *Anales de Botanica Agricola* 2: 16-19.

Ojiambo, P.S., Ayiecho, P.O. and Nyabundi, J.O. (1999). Effect of plant age on sesame infection by Alternaria leaf spot. *African Crop Science Journal* 7: 91-96.

Ojiambo, P.S., Ayiecho, P.O., Narla, R.D. and Mibey, R.K. (2000). Tolerance level of *Alternaria sesami* and the effect of seed infection on yield of sesame in Kenya. *Experimental Agriculture* 36: 335-342.

Pandya, R.K., Triphathi, M.L., Reetu Singh and Singh, R. (2000). Change in mineral contents in sesame leaves infected by *Pseudomonas syringae* pv. *sesami* (Malkoff). *Crop Research-* Hisar 20: 140-144.

Parani, M., Singh, K.N., Rangasamy, S.R.S. and Ramalingam, R.S. (1996). A study on mechanism of phyllody disease resistance in *Sesame alatum* Thonn. *Current Science* 70: 86-89.

Pathirana, R. (1992). Gamma ray induced field tolerance to Phytophthora blight in sesame. *Plant Breeding* 108: 314-319.

Phookan, A.K. and Hazarika, R. (1994). Damping off of Sesame caused by *Xanthomonas campestris* pv. *sesami* in Assam. *Indian Phytopathology*. 47: 249-250.

Poswal, M.A.T. and Misari, S.M. (1994). Resistance of sesame cultivars to Cercospora leaf spot induced by *Cercospora sesami* Zimin. *Discovery and Innovation* 6: 66-70.

Rajpurohit, T.S. (1981). Morphology and taxonomy of sesamum Alternaria. Madras Agricultural Journal. 68(10): 696-697.

Rajpurohit,T.S. and Prasad, N. (1982). Effect of different media, temperature and levels of pH on the growth and sporulation of *Alternaria sesami.Geobios* Vol.9(3)142-144.

Rajpurohit, T.S. and Prasad, N. (1982). Production of pectinolytic enzymes by *Alternaria sesami* in vitro. *Indian Journal of Mycology and Plant Pathology*.Vol 12(2): 220-221.

Rajpurohit, T.S.N. Prasad and Gemawat, P.D. (1984). Evaluation of fungicides against *Alternaria sesami* in vitro. Pesticides special issue on fungicides Jan.: 57.

Rajpurohit, T.S. (1993). Occurrence, varietal reaction and chemical control of new powdery mildew (*Erysiphe orontii* Cast) of sesame. *Indian Journal of Mycology and Plant Pathology*. 23: 207-209.

Rajpurohit,T.S. (1997). Management of Macrophomina stem and root rot of sesame through fungicidal seed treatment and varietal resistance *Indian Journal of Mycology and Plant Pathology*. 27: 98.

Rajpurohit,T.S. (1998). Integrated management of Phytophthora blight, Cercospora leaf spot and Alternaria leaf spot of sesame. *Indian Journal of Mycology and Plant Pathology*. 28: 68.

Rajpurohit, T.S. (1999). *Trichoderma viride* Biocontrol agent effective against Macrophomina stem and root rot of sesame. 4th Agricultural Science Congress, Feb. 21-24, Jaipur. Abstract P.297.

Rajpurohit,T.S and Solanki, Z.S. (2002). Identification of resistance sources of Alternaria leaf spot of sesame. *Asian congress of Mycology and Plant Pathology*. Oct. 1-4, University of Mysore.Page : 207.

Rajpurohit, T.S., Solanki, Z.S. and.Ahuja, D.B. (2003). Screening of promising sesame genotypes against stem and root rot (*Macrophomina phaseolina*) under sick plot. National Seminar on Strees management in oilseeds for attaining self reliance in vegetable oils, Indian Society of Oilseeds Research, DOR Hyderabad. 28-30, Jan. 2003.

Rajpurohit, T.S. (2002). Influenced of intercropping and mixed cropping sesame with pearl millet, greengram and mothbean on the incidence of stem and root (*Macrophomina phaseolina*) of sesame *Sesamum* and *Safflower News letter*. No.-17: 40-41.

Rajpurohit, T.S. (2003). Studies on fungicides plant products and bioagents as spray in sesame disease management. *Journal of Eco-physiology* 6(3-4): 87-88.

Rajpurohit, T.S. (2004). Studies on plant products and new chemicals as spray against Alternaria blight of sesame. Geobios, 31: 64-65.

Rajpurohit, T.S. (2004). Efficacy of foliar spray on incidence of diseases of sesame. *Sesamum* and *Safflower News letter*. No.-19: 67-69.

Rajpurohit, T.S. (2004). Evaluation of plant products and fungicides and insecticides against Alternaria blight, Phyllody and leaf curl diseases of sesame. Indian *Phytopathology*. 57: 340-341.

Rajpurohit, T.S. and Bishnoi, H.R. (2004). Integrated disease management in sesame (*Sesamum indicum* L.). Paper presented in 26 Annual Conf. and Symposium. ISMPP. Goa University, Goa, 7-9 Abstract. pp. 144-145.

Rajpurohit, T.S. (2005). Seed treatment and Bioagent for the management of stem and root rot of sesame. *Sesamum* and *Safflower* News letter. (Under publication)

Ram, T., Ansari, M.M. and Gangwar, B. (1993). Evaluation of sesame varieties in rainfed conditions for better yield and resistance to diseases in a rice based cropping system. In Proceedings of Symposium on farming systems for sustained productivity in humid tropics. Ed. Rao, G.C., Central Agricultural Research Institute, Port Blair. pp. 119-122.

Rao, N.R. and Vijayalakshmi, M. (2000). Studies on *Alternaria sesami* pathogenic to sesame. *Microbiological Research* 155: 129-131.

Rao, R. and Pavgi, M.S. (1975). Two sclerotial diseases of crops in Varanasi. *Indian Phytopathology*. 28: 401-404.

Sabet, K.A. and Dowson, W.J. (1960). Bacterial leaf spot of sesame (*Sesame orientale* D.). *Phytopath. Z.* 37: 252-258.

Saharan, G.S. and Chand, J.N. (1988). Tilhani Faslon ke Rog. Commission for Scientific and Technical Terminology, Government of India, New Delhi. pp. 47-83.

Siddarmaiah, A.L., Kulkarni, S. and Basavarajaiah, A.B. (1979). Occurrence of a new collar rot disease of Niger (*Guizotia abyssinica* Cass.) in India. *Current Science* 18: 174.

Siddarmaiah, A.L., Desai, S.A., Hegde, R.K. and Kulkarni, S. (1980a). Seed mycoflora of *Guizotia abyssinica* Cass. and their control. *Mysore Journal of Agricultural Sciences*. 14: 334-336.

Siddarmaiah, A.L., Desai, S.A., Hegde, R.K. (1980b). Effect of seed mycoflora of Niger on seed germination and oil content under different moisture levels. *Mysore Journal of Agricultural Sciences*. 14: 497-499.

Sasikumar, B. and Sardana, S. (1988). Siping borok (Jhum til). Source of some valuable genes. *Journal of Oilseeds Research* 5: 183-184.

Selim, A.K., Serry, M.S., Omran, A.O. and Satour, M.M. (1976a). Breeding for disease resistance in Sesame, *Sesamum indicum* L. II. Inheritance of resistance to root and stem rot disease caused by *Sclerotium bataticola* Taub. *Egyptian Journal of Phytopathology* 8: 15-18.

Selim, A.K., Serry, M.S., Satour, M.M. and Al Ahmar, B.A. (1976b). Breedidng for disease resistance in sesame, *Sesamum indicum* L. III Inheritance of resistance to *Fusarium oxysporum* Schlecht. *Egyptian Journal of Phytopathology* 8: 19-24.

Shadakshari, Y.G., Virupakshappa, K. and Siddaramaiah, A.C. (1989). Reactions of Sesame genotypes against powdery mildew and bacterial leaf spot. *Current Research* 18: 4-5.

Shreemali, J.L. (1978). Two new species of Phoma from India. *Indian Journal of Mycology and Plant Pathology*. 8: 220-221.

Shukla, B.N. and Vyas, S.C. (1997). Two new diseases of Sesame (*Sesamum indicum* L.). J.N.K.V.V. Research Journal, 11: 142-143.

Singh, B.P., Shukla, B.N. and Kaushal, P.K. (1976). Evaluation of Sesame varieties for their susceptibility to *Phytophthora parasitica* Dastur at Jabalpur, M.P. J.N.K.V.V. Research Journal, 10: 76-77.

Singh, T. and Singh, D. (1990). Incidence of *Macrophomina phaseolina* (*Tassi*) Goid in sesame seeds of Rajasthan. *Indian Journal of Mycology and Plant Pathology*. 20: 111-114.

Tripathi, S.K., Tiwari, K.L. and Patel, R.P. (1986). Assessment of stability in resistance to Cercospora leaf spot in Sesame. *Agricultural Science Digest Karnal*, 16: 233-236.

Urdaneta, U.R. and Mazzani, B. (1976). Varietal differences in susceptibility to bacteriosis in sesame (*Sesamum indicum L.*) in Venezuela. Ag*ronomia Tropical* 26: 311-320.

Variar, M., Pavgi, M.S. and Variar, M. (1981). Varietal reaction of sesame to Synchytrium gall disease. *Indian Phytopathology*. 34: 430-433.

Verma, M.L. and Bajpai, R.P. (2001). Effect of bio-inoculants on phytopthora blight and Macrophomina root/stem rot of Sesame (*Sesamum indicum* L.). National Symposium on plant protection strategies for sustainable Horticulture Society at Plant Protection Science. SKAUT, Jammu, 12-13 Oct. pp: 113.

Vyas, S.C., Prasad, K.V.V. and Khare, M.N. (1981). Diseases of Sesame and Niger and Their Control. Department of Plant Pathology, J.N.K.V.V, Jabalpur, M.P. Bull. pp.16.

Wasnikar, A.R., Khare, M.N., Deshmukh, M.R. and Yasin, M. (1991). Bacterial blight of sesame. *Current Research* 20: 183.

Yirgu, D. (1964a). Some diseases of *Guizotia abyssinica* in Ethiopia. *Plant Disease Reporter* 48: 672.

Yirgu, D. (1964b). *Xanthomonas guizotia* sp. nov. on *Guizotia abyssinica*. *Phytopathology* 54: 1490-1491.

Microbial Biotechnology and Ecology (2011) *Pages* **293–302**
Editors: **Deepak Vyas, G.S. Paliwal, P.K.Khare and R.K Gupta**
Published by: **DAYA PUBLISHING HOUSE, NEW DELHI**

Chapter 17

Integrated Disease Management of *Piper betle*

☆ *Deepak Vyas*[1]* *and Rajesh Yadav*[2]

[1]*Lab of Microbial Technology and Plant Pathology*
Department of Botany, Dr. H.S. Gour University, Sagar – 470 003, M.P.
[2]*ABR PG College, Anpra, Sonbhadra, U.P.*

ABSTRACT

The present investigation deals with the study on disease management of Pan caused by *Phytophthora parasitica* var. *piperina*. The result reveals that when the different control agents such as Ofloxacin (antibiotic), Calc. sulph (homoeodrug), Fytalon (fungicide), *Allium sativum* (botanical), *Trichoderma* sp. (biocontrol agent), *Microcystis* cell extract (Algal bloom), are used in higher concentration, they all shows complete killing of pathogen but when they used in lower concentration they show variable inhibitory effect against the pathogen. Different combination of control agents under *in vitro* and *in vivo* condition showed variation in the inhibitory action against the pathogen. When 100 ppm of ofloxacin combined with 10 per cent plant extract of *Allivum sativum* produces best result and shows 100 per cent killing of pathogen *in vitro* and *in vivo* condition.

Introduction

The Concept of integrated disease management (IDM), which is derived from integrated pest management, is being considered quite important in relation to disease management. Integrated pest management; aims at keeping population below the level of economic injury using best combination of various components of management. FAO has defined IPM as a pest management system that in the context of associated environment and the population dynamics of the pest species, utilizes all suitable

* E-mail: dvyas64@yahoo.co.in.

techniques and methods, in as compatible manner as possible and maintains the pest populations at levels below those causing economic injury. This new concept actually, describes what the farmers were doing before chemical pesticides came in vogue. In IPM, various components of management are tailored so as to get cost-effective control without any adverse effect on environment and ecosystem. IPM or IDM relied more on non-toxic means but does not intend to keep pesticides at bay from the management system; rather ecosystem advocates minimal or requisite use of pesticides as may be appropriate to make disease management not only effective but non-hazardous and environment friendly. IDM or IPM thus rules out total reliance on pesticides, which by and large is used as the strongest weapon against pests.

An effective IDM system aims at minimising crop losses due to diseases through efficient management of soil, water, nutrient and plant by integrating cultural and agronomic practices, chemical application, host plant resistance, biological agents and by the application of physical, mechanical and regulatory control methods.

Integrated disease management (IDM) as practiced in the developed countries is a sophisticated technology achieving a computerised on line service. Practical IDM methods have been suggested for less developed countries or traditional agro-ecosystems. It is based on efficacy, ecology and economics employing routine and supervisory practices with minimum use of agrochemicals being backed by epidemiology and management science without using computers. Disease management has to be integrated with total crop management.

Disease management strategy largely depends upon the stage/form of pathogen responsible for carrying over a particular disease from one season to another and role of various agencies such as soil, stubbles and crop debris play important role in the spread of pathogen. Thus, to evolve a suitable and efficient system for disease control, prerequisite is availability/generation of information on various aspects such as diagnosis, measurement and appraisal *i.e.* economic significance epidemiology and influence of cultural practices on intensity of disease initiation and development. Several workers have reported the integrated management of various crop diseases (Kumar and Dubey, 2001; Dohroo, 1995; Vyas, 1994; Vadhera *et al.*, 2001; Koshy *et al.*, 2002 and Dubey, 2002). Srivastava (2000) recommended integrated plant disease management for sustainable agriculture. According to Ahuja and Kandhari (2000) integrated disease management has been adopted in rice crop.

Over the years people have been trying to control pan disease by using single drug (fungicide, homoeopathic drug, antibiotic, plant extract, etc.) against test pathogen but recurrence of diseases again and again reflects that no single drug can control the disease until and unless it is used in high concentration. The existing methods for disease control of pan mainly aim for the management of individual disease through different practices. Cultural practice/chemical and plant products together and their use for control measures come under IDM. Though not impossible but cumbersome is the task to combine high level of resistance in a genotype with desirable attributes. Therefore it would be worth-while to integrate available compatible control practices to develop viable and economic management strategies against diseases particularly for the genotype which possess desirable attributes with high yield and low level of resistance to diseases.

The aim of present study is to follow the integrated approach or combination of various drugs and other agents. Thus the attempts have been made to control the leaf rot of pan caused by *P. parasitica* var. *piperina* by integrated disease management system.

Materials and Methods

In vitro Effect of Various Combinations on Mycelial Growth of *P. parasitica* var. *piperina*

In the present study the drugs and organisms were selected and tested on the basis of their earlier good performance. These include Ofloxacin, Calc sulph, Fytalon, extract of *Allium sativum*, biocontrol agents *Trichoderma* sp. and *Microcystis* cell extract (CELk). The selected drugs and organisms were taken into different combination *viz.* double and triple and their approx LD-50 doses were selected for this study. The different combinations of Ofloxacin (100 ppm), Calc sulph (30 potency), Fytalon (200 ppm), *Allium sativum* (10 per cent), CELk (100 mgl^{-1}) and *Trichoderma* sp. were taken. For double combination, 1:1 v/v ratio was used and is case of *Trichoderma* sp. dual culture method was used. In triple combination 1:1:1 v/v/v ratio was followed. The various combination of drugs mixed in oat meal agar medium before pouring except *Trichoderma* sp. Then plates were inoculated with disc of test pathogen and *Trichoderma* sp. Plates were incubated at 27±1°C for 144 hours. Plates without having any drug served as control. Each experiment was designed in triplicate form. Data were taken after 144 hours. From this, mycelial growth and percent inhibition of test pathogen were calculated.

The Effect of Various Combinations on the Leaf Rot of Pan

Healthy betelvine leaves were collected from natural habitat (bareja) and washed with running tap water and followed by sterilized distilled water. Then leaves were dipped in each combination for an hour, these were inoculated with disc of test pathogen. Before inoculation leaves were injured with sterilized needle at centre of leaves. Each experiment was performed in triplicate with suitable control. In case of *Trichoderma* sp. suspension was prepared and mixed with double combination in 1:1 v/v ratio and triple combination in 1:1:1 v/v/v ratio respectively. Then plates were incubated at 27±1°C for 144 hours. The rotten area of leaf was measured and from this growth and percent inhibition were recorded.

Result

Table 17.1 shows that drug Ofloxacin caused 33.4 per cent inhibition of mycelial growth at 100 ppm concentration; however 200 ppm concentration of this drug caused complete killing of the test pathogen. Calc sulph caused 37.5 per cent reduction of mycelial growth at 30 potency and it's 200 potency was found to suppress the growth of test pathogen completely. Fytalon, a fungicides at 200 ppm concentration caused 54.5 per cent inhibition however its 500 ppm caused complete suppression of mycelial growth of test pathogen. Extract of *A. sativum* caused 80 per cent reduction in mycelial growth at 10 per cent concentration and at 20 per cent concentration it inhibited the growth of test pathogen completely. It was found that fungal bio-agent *Trichoderma* spp. reduced 66.7 per cent mycelial growth when used as dual culture. Cyanobacterial extract CELk showed 51.2 per cent inhibition of mycelial growth at 100 mgl^{-1} whereas the complete killing of fungal growth of test pathogen was observed at 400 mgl^{-1} concentration.

Table 17.2 showed that Ofloxacin + *A. sativum* combination cause complete suppression of mycelial growth under *in vitro* and *in vivo* condition followed by Fytalon + CELk combination which caused 93.1 per cent reduction and 88.4 per cent reduction under both *in vitro* and *in vivo* condition respectively. Combination of *Trichoderma spp.* + Calc sulph caused 77.8 per cent reduction in mycelial growth *in vitro* and 80.7 per cent reduction *in vivo* condition. Trichoderma sp. + Fytalon combination showed 66.6 per cent inhibition of mycelial growth under *in vitro* condition and 65.4 per cent inhibition under *in vivo* condition.

Table 17.1: Comparison of Highly Effective Control Agents of *P. parasitica* var. *piperina*

Treatments	*LD-50*		*LD*	
	Inhibition (%)	*Concentration*	*Inhibition (%)*	*Concentration*
Ofloxacin	33.4	100	100	200 ppm
Calc. sulph	37.5	30 potency	100	200 potency
Fytalon	54.5	200 ppm	100	500 ppm
A. sativum	80.0	10 per cent	100	20 per cent
Trichoderma spp.	66.7	-	-	-
CELk	51.2	100 mgl^{-1}	100	400 mgl^{-1}

Table 17.2: Integrated Effect of Different Control Agents Against *P. parasitica* var. *piperina*

Treatments	*In vitro*		*In vivo*	
	Growth (mm)	*Inhibition (%)*	*Growth (mm)*	*Reduction (%)*
Control	90	00	52	00
Ofloxacin + *A. sativum* (100 ppm) (10 per cent)	00	100	00	100
Fytalon + CELk (200 ppm) (100 mgl^{-1})	06	93.1	06	88.4
Trichoderma sp. + Calc sulph (30 potency)	20	77.8	10	80.7
Trichoderma sp. + Fytalon (200 ppm)	30	66.6	18	65.4
CD (P = 0.05)	ND	1.35	ND	1.77

ND: Not detected.

Table 17.3 shows that combination of Ofloxacin and Fytalon caused 93.1 per cent and 88 per cent inhibition of mycelial growth of test pathogen under *in vitro* and *in vivo* condition respectively, and this was followed by the combination of Ofloxacin + CELk which caused 91.2 per cent *in vitro* and 84 per cent reduction *in vivo* condition. Combination of *A. sativum* and *Trichoderma* sp. caused 88.9 per cent reduction of mycelial growth under *in vitro* and 80 per cent inhibition under *in vivo* condition. Combination of CELk and Calc sulph caused 75.5 per cent and 70 per cent reduction of mycelial growth of test pathogen under in vitro and *in vivo* conditions.

Table 17.4 shows that Fytalon and *A. sativum* combination caused complete suppression of mycelial growth under *in vitro* condition. Interestingly the same combination of the drugs was not so much effective and caused only 88.4 per cent reduction of mycelial growth of test pathogen under *in vivo* condition. This is followed by the combination of *Trichoderma* sp. and CELk which caused 73.4 per cent reduction of mycelial growth under *in vitro* and 71.2 per cent under *in vivo* condition. When Fytalon used with Calc sulph it caused 71.2 per cent reduction of mycelial growth under *in vitro* condition and 65.4 per cent reduction *in vivo* condition. The combination of *Trichoderma* sp. + Ofloxacin exhibited 71.2 per cent and 69.2 per cent reduction under *in vitro* and *in vivo* conditions respectively.

Combination of Ofloxacin and Calc sulph was found less effective and caused only 62.2 per cent and 65.4 per cent inhibition of growth under *in vitro* and *in vivo* condition respectively.

Table 17.3: Integrated Effect of Different Control Agents against *P. parasitica* var. *piperina*

Treatments	*In vitro*		*In vivo*	
	Growth (mm)	*Inhibition (%)*	*Growth (mm)*	*Reduction (%)*
Control	90	00	50	00
Ofloxacin + Fytalon (100 ppm) (200 ppm)	06	93.1	06	88
A. sativum + *Trichoderma* sp. (10 per cent)	10	88.9	10	80
CELk + Calc sulph ($100\ mgl^{-1}$) (30 potency)	22	75.5	15	70
Ofloxacin + CELk (100 ppm) ($100\ mgl^{-1}$)	08	91.2	08	84
CD (P = 0.05)	ND	3.06	ND	3.95

ND: Not detected.

Table 17.4: Integrated Effect of Different Control Agents against *P. parasitica* var. *piperina*

Treatments	*In vitro*		*In vivo*	
	Growth (mm)	*Inhibition (%)*	*Growth (mm)*	*Reduction (%)*
Control	90	00	52	00
Fytalon + *A. sativum* (200 ppm) (10 per cent)	00	100	06	88.4
Ofloxacin + Calc sulph (100 ppm) (30 potency)	34	62.2	18	65.4
Trichoderma spp + CELk ($100\ mgl^{-1}$)	24	73.4	15	71.2
Fytalon + Calc sulph (200 ppm) (30 potency)	26	71.2	18	65.4
Trichoderma spp + Ofloxacin (100 ppm)	26	71.2	16	69.2
CD (P = 0.05)	ND	2.47	ND	3.82

ND: Not detected.

Table 17.5 shows that combination of Ofloxacin, Fytalon and Calc sulph proved highly effective among the other tested combinations. It caused 93.1 per cent reduction in mycelial growth under *in vitro* and 88 per cent under *in vivo* condition. This was followed by the combination of *A. sativum* + *Trichoderma* sp. + Calc sulph which caused 88.9 per cent and 84 per cent of mycelial growth of test pathogen under *in vitro* and *in vivo* condition respectively. Combination of *A. sativum* + CELk + Ofloxacin was also found very inhibitory causing 86.6 per cent and 80 per cent reduction of mycelial growth under *in vitro* and *in vivo* condition respectively. The combination of *A. sativum* + CELk +

Trichoderma sp. appeared to be less effective and caused 77.8 per cent reduction in mycelial growth under *in vitro* condition and 70 per cent under *in vivo* condition.

Table 17.5: Integrated Effect of Different Control Agents against *P. parasitica* var. *piperina*

Treatments	*In vitro*		*In vivo*	
	Growth (mm)	*Inhibition (%)*	*Growth (mm)*	*Reduction (%)*
Control	90	00	50	00
A. sativum and Trichoderma spp. + Calc sulph (10 per cent) (30 potency)	10	88.9	08	84
Ofloxacin + Fytalon + Calc sulph (100 ppm) (200 ppm) (30 potency)	06	93.1	06	88
A. sativum + CELk + Ofloxacin (10 per cent) (100 mgl^{-1}) (100 ppm)	12	86.6	10	80
A. sativum + CELk + *Trichoderma* sp. (10 per cent) (100 mgl^{-1})	20	77.8	15	70
CD (P = 0.05)	ND	2.03	ND	4.47

ND: Not detected.

For comparative study of control measures used against test pathogen are presented in Table 17.6.

Table 17.6: Comparative Analysis of Control Measures Used against *P. parasitica* var. *piperina*

Source	*Easily Available (M/L)*	*Eco-friendly (Yes/No)*	*Effective (Yes/No)*	*Economically Viable (Yes/No)*
ANTIBIOTICS				
Acyclovir	M,L	Yes	Yes	Yes
Amoxycillin	M,L	Yes	No	Yes
Ampicillin	M,L	Yes	Yes	Yes
Azithromycin	M,L	Yes	No	No
Cefadroxil	M,L	Yes	No	Yes
Cefixime	M,L	Yes	No	No
Cefpodoxime	M,L	Yes	Yes	No
Cefuroxime	M,L	Yes	Yes	No
Cephalexin	M,L	Yes	No	Yes
Clarithromycin	M,L	Yes	Yes	No
Clomiphene	M,L	Yes	Yes	Yes
Cloxacillin	M,L	Yes	Yes	Yes
Ofloxacin	M,L	Yes	Yes	Yes
Roxithromycin	M,L	Yes	No	Yes
Sparfloxacin	M,L	Yes	Yes	Yes
Amoxycillin + Clavulanate	M,L	Yes	No	No

Contd...

Table 17.6–Contd...

Source	*Easily Available (M/L)*	*Eco-friendly (Yes/No)*	*Effective (Yes/No)*	*Economically Viable (Yes/No)*
Amoxycillin + Carbocisteine	M,L	Yes	Yes	Yes
Ampicillin + Cloxacillin	M,L	Yes	No	Yes
Ciprofloxacin + Tinidazole	M,L	Yes	Yes	Yes
Ofloxacin + Tinidazole	M,L	Yes	Yes	Yes
FUNGICIDES				
Thiram	M,L	No	No	Yes
Bavistin	M,L	No	No	No
Fytalon	M,L	No	Yes	No
HOMOEOPATHIC DRUGS				
Kali Sulph	M,L	Yes	No	Yes
Calc Sulph	M,L	Yes	Yes	Yes
Nat Sulph	M,L	Yes	Yes	Yes
PLANT EXTRACTS				
Allium sativum	L	Yes	Yes	Yes
Allium cepa	L	Yes	Yes	Yes
Azadirechta indica	L	Yes	Yes	Yes
Cinnamomum camphora	L	Yes	Yes	Yes
Lantana camara	L	Yes	Yes	Yes
Mentha spicata	L	Yes	No	Yes
Ocimum sanctum	L	Yes	Yes	Yes
Zingiber officinale	L	Yes	Yes	Yes
BIOCONTROL AGENTS				
Trichoderma sp.	M,L	Yes	Yes	Yes
Penicillium sp.	L	Yes	No	Yes
Aspergillus sp.	L	Yes	No	Yes
MICROCYSTIS EXTRACT				
CELk	L	Yes	Yes	Yes
CELb	L	Yes	No	Yes

L: Lab; M: Market.

Discussion

To control the leaf rot of pan caused by *P. parasitica* var. *piperina* many control measure such as antibiotics, fungicides, homoeopathic drugs, plant extracts, biocontrol agents have been tried. For the purpose of IDM the best drug, plant extract, bio agents which were showing greater inhibitory activity against test pathogen were selected. Drugs, fungicides, plants extracts and bioagents were randomly selected for preparing different combinations. For avoiding unnecessary use of fungicides and antibiotics only those concentrations of these drugs were choosen which had shown 50 per cent reduction of mycelial growth when used separately.

Results obtained from the present study suggest that when chemical agents such as fungicide Fytalon was used with *A. sativum* in combination, the complete killing of the test pathogen was seen under *in vitro* condition. *A. sativum* contains allylopropyl disulphide, diallyl disulphide and two more sulphur compound including which are known to be fungitoxic in nature, allicin (Chopra *et al.*, 1956, 1969). Fytalon is a copper oxychloride and copper compounds are well known antifungal agents. Chandel and Katoch (2001) have also reported chemical control of *Fusarium oxysporum*, an incitant of carnation wilt with Fytalon. Here it appears that both fungicide and plant extract caused synergestic effect due to which test pathogen could not grow. Under *in vivo* condition however, the combination caused 88.4 per cent reduction which may also be considered a quite significant. The differential behaviour of the Fytalon and *A. sativum* under *in vitro* and *in vivo* condition is the subject of further work. Management of Tikka leaf spot of ground nut by spraying potash and Neem seed extract has been reported by Chandrasekar *et al.* (1994).

Almost similar results were obtained when Ofloxacin and *A. sativum* were used together, the only difference here was that this combination caused complete inhibition under both *in vitro* and *in vivo* conditions.

Combined effect of Ofloxacin and Fytalon showed 93.1 per cent and 88 per cent reduction of mycelial growth of test pathogen respectively. Since both antibiotic and fungicide were found to be more effective in concentration, it appears as if both induced synergestic effect against test pathogen.

However, when Fytalon was used with Calc sulph no satisfactory result could be obtained. The poor inhibitory action of this combination might be due to lower concentration of the drug or it might be quite possible that Calc sulph becomes ineffective when used in combination.

Interestingly when CELk was used in combination with Fytalon and antibiotic it reduced more or less 90 per cent mycelial growth under *in vitro* and *in vivo* conditions but when it was used in combination with Calc sulph its inhibitory activity was affected adversely and it could cause only 75 per cent reduction of the mycelial growth of test pathogen.

When *Trichoderma* sp. was used with Calc sulph it caused only 77.8 per cent and 80.7 per cent and when it was used with CELk it caused 73.4 per cent and 71.2 per cent reduction in mycelial growth under *in vitro* and *in vivo* conditions respectively. Almost similar results were obtained when *Trichoderma* sp. was used with Ofloxacin showing 71.2 per cent and 69.2 per cent reduction; when used with Fytalon it caused 66.6 per cent and 65.4 per cent reduction under both *in vitro* and *in vivo* conditions. The lesser efficacy of this combination may be attributed to the detrimental affect of all the drugs on *Trichoderma* sp. resulting poor growth of the fungal antagonist. Klassen (1981) observed the opposite result when he used *Trichoderma* sp. with (PCNB) fungicide. Pant and Mukhopadhyay (2001) reported the maximum control of seed and seedling rot complex of soyabean by using *T. harzianum* and Vitavax (fungicide). According to Sharma and Mishra (1995) *Trichoderma* sp. gained considerable importance either alone or when integrated with lower dose of fungicides for the management of soil-born plant pathogens. Dubey (2002) also reported the similar result with the combination of Thiram and *T. viride* against collar rot of french bean.

In contrast, combination of plant extract of *A. sativum* and *Trichoderma* sp. was found more effective by causing 88.9 per cent and 80 per cent reduction in the mycelial growth of test pathogen in both *in vitro* and *in vivo* conditions. This is an interesting finding because it would be expected that presence of many sulphur containing substances in *A. sativum* will inhibit the growth of *Trichoderma* spp. and it would not cause significant harm to *Phytophthora* but the result of present study suggest that the plant extract of *A. sativum* was so powerful that it not only restricted the growth of *Trichoderma* but also reduced the growth of test pathogen. Since this combination is a purely of origin. It is also a ecofriendly and highly effective. Use of this combination certainly reduces he unnecessary load of chemical drugs, which are not only hazardous to plants but also to human beings.

No triple combination was found to cause complete reduction of the growth of *Phytophthora*. Among he tested combinations best effective against test pathogen was Ofloxacin + Fytalon + Calc sulph which caused 93 per cent and 88 per cent reduction under both conditions, then followed by *A. sativum* and *Trichoderma* sp. and Calc sulph 88.9 per cent and 84 per cent suppression of pathogen; and least effective combination was *A. sativum* + CELk + Calc Sulph with 77.8 per cent and 70 per cent reduction of mycelial growth under *in vitro* and *in vivo* conditions. As such it may be summarised from these results that combination of only two, one chemical control agent and one plant extract are two drugs or two botanical agents were more effective measure for IDM. Thus on the basis of these results it may be we recommended that Ofloxacin + *A. sativum*, Fytalon + *A. sativum*, Fytalon + CELk and *A. sativum* + *Trichoderma sp.* may be successfully used for the control of betelvine *Phytophthora*. The present findings are supported by other workers who have also achieved success on integration of biological agents with fungicides to get significantly higher disease control in several crops than that obtained by either biocontrol agent or fungicide alone (Henis *et al.*, 1978; Upadhyay and Mukhopadhyay, 1983; Mukhopadhyay and Kaur, 1990; Sawant and Mukhopadhyay, 1990; Vyas, 1994; Dubey *et al.*, 1996; Dubey, 1997). Obviously if the disease can be controlled by two drugs in their low concentration then there is not need of using higher concentration. This is the only way by which practical viability of IDM can be achieved. This practice will help to avoid the anticipated environmental pollution due to the sole use of chemicals for disease management and also will nullify the probable development of fungicide resistant strains in pathogen if any, arising out of the repeated use of some fungicides for disease management.

References

Ahuja, S.C. and Kandhari, J. (2000). Integrated disease management in Rice crop. In: *Glimpses in Plant Sciences* (Eds. Aneja, K.R., Charaya, M.U., Aggarwal, A., Hans, D.K. and Khan, S.A.) Pragati Prakashan, Meerut. 153-160.

Chandel, S.S. and Katoch, R. (2001). Chemical control of *Fusarium oxysporum f.sp. dianthi* an incitant of carnation wilt. *Indian J. Microbiol.* 41: 135-137.

Chandrasekar, V., Narayanaswami, R. and Ramabardran, R. (1994). Effect of foliar spray of potash and neem seed extract on the Tikka leaf spot of groundnut. *Indian Phytopatho.* 47: 188-189.

Chopra, R.N., Nayar, S.L. and Chopra, I.C. (1956). *Glossary of Indian Medicinal Plants.* Publications and Information Directorate. C.S.I.R. New Delhi.

Correa, F. and Zeigler, R. (1991). Integrated management of rice blast disease (*P. oryzae*). *Fitopathologia Colombiana.* 15: 35-37.

Dohroo, N.P. (1995). Integrated management of yellows of ginger. *Indian Phytopath.* 48: 90-92.

Dubey, S.C. (1997). Biological control of web blight of groundnut caused by *Thanatephorus cucumeris. Golden Jubilee International Conference of IPS* at IARI, New Delhi. p. 194.

Dubey, S.C. (2002). Bio-agent based integrated management of collar rot of french bean. *Indian Phytopath.* 55: 230-231.

Ghewande, M.P., Desai, S., Prem Narayan and Ingle, A.P. (1991). An integrated approach to the management of foliar diseases of groundnut (*Arachis hypogea L.*). *Poster Abstract, Second Inter. Groundnut Workshop*, ICRISAT, Patancheru, A.P. India.

Henis, Y., Ghaffar, A. and Baker, R. (1978). Integrated control of *Rhizoctonia solani* of radish: Effect of successive plantings, PCNB and *Trichoderma harzianum* on pathogen and disease. *Phytopathol.* 69: 1164-1169.

Hooda, K.S. and Srivastava, M.P. (1995). Efficacy of different fungitoxicant and neem products in the control of rice blast. *Pestology*. 19: 41-44.

Khurana, S.M.P., Garg, I.D. Singh, B.P. and Gadewar, A.V. (1998). Major disease of potato and their management. *Integrated pest and disease management* (Eds.) Upadhyay, R.K., Mukerji, K.G., Chamola, B.P., and Dubey, O.P.. APH Publishing Corporation, New Delhi. 11-64.

Koshy, P.K., Kumar, Y., Jayasree, D., Joseph, V. and Sosamma, V.K. (2002). Integrated management of leaf rot diseases and insect pests on coconut. *Indian Phytopath*. 55: 545-56.

Kumar, D. and Dubey, S.C. (2001). Management of collar rot of pea by the integration of biological and chemical methods. *Indian Phytopath*. 54: 62-66.

Mukhopadhyay, A.N. and Kaur, N.P. (1990). Biological control of chickpea wilt complex by *Trichoderma harzianum Proc. Third Intr. Conf. on plant protection in the tropics*. Malaysia, 1990.

Mukhopadhyay, A.N., Shrestha, S.M. and Mukherjee, P.K. (1992). Biological seed treatment for control of soil born plant pathogens. *FAO Plant Prot. Bull*. 40: 21-30.

Pun, K.B., Doraiswamy, S. and Jeyarajan, R. (2000). Screening of virus inhibitory chemicals and neem products against okra yellow vein mosaic virus. *Indian Phytopath*. 53: 95-96.

Rajappan, K., Ramaraj, B. and Natarajan, S. (2000). Field evaluation of fungicides and neem formulations against bacterial blight of French bean. *Indian Phytopath*. 53: 303-304.

Sharma, S.D. and Mishra, A. (1995). Tolerance of *Trichoderma harzianum* to agrochemicals. *Indian J. Mycol. Pl. Pathol.* 25: 129.

Srivastava, M.P. (2000). Integrated plant disease management for sustainable agriculture. *Glimpses in Plant Sciences* (Eds.) Aneja, K.R., Charya, M.U., Aggarwal, A. Hans, D.K. and Khan, S.A. Pragati Prakashan, Meerut. 77-81.

Tondon, J.P. (1998). Research development and management for production of wheat. *Integrated pest and disease management* (Eds.) Upadhyay, R.K., Mukerji, K.G., Chamola, B.P. and Dubey, O.P. A.P.H. Publishing Corporation, New Delhi. 131-144.

Upadhyay, J.P. and Mukhopadhyay, A.N. (1983). Effect of non volatile and volatile antibiotics of *Trichoderma harzianum* on growth of *Sclerotium rolfsii*. *Indian J. Mycol. Pl. Pathol*. 13: 232-233.

Vadhera, I., Tiwari, S.P. and Shukla, B.N. (2001). Estimation of yield losses in tomato and integrated management of root-knot nematode. *Indian Phytopath*. 54: 495-496.

Vyas, S.C. (1994). Integrated biological and chemical control of dry root rot of Soyabean Indian *J. Mycol. Pl. Pathol*. 24: 132-134.

Microbial Biotechnology and Ecology (2011) *Pages* **303–319**
Editors: **Deepak Vyas, G.S. Paliwal, P.K.Khare and R.K Gupta**
Published by: **DAYA PUBLISHING HOUSE, NEW DELHI**

Chapter 18

Thermophilic Fungi: Occurrence, Characteristics and Enzymatic Potential

☆ *Naveen Kango**

Department of Applied Microbiology and Biotechnology
Dr. H.S. Gour University, Sagar – 470 003, M.P.

ABSTRACT

A heterogeneous group of high temperature loving molds is commonly referred as thermophilic fungi. The biotechnological potential of thermophilic molds stems from their inherent ability to utilize complex organic matter including lignocelluloses for their growth by virtue of their extracellular thermostable hydrolytic enzymes. About 30 such fungal species that infringe the upper temperature limit of eukaryotes deserve special attention. These fungi are naturally bestowed with the ability to produce several thermostable bioactive molecules. Thermostable enzymes are of special interest in industries as higher operational temperature allows better solubility of the substrate and undesirable contaminations are avoided. The present article takes a glance at the occurrence, properties and enzymatic potential of this interesting group of molds.

Introduction

Temperature is one of the key environmental variables that profoundly affects survival, growth and distribution of microorganisms in nature. Hence, temperature relationship of microorganisms is an important criterion used for their assortment to different groups namely mesophiles, psychrophiles and thermophiles. Most microorganisms are mesophilic and grow in the temperature range of 10-

* E-mail: nkango@gmail.com

45°C. Those growing at lower temperatures between 0-20°C are referred as psychrophiles while thermophilic microorganisms have ability to survive and proliferate in the habitats of high temperature regimes (45-100°C) such as hot springs, composts, geothermal soils, storages, bird's nests, woodchip piles etc. Extreme thermophily is witnessed in eubacteria or archaea, some of which are able to grow near or above 100°C in thermal springs, solfatara fields, or hydrothermal vents (Tansey and Brock, 1978; Brock, 1995; Blochl *et al.*, 1997). Among fungi, only a few species have the ability to thrive at temperatures between 45-55°C. Such fungi comprise thermophilic and thermotolerant forms, which are arbitrarily distinguished on the basis of their minimum and maximum temperature of growth (Cooney and Emerson, 1964). Thermophilic fungi have a growth temperature maximum at or above 50°C and a minimum at or above 20°C while thermotolerant fungi are distinguished from the above on the basis of their lower temperature limit of growth, allowing them to grow below 20°C. Considering that the vast majority of eukaryotes cannot survive prolonged exposure to temperatures above 45°C, the ability of some fungal species to infringe this upper temperature limit is an important phenomenon (Alexandrov, 1977; Maheshwari *et al.*, 2000). Thermotolerant fungi are the chief components of the microflora that develops in heaped masses of plant material, piles of agricultural and forestry products, and other accumulations of organic matter wherein the self-heating, aerobic and organically rich conditions provide a conducive atmosphere for the development of thermophilic mycoflora in succession (Allen and Emerson, 1949; Cooney and Emerson, 1964). High temperature loving fungi constitute a

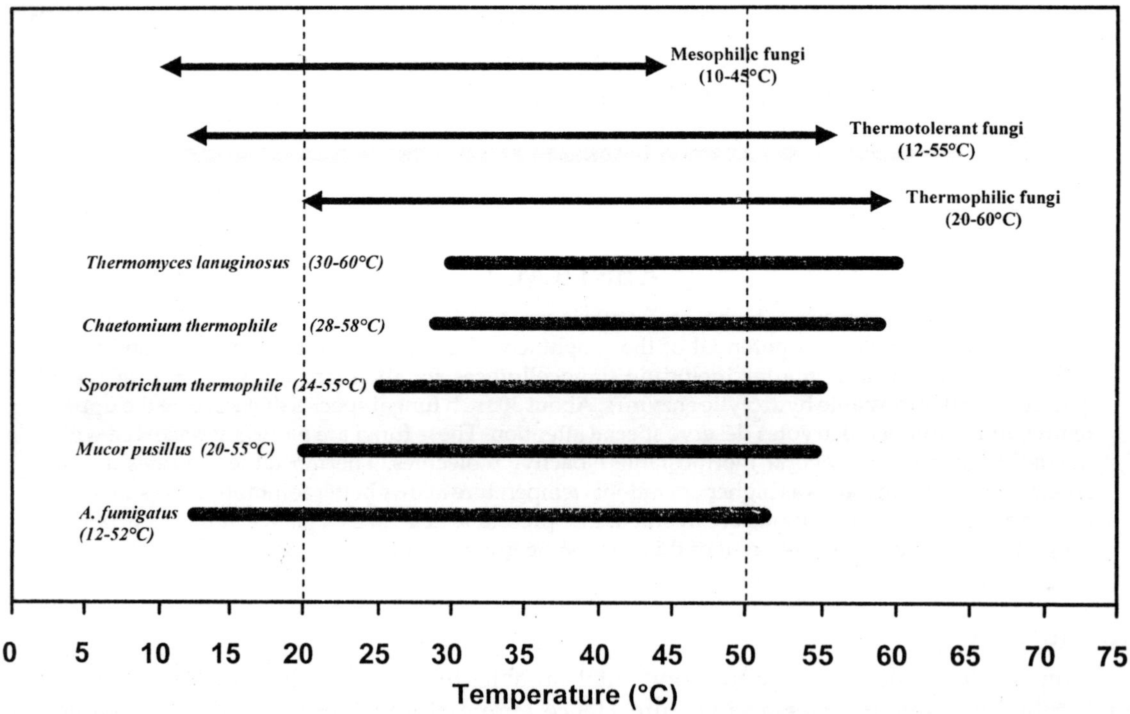

Figure 18.1: Range of Growth Temperature of Various Groups of Fungi and Some Typical Thermophilic and Thermotolerant (*A. fumigatus*) Forms (Modified from Cooney and Emerson, 1964).

heterogeneous group of moulds consisting of a large number of morphologically and physiologically different forms including representatives from zygomycetes, ascomycetes, basidiomycetes and fungi imperfecti, having in common certain biochemical and ecological characteristics associated with their ability to grow at elevated temperatures (Mouchacca, 1997). Figure 18.1 shows range of temperature for growth of mesophilic, thermophilic and thermotolerant fungi. True thermophilic molds fail to grow below 20°C while *A. fumigatus*, being thermotolerant, is able to grow at 12°C. Over the years different workers have proposed terminologies and definitions to express temperature relations of high temperature loving fungi. Some of these are as follows:

Thermophilic Fungi

La Touche (1950) suggested this term to designate those fungi that can grow and reproduce between 40-50°C. Cooney and Emerson (1964) used this term for fungi that have maximum growth temperature at or above 50°C and minimum at or above 20°C while Crisan (1973) considered all those forms growing above 40°C as thermophilic. Abdel-Hafez *et al.* (1983) included all those fungi in this category that could grow in temperature range of 45-60°C with optimum growth at 45°C.

Thermotolerant Fungi

This term was used by Cooney and Emerson (1964) for fungi with maximum temperature limit for growth at or above 50°C but could grow below 20°C.

Transition Thermophiles

Apinis and Pugh (1967) applied this term to the fungi that have ability to grow below 20°C but can also withstand high temperatures approaching 40°C.

Eurythermal Fungi

These include fungi that have 50°C as optimum growth temperature while their minimum lies between 28-40°C (Henssen, 1957).

Sterothermal Fungi

Moulds that have an optimum growth temperature at 50°C but can also grow well at higher temperatures (Henssen, 1957).

Thermophilous Fungi

Johri (1980) while enumerating fungi living at high temperature has used this term to include both thermophilic and transitional mesophilic moulds.

Although breaching of defined temperature limits, specially with changing nutritional conditions, has been observed in many cases, the definition of thermophilic moulds as given by Cooney and Emerson (1964) provides a good working basis and holds appropriate to most fungi thriving at high temperatures (Cole, 1980).

Prevalence of these moulds is anticipated in habitats where temperature regime approaches their optimum and higher *i.e.* 45-60°C. Early reports of their occurrence were made from composts wherein the daily temperature during the peak heating hours reaches 65-70°C (La Touche, 1950; Fergus, 1964; Fergus and Sinden, 1969; Chang and Hudson, 1967; Hayes, 1969; Jodice *et al.*, 1975; Moubasher *et al.*, 1982; Satyanarayana and Johri, 1983). Stored grains likewise provide an environment where internal heating leads to regular appearance of thermophilic moulds (Flanningan, 1969, Mulinge and Apinis, 1969; Wallace *et al.*, 1986; Jain, 1999).

Occurrence of Thermophilic Fungi

Isolation of thermophiles from soil was reported as early as 1939 and it still remains to be an ideal niche for thermophilic moulds (Waksman and Cordon, 1939; Waksman *et al.*, 1939). Sun-heated soils remain insolated during day hours and this causes rise in its temperature. Tansey and Jack (1977) have reported isolation of thermophilic moulds from sun heated soils whereas other soils including hot spring soils (Tansey and Brock, 1971; Tansey and Brock, 1972), tropical and subtropical soils (Eggins *et al.*, 1972; Udagawa *et al.*, 1973; Morgan-Jones, 1974, Tansey and Jack, 1977; Rajasekaran and Maheshwari, 1993), coal mine soils (Thakre and Johri, 1976), pasture soils (Eggins and Malik, 1969) and those of temperate and cold regions are also known to lodge a good number of these moulds (Ellis, 1980).

Bird nests act as incubators providing ambient temperature for thermophilic moulds (Apinis and Pugh, 1967; Tansey, 1973; Satyanarayana *et al.*, 1977). Stored peat and oil palm kernels have also been reported to harbor a variety of these moulds (Eggins and Coursey, 1964; Kuster and Locci, 1964). Retting guayule (Cooney and Emerson, 1964), industrial wood chip piles (Hajny, 1966; Ofosu-Asiedu and Smith, 1973; Greaves, 1971), moldy hay (Gregory *et al.*, 1963), sugarcane bagasse (Lacey, 1974), litter (Johri, 1980), saw dust (Singhai and Jain, 1992) etc. are certain more habitats in which thermophilic moulds proliferate and play important role. Kango *et al.* (2003a) have reported occurrence of 48 fungal forms belonging to 13 different genera and 18 species of thermophilic and thermotolerant fungi from soil, wheat straw, compost, wood chips, litter, bird nesting materials and storage seeds.

The ubiquitous nature of thermophilic moulds is well established as evidenced by numerous reports of their occurrence from different parts of the world. Subrahmanyam (1999) has reviewed the ecology of thermophilic moulds and has given distribution of more than ninety species belonging to forty-two genera of such moulds over 18 countries including Africa, Antarctica and sub-Antarctica, Australia, Germany, Poland, India, Indonesia, Italy, Japan, The Netherlands, New Zealand, Malaysia, Pakistan, Republic of Honduras, Saudi Arabia, United Kingdom, USA and Russia. Amongst documented species, a maximum of 52 have been found in India and it seems to have more diversity of these moulds in comparison to other countries. Other countries having relatively more accounts of isolation of different thermophilic moulds include USA, UK and Africa with reports of harboring a total of 37, 33 and 25 species, respectively. *Aspergillus fumigatus, Chaetomium thermophile, Humicola insolens, Malbranchea pulchella* var. *sulfurea, Mucor pusillus, Sporotrichum thermophile, Talaromyces emersonii, Thermoascus aurantiacus, Thermomyces lanuginosus* are relatively more prevalent thermophilic moulds than others (Subrahmanyam, 1999). On behalf of high temperature requirement of these moulds it is anticipated that they must be more prevalent in tropics. However, several records of thermophilic fungi from colder areas including Arctic and Antarctic regions indicate their surprisingly successful survival in extreme cold. Moreover, soil samples from Brazil, Honduras and Costa Rica and those of USA and UK are reported to contain more or less similar thermophilic mycoflora.

Growth Requirements

Most of the thermophilic fungi can grow on simple media containing carbon and nitrogen sources with few mineral salts (Cooney and Emerson, 1964; Rosenberg, 1975; Deploey, 1976; Sahm and

Chapman, 1976; Satyanarayana and Johri, 1984) while many of these have been found to grow on a wide range of carbon sources including simple sugars, organic acids and complex polymers such as starch, cellulose, hemicellulose, lignin and pectin (Subrahmanyam, 1980; Satyanarayana and Johri, 1984; Satyanarayana *et al.*, 1992; Gupta and Maheshwari, 1985). They proliferate in decaying organic matter during high-temperature phase succeeding mesophilic generation (Hedger and Hudson, 1974). At this point of time, initially available simple C-sources are depleted and the C-source available for the growth of thermophilic fungi is mainly polysaccharide component of lignocellulosic biomass. Interestingly, many compost fungi such as *Thermomyces lanuginosus*, *Talaromyces duponti*, *Mucor pusillus*, *Melanocarpus albomyces* etc. are reported to be non-cellulolytic. These species grow commensally by utilizing sugars produced by cellulolytic partners *e.g. Chaetomium thermophile* (Hedger and Hudson, 1974). Some of thermophilic fungi, namely *T. lanuginosus*, *Humicola insolens* etc. rapidly utilize xylan as an accessible C-source (Chang, 1967) while some cellulolytic thermophiles such as *Sporotrichum thermophile* could utilize cellulose as efficiently as glucose (Gaikwad and Maheshwari, 1994). Concurrent utilization of mixed C-substrates has also been indicated in *T. lanuginosus* and *Penicillium duponti* (Maheshwari and Balasubramanyam, 1988).

Simple N-sources such as nitrates and asparagine support good growth of these moulds (Satyanarayana and Johri, 1984) while protein hydrolysis by growing cultures of some thermophilic moulds, as monitored by pulse-labeled protein, has been reported to be twice that of mesophilic fungi suggesting high turnover of soluble protein in cultures of thermophilic fungi (Rajasekaran and Maheshwari, 1990).

Thermophilic fungi inhabiting different environmental conditions in different habitats may show a great variation in their temperature requirement for their optimum growth. Whether different geographical isolates of same fungal species inhabiting different habitats possess same temperature growth relations is a subject of active interest and enquiry. It has been suggested that subtropical strains of a particular fungus, may have higher optimum growth temperatures than strains from temperate regions and hence geographical isolates of the same species may differ in their temperature-growth relationships and possibly other physiological aspects (Maheshwari *et al.*, 2000). Thermostability in enzymes of thermophilic moulds is an important characteristic and most of them have optimal activity between 40-55°C (Liu *et al.*, 1974; Somkuti and Babel, 1968a, b; Ogundero, 1980; Craveri *et al.*, 1974). Enzymes with carbohydrate moiety (glycoproteins) *e.g.* xylanases and cellulases have been found to have temperature optima in the range of 50-70°C (Matsuo *et al.*, 1977; Margaritis and Creese, 1981; Vandamme *et al.*, 1982) and removal of carbohydrate residue reduces their thermal stability suggesting important role played by carbohydrate moiety in stabilizing their conformation at elevated temperatures.

Enzymatic Potential

The desire of identifying fungi capable of secreting high levels of thermostable enzymes with high temperature optima and a long shelf life has stimulated many workers to explore thermophilic moulds for industrial usage. The adaptability of thermophilic moulds to various habitats heterogeneous in terms of temperature, nutrients, gases, water activity (a_w) and saprophytic microbiota indicates their potentiality desirable in industrial strains. Thermophilic moulds possess a comprehensive range of extracellular enzymes as evidenced by their growth in plant organic matter where they degrade complex polymers such as starch, protein, pectin, hemicellulose and lignin (Emerson, 1968; Johri and Satyanarayana, 1986; Maheshwari *et al.*, 2000).

Several enzymes from these moulds including protease, lipase, α-amylase, glucoamylase, cellulase, cellobiose dehydrogenase, glucuronidase, pectinase, lactase, laccase, phytase have been found to exhibit better thermal stabilities as compared to their mesophilic counterparts (Berka *et al.*, 1997; 1998). However, their thermostability is much lower to that of enzymes from hyperthermophilic bacteria (Adams, 1993) but the optimal temperature of enzymes from hyperthermophilic bacteria is too high to suit realistic operational temperature range in industries *i.e.* 50-65°C (Maheshwari *et al.*, 2000).

Proteases have long been used in food, dairy, detergent and leather industries. A variety of thermophilic moulds have been reported to produce thermostable proteases. These include *Mucor pusillus* (Arima *et al.*, 1967; Somkuti and Babel, 1968a), *M. miehei* (Ottesen and Rickert, 1970), *Penicillium duponti* (Hashimoto *et al.*, 1972), *Malbranchea pulchella* var. *sulfurea* (Ong and Gaucher, 1976) and *Humicola lanuginosa* (Shenolikar and Stevenson, 1982).

Lipases (EC 3.1.1.3) catalyzing hydrolysis of triacylglycerides are also used in laundry detergents where they are expected to remain active in presence of surfactants at 30-60°C. These have been characterized from *Mucor pusillus, Humicola lanuginosa, Rhizomucor miehei* (Somkuti and Babel, 1968b; Arima *et al.*, 1972; Huge-Jensen *et al.*, 1989). *H. lanuginosa* lipase (Lipolase, Novo industries) is being used in detergent formulations in conjunction with other microbial enzymes. A recombinant plasmid containing the cDNA for *R. miehei* lipase was successfully expressed in *Aspergillus oryzae* and large amounts of this enzyme without any alteration in thermal properties could be produced (Huge-Jensen *et al.*, 1989). Similarly, *H. lanuginosa* lipase gene has also been cloned in *A. oryzae* (Arima *et al.*, 1972).

Enzymatic hydrolysis of glycosidic linkages of starch using amylases finds main application in food and textile industries. α-Amylase (EC 3.2.1.1) have been produced from several thermophilic moulds *viz. Rhizomucor pusillus, H. lanuginosa* (Adams, 1994), *Mucor pusillus, M. miehei* (Adams and Deploey, 1976), *Humicola* sp. (Barnett and Fergus, 1971), *Talaromyces emersonii* (Bunni *et al.*, 1989), *Thermoascus aurantiacus* (Jaychandran and Ramabadran, 1970) and *Myceliophthora thermophila* D14 (Sadhukhan *et al.*, 1992) while strains of *T. lanuginosus* have been shown to produce both α-amylase and glucoamylase (EC 3.2.1.3) (Rao *et al.*, 1979; Arnesen *et al.*, 1998).

Enzymatic conversion of cellulose into glucose using fungal cellulases (EC 3.2.1.4) has multifarious prospects since glucose is the most readily utilized source of energy and carbon. Several thermophilic moulds used as source of this enzyme include *Sporotrichum thermophile* (Coutts and Smith, 1976), *Talaromyces emersonii* (Folan and Coughlan, 1978), *Thermoascus aurantiacus* (Kawamori *et al.*, 1987), *Humicola insolens* (Hayashida and Yoshioka, 1980) and *Humicola grisea* var. *thermoidea* (Yoshioka *et al.*, 1982). Pectin hydrolyzing enzyme, polygalactouronase (EC 3.2.1.15) cleaves α-1,4 linkages of pectin. Among all pectinases, polygalactouronase of *T. aurantiacus* has been found to be most stable exhibiting a temperature optima of 65°C and a half life of 20 minutes at 55°C (Inamdar, 1987).

Laccase (EC 1.10.3.2) activity is believed to be significant in polymerization of phenolic substrates to humic substances. Laccase catalyze oxidation of phenolic compounds coupled with reduction of oxygen to water. Among thermophilic moulds laccase has been characterized from *Myceliophthora thermophila* (Berka *et al.*, 1997) and *Chaetomium thermophilum* (Chefetz *et al.*, 1998). The gene encoding laccase of *M. thermophila* was cloned and expressed in mesophilic *A. oryzae*. Laccase from *C. thermophilium* was found to have half-life of 12 hours at 50°C and was active over broad pH range of 5-12.

Table 18.1: Characteristics of Some Industrially Important Thermostable Glycosidases Produced by Thermophilic Fungi

Thermophilic Fungi		*M(r) kDa*	*Optimum*		*Reference*
			pH	*T (°C)*	
		Endo-glucanase (EC 3.2.1.4)			
Chaetomium thermophile var. coprophile		36	6.0	60	Ganju *et al.*, 1990.
Humicola grisea var. thermoidea		63	5.0	–	Hayashida and Mo, 1986
Myceliophthora thermophila		100	4.8	65	Roy *et al.*, 1990
Thermoascus aurantiacus		32	2.9	76	Khandke *et al.*, 1989
		β-glucosidase (EC 3.2.1.21)			
Sporotrichum thermophile		240	5.4	65	Bhat and Bhat, 1997
Humicola grisea var. thermoidea		156	4.0-4.5	60	Filho, 1996
Humicola insolens		250	5.0	50	Hayashida *et al.*, 1988
Humicola lanuginosa		110	4.5	60	Anand and Vithayathil, 1989
Thermoascus aurantiacus		15.1	4.2	71	Khandke *et al.*, 1989
		α-amylase (EC 3.2.1.1)			
Thermomyces lanuginosus			4.6-5.2	60	Jensen and Olsen, 1992
Scytalidium sp.		87	6.5	50	Odibo *et al.*, 1992
Rhizopus		64	4.0-6.0	60-65	Siqueria *et al.*, 1997
		Glucoamylase (EC 3.2.1.3)			
Humicola lanuginosa		–	4.9-6.6	65, 70	Taylor *et al.*, 1978
Thermomyces lanuginosus		37	–	70	Rao *et al.*, 1981
		Xylanase (EC 3.2.1.8)			
Humicola insolens		6, 21	6-6.5	55-60	Rana *et al.*, 1996
Malbranchea pulchella var. *sulfurea* No. 48		--	6-6.5	70	Matsuo and Yasui, 1985
Melanocarpus albomyces		24, 38	5.6, 6.6	65	Prabhu and Maheshwari, 1999
Paecilomyces variotii		20	4	50	Kelly *et al.*, 1989
Thermomyces lanuginosus SSBP		23.6	6.5	70-75	Lin *et al.*, 1999
Thermomyces lanuginosus		25.5	7	60-70	Cesar and Mrsa, 1996
Thermomyces lanuginosus		32	5.1	80	Gomes *et al.*, 1993
Chaetomium thermophile	Xyl I	26	4.8-6.0	70	Ganju *et al.*, 1989
	Xyl II	7	5.4-6.0	60	

Spurt of interest in identifying thermostable xylanases for biobleaching of pulp in paper industries to reduce the burden of residual organochloro compounds due to chemical bleaching has lead many workers to investigate thermophilic moulds (Kango and Jain, 2005). Xylanolytic enzymes include endoxylanase, β-xylosidase, arabino-furanosidase, acetyl xylan esterase, glucuronidase and exo-xylanase. Among these, endoxylanase (EC 3.2.1.8) is understood to be the chief enzyme hydrolyzing the backbone of xylan macromolecule and therefore plays a central role in xylan depolymerization. Applications of xylanases include saccharification of lignocelluloses, clarification of fruit juices and wines, enhancement of quality and nutritional value of feed, fibre upgradation etc. (Kango, 2007). Biobleaching using microbial xylanases that replaces toxic chlorine has been realized as a major application of these enzymes. Development of an efficient process for xylan hydrolysis using microbial xylanases opens up novel prospects for xylan utilization where the hydrolysate can be utilized for the production of fuel, SCP and other value-added products. *Humicola insolens*, a thermophilic mould, secreted two xylanases of molecular weight equal to 6 kDa and 21 kDa (Dusterhoft *et al.*, 1997). Similarly, *H. grisea* var. *thermoidea* also secreted two xylanases having different low molecular weights *i.e.* 13 kDa and 73 kDa (Rana *et al.*, 1996). Two different strains of *Melanocarpus albomyces* have been shown to produce thermostable xylanases with distinct characteristics (Prabhu and Maheshwari, 1999; Saraswat and Bisaria, 2000). Strains of *Thermomyces lanuginosus*, a potent xylanolytic fungi elaborating high levels of thermostable xylanase have been shown to produce single form of xylanase in the range of 20-30 kDa (Lin *et al.*, 1999) while *T. lanuginosus* BS1 strain has been reported to secrete two xylanases of 25 kDa and 54 kDa (Chadha *et al.*, 1999). *E. nidulans* NK-62 and *T. lanuginosus* NK-2 have been reported to produce higher titres of thermostable xylanase, 362 U/ml and 1418 U/ml, respectively utilizing wheat bran (Kango *et al.*, 2003b; Kango *et al.*, 2006).

Along with the above said secretory enzymes, many useful intracellular enzymes have also been noticed in thermophilic moulds. Characterization of intracellular pentose phosphate pathway enzyme, glucose-6-phosphate (EC 1.1.1.49) has been done from *Penicillium duponti* (Broad and Shepherd, 1970) while malate dehydrogenase (TCA cycle) could be isolated from *H. lanuginosa* and *M. pusillus* (Yang *et al.*, 1997). Intracellular trehalose (EC 3.2.1.28) acts upon trehalose, which is believed to provide stability to membranes and enzymes against drying and thermal denaturation (Crowe *et al.*, 1984; Colaco *et al.*, 1992), has been isolated from *T. lanuginosus* (Prasad and Maheshwari, 1978), *H. grisea* var. *thermoidea* (Cardello *et al.*, 1994) and *Scytalidium thermophilum* (Kadowaki *et al.*, 1996). Endocellular invertase and phosphodiesterase have also been characterized from thermophilic *T. lanuginosus* and *T. duponti*, respectively (Fontana *et al.*, 1974; Maheshwari *et al.*, 1983). Other intracellular enzymes of thermophilic moulds include aminopeptidase (Chapuis and Zuber, 1970), ATP sulfurylase (Renosto *et al.*, 1985), galactosidases (Arnaud *et al.*, 1976; Fischer *et al.*, 1995), β-D-xylosidase (Almeida *et al.*, 1995) etc.

Enzymes having high temperature optima and thermostability, as in case of thermophilic moulds, are better suited to industrial applications. This fact is substantiated by a number of available patents disclosing classified applications of enzymes from thermophilic moulds (Fangerstrom *et al.*, 1997; Bruins *et al.*, 2001). Looking at their ubiquity, ease of isolation and culture, minimal growth requirements and broad spectra of thermostable enzymes, thermophilic moulds appear to be good candidates for potential biotechnologies.

References

Abdel-Hafez, S.I.I., Aboel-Hafez, A.H. and Abdel-Kader, M.I.A. (1983). Composition of fungal flora of Syrian-soils 4. Thermophilic fungi. *Mycopathol.* 81: 177-182.

Adams, M. W. W. (1993). Enzymes and proteins from organisms that grow near and above 100°C. *Annu. Rev. Microbiol.* 47: 627-658.

Adams, P. R. (1994). Extracellular amylase activities of *Rhizomucor pusillus* and *Humicola lanuginosa* at initial stages of growth. *Mycopathol.* 128: 139-141.

Adams, P. R. and Deploey, J.J. (1976). Amylase production by *Mucor miehei* and *M. pusillus*. *Mycologia* 68: 934-938.

Alexandrov, V. Y. (1977). *Cells, Molecules and Temperature*. Translated from the Russian by V. A. Bernstam. Springer-Verlag KG, Berlin, Germany.

Allen, P. J., and Emerson, R. (1949). Guayule rubber, microbiological improvement by shrub retting. *Ind. Eng. Chem.* 41: 346-365.

Almeida, E. M. de, Lourdes, M. de, Polizeli, T. M., Terenzi, H. F. and Jorge, J. A.(1995). Purification and biochemical characterization of β-xylosidase from *Humicola grisea* var. *thermoidea*. *FEMS Microbiol. Lett.* 130: 171-176.

Anand, L. and Vithayathil, P. J. (1989). Purification and properties of β-glucosidase from a thermophilic fungus *Humicola lanuginosa* (Griffon and Maublanc) Bunce. *J. Ferment. Bioeng.* 67: 380-386.

Apinis, A.E. and Pugh, G. J. F. (1967). Thermophilic fungi of bird's nest. *Mycopathol. Mycol. Appl.* 33: 1-9.

Arima, K., Iwasaki, S. and Tamura, G. (1967). Milk clotting enzyme from microorganisms I. Screening test and the identification of the potent fungus. *Agric. Biol. Chem.* 31: 540-545.

Arima, K., Liu, W.H. and Beppu, T. (1972). Studies on the lipase of thermophilic fungus *Humicola lanuginosa*. *Agric. Biol. Chem.* 36: 893-895.

Arnaud, N., Bush, D. A. and Horisberger, M. (1976). Study of an intracellular α-galactosidase from the thermophilic fungus *Penicillium duponti*. *Biotechnol. Bioeng.* 18: 581-585.

Arnesen, S., Eriksen, S.H., Olsen, J. and Jensen, B. (1998). Increased production of α-amylase from *Thermomyces lanuginosus* by the addition of Tween 80. *Enzyme Microb. Technol.* 23: 249-252.

Barnett, E. A. and Fergus, C. L. (1971). The relation of extracellular amylase, mycelium, and time, in some thermophilic and mesophilic *Humicola* species. *Mycopathol. Mycol. Appl.* 44: 131-141.

Berka, R. M., Schneider, P., Golightly, E. J., Brown, S. H., Madden, M., Brown, K. M., Halkier, T., Mondorf, K. and Xu, F. (1997). Characterization of the gene encoding an extracellular laccase of *Myceliophthora thermophila* and analysis of the recombinant enzyme expressed in *Aspergillus oryzae*. *Appl. Environ. Microbiol.* 63:3151–3157.

Berka, R. M., Rey, M. W., Brown, K. M., Byun, T. and Klotz, A. V. (1998). Molecular characterization and expression of a phytase gene from the thermophilic fungus *Thermomyces lanuginosus*. *Appl. Environ. Microbiol.* 64: 4423-4427.

Bhat, M. K., and Bhat, S. 1997. Cellulose degrading enzymes and their potential industrial applications. *Biotechnol. Adv.* 15: 583-620.

Blochl, E., Rachel, R., Burggraf, S., Hafenbradl, D., Jannasch, H. W. and Stetter, K. O. (1997). *Pyrolobus fumarii*, gen. and sp. nov., represents a novel group of archaea, extending the upper temperature limit for life to 113°C. *Extremophiles* 1: 14-21.

Broad, T. E. and Shepherd, M. G. (1970). Purification and properties of glucose-6-phosphate dehydrogenase from the thermophilic fungus *Penicillium duponti. Biochim. Biophys. Acta* 198: 407-414.

Brock, T. D. (1995). The road to Yellowstone and beyond. *Annu. Rev. Microbiol.* 49: 1-28.

Bruins, M.E., Janssen, A.E.M. and Boom, R.M. (2001). Thermozymes and their applications-A review of recent literature and patents. *Appl. Biochem. Biotechnol.* 90: 155-186.

Bunni, L., McHale, L. and McHale, A. P. (1989). Production, isolation and partial characterization of an amylase system produced by *Talaromyces emersonii* CBS 814.70. *Enzyme Microb. Technol.* 11: 370-375.

Cardello, L., Terenzi, H. F. and Jorge, J. A. (1994). A cytosolic trehalase from the thermophilic fungus *Humicola grisea* var. *thermoidea. Microbiology* 140:1671-1677.

Cesar, T. and Mrsa, V. (1996). Purification and properties of the xylanase produced by *Thermomyces lanuginosus. Enzyme Microb. Technol.* 19: 289-296.

Chadha, B.S., Jaswinder, K., Rubinder, K., Saini, H.S. and Singh, S. (1999). Xylanase production by *Thermomyces lanuginosus* wild and mutant strains. *World J. Microbiol. Biotechnol.* 15: 217-221.

Chang, Y. (1967). The fungi of wheat straw compost. II. Biochemical and physiological studies. *Trans. Br. Mycol. Soc.* 50:667–677.

Chang, Y. and Hudson, H.J. (1967). Fungi of wheat straw compost I. Ecological studies. *Trans. Br. Mycol. Soc.* 50: 649-666.

Chapuis, R. and Zuber, H. (1970). Thermophilic aminopeptidases: AP I from *Talaromyces duponti. Methods Enzymol.* 19: 552-555.

Chefetz, B., Chen, Y. and Hadar, Y. (1998). Purification and characterization of laccase from *Chaetomium thermophilium* and its role in humification. *Appl. Environ. Microbiol.* 64: 3175-3179.

Colaco, C., Sen, S., Thangavelu, M., Pinder, S. and Roser, B. (1992). Extraordinary stability of enzymes dried in trehalose: simplified molecular biology. *Biotechnol.* 10: 1007-1011.

Cole, A.L.J. (1980). Thermophilic fungi. In: *Recent advances in Biology of Microorganisms* (Bilgrami, K.S. and Vyas, K.M., Eds.) Bishen Singh Mahendrapal singh, Dehradun, pp. 113-118.

Cooney, D. G. and Emerson, R. (1964). *Thermophilic Fungi: An Account of Their Biology, Activities and Classification*. W. H. Freeman and Company, San Francisco, p188.

Coutts, A. D. and Smith, R. E. (1976). Factors influencing the production of cellulases by *Sporotrichum thermophile. Appl. Environ. Microbiol.* 31: 819-825.

Craveri, R., Manachini, P.L. and Aragozini, F. (1974). Thermostable and alkaline lipolytic and proteolytic enzymes produced by a thermophilic mold. *Mycopathol. Mycol. Appl.* 54: 93-204.

Crisan, E.V. (1973). Current concepts of thermophilism and thermophilic fungi. *Mycologia* 63: 1171-1198.

Crowe, J. H. Crowe, L. M. and Chapman, D. (1984). Preservation of membranes in anhydrobiotic organisms: the role of trehalose. *Science* 223: 701-703.

Deploey, J.J. (1976). Carbohydrate nutrition of *Mucor miehei* and *Mucor pusillus*. *Mycologia* 68: 190-194.

Dusterhoft, E.M., Linssen, V.A.J.M., Voragen, A.G.J. and Beldman, G. (1999). Purification, characterization and properties of two xylanases from *Humicola insolens*. *Enzyme Microb. Technol.* 20: 437-445.

Eggins, H.O.W. and Coursey, D.G. (1964). Thermophilic fungi associated with Nigerian oil palm produce. *Nature* 203: 1081-1084.

Eggins, H.O.W. and Malik, K.A. (1969). The occurrence of thermophilic and cellulolytic fungi in pasture land soil. *Antonie Van Leeuvenhoek* 354 : 178-184.

Eggins, H.O.W., Szilvinyi, A. and Allsopp, D. (1972). The isolation of actively growing thermophilic fungi from insolated soils. *Int. Biodet. Bull.* 8: 53-58.

Ellis, D.H. (1980). Thermophilous fungi isolated from some Antarctic and sub-Antarctic soils. *Mycologia* 72: 1033-1036.

Emerson, R. (1968). Thermophilic fungi. In: *The Fungi*, Vol 3. (Anisworth, G.C. and Sussman, A.S., Eds.), Academic Press, New York. pp. 105-128.

Fangerstrom, R., Paloheimo, M., Lantto, R., Lahtinen, T. and Suominen, P. (1997). *Novel xylanases and their uses*. (Assignees: Primalco, Finland) Patent No. WO 9722691.

Fergus, C.L. (1964). Thermophilic and thermotolerant moulds in mushroom compost during peak heating. *Mycologia* 56: 267-284.

Fergus, C.L. and Sinden, J.W. (1969). A new thermophilic fungus from mushroom compost *Thielavia thermophila* sp. Nov. *Can. J. Bot.* 47: 1635.

Filho, E. X. F. (1996). Purification and characterization of a β-glucosidase from solid-state cultures of *Humicola grisea* var. *thermoidea*. *Can. J. Microbiol.* 42:1-5.

Fischer, L., Scheckermann, C. and Wagner, F. (1995). Purification and characterization of a thermotolerant β-galactosidase from *Thermomyces lanuginosus*. *Appl. Environ. Microbiol.* 61: 1497-1501.

Flanningan, B. (1969). Microflora of barley grains. *Trans. Br. Mycol. Soc.* 53: 371-379.

Folan, M. A. and Coughlan, M. P. (1978). The cellulase complex in the culture filtrate of the thermophilic fungus, *Talaromyces emersonii*. *Int. J. Biochem.* 9: 717-722.

Fontana, A., Boccu', E. and Veronese, F. M. (1974). Purification of a phosphodiesterase from the thermophilic fungus *Talaromyces duponti*. Hoppe-Seyler's Z. *Physiol. Chem*. 355: 1436-1442.

Gaikwad, J. S. and Maheshwari, R. (1994). Localization and release of b-glucosidase in the thermophilic and cellulolytic fungus, *Sporotrichum thermophile. Exp. Mycol.* 18: 300-310.

Ganju, R. K., Vithayathil, P. J. and Murthy, S. K. (1989). Purification and characterization of two xylanases from *Chaetomium thermophile* var. coprophile. *Can. J. Microbiol.* 35: 836-842.

Ganju, R. K., Murthy, S. K. and Vithayathil, P. J. (1990). Purification and functional characteristics of an endocellulase from *Chaetomium thermophile* var. *coprophile. Carbohydr. Res.* 197: 245-255.

Gomes, J., H. Purkarthofer, M. Hayn, J. Kapplmuller, M. S. and Steiner W. (1993). Production of a high level of cellulase-free xylanase by the thermophilic fungus *Thermomyces lanuginosus* in laboratory and pilot scales using lignocellulosic materials. *Appl. Microbiol. Biotechnol.* 39: 700-707.

Greaves, H. (1971). Biodeterioration of tropical hardwood chips in outdoor storage. *TAPPI* 54: 1128-1133.

Gregory, P.H., Lucey, M.E., Festenstein, G.N. and Skinner, F.A. (1963). Microbiological and biochemical changes during molding of hay. *J. Gen. Microbiol.* 33: 147-174.

Gupta, S. D. and Maheshwari, R. (1985). Is organic acid required for nutrition of thermophilic fungi? *Arch. Microbiol.* 141: 164-169.

Hajny, G.J. (1966). Outside storage of pulp wood chips, a review and bibliography. *TAPPI*. 49: 97-109.

Hashimoto, H., Iwaasa, T. and Yokotsuka, T. (1972). Thermostable acid protease produced by *Penicillium duponti* K1014, a true thermophilic fungus newly isolated from compost. *Appl. Microbiol.* 24: 986-992.

Hayashida, S. and Yoshioka, H. (1980). Production and purification of thermostable cellulases from *Humicola insolens* YH-8. *Agric. Biol. Chem.* 44: 1721-1728.

Hayashida, S. and Mo., K. (1986). Production and characteristics of aviceldisintegrating endoglucanase from a protease-negative *Humicola grisea* var. *thermoidea* mutant. *Appl. Environ. Microbiol.* 51:1041-1046.

Hayashida, S., K. Ohta, and Mo. K. (1988). Cellulases of *Humicola insolens* and *Humicola grisea. Methods Enzymol.* 160: 323-332.

Hayes, W. A. (1969). Microbiological changes in composting wheat straw horse manure mixtures. *Mush. Sci.* 7: 173-186.

Hedger, J. N. and Hudson, H. J. (1974). Nutritional studies of *Thermomyces lanuginosus* from wheat straw compost. Trans. *Br. Mycol. Soc.* 62: 129-143.

Henssen, A. (1957). Uber die Bedeutung der thermophilen Mikro-organismen fur die Zersetzung des Stallmistes. *Arch. Mikrobiol.* 27: 63-81.

Huge-Jensen, B., Andreasen, F., Christensen, T., Christensen, M. Thim, L. and Boel, E. (1989). *Rhizomucor miehei* triglyceride lipase is processed and secreted from transformed *Aspergillus oryzae*. *Lipids* 24: 781-785.

Inamdar, A. (1987). Polygalacturonase from *Thermoascus aurantiacus*: isolation and functional characteristics. *Ph.D. thesis*. Indian Institute of Science, Bangalore, India.

Jain, P.C. (1999). Spoilage of stored products. In: *Thermophilic Moulds in Biotechnology* (B.N. Johri, T. Satyanarayana and J. Olsen, Eds.) Kluwer Academic publishers, Dordrecht, pp. 289-316.

Jaychandran, S. and Ramabadran, R. (1970). Production of amylase by *Thermoascus aurantiacus*. *Indian J. Exp. Biol.* 8: 344.

Jensen, B., and Olsen, J. (1992). Physicochemical properties of a purified alpha-amylase from the thermophilic fungus *Thermomyces lanuginosus*. *Enzyme Microb. Technol.* 14: 112-116.

Jodice, F.R., Jole, C.S., Fiusello, N., France, O. and Cantini, G. (1975). Thermophilic fungi I. Contribution to their isolation, metabolism and their ability to degrade organic matter. *Allonia* 20: 53-74.

Johri, B.N. (1980). Biology of thermophilous fungi. In: *Recent Advances in Biology of Microorganisms* (Bilgrami, K.S. and Vyas, K.M., Eds.) Bishen Singh Mahendrapal singh, Dehradun, pp. 265-277.

Johri, B.N. and Satyanarayana, T. (1986). Thermophilic moulds- Perspective in basic and applied research. *Indian Rev. Life Sci.* 6: 75-100.

Kadowaki, M. K., Polizeli, M. L. T. M., Terenzi, H. F. and Jorge, J. A. (1996). Characterization of trehalose activities from the thermophilic fungus *Scytalidium thermophilum*. *Biochim. Biophys. Acta* 1291: 199-205.

Kango, Naveen (2007) Lignocellulose Biotechnology: Harnessing renewable plant materials for energy and chemicals. *Everyman's Science*, XLII (4) 194-197.

Kango, Naveen, Agrawal, S.C. and Jain, P.C. (2006). Production and properties of thermostable xylanase by *Thermomyces lanuginosus* NK-2 grown on lignocelluloses. *Biotechnol.* 5(2): 148-152.

Kango, Naveen and Jain, P.C. (2005). Production and application of fungal xylanases. In: *Fungi: Diversity and Biotechnology* (Rai, M.K. and Deshmukh S.K., Eds.), Scientific Publishers, New Delhi. pp. 251-281. ISBN 81-7233-403-6.

Kango, Naveen, Agrawal, S.C. and Jain, P.C. (2003a). Isolation of thermophilic fungi from soil and decomposing organic matter. *Asian Jr. of Microbiol., Biotechnol. Env. Sci.* 5(4): 445-447.

Kango, Naveen, Agrawal, S.C. and Jain, P.C. (2003b). Production of xylanase by *Emericella nidulans* NK-62 on low-value lignocellulosic substrates. *World J. Microbiol. Biotechnol.* 19 (7): 691-694.

Kawamori, M., Takayama, I. K. and Takasawa, S. (1987). Production of cellulases by a thermophilic fungus *Thermoascus aurantiacus* A-131. *Agric. Biol. Chem.* 51: 647-654.

Kelly, C.T., O'Mahony, M.R. and Fogarty, W.M. (1989). Extracellular xylanolytic enzymes of *Paecilomyces variotii*. *Biotechnol. Lett.* 9: 125-130.

Khandke, K. M., P. J. Vithayathil, and S. K. Murthy. (1989). Purification of xylanase, β-glucosidase, endocellulase, and exocellulase from a thermophilic fungus, *Thermoascus aurantiacus*. *Arch. Biochem. Biophys*. 274: 491-500.

Kuster, E. and Locci, R. (1964). Studies on peat and peat microorganisms II. Occurrence of thermophilic fungi in peat. *Arch. Microbiol*. 48: 319-324.

La Touche, C. J. (1950). On thermophilic species of *Chaetomium*. *Trans. Br. Mycol. Soc*. 33: 94-104.

Lacey, J. (1974). Moulding of sugarcane bagasse and its prevention. *Ann. Appl. Biol*. 76: 63-76.

Lin, J., Ndlovu, L.M., Singh, S. and Pillay, B. (1999). Purification and biochemical characteristics of β-D-xylanase from a thermophilic fungus, *Thermomyces lanuginosus*-SSBP. Biotechnol. *Appl. Biochem*. 30: 73-79.

Liu, W.H., Beppu, T. and Arima, K. (1974). Cultural conditions and some properties of lipase of *Humicola lanuginosa* S-38. *Agr. Biol. Chem*. 36: 1919-1924.

Maheshwari, R. and Balasubramanyam, P.V. (1988). Simultaneous utilization of glucose and sucrose by thermophilic fungi. *J. Bacteriol*. 170: 3274-3280.

Maheshwari, R., Balasubramanyam, P. V. and Palanivelu, P. (1983). Distinctive behaviour of invertase in a thermophilic fungus, *Thermomyces lanuginosus*. *Arch. Microbiol*. 134: 255-260.

Maheshwari, R., Bharadwaj, G. and Bhat, M. K. (2000). Thermophilic fungi: their physiology and enzymes. *Microbiol. Mol. Biol. Rev*. 64: 461-488.

Margaritis, A. and Creese, E. (1981). Thermal stability characteristics of cellulase enzymes of *Sporotrichum thermophile*. *Biotechnol. Lett*. 3: 471-476.

Matsuo, M., and Yasui, T. (1985). Properties of xylanase of *Malbranchea pulchella* var. *sulfurea* no. 48. *Agric. Biol. Chem*. 49:839-841.

Matsuo, M., Yasui, T. and Kobayashi, T. (1977). Purification and some properties of β-glucosidase from *Malbranchea pulchella* var. *sulfurea* no. 48. *Agric. Biol. Chem*. 41: 1593-1599.

Morgan-Jones, G. (1974). Notes on hyphomycetes V. A new thermophilic species of *Acremonium*. *Can. J. Bot*. 52: 429-431.

Moubasher, A.H. Aboel-Hafez, S.I.I., Aboelfattah, M.M. and Moharrarh, A.M. (1982). Fungi of wheat and bean straw composts 2. Thermophilic fungi. *Mycopathol*. 84: 61-72.

Mouchacca, J. (1997). Thermophilic fungi: biodiversity and taxonomic status. Cryptogam. *Mycol*. 18: 19-69.

Mulinge, S.K. and Apinis, A.E. (1969). Occurrence of thermophilic fungi in stored moist barley grain. *Trans. Br. Mycol. Soc*. 53: 361-370.

Odibo, F.J.C., Okafor, N., Tom, M.U. and Oyeka, C.A. (1992). Purification and characterization of β-amylase from *Hendersonula toruloidea*. *Starch* 44: 192-195.

Ofosu-Asideu, A. and Smith, R. S. (1973). Degradation of soft wood by thermophilic and thermotolerant fungi. *Mycologia* 65: 240-244.

Ogundero, V.W. (1980). Lipase activity of thermophilic fungi from mouldy groundnuts in Nigeria. *Mycologia* 72: 118-126.

Ong, P. S. and Gaucher, G. M. (1976). Production, purification and characterization of thermomycolase, the extracellular serine protease of the thermophilic fungus *Malbranchea pulchella* var. sulfurea. *Can. J. Microbiol.* 22: 165-176.

Ottesen, M. and Rickert, W. (1970). The isolation and partial characterization of an acid protease produced by *Mucor miehei*. C. R. *Trav. Lab. Carlsberg* 37: 301-325.

Prabhu, K. A., and Maheshwari, R. (1999). Biochemical properties of xylanases from a thermophilic fungus, *Melanocarpus albomyces*, and their action on plant cell walls. *J. Biosci.* 24: 461-470.

Prasad, A. R. S. and Maheshwari, R. (1978). Growth and trehalase activity in the thermophilic fungus *Thermomyces lanuginosus*. *Proc. Indian Acad. Sci.* 87B: 231-241.

Rajasekaran, A. K. and Maheshwari, R. (1990). Growth kinetics and intracellular protein breakdown in mesophilic and thermophilic fungi. *Indian J. Exp. Biol.* 28: 46-51.

Rajasekaran, A. K. and Maheshwari, R. (1993). Thermophilic fungi: an assessment of their potential for growth in soil. *J. Biosci.* 18: 345-354.

Rana, B.K., Johri, B.N. and Thakur, I.S. (1996). Formation and activities of xylan-hydrolysing enzymes of *Humicola grisea* var. *thermoidea*. *World J. Microbiol. Biotechnol.* 12: 12-15.

Rao, V. B., Maheshwari, R., Sastry, N. V. S. and Subba Rao, P. V. (1979). A thermostable glucoamylase from the thermophilic fungus *Thermomyces lanuginosus*. *Curr. Sci.* 48: 113-115.

Rao, V. B., Sastri, N. V. S. and Subba Rao, P. V. (1981). Purification and characterization of a thermostable glucoamylase from the thermophilic fungus *Thermomyces lanuginosus*. *Biochem. J.* 193: 379-387.

Renosto, F., Schultz, T., Re, E. Mazer, J., Chandler, C. J., Barron, A. and Segel, H. (1985). Comparative stability and catalytic and chemical propertiesm of the sulfate-activating enzymes from *Penicillium chrysogenum* (mesophile) and *Penicillium duponti* (thermophile). *J. Bacteriol.* 164: 674-683.

Rosenberg, S. L. (1975). Temperature and pH optima for 21 species of thermophilic and thermotolerant fungi. *Can. J. Microbiol.* 21: 1535-1540.

Roy, S. K., Dey, S. K., Raha, S. K. and Chakrabarty S. L. (1990). Purification and properties of an extracellular endoglucanase from *Myceliophthora thermophila* D-14 (ATCC 48104). *J. Gen. Microbiol.* 136: 1967-1971.

Sadhukhan, R., Roy, S. K., Raha, S. K., Manna, S. and Chakrabarty, S. L. (1992). Induction and regulation of a-amylase synthesis in a cellulolytic thermophilic fungus *Myceliophthora thermophila* D14 (ATCC 48104). *Indian J. Exp. Biol.* 30: 482-486.

Sahm, D.F. and Chapman, E.S. (1976) Effects of carbon sources on *Papulospora thermophila* at various temperatures. *Mycologia* 68: 168-174.

Saraswat, V. and Bisaria, V.S. (2000). Purification, characterization and substrate specificities of xylanase isoenzymes from *Melanocarpus* albomyces IIS 68. *Biosci. Biotechnol. Biochem.* 64: 1173-1180.

Satyanarayana, T. and Johri, B.N. (1983). Extracellular protease of thermophilic fungi isolated from decomposing paddy straw. *Trop. Plant. Sci. Res.* 1: 137-140.

Satyanarayana, T. and Johri, B.N. (1984). Ecology of thermophilic fungi. In. *Progress in Microbial Ecology* (Mukerjee *et al.*, Eds.) Print House, Lucknow pp. 349-361.

Satyanarayana, T., Johri, B.N. and Klein, J. (1992). Biotechnological potential of thermophilic fungi. In: *Handbook of Applied Mycology*, Vol.4 (Arora, D.K., Elander, R.P. and Mukerji, K.G., Eds.) Marcell Dekker, New York, pp. 729-761.

Satyanarayana, T., Johri, B.N. and Saksena, S.B. (1977). Seasonal variation in mycoflora of nesting materials of birds with special reference to thermophilic fungi. *Trans. Br. Mycol. Soc.* 68: 307-309.

Shenolikar, S. and Stevenson, K. J. (1982). Purification and partial characterization of a thiol proteinase from the thermophilic fungus *Humicola lanuginosa*. *Biochem. J.* 205: 147-152.

Singhai, S. and Jain, P.C. (1992). Thermophilous fungi from decomposing organic matter: compost and sawdust. *Geobios New Reports*. 11: 41-44.

Siqueira, E.M.D., Mizuta, K. and Giglio, J..R. (1979). α-amylase of *Pycnoporus sanguineus*. *Mycol. Res.* 101: 188-190.

Somkuti, G. A. and Babel, F. J. (1968a). Purification and properties of *Mucor pusillus* acid protease. *J. Bacteriol.* 95: 1407-1414.

Somkuti, G.A. and Babel, F.J. (1968b). Lipase activity of *Mucor pusillus*. *Appl. Microbiol.* 16: 617-619.

Subrahmanyam, A. (1980). Studies on *Thermoascus aurantiacus*. *Acta Mycologia* 16: 121-131.

Subrahmanyam, A. (1999). Ecology and Distribution. In: *Thermophilic Moulds in Biotechnology*. (Johri, B.N. Satyanarayana, T. and Olsen, J., Eds.), Kluwer Academic Publishers, Dordrecht, pp. 13-42.

Tansey, M. R. and Brock, T. D. (1972). The upper temperature limit for eukaryotic organisms. *Proc. Natl. Acad. Sci. USA* 69: 2426-2428.

Tansey, M. R. and Brock, T. D. (1978). Microbial life at high temperatures: ecological aspects. In: *Microbial Life in Extreme Environments* (Kushner, D. J., Ed.), Academic Press Ltd., London, pp. 159-216.

Tansey, M.R. (1973). Isolation of thermophilic fungi from alligator nesting material. *Mycologia* 65: 594-601.

Tansey, M.R. and Brock, T.D. (1971). Isolation of thermophilic and thermotolerant fungi from hot spring effluents and thermal soils of yellow stone national park. *Bacteriol. Proc.* p 36.

Tansey, M.R. and Jack, M.A. (1977). Growth of thermophilic fungi in soil *in situ* and *in vitro*. *Mycologia.* 69: 563-578.

Taylor, P. M., Napier, E. J. and Fleming I. D.(1978). Some properties of glucoamylase produced by the thermophilic fungus *Humicola lanuginosa*. *Carbohydrate Research* 61: 301-308.

Thakre, R. P. and Johri, B. N. (1976). Occurrence of thermophilic fungi in coalmine soils of Madhya Pradesh. *Curr. Sci.* 45: 271-273.

Udagawa, S., Furuya, K. and Horie, Y. (1973). Notes on some ascomycetous micro fungi from soil. *Bull. Natl. Sci. Mus.* (Tokyo), 16: 503-520.

Vandamme, E.J., Logghe, J.M. and Geeraerts, H.A.M. (1982). Cellulase activity of a thermophilic *Aspergillus fumigatus* (Fresinius). *J. Chem. Technol. Biotechnol.* 32: 968-974.

Waksman, S.A., and Cordon, T.C. (1939). Thermophilic decomposition of plant residues in composts by pure and mixed culture of microorganisms. *Soil Sci.* 47: 217-225.

Waksman, S.A., Umbreit, W.W. and Cordon, T.C. (1939). Thermophilic actinomycetes and fungi in soils and in composts. *Soil Sci.* 47: 37-61.

Wallace, H.A., Sinha, R.N. and Mills, J.T. (1986). Fungi associated with small wheat bulks during prolonged storage. *Can. J. Bot.* 54: 1332-1343.

Wyss, M., Brugger, R., Kronenberger, A., Re´my, R., Fimbel, R., Osterhelt, G. Lehmann, M. and Van Loon, A. P. G. M. (1999). Biochemical characterization of fungal phytases (myo-inositol hexakisphosphate phosphohydrolase): catalytic properties. *Appl. Environ. Microbiol.* 65: 367-373.

Yang, J., Teplyakov, A. and Quail, J. W. (1997). Crystal structure of the aspartic proteinase from *Rhizomucor miehei* at 2.15 Å resolution. *J. Mol. Biol.* 268: 449-459.

Yoshioka, H., Anraku, S.I. and Hayashida, S. (1982). Production and purification of a novel type of CMCase from *Humicola grisea* var. *thermoidea* YH-78. *Agric. Biol. Chem.* 46: 75-82.

Microbial Biotechnology and Ecology (2011) Pages 320–343
Editors: Deepak Vyas, G.S. Paliwal, P.K.Khare and R.K Gupta
Published by: DAYA PUBLISHING HOUSE, NEW DELHI

Chapter 19

Biodiversity and Conservation of Mushroom in Chhattisgarh Region

☆ *M.P. Thakur*, C.S. Shukla and V.K. Yadav*

Department of Plant Pathology,
Indira Gandhi Agricultural University, Raipur – 492 006, Chhattisgarh

ABSTRACT

Chhattisgarh State is divided into three agroclimatic zone (Plains of Chhattisgarh, Bastar Plateau, Northern Hill Region) and occupied 135191 sq. km area of which 43.8 per cent area is under forest. More than 80 per cent of the population is engaged in agricultural and forest activities. Among these activities mushrooms are one of the choicest item which are collected and consumed by local people during rainy season. It constitutes an important source of their income. They sold mushroom in local market hats / towns. The survey was conducted for occurrence of naturally growing mushrooms in all districts of Chhattisgarh. The survey for naturally growing mushroom flora was initiated in 1988 and continued till date. In all, 78 mushrooms were collected during rainy season and identified. Among the collected mushrooms, 33 edible, 22 were as non edible, 2 medicinal, 2 poisonous and 19 were unidentified. Maximum number of mushrooms were collected from Bastar Plateau and most of them were consumed by tribals. The most frequent mushrooms observed from different places of Chhattisgarh were *Volvariella volvacea*, *Cantherellus* spp., *Tuber* sp., *Rusula* sp. and *Termitomyces* spp. Some mushrooms like *Termitomyces* spp., *Rusulla* spp., *Volvariella* like and *Ganoderma lucidum* were used as medicine to reduce labour pain, headache, constipation and healing of wounds respectively.

* E-mail: mp_thakur@yahoo.com

Introduction

Chhattisgarh State has 135191 sq. km area, which accounts 4.11 per cent of the total geographical area of the country. The state has 43.8 per cent of total geographical area covered by forest and 35 per cent area is under cultivated land. It is located between 17° 48 N and 24° 5 N latitude and 80° 15–84° 20E longitude. It is touched with the boundaries of Uttar Pradesh and Jharkhand in North, Orissa in East, Andhra Pradesh in South and M. P., Maharastra in West. Administratively, it is divided into 146 blocks, 97 Tehsils with 16 Districts. According to Census Report of 2001, the total population is about 20795956 of which 16622627 lives in villages and rest in Urban areas. Near about 80 per cent of the population is engaged in agricultural activities. The population is dominated by Schedule Tribe (32 per cent), Schedule Caste (12 per cent) and backward classes. Thus, the weaker section constitute about 44 per cent of the total population. The population density is 154 per sq. km. The state has 28 per cent area under irrigation and paddy is the main crop grown in 80 per cent of the area as rainfed.

The climatic conditions prevailing in Chhattisgarh made the natural habitat for a large number of mushroom flora. The detailed information for the presence of natural agaric flora in different agroclimatic zones of Chhattisgarh was started with the inception of All India Coordinated Mushroom Improvement Project in 1988 at IGAU, Raipur. Prior to 1988, no systematic efforts were made for collection, identification and preservation of fleshy fungi of this state. But, thereafter, the systematic work was started to explore the kind of wild mushroom flora available and exploit their production potential in a commercial scale.

Agroclimatic Zones

Chhattisgarh is divided into 3 agroclimatic zones *i.e.* Plains of Chhattisgarh, Bastar Plateau, Northern Hill Region (Table 19.1). About 74 per cent area of Chhattisgarh Plains, 97 per cent of Bastar Plateau and 95 per cent in Northern Hills are rainfed (Table 19.2). Forest area in Bastar Plateau is highest followed by Northern Hills and Chhattisgarh Plains. The main source of irrigation is canals. The cropping intensity of the region is 121 per cent.

Table 19.1: Area, Classification and Rainfall in Chhattisgarh

Months	*Raipur*				*Jagdalpur*				*Ambikapur*			
	Max. Temp.	*Min. Temp.*	*Rain-fall*	*Relative Humidity*	*Max. Temp.*	*Min. Temp.*	*Rain-fall*	*Relative Humidity*	*Max. Temp.*	*Min. Temp.*	*Rain-fall*	*Relative Humidity*
January	26.8	11.9	026.0	92	25.2	10.8	38.3	94	24	8	28	85
February	28.9	12.3	031.0	88	28.1	12.1	13.6	92	28	11	22	79
March	36.6	17.7	011.6	74	35.0	17.1	6.4	89	31	14	55	63
April	39.9	23.4	004.2	57	36.0	22.9	43.5	84	34	22	26	50
May	41.1	27.9	008.9	48	35.6	23.0	85.6	73	39	23	31.5	50
June	36.0	25.9	252.2	74	30.0	22.3	351.8	84	35	23	300	79
July	31.8	24.3	379.8	92	27.6	22.2	356.3	91	32	21	455	90
August	29.1	24.4	156.6	92	25.9	21.8	308.6	93	32	21	349	89
September	32.2	24.4	136.4	92	29.6	21.9	258.2	93	30	19	310	89
October	30.8	19.3	020.2	93	29.3	19.0	151.4	93	30	19	50	88
November	30.2	13.8	000.0	90	27.1	12.2	7.6	92	28	10	25	87
December	28.1	10.9	000.0	90	26.8	8.0	0.0	94	26	8	22	85

Table 19.2: Details of Agroclimatic Zones of Chhattisgarh

Agroclimatic Zone	*Districts*	*Geographical Area*	*Area Under Forest (%)*	*Rainfall*	*Soil Type*
Plains of Chhattisgarh	Raipur, Mahasamund, Dhamtari, Durg, Rajnandgaon, Kawardha, Kanker, Bilaspur, Janjgir, Korba and Raigarh	7493.781 sq km	33.3	900-1000 mm	Antisol (Bhata) Alfisol (Dorsa) Vertisol (Kanhar) Inceptisol (Matasi)
Bastar Plateau Zone	Bastar, Dantewada	3262.626 sq km	65.2	1200-1600 mm	Mollisoils
Northern Hill Region	Sarguja, Korea, Jashpur	2846.954 sq km	43.8	1200-1400 mm	Vertisols Alfisols

Fleshy Fungi of Chhattisgarh

The work on survey, collection, isolation and preservation of naturally growing fleshy fungi during monsoon season was initiated in 1988. The local people of Chhattisgarh do collect and consume some edible mushrooms as well as bring them to the local market for sale during July to October. According to Kumar and Agrawal (1989) nearly 50 tons of naturally growing edible fleshy fungi is collected from Bastar alone during monsoon season. A survey for the presence of naturally occurring fleshy fungi was conducted from 1989 to 1995 (Anonymous, 1989-1995). Shukla *et al.* (2001) surveyed the forest area of Chhattisgarh and collected edible and wild mushrooms. Thakur *et al.* (2005) reported enormous wild mushroom flora existing in Chhattisgarh. They isolated, identified and characterized some of the mushroom flora. Some wild edible mycorrhizal mushrooms of Chhattisgarh were also reported by Shukla (2005). Shukla *et al.* (2001) collected various species of mushroom from Bilaspur division and reported that some mushrooms have antitumour, antibacterial, antifungal and antiviral properties. The species of *Agaricus, Russula, Termitomyces* are being used for treatment of goitre, wound and small pox respectively. During survey, most commonly available mushrooms encountered were *Russula* spp., *Volvariella* spp., *Termitomyces* spp., *Tuber* spp. and *Cantherellus* spp. (Annonymous, 2002, 2003, 2004, Sharma and Samota, 2006). The fleshy fungi collected from different areas of Chhattisgarh with respect to habitat, season, distribution and ethnic information is presented. Most of these fungi were identified using standard monographs (Purkayastha and Chandra, 1985; Pacioni, 1985; Rinaldi and Tyndalo, 1974; Pegler and Spooner, 1997). The details are as under:

Agrocybe erebia

Habit and habitat : Sporophore in groups on ground

Season : Rainy season (June–October)

Distribution : Achanakamar sanctuary area of Bilaspur

Edibility : Non edible

***Agrocybe* sp.**

Habit and habitat : Sporophore in groups or scattered in soil

Season : Rainy season (June–October)

Distribution : From different places of Bilaspur

Edibility : Non edible

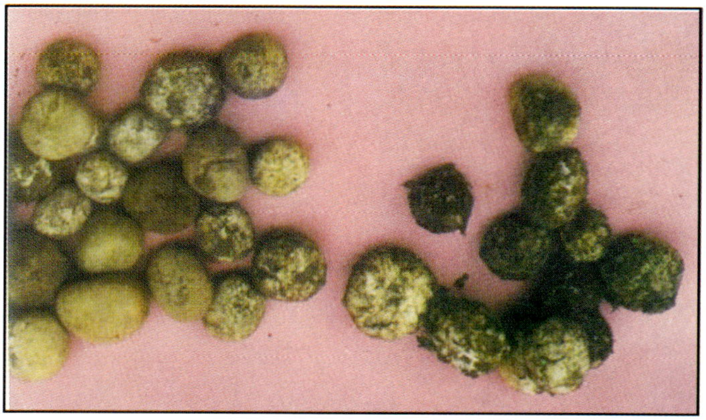

***Tuber* sp.**

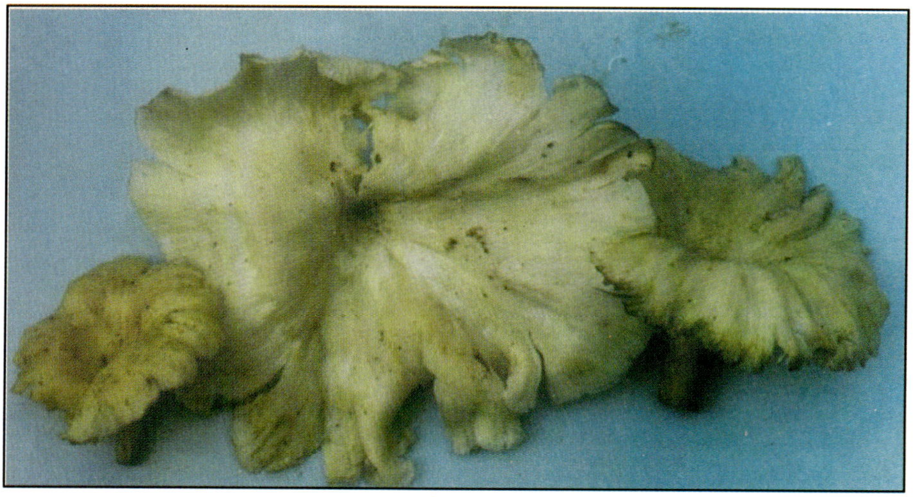

***Cantherellus* sp.**

***Russula* sp.**

Pleurotus florida

***Amanita* sp.**

***Coprinus* sp.**

Termitomyces microcarpus

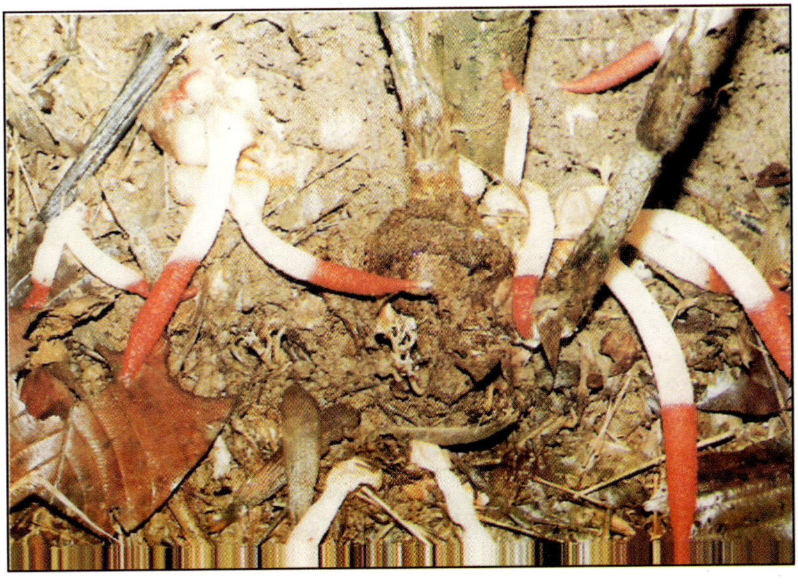

***Mutinus* sp.**

***Ramaria* sp.**

***Termitomyces* sp.**

Amanitopsis

Habit and habitat : Growing singly in scattered on leaf litter

Season : Rainy season (August)

Distribution : Barnavapara

Edibility : Edible but consumed by local people

Amanita citrina

Habit and habitat : Under shrubs

Season : Rainy season (June–October)

Distribution : Collected from different places of Bastar forest

Edibility : Poisonous

***Amanita* sp.**

Habit and habitat : Sporophore occurred in scattered on the soil under shrubs

Season : Rainy season (June–October)

Distribution : Collected from different places of Bastar forest

Edibility : Not known

Armillaria mellea

Habit and habitat : Habit and habitat- sporophore growing on deciduous stumps of dead trunk of tree

Season : Rainy season (June–October)

Distribution : Raipur and Bilaspur

Edibility : Non edible

***Astreatus hygrometricus* (Boda)**

Habit and habitat : Scattered on the soil, Hypogeal mycorrhizal association with root of sal tree

Season : Rainy season (June–October)

Distribution : All the forest area of Chhattisgarh

Edibility : Excellent

Auricularia polytricha

Habit and habitat : Solitary or gregarious on dead wood and logs in all region of Chhattisgarh

Season : Rainy season (June–October)

Distribution : All region of Chhattisgarh

Edibility : Edible

***Auricularia* sp.**

Habit and habitat : On dead branches of mango, Custard apple, subabool, gulmohar

Season : Rainy season (June–October)

Distribution : Raipur

Edibility : Edible

***Boletus* sp.**

Habit and habitat	:	Sporophore solitary on sandy soil under dense shrubs on leaf litter
Season	:	After good monsoon rains
Distribution	:	In the area of Uddanti and Barnawapara Wild life Sanctuary of Raipur
Edibility	:	Edible but not consumed by local people

Bovista appendicellata

Habit and habitat	:	Sporophore scattered on leaf litter under the shed of broad leaves trees
Season	:	Rainy season (June–October)
Distribution	:	Bilaspur
Edibility	:	Not known

***Calocybe* sp.**

Habit and habitat	:	Sporophore solitary in soil under broad leaves trees
Season	:	Rainy season (June–October)
Distribution	:	Restricted in few localities of Ambikapur forest
Edibility	:	Edible but not consumed by the local people

Clavatia cyantiformis

Habit and habitat	:	Solitary or in groups on leaf litter of shrubs
Season	:	Rainy season (June–October)
Distribution	:	Forest area of Korea District
Edibility	:	Edible but not consumed by local people

***Cantherellus* sp. (Bans phutu)**

Habit and habitat	:	Sporophore in groups, have with mycorrhizal association with bamboo trees
Season	:	Rainy season during August month near Raksh Bandhan
Distribution	:	In all bamboo growing areas of Chhattisgarh
Edibility	:	Excellent and sold in local market @ 60-80/kg

Clavaria fumosa

Habit and habitat	:	Sporophore grows in groups on the ground on fallen branches
Season	:	Rainy season
Distribution	:	Sitanadi Wild Life Sanctuary
Edibility	:	Not Known

***Clavaria* sp.**

Habit and habitat	:	On wood in scattered
Season	:	Rainy season
Distribution	:	Tirathgarh, Narayanpur
Edibility	:	Not edible

Clitocybe gestrpoa

Habit and habitat	:	Sporophore grows in groups under the shrubs on leaf litter
Season	:	Rainy season
Distribution	:	Tirathgarh, Chitrakut and other places of Bastar
Edibility	:	Poisonous

***Coprinus comatus* (Shaggy cap)**

Habit and habitat	:	Gregarious in grassland and road side
Season	:	Rainy season
Distribution	:	In all districts of Chhattisgarh
Edibility	:	Edible but rarely consumed by local people

***Coprinus* sp. (Ink cap)**

Habit and habitat	:	Sporophore grows solitary or in groups
Season	:	Rainy season
Distribution	:	Cow dung, damp soil, rotten paddy straw in heaps
Edibility	:	Not edible but people in Korea district do eat it before pileus formation.

Coriolus versicolor

Habit and habitat	:	On dead wood
Season	:	Rainy season
Distribution	:	Narayanpur
Edibility	:	Medicinal or therapeutic value

***Crepodorus* sp.**

Habit and habitat	:	Sporophore grows solitary or in groups on dead wood of broad leaves trees
Season	:	Rainy season during August
Distribution	:	Teerathgarh
Edibility	:	Not known

Cyathus limbatus

Habit and habitat	:	Sporophore grows solitary or in groups in open grass and same times in the soil
Season	:	Rainy season
Edibility	:	Not known

Daldinia concentrica

Habit and habitat	:	Dead wood logs or live plants
Season	:	Rainy season
Distribution	:	Teerathgarh, Narayanpur forest area
Edibility	:	Not known

***Flamulina* sp.**

Habit and habitat : On dead wood, mainly stumps
Season : Rainy season
Distribution : Teerathgarh, Barnawapara, Khajuri Nala and Uddanti Sanctuary
Edibility : Edible

Ganoderma applanatum

Habit and habitat : Sporophore grows solitary or in groups on dead trunk of broad leaves tree
Season : Rainy season
Distribution : All forest area of Chhattisgarh
Edibility : Not edible

Ganoderma lucidum

Habit and habitat : Sporophore grows solitary or in groups on dead trunk of broad leaves tree
Season : July- October
Distribution : All forest area of Chhattisgarh, mainly on live/dead trunk of subabul, gulmohar, babool
Edibility : Not edible

Ganoderma suiji

Habit and habitat : Sporophore grows solitary or in groups
Season : Rainy season
Distribution : Ambikapur, Bastar and Bilaspur forest area
Edibility : Not edible

***Geastrum fimbriatum* (Earth stars)**

Habit and habitat : Sporophore grows solitary or in groups in soil
Season : Rainy season
Distribution : Labhandi research farm
Edibility : Not edible

Hygrocybe marchee

Habit and habitat : Sporophore scattered on grass land
Season : Rainy season
Distribution : Churra, Gariabandh and Manpur area of Raipur
Edibility : Not known

Hypholoma udum

Habit and habitat : Sporophore grows in groups on the trunk of deciduous tree
Season : Rainy season
Distribution : Kevnchi, Lormi and Achanakmar forest
Edibility : Not known

Hypoxylon fragiformae

Habit and habitat : Normally gregarious on rooting beech, grow on wood, crust like stromata

Season : Rainy season

Distribution : Manpur, Sihava, Gariabandh forest

Edibility : Not edible

Laccaria laccata

Habit and habitat : From soil of mixed forest, sporophore in groups, singly or scattered

Season : Rainy season

Distribution : Chitrakut, Teeratgarh, Narayanpur

Edibility : Edible but not consumed by local people

Lactarius aceverimus

Habit and habitat : Solitary or in groups under broad leaves trees

Season : Rainy season

Distribution : Narayanpur

Edibility : Not known

Lactarius vellereus

Habit and habitat : Sporophore grows singly in scattered form under shrubs of broad leaves

Season : Rainy season

Distribution : Narayanpur

Edibility : Not known

***Lactarius* sp.**

Habit and habitat : Grows singly on leaf litter

Season : Rainy season

Distribution : Narayanpur

Edibility : Not known

***Lentinus* sp.**

Habit and habitat : Sporophore in bunches on dead wood logs.

Season : Rainy season

Distribution : Korea

Edibility : Edible but not consumable by local peoples

Lepista nuda

Habit and habitat : In humus rich ground under broad leaves shrubs

Season : Rainy season

Distribution : Achanakmar (Bilaspur)

Edibility : Good

Lycoperdon maximum

Habit and habitat : Sporophore solitary or scattered in grass land

Season : Rainy season

Distribution : Sandy loam soil

Edibility : Edible

Marasmius oreades

Habit and habitat : Sporophore gregarious in grass land

Season : Rainy season

Distribution : Bastar forest

Edibility : Edible

Melanoleuca alboflavida

Habit and habitat : Sporophore grows in soil or in leaf litter

Season : Rainy season

Distribution : Barnawa forest area

Edibility : Not edible

Microglossum viridae

Habit and habitat : On the ground in grass or woods

Season : Rainy season

Distribution : Narayanpur

Edibility : Not edible

***Mutinus* sp.**

Habit and habitat : Sporophore in group under shade of old bamboo trees

Distribution : Bamboo rhizomes near Bilaspur area

Edibility : Inedible

Panaiolima foenislcic

Habit and habitat : Gregarious in grassy areas

Season : Rainy season

Distribution : Achanakmar, Korea, Bilaspur

Edibility : Not edible

***Phallorina* sp.**

Habit and habitat : Sporophore solitary or in groups on soil under the trees

Season : Rainy season

Distribution : Near guest house of Guru Ghasidas University, Bilaspur

Edibility : Edible but not consume by local people

Termitomyces heimii

***Polyporus* sp.**

***Tricholoma* sp.**

Coriolus versicolor

***Lactarius* sp.**

Ganoderma lucidum

Auricularia polytricha

Pleurotus florida (Pihri)

Habit and habitat	:	Sporophore grows solitary or in groups on mango logs
Season	:	Rainy season
Distribution	:	Narayanpur area
Edibility	:	Edible

Pleurotus sajor caju

Habit and habitat	:	Sporophore in groups on dead wood logs
Season	:	Rainy season
Distribution	:	Branches of mango trees
Edibility	:	Good

Pleurotus stramonius

Habit and habitat	:	Sporophore in groups on dead branches of Mango
Season	:	Rainy season

Schizophyllum commune

Distribution	:	House of C. S. Shukla
Edibility		Good

Pleurotus sp.

Habit and habitat	:	Sporophore in groups on dry wood logs of many trees
Season	:	Rainy season
Distribution	:	Bilaspur
Edibility		Good

Podbrella microcarpa

Habit and habitat	:	In soil in singly or groups
Season	:	Rainy season
Distribution	:	College campus Nandghat Bilaspur

***Podaxis* sp.**

Habit and habitat	:	Sporophore solitary and grows on sandy soil
Season	:	Rainy season
Distribution	:	Tarra village
Edibility	:	Edible but not consumed by local people

Podoscypha mutzonata

Habit and habitat	:	Sandy loam in singly
Season	:	Rainy season
Distribution	:	Nilanjee
Edibility	:	Non edible

Polyporus varius

Habit and habitat	:	Sporophore in groups on dead wood
Season	:	Rainy season
Distribution	:	All districts of Chhattisgarh
Edibility	:	Not known

Polyporus elegans

Habit and habitat	:	On dead wood
Season	:	All districts of Chhattisgarh
Distribution	:	Bilaspur
Edibility	:	Not edible

Ramaria formosa

Habit and habitat	:	On wood logs of many trees
Season	:	Rainy season
Distribution	:	Narayanpur
Edibility	:	Not edible

Russula delica

Habit and habitat	:	Sporophore grows solitary or in groups and occur on shades of many forest shrubs
Season	:	Rainy season
Distribution	:	All districts of Chhattisgarh
Edibility	:	Good

Russula lepida

Habit and habitat	:	Under broad leaves trees in soil, scattered
Season	:	Rainy season
Distribution	:	All places of Chhattisgarh
Edibility	:	Edible

Russula aeruginla

Habit and habitat	:	In groups under sal forest
Season	:	Rainy season
Distribution	:	All places of Chhattisgarh
Edibility	:	Edible

Russula atropurpurea

Habit and habitat	:	Under broad leaves trees specialy in acidic soil
Season	:	Rainy season
Distribution	:	Bastar, Bilaspur
Edibility	:	Edible

Russula mairei

Habit and habitat	:	Under shrubs or mixed forest
Season	:	Rainy season
Distribution	:	Achanakma, Barnawapara
Edibility	:	Not edible

Russula virginca

Habit and habitat	:	Under shrubs or mixed forest
Season	:	Rainy season
Distribution	:	Achanakma, Barnawapara
Edibility	:	Edible

Rozites caperata

Habit and habitat	:	Under broad leaves on leaf litter
Season	:	Rainy season
Distribution	:	Korea, Manendragarh
Edibility	:	Not edible

***Scleroderma verucosum* (Dhela phutu)**

Habit and habitat	:	Sporophore grows solitary or in groups and occur in laterite soil
Season	:	Rainy season
Distribution	:	Labhandi farm
Edibility	:	Good

***Scleroderma* sp.**

Habit and habitat	:	Sporophore grows singly or gregarious in the ground
Season	:	Rainy season
Distribution	:	Bastar
Edibility	:	Good

***Stereum* sp.**

Habit and habitat	:	Sporophore on dead wood of broad leaves trees
Season	:	Rainy season
Distribution	:	Korea, Ambikapur, Pendra, Marvahi
Edibility	:	Not edible

Strophoria semiglobata

Habit and habitat	:	Sporophore grows on soil and either solitary or gregarious
Season	:	Rainy season
Distribution	:	Bilaspur, Ratanpur, Pali, Kathghora forest
Edibility	:	Not known

Schizophyllum commune

Habit and habitat		On dead wood
Season		Rainy season
Distribution		Mahasamund
Edibility		Medicinal or therapeutic uses

***Termitomyces clypeatus* (Patera phutu)**

Habit and habitat	:	Sporophore solitary or scattered on the ground
Season	:	Collected from laterite and bunds of ricefield soil after onset of good monsoon
Distribution	:	Associated with termite mounds
Edibility	:	Excellent

***Termitomyces heinii* (Bhado phutu)**

Habit and habitat	:	Sporophore gregarious and grows in large numbers in forest in undisturbed ground. Local people collected for consumption's and sold in local market
Season	:	In the month of September
Distribution	:	All agroclimatic zone of Chhattisgarh
Edibility	:	Excellent

***Termitomyces microcarpus* (Kanki phutu)**

Habit and habitat	:	Sporophore grows in large numbers on the termite mounds
Season	:	After onset of good monsoon
Distribution	:	All agroclimatic zone of Chhattisgarh
Edibility	:	Excellent

***Termitomyces robustus* (Dashara phutu)**

Habit and habitat	:	Sporophore scattered or groups in undisturbed forest area on the ground
Season	:	During September to October

Distribution : All agroclimatic zone of Chhattisgarh

Edibility : Excellent

***Tricholoma* sp. (Hardola phutu)**

Habit and habitat : Sporophore solitary or group in under the shade of broad leaves tree

Season : During Rainy season in August and September

Distribution : Barnawapara Sanctuary, Achanakmar and forest area of Bastar Plateau

Edibility : Good

***Tuber* sp. (Safed boda)**

Habit and habitat : Sporophore scattered under soil in mixed forest area

Season : During September to October

Distribution : Collected from all forest of state

Edibility : Excellent

Volvaria umbonata

Habit and habitat : Sporophore scattered on the ground

Season : Rainy season

Distribution : Bastar

Edibility : Not known

***Volvariella volvacea* (Paira phutu)**

Habit and habitat : Sporophore grows solitary or scattered on paddy straw leaf

Season : All the localities of Chhattisgarh state. Local people collected and sold in local market

Distribution : All districts of Chhattisgarh

Edibility : Excellent

***Volvariella* like**

Habit and habitat : Sporophore scattered in leaf litter on living trees

Season : After onset of good rains in the during August

Distribution : Dense forest area of Pali and Katghora

Edibility : Good

***Volvariella* sp.**

Habit and habitat : Sporophore scattered or groups on rotten paddy straw

Season : During September to October

Distribution : All Agroclimatic Zone of Chhattisgarh

Edibility : Excellent

***Xylaria* sp.**

Habit and habitat	Sandy loam in groups
Season	Rainy season
Distribution	India
Edibility	Non edible

Xylaria polymorpha

Habit and habitat	Sandy loam in groups
Season	Rainy season
Distribution	Nilanjee
Edibility	Non edible

Future Strategies

- ☆ Intensive survey work needs to be carried out to collect wild and edible mushroom flora from different districts of Chhattisgarh and these should be conserved *in situ* and *ex situ*.
- ☆ Evaluation of locally available mushroom germplasm for cultivation, yield potential, insect pests and diseases resistance, nutritional values, medicinal values, export etc.
- ☆ Multidisciplinary research is required to characterize and study the wild mushroom flora.
- ☆ A need was felt to monitor and map the mushroom diversity of forest ecosystem using Satellite Remote Sensing, GPS and GIS techniques.
- ☆ Isolation, purification, identification and characterization of endangered mushroom and other mushroom species by molecular approaches, community participation and their cataloguing.
- ☆ An efficient, cheap solar dryer with electrical backup needs to be developed for drying of fresh mushrooms.
- ☆ A need was felt to develop different varieties of mushrooms suiting to taste of people and prevailing agroclimatic conditions of Chhattisgarh.
- ☆ An awareness is to be created to grow mycorrhizal mushroom species either by forest departments or by tribals in forest areas in order to conserve these species from deforestation or environmental degradation.
- ☆ To compile ethnic information/traditional knowledge from tribals regarding nutritional and medicinal values of mushrooms.
- ☆ Efforts are to be made to exploit the medicinal values of wild mushrooms (*Ganoderma lucidum*) commonly present in the forest areas of Chhattisgarh.
- ☆ The potential of medicinal mushroom in India is untapped and need to evolve into a successful biotechnological industry for the benefit of the mankind.

References

Annonymous. (1990). *Survey, Collection and identification of Mushrooms in Chhattisgarh*. Annual report of AICMIP centre Raipur.

Annonymous. (1991). *Survey, Collection and Identification of Mushrooms in Chhattisgarh.* Annual report of AICMIP centre Raipur.

Annonymous. (1992). *Survey, Collection and Identification of Mushrooms in Chhattisgarh.* Annual report of AICMIP centre Raipur.

Annonymous. (1993). *Survey, Collection and Identification of Mushrooms in Chhattisgarh.* Annual report of AICMIP centre Raipur.

Annonymous. (1994). *Survey, Collection and Identification of Mushrooms in Chhattisgarh.* Annual report of AICMIP centre Raipur.

Annonymous. (1995). *Survey, Collection and Identification of Mushrooms in Chhattisgarh.* Annual report of AICMIP centre Raipur.

Annonymous. (2002). *Survey of Fleshy Fungi of Chhattisgarh.* Annual report of AICMIP centre Raipur.

Annonymous. (2003). *Survey of Fleshy Fungi of Chhattisgarh.* Annual report of AICMIP centre Raipur.

Annonymous. (2004). *Survey of Fleshy Fungi of Chhattisgarh.* Annual report of AICMIP Centre Raipur.

Annonymous (1989). *Survey, Collection and Identification of Mushrooms in Chhattisgarh.* Annual Report of AICMIP Centre, Raipur.

Kumar, S.M. and Agrawal, K.C. (1999). Mushroom a supplement to diet in Chhattisgarh. Paper presented at symposium on Chhattisgarh Agricultural system- Accomplishment, present status and future strategies. Held at IGKV Raipur from 20-23rd January, 1989.

Pacioni, G. (1985). *The Macdonand–Encyclopedia of Mushrooms and Toadstools.* Mac donald and company limited London Sydnee. p. 511.

Pegler, D. and Spooner, B. (1997). *The mushroom identifier.* Leefung- Ascoprinters Limited China. P. 144.

Purkyastha, R. P. and Chandra, A. (1985). *Manual of Indian Edible Mushrooms.* Jagmandar Book Agency, New Delhi. p. 316.

Ramsbottom, J. (1989). *Mushroms and Toadstools.* Willam collins sons and Company Ltd. London. p. 382.

Rinaldi, A. and Tyndalo, V. (1974). *The Complete Book of Mushrooms.* Crown Publishers Inc. Crescent Books, New York. p. 430.

Sharma, S.S. and Samota, R.K. (2006). Tapping into India's mushroom biodiversity: identification, conservation and domestication of wild mushroom flora. In : Compendium of lectures of ICAR short course on " *Emerging Areas in Mushroom Diversity, Production and Postharvest Developments*" at IGAU, Raipur from 9th to 18th January, 2006. pp. 69-84.

Shukla, C.S., Khatri, A.K. and Thakur, M.P. (2002). Fleshy fungi from Chhattisgarh. *Paper presented in National Seminar on Role of Antimicrobials on Sustainable Horticulture* held at IGAU Raipur on 28th January. p.70.

Shukla, R.V., Tiwari, M., and Tripathi, A. (2005). Mushrooms and tribals in sal forest of Chhattisgarh. In National Mushroom Workshop on "*Awareness Creation on Biodiversity and Conservation of Mushrooms*" organized by IGAU, Raipur from December 1-2, 2005. p 5.

Shukla, R.V. (2005). Wild edible mycorrhizal mushrooms of Chhattisgarh. In: National Mushroom Workshop on *"Awareness Creation on Biodiversity and Conservation of Mushrooms"* organized by IGAU, Raipur from December 1-2, 2005. p 4.

Thakur, M.P., Shukla, C.S. and Yadav, V.K., 2005. Mushroom wealth of C. G., their conservation and exploitation. In National Mushroom Workshop on *"Awareness Creation on Biodiversity and Conservation of Mushrooms"* organized by IGAU, Raipur from December 1-2, 2005, p.3.

Microbial Biotechnology and Ecology (2011) *Pages* 344–352
Editors: Deepak Vyas, G.S. Paliwal, P.K.Khare and R.K Gupta
Published by: DAYA PUBLISHING HOUSE, NEW DELHI

Chapter 20

Mushrooms: A Nutritive Food for Human Beings

☆ *Poonam Dehariya*, Anjuli Chaubey and Deepak Vyas*

Lab of Microbial Technology and Plant Pathology,
Department of Botany, Dr. H.S. Gour University, Sagar – 470 003, M.P.

Introduction

Mushrooms have played an important role in the diet of many people for thousands of years. Ancient Egyptian hieroglyphics reveal that mushrooms were thought to bring immortality. Many later cultures believed that eating mushrooms could endow them with super-human strength. Give them clairvoyance in locating lost objects and lead the soul to reside with the gods (Ingramm, 2002). Mushrooms were believed by the Greeks to give strength to the warriors while Romans regarded them as the 'Food for the Gods'. The Chinese prized mushrooms as health food, the 'elixir of life'. The mushrooms were worshipped and consumed as hallucinogens by the Mexican Indians.

In the era of 'healthy eating' by cutting down the calories, saturated fat and cholesterol, mushrooms were bound to attract the attention. Mushrooms have long been considered, especially in the orient, to have medicinal values (King, 1993). Due to unique chemical composition, mushrooms are suited to the specific groups suffering with some ailments. As a low calorie-high protein diet with almost no starch and sugars mushrooms are the 'delight of the diabetic'. Due to high K: Na ratio, few calories and low fat, mushrooms are the choice of the dietician for those with obecity, hypertension and atherosclerosis (heart disease). Alkaline ash and high fibre contents make them suitable for those suffering from hyperacidity and constipation (Rai and Sohi, 1988).

* E-mail: poonam.dehariya@yahoo.com

Mushrooms are good source of high quality proteins and are rich in vitamins and minerals. They contain 20-35 per cent protein which is higher than in vegetables and fruits and is a superior quality. They are rich in lysine and tryptophan the two essential amino acids that are deficient in cereals. They contain good amount of vitamin C and B complex (thiamine, riboflavin and niacin), potassium, phosphorus and sodium, also low but available form of iron. In mushrooms, potassium, sodium ratio is very high which is ideal for patients of hypertension. They are low caloric food with very little fat and sugars and without starch and cholesterol (Pathak, Yadav and Gour, 2000).

Nutraceuticals, Nutriceuticals and Pharmaceuticals

The term Nutraceutical is proposed by DeFelice in 1979 and quoted by Brower in 1988 and reported by Chang and Miles (1992). These are often referred as phytochemicals or functional foods. These are natural, bioactive, chemical compounds that have health-promoting, disease-preventing or medicinal properties.

Chang and Buswell in 1996 have coined the term 'mushroom nutriceuticals', is a refined/partially defined extractive which is consumed in form of capsules or tablets as a dietary supplement (not a food) and which has potential therapeutic applications increases resistance to diseases and causes regression in disease state.

Pharmaceuticals are defined chemical preparations, specifications for which are listed in a pharmacopoeia and are mainly used therapeutically; they have high potency, chances of overdose and toxicity are there thus needing regulation by the authorities.

Table 20.1

	Composition	*Usage/Activity*	*Mode of Delivery*	*Accessibility*
Nutraceutical	Modified/enriched food material	Maintenance of healthy diet low potency	Consumed as part of normal diet	Readily available from general retail outlets
Nutriceutical	Refined/partially defined food extractives	Dietary supplements, Medicinal properties, use primarily prophylactic but with potential therapeutic usage	Capsules/tablet low toxicity little scope for overdose	Comparable with over the counter medicines
Pharmaceutical	Almost invariably a defined chemical preparation with medicinal properties. Specification listed in a pharmacopoeia	Specific medical treatment, mostly therapeutic usage.	Oral administratio/ inhalation/topical/ injection High potency, Toxic potential	Range from over the counter medicines to strictly controlled dangerous drugs

Nutritive Contents of Mushrooms

Protein

Protein is one of the most important food factors. The sufficiency of protein in a diet is an important measure of its adequacy and quality. Mushrooms due to high quantity and quality of protein have been recognized by the Food and Agriculture Organization as food contribution to the protein nutrition of the countiers depending largely on cereals (Geda and Joshi, 2006).

On a dry weight basis, mushrooms normally contain 19-35 per cent protein, as compared to 7.3 per cent in rice,13.2 per cent in wheat, 39.1 per cent in soybean, and 25.2 per cent in milk. Therefore in amount of crude protein, mushrooms rank below most animal meats but well above most other foods, including milk, which is an animal product. Jandaik (1974) from Solan reported 33.5 per cent protein in *P. sajor-caju*. Protein content of mushrooms has also been reported to vary from to flush (Crisan and Sands, 1978) and also with the cultivation substrate (Bano and Rajarathanam, 1982). Chang and Miles (1989) have suggested that total protein content of 3.5-4 per cent in commercial edible mushrooms may be more representative. Quality of mushroom protein is far superior to the vegetables proteins, amino acids (Crisan and Sands, 1978; Bano and Rajarathanam, 1982; Chang and Miles, 1989).

Essential amino acids-proteins are made up from over 20 amino acids in varying amounts, they are quantitatively different. The human body can convert some of these amino acids into others, but there are nine essential amino acids (lysine, methionine, tryptophane, threonine, valine, leucine, histidine, and phenylalanine). Table 20.3 shows that for all mushrooms listed in table, the most abundant essential amino acid is lysine, and the lowest levels among the essential amino acids are those of tryptophane and methionine.

Pleurotus flabelltus (Bano *et al.*, 1963) and *P. sajor-caju* (Jandaik and Kapoor, 1976) have been reported to contain seventeen to eighteen amino acids respectively including all the essential amino acids as well as most commonly occurring non-essential amino acids and amides.

Table 20..2: Essential Amino Acid Comparison of Edible Mushrooms

Amino Acids	*Agaricus bisporus*	*Lentinula edodes*	*Pleurotus ostreatus*	*Pleurotus sajor-caju*	*Volvariella volvacea*	*Hen's egg*
Leucine	7.5	7.9	6.8	7	4.5	8.8
Isoleucine	4.5	4.9	4.2	4.4	3.4	6.6
Valine	2.5	3.7	5.1	5.3	5.4	7.3
Tryptophan	2	nd	1.3	1.2	1.5	1.6
Lysine	9.1	3.9	4.5	5.7	7.1	6.4
Threonine	5.5	5.9	4.6	5	3.5	5.1
Phenylalnine	4.2	5.9	3.7	5	2.6	5.8
Methionine	0.9	1.9	1.5	1.8	1.1	3.1
Histidine	2.7	1.9	1.7	2.2	3.8	2.4
Total essential amino acids	38.9	36	33.4	37.6	32.9	47.1

Source: Chang and Miles, 2004.

Carbohydrate and Fiber

According to Crisan and Sands (1978), pentose, methylpentoses, hexoses, as well as disaccharides, amino sugars, sugar alcohols, and sugar acids are constituents of mushroom carbohydrate. *Pleurotus* species contain carbohydrates, ranging from 46.6 to 81.8 per cent (as compared to 60 per cent in *A. bisporus* on a dry weight basis). Fiber is considered to be an important ingredient in a balanced and healthy diet. The fiber content ranges from 7.4 to 27.6 per cent in *Pleurotus* species (as compared to 10.4 per cent in *A. bisporus*, and 4 to 20 per cent in *V. volvacea*).

Minerals

Mushrooms are good source of minerals. In general, mushrooms contain significant quantities of phosphorus, potassium, and copper, *Agaricus bisporus* being especially high in these minerals. A serving of white mushrooms has more potassium than a tomato.

Mushrooms are rich in minerals as may be evident from their ash content 8-10 per cent in *A. bosporus*, 5-15 per cent in *Pleurotus* spp., 11.5 per cent in *V. volvacea* and 7.0 per cent in *Lentinus edodes* mineral of higher content is potassium (45 per cent of total ash content) followed by phosphorus, sodium, magnesium and calcium which together constitute about 56-70 per cent of total ash content (Chang and Miles, 1989). Potassium and phosphorus are the main constituents of ash of *Pleurotus* species also (Bano and Rajarathanam, 1982) iron are present in appreciable amount in *Pleurotus* species.

Vitamins

There are many essential vitamins required daily in the diet. The fruit body of a mushroom is an excellent source of B-complex vitamins including riboflavin (B 2), niacin, pantothenic acid, thiamine (B1) biotin, folate and vitamin B12. Mushrooms are a particularly rich source of riboflavin. Mushrooms are unique in that they contain Vitamin B12, something that vegetables can't produce at all (Ingramm, 2002).

Riboflavin content was 5 mg in *A. bosporus*, 4.9mg in *L. edodes* and 1.63 to 2.98 mg in *V. volvacea* (Chang and Miles, 1989). Folic acid and vitamin B12, which are generally absent in plant foods, are present in mushrooms, although in small quantities. Mushroom also contain vitamin C (Crisan and Sands, 1978; Bano and Rajarathanam, 1982). On fresh weight basis *A. bisporus* contains about 3 mg ascorbic acid (Beelman and Edwards, 1989).

Fat

In general the crude fat of mushrooms has representatives of all classes of lipid compounds including free fatty acids, monoglycerides, diglycerides, triglycerides, sterols, sterol esters, and phospholipids.The values for saponifiable lipid range from 78.1 per cent in *Auricularia auricula* to 58.8 per cent in *Volvariella vovacea*. The low percentage of saponifiable lipid found in *Volvariella volvacea* is mainly due to the presence of unusually high contents of provitamin D2 and ergosterol. Saturated fatty acid found in *Volvariella volvacea* is 14.6 per cent, *Lentinula edodes* 19.9 per cent, *Agaricus bisporus* 19.5 per cent, *Pleurotus sajor caju* 20.7 per cent, *Auricularea auricula* 25.8 per cent, *Tremella fusiformis* 22.8 per cent.

The unsaturated fatty acid found in *Volvariella volvacea* is 85.4 per cent, *Lentinula edodes* 80.1 per cent, *Agaricus bisporus* 19.5 per cent, *Pleurotus sajor-caju* 20.7 per cent, *Auricularea auricula* 25.8 per cent, *Tremella fusiformis* 22.8 per cent (Chang and Miles, 2004).

Moisture

Generally the moisture of fresh mushrooms ranges from 85-95 per cent on an average; the fresh mushroom contains approximately 90 per cent moisture. But it varies from 78.3-90.5 per cent in *Agaricus bisporus*, 89.7 per cent in *Agaricus campestris*, 89.1 per cent, *Auricularia* sp., 87.3 per cent in *Boletus edodes*, 89.2 per cent in *Flammulina velutipes*, 90.0-91.8 per cent in *Lentinula edodes*, 92.2 per cent in *Pleurotus eous*, 91.5 per cent in *Pleurotus florida*, 73.7-90.8 per cent *Pleurotus ostreatus*, 90.1 per cent in *Pleurotus sajor-caju*, 90.4 per cent in *Volvariella diplasia*, 89.1 per cent in *Volvariella volvace* (Chang and Miles, 2004)

Table 20.3: The Nutritive Content of some Mushrooms

	Agaricus bisporus	*Lentinus edodes*	*Pleurotus sajor-caju*	*Pleurotus ostreatus*
Calories	355	356	300	360
Protein g/100g	34.44	32.93	26.6	27.25
Fat g/100g	3.10	3.73	2.0	2.75
Carbohydrate g/100g	47.38	47.60	50.7	52.76
Dietary fiber g/100g	20.90	28.90	13.3	43.80
Vitamin A 1U/100g	0	0	–	0
Thiamine (B1) mg/100g	0.27	0.25	–	0.26
Riboflavin (B2 mg/100g	4.13	2.30	–	2.45
Niacin (B3) mg/100g	69.20	20.40	–	65.80
Pantothenic acid (B5)mg/100	12.70	11.6	–	33.20
Vitamin C mg/100g	0	0	–	0
Vitamin D 1U/100g	235	110	–	136
Calcium mg/100g	23	23	20	5
Copper mg/100g	4.33	1.23	12.2	1.61
Iron mg/100g	2.1	5.5	12.4	11
Potassium mg/100g	4500	2700	3260	4600
Sodium mg/100g	52	18	60	13
Selenium mg/100g	0.415	0.076	–	0.175

Source: Stemet, 2005.

Nutritional Index

Crisan and Sands in 1987 proposed the use of a nutritional index calculated as follows.

$$\text{Nutritional index} = \frac{(\text{EAA Index} \times \text{Percentage protein})}{100}$$

The essential amino acid index (EAA) is measured according to dietary protein in terms of an essential amino acid pattern based on known adult human dietary requirements. The amino acid score (chemical score) is the amount of the most limiting amino acid in the food protein expressed as a percentage of the same amino acid present in the reference protein. The EAA Indexes, amino acid scores, and nutritional indexes for various mushrooms and other foods are shown in Table 20.4. In terms of EAA Indexes and amino acid scores, the most nutritive mushrooms rank alongside meat and milk in potential nutritive value and considerably higher than most legumes and vegetables. The least nutritive mushrooms rank appreciably lower but are still comparable to some common vegetables (Chang and Miles, 2004).

Essential Amino Acid index (EAA Index) was proposed by Oser (1951) reported by Chang and Miles (2004) to evaluate the quality of dietary protein in terms of the ratio of the essential amino acids contained in a food, relative to the essential amino acid content of a highly nutritive reference protein.

Oser (1951) has used whole egg protein as his reference protein.

$$\text{EAA Index} = n\sqrt{\frac{\text{Lysine P} \quad \text{Tryptophane P} \quad \text{Histidine P}}{\text{Lysine S} \quad \text{Tryptophane S} \quad \text{Histidine S}}}$$

P: Food protein; S: Standard (egg); n: The number of amino acids.

Chemical Score or Amino Acid Score is another estimate of nutritive value used by some investigators (FAO, 1968). The Chemical Score is based on the amount of the most limiting amino acid present in the reference protein (egg) and is calculated using the following formula (FAO, 1968).

$$\text{Chemical Score} = \frac{\text{Grams essential amino acid in test protein}}{\text{Total essential acids in test protein}} \times \frac{\text{Total essential amino acids in egg}}{\text{Grams essential amino acids in egg}} \times 100$$

The Chemical Score reflects the 'percentage of adequacy' of the protein and approximates the probable efficiency of utilization of test protein in children (FAO, 1993). By calculating a factor equal to 100 times the reciprocal of the amino acids scores one can determine the present in the reference protein (Crisan and Sands, 1978).

The Chemical Score is also calculated by using FAO reference protein (FAO, 1973) with the help of the following formula.

$$\text{Amino Acid Score} = \frac{\text{Mg of amino acid in 1 g test protein}}{\text{Mg of amino acid in 1 g - reference protein}} \times 100$$

Table 20.4: Comparison of Nutritive Value of Mushrooms with Various Foods

Essential Amino Acid Indexes		*Amino Acid Sources*		*Nutritional Indexes*	
100	Pork; chicken; beef	100	Pork	59	Chicken
99	Milk	98	Chicken; beef	43	Beef
98	Mushrooms (high)	91	Milk	35	Pork
91	Potatoes; kidney beans	89	Mushrooms (high)	31	Soybeans
88	Corn	63	Cabbage	28	Mushrooms (high)
86	Cucumbers	59	Potatoes	26	Spinach
79	Peanuts	50	Corn	25	Milk
76	Spinach; soybeans	46	Kidney beans	21	Kidney beans
72	Mushrooms (low); cabbage	42	Cucumbers	20	Peanuts
69	Turnips	33	Turnips	17	Cabbage
53	Carrots	32	Mushrooms (low)	14	Cucumbers
44	Tomatoes	31	Carrots	11	Corn
		28	Spinach	10	Turnips
		23	Soybeans	9	Potatoes
		18	Tomatoes	8	Tomatoes
				6	Carrots
				5	Mushrooms (low)

Source: Chang and Miles, 2004.

Delicious Recipies of Mushrooms

Mushroom Pakoras

Ingredients

- ☆ Small size mushroom (only the pileups part, 200g)
- ☆ Onion -100g
- ☆ Basan -100g
- ☆ Baking powder-1 tsp.
- ☆ Chilli powder -1/2 tsp.
- ☆ Egg -1
- ☆ Salt to taste
- ☆ Mustard oil/butter-1 tsp.

Method

- ☆ Beat the eggs and add flour, salt, chilli powder, baking powder.
- ☆ Make dough of 25g each. Make a hole on a dough.
- ☆ Stuff the hole with mushroom.
- ☆ Roll the doughs in bread crumbs.
- ☆ Fry the doughs with oil or butter till they become brown.

Button Mushroom fry

Ingredients

- ☆ 100gms button mushroom
- ☆ 1 tsp. chilli powder
- ☆ 1/3 tsp. salt
- ☆ 2 gm cumin seeds
- ☆ Oil required quality.

Method

- ☆ Wash mushroom and place it on Blotting/Old news paper.
- ☆ Heat oil on a tawa. When the oil heats up add the mushrooms.
- ☆ Fry the mushrooms for 2 minutes. Add all the other ingredients and fry for 3 minutes.

Mushroom Sandwich

Ingredients

- ☆ Mushroom-100gm
- ☆ Onion-2 small bulbs
- ☆ Tomato-1
- ☆ Green coriander leaves
- ☆ Green-chilli-2

- ☆ Red chilli powder as required.
- ☆ Salt to taste
- ☆ Butter-25 gm.

Method

- ☆ Wash mushrooms and cut into pieces.
- ☆ Cut onion into small pieces.
- ☆ Put the pieces of onion and mushroom on a pan, add butter and heat gently.
- ☆ In a separate pan, put tomato, pieces of green chilli, green coriander leaves.
- ☆ Add salt and other spices, fry for 10 minutes.
- ☆ Put baked mushroom with onion pieces in the pan.
- ☆ Add few drops of Onion juice and heat gently with butter for 2-3 minutes.
- ☆ Remove from the fire.
- ☆ Put the prepared material into the pieces of bread.

Mushroom Starter

Ingredients

- ☆ Button mushroom-1 packet
- ☆ Butter-250gm.
- ☆ Parsley 1 bunch chopped finely
- ☆ Bread pieces.
- ☆ 1 tps of Garlic paste.

Method

- ☆ Mix butter with finely chopped parsley and garlic paste, remove stem from the mushrooms.
- ☆ Fill the mushroom caps with the herbed butter, in a oven proof dish and grill for 15 min or till the mushrooms are nicely coated with the butter.
- ☆ Serve hot with small bread pieces.

Mushroom Soup

Ingredients

- ☆ Mushroom-250 gm.
- ☆ Butter -25 gm.
- ☆ White flour-one tsp
- ☆ Black peeper-1/2 tsp
- ☆ Ground spices (Chilli, black pepper, cardamom etc.)-3g
- ☆ Milk 1-2 litre

Method

- ☆ Wash and cut mushrooms into very small pieces.

- Put pressure to remove water from mushrooms
- Fry mushrooms with butter in a pan.
- Add milk and boil for 5-7 minutes.
- A suspension of water and flour is added to the milk for make it thick.
- Boil the liquid for 5-10 minutes.
- Add the spices and salt to make it more tasty.

References

Bano, Z., Srivastava, K. S., and Srivastava, H.C. (1963). Amino acid composition of the protein from a mushroom. (*Pleurotus* sp.). (Later on identified as *Pleurotus flabellatus.*). *Appl. Microbiol.*, 11(12): 363-365.

Bano, Z. and Rajarathanam, S. (1982). *Pleurotus* mushroom as nutrition food. In: *Tropical Mushrooms-Biological Nature and Cultivation Methods* (Chang, S. T. and Quimio, T. H., Eds). The Chinese University Press, Hong Kong. 363-382.

Beelman, R.B. and Edwards, C.G. (1989). Variability in the composition and nutritional value on the Cultivated Mushroom, *Agaricus bisporus*. *Mushroom News*, 37: 17-26.

Chang, S. T. and Miles, P.G. (1989). *Edible Mushrooms and their Cultivation*. CRC Press, Boca Raton, pp. 345.

Chang, S.T. and Miles, P.G. (2004). *Mushrooms: Cultivation, Nutritional Value, Medicinal Effect and Environmental Impact*, 2nd edn. CRC Press, p. 28-30.

Chang, S.T. and Buswell, J.A. (1996). Mushroom nutriceuticals. *World J. Microbiol. Biotechnol.*, 12: 473-476.

Crisan, E. V. and Sands, A. (1978). Nutritional value. Un: *The Biology and Cultivation of Edible Mushrooms*, Chang, S.T. and Hayes, W. A., Eds., Academic Press, New York, 137-168.

Food and Agriculture Organisation (FAO) (1968). *Amino acid content of Foods and Biological Data on Proteins*. Food Agric. Organ., U.N., Rome.

Geda, A.K. and Joshi, P. K. (2006). Nutritional qualities of Mushrooms. *Compandium of Lectures-Emerging Areas in Mushroom Diversity, Production and postharvest Developments*, pp. 53-68.

Ingramm, S. (2002). The real nutritional value of fungi. NUTRITIONAL _VALUE.htm.

Jandaik, C.L. and Kapoor, J.N. (1976). Studies on *Pleurotus sajor-caju* (Fr) Singer. *Mushroom Sci.*, 9: 667-672.

King, T.A. (1993). Mushrooms: The ultimate health food but little research in US to prove it. *Mushroom News*, 41(2): 46-29.

Pathak, V. N., Yadav, N. and Gour, M. (2000). *Mushroom Production and Processing Technology*. Agrobios, India.

Rai, R.D. (1997). Medicinal mushrooms. *Advances in Mushroom Biology and Production*, (Rai, Dhar and Verma, eds.) MSI Solan.

Rai, R.D. and Sohi, H.S. (1988). How protein rich are mushrooms. *Indian Horticulture*, 33(2): 2-3.

Stemets, P. (2005). Notes on Nutritional properties of culinary-medicinal mushrooms. *International Journal of Medicinal Mushrooms*, 7: 103-110.

Microbial Biotechnology and Ecology (2011) Pages 353–376
Editors: Deepak Vyas, G.S. Paliwal, P.K.Khare and R.K Gupta
Published by: DAYA PUBLISHING HOUSE, NEW DELHI

Chapter 21

Mushroom: The Beneficial Fungi

☆ *Anjuli Choubey[1]*, Deepak Vyas[1], Poonam Dehariya[1], A.K. Pandey[2] and D.C. Atri[1]*

[1]*Lab of Microbial Technology and Plant Pathology, Department of Botany, Dr. H.S. Gour University, Sagar – 470 003, M.P.*
[2]*Chairman, Madhya Pradesh Private University Commission, Bhopal, M.P.*

ABSTRACT

Mushrooms are the fruiting bodies of macrofungi. They include both edible/medicinal and poisonous species. Mushroom farming is both science and art. The science is developed through research, the art is perfected through curiosity and practical experience. Present review article has studied under following heads:

☆ First of all ecological description and taxonomic position of the mushrooms have been described on the basis of their spore production method and their properties.

☆ Second, cultivation technology of some mushrooms have been described. Mushroom cultivation is friendly to the environment and helpful in bioconversion/bioremediation of environment adulterants and maintenance of balanced ecosystem.

☆ Third, the major attributes of mushrooms, their medicinal properties, has also been drawn to our attention for study *e.g.* immunomodulatory, antitumour activity, hypotensive etc.

☆ Forth heading highlights the nutraceutical value of the mushrooms because they are rich in crude fibre and protein. Infact, mushrooms also contain low fat, low calories and good vitamins.

☆ Last step of this article is marketing of the mushrooms, which is equally important with the production of the crop. The development of a crop like mushroom depends on its efficient

* E-mail: anjuli83@yahoo.co.in

marketing. The efficiency of mushroom marketing is crucial in determining the profit of the producer on one hand and the level of satisfaction of the consumer on the other. So it is important to throw some light on this aspect with specific objective: to examine the existing marketing pattern of mushroom strategies in order to improve the marketing of mushroom.

Biodiversity of Mushroom

Everyone has had some encounter with mushrooms whether on a walk through the woods, strolling through the park or just finding them in our lawns and gardens. Mushrooms are phenomenon of the nature that is hard to overlook. Mushrooms have been objects of much curiosity and speculation since times immemorial. They are an important component of the forest ecosystem. Their edibility, poisonous nature, psychotropic properties, mycorrhizal and parasitic association with the forest trees make them economically important and interesting to study (Lakhanal, 1996).

India is one of the top ten megadiversity centres and has ample species of wild mushrooms which occur during rainy season. There are at least 12,000 species of fungi that can be considered as mushrooms with at least 2,000 species showing various degrees of edibility (Chang, 1999). Furthermore, over 200 species of mushroom have been collected from the wild and utilized for various traditional medicinal purpose in the far east. Mushrooms are a macrofungi belong to ascomycotina and basidiomycotina. They are achlorophyllous and saprophytic. They absorb their nutrients with the help of very fine thread like structures called mycelium which penetrate into substrate and forms fruiting body that is called mushroom. The fruiting body of the mushroom may be fleshy, subfleshy or sometimes leathery and umbrella like or the various other shapes, size and colour. Generally mushrooms consists a pileus or cap, a long stalk or stipe and gills but some have additional structures like a veil, or annulus, a cup or volva.

Table 21.1: Ascomycotina

Sl.No.	*Group*	*Botanical Name*	*Common Name*	*Morphological Description*	*Ecology*
1.	Cup shaped mushroom	*Aleuria aurantia*	(Orange peel)	Bright orange 2-10 cm wide shallow saucer shaped cup made of brittle flesh. Its outer surface is lighter and often slightly hairy with the inner surface a smooth bright orange that fades to an orange yellow with age.	A very common mushroom found in disturbed areas. They are found in groups or clusters in hard soils on roadsides and in newly disturbed soil.
		Chlorociboria aeruginascens	(Black rubber cup)	A very small (usually to 0.5 cm) blue green cup that leaves a green stain on the rotting logs and barkless wood it grows on.	It is found from june to November on rotting logs and barkless wood.
2.	Morels	*Gyromitra infula*	(Saddle shaped false morel)	This mushroom has a 3-10 cm wide cap that is brownish and saddle shaped with interior chamber and brittle flesh and has 1-5 cm. long stalk is 2 cm thick and whitish or buff in colour. It is hollow and has no ridges.	Found on rotten wood or on the ground in woody debris.

Contd...

Table 21.1–Contd...

Sl.No.	Group	Botanical Name	Common Name	Morphological Description	Ecology
		Helvella lacunose	(Fluted black helvella)	A dark saddle shaped 1-5 cm wide cap that is lobed and wrinkled above but smooth below. It is dark brown or grey to grey-black in colour. It has a 4-10 cm. long deeply fluted or ribbed stalk with interior chambers. It is greyish in color.	It is found on the ground or on decaying wood in coniferous and deciduous woods. It found from September-November.

Table 21.2: Basidiomycotina

Sl.No.	Group	Botanical Name	Common Name	Morphological Description	Ecology
1.	Toad stools (All fleshy gill bearing fungi)	*Amanita phalloides*	Death cap	Pileus 3-3.5 inches across, rounded at first them more or less flattened, variable in color, but usually greenish or yellowish-olive and streaked with fine, darker fibrils in early stages a few large white patches of the volva adhere to the cap.	Grows in mixed deciduous woodland particularly in association with oak. The season is early summer to quite autumn.
		Volvariella volvacea	Paddy straw mushroom	It is a dark species of warm countries, which is occasionally found on decaying vegetable matter in hot houses.	Grows solitary or gregarious on rotten paddy straw mushroom. It found from April–September.
		Lepiota procera	Parasol mushroom	Pileus upto 8 inches across, dull brown or greyish brown, covered with shaggy scales, at first somewhat rounded or oval, so that the young plant has the appearance of a drum stick, later becoming bell shaped and finally expanded and with a prominent umbo, the margin thin, usually turning downwards and fringed.	Grows in small troops among leaf litter or short grass often at the edge of woods.
		Agaricus bisporus	White button mushroom	The fruiting body of this mushroom has a stipe and an umbrella shaped cap. on the under side of the pileus gills are present. When the fruiting body is young the gills are enclosed by a membrane that extends from	It grows on compost heaps, in garden waste and beside roads. Occasionally on the edges of hedges.

Contd...

Table 21.2–Contd...

Sl.No.	Group	Botanical Name	Common Name	Morphological Description	Ecology
				the margin of the pileus to the stipe. This membrane is called veil. It is torn around the margin of the pileus. When the pileus expands but it remains as a ring called annulus.	
2.	Bracket fungi	*Ganoderma applanatum*	Artist's conk	A woody often horney, flat to hoof shaped, grey to brownish, stalkless cap with a white pore surface that bruises readily, hence the other common name is "Artist's conk" because a drawing can be made on the surface. The cap is 5-50 cm wide with 0.5-5cm thick brown flesh.	Grows year round on dead wood, particularly deciduous, but has been found on conifers and living trees as well.
3.	Jelly fungi	*Tremella mesenterica*	Witches butter	2.5 cm. wide globular or brain like or lobed, orange yellow to clear yellow in colour, turns orange red on drying.This one looks a lot like Orange jelly (*Dacrymyces palamatus*) but it lacks a stem.	It grows year round on deciduous wood.
4.	Polypores	*Tramates versicolor*	Turkey tail	An overlapping or rosette forming stalkless polypore with thin leathery kidney or spoon shaped, multicolored caps that may be flat or wavy silky, hairy or velvety with alternate smooth zones.	Found on dead woods or wounds in living trees
		Pleurotus ostreatus	Oyster mushroom	A broad fleshy, white, grey or brown cap with broad whitish or yellow tinged gilles. Cap is 15-20 cm. wide, oyster shaped semicircular to elongated margin lobe is wavy, pleasant odour, thick, white flesh, stalk is short or absent.	Usually grows in clusters on dead tree trunks or branches and rarely on living trees. It grows year round in favourable condition.
		Pycnoporus cinnabarinus	*Cinnabar red polypore*	A 2-12 cm. wide semicircular or kidney shaped flattened stalkless polypore. The cap is cinnabar-red to orange, rough and slightly hairy but with age becomes warty and wrinkled.	Found in all seasons on dead deciduous and occasionally aged coniferous wood.

Contd...

Table 21.2–Contd...

Sl.No.	Group	Botanical Name	Common Name	Morphological Description	Ecology
5.	Coral fungi	*Clavularia cristata*	Crested coral	A many branched whitish, crested coral with characteristic tooth like jagged tips color is whitish but becomes yellowish to ochare with age. It grows 2-3 cm wide and about 2-8 cm high.	Found growing on the ground in wooded areas from August to November.
		Lentaria bysiseda	Cotton based coral	A creamy or pinkish tan coloured branched coral mushroom with characteristic pliant-tough flesh. It grows 1-6 cm. high and about as wide, it has short and thick stalk rising out of hairy white mycelial patch, though the stalk is not always present.	This is very common coral mushroom on twigs, leaves and cones or rotten wood of both coniferous and deciduous trees. It can be found in season anywhere from August–November depending on the conditions.
6.	Puff balls	*Lycoperdon perlatum*	Gem studded puffball	A whitish tan coloured round or turban shaped mushroom with an stalk like base. It is covered with detachable long and short spines. It develops a pore in the center and has white flesh that becomes greenish-brown with the age.	Often found growing in moist lowland forests sometimes on rotten wood.
		Pseudohydnum gelatinosum	Jelly tooth	A whitish translucent tongue like jelly with gelatinous flesh, occasionally with brownish or grayish tones. It has tooth like white spines on its underside.	Found on rotting woods in coniferous or mixed forest.
7.	Slime fungi	*Lycogala epidendrum*	Wolf's milk slime	A pinkish or reddish massed that produces small (0.3–0.5 cm wide and high) rounded cushions that may be pink, orange, grey or brownish and exude a pinkish paste when squeezed.	A common mushroom from June-November found scattered or clustered on dead logs
		Physarum polycephalum	Yellow slime	A sheet like yellowish expanding and contracting slimy mass roundish heads on unbranched stalks.	Found from June-October on dead wood and fleshy fungi.

Table 21.3: Edible Mushrooms

Sl.No.	Common Name	Botanical Name
1.	Button mushroom or champignon	*Agaricus bisporus*
2.	Oyster mushroom	*Pleurotus ostreatus*
3.	Pig mushroom or stone mushroom	*Boletus edulis*

Contd...

Table 21.3–Contd...

Sl.No.	Common Name	Botanical Name
4.	Cauliflower mushroom	*Sparassis crispa*
5.	Birch bolete	*Leccinum scabrum*
6.	Saffron milk	*Lactarius deliciosus*
7.	Sweet tooth fungus	*Hydnum repandum*
8.	Chicken mushroom	*Laetiporous sulphureus*
9.	Shimji or beech mushroom	*Hysizygus tessulatus*
10.	Snow fungus	*Tremella fuciformins*
11.	Tooth fungus	*Hericium erinaceus*
12.	Tree ear fungus	*Auricularia polytincha*
13.	Winter mushroom or Enokitake	*Flammulina velutipes*
14.	Chanterelle or Girolle	*Cantharellus cibarius*
15.	Pig's ear gomphus	*Gomphus clavatus*
16.	Wood blewit	*Lepista nuda*
17.	Golden-spotted waxy cap	*Hygrophorus chrysodons*
18.	Trumpet clitocybe	*Clitocybe geotropa*
19.	Silky grey tricholoma	*Tricholoma portentosum*
20.	Wood wooly foot	*Collybia peronata*

Table 21.4: Poisonous Mushrooms

Sl.No.	Common Name	Botanical Name
1.	Herald of the winter	*Hygrophorus hypothejus*
2.	Common funnel cap	*Clitocybe infundibuliformis*
3.	Yellow brown tricholoma	*Tricholoma fulvum*
4.	Blackening lyophyllum	*Lyophyllum fumato fotens*
5.	Collybia confluens	*Tufted collybia*
6.	The stainer	*Mycena crocata*
7.	Tall amanita	*Amanita excelsa*
8.	Garlic mushroom	*Marasmius alliaceus*
9.	Red gilled agaric	*Melanophyllum echinatum*
10.	Safron parasol	*Cystoderma aminthinum*
11.	Glistening ink-cap	*Coprinus micaceus*
12.	Fringed crumble cap	*Psathyrella candolleana*
13.	Scaly crumble cap	*Psathyrella squamosa*
14.	Verdigris agaric	*Stropharia aeruginosa*
15.	Scaly psilocybe	*Psilocybe squamosa*
16.	Fruity cortinarius	*Cortinarius amoenolens*

Contd...

Table 21.4–Contd...

Sl.No.	Common Name	Botanical Name
17.	Slimy banded	*Cortinarius collinitus*
18.	Fleecy milk cap	*Lactarius vellereus*
19.	Golden false pholiota	*Phaeolepiota aurea*
20.	Blood red Cortinarius	*Dermocybe sanguinea*

Table 21.5: Fungi with Conflicting Reports on Edibility

Binomial	Notes
Agaricus arvensis	Reported mostly as edible and eaten in some parts of the world; also said to be a gastrointestinal irritant (Lincoff and Mitchel, 1977).
Agaricus semotus	Said to be edible from Hong Kong (Chang and Mao, 1995).
Amanita flavoconia	Conflicting accounts from Mexico: one report says it is edible, the other that it is poisonous.
Coprinus atramentarius	Edible if eaten in the absence of alcohol; this produces an unpleasant effect if imbibed at the same time, hence remarks that it is potentially poisonous (Lincoff and Mitchel, 1977).
Gyromitra esculenta	In Finland it is a delicacy (Härkönen, 1998) and it is also widely eaten in the Russian Federation and neighbouring regions. In other countries it is said to be poisonous and can kill when raw (Hall *et al.*, 1998a). The toxic properties are mitigated by suitable preparation prior to eating.
Helvella lacunosa	Widely eaten but also reported as toxic if eaten raw (Lincoff and Mitchel, 1977).
Lampteromyces japonicus	A common cause of poisoning in Japan (Hall *et al.*, 1998a) but also has medicinal properties (Hobbs, 1995).
Morchella esculenta	Like other morels said to be poisonous if eaten raw (Lincoff and Mitchel, 1977). Edible and good when cooked.
Tricholoma pessundatum	Edible in Hong Kong (Chang and Mao, 1995) but *T. pessundatum* var. *montanum* reported as poisonous elsewhere (Lincoff and Mitchel, 1977).
Tricholoma sulphureum	All records say it is poisonous apart from an account from India that says it is edible (Purkayastha and Chandra, 1985).

Cultivation Technology of Mushroom

Mushroom cultivation is completely different from growing green plants. Mushroom do not contain chlorophyll and therefore depend on other plant material (the substrate) for their food. A mushroom cultivar should be thoroughly familiar with the life cycle of fungi. Mushrooms can also be grown in lawns, polytonal, vegetable garden, woodlands and in permaculture design (Edwards, 2000).

Mushrooms can be cultivated through a variety of methods. Some methods are extremely simple and demand little or no technical expertise on other hand, cultivations which require aspects of sterile handling technology are much more technically demanding (Chang and Miles, 1989). Mushroom cultivation involves several different operations.

Each of which must be performed properly if the enterprise is to be successful. Failure of any phase will result in a decreased harvest or total loss.

Steps for Mushroom Cultivation

1. Selection of mushroom spores or strains
2. Maintenance of mycelial culture
3. Development of spawn/inoculum
4. Preparation of growing medium
5. Spawn inoculation and colonization of substrate

Mushroom can easily grow on different substrates for instances on domestic, industrial or agricultural wastes, such as paddy straw, cotton waste, jute waste, palm bunch waste, banana leaves, sugarcane waste, water hyacinth, paper waste, sawdust, wood etc.

Cultivation Technology of *Pleurotus* Mushroom

Common Name–Oyster mushroom

Indian Name–Dhingri

Substrate

Wheat/paddy straw, soybean straw or other agricultural wastes can be used as a basal substrate. The straw should be fresh, dried and stored in proper conditions. The substrate should be free from diseases, contaminants and other plant material. For substrate preparation, wheat straw is chemically sterilized/soaked in water 18-24 hours and dipped in hot water for 2-3 hours at 60°C. Excess water is drained off. Polythene bags may be used as containers.

Supplementation

For substrate supplementation, supplements rich in N_2 are added. Soybean meal, Gram powder, Ground nut powder may be used as supplements.

Spawning

For spawning, 18-20 day old spawn is used, very old spawn develops clumps and results into proper mycelial growth. Spawning is done in 3 layers with 40 per cent of grain spawn. The straw are taken in polythene bags and seeded with wheat grain spawn. The spawn layer is covered with wet straw. Two per cent vent area recommended for polythene bags to prevent spoilage of the substrate.

Fruiting

After19 days of spawning, mycelia run through the substrate now polythene bags is removed. Watering is done almost every day. Pinheads are generally appeared after 23 days of spawning. Some of these develop into mature fruit bodies within another 3-4 days. These are picked up and sold in the markets either in fresh conditions or after drying.

Harvesting

First flush occurs 3-4 weeks of spawning. Usually matured ones are picked up gently without disturbing the smaller ones. After first harvest the bed are kept as such for another one week or so and are watered as and when necessary in order to provide appropriate moisture. Yield of second flush is much less than that of first one.

Risk Factors

Tricholoma spp., *Aspergilus* spp., and *Rhizopus* spp. are commonly observed on the substrate during spawn run. These can be avoided by good hygeine.

Cultivation Technology of *Calocybe* Mushroom

Common Name–White milky mushroom

Indian Name–Dudhchatta, Kuduk

Other Name–Saint George's mushroom

Substrate

Wheat/paddy straw can be used as a basal substrate. The straw should be fresh, dried and stored in proper conditions. The substrate should be free from diseases, contaminants and other plant material. For substrate preparation, wheat straw is chemically sterilized/soaked in boiled water for 2-3 hrs. And used for seeding the beds. Polythene bags may be used as containers. Two percent vent area recommended for polythene bags to prevent spoilage of the substrate.

Supplementation

For substrate supplementation, supplements rich in N_2 are added. Soybean meal, may be used as supplements.

Spawning

For spawning, 18-20 day old spawn is used. Very old spawn develops clumps and results into proper mycelial growth. Spawning is done in 3 layers with 40 per cent of grain spawn.

Casing

After completion of mycelial growth the casing is done with two year old cow dung patties of size (1-2") or garden soil, coirdust and farm yard manure (FYM). The thickness of casing should be kept 1".

Fruiting

After 10-12 days of casing, fruit primordia are formed with in 5-6 days, mature fruit bodies of the mushroom become ready for harvest. The bags should not be opened before the mycelial growth completion. The bags can be arranged on alluminium racks at a distance of 5-10cm. The cropping room should be properly ventilated and adequately lighted.

Risk Factors

Due to poor hygenic conditions, many competitor moulds and fungi may develop and hamper the mycelial growth. *Tricholoma* spp., *Aspergilus* spp., and *Rhizopus* spp. are commonly observed on the substrate during spawn run. These can be avoided by good hygeine.The ink caps (*Coprinus* spp.) in the spawned bags can be removed by hand picking or by reducing the humidity of cropping room below 90 per cent. The disease can be reduced by avoiding excess moisture and providing proper aeration.

Harvesting

The right stage for picking can be judged by the size of fruit body. The fruit bodies should be harvested before spore release by twisting so that the stubs are not left on the straw. It is better to pick all the mushroom in the bag at a time and then irrigate them.

The spent straw from *Calocybe indica* can be used for biogas production. The spent substrate can also be used as manure.

Cultivation Technology of *Lentinus edodes*

Common Name–Shittake

Indian Name–King or monarch of mushroom

Other Name–Xiang gu

Substrates

There is good development and high production only with rich sources of wood. The substrate usually comprises a mixture of wood chips, cereal bran and millet. Wood chips from hardwood species are usually preferred, although a small proportion of Softwood species may be mixed with the hardwoods. A formula of sawdust (80 per cent), wheat bran (10 per cent) can also be used. After mixing the drying ingredients by hand or with a mechanical mixer, water is added to the mixture so that the final moisture content of the substrate is between 55 and 68 per cent. Depending on the capacity of the sawdust to absorb water. The ingredients are then packed into autoclavable polypropylene or high density polyethylene bags. After the bags have been filled (1.5 to 4 kg wet wt.) The neck of the bag is fitted with a plastic collar and plugged with a cottonwool stopper.

After filling the bags with either noncompacted or compacted nutrient supplemented sawdust, the substrate is heat–treated to kill unwanted competitive microorganisms. Heat treatment may carried out in an autoclave at 121°C for 1 hr or in a brick or cement lined tower at 90-95°C for 5-7 hr.

Supplementation

Some growers use feather meal, mineral and vitamin supplements.

Spawning

After cooling, the substrate is inoculated with spawn. Inoculation is usually performed in a laminar flow. Spawn run (the time required for the mycelium to colonize the substrate) may range from 18-30 days. During this period, the bags are usually incubated in a 4hr/20hr light/dark cycle at 23-25°C. The substrate may lose 5-10 per cent of its dry weight and may subsequently have a higher moisture content. With longer spawn runs (over 40 days), The surface of the colonized substrate may begin to turn brown. Some exudates may be produced and drain to the bottom of the plastic bag. At the end of the spawn run the plastic bag is cut away and the synthetic logs are subjected to production conditions. The wood chips are held together by the fungal mycelium. In some cases, Portions of the bag may be left on the synthetic logs to retain moisture. Misting be provided daily by overhead nozzles controlled automatically. Relative humidity is often maintained at 85 per cent. Temperature range of 14-16°C optimum for most lines of Shiitake.

Fruiting

Fruiting is near when rings of white mycelia appear on the bags.

Harvesting

Mushroom should be harvested when their caps are about two-third open. Mushrooms may be stored in a cardboard box for up to a month at 36-41°F.

Mushrooms may also be dried whole or sliced before marketing.

Risk Factors

Many competitor moulds, pests, insects may be develop due to poor hygenic conditions. In which *Hypoxylon coccineum, Merulius tremellosus, Monilia* spp., *Poria vaporaria, Achorute* spp., *Lepidocyrtus spp.* are included.

Cultivation Technology of *Ganoderma lucidum*

Common Name–Lingzhi

Indian Name–Reishi

Other Name–Mannentake

Reishi often is associated with health and recuperation, longevity, wisdom, and happiness (Stamets 1990, 1993). It is believed that certain triterpenes and polysaccharides may account for the multiple activities of Reshi. Thus, considerable time and effort has gone into the isolation and characterization of these compounds. About 100 triterpenes have been isolated from either the fruitbodies or mycelium but only a few have been tested for bio-activity (Mizuno *et al.*, 1995).

Substrate

Hardwood sawdust like Oak (*Quercus* spp.), Maple (*Acer* spp.) etc. is the most popular basal ingredient used in synthetic formulation of substrate used to produce reishi (Royse, 1997; Tong and Cheh, 1990; Soto-vela Zco *et al.*, 2002). A very fine grade of saw dust is generally used for cultivation. Other basal ingredients that may be used include wheat straw corn cobs or mixtures paddy straw, bagasse, mixture of saw dust and wheat, paddy etc.

Supplementation

Supplements such as wheat bran, rice bran, millet, rye, corn etc. are added to the mix into the main ingredient. There supplements serve as nutrients to provide an optimum growing medium.

Spawning

For spawning, grain spawn 30-40 g/kg substrate and sawdust spawn 60-80g/kg substrate is used. The inoculated bags are kept in growing houses for approximately 1-1½ months for spawn run. During spawn run, bag should be kept under low light intensity and high humidity with little ventilation. When the spawn run is completed within 30-35 days, the polythene bag cover is removed and the substrate blocks are exposed.

Fruiting

Formulation of young mushroom buttons. 2-expanding (opening) of mushroom caps, colour at the margin is whitish till it grows but turned to red when the fruit body is matured it takes about 20-25 days from the button stage.

Harvesting

Harvesting is done when the edge of mushroom turned to red colour. Harvesting is done by picking the mushroom and holding the mushroom stalk and gently twisting it from the substrate block. Reishi mushroom can be dried. Good yield would be 300 gm reishi per kg dry substrate.

The advantages of mushroom cultivation can be summarized as:

1. Wastes such as cereal straws are largely burnt by the farmers, which causes air pollution. However, these raw materials can actually be used for the cultivation of mushrooms. This kind of bioconversion exercise can greatly reduce environmental pollution.
2. Mushroom cultivation can be a labour intensive activity. Therefore, it will serve as means of generating employment, particularly for rural women and youths in order to raise their social status. It will also provide additional work for the farmers during winter months when the farming schedule is light.
3. It will provide the people with an additional vegetable of high quality, and enrich the diet with high quality proteins, minerals and vitamins which can be of direct benefit to the human health and fitness. The extractable bioactive compounds from medicinal mushrooms would enhance human's immune systems and improve their quality of life.
4. Mushroom is a cash crop. The harvested fruiting bodies can be sold in local markets for additional family income or exported for an important source of foreign exchange that will definitely improve the economic standards of the people

5. Some warm mushrooms, *e.g. Volvariella volvacea* (Straw mushrooms) and *Pleurotus sajor-caju* (Oyster mushrooms) are relatively fast growing organisms and can be harvested in 3 to 4 weeks after spawning. It is a short return agricultural business and can be harvested in 3 to 4 weeks after spawning. It is a short return agricultural business and can be of immediate benefit to the community.

Therapeutic Properties of Mushrooms

There are at least 2000 species of fungi that can be considered as mushroom in which at present 270 species of mushroom that are known to have various therapeutic properties (Ying *et al.*, 1987). The practice of using fungi, especially mushrooms, in chinese herbal medicines has been recordes of the *"Materia medica"*. The earliest book on medicinal materials in China, the *"Shen Noug's Herbal"* (Shen Noug Pen T's ao Jing) (100-200AD) recorded the medicinal effects of several mushrooms including *Ganoderma lucidum, Poria cochos, Tremella fusiformis* and others medicinal mushrooms have become even more widely used as traditional medicinal ingredients for the treatment of various diseases and related health problems largely due to the increased ability to produce the mushrooms by artificial method.

Higher basidiomycetes mushrooms have been used in folk medicine throughout the world since ancient time (Wasser and Weis, 1999). Far from being an outdated curiosity, medicinal mushrooms earn ever growing popularity, stimulated by a high number of scientific studies. The discovery of antibiotics has convinced the public that fungi could produce particularly active substances and that the ancient traditions could have a part of truth. Mushrooms are known to possess all essential amino acids, minerals, vitamins etc., in adequate quantities and low in sugars and such are selective medicinal food for diabetics. Probably this might have captured the attention of F.A.O. for recommending mushrooms as a food especially to the under developed countries where the protein malnutrition has become a real threat to human health.

Mushrooms in Ayurveda

In ayurveda it has been mentioned that mushrooms are body builders, energisers and they cure stomach problems. A special preparation recommended for consumption to tuberculosis patient is mushrooms boiled in milk with sugar. Even though references are lacking in literature, it is a well known belief that "Nilamanga" (*Sclerotium stipipatum*) a rare underground mushroom is supposed to be a valuable natural medicine for a number of common diseases. Ayurvedic medical practitioners accept this as an incomparable medicine for a number of body disorders. Nilamanga are just like a mango in their shape and size and are found underground, usually associated with old termite nests under the soils. They appear singly or in bunches with small stalks. Inner flesh is white and soft when young at maturity it become hard with a thick black outer coating.

It has been reported that a boiled water extract of Nilamanga if consumed will prevent the body dehydration. Boiled Nilamanga extract in *sesamum* oil is a very good medicine for the internal application to cure ear pain in children. A mixture of crushed Nilamanga in butter is very congenial for external application for faster curing of body abscissions. Similarly, extract of crushed "Nilamanga" is reported to be a good remedy for jaundice than any other medicines (Nair *et al.*, 1986).

Mushrooms have various immunological and anticancer properties (Mizuno *et al.*, 1995). They also offer other potentially important therapeutic properties including antioxidant, antihypertensive, cholesterol lowering, liver protection, antifibrotic, anti-inflammatory, antidiabetic, anti-inflammatory, antiviral and antimicrobial (Lakhanpal *et al.*, 2005). Many pharmaceutical companies are viewing the medicinal mushrooms as a rich source of innovative biomedical molecules. Many polysaccharide bound proteins produced by Basidiomycetae fungi have classified as anti-tumour chemicals by the U.S. National Cancer Institute.

Antitumoral

A number of mushroom compounds capable of modifying the biological response of the host by stimulation of the immune system provoking therapeutic effects like inhibition of tumour growth and relief in immunodeficient diseases such as AIDS. Japanese researchers in the 1970's, at a French convention, presented a new type of antitumoral molecule found in a mushroom. This compound, a polysaccharide, does not act on the tumour by direct cytotoxicity like already known substances but acts by stimulation of the immune system (Takashi, 1995). Since then several other substances have been discovered. Several immunotherapeutic drugs have been commercialized in Japan, among them PSK (Krestin) extracted from *Trametes versicolor* mycelium marketed in 1977 to treat digestive organ, lung and breast cancer. About 10 years later was commercialized a similar product, PSP. Lentinan from the *Lentinus edodes* fruit body commercialized in 1985 for stomach cancer, SPG (*Schizophyllan*) from *Schizophyllum commune* put on market 1986 against cervical cancer.

Hypocholesterolemic

The major cause of mortality world is cardiovascular disease. One of the preventive measures against these diseases is to reduce the cholesterol level in the blood. This can be done by diet or by a therapeutical agent, like lovastanin (mevinoline), among Basidiomycetes, *Pleurotus ostreatus* has shown the highest lovastatine content. Other species are active by lowering blood cholesterol *i.e.Pleurotus cornuciopiae, P.eryngii, P. sapidus, Lentinus edodes, Polyporus confluens, Auricularia polytricha, A. auricular-judae, Flammulina velutipes* and *Ganoderma lucidum* (Wasser and Weis, 1999).

Anti-inflammatory Effects

More than half of the people over 65 years old suffer from arthritis. Drugs on the market (aspirin, steroids) are of limited efficiency, because of potential secondary effects. In reason of the long traditional use of *Ganoderma lucidum* in Asia for rheumatism and stiff articulations, several studies have been done on this mushroom seeking a non steroidic drug. The possibility seems to be very promising on developing on an antiinflammtory drug from *Ganoderma* extracts. Some heterogalacton–protein complexes with anti inflammatory and antiallergic functions in *Ganoderma lucidum*. According to Wasser and Weis (1999) anti inflammatory effects can be found in other mushroom products from *Tremella fusiformis, Schizophillum commune, Lentinus edodes, Flammulina velutipes, Marasmius androsaceus.*

Antiviral

Diverse basidiomecetous substances are effective against viral infections including AIDS. These are mostly substances which mobilize the body's humoral defences (Wasser and Weis, 1999) for instances, Lentinane extracted from *Lentinus edodes* suppressed the surface expression of human immunodeficiency virus on T cells when dosed in combination with ayodothymidine (Tochikura *et al.*, 1987). It was also active against vesicular stomatitis virus in mice, encephalitis virus, abelson virus and adenovirus type 12. Other mushrooms like *Grifola frondosa, Tsramates versicolor, Ganoderma lucidum* have antiviral effects too (Wasser and Weis, 1999).

Hypoglycemic

Diabetes is a disease due to the lack of regulation of the blood sugar level due to the malfunction or non function of specialized pancreatic cells. Certain mushrooms could have a positive effect on early stage of type 2 diabetes. For instances Ganoderanes isolated from *Ganoderma lucidum* showed hypoglycemic activity in mice (Hikino *et al.*, 1989). According to Lo *et al.* (2005) *Tremella mesenterica* might be developed as a potential oral hypoglycaemic agent or functional food for diabetic patients and for persons with high risk for diabetes mellitus. *Lentinus edodes* and *calocybe georgii* are two other examples of mushrooms showing hypoglycaemic activity.

Hypotensive

Certain compounds can maintain blood pressure at a normal level (Kabir *et al.*, 1988) showed the effect on blood pressure of a powder prepared from *Ganoderma lucidum* fruit bodies on hypertensive rats. *Grifola frondosa, Lentinus edodes, Auricularia auricular-judae* and *Armellaria mellea* are other mushrooms lowering blood pressure (Wasser and Weis, 1999).

Antifungal

Ecologically the production of antifungal substance is almost surely due to the constant need of the organisms to compete against other fungi. There are indeed a lot of observations of aggressive action of one species towards another. It was 1923 that Falck first recognized such an activity in *Sparassis crispa* (Delmas, 1989).

Miscellanous Activities

Other kinds of activities can be found in Basidiomycetes, *i.e.* Cardiovascular (*i.e. Lentinus edodes, Auricularia auricular-judae, Armellaria mellea, Grifola frondosa, Volvariella volvacea*), Antiallergic (*i.e. Agaricus compestris*), Hepatoprotective (*i.e. Schizophyllum commune, Dendropolyporus umbellatus, Tramates Versicolor, Ganoderma lucidum, Lentinus edodes*).

Table 21.6: Medicinal Mushrooms

Sl.No.	Species	Common Name	Medicinal Value	Key Active Constituents
1.	*Auricularia dimidate*	Jew's ear or wooden ear or tree ear	1. Pile treatment and lowers blood cholesterol 2. Stopping pain and bleeding. 3. Treatment in haemorrhoids and excessive uterine bleeding (ying *et al.*, 1987)	Polysaccharides
2.	*Pleurotus ostreatus*	Oyster mushroom or mushroom of the flower heaven	1. Anticancerous activity 2. Immunomodulating effects. 3. Antiviral, antibiotic and anti-inflammatory activities. 4. Cholestrol lowering ability.	Beta glucans (antitumorous) Lovastatin (Cholestrol lowering)
3.	*Ganoderma lucidum*	Reishi or mannentake or lingzhi or ling chu or ling chin or mushroom of immortality.	1. Treatment of hepatopathy and chronic hepatitis. 2. Treatment of hypertension, arthritis and insomnia. 3. Treatment in asthma, gastric ulcer and bronchitis. 4. Reduce B.P., blood cholesterol and blood sugar levels as well as inhibit platelet aggregations.	Beta and hetro-B-glucans (antitumor, immunostimulating) Lingzhi-8-protein (antiallergenic, immunomodulating). Ganodermatic acids-Triterpenes (antiallergenic agents, cholesterol and B.P. reducing).

Contd...

Table 21.6–Contd...

Sl.No.	*Species*	*Common Name*	*Medicinal Value*	*Key Active Constituents*
4.	*Lentinus edodes*	Shittake or xiang gu or fragrant mushroom.	1. Treating high blood pressure and lowering blood cholesterol 2. Strenthening health. 3. Anticancerous activity. 4. Enhancing the immune system.	B-D-glucan (lentinan) Heteroglucan protein (LEM). Eritadenine (cholesterol reducing) Ergosterol (Provitamin D-2) RNA fractions (antiviral-nucleic acids)
5.	*Hericium erinaceus*	Monkey head mushroom or hedgehog mushroom	1. Beneficial for heart, Kidney, liver, lung and spleen. 2. To promote digestion and general vigour, strength and general nutrition. 3. The polysaccharide from this mushroom have cytostatic effects on gastric oesophageal hepatic and skin cancers (Mizuno, 1999.)	Beta-D-Glucans (antitumerous) Ergosterol (provitamin D) Cyathane derivatives (nerve growth stimulators)
6.	*Grifola frondosa*	Hen of the woods or sheep's head orfan shaped mushroom	1. Anticancer activity (Hishida *et al.,* 1988, Kurashige *et al.,* 1997) 2. Immunological enhancement together with property of anti HIV, antihypertension, antidiabetic and antiobesity (Zhunang and mizuno,1999) 3. Treatment of flu and common infection (Bacteria and Virus) 4. In diabetes mellitus 5. In hypercholesterolemia 6. In urinary tract infections	1,3, and 1,6 Beta glucans (antitumor and immuno-modulating) commercial product 'Grifolan'.
7.	*Tremella mesenterica*	White auricularia or white jelly fungus	1. Act as immune tonic. 2. For treating debility and exhaustion together with many other ailments including skin 3. Special significance in cosmetic industry.	Acidic polysaccharide-Glucuronoxylomannen (Antitumor, Immunostimulatory, antidiabetic, skin enhancing.)
8.	*Trametes* (*Coriolus*) *versicolor*	Turkey tail	1. Act directly on tumor cells as well as directly in the host to boost cellular immunity (Tsukagoshi,1984) 2. Antiviral activity through stimulation of interferon production. 3. Inhibit HIV-1 (Collins and Ng, 1977).	PSK or "Krestin"- a water soluble protein bound polysaccharide and PSP (direct act on tumour cells)
9.	*Poria cocos*	Wolfiporia cocos	1. Treatment of viral hepatitis. 2. Anticancer activity. 3. Immunomodulatory activities.	Polysaccharide-spachyman and packymaran (antitumour and immunomodulating). Tetracyclic triterpenes (immunomodulation and antiviral).

Contd...

Table 21.6–Contd...

Sl.No.	Species	Common Name	Medicinal Value	Key Active Constituents
10.	*Phellinus linteus*	Meshimakobu or "song gen"	1. "Worked as a miracle medicine" refreshing the human body and prolonging longevity (Ying *et al.*, 1987). 2. Tumour proliferation suppression (Mizuno, 2000). 3. Effects on digestive system cancers *e.g.* gastric, duodenal, colon, rectal and also liver cancer. 4. To treat liver diseases and gastric ulcers. 5. Inhibit the growth of cancers	Beta-glucans marketed as meshima capsules.
11.	*Flammulina velutipes*	Enokitake	1. To treat liver diseases and gastric ulcers. 2. Inhibit the growth of cancers.	Beta glycoprotein-proflamin
12.	*Calocybe indica*	Dudhchhatta	1. Antidiabetic activity	*P. anisaldehyde*
13.	*Agaricus bisporus*	Button mushroom	1. Immune enhancing 2. Antitumour properties.	Proteoglucans Beta (1-6)-D-glucan

Nutritional Value of Mushroom

The greatest difficulty in feeding man is to supply a sufficient quantity of the body-building material – protein. The other three nutritional categories are: the source of energy food-carbohydrates and fats; accessory food factors– vitamins; and inorganic compounds which are indispensable to good health. Of course, water, too, is essential. The moisture content of fresh mushrooms varies within the range of 70–95 per cent depending upon the harvest time and environmental conditions, whereas it is about 10–13 per cent in dried mushrooms. The protein content of the cultivated species ranges from 1.75 to 5.9 per cent of their fresh weight. It has been estimated that an average value of 3.5 to 4.0 per cent would be more representative. This means that the protein content of edible mushrooms in general, is about twice that of onion (1.4 per cent) and cabbage (1.4 per cent), and four times and 12 times those of oranges (1.0 per cent) and apples (0.3 per cent), respectively. In comparison, the protein content of common meats is as follows: pork, 9-1629 per cent; beef, 12-20 per cent; chicken, 18-20 per cent; fish, 18-20 per cent; and milk, 2.9-3.3 per cent. On a dry weight basis, mushrooms normally contain 19-35 per cent protein, as compared to 7.3 per cent in rice, 12.7 per cent in wheat, 38.1 per cent in soybean, and 9.4 per cent in corn. Therefore, in terms of the amount of crude protein, mushrooms rank below animal meats, but well above most other foods, including milk, which is an animal product (Chang and Miles 1989).

Quantitative data relating to the nutritive value of mushrooms is sparse. In the absence of feeding trials, alternative methods have been used to determine or predict the nutritional value of foods based on their content of essential amino acids (Crimson and Sands 1978). The Essential Amino Acid Index (EAA. Index) rates dietary protein in terms of an essential amino acid pattern based on known adult human dietary requirements. The Amino Acid Score (Chemical Score) is the amount of the most limiting amino acid in the food protein expressed as a percentage of the same amino acid present in the reference protein. In an attempt to resolve the difficulties inherent in comparisons between those mushrooms containing small amounts of high quality protein with those containing larger amounts of a protein of lesser nutritional quality, (Crisan and Sands, 1978) proposed the use of a Nutritional Index calculated as:

$$\text{Nutritional index} = \frac{(\text{EAA Index} \times \text{Percentage protein})}{100}$$

Table 21.7: Proximate Composition (Percent fresh weight of cultivated mushrooms)

Mushroom/Vegetable	*Moisture*	*Protein*	*Fat*	*Carbohydrate*	*Fibre*	*Ash*	*Calory*
Agaricus bisporus	90.1	2.9	0.3	5.0	0.9	0.8	36
Pleurotus sajor-caju	90.2	2.5	0.2	5.2	1.3	0.6	35
Cabbage	91.9	1.8	0.1	4.6	1.0	0.6	27
Cauliflower	90.8	2.6	0.4	4.0	1.2	1.0	30
Potato	74.7	1.6	0.1	22.6	0.4	0.6	97

Table 21.8: Essential Amino Acids in Edible Mushroom

Amino Acid	*White Button Mushroom*	*Oyster Mushroom*	*White Milky Mushroom*
Valine	1.63	4.7	2.48
Alanine	2.4	–	1.6
Argenine	1.90	6.7	trace
Aspartic acid	3.14	–	1.00
Metheonine	0.39	1.2	–
Histidine	0.64	2.0	2.4
Tryptophane	3.94	0.89	–
Lysine	1.62	5.0	5.56
Threonine	1.48	4.2	0.52
Phenel alanine	1.55	2.0	–
Cystine	–	–	–
Glutamic acid	7.06	–	2.12
Glycine	1.2	–	10.80
Leucine	2.16	4.4	–
Isoleucine	1.28	5.6	–
References	Hayes and Haddad (1976)	Bano *et al.* (1963)	Purkayastha and Nayak (1981)

The EAA Indices, the Amino Acid Scores, and Nutritional Indices for various mushrooms and other foods, are reported by (Crisan and Sands 1978). EAA Indices and Amino Acid Scores of the most nutritive mushrooms (highest values) rank in potential nutritive value with those of meat and milk, and are significantly higher than those for most legumes and vegetables. The least nutritive mushrooms rank appreciably lower but are still comparable to some of our common vegetables. In addition to their good proteins, mushrooms are a relatively good source of the following individual nutrients: fat, phosphorus, iron, and vitamins including thiamine, riboflavin, ascorbic acid, ergosterine and niacin. They are low in calories, carbohydrates and calcium. It has also been reported that a total lipid content varying between 0.6 and 3.1 per cent of the dry weight, is found in the commonly cultivated mushrooms. At least 72 per cent of the total fatty acids are found to be unsaturated in all the four tested mushrooms (Huang, *et al.*, 1985). It should be noted that unsaturated fatty acids are essential and significant in our diet and to our health. In recent years, there has been a trend toward discovering ways of treating mushrooms so as to give them added value. For example, Wermer and Beelman (2002) have reported on growing mushrooms enriched in selenium. By adding sodium selenite to compost over a range of 30-300 parts per million, they found that the mushrooms increasingly absorbed selenium according to the amount in the compost, so that it is possible to grow mushrooms containing a desired concentration. Selenium is an essential micronutrient that has generated much recent interest in nutritional and medical research–and, more recently, within the food industry (Beelman and Royse, 2006). Selenium has numerous physiological functions, but is best known as necessary 30 cofactor for the glutathione peroxidase enzyme system. This system is responsible for removing free radicals from the body, thus reducing oxidative damage. The desirability of a food product does not necessarily bear any correlation to its nutritional value. However, its appearance, taste, and aroma, sometimes can stimulate one's appetite (preference). In addition to nutritional value, mushrooms have some unique colour, taste, aroma, and texture characteristics, which attract their consumption.

Marketing of Mushrooms

Marketing is the most important consideration for the success of any firm as far as the mushroom's marketing are concerned, if we can't sell our mushrooms at a price that ensures a reasonable profit margin, we don't want to invest in this enterprise. So it is necessary that a good cultivar of mushroom should be well familiar with their marketing management.

"Marketing management is a total system of business activities designed to plan, price, promotion, and distribute want to satisfying products, services and ideas to target markets in order to achieve organizational objectives" (Kotler, 1999).

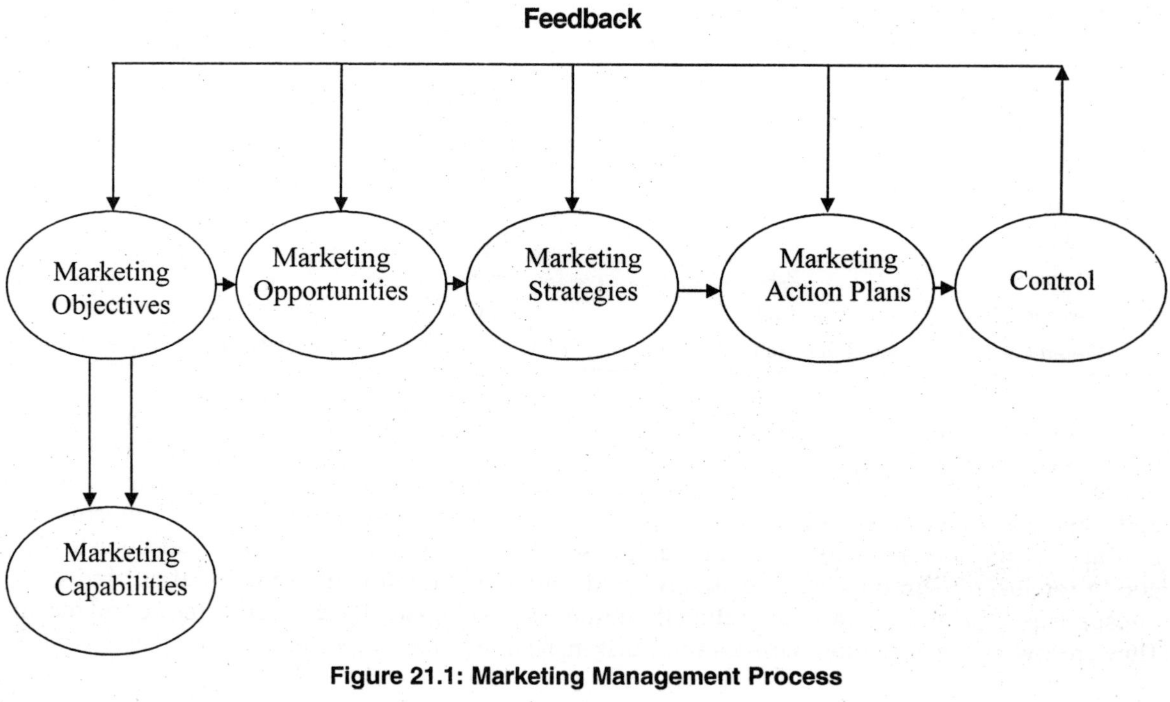

Figure 21.1: Marketing Management Process

Objectives of Marketing Management

- The size, extent and future potential of market.
- The factors governing the purchasing decisions of the consumers.
- The collection of the data about users opinion about the idealness of the product and use this as pointer to the future trend.
- The investigations about the possible new application as of the product.
- The judgement about the threat and its extent of competition and
- The development of optimum effective methods.

These are some techniques which are needed for management of mushroom marketing

- Intellectual skill
- Communication skill
- Organization understanding skill
- Decision making skill
- Motivation skill

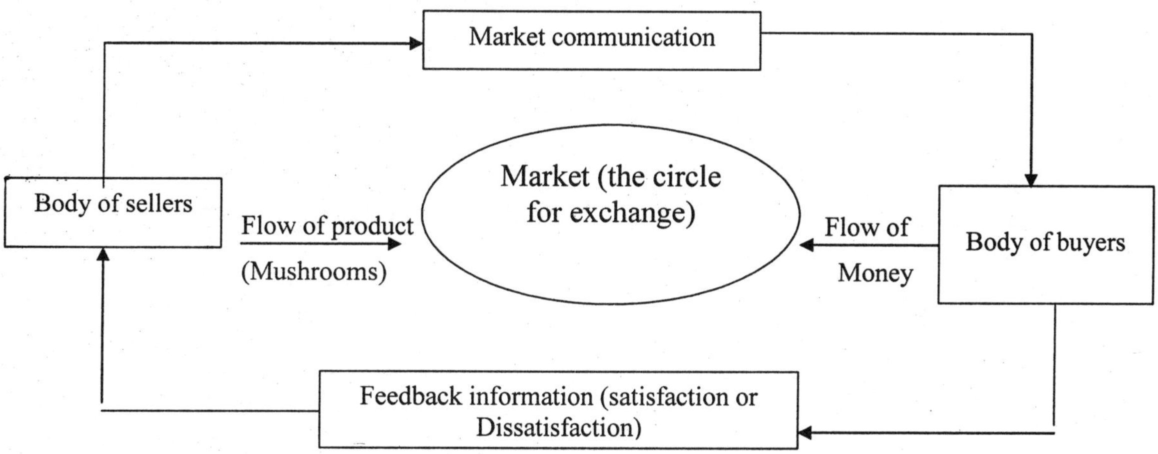

Figure 21.2: A Simple Marketing System

Need of the Mushroom's Marketing

Mushrooms have various nutraceuticals, pharmaceutical, cosmaceutical and medicinal properties. But a number of people of our country are unknown with their health increasing properties. So the main target of mushroom's marketing is may enable to popularize the mushrooms as potential food and medicines. Besides this, their cultivation may be helpful to provide employment and management of organic waste.

Marketing Strategy

Marketing strategy is the set of objectives, policies and rules that guide the firm's marketing efforts. Marketing mix, market segmentation, and distribution strategy are included in it.

Marketing Mix

Marketing mix is a set of numerous tools that the firm or marketers use to elicit desired responses from their target market. These tools classified by McCarthy into four broad groups that he called the 4P's of marketing.

These elements of marketing mix (4P's) discussed below.

Product

It is a bundle of satisfactions, fulfilling the number of needs of targeted customers. The variables are quality, features, brand name, packaging, size etc.

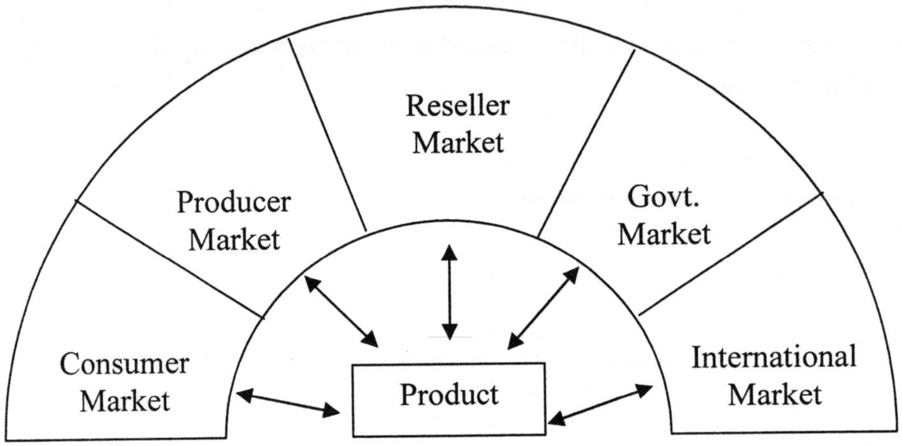

Price Activities

It is the amount which the customer a pay for the product. The deciding factors are target customers, cost, competition etc.

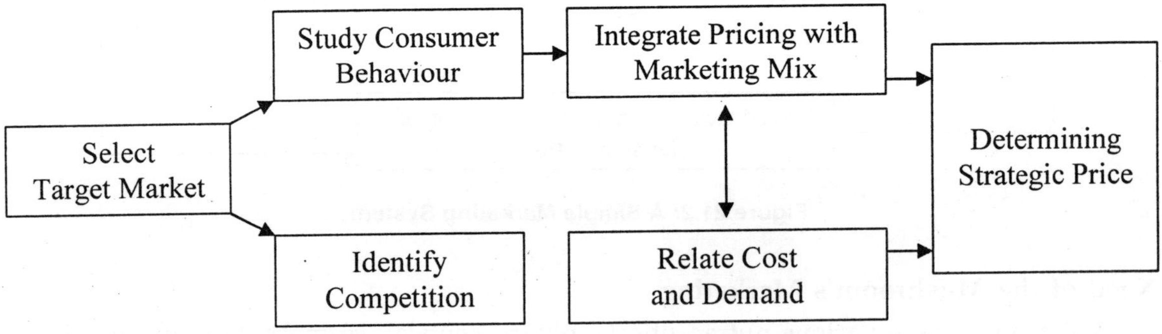

Promotional Activities

Advertisement is the major promotional activity industry. The various sources are sign boards, news papers and periodicals, sponsored programme on radio and T.V.

Place

Place of availability like cooperative stores, nearby shop and general stores in main markets.

R. lanterborn opines that 4P's represent the sellers view of marketing tools to influence buyers. From a buyer point of view, each marketing tool is designed to deliver a customer benefit. He suggest 4P's corresponds to 4C's *e.g.*

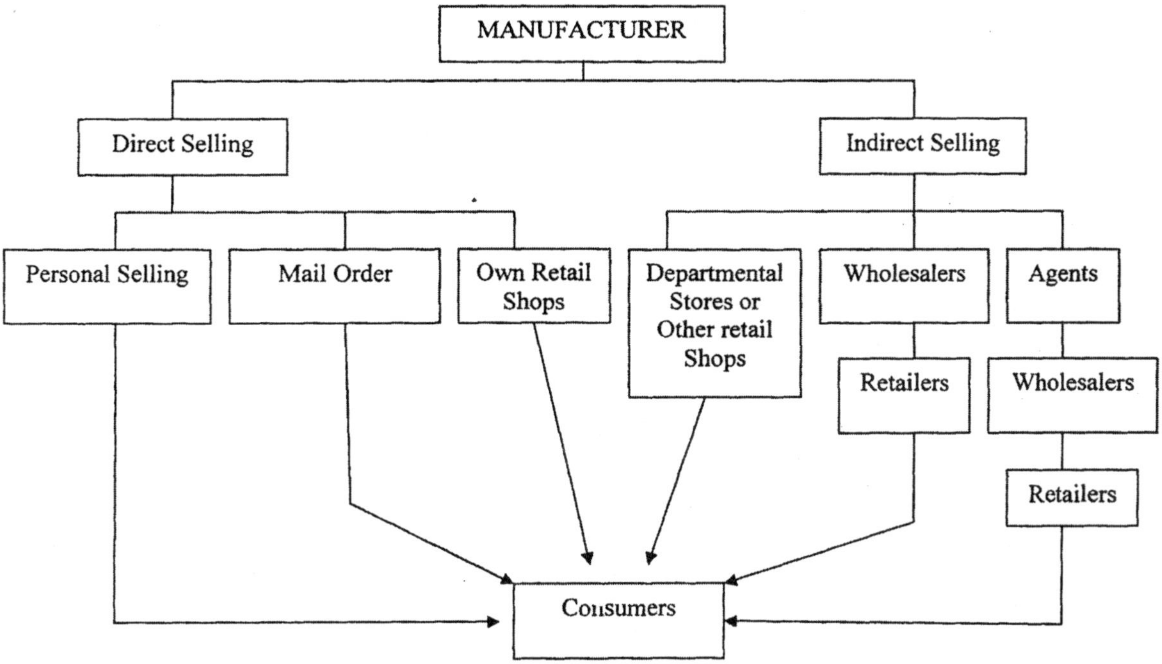

Four Ps	*Four Cs*
Product	Customer solution
Price	Customer cost
Place	Convenience
Promotion	Communication

Market Segmentation

"Market segmentation is the process of identifying group of buyers with different buying desires or requirements."Their conception lies on various base, important one's are listed below.

- ☆ Demographic basis like age, family size, education, occupation etc.
- ☆ Social class source of income, type of house, residential area etc.
- ☆ Family life cycle young, married/unmarried family size, age of youngest child, number of children in family etc.
- ☆ Psycholographic basis mode of living etc.
- ☆ Others like brand loyalty etc.

Goals of the Mushroom Marketing

- ☆ Maximize consumption of the mushrooms.
- ☆ Maximize satisfaction of mushroom's consumers.
- ☆ Maximize choice.
- ☆ Maximize life quality.

References

Adhikari, M.K. and Durrieu, G. (1996). Ethnomycologie nepalise. *Bull soc. Myc. France.* 112: 31-41.

Beelman, R.B., and Royse, D.J. (2006). Selenium enrichment of *Pleurotus cornucopiae* (paulet) Rolland and *Grifola frondosa* (Dicks: fr) S.F.Gray mushrooms. *Int. J. Med. Mush.* 8: 77-80.

Beyer, D. (2005). Spent mushroom substrate (SMS) research in the U.S. AMGA *J. Summer Issue* 31-32.

Chang, S.T. and Buswell, J.A. (1999).*Ganoderma lucidum* (Curt: fr), P.Karst. (Aphyllophoranymycetideae) a mushrooming medicinal mushroom. *Int. J. of Med. Mush.*1: 139-146.

Chang, S.T. and Miles (1989). Recycling of solid wastes with emphasis on organic wastes. *Green Productivity*, March issue 23-26.

Chang, S.T. (1999).World production of cultivated edible and medicinal mushroom in 1997 with emphasis on *Lentinus edodes* (Berk) Sing in China. *Int. J. Med. Mushrooms.*1 (4), 153-167.

Chang, S.T. and Miles, P.G. (2004). *Mushrooms cultivation, Nutritional value, Medicinal effect and Environmental impact*. Second edition.CRC press LLC, Florida.

Chaubey, A., Dehariya, P., Vyas, D. and Jain, A. (2010). Solid waste management by oyster mushroom cultivation. *Biozone International Journal of Life Science*, Vo. 2(1 and 2): 369–372.

Chaubey, A., Dehariya, P. and Vyas, D. (2010). Yield performance of *Calocybe indica* on conventional and non-conventional substrates. *J. of Mycol. and Pl. Pathol.*, 40(2): 176–178.

Collins, R.A., and Ng., T.B. (1987). Polysaccharopeptide from *Coriolus versicolor* has potential for use against human immunodeficiency virus type 1 infections. *Life science* 60, 383-387.

Crisan, E.V. and Sands, A. (1978). Nutritional value. In *"The biology and cultivation of edible mushroom"* edited to S.T.Chang and W.A. Hayes. Academic Press, New York.

Dehariya, P., Chaubey, A., Wagay, J.A. and Vyas, D. (2010). Wild mushrooms of Patharia forest of Sagar. *J. of Mycol. and Pl. Pathol.*, 40(2): 179–186.

Delmas, J. (1989). Les champignons et leur culture. Ed. Flammarion, Paris, 689 pp.

Edwards, R. (2000). The missing link? Mushrooms in permaculture. *Permaculture magazine.* No. 25.p. 37-39.

Hayes, W.A. and Haddad N. (1976). The food value of cultivated mushroom and its importance to the mushroom industry. *The Mushroom Journal* 40: 104-110.

Hikimo, H., Ishiyama, M., Suzuki, Y. and Konno, C. (1989). Mechanisms of hypoglycaemic activity of ganoderan B: a glycan of *Ganoderma lucidum* fruitbodies. *Planta Medica* 55, 423-428.

Hikino,H., and Mizuno,T. (1989). Hypoglycemic actions of some heteroglycans of *Ganoderma lucidum* fruit bodies. *Planta Med Aug.*, 55 (4): 385.

Hobbs,C. (1995). *Medicinal Mushrooms: An Exploration of Tradition, Healing and Culture.* Santa cruz, Botanica press.

Huang, B.H., Yung, K.H., Chang, S.T. (1985). The sterol composition of *volvariella volvacea* and other edible mushrooms. *Mycologia* 77 : 959-963.

Jain, A.K. (2005). *Thesis* on Cultivation technology of *Pleurotus* sp. with special reference to marketing potential in Sagar region.

Kabir,Y., Kimura, S.and Tamura, T. (1988). Dietary effect of *Ganoderma lucidum* mushroom on blood pressure and lipid levelsin spontaneously hypertensive rats (SHR). *J. Nutr. Sci. Vitaminol* (Tokyo). Aug., 34(4): 433-438.

Kotler, P. (1999). *Marketing Management.* The millennium edition. Printice Hall, India, New Delhi.

Kurashige, S., Akuzawa, Y.and Endo, F. (1997). Effect of *Lentinus edodes, Grifola frondosa, and Pleurotus ostreatus* administration on cancer outbreak, and activities of macrophages and immuno pharmacol. *immunotoxicol.* Maym 19(2): 175-183.

Lakhanpal, T. N. (1996). *Mushrooms of India Boletaceae* (eds. Mukherji, K.G.). A.P.H. publishing corporation.Vol. 1.

Lauterborn, Robert (1990). New marketing litany : 4Ps passwords take over, Advertising age,1-26.

Lincoff, Gary, Mitchel, (1977). *Toxic and hallucinogic mushrooms poisoning : A handbook for physians and mushroom hunters.*

Mizuno, T., Satio, H., Nishitoba and kowagishi, H. (1995). Antitumor–active substances from mushrooms. *Food Reviews Internationals* 11: 23-61.

Nair, M.C., and Bhavani Devi, S. (1986). Collection and identification of Agaricales. In beneficial fungi and their utilization. (Eds. Nair, M.C. and Balakrishnan, S.).pp. 25-27.

Noble, R. (2005). Spent mushroom substrate an alternative use. *AMGA J. Summer Issue,* pp. 33-35.

Pathak, V. N., Yadav, N., Gour, M. (1998). *Mushroom Production and Processing Technology.* Agrobios, India.

Purkayastha, R.P.and Nayak, D. (1981). Development of cultivation method and analyses of protein of a promising edible mushroom *Calocybe indica* P and C. *Mushroom Sci.,* 9: 697-713.

Purkayastha, R.P. and Chandra, A. (1985). *Manual of Indian Mushrooms.* 300 pp. New Delhi, India, Today and Tomorrow's printers and Publishers.

Royse, D.J. (1997). Speciality Mushrooms : Consumption, Production and Cultivation. Revista Maxicane *de micologia,* 13 : 1-11.

Soto-velazco, C., Lopez, M.C., Vazquez, E. and Alvarez, I. (2002). Cultivation of *Ganoderma Lucidum* and its effect on the production of lymphocytes in : *Mushroom Biology and Mushroom Products,* Sanchez, J.E., Muerta, G. and Mohtiel, E., Eds., Universities Autonoma Del Estado De Morelos, Mexico, 379-382.

Stamets, P, and Yao, C.C.W. (1998). Mycomedicinals: An informational booklet on medicinal mushrooms. *Mycomedia,* Olympia. 46 pp.

Stamets, P. (1990). A discussion on the cultivation of *Ganoderma lucidum* (Curtis: Fr.) Kar. The reishi of ling zhi mushroom of immortality. *McIlvainea* 9(2): 40-50.

Stamets, P. (2005). *Mycelium running : How mushroom can help save the world.* Ten speed press, Berkeley and Toronto, 574 pp.

Stamets, P. (1993). *Growing Gourmets and Medicinal Mushrooms.* Ted Speen Press. Hong Kong, 551 P.

Stamets, P. (2000). *Growing Gourmets and Medicinal Mushrooms.* Ten Speed Press, Berkely and Toronto, 339 pp.

Thakur, M.P., Shukla, C.S., Saxena, R.,Yadav,V. (2005). Emerging areas in mushroom diversity, production and postharvest developments (compendium of lectures). *IGAU* Raipur.

Tochikura, T.S., Nakashima, H., Kaneko,Y., Kobayashi,. N.(1987). Suppression of human immuno deficiency virus replication by 3-Azido-dexot Hymidine in various human haematopoietic cell lines in vitro. Augmentation by the effect of lentinan. *J. Cancer Res.* (Gann), p. 578-583.

Tong, C. and Z. Chen. (1990). Cultivation of *Ganoderma Lucidum* in Malysia, *Mush J. Tropics,* 10 : 27-30.

Tsukagoshi, N., Sasaki, T., Takemura, T., Ihara, H., Idota, Y., Yamagota, H. and Udaka, S. (1984). Efficient synthesis and secretion of a thermophillic alpha-amylase by protein producing *Bacillus brevis* 47 carrying the Bacillus stearothermophillus amylase gene. *J. Bacteriol.* 164 (3) : 1182-1187.

Tsukagoshi, S. (1984). Krestin (PSK). *Cancer Treatment Reviews* 11,131-155.

Vyas, D., Chaubey, A., Dehariya, P., Wagay, J.A., Bajpay, A., Lodhwal, R.K. (2009). Mushroom cultivation a tool for agripreneurship and rural management. In: *Perspectives on Agripreneurship and Rural Development* (Eds.) Prof. R.K. Pandey, Prof. P.S. Tripathi, Dr. S.K. Dubey, Dr. Ashish Bajpai. Faculty of Studies, BHU, pp. 463–473.

Wang, H.X. and Ng, T.B. (2001). Isolation of pleuriregin, a novel ribosome-inactivating protein from fresh sclerotia of the edible mushroom (*Pleurotus tuber-regium*). Sclerotia. *Int. J. of Biochem. Cell. Biol.,* 33(5): 483-490.

Wasser, S.P. and Weis, A.L. (1999). Therapeutic effects of substances occurring in higher basidiomycetes mushrooms : a modern perspective. *Crit. Rev. Immunol.,* 19(1): 65-96.

Ying, J., Mao.X., Ma, Q., Zhong,Y. and Wen, H. (1987). Icons of medicinal fungi from China. *Science Press,* Beijing, China.

Zhuang, C. and Mizuno,T. (1999). Biological responses from *Grifola frondosa* (Dick. fr) S..F. gray-maitake. (Aphyllophoranymycetideae). *Int. J. of Med. Mush.* 1: 317-324.

Vyas, D., Jain, A.K., Bajpei, A. (2007). Marketing potential of *Pleurotus* mushroom and management strategies in Sagar region. *J. Bot. Soc. Univ.* Sagar.

Microbial Biotechnology and Ecology (2011) *Pages* **377–390**
Editors: **Deepak Vyas, G.S. Paliwal, P.K.Khare and R.K Gupta**
Published by: **DAYA PUBLISHING HOUSE, NEW DELHI**

Chapter 22

Trichoderma Species: The History and Evolution of Current Concepts of Biological Control

☆ ***Pradeep Kumar Singh*[1]* *and Deepak Vyas*[2]**

[1]*Government Naveen College, Janakpur, Koriya, CG*
[2]*Department of Botany, Dr. Hari Singh Gour University, Sagar, M.P.*

Introduction

Traditional methods used to protect crops from diseases have been largely based on the use of chemical pesticides. Applications of fungicides and fumigants can have drastic effects on the environment and consumer, and are often applied in greater quantities than herbicides and insecticides in agricultural production. Chemical methods are not economical in the long run because they pollute the atmosphere, damage the environment, leave harmful residues, and can lead to the development of resistant strains among the target organisms with repeated use (Naseby *et al.*, 2000). A reduction or elimination of synthetic pesticide applications in agriculture is highly desirable. One of the most promising means to achieve this goal is by the use of new tools based on biocontrol agents (BCAs) for disease control alone, or to integrate with reduced doses of chemicals in the control of plant pathogens resulting in minimal impact of the chemicals on the environment (Chet and Inbar, 1994; Harman and Kubicek, 1998). To date, a number of BCAs have been registered and are available as commercial products, including strains belonging to bacterial genera such as *Agrobacterium, Pseudomonas, Streptomyces* and *Bacillus*, and fungal genera such as *Gliocladium, Trichoderma, Ampelomyces, Candida* and *Coniothyrium*.

---

* E-mail: pksbotany@gmail.com

Trichoderma spp. are among the most frequently isolated soil fungi and present in plant root ecosystems (Harman *et al.*, 2004). These fungi are opportunistic, avirulent plant symbionts, and function as parasites and antagonists of many phytopathogenic fungi, thus protecting plants from disease. So far, *Trichoderma* spp. are among the most studied fungal BCAs and commercially marketed as biopesticides, biofertilizers and soil amendments (Harman, 2000; Harman *et al.*, 2004; Lorito *et al.*, 2004). Depending upon the strain, the use of *Trichoderma* in agriculture can provide numerous advantages: (i) colonization of the rhizosphere by the BCA ("rhizosphere competence") allowing rapid establishment within the stable microbial communities in the rhizosphere; (ii) control of pathogenic and competitive/deleterious microflora by using a variety of mechanisms; (iii) improvement of the plant health and (iv) stimulation of root growth (Harman *et al.*, 2004).

This review presents a compilation of the most recent advances in understanding the mechanisms involved in the interaction of *Trichoderma* spp. with phytopathogenic fungi and plants. We emphasize the biological and biochemical aspects of this topic, with particular attention paid to the molecular factors involved in the natural crosstalk occurring in soil and root environment. A better understanding of the principles regulating the interaction between fungal pathogens, host plants, and BCAs such as *Trichoderma* would enhance the practical application of these beneficial microorganisms for plant disease control.

Trichoderma–Pathogen Interaction

Trichoderma (teleomorph *Hypocrea*) is a genus of asexual fungi found in the soils of all climatic zones. *Trichoderma* is a secondary opportunistic invader, a fast growing fungus, a strong spore producer, a source of cell wall degrading enzymes (CWDEs: cellulases, chitinases, glucanases, etc.), and an important antibiotic producer. Numerous strains of this genus are 'rhizosphere competent' and are able to degrade hydrocarbons, chlorophenolic compounds, polysaccharides and the xenobiotic pesticides used in agriculture (Harman and Kubicek, 1998; Harman *et al.*, 2004). The main biocontrol mechanisms that *Trichoderma* utilizes in direct confrontation with fungal pathogens are mycoparasitism (Papavizas, 1985; Harman and Kubicek, 1998; Howell, 2003) and antibiosis (Howell, 1998; Sivasithamparam and Ghisalberti, 1998).

Mycoparasitism and Lytic Enzymes

The complex process of mycoparasitism consists of several events, including recognition of the host, attack and subsequent penetration and killing. During this process *Trichoderma* secretes CWDEs that hydrolyze the cell wall of the host fungus, subsequently releasing oligomers from the pathogen cell wall (Kubicek *et al.*, 2001; Howell 2003; Woo *et al.*, 2006). It is believed that *Trichoderma* secretes hydrolytic enzymes at a constitutive level and detects the presence of another fungus by sensing the molecules released from the host by enzymatic degradation (Harman *et al.*, 2004; Lorito *et al.*, 2006; Woo and Lorito, 2007).

The molecular biology of the mycoparasitic interaction between pathogen and antagonist has been studied in detail. The factors activating the biocontrol gene cascade in *T. atroviride* strain P1 mutants containing the green fluorescent protein (*gfp*) or glucose oxidase (*gox*) gene reporter systems controlled by different inducible promoters (*i.e.* from the exochitinase *nag1* gene or the endochitinase *ech42* gene) have been evaluated. Interestingly, the expression of these genes involved in mycoparasitism was induced by the digestion products obtained after treatments of fungal cell walls and colloidal chitin with purified CWDEs or fungal culture filtrates. LC/MS–MS analysis revealed that these novel mycoparasitism related inducers have an oligosaccharide structure (Woo *et al.*, 2004). Recently, the role of *Trichoderma* ABC transporters in both mycoparasitism and nutritional uptake by *Trichoderma* has been investigated (Singh and Vyas, 2008). Unpublished but convincing data demonstrated that culture filtrates or mycelia of numerous plant pathogens induced the expression of specific *T. atroviride* ABC transporter genes, such as *tabc2*. This conclusion was confirmed by producing and analyzing knock-out mutants that showed a slower growth on different culture media or in presence of different fungal pathogens (*Botrytis cinerea, Rhizoctonia solani* and *Pythium ultimum*).

The antifungal arsenal of *Trichoderma* spp. includes a great variety of lytic enzymes (Lorito, 1998; Lorito *et al.*, 1994a, 1996a), most of which play a great role in biocontrol (Harman and Kubicek, 1998; Baek *et al.*, 1999; Carsolio *et al.*, 1999; Woo *et al.*, 1999; Zeilinger *et al.*, 1999; Kullnig *et al.*, 2000; Kubicek *et al.*, 2001). Many CWDEs from different *Trichoderma* strains have been purified and characterized (Lorito, 1998). Interestingly, when tested alone or in combinations, the purified proteins showed antifungal activity towards a broad spectrum of fungal pathogens (*i.e.* species of *Rhizoctonia, Fusarium, Alternaria, Ustilago, Venturia* and *Colletotrichum*, as well as fungus like organisms such as the Oomycetes *Pythium* and *Phytophthora* which lack chitin in their cell walls) (Tronsmo, 1991; Lorito *et al.*, 1993, 1994a).

The direct application of anti-microbial compounds produced by fungal BCAs, instead of the whole "live" organisms, has numerous advantages in industry and agriculture, and may be more amenable to public opinion because of the inability of the agent to reproduce and spread. The selective production of active compounds may be performed by modifying the growth conditions, *i.e.* type and composition of culture medium, temperature of incubation and pH etc. (Lorito and Scala, 1999; Woo and Lorito, 2007). The presence of different carbon sources, such as mono- or polysaccharides, colloidal chitin, or fungal tissues, has been shown to induce the secretion of CWDEs (Mach *et al.*, 1999). Enhanced anti-fungal activity can be obtained by the combined application of *Trichoderma* enzymes with the fungus, different classes of synthetic fungicides, and in particular with compounds that affect the integrity of the cell membrane (Lorito *et al.*, 1994b, 1996a). Moreover, purified mixes of CWDEs with different lytic activities showed improved antifungal effects against various plant pathogens, sometimes comparable to those obtained by using synthetic pesticides alone (Lorito *et al.*, 1994b, 1996a; Baek *et al.*, 1999; Carsolio *et al.*, 1999).

Antibiosis and Secondary Metabolites

Trichoderma produces a plethora of secondary metabolites with biological activity (Ghisalberti and Sivasithamparam, 1991; Sivasithamparam and Ghisalberti, 1998). The term "secondary metabolite" includes a heterogeneous group of chemically different natural compounds possibly related to survival functions for the producing organism, such as competition against other micro- and macroorganisms, symbiosis, metal transport, differentiation etc. (Demain and Fang, 2000). Included in this group are antibiotics, which are natural products able to inhibit microbial growth. Antibiotic production is often well correlated with biocontrol ability, and the application of purified antibiotics was found to show effects on the host pathogen similar to those obtained by using the corresponding living microbe. Ghisalberti *et al.* (1990) demonstrated that the biocontrol efficacy of *Trichoderma harzianum* isolates against *Gaeumannomyces graminis* var. *tritici* is related to the production of pyrone-like antibiotics.

The production of secondary metabolites by *Trichoderma* spp. is strain dependent and includes antifungal substances belonging to a variety of classes of chemical compounds. They were classified by Ghisalberti and Sivasithamparam (1991) into three categories: (i) volatile antibiotics, *i.e.* 6-pentyl-a-pyrone (6PP) and most of the isocyanide derivates; (ii) water-soluble compounds, *i.e.* heptelidic acid or koningic acid; (iii) peptaibols, which are linear oligopeptides of 12–22 amino acids rich in a-aminoisobutyric acid, N-acetylated at the N-terminus and containing an amino alcohol (Pheol or Trpol) at the C-terminus (Le Doan *et al.*, 1986; Rebuffat *et al.*, 1989).

Recently, we isolated and characterized the main secondary metabolites obtained from culture filtrates of two commercial *T. harzianum* strains (T22 and T39), and their production during the antagonistic interaction with the pathogen *R. solani* was also investigated (Vinale *et al.*, 2006). The chemical structures of *Trichoderma* antibiotics may suggest two different mechanisms of action. The production of low molecular weight, non-polar, volatile compounds (*i.e.* 6PP) results in a high concentration of antibiotics in the soil environment that have a relatively long distance range of influence on the microbial community. On the contrary, a short distance effect may be due to the polar antibiotics and peptaibols acting in close proximity to the producing hyphae. Lorito *et al.* (1996b) demonstrated that peptaibols inhibited b-glucan synthase activity in the host fungus, while acting synergistically with *T. harzianum* b-glucanases. The inhibition of glucan synthase prevented the reconstruction of the pathogen cell wall, thus facilitating the disruptive action of b-glucanases. The synergism existing between enzymes and polar antibiotics is strictly related to their mechanism of action (Schirmbock *et al.*, 1994; Lorito *et al.*, 1996a,b; Fogliano *et al.*, 2002). Although the role and the effects of peptaibols are clear, the mode of action of other *Trichoderma* secondary metabolites (*i.e.* pyrones), and their possible synergisms with other compounds have not yet been elucidated (Claydon *et al.*, 1987; Serrano-Carreon *et al.*, 1993; Howell, 1998).

According to the secondary metabolite produced, Howell *et al.* (1993) divided strains of *T. virens* into two groups: the "Q" strains able to produce the antibiotic gliotoxin and the "P" strains that produce a related compound, gliovirin, instead of gliotoxin (Howell and Stipanovic, 1983). Gliotoxin has a broad spectrum of antibiotic activity, while gliovirin is a specific potent inhibitor of Oomycetes and its production was positively correlated with biocontrol efficacy of "P" group strains to control *Pythium* damping-off of cotton (Chet *et al.*, 1997; Howell, 1998). On substrates with high C/N ratios, both "P" and "Q" strains of *T. virens* produce a phytotoxin similar to viridin, that is called viridiol. The viridiol-producing strains may be applied to surface soil as bio-herbicide for weeds, where they do not affect the crop plant that is planted in the treated soil (Howell, 2006). Other observations indicated that the biological control of pre-emergence damping-off by *T. virens* could be also related to its ability to degrade seed-emitted compounds that stimulate pathogen propagule germination (Howell, 2002). On the other hand, the induction of plant defence responses by some strains of *T. virens* plays a pivotal role in successful disease control of *R. solani* on cotton (Howell *et al.*, 2000). In a recent study, Howell and Puckhaber (2005) indicated that "P" strains unable to induce the production of phytoalexins in cotton were ineffective as BCAs and pathogenic to susceptible cultivars. Conversely, "Q" strains inducing high levels of phytoalexin synthesis showed improved biocontrol efficacy and were not pathogenic to cotton roots. Phytoalexin synthesis in cotton is elicited by a protein produced by *T. virens* (Hanson and Howell, 2004), but the exact biochemical processes involved are not yet understood.

Competition with Pathogens and Soil Microbial Community

Competition for carbon, nitrogen and other growth factors, together with competition for space or specific infection sites, may be also used by the BCA to control plant pathogens. *T. harzianum* is able to control *B. cinerea* on grapes by colonizing blossom tissue and excluding the pathogen from its infection site (Gullino, 1992). Sivan and Chet (1989) demonstrated that competition for nutrients is the major mechanism used by *T. harzianum* to control *F. oxysporum* f. sp. *melonis*. Moreover, *Trichoderma* has a strong capacity to mobilize and take up soil nutrients, thus making it more efficient and competitive than many other soil microbes (Benytez *et al.*, 2004).

The biotic components of the soil environment have relevant effects on the biocontrol activity of *Trichoderma* against plant pathogens. Bae and Knudsen (2005), by using a Gfp-tagged mutant, showed that higher levels of microbial soil biomass induced a shift from hyphal growth to sporulation in *T. harzianum*, thus reducing its biocontrol efficacy. This effect may be associated with a phenomenon known as "soil fungistasis", which is largely dependent on the soil microbial community composition (de Boer *et al.*, 2003). In particular, the production of antibiotic compounds and the presence of bacteria belonging to the genus *Pseudomonas* seem to be essential for the development of this phenomenon. In this context a detailed study of the metabolites produced by microorganisms present in the soil environment should be performed in order to avoid the suppression of BCAs.

Trichoderma–Plant Interaction

In addition to the beneficial effects that occur in direct interactions with plant disease agents, some *Trichoderma* species are also able to colonize root surfaces and cause substantial changes in plant metabolism (Harman *et al.*, 2004). It is well documented that some strains promote plant growth, increase nutrient availability, improve crop production and enhance disease resistance (Harman *et al.*, 2004).

Plant Root Colonization

The physical interaction between *Trichoderma* and the plant was observed by electron microscopy to be limited to the first few cell layers of plant epidermis and root outer cortex (Yedidia *et al.*, 1999). The hyphae of the BCA penetrate the root cortex but the colonization by *Trichoderma* is stopped, probably by the deposition of callose barriers by the surrounding plant tissues (Yedidia *et al.*, 1999). It appears that this interaction evolves into a symbiotic rather than a parasitic relationship between the fungus and the plant, whereby the fungus occupies a nutritional niche and the plant is protected from disease. A very active, direct molecular cross-talk occurs between the fungus and the plant. Elicitors from *Trichoderma* activate the expression of genes involved in the plant defence response system, and promote the growth of the plant, root system and nutrient availability. This effect in turn augments the zone for colonization and the nutrients available for the biocontrol fungus, subsequently increasing the overall antagonism to plant pathogens (Yedidia *et al.*, 2003; Hanson and Howell, 2004; Harman *et al.*, 2004).

Plant Growth Promotion

Many BCAs, such as fungi, bacteria and viruses, are not only able to control the pathogens that cause plant disease, but are also able to promote plant growth and development. In greenhouse and field trials, the ability of *T. harzianum* T22 and *T. atroviride* P1 to improve the growth of lettuce, tomato and pepper plants under field conditions was investigated (Vinale *et al.*, 2004). Crop productivity was increased up to 300 per cent, as determined by comparing the treated plots with the untreated controls and measuring fresh/dry root and above ground biomass weights, height of plants, number of leaves and fruits. This study also demonstrated the compatibility of *T. harzianum* T22 and *T. atroviride* P1 with pesticides conventionally used in organic farming by monitoring the effect on mycelia growth in both liquid and solid media. Results indicated a high level of tolerance by *Trichoderma* strains to concentrations of copper oxychloride varying from 0.1 up to 5mM (Vinale *et al.*, 2004). These positive effects of *Trichoderma* may be obtained with different plant species, thus the genetic base of such interactions seems not to be predominant. Conversely, at least in maize the plant growth promotion effect is genotype specific and some inbreds respond negatively to different strains (Harman, 2006).

A yield increase was also observed when plant seeds were exposed to *Trichoderma* conidia that were separated from them by cellophane, suggesting that *Trichoderma* metabolites can influence the plant growth (Benitez *et al.*, 2004). On the other hand, only a few reports deal with the ability of antagonistic fungal strains to produce compounds acting as growth promoting factors. Cutler *et al.* (1986, 1989) reported the isolation, identification and biological activity of secondary metabolites produced by *T. koningii* (koninginin A) and *T. harzianum* (6-pentyl-apyrone) that acted as plant growth regulators. Both metabolites significantly inhibited the growth of etiolated wheat coleoptiles at a relatively high concentration (10^{-3}M), but no effect was registered at lower doses (range from 10^{-4} to 10^{-3} M). It is hypothesized that such Trichoderma secondary metabolites may act as auxin-like compounds, which typically have an optimum activity between at 10^{-5} and 10^{-6}M while having an inhibitory effect at higher concentrations (Thimann, 1937; Cleland, 1972; Brenner, 1981), and/or are involved in the production of auxin inducers. The dose–effect response of such compounds on plant growth and development requires further investigation. *Trichoderma* spp. also produce organic acids, such as gluconic, citric or fumaric acids, that decrease soil pH and permit the solubilization of phosphates, micronutrients and mineral cations like iron, manganese and magnesium, useful for plant metabolism (Benitez *et al.*, 2004; Harman *et al.*, 2004).

Induction of Plant Defence Responses

The induction of plant defence responses mediated by the antagonistic fungus has been well documented (De Meyer *et al.*, 1998; Yedidia *et al.*, 1999; Hanson and Howell, 2004; Harman *et al.*, 2004). Various plants, both mono- and dicotyledonous species, showed increased resistance to pathogen attack when pre-treated with *Trichoderma* (Harman *et al.*, 2004; Singh *et al.*, 2007). Plant colonization by *Trichoderma* spp. reduced disease caused by one or more different pathogens, at the site of inoculation (induced localized acquired resistance, LAR), as well as when the biocontrol fungus was inoculated at different times or sites than that of the pathogen (induced systemic resistance or ISR).

The induction of plant resistance by colonization with some *Trichoderma* species is similar to that elicited by rhizobacteria, which enhance the defence system but do not involve the production of pathogenesis-related proteins (PR proteins) (Van Loon *et al.*, 1998; Stacey and Keen, 1999; Harman *et al.*, 2004). In a recent work Alfano and co-workers (2007) investigated at a molecular level the plant genes involved in *T. hamatum* 382 resistance inductions by using a high-density oligonucleotide microarray approach. Interestingly, *Trichoderma*-induced genes were associated with biotic or abiotic stresses, as well as RNA, DNA, and protein metabolism. In particular, genes that codify for extensin and extensin-like proteins were found to be induced by the BCA, but not those codifying for proteins belonging to the PR-5 family (thaumatin-like proteins), which are considered the main molecular markers of SAR.

During the interaction of *Trichoderma* with the plant, different classes of metabolites may act as elicitors or resistance inducers (Harman *et al.*, 2004; Woo *et al.*, 2006; Woo and Lorito, 2007). These molecules include: (i) proteins with enzymatic activity, such as xylanase (Lotan and Fluhr, 1990); (ii) avirulence-like gene products able to induce defence reactions in plants (Woo *et al.*, 2004); (iii) low-molecular-weight compounds released from fungal or plant cell walls by the activity of *Trichoderma* enzymes (Harman *et al.*, 2004; Woo *et al.*, 2006; Woo and Lorito, 2007). Some of the low-molecular-weight degradation products released from fungal cell walls were purified and characterized, and found to consist of short oligosaccharides comprised of two types of monomers, with and without an amino acid residue (Woo *et al.*, 2006; Woo and Lorito, 2007). These compounds elicited a reaction in the plant when applied to leaves or when injected into root or leaf tissues. Further, they also stimulated the biocontrol ability of *Trichoderma* by activating the mycoparasitic gene expression cascade. Recently, Djonovic *et al.* (2006) identified a small protein (Sm1) elicitor secreted by T. *virens*, and demonstrated its involvement in the activation of plant defence mechanisms and the induction of systemic resistance. In addition to their innate antimicrobial effect, their action may also stimulate the biological activity of resident antagonistic microbial populations or introduced *Trichoderma* strains, and promote an ISR effect in the plant. Other secondary metabolites, like peptaibols, may act as elicitors of plant defence mechanisms against pathogens. In fact, application of peptaibols activated a defence response in tobacco plants (Bený´ tez *et al.*, 2004; Viterbo *et al.*, 2006, personal communication). A peptaibol synthetase from *T. virens* was purified (Wiest *et al.*, 2002), and the achieved cloning of the corresponding gene will facilitate an understanding of the role of this class of compounds in plant defence response.

Influence of Soil Environment on *Trichoderma*–Plant Interaction

The activities of BCAs are also affected by the presence of organic nutrients in soil (Hoitink and Boehm, 1999). Organic matter composition and the associated biotic and abiotic environment can affect the activities of *Trichoderma*, especially in relation to the conduciveness/receptivity of the soil to the strain (Simon and Sivasithamparam, 1989; Wakelin *et al.*, 1999). So far, composts represent an optimal substrate for BCAs, thus encouraging their establishment into the soil environment (Hoitink and Boehm, 1999; Leandro *et al.*, 2007). The mechanisms of action used by *Trichoderma* (competition, antibiosis, parasitism and systemic-induced resistance) are influenced by concentration and availability of nutrients (carbohydrates in lignocellulosic substances, chitin, lipids, etc.) within the soil organic matter (Hoitink *et al.*, 2006). Krause *et al.* (2001) demonstrated that *T. hamatum* inoculation of potting mix with a high microbial capacity, which supported high populations of BCAs, significantly reduced the severity of *Rhizoctonia* damping-off of radish or *Rhizoctonia* crown and root rot of poinsettia. Moreover, *T. hamatum* inoculated into the compost amended potting mix induced systemic acquired resistance on cucumber, reducing the severity of *Phytophthora* leaf blight (Khan *et al.*, 2004). This induction was more effective on plants grown in compost-amended media when compared to lower microbe carrying capacity sphagnum peat media (Hoitink *et al.*, 2006). A better understanding of the activities of *Trichoderma* strains in plant growth media high in organic matter could also help to select strains suitable for multiple acre field conditions associated with stubble retention practices and/or organic farming which are becoming increasingly popular world-wide.

The Three-Way Interaction

Trichoderma–Plant–Pathogen

The three-way interactions involving *Trichoderma*, plant and fungal pathogen have received less attention in comparison to the "simple" two-partner systems (*i.e.* plant–pathogen, plant–antagonist or pathogen–antagonist). There are obvious difficulties in studying such a complex system even if it is reproduced *in vitro*, although it better simulates the natural interactions occurring in soil agro-ecosystems. Recent studies have investigated some of the morphological or molecular aspects involved in plant–pathogen–antagonist interactions by using novel methods such as proteomics (Marra *et al.*, 2006) and gene reporter systems (Lu *et al.*, 2004). The molecular cross-talk taking place during three-way interactions requires experiments that investigate the changes in gene expression occurring in each partner involved, singly and subsequently in all possible combinations. Further, an *in situ* analysis of the compounds implicated when plants are exposed concurrently to different beneficial and/or pathogenic microorganisms could be performed.

Marra *et al.* (2006) studied the three-way interactions of *Trichoderma* with plant and different fungal pathogens by using a proteomic approach in order to analyze the differential proteins produced. Proteins were identified and characterized by using tryptic digestion, mass spectrometry (MS) and in silico analysis. Results indicated that in the plant proteome-specific PR proteins and other disease related factors (*i.e.* potential resistance genes) may regulate the three-way interaction, and that the presence of the antagonist modifies quantitatively and qualitatively the plant response to a pathogen attack. In some cases, the antagonistic fungus reduced production of some defence proteins, but resulted in a higher accumulation of others. These observations suggest that the plant response to a specific BCA depends upon each of the three partners involved. On the microbial side, many differential proteins obtained from the *T. atroviride* interaction proteome showed interesting homologies to those of a fungal hydrophobin and ABC transporters. Virulence factors, like cyclophilins, were also up-regulated in the pathogen proteome during the interaction with the plant alone, as well as with the antagonist.

Gfp-tagged mutants of *T. atroviride* were used to study the *in situ Trichoderma*–plant–pathogen interaction by using different promoters of biocontrol-related genes to drive the expression of the living producer (Lu *et al.*, 2004). In particular, induction of *Trichoderma* genes encoding for different CWDEs in the presence of the soil-borne pathogens *R. solani* and *P. ultimum* was monitored by confocal and fluorescence microscopy. During the three way interaction the transformants were activated by the presence of the host fungal pathogen and purified colloidal chitin chitoligomers, and appeared to fluoresce during the early phases of contact. This approach allowed for the first time a direct visualization of the mycoparasitic gene expression cascade *in vivo*. The authors suggested that specific compounds released by the host cell walls were actively involved in mycoparasitism induction. In addition, the involvement of *T. atroviride* endo- and exochitinases (*nag1* and *chit42*) in the mycoparasitic process other than in the simple host hyphae degradation was also demonstrated.

Further understanding of the mechanisms operating in the interaction between plant and microbes in the soil communities could encourage development of new powerful biotechnologies, useful in the management of fungal diseases and in the improvement of crop production yields.

Conclusions

The success of biocontrol agents is dependant upon the complex interactions that these beneficial microbes establish with pathogens and plants in the soil ecosystem. A better understanding of these processes and of the molecular cross-talk occurring among the participants will not only result in the application of safer and less expensive methods to protect plants and increase crop yield, but also will extend our knowledge of how a disease process develops. Recent advances in modern techniques such as proteomics and metabolomics could provide novel information about the complex tripartite interactions, in particular about the ability of *Trichoderma* to sense the environment, the plant and the microbial community. However, it is clear that different approaches, *i.e.* genetic, molecular, biochemical and ecological, should be integrated to conduct future studies in biocontrol research and development of new technologies. In particular, a modern and more effective use of beneficial microbes such as *Trichoderma* should take into account an actual understanding of the biology and the interaction capabilities of these agents, starting with the implementation of new strain selection protocols that consider the multiple beneficial effects exerted on the colonized plant. Genetic manipulation offers novel opportunity to achieve improved biocontrol efficacy. Brunner *et al.* (2005) by over expressing a glucose oxidase gene from *Aspergillus niger* in *T. atroviride* strain P1 obtained mutants able to control fungal pathogens and induce plant systemic resistance better than the wild-type strain. Finally, the information gathered by fundamental and applied studies conducted with a "wide-view" approach may allow us to overcome in the future, at least in some applications, the difficulties associated with use of living microbes. This can be done by introducing new biopesticides and biofertilizers, *i.e.* based on the metabolites or bioactive compounds responsible for the desired beneficial effects on crops (Singh and Vyas, 2007). The application of *Trichoderma* metabolites for crop protection, such as the host defence inducers and antibiotics, may become a reality in the near future, as they can be produced cheaply in large quantities on an industrial scale, easily separated from the fungal biomass, dried and formulated for spray or drench applications. In the meantime, further experiments should be performed to better understand the mechanisms of action of *Trichoderma* secondary metabolites and their possible synergisms with other compounds used in agriculture.

References

Alfano G, Lewis I, Cakir C, Bos JIB, Miller SA, Madden LV, Kamoun S, Hoitink HAJ. (2007). Systemic modulation of gene expression in tomato by *Trichoderma hamatum* 382. *Phytopathology* 97: 429–437.

Bae YS, Knudsen GR. (2005). Soil microbial biomass influence on growth and biocontrol efficacy of *Trichoderma harzianum*. *Biological Control* 32: 236–242.

Baek JM, Howell CR, Kenerley CM. (1999). The role of an extracellular chitinase from *Trichoderma virens* Gv29-8 in the biocontrol of *Rhizoctonia solani*. *Current genetics* 35: 41–50.

Benytez T, Rincon AM, Limon MC, Codon AC. (2004). Biocontrol mechanisms of *Trichoderma* strains. *International Microbiology* 7: 249–260.

Brenner ML. (1981). Modern methods for plant growth substance analysis. *Annual Review of Plant Physiology* 32: 511–538.

Brunner K, Zeilinger S, Ciliento R, Woo SL, Lorito M, Kubicek CP, Mach RL. (2005). Genetic improvement of a fungal biocontrol agent to enhance both antagonism and induction of plant systemic disease resistance. *Applied Environmental Microbiology* 71: 3959–3965.

Carsolio C, Benhamou N, Haran S, Cortes C, Gutierrez A, Chet I, Herrera-Estrella A. (1999). Role of the *Trichoderma harzianum* endochitinase gene, *ech42*, in mycoparasitism. *Applied Environmental Microbiology* 65: 929–935.

Chet I, Inbar J, Hadar I. (1997). Fungal antagonists and mycoparasites. In: Wicklow, D.T., Soderstrom, B. (Eds.), *The Mycota* IV: *Environmental and Microbial Relationships*. Springer, Berlin, pp. 165–184.

Chet I, Inbar J. (1994). Biological control of fungal pathogens. *Applied Biochemistry and Biotechnology* 48: 37–43.

Claydon N, Allan M, Hanson JR, Avent AG. (1987). Antifungal alkyl pyrones of *Trichoderma harzianum*. *Transactions of the British Mycological Society* 88: 503–513.

Cleland R. (1972). The dosage–response curve for auxin-induced cell elongation: a re-evaluation. *Planta* 104: 1–9.

Cutler HG, Cox RH, Crumley FG, Cole PD. (1986). 6-Pentyl-apyrone from *Trichoderma harzianum*: its plant growth inhibitory and antimicrobial properties. *Agricultural and Biological Chemistry* 50: 2943–2945.

Cutler HG, Himmetsbach DS, Arrendale RF, Cole PD, Cox RH. (1989). Koninginin A: a novel plant regulator from *Trichoderma koningii*. *Agricultural and Biological Chemistry* 53: 2605–2611.

de Boer W, Verheggen P, Klein GPJ, Kowalchuk GA, van Veen JA. (2003). Microbial community composition affects soil fungistasis. *Applied Environmental Microbiology* 69: 835–844.

De Meyer G, Bigirimana J, Elad Y, Hofte M. (1998). Induced systemic resistance in *Trichoderma harzianum* T39 biocontrol of *Botrytis cinerea*. *European Journal of Plant Pathology* 104: 279–286.

Demain AL, Fang A. (2000). The natural functions of secondary metabolites. *Advances in Biochemical Engineering/Biotechnology* 69: 1–39.

Djonovic S, Pozo MJ, Dangott LJ, Howell CR, Kenerley CM. (2006). *Sm1*, a proteinaceous elicitor secreted by the biocontrol fungus *Trichoderma virens* induces plant defense responses and systemic resistance. *Molecular Plant–Microbe Interaction* 19: 838–853.

Fogliano V, Ballio A, Gallo M, Woo SL, Scala F, Lorito M. (2002). Pseudomonas lipodepsipeptides and fungal cell wall-degrading enzymes act synergistically in biological control. *Molecular Plant–Microbe Interaction* 15: 323–333.

Ghisalberti EL, Narbey MJ, Dewan MM, Sivasithamparam K. (1990). Variability among strains of *Trichoderma harzianum* in their ability to reduce take-all and to produce pyrones. *Plant and Soil* 121: 287–291.

Ghisalberti EL, Sivasithamparam K. (1991). Antifungal antibiotics produced by *Trichoderma* spp. *Soil Biology and Biochemistry* 23: 1011–1020.

Gullino ML. (1992). Control of Botrytis rot of grapes and vegetables with *Trichoderma* spp. *In*: Tjamos EC, Papavizas, GC, Cook RJ. (Eds.), *Biological Control of Plant Diseases, Progress and Challenges for the Future*. Plenum Press, New York, pp. 125–132.

Hanson LE, Howell CR. (2004). Elicitors of plant defense responses from biocontrol strains of *Trichoderma virens*. *Phytopathology* 94: 171–176.

Harman GE, Howell CR, Viterbo A, Chet I, Lorito M. (2004). *Trichoderma* species-opportunistic, avirulent plant symbionts. *Nature Review Microbiology* 2: 43–56.

Harman GE, Kubicek CP. (1998). *Trichoderma* and *Gliocladium*. Taylor and Francis, London.

Harman GE. (2000). Myths and dogmas of biocontrol: changes in perceptions derived from research on *Trichoderma harzianum* T-22. *Plant Disease* 84: 377–393.

Harman GE. (2006). Overview of mechanisms and uses of *Trichoderma* spp. *Phytopathology* 96: 190–194.

Hoitink HAJ, Boehm MJ. (1999). Biocontrol within the context of soil microbial communities: a substrate-dependent phenomenon. *Annual Review of Phytopathology* 37: 427–446.

Hoitink HAJ, Madden LV, Dorrance AE. (2006). Systemic resistance induced by *Trichoderma* spp.: interactions between the host, the pathogen, the biocontrol agent, and soil organic matter quality. *Phytopathology* 96: 186–189.

Howell CR, Hanson EL, Stipanovic RD, Puckhaber LS. (2000). Induction of terpenoid synthesis in cotton roots and control of *Rhizoctonia solani* by seed treatment with *Trichoderma virens*. *Phytopathology* 90: 248–252.

Howell CR, Puckhaber LS. (2005). A study of the characteristics of "P" and "Q" strains of *Trichoderma virens* to account for differences in biological control efficacy against cotton seedling diseases. *Biological Control* 33: 217–222.

Howell CR, Stipanovic RD, Lumsden RD. (1993). Antibiotic production by strains of *Gliocladium virens* and its relation to the biocontrol of cotton seedling diseases. *Biocontrol Science and Technology* 3: 435–441.

Howell CR, Stipanovic RD. (1983). Gliovirin, a new antibiotic from *Gliocladium virens*, and its role in the biological control of *Pythium ultimum*. *Canadian Journal of Microbiology* 29: 321–324.

Howell CR. (1998). The role of antibiosis in biocontrol. In: Harman GE. Kubicek CP. (Eds.), *Trichoderma and Gliocladium. Enzymes, Biological Control and Commercial Application*, vol. 2. Taylor and Francis Ltd., London, pp. 173–183.

Howell CR. (2002). Cotton seedling pre-emergence damping-off incited by *Rhizopus oryzae* and *Pythium* spp. and its biological control with *Trichoderma* spp. *Phytopathology* 92: 177–180.

Howell CR. (2003). Mechanisms employed by *Trichoderma* species in the biological control of plant diseases: the history and evolution of current concepts. *Plant Disease* 87: 4–10.

Howell CR. (2006). Understanding the mechanisms employed by *Trichoderma virens* to effect biological control of cotton diseases. *Phytopathology* 96: 178–180.

Khan J, Ooka JJ, Miller SA, Madden LV, Hoitink HAJ. (2004). Systemic resistance induced by *Trichoderma hamatum* 382 in cucumber against *Phytophthora* crown rot and leaf blight. *Plant Disease* 88: 280–286.

Krause MS, Madden LV, Hoitink HAJ. (2001). Effect of potting mix microbial carrying capacity on biological control of *Rhizoctonia* damping-off of radish and *Rhizoctonia* crown and root rot of poinsettia. *Phytopathology* 91: 1116–1123.

Kubicek CP, Mach RL, Peterbauer CK, Lorito M. (2001). *Trichoderma*: from genes to biocontrol. *Journal of Plant Pathology* 83: 11–23.

Kullnig C, Mach RL, Lorito M, Kubicek CP. (2000). Enzyme diffusion from *Trichoderma atroviride* (*T. harzianum* P1) to *Rhizoctonia solani* is a prerequisite for triggering of *Trichoderma ech42* gene expression before mycoparasitic contact. *Applied Environmental Microbiology* 66: 2232–2234.

Le Doan T, El-Hajii M, Rebuffat S, Rajeswari MR, Bodo B. (1986). Fluorescein studies on the interaction of trichorzianine A IIIc with model membranes. *Biochimica et Biophysica Acta* 858: 1–5.

Leandro LFS, Guzman T, Ferguson LM, Fernandez GE, Louws FJ. (2007). Population dynamics of *Trichoderma* in fumigated and compost amended soil and on strawberry roots. *Applied Soil Ecology* 35: 237–246.

Lorito M, Farkas V, Rebuffat S, Bodo B, Kubicek CP. (1996b). Cell wall synthesis is a major target of mycoparasitic antagonism by *Trichoderma harzianum*. *Journal of Bacteriology* 178: 6382–6385.

Lorito M, Harman GE, Hayes CK, Broadway RM, Tronsmo A, Woo SL, Di Pietro A. (1993). Chitinolytic enzymes produced by *Trichoderma harzianum*: antifungal activity of purified endochitinase and chitobiosidase. *Phytopathology* 83: 302–307.

Lorito M, Hayes CK, Di Pietro A, Woo SL, Harman GE. (1994a). Purification, characterization and synergistic activity of a glucan 1,3-b-glucosidase and an N-acetyl-b-glucosaminidase from *Trichoderma harzianum*. *Phytopathology* 84: 398–405.

Lorito M, Peterbauer C, Hayes CK, Harman GE. (1994b). Synergistic interaction between fungal cell wall degrading enzymes and different antifungal compounds enhances inhibition of spore germination. *Microbiology* 140: 623–629.

Lorito M, Scala F. (1999). Microbial genes expressed in transgenic plants to improve disease resistance. *Journal of Plant Pathology* 81: 73–88.

Lorito M, Woo SL, DAmbrosio M, Harman GE, Hayes CK, Kubicek CP, Scala F. (1996a). Synergistic interaction between cell wall degrading enzymes and membrane affecting compounds. *Molecular Plant–Microbe Interaction* 9: 206–213.

Lorito M, Woo SL, Iaccarino M, Scala F. (2006). Microrganismi antagonisti. In: Iaccarino M. (Ed.), *Microrganismi Benefici per le Piante*. Idelson-Gnocchi s.r.l., Napoli, Italia, pp. 146–175.

Lorito M, Woo SL, Scala F. (2004). Le biotecnologie utili alla difesa sostenibile delle piante: i funghi. *Agroindustria* 3: 181–195.

Lorito M. (1998). Chitinolytic enzymes and their genes. In: Harman GE, Kubicek CP. (Eds.), *Trichoderma and Gliocladium*, Vol. 2. Taylor and Francis Ltd., London, pp. 73–99.

Lotan T, Fluhr R. (1990). Xylanase, a novel elicitor of pathogenesisrelated proteins in tobacco, uses a nonethylene pathway for induction. *Plant Physiology* 93: 811–817.

Lu Z, Tombolini R, Woo SL, Zeilinger S, Lorito M, Jansson JK. (2004). *In vivo* study of *Trichoderma*–pathogen–plant interactions with constitutive and inducible *gfp* reporter systems. *Applied Environmental Microbiology* 70: 3073–3081.

Mach RL, Peterbauer CK, Payer K, Jaksits S, Woo SL, Zeilinger S, Kullnig CM, Lorito M, Kubicek CP. (1999). Expression of two major chitinase genes of *Trichoderma* atroviride (*T. harzianum* P1) is triggered by different regulatory signals. *Applied Environmental Microbiology* 65: 1858–1863.

Marra R, Ambrosino P, Carbone V, Vinale F, Woo SL, Ruocco M, Ciliento R, Lanzuise S, Ferraioli S, Soriente I, Gigante S, Turra D, Fogliano V, Scala F, Lorito M. (2006). Study of the three-way interaction between *Trichoderma* atroviride, plant and fungal pathogens by using a proteomic approach. *Current Genetics* 50: 307–321.

Naseby DC, Pascual JA, Lynch JM. (2000). Effect of biocontrol strains of *Trichoderma* on plant growth, *Pythium ultimum* population, soil microbial communities and soil enzyme activities. *Journal of Applied Microbiology* 88: 161–169.

Papavizas GC. (1985). *Trichoderma* and *Gliocladium*: biology, ecology, and potential for biocontrol. *Annual Review Phytopathology* 23: 23–54.

Rebuffat S, El Hajji M, Hennig P, Davoust D, Bodo B. (1989). Isolation, sequence and conformation of seven trichorzianines B from *Trichoderma harzianum*. *International Journal of Peptide and Protein Research* 34: 200–210.

Schirmbock M, Lorito M, Wang YL, Hayes CK, Arisan-Atac I, Scala F, Harman GE, Kubicek CP. (1994). Parallel formation and synergism of hydrolytic enzymes and peptaibol antibiotics, molecular mechanisms involved in the antagonistic action of *Trichoderma harzianum* against phytopathogenic fungi. *Applied Environmental Microbiology* 60: 4364–4370.

Serrano-Carreon L, Hathout Y, Bensoussan M, Belin JM. (1993). Metabolism of linoleic acid and 6-pentyl-a-pyrone biosynthesis by *Trichoderma* species. *Applied and Environmental Microbiology* 59: 2945–2950.

Simon A, Sivasithamparam K. (1989). Pathogen suppression: A case study in biological suppression of *Gaeumannomyces graminis* var. *tritici* in soil. *Soil Biology and Biochemistry* 21: 331–337.

Singh PK, Mishra MK, Vyas D. (2007). Efficacy of *Trichoderma* spp. strains in the control of Fusarium wilt of tomato. *Journal of Mycology and Plant Pathology* 37: 105–107.

Singh PK, Vyas D. (2007). Efficacy of *Trichoderma* sp. in control of wilt of tomato. *Journal of Basic and Applied Mycology* 6: 93–95.

Singh PK, Vyas D. (2008). Effects of chemical fertilizers on the growth of *Trichoderma* spp. in soil. *Journal of Botanical Society of University Sagar*, 42: 82-86.

Sivan A, Chet I. (1989). The possible role of competition between *Trichoderma harzianum* and *Fusarium oxysporum* on rhizosphere colonization. *Phytopathology* 79: 198–203.

Sivasithamparam K, Ghisalberti EL. (1998). Secondary metabolism in *Trichoderma* and *Gliocladium*. In: Harman GE, Kubicek CP. (Eds.), *Trichoderma and Gliocladium*, Vol. 1. Taylor and Francis Ltd., London, pp. 139–191.

Stacey G, Keen NT. (Eds.) 1999). *Plant–Microbe Interactions*, vol. 4. APS Press, St. Paul.

Thimann KV. (1937). On the nature of inhibitions caused by auxin. *American Journal of Botany* 24: 407–412.

Tronsmo A. (1991). Biological and integrated controls of *Botrytis cinerea* on apple with *Trichoderma harzianum*. *Biological Control* 1: 59–62.

van Loon LC, Bakker PAHM, Pieterse CMJ. (1998). Systemic resistance induced by rhizobacteria. *Annual Review of Phytopathology* 36: 453–483.

Vinale F, DAmbrosio G, Abadi K, Scala F, Marra R, Turra D, Woo SL, Lorito M. (2004). Application of *Trichoderma harzianum* (T22) and *Trichoderma atroviride* (P1) as plant growth promoters, and their compatibility with copper oxychloride. *Journal of Zhejiang University Science* 30: 2–8.

Vinale F, Marra R, Scala F, Ghisalberti EL, Lorito M, Sivasithamparam K. (2006). Major secondary metabolites produced by two commercial *Trichoderma* strains active against different phytopathogens. *Letters in Applied Microbiology* 43: 143–148.

Viterbo A, Shoresh M, Brotman Y, Chet I. (2006). Towards understanding the molecular basis for induced resistance in the *Trichoderma*–plant interaction. In: Proceedings of the Ninth International workshop on *Trichoderma* and *Gliocladium*. Vienna, Austria, April 6–8, KS6.

Wakelin SA, Sivasithamparam K, Cole ALJ, Skipp RA. (1999). Saprophytic growth in soil of a strain of *Trichoderma koningii*. *New Zealand Journal of Agricultural Research* 42, 337–345.

Wiest A, Grzegorski D, Xu B, Goulard C, Le Doan S, Ebbole DJ, Bodo B, Kenerley C. (2002). Identification of peptaibols from *Trichoderma virens* and cloning of a peptaibol synthetase. *Journal of Biological Chemistry* 277: 20862–20868.

Woo SL, Donzelli B, Scala F, Mach R, Harman GE, Kubicek CP, Del Sorbo G, Lorito M. (1999). Disruption of the *ech42* (endochitinase encoding) gene affect biocontrol activity in *Trichoderma harzianum* P1. *Molecular Plant–Microbe Interaction* 12: 419–429.

Woo SL, Formisano E, Fogliano V, Cosenza C, Mauro A, Turra D, Soriente I, Ferraioli S, Scala F, Lorito M. (2004). Factors that contribute to the mycoparasitism stimulus in *Trichoderma atroviride* strain P1. *Journal of Zhejiang University Science* 30: 421.

Woo SL, Lorito M. (2007). Exploiting the interactions between fungal antagonists, pathogens and the plant for biocontrol. In: Vurro M, Gressel J. (Eds.), *Novel Biotechnologies for Biocontrol Agent Enhancement and Management*. IOS, Springer Press, Amsterdam, the Netherlands, pp. 107–130.

Woo SL, Scala F, Ruocco M, Lorito M. (2006). The molecular biology of the interactions between *Trichoderma* spp., phytopathogenic fungi, and plants. *Phytopathology* 96: 181–185.

Yedidia I, Benhamou N, Chet I. (1999). Induction of defence responses in cucumber plants (*Cucumis sativus* L.) by the biocontrol agent *Trichoderma harzianum*. *Applied Environmental Microbiology* 65: 1061–1070.

Yedidia I, Shoresh M, Kerem Z, Benhamou N, Kapulnik Y, Chet I. (2003). Concomitant induction of systemic resistance to Pseudomonas syringae pv. lachrymans in cucumber by *Trichoderma asperellum* (T-203) and accumulation of phytoalexins. *Applied Environmental Microbiology* 69: 7343–7353.

Zeilinger S, Galhaup C, Payer K, Woo SL, Mach RL, Fekete C, Lorito M, Kubicek CP. (1999). Chitinase gene expression during mycoparasitic interaction of *Trichoderma harzianum* with its host. *Fungal Genetics and Biology* 26: 131–140.

Microbial Biotechnology and Ecology

Volume-2

Microbial Biotechnology and Ecology

Volume-2

Deepak Vyas
G.S. Paliwal
P.K. Khare
R.K. Gupta

2024

Daya Publishing House®
A Division of
Astral International Pvt. Ltd.
New Delhi – 110 002

© EDITORS

First Published, 2011

Reprinted, 2024

ISBN: 978-81-7035-694-3 (HB)

Disclaimer:

Every possible effort has been made to ensure that the information contained in this book is accurate at the time of going to press, and the publisher and author cannot accept responsibility for any errors or omissions, however caused. No responsibility for loss or damage occasioned to any person acting, or refraining from action, as a result of the material in this publication can be accepted by the editor, the publisher or the author. The Publisher is not associated with any product or vendor mentioned in the book. The contents of this work are intended to further general scientific research, understanding and discussion only. Readers should consult with a specialist where appropriate.

Every effort has been made to trace the owners of copyright material used in this book, if any. The author and the publisher will be grateful for any omission brought to their notice for acknowledgment in the future editions of the book.

All Rights reserved under International Copyright Conventions. No part of this publication may be reproduced, stored in a retrieval system, or transmitted in any form or by any means, electronic, mechanical, photocopying, recording or otherwise without the prior written consent of the publisher and the copyright owner.

Published by : **Daya Publishing House®**
A Division of
Astral International Pvt. Ltd.
– ISO 9001:2015 Certified Company –
4736/23, Ansari Road, Darya Ganj,
New Delhi-110 002
Ph. 011-43549197, 23278134
E-mail: info@astralint.com
Website: www.astralint.com

Printed at : Replika Press Pvt. Ltd.

– Biography –
Late Prof. K.M. Vyas

Born on 19th September, 1942 in village Dharoan then in district Hamirpur and presently in district Mahoba (UP) in the family of Smt Radhika Rani and Sri Rooplal Vyas. Krishna Murari had his primary education in village and for the secondary education he moved to a Tehsil place in Charkhari. He obtained BSc from Maharaja College, Chhatarpur and MSc (Botany) degree from Sagour University, Sagar in 1963. Young Krishna Murari joined as a lecturer in the Botany Department then Sagour University, Sagour now Dr. H.S.G. V.V., Sagar (M.P.). Simultaneously joined PhD programme under the supervision of Prof SB Saksena (FNA), a renowned mycologist of that time. He did his PhD on "*Studies on the Ecology and Physiology of Soil Microorganisms in Relation to Soil Borne Plant Diseases*". For post doctoral work (1973-74), he went to Department of Biochemistry, University of Hull, England (UK). He worked on "*Production and Activity of Cell Wall Degrading Enzymes by Late Blight Fungus– Phytopthora infestens*" with Prof J Friend.

Foreign Visits

Visited and delivered lectures at various important Research Institutions and Universities of England, Scotland, Belgium and Holland, Imperial College of Science and Technology, London, Long Ashton Research Station, Bristol, University of Glasgow, International Agriculture Institute, Wageninghe, Phytopathologic Research Institute Baarn, Holland.

Outstanding Research Contributions

1. Successful control of *Phytophthora* diseases of pan by streptomycin and other antibiotics.
2. Antibiotics as a new group of plant growth regulators.
3. Control of mosquitoes by certain isolates of bacteria and fungi.
4. Fungal pathogenesis of plants with special reference of cell wall degrading enzymes and phenolics in fungal pathogenesis.

5. *Alectra parasitica*: a traditional drug for the treatment of leprosy and other skin diseases.
6. Fungal respiration with special reference to mycelial respiration of plant pathogenic fungi.
7. Rhizosphere, rhizoplane, phyllosphere and phylloplane microflora and their role in plant diseases.

Professor and Head

In August 1984 he became a professor and in the month of October he became Head of the Department. As soon as he assumed the duty of HOD he brought DRS programme from UGC to strengthen the teaching and research facilities. During the period of 17 years of headship he also brought DSA phase I and II, and laid a stone for Cosist (Assistant programme) under SAP programme of UGC.

Prof Vyas established a strong centre of research in Mycology and Microbiology at Sagar. Where he undertook and successfully completed various research projects funding by UGC, CSIR, ICMR etc. He has published his work in various journals of National and International repute. He has edited book "*Recent Advances in Microbiology*". More than 40 students have obtained PhD under his supervision. Many of his students are actively engaged in research work and are occupying prestigious position in country and abroad.

During his tenure department grow, and many learned people from country and abroad visited the department and deliver their talks. He organized many seminars/ symposia/ workshops and refresher courses. Beside his scientific tamper and aptitude he has got administrative skills and proved himself an administrator in various capacities.

Administrative Experiences

1. As Head of the Botany Department. Dr. H.S.G. V.V., Sagar: 17 years (November 1984 to October 2001).
2. As Dean Faculty of Life Science: 4 years (two terms: 1988-90 and 1996-98).
3. As University Proctor: 6 Years (1986-1982).
4. As Chairman Board of Studies in Botany: 6 years (two terms: 1984-87 and 1994-97).
5. As Chief Coordinator of University Examinations: (three terms: 1985-86, 1988-89, and 1994-1995).
6. As Coordinator UGC NET- Exam, December 1996, June 1997, December 1997 and June 1998.
7. As Member of Academic Council Dr. H.S. G. V.V., Sagar: 10 years.
8. As Member of Executive Council; As Dean Faculty of Life Science (1988-90) and as Senior Professor (1996-99).
9. As Member of Central Board of Studies in Botany, Bhopal (MP): 10 years.
10. As Vice Chancellor of Dr. H.S.G. V.V., Sagar on various occasions/ day (during short term leave vacancy).
11. As Programme Coordinator of DRS/ DSA phase I and II programme of UGC in Botany Dr. H.S.G. V.V., Sagar.
12. As Member, Board of Studies, RDC and Selection Committees of various Universities and Colleges: 14 years.
13. As Professor Incharge of Development, Dr. H.S.G. V.V., Sagar: 3 years.

14. As Chairman, Organizing Committee of various symposia and seminars.
15. As Coordinator of Refresher Course in Botany: (1995-1998).
16. As Director Development in Botany: 1 October 2001–30 April 2004.

He was retired on 30th April 2004. He was an amazing personality with strong will power. His fight against Hepatitis B was amazing. After obtaining superannuation from the university he joined as a Principal at Pt. Vasudev Prasad Tiwari Degree College, Jhansi. He worked for the welfare of humanity under the umbrella of Bundelkhand Janseva Samiti, Jhansi.

Prof. Vyas was a great source of inspiration for his students, family members. His untimely death created a vacuum not only in scientific community but also in the family. It is said that there is a woman behind every great man. In this case it is Mrs Urmila Vyas, since 1958 when she was married to Prof KM Vyas. She has sacrificed all her comforts and has taken over all the household responsibilities so that Prof KM Vyas could be free to devote all his time to academics. Any visitor to their home was deeply impressed by their modesty, simplicity and warmth of their hospitability.

The Vyas are a well knit family. They have well educated children. One son and three daughters. Son is Dr Deepak Vyas presently serving in the same Department (Botany), Dr. H.S.G. V.V., Sagar married with Mrs. Nisha Vyas who is lecturer in college and have two sons Siddhant and Vedant studying in school. Among the daughters, elder daughter is Dr Jyoti married with with Dr Sanjay Chaubey, having their own hospital at Jhansi and having two sons Rohan and Kartik, then Mrs Arti married with Sunil Sharma who is working as an area manager in private firm. Arti is lecturer in private college at Bhopal have two kids Shivani and Shreyansh. Younger daughter Mrs Bharti is married with Dr Anand Sharma. Who is working at Cambridge Hospital, London, England.

List of Publications

1. Vyas, A. D. Vyas and K.M. Vyas (2006). Microbial cellulases for Industrial application. Every man's Science XL (6) 411-416.
2. Sharma, C.D., A.N. Rai and K.M. Vyas (2006). An undescribed fungal genus of the forest flora of Madhya Pradesh J. Bot. Soc. Univ. Sagar, 41 : 94-98.
3. Vyas, A., A. Jain, D. Vyas and K. M. Vyas (2005). Production and partial characterization of cellulases by two *Aspergillus* spp. on pretreated low value ligno-cellulases. Proc. Natl. Acad. Sci. India, 75(B) p. 1-7.
4. Vyas, Ashish, D. Vyas and K.M. Vyas (2005). Biodiversity conservation through integrated participatory management J. Bot. Soc. Univ. Sagar, 40 : 178-189.
5. Dwivedi, O. P., R. K. Yadav, D. Vyas and K. M. Vyas (2004). Role of Potassium in the occurrence of VAM spores in the rhizosphre soils of Lantana species. In: Microbiology and Biotechnology for sustainable development. Ed. P.C. Jain, Pp. 248–253. CBS Publishers, New Delhi.
6. Yadav, R.K., O.P.Dwivedi, D. Vyas and K. M. Vyas (2004). Antifungal effect of *Microcystis* against Beetalvine *Phytophthora.* J. Basic and applied Myco. 3(1+2) : 105–107.
7. Vyas Ashish, Vyas Deepak and Vyas K.M. (2004). Production and optimization of cellulases on pretreated groundnut shell by *Aspergillus tenuis* AV49. Journal of Scientific and Industrial Research Vol. 64, Pp. 281-286.
8. Vyas, A., D. Vyas and K.M. Vyas (2004). Cellulolytic strains of *Aspergillu* and *Penicillium*. J. Bot.Soc.Univ. Sagar 39 66-71.

9. Yadav, R.K., O.P. Dwivedi, D. Vyas and K.M. Vyas (2003). Management of *Phytophthora* disease of betel vine by plant extracts J. Basic and Applied Mycol. 2 (1) : 79-82.
10. Agrawal, R., A.N. Rai and K.M. Vyas (2003). A new species of *Tretospora* from India. J. Basic and Applied Mycol. 2 (1) 48-49.
11. Soni, Anuradha, D. Vyas and K.M. Vyas (2003). Effect of VAM fungi on the growth and productivity of Soybean (*Glycine max.*) J. Basic and Applied Mycol. 2 (1) : 40-41.
12. Vyas, Ashish, D. Vyas and K.M. Vyas (2003). Screening of extra-cellular cellulase producing fungi from different lignocellulosic wastes. J. Basic and Applied Mycol. 2 (1) : 14-16.
13. Khare, N. and Vyas K.M. (2003). Effect of nitrogen sources on the production of pectolytic and cellulytic enzymes by *Rhizopus nodosus* and *Phytophthora nicotiana* J. Bot Soc. Univ. Sagar 36 : 87-94.
14. Dwivedi, O.P., R.K. Yadav, D. Vyas and K.M. Vyas (2002). Distribution of VAM fungi in the rhizosphere soils of Betel vine J. Phytopathol. 56 (2) : 228-229.
15. Dwivedi, O.P., Yadav, R.K., Vyas, D. and Vyas, K.M. (2002). Vesicular-Arbuscular Mycorrhizal Association with *Lantana* sp. In Sagar University Campus; in Different Disordered Systems INDIAS 2000. (eds) K. Fukawa, J.T. Watson, S.N. Saxena and S.K. Shrivastava, pp. 154-157. INDIAS Publication, Allahabad.
16. Vyas, D., O.P. Dwevedi, R.K. Yadav and K.M. Vyas (2002). Arbuscular Diversity of VAM Fungi. In frontiers of fungal diversity in India. (Eds.) G.P. Rao, C. Manoharachari, D.J. Bhat, R.C. Rajak and T.N. Lakhanpal pp. 873-889. International Book Distributing Company Lucknow, India.
17. Singh, N., Vyas, D., Gupta, U.S. and Vyas, K.M (1999). Physico-chemical properties of Sagar lake with special reference to *Microcystis* bloom in the proceeding of the New Millenium Conference on Retrospect of Indian Research on Environmental Pollution : Focus 21st Century, 38.
18. Sharma, C.D., Gadpandey, K.K., Firdousi, S.A., Rai, A.N. and Vyas, K.M. (1998). Three New species of *Cladosporium* from Madhya Pradesh, India. Indian Phytopath. 51 (2): 152-160.
19. Chaursia, J.P., Chaurasia, P. and Vyas, K.M. (1997). Effect of *Ocimum sanctum* leaf extract on betelvine *Phytophthora* disease; An ecofriendly new approach. Environment pollution in 21 Century (Ed). V.S. Bais. Narendra Publishing House, Delhi, p. 46-51.
20. Chaurasia, J.P. and Vyas, K.M (1997). *In vivo* evaluation of some homoeopathic drugs against betelvine *Phytophthora* disease, Indian Phytopath. 4:148-153.
21. Vyas, K.M., Chaurasia, J.P. and Mishra, A.S. (1996). Effect of hepar sulphate on betelvine *Phytopathora* disease. Madhya Bharti Journal, Dr HS Gour Vishwavidyalaya, Sagar, India 36A 40B: 65-71.
22. Vyas, K.M. and Firdousi, S.A. (1996). Leaf blight of *Terminalia tomentosa.* Indian Phytopathology (Accepted).
23. Vyas, K.M., Gadpandey, K.K., Sharma, C.D., Firdousi, S.A. and Rai, A.N. (1996). Two new Hyphomycetes from India. Mycological Research (Accepted).
24. Vyas, K.M., Sharma, C.D., Gadpandey, K.K., Rai, A.N. and Firdousi, S.A. (1996). Three new species of *Cladosporium* from Madhya Pradesh. Indian Phytopathological Society IARI, New Delhi (In press).

25. Vyas KM, Dubey MP and Sharma RS (1995). Low cost agronomic practice for soybean (*Glycine max*) cultivation in Madhya Pradesh. Indian Journal of Agriculture Sciences 65(10):743-745.
26. Vyas KM, Kumar R and Rai AN (1995). Two new *Alternaria* species from Uttar Pradesh. Acta Botanica Indica 23 (1): 151-153.
27. Vyas KM and Firdousi SA (1994). A new leaf spot disease of *Casearia tomentosa* caused by *Asteromella* species. Acta Botanica Indica 20: 120-121.
28. Vyas KM and Saxena AP (1993). Antimicrobial activity of *Alectra parasitica*, A rice var. *chitrakutensis* rau. Journal of Economic and Taxonomic Botany 17 (1) 55-59.
29. Vyas KM, Firdousi SA and Rai AN (1993). A new species of *Stenella* from India. Acta Botanica Indica 25: 131-132.
30. Vyas KM and Firdousi SA (1993). A new spot of *Miliusa tomentosa* Roxb. Acta Botanica Indica 21 (1): 124.
31. Vyas KM and Mehta A (1992). Degradation of plant tissues by fungal enzymes. In: Dubey NK and Sharma PD, Rastogi Publications, Meerut, India.
32. Vyas KM, Firdousi SA and Rai AN (1992). A new species of *Pseudocercospora* from India. Indian Phytopathology 45(4): 449-451.
33. Vyas KM, Firdousi SA and Rai AN (1992). A new species of *Pseudocercospora* from India. Acta Botanica Indica.
34. Vyas KM, Firdousi SA and Rai AN (1992). *Mycovellosiella adinae* Sp nov. from India. Indian Phytopathology 45(4): 451-452.
35. Vyas KM (1990). A new leaf spot disease of *Ficus begalensis* caused by *Coniothyrium olivaceum* from India. Acta Botanica Indica 1: 150.
36. Vyas KM, Firdousi SA and Rai NA (1990). *Pseudocercospora gymnosporinae* sp. nov. India. Mycological Research (UK in press).
37. Vyas KM, Firdousi SA and Rai NA (1990). A new host record of *Thielvia appendiculata* from India. Indian Phytopath 43(1): 124.
38. Vyas KM, Firdousi SA and Rai NA (1990). A new leaf spot disease of *Eleodendron glaucum* from India. Indian Phytopath 43(1): 122.
39. Vyas KM and Gupta DK (1989). Efficacy of *Bacillus subtilis* against Mosquito larvae (*Anophelis culicifacies*). Zeitschrift Fur Angewandte Zoologie (German Journal for Applied Zoology) 85-91.
40. Vyas KM and Chile SK (1989). Metabolism of ascorbic acid and *Piper betle* leaves under *Phytopathora* pathogenesis. Perspective in Mycological Research 1: 143-150.
41. Vyas KM and Tripathi HSS (1986). Production of toxin by *Xanthomaonas compestris* pv *betticola* causing leaf spot disease of *Piper betle* L. Acta Botanica Indica 14: 220-225.
42. Vyas KM and Saxena NK (1986). Microbial synthesis of 11-deoxy corticosteron from progesterone by a nutant of *Rhizophus nodosus* (Abst.) XIV International Congress of Microbiology: Menchester, England 7-13 September.
43. Vyas KM and Chile SK (1985). Electrolyte leakage from *Piper betle* leaves after infection with *Phytopathora parasitica* var *piperina*. Indian Phytopathology 38 (3): 529-530.

44. Vyas KM and Tripathi HSS (1985). Production of cellulolytic enzymes by *Xanthomonas compestris* pv *betticola*: *in vivo* and *in vitro*. Bull Bot. Soc. 33.

45. Vyas KM, Tripathi HSS, Saxena NK and Sadique Y (1985). Efficacy of hydroxyl stilbamidine isethionate USP against *Phytopathora parasitica* var *piperina*. Phytopathora Newsletters 13: 32-33.

46. Vyas KM, Saxena NK, Tripathi HSS and Sadique Y (1985). Complete control of *Phytopathora parasitica* var *piperina* by S-flurocytosine. Phytopathora News Letters 13: 30-31.

47. Vyas KM, Dubey GL and Dubey O (1985). Physico-chemical properties of linseed oil as affected by various seed borne fungi. Phytopath. z. 113: 66-70.

48. Vyas KM and Chile SK (1984). Efficacy of *Vinca rosea* extracts against protease from human pathogenic strain of *Trichophytom rubrum* sab. Hindustan Antibiotics Bull. 26: 114-116.

49. Vyas KM, Kher AK and Saxena SB (1984). Persistence of various fungicides in soil. Indian Phytopath 27(3): 574-576.

50. Vyas KM and Shrivastava A (1984). Effect of cholesterol on the toxicity of certain antibiotics against betelvine *Phytopathora*. Phytopathora News Letter 12 pp. 33.

51. Vyas KM, Soni, Saxena NK and Tripathi HSS (1984). Decline in the total reducing sugar content during the postharvest infection of *Lycopersicon lycopersicum* L. Karsgen fruits by *Bacillus fructodrestruens*. Bull Bot. Soc. 30-31: 69-71.

52. Vyas KM, Saxena NK and Tripathi HSS (1984). Cationic degradation of Streptomycine and Kanemycine by *Aspergillus niger* in broth cultures during fragmentation in shade flask. Hindustan Antibiotics Bulletin 26 (3 and 4) pp. 96-101.

53. Chile SK, Vyas KM and Singh R (1984). Induced resistance in *Piper betle* leaves in relation to *Phytopathora parasitica* var. *piperina*. Indian Phytopath. 37: 719-721.

54. Chile SK and Vyas KM (1984). Inhibitors of certain enzymes of *Botrydiplodia theobromae* in *Piper betle* leaf extract. Hindustan Antibiotics Bull. 26: 27-32.

55. Vyas KM and Chile SK (1984). Respiratory response of betlevine *Phytopathora* induced by various volatile oils. Jr Mycology Plant Path.

56. Vyas KM, Saxena NK and Tripathi HSS (1984). Accumulation of D.L. 3,4 dihydroxy-phenyalanine during the soft rot of *Lycopersicon esculentum* fruits caused by *Bacillus fructodestruens*. Abst. 36th Ann. Meeting Indian Phytopath (Hissar) 37: 401.

57. Vyas KM, Saxena NK and Tripathi HSS (1984). Postharvest accumulation of vitamin C induced by *Bacillus fructodestruens* in the *Lycopersicon esculentum* fruits. Abst. 36th Ann. Meeting Indian Phytopath, Hissar and paper published in Nat. Acad. Sci. Letters 7(4): 109-110.

58. Vyas KM and Shrivastava A (1984). Chemical control of betelvine *Phytopathora*. Abst. Symp. Chemical Control of Plant Disease. Indian Phytopath, Hissar.

59. Vyas KM and Singh R (1984). Nitrogen nutrition of *Phytopathora parasitica* var *piperina* (Dast) causing foot and leaf rot disease of *Piper betle* Linn. Proc. 71st Ind. Sci. Cong. Part III Abstract, 57: 26.

60. Vyas KM and Atri DC (1983). Pectolytic and cellulolytic activity during pathogenesis caused by *Botryodiplodia theobromae* Pat. In *Pyrus pyrifolia* Burm. fruits. Abst National Acad. Sci. (Biol.) Abst. No. 257 pp. 94.

61. Vyas KM and Chile SK (1983). Free amino acids in relation to *Phytopthora* leaf rot pathogenesis of Pan. Indian Phytopath 36(4): 626-628.

62. Vyas KM and Chile SK (1983). Dehydrogenase activity in *Piper betle* L. leaves infected with *Phytopathora parasitica* var. *piperina*. Indian Phytopath 36(4): 613-617.

63. Vyas KM and Sxena N (1983). Effect of antibiotics on certain cell wall degrading enzymes of *Rhizopus nodosus* and *Phytopathora nitotianae*. Hindustan Antibiotics Bulletin 25 (1 and 2) pp. 21-24.

64. Vyas KM and Shrivastava A (1983). In vitro efficacy of tetracycline against betelvine *Phytopathora*. Hindustan Antibiotics Bulletin 25 (1 and 2) pp. 15-17.

65. Vyas KM and Saxena AP (1983). Ethnobotany of Dhasan valley. Jr. of Econ. Tax. Bot., Jodhpur.

66. Vyas KM and Saxena AP (1983). Traditional treatment of leprosy and leucoderma by tribals from Bundelkhand UP. Bull. of Medico Ethn. Bot. Res. CCRAS, New Delhi.

67. Vyas KM and Thakur MS (1983). Production of plant growth regulators by some *Fusarium* species. Folia Microbial. 28: 124-129.

68. Vyas KM, Gupta DK and Shrivastava A (1983). In vitro antibacterial properties of some fixed oils. Proc 70th Ind. Sci Cong. Assoc 29.

69. Vyas, Chile SK and Shrivastava A (1982). Efficacy of some growth hormones on sporulation of betelvine *Phytopathora*. Bull. Bot. Soc. Univ. Saugar 29: 25-27.

70. Chile SK and Vyas KM (1982). Efficiency of carboxin in relation to *Phytopathora* leaf rot of *Piper betle*. Abstract 1st National Symposium of Synthetic Fungicides in Plant Disease Control in IARI, June 1982.

71. Chile SK, Vyas KM and Vyas RS (1982). Effect of culture filtrate of various fungi on sporulation of betelvine *Phytopathora*. Hind. Ant. Bull. 23: 25-26.

72. Chile SK and Vyas KM (1982). Effect of culture filtrate of various fungi on acidative enzymes of betelvine *Phytopathora*. Hind. Ant. Bull. 24: 26-28.

73. Vyas KM, Gupta DK and Chile SK (1982). A fruit rot disease of Momardiaca. Indian Phytopath 35(2): 351.

74. Vyas KM and Gupta DK (1982). Biological control of mosquito. IX Annual Conference of Indian Society of Human Genetics, Bhopal 58.

75. Vyas KM and Gupta DK (1982). Effect of *Aspergillus* mycotoxin on *Culex* larvae. Proc 69th Ind. Sci. Cong. Asso. Abstract 61.

76. Vyas KM and Thomas T (1982). On the abundance of *Trichoderma* species in the soil of Andaman and Nicobar Islands. Proc 69th Ind. Sci. Cong. Asso. Abstract 59.

77. Vyas KM and Saxena N (1982). Pectolytic enzymes of *Rhizopus nodosus* and *Phytopathora nictianae* in relation to electrolyte leakage and tissue maceration. Proc 69th Ind. Sci. Cong. Asso. Abstract 44.

78. Vyas KM and Saxena AP (1982). In vitro antimicrobial activity of some medicinal plants against human pathogens. Proc 69th Ind. Sci. Cong. Asso. Pt. III: 43-44.

79. Vyas KM, Singh R and Chile SK (1982). Nitrogen nutrition of *Phytopathora parasitica* var *piperina* (Dast) Causing foot rot and leaf rot diseases of *Piper betle* Linn. Jur. Biol. Res.

80. Vyas KM (1982). Peroxiodase and catalases in relation to betelvine. Phytopathora News Letters, USA 10: 8.

81. Vyas KM and Chile SK (1982). Breakdown of disease resistance in betelvine *Phytopathora* system by vitamin C. Phytopathora News Letters, USA 10: 6-7.

82. Vyas KM and Thakur MS (1982). Plant growth regulatory metabolites produced by fungi. Abstract Symp. Fermented Foods, Food Contaminants, Biofertlizers and Bioenergy pp. 54.

83. Vyas KM and Thoimas T (1982). *Gliocladium virens*- A new report from India. Geobios News reports 1: 45.

84. Vyas KM and Saxena AP (1981). *Martynia annua* Linn. A drug for Asthma, Itch and Eczema. Bull of Medico. Ethno Bot. Res. CCRAS, New Delhi II (3): 427- 429.

85. Vyas KM (1981). Problem and progress in *Phytopathora* diseases of Pan group discussion on improvement of betelvine cultivation. NBRI, Lucknow p 92-93.

86. Vyas KM, Thomas T and Gupta DK (1981). *Gilmaniella subornate*: A new report from India. Bull. Bot. Soc. Univ. of Saugar 28: 38.

87. Vyas KM, Thomas T and Gupta DK (1981). *Trichoderma pseudo-konningii* Rifai: A new report from India. Bull. Bot. Soc. Univ. of Saugar 28: 28.

88. Vyas KM (1981). Effect of culture filtrates of various fungi on sporulation of betelvine *Phytopathora*. Hind. Antibiot. Bot. 23: 25-26.

89. Vyas KM, Chourasia SC and Saxena SB (1981). Management of *Phytopathora* diseases of Pan (*Piper betle*). Abst. 3rd International Symposium on Plant Pathology (IPS) pp. 253.

90. Chile SK and Vyas KM (1981). Polyphenol oxidase in relation to betelvine *Phytopathora* pathogenesis of Pan. Phytopathora News Letters, USA 9: 40-41.

91. Vyas KM and Saxena AP (1981). Medicinal plants of Bundelkhand I.P preliminary ethnobotanical survey of Banda district UP. Proc. 68th Ind. Sci. Cong. Pt III (Abst): 147.

92. Vyas KM and Saxena AP (1981). Flora of Bundelkhand II. Hydrophytic tract of Ken-Baghain tract (Banda district UP). Ind. Soc. Weed. Sci. Abst: 40-41.

93. Vyas KM and Saxena AP (1981). Ethnobotanical records on infectious diseases from tribals of Banda district (UP). Ind. Soc. Weed. Sci. Abst: 40-41.

94. Vyas KM and Gupta DK (1981). Biological control of mosquitoes: I Effect of culture filtrate of certain *Aspergilli* on Anopheles larvae. Proc. 51 Session Nat. Acad. Sci.: 51.

95. Vyas KM and Chile SK (1981). Effect of the age of betel vine leaves on phenolics and *Phytopathora* rot. Ind. Sci. Cong.

96. Chile SK, Chourasia and Vyas KM (1981). Role of cuttings on the growth of the betel vine *Phytopathora*. Phytopathora News Letters, USA 8: 31.

97. Vyas KM and Chile SK (1980). Role of polyphenol oxidase in disease resistance with special reference to the betel vine *Phytopathora*. Symp. Plant Disease Problem at PAU, Ludhiana (Punjab) Abstract IIX.

98. Vyas KM, Chile SK and Saraf M (1980). Effect of some fungicides on sporulation of different isolates of betel vine *Phytopathora*. Bull. Bot. Univ. of Saugar, Sagar (MP) 27 pp. 16-18.

99. Vyas KM, Chourasia SC and Chile SK (1980). Role of infested soil and cuttings on the spread of betel vine *Phytopathora*. Phytopathora News Letters, USA 8: 28.

100. Vyas KM and Chile SK (1980). Role of infested soil and cuttings on the spread of betel vine *Phytopathora*. Symp. on Phytopathora Diseases of Tropical Plants (Calicut) pp. 22.

101. Vyas KM and Chile SK (1980). Varietal reaction of *Piper betle* leaves against different isolates of betel vine *Phytopathora*. Phytopathora News Letters, USA 8: 29.

102. Vyas KM, Chile SK and Vyas RS (1980). Two new diseases of cultivated plants. Indian Phytopath. 33(3): 493-495.

103. Vyas KM and Gupta M (1979). Studies on cell wall degrading enzymes in relation to banana fruit rot. Indian Phytopath. 32: 456- 457.

104. Vyas KM, Raghvan U and Saxena SB (1979). Studies on amino acid composition of mycelia of 14 of *Botryodiplodia theobromae* Pat. Bull. Bot. Soc. Univ. Saugar, 23-24: 21-25.

105. Vyas KM and Thakur MS (1979). Production of indole auxins by *Curvularia* spp. Abst. 66th Sess. Ind. Sci. Cong. p. 38.

106. Vyas KM and Dubey GL (1979). Physico-chemical properties of linseed oil as affected by various seed borne fungi. Abst. 66th Sess. Ind. Sci. Cong. p. 15.

107. Vyas KM and Saxena AP (1979). Flora of Bundelkhand I, A contribution to grasses of Banda and Hamirpur districts (UP). Bull. Bot. Soc. Univ. of Saugar, Sagar 25-26: 73-78.

108. Vyas KM and Chourasia SC (1979). Efficacy of certain antibiotics against betel vine *Phytopathora*. Phytopathora News Letters, USA 7: 28.

109. Vyas KM and Chourasia SC (1979). Leaf spot, a new disease of bakoli (*Mimusops elengi*) in Sagar, India 63: 806.

110. Vyas KM and Chourasia SC (1979). Oxidative metabolism under *Phytopathora* leaf rot pathogenesis of Pan. Phytopathora News Letters, USA 7: 27.

111. Vyas KM (1978). Effects of fungicides on the growth of *Phytopathora parasitica* var. *piperina* causing foot rot and leaf rot diseases of Pan. Proc. Nat. Acad. Sc. (Allahabad) 47 (B) III: 141-144.

112. Vyas KM and Chourasia SC (1978). Formation of cospores in vivo by *Phytopathora parasitica* var. *piperina*. Phytopathora News Letters USA, Feb. 1978, No. 6 pp. 42-43.

113. Vyas KM and Chourasia SC (1978). Leaf surface mycoflora in relation to *Phytopathora* leaf rot on Pan (*Piper betle*). Phytopathora News Letters USA, Feb. 1978, No. 6 pp. 40-41.

114. Vyas KM and Chourasia SC (1978). Host parasite interactions in betelvine *Phytopathora* varietal response to leaf rot pathogenesis. Phytopathora News Letters USA, Feb. 1978, No. 6 pp. 38-39.

115. Vyas KM (1978). Effect of fungal metabolites on the growth ot morphogenesis of crop plants. Abst. Ibid. pp. 43.

116. Vyas KM (1978). Studies on the molds and mycotoxins with special reference to mycoflora associated with the grains of jowar. Abst. All India Symposium on Physiology of Parasitism Jabalpur. pp. 20.

117. Vyas KM, Kazmi SM and Soni NK (1978). Studies on the antifungal properties of mamalian urine. Agriculture and Agro Industries Jur.

118. Chourasia SC, Vyas KM and Pathak RK (1977). In vitro inhibitory effects of certain fungicides in the growth of *Phytopathora parasitica* var. *piperina* causing foot rot and leaf rot diseases of Pan (*Piper betel* Linn.) Proc. Nat. Acad. Sci. Ind. 47(B) III: 141-144.

119. Vyas KM and Chourasia SC (1977). Effect of fungicides on the oxidative metabolism of *Phytopathora parasitica* var. *piperina* (Dast.) causing foot rot and leaf rot diseases of Pan (*Piper*

betel) in physiology of microorganism edited by KS Bilgrami Todays and Tomorrow Printers and Publishers, New Delhi 11-16.

120. Vyas KM and Jain SK (1977). Production of auxin like plant growth regulatory metabolites by soil fungi. In: physiology of microorganisms (Edited by KS Bilgrami), Todays and Tomorrow Printers and Publishers, New Delhi 331-340.
121. Vyas KM, Thind TS and Prakash V (1977). Effect of some antibiotics on the germination of corainder seeds. Ind Jr. of Exp. Biol. Vol. 15, No. 3, March 77 pp. 247-248.
122. Vyas KM and Chourasia SC (1977). In vitro effect of some volatile oils against *Phytopathora parasitica* var. *piperina*. Jr. Res. Ind. Med. Yoga and Homoeo. 12:3. 139-142.
123. Vyas KM and Chourasia SC (1977). The influence of age and starvation on endogenous respiration of mycelium of *Phytopathora parasitica* var. *piperina*. Acta Botanica Indica 5: 54-57, 77.
124. Vyas KM (1976). Respiratory characterstics of plant pathogenic fungi, *Ibid*.
125. Vyas KM (1976). Production of cell wall degrading enzymes by *Botryodiplodia theobromae, Ibid*.
126. Vyas KM (1976). Microbial metabolites and plant growth promoting activity of some antifungal antibiotic compounds, *Ibid*.
127. Vyas KM (1976). Microbial metabolites and plant growth production of indole auxins by soil fungi, *Ibid*.
128. Vyas KM (1976). Effect of fungicides on the oxidative metabolism of *Phytopathora parasitica* var. *piperina* (Dast) causing foot rot and leaf rot of Pan (*Piper betle*). In: Physiology of Microorganisms (Edited by KS Bilgrami), Today and Tomorrow Publications pp. 11-16.
129. Vyas KM (1976). Interactions of soil microorganisms with *Phytopathora parasitica* var. *piperina*. Abst. Symp. on Physiology of Microorganisms held at Bhagalpur Feb. 1976.
130. Vyas KM and Chourasia SC (1976). Activity of some volatile oils against *Phytopathora parasitica* var. *piperina*. Indian Drugs Jan 9-15.
131. Vyas KM and Chourasia SC (1976). Influence of culture media on fungal respiration. Hind. Antibiot. Bull. Vol. 18: 3 and 4.
132. Vyas KM and Mehta P (1975). Effect of plat growth regulators on production and activity of cellulase. Ind. J. Exptl. Biol. Vol 14, 206-208.
133. Vyas KM and Chourasia SC (1975). Studies on leaf rot of Pan (*Piper betle*) varietal factors in relation to disease resistance. Symp. on Plant Pathlogical problems held at Udaipur 5(1): 10-11.
134. Vyas KM, Mehta P and Saxena SB (1975). Pathological studies of fruit rot diseases of tamato caused by *Alternaria solani* and *A. tenuis*. Indian Phytopath. Vol. XXVIII No 2, 247-252.
135. Vyas KM, Mehta P and Saxena SB (1975). Metabolic changes during pathogenesis of tomato fruit rot diseases. Indian Phytopath. Vol. XXVIII No 2, 253-255.
136. Vyas KM and Soni NK (1975). Effect of antibiotics and fungicides on the mycelial respiration and growth of *Rhizoctonia capsici*. Ind. Jr. Exptl. Biol. Vol. 13, pp. 216-218.
137. Vyas KM, Mehta P and Saxena SB (1975). Production of pectolytic enzymes by *Alternaria* sp. of various culture media. J. Ind. Bot. Soc. 200-206.

138. Vyas KM and Soni NK (1975). Responses of *Rhizotonia solani* Kuhn and *Colletotrichum capsici* (Syd.) Butler and Bisby to various chemical substances. Madhya Bharti Univ. Sagar, Vol. 22-23 pp. 43-48.

139. Vyas KM, Mehta P and Saxena SB (1975). Effect of native carbon sources and pH on the cellulolytic enzymes of *Alternaria solani* and *A. tenuis*. Science and Culture Vol. 41 pp. 400-402.

140. Vyas KM and Chourasia SC (1975). Effect of vitamins on the mycelial growth of *Phytopathora parasitica* var. *piperina*. Madhya Bharti, vol. 22 and 23 pp. 63-65.

141. Vyas KM and Soni NK (1975). Respiratory response of *Gleosporium papayae* induced by various substances. Proc. Ind. Sci. Acad. Vol. 41, Part B. No. 5 pp. 458-461.

142. Vyas KM, Saxena RK and Saxena SB (1974). Physiology of *Tectona grandis* leaves infected with *Uncinula tectonae*. J. Indian Bot. Soc. 53: 256-270.

143. Vyas KM, Jain SK and Raghvan P (1974). Microbial metabolites and plant growth: Plant growth regulatory activity of culture filtrates of soil microorganisms and some antibiotic compounds. Abst. Symp. on Biology of Soil Microorganisms, Sagar pp. 35-36.

144. Vyas KM, Mehta P and Saxena SB (1974). Metabolic changes during pathogenesis of fruit rot diseases of tomato. MP Vigyan Acad. Bhopal Abst. pp. 7-8.

145. Vyas KM, Saxena SB and Soni NK (1974). Effect of antibiotics and fungicides on endogenous mycelial respiration and growth of *Rhizoctonia solani* Kuhn. and *Colletotrichum capsici* (Syd.) Butler and Bisby. MP Vigyan Acad. Bhopal Abst. pp. 5-6.

146. Vyas KM, Mehta P and Saxena SB (1974). Production of pectolytic and cellulolytic enzymes by *Alternaria* sp. during pathogenesis of tomato fruits. Hind. Antibiot. Bull. Vol. 16, No. 4 pp. 20-214.

147. Vyas KM, Mehta P and Saxena SB (1974). Effect of native carbon sources and pH on the pectolytic enzymes of *Alternaria* sp. during pathogenesis of tomato fruits. Hind. Antibiot. Bull. Vol. 40 Part B, No. 4 pp. 433-439.

148. Vyas KM, Pandey P and Soni NK (1974). Effect of antibiotics on endogenous mycelial respiration and growth of some plant pathogenic fungi. Hind. Antibiot. Bull. Vol. 16: 199-201.

149. Vyas KM and Soni NK (1973). Effect of the age of culture and starvation period on mycelial respiration. Bull. Bot. Soc., Univ. Saugar, Vol. 20 pp. 17-20.

150. Vyas KM and Saxena SB (1973). Studies on mycelial respiration of *Sclerotium rolfsii*. Proc. Ind. Nat. Acad. Vol. 39 Part B No. 5 569-575.

151. Vyas KM, Raghvan P and Jain SK (1973). Auxins like activity of the antibiotic aureofungin. Hind. Antibiot. Bull. Vol. 16 No 1 pp. 29-31.

152. Vyas KM and Jain SK (1973). Production of auxins by microorganisms. Hind. Antibiot. Bull. Vol. 11, pp 217-219.

153. Vyas KM and Soni NK (1973). Respiratory and growth responses of *Fusarium oxysporum* Sacc. induced by various substances. Ind J. Exptl. Biol. Vol. 11 pp. 217-219.

154. Vyas KM, Chourasia SC and Soni NK (1973). Efficiency of certain antibiotics against *Phytopathora parasitica* var. *piperina* causing leaf rot of Pan. Hind. Antibiot. Bull. Vol. 16 No. 1 pp. 4-8.

155. Vyas KM and Soni NK (1973). Effect of antifungal antibiotics on endogenous mycelial respiration of *Rhizoctonia solani*. Hind. Antibiot. Bull. Vol. 15, pp. 160-163.

156. Vyas KM, Raghavan P and Jain SK (1973). Synergistic interaction of cloremphenicol and IAA on the growth of *Avena* coleoptile sections. Hind. Antibiot. Bull. Vol. 15, pp. 81-83.

157. Vyas KM and Mishra AS (1972). Production of P-proline by soil microorganisms. Hind. Antibiot. Bull. Vol. 15 No 1 and 2 pp. 30-33.

158. Vyas KM, Soni NK and Saxena SB (1972). Effect of culture of various fungi on the mycelial respiration of *Fusarium oxysporum* Sacc. Hind. Antibiot. Bull. Vol. 14, pp. 179-180.

159. Vyas KM, Saxena SB and Jain BK (1972). Respiratory characteristics of some soil borne plant pathogenic fungi. J. Ind. Bot. Soc. 51: 92-96.

160. Vyas KM and Mehta P (1972). Control of fruit rot of tomatoes by phenolics and fungicides. Bull. Bot. Soc., Univ. Saugar, Vol. 19 pp. 35-37.

161. Vyas KM and Dubey GL (1972). Effect of plant growth regulators on the growth of *Sclerotium rolfsii*. Bull. Bot. Soc., Univ. Saugar, Vol. 19 pp. 35-37.

162. Vyas KM, Saxena SB and Soni SK (1972). Effect of plant growth regulators on oxygen uptake and growth of *Rhizoctonia solani* Kuhn and *Colletotrihum capsici* (Syd.) Butler and Bisby. Bull. Bot. Soc., Univ. Saugar, Vol. 19 pp. 24-27.

163. Vyas KM and Mehta P (1970-71). Production of cellulose and mycelial growth of *Alternaria solani* and *A. tenuis*. Bull. Bot. Soc., Univ. Saugar, Vol. 17 and 18 pp. 71-72.

164. Vyas KM and Soni NK (1970-71). Respiratory characteristics of mycelium of *Gleosporium papaya* P. Henn. Bull. Bot. Soc., Univ. Saugar, Vol. 17 and 18 pp. 48-50.

165. Vyas KM, Soni NK and Mishra AS (1970-71). Respiratory characteristics of *Gleosporium* infected papaya fruits in relation to disease development. Bull. Bot. Soc., Univ. Saugar, Vol. 17 and 18 pp. 48-50.

166. Vyas KM, Saxena SB and Atri DC (1970-71). Production of amino acids by soil fungi. Bull. Bot. Soc., Univ. Saugar, Vol. 17 and 18 pp. 31-34.

167. Vyas KM and Saxena SB (1962-64). The wood decaying fungi of Sagar. Madhya Bharti, Vol. 11-13 pp. 15-28.

List of Ph.D. Awarded Under the Supervision of Prof. K.M. Vyas

1. Mehta, Pradeep 1974. "Biochemical Investigations of Fruit-Rot Diseases of Tomatoes."
2. Jain, Satish Kumar 1975. "Studies on the Microbial Production of Auxins".
3. Chourasia, Shiv Charan 1976. "Studies on the Foot-Rot and Leaf-Rot Diseases of Pan (*Piper betle* L.) with Special Reference to Pathogenesis and Control Measures".
4. Raghavan (alias) Srinivasan, Parvati 1976. "Physiological Studies on the Effect of Antibiotics on Plants with Special Reference to their Growth Regulatory Effects"
5. Mishra, Jayendra Kumar 1979. Biochemical Studies on Some *Botryodiplodia* Rot with Special Reference to Cell-Wall Degrading Enzymes".
6. Arti, Dinesh Chandra 1980. "Study of Fruit-Rot Disease of *Pyrus pyrifolia* Caused by *Botryodiplodia theobromae*".
7. Dubey, Guljari Lal 1980."Fungal Succession in Relation to the Bio-deterioration of Linseed".

8. Thakur, Munna Singh 1981. "Studies on the Metabolities of Some Microorganisms and their Effects on Plant Growth".
9. Jain, Maya (nee) Singhai, Maya 1982. "Studies on Plant Growth Regulatory Activities of Certain Antibiotics".
10. Saxena, Neeraj 1982. "Studies on Cell Wall Degrading Enzymes in Relation to Fungal Pathogenesis".
11. Thomas, Thankamma 1982. "Studies on the Soil Fungi of the Andaman and Nicobar Islands".
12. Chiele, Satish Kumar 1983. "Studies on Host–Parasite Relationship with Special Reference to Betel–Vine *Phytopathora*".
13. Saxena, Anand Prakash 1983. "Studies on Some Medicinal Plants of Bundelkhand Region with Special Reference to their Antimicrobial Activity".
14. Gupta, Deepak Kumar 1984. "Studies on the Microbiological Control of Mosquito, Causing Malaria".
15. Shrivastava, Anjali 1985. "Studies on the Efficacy of Certain Antibiotics Against Betel-Vine *Phytopathora*".
16. Katare, Daya Shanker 1987. "Genetic Analysis of Metric and Quality Attributes in Tomato".
17. Saxena, Nitin Kumar 1987. "Microbial Transformation of Steroids".
18. Tripathi, Hari Har Sharran 1987. "Study on the Production and Activity of Toxic Metabolites by Certain Bacterial and Fungal Pathogens of *Piper betle* L".
19. Sadiq, Yasmin 1989. "Studies on the Role of Phenolic Compounds in Disease Resistance in Plants (with special reference to Betel-vine *Phytopathora*)".
20. Shrama, Pratap Bhanu 1991. "Study on the Effect of Cultural and Chemical Control Systems of Kharif Weeds in Tawa Command Area".
21. Firdousi, Shakeel Ahamad 1992. "Studies on the Fungal Disease of Some Important Forest Plant with Special Reference to the Foliage Diseases".
22. Sediq, Roohi 1992. "Studies on the Effect of Certain Factors on the Forest Vegetation of Sagar".
23. Tewari, Prem Narayan 1992. "Studies on Yield, Oil and Protein Content of Soybean as Affected by Fertility Levels, Different Stages of Sowing and Plant Density".
24. Malaiya, Ku. Seema 1992. "Antimicrobial Evaluation of Certain Medicinal Plants of Ethnobotanical-Importance".
25. Dubey, Mahadev Prasad 1993. "Effect of Soil Tilth Seed Inoculation and Sowing Methods on Growth and Yield of Soybean".
26. Goswami, Ravi Prakash 1994. "Studies on the Effect of Volatile Fungistatic Factors on Fungal Morphogenesis".
27. Dixit, Ashwini Kumar 1995. "Ethno-medicinal, Photochemical and Anti-microbial Studies of Some Medicinal Plants of Sagar District".
28. Chourasia, J.P. 1995. "Studies on the Management of Betalvine *Phytopathora* Diseases in Sagar".
29. Sharma, C.D. 1995. "Folicolus Fungal from the Forest Flora of M.P. with Special Reference to Deuteromycotina".

30. Goswami, Sita Ram 1995. "Response of Different Levels of Phosphetic Fertilizers and their Methods of Application on Growth Parameters Productivity and Quality of Soybean".
31. Malaiya, Manju 1998. "Studies on Antimicrobial Activity of Some Indian Traditional Remedies (Ethnomedicinal plants)".
32. Mishra, Avanish 1998. "Studies on the Biology of Some Ethno-medicinal Plants of Sagar Region".
33. Kaushal, A.K. 1998. "Role of Irrigation Schedule on Productivity of Sunflower".
34. Sthapak Smt. Madhu 2000. "Studies on the Fungal Diseases of Potato Under Storage Conditions at Sagar".
35. Nema Ku. Shushma 2000. "Studies on Some *Phytopathora* Species from Central India with Special Reference to their Reproduction".
36. Dwivedi, O.P 2003. 'Studies on Soil Microorganisms with Special Reference to Vesicular-Arbuscular Mycorrhizal (VAM) Fungal Association with Wheat Crop of Sagar Region".
37. Yadav, Rajesh 2003. "Studies on Integrated Management of Betelvine *Phytopathora*".
38. Agrawal, Roop Shikha 2004. "Studies on Fungal Diseases of Forest Plants of Mandla Region".
39. Jain, Ashish 2005. "Studies on Mushroom Cultivation with Special Reference to *Pleurotus* Species and their Marketing Potential in Sagar Region".
40. Vyas, Ashish 2005. "Studies on Microbial Biodiversity with Special Reference to their Cellulolytic Activity".
41. Soni Anuradha 2006. Studies on occurrence of VAM Fungi in leguminous crops of sagar.

Acknowledgements

A journey into unknown territory, this is how I felt at the onset of this task. The work undertaken has made me experience many ups and downs throughout the duration of compilation. Although I had been living with it days and nights but there are many people who have knowingly or unknowingly figured a major role in its completion. Although words are not sufficient to express my feelings but still I shall try to comprehend as far as I can and the feelings behind are really heartfelt.

This book is presented with obligation to the under mentioned with full acceptance of its limitations.

I thanks my Shredye Guru Ji Pt Omkar Dev Ji, whose wishes enabled me to reach this mile stone who showed me path to reach the destination and responsible for how to thank god.

I note down here with humility and pride that I am fortunate enough in having support of esteemed personalities like Prof. H.D. Kumar (FNA), BHU, Varanasi, Prof. D.P. Tiwari, Ex Head, Department of Botany, Govt. Science P.G. College, Jabalpur, Prof. C. Manoharachari (Emeritus Prof.), Department of Botany, Osmania University, Hyderabad, Prof. Anupam Dixit, Department of Botany, Allahabad University, Prof. S.P. Adhikari, Department of Biotechnology, Visva Bharti University, Shantiniketan (West Bengal), Prof K.L. Tiwari, Ex Head, Department of Biotechnology, Pt R.S. University, Raipur, Prof G.S. Paliwal (Emeritus Prof.) Senior Consultant National Afforestation and Eco-development Board, New Delhi, Prof H.K. Goswami, Ex Head, Department of Genetics, Barkatullah University, Bhopal and Prof. H.C. Lakshman, Head, Department of Botany, Karnataka University, Dharwad.

I am happy to remember my teachers who happened to be students of my father late Prof. K.M. Vyas, therefore I gratefully acknowledges Prof A.D. Adoni, Prof. T.R. Sahu, Prof. N.K. Soni, Prof. D.C. Atri, Prof. A.K. Kandya, Prof. P. Mehta, Prof. P.K. Khare, Prof. A.N. Rai, Prof. (Smt.) J. Dubey, Dr. A.S. Mishra, Dr. S.K. Yadav, Dr. N.P. Bhalla and my colleague Dr. A. Mehta, Dr. A. Biswas, Miss Poonam Dehariya and Dr. Ashok Shukla for their necessary help whenever I was in need.

I express sincere gratitude to all the authors who have contributed for this volume.

It gives me immense pleasure in acknowledging with sincere gratitude, the enthusiastic co-operation received from my students Dr. O.P. Dwivedi, Dr. Rajesh Yadav, Dr. Ashish Vyas, Dr. Anuradha Soni, Dr. Ashish Jain, Dr. Archana Dubey, Dr. Mahendra Mishra, Dr. Pradeep Singh, Dr. Prashant Soni, Dr. H.C. Gena, Pramod Richhariya, Meenakshi Mishra, Anjuli Chaubey, Javed Ahmad Wagay and Mrs. Anuradha Shukla.

I express a deep valuation to all the non-teaching staff especially Dr. Pradeep Tiwari, K.K. Dixit, J.P. Chaubey, P. Gujre and Shailesh Katare of the department for their time to time help and co-operations.

It is beyond the capacity of all the words to express my respect and profound affection for my mother Mrs. Urmila Vyas. Today whatever I am that is only by virtue of her love, care, affection and blessings. At this juncture I can not forget my wife Mrs. Nisha and children Sidhhant and Vedant and pain taken and sacrifices made by them can not be summed up. They remained as a source of constant inspiration; I bow down before their love and affection towards me. The love and affection extended by my sisters Dr. Jyoti, Arti and Bharti, and their family is memorable.

Last, but certainly not the least, I must thanks to Mr. Nafees of Perfect Computers for designing this cover page in a careful and elegant manner, which is miraculous. I also thank Mr. Anil Mittal of Daya Publishing House for publishing the volume a scientific tribute to my father.

Deepak Vyas

H.D. Kumar

Ph.D. (London). FASc. FNASc, FNA

Formerly Professor of Botany,
Dean, Faculty of Science, and Coordinator,
Biotechnology Programme
Banaras Hindu University

Phone (Res.): (0542) 2315180

214, Saketnagar, Naria,
VARANASi – 221 005

Foreword

Microbes, the real masters of life on Earth, are both a boon and bane for the planet. Our microbial friends live within our body, protecting us from disease, aiding digestion, and preserving immunity. On the other side, some of the deadliest diseases are caused by viruses and pathogenic bacteria. Whereas beneficial fungi are an asset to the industrialist and the environment, pathogenic fungi cause havoc and can devastate cereal crops and fruit trees.

The present volume is a Festschrift in honour of my long-term friend, late Prof. K.M. Vyas, a reputed mycologist with strong interest in microbiology. This volume includes some 50 chapters including 23 on microbiology, 13 on biotechnology and 14 on ecology. These chapters are authored by senior professors and younger scientists. The editors are practising researchers and I have pleasure in commending this Festschrift to all those interested in the exciting fields of mycology, microbiology and general biology.

H.D. Kumar

Preface

With the advent of new technologies, the knowledge of science is expanding in new dimensions every day. The frequent review of the state of our knowledge not only to keep ourselves up to date but also to provide over the various strategies ought to be adopted for further revolution of science in all its sphere.

As one know that presently existing life form are evolve from the microorganisms during the course of evolution. These microorganisms can adapt to virtually any physio-chemical environment and are found everywhere.

Despite their huge importance, the fact is that much of what we know about and microbial processes is based on the work done on only a harmful of species-weed like species which can be readily grown in laboratory culture. The vast majority of microorganisms found in nature (Perhaps over 95 per cent) have not so far been cultured but earnest attempts are under ways to grow as many of them as possible in culture.

One most striking property of many microorganisms is the highly promiscuous mode of gene transfer among unrelated or remotely related taxa. Such horizontal (lateral) gene transfers have been the subject of intense research activity in the past few years, in view their strong bearing on the applicability of the biological species concept and the Linnaean system of binominal nomenclature to prokaryotes.

Technological advancement has revolutionized and transform microbiology with biotechnology and created exciting opportunities for agriculture, medicine, industry and environment remediation the organism involved in these mechanisms mostly belongs to bacteria, fungi, cyanobacteria etc.

To keep pace with the growing knowledge in every field of science, our agenda for the coming few years should be to identify the key organizational and genetic principles underlying various morphogenetic processes in both prokaryotes and eukaryotes, while also pursuing the laborious elucidation of the molecular network that control the organizational of living matter.

Considering the importance of microorganism and their biotechnological processes are the major contents of the monograph. Beside microbiology and biotechnology few chapters on ecology are included.

Editors

Contents

— Volume 2 —

— BIOTECHNOLOGY —

List of Editors & Contributors

Editors

Dr. Deepak Vyas
Assistant Professor, Department of Botany, Dr. H.S.G. University, Sagar – 470 003, M.P.

Prof. G.S. Paliwal
Senior Consultant, Regional Centre of the National Afforestation and Eco-development Board, Ministry of Environment and Forests, Government of India, New Delhi

P.K. Khare
Professor, Department of Botany, Dr. H.S.G. University, Sagar – 470 003, M.P.

Dr. R. K Gupta
Assistant Professor, Department of Botany, Govt. P.G. College, Rhishikesh

Associate Editors

Prof. A.K. Pandey
Chairman, M.P. Govt. Private University Commission, Bhopal, M.P.

Dr. Jamaludin
Emeritus Prof. Department of Biological Sciences, R.D. University Jabalpur, M.P.

Dr. Neeraj Khare
Department of Botany, Dr. H.S.G. University, Sagar – 470 003, M.P.

Contributors

Adhikary, S.P.
Department of Biotechnology, Vishwa-Bharti University, Shantiniketan, West Bengal

Agrawal, Pooja
Department of Applied Microbiology and Biotechnology, Dr. H.S.G. University, Sagar – 470 003, M.P.

Akhtar, Nasim
Department of Biotechnology, GITAM University, Gandhi Nagar Campus, Rushikonda, Vishakapatnam – 530 045

Ashu
Department of Botany, Sahu Jain PG College, Najibabad – 246 763, U.P.

Atri, D.C.
Department of Botany, Dr. H.S.G. University, Sagar – 470 003, M.P.

Banerjee, Meenakshi
Laboratory of Algal Biotechnology, Department of Bioscience, Barkatullah University, Bhopal, M.P.

Behera, R.K.
Department of Botany, Berhampur University, Berhampur – 760 007, Orissa

Bharose, Ram
School of Forestry and Environment, SHIATS-Deemed University, Allahabad

Bhatt, R.K.
Indian Grassland and Fodder Research Institute, Jhansi – 284 003, U.P.

Bhattacharya, Abhishek
Bacteriology Lab, Department of Biological Sciences, Rani Durgavati University, Jabalpur, M.P.

Biradar, D.P.
Department of Agronomy, UAS, Dharwad – 580 005

Byatanal, Mahesh B.
Post Graduate Department and Research in Botany (Microbiology Lab), Karnataka University, Dharwad – 580 003

Chakravarty, Neha
National Research Centre for Agroforestry, Jhansi – 284 003, U.P.

Chaubey, Anjuli
Lab of Microbial Technology and Plant Pathology, Department of Botany, Dr. H.S.G. University, Sagar – 470 003, M.P.

Chauksay, P.
Department of Botany, Dr. H.S.G. University, Sagar – 470 003, M.P.

Chaurasia, Bhaskar
Department of Botany, Dr. H.S.G. University, Sagar – 470 003, M.P.

Chouhan, A.K.S.
Department of Botany, Dr. H.S.G. University, Sagar – 470 003, M.P.

David, Arun A.
School of Forestry and Environment, SHIATS-Deemed University, Allahabad

Dehariya, Poonam
Lab of Microbial Technology and Plant Pathology, Department of Botany, Dr. H.S.G. University, Sagar – 470 003, M.P.

Dikshit, Anupam
Biological Product Laboratory, Department of Botany, University of Allahabad, Allahabad –211 002

Diwedi, O.P.
Lab of Microbial Technology and Plant Pathology, Department of Botany, Dr. H.S.G. University, Sagar – 470 003, M.P.

Goswami, H.K.
24, Kaushalnagar, PO Misrod, Bhopal – 462 047, M.P.

Gupta, Pushpa
National JALMA Institute for Leprosy and Other Mycobacterial Diseases, Agra – 282 001

Gupta, Rajan Kumar
Department of Botany, Govt. P.G. College, Rishikesh, Uttarakhand

Gupta, U.S.
Department of Zoology, Dr. H.S.G. University, Sagar – 470 003, M.P.

Gupta, Umesh Dutta
National JALMA Institute for Leprosy and Other Mycobacterial Diseases, Agra – 282 001

Guru, S.D.
Department of Botany, Ranchi Women's College, Ranchi, Jharkhand

Hegde, Ganesh R.
Department of Botany, Sri Krishnadevaraya University, Anantapur – 515 0003

Jadhav, S.K.
School of Studies in Biotechnology, Pt. Ravishankar Shukla University, Raipur – 492 010,

Jain, Arun K.
Department of Anthropology and Tribal Development, GG University, Bilaspur, C.G.

Jain, Devendra
School of Life Sciences, Jaipur National University, Jaipur – 302 025, Rajasthan

Jamaluddin
Ex Group Coordinator and Director, Tropical Forest Research Institute, Jabalpur, M.P.

Jha, Anuradha
National Research Centre for Agroforestry, Jhansi – 284 003, U.P.

Kamalvanshi, Madhavi
National Research Centre for Agroforestry, Jhansi – 284 003, U.P.

Kamran, A.
Biological Product Laboratory, Department of Botany, University of Allahabad, Allahabad –211 002

Kango, Naveen
Department of Applied Microbiology and Biotechnology, Dr. H.S.G. University, Sagar – 470 003, M.P.

Karwa, Alka
Department of Biotechnology, S.G.B. Amravati University, Amravati – 444 602

Kaushik, B.D.
Division of Biotechnology, Netaji Subash Institute of Technology, Sector 3, Dwarka, New Delhi – 110 078

Khare, M.N.
Agricultural Research Station, Mandor, Jodhpur, Rajasthan (Ex. Dean and Professor Emeritus, Plant Pathology)

Khare, P.K.
Department of Botany, Dr. H.S.G. University, Sagar – 470 003, M.P.

Kumar, Anil
Principal Scientist (Plant Pathologist), National Research Centre for Agroforestry, Jhansi – 284 003, U.P.

Kumar, Mukesh
Department of Botany, Sahu Jain PG College, Najibabad – 246 763, U.P.

Kunjam, S.
School of Studies in Biotechnology, Pt. Ravishankar Shukla University, Raipur – 492 010,

Lakshman, H.C.
Department of Botany, Microbiology Laboratory, Karnataka University, Dharwad – 580 003, Karnataka

Maya, C.
Department of Botany, Bangalore University, J.B. Campus, Bangalore – 560 056

Mehta, A.
Department of Botany, Dr. H.S.G. University, Sagar – 470 003, M.P.

Mehta, P.
Department of Botany, Dr. H.S.G. University, Sagar – 470 003, M.P.

Mishra, A.K.
Indian Grassland and Fodder Research Institute, Jhansi – 284 003, U.P.

Mishra, P.
Biological Product Laboratory, Department of Botany, University of Allahabad, Allahabad –211 002

Mishra, Arun Kumar
Laboratory of Microbial Genetics, Department of Botany, Banaras Hindu University, Varanasi – 221 005

Mishra, R.K.
Biological Product Laboratory, Department of Botany, University of Allahabad, Allahabad –211 002

Nema, Sushma
Senior Scientist, Plant Pathology, AICRP (S&N) J.N.K.V.V., Jabalpur

Pandey, A.K.
Chairman, M.P. Council Pvt University Ltd., Bhopal

Pandey, J.
Department of Botany, Banaras Hindu University, Varanasi – 221 005, U.P.

Pandey, R.K.
Biological Product Laboratory, Department of Botany, University of Allahabad, Allahabad – 211 002

Pandey, Usha
Faculty of Science and Technology, MG Kashividya Pith, Varanasi – 221 005, U.P.

Panigrahi, A.K.
Department of Botany, Berhampur University, Berhampur – 760 007, Orissa

Pataik, Anita R.
Department of Botany, Berhampur University, Berhampur – 760 007, Orissa

Pujari, Rajesh
Bacteriology Lab, Department of Biological Sciences, Rani Durgavati University, Jabalpur, M.P.

Rai, A.N.
Department of Botany, Dr. H.S.G. University, Sagar – 470 003, M.P.

Rai, Mahendra
Department of Biotechnology, S.G.B. Amravati University, Amravati – 444 602

Rajpurohit, T.S.
Pathologist, Oilseeds (Rajasthan Agricultural University)

Richhariya, Pramod K.
Lab of Microbial Technology and Plant Pathology, Department of Botany, Dr. H.S.G. University, Sagar – 470 003, M.P.

Rishi, Narayan
School of Virology, Amity University, Noida, U.P.

Rishi, Sushma
School of Virology, Amity University, Noida, U.P.

Roy, A.K.
Department of Botany, T.M. Bhagalpur University, Bhagalpur – 812 007

Sabannavar, Shweta J.
Post Graduate Department and Research in Botany (Microbiology Lab), Karnataka University, Dharwad – 580 003

Sahu, Alka
Department of Botany, Berhampur University, Berhampur – 760 007, Orissa

Satya
Lichenology Laboratory, National Botanical Research Institute (CSIR), Rana Pratap Marg, Lucknow – 226 001, U.P.

Sethi, S.K.
PG Department of Biotechnology, Utkal University, Bhubaneswar – 751 004, Orissa

Sharma, A.N.
Department of Anthropology, Dr. H.S.G. University, Sagar – 470 003, M.P.

Sharma, Anjana
Bacteriology Lab, Department of Biological Sciences, Rani Durgavati University, Jabalpur, M.P.

Shukla, Ashok
Lab of Microbial Technology and Plant Pathology, Department of Botany, Dr. H.S.G. University, Sagar – 470 003, M.P.

Shukla, C.S.
Department of Plant Pathology, Indira Gandhi Agricultural University, Raipur – 492 006, Chhattisgarh

Singh, A.
Laboratory of Microbial Genetics, Department of Botany, Banaras Hindu University, Varanasi – 221 005

Singh, A.N.
Department of Botany, T.M. Bhagalpur University, Bhagalpur – 812 007

Singh, Abhijeet
School of Life Sciences, Jaipur National University, Jaipur – 302 025, Rajasthan

Singh, P.
Laboratory of Microbial Genetics, Department of Botany, Banaras Hindu University, Varanasi – 221 005

Singh, P.K.
Department of Botany, Dr. H.S.G. University, Sagar – 470 003, M.P.

Singh, Siddhartha
Department of Botany, Dr. H.S.G. University, Sagar – 470 003, M.P.

Singh, Surendra
Centre of Advanced Study in Botany, Banaras Hindu University, Varanasi – 221 005, U.P.

Singh, Vijendra
Institute of Microbial Technology, Chandigrah – 160 036

Singh. S.S.
Laboratory of Microbial Genetics, Department of Botany, Banaras Hindu University, Varanasi – 221 005

Soni, Prashant
Department of Botany, Dr. H.S.G. University, Sagar – 470 003, M.P.

Srivastava, A.
Laboratory of Microbial Genetics, Department of Botany, Banaras Hindu University, Varanasi – 221 005

Suresh, B.
Department of Applied Genetics, Karnatak University, Dharwad – 580 003

Tamphasana, R.K.
Department of Applied Geology, Dr. H.S.G. University, Sagar – 470 003, M.P.

Thakur, A.S.
Department of Botany, Govt College, Khurai, Sagar – 470 117, M.P.

Thakur, M.P.
Department of Plant Pathology, Indira Gandhi Agricultural University, Raipur – 492 006, Chhattisgarh

Thangadurai, D.
Department of Botany, Karnatak University, Dharwad – 580 003

Thomas, H.
Department of Applied Geology, Dr. H.S.G. University, Sagar – 470 003, M.P.

Thomas, T.
School of Forestry and Environment, SHIATS-Deemed University, Allahabad

Tiwari, H.S.
Indian Grassland and Fodder Research Institute, Jhansi – 284 003, U.P.

Tiwari, K.L.
School of Studies in Biotechnology, Pt. Ravishankar Shukla University, Raipur – 492 010, C.G.

Tripathi, V.D.
National Research Centre for Agroforestry, Jhansi – 284 003, U.P.

Upadhyay, Mukesh K.
School of Life Sciences, Jaipur National University, Jaipur – 302 025, Rajasthan

Upadhyay, Saurabh
School of Forestry and Environment, SHIATS-Deemed University, Allahabad

Upreti, D.K.
Lichenology Laboratory, National Botanical Research Institute (CSIR), Rana Pratap Marg, Lucknow – 226 001, U.P.

Valho, MAA Pinheirho De Car
ISO Plexis Germplasm Bank, Department of Biology, University of Madeira, Funchal 9000-390, Portugal

Vasathaiah, Hemanth K.N.
Center for Viticulture and Small Fruit Research, Florida A and M University, 6505 Mahan Drive, Tallahassee, Florida 32317, USA

Verma, Naveen
Department of Botany, Dr. H.S.G. University, Sagar – 470 003, M.P.

Verma, Vidhi
Laboratory of Algal Biotechnology, Department of Bioscience, Barkatullah University, Bhopal, M.P.

Vyas, Ashish
Division of Biotechnology, Lovely Professional University, Ludhiana (Punjab)

Vyas, Deepak
Lab of Microbial Technology and Plant Pathology, Department of Botany, Dr. H.S.G. University, Sagar – 470 003, M.P.

Wagay, Javed Ahmad
Lab of Microbial Technology and Plant Pathology, Department of Botany, Dr. H.S.G. University, Sagar – 470 003, M.P.

Yadav, Rajesh
ABR PG College, Anpra, Sonbadhra, U.P.

Yadav, S.K.
Department of Botany, Dr. H.S.G. University, Sagar – 470 003, M.P.

Yadav, V.K.
Department of Plant Pathology, Indira Gandhi Agricultural University, Raipur – 492 006, Chhattisgarh

Biotechnology

Microbial Biotechnology and Ecology (2011) *Pages* 393–438
Editors: Deepak Vyas, G.S. Paliwal, P.K.Khare and R.K Gupta
Published by: DAYA PUBLISHING HOUSE, NEW DELHI

Chapter 23

Fungitoxic Potential of Essential Oil and Extracts from Higher Plants in the Management of Postharvest Deterioration: A Review

☆ *Anupam Dikshit*, Ravindra Kumar Pandey, Rohit Kumar Mishra, Piyush Mishra and Ahsan Kamran*

Biological Product Laboratory, Department of Botany, University of Allahabad, Allahabad – 211 002

ABSTRACT

Postharvest losses of stored products are enormous due to fungal deteriorations. Although, there are number of synthetic fungicides available in the market for checking these deteriorations, they have several side effects such as high toxicity, long degradation periods, their concentration in food chain, chronic poisoning through the continuous intake of small quantities, development of new races of pathogens, non-biodegradable nature and exhaustible source. The history of pesticide development has been instructive to us in term of benefits derived as well as the hazard, which accompany the indiscriminate use of these poisonous chemicals. Therefore, we must search the new sources of pesticides, which in addition to their efficiency must be safe and selective to target pests or pathogen. Plants are very rich sources of bioactive chemicals such as phenolics, polyphenols, quinones, flavones, flavonoids, flavonols, tannins, cumarins, terpenoids, lectins and polypeptides. The essential oil fraction of botanicals is often the most inhibitory chemical to growth and survival

* E-mail: anupambplau@rediffmail.com

of microorganisms. They are highly enriched with terpenoids. Further they are volatile, biodegradable, and environmentally safe and are easily available in local environment. Therefore, present communication reviews the work done in investigating the fungitoxic potential of essential oils and extracts from higher plants in fungal deterioration of stored products.

Keywrds: *Myotoxins, Fungitoxic, Biodeterioration.*

Introduction

It is estimated that between 60 to 80 per cent of all grains produced in the tropics are stored by farmers. For small farmers, the main purpose of storing grains is to ensure household food supplies. It also provides a form of saving to cover future cash needs through sale or for barter or gift exchange. Small quantities of grains are also stored for seed. Farmers who produce surplus may also store grain for sale later to take advantage of seasonal price rise. A traditional farm storage system provides maximum protection against deterioration of commodity by inclement weather and pests. Farm storage system have evolved over a long periods to satisfy these requirements. Most are well adapted to their environment and losses are generally low, often below 5 per cent of grain weight over a storage season (Tyler and Boxall, 1984). However, for resource poor farmers, even losses of this magnitude have important implications for food security. In addition to storage of food grains at farmer's level, the governments of different countries also procure and store them for reasons of food security for its growing population. There was spiral increase in populations of several countries including India. In case of India, it has crossed one billion marks in 2000 A.D. and it will give India the dubious distinction of being the most populated country in the world by 2025 A.D. Therefore, the challenge of feeding ever-growing population of human being shall be very difficult task. We can not meet such a challenge with the increase in food production alone but their protection from deterioration caused by fungi and other pests during storage have to be given due emphasis (Shukla, 1997).

In developing countries, fungal pests cause the greatest losses during storage to cereals and other durable commodities. Deteriorations due to fungi are due to unhygienic conditions of storage and this in turn is associated with initial high moisture content of the stored products or absorption of moisture during storage due to defects in the storage system (Stinson *et al.*, 1980). There are several storage fungal pests that cause deteriorations in stored commodities. The most common among them are– *Aspergillus flavus, A. niger, A. clavatus, A. terreus, A. versicolor, A. candidus, Alternaria alternata, Curvularia lunata, Cladosporium cladosporoides, C. herbarum, Epicoccum nigrum, Emericella nidulans, Emericella rugulosa, Fusarium acuminatum, F. moniliforme, Mucor hiemalis, Penicillium citrinum, P. chrysogenum, P. expansum, P. funiculosum, P. italicum, Rhizopus arrhizus, Rhizopus nigricans, Syncephalastrum racemosum* etc. Several of these fungal pests such as species of *Aspergillus, Penicillium, Fusarium, Alternaria* and *Cladosporium* etc. not only bring about deterioration in the quality and quantity of agricultural produce in storage and transit but also create health hazards in animals and human beings, by producing toxic metabolites in the form of mycotoxins in the stored commodities. These organisms are capable of growing under a diversity of conditions of moisture, pH and temperature. If the mould growth occurs, there is always the concomitant possibility of mycotoxin production (Zohri and Abdel-Gawad, 1993).

Mycotoxins are dangerous chemical that cause several complications in the body. They are carcinogenic, hepatoxic, nephrotoxic and teratogenic. Extreme toxicity of mycotoxins lies in the fact that they are extremely stable and dangerous in minute quantities. Further, once formed they cannot be removed from the commodity concerned by processing or removal of visible mould growth. They are heat stable, so they cannot be destroyed by cooking. Since mycotoxins are extremely toxic, regulatory and industry guidelines limit are set at very low levels. In developing countries often the good quality products are exported while substandard produce unacceptable to foreign buyers (because they exceed regulatory limits for mycotoxin content) is sold to the domestic market (Dawson, 1991). Therefore, the mycotoxin contamination of food and feeds is not a particular problem to the developed world, although heavy economic costs are incurred in ensuring low concentrations of mycotoxins (Mannon and Johanson, 1985). In poorer developing countries, such contamination has more serious consequences, affecting agricultural economies, reducing annual production and good quality exports and seriously affecting the health of the population.

To control fungal deterioration of agricultural produce, many organic and inorganic fungicides have been developed and are being used. The use of many of these has, however, been restricted due to their undesirable side effects such as a high and acute toxicity, the long degradation periods, their concentration in food chain, the suspected dangers of chronic poisoning through the continuous intake of small quantities. Besides, due to development of new races of pathogens, many of these fungicides are gradually becoming out of date (Dikshit, 1980). Furthermore, the sources of these synthetic fungicides are largely petrochemicals, which are exhaustible. Therefore, hunt for inexhaustible sources of such chemotherapeutants is highly desirable. According to Brandes (1967) much of our efforts are being wasted in routine testing of the standard fungicides, when there is a pressing need to investigate new sources of effective fungicides.

Fungitoxic Potential of Higher Plants

Green plants appear to be the reservoir of effective chemotherapeutants and can provide reversible source of useful pesticides (Swaminathan, 1978). Tropical floras, in contrast to their temperate zone counterparts, have developed a more efficient and varied defence mechanism because of the far severe conditions for survival. They, thus provide a rich and intriguing source for isolating natural secondary plant metabolites, which exhibit interesting antimicrobial properties. Although only some 15,000 secondary plant metabolites have been chemically identified, their total number may exceed 4,000,000 (Saxena, 1993). They are vast cornucopia of defense chemicals. Recent reports on the possibility of use of higher plants and their constituents have indicated their usefulness in providing fungicides which are largely non-phytotoxic, more systemic and easily biodegradable (Fawcett and Spencer, 1969; Beye, 1978). They are sustainable, can be continuously propagated year after year and do not have any negative impact on the environment as long as care is taken to avoid the propagation of plants from foreign ecosystems which might, therefore, become established as weeds.

Farmers use plants as storage protectants since long times. Further, where plants are used as storage protectants, they are almost always applied to control insect pests. This is reflected in volumes of research directed to identifying insecticidal or insect repellent plants and plant extracts. Nevertheless, some work has been undertaken to determine whether plants can control storage fungi. Most workers have investigated the properties of spices as inhibiting agents of mycelial growth of *Aspergillus* species and of its toxin production. *Syzygium aromaticum* (cloves) have been found to be particularly effective, often completely inhibiting both fungal growth and toxin production (Hitokoto *et al.*, 1980; Mabrouk and El-Shayeb, 1980). Many commercially available spices and herbs, turmeric, basil, marjoram, anise, cumin and coriander are able to completely inhibit toxin production, but only partially inhibit fungal mycelial growth (Hitokoto *et al.*, 1980). Aqueous extracts of weeds and medicinal plants have also been found to inhibit toxin production by *Aspergillus flavus*. These include *Ricinus communis, Arnebia nobilis* and *Nicotiana plumbaginifolia*. (Bilgrami *et al.*, 1980). Other fungi, *Fusarium solani, F. phaseoli* and *Verticillium albo-atrum*, have been shown to be susceptsible to tannins extracted from bark of various trees, including chestnut and wattle (Lewis *et al.*, 1967). In addition to above several other fungi, have been shown to be susceptible to essential oils extracted from higher plants (angiosperms and gymnosperms). A recent compilation of work on fungitoxicity of higher plants against dominant storage fungi and a few other fungi have been made as under:

Abutilon indicum (Malvaceae)

Jain *et al.* (1979) reported the toxicity of oil isolated from leaves against *Alternaria robi, Aspergillus flavus, A. fumigatus, A. nidulans, A. niger, Cladosporium hebarum, Fusarium moniliforme, Helminthosporium* sp. *Penicillium digitatum, P. italicum, Rhizopus nigricans* and *Trichothecium roseum.*

Acorus calamus (Araceae)

Samuel (1993) isolated the oil from the rhizome and reported its toxicity against *Fusarium moniliforme* and *Helminthosporium oryzae* at a minimum dose of 500 ppm. The oil possessed fungistatic activity at MIC but turned fungicidal at 4000-ppm conc. The active principle isolated from the oil was recognized as Aṣarone. The volatile oil at its MIC was reported to possess toxicity against *Aspergillus flavus, A. nidulans, Cladosporium herbarurn, Drechslera sativa, Fusarium oxysporum, Macrophomina phaseolina, Rhizoctonia solani* and *Sclerotium rolfsii.* It did not show any phytotoxicity on seed germination and seedling growth of *Oryza sativa* and *Hordeum vulgare.* Siddiqui *et al.* (2003) reported the fungitoxicity of the essential oil against *Helminthosporium oryzae* and *Fusarium moniliforme* at a minimum dose of 500 ppm (0.5×10^3ml/l), and it was fungistatic in nature.

Adenocalymma allicea (Bignoniaceae)

Chaturvedi (1979) reported 0.3 per cent yield of water-soluble essential oil containing 10 major and 35 minor components. The oil inhibited the mycelial growth of *Alternaria raphani, A. solani, Cephalosporium sacchari, Cladosporium cladosporioides, Colletotrichum lindemutheanum, C. truncatum, Fusarium moniliforme, F. solani, Paecilomyces* sp. *Rhizoctonia bataticola, R. solani* and *Talaromyces* sp. at 500 ppm (0.5×10^3ml/l). The oil killed *Helminthosporium oryzae* within 30 minutes (by contact). The antifungal factor was thermostable and retained its toxicity up to 21 days of storage. The oil controlled leaf spot disease in paddy without showing any adverse side effects on the host plant (Chaturvedi *et al.*, 1987).

Aegle mormelos (Rutaceae)

Rao and Joseph (1971) found that the oil was toxic to *Curvularia lunata, Fusarium moniliforme, Helminthosporium sacchari, Physalospora tucumanensis* and *Sclerotium rolfsii.* The leaves yielded 0.4 per cent oil, which killed 40 mycelial discs (5 mm. in diameter) of the test fungus *Rhizoctonia solani* within 2 hours (by contact) and completely inhibited the mycelial growth of *Helminthosporium corbonum* at 0.3 per cent (3.0×10^3 ml/l) dose (Srivastava, 1981). The thermostable fungitoxic factor of the oil retained its toxicity upto 105 days in storage at room temperature and was non phytotoxic to seed germination of *Vigna mungo* (Renu *et al.*, 1986).

Ageratum conyzoides (Asteraceae)

Rao (1976a) reported that the leaf oil was toxic to *Colletotrichum capsici, Helminthosporium oryzae, H. turcicum, Penicillium chrysogenum, P. javanicum* and *Pyricularia setariae.* It was also found to be toxic against *Alternaria helianthi, Fusarium moniliforme, F. solani, Pyricularia oryzae, Pythium vexans, Rhizoctonia bataticola* and *R. solani* (Sharma *et al.*, 1978). Chandra and Dikshit (1981) recorded fungistatic nature of the oil at 0.2 per cent (2.0×10^3 ml/l) concentration against *Absidia ramosa, Alternaria alternata, A. citri,*

A. raphani, Aspergillus flavipes, A. fumigatus, A. oryzae, A. quadrilineatus, A. ruber, A sydowi, A. tamarii, A. terreus, Botryodiplodia theobromae, B. palmicolum, Botrytis cinarea, Cladosporium cladosporioides, Colletotrichum falcatum, C. gloeosporioides, Cunninghamella elegans, Curvularia lunata, Fusarium oxysporum, Geotrichum candidum, Gloeosporium fructigenum, Penicillium funiculosum, P. expansum, Rhizopus arrhizus and *Tricothecium roseum.* Chandra (1984) reported that the oil controlled the blue mould rot during storage of oranges, caused by *Penicillium italicum* at 0.1 per cent (1.0×10^3 ml/l) concentration. Singh *et al.* (1986a) reported that the mycelial disc of *Epidermophyton floccosum, Microsporum canis* and *Trichophyton mentagrophytes* were completely killed when dipped in 4000-ppm (4.0×10^3ml/l) dose of the oil. The isolated de-methoxy ageratochromene, the antifungal principle of the oil, which was found toxic at 1000 ppm (1.0×10^3ml/l) against the dermatophytes tested.

Ageratum haustonianum (Asteraceae)

The leaf oil was found toxic at 100-ppm (0.1×10^3 ml/l) concentration, possessing fungistatic nature against *Blastomyces dermatitides, Epidermophyton floccosum, Histoplasma capsulatum, Trichophyton mentagrophytes, T. simii, T. terrestre, T. tonsurans, T. verrucosum* and *T. violaceum.* However, it killed *Microsporum gypseum* with in one second at 300-ppm (0.3×10^3 ml/l) dose by contact (Pandey *et al.*, 1983a). Pandey *et al.* (1983b) reported that the fungitoxic factor of the oil was thermostable and fungistatic against *Fusarium lateritium* at 0.05 per cent (0.5×10^3ml/l) concentration.

Aglaja odorata (Meliaceae)

Kakrani and Nair (1982) reported that the seed oil possessed toxicity against *Rhizopus oryzae.*

Allium sativum (Liliaceae)

Skinner (1955) reviewed in detail the antimicrobial spectrum of the volatile as well as non-volatile component of the plant and its active principle, allicin. Further publication of the Maruzzella and Liquroi (1958), Maruzzella and Baiter (1959), Maruzella *et al.* (1960), Thind and Dahiya (1977), Petricic *et al.* (1978), Singh and Singh (1981), and Singh and Deshmukhi (1984) revealed that the oil possessed toxicity against a wide range of fungi including *Absidia spinosa, Alternaria alternata, A. tenuis, Aspergillus flavus, A. fumigatus, A. japonicus, A. nanus, A. nidulans, A. sydowi, Candida albicans, C. monosa, Cephalosporium sacchari, Claviceps purpurea, Curvularia lunata, Erysiphe polygoni, Fusarium nivale, F. solani, Helminthosporium gramineum, Lentinus lepideus; Lenzites trabea, Polyporus versicolor, Trichoderma viride, Microsporum gypseum* and *Verticillium terrestrae.*

Allium cepa (Liliaceae)

The oil has been reported fungitoxic against many fungi *viz. Alternaria alternata, Aspergillus niger, Botrytis allii, Claviceps purpurea Chrysosporium tropicum, Keratinophyton terreum, Malbranchea pulchela, Microsporum gypseum, Phytophthora infestans* and *Verticillium tenuipes* (Meruzzella and Baiter, 1959; Korableva and Patapova, 1966; Jain and Agarwal, 1978; Singh and Deshmukhi, 1984).

Alpinia galanga (Zingiberaceae)

The oil extracted from rhizome was toxic at 4000 ppm (4.0×10^3ml/l) concentration to *Absidia ramose, Alternaria alternata, A. raphani, A. solani, Aspergillus fumigatus, A. versicolor, Cephalosporium sacchari, Colletotrichum capsici, C. gloeosporioides, C. tinosporae, Epicoccum roseum, Fusarium moniliforme, F. oxysporium, F. solani, Helminthosporium carbonum, H. gramineum, H. maydis, H. turcicum, H. oryzae, Pestalotia versicolor, Pyricularia oryzae, Talaromyces* sp. and *Trichoderma viride* (Tripathi, 1980). The active principle of the oil was thermolabile and it got destroyed during autoclaving but remained unaffected by temperature upto 100°C and by storage upto 91 days at room temperature. Tripathi *et al.* (1983) during preliminary *in-vivo* trials observed that the oil prevented the appearance of symptoms of leaf spot disease in paddy without affecting the germination and seedling growth of host plant. Janseen and Scheffer (1985) isolated acetoxychavicol, the active principle of the oil and reported its fungitoxicity at 250 mg/ml doses against *Aspergillus niger, Candida albicans, Epidermophyton floccosum, Penicillium expansum, Trichophyton mentagrophytes, T. rubrum* and *T. concentricum.*

Alseodaphne keanamia (Lauraceae)

The leaves and bark extracts of plant were screened for antifungal activities. Various pathogens were used *viz. Aspergillus niger, A. flavus, Trichoderma* sp. and *Cladosporium* sp. (Versha Parcha *et al.*, 2003).

Ammoides pusilla (Apiaceae)

Oil from aerial parts showed antimicrobial activity against eight strains of bacteria, several fungi and yeast such as *Aspergillus niger* and *Candida albicans.* GC and GC/MS of oil showed 46 constituents among which thymol (44.5 per cent), γ-terpinene (32.9 per cent) and p- cymene (13.5 per cent) were the chief. (Laouer *et al.*, 2003)

Amomum subulatum (Zingiberaceae)

Kumar (1992) reported that the volatile oil was toxic to *Aspergillus flavus, A. ochraceus, Fusarium oxysporum* and *Penicillium citrinum* at a conc. of 1000 ppm. The nature of fungotoxicity was static at MIC but turned fungicidal at a conc. of 8000 ppm.

Anaphalis controrta (Asteraceae)

Saxena *et al.* (1984) found that the leaf oil was toxic to *Aspergillus niger, Microsporum gypseum* and *Penicillium notatum.*

Anethum graveolens (Apiaceae)

Rao and Joseph (1971) reported that seed oil was toxic to *Candida albicans, Epidermophyton floccosum, Trichophyton mentagrophytes* and *T. rubrum.* Later, it was reported toxic to species of *Aspergillus, Curvularia* and *Penicillium* (Chaursia and Kher 1978a, b); *Alternaria* sp., *Aspergillus candidus, A. flavus, A. fumigatus, A. nidulans, A. niger, Cladosporium herbarium, Cunninghamella echinulata, Fusarium monilliforme, F. oxysporum, Helminthosporium sacchari, Microsporum cocci, M. gypseum, Mucor mucedo, Penicillium digitatum, Rhizopus* sp., *Trichophyton mentagrophytes, T. rubrum* and *Trichothecium roseum* (Sharma and Singh, 1979b) and *Alternaria tenuis, A. fumigatus, Candida albicans, C. tropicallis, Diplodia natalensis, Keratinomyces ajelloi, Microsporum gypseum, Penicillium digitatum, P. italicum, Trichophyton equinum* and *T. rubrum* (Narsimha Rao and Subba Rao, 1982; Arora and Pandey, 1984 a.b; Saksena, 1984). The oil showed fungistatic nature at 1000 ppm (1.0×10^3 ml/l) dose, against *Alternaria alternata, Aspergillus flavus, A. awamori, A. niger, A. ochraceus, A. parasiticus, A. sydowi, A. terreus, Candida tropicalis, Cladosporium herbarum, Epidermophyton floccosum, Fusarium moniliforme, F. oxysporum, Microsporum gypseum, Nigrospora oryzae, Penicillium chrysogenum, P. citrinum, Rhizopus nigricans* and *Trichophyton mentagrophytes* (Dubey, 1986). The toxic principle of the oil *i.e.* Anethol, proved non phytotoxic to seed germination and seedling growth of *Pisum sativum* and was not affected by temperature as well as by long storage duration at room temperature (Dubey, 1986).

Angelica archenglica (Apiaceae)

Korta and Starzyk (1963) reported toxicity of the oil against *Candida albicans, C. krusci, C. parakrusei, C. pelliculosa, C. silosis, C. stellatoidea* and *C. tropicalis*. Later, its toxicity was reported against *Nannizzia fulva, N. gypsea* and *N. incurvata* (Gautam *et al.*, 1980).

Apium graveolens (Apiaceae)

Rao and Joseph (1971), Sharma and Singh (1979 b) and Narsimha Rao and Subba Rao (1982) reported that the oil from seeds produced marked inhibitory zones in cultures seeded with spores of *Candida albicans, Epicoccum floccosum, Epidermotophyton floccosum, Trichophyton mentagrophytes. T. rubrum, Alternaria* sp., *Aspergillus candidus, A. flavus, A. fumigatus, A. nidulans, A. niger, Cladosporium herbarum, Cunninghamella echinulata, Fusarium moniliforme, F. oxysporum, Helminthosporium sacchari, Microsporum cocci, M. gypseum, Mucor mucedo, Penicillium digitatum, Rhizopus* sp. and *Trichothecium roseum.*

Arnica latifolia (Asteraceae)

Singh (1976) reported the toxicity of the oil against *Candida albicans, Cephalosporium sacchari, Ceratocystis paradoxa, Curvularia lunata, Fusarium moniliforme, Helminthosporium sacchari, Physalospora tucumanensis, Trichophyton rubrum* and *Sclerotium rolfsii.*

Artemisia absinthium (Asteraceae)

Kaul *et al.* (1976) reported the toxicity of oil at 1:1000 (1.0×10^3 ml/l) dilution against *Candida albicans, C. utilis* and *Aspergillus niger.*

Artemisia capillaris (Asteraceae)

Ikenaka *et al.* (1956) reported that the oil was toxic to *Alternaria kikuchiana, Aspergillus awamori, A. niger, A. oryzae, Gibberella fujikuroi, G. bataticola, G. sanbinetii, Colletotrichum miyabeanus, Penicillium chrysogenum, Pyricularia oryzae* and *Rhizopus javanicus.* Imai *et al.* (1956) isolated Capillin the antifungal factor of the oil.

Artemisia cina (Asteraceae)

Vichkanova *et al.* (1972) reported the toxicity of the oil against *Candida albicans, Microsporum lanceoletum* and *M. gypseum.*

Artemisia martima (Asteraceae)

Dikshit and Husain (1984) reported the toxicity of the oil against *Microsporum gypseum, Trichophyton equinum* and *T. rubrum.*

Artemisia vestita (Asteraceae)

The toxicity of the oil at 1:1000 (1.0×10^3 ml/l) dilutions against several dermatophytes was reported by Kaul *et al.* (1976). Gautam *et al.* (1980) reported the toxicity of this oil against *Nannizzia fulva, N. gypsia* and *N. incurvata.*

Azadirachta indica (Meliaceae)

Kher and Chaurasia (1977), Thind and Dahiya (1977) and Jain and Agrawal (1978) reported the toxicity of the oil against some kerationophylic fungi *viz. Verticillium tenuipes, Keratinophyton terreum, Malbrachea pulchella* and *Chrysosporium tropicum.* Later, Singh and Singh (1981) and Dwivedi and Dubey (1986) reported the complete inhibition of the conidial and sclerotial germination of *Erysiphe polygoni* and *Macrophomina phaseoli* by the volatile vapour of the oil.

Blumea membranacea (Asteraceae)

Zutshi and Mehta (1977) and Sharma *et al.* (1978) reported the toxicity of the oil against *Alternaria helianthi, Fusarium moniliforme, Helminthosporium oryzae, H. turcicum, Pyricularia setariae* and *Rhizoctonia solani.* Later, its toxicity was reported against *Aspergillus luchuensis, A. sydowi* and *Cladosporium cladosporioides.*

Bothriocloa perrtusa (Poaceae)

Jain *et al.* (1979) reported that the oil was inhibitory to *Alternaria robi, Aspergillus flavus, A. fumigatus, A. niger, A. nidulans, Cladosporium herbarum, Rhizopus nigricans* and *Trichothecium roseum.*

Boswellia serrata (Burseraceae)

Garg (1974) repoıted that the oil possessed toxicity against *Curvularia geniculata, C. pallescens, Fusarium culmorum, F. oxysporum* and *Pythium aphanidermatum.* Girgune *et al.* (1978) reported its inhibitory activity against *Alternaria helianthi, Fusarium solani, Helminthosporium oryzae, H. turcicum* and *Rhizoctonia solani* and found that the oil contains α-thuzene, α-pinene and α-phellendrene.

Caesulia axillaris (Asteraceae)

The antimicrobial activity of the oil was first reported by Zutshi *et al.* (1975) and later confirmed by Geda *et al.* (1978). The oil was fungicidal to *Helminthosporium oryzae* at 0.5 per cent (5.0×10^3ml/l) concentration and it controlled leaf spot disease of paddy at 0.025 per cent (0.25×10^3 ml/l) concentration with no side effects on the host plant (Pandey *et al.*, 1982a). The antifungal factor of the oil was thermostable, durable up to 180 days of storage and could bear increased inoculum density (Pandey *et al.*, 1982).

Caesalpinia sappan (Caesalpiniaceae)

The toxicity of the oil against *Aspergillus nidulans* was reported by Yadav *et al.* (1978).

Calamus sp. (Arecaceae)

Arora and Pandey (1984a) reported toxicity of the oil against *Alternaria tenuis, Diplodia natalensis, Penicillium digitatum* and *P. italicum.* The oil was used as preservative to prevent fungal deterioration of fruits of *Citrus reticulata* during storage (Arora and Pandey 1984b).

Calamintha nepeta (Lamiaceae)

In-vitro microbicidal properties of the essential oil of *Calamintha nepeta* were studied on *Aspergillus niger, A. parasiticus, Rhizopus oryzae, Rhizoctonia oryzae-sativae, Candida albicans* and *Colletotrichum musae.* The essential oil showed the highest activity against *R. oryzae-sativae* and *C. albicans.* (Tajo *et al.*, 1999).

Callistemon lanceolatus (Myrtaceae)

Dwivedi *et al.* (1990) tested the oil against *Macrophomina phaseolina* and found to be toxic at 500, 1000 and 2000 ppm (0.5×10^3 ml/l, 1.0×10^3 ml/l, 2.0×10^3 ml/l doses.

Capsicum annuum (Solanaceae)

The toxicity of the oil was tested against *Aspergillus niger* by Capek (1955).

Carum balbocastanum (Apiaceae)

The toxicity of the oil against *Aspergillus fumigatus, Curvularia lunata* and *Trichoderma viride* was reported by Nigam and Rao (1977).

Carum carvi (Apiaceae)

Slavenas (1963) reported that the oil from fruits possessed toxicity against *Aspergillus niger*. Do Vgich (1971) recorded the toxic nature of the oil against *Aspergillus oryzae, Alternaria tenuis* and *Penicillium chrysogenum*. The oil was later reported to possess toxicity against *Aspergillus flavus, A. fumigatus, A. versicolor, Fusarium* sp., *Trichoderma viride* and *Curvularia* sp. (Nigam and Rao, 1977). *Aspergillus candidus, A. nidulans, Cladosporium herbarum, Cunninghmella echinulata, Fusarium moniliforme, F. oxysporum, Helminthosporium sacchari, Microsporum cocci, M. gypseum, Mucor mucedo, Penicillium digitatum, Rhizopus* sp., *Trichophyton mentagrophytes, T. rubrum* and *Trichothecium roseum* (Sharma and Singh, 1979b) as well as *Candida tropicalis* and *Rhizopus oryzae* (Narayanan *et al.*, 1980a).

Carum roxburghianum (Apiaceae)

Nigam and Rao (1977) observed that the oil produced marked inhibitory zone in cultures seeded with spores of *Aspergillus flavus, A. fumigatus, A. oryzae, Alternaria tenuis, Curvularia* sp. and *Trichoderma viride*.

Cassia sp. (Caesalpiniaceae)

Maruzzela and Liquori (1958) and Maruzzella *et al.* (1960) reported the toxicity of the oil against *Lentinus lepideus, Lenzites trabea* and *Polyporus versicolor*.

Cedrus deodara (Pinaceae)

Dikshit *et al.* (1979) isolated 2.72 per cent oil from root wood and reported that the oil was toxic to *Helminthosporium oryzae* at 3000 ppm (3.0×10^3 ml/l) concentration. Dikshit (1980) reported that the thermostable fungitoxic factor of the oil retained its toxicity during storage up to 600 days and sustained heavy inoculum density but got destroyed by autoclaving when kept at 20 lbs per square inch pressure for 20 minutes. The oil at 3000 ppm (3.0×10^3 ml/l) dose inhibited the growth of *Absidia romosa, Alternaria brasicae, A. raphani, A. solani, Aspergillus carneus, A. flavipes, A. fumigatus, A. niger, A. versicolor, Cephalosporium sacchari, Cladosporiun* sp., *Colletotrichum falcaturn, C. lindemuthianum, C. tinosporae, Cornespora casicola, Fusarium moniliforme, F. solani, Pacilomyces* sp., *Pyricularia oryzae, Talaromyces* sp. and *Trichoderma viride*. The oil killed the test fungus *H. oryzae* in 3 hr. by contact at 3.0×10^3 ml/l concentration in 90 min at 5×10^3 ml/l, while the contact of pure oil killed the fungus in seconds. Dikshit and Dixit (1982) demonstrated that the oil successfully prevented the appearance of symptoms of leaf spot disease in paddy without any adverse effect on seed germination and general growth of the host plant (Dikshit *et al.*, 1982). Dikshit *et al.* (1983) reported that the oil at 5.0×10^3 ml/l dose controlled the fungal deterioration of fruits of *Foeniculum vulgare* and *Coriandrum sativum* during storage.

Centella asiatica (Apiaceae)

Minija and Thoppil (2003) studied the antimicrobial activity of oil on six fungi *viz. Aspergillus niger, A. parasiticus, Rhizopus oryzae, Candida albicans, Fusarium solani* and *Colletotrichum musae*. The essential oil showed remarkable activity against *A. niger, R. oryzae, F. solani* and *C. albicans*.

Cestrum diurnum (Solanaceae)

Renu *et al.* (1980) reported that the leaves yielded 0.22 per cent water soluble oil which exhibited toxicity against *Absidia ramosa, Alternaria alternata, A. brassicae, A. raphani, A. solani, Aspergillus flavipes,*

A. flavus, A. fumigatus, A. terreus, C. cladosporioides, Colletotrichum capsici, C. falcatum. C. gloeosporioides, C. tinosporae, Curvularia lunata, Epiccoccum sp., *Fusarium moniliforme, F. oxysporum, F. solani, Helminthosporium carbonum, H. gramineum, H. maydis, H. oryzae, H. turcicum, Pacilomyces* sp., *Pyricularia oryzae, Penicillium decumbens, Pestalotia versicolor, Phoma* sp., *Rhizoctonia batalicola, Talaromyces* sp., *Trichoderma viride* and *Trichothecium roseum* at 0.7 per cent (7.0×10^3ml/l) concentration and killed the test fungus, *Rhizoctonia solani* in 12 hrs by contact. The fungitoxic factor of the oil sustained heavy inoculum density and remained unaffected by temperature treatment and storage upto 400 days. The seeds of *Vigna mungo* when soaked in 0.7 per cent (7.0×10^3 ml/l) dilution of the oil for 12 hrs. did not show any adverse effect on their germination and seedling growth (Srivastava, 1981).

Chenopodium ambrosioides (Chenopodiaceae)

Kumar, *et, al.* (2007). extracted essential oil from the leaves of *Chenopodium ambrosioides* Linn. (Chenopodiaceae) and tested against the aflatoxigenic strain of test fungus *Aspergillus flavus* Link. The oil completely inhibited the mycelial growth at100 µg/ml. The oil exhibited broad fungitoxic spectrum against *Aspergillus niger, Aspergillus fumigatus, Botryodiplodia theobromae, Fusarium oxysporum, Sclerotium rolfsii, Macrophomina phaseolina, Cladosporium cladosporioides, Helminthosporium oryzae* and *Pythium debaryanum* at 100 µg/ml. Kishore *et al.* (1982b) reported that the oil from inflorescence was fungicidal to *Absidia romosa, Alternaria alternata, Aspergillus flavus, A. fumigatus, A. nidulans, A. niger, A. versicolor, Cladospoium cladosporioides, Chaetomium globosum, Colletotrichum capsici, C. falcutum, C. gloeosporioides, Curvularia lunata, Fusarium moniliforme, F. oxysporum, Helminthosporium maydis, Pacilomyces* sp., *Penicillium italicum* and, *Rhizopus arrhizus* at 1000 ppm (1.0×10^3 ml/l) concentration.

Chloroxylon swietenia (Rutaceae)

Garg and Oswal (1982) reported the toxicity of the oil at 1 : 50 (20×10^3 ml/l) dilution against *Aspergillus oryzae, A. terreus* and *Candida albicans.*

Cicer arietinum (Leguminosae)

Rao and Joseph (1971) reported that the oil from the seed husk was toxic to *Cephalosporium sacchari, Ceratocystis paradoxa, Curvularia lunata, Fusarium moniliformae, Helminthosporium sacchari, Physalospora tucumanesis, Rhizopus arrhizus* and *Sclerotium rolfsii.*

Cinnamomum sp. (Lauraceae)

The bark oil was reported to be toxic against *Aspergillus niger* by Capek (1955). Later Mukherjee (1974) reported that oil completely prevented the sporulation in *Alternaria solani, Curvularia lunata, Fusarium solani* and *Helminthosporium oryzae.*

Cinnamomum camphora (Lauraceae)

Suri and Thind (1979) reported the toxicity of the oil against *Aspergillus fumigatus* and *Penicillium notatum.* Tiwari *et al.* (1988) found that the oil at 1.0×10^3ml/l effectively reduced the decay due to *Aspergillus niger* of onion during storage. Dwivedi *et al.* (1990) tested the oil against *Macrophomina phaseolina* and found to be toxic at 0.5×10^3 ml/l, 1.0×10^3 ml/l and 2.0×10^3 ml/l doses. This fungus causes root rot disease of crops.

Cinnamomum tamala (Lauraceae)

Girgune *et al.* (1978) reported that the bark oil which contained α-phellendrene was toxic to *Alternaria helianthi, Fusarium moniliforme, F. solani, Helminthosporium oryzae, Pythium vexans, Rhizoctonia bataticoia* and *R. solani.*

Cinnamomum zeylanicum (Lauraceae)

Chaurasia and Kher (1978a, b) reported that the bark oil possess fungitoxicily. Jain and Agrawal (1978) reported toxicity against *Chrysosporium tropicum, Keratinophyton terreum, Malbranchea pulchella* and *Verticillium tenuipes.* Suri and Thind (1979) reported toxicity against pathogen *viz. Keratinophyton terreum, Malbranchea pulchella* and *Trichophyton mentogrephytes.* Narayan *et al.* (1980 b) reported that the oil at 1:64 dilution inhibited the mycelial growth of *Aspergillus fumigatus, A. niger, Candida troipicalis, Fusarium* sp. and *Rhizopus oryzae* whereas Saxena (1989) reported the toxicity of the bark oil against *Candida albicans, Keratinomyces ajelloi, Microsporum gypseum, Trichophyton equinum* and *T. rubrum.* Mishra *et al.* (1989) isolated the essential oil from leaves and reported its toxicity against *Aspergillus flavus* at 2.0×10^3 ml/l dose. Tiwari *et al.* (1994) reported that the vapour emitted by bark completely inhibited the mycelial growth of *A. flavus* and *A. niger.* Tiwari *et al.* (1998) found that the bark oil of *Cinnamomum zeylanicum* having the insect repellent activity against the pulse bettle, *Callosobrachus chinensis* L. by olfactometer. The minimum doses of repellency and LC 50 were found to be 0.002 per cent and 0.004 per cent respectively. Besides, the insect repellent potency of the oil was found to be thermostable and posses long shelf life at its absolute repellency concentration (0.004 per cent). Thus the oil of *Cinnamomum* demonstrated its activity as potent insecticides which may prove its usefulness as biodegradable, indigenous and herbal preservatives for safe storage of gram meant for human consumption.

Citrus aurantifolia (Rutaceae)

Gupta and Singh (1982) reported the toxicity of leaf oil against many fungi *viz. Aspergillus fumigatus, A. nidulans, A. niger, Cunninghamella echinulata, Cladosporium herbarum, Fusarium moniliforme, F. oxysporum* and *Rhizopus* sp. Narsimha Rao and Subba Rao (1982) reported the toxicity of the oil against human pathogenic fungi *viz. Candida albicans, Epidermophyton floccosum, Trichophyton mentagrophytes* and *T. rubrum.*

Citrus aurantifolium (Rutaceae)

Rao and Joseph (1971) reported that the oil was toxic to *Candida albicans, Cephalosporium sacchari, Ceratocystis paradoxa, Curvularia lunata, Epidermophyton floccosum, Fusarium moniliforme, Helminthosporium sacchari, Physalospora tucumanesis, Sclerotium rolfsii, Trichophyton mentagrophytes* and *T. rubrum.*

Cinnamomum bergamia (Rutaceae)

Ceruti *et al.* (1982) reported the toxicity of the oil against *Aspergillus flavus, A. fumigatus, Fusarium oxysporum, Penicillium digitatum* and *Trichophyton mentagrophytes.*

Citrus sinensis (Rutaceae)

The oil from epicarp exhibited strong fungitoxicity at 0.5 per cent concentration against *A. flavus, P. italicum* and *Alternaria alternata* as a contact toxicant. The oil at the same concentration inhibited heavy doses of inocula and exhibited quick killing action. The pesticidal action of the oil was thermostable upto 80°C and lasted even upto 24 months. The oil was broad spectrum (Shukla *et al.*, 2000).

Citrus lemon (Rutaceae)

de Souza *et al.* (2005) noted that oil and the phytochemicals citral, eugenol and mircene showed significant antimould activity. Among the products that evidenced the antimould activity citral and eugenol showed the lowest minimum inhibitory concentration, which was 1 per cent and 4 per cent respectively. The mould strains assayed are *Fusarium* sp. *Rhizopus* sp. *Aspergillus flavus, A. niger* and *Penicillium* sp.

Citrus medica (Rutaceae)

Essien, *et al.* (2008) evaluated the in vitro antifungal effect of the essential oil of *Citrus medica* L. on storage fungi of *Arachis hypogea* L. stored for 6 months using the disc diffusion agar method. The oil exhibited a wide spectrum of fungitoxicity, inhibiting all 14 fungus species tested. Thus, the oil can be exploited as a fumigant against storage fungi for the preservation of stored legume seeds due to its wide range of activity, non-phytotoxicity, and long-term persistence of fungitoxicity. Dubey (1981) isolated 0.5 per cent oil from the leaves and found it to be a mixture of 3 major and 10 minor components. The oil was fungistatic at 2000 ppm (2.0×10^3 ml/l) concentration against the test fungi, *Aspergillus flavus* and *A. versicolor*. The fungitoxic factor of the oil was susceptible to temperature treatment beyond 80°C but retained its toxicity up to 380 days during storage at room temperature. Dubey *et al.* (1982 a, b) reported that the oil at 2000 ppm (2.0×10^3 ml/l) dose, inhibited the mycelial growth of the *Acrophialospora fusispora, A. racemosa, Acremonium* sp., *Alternaria alternata, A. solani, Aspergillus nidulans, A. niger, A. flavipes, A. fumigatus, A. rubber, Chaetomium* sp., *Cladosporium cladosporioides, Colletotrichum capsici, Curvularia lunata, Paecilomyces varioti, Penicillium pinophilum, Rhizopus* sp. and *Trichoderma viride*. The vapour of the oil also controlled fungal deteriorations of the seeds of *Coriandrum sativum* and *Foeniculum vulgare* at 4000-ppm (4.0×10^3 ml/l) concentration up to 120 days in storage, without showing any adverse effect on their germination and seedling growth. Mishra *et al.* (1989) isolated the essential oil from leaves and reported its toxicity against *Aspergillus flavus* at 2000-ppm (2.0×10^3 ml/l) dose.

Citrus reticulata (Rutaceae)

The germicidal property of the leaf oil was first reported by Murdoch and Allen (1960). They reported that the oil was toxic to species of *Aspergillus* and *Torula*. Baiter (1971) reported that the fungitoxicity of the oil against *Candida albicans* and *Trichoderma lignorum* was due to its adverse effect on respiration. Devis and Smooth (1972) reported that the oil from fruits peels inhibited the spore germination of *P. digitatum*. Alderman and Marth (1976) reported that the oil inhibited the growth as well as aflatoxin production in *Aspergillus parasiticus*.

Citrus sinensis (Rutaceae)

Singh *et al.* (1978b) reported toxicity of the oil against *Nannizzia incurvata* (+), *N. incurvata* (–), *Botryotrichum kerotinophyllum* and *Malbranchea aurantica*.

Clerodendrum inerme (Verbinaceae)

Singh *et al.* (1979b) reported the toxicity of the oil against *Alternaria* sp., *Aspergillus candidus, A. flavus, A. fumigatus, A. nidulans, A. niger, Cladosporium herbarum, Cunninghamella echinulata, Fusarium moniliforme, F. oxysporum, Helminthosporium sachhari, Microsporum cocci, M. gypseum, Mucor mucedo, Penicillium digitatum, Rhizopus* sp., *Trichophyton mentagrophytes, T. rubrum* and *Trichothecium roseum*.

Coleus ambonicus (Lamiaceae)

Srivastava (1990) reported its toxicity against *Aspergillus flavus*, and *A. niger* at 1.0×10^3 ml/l concentration. The nature of toxicity was fungistatic at MIC but turned fungicidal at 0.3×10^3 ml/l concentration. Besides, the essential oil also inhibited the mycelial growth of 33 fungal species at its MIC.

Coriandrum sativum (Apiaceae)

Capek (1955) and Slavenas (1963) reported that the oil was toxic to *Aspergillus niger*. Inhibitory effect of the oil against *Alternaria tenuis, Aspergillus oryzae* and *Penicillium chrysogenum* was demonstrated by Do Vgich (1971). Sharma and Singh (1979b) reported that the oil was toxic to *Alternaria* sp., *Aspergillus candidus, A. flavus, A. fumigatus, A. nidulans, Cladosporium herbarum, Cunninghamella echinulata, Fusarium moniliforme, F. oxysporum, Helminthosporium sachhari, Microsporum cocci, M. gypseum, Mucor mucedo, Penicillium digitatum, Rhizopus* sp., *Trichophyton mentagrophytes, T. rubrum* and *Trichothecium roseum*. The oil was studied by Garg and Siddiqui (1992) *in vitro* for its antifungal activity against 18 fungal organisms using "Filter Paper Disc Agar Diffusion Technique". The strong antifungal activity against *Alternaria alternata, Curvularia lunata, Pestalotia psidi, Phytopthora parasitica, Trichoderma viride* and *Collectotrichum capsici* can be exploited.

Cuminum cyminum (Apiaceae)

Nigam and Rao (1977) reported that the oil produced inhibitory zone in culture seeded with *Aspergillus fumigatus, Fusarium tenuis,* and *Trichoderma viride* but was ineffective against *Aspergillus flavus, A. versicolor, A. oryzae, Curvularia* sp., *Penicillium javanicum* and *P. funiculosum*. Sharma and Singh (1979b) using filter paper disc technique reported that *Alternaria* sp., *Aspergillus candidus, A. flavus, A. nidulans, A. niger, Cladosporium herbarum, Cunninghamella echinulata, Fusarium moniliforme, F. oxysporum, Helminthosporium sacchari, Microsporum cocci, M. gypseum, Mucor mucedo, Penicillium digitatum, Rhizopus* sp. *Trichophyton mentagrophytes. T. rubrum* and *Trichothecium roseum* were susceptible to the oil. Saksena (1984) reported that the fungitoxic efficacy of the oil against *Aspergillus fumigatus, Candida albicans, C. tropicalis, Keratinomyces ajelloi, Microsporum gypseum, Trichophyton equinum* and *T. rubrum* was superior to griseofulvin. Singh and Upadhyay (1991) also reported that the oil was strongly fungitoxic against *Aspergillus flavus* and *A. niger*. Oil from seeds of *Cuminum cyminum* contained more than 60 compound principal among them were cumin aldehyde (36 per cent), β-pinene (19.3 per cent), p-cymine (18.4 per cent) and γ-terpinene (15.3 per cent). Antimicrobial testing showed high activity against *Aspergillus niger* and yeast *Saccharomyces cereviceae* and *Candida albicans*. (Jirovetz *et al.*, 2005)

Curcuma amada (Zingiberaceae)

Rao and Joseph (1971) reported that rhizome oil was toxic to *Cephalosporium sacchari, Ceratocystis paradoxa, Curvularia lunata, Fusarium moniliforme, Helminthosporium sacchari, Physalospora tucumanensis* and *Sclerotium rolfsii*. Banerjee and Nigam (1978a) and Mahajan *et al.* (1978) reported antifungal activity of the oil against *Aspergillus niger, A. flavus, Curvularia oryzae, Helminthosporium oryzae, Microsporum gypseum, Penicillium javanicum, P. liliacinum, Pestalotia lapagericola, Trichoderma viride* and *Trichophyton mentagrophytes*.

Curcuma angustifolia (Zingiberaceae)

Banerjee and Nigam (1977) reported that 0.1 ml of 1 : 1000 dilution of the oil produced significant inhibitory zone in cultures of *Aspergillus flavus, A. niger, Curvularia* sp., *P. lilacinum, Pestalotia lapagericola, Trichoderma viride* and *Trichophyton mentagrophytes*.

Curcuma cassia (Zingiberaceae)

Garg and Jain (1998) reported the toxicity of the rhizome oil against *A. niger* and *A. fumigatus*.

Curcuma aromatica (Zingiberaceae)

Toxicity of the rhizome oil was reported against *Aspergillus flavus, A. niger, Curvularia oryzae, Helminthosporium oryzae, Microsporum gypseum, Penicillium javanicum, P. lilacinum, Pestalotia lapagericola, Trichoderma viride* and *Trichophyton mentagrophytes*. (Rao, 1976b and Banerjee *et al.*, 1978b).

Curcuma longa (Zingiberaceae)

The rhizome oil was reported to possess toxicity against *Aspergillus flauvs, A. niger, Curvularia oryzae, Helminthosporium oryzae, Microsporum gypseum, Penicillium javanicum, P. lilacinum, Pestalotia lapagericola, Trichoderma viride* and *Trichophyton mentagrophytes* (Banerjee and Nigam, 1978a, b). Chandi *et al.* (1999) reported that the turmeric leaf and rhizome oils were fungicidal against *Candida albicans* MTCC-183. Leaf oil showed better activities in comparison to rhizome oil. Leaf oil killed the test pathogen with in 1-15 minute. The activities were well compared with standard antifungal drugs. The essential oil from the rhizomes of *Curcuma longa* has been studied for antifungal activity against five fungi *viz. Aspergillus niger, A. fumigatus, Curvularia lunata, Fusarium psidi* and *Rhizopus oryzae* (Garg and Jain, 2003). Gurdip Singh *et al.* (2003) reported that the rhizome oil was toxic to several bacteria and is better antifungal as compared to synthetic fungicides. The oil has no adverse effect on seed germination and seedling growth of some common agricultural crops. The fungitoxic spectrum shows that oil has strong fungitoxic activity (100 per cent mycelial inhibition) against *Colletotrichum falcatum, Fusarium moniliforme,* and *Helminthosporium oryzae* at 1000-ppm concentration and against *Aspergillus niger, Curvularia pallescens, Fusarium oxysporum* and *Periconia atropurpurea* at 2000 ppm concentration. Mycelial growth of *A. flavus, Curvularia lunata* and *Rhizoctonia solani* was completely inhibited at 3000-ppm concentration.

Curcuma zedoaria (Zingiberaceae)

Rao and Joseph (1971) reported the toxicity of the oil against *Cephalosporium sacchari, Ceratocystis paradoxa, Curvularia lunata, Fusarium moniliforme, Helminthosporium sacchari, Physalospora tucumanensis* and *Sclerotium rolfsii*. Mishra and Dixit (1978), Banerjee and Nigam (1978b) and Banerjee *et al.* (1978b) reported toxicity of the oil against *Alternaria tenuis, Aspergillus flavus, A. niger, Curvularia oryzae, Fusarium nivale, Helminthosporium oryzae, Microsporum gypseum, Penicillium lilacinum, P. javanicum, Pestalotia lapagericola, Trichoderma viride* and *Trichophyton mentagrophytes*. Narsimha Rao and Subba Rao (1982) further reported the antifungal spectrum of the oil against many pathogenic fungi including *Candida albicans, Epidermophyton floccosum* and *Trichophyton rubrum*.

Cydonia oblonga (Rosaceae)

Shikhaliev (1974) reported that the oil from fruits was toxic to *Fusarium avenaceum*.

Cymbopogon citratus (Poaceae)

The toxicity of the oil against *Microsporum gypseum* and *Penicillium notatum* was reported by Sawhney *et al.* (1977). Arora and Pandey (1977) reported that the oil was not only toxic to *Penicillium italicum* but also controlled blue mould decay of citrus fruits during storage. Handique and Singh (1990) extracted essential oil from a mutant strain M-81 of lemon grass and tested against *Rhizoctonia solani, Sclerotium rolfsii* and *Sclerotinia sclerotiorum* at 100 ppm (0.1×10^3 ml/l). The growth of *Rhizoctonia solani* and *Sclerotinia sclerotiorum* was decreased by 67 per cent and 40 per cent respectively whereas it was ineffective against *Sclerotium rolfsii*. Dwivedi *et al.* (1990) tested the oil against *Macrophomina phaseolina* and found to be toxic at 500, 1000 and 2000 ppm (0.5×10^3 ml/l, 1.0×10^3 ml/l and 2.0×10^3 ml/l) doses. Oil and leaf powder of *Cymbopogon citratus* (lemon grass) at 0.1 and 0.25 per cent (v/w) and ground leaves at 10 per cent (w/w) significantly reduced deterioration and aflatoxin production in shelled melon seeds inoculated with toxigenic *Aspergillus flavus, A. niger, A. tamarii* and *Penicillium citrinum* (Bankole and Joda, 2003).

Cymbopogon flexuosus (Poaceae)

Saksena (1984) reported the toxicity of the oil agaisnt *Aspergillus fumigatus, Candida albicans, C. tropicalis, Keratinomyces ajelloi, Microsporum gypseum, Trichophyton equinum* and *T. rubrum*. The oil from leaves was strongly toxic against following storage fungal pests *viz. Aspergillus flavus, Penicillium italicum* and *Alternaria alternata*. The oil showed broad spectrum, inhibited heavy doses of inocula, thermostable and toxicity persisted for at least 12 months (Shukla *et al.*, 2000). The minimum bioactive concentration with fungicidal action of the oil was found to be 0.2 µl/ml for *Alternaria alternata*, 0.4 µl/ml for *A. flavus, A. fumigatus, A. parasiticus, Cladosporium cladosporioides, P. italicum, P. digitatum* and a number of other fungi, and 0.5 µl/ml for *Botrytis cyneria* and *Helminthosporium oryzae*. The bioactivity of oil was broad spectrum, thermostable and lasted upto 48 months. The oil was used for *in vivo* controlling postharvest spoilage of *Malus pumilo* (Shahi *et al.*, 2002). de Souza *et al.* (2005) noted that oil from *Cymbopogon citratus* and the phytochemicals citral, eugenol and mircene showed significant antimould activity. Among the products that evidenced the antimould activity citral and eugenol showed the lowest minimum inhibitory concentration, which was 1 per cent and 4 per cent respectively. The mould strains assayed are *Fusarium* sp. *Rhizopus* sp. *Aspergillus flavus, A. niger* and *Penicillium* sp.

Cymbopogon martinii (Poaceae)

Kala *et al.* (1984) reported that the oil from shoots completely prevented the growth and production of hepatocarcinogenic aflatoxin by *Aspergillus parasiticus*. Jeyalakshmi *et al.* (1997) evaluated the oil against chili fruit rot caused by *Colletotrichum capsici* (syd) Butler and Bisby at different concentration *viz.* 0.1, 0.075, 0.05, 0.025, 0.01 and 0.005 percent *in-vitro* following poisoned food technique and spore germination assay. It exerted cent percent inhibition on mycelial growth and spore germination even at 0.025 per cent concentration. Because of its strong fungitoxic activity it may be exploited for the management of chili fruit rot disease. Babu *et al.* (2000) to test the antifungal properties of palma rosa [*Cymbopogan martinii* (Roxb.) Watson] oil used it against *Alternaria solani*, the tomato leaf blight pathogen. Palma rosa oil inhibited the mycelial growth and spore germination of *A. solani* at 0.3 per cent concentration. Mycelial growth and spore germination were observed at very low concentration of the oil (below 0.2 per cent).

Cymbopogon nardus (Poaceae)

The growth and production of hepatocarcinogenic aflatoxin by *Aspergillus parasiticus* was reduced when cultures were supplemented with graded concentration of the oil (Kala *et al.*, 1984).

Cymbopogon winteriunus (Poaceae)

Sawhney *et al.* (1977) reported that the oil possessed toxicity against *Microsporum gypseum* and *Penicillium notatum*.

Cyperus scariosus (Cyperaceae)

Lahariya and Rao (1979) reported that the oil was toxic to *Aspergillus fumigatus, A. niger, Chrysosporium tropicum, Malbranchea pulchella, Microsporum gypseum* and *Penicillium lilacinum*. Dikshit and Hussain (1984) reported inhibitory effect of the oil against *Trichsophyton equinum* and *T. rubrum*.

Dalbergia sissoo (Fabaceae)

Jain *et al.* (1979) reported toxicity of the oil against *Alternaria solani, Aspergillus flavus, A. fumigatus, A. nidulans, A. niger, Cladosporium herbarum, Cunninghamella* sp., *Fusarium moniliforme, Helminthosporium* sp., *Penicillium digitatum, P. italicum, Rhizopus nigricans* and *Trichothecium roseum.*

Daucus carota (Apiaceae)

Chaurasia and Kher (1978a, b) and Grover and Rao (1978a) reported that the oil was toxic to species of *Aspergillus, Curvularia* and *Penicillium.*

Dinnetia tripetala (pepperfruit)

Essential oil and phenolic extracts inhibited growth of several food borne microorganisms including sp. of *Penicillium* and *Aspergillus* (Ejechi and Akpomedaye, 2004).

Dysoxylum malabaricum (Meliaceae)

Parcha *et al.* (2003) tested for antimicrobial activity of essential oil from the leaves of *Dysoxylum malabaricum* against *Aspergillus niger, A. flavus, Penicillium* sp. and *Candida albicans.* The oil was found effective in inhibiting the growth of all the organisms. Maximum inhibition was found in case of *C. albicans.*

Eucalyptus sp. (Myrtaceae)

The effectiveness of *Eucalyptus* oil to control the black mould of onion bulbs was observed by Tiwari *et al.* (1988) and it was found that the oil at 1.0×10^3 ml/l concentration effectively reduced the decay as well as the sprouting of stored onions. Oil showed inhibitory effect on the mycelial growth of three soil borne plant pathogen *viz., Sclerotium rolfsii, Rhizoctonia solani* and *Sclerotinia rolfsii* (Handique and Singh, 1989). Suri and Thind (1979) reported that the *Eucalyptus camaldulensis* oil was toxic to *Aspergillus fumigatus, Penicillium notation* and some human pathogens. Kaul and Nigam (1977) and Sampurna and Nigam (1979) reported that the *E. citriodora* oil was toxic to *Aspergillus niger, Candida albicans, C. utilis, Keratinophyton terreus, Malbranchea pulchella* and *Trichophyton mentagrophytes.* Suri *et al.* (1979) recorded that the pure oil of *E. citriodora* controlled fruit rot of apple caused by *Clathridium corticola* and 1: 1000 dilution of the same oil completely checked the growth of *A. niger, C. corticola* and *Microsporum gypseum.* The toxicity of the *E. citriodora* oil against *Aspergillus flavus, A. fumigatus, Fusarium oxysporum* and *Penicillium digitatum* was also reported by Ceruti *et al.* (1982). Haerdtl (1962) reported the fungitoxicity of the volatile constituent from *E. gennii* against *Aspergillus niger.* Egawa *et al.* (1977) isolated gallic acid and unidentified isomers of benzene from the volatile constituent of the same plant and reported that the compounds were inhibitory to conidial germination of *Cochliobolus benaus.* However, Arora and Pandey (1984a, b) recorded the toxicity of the oil against *Alternaria tenuis, Diplodia metalensis* and *Penicillium digitatum.* Ceruti *et al.* (1982) reported the toxicity of the *E. globulus* oil against *Aspergillus flavus, A. fumigatus, Fusarium oxysporum, Penicillium digitatum* and *Trichophyton mentagrophytes.* Benjinali *et al.* (1984) reported that the oil possessed broad antifungal spectrum and Dwivedi and Dubey (1986) observed that the oil was inhibitory to sclerotial germination of *Macrophomina phaseoli.* Ceruti *et al.* (1982) recorded the toxicity of the *E. piperita* oil against *Aspergillus flavus, A. fumigatus, Fusarium oxysporum,* and *Penicillium digitatum* and *Tichophyton mentagrophytes.* Shahi *et al.* (1997) noted that essential oils from *Eucalyptus citriodora, E. dalarympleana, E. laveopinea, E. pauciflora* at 1000 ppm showed complete inhibition of the storage spoilage test fungus *Penicillium italicum.* The oil of *E. dalarympleana* and *E. laveopinea* showed fungistatic toxicity against the test fungus at 3000-ppm but the oil of *E. laveopinea* showed partial inhibition at 3000 ppm. The oil of *E. citriodora* at 1000-ppm exhibited fungicidal nature and withstood heavy inoculum. Shahi *et al.* (2002) noted that essential oil extracted from leaves of *Eucalyptus pauciflora* showed toxicity against fungal pests *viz.*, sp. of *Alternaria, Aspergillus, Penicillium,* and *Rhizopus.* The effective pesticidal concentration (EPC) of the oil was found to be 0.3×10^3, 0.4×10^3, 0.5×10^3 and 0.6×10^3ml/l respectively inhibiting the mycelial growth of the tested pests completely.

Eugenia bracteata (Myrtaceae)

Rao and Joseph (1971) recorded that the oil was toxic to *Cephalosporium sachhari, Ceratocystis paradoxa, Curvularia lunata, Fusarium moniliforme, Helminthosporium sacchari, Physalospora tucumanensis* and *Sclerotium rolfsii.*

Eugenia caryophylata (Myrtaceae)

Clove oleoresin at 0.2 to 0.8 per cent (v/v) was studied against the test microorganisms such as *Candida albicans, Pencillium citrinum, Aspergillus niger* and *Trichophyton mentagrophytes.* It was highly effective against *T. mentagrophytes* and *Candida albicans. P. citrinum* and *A. niger* were relatively more resistant. Clove oleoresin was first dispersed in sugar solution and then used for antifungal testing (Nunez *et al.*, 2001)

Eugenia heyneana (Myrtaceae)

Rao and Joseph (1971) recorded the toxicity of the oil against *Ceratocystis paradoxa, Curvularia lunata, Helminthosporium sacchari* and *Physalospora tucumanensis.* Garg (1974) while testing the range of antifungal activity reported that the oil was toxic to *Curvularia geniculata, C. cappescens, Fusarium culmorum, F. oxysporum, Helminthosporium* sp. and *Pythium aphanidermatum.* Narsimha Rao and Subba Rao (1982) recorded toxicity of the oil against *Candida albicans, Epidermophyton floccosum, Trichophyton mentagrophytes* and *T. rubrum.*

Eupatorium ayapana (Asteraceae)

Chaurasia and Kher (1978a, b) reported toxicity of the oil against species of *Aspergillus, Curvularia and Penicillium.* Sharma and Singh (1979b) recorded toxicity of the oil against *Alternaria* sp., *Aspergillus candidus, A. flavus, A. fumigatus, A. nidulans, A. niger, Cladosporium herbarum, Cunninghamella echinulata, Fusarium moniliforme, F. oxysporum, Helminthosporium sacchari, Microsporum cocci, M. gypseum, Mucor mucedo, Penicillium digitatum, Rhizopus* sp., *Trichophyton mentagrophytes, T. rubrum* and *Trichothecium roseum.*

Eupatorium cannabinum (Asteraceae)

Dubey, *et al.* (2007) extracted essential oil of *Eupatorium cannabinum,* it was found to be fungitoxic in nature against *Botryodiplodia theobromae* and *Colletotrichum gloeosporioides* causing stem end rot disease and anthracnose disease in mango respectively. Dwivedi *et al.* (1990) tested the oil against *Macrophomina phaseolina* and found to be toxic at 500, 1000 and 2000 ppm (0.5×10^3 ml/l, 1.0×10^3 ml/l and 2.0×10^3 ml/l) doses.

Eupatorium capillifolium (Asteraceae)

Chandra *et al.* (1982c) reported that the leaves yielded 1 per cent of the oil which was fungitoxic to *Absidia ramose, Alternaria raphani, Aspergillus awamori, A. sydowi, Colletotrichum capsici, C. gloeosporioides, Curvularia lunata, Helminthosporium oryzae* and *Rhizoctonia solani* at 1.0×10^3 ml/l dose. They further reported that the fungitoxicity of the oil was enhanced at pH 7 and pH 9.

Eupatorium triplinerve (Asteraceae)

Garg (1974) reported that the leaf oil was toxic to *Aspergillus candidus, A. fumigatus, A. nidulans, A. terreus, Curvularia geniculata, C. pallescens, Fusarium culmorum, F. oxysporum, Paecilomyces* sp., *Penicillium chrysogenum, Helminthosporium* sp., *Pythium aphanidermatum, Trichoderma viride* and *Trichurus* sp.

Feronia elephantum (Rutaceae)

Sharma *et al.* (1978) reported that the oil, which contained estragol, was toxic to *Alternaria helianthi, Colletotrichum capsici, Fusarium moniliforme, F. solani, Helminthosporium oryzae, H. turcicum, Pythium vexans, Pyricularia setariae, Rhizoctonia bataticola* and. *R. solani.* Zutshi *et al.* (1975), Zutshi and Mehta (1977) and Gupta and Singh (1982) reported that the oil possessed toxicity against *Aspergillus fumigatus, A. niger, A. nidulans, Cladosporium herbarum, Cunninghamella echinulata, Fusarium moniliforme, F. oxysporum* and *Rhizopus* sp. Dubey (1993) reported the toxicity of the oil against *Aspergillus flavus* and *A. niger* at 1.0×10^3 ml/l doses and was fungicidal.

Ferula jaeschkeana (Apiaceae)

Gautam *et al.* (1990) reported that the oil was toxic to *Nannizzia fulva, N. gypsea* and *N. incurvata.*

Foeniculum vulgare (Apiaceae)

Ozcan *et al.* (2006) examined the chemical composition of the flower and unripe and ripe fruits from fennel (bitter) (*Foeniculum vulgare* ssp. *piperitum*) by gas chromatography and gas chromatography-mass spectrometry. The main identified components of the flower and unripe and ripe fruit oils were estragole (53.08 per cent, 56.11 per cent, and 61.08 per cent), fenchone (13.53 per cent, 19.18 per cent, and 23.46 per cent), and α-phellandrene (5.77 per cent, 3.30 per cent, and 0.72 per cent), respectively. The oils exerted varying levels of antifungal effects on the experimental mycelial growth of *Alternaria alternata, Fusarium oxysporum,* and *Rhizoctonia solani.* The 40 ppm concentrations of fennel oils showed inhibitory effect against mycelial growth of *A. alternata,* whereas 10 ppm levels were ineffective. Capek (1956) reported toxicity of the oil against *Aspergillus flavus.* Later Do Vgich (1971) reported the toxicity against *Alternaria tenuis, Aspergillus oryzae* and *Penicillium chrysogenum.* Sharma and Singh (1979b) reported the toxicity against *Alternaria* sp., *Aspergillus candidus, A. flavus, A. fumigatus, A. nidulans, A. niger, Cladosporium herbarum, Caninghamella echinulata, Fusarium moniliforme, F. oxysporum, Helminthosporium sacchari, Microsporum cocci, M. gypseum, Mucor mucedo, Penicillium digitatum, Rhizopus* sp., *Trichophyton mentagrophytes, T. rubrum, Trichothecium roseum.* However, Narayan *et al.* (1980a) reported the toxicity of the oil against *A. fumigatus, A. niger, Candida tropicalis* and *Rhizopus oryzae.* Shukla (1987) reported that the fungitoxic factor of the oil was thermostable and remained active during storage upto 365 days. The antifungal efficacy of the oil was not affected when the inoculum density of *Penicillium italicum* was increased. Shukla and Tripathi (1987a) reported that the oil at 1.5×10^3 ml/l completely inhibited the mycelial growth of *Alternaria alternata, A. tenuissima, Aspergillus awamori, A. ochraceus, A. parasiticus, A. sydowi, A. tamarii, A. terreus, Botryodiplodia theobromae, Colletotrichum capsici, Curvularia lunata, C. pallascens, Drechslera hawaiiensis, D. oryzae, Epicoccum nigrum, Fusarium acuminatum, F. equiseti, F. semitectum, F. udum, Macrophomina phaseolina, Mwemnoniella echinata, Nigrospora oryzae, Penicillium chrysogenum, P. citrinum* and *Rhizopus nigricans* and enhanced the seed germination *of Arachis hypogea.*

Geranium sp. (Geraniaceae)

Sampurna and Nigam (1979) and Saksena (1984) reported that the oil was toxic to *Aspergillus fumigatus, Candida albicans, C. tropicalis, Keratinophyton terreum, Keratinomyces ajelloi, Microsporum gypseum, Trichophyton rubrum* and *T. equinum.*

Glossocardia bosvallia (Asteraceae)

Pathak and Dixit (1984) reported toxicity of the oil against *Aspergillus niger, Botryodiplodia theobromae, Botryotrichum keratinophyllum, Chrysosporium tropicum, Microsporum gypseum, Malbranchea pulchella, Phytophthora parasitica* var. *piperina* and *Rhizopus nodosus.*

Hedychium spicatum (Zingiberaceae)

Samuel and Tripathi (1994) reported that the essential oil extracted from the rhizome completely inhibited the mycelial growth of *Helminthosporium oryzae* and *Fusarium moniliforme* at a minimum dose of 1000 ppm. The nature of toxicity against the test fungi at this concentration was fungistatic but turned fungicidal at 5000 ppm dose. Temperature treatment upto 120°C and storage up to 365 days (at 28 + 2°C) did not alter the fungistatic efficacy of the oil. The oil exhibited broad fungitoxic spectrum and proved non-phytotoxic to seed germination and seedling growth of *Oryza sativa* and *Hordeum vulgare.*

Helichrysum italicum (Asteraceae)

Chirkina and Osipova (1974) reported that the oil obtained from flowers showed the presence of eugenol, furfural, geroniol, neusol and pinene and inhibited the growth of *Candida albicans.*

Hypericum patulum (Hyperaceae)

Kumar (1992) reported the toxicity of the oil against *Aspergillus flavus, A. ochraceus,* and *Fusarium oxysporum* at 1.2×10^3ml/l. The oil was fungistatic at its MIC and remained static even upto 10×10^3 ml/l conc. of the oil.

Hyptis suaveolens (Lamiaceae)

Pandey *et al.* (1982c) recovered 0.22 per cent oil from the leaves and reported its toxicity at 4.0×10^3 ml/l concentration against *Aspergillus flavus, A. rubrum, A. versicolor, Chaetomium globosum, Cladosporium cladosporioides, Fusarium oxysporum, Helminthosporium oryzae, Microsporum gypseum, Monilia sitophila, Pyricularia oryzae, Rizopus arrhizus, Trichophyton equinum, T. mentagrophytes* and *T. rubrum.* They further reported that the thermostable antifungal factor of the oil was most effective at pH 9 and remained unaffected by autoclaving and storage upto 180 days. They also observed that the oil successfully prevented the leaf spot disease of paddy without showing any toxicity to the host plant. However, Singh *et al.* (1983a) reported that the oil at 1.0×10^3 ml/l concentration completely inhibited the mycelial growth of *Epidermophyton floccosum, Microsporum canis* and *Trichophyton mentagrophytes.*

Hyssopus officinalis (Lamiaceae)

Do Vgich (1971) reported the toxicity of the oil against *Alternaria tenuis, Aspergillus oryzae* and *Penicillium chrysogenum.*

Inula cuspidata (Asteraceae)

Chauhan and Saxena (1985) reported that the oil was toxic to *Aspergillus flavus, A. fumigatus, A. glaucus, A. nidulans* and *A. terreus.*

Inula racemosa (Asteraceae)

Mishra and Dixit (1978) reported that the oil possessed toxicity against *Colletotrichum capsici* and *Fusarium moniliforme.* Later Mishra *et al.* (1979a) recorded toxicity of the oil against *Alternaria helianthi, Fusarium solani, Helminthosporium turcicum, H. oryzae, Pyricularia setariae, Pythium vexans, Rhizoctonia bataticola* and *R. solani.*

Juniperus communis (Pinaceae)

The toxicity of the oil against *Aspergillus niger* was reported by Slavenas and Razinskaite (1962) and Slavenas (1963). Dwivedi *et al.* (1990) tested the oil against *Macrophomina phaseolina* and found to be toxic at 500, 1000 and 2000 ppm (0.5 × 10^3 ml/l, 1.0 × 10^3 ml/l, 2.0 × 10^3ml/l). This fungus causes root rot disease of crops.

Juniperus macropoda (Pinaceae)

Singh *et al.* (1978b) and Mishra and Chauhan (1984) reported that the oil possessed toxicity against *Botryotrichum keratinophyllum, Epidermophyton floccosum, Malbranchea aurantica, Nannizzia incurvata* (+), *N. incurvata* (–) and *Trichophyton rubrum.*

Juniperus verginiana (Pinaceae)

Mall *et al.* (1985) reported that the wood oil was fungistatic to *Candida pseudotropicalis, Epidermophyton floccosum, Microsporum gypseum, Sporothrix schenekii* and *Trichophyton rubrum* at 1.0 × 10^3 ml/l dose but became fungicidal at 6.0 × 10^3 ml/l.

Lantana camara (Verbinaceae)

Avadhoot and Verma (1978) reported that the seed oil was toxic to *Candida albicans, Curvularia lunata* and *Fusarium oxysporum.* Sharma and Singh (1979b) reported the toxicity of the oil against *Alternaria* sp., *Aspergillus candidus, A. flavus, A. fumigatus, A. nidulans, A. niger, Cladosporium herbarum, Cunninghamella echinulata, Fusarium moniliforme, Helminthosporium sacchari, Microsporum cocci, M. gypseum, Mucor mucedo, Penicillium digitatum, Rhizopus* sp., *Trichophyton mentagrophytes, T. rubrum* and *Trichothecium roseum.*

Lantana indica (Verbinaceae)

Rao and Joseph (1971) and Narsimha Rao and Subba Rao (1982) reported that the oil was toxic to *Candida albicans, Epidermophyton floccosum, Trichophyton mentagrophytes* and *T. rubrum.*

Laurus nobilis (Lauraceae)

The toxicity of the oil against *Aspergillus niger* was reported by Capek (1955).

Lepidogathis hyalina (Acanthaceae)

The essential oil of *L. hyalina* inhibited the mycelial growth of *Aspergillus flavus, A. niger* and some bacteria (Yadava, 1986).

Lapidium ruderale (Brassicaceae)

Singh *et al.* (1984) reported that the oil from growing tips was toxic to *Aspergillus flavus* at 1.0 × 10^3 ml/l dose.

Leucus aspera (Lamiaceae)

Rao and Joseph (1971) and Narsimha Rao and Subba Rao (1982) reported the toxicity of the oil against *Candida albicans, Epidermophyton floccosum, Trichophyton mentagrophytes* and *T. rubrum.*

Lippia alba (Verbinaceae)

Dwivedi *et al.* (1990) tested the oil against *Macrophomina phaseolina* and found to be toxic at 500, 1000 and 2000 ppm. This fungus causes root rot of crops. de Souza *et al.* (2005) noted that oil from

Lippia alba, Lippia microphylla and the phytochemicals citral, eugenol and mircene showed significant antimould activity. Among the products that evidenced the antimould activity citral and eugenol showed the lowest minimum inhibitory concentration, which was 1 per cent and 4 per cent respectively. The mould strains assayed are *Fusarium* sp., *Rhizopus* sp., *Aspergillus flavus, A. niger* and *Penicillium* sp.

Lippia javanica (Verbinaceae)

Leaves of Bangladeshi *Lippia javanica* gave essential oil content of 0.15 per cent. Thirty-three compounds comprising 99.3 per cent of the oil were identified by GC/MS. The major constituents were geraniol (29 per cent), neral (19.3 per cent), isopropenyl acetophenone (15.9 per cent) and linalool (10.6 per cent). The oil was tested for fungitoxic properties against six pathogens. The minimum concentration for absolute inhibition of the *Alternaria alternata, Fusarium equiseti, Curvularia lunata, Colletotrichum corchori* was found to be 250 ppm and *Macrophomina phaseolina, Botrydiplodia theobromae* were found to be 750 ppm and 1000 ppm respectively (Chowdhury *et al.*, 2003).

Majorana hortensis (Lamiaceae)

Yadav and Saini (1991) reported its antimicrobial activity against *Aspergillus fumigatus* and *A. niger*.

Melaleuca leucadendron (Myrtaceae)

Dubey *et al.* (1983 b) reported that the leaf oil was fungistatic to *Colletotrichum capsici, Curvularia lunata, Helminthosporium oryzae, Pyricularia oryzae, Pythium aphanidermatum, P. debaryanum, Rhizoctonia solani* and *Syncephalastrum* sp. at 0.05×10^3 ml/l but at 0.5×10^3 ml/l it turned fungicidal. However, at 0.5×10^3 ml/l the oil was proved to have no adverse effect on seed germination and seedling growth of *Phaseolus aureus*. Mishra *et al.* (1989) isolated the essential oil from leaves and reported its toxicity against *Aspergillus flavus* at 2.0×10^3ml/l dose.

Melissa officinalis (Lamiaceae)

Dikshit and Hussain (1984) reported the toxicity of the oil against *Microsporum gypseum, Trichophyton equinum* and *T. rubrum*.

Melissa parviflora (Lamiaceae)

The toxicity of the oil against *Colletotrichum* sp. was reported by Lakshmi and Rao (1978).

Mentha arvensis (Lamiaceae)

Kumar, *et al.* (2007) extracted essential oils from *Mentha arvensis*, it was found to be effective against strains of *A. flavus* and completely stopped the radial mycelial growth of *A. flavus* at 0.10 mg mL^{-1}. It was found to be superior over the synthetic fungicides tested and showed a broad fungitoxic spectrum against *A. niger, A. fumigatus, Botryodiplodia theobromae, Cladosporium cladosporioides, Fusarium oxysporum, Helminthosporium oryzae, Macrophomina phaseolina* and *Sclerotium rolfsii* at 0.10 mg mL^{-1}. Sanyal and Verma (1969) reported that the oil was fungitoxic. Later Arora and Pandey (1977, 1984a,b) reported the toxicity of the oil against *Alternaria tenuis, Diplodia metalensis, Penicillium digitatum* and *P. italicum*. Chaurasia and Kher (1978a, b) and Dikshit *et al.* (1979) noted the toxicity of oil against *Aspergillus* sp., *Curvularia lunata*, and *Helminthosporium oryzae*. While Gautam *et al.* (1980) noted the toxicity of oil against dermatophytes *viz. Nannizia fulva, N. gypsea, N. incurvata*. Ceruti *et al.* (1982) reported the toxicity of the oil against *Aspergillus flavus, A. fumigatus, Fusarium oxysporum* and *Trichophyton mentagrophytes*. Singh *et al.* (1983a) recorded that the oil at 2.0×10^3 ml/l was fungicidal to *Absidia ramose, Alternaria brassicae, A. raphani, A. solani, Aspergillus niger, A. lindemuthianum, C. tinosporae, Fusarium monoliforme, F. solani, Paecilomyces fusisporus, Talaromyces* sp. and *Trichoderma viride*. Kala *et al.* (1984) reported that the oil significantly reduced aflatoxin production in *Aspergillus parasiticus*. Arora and Pandey (1984 b) reported that the oil can be successfully exploited for control of blue mould decay in citrus fruits during storage.

Mentha citrata (Lamiaceae)

Sawhney *et al.* (1977) noted the toxicity of the oil against *Microsporum gypseum* and *Penicillium notatum*.

Mentha lavanduliodora (Lamiaceae)

Ceruti *et al.* (1982) reported the oil was toxic to *Aspergillus flavus, A. fumigatus, Fusarium oxysporum, Penicillium digitatum* and *Trichophyton mentagrophytes*.

Mentha piperita (Lamiaceae)

The fungitoxicity of the oil against *Fusarium moniliforme* and *Penicillium luteum* was reported by Makorova (1961) and Kovacs (1964). Gautam *et al.* (1980) reported toxicity of the oil against some dermetaphytes *viz., Nannizzia fulva, N. gypseum* and *N. incurvata*.

Murraya exotica (Rutaceae)

The toxicity of the leaf oil against *Alternaria alternaria, A. solani, Aspergillus flavus, A. fumigatus, A. nidulans, A. niger, Cladosoporium herbarum, Cunninghamella* sp., *Fusarium moniliforme, Helminthosporium* sp., *Penicillium digitatum, P. italicum, Rhizopus nigricans* and *Trichothecium roseum* was first reported by Jain *et al.* (1979). Later Gupta and Singh (1982) reported the toxicity against *Cunninghamella echinulata, Fusarium moniliforme, F. oxysporum* and *Rhizopus* sp.

Murraya paniculata (Rutaceae)

Garg (1974) reported that the oil was inhibitory to *Curvularia geniculata, C. pallescens, Fusarium culmorum, F. oxysporum, Helminthosporium* sp. and *Pythium aphanidermatum*.

Myristica fragrans (Myristicaceae)

Srivastava (1990) reported that the essential oil of *M. fragrans* was toxic to *Aspergillus flavus* and *A. niger* at 2.0×10^3ml/l and 1.0×10^3ml/l doses respectively. The nature of fungitoxicity was found to be fungistatic even at a higher concentration of 15×10^3 ml/l. However, the oil also proved fungitoxic to 33 fungal species.

Nardostachys jatamansi (Valerianaceae)

Girgune *et al.* (1978) reported toxicity of the oil against *Alternaria helianthi, Colletotrichum capsici, Fusarium solani, Helminthosporium oryzae, H. turcicum, Pyricularia setariae, Pythium vexans, Rhizoctonia bataticola* and *R. solani*. Mishra *et al.* (1995) reported that the oil at 1.0×10^3 ml/l conc. inhibited the growth of *Aspergillus flavus, A. niger* and *Fusarium oxysporum*. The oil also proved effective against six other fungi at the same conc. and did not adversely affect seed germination and seedling growth of *C. arietinum* and *Cajanus cajan*.

Nepeta cataria (Lamiaceae)

Makorova (1961) reported the toxicity of the seed oil against *Aspergillus flavus, A. versicolor, Fusarium tenuis* and *Trichoderma viride*. Rao and Nigam (1978) isolated d-limonene, p-cymene, citronellyl acetate, carvone and d-citroneual from oil and reported that each component possessed fungitoxicity. Agarwal *et al.* (1979) reported that the oil besides being anti-helminthic also possessed toxicity against *Aspergillus flavus, A. niger, Curvularia lunata, C. oryzae, Microsporum gypseum, Penicillium chrysogenum, P. javanicum* and *Trichoderma viride* either at 1:10 or at 1:1000 dilutions. Rathee *et al.* (1982) reported that carvone was antifungal factor of the oil and was toxic to *Alternaria helianthi, Colletotrichum capsici, Fusarium moniliforme, F. solani, Helminthosporium oryzae, H. turcicum, Pyricularia setariae, Pythium vexans, Rhizoctonia bataticola* and *R. solani.*

Ocimum adscendens (Lamiaceae)

Asthana and Singh (1981) isolated oil from leaves and reported its fungicidal toxicity against *Aspergillus flavus* and *A. niger* at 4.0×10^3 ml/l and 0.2×10^3 ml/l concentration respectively. Asthana *et al.* (1982) reported the toxicity against *Helminthosporium oryzae* and reported that the oil controlled fungal deterioration of *Vigna radiata* during storage without affecting its seed germination and seedling growth. Asthana (1984) recorded that the oil at 0.4×10^3 ml/l conc. completely inhibited the growth of *Alternaria alternata, A. citri, Aspergillus awamori, A. flavipes, A. fumigatus, A. nidulans, A. ochraceus, A. sydowi, A. terreus, Chaetomium globosum, Cladosporium cladosporioides, Colletotrichum capsici, C. gloeosporioides, Gleosporium fructigenum* and *Trichothecium roseum* and its fungitoxic factor was not affected by temperature treatment up to 100°C and storage up to 360 days.

Ocimum basilicum (Lamiaceae)

Herath and Abeywickrama (2008) isolated essential oil of *Ocimum basilicum* most effective chemical component is eugenol inhibiting conidial germination of *Colletotrichum musae* and *Fusarium proliferatum in vitro.* Do Vgich (1971) reported toxicity of the oil against *Alternaria tenuis, Aspergillus oryzae* and *Penicillium chrysogenum.* Kaul and Nigam (1977) and Sawhney *et al.* (1977) recorded the toxicity against *Aspergillus niger, Candida albicans, C. utilis, Microsporum gypseum* and *Penicillium notatum.* Lahariya and Rao (1979) reported that the oil was toxic to *Aspergillus fumigatus, A. niger, Chrysosporium tropicum, Malbranchea pulchella, Microsporum gypseum* and *Penicillium lilacinum* while inhibition of *Diplodia metalensis, Penicillium digitatum* and *P. italicum* was reported by Arora and Pandey (1984a, b). Reuveni *et al.* (1984) reported that cineole, linalool, methyl chavicol and eugenol were probably the active constituents of the oil which completely inhibited the growth of *D. metalensis* and *Fusarium oxysporum.* Dube *et al.* (1989) reported that the essential oil completely inhibited the mycelial growth of 22 species of fungi including mycotoxin producing strains of *Aspergillus flavus* and *A. parasiticus* at 1.5×10^3 ml/l concentration. It also exhibited significant insect repellent activity against *Allacophora foveicollis.*

Ocimum canum (Lamiaceae)

Rao and Joseph (1971) reported that the leaf oil was toxic to *Candida albicans, Cephalosporium sacchari, Ceratocystis paradoxa, Curvularia lunata, Epidermophyton floccosum, Fusarium moniliforme, Helminthosporium sacchari, Physalospora tucumanensis, Sclerotium rolfsii, Trichophyton mentagrophytes* and *T. rubrum.* Jain and Agarwal (1978) observed presence of methyl heptone, eugenol and geraniol in the oil, which was toxic to *Chrysosporium tropicum, Keratinophyton tenuipes, Malbranchea pulchella* and *Verticillium tenuipes.* Bhargava *et al.* (1979) isolated 1.4 per cent of the oil from leaves and recorded its toxicity against *Aspergillus flavus* and *A. versicolor* at 3.0×10^3 ml/l. Bhargava *et al.* (1981) reported fungistatic nature of the oil at 3.0×10^3 ml/l which become fungicidal at 6.0×10^3 ml/l concentration. Dubey (1981) noted the toxicity of the thermostable fungitoxic factor, which remained active during storage up to 390 days. Narsimha Rao and Subba Rao (1982) reported that the oil inhibited the growth of *E. floccosum.* Dubey *et al.* (1983 a) reported that the oil successfully prevented fungal deterioration of

Foeniculum vulgare and *Coriandrum sativum* during storage without showing any phytotoxic effect on the treated commodities. Mishra *et al.* (1989) isolated the essential oil from leaves and reported its toxicity against *Aspergillus flavus* at 2000 ppm (2.0×10^3 ml/l). Thakur *et al.* (1989) isolated linalool from the *Ocinum canum* oil and tested against *Rhizoctonia solani* and *Fusarium solani.* It was found to be strongly fungitoxic to the test fungi.

Ocimum citriodora (Lamiaceae)

Sawhney *et al.* (1977) reported the toxicity of the oil against *Microsporum gypseum* and *Penicillium notatum.*

Ocimum gratissimum (Lamiaceae)

The toxicity of the oil against *Microsporum gypseum* and *Penicillium notatum* was reported by Sawhney *et al.* (1977) and Grover and Rao (1977). Singh *et al.* (1983a) reported that the oil at 1.0×10^3 ml/l completely inhibited the growth of *Epidermophyton floccosum, Microsporum canis* and *Trichophyton mentagrophytes.* Tripathi *et al.* (1985) reported the toxicity against *Alternaria alternata, Colletotrichum capsici* and *Sclerotium rolfsii* at 0.25×10^3 ml/l, 0.5×10^3 ml/l conc. respectively. They further reported that the oil possessed fungistatic nature at lower doses but became fungicidal at higher doses and when chemically analyzed yielded myrcene, b-pinene, p-cymene, linalyl acetate, terpineol, terpinyl acetate, linalool, geranial, citroneollol and eugenol each exhibiting toxicity against *A. alternata, C. capsici* and *S. rolfsii* at different doses. Dixit (1986) isolated eugenol from the oil and reported that the oil as well as eugenol can be used to prevent bakery bread from fungal deterioration during storage. Mishra *et al.* (1989) isolated the essential oil from leaves and reported its toxicity against *Aspergillus flavus* at 2000 ppm. Thakur *et al.* (1989) isolated the main constituent eugenol from the *Ocimum* oil. 0.1 per cent of eugenol inhibited the radial growth of *Sclerotium rolfsii, Rhizoctonia* sp. and *Alternaria alternata. O. gratissimum* was found to exhibit absolute toxicity against the tested fungi *A. flavus* and *A. niger* (Sharma, 2000).

Ocimum killimandscharicum (Lamiaceae)

Suri and Thind (1979) reported the toxicity of the oil against *Aspergillus fumigatus* and *Penicillium notatum.*

Ocimum sanctum (Lamiaceae)

Oil from leaves showed absolute toxicity against *Aspergillus flavus* but was moderately active against *A. niger* (Sharma, 2000).

Ocimum thesiforum (Lamiaceae)

Rao and Joseph (1971) reported the toxicity of the oil against *Cephalosporium sacchari, Ceratocystis paradoxa, Curvularia lunata, Fusarium moniliforme, Helminthosporium sacchari, Physalospora tucumanensis* and *Sclerotium rolfsii.*

Ocimum viridae (Lamiaceae)

Thakur *et al.* (1989) isolated thymol from the oil. 0.1 per cent of thymol inhibited the growth of *Rhizoctonia solani* and *Sclerotium rolfsii.*

Oenanthe javanica (Apiaceae)

Sharma and Singh (1979 a) reported the toxicity of the oil against *Alternaria* sp., *Aspergillus candidus, A. flavus, A. fumigatus, A. niger, Cladosporium herbarum, Cunninghamella echinulata, Fusarium moniliforme,*

F. oxysporum, Helminthosporium sacchari, Microsporum cocci, M. gypseum, Mucor mucedo, Penicillium digitatum, Rhizpus sp., *Trichophyton mentagrophytes, T. rubrum* and *Trichothecium roseum.*

Oenanthe marjorana (Apiaceae)

Capek (1955) recorded toxicity of the oil against *Aspergillus niger.* Maruzzella and Liguroi (1958) and Maruzzella *et al.* (1960) noted that the oil was inhibitory to *Lentinus lepideus, Lenzites trabea* and *Polyporus versicolor.*

Oenanthe vulgaris (Apiaceae)

Korta and Staryzk (1963) recorded mycostatic action of the oil against *Candida albicans, C. tropicalis, C. parakrusei, C. krusei, C. pelliculose, C. stellatoidea* and *C. silosis.*

Parthenium hysterophorus (Asteraceae)

Sharma and Singh (1979) reported that the oil completely inhibited the growth of *Alternaria* sp., *Aspergillus candidus, A. flavus, A. fumigatus, A. nidulans, A. niger, Cladosporium herbarum, Cunninghamella echinulata, Fusarium moniliforme, F. oxysporum, Helminthosporium sacchari, Microsporum cocci, M. gypseum, Mucor mucedo, Penicillium digitatum, Rhizopus* sp., *Trichophyton mentagrophytes, T. rubrum* and *Trichothecium roseum.*

Pastinaca sativa (Apiaceae)

Do Vgich (1971) reported toxicity of the oil against *Alternaria tenuis, Aspergillus oryzae* and *Penicillium chrysogenum.*

Piperomia pellucida (Piperaceae)

Singh *et al.* (1980) reported that the leaf oil completely inhibited the mycelial growth of *Absidia ramosa, Aspergillus fumigatus, A. niger, A. versicolor, Cephalosporium sacchari, Chaetomium globosum, Cladosporium cladosporioides, Colletotrichum falcatum, C. gloeosporioides, Curvularia lunata, Helminthosporium carbonum, H. gramineum, H. maydis, H. oryzae, H. sturcicum, Trichoderma viride* and *Trichothecium roseum* at 2.0×10^3 ml/l and the fungitoxic factor of the oil was thermostable as well as durable during storage up to 150 days. Later, Singh *et al.* (1983b) noted that the oil sustained heavy inoculum density and killed the test fungus, *H. oryzae* within 15 minutes (by contacts). They further observed that the oil successfully prevented the leaf spot disease in paddy without affecting the germination, seedling growth and morphology of the host plant.

Petroselinum crispum (Apiaceae)

Crisan and Hodisan (1975) reported the toxicity of the oil against *Alternaria alternata.*

Peucedanum oreoselnum (Apiaceae)

Crisan and Hodisan (1975) reported the toxicity of the oil against *Alternaria alternata.*

Phytolacca americana (Phytolaccaceae)

Vichkanova *et al.* (1972) reported the toxicity of the oil against *Candida albicans, Microsporum gypseum* and *M. lanantum.*

Pimpinella anisum (Apiaceae)

Capek (1956) reported the toxicity of the oil against *Aspergillus niger.* Shukla and Tripathi (1987b) recorded that the oil at 1.0×10^3 ml/l completely inhibited the mycelial growth of *Aspergillus flavus* and

Penicillium italicum and antifungal factor of the oil remained unaffected by temperature treatment up to 100°C and storage up to 365 days. They isolated anethole as antifungal principle of the oil and reported its fungistatic nature of toxicity at 0.6×10^3 ml/l against *Alternaria alternata, A. tenuis, Aspergillus awamori, A. fumigatus, A. parasiticus, A. sydowi, A. tamarii, A. terreus, Colletotrichum capsici, Curvularia lunata, C. pallascens, Epicoccum nigrum, Fusarium acuminatum, F. equiseti, F. moniliforme, F. oxysporum, F. semitectum, F. udum, Macrophomina phaseolina, Nigrospora oryzae, Penicillium chrysogenum, P. citrinum* and *Rhizopus nigricans.* Seed dressing with anethole at 5.0×10^3 ml/l completely prevented the fungal deterioration of groundnut during storage without affecting its germination and seedling growth (Shukla, 1987).

Pinus pondorosa (Pinaceae)

Cobb *et al.* (1968) detected the presence of comphene, carvene, limonene, myrcene, oleoresin, phellendrene and pinene in the oil which possessed toxicity against *Ceratocystis annosus, C. minor, C. pilifera, C. schrenkiana* and *Fames annosus.*

Pinus sylvestris (Pinaceae)

Melvin and Krupa (1971) and Krupa and Nylund (1972) reported that the oil contained some volatile organic constituents, which were toxic to *Fomes annosus* and *Phytophthora cinnamoni* and also to the mycorrhizal fungi of the plant from which the oil was isolated.

Piper betel (Piperaceae)

Dubey (1986) reported fungistatic nature of the oil against *Aspergillus flavus* at 0.5×10^3 ml/l dose, which remained unaffected by autoclaving, temperature treatment, storage and heavy inoculum density. He further reported that the oil showed no adverse effect on seed germination and seedling growth of *Pisum sativum.* Dubey and Tripathi (1987) isolated eugenol as antifungal principle of the oil and reported its fungistatic nature at 0.2×10^3 ml/l dose against *Alternaria alternata, Aspergillus awamori, A. fumigatus, A. parasiticus, A. sydowi, A. terreus, Botrydiplodia theobromae, Curvularia lunata, Nigrospora oryzae, Penicillium citrinum* and *Rhizopus nigricans.*

Piper nigrum (Piperaceae)

Chaurasia and Kher (1978 a, b) reported that the oil possessed toxicity against *Curvularia lunata* and species of *Aspergillus, Penicillium* and *Rhizopus.*

Pogostemon plectranthoides (Lamiaceae)

Kaul and Nigam (1977) recovered 0.5 per cent oil from leaves and reported its toxicity against *Aspergillus niger, Candida albicans* and *C. utilis* at 1:1000 dilution.

Polyalthia longifolia (Annonaceae)

Narsimha Rao and Subba Rao (1982) recorded toxicity of the oil against *Candida albicans, Epidermophyton floccosum, Trichophyton mentagrophytes* and *T. rubrum.*

Premna tomentosa (Verbinaceae)

Rao and Joseph (1971) reported the toxicity of the oil against *Cephalosporium sacchari, Ceratocystis paradoxa, Curvularia lunata, Fusarium moniliforme, Helminthosporium sacchari, Physalospora tucumanensis* and *Sclerotium rolfsii.*

Prunus amygdalus (Rosaceae)

Singh (1976) reported the toxicity of the oil against *Colletotrichum falcatum, Helminthosporium oryzae* and *Pyricularia oryzae.*

Psidium guajava (Myrtaceae)

Nigam and Rao (1977) reported the toxicity of the oil against *Aspergillus flavus, A. fumigatus, A. oryzae, A. versicolor, Curvularia* sp., *Fusarium tenuis, Penicillium funiculosum, P. javanicum* and *Trichoderma viride.* Kamran (2008) tested the oil against *Aspergillus fumigatus, A. niger, A. ochraceous, Penicillium expansum* and *Alternaria alternata.*

Psoralea corylifolia (Leguminosae)

Grovers and Rao (1979) and Sharma and Singh (1979 b) noted toxicity of the oil against *Alternaria* sp., *Aspergillus candidus, A. flavus, A. fumigatus, A. nidulans, A. niger, Cladosporium herbarum, Cunninghamella echinulata, Fusarium moniliforme, F. oxysporum, Helminthosporium sacchari, Microsporum cocci, M. gypseum, Mucor mucedo, Penicillium digitatum, Rhizopus stolonifer, Trichophyton mentagrophytes, T. rubrum* and *Trichothecium roseum.*

Putranjiva roxburghii (Euphorbiaceae)

Tripathi and Kumar (2007) collected samples of peanut seeds from stores in eastern Uttar Pradesh, India and examined for their associated mycoflora and insects. The oil of *Putranjiva roxburghii* exhibited the greatest toxicity. The oil was found to be fungicidal and thermostable at its minimum inhibitory concentration (MIC) of 400 ppm evaluated against the dominant fungi, *Aspergillus flavus* and *Aspergillus niger*, as well as *Trogoderma granarium*. Kumar and Tripathi (2003) reported the fungitoxic properties of the oil against *Aspergillus flavus* and *A. niger* and insect *Trogoderma granarium* causing heavy deterioration in fresh seeds of ground nut during storage.

Raphanus sativus (Brassicaceae)

The seed oil was reported to possess toxicity at 1:1000000 dilutions against *Alternaria brassicae, Fusarium avenaceum* and *Phoma* sp. (Nehrash, 1961). Sehgal (1961) reported that the fungitoxic property of members of Brassicaceae was due to presence of isocyanates.

Randia dumentorium (Rubiaceae)

Mishra *et al.* (1979 b) reported the toxicity of the oil against *Colletotrichum capsici* and *Fusarium solani.*

Rosa indica (Rosaceae)

Maruzzella (1963) reported that the oil used in perfumery was toxic to *Claviceps graminicola, C. purpurea, Colletotrichum graminicola, Gibberella fujikuroi, G. zeae* and *Helminthosporium turcicum.*

Salvia officinalis (Lamiaceae)

Brieskorm (1950) and Crisan and Hodisan (1975) reported that the oil was toxic to *Mucor mucedo* and *Penicillium glaucum.*

Santalum album (Santalaceae)

Singh (1976) reported the toxicity of the oil against *Colletotrichum falcatum, Helminthosporium oryzae,* and *Pyricularia oryzae.* Dikshit and Hussain (1984) recorded toxicity against *Microsporum gypseum, Trichophyton equinum* and *T. rubrum.*

Satureja delmatica (Lamiaceae)

Korta and Staryzk (1963) reported that the oil possessed toxicity against *Candida albicans, C. krusei, C. parakrusei, C. pelliculosa, C. silosis, C. stellatoidea* and *C. tropicalis.*

Satureja hortensis (Lamiaceae)

The oil possessed toxicity against *Candida albicans, C. krusei, C. parakrusei, C. pelliculosa, C. stellatoidea, C. tropicalis* (Korta and Staryzk, 1963), and *Aspergillus oryzae, Alternaria tenuis* and *Penicillium chrysogenum* (Do Vgich, 1971).

Selinum tenufolium (Apiaceae)

Murari and Mathela (1978) and Mathela and Sinha (1978) reported the toxicity of the oil against many fungi.

Sequoia sempervirens (Taxodjaceae)

Hital *et al.* (1977) reported the toxicity of the oil against different species of *Aspergillus.*

Seseli indicum (Apiaceae)

Dwivedi *et al.* (1990) tested the oil against *Macrophomina phaseolina* and found to be toxic at 0.5×10^3 ml/l, 1.0×10^3 ml/l and 2.0×10^3 ml/l doses. This fungus causes root rot disease of crops. Chaturvedi and Tripathi (1989) reported the toxicity of the oil against *Aspergillus niger, A. flavus* and *Fusarium oxysporum* at 8.0×10^3 ml/l concentration.

Sassafras officinalis (Lauraceae)

Singh *et al.* (1978b) reported toxicity of the oil against *Botryotrichum keratinophyllum, Malbranchea aurantica, Nannizzia incurvata* (+) and *N. incurvata* (–).

Skimmia laureda (Rutaceae)

Sampurna and Nigam (1979) reported that the oil possessed toxicity against *Keratinophyton terreum, Malbranchea pulchella* and *Trichophyton mentagrophytes.*

Sphaeranthus indicus (Asteraceae)

Rao and Joseph (1971) reported that the oil was toxic to *Cephalosporium sacchari, Ceratocystis paradoxa, Curvularia lunata, Fusarium moniliforme, Helminthosporium sacchari, Physalospora tucumanensis* and *Sclerotium rolfsii.*

Syzygium aromaticum (Myrtaceae)

Capek (1955) reported the toxicity of the oil against *Aspergillus niger*. However, the oil was also inhibitory to 8 species of keratinophilic fungi (Thind and Dahiya, 1977). Khan *et al.* (1994) found that the flower bud extract of *Syzygium aromaticum* exhibited absolute toxicity against *Aspergillus niger* and *Curvularia ovoidea,* inhibiting the mycelial growth completely.

Syzygium caryophyllata (Myrtaceae)

Singh (1976) reported the toxicity of the oil against *Colletotrichum falcatum, Helminthosporium oryzae* and *Pyricularia oryzae.*

Syzygium cumini (Myrtaceae)

The oil possessed toxicity against *Cephalosporium sacchari, Ceratocystis paradoxa, Curvularia lunata, Fusarium moniliforme, Helminthosporium sacchari, Physalospora tucumanensis* and *Sclerotium rclfsii* (Rao and Joseph, 1971 and Grover and Rao, 1978b).

Thuja orientalis (Cupressaceae)

Nigam and Rao (1977) reported that the oil was toxic to *Aspergills flavus, A. fumigatus, A. oryzae, A. versicolor, Curvularia* sp., *Fusarium tenuis, Penicillium funiculosum, P. javanicum* and *Trichophyton viride.*

Thujopsis dalabrata (Cupressaceae)

Uchida (1928) reported toxicity of the oil against *Merulius lacrymans* and *Polyporus lacrymans.* Later, Okazaki and Homma (1953) observed that the fungitoxicity of the oil against *Achorion* sp. and species of *Epidermophyton* and *Trichophyton* was due to the presence of kinokitiol, 1-rhodinic acid and thujopsol.

Thymus vulgaris (Lamiaceae)

The toxicity of the oil against *Aspergillus niger* was first reported by Capek (1955). Later Do Vgich (1971) reported toxicity against *Alternaria tenuis, Aspergillus oryzae* and *Penicillium chrysogenum,* and Ceruti *et al.* (1982) against *Aspergillus flavus, A. fumigatus, Fusarium oxysporum, Penicillium digitatum* and *Trichophyton mentagrophytes.*

Thymus serpyllum (Lamiaceae)

Oil extracted from dried, crushed flowering plants showed antifungal properties against *A. flavus, A. awamori, A. niger, A. foetidus* and *A. oryzae.* It also inhibited all the three stages of asexual reproduction *i.e.* spore germination, mycelial growth and spore formation (Rahman *et al.*, 2003)

Trachyspermum ammi (Apiaceae)

Nigam and Rao (1977) reported that the oil was toxic to *Fusarium tenuis.* Sharma and Singh (1979b) and Narayan *et al.* (1980a) recorded that the oil was inhibitory to *Alternaria* sp., *Aspergillus candidus, A. flavus, A. fumigatus, A. nidulans, A. niger, Candida tropicalis, Cladosporium herbarum, Cunninghamella echinulata, Fusarium moniliforme, F. oxysporum, Helminthosporium sacchari, Microsporum canis, M. gypseum, Mucor mucedo, Penicillium digitatum, Rhizopus oryzae, Trichophyton mentagrophytes, T. rubrum* and *Trichothecium roseum.* Singh *et al.* (1986b) reported toxicity of the dethymolysed oil against *Alternaria brassicae, A. solani, Cephalosporium sacchari, Cladosporium cladosporioides, Colletotrichum falcatum, C. lindemuthianum, C. tinosporae, Cornespora casicola, Fusarium solani, Helminthosporium oryzae, Paecilomyces* sp., *Talaromyces* sp. and *Trichoderma viride.* Dikshit and Hussain (1984) reported toxicity against *Microsporum gypseum, Trichophyton equinum* and *T. rubrum,* and Saksena (1984) against *Candida albicans, C. tropicalis* and *Keratinomyces ajelloi.* Singh *et al.* (1986b) reported that the oil at 900 ppm (0.9×10^3 ml/l) dilution killed *Epidermophyton floccosum, Microsporum canis* and *Trichophyton mentagrophytes* within 20 seconds (by contact). Tripathi *et al.* (1986) reported fungicidal nature of the oil against *Alternaria alternata, A. solani, Aspergillus flavipes, A. flavus, A. niger, A. ochraceus, A. sydowi, A. tamarii, A. terreus, A. versicolor, Botrytis cinerea, Colletotrichum capsici, C. falcatum, Curvularia lunata, C. pallescens, Chaetomium globosum, Fusarium semitectum, Monilia pruinosa* and *Penicillium funiculosum* at 800 ppm (0.8×10^3 ml/l). They further added that the fungitoxic factor remained unaffected by temperature treatment up to 100°C and storage at room temperature up to 365 days. The seeds of

Arachis hypogea when dressed at 5000 ppm (5.0×10^3 ml/l) of the oil and stored for 12 months showed no fungal contamination indicating the grain protection efficacy of the oil. Thymol and p-cymene were isolated as antifungal principles of the oil and reported to possess toxicity against *Aspergillus flavus* and *A. niger* at 1.0×10^3ml/l concentration (Tripathi *et al.*, 1986).

Valeriana wallichii (Valerianaceae)

Suri and Thind (1979) reported the toxicity of the oil against *Aspergillus fumigatus* and *Penicillium notatum.*

Vitex negundo (Verbenaceae)

Uppalpati and Rao (1979) reported the toxicity of the oil against species of *Colletotrichum, Fusarium, Helminthosporium* and *Trichoderma.* Dwivedi *et al.* (1990) tested the oil against *Macrophomina phaseolina* and found to be toxic at 500, 1000 and 2000 ppm (0.5×10^3 ml/l, 1.0×10^3 ml/l, 2.0×10^3 ml/l). This fungus causes root rot of crops.

Vitiveria zizanoides (Poaceae)

Singh *et al.* (1978b) reported the fungitoxicity of the oil against *Botryotrichum keratinophyllum, Malbranchea aurantica, Nannizzia incurvata* (+) and *N. incurvata* (–). Dikshit and Hussain (1984) reported toxicity against *Microsporum gypseum, Trichophyton equinum* and *T. rubrum.*

Zanthoxylum alatum (Rutaceae)

The essential oil recovered from the fruits, at 2.0×10^3 ml/l dose was fungistatic to 24 fungi, including the aflatoxin producing strains of *Aspergillus flavus* and *A. parasiticus.* (Dubey *et al.*, 1990).

Zanthoxylum avicennae (Rutaceae)

Chen *et al.* (1990) extracted essential oil by steam distillation. Separation of essential oil yielded 27 compounds, out of which 22 were identified using Gas Chromatography. Citral, 1-octanol, 4-methyl-1-6 acetohexanol and linalool showed anti-mildew activity.

Zingiber officinale (Zingiberaceae)

Sharma and Singh (1979b) reported toxicity of the oil against *Alternaria* sp., *Aspergillus candidus, A. flavus, A. fumigatus, A. nidulans, A. niger, Cladosporium herbarum, Cunninghamella echinulata, Fusarium moniliforme, F. oxysporum, Helminthosporium sacchari, Microsporum cocci, M. gypseum, Mucor mucedo, Penicillium digitatum, Rhizopus* sp., *Trichophyton mentagrophytes, T. rubrum* and *Trichothecium roseum.* Singh and Singh (1981) recorded complete inhibition of conidial germination of *Erysiphe polygoni.* Singh *et al.* (1983b) isolated citral (3, 7 dimethyl 1-2, 6-octadieneal) as antifungal principle of the oil and reported its toxicity against *Fusarium nivale.* Singh *et al.* (1999) studied the activity of the aqueous extract of Zingiber *officinalis* on *Sclerotinia sclerotiorum* which causes stem rot of gram and reported that growth and sclerotium formation were completely inhibited at 80,000 ppm and above.

Efficacy and Applications

Several hundred-research papers are published each year on the antimicrobial activity or other functional activity of botanicals from higher plants, and a complete review of all of them is beyond the scope of this article. The most excellent ones are given above. A brief discussion on efficacy and application of antimicrobial botanicals from higher plants is crucial. The antimicrobial botanicals, which have the potential to be used as storage protectants, can be divided into several useful categories, including phenolics, polyphenols, quinones, flavones, flavonoids, flavonols, tannins, cumarins, terpenoids, lectins and polypeptides (Cowan, 1999). Many herbs, such as thyme, contain multiple active compounds, which represent different chemical families. The essential oil fraction of botanicals is often the most inhibitory chemical fraction to growth and survival of microorganisms. Essential oils are highly enriched with terpenoids. Examples of herbs and spices containing terpenoids which have been shown to have antimicrobial activity include allspice, basil, bay, burdock, cinnamon, paprika, chilli pepper, clove, eucalyptus, dill, gotu kola, grape fruit seed extract, horseradish, lemon verbena, oregano, paod' arco, papaya, peppermint, rosemary, savory, sweet flag, tansy, tarragon, thyme, turmeric, valerian and willow (Duke, 1985; Cate, 2000). The other major chemical group found in plants, which has been frequently reported to have antimicrobial and antifungal activity, is the sulfoxide/ isothiocyanate family, which includes onion, garlic, mustard and members of the *Brassica* family. Approximately 30 per cent of essential oils, which have been examined, are inhibitory to bacteria, and more than 60 per cent of essential oil derivatives have been shown to be inhibitory to fungi (Chaurasia and Vyas, 1977; Cowan, 1999).

The mechanism of action for the antimicrobial activity of botanical storage protectants is not fully understood. However, terpenoids and phenolics are thought to exert inhibitory action against microorganisms by membrane disruption (Cichewick and Thorpe, 1996; Lambert *et al.*, 2001; Schultes, 1978). Simple phenols and flavonoids appear to inhibit growth by binding to biochemicals essential for metabolism (Peres *et al.*, 1997). Both coumarins and alkaloids are thought to inhibit growth of microorganisms at the genetic level (Hoult *et al.*, 1996; Rahman and Chaudhary, 1995).

Although numerous studies have been done *in vivo* to evaluate the antimicrobial activity of botanicals, only a few studies have been done with stored products for preventing or controlling mould growth. Inhibition of fungal growth on coriander and fennel seeds dressed with 0.5 per cent concentration of *Cedrus* oil has been reported (Dikshit 1980). Seeds of coriander showed good result when fumigated with essential oil from *Citrus media* and *Ocimum canum* at their MIC (Dubey *et al.*, 1983). *In vivo*, application *Cymbopogon citratus* oil showed that growth of *Aspergillus flavus* was greatly checked (Mishra *et al.*, 1994). Seeds of wheat and groundnut fumigated with oil from *Eucalyptus citriodora* showed excellent result (Shukla, 1997). Oil and leaf powder of *Cymbopogon citratus* significantly reduced deterioration and aflatoxin production in shelled melon seeds inoculated with toxigenic *Aspergillus flavus*, *A. niger*, *A. tamarii* and *Penicillium citrinum* (Bankole and Joda, 2003). Use of *Trachyspermum ammi* oil inhibits growth of dominant storage fungi such as species of *Aspergillus*, *Penicillium*, and *Alternaria* etc. *in-vivo* condition.

These studies show that some botanicals have the potential to be effective storage protectant although product development to optimize functionality and flavour will be challenging. More studies are needed on applications of botanicals from higher plants in storage protection to fully understand how best to optimize their use.

Safety

Use of many plants in storage protection is commonly regarded as safe but some plants contain noxious compounds, which may render them unsafe for both animals and humans to consume. Toxicology axiom."The dose makes the poison" also apply in some cases. It means that a substance that is safely consumed in the diet at low levels may be unsafe if consumed at a higher level in the diet. Therefore, the data demonstrating that the botanical is safe when consumed at the higher level is needed. And it is virtually difficult to find typical toxicological data such as Acceptable Daily Intake and No Effect Level. Further, there may be unusual sensitivities of some parts of the population to specific herbal compounds or strong aromatic ingredients. Therefore, while using botanicals in storage protection these points should be duly emphasized to avoid negative nutritional or health consequences.

A Sure Future for Botanical Storage Protectants

We are currently limited to a rather small number of antimicrobial agents, which have been used for many years with little expansion, and there is a real need to expand the list of storage protectants, which can be used to ensure safety and quality of stored products. These systems may have synergistic or additive uses with one another or may also be used with conventional antimicrobial compounds. The future of naturally occuring antimicrobial system seems sure, as new storage protection systems are being rapidly developed and used in a variety of storage products.

Acknowledgements

The authors are thankful to the Head, Department of Botany, University of Allahabad, Allahabad for providing the facilities and to the Council of Scientific and Industrial Research, New Delhi for providing financial assistance.

References

Agarwal, R., Kharya, M.D. and Srivastava, R. (1979). Antimicrobial and antihelmintic activities of essential oil of *Nigella sativa* L. Family Ranunculaceae (Kalajira). *Indian J. Expl. Biol.* 17(11): 1264-1265.

Alderman, G.G. and Morth. E.H. (1976). Inhibition of growth and aflatoxin production of *Aspergillus parasiticus* by *Citrus* oils. *Z. Lebensm. Unters, Forsch.* 160: 353-358.

Arora, R. and Pandey, G.N. (1977). The application of essential oils and their isolates for blue mould decay control in *Citrus reticulata. J. Food Sci. Technol.* 14: 15-16.

Arora, R. and Pandey, G.N. (1984). Effect of essential oils on *Citrus* decay pathogens. *Biol. Mem.* 9(1): 69-72.

Arora, R. and Pandey, G.N. (1984). Application of essential oil and their isolates as preservative of *Citrus reticulata. Biol. Mem.* 9(1): 98-104.

Asthana, A. (1984). Studies on volatile activity of some higher plants against fungal deterioration of *Vigna radiata* (L.) Wilczak during storage. *Ph.D. Thesis*, University of Gorakhpur, Gorakhpur, India.

Asthana, A. and Singh, A.K. (1981). Fungitoxic properties of essential of *Ocimum adscendens. J. Indian Bot. Soc.* (Suppl.). 60:13.

Asthana, A., Chandra, H., Dikshit, A. and Dixit S.N. (1982). Volatile fungitoxicants from leaves of some higher plants against *Helminthosporium oryzae. Z. Pflanzen Krankheiter Pflanzens Chutz.* 89:475-479.

Avadhoot, Y. and Verma, K.C. (1978). Antimicrobial activity of essential oil of seeds of *Lantana camara* var. *aculeata* Linn. *Indian Drugs Pharm.* 13: 41-42.

Babu, S., Seetharaman, K., Nandakumar, R. and Johnson. I. (2000). *In vitro* antifungal activity of Palmarosa (*Cymbopogon martinii* (Roxb.) Watson) oil against *Alternaria solani. Indian Perfumer* 44 (4): 275-277.

Baiter, J. (1971). Effect of *Citrus* oil on respiration of *Trichoderma lignorum. Z. Labensm. Unters-Forsch* 169: 22-25.

Banerjee, A. and Nigam, S.S. (1977). Antifungal activity of the essential oil of *Curcumu angustifolia. Indian J. Pharm.* 39: 143.

Banerjee, A. and Nigam, S.S. (1978). Antimicrobial efficacy of the essential oil of *Curcuma longa. Indian J. Med. Res.* 68: 864-866.

Banerjee, A. and Nigam, S.S. (1978). Antifungal efficiency of the essential oils derived from the various species of the genus *Curcuma* Linn. *J. Res. Indian Med. Yoga Homeo.* 13: 63-70.

Bankole, S.A. and Joda, A.O. (2003). Effect of lemon grass oil (*Cymbopogon citratus* Stapf.) powder and essential oil on mould deterioration and aflatoxin contamination of melon seeds (*Colocynthis citrullus* L.). *E-mail: sabankole@juno.com*

Benjinali, B., Tanaqui–Elaraki, A., Ayadi, A. and Ihlal, M. (1984). Method to study antimicrobial effects of essential oils. Application to the antifungal activity of six morocean essences. *Journal Food Protection.* 47(10): 748-752.

Beye, F. (1978). Insecticides from Vegetable Kingdom. *Plant Research and Development* 7: 13-31.

Bhargava, K.S., Dixit, S.N., Dubey, N.K. and Tripathi, R.D. (1979). Fungitoxic properties of *Ocimum canum. J. Indian Bot. Soc. (Suppl.)* 58: 20.

Bhargava, K.S., Dixit, S.N., Dubey N.K. and Tripathi R.D. (1981). Fungitoxic properties *of Ocimum canum. J. Indian Bot. Soc.* 60: 24-27.

Bigrami, S.K., Misra, R.S., Sinha, K.K and Singh, P. (1980). Effect of some wild and medicinal plant extracts on aflatoxin production and growth of *Aspergillus flavus* in liquid culture. *Indian Botanical Society* 59: 123-126.

Bishop, C.D. and Thornton, I.B. (1997). Evaluation of antifungal activities of essential oils of *Monarda citriodora* var. *citriodora* and *Melaleuca alternifolia* on postharvest pathogens. *J. Essential Oil Research,* USA. 9 (1): 77-82.

Brandies, G.A. (1967). Commercial development of fungicides (Discussion): 246-247. In Holten *et al.* (eds.) Plant Pathology problems and progress, 1908-1958. *Indian University Press,* Allahabad, India.

Brieskorm, C.M. (1950). Active ingradients of *Salvia offinalis* VI. Bactericidal activity of oil. *Dage Arch. Pharma.* 283: 33-35.

Capek, A. (1955). Phytoncides of species. *Prumsyl Potranin* 6: 433-435.

Capek, A. (1956). Phytoncides of anise and fennel. *Prumsyl Potranin* 7: 260-262.

Cate, M. (2000). Antimicrobial and Toxicological characteristics of commercial herbal extracts and the antimicrobial efficacy of herbs in marinated chicken. *M.S. thesis* directed by F. A. Draughon, May, Univ. of Tennessee, Knoxville.

Ceruti, A., Sacco, T. and Vinai. A. (1982). Action of some essential oils on fungal growth, *Allionia* 25: 5-8.

Chandi, C.R., Dash, S.K., Mishra, R.K. and Charyulu, J.K. (1999). *In-vitro* evaluation of antimycotic activity of turmeric (*Curcuma longa* L.) essential oils against *Candida albicans* and *Cryptococcus neoformans. Indian Perfumer* 43(4): 172-178.

Chandra, H. (1984). Evaluation of some higher plants for their volatile activity against blue mould rots of orange. *Ph.D. thesis*, University of Gorakhpur, Gorakhpur, India.

Chandra, H. and Dikshit, A. (1981). Volatile Fungitoxicant from the leaves of *Ageratum conyzoides* against *Colletotrichum capsici* and *Penicillium italicum*. *J. Indian Bot. Soc.* (*Suppl.*) 60:13.

Chandra, H., Asthana, A., Tripathi, R.D. and Dixit S.N. (1980). Fungitoxicity in the vapours from the fruits of *Cinnamomum cecidodaphne*. *Jour. Ind. Bot. Soc.* (*suppl.*) 59:19.

Chandra, H., Pandey, D.K., Tripathi, N.N. and Dixit, S.N. (1982). Toxicity of the essential oil of *Eupatorium capillifolium* against *Helminthosporium oryzae*. *Indian Perfumer* 26 (3-4): 231-236.

Chaturvedi, R. (1979). Evaluation of higher plants for their fungitoxicity against *Helminthosporium oryzae*. *Ph.D. Thesis*, University of Gorakhpur, Gorakhpur, India.

Chaturvedi, R., Dikshit, A and Dixit, S.N. (1987). *Adenocalyma allicea*. A new source of natural fungitoxicant. *Tropical Agriculture* 64(4): 318-322.

Chaturvedi, R.V. and Tripathi, S.C. (1989). Fungitoxic, physiochemical and phytotoxic properties of essential oil of *Seseli indicum*. *W and A Journal of Phytopathology* 124 (1-4): 316-322.

Chaurasia, S.C. and Vyas, K.K. (1977). *In vitro* effect of some volatile oil against *Phytophthora parasitica* var. *piperina*. *J. Res. Indian Med. Yoga Homeoath*. 24-26.

Chaurasia, S.C. and Kher, A. (1978). Inhibitory effects of essential oils of three medicinal plants against various fungi. *Indian Drugs Pharm.* 13: 7-9.

Chaurasia, S.C. and Kher, A. (1978). Activity of essential oils of three medicinal plants against various pathogenic and non-pathogenic fungi. *Indian J. Hosp. Pharm.*15: 139-141.

Chawdhary, J.U., Yusuf. M., Jaripa Begum, Sultana, S.A. and Husain, M.M. (2003). Composition and Fungitoxic properties of the essential oil of *Lippia javanica* leaves. *Indian perfumer* 47(4): 385-388.

Chen. S.F., Zhu, L.P., Lu, B.Y., Yu, K.K., Lin, Z.S. (1990). Studies on essential oil from *Zanthoxylem avicennae* (Lam.) D.C. and its antimildew activity. *Acta Botanica Sinica* 31(1): 49-53.

Cichewick, R.H. and Thorpe, P.A. (1996). The antimicrobial properties of chile pepper (*Capsicum* species) and their use in Mayan medicine. *J. Ethanopharmacol.* 52: 61-70.

Chirkina, N.N. and Osipova, E.A. (1974). Antimicrobial properties of essential oil from everlasting flower *Helichrysum italicum* cultivated in Crimea. *Biol. Nauki* 17: 86-89.

Cobb Jr. F.W., Kristie, M., Zavarin, E. and Barber Jr. H.W. (1968). Inhibitory effects of volatile olioresin component on *Fomes annosus* and four *Ceratocystis* sp. *Phytopath.* 58: 1327-1335.

Cowan, M.M. (1999). Plant products as antimicrobial agents. *Clin. Microbiol. Rev.* 12: 564-582.

Crisan, A. and Hodisan, V. (1975). Investigations on fungistatic and flungicidal action of some extracts and volatile oils from medicinal plants. *Contributii Botanica* 171-179.

Davis, P.L. and Smoot, J.L. (1972). Germination of *Penicillium digitatum* spores as affected by solutions of volatile components of *Citrus* fruits. *Phytopath.* 62: 488-489.

Dawson, R.J. (1991). A global view of the mycotoxin problem. In: Champ B.R., Higley E., Hocking, A.D., Pitt, J.I. (eds.) *Fungi and mycotoxin in stored products*. Canberra: ACIAR Proc.36: 22-28.

de Souza, E.L., de Oliveira Lima, E., de Luna, K.R., de Sousa, F. and de Sousa, C.P. (2005). Inhibitory action of some essential oils and phytochemicals on the growth of various mould isolated from foods. *Brazilian Arch. Biology and Technology* 48(2): 245-250.

Dikshit, A. (1980). Fungitoxic evaluation of some plants and active product (Cedrus oil). *Ph.D. Thesis,* University of Gorakhpur, Gorakhpur, India.

Dikshit, A., Dubey, N.K., Tripathi, N.N., Tripathi, R.D. and Dixit, S.N. (1982). Control of fungal deterioration by some species during storage by the essential oil of *Cedrus deodara, A supplement to Annals of Applied Biology* 100: 56-57.

Dikshit, A. Singh, A.K., Tripathi, R.D. and Dixit, S.N. (1978). The volatile fungitoxic activity of *Cedrus deodara. Proc. Symp. Plant Disease Problems,* Jaipur, 47.

Dikshit, A., Naqvi, A.A. and Hussain, A. (1986). *Schinus molle*: A new source of fungitoxicant. *Appl. Environ. Microbiol.* 15: 1085-1088.

Dikshit, A., Singh, A.K., Tripathi, R.D. and Dixit, S.N. (1979). Fungitoxic and phytotoxic properties of some essential oils. *Bio. Bull. Indian* 1: 45-51.

Dikshit, A. and Hussain, A. (1984). Antifungal action of some essential oils against animal pathogens. *Fitoterapia* 55 (3): 171-176.

Dikshit, A., Dubey, N.K., Tripathi, N.N., Dixit, S.N. (1983). *Cedrus* oil- A promising storage fungitoxicant. *J. Stored Prod. Res.* 19(4): 159-162.

Dixit, K. (1986). Evaluation of some higher plants and their products for preservation of some food stuffs against fungal deterioration during storage. *Ph. D. Thesis.* University of Gorakhpur, Gorakhpur, India.

Dovgich, N.A. (1971). Antimicrobial effect of essential oils. *Microbiol. Zh.* 33: 253-259.

Dube, P. and Tripathi, S.C. (1987). Studies on antifungal, physico-chemical and phytotoxic properties of essential oil of *Piper betel. Pflan. Und. Pflanzen. Shutz.* 94: 235-242.

Dubey, R.K., Kumar, R. and Dubey, N.K. (2007). Evaluation of *Eupatorium cannabinum* Linn. oil in enhancement of shelf life of mango fruits from fungal rotting. *World J Microbiol Biotechnol.* 23:467–473.

Dube, S., Kumar, A. and Tripathi, S.C. (1990). Antifungal and insect repellant activity of essential oil of *Zanthoxylum alatum. Annals of Botany* 65:75-76.

Dube, S., Upadhyay, P.D. and Tripathi, S.C. (1989). Antifungal, physicochemical and insect repellant activity of essential oil of *Ocimum basilicum. Canadian J. Bot.*67: 2085-2087.

Dubey, N.K. (1981). Studies on volatile activity of some higher plants against fungi causing seed deterioration in storage. *Ph.D. Thesis,* Univ. of Gorakhpur, Gorakhpur, India.

Dubey, N.K., Bhargava, K.S. and Dixit, S.N. (1983). Protection of some stored food commodities from fungi by essential oils *of Ocimum canum* and *Citrus medica. Indian J. Tropical Plant Disease* 1: 177-179.

Dubey, N.K., Kishore, N. and Singh, S.K. (1983). Antifungal properties of the volatile fraction of *Melaleuca leucadendron. Trop. Agric.* (Trinidad) 60 (3): 227-228.

Dubey, N.K., Kishore, N., Tripathi, N.N., Tripathi R.D. and Dixit S.N. (1982). Fungitoxicity of the essential oil of *Citrus medica* against storage fungi. *Suppl. Ann. Appl. Biol.* 100: 58-59.

Dubey, N.K., Dixit, S.N. and Bhargava, K.S. (1982). Evaluation of leaf extracts of higher plants against some storage fungi. *J. Indian Bot. Soc.* 5: 20-22.

Dubey, N.K., Kishore, N., Tripathi, N.N., Tripathi, R.D. and Dixit, S.N. (1981). Fungitoxicity of the essential oils of *Ocimum canum* against *Aspergillus flavus* and *A. vesicolor*. *Indian Perfumer* 25 (2): 1-5.

Dubey, P. (1986). Studies on antifungal activity of some essential oils. *Ph.D. Thesis.* University of Gorakhpur, Gorakhpur, India.

Duke, P.J. (1985). "Hand book of Medicinal Herbs." *CRC Press, Inc.* Boca Raton, Fla.

Dwivedi, R.S. (1993). Biopesticidal plant protection for substainable crop productivity. Presidential address. *Botany section, 80th Indian Science Congress Association.* Goa 1-35.

Dwivedi, R.S. and Dubey, R.C. (1986). Effect of volatile fractions of two medicinal plants on germination of *Macrophomina phaseolina* Sclerotia. *Trans. Brit. Mycol. Soc.* 87(2): 326-328.

Dwivedi, S.K., Kishore, N., Dwivedi, S.K. (1990). Fungitoxicity of some essential oils against *Macrophomina phaseolina. Indian Perfumer* 34(1): 20-21.

Dwivedi, S.N. and Shukla T.N. (1990). Mycoflora of gram seeds in different agroclimatic regions and their pathology. *Indian Phytopathology* 43: 98-99

Egawa, A., Tsulsui, D., Talsuyama, R. and Halta, T. (1977). Antifungal substances found in leaves of *Eucalyptus* sp. *Experientia* 33: 889–890.

Ejechi, B.O. and Akpomedaye, D.E. (2005). Activity of essential oil and phenolic acid extracts of Pepperfruits (*Dennetia tripetala* G. Baker, Anonaceae) against some food-borne microorganisms. *African J. Biotechnology* 4 (3): 258-261.

Essien, P.E., Essien, J.P., Basil, N. and Ebong, G.A. (2008). Physicochemical Properties and Fungitoxicity of the Essential Oil of *Citrus medica* L. against Groundnut Storage Fungi. *Turk J Bot.* 32, 161-164.

Fawcett, C.H. and Spencer, D.M. (1969). Natural antifungal compounds: 637-669. In Torgeson, D.C. (eds.) Fungicides an Advance Treatise Vol. II. *Academic Press,* New York and London.

Garg, S.C. (1974). Antifungal activity of some essential oils. *Indian J. Pharm.* 36: 46-47.

Garg, S.C. and Jain, R.K. (1998). Antimicrobial efficacy of essential oil from *Curcuma cassia. Indian J. Microbiology* 38: 169-170.

Garg, S.C. and Jain, R.K. (2003). Antimcrobial activity of the essential oil of *Curcuma longa, Ind. Perfumer* 47(2): 199-202.

Garg, S.C. and Oswal, V.B. (1982). *In-vitro* antifungal activity of the essential oil of *Chloroxylon swietenia* Roxb. Coram. *Indian Drugs* 19(5): 189-190.

Garg, S.C., Siddiqui IV. (1992). *In-vitro* antifungal activity of the essential oil of *Coriandrum sativum* Linn. *J. Research and Education In: Indian medicine* 11(3): 11-13.

Gautam, M.P., Jain, P.C. and Singh, K.V. (1980). Activity of some essential oils against dermatophytes. *Indian Drugs* 17(9): 269-270.

Geda, A. and Bokadia, M.M. (1979). Antifungal activity of the essential oil of *Blumea membranacea. Ind. Drugs Pharma.* 14: 21-22.

Geda, A., Saxena, S.P. and Bokadia, M.M. (1978). *In-vitro* antimicrobial efficiency of essential oil against human pathogenic bacteria and phytopathogenic fungi. *Indian Drugs* 15: 253-255.

Girgune, J.B., Jain, N.K. and Garg, B.D. (1978). Antifungal activity of some essential oils. *Indian Drugs* 16: 24-26.

Girgune, J.B., Jain, N.K. and Garg, B.D. (1978). Antimicrobial activity of the essential oil from *Leucas cilliata* (Labiatae). *Indian Drugs Pharm. Ind.* 13: 39-40.

Grover, G.S. and Rao, J.T. (1977). Investigation on the anti-microbial efficiency of essential oils of *Ocimum sanctum* and *O. gratissimum* L. *Perfumer U. Kosmet* 58: 326-328.

Grover, G.S. and Rao, J.T. (1978). *In-vitro* antimicrobial studies of essential oil of *Eugenia jambolana. Indian Drugs* 15: 143-144.

Grover, G.S. and Rao, J.T. (1979). *In-vitro* antimicrobial studies of the essential oil of *Psoralea corylifolia. Indian Perfumer* 23: 135-137.

Grover, G.S., Rao, J.T. (1978). *In-vitro* antimicrobial studies of essential oil of *Daucus carota. Indian Drugs Pharm.* India 13: 39-40.

Gupta, C. and Singh, V.P. (1982). *In-vitro* antifungal effect of the essential oils of some medicinal plants. *Science and culture* 48(2): 441-443.

Haerdtl, H. (1962). Fungicidal properties of oils and solid filters. *Deuts. and pothek,* Ztd. 102: 447-448.

Handique, A.K. and Singh, H.B. (1990). Antifungal action of Lemon grass oil on some soil-borne plant pathogens. *Indian Perfumer* 34(3): 232-234.

Handique, A.K. and Singh, H.B. (1989). Antifungal activity of *Eucalyptus* oil on some soil-borne plant pathogens. *Ind. Drugs* 26(6): 274-276.

Herath, H. and Abeywickrama, K. (2008). *In-vitro* application of selected essential oils and their major components in controlling fungal pathogens of crown rot in Embul banana (*Musa acuminate*-AAB) *International Journal of Food Science and Technology*. 43, 440–447.

Hital, S.H., Zedan, H.H., Haggag, M.Y. and Saliman, P.M. (1977). Volatile of *Sequoia semipervirens* Engl. *Egypt. J. Pharm. Sci.*18: 85-91.

Hitokoto, H., Morozumi, S., Wauke, T., Sakai, S. and Kurata, H. (1980). Inhibitory effects of spices on growth and toxin production of toxigenic fungi. *Applied and Environmental Microbiology* 39: 818-822.

Hoult, J.R.S. and Paya. M. (1996). Pharmacological and biochemical action of simple coumarins: Natural products with therapeutic potential. *Gen. Pharmacol.* 27: 713-722.

Ikenaka, T., Ikeda, N., Taneka, K. and Sugawara, S. (1956). Studies on the essential oil of *Artemisia capillaris* Thumb. I. Antifungal activity of essential oil (4). *Bayer Synp.* 5: 325-343.

Imai, K., Ikeda, N., Tanaka, K. and Sugawara, S. (1956). Studies on the essential oil *of Artemisia capillaris* Thumb. I. Antifungal activity of the essential oil (2) Isolation of the antifungal principle. *J. Pharm. Soc. Jap.*76: 400-406.

Jain, P.C. and Agarwal, S.C. (1978). Notes on the activity of some odoriferous organic compounds against some keratinophillic fungi. *Trans. Mycol. Soc. Jap.* 19: 197-200.

Jain, P.K., Charia, A.K., Sharma, S.K. and Bokadia, M.M. (1979). Study of some essential oils for their antifungal activities. *Indian Drugs* 16: 122-123.

Janseen, A.M. and Scheffer, J.J.C. (1985). Acetoxychavicol acetate an antifungal component of *Alpinia galanga. Plant Medica* 6: 507-511.

Jeyalakshmi, C. and Seetharaman, K. (1997). Antifungal properties of palmarosa oil against *Colletotrichum capsici* causing fruit rot of chilli. *Indian Perfumer* 41(3): 106-108.

Jham, G.N., Dhingra, O.D., Jardim, C.M. and Valente, V.M.M. (2005). Identification of the major fungitoxic component of cinnamon bark oil. *Fitopatol. Bras.* 30 (4) July / Aug. 2005.

Jirovetz, L., Buchbauer, G., Stoyanova, A.S., Georgiev, E.V., and Damianova S.T. (2005). Composition, quality control and antimicrobial activity of the essential of cumin (*Cuminum cyminum*) seeds from Bulgaria that had been stored for up to 26 years. *J. Food Science and Technology* 40(3): 305.

Kakrani, H.K. and Nair, G.V. (1982). Antibacterial and antifungal activity of volatile oil from the seeds of *Aglaca odoratissima. Fitoterapia* 53(4): 107-109.

Kala, P.K., Tripathi, R.K., Gupta, K.C. and Singh, A.K. (1984). Effect of some essential oil on growth and aflatoxin production by *Aspergillus parasiticus* in stored grains. *Pesticides* 18: 43-46.

Kamran, A. (2008). Studies on antifungal action of some biochemotherapeutants. *Ph.D. Thesis,* University of Allahabad, Allahabad, India.

Kaul, V.K., and Nigam, S.S. (1977). Antibacterial and antifungal studies of some essential oils. *J. Res. Indian Med. Yoga Homeo* 12: 132-135.

Kaul, V.K., Nigam, S.S. and Dhar, K.L. (1976). Antimicrobiol activities of essential oils of *Artemisia absinthium* Linn., *A. vestita* Wall and *A. vulgaris* Linn. *Indian J. Pharm.* 38: 21.

Khan, S.A. and Tripathi, N.N. (1994). Volatile antifungal activity of some higher plants against storage fungi. *Neo Botanica* 2: 111-119.

Kher, A. and Chaurasia, S.C. (1977). Antifungal activity of essential oils of three medicinal plants, *Indian Drugs* 15: 41-42.

Kishore, N., Srivastava, O.P., Dubey N.K. and Dixit, S.N. (1982). Evaluation of the essential oils from the inflorescence of *Chenopodium ambrosioides* L. against *Rhizoctonia solani. Indian Perfumer* 26(2): 228-230.

Korableva, N.P. and Patapora, L.M. (1966). Role of essential oil in resistance of onion to microorganism. *Biokhim Osn. Zashch. Rost. Acad. Nauk. S.S.R.* Inst. Biochem. 79-88.

Korta, J. and Starzyk, J. (1963). Investigation on antibiotic properties of essential oils of certain species of Umbelliferae. *Acta Botanica Cracoviensia* 6: 149-160.

Kovacs, G. (1964). Studies on antibiotic substances from higher plants with special reference to their plant pathological importance. *K. Vet. Hjsk. Arsskr.* 47-92.

Krupa, S. and Nylund, *I.E.* (1971). Ectomycorrhizae of Pine III. Growth inhibition of two root pathogenic fungi by volatile organic constituents of ectomycorrhizal root systems of *Pinus sylvestris. Eur. J. For. Path.* 2: 88-94.

Kumar, A. (1992). Evaluation of some essential oils against mycotoxin producing fungi deteriorating pulses during storage. *Ph.D. Thesis,* University of Gorakhpur, Gorakhpur, India.

Kumar, N. and Tripathi, N.N. (2003). *Putranjiva* oil an ecofriendly herbal bio-pesticide. *Indian Phytopath.* 56 (3): 352.

Kumar, R., Dubey, N.K., Tiwari, O.P., Tripathi, Y.B. and Sinha, K.K. (2007). Evaluation of some essential oils as botanical fungitoxicants for the protection of stored food commodities from fungal infestation. *Journal of the Science of Food and Agriculture.* 87(9) p 1734-1743.

Kumar, R., Mishra, A.K., Dubey, N.K. and Tripathi, Y.B. (2007). Evaluation of *Chenopodium ambrosioides* oil as a potential source of antifungal, antiaflatoxigenic and antioxidant activity. *Int. J. Food Microbiol.* 115(2):159-64.

Lahariya, A.K. and Rao, J.T. (1979). *In-vitro* antimicrobial studies of the essential oil of *Cyperus scariosus* and *Ocimum basilicum*. *Indian Drugs* 16(7): 150-152.

Lakshmi, U.S. and Rao, J.T. (1978). Antimicrobial efficacy of the essential oil of *Melissa parviflora*. *Indian Drugs* 16(3): 69.

Lambert, R.J.W., Skandamis, P.N., Coote, P.J. and Nychas, G.J.E. (2001). A study of the minimum inhibitory concentration and mode of action of oregano essential oil, thymol and carvacrol, *J. Appl. Microbiol.* 91: 453-562.

Laouer, H., Zerrouq, M.M., Sahli, F., and Chaker, A.N. (March/April 2003). Composition and antimicrobial activity of *Ammodes pusilla* (Brot.) Breistr. essential oil. *J. Essential Oil Research.*

Lewis, J.A. and Papavizas, G.C. (1967). Effects of tannins on spore germination and growth of *Fusarium solani* f. *phaseoli* and *Verticillium albo-atrum*. *Canadian J. Microbiology* 13: 1655-1661.

Mabrouk, S.S. and El-Shayeb, N.M.A. (1980). Inhibition of aflatoxin formation by some spices. *Lebensm. Unters. Forsch* 171: 344-347.

Mahajan, V., Omray, A.K. and Verma, K.C. (1978). Antimicrobial activity of essential oil from rhizome of *Curcuma* sp. *Indian Perfumer* 22(4): 293-295.

Mall, H.V., Asthana, A., Dubey, N.K. and Dixit, S.N. (1985). Toxicily of Cedar wood oil against some dermatophytes. *Indian Drugs* 22: 296- 298.

Mannon, J., Johanson, E. (1985). Fungi down on the farm. *New Scient* 28:12-16.

Marruzzella, J.C. (1963). The effect of perfume oils on the growth of phytopathogenic fungi. *Plant Dis. Rep.* 47: 756-757.

Maruzzela, J.C., Baiter, J. and Katz, A. (1959). Further studies on the action of perfume oil vapours on microorganism. *Perfume Essent. Oil Res.* 50: 955-957.

Maruzzella, J.C. and Baiter, J. (1959). The action of essential oils on phytopathogenic fungi. *Plant Dis. Rep.* 43 (11): 1143-1147.

Maruzzella, J.C. and Liguroi, L. (1958). The *In-vitro* antifingal activity of essential oils. *J. Amer. Pharm. Ass. Sci. Ed.* 47: 250-254.

Maruzzella, J.C., Sorandis, D.A., Sorandis, J.B. and Grabon, G. (1960). Action of odoriferous organic chemicals and essential oil on wood destroying fungi. *Plant Dis. Rep.* 44: 789-792.

Mathela, C.S. and Sinha, G.K. (1978). Antibacterial and antifungal study of some indigenous essential oils. *J. Res. Indian Med. Yoga Homeo.* 13(3): 122-124.

Melvin, E. and Krupa, S. (1971). Studies on ectomycorrhizae of Pine III. Growth inhibition of mycorrhizal fungi by volatile organic constituents *of Pinns sylvestris* (Scots Pine) roots. *Physiol. Plant* 25: 337-340.

Miniza, J. and Thoppil. *I.E.* (2003). Antimicrobial activity of *Centella asiatica* (L.) Urb. essential oil. *Indian Perfumer* 47(2): 179-181.

Mishra, A. (1993). Evaluation of some essential oils for the control of fungi producing aflatoxin in pulses during storage. *Ph.D. Thesis*, Awadh University, Faizabad, India.

Mishra, A.K. and Dubey, N.K (1990). Fungitoxic properties of *Prunus persica* oil. *Hindustan Antibiot. Bull.* 32(3-4): 91-93.

Mishra, A.K. and Dubey, N.K. (1994). Evaluation of some essential oils for their toxicity against fungi causing deterioration of stored food commodities. *Applied and Environmental Microbiology* 60: 1101-1105.

Misra, D., Mishra, M. and Tiwari, S.N. (1997). Toxic effect of volatiles from *Callistemon lanceolatus* on six fungal pathogens of rice. *Indian Phytopath.* 50 (1): 103-105.

Mishra, A.K., Dwivedi, S.K. and Kishore, N. (1989). Antifungal activity of some essential oil. In *Nat. Aca. Sci. Letters* 12(10): 33-36.

Mishra, D., Chaturvedi, R.V. and Tripathi, S.C. (1995). The fungitoxic effect of the essential oil of the herb *Nardostachys jatamansi* D.C. *Tropical Agri.* 72(1): 48-52.

Mishra, P. and Chauhan, C.S. (1984). Antimicrobial studies of the essential oil of the berries of *Juniperus macropoda* Boiss. *Hindustan Antibio. Bull.* 25(182): 38-40.

Mishra, S.B. and Dixit, S.N. (1978). Antifungal properties of leaf extract of *Ranunculus scleratus* L. *Experientia* 34: 1442.

Mishra, S.H., Chaturvedi, S.C. and Dixit, V.K. (1979). Antimicrobial activity of essential oil of *Inula racemosa. Indian Drugs* 16(7): 141-143.

Mishra. S.H., Chaturvedi, S.C. and Dixit, V.K. (1979). Anti-microbial activity of essential oil of *Randia dumentorium* Lank. *Indian Perfumer* 23(1): 59-60.

Misra, S.B. and Dixit, S.N. (1978). Fungitoxic activity of volatile fraction of three plants. *Indian Perfumer* 22: 61-62.

Mokurova, V.A. (1961). The action of antimicrobial substances from higher plants on the causal agents of maize disease. *Nauk. Dokl. Vysk. Shkol. Biol. Sci.* 4: 152-154.

Mukherjee, N. (1974). Antifungal activities of some oils and detergents. I. Action on spore germination, growth and sporulation of some plant pathogenic fungi. *Z. pflazenkrank, Pflanzen Schutz.* 81: 468-471.

Murari, N.D. and Mathela, C.S. (1978). Antifungal studies of some indigenous essential oils. *Indian Perfumer* 22(4): 291-292.

Murdoch, D.I. and Allen, W.E. (1960). Germicidal effect of orange peel oil and d-limonene in water and orange juice. I. Fungicidal properties against yeast. *Food Techol.* 14: 441-445.

Narayan, V., Rao, K.K. and Giridhar, R. (1980). Antimicrobial activity of essential oil of *Cinnamomum zeylanicum* Bl. *Indian Drugs* 17(11): 360-362.

Narayan, V., Rao, K.K. and Giridhar, R. (1980). The *in vitro* efficiency of essential oils of some umbelliferous plants. *Indian Drugs* 17(12): 394-396.

Narsimha Rao, B.G.V. and Subba Rao, P. (1982). The efficacy of some essential oils on pathogenic fungi. *Flavour Industry* 3: 368.

Nehrash, A.K. (1961). The antimicrobial properties of cultivated radish I. The antimicrobial activity of extracts and essential oil from cultivated and wild radish. *J. Microbiol.* Kiew, 23: 32-37.

Nigam, S.S. and Rao, T.S.S. (1977). Antimicrobial efficacy of some Indian essential oils. In Kapoor, L.D. and Ramkrishanan, 1977 (eds.) *Advances in essential oil industry*. Today and Tomorrow's Printers and Publishers, New Delhi, India. 177-180.

Nunez, L., Aquino, M.D. and Chirife, J. (Apr./Jun. 2001). Antifungal properties of Clove oil (*Eugenia caryophylata*) in sugar solution. *Braz. J. Microbiol.* 32(2).

Okazaki, K. and Homma, A. (1953) Antibacterial activity of higher plants XXVI. Antimicrobial effect of essential oil, I. Oil of *Thujopsis dolabrata* and its components. *J. Pharm. Soc.* Japan, 74: 174-176.

Ozcan, M.M., Chalchat, J.C., Arslan, D., Ates, A. and Unver, A. (2006). Comparative essential oil composition and antifungal effect of bitter fennel (*Foeniculum vulgare* ssp. *piperitum*) fruit oils obtained during different vegetation. *J Med Food* 9 (4) 552–561.

Palanisami, A., Manickam, A. and Neelakantan, S. (1990). Fungal flora and aflatoxin production in relation to postharvest practices in groundnut. *Madras Agricultural Journal* 77(1): 26-31.

Pandey, D.K. Chandra, H., Tripathi, N.N. and Dixit, S.N. (1983). Mycotoxicity in leaves of some higher plants with special reference to that of *Ageratum houstonianum*. *Mil. Mykosen.* 26(11): 565-573.

Pandey, D.K., Chandra, H. and Tripathi, N.N. (1982). Volatile Fungitoxic activity in higher plants with special reference to that of *Callistemon lanceolatus* D.C. *Phytopath. Z.* 105: 175-182.

Pandey, D.K., Chandra, H., Tripathi, N.N., and Dixit, S.N. (1983). Toxicity of the essential oil of *Ageratum houstonianum* against *Fusarium lateritium*. *Beitr. Biol. Pflanz.* 58(1): 115-122.

Pandey, D.K., Tripathi, N.N., Tripathi, R.D. and Dixit S.N. (1982). Fungitoxic and phytotoxic properties of the essential oil of *Caesulia axillaris* Roxb. (Campositae) *Angew. Botanik* 56: 259-267.

Pandey, D.K., Tripathi, N.N., Tripathi, R.D. and Dixit, S.N. (1982). Fungitoxic and phytotoxic properties of the essential oil of *Hyptis suaveolens*. *J. Plant Disease and Protection* 89(6): 344-349.

Parcha, V., Gahlot, M., Kumar, P. and Bishwas, S.A.S. (2003). Antimicrobial activity of *Alseodaphne keanamia* extracts. *Ind. Perfumer* 47(1): 87-90.

Parcha, V., Gahlot, M. and Bajaj, B.K. (2003). Antimicrobial activity of essential oil of *Dysoxylum malabaricum*. *Indian Perfumer* 47 (2): 183-187.

Pathak, A.K. and Dixit, V.K. (1984). Antifungal and antibacterial studies on essential oil of *Glossocardia bosvallia*. *Indian Perfumer* 28(2): 91-94.

Peres, M.T., Monache, F.D., Cruz; A.B., Pizzolatti, M.G. and Yunes, R.A. (1997). Chemical composition and antimicrobial activity of *Croton urucurana* Baillon (Euphorbiaceae). *J. Ethnopharmacol.* 56: 223-226.

Petricic, J., Kupinic, M. and Lulie, B. (1978). Garlic (*Allium sativum* Linn.) Antifungal effects of some component of volatile oil. *Acta. Pharm. Jugosal.* 28: 41-43.

Rahman, M.U., Gul, S. (May-June 2003). Mycotoxic effect of *Thymus serpyllum* oil on the asexual reproduction of *Aspergillus* species. *J. Essential Oil Research.*

Rahman, A. and Chaudhary, M.I. (1995). Diterpinoid and steroidal alkaloids. *Nat. Prod. Rep.* 12: 361-379.

Rao, B.G.V.N. and Joseph, P.L. (1971). Activity of some essential oils towards phytopathogenic fungi. *Riechest Aromas Koir* 21: 405-410.

Rao J.T. (1976). The *in vitro* antimicrobial effects of the essential oil of *Ageratum conyzoides*. *Riecht. Aromen. Koerperf* 26: 50.

Rao, J.T. (1976). Antifungal activity of some essential oil of *Curcuma aromatica*. *Ind. J. Pharma.*38: 53-54.

Rao, T.S.S. and Nigam, S.S. (1978). Chemical and antimicrobial examination of the essential oil from the seeds of *Nigella sativa* var. *aromatica*. Abstracts of papers presented at the seminar on "*Advances in Aromatic plants and Aroma chemicals*" by Essential oil Association of India held at Bombay, 8-9th April, Session-I, ab-18.

Rathee, P.S., Mishra, S.M. and Kousal, R. (1982). Antimicrobial activity of essential oil, fixed oil and unsapioniafiable matter of *Nigella saliva* L. *Indian J. Pharm. Sci.* 44(1): 8-10.

Renu, Mall, H.V., Dubey, N.K. and Dixit, S.N. (1986). Mycotoxic properties of the essential oil of *Aegle mermelos* Corr. *Beitr. Biol. Pflanzen Go.* 325-331.

Renu, Tripathi, R.D. and Dixit, S.N. (1980). Fungitoxic properties of *Cestrum diurnum*. *Naturwiss* 67: 150.

Renuveni, R., Fleischer, A. and Putieysky, E. (1984). Fungistatic activity of essential oil from *Ocimum basillicum* Chemotypes. *Phytopath. Z.* 110(1): 20-22.

Saksena, N.K. (1984). Comparative evaluation of some essential oils for their antifungal action some dermatophytes. *Indian Perfumer* 28(1): 35-37.

Sampurna, T. and Nigam, S.S. (1979). Efficacy of some Indian oils against *Trichophyton mentagrophytes, Malbranchea pulchella* and *Keratinophyton terreum*, at different conc. *Indian Drugs Pharm.* India 14(4): 20-30.

Samuel, C.O. (1993). Evaluation of essential oils for their toxicity against some seed-borne pathogen. *Ph.D. Thesis*, University of Gorakhpur, Gorakhpur, India.

Samuel, C.O. and Tripathi, S.C. (1994). Fungitoxic properties of essential oil of *Hedychium spicatum*. *Indian Perfumer* 38(4): 105-111.

Sanyal, A. and Verma, K.C. (1969). *In-vitro* antibacterial and antifungal activity of *Mentha arvensis* var. *piperescenes* oil obtained from different sources. *Indian J. Microbiol.* 9: 23-24.

Sawhney, S.S., Suri, A.K. and Thind, T.S. (1977). Antimicrobial efficacy of some essential oil *in vitro*. *Indian Drugs* 15: 30-32.

Saxena, A.R. (1980). Evaluation of higher plants for their fungitoxic properties, *Ph.D. Thesis*, University of Gorakhpur, Gorakhpur, India.

Saxena, V.K., Sahai, A. and Sandhya, G. (1984). Studies on antimicrobial efficacy of essential oil of the leaves of *Anaphalis centorta*. *India Perfumer* 28 (3-4): 177-178.

Saxena, R.C. (1993). Neem as a source of natural insecticides: An Update-*Proc. Symp. on Botanical Pesticides* at Central Tobacco Research Institute, Rajamundry.

Schultes, R.E. (1978). The kingdom of plants. In "Medicines from the Earth," ed. W.A.R.Thompson. 208. McGraw-Hill Book Co., New York.

Sehgal, J.M. (1961). Antimicrobial substances from flowering plants. I. Antifungal substances. *Hindustan Antibiotics. Bull.* 4: 3-29.

Shahi, S., Shukla, A.C., Dikshit, S. and Dikshit, A. (1997). Modified spore germination inhibition technique for evaluation of candidate fungitoxicant (*Eucalyptus* sp.) H. W. Dehne *et al.* eds. *Proc. 4th Int. Symp. Diagnosis and Identification of Plant Pathogens*, Kluwer Academic Publishers, Netherlands, 253-263.

Shahi, S.K., Shukla A.C., Patra, M. and Dikshit, A. (2002). Use of essential oil as botanical-pesticide against postharvest spoilage in fruits, *Malus pumilo*. *J. BioControl*, Kluwer Academic Publishers. 00: 1-10.

Shahi, S.K., Shukla A.C. and Dikshit, A. (2002). *Eucalyptus pauciflora*: A potential source of ecofriendly pesticide, PROTECTON. *Proc. National Seminar on Plant Biotechnology for Sustainable Hill Agriculture.* (eds.) Kumar, N., Negi, P.S. and Singh, N.K. 149-152.

Sharma, N. (2001). Preservation of dried fruits and nuts from biodeterioration by natural plant volatiles. Donahaye, E. J. *et al.*, eds. (2001) *Proc. Int. Conf. CAF in stored Products*, Executive Printing Services, Clovis California, USA, 195-208.

Sharma, G.P., Jain, N.K. and Garg, B.D. (1978). Antifungal activity of some essential oils, *Indian Drugs* 16: 21-23.

Sharma, S.K. and Singh, V.P. (1979). Antifungal activity of the essential oil of *Oenanthe javanica*. *Indian Drugs* 17: 289-291.

Sharma, S.K. and Singh, V.P. (1979). Antifungal activity of some essential oils. *Indian Drugs and Pharm. Ind.* 14: 3-6.

Sharma, A. and Gautam, M.P. (1977). Investigation on the antifungal activity of volatile oil derived from *Nepeta hindostana* (Roth) Hains. *Indian Drugs Pharm. Ind.*12: 33-34.

Sharma, G.P., Jain, N.K. and Garg, B.D. (1978). Antifungal activity of some essential oils. *Indian Drugs* 16: 21-23.

Shukla, A.C. (1997). Fungitoxic studies of some aromatic plants against storage fungi, *Ph D Theis*, University of Allahabad, Allahabad, India.

Shukla, A.C., Shahi, S.K. and Dikshit, A. (1997). Fungicidal toxicity of the essential oil of *Cymbopogon flexuosus* against some storage fungi. *Proc. 49th Annual meeting and National Symposium on emerging issues in Plant Pathology*, Jabalpur. (abs.) 20.

Shukla, A.C., Shahi, S.K. and Dikshit, A. (2000). Epicarp of *Citrus sinensis*: A Potential source of natural pesticide, *Indian Phytopathology* 53(3): 318-322.

Shukla, H.S. (1987). Studies on some volatile fungitoxic constituents of some higher plants. *Ph.D. Thesis*, University of Gorakhpur, Gorakhpur.

Siddiqui, N., Samuel, C.O. and Tripathi, S.C. (2003). Fungitoxic properties of essential oil of *Acorus calamus*. *Indian Phytopath.* 56(3): 355.

Singh, A.K. (1980). Antifungal activity of volatile fractions of some higher plants. *Ph. D. Thesis*, University of Gorakhpur, Gorakhpur, India.

Singh, A.K., Dikshit, A., Sharma, M.L. and Dixit, S.N. (1980). Fungitoxic activity of some essential oils. *Eco. Bot.* 34(2): 186-190.

Singh, A.K., Dikshit, A. and Dixit, S.N. (1983). Fungitoxic properties of essential oil of *Mentha arvensis* var. *piperascens*. *Perfumer and Flavorist* 8: 55-58.

.Singh, A.K., Dikshit, A. and Dixit, S.N. (1983). Antifungal studies of *Peperomia pellucida*. *Beitr. Biol. Pflanzen* 58: 357-368.

Singh, A.K., Tripathi, S.C. and Dixit, S.N. (1978). Fungitoxicity of volatile fraction of some angiospermic plants. *Proc. Symp. Environ. Sci. and Human welfare*, Ujjain, India, 13.

Singh, G., Singh, B.S. and Kumar, B.V.A. (1978). Antimicrobial activity of the essential oil of four medicinal plants against some keratinophilic fungi. *Indian Drugs* 16: 44-45.

Singh, G., Upadhyay, R.K. (1991). Fungitoxic activity of cumaldehyde main constituents of the *Cuminum cyminum* oil. *Fitoterapia* 62(1): 87.

Singh, G., Upadhyaya, R.K., Narayanan, C.S., Padmkumar, K.P. and Rao, G.P. (1993). Chemical and fungitoxic investigations on the essential oil of *Citrus sinensis* (L) Pers, *Zeitschriftfuer Pflanzenkrankheiten-und-Pflanzenschutz*, Germany. 100 (1): 69-74.

Singh, H.B. and Singh, U.P. (1981). Effect of volatile of some plant extracts and their oil on conidia of *Erysiphae polygoni*. DC. *Aust. Plant. Path.*10 (4): 66-67.

Singh, J. and Tripathi, N.N. (1999). Inhibition of storage fungi of blackgram. (*Vigna mungo* L.) by some essential oils. *Flavour Fragr. J.* 14 (1-4).

Singh, K.V. and Deshmukhi, S.K. (1984). Volatile constituents from members of Liliaceae and spore germination of *Microsporum gypseum* Comples. *Fitoterpia* 55(5): 297.

Singh, L. (1976). Effect of some oil on *in vitro* growth of three phytopathogenic fungi. *J. Indian Bot. Soc.* 55: 186-189.

Singh, R.B., Singh, J. and Singh, H.B. (1999). Effect of ginger (*Zingiber officinale* Roscoe) extracts on growth and sclerotium (Lib) de Bary. *Indian Perfumer* 43(2): 60-63.

Singh, S.K., Dubey, N.K., Tripathi, S.C. and Singh, S.K. (1984). Fungitoxicity of some essential oils against *Aspergillus flavus*. *Indian Perfumer* 28(3-4): 164-166.

Singh, S.P., Dubey, P. and Tripathi, S.C. (1986). Antifungal properties of essential oil of *Ageratum conyzoides L. Nat. Acad. Sci. Lett.* 9(4): 97-99.

Singh, S.P., Dubey, P. and Tripathi, S.C. (1986). Fungitoxic properties of the essential oil of *Trachyspermum ammi*. *Sprague–Mykosen* 29(1): 37-40.

Singh, Y., Tripathi, R.D., Tripathi, N.N. and Dixit, S.N. (1983). The isolation and properties of fungitoxic principle from *Zingiber officinale*. *India J. Pl. Pathol.*1: 89-96.

Skinner. F.A. (1955). Antibiotics, In: *Modern method of plant Analysis Vol. VII*, Peach. K. and Tracey, M.N. (eds.) Springer I Verlag, Berlin, 626-726.

Slavenas, J. (1963). Physiological role of essential oils. *Obschch Zakonomer Rosta. Razu. Rast. Dokl. Nauk. Konz. Vilnyus* 313-338.

Slavenas. J. and Razinskaite. D. (1962). Some studies of phytoncidal substances and *Juniperus* oil from common Juniper Leit. *T.S.R. Moks. Akad. Darbai, Er. C.* 4(2): 63-74.

Srivastava, L. (1990). Fungitoxic evaluation of volatile constituents of some higher plants against fungi infesting dry fruits during storage, *Ph.D. Thesis*, Gorakhpur University, Gorakhpur.

Srivastava, R. (1981). Evaluation of higher plants for their fungitoxicity against *Rhizoctonia solani* Kuhn. *Ph.D. Thesis*, University of Gorakhpur, Gorakhpur, India.

Stinson, E.E., Billis, D.D., Osman, S.F., Sciliano, J., Ceponis, M.J., Heisler, E.G. (1980). Mycotoxin production by *Alternaria* species grown on apples, tomatoes and blueberries. *Agric. Food Chem.* 28: 960-963.

Suri, R.K. and Thind, T.S. (1979). *In-vitro* antifungal efficacy of some essential oils. *East Pharm.* 22(257): 109-110.

Suri, R.K., Nigam, S.S. and Thind, T.A. (1979). *In-vitro* antimicrobial efficacy of essential oil of *Eucalyptus citriodora. Indian Drugs* 14(3): 35-37.

Swaminathan, M.S. (1978). Inaugural Address *First Bot. Conference,* Meerut, India: 1-31.

Thind, T.S. and Dahiya, M.S. (1976). Inhibitory effect of essential oils of some medicinal plants against soil inhabiting dermatophytes. *Indian Drugs* 14: 17-20.

Thind, T.S. and Dahiya, M.S. (1977). Inhibitory effect of essential oil of four medicinal plants against some keratinophillic fungi. *East Pharm* 20: 147-148.

Thakur, R.N., Singh, P. and Khosla, M.K. (1989). *In-vitro* studies on antifungal activity of some aromatic oils. *Indian Perfumer* 33(4): 257- 260.

Tiwari, R., Dixit, K. and Dixit, S.N. (1994). Studies on fungitoxic properties of essential oil of *Cinnamomum zeylanicum* Breyn. *Indian Perfumer* 38 (3): 98-104.

Tiwari, R., Dixit, V. and Behal, K. (1998). Insect repellent activity of the essential oil of *Cinnamomum zeylanicum* Breyn. bark. *Indian Perfumer* 42(2): 68-72.

Tiwari, R., Mishra, D.N. and Upadhyay, P.S. (1988). Efficacy of some plant volatiles for the control of black mold of onion caused by *Aspergillus niger* Van Tiegh. during storage. *L. Nat. Acad. Sci.* 2(11): 345-347.

Tripathi, N.N. (1980). Fungitoxicity in some higher plants. *Ph. D. Thesis,* University of Gorakhpur, Gorakhpur, India.

Tripathi, N.N., Dubey, N.K., Dikshit, A., Tripathi, R.D. and Dixit, S.N. (1983). Fungitoxic properties of *Alpinia galanga* oil. *Trop. Sci. Res.* 1(1): 49-52.

Tripathi, R.D., Banerji, R., Sharma, M.L., Balasubramanyam, V.R. and Nigam, S.K. (1985). Toxicity of essential oil from the new strain of *Ocimum gratissimum* (Clocimum) against Betilvine Pathogenic Fungi. *Agric. Biol. Chem.* 49(8): 2277-2282.

Tripathi, S.C., Singh, S.P. and Dube, S. (1986). Studies on antifungal properties of *Trachyspermum ammi* (L.) Sprague. *Phytopath. Z.* 116: 113-120.

Tripathi, N.N. and Kumar, N. (2007). *Putranjiva roxburghii* oil-A potential herbal preservative for peanuts during storage. *Journal of Stored Product Research.* 43(4) p 435-442.

Tyler, P.S. and Boxall, R.A. (1984). Postharvest loss reduction Programmes: A decade of activities:- What consequences? *Tropical Stored Products Information* 23: 13-28.

Uchida, S. (1928). In essential oils of leaves of "Hibs" wood and its relation to the resistance of the wood to destruction by fungi. *J. Soci. Chem.* India, 31: 501-508.

Uppalpati, L. and Rao, J.T. (1979). Antimicrobial properties of the essential oil of *Vitex negundo* Linn. *Indian Drugs Pharm.* 14: 31-33.

Vichkanova, S.A., Adeqina, V.V. and Izosimova, S.B. (1972). Essential oil as a source of new antifungal preparations. *Fintontsidy mater–Soveshch,* Naukova Dumka, Kiev.123-126.

Yadava, R.N. (1986). *In-vitro* antimicrobial studies on the essential oil of *Lepidagathis hylina* Nees. *Indian Perfumer* 30(4): 481-482.

Yadava, R.N., Saxena, V.K. and Nigam, S.S. (1978). Antimicrobial activity of the essential oil of *Caesalpinia Sappan* L. *Indian Perfumer* 22: 73- 75.

Zohri, A.A., Abdel-Gawad, K.M. (1993). Survey of mycoflora and mycotoxins of some dry fruit in Egypt. *Basic Microbiol.* 33: 279-288.

Zutshi, S.K. and Mehta, S.C. (1977). Screening of some essential oils for antifungal properties. *Indian Drugs and Pharma.* 12: 13-14.

Zutshi, S.K., Joshi, S.K. and Bokadia, M.M. (1975). Antimicrobial activity of some Indian essential oils. *Indian J. Pharm.* 37: 129–130.

Microbial Biotechnology and Ecology (2011) *Pages* **439–451**
Editors: **Deepak Vyas, G.S. Paliwal, P.K.Khare and R.K Gupta**
Published by: **DAYA PUBLISHING HOUSE, NEW DELHI**

Chapter 24

Carbonic Anhydrase: An Ancient Enzyme with Multidimensional Roles

☆ *Anjana Sharma, Rajesh Pujari and Abhishek Bhattacharya*

Bacteriology Lab, Department of Biological Sciences, Rani Durgavati University, Jabalpur, M.P.

ABSTRACT

Carbonic anhydrases catalyze the reversible hydration of CO_2 and are ubiquitous in highly evolved eukaryotes and in single cell prokaryotes. The existence of the enzyme in five classes and its identification in thermophillic species from the Archaea domain indicates the systematic evolution along with metabolic and phylogenetic diversity of the enzyme. The enzyme plays an important role in the maintaining the ionic concentration and pH; interconversion of CO_2 and HCO_3^-; homeostasis; photosynthesis and inorganic ion transport. Recently new roles related to gluconeogenesis and pathogenesis has also been assigned to this multifaceted enzyme. Thus in nut shell the enzyme has been envisaged to provide an enroute into the Biochemistry, Physiology and Pathology of individual organism.

Introduction

Carbonic anhydrases (CA) (Carbonate hydrolases; E. C. 4.2.1.1.) has long been one of the most well characterized zinc metalloenzymes that fall into three distinct evolutionary and structural classes *i.e.* α- (alpha), β- (beta) and γ- (gamma) however, recently another two classes has been investigated *i.e.* δ- (delta) and ε- (epsilon) class CA. Although α-class enzymes, particularly mammalian CA II have been the subject of extensive structural studies but for the β-class, consisting of a wide variety of prokaryotic and plant chloroplast CAs, the structural data is quite limited.

Physiologists predicted the existence of CA almost 70 years ago that plays a lead role in catalyzing as well as overcoming the slow-rate conversion between carbon dioxide and bicarbonate. Thereafter, for the first time carbonic anhydrase was purified from erythrocytes (Meldrum and Roughton, 1933; Stadie and O' Brien, 1933) and was further identified, belonging to α-class containing a zinc ion (Keilin and Mann, 1940). The first report of bacterial CA was presented in 1963 by Veitch and Blankenship. A large number of α-CA isoenzymes, primarily from mammalian species have subsequently been identified and characterized revealing different aspects of their molecular structures (Supuran, 2008). The reaction catalyzed, $CO_2 + H_2O \longleftrightarrow HCO_3^- + H^+$, the interconversion of carbon dioxide and bicarbonate, is of fundamental physiological and biochemical importance, relating in various *in vivo* context to pH or ionic balance and metabolic flux (Henry, 1996). Besides these, innumerable number of function makes the enzyme highly significant either in the perspectives of physiology or pathology and therapeutic concerns. Therefore, present review discusses and unravels the enigmatic roles of CA in lieu of different type of functions and existence in different forms in various living sources.

Carbonic Anhydrase: Ubiquitous Nature of the Enzyme

The ubiquitously expressed zinc metalloenzymes, present in prokaryotes and eukaryotes being encoded by three distinct, evolutionarily unrelated gene families: α-CAs (prevalent in vertebrates, bacteria, algae and cytoplasm of green plants), the β-CAs (predominant in bacteria, algae and chloroplasts of both mono as well as dicotyledons), the γ-CAs (mainly in archaea and some bacteria) and another class, the δ-CA was found to be present in marine diatoms, respectively (Supuran, 2008). The comparison of sequences and crystal structures of mammalian and plant CAs demonstrates that they evolved independently and therefore designated as α- and β-class respectively however an additional independently evolved γ-class was reported by Alber and Ferry (1994) for which the phylogenetic analyses predict an ancient origin (Smith *et al.*, 1999).

The α-class is the best characterized class with 16 isozymes identified in mammals with different catalytic activity, subcellular localization and tissue distribution (Supuran, 2008) (Table 24.1). Several isozymes were implicated in various disease states for which treatment frequently involves the application of sulfonamides that inhibit CA activity (Supuran *et al.*, 2008). Prokaryotic α-class enzymes have been compared with other two classes.

Though less study has been made for β-class CA but the first crystal structure for any β-class CA was reported by Mitsuhashi *et al.* (2000). β-class CA, as presumed earlier belonged to higher plants only but were later isolated from a variety of algae (Yagawa *et al.*, 1987; Hiltonen *et al.*, 1995; Eriksson *et al.*, 1996) and found to be widely distributed in bacteria and archaea domain (Smith *et al.*, 1999). The characterization of enzymes from the β-class revealed sharp differences from the other three classes. The α–class and γ–class enzymes are strictly monomers and trimers, respectively however, members of the β-class are dimers, tetramers, hexamers and octamers which suggested a dimer as the basic building block (Kimber and Pai, 2000). Phylogenetic analyses indicated that the β-class is more diverse than the other two classes (Smith *et al.*, 1999). Sequence alignment indicated that only 5 residues (three zinc ligand, one aspartate and one arginine residue) were completely conserved (Smith *et al.*, 2000). Plant sequences were found to form two monophyletic clades representing dicot and monocot plants (Smith *et al.*, 1999). Carbonic anhydrase obtained from *Pisum sativum* was found to be a dimer of homodimers whereas CA obtained from algae was a homodimer in which the monomer was found to be composed of two internally repeated structures each with an active site (Tripp *et al.*, 2001).

Gamma-class CA was thought to have evolved between 3.0 and 4.5 billion years ago (Smith *et al.*, 1999) and therefore preceded evolution of the α-class at 200-300 million years ago (Hewett-Emmett and Tashian, 1996; Jiang and Gupta, 1999). The only γ-class enzyme characterized was "Cam" from the archaeon *Methanosarcina thermophila* (Alber and Ferry, 1994). Cam was found to be a homodimer that adopts a left-handed parallel β-helical fold (Kisker *et al.*, 1996) and heterologously produced in *E. coli* at high levels to yield a zinc enzyme (Alber and Ferry, 1996). However, iron and cobalt-substituted forms exhibit greater CO_2 hydration rates than the zinc enzyme (Alber *et al.*, 1999) and it was thought that possibly Cam functions in *M. thermophila* using a different transition metal than zinc (Tripp *et al.*, 2001).

Table 24.1: Distribution of Various Human CA Isoforms with Different Catalytic Activities

Isozyme	*Kcal/Km (M^{-1} s^{-1})*	*Subcellular Localization*	*Tissue/Organ localization*
hCAI	5.0×10^7	Cytosol	Erythrocytes, GI tract
hCAII	1.5×10^8	Cytosol	Erythrocytes, GI tract, eye, bone osteoclasts, kidney, lung, testis, brain
hCAIII	3.0×10^5	Cytosol	Skeletal muscle, adipocytes
hCAIV	5.1×10^7	Membrane-bound	Kidney, lung, pancreas, brain, capillaries, colon, heart muscle
hCAVA	2.9×10^7	Mitochondria	Liver
hCAVB	9.8×10^7	Mitochondria	Heart and skeletal muscle, pancreas, kidney, spinal cord, GI tract
hCAVI	4.9×10^7	Secreted (saliva, milk)	Salivary and mammary glands
hCAVII	8.3×10^7	Cytosol	CNS
hCAVIII	–	Cytosol	CNS
hCAIX	5.5×10^7	Transmembrane	Tumours, GI mucosa
hCAX	–	Cytosol	CNS
hCAXI	–	Cytosol	CNS
hCAXII	3.5×10^7	Transmembrane	Renal, intestinal, reproductive epithelia, eye, tumours
hCAXIII	4.3×10^7	Cytosol	Kidney, brain, lung, gut, reproductive tract
hCAXIV	3.9×10^7	Transmembrane	Kidney, brain, liver
hCAXV	9.6×10^7	Membrane-bound	Kidney

Francois Morel and co-workers have reported the purification of a 27-kDa monomeric CA called TWCA1, from marine diatom *Thalassiosira weissflogii* (Roberts *et al.*, 1997) that was classified under a separate class called δ-class. The catalytic zinc ion was shown by X-ray absorption spectroscopy to be coordinated by three histidines and a water molecule, similar to the active sites of α-class and γ-class CAs and the near-edge spectra revealed that the active site geometry is similar to that of α-class enzymes (Cox *et al.*, 2000). The existence of a water molecule as a fourth ligand suggested that δ-class CA also follows the zinc hydroxide mechanism of the other three classes (Tripp *et al.*, 2001).

Although TWCA1 has biochemical properties similar to the three known classes of CA however the deduced sequence of the gene encoding TWCA1 revealed no significant identity to the three classes. Therefore TWCA1 is the prototype for a fourth class of CA that has been designated as the δ–class (Tripp *et al.*, 2001). During the conditions in which the level of TWCA1 was found low in *T. weissflogii*, in that case a 43-kDa cadmium-specific CA has been expressed (Lane and Morel, 2000). The sequence of the gene encoding δ–CA has not yet been reported (Tripp *et al.*, 2001).

Carbonic anydrases are widespread enzymes, present in mammals in 16 different isoforms (Supuran, 2008). Some of these ioszymes are cytosolic (CA I, II, III, VII and XIII) others are membrane-bound (CA IV, IX, XII, and XIV), CA V is mitochondrial by origin and CA VI is secreted in the saliva and milk. Three cytosolic a catalytic forms of CA *i.e.* CA related proteins (CARP) are also known (CARPVIII, CARPX and CARPXI). The catalytically isoforms, which play important physiological and pathological functions are strongly inhibited by aromatic and heterocyclic sulfonamides.

Under the name of CA, Meldrum and Roughton (1933) were the first to describe a new enzyme present in R.B.Cs. They purified the enzyme further and gave a detailed account of it's properties and the kinetics of the enzyme. Isoenzymes of CA later isolated not only from mammalian blood but also from other tissues of dog like kidney (Byvoet and Gotti, 1967), prostrate, uterus (McIntosh, 1969; 1970), stomach, colon, caecum (Carter and Parsons, 1970a; 1970b; 1971) and rumen epithelial tissue (Carter, 1971). Later different forms of the enzyme have been isolated from erythrocytes of man, green monkey, baboon, horse, pig, guinea pig, rat, blue-white dolphin and yellowfin tuna (Lindskog *et al.*, 1971; Anderson *et al.*, 1972; Laurentt-Tabusse *et al.*, 1972). CA has also been isolated from erythrocytes of sheep (Tanis *et al.*, 1970). The major form of CA from dog erythrocytes has been isolated and characterized by Soliman and Klub (1974). CA XII is a membrane glycoprotein with an active extracellular CA domain that overexpressed on cell surfaces of certain renal cell cancers (Ulmasov *et al.*, 2000). The last mammalian α-CA isoform (CAXV) has now been isolated, purified and characterized along with representatives of the α-, β-, γ-, and δ-classes from various pathogenic organisms from across the phylogeny.

During the past few decades, considerable efforts have been made on investigations of the properties and function of the enzyme carbonic anhydrase from microbial sources. Progress reviewed by Lindskog *et al.* (1971) and Maren (1967) paved the way to achieve knowledge of the kinetic behaviour of the enzyme as well as structural features. The presence of CA activity in bacteria was firstly reported by Veitch and Blankenship (1963) who measured significant activities in strains of several species of the genus *Neisseria* and in a strain of *Streptococcus salivaria*. The highest activity found was in *Nesseria sicca*, strain 6021. Gill *et al.* (1984) observed an intracellular CA but not periplasmic in *Rhodospirillum rubrum* however, the biological function of this enzyme has not yet been investigated. Guilloton *et al.* (1992) established the existence of CA in *E. coli* and identified that the enzyme is being encoded by *cyn operon*. Later Alber and Ferry (1994), purified and characterized CA from *Methanosarcina thermophila* grown in acetate and was the first report on CA from the archaeal domain. The enzyme represented a prototype of the γ-class of CAs. Methanogenic anaerobes have been the largest group within archaea, a domain which has been phylogenetically closer to the eukarya than to the bacteria (Woese *et al.*, 1990).

Later Braus-Stromeyer *et al.* (1997) detected the prevalence of CA activity in *Acetobacterium woodii, Acetohalobium arabaticum, Clostridium formicoaceticum, Sporomusa silvacetica, Moorella thermoautotrophica, Moorella thermoacetica subsp. "pratumsolum", Sporomusatermitida* and *Thermoanaerobacter kivui. A. woodii* was found to exhibit highest CA specific activity (approximately 14 U mg of protein^{-1}) of all the microbes represented till date. In 1999, Smith and Ferry reported a plant type β-class CA in the thermophilic methanoarchaeon *Methanobacterium thermoautotrophicum* reflecting no esterase activity however the enzyme had a KCal value of 1.7×10^4 S^{-1}. The results revealed that the β-class CA extends not only into the archaea domain but also into the thermophilic prokaryotes. *Rhodopseudomonas palustris*, a purple non-sulfur bacterium behaved surprisingly as it was observed that the organism exhibited CA activity when grown anaerobically but not aerobically (Puskas *et al.*, 2000).

Enigmatic Role of Carbonic Anhydrase

CAs were shown to be present in a multitude of prokaryotes, where it plays many important functions such as respiration, transport of CO_2 and photosynthesis (Supuran, 2008). Therefore, CA is widely distributed in nature and participates in various functions which includes facilitated diffusion of CO_2, interconversion of CO_2 and HCO_3^- during photosynthesis and biosynthetic processes, pH homeostasis and ion transport (Tashian, 1990; Coleman, 1991). Methanogenic anaerobes exhibit CA activity whereby they obtain energy for growth by producing methane from the reduction of CO_2 with H_2 or by the conversion of methyl groups of acetate, methanol or methanolamines to methane (Boone *et al.*, 1993) and also has been proposed that CA is required for an CH_3CO^{2-}/HCO_3^- antiport system.

In addition to the physiological reactions CAs catalyze a variety of other reactions (Figure 24.1) such as: hydration of cyanate to carbamic acid or of cyanamide to urea; aldehyde hydration to gem-diols; hydrolysis of carboxylic or sulfonic esters as well as other hydrolytic processes (Brig *et al.*, 1999; Guerri *et al.*, 2000; Supuran and Scozzafava, 2000; Supuran *et al.*, 2003).

The Carbonic anhydrase role was also found to be involved in many other biochemical and pathological processes which includes electrolyte secretion in various tissues and organs, biosynthetic reactions such as gluconeogenesis, lipogenesis and ureagenesis, bone resorption, calcification and tumorigenicity (Supuran and Scozzafava, 2000; 2002). Being involved in pathological basis for a number of diseases, the enzyme has been an important therapeutic targets with the potential to be inhibited to treat a range of disorders including oedema, glaucoma, obesity, cancer, epilepsy and osteoporosis (Supuran, 2008). There are atleast 25 clinically used drugs have been reported to possess significant CA inhibitory properties to treat several diseases *viz.* acetazolamide, methazolamide,

$$O{=}C{=}O + H_2O \longleftrightarrow HCO_3^- + H^+$$

$$O{=}C{=}NH + H_2O \longleftrightarrow H_2NCOOH$$

$$HN{=}C{=}NH + H_2O \longleftrightarrow H_2NCONH_2$$

$$RCHO + H_2O \longleftrightarrow RCH(OH_2)^{\cdot}$$

$$RCOOAr + H_2O \longleftrightarrow RCOOH + ArOH$$

$$RSO_3Ar + H_2O \longleftrightarrow RSO_3H + ArOH$$

$$ArF + H_2O \longleftrightarrow HF + ArOH \quad (Ar = 2,4\text{-dinitrophenyl})$$

$$PhCH_2OCOCl + H_2O \longleftrightarrow PhCH_2OH + CO_2 + HCl$$

$$RSO_2Cl + H_2O \longleftrightarrow RSO_3H + HCl \quad (R = Me; Ph)$$

Figure 24.1: Reactions Catalyzed by α-Class Carbonic Anhydrases

ethoxzolamide, sulthiame, dichlorophenamide, dorzolamide, brinzolamide, indisulam, topiramate, zonisamide, sulpiride, COUMATE, EMATE, celecoxib, valdecoxib and saccharin (Table 24.2).

Table 24.2: Biomedical Applications of Inhibitors of Carbonic Anhydrase Isoforms

Sl.No.	Compound	CA Isoforms (as therapeutic targets)	Application
1.	Acetazolamide	CAII, VA,VB, VII, VI, IX, XII, XIII and XIV	Nor-mercurial diuretic and glaucoma treatment
2.	Methazolamide	-do-	Qedema treatment and glaucoma treatment
3.	Ethoxzolamide	-do-	-do-
4.	Sulthiame	-do-	Glaucoma treatment
5.	Dichlorophenamide	-do-	Qedema treatment
6.	Dorzolamide	-do-	Glaucoma treatment
7.	Brinzolamide	-do-	-do-
8.	Indisulam	-do-	Treatment of cancer (antitumoric)
9.	Topiramate	CAII, VII, VA, VB, VI, XII and XIII	Antiepileptic
10.	Zonisamide[a]	CAII, VA, VB, IX	Antiepileptic, treatment for neuropathic pain, bipolar disorder, migraine, obesity, eating disorders, parkinsons disease
11.	Sulpiride[a]	CAVI, XII	Antipsychotic
12.	COUMATE	CAIX and XII*	Anticancer
13.	EMATE	CAIX and XII*	-do-
14.	Celecoxib	CAII	Antinflammatory
15.	Valdecoxib	CAXII and CAII	
16.	Saccharin	CA VII[a]	
17.	Fluorescent sulphonamide derivative	CAIX	Anticancer
18.	Pyridinium derivative	-do-	-do-
19.	Chlorothiazide and Hydrochlorothiazide	CAII, IV, XII and XIV	Diuretics
20.	Quinethazone	CAII, CAIV, XII and CAXIV	-do-
21.	Metolazone	-do-	-do-
22.	Chlorthalidone	-do-	-do-
23.	Indapamide	-do-	-do-
24.	Furosemide	-do-	-do-
25.	Bumetamide	-do-	-do-

a: Low affinity to other isoforms.

Furthermore, evidences have been laid down that various forms of organisms such as protozoa, fungi and bacteria were found to exhibit virulence and growth as a parasite due to carbonic anhydrase activity and therefore the aspect has emerged as a new and recent directive for research. *Plasmodium falciparum* (Krungkrai *et al.,* 2005), *Helicobacter pylori* (Nishimori *et al.,* 2006; 2007), *Mycobacterium tuberculosis* (Suarez Covarrubias *et al.,* 2005), *Candida albicans* (Klengel, 2005) and *Cryptococcus neoformans* (Morgensen *et al.,* 2006; Bahn *et al.,* 2005) are some of the organisms that have been well characterized in terms of their pathogenicity due to CA activity. Later it has been proved that the class of CA prevalent in the organism is critical for the growth or virulence of these pathogens and infact for their different degrees of resistance to the currently available drugs. An α-class CA (pfCA) is prevalent in *P. falciparum*. Possess catalytic properties that are distinct to that of the human host enzymes CA I and CA II and plays a major role in causing malaria. On the other hand *H. pylori*, organism responsible for different gastric mucosal lesions, including gastritis, stomach ulcers and gastric cancer (Nishimori *et al.,* 2006; 2007) exhibits both α- and β-class CAs called hpαCA and hpβCA, respectively. Both the classes of CAs are essential for the survival of the pathogen and plays crucial role in acid acclimatization of the pathogen within the human stomach (Supuran, 2008). Concrete evidence was put forth when it was revealed that inhibition of the two enzymes leads to the death of the bacteria and a possible eradication of *H. pylori* from the stomach thereby making the enzyme, a druggable target (Nishimori *et al.,* 2006; 2007). Another remarkable aspect was revealed by Suarez Covarrubias *et al.* (2005) when it was established that *M. tuberculosis Rv3588c* gene required for *in vivo* growth of the pathogen and the gene encodes for a β-class CA with an activity that is highly dependent on pH, being active at pH 8.4 but not at pH 7.5. As identified by Muhlschlegel's and Heitman's groups (2005), the signaling role of CAs in *C. albicans* and *C. neoformans* was evident proving that CA is important for the virulence of these two fungal pathogens.

Valdivia and Falkow (1999), investigated the inevitable role of *mig-5*, a gene encoding a putative β-class CA in facultative intracellular pathogen *Salmonella typhimurium* and was found that the gene get induced 24-fold upon infection in host macrophages and it was necessary for bacterial survival within the host. Sharma *et al.* (2008) investigated the prevalence of Carbonic anhydrase in diversified genera that includes *Enterobacter, Aeromonas, Shigella* and *Klebsiella* and observed that all the organisms exhibited CA activity irrespective of their magnitude of production, low or high. Noticeably, it was observed that high level of CA activity was more predominant in the organisms, those which were urease positive and are opportunistic pathogen thereby indicating that CA might play a role in their pathogenesis or microbial virulence, results in accordance to the earlier reports of Smith *et al.* (1999).

Multidimensional Properties of the Enzyme

Enzymes, the regulatory bodies within an organism has been represented by their known specificity towards substrate and their typified allotted function within a biological system and in nature all enzyme do follow the natural rule however, Carbonic anhydrase rules as an exception that deviates from the nature's rule in the sense that it has multiple functions to carry out and that which are not at all correlated with each other and the enzyme is meant for several diverse metabolic pathways as mentioned earlier and hence known for its multidimensional properties. It performs different functions under different physiological conditions. The same enigmatic behaviour is also evident from the reports of Liljas and Laurberg (2000). The enzyme belongs to different classes that evolved independently and have no sequence homology. The amino acid sequences are strikingly different however inspite of the differences in sequence, all forms of CA have an essential zinc ion in the active site.

The active sites show essential features of remarkable similarity though structurally there exists a wide variations. All of the CAs are completely different from one another at the level of their tertiary and quarternary structures (Figure 24.2). Gamma-class CA exhibits a primitive β-helix structure where short repeated amino acid sequence produces the main part of the protein fold whereas the α- and β-CAs have folds that are constructed by different arrangements of α-helices and β-strands.

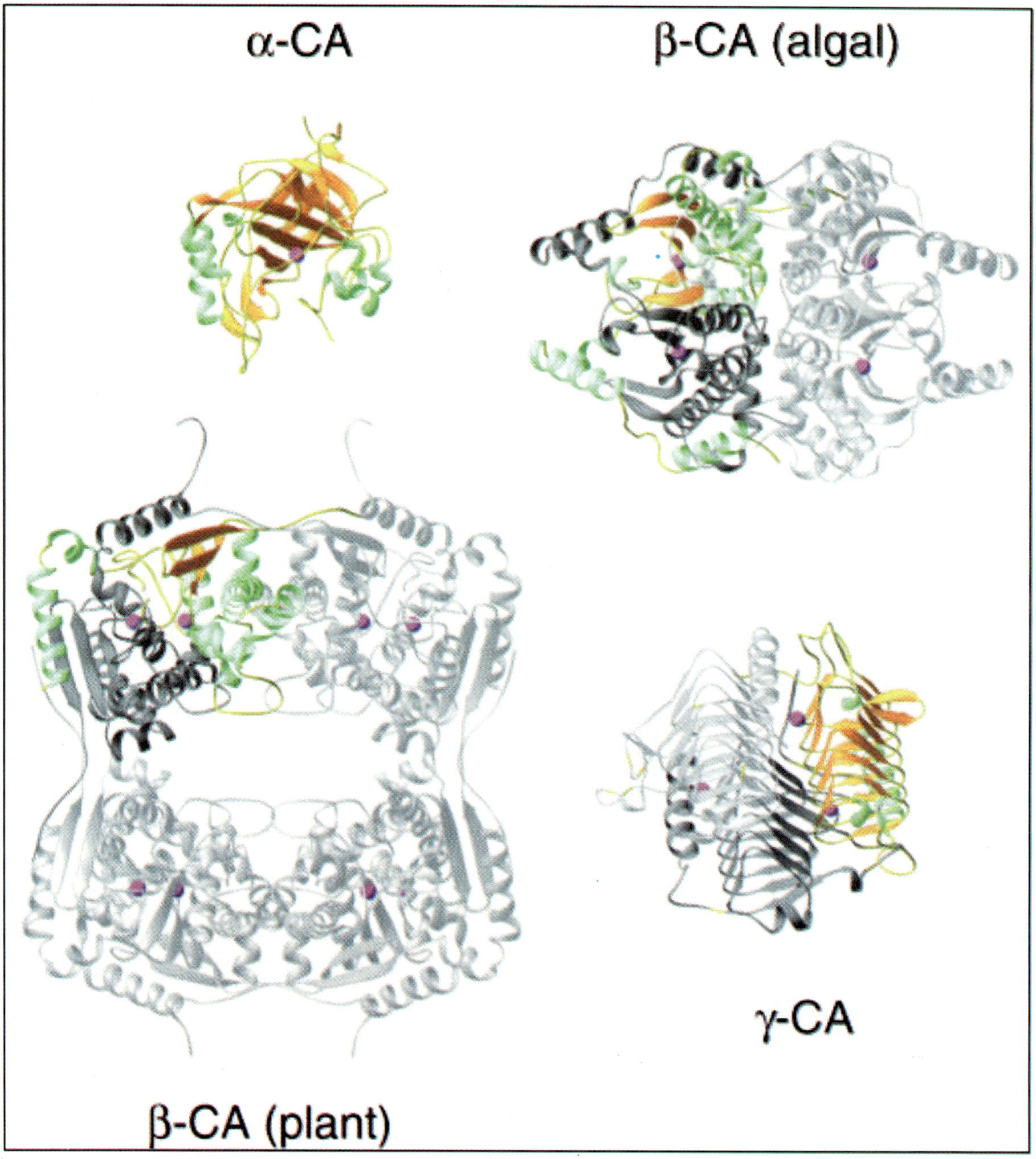

Figure 24.2: The Tertiary Structures of the Three Forms of Carbonic Anhydrase

Differences in quarternary structure exist even within the class of β-CA (Figure-2) where algal β-CA is a dimer with internal repeat and on the same hand, plant β-CA is an octamer which does not based on 4-fold but on repeated 2-fold symmetry.

Therefore, within the same class remarkable dissimilarity exist though the enzyme is same and performing the same function. The active site retains a zinc ion in common however, the zinc ion ligands are completely different and hence the essential residues of the active sites differ as well (Figure 24.3). More striking difference is that β-CAs from plant has an acetate ion bound to the zinc while the acetate ion is being replaced by aspartate in the algal β-CA.

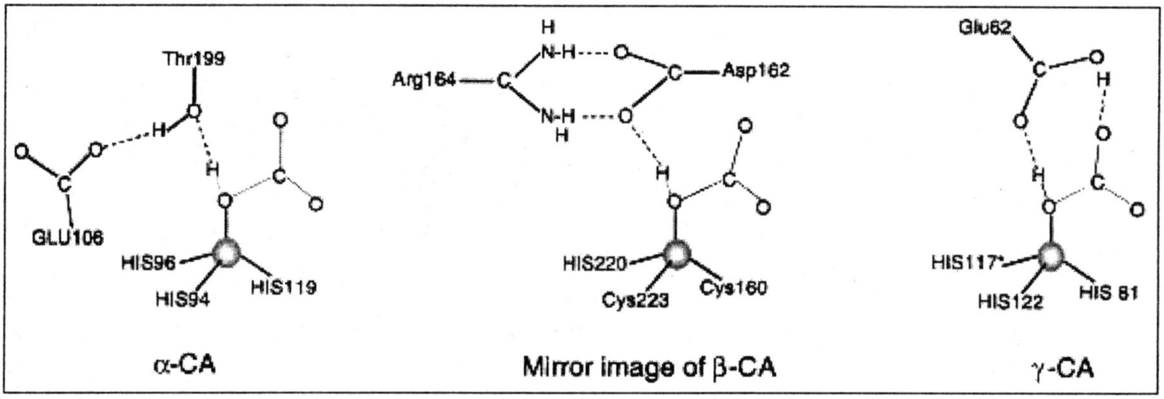

Figure 24.3: Schematic Representations of the Active Sites of the Three Forms of Carbonic Anhydrase

Inspite of the differences, these enzymes catalyzes reaction in common where the zinc ion activates a water molecule that reacts with CO_2 or destabilizes bicarbonate in the reverse reaction. Therefore, CA illustrates a detail enigmatic account that being the same enzyme it strongly holds a versatile nature.

Highlighting the genomics account and expression of the enzyme, Fukuzawa *et al.* (1990) investigated that *Chlamydomonas reinhardtii*, an alga encodes two different CAs within the same genome as gene *CAH1* and *CAH2* and are being expressed and regulated under different physiological conditions. The 5′ upstream gene *CAH1* encodes a major periplasmic CA and its level of expression is rapidly induced under low CO_2 level in the presence of light while the 3′ downstream gene *CAH2* encodes another periplasmic isozyme induced at high CO_2 level *i.e.* 1-5 per cent CO_2.

In regards to the cellular localization the enzyme exhibit a great diversity as the enzyme have different destination in different model systems such as either extracellular or intracellular. When intracellular it could be cytosolic, periplasmic, membrane bound and transmembranal and secreted outside a cell when extracellular by nature.

Carbonic anhydrase enzyme has been a strong example and representative of convergent evolution and hence though differ distinctly still has its unique features to correlate. It has been of significant interest to denote that the enzyme has evolved three times completely in an independent way (Liljas and Laurberg, 2000) thereby generating enzymes in several different isoforms that appears to be different superficially but are actually posssess great underlying similarities.

Ferry *et al.* (1999) detected the prevalence of β-class and γ-class carbonic anhydrases in methanoarachaea and established that carbonic anhydrases are widely distributed in prokaryotes and that the β-class and γ-class carbonic anhydrases are predominant. They also documented that these two classes of enzymes are of ancient origin. On the other hand presence of both β-class and γ-class carbonic anhydrases in thermophilic chemolithoautotrophs has been established suggesting that ancient carbon dioxide fixation pathways are dependent on carbonic anhydrase for efficient Carbondioxide fixation. However, Alber and Ferry (1996) purified the carbonic anhydrase enzyme from the archaeon methanosarcina thermophila which was noticeably found to be distinct from the α- and β- class, and is the prototype of a different class *i.e.* γ-class.

Ferry *et al.* (1999) found that β-class carbonic anhydrases occur not only in species of archaea domain, but also in thermophilic prokaryotes with an obligate chemilithoautotrophic mode of growth, and are among the species that represent some of the deepest branches of the universal tree of life. Hewwett-Emmett and Tashain (1996) proposed about the α-class carbonic anhydrase so that the evolution of this class has been from a common ancestral gene 0.5-0.6 billion years ago and the consistency has been followed with the evidence that few prokaryotic genomes encoding α-class carbonic anhydrases. However, Ferry *et al.* (1999) claimed that the γ-class carbonic anhydrases has evolved approximately 4.2 billion years ago, with the major cluster of bacterial and archaeal carbonic anhydrases thereby supporting that the carbonic anhydrases must be extremely ancient enzymes, existing before the divergence of archaea and bacteria domains.

Additional support for an ancient origin of carbonic anhydrase is the presence of the β-class and γ-classes in thermophiles representing deep branches of the universal tree of life that includes organisms like *C. hydrogenoformans, M. thermoautotrophicum, Methanococcus janaaschii, Sulfolobus solfataricus, P. furiosus, Pyrococcus horikoshii* etc. (Ferry *et al.*, 1999).

Evidence against prevalence of carbonic anhydrase was obtained from freshwater, marine, mesophilic, thermophilic, aerobic, anaerobic, pathogenic, symbiotic, methylotrophic, acetotrophic, methanogenic, acetogenic, autotrophic, heterotrophic and photosynthetic species (Ferry *et al.*, 1999). Prokaryotic carbonic anhydrases also have medical significance as are represented by several pathogenic species, suggesting the possibility that carbonic anhydrase exhibit some role and is required for microbial virulence. Evidence in support of this has been documented by Valdivia and Falkow (1997) for a gram-negative, facultative intracellular pathogen *S. typhimurium*. Expression of mig-5, a gene encoding a putative β-class carbonic anhydrase has been induced upto 24 fold upon infection in host macrophages and it is necessary for bacterial survival within the host.

Eriksson *et al.*, 1996 reported the prevalence of carbonic anhydrase in the green unicellular alga, *Chlamydomonas reinhardtii* when grown under Carbondioxide–limiting conditions and further the enzyme was characterized as 22-kDa polypeptide in *C. reinhardtii*, one more additional feature to the molecular nature of carbonic anhydrase with respect to the usual molecular weight of carbonic anhydrase that ranges between 29kDa–35kDa. Localization study of the enzyme in *C. reinhardtii* revealed that the enzyme was localized and induced within the mitochondria by a lowering of the partial pressure of Carbondioxide in the growth medium. The study brought-in a significant role of carbonic anhydrase in adaptation of the alga to low Carbondioxide conditions and thus provides the basis for experimental studies of mitochondrial carbonic anhydrases, which has been a rarely localized enzyme in living systems other than mammalian isoforms hCA VA and hCAVB.

Soltes-Rak *et al.*, 1997 identified and characterized Carbonic anhydrase from two disparate cyanobacteria, *Anabaena* sp. strain PCC 7120 and *Synechococcus* sp. strain PCC 7942 and was found out that the enzyme resembled to be of vertebrate-type. Carbonic anhydrase and that they revealed significant sequence similarity to human CA I and CA II as well as of *Chlamydomonas reinhardtii*. In addition, the similarity of the active-site structure of the cyanobacterial Carbonic anhydrase indicated that it could be an important tool for structural studies on amino acid substitutions resulting in the evolution of the highly efficient vertebrate CA II isoforms.

Marine diatom, *Phaeodactylum tricornutum* exhibited a single intracellular Carbonic anhydrase under different conditions *viz.* air-grown, and at reduced levels, in high Carbondioxide grown cells of the alga as per the study made by Satoh *et al.*, 2001. However, no external Carbonic anhydrase activity was detected irrespective of growth under cardondioxide conditions. Acetazolamide (2mM), a strong and specific inhibitor of Carbonic anhydrases had little effect on the affinity for dissolved inorganic carbon, thereby suggesting that internal Carbonic anhydrase was crucial for the operation of a carbon concentrating mechanism in *P. tricornutum*. Their study concluded that intracellular Carbonic anhydrase appeared to play a major role in acquiring and supplying substrate for photosynthesis in marine diatoms. Later *P. tricornutum* Carbonic anhydrase was described during the study to be the first β-class Carbonic anhydrase in a diatom species.

The other diatom Carbonic anhydrase isolated from *T. weissflogii* (TWCA1), showed no homology to the known algal Carbonic anhydrases and therefore a fundamental difference of carbon acquisition among algal phyla was proposed by Roberts *et al.* (1997).

Xia and Gao (2005) detected the prevalence of Carbonic anhydrase and it's activity from fresh green algae *Chlorella pyrenoidosa* chick and *C. reinhardtii* Dang cells when grown under different Carbondioxide concentrations within a range of 3-186 μmol/L. They observed the prevalence of both intracellualar and extracellular Carbonic anhydrase through activity and furthermore they concluded that the extracellular Carbonic anhydrase can be induced by a change in environmental Carbondioxide concentration and it was presumed to play an important role in photosynthesis by maintaining the inorganic carbon species equilibrium at the site of transport or diffusion into cells however with the increased Carbondioxide concentration in the medium both intracellular and extracellular Carbonic anhydrase activity exhibited a significant decline but to varying extents. And thus it implied that the susceptibility of Carbonic anhydrase activity to Carbondioxide enrichment was species dependent.

Intracellular Carbonic anhydrase in *Chlorella* species and *C. reinhardtii* was found to be of two types: (*i*) a soluble Carbonic anhydrase localized within the Cytosol as reported by Xia and Gao (2005) and (*ii*) an insoluble membrane-bound Carbonic anhydrase associated with the chloroplast however in contrast to this Sultemeyer (1998) earlier reported an intracellular Carbonic anhydrase in mitochondria during acclimation to low Carbondioxide concentrations in *C. reinhardtii*.

Concluding Remarks

Literature reviewed in this report is with an intention to putforth the uncommon role (with respect to nature, structure, function, expression level and evolution) of a common enzyme (being ubiquitous) in nature. A single biological derivative encompassing a number of functionality has reached a new height exceptionally in the field of biochemistry, physiology and pathology. The same molecule has gained a great foot hold now in terms of applied biotechnology either in regards to the approach of medical diagnosis on one hand and carbondioxide sequestration on the other hand whereby researchers foresee an ultimate alternative towards disease cure and eradicating the problem of global warming through exploitation of its CO_2 hydration activity, respectively. These facets let us assume the multidimensionality in total for the enzyme Carbonic anhydrase.

The enzyme seeks due importance and recognition being a novel entity and lays down path to unravel several other biological hurdles for researchers in the time to come. Further detail understanding is deemed necessary and the authors are paving the way for deep interaction of the molecule in biochemical and physiological perspectives to shape-up the dynamicity of the molecule.

References

Alber BE and Ferry JG. (1994). *Proc Natl Acad Sci USA* 91: 6909–6913.

Anderson B Nyman PO and Strid L. (1972). *Biochem Biophysics Res Commun* 48: 670-677.

Boone DR Whitman WB and Rouveiere P. (1993). in *Methanogenesis*, ed. Ferry JG (Chapman and Hall, London) pg 35-80.

Braus-Stromeyer S, Schnappauf G, Braus GH, Gobner A and Drake HL. (1997). *J Bacteriol* 179: 7197–7200.

Brigarti F Manganin S Scozzafava Vernaglione G and Supuran CT. (1999). *J Biol Inorg Chem* 4: 528-536.

Byvoet P and Gotti A. (1967). *Mol Pharm* 3: 142-152.

Carter MJ. (1971). *Biochem Biophysics Acta* 235: 222-236.

Carter MJ and Parsons DS. (1970 a). *Biochem biophysics Acta* 206: 190-192.

Carter MJ and Parsons DS. (1971). *J Physiol* 215: 71-94.

Carter MJ and Parsons DS. (1970 b). *J Biol Chem* 120: 797-808.

Coleman JR. (1991). 14: 861-867.

Cronk JD, Endrizzi JA, Cronk M and Zhang KYJ. (2001). *Protein Science* 10 : 911-922.

Errikson M Karlsson J Ramazanov Z Gardershrom P and Sameulsson G. (1996). *Proct Natl Acad Sci.* 93: 12031-12034.

Gill SR, Fedorka PJ, Tweten RK and Sleeper BP. (1984). *Arch Microbiol* 138: 113–118.

Henry RP. (1996). *Annu Rev Physiol* 58 : 523-538.

Hewett-Emmett D and Tashian RE. (1996). *J Bacteriol* 178: 3270-3274.

Hiltonen T Karlsson J Clarke AK and Sameulesson G. (1995). *Planta* 195 : 345-351.

Kimber MS and Pai EF. (2000). *EMBO* 19: 1407-1418.

Laurent-Tabuse G Giraud N and Limozin n. (1972). *Biochem Biophysics Res Commun* 54: 809-821.

Lindskog S Hederson LE Kaman KK Lijlas A Nyman PO and Strandberg B. (1971). *The Enzymes*, vol V (Boyer PO). 3rd edn pp. 587-665 Academic Press, New York.

Maren TH (1967). *Physiol Rev* 47: 595.

McIntosh JEA. (1969). *J Biol Chem* 114: 463-476.

McIntosh JEA. (1970). *J Biol Chem* 120: 299-310.

Meldrum NO and Roughton FJW. (1933). *J Physiol* 80 : 113-141.

Mitsuhashi S Mizurhima T yamashita E Yamato m and Trukihara T. (2000). *J Biol Chem* 275 : 5521-5526.

Pushkas LG, Inui M, Zahan K and Yukawa H. (2000). *Microbiology* 146: 2957–2966.

Roberts SB lane TW Morel FMM. (1997). *J Phycol* 33: 845-850.

Satoh D Hiraoka Y Colman B and Matsuda Y. (2001). *Plant Physiol* 126: 1459-1470.

Smith KS Zakubzick C, Whittam TS and Ferry JG. (1999). *Proc Natl Acad Sci USA* 96: 15184–15189.

Soltes-Rak E mulligan ME and Coleman JR. (1997). *J Bacteriol* 179: 769-774.

Stadie WC and O' Brien H. (1933). *J Biol Chem* 103 : 521-529.

Sultemeyer D. (1998). *Can J Biol* 76: 962-972.

Supuran CT. (2008). *Current Pharmaceutical Design* 14 : 603-614.

Supuran CT and Scozzofava A. (2000). *Exp Opin Ther* 10: 575-600.

Supuran CT Scozzofava A Vernaglione G and Supuran CT. (1999). *J Biol Inorg Chem* 4: 528-536.

Tashian RE. (1990). *Biossays* 10: 186-192.

Tripp BC Smith K and Ferry JG. (2001). 276: 46615-48618.

Ulmasov B Waheed A Shah GN Grubb HJ Sly WS and Silverman DN. (2000). 97: 14212-14217.

Valdivia RH and Falkow S. (1997). *Science* 277: 2007–2011.

Veitch FP and Blakenship LC. (1963). Carbonic anhydrase in bacteria. *Nature* 197: 76-77.

Woere CR Kandler O and Wheelis ML. (1990). Towards a natural system of organisms: proposals for the domains Archea, Bacteria and Eukarya. *Proc Natl Acad Sci USA* 87: 4576-4579.

Xia J and Gao K. (2005). Impacts of elevated CO_2 concentration on biochemical composition, carbonic anhydrase and nitrate reductase activity of freshwater green algae. *J Integrative Plant Biology* 47(6): 668-675.

Yagawa Y, Shiraiwa Y and Miyachi S. (1984). Carbonic anhydrase from the blue green algae (cyanobacterium). *Anabaena variabilis. Plant Cell Physiol* 25: 775–783.

Microbial Biotechnology and Ecology (2011) *Pages* **452–471**
Editors: **Deepak Vyas, G.S. Paliwal, P.K.Khare and R.K Gupta**
Published by: **DAYA PUBLISHING HOUSE, NEW DELHI**

Chapter 25

Pharmacological Potential of Mushrooms

☆ *Alka Karwa and Mahendra Rai*

Department of Biotechnology, S.G.B Amravati University, Amravati – 444 602

ABSTRACT

Biodiversity of mushrooms in India reveals a wide spectrum of culinary and pharmacologically important species that shows a wide spectrum of bioactivity. These are traditionally being utilized and modern techniques of molecular biology are continuously being worked for the same. Certain activities of mushrooms like immunomodulator, antitumor and anti HIV effects are being exploited increasingly. Mushrooms after all are now no longer mysterious: they may inspire excitement, fascination and even wonder. Nature's oldest art, the food of the Gods, the mushroom promise to be the next marvel after the green revolution.

Introduction

Mushrooms have always been mysterious. Their extraordinary biology, history, appearance, flavors, folklore, cuisine, gathering and more, adds to the mystery. Mushrooms are the invaluable link in the energy chain whose enzymes break the tough lignin of the wood and scavenge the organic material found in the forest floor. Their role is as nature's trash burners. There are thousands of regularly harvested edible fungi in the world, in addition to literally hundreds of thousands of other edible species. Some species are highly priced because they cannot be cultivated and are often harvested from natural settings.

There is an enormous development in the food industry and mushroom production in particular in the past few decades. Mushrooms comprise an unlimited source of neutraceutical and pharmacological products with immunomodulator potential. Medicinal mycology is largely practiced in China, Japan, and to some extent in India and some Eastern countries since time immemorial. The power of mushrooms is considered divine. The multifaceted uses of mushrooms as food, medicine, feed, dyes and anti aging device besides ecological balance are continuously being exploited in different parts of the globe (Rai and Tidke, 2005). Most of the identified mushrooms have important salutary effects on health and in treating diseases.

Increasingly, many varieties of mushroom are becoming popular, and these offer a wider range of tastes and textures. Many are cultivated versions of wild mushrooms, now called specialty mushrooms, while others are varieties that haven't yet been domesticated and so must be picked in the wild and, therefore, tend to be rather expensive. Many specialty mushrooms are available in dried form; some are canned or jarred, as well as fresh.

The first reference of occurrence of basidiomycetes was recorded by Montagne in 1842 (Sathe, 1979). Manjula, (1983) later on compiled 'A revised list of agaricoid and boletoid basidiomycetes from India and Nepal'. Biodiversity and occurrence of basidiomycetes in India was studied by various workers (Sathe and Deshpande, 1979, Natrajan and Raman, 1983, Purkayastha and Chandra, 1985, Verma *et al.*, 1987, Lakhanpal 1995, Pradeep *et al.*, 1998, Vrinda *et al.*, 1999 and Atri *et al.*, 2000).

Culinary Awareness and Acceptance

Mushroom is frequently mentioned as an alternative resource for food. The modern mushroom culture produces more proteins per unit area of land than by any other form of agriculture. Its use as food is probably as old as cultivation itself. But the nutritive value of mushroom was recognized only in recent times. Not usually thought of as a particularly good source of nutrients, mushrooms actually rank rather high in nutritive value. They contain a substantial amount of B vitamins, selenium, copper, and some other trace minerals. Mushrooms are low in calories (a cup of raw mushrooms has about 20 calories). Moreover, researchers have discovered antibacterial and other medicinal components, including anti-tumor compounds called triterpenoids in many mushrooms.

A large population of India is vegetarian. With human population billowing as rapidly as it is, conventional agriculture is unlikely to keep pace with food requirements. Alternative sources of energy are required if serious food crisis is to be averted. It is here that mushrooms provide a rich addition to the diet. Their nutritional value lies between that of meat and vegetables. As supplements to vegetarian diets, mushrooms have been called 'vegetable meat', both for their nutritive value and the culinary sophistication. But they do add variety, flavor, along with supplemental protein and vitamins to otherwise inadequate diets. The rarity of using mushrooms as food, the apparently mysterious source of their origin, their texture, and their flavors, combined perhaps with the culinary skill of expert chefs lends it the suspicion of being meat. And this is the reason mushrooms are not yet widely served as food in this country. But gradually the concept of mushroom dishes is definitely catching on. In a non-metropolis one may not get the variety and freshness in different types of mushrooms as in the major cities. But canned mushrooms are available. If freshness is not a criterion, in these days of communication revolution, accessibility is no longer a constraint. Once the mushrooms are used extensively, the demand would increase and the local cultivation and the supply would naturally keep pace.

The number of mushrooms on Earth is estimated at 140,000, yet perhaps only 10 per cent *i.e.* approximately 14,000 named species are known (Wasser and Didukh, 2005) and out of these known species of mushrooms, it was thought that 700 are edible and more than 200 may have medicinal value, although only a small percentage of these mushrooms are available to the consumer (Kendriok, 1985). Tribal members subsist to a great extent on nontraditional food sources, especially wild edible mushrooms, which grow abundantly during monsoon and post-monsoon seasons and are sometimes collected in large quantities for sale in urban markets (Rajak and Rai, 2005). In India more than 200 species of edible mushrooms have already been listed. The artificial cultivation of a few edible mushrooms ensures that mushrooms sold are truly edible. Therefore, there is a need to develop newer varieties of mushrooms by the scientific community and introduce to the population that is interested in consumption of mushrooms due to its culinary and medicinal properties. This could serve them with the improved mushrooms from the biologists' bench.

The health conscious people are turning towards mushroom to a greater extent. The low calorie content and high fiber of mushrooms, as well as lack of sugar and fat when fresh, is a boon for persons on a diet: the obese, the hypertensive and the diabetic. They fill the stomach and give a feeling of satiety; the high fiber slows down absorption of food from the intestines. One can say that mushrooms have now entering mainstream food in India.

Mushrooms are consumed as an energy rich food due to their peculiar property of containing rich proteins, vitamins, and minerals. They are superior to many vegetables and beans in their nutritive value. Mushrooms contain all essential amino acids. The digestibility coefficient of mushroom is also high, about 87 per cent. Their fat content is less than 0.3 per cent. It is less as compared to the fat of meat, egg, and milk. The fat in mushroom is rich in ergosterol, which is involved in the biosynthesis of Vitamin- D in human body. It also contains high proportion of thiamin, riboflavin, niacin, and ascorbic acid. They are also rich in minerals such as Ca, P, Fe, Na, and K. Selenium, an essential mineral, works closely with Vitamin E to produce antioxidants that neutralize the cell-damaging "free radicals" that can increase the risk of cancer and other diseases of aging. It plays an important role in the immune system, the thyroid system and the male reproductive system. *Potassium* is a mineral a human body just can't do without. It helps maintain normal heart rhythm, fluid balance, muscle and nerve function. The U.S. Food and Drug Administration recently gave a nod to its disease-fighting capability by stating: "Diets containing foods that are good sources of potassium and low in sodium may reduce the risk of high blood pressure and stroke." Some mushrooms supply more potassium than foods better known as sources for this mineral. *Copper* is another essential mineral and mushrooms are a good source. We're all aware of iron's role in making red blood cells and delivering oxygen to every part of the body. But the key is that iron can't do its job without copper.

As they have high nutritional value, FAO has recommended dried mushroom as a source of protein (3-4g/100g) in human diet. They are cheaper in price than any other nutritive diet in the market. Thus functional mushrooms deserve further serious investigations. The Indian organic foods industry is expected to witness a boom period. There is a huge export opportunity for the Indian organic farming industry both in niche markets, as well as mainstream markets world over.

Many mushrooms are not strictly pharmaceutical products (real medicines) but represent a novel class of dietary supplements (DSs), functional foods, nutriceutical, nutraceutical, phytochemical, mycochemical, biochemopreventives and designer foods (Chang and Buswell, 1999, 2003; Ziesel, 1999; Wasser *et al.*, 2000 a, b, 2001; Chang and Miles, 2004 and Wasser *et al.*, 2004). DSs are ingredients obtained from foods, plants and mushrooms that are taken without further modification separately from foods for their presumed health-enhancing benefits (McNamara, 1999). Mushrooms are a nutritionally functional food and a rich source of physiologically beneficial medicines.

Mushrooms are valuable health food–low in calories, high in vegetable proteins, iron, zinc, chitin, fiber, vitamins and minerals. Mushrooms also have a long history of use in Traditional Chinese Medicine (Wasser *et al.*, 2004). Their legendary effects on promoting good health and vitality and increasing body's adaptive abilities have been supported by recent studies (Wasser and Didukh, 2005). These studies suggest that Mushrooms are probiotic–they help our body strengthen itself and fight off illness by maintaining physiological homeostasis–restoring our bodies balance and natural resistance to disease. The compounds they contain have been classified as Host Defense Potentiators (HDP) which can have immune system enhancement properties (boost immune system). That is one of the reasons they are currently used as adjuncts to cancer treatments in Japan and China.

The increasing interest in the traditional remedies for various physiological disorders and the recognition of numerous biological activities of mushroom products have led to the coining of the term *mushroom* nutriceutical (Chang and Buswell, 1996, 1999, 2003; Chang, 1999; Wasser *et al.*, 2000 a, b, 2001; Chang and Miles, 2004), not to be confused with *nutraceuticals/functional foods and pharmaceuticals.*

A mushroom nutraceutical is a refined or partially refined extract or dried biomass from either the mycelium or the fruit body of a mushroom, is consumed in the form of capsules or tablets as a DSs and not as a food, and has potential therapeutic applications. Regular intake may enhance the immune response of the human body thereby increasing resistance to disease and in some cases causing regression of the disease state. Acting as immunopotentiators, mushroom preparations modify host biological responses (Hobbs, 1995; Wasser and Weis, 1999a, b; Wasser, 2002) and are also known as *Biological response modifiers* (BRMs).

Several types of culinary and therapeutic formulations of mushrooms available for consumption are like commercially cultivated fruit body powders, alcohol and hot water extracts and their mixtures, dried and pulverized preparations of mycelium, with mushroom primordia, or biomass extracted from the fermentation tank, or naturally grown dried mushroom fruit bodies in capsule or tablet form. Many mushrooms or mushroom preparations traditionally taken as treatments for specific conditions are now marketed for use as prophylactic agents. However, mushrooms are easily vegetatively propagated and thus keep to one clone (Rai *et al.*, 2005).

The safety advantages of using mushroom-based DSs as opposed to herbal preparations are (1) the overwhelming majority of mushrooms used for production of DSs are cultivated commercially and not gathered in the wild, this provides very good chances of proper identification, pre and unadulterated products. It is also genetically uniform. (2) Mushrooms are easily propagated vegetatively and thus are clones. The mycelium can be stored for a long time and the genetic and biochemical consistency may be checked after considerable time.

Pharmacological Potential

In the second half of the 20th century, mushroom-producing technologies have grown enormously. In 2002 the value of world mushroom production was estimated at about $20 billion USD, which is almost the same as the value of coffee production (Chang, 1999, 2001, Wasser *et al.*, 2000a, b, 2001., Wasser 2002, Chang and Buswell, 2003, Chang and Miles, 2004).

Reports indicate the potential of mushrooms as life-saving drugs antibiotics and biologically active substance. It has various industrial uses. It is tempting to say that many mushrooms known today to have medicinal benefits, which arose from observations of their uses in China and Japan, are objects of scientific studies that continue to determine the active components in such mushrooms (Doris *et al.*, 2005). Whereas the documentation of this knowledge is important for the understanding of the sociocultural practices of the Indians, it is equally good for science, because detailed scientific studies of mushroom uses and practices of these people can lead to the discovery of new drugs and nutraceuticals from Indian mushrooms.

A purified compound is very different from the whole mushroom, however, and the inevitable questions that follow are whether eating the whole mushroom has preventive or therapeutic value, and if so, how much mushroom should be consumed and in what form. For shiitake, researchers found that the powdered mushroom fruit bodies given to rats as 10-20 per cent of their diet inhibited transplanted tumors, and small studies have demonstrated lipid-lowering effects consumption of 9 g of dried mushrooms or 90 g of fresh mushrooms (Wasser, 2005).

The content and potency of bioactive ingredients may differ depending on how the mushroom is prepared and ingested. For example, the anticarcinogenic thioproline content varies from undetectable amounts in fresh shiitake, to 134 mg/100 g of dried shiitake, to 843 mg/100 g of boiled shiitake. As is the case for most plants and herbs, the specific strain, growing conditions, and other environmental factors will also significantly affect the taste, form, and substance of the mushroom, as well as affect the bioactive content (Nanba, 1967., Mori *et al.*, 1976., Susuki and Oshirna, 1976 and Royse and Bahler, 1996). Antitumor and immmunostimulating polysaccharides of higher basiomycetes are shown in Table 25.1.

Table 25.1: Antitumor and Immmunostimulating Polysaccharides of Higher Basidiomycetes

Polysaccharides	*Species*	*References*
Glucans		
α-(1→3)-glucan	*Armillariella tabescens*	Kiho *et al.*, 1992a
Linear α-(1→3)-glucan	*Amanita muscaria* *Agrocybe aegerita*	Kiho *et al.*, 1994 Yoshida *et al.*, 1996
α(1→4)-; β-(1→6)-glucan	*Agaricus blazei*	Fujimiya *et al.*, 1998b
α-(1→6)-; α(1→4)-glucan A	*Agaricus blazei*	Mizuno *et al.*, 1990a
β-(1→6)-glucan	*Lycophyllum decastes* *Armilalriella tabescens*	Ukawa *et al.*, 2000 Kiho *et al.*, 1992a
β-(1→6)-; β-(1→3)-glucan	*Agaricus blazei*	Mizuno *et al.*, 1990a
	Grifola frondosa	Nanba *et al.*, 1987
β-(1→6)-; α-(1→3)-glucan	*Agaricus blazei*	Mizuno *et al.*, 1990a
β-(1→3)-glucoronoglucan	*Ganoderma lucidum*	Saito *et al.*, 1989
Mannoxyloglucan	*Grifola frondosa*	Mizuno *et al.*, 1986
Galactoxyloglucan	*Hericium erinaceus*	Mizuno. 1999b
Xyloglucan	*Grifola frondosa*	Mizuno *et al.*, 1986
	Polyporus confluence	Mizuno *et al.*, 1992
	Pleurotus pulmonarius	Zhuang *et al.*, 1993
Xylogalactoglucan	*Inonotus obliquus*	Mizuno *et al.*, 1990a
Mannogalactoglucan	*Pleurotus pulmonarius* *Pleurotus Cornucopiae* *Ganoderma lucidum* *Agaricus blazei*	Gurtirrez *et al.*, 1996 Kim *et al.*, 1994 Cho *et al.*, 1999
Galactomannoglucan	*Flammulina velutipes* *Hohenbuehelia serotina* *Leucopaxillus giganteus*	Ikekawa *et al.*, 1982 Mizuno *et al.*, 1994 Mizuno 1995a

Contd...

Table 25.1–Contd...

Polysaccharides	Species	References
Arabinoglucan	*Ganoderma tsugae*	Zhang *et al.*, 1994b
Riboglucan	*Agaricus blazei*	Cho *et al.*, 1999
Glycans		
Arabinogalactan	*Pleurotus citrinopileatus*	Zhang *et al.*, 1994a
Glucogalactan	*Ganoderma tsugae*	Wang *et al.*, 1993
Fucogalactan	*Sarcodon asparatus*	Mizuno 2000
α-(1→6)-mannofuco-galactan	*Fomitella fraxinea*	Cho *et al.*, 1998
Fucomannogalactan	*Dictyophora indusiata*	Hara *et al.*, 1993
Mannoaglactofucan	*Grifola frondosa*	Zhuang *et al.*, 1993
Xylan	*Hericium erinaceus*	Mizuno. 1999b
Glucoxylan	*Hericium erinaceus*	Mizuno. 1999b
	Pleurotus pulmonarius	Zhuang *et al.*, 1993
Mannoglucoxylan	*Hericium erinaceus*	Mizuno. 1999b
α-(1→3)-mannan	*Dictyophora indusiata*	Ukai *et al.*, 1983
Glucomannan	*Agaricus blazei*	Hikichi *et al.*, 1999
β-(1→2)-; β-(1→3)-gluco-mannan	*Agaricus blazei*	Tsuchida *et al.*, 2001 Mizuno *et al.*, 1999b
Galactoglucomannan	*Lentinus edodes*	Fujii *et al.*, 1979

Source: Reshetnikov *et al.*, 2001.

It is reported that mushrooms have significant pharmacological effects of homeostasis, regulation of biorhythm, curing of various diseases, and prevention and improvement from life threatening diseases such as cancer, cerebral stroke, and cardiac problems. At the same time it is also being confirmed that mushrooms have effective substances for decreasing the blood cholesterol and possess antihypertension, antihyperlipemia, antidiabetes, antitumor, immunomodulating and other qualities (Chang and Miles, 2004).

Very few agents are effective for both treatment and prevention of disease, but research has shown that certain mushrooms can be used in this way. The fungus *Coriolus versicolor* is found almost worldwide; however, its bioactivity varies depending on the habitat in which it grows (Monro, 2003). To eliminate these variations, Gourmet Mushrooms, Inc. (Sebastopol, California) established the CV-OH1 strain a few years ago. This strain demonstrates rapid and aggressive colonization and far outmatches other isolates in bioactivity and vigor. A mother culture of strain CV-OH1 has been developed and maintained to produce spawn. In-house checks confirm that the mother culture does not alter from the original isolate. The production process involves the inoculation of sterile organic edible grain with spawn from the mother culture. The fungus is allowed to completely colonize the growth medium aseptically. At the correct stage of development–at maximum bioavailability–the living biomass is aseptically air-dried, granulated, and tested microbiologically.

Besides the culinary magics of mushrooms, this review article emphasises the pharmacological properties explored, practiced and accepted widely, of three globally popular mushrooms *viz.*, Lentinus, Ganoderma and Grifola.

Shiitake

At present shiitake is one of the five most cultivated edible mushrooms in the world (Chang, 1999 and Wasser 2005). These mushrooms have excellent nutritional value with water soluble polysaccharides. The mushroom exhibits anticarcinogenic (Chihara *et al.*, 1987) and antitumor (Smith *et al.*, 2002.,Yap and Ng, 2003 and Moriyama 1981) effects, immunomodulating effects (Wasser and Weis 1997 and Nagai and Ng, 2003), cardiovascular effects (Hobbs, 1995., Wasser and Weis 1997., Hobbs, 2000 and Mizuno, 1996), hepatoprotective effects (Przbylowicz and Donoghue, 1990, and Chang, 1999), antiviral, antibacterial and antiparasistic (Wasser and Weis,1997., Hobbs, 2000., Smith *et al.*, 2002 and Ngai and Ng, 2003) effects as well as antifungal activity (Nagai and Ng., 2003).

In 1969, Ikekawa *et al.*, noted that a water extract of shiitake fruiting bodies could inhibit transplanted tumors in mice. Around the same time, Chihara *et al.* (1987) isolated an antitumor polysaccharide from shiitake and named it lentinan. The molecular formula for lentinan is $(C_6H_{10}O_5)n$, and the mean molecular mass is 500 kDa. Lentinan is a D-glucan as shown by electrophoresis, ultracentrifugation, and other instrumental analyses. Its structure was confirmed as (1, 3)-D-glucopyranan with a branched chain of (1,6)-monoglycosyl, showing a right-handed triple helix. Lentinan has been found to activate macrophages, T lymphocytes, and other immune effector cells that modulate the release of cytokines, which may in turn account for its indirect antitumor and antimicrobial properties (Ikekawa *et al.*, 1969., Chihara *et al.*, 1970., Sasaki and Takasuka., 1976; Bluhm and Sarco, 1977 and Hamuro and Chihara, 1985).

Besides the well-studied above mentioned compounds, other potentially beneficial compounds have been found in shiitake. One example is thioproline, which blocks the formation of carcinogenic N-nitroso compounds (Kurashirna *et al.*, 1990).

Ganoderma lucidum

The virtues of another mushroom *Ganoderma lucidum* extracts, handed down from generation to generation, include it as a "cancer cure" and a symbol of happy augury, good fortune, good health, longevity, and even immortality (Wasser, 2005). Medicinal uses of *G. lucidum* in ancient Far East countries included the treatment of neurasthenia, debility from prolonged illness, insomnia, anorexia, dizziness, chronic hepatitis, hypercholesterolemia, mushroom poisoning (antidote), coronary heart disease, hypertension, prevention of altitude sickness, treatment of deficiency fatigue, carcinoma, and bronchial cough in the elderly (Natalia *et al.*, 2005). According to Traditional Chinese Medicine (TCM) different types of *G. lucidum* have different tastes and thus affect different organs. Based on their colors six different types of *G. lucidum* have been classified each with different uses (Table 25.2). The fruiting body, mycelia, and spores of *G. lucidum* contain approximately 400 different bioactive compounds, nucleotides, sterols, steroids, fatty acids, proteins/peptides, and trace elements (Wasser, 2005). It also contains oleic acid, cyclo-octasulfur, an ergosterol peroxide (5, 8-epidioxy-ergosta-6, 22-dien-3-ol) and cerebrosides. Polysaccharides (β-D-glucans, heteropolysaccharides, and glycoprotein) isolated from *G. lucidum* demonstrated antitumor activity against Sarcoma 10 in mice (Doris *et al.*, 2005). Triterpenoids like ganoderic acids T-Z showed cytotoxic activity *in vitro* on hepatoma cells the polysaccharide-mediated potentiation of immune function is thought to be the major mechanism of antitumor action by *G. lucidum*.

Ganoderma has been used as folk medicine since ancient times and it is a popular health food frequently promoted as a cancer cure. *Ganoderma* has been recognized traditionally and scientifically as potentially useful in the treatment of cancer, (Chang, 2001).

Table 25.2: *Ganoderma lucidum* and their Medicinal Properties

Color	Taste	Japanese name	Use
Blue	Sour	Aoshiba	Improves eyesight and liver function, calms nerves
Red[a]	Bitter	Akashiba	Aids internal organs, improves memory, enhances vitality
Yellow	Sweet	Kishiba	Strengthens spleen function, calms the spirit
White	Hot/pungent	Shiroshiba	Improves lung function, gives courage and strong will
Black	Salty	Kuroshiba	Protects kidney
Purple	Sweet	Murasakishiba	Enhances function of ears, joints, muscles, helps complexion

[a] The red colored variety of *G. lucidum* is generally regarded as the most potent and medicinal (Wasser, 2005).

Immunostimulation with Ganoderma Polysaccharides

Nakashima *et al.* (1979) demonstrated T-cell activation by orally administered *Ganoderma* polysaccharides (GP). Subsequently, activation of immune function including the activation of cytotoxic NK cells (Won *et al.*, 1989), and stimulation of cytokines (Chang *et al.* 1988) has also been reported with GP. The B-1, 3 glucan polysaccharide fraction in particular, has been demonstrated to have a broad-based adjuvant stimulatory activity on T-cells and macrophages, and lead to the enhance production of IL-I (Jia *et al.*, 1994), IL-2 (Chang *et al.*, 1988, Zhang *et al.*, 1 993a), and TNF (Zhang *et al.*, 1993b) which have been reported to play a role in antitumor immune surveillance (Lattime and Stutman 1991). Of particular interest is the demonstrated activity of a crude Ganoderma extract on NK cell activity *in vitro* (Zhang *et al.*, 1994) and the effectiveness of a water soluble GP fraction derived from mycelium given by intraperitoneal, intravenous, or oral route in enhancing splenic NK activity in normal mice and restoring depressed NK cytotoxicity in tumor-bearing mice (Won *et al.*, 1989). Other studies demonstrated the ability of GP in activating macrophages (Zhang *et al.*, 1994) and in markedly enhancing the cytotoxicity of CTL (Lei, 1992), both established tumoricidal effector pathways of the host immune system.

Potential of GP Immunostimulation for Prevention of Cancer

Ganoderma has been recognized traditionally and scientifically as potentially useful in the treatment of cancer (Chang 1993). Although triterpenoids and other fractions of *Ganoderma* may have direct cytotoxic activity (Toth *et al.*, 1983, Lin *et al.*, 1997), some evidence supports the notion that major anti-tumor efficacy of *Ganoderma* is host dependent and mediated through GP immune stimulation (Xia *et al.*, 1989, Mizuno 1995).

Ikekawa *et al.* (1968) first reported on the efficacy of soluble extracts from *Ganoderma* in inhibiting transplanted Sarcoma 180 in mice. This host-dependent antitumor activity has been subsequently confirmed to be from the polysaccharide fractions of *Ganoderma* (Sasaki *et al.*, 1911). Multiple similar studies subsequently confirms this observation and antitumor efficacy of *Ganoderma* has been demonstrated from various species, at different stages of growth and using different solvents for extraction and different routes of administration. Antitumor activity has been demonstrated *in vitro* as well as in syngeneic tumor systems in animals. However, no human trials of *Ganoderma* against cancer in peer reviewed journals nor any controlled clinical trials in humans have yet been conducted or published. Since it is now increasingly clear that immunostimulatory bioactivity from most beta-glucan based compounds function via a similar beta-glucan receptor (Czop 1985), it has been possible to hypothesize that *Ganoderma* polysaccharides should function similarly (Chang, 1996). Data from such bioactively comparable compounds all suggest improved quality of life or survival for cancer patients may be possible with *Ganoderma* supplementation.

Standardization in bioactive polysaccharide content and dosages will be necessary to assure its rational use, and clinical trials in selected cancers with defined endpoints will confirm its efficacy. Ultimately, one can believe that integration of the natural, complementary modalities and conventional modern medicine will lead to healing and the saving of lives. The treatment of various ailments through Ganotherapy has been initiated very recently in India and is now under experiment (Rahman *et al.*, 2005). The efforts were intended for promoting alternative medication for curing and maintaining health and at the same time creating social acceptance, adaptation, and affordability of this treatment among a greater segment of people. And it has been found that 80 per cent of the patients were cured and are now leading a better quality of daily life. It may be urged and stressed that a wider scope of clinical research work on Ganotherapy in the country is needed.

Grifola frondosa

Much research has reported that *Grifola frondosa* and its extracts showed remarkable antitumor activities by activating the cellular immune system, enhancing the production of cytokines, or inducing apoptosis. Kawagishi *et al.*, 1990 isolated an N-aceylgalactosamine-specific lectin from fresh fruiting bodies of *G. frondosa.*

Antitumor activity of *G. frondosa* polysaccharide, Grifron D-fraction (GD), on human prostate cancer PC-3 cells *in vitro* was exhibited after an investigation by Fullerton *et al.*, in 2000. Earlier, the anti-HIV activity of *G. frondosa* extract has been recognized by the Japan National Institute of Health (Ishikawa, 1991) and confirmed by the US National Cancer Institute (Screen 1992) that *G. frondosa* extract was able to prevent as much as 97 per cent of HIV-infected T-helper lymphocytes from being destroyed *in vitro*. Kabir *et al.* (1987) and Ohtsuru *et al.* (1999) reported that *G. frondosa* powder lowers the blood pressure in spontaneously hypertensive rats. Whereas in an antidiabetic report diabetic levels of blood glucose were significantly decreased in the rats fed *G. frondosa* (Horio *et al.*, 1995). Kubo *et al.* (1997) reported that *G. frondosa* helps to improve the lipid metabolism, because it inhibits both liver lipid and serum lipid and antiobesity effect of *G. frondosa* was studied by Ohtsuru *et al.*, in 1999. Wu (1994) reported that a proteoglucan from *G. frondosa* was clinically used to treat chronic hepatitis B. Whereas Ochiai (2003) reported that a proteoglucan extracted from *G. frondosa* strongly inhibited the growth of influenza virus.

Adachi *et al.* (1989) and Zhuang *et al.* (1994) reported that the modification of some polysaccharide fractions from the fruiting body or mycelium of *G. frondosa* by Smith degradation and formic acid degradation may increase their antitumor activities (Zhuang and Wasser, 2004).

Need of Molecular Characterization

The vagaries of classical morphological characterization make identification the weak link in research. The use of molecular methods has potential to give greater objectivity to approaches to characterization of species and strains, enabling to look past the influence of habitat on morphology. Mushroom classification and taxonomy is still not an exact science, molecular techniques will help but definitive tools for objective characterization are still a long way off (different methods giving convergent results or comparable ease of utilization).

Improvements can be made by exploiting diversity in wild strains or by using directed molecular breeding. The mushroom genetics program comprises: improvement and exploitation of transformation; molecular analysis of inter- and intra-species relatedness; characterization of developmental genes.

Now-a-days ITS sequencing is being used for identification of species, to establish species relationships within *Agaricus* and develop molecular phylogenetics for the genus (Singh *et al.*, 2006). RAPD markers are also being used as a tool to measure genetic diversity and to aid the analysis of mushroom breeding systems. It's because of these sophisticated techniques of biotechnology that the genetic diversity or any threat to any species can be identified and worked out. Like in *Agaricus*, restricted genetic diversity in cultivated lines represents a risk factor and breeding is needed to incorporate greater variability from wild types.

Analysis of commercial strains of two edible mushrooms, *Pleurotus ostreatus* and *Agaricus bisporus*, using PCR and isozyme electrophoresis techniques lead to differentiate groups of genetically similar and distant strains. Among the commercial strains of *P. ostreatus*, the level of genetic variation was higher suggesting a broader genetic basis employed in breeding of this mushroom. On the other hand, the cultivars and hybrids of *A. bisporus* showed a higher level of homology. The isozyme markers (nonspecific esterase, leucinaminopeptidase, and phosphoglucoisomerase) are recommended for identification of the commercial strains of edible mushrooms (Shnyreva *et al.*, 2003).

Genetic Resources

The collection, molecular characterization and germplasm conservation of wild types becomes critically essential to assure availability of breeding material for research and commercially cultivated mushrooms.

In developing countries, natural mushroom genetic diversity combined with the application/ adaptation of presently available cultivation technologies offers potential for improving productivity of small-scale mushroom growing. Potential advantages that can be obtained from better international coordination in mushroom germplasm include:

- ✰ Generation of better/more complete knowledge of habitat/organism diversity
- ✰ Opportunities to develop appropriate protocols for collecting, identifying and conserving mushroom germplasm
- ✰ Strategic use of scarce resources
- ✰ International collaboration facilitated
- ✰ Opportunities for training for scientists as well as growers.

Challenges

Mushroom genetic resources can contribute significantly to the solution of food supply problems. There is a need for worldwide agreement on how and where to store mushroom germplasm in order to ensure there can be a future for breeding edible mushrooms. Demonstration of serious coordinated pooling efforts on the part of research institutions could be helpful in encouraging industry players and researchers to appreciate the interest for them in participating and helping finance mushroom germplasm collection and conservation work.

Development of high temperature cultivated strains is a major objective in many warm countries and germplasm collection efforts need to be oriented to identifying useful characteristics in wild types. A crash program from time to time to document, collect and conserve wild edible and useful mushroom germplasm is needed in developing countries, especially in India, tapping indigenous knowledge as much as possible.

Research in several Asian countries can collaboratively and usefully focus on identification and domestication of local taxa/strains, with the aim of leading to higher diversity within cultivated mushrooms and reduced vulnerability of the sector. There is a need for training in methods of identification, conservation characterization and utilization of mushroom genetic resources, including training in preparation of high quality spawn.

Strategy for Conservation

Ex situ mushroom germplasm collections–including also associated characterization and evaluation functions–need to be correlated with research and training activities. In India, it is important to harness local peoples' knowledge of edible/useful mushroom habitat and characteristics, and establish local human capacity to survey, characterize and conserve useful mushrooms.

Storage techniques for mushroom strain isolates and cultures needs further work to develop low-cost, feasible, reliable protocols and alternatives that minimize risk of stress-induced degradation. The genomic DNA sequencing can be another advanced technique for germplasm conservation with minimized risk. The group's attention should be focused on macro fungi, in terms of collection, conservation, and characterization of genetic resources. Mushroom genetic resource conservation strategies need elements of cooperation, coordination and strategic planning.

Cooperation among different laboratories is needed to develop molecular techniques as a tool to help species/strain identification and characterization. Responsibility for mushroom genetic resource conservation strategies starts with governments; *ex situ* conservation through mushroom collections at research institutions in each country should be supported by governments and links established with other banks within and among regions. It is necessary and opportune to establish concrete strategic objectives and working arrangements for achieving them first; then address funding issues accordingly.

Mycorrhizal mushrooms are another group of medicinal mushrooms with high therapeutic values and more preferable flavor. Like in other developed countries dedicated task force on research and development of mycorrhizal mushrooms also should be established. In terms of strategic objectives and technical activities to be pursued, the original terms of reference for the mushroom germplasm science network should be retained and updated, *viz.*

- ☆ Collection, characterization and utilization studies on species and strains of edible mushrooms and related fungal organisms.
- ☆ Development of conservation methodologies for mushroom germplasm to protect against degradation or mutation.
- ☆ Development of molecular techniques as tools (to be applied in association with morphological/taxonomic approaches) for characterization and as markers for classification and certification of mushroom strains.
- ☆ Data base establishment and maintenance.
- ☆ It can also be emphasized that the network should promote close cooperation between taxonomists and molecular biologists for identification of macro fungi.

Yet, there are a far greater number of wild edible mushrooms and related fleshy fungi that have not been subject to cultivation. The growing interest in gourmet or exotic mushrooms is the direct result of research. In Asia and in Europe, many species of mushrooms have been eaten and enjoyed for a long period of time. Therefore, the industry has a continuing need for new and improved varieties with well-defined characteristics. International collaborative research and development activity is being encouraged and welcomed on an urgent basis to continue clinical research.

Acknowledgements

The authors are thankful to Department of Science and Technology, New Delhi for providing financial assistance for the mushroom project.

References

Adachi Y., Ohno N., Ohsawa M., Sato K., Oikawa S and Yadomae T. (1989). Physicochemical properties and antitumor activities of chemically modified derivatives of antitumor glucan "Grifolan LE" from *Grifola frondosa*. *Chem Pharm Bull*, 37, 7, 1838-1843.

Anke, T., Oberwinkler, F., Steglich, W. and Schramm, G. (1977) The Strobilurins- new antifungal antibiotics from the basidiomycete *Strobilurus tenacellus* (Pers. ex. Fr.) Sing. *J. Antibiotics*. 30: 806-810.

Anke, T., Werle, A., Bross, M. and Steglich, W. (1990) Antibiotics from basidiomycetes XXXIII, Oudemansin X, a new antifungal *E-?*-Methoxyacrylate from *Oudemansiella radicata* (Rehlan EX FR.) Sing. *J. Antibiotics* 43: 1010-1011.

Anke, T.,Werle, A., Zapf, S., Velten, R. and Steglich, W. (1995) Favolon, a new antifungal triterpenoid from a *Favolaschia* Species. *J. Antibiotics* 48: 725-726.

Arnone, A., Cardillo R., Nasini, G., and Pava, O.V.D. (1993). Cyathiformines A-D, new chorismate-derived metabolites from the fungus *Clitocybe cyathifomis*. *Tetrahedron* 49: 7251–7258.

Arnone, A., Brambilla, U., Nasini, G., and Pava, O.V.D. (1995). Isolation and structure elucidation of Tsugicolines A-D, novel protoilludane sesquiterpines from *Laurilia tsugicola*. *Tetrahedron*. 51: 13357-13364

Arnone, A., Capelli, S., Nasini,G., Meille, S. V., and Pava, O. V. D. (1996) Secondary mould metabolites II Structure elucidation of diatretol–A new Diketopiperzine metabolite from the fungus *Clitocybe diatreata*. Liebigs Ann. 1875-1877.

Atri N.S., Kaur A and Saini S.S. (2000). Taxonomic studies on agarics from Punjab plains. *Indian J. Mush*. 18, 6-14.

Atsumi, S., Umezawa, K., Iinuma, H., Naganawa, H., Nakamura, H.,Iitaka, Y. and Takeuchi, T. (1990). Production, Isolation and Structure Determination of A Novel β- Glucosidase inhibitor, cyclophellitol, from *Phellinus* sp. *J.Antibiotics* 43: 49-53.

Bluhm T.L. and Sarco A. (1977). The triple helical structure of lentinan, β-(1 -3)-D-glucan. *Can J Chem*; 55: 293-9-9

Chang R. (1993). Limitations and potential applications of Ganoderma and related fungal polyglycans in clinical ontology. *First international conference on mushroom biology and mushroom products*: 96.

Chang R. (1996). Functional properties of edible mushrooms, *Nutr Rev*, 54, 11, S91-S93.

Chang S. T and Buswell J. A. (1996). Mushroom nutriceuticals. *World J Microb Biotech*, 12. 473-476.

Chang S. T. (1999). World production of cultivated edible and medicinal mushrooms in 1997 with emphasis on *Lentinus edodes* (Berk.) Sing. In China. *Int. J. Med. Mushr*. 1, 387-409.

Chang S. T. (2001). A 40-year journey through bioconversion of lignocellulosic wastes to mushrooms and dietary supplements. *Int. J. med. Mushr*, 3 299-311.

Chang S. T. and Buswell J. A. (2003). Medicinal mushrooms- a prominent source of nutriceuticals for the 21st century. *Curr Topics Nutraceut res*, 1, 257-280.

Chang S. T. and Miles Ph. G. (2004). Mushrooms. Cultivation, nutritional value, medicinal effect and environmental impact. *CRC Press*, Boca Raton, FL, 451pp.

Chang S.T. and Buswell J. A. (1999). *Ganoderma lucidum* (W.Curt.: Fr.) P. Karst. (Aphyllophoromycetideae)- a mushrooming medicinal mushroom. *Int J Med Mushr*, 1,139-146.

Chang, R. (1993). Limitations and Potential applications of Ganoderma and related fungal polyglycans in clinical ontology; *First International Conference on Mushroom Biology and Mushroom products*: 96

Chihara G., Hamuro J., Maeda Y., Arai Y. and Fukuoka F. (1970). Fractionation and purification of the polysaccharides with marked antitumor activity especially Lentinan, from *Lentinus edodes* (Berk.). Sing., an edible mushroom. *Cancer Res* 30: 2776-81.

Chihara, G., Hamuro, J., Maeda, Y. Y., Shiio, T., Suga, T., Takasuka, N and Sasaki T. (1987). Antitumor and metastasis- inhibitory activities of lentinan as an immuno-modulator: an overview. *Cancer Detect. Rev. Supp.* 1, 423-443.

Cho, S. M., Koshino, H., Yu, S. H. and Yoo, I. D. (1998). A mannufucogalavtan, fomitellan A, with mitogenic effect from fruit bodies of *Fomitella fraxinea* (Imaz.). *Carbohyd. Polym* 37: 31-18.

Cho, S. M., Park, J. S., Kim, K. P., Cha, D. Y. and Yoo, I. D. (1999). Chemical features and purification of immunostimulating polysaccharides from fruit bodies of *Agaricus blazei*. *Korean J. Mycol.* 27: 170-174.

Cuellar, M. J. Giner, R. M., Recio, M. C., Just M. J., Manez, S. and Rios, J. L. (1996). Two Fungal lanostane derivative as Phospholipase A2 Inhibitors. *J. Nat. Prod.* 59: 977 -979.

Czop, J.K., Austen, K.F. (1985). A beta-Glucan inhibitable receptor on human monocytes, *J. Immunol.* 134, 2588-593.

Daferner, M., Anke, T., Hellwig, V., Steglich, W. and Sterner, O. (1998). Strobilurin M, Tetrachloropyrocatechol and Tetrachloropyrocatechol methyl ether: new antibiotics from *Mycena* species. *J. Antibiotics* 51: 816-822.

Doris O. Osagualɛkhor and J. A. Okhuoya (2005). Sociocultural and ethnomycological uses of mushroom among the Esan people of Nigeria. *Intl J. of Med. Mush.* 3(7)–442-443.

Dufresne, C., Young, K., Pelaez, F., Val, A.G.D., Valentino, D., Graham, A., Platas, G., Bernard, A. and Zink, D. (1997). Illudinic acid, a novel Illudane sesquiterpine antibiotic. *J Nat. Prod.* 60: 188 -190.

Engler, M., Anke, T., Klostermeyer, D. and Steglich, W. (1995). Hydroxystrobilurin A, a new antifungal *E*-β-Methoxyacrylate from a *Pterula* species.*J.Antibiotics* 48: 884 -885.

Erkel, G., Becker, U., Anke, T. and Sterner, O. (1996). Nidulal, a novel inducer of differentiaton of Human Promyelocytic Leukemia Cells from *Nidula candida*. *J.Antibiotics* 49: 1189-1195.

Fujii, T., Ishida, N., Maeda, H., Mizuni, I. and Suzuki, F. (1979). KS-2-A. *Us Patent 4143780*, 8 July 1979.

Fujimiya, Y., Suzuki, Y., Oshima, K. I., Kobori, H., Moriguchi, K., Nakashima, H., Matumoto, Y., Takahara, S., Ebina, T. and Katakura, R.. (1998b). Selective tumoricidal effect of soluble proeoglucan extracted from the basidiomycete, *Agaricus blazei* Murrill, mediated via natural killer cell activation and apoptosis. *Cancer Immunol Immmunother* 46: 147-159.

Fullerton S.A., Smadi A.A., Tortorelis D.G., Choudhury M.S., Mallouh C., Tazaki H. and Konno S. (2000). Induction of apoptosis in human prostatic cancer cells with Beta-glucan (Maitake mushroom polysaccharide). *Molec Urol*, 4, 1, 7-13.

Gan, K. H., Fann, Y-F., Hsu, S-H., Kou, K-W. and Lin, C-N. (1998). Mediation of the cytotoxicity of lanostanoids and steroids of *Ganoderma tsugae* through apoptosis and cell cycle. *J Nat. Prod.* 61: 485-487.

Gutierrez, A., Prito, A., and Martinez, A, T. (1996). Structural characterization of extracellular polysaccharides produced by fungi from the genus *Pleurotus*. *Carbohydr Res* 281: 143-154.

Hamuro J. and Chihara G. Lentinan, (1985). A T-cell oriented immunopotentiator: its experimental and clinical applications and possible mechanism of immune modulation. In: Fenichel RL, Chirigos MA, eds. *Immune Modulation Agents and Their Mechanisms*. New York: Marcel Dekker, 409-36.

Hautzel, R., Anke, H. and Sheldrick, W.S. (1990). Mycenon, A new metabolite from a *Mycena* species TA 87202 (Basidiomycetes) as an Inhibitor of Isocitrate Lyase. *J.Antibiotics*. 43: 1240-1244.

Heim, J., Anke, T., Mocek, U., Steffan, B. and Steglich, W. (1988). Antibiotics from Basidiomycetes XXIX: Pilatin, A new antibiotically active marasmane derivative from cultures of *Flagelloscypha pilatii* Agerer. *J.Antibiotics* 41: 1752-1757.

Hikichi, M., Hiroe, E., and Okubo, S. (1999). Protein polysacchride 0041. *European Patent 0939082*, 9 January 1999.

Hobbs Ch. (1995). Medicinal mushrooms: an exploration of tradition, healing and culture. *Botanica Press*, Santa Cruz, CA, 251.

Hobbs, Ch. (2000). Medicinal value of *Lentinus edodes* (Berk.) Sing. A literature review. *Int. J. Med. Mushr*. 2, 287-302.

Horio H. and Ohtsuru M. (1995). Maitake (*Grifola frondosa*) improves glucose tolerance of experimental diabetic rats. *J Nutr Sci Vitaminol*, 47, 57-63.

Ikekawa T., Uehara N., Maeda Y., Nankinishi M. and Fukuoka F. (1969) Antitumor activity of aqueous extracts of edible mushrooms. *Cancer Res*; 29: 734-5-5

Ikekawa, T., Ikeda, Y., Yoshida, Y., Nikanishi, M., Yokoyama, E., and Yamazaki, E. (1982). Antitumor polysaccharides of *Flamulina velutipes* 2. The structure of EA-3 and further purification of EA-5. *J Pharmacobiol Dyn* 5: 576-581.

Ishikawa K. (1991). Anti-HIV activity in cytopathic effect of proteglucan extracted from Maitake mushroom. *Japan National Institute of Health*, January 23, 1991.

Jia, Y.F., Zhou, XB. Meng, H., and Zhang, (1993). Effects of Ling-Zhi on hemopoietic system in mice–immunopharmacological study (11). In *The research on Ganoderma (part I)*. Zhu S. and Mori M. (eds). Shanghai Med. U. Press, Shanghai, P. 284-288.

Kabir T., Yamaguchi M. and Kimura S. (1987). Effect of Shiitake (*Lentinus edodes*) and Maitake (*Grifola frondosa*) mushrooms on blood pressure and plasma lipids of spontaneously hypertensive rats. *J Nutri Sci Vitaminol*, 33, 341-346.

Kawagishi H., Nomura A., Mizuno T., Kimura A., and Chiba S. (1990). Isolation and characterization of lectin from *Grifola frondosa* fruiting bodies. *Biochem Biophys Acta*, 1034, 247-252.

Kendriok B. (1985). *The Fifth Kingdom*. Waterloo: Mycologue Publication.

Kiho, T., Nagai, Y. S., Sakushima. M. and Ukai, S. (1992a). Polysaccharides in fungi. XXIX. Structural features of two antitumor polysaccharides from the fruiting bodies of *Armillariella tabescence Chem Pharm Bull* (Tokyo) 40: 2212-2214.

Kiho, T., Yoshida, I., Katsuragava, M., Sakushima, M., Usiu, S. and Ukai, S. (1994). Polysaccharides in fungi. XXXIV. Polysaccharides from the fruiting of *Amanita muscaria* and the antitumor activity of its carboxymethylated product. *Biol Pharm Bull* 17: 1460-1462.

Kim, B. K., Kim H.W. and Choi, E.C. (1994). Anti-HIV effects of *Ganoderma lucidum*. In: *Ganoderma*: Systematics, Phytopathology and Pharmacology: Proceedings of Contributed Symposium 59 A, B. 5th International Mycological Congress. Vancouver.

Kubo K. and Nanba H. (1997). Anti-hyperliposis effect of Maitake fruit body (*Grifola frondosa*). *Biol Pharm Bull*, 20, 7, 781-785.

Kubo, I., Kim, M., Wood, W.F. and Naoki, H. (1986). Clitocine, A New Insecticidal Nucleoside from the mushroom *Clitocybe inversa*. *Tetrahedron Letters* 27: 4277-4280.

Kuhnt, D., Anke, T., Besl, H., Bross, M., Herrmann, R., Mocek, U., Steffan, B., and Steglich, W. (1990). Antibiotics from basidiomycetes XXXVII. New inhibitors of cholesterol biosynthesis from cultures of *Xerula melanotricha* Dorfelt. *J. Antibiotics* 43: 1413-1420.

Kurashirna Y., Tsuda M. and Sugimura T. (1990). Marked formation of thiazolidine-4-carboxylic acid, an effective nitrite trapping agent *in vivo*, on boiling of dried shiitake mushroom (*Lentinus edodes*). *J Agric Food Chem* 38: 1945-9-9.

Kuschel, A., Anke, T., Velten, R., Klostermeyer, D., Steglich, W. and Konig, B. (1994). The Mniopetals, New Inhibitors of reverse transcriptases from a *Mniopetalum* species (Basidiomycetes). *J.Antibiotics* 47: 733-739.

Lakhanpal T.N.(1995). Mushroom flora of North West Himalayas. In *Advances in horticulture 13, Mushrooms* (Chandra K.L. and Sharma S.R. Eds.) MPH New Delhi India 351-373.

Lauer, U., Anke, T., Sheldrick, W.,Scherer, A. and Steglich, W. (1989). Antibiotics from basidiomycetes XXXI Aleurodiscal: An Antifungal Sesterterpenoid from *Aleurodiscus mirabilis*. (Berk. and Curt) Hohn. *J. Antibiotics* 42: 875-882.

Lee, I-K., Yun, B-S., Cho, S-M., Kim, W-G., Kim, J-P., Ryoo, I-J., Koshino, H. and Yoo, I-D. (1996). Betulinans A and B, Two Benzoquinone compounds from *Lenzites betulina*. *J. Nat. Prod*. 59: 1090-1092.

Lei J, (1992). Pharmacological study on Cordyceps sinensis (Berk.) Sacc. and ze-e Cordyceps *Chung Kuo Chung Yao Tsa Chih* 17(6), 364-366.

Lin WH, (1997). Dimerization of the N-terminal amphipathic alpha-helix domain of the fungal immunomodulatory protein from *Ganoderma tsugae* (Fip-gts) defined by a yeast two-hybrid system and site-directed mutagenesis. *J Biol Chem* 272(32), 20044-20048.

Manjula B. (1983). A revised list of agricoid and boletoid basidiomycetes from India and Nepal. *Proc. Indian Acad. Sci.* (Plant Sci) 92: 81-213.

Mc Namara H. S. (1999). Dietary supplement legislation. In: *Botonical Medicine*. Eshkinazi D., ed., Mary Ann Liebert, Larchmont, NY. 199-204.

Mizuno T, (1992). Antitumor-active polysaccharides isolated from the fruiting body of *Hericium erinaceum*, an edible and medicinal mushroom called yamabushitake or houtou. *Biosci Biotechnol Biochem* 56(2), 347-348.

Mizuno T. (1995). Bioactive molecules of mushrooms- food, function and medicinal effect of mushroom fungi. *Food RevInt*, 11. 7-21.

Mizuno, M., Minato, K-I, Ito H., Kawade M., Terai H., and Tsuchida H. (1999). Anti-tumor polysaccharide from the mycelium of liquid cultured Agaricus *blazei* Mill. *Biochem. Mol. Biol. Int.* 47: 707-714.

Mizuno, T. (1995a). Shiitake, *Lentinus edodes*: functional properties for medicinal and food purposes. *Food Rev. Int.*, 11(1): 111-128.

Mizuno, T. (1996). A development of antitumor polysaccharides from mushroom fungi. *Food Ingred. J. Jpn.* 167, 69-85.

Mizuno, T. (1999b). Bioactive substances in *Hericeum erinaceus* (Bull.: Fr.) Pres. (Yamabushitake), and its medicinal utilization. *Int J Med Mushr*, 1: 105-119.

Mizuno, T. (2000). Development of an antitumor biological response modifier from *Phellinus linteus* (Berk. et Curt.) Teng (Aphyllophoromycetedeae) *Int J Med Mushr*, 2: 21-33.

Mizuno, T. Ma, Y., Ito, H. and Suzuki, C. (1994). Water insoluble polysaccharides originated in mushroom its production, and antitumor agent mainly comprising the polysaccharide. *Japanese Patent 06-080703*. 22 March 1994.

Mizuno, T., Hagiwara, T., Nakamura, T., Ito, H., Shimura, K., Sumiya, T. and Asakura A. (1990a). Antitumor activity and some properties of water soluble polysaccharides from "Himematsutake" the fruiting body of *Agaricus blazei* Murrill. *Agric Biol Chem.* 54: 2889-2896.

Mizuno, T., Ohsava, K., Hagiwara, N. and Kuboyama, R. (1986). Fractionation and characterization of antitumor polysaccharides from Maitake, *Grifola frondosa*. *Agric Biol Chem* 50: 1679-1688.

Monro. J. A. (2003). Treatment of cancer with mushroom products. *Archives of environmental health.*

Mori K., Toyomasu T., Nanba H. and Kuroda H. (1987). Antitumor activities of edible mushrooms by oral administration. In: Wuest PJ, Royse DJ, Beelman RB, eds. *Proceedings of the International Symposium on Scientific and Technical Aspects of Cultivating Edible Fungi.* Amsterdam: Elsevier, 1-6.

Morita, Y., Hayashi, Y., Sumi, Y., Kodaira, A., and Sibata, H. (1995). Haploporic acid A, a novel dimeric drimane sesquiterpenoid from the basidiomycete *Haploporus odorus*. *Biosci. Biotech. Biochem.* 59: 2008-2009.

Moriyama, M. (1981). Antitumor effect of polysaccharide lentinan on transplanted ascites hepatoma-134 in C3H/He mice. In *Manipulation of host defence mechanisms*, Intl Congress Series 576. Excerpta Media: Amsterdam.

Nagai, P. H. and Ng, T. B. (2003). Lentin, a novel and potent antifungal protein from shiitake mushroom with inhibitory effects on activity of human immunodeficiency virus-1 reverse transcriptase and proliferation of leukemia cells. *Life Sci.* 73, 3363-3374.

Nanba H. (1967). Antitumor action of Shiitake (*Lentinus edodes*) fruit bodies orally administered to mice. *Chem Pharm Bull.* 35: 2459-6-6.

Natalia L., Poyedinok, A. S., Buchalo, O. V., Efremenkova and A. M. Negriyko. (2005). The light factor in biotechnology cultivation of medicinal mushrooms. *Intl J. of Med. Mush.* 3(7)- 448-449.

Natrajan K. and Raman, N. (1983). South Indian agaricales: Today and tomorrow. Printers and Publishers New Delhi 204.

Ochiai H. (2003). The effect of *Grifola frondosa* on influenza virus. In: *The abstract at the 16th annual meeting of the Japanese Biotherapy Association.*

Ohtsuru M., Horio H., Masui H. and Takada I. (1999). Effects of administration of *Grifola frondosa* on blood pressure and body weight in spontaneously hypertensive rats. *Nippon Shokuhin kagaku Kogaku Kaishi,* 46. 12, 806-814.

Pradeep C.K., Vrinda K.B., Mathew, S. and Abraham T.K. (1998). The genus Volvariella in Kerala state, India. *Mushroom Res.* 7, 53-62.

Przbylowicz P. and Donoghue J.(1990). *Shiitake Growers Handbook: The art and Science of mushroom cultivation.* Hunt Publ.Co.: Dubugue, Kendall, 199.

Purkayastha R.P. and Chandra, A. (1985). *Manual of Indian edible mushrooms.* Jagmander Book Agency Deshbandhu Gupta Road New Delhi 267.

Rahman H., Rahman S., Rahman A. and Rahman, K. (2005) Dissemination of case studies on Ganotherapy. *Intl J. of Med. Mush.* 3(7)–456.

Rai M. K. and Tidke G. (2005). Biotechnological potential of mushrooms: drug and dye production. *Intl J. of Med. Mush.* 3(7)- 457.

Rai, M K. (1997). Wild edible mushrooms of tribal areas of Seoni District, Madhya Pradesh, *Mushr. Res.,* 6(2): 107-108.

Rai M. K., Tidke G. and Wasser S. P. (2005). Therapeutic potential of mushrooms. *Natural product radiance.* 246-257.

Razak R. C and Rai M. K. (2005). Edible tribal mushroom resources of Central India and their ethnological aspects. *Intl J. of Med. Mush.* 3(7)- 458.

Reshetnikov S. v., Wasser S. P. and Tan, K.K. (2001). Higher Basidiomycetes as a source of antitumor and immumomodulating polysaccharides (Review). *Int J Med Mushr.* 3, 361-394.

Royse D.J. and Bahler C.C. (1996). Effects of genotype spawn run time and substrate formulation on biological efficiency of shiitake. *Appl Environ Microbial.* 52: 1425-7.

Saito, K., Nishijima, M. and Miyazaki, T. (1989). Studies on fungal polysacchrides: XXXV. Sructural analysis of an acidic polysacchride from *Ganoderma lucidum. Chem Pharm Bull* (Tokyo) 37: 3134-3136.

Saito, T., Akoi, F., Hirai, H., Inagaki, T., Matsunaga, Y., Sakakibara, T., Sakemi, T. S., Suzuki, Y., Watanabe, S., Suga, O., Sujaku, T., Smogowicz, A. A., Truesdell, S.J., Wong, J.W., Nagahisa, A., Kojima, Y., and Kohjima, N. (1998). Erinacine E as a kappa opioid receptor agonist and its new analogs from a basidiomycete, *Hericium ramosum. J. Antibiotics* 51: 983-990.

Sasaki T and Takasuka N. (1976). Further study of the structure of lentinan, an antitumour polysaccharide from *Lentinus edodes. Carbohydr Res.* 47: 99-104.

Sathe A. V. (1979). Agaricology in India-A review of work on Indian Agaricales. *Biovigyanam.* 5: 125-130.

Sathe A. V. and Deshpande S. (1979). Agaricales of Maharashtra In *Advances in Mycology and Plant pathology* (Chattopadhyay S.B. and Samajpati N. eds) 82-88.

Screen A. (1992). Developmental therapeutics program *in-vitro* testing results. National Cancer Institute, January 1992.

Shnyreva A.V., Belokon Y.S. and Belokon M.M. (2003). Differentiation between Commercial Strains of Oyster and Button Mushrooms Using Molecular Markers. *Russian J Genetics*, 39, 11, 1232-1239(8).

Singh S.K., Sharma V.P., Sharma S.R., Satish Kumar and Mugdha Tiwari. (2006). Molecular characterization of Trichoderma taxa causing green mould disease in edible mushrooms. *Current Sci.* 90, 3, 427-431.

Smith, J., Rowan N. and Sulivan, R. (2002). Medicinal mushrooms: Their therapeutic properties and current medical usage with special emphasis on cancer treatment; *special report commissioned by cancer research UK*. The University of Strathelyde in Glasgow, 256.

Stadler, M., Anke, H. and Sterner, O. (1994). New nematicidal and antimicrobial compounds from the basidiomycete *Cheimonophyllum candidissimum* (Berk and Curt) Sing. *J.Antibiotics*. 47: 1284-1289.

Sundin, A., Anke, H., Bergquist, K-E., Mayer, A., Sheldrick, W. S., Stadler, M. and Sterner, O. (1993). The structure determination of Panellon and Panellol, two 14-Noreudesmanes Isolated from *Resupinatus leightonii*. *Tetrahedron* 49: 7519-7524.

Susuki S and Oshirna S. (1976). Influence of Shi-Ta-Ke (*Lentinus edodes*) on human serum cholesterol. *Mushroom Sci* 1 976; 9 (Part 1): 463-7.

Tanimoto,T., Onodera, K., Hosoya, T., Takamatsu, Y., Kinoshita, T., Tago, K., Kogen, H., Fujioka, T., Hamano, K. and Tsujita, Y. (1996). Schizostatin, a Novel Squalene Synthase Inhibitor Produced by the Mushroom, *Schizophyllum commune*. *J. Antibiotics* 49: 617-623.

Toth J.O., Luu B. and Ourission G. (1983). Les acides ganoderiques T a Z: triterpenes cytoxiques de *Ganoderma leucidum* (Polyporaceae). *Tetrahedron Lett.* 24 1081-1084.

Tsuchida, H., Mizuno, M., Taniguchi, Y., Ito, H., Kawade M. and Akasaka K. (2001). Glucomannan separated from Agaricus blazei mushroom culture and antitumor agent containing as an active ingredient. *Japanese Patent 11-080206*, 26 March 2001.

Ukai, S., Kohi, T., Hara, C., Kuruma, I. and Tanaker, Y. (1983). Polysaccharides in fungi: XIV. Antiinflammatory effect of the polysaccharides from the fruit bodies of several fungi. *Journal of Pharmacology Dynamics* 6, 983-990.

Verma R.N., Singh G.B. and Bilgramy K.S. (1987). Fleshy fungal flora of NEH India-I. Manipur and Meghalaya. In *Indian Mushroom Science*: II (Kaul T.N. and Kapoor B.M. Eds.) RRL CSIR Jammu Tawi 414-421.

Vrinda K.B. Pradeep C.K., Mathew S and Abraham T.K. (1999). Agaricales from Western Ghats VII. *Mushroom res.* 8, 9-12.

Wasser S. P. (2002). Medicinal mushrooms as a source of antitumor and immunomodulating polysaccharides. *Appl Microb Biotechnol*, 60, 258-274.

Wasser S. P. (2005). Shiitake (*Lentinus edodes*). Encyclopedia of dietary supplements DOI: 10.1081/E-EDS-120024880.

Wasser, S.P. and Weis A. L. (1999a). Medicinal properties of substances occurring in higher basidiomycetes mushrooms: a modern perspective. *Crit Rev Immunol,* 1965-96.

Wasser, S.P. and Weis A. L. (1999b). Therapeutic effects of substances occurring in higher basidiomycetes mushrooms: current perspectives (Review). *Int J Med Mushr,* 1: 31-62.

Wasser, S.P., Nevo E., Sokolov D., and Timor-Timenetsky M. (2000b). The regulation of dietary supplements from medicinal mushrooms. In: *Science and cultivation of edible fungi.van Griensven,* (ed.) Balkema, Rotterdam. 789-801.

Wasser, S. P., Nevo E., Sokolov D., Reshetnikov S., and Timor-Timenetsky M. (2000a). Dietary supplements from medicinal mushrooms: diversity of types and variety of regulations. *Int. J. Med. Mushr,* 2, 1-19.

Wasser, S. P., Reshetnikov S. V., Solomko E. F., Buchalo A. S and Nevo E. (2002). For higher basidiomycetes mushrooms grown as biomass in submerged culture. *US patent no. 6,372,462 B2,* Publ. Apr. 16, 2002.

Wasser, S. P., Sokolov D., Nevo E., Reshetnikov S., and Timor-Timenetsky M. (2001). Dietary supplements from medicinal mushrooms: how we are going to ensure their quality and safety. *Int. J. Med. Mushr.* 3, 94.

Wasser, S.P. and Didukh M.Y. (2005). Culinary-medicinal higher basidiomycete mushrooms as a prominent source of dietary supplements and drugs for the 21st century. *Fifth international conference on mushroom biology and mushroom products.* 12.20-34.

Wasser, S.P., Didukh M.Y and Nevo E. (2004). Dietary supplements from culinary-medicinal mushrooms: A variety of regulations and safety concerns for the 21st century. *Int J Med Mushr,* 6, 231-248.

Wasser, S. P. (2005). Reishi or Ling Zhi (*Ganoderma lucidum*). *Encyclopedia of dietary supplements DOI*: 10.1081/E-EDS-120022119.

Wasser, S. P., Didukh, M. Y. and Nevo, E. (2004). Dietary supplements from culinary medicinal mushrooms: A variety of regulations and safety concerns for the 21st century. *International J. Med. Mush,* 6- 231-248.

Wasser, S. P., Weis, A. L. (1997). Meidcinal mushrooms. *Lentinus edodes* (Berk) Singer, Nevo, E., Ed.; Peledfus Publ. House: Haifa, Israel. 95.

Weber, W., Anke, T., Steffan, B. and Steglich, W. (1989). Antibiotics from Basidiomycetes XXXII Strobilurin E: A New Cytostatic and Antifungal (E)–Methoxyacrylate Antibiotic from *Crepidotus fulvotomentosus* Peck. *J.Antibiotics* 43: 207-212.

Won, S.J., Lee, SS., Ke, YH., Lin, MT. (1989). Enhancement of splenic NK cytotoxic activity by the extracts of *Ganoderma lucidum* mycelium in mice. *J Biomed Lab Sci* 2: 201-213.

Wood, K.A., Kau, D-A., Wrigley, S.K., Beneyto, R., Renno, D. V., Ainsworth, A.M., Penn, J., Hill, D., Killacky, J. and Depledge, P. (1996). Novel -Methoxyacrylates of the 9- Methoxystrobilurin and Oudemansin Classes Produced by the Basiodiomycete *Favolaschia pustulosa. J.Nat. Prod.* 59: 646-649.

Wu S. (1994). Therapeutic effect of Grifola polysaccharides in chronic hepatitis B. In: *the abstracts at International symposium on production and products of Lentinus Mushroom,* Nov 1-3, 1994. China.

Xia, E. and Q. Chen. (1989). Isolation, analysis and biological activities of the polysaccharide of *Tremella Fuciformis*. *ACTA Mycol. Sin.* 7: 166-174.

Yap, A. T., Ng, M. L. (2003). Immunopotentiating properties of lentinan (1-3)-B-D-glucan extracted from culinary- medicinal shiitake mushroom *Lentinus edodes* (Berk) Singer (Agaricomycetideae). *Int. J. Med. Mush.* 5, 352-372.

Yoshida, I., Kiho, T., Usai, S., Sakushima M. and Ukai, S. (1996). Polysacchrides in funji. XXXVII. Immunomodulating activities of carboxymethylated derivatives of linear (1→ 30-alpha-glucans extracted from the fruiting bodies of *Agrocybe cylindrical* and *Amanita muscaria*. *Biol. Pharm. Bull* 19: 114-121.

Zhang J, *et al.* (1994). Antitumor polysaccharides from a Chinese mushroom, "yuhuangmo," the fruiting body of *Pleurotus citrinopileatus*. *Biosci Biotechnol Biochem.*; 58(7): 1195-1201.

Zhang LX, (1993b). Effect of Japanese *Ganoderma Lucidum* on production of interleukin-2 from *Murine splenocytes Chung Kuo Chung Hsi I Chieh Ho Tsa Chih* 13(10), 613-615.

Zhang, J., Wang, g., Li, H., Zhuang, C., Mizuno, T., Ito H., Suzuki, C., Okamoto, H. and Li, J. (1994a). Antitumor Polysaccharides from a Chinese mushroom "Yuhuahgmo" the fruiting body of *Pleurotus citrinopiliatus*. *Biosci Biotechnol Biochem* 58: 1195-1201.

Zhang, J., Wang, g., Li, H., Zhuang, C., Mizuno, T., Ito H., Suzuki, C., Okamoto, H. and Li, J. (1994b). Antitumor activity of protein containing glycans from the Chinese mushroom Songshan Linzhi, *Ganoderma tsugae*, mycelium. *Biosci Biotechnol Biochem* 58: 1202-1205.

Zhang, L. and M. Yu, (1993a). Influence of ling zhi on natural killer cells-Immunopharmacological study (5). From *The research on Ganoderma lucidum (part one)*. Shanghai: Shanghai Medical University Press, pp. 246-253.

Zhuang C., Mizuno T., Ito H and Shimura K. (1994). Chemical modification and antitumor activity of polysaccharides from the mycelium of liquid cultured *Grifola frondosa*. *Nippon Shokuhin Kogyo Gakkaishi*, 41, 10, 733-740.

Zhuang Cun and Wasser S.P. (2004). Medicinal Value of culinary- medicinal Maitake mushroom *Grifola frondosa* (Dicks.: Fr.) S.F. Gray (Aphyllophoromycetodeae). Review. *Int J Med Mushr*, 6, 287-313.

Zhuang, C., Mizuno, T., Shimada, A., Ito, H., Suzuki, C., Mayuzumi, Y., Okamoto, H., Ma, Y. and Li, J. (1993). Antitumor Protein-containing Polysaccharides from a Chinese mushroom Fengweigu or Houbitake, *Pleurotus sajor-caju* (Fr.) Sing *Biosci Biotechnol Biochem* 57: 901-906.

Ziesel S. M. (1999). Regulation of nutraceuticals. *Science*. 285: 1853-1855.

Microbial Biotechnology and Ecology (2011) *Pages* **472–483**
Editors: **Deepak Vyas, G.S. Paliwal, P.K.Khare and R.K Gupta**
Published by: **DAYA PUBLISHING HOUSE, NEW DELHI**

Chapter 26

Significance of GM Crops in Modern Agriculture

☆ *Shweta J. Sabannavar*[1], *H.C. Lakshman*[1], *Mahesh B. Byatanal*[1] *and C. Maya*[2]

[1]*Post Graduate Department and Research in Botany (Microbiology Lab), Karnataka University, Dharwad – 580 003, Karnataka*
[2]*Department of Botany, Bangalore University, JB Campus, Bangalore – 560 056*

Introduction

Crop improvement was greatly accelerated in the 20th century by the rediscovery of the laws of inheritance. It was by Gregor Johan Mendel in 1865, and after scientists established that the chromosomes contain the information for specific traits encoded in DNA. The rise of molecular biology, starting in the 1950s with the discovery of DNA's structure, led to a series of further discoveries showing how traits are encoded as proteins in the cell and the plants. This understanding and new techniques that allow scientists to manipulates DNA in the laboratory, are revolutionizing plant breeding.

Genetic engineering depends on the fact that the same molecule, DNA, makes up the genes of all organisms. This means that a gene from a fungus (made of DNA) can be inserted in a rice plant's chromosome (also made of DNA) and, with luck and experimental skill, can be induced to function. The ability to transform crops with genes from any other organism opens up the entire community of life as the source for new genes for a crop. Although the process of genetic modification started long ago, only crops that are genetically engineered are referred to as genetically modified (GM) crops. Some writers prefer the term genetically enhanced or improved.

There are several advantages of GM technology over other methods. First and foremost, it is specific: Unlike crosses where many genes are transferred during sexual reproduction, here the scientist transfers only a single gene (or a few genes) to the crop plants. Second, the method is fast; as in any micropropagation technique, many plants can be generated quickly. Finally, the genes themselves are an advantage. Unlike any other breeding methods, plant biotechnology allows a scientist to put virtually any gene from any organism into a crop plant. In practice, scientists create a large number of transformed crop lines for a specific gene and then carry all these lines through to many rounds of conventional plant breeding, to ensure that the introduced characteristic. Thus GM technology does not eliminate plant breeding but supplements it just like other laboratory-based methods.

There is an urgent need to develop techniques in modified crops, which are sustainable from environmental production and socioeconomic points of view. This is to guarantee the sufficient food production in the coming decades. It is beyond critical because modern agricultural production throughout the world does not appear to be sustainable in the long-term FAO (1980), seems to appropriate in this contest.

But while GM crops are now being used widely by farmers in the USA, consumers in the European Union and Japan have reacted strongly against them. Although this has slowed the rate at which GM crops and foods are being introduced, the biotech industry is continuing to push ahead with them. This briefing explains why people should still be concerned about GM crops.

Genetically modified (GM) foods are foods produced from genetically modified organisms (GMO) that had their DNA altered through genetic engineering. GM foods were first put on the market in the early 1990s. The most common modified foods are derived from plants: soybean, corn, canola and cotton seed oil (Maessen, 1997).

Controversies surrounding GM foods and crops commonly focus on human and environmental safety, labeling and consumer choice, intellectual property rights, ethics, food security, poverty reduction and environmental conservation.

GM Foods Currently Used in Development Countries

Currently, there are number of foods which exist in a genetically modified version shown in Table 26.1.

Currently Known GM Crops

Between 1995 and 2005, the total surface area of land cultivated with GMOs had increased, by a factor of 50, from 17,000 km^2 (4.2 million acres) to 900,000 km^2 (222 million acres), of which 55 per cent were Brazil.

Although most GM crops are grown in North America, in recent years there has been rapid growth in the area sown in developing countries. For instance in 2005 the largest increase in crop area planted to GM crops (soybeans) were in Brazil (94,000 km^2 in 2005 versus 50,000 km^2 in 2004.) (Domingo, 2000). There has also been rapid and continuing expansion of GM cotton varieties in India since 2002 (Cotton is a major source of vegetable cooking oil and animal feed.) It is predicted that in 2006/7 32,000 km^2 of GM cotton will be harvested in India (up more than 100 per cent from the previous season). Indian national average cotton yields of GM cotton were seven times lower in 2002, because the parental cotton plant used in the genetic engineered was not well suited to the climate of India and failed. The publicity given to transgenic trait Bt insect resistance has encouraged the adoption of better performing hybrid cotton varieties, and the Bt trait has substantially reduced losses to insect predation. Economic and environmental benefits of GM cotton in India to the individual farmer have been documented (May, Sir R; Halford NCordell and Halford 1999).

Table 26.1: Sowing Genetically Modified Common Food Products Available in Developing Countries

Food	*Properties of the Genetically Modified Variety*	*Trade Name and the company which Produced Initial Version*	*Specific Genetic Modification*	*Percent of Genetically Modified Version Occupies of the Given Food' Agriculture in USA*	*Percent this GMF takes up of agriculture compared to non-GMF version, globally*
Soybeans	Resistant to herbicide	Roundup Ready, Monsanto	Herbicide resistant	TBA	TBA
Corn	Resistance to certain pesticides (tolerating crop spray this way a farmer can use amounts of pesticides which would normally kill the plant, without harming it)	TBA	New gene added/transferred into plant genome	TBA	TBA
Cotton	Pest-resistant cotton	TBA	New gene added/transferred into plant genome	TBA	TBA
Tomatoes	Variety that does not rot (degrade) as fast–the genetically modified tomatos do not produce a substance that normally causes tomatoes to rot	*E.g.* FlavrSavr	First genetically modified tomatoes contained genes that made them resistant to antibiotics. After concern from doctors and the medical community, tomatoes are now genetically modified in an alternative way	TBA	TBA
Rapeseed plants	Resistance to certain pesticides	TBA	New gene added/transferred into plants genome	TBA	TBA
Sugar cane	Resistance to certain pesticides (tolerating crop spray)	TBA	New gene added/transferred into plants genome	TBA	TBA
Sweet corn	Produces its own insecticide (a toxin to insects, so insects, so insect attacks are less likely	*Bt* corn	Insect-killing gene added to the plant. The gene comes from the bacteria *Bacillus thuringiensis*	TBA	TBA
Rice	Genetically modified to contain high amount of Vitamin A (beta carotene)	Golden rice	Three new genes implanted; two from daffodils and the third from a bacterium	TBA	TBA

In 2003, countries that grew 99 per cent of the global transgenic crops were the United States (63 per cent), Argentina (21 per cent), Canada (6 per cent), Brazil (4 per cent), China (4 per cent), and South Africa (1 per cent) (Thompson, 1990). The Grocery Manufacturers of America estimate that 75 per cent of all processed foods in the U.S. contain a GM ingredient (USDA 2006). In particular, Bt corn, which produces the pesticide within the plant itself is widely grown, as are soybeans genetically designed to tolerate glyphosate herbicides. These constitute "input-traits" are aimed to financially benefit the producers, have indirect environmental benefits marginal cost benefits to consumers.

In the US, by 2006, 89 per cent of the planted area of soybeans, 83 per cent of cotton, and 61 per cent maize was genetically modified varieties. Genetically modified soybeans carried herbicide tolerant traits only, but maize and cotton carried both herbicide tolerance and insect protection traits (the latter largely the *Bacillus thuringiensis* Bt insecticidal protein). In the period 2002 to 2006, there were significant increases in the area planted to Bt protected cotton and maize, and herbicide tolerant maize also increased in sown area (Simpson, 1999).

However, several studies have found that genetically modified varieties of plants do not produce higher yields than normal plants.

Transgenic Plants Function as Bioreactor

Plants are easy to grow and can generate considerable biomass. With these features in mind, research has been carried out to determine whether transgenic plants can be used for the production of commercial proteins and chemicals.

Through genetic engineering techniques, compounds of commercial interest that were previously produced only from exotic sources of plants, animal and microbial origin is now produced in domesticated crops. Transgenic plants can act as living bioreactors for the inexpensive production of chemicals and pharmaceuticals, and this is also known as *molecular farming* (Table 26.2). It has been possible to produce carbohydrates, fatty acids, polypeptides, vaccines, industrial enzymes and biodegradable plastics using the molecular farming approach.

GM Crops in Carbohydrates Improvements

Starch is one of the components of plant cells. Tailor-made starches with reduced level o' amylose or increased amount of starch have been developed in potato. Starch precursors have also been rerouted into the biosynthetic pathways of other storage carbohydrates. Nonfructan storing tobacco and potato plants have been transformed with fructosyl trans-erase gene from *Bacillus subtilis* to accumulate fructan. Production of novel compounds has also been attempted through genetic engineering. Tobacco plants have been transformed with the mannitol-1-phosphate dehydrogenase gene (mf/D) from *E. coli*. These transgenics synthesized more mannitol and were also tolerant to high salinity (Tarczynski *et al.*, 1993). Similarly myo-inositol o-methyl transferase gene from *Mesembryanthemum crystallimum* (ice plant) has been incorporated into tobacco to produce pinnitol a myoinositol derived cyclic sugar alcohol. Trehalose is an additive that improves processed and dried food by making them astier. The plant biotechnological companies MOGEN (Leiden, The Netherlands) and Calgene (Davis, USA) have reported the synthesis of small quantities of trehalose in transgenic tobacco plants.

GM Crops in Lipids

Monounsaturated fatty acid level in plants is desirable so that the nutritional value of oil is enhanced. When a rat desaturase gene was introduced into tobacco, level of palmitoleic acid (16:1) and oleic acid (18:1) levels increased (Grayburn *et al.*, 1992). For the manufacture of detergents and for specialized nutritional applications, thioesterase gene from California bay tree was incorporated into *Arabidopsis* for the accumulation of medium chain 12-carbon fatty acid as storage lipids (Voelker *et al.*, 1992). The first nonfood products of plant bioengineering in commercial production is a genetically engineered rape seed plant variety (Laurical) that was modified to produce lauric acid, a 12-carbon fatty acid used to make soaps and detergents (Table 26.3).

Table 26.2: Some of the Therapeutic Agents Produced in Transgenic Plants

Protein	*Plant(s)*	*Application*
Human protein C	Tobacco	Anticoagulant
Human hirudin variant 2	Tobacco, canola, Ethiopian mustard	Anticoagulant
Human granulocyte-macrophage colony-stimulating factor	Tobacco	Neutropenia
Human erythropoietin	Tobacco	Anemia
Human enkephalins	Thale cress, canola	Antihyperanalgesic by opiate activity
Human epidermal growth factor	Tobacco	Wound repair/control of cell proliferation
Human α-interferon	Rice, turnip	Hepatitis C and B
Human serum albumin	Potato, tobacco	Liver cirrhosis
Human hemoglobin	Tobacco	Blood substitute
Human homotrimeric collagen-I	Tobacco	Collagen synthesis
Human al-antitrypsin	Rice	Cystic fibrosis, liver disease, hemorrhage
Human growth hormone	Tobacco	Dwarfism, wound healing
Human aprotinin	Corn	Tryp: in inhibitor for transplantation surgery
Angiotensin-1-converting enzyme	Tobacco, tomato	Hypertension
α-Tricosanthin	Tobacco	HIV therapy
Glucocerebrosidase	Tobacco	Gaucher disease

GM Crops in Protein Quality

Considerable progress has also been made in the improvement of nutritional quality of legume storage protein by expression of the gene responsible for a methionine-rich protein. The Brazil nut (*Bertholletia excelsa*) seed protein, 2S albumin of Brazil nut under the promoter of phaseolin gene was introduced into tobacco. The accumulation of 2S albumin resulted in a 30 per cent increase in the methionine content in the transgenic tobacco seed (Altenbach *etal.*, 1989). Similarly, expression and accumulation of the Brazil nut 2S albumin protein in transgenic *Arabidopsis*, rapeseed, soybean, French bean\ seeds, and potato have also been reported. Unfortunately, the Brazil nut 2S albumin was found to be allergenic (Nordlee *et al.*, 1996), which raised the safety concern of development of transgenic plants expressing this protein. Sunflower seed albumin, rich in methionine, is not an allergen and has been used for development of transgenic plants, the sunflower albumin gene under the control of the pea vicilin gene promoter has been transferred to lupin, and the, methionine content of seed protein of transgenic plants increased by twofold. Rat feeding experiments demonstrated the superior nutritive value of transgenic lupin (Molvig *et al.*, 1997). Another seed storage protein gene (*Ama1*) isolated from

Table 26.3: The Use of Plants as Bioreactors for the Production of Lipids, Carbohydrates and Proteins

Compound	*Origin of Gene(s)*	*Application*	*Plant Species Used*
	Lipids		
Medium chain fatty acids	California bay tree (*Umbellularia californica*)–Thioesterase	Food, detergent, industrial	Oilseed rape
Mono-unsaturated fatty	Rat-desaturase	Food	Tobacco
	Acids		
Poly-hydroxybutyric acid	*Alcaligenes vutrophus*	Biodegradable plastics	*Arabidopsis,* oilseed rape, soybean.
Saturated fatty acids	*Brassica rapa*	Food, confectioneries	Oilseed rape
γ-Linolenic acid	Desaturase (*Synechocystis*); *Mortierella alpina*	For reducing coronary diseases	Tobacco; canola
Ricinoleic acid	Castor oleate hydroxylase	Nylon paints, varnishes, lubricants	*Arabidopsis*
	Carbohydrates		
Amylose-free starch	*Solarium tuberosutn* (GBSS)	Food, industrial	Potato
Cyclodextrins	*Klebsiella* pnei/mon/ae-Cyclodextrin glucosyl transferase	Food, pharmaceutical	Potato
Fructans	*Bacillus subtilis-F*\jctosyl transferase	Food, Industrial	Tobacco, potato
Increased amount of	*E. coli* (*glgCiG*)	Food, industrial	Potato
	Starch		
Trehalose	*E. coli*	Food stabilizer	Tobacco
Sucrose	Maize sucrose phosphate synthase (SPS)	Sweetner	Tomato
	Proteins		
Alpha-trichsantin	Chinese medicinal plant	Inhibition of HIV replication	*Nicotiana benthamiana*
Angiotensin converting enzyme inhibitor	Milk	Anti-hypersensitive effect	Tobacco, tomato
Antigens I	Bacteria, viruses	Orally administered vaccines	Tobacco, tomato/ potato, lettuce/
Antigens	Pathogens	Subunit vaccine	Tobacco/
Enkephalin	Human	Opiate activity	Oilseed rape, *Arabidopsis*
Hirudin	Synthetic	Thrombin inhibitor	Oilseed rape

Table adapted from Goddjin and Pen *et al.,* 1997). In transgenic plants, sucrose level increased with a concomitant reduction in starch.

Amaranthus is a good candidate for introduction into crop plants as the protein has well-balanced amino acid composition (Raina and Datta, 1992). This gene has been introduced into potato, the level of accumulation of *Ama1* was found to be 0.3 per cent of total soluble protein (Chakraborty *et al.*, 1998). The soybean glycinine gene has been introduced into rice to increase its protein content and to make it more digestible. In addition, some important amino acids, such as lysine, lacking in quantity in normal rice have been replinshed (Momma *et al.*, 1999).

GM Crops in Vitamin and Mineral

Vitamin A content in the diet plays a major role in preventing disorders such as color blindness. Rice, the staple food for billions of people, does not contain any provitamin A in the endosperm. Immature rice endosperm is capable of synthesizing the early intermediate geranyl geranyl diphosphate (GGPP), which can be used to produce the uncolored carotene phytoene by expressing the enzyme phytoene synthase in rice endosperm. The synthesis of ^-carotene requires three additional plant enzymes: phytoene desaturase and x-carotene desaturase, each catalyzing the introduction of two double bonds, and lycopene β-cyclase, encoded by the *lcy*[l] gene. Ye *et al.* (2000) used the *Agrobacterium*-mediated transformation method in rice to introduce the entire p-carotene biosynthetic pathway into the rice endosperm. Plant phytoene synthase (*phy*) gene originated from daffodil {*Narcissus pseudonarcissus*), bacterial phytoene desaturase (*Gt1*) originating from *Erwinia uredovora* placed under the control of endosperm-specific glutelin (Gfi) and the constitutive CaMV35S promoter respectively. The 'phytoene synthase cDNA contained a 5'-sequence coding for a functional transit peptide, and the *crtl* gene contained the transit peptide (fp) sequence of the pea Rubisco small subunit. These two direct the formation of lycopene in the endospeum plastids, the site of geranyl geranyl diphosphate formation. A third gene, lycopene β-cyclase, from *Narcissus pseudonarcissus,* controlled by the rice glutelin promoter; which carried a functional transit peptide allowing piastid import and selectable marker *aphlV*, were introduced. The resulting transgenic rice plants produce sufficient p-carotene converted to vitamin A and rice has been named as "Golden rice".

GM Crops in Commercial Transgenic Crops

Transgenic plant products have now been released as commercial varieties in a range of crops (Table 26.4). By all accounts, 1996 was a critical year in the history of plant biotechnology.

For the first time, farmers planted large areas under transgenic plants. It was in 1995 that potato varieties expressing Bt crystal protein toxic to Colorado potato beetle were made commercially available. In the following year, cotton varieties expressing the *Bt* crystal protein toxic to a number of lepidopteran pests, including the tobacco budworm and cotton bollworm, were introduced. At the same time, Bt corn hybrids with improved resistance to European cornborer became available to the farmers. All these commercial introductions have taken place in USA. *Bt* cotton varieties carrying the same technology as used in USA are being introduced in Australia. Transgenics for herbicide tolerance was the dominant trait in 1997. Transgenics have now been released for various traits from virus resistance to quality and for industrial use.

Transgenics for Quality and Improved Storage

The first approval for commercial sale of a food product was a transgenic tomato "Flavr Savr" with delayed ripening, developed by Calgene, USA in 1994. Improved storage or long shelf life of tomato that is suitable for food processing can be developed by two approaches using (i) antisense RNA technology or (ii) using gene for 1-aminocyclopropane-1-carboxylic acid (ACC; deaminase, which degrades ACC to ethylene. Calgene used antisense RNA against polygalac turonase (PG) enzyme encoding gene. On the basis of PG gene sequence, an antisense PG gene was constructed and tomato plants were transformed. These transgenic tomato plants produced both sense and antisense mRNA for the PG gene resulting in RNA-RNA pairing. This would result in nonproduction of PG gene product, and thus preventing the attack of PG gene upon pectin in the cell wall of ripening fruit and thereby preventing softening of fruit.

Table 26.4: Commercial Releases of Transgenic Plant Varieties

Trait	*Crop*	*Product Bame*	*Organization*
		Insect resistance	
Bt	Cotton	Bollguard	Monsanto Co.
Bt	Cotton	-	Calgene
Bt	Potato	Newleaf	Monsanto Co.
Bt	Maize	Yield Guard	Monsanto Co.
Bt	Maize	Maximizer	Ciba Seeds
Bt	Maize	-	Novartis, Mycogen, Pioneer Hi-Bred
Virus resistance	Squash	Freedom II	Asgrow Seeds, Inc.
Virus resistance	Tobacco, tomato, capsicum	-	(China)
Virus resistance	Cantaloupe	-	Asgrow Seeds, Inc.
Virus resistance	Papaya	Rainbow	-
		Herbicide resistance	
Glufosinate	Maize	Liberty Link	AgrEvo Canada inc.
Glufosinate		Liberty Link	AgrEvo USA Co.
Sethoxydim		-	BASF
Glufosinate		-	DeKalb Genetics Corp
Glyphosate		Roundup Ready	Monsanto Co.
Glufosinate		-	Plant Genetic Systems V
Glufosinate	Oilseed rape, Canola	Innovator	AgrEvo Canada Inc.
Glufosinate		"	AgrEvo USA Co.
Glyphosate		Roundup Ready	Monsanto Co.
Glufosinate		–	Plant Genetic Systems
Bromoxynil		–	Rhone-Poulene
Bromoxynil	Cotton	BXN	Calgene *rz*
Sulfonylurea		-	DuPont
Glyphosate		Roundup Ready	Monsanto Co.
Glyphosate	Soybean	Roundup Ready	Monsanto Co.
Sulfonyl urea	Flax	Triffid	Univ. of Saskatchewan
Glufosinate	Chicory	-	Bejo-Zaden

Contd...

Table 26.4–Contd...

Trait	Crop	Product Bame	Organization
		Quality	
Vine ripened, shelf life	Tomato	Flavr Savr	Calgene
Vine ripened, shelf life	Tomato	Endless Summer	DNA Plant Technology -
Delayed softening, paste Consistency	Tomato	-	Zeneca
Modified starch	Potato		AVEBE
Increased vase life	Carnation		Florigene
Modified flower color	Carnation	-	Florigene
Oil characteristics	Canola	Laurical	Calgene
Male sterility hybrid system	Canola	-	Plant Genetics Systems

More than 4500 GM plant varieties have been tested in the United States, and 50 have been approved for unlimited release. Researchers have been transferred genes to mycorrhizal symbionts of plant roots to confer resistance to pathogens and increase efficiency (Hiremath and Podila, 2000).

Most GM plants target plant protection, including herbicide resistance, insect resistance and disease resistance. There are also many modifications to plant quality including nutrition and chemical content, timing of ripening, increased photosynthesis, and stress tolerance. Plants are also being designed to produce pharmaceuticals, industrial proteins, and biodegradable plastics. Others will remove heavy metals from contaminated soils (Krimsky *et al.*, 1996; Hapgood, 2000).

But regulations will be difficult to enforce. If anybody can produce GM plants, many will. There is no way that every new product can be thoroughly tested for health and environmental safety. Patents can be pirated. Products may be introduced with disclosure of their genetic identity. GM crops produced in foreign countries may cross our borders, and vice-versa. Careless application of marketing rules will lead to debacles such as the recent contamination of taco shells by Starlink corn not approved for human consumption.

Developing nations see great hop in GM plants to help solve their problems. A good example is China (Altman and Watanabe). Chinese scientists have already inserted virus coat proteins into tobacco, tomato, potato, green pepper, chrysanthemum, and rice. The virus coat proteins induce some as-yet unknown mechanism of immunity against virus infection in the host plants. The Chinese have also inoculated more than a million hectares of rice and soybean with recombinant strains of *Rhizobium* bacteria that fix nitrogen more efficiently. Chinese scientists claim a 5-10 per cent increase in yield. Increase cadmium uptake and tolerance, and plan to use the weeds to clean up pollution. They are also working on saline and drought resistance, starch and oil content, and increased protein storage. In area where Chinese growers were spraying pesticides on cotton an average of 15 to 20 times per season, introduction of *Bt* cotton reduced the sprays to near zero in may cases (Cline, Western Farm Press, 2000).

In 1996 the International Food Biotechnology Council convened a panel of allergy experts to develop a standard protocol. First, if transgenes are from a known source of allergy, standard serological and skin prick tests are applied to the new product to check for the known allergen. If the transgenes are from a novel source, they are analyzed for amino acid sequences similar to known allergens, and if some are found, serum antibody tests using the known allergen can be test for reaction to the novel protein. If the researchers find no homologies to known allergens, the novel proteins are submitted to simulated digestion experiments. If they are not digested quickly enough, further tests are needed before marketing or regulatory approval (Lehrer, Samuel, 1999).

Another recent finding is that both plant and animal chromosomes contain large amounts of ancient virus DNA that has become inactivated and lies dormant with the host genome. Ho and her colleagues fear that recombination with CaMV promoter could reactivate some of these provirus segments to form virulent new plant viruses. CaMV (Cauliflower Mosaiç Viral) is related to human hepatitis B virus and retroviruses such as HIV, and some critics hint that CaMV could even recombine to form new virus strains pathogenic to humans, a very scary thought to the public. Of course, GM defenders have rebuttals to all these theories by the critics (Ho *et al.*, 1999).

What are the Concerns About Growing GM Crops?

Farmland is already in severe decline because of intensive, chemical farming. For example, plants which were considered to be arable weeds 40 years ago are now listed as rare or scarce and some are endangered species (Wilson *et al.*, 1990). There is widespread concern that the use of GM herbicide tolerant crops could make this worse.

GM herbicide tolerant crops allow farmers to apply 'broad spectrum' weed killers to their field, which kill all other plants. There is concern that this will continue the decline of farmland wildlife because the use of these GM crops could lead to the removal of weeds from all crops in the normal arable rotation. "Untested introduction of GM crops could be the final blow for such species as the skylark, corn bunting and the linnet, as the seeds an insects on which they feed disappear" (Domingo, 2000).

The GM oilseed rape supplied by Aventis for the farm scale trials in England and Scotland was found to be contaminated with an unauthorized GM variety containing antibiotic resistance markers. The contaminated seed had been plants at more than 20 sites since 1999.

As can be seen, GM contamination is already a problem. The long distance that viable pollen can travel mean that the separation distances between GM and other crops would have to be very large to be effective.

Most GM crops being grown at the moment are destined for markets in rich countries. Such products will not help to feed the poor and hungry of the world. The majority of GM crops are designed for use in intensive farming systems, with single crop in large fields requiring heavy use of chemical inputs. Many farmers in developing countries are small scale, growing many different crops and they often cannot afford the chemicals needed.

Scientists have already raised similar concerns about the feeding of GM in high quantities to both farm animals and babies. Scientists working for the Zambian Government which rejected GM food aid stated that "While it is often said that GM maize is consumed by millions of Americans, it was noted that it is eaten in highly processed form and is not a staple food in the USA. In Zambia maize is the staple food and is usually the only carbohydrate source" (Guest, 1992).

References

Adoption of Genetically Engineered Crops in the U.S. *USDA ERS* July 14, 2006.

Altenbach, S.B., Pearson, K. W., Maker, G., Staraci, L. C., and Sun, S.S.M. (1989) Enhancement of methionine content of seed proteins by the expression of a chimeric gene encoding a methionine rich protein in transgenic plants. *Plant Mol. Biol.* (13): 513-522.

Altman, David w., and Kazuo N. Watanabe (ed.) *Plant biotechnology transfer to developing countries.* R. G. Landes Co. Austin, Texas.

Chakraborty, N., Chakraborty, S., Kesarwan, M., Azam, M and Datta, A. (1998) *Proceedings of IUBS symposium*, Taipei, Taiwan, 1997, pp. 125-131.

Cline, Harry, (2000) *Ag. Biotech quizzing aplenty.* Western Farm Press (Sept, 2, 2000).

Domingo, J. L. (2000) Health risks of GM foods: many opinions but few data. *Science* (288): 1748-9.

Goddjin, O.J.M, and Pen, J (1995) Plants as bioreactors. *TIBTECH* (13): 379-387.

Grayburn, W.S., Collius, G. B. and Hildebrand. D. F. (1992) *Biotechnology,* (10): 675-678.

Guest, G. (1992) *Response to FDA Draft Federal Register Notice on Food Biotechnology.*

Halfore Ncordell, Halford. N (2003). *Genetically modified crops.* Imperial college Press. ISBN 1-86094-353-5.

Hapgood Fred. (2000). Garage biotech is here, or just around the corner. Will genetic modification for fun and profit become a homegrown industry? *Civilization* (April/May 2000): 45-51.

Hiremath, Shivanand, and Gopi Podila. (2000). Development of genetically engineered mycorrhical fungi for biological control. In Podila, Gopi and David Douds, Jr. (ed.) *Current Advanes in Mycorrhizae Research.* APS Press, St. Paul, Minnesota.

Ho, Mae-Wan, Angela Ryan, and Joe Cummins (1999). Cauliflower mosaic Viral Promoter–a recipe for disaster? *Microbial Ecology in Health and Disease* II (4).

Krimsky, Sheldon, and Roger Wrubel. (1996). *Agricultural biotechnology and the environment: Science, policy and social issues.* Univ Illinois Press, Urbana, Illinois.

Lehrer, Samuel, B. (1999). The Potential Health Risks of Genetically Modified Organisms: How can Allergens be Assessed and Minimized? *CGIAR/NAS Biotechnology Conference–October 21-22, 1999.* Univ. Illinois Press, Urbana.

Maessen, G.D.F. (1997). *Genomic stability and stability of expression in genetically modified plants.*

May, Sir. R. (1999) Quote from Sunday Express 14th November 1999.

Mitchell C. Tarazynski, Richard G. Jensen, and Hans J. Bhnert (1993). Stress Protection of transgenic Tobacco by Production of the Osmlyte Mannitol *Science,* (239): 508-510.

Molvig, L., Tabe, L. M., Eggum, B.O., Moore, A.E., Craig, S., Spencer, D. and Higgins, T. J.V. (1997) Enhanced methionine levels and increased nutritive value of seeds of transgenic Lupin (*Lupinus angustifolius*) expressing a sunflower seed albumin gene. *Proc. Natl. Molecular cloning of a gene encoding a seed specific protein with nutritionally balanced amino acid composition from Amaranthus.* Proc. Natl. Acad. Sci. USA (89): 11774-11778.

Momma, K., Hashimoto, W., Ozama, S., Kawai, S., Katsube, T., Takaiwa, F., Kitos, M., Utsumi, S., and Murata, K. (1999) Quality and Safety evaluation of genetically engineered rice with soybean glycinin: analyses of the grain composition and digestibility of glycinin in transgenic rice. *Biosci. Biotechnol. Biochem.* (63): 314-318.

Nordlee, J. A., Taylor, S.L., Townsend, J. A., Thomas, L. A., and Bush, R. K. (1996) Identification of a Brazil nut alhrgen in transgenic soybeans. *New Eng. J. Med.* (339): 688-692.

Raina, A., and Datta, A. (1992) Molecular cloning of a gene encoding a seed specific protein with nutritionally balanced amino acid composition from Amaranthus. *Proc. Natl. Acad. Sci. USA.* (89): 11774-11778.

Simpson, EC *et al.* (1999) Gene flow in genetically modified herbicide tolerant oilseed rape (*Brassicanapus*) in the UK. Gene flow and agriculture: Relevance for Transgenic crops. 1999 *BCC Symposium Proceedings* No.72 pp. 72-81.

Voelker, T. A., Worrell, A. C., Anderson, L., Bleinbaum, J., Fane, C., Hawkins, D.J., Radke, S.E and Davis, H. M. (1992) Fatty acid biosynthesis redirected to medium chains in trangenic oil seed plants. *Science*, (257): 872-874.

Wilson, P.J, Boatman, N.D. and Edwards P.J. (1990) Strategies for the conservation of endangered arable weeds in Great Britain. *Proceedings of the European Weed Research Society Symposium 1990*: Integrated Weed Management in Cereals pp. 93-100. *European Weed Research Society*, Helsinki.

Ye, S., Al-babili, s., Kloeti, A., Zhang, J., Lucca, P., Beyer, P. and Potrykus, I. (2000) Engineering the provitan in A (Beta Carotene) biosynthetic pathway into (Carotenoid free) rich endosperm. *Science* (287): 303-305.

Microbial Biotechnology and Ecology (2011) *Pages* **484–500**
Editors: **Deepak Vyas, G.S. Paliwal, P.K.Khare and R.K Gupta**
Published by: **DAYA PUBLISHING HOUSE, NEW DELHI**

Chapter 27

Plant DNA Banking for Biodiversity Assessment and Conservation Research

☆ *D. Thangadurai[1], M.A.A. Pinheirho De Car Valho[2], D.P. Biradar[3], Hemanth K.N. Vasathaiah[4], B. Suresh[5] and Ganesh R. Hegde[6]*

[1]Department of Botany, Karnatak University, Dharwad – 580 003
[2]ISO Plexis Germplasm Bank, Department of Biology, University of Madeira, Funchal 9000-390, Portugal
[3]Department of Agronomy, UAS, Dharwad – 580 005
[4]Center for Viticulture and Small Fruit Research, Florida A and M University, 6505 Mahan Drive, Tallahassee, Florida 32317, USA
[5]Department of Applied Genetics, Karnatak University, Dharwad – 580 003
[6]Department of Botany, Sri Krishnadevaraya University, Anantapur – 515 003

Introduction

The sequencing of plant genomes opened new frontiers in conservation of their biodiversity and genetic improvement. The advances in gene isolation and sequencing in many plant species allows to envisage that within a few years, gene-bank curators may replace their large cold stores of seeds with plant DNA bank that will be electronically stored (Deutsche Forschungsgemeinschaft, 2008). The characterization of plant genomes will ultimately create a true DNA bank, which should possess a large and accessible gene inventory of today's non-characterized plant gene pools (Willis and van Wyk, 2006). Of course, seed banks of comprehensively investigated stocks should remain because geneticists and plant breeders, the main users of gene banks, will need this germplasm for their work (Klenk, 2008). Genomics may accelerate the utilization of candidate genes available at these gene banks through transformation without barriers across plant species or other living kingdoms (de Vicente and Andersson, 2006). Nonetheless, genetic engineering should be seen as one of the methods of plant breeding that permits the direct alteration and re-building of a crop population. 'Shutting-off' genes coding for undesired characteristics may be another application of transgenics in crop improvement (de Vicente *et al.*, 2006; Moore and Ming, 2008).

There are many areas of basic and strategic research in plant breeding and genetics that are being facilitated by marker-aided analysis (Paterson, 1996). With molecular markers, plant biologists are reviewing crop evolution and gathering new knowledge. Such information should be incorporated into genetic enhancement programmes, especially those with an evolutionary breeding scheme. Likewise, plant ideotypes for each crop should drive the work of plant breeders (Jan Engels and Visser, 2003). Specific plant morphotypes have been defined in rice and wheat based on accumulated knowledge of crop physiology and crop protection (de Vicente and Andersson, 2006). The needed characteristics required to develop improved plant prototypes ensuing from such a 'virtual breeding' approach may be available in gene banks of the crop or in those of other species. Otherwise, breeders may obtain novel transgenes to develop the required ideotype (Zetzsche and Gemeinholzer, 2007).

Nowadays, the finding of new genes that add value to agricultural products seems to be very important in the private agri-business (Dunwell, 2000). Unique gene databases are being assembled by the industry with the massive amount of data generated by genomics research (Jauhar, 2006). A new term 'bioresource' was coined recently to refer to a fast and effective licensed technology of pinpointing genes. With this method, a 'benign' virus infects a plant with a specific gene that allows researchers to observe directly its phenotype (Henry, 2006). Bioresource replaces the standard time-consuming approach of first mapping a gene to subsequently determine its exact function (Sharma *et al.*, 2000). Gene identification in DNA libraries coupled with bioresource technology and an enhanced ability to put genes into plants will be routine for improving crops in the next decade (Li *et al.*, 2001).

Plant DNA banks played a crucial role in restoring crop diversity lost during emergencies in Cambodia, across Africa and in the Middle East. These instances highlight the value and importance of accurate characterization and documentation of DNA bank accessions, so that the right materials can be found quickly and targeted appropriately. Several crops are more difficult to conserve (de Vicente and Andersson, 2006). Potatoes, yams and cassava, for example, are not generally propagated by seed but are grown from tubers or other plant parts. The storage organs of such crops may be kept under cool conditions and at an appropriate humidity (Henry, 2006). But they take up much more space than seed, and must be regenerated (replanted) much more frequently, making it considerably more expensive to conserve such crops than those that can be stored as seeds (de Vicente and Andersson, 2006). Alternatively, root and tuber crops, as well as crops that do not produce seed at all (such as banana and plantain), or that produce seed that dies when it is dried and cooled (such as many tropical fruit species), can be maintained as living collections of plants growing in the field; these are called field gene banks (ETC, 2002; Johnson, 2008). However, such collections are very susceptible to natural hazards–pests, diseases and adverse weather–and it is desirable to also maintain them in more protected conditions (Frankel *et al.*, 1995; Smale, 2000). Thus collections of crops that cannot be stored as seed, are often maintained as DNA. But as the technical complexity of conservation increases, so does the cost (Virchow, 2003).

Modern agriculture entirely depends on the genetic and population diversity of a particular crop. If a new rice disease appears, researchers can search the DNA for resistant varieties. The alleles required to make rice more tolerant of drought, for example, exist within the varieties contained in the collection (Mc Cullum *et al.*, 2003; Purty *et al.*, 2008). The DNA bank contains the genetic diversity helps in identifying such germplasm for genetic improvement purposes. Hence, this article summarizes information on different strategies to plant DNA bank (Henry, 2006).

Diversity and Distribution of Plant Genetic Resources

To exercise national sovereignty over its genetic resources, documentation and characterization of biodiversity based on various approaches, is a prerequisite for any country on the globe. In most instances, genetic diversity is the key to developing crop varieties from wild species. Variability of genetic traits allows some plants in a population to adapt to particular conditions and climates, to be more palatable to humans, or to be more resistant to attacks by pests and disease. Seeds selected from genetically diverse, wild populations performed with varied success in subsequent years. Regional varieties developed as seeds were continuously grown and selected from the same location. When humans began transporting seeds around the globe, regional varieties emerged in diverse regions from diverse cultures (de Vicente and Andersson, 2006). Today for example, we have regional varieties of chiles from Thailand, China, Europe, and even our local improved variety from Araku Valley, all of which derived from a common ancestral gene pool in South America. Human agricultural influence increased genetic diversity within a species by moving it from region to region and propagating it from year to year. Until the twentieth century, it was farmers working with open-pollinated crops who were the plant breeders (Henry, 2006).

During the last century, much has changed with the discovery of genetics and the development of the hybrid seed industry (Griliches, 1991). Hybrids (made from crossing two inbred, open-pollinated varieties) were revolutionary, exhibiting the extreme uniformity advantageous to mechanized farming. Additionally, in certain crops, hybrid varieties exhibited increased vigor compared to some open-pollinates (Peferoen, 1997). Hybrids, however, not only appear uniform, but are genetically extremely uniform, further limiting crop genetic diversity. The first hybrid seed sold to the public was a corn variety developed by Pioneer Seed in the 1920's (Griliches, 1991). Its success came, in part, because corn is one of the crops that benefit most from hybridization (Morrisa *et al.*, 1998). However, due to its success, farmers were quick to adopt hybrid varieties of other crop species that may not have exhibited such improvement. Throughout the 1930s and 40s, hybrids swept the marketplace. As farmers shifted from saving seed to buying from 'professionals', the culture of seed saving dwindled (Pray 2003).

Today, genetic diversity is further reduced by the modern biotechnology industry (Komen and Persley, 1993; Deeds *et al.*, 1997). Rapid reproduction of tissue cultures creates fields of genetically identical plants (de Carolis and Deeds, 1999). The insertion of genetic material in crops places the same genes into each plant. In addition, genetically modified genes pose a threat to our crop species' wild progenitors growing within their regions of origin (Uzogara, 2000). Currently, there is a heated controversy over whether genetically modified genes have been found in land race and wild species of corn growing in Oaxaca, Mexico, modern corn's genetic birthplace (Giesecke, 2000).

The concern is a scrubby wild plant could hold the key to agricultural improvements of enormous value to struggling farmers and the nutrient-deprived poor (Dale, 1992). For example, wild relatives of crops such as the brassicas have great potential to assist in crop improvement. The wild relatives of crops, which include crop ancestors as well as other species more or less closely related to them, have been an important part of crop improvement programme for over 60 years (Raybould and Gray, 1993). Genes from wild relatives have provided crops with resistance to pests and diseases, enhanced tolerance to abiotic stresses and have made possible the production of hybrid varieties and improved quality traits such as protein content (de Vicente and Andersson, 2006). Some successful examples of the use of wild relatives in crop improvement include the introduction of resistance to potato late blight (the disease that caused the Irish famine of the 1800s); the production of rice hybrids, currently planted on 45 per cent of China's rice-growing area; and increased soluble solid content in tomatoes, a trait worth US$240 million per year (Xiao *et al.*, 1998).

A recent study of the presence of genes from crop wild relatives in released varieties of 16 mandate crops of the Future Harvest Centres highlights the successful use of wild relatives in breeding programmes (Hilder and Boulter, 1999; Johnson *et al.*, 2003; Bragdon *et al.*, 2005). While breeding efforts continue to emphasize the use of pest and disease resistance genes, a wider range of characteristics are being introduced from wild relatives than in the past, such as drought tolerance, ability to cope with high or low temperatures, enhanced yield and quality traits such as higher levels of iron (Pasternak, 1987; Plucknett, 1987; Giri and Laxmi, 2000; Smith and Dilday, 2002). Nevertheless, crop wild relatives are still under exploited. Recent advances in molecular technologies and improved interspecies crossing procedures have overcome many genetic and agronomic barriers to using wild relatives in breeding programmes (Jarvis and Hodgkin, 2002). It is hoped that this development will foster greater recognition of the importance and value of using wild relatives in crop improvement (Stewart *et al.*, 2003).

Global Loss of Plant Genetic Resources

The world's biota is under siege with a human population nearing 6 billion places unprecedented pressures on the biosphere (Altieri and Merrick, 1987). Approximately 95 per cent of the terrestrial surface is now occupied by human settlements or ecosystems managed for food and materials production. As a result, natural ecosystems are destroyed and fragmented, species are destroyed or doomed, and global genetic diversity is diminished (Cohen *et al.*, 1991). The scale of this destruction directly threatens human civilization, which is dependent in numerous ways on biological diversity. The enormity of the crisis has stimulated international efforts to conserve biological diversity and to ensure sustainable use of its components (Ellstrand and Elam, 1993).

The world is facing a biodiversity crisis of unprecedented magnitude. Unless there are radical changes in human behavior, we can expect significant and alarming species loss in the near future. For a few reasons, DNA banks have great potential as tools for biodiversity conservation, yet few have been established (Francisco-Ortega *et al.*, 2000). One reason to conserve crop diversity in DNA banks is that it is under threat elsewhere. Habitats continue to be destroyed and with the habitats go the plants. One of the main threats to crop diversity is modern agriculture. As new varieties are taken up by farmers because they offer genuine benefits, they may displace the diversity that was there before (de Vicente and Andersson, 2006). This is especially ironic because all plant breeding is built on existing crop diversity, which makes it imperative that this diversity be conserved and remain available somewhere (Fowler and Hodgkin, 2004). Crop diversity, once lost, is impossible to replace (Henry, 2006).

DNA: The Molecule of Heredity

DNA is a very long, threadlike macromolecule made up of large number of deoxy ribonucleotides, each composed of a base, a sugar and a phosphate group. The bases of DNA molecules carry genetic information, whereas their sugar and phosphate group performs a structural role. This hereditary molecule in recent times can be used to find answers to many unresolved biological questions (Avise *et al.*, 1987). In any given eukaryotic organism, this can be isolated, purified, quantified and maintained for many hundreds of years under controlled conditions (Avise 1991).

The Concept of DNA Banking

The accelerated use of genomic technology in biological research portends a changing pattern in the use of genetic resource collections and fuels the desire among researchers to establish DNA banks. Also, DNA banks are very fundamental in organizing a comprehensive 'Encyclopaedia of Life', as envisioned by E.O. Wilson (2003). DNA sequencing of herbarium sheet specimens, cell cultures and germplasm offers a new approach to studies of gene function, evolution, taxonomy and epidemiology among plants. A catalog of unique genetic sequences 'DNA barcodes' can conceptually unite diverse assemblages of repositories under a common registry of sequence accessions (Armstrong and Ball, 2005; Brower, 2006; Rubinoff, 2006). The challenge for DNA repositories of today is to provide a product that is amenable for uses anticipated in the future. Long-term needs will have a decisive impact on operational organization and repository design and a major limitation involves specimen quality and shelf-life (de Vicente and Andersson, 2006).

For many species that are difficult to conserve by conventional means either as seeds or vegetatively or that are highly threatened in the wild, DNA storage may provide the ultimate way to conserve the genetic diversity of these species and their populations in the short term, until effective methods can be developed (Adams, 1997). As progress in biotechnology advances, numerous techniques are in hand now for conserving germplasm, each with their own advantages and disadvantages (Maxted *et al.*, 1997; Maunder *et al.*, 2004). With the recent developments of Marker Assisted Selection and DNA Barconding of plant genetic resources, DNA extraction and storage are regarded as one of the emerging *ex situ* techniques for plant germplasm conservation, management and further future utilization (Ribaut and Hoisington, 1998).

Loosely, it has been argued that DNA banking is an unrealistic option for germplasm conservation given that DNA cannot resurrect extinct species (Maunder *et al.*, 2004); of course, the DNA extracted from those herbarium specimens of extinct species may provide ample evidences to speculate the speciation and evolutionary significance of such plants. In recent times, DNA is the most frequent and easiest forms of germplasm requested from genebanks for research purpose (de Vicente, 2004) especially for molecular, phylogenetic and systematic research in plants (Henry, 2006).

Approaches of *in situ* conservation have been applied most frequently to cultivated plants, landraces and weedy forms, and not so much for the wild relatives of crops (Kennedy, 2004). Collecting and maintaining representative samples of such wild plant populations in the DNA is crucial, though there has been much debate on how much to collect and from how many populations to ensure the maximum of genetic diversity (Brown and Briggs, 1991; Neel and Cummings, 2003). Today, there has been an increasing global concern towards biodiversity assessment and conservation for future generations (Orr, 2003). There are many approaches exclusively available for this purpose, including *in situ*, *ex situ* and combinations of both techniques. Moreover, when compared to animals, plant conservation activities are inexpensive and less time-consuming. In the late 1980s, efforts were made to encourage groups and institutions to establish repositories of genomic DNA in an attempt to complement other conservation efforts. By 1992, thirty one institutions were participating in a network of DNA banks. In addition, a number of international, national and regional DNA banks have now recently been established (Adams, 1997).

With the advent of traditional plant conservation activities such as the management of *ex situ* living collections in seed banks and botanic gardens, and specimen collections in herbaria and museums, it becomes as easy as to extract and store DNA from such collections for various purposes. It has been estimated in the past that the genetic variation in different plant species is in the range of 1 million to 10 million base pairs and that a DNA bank holding such 10 million species on earth (Mattick *et al.*, 1992). It means that, only very little effort is needed to collect and document such DNA as a genetic resource (Ryder *et al.*, 2000).

A DNA bank is a particular type of genetic resource bank that preserves and distributes the DNA molecule and provides associated information. Existing DNA banks usually support agricultural applications, but may also be necessary to support genomic projects in all species. Material stored at the bank is made available through research groups and individuals making donations of plant material and/or DNA. The DNA is stored at –80°C with the equivalent voucher stored within a herbarium. For example, databased information on DNA stocks and associated data on the source plant species is available as given in the Table 27.1. DNA from plants provides endless information about each species. Researchers that can take advantage of these facilities include phylogenetic, phylogeographic and population studies as well as gene discovery and marker development (Hodkinson *et al.*, 2007).

Table 27.1: Databased Information on DNA Stocks and Associated Data on the Source Plant Species Available at Fairchild Tropical Botanic Garden, Florida, USA for Phoenix DNA Samples

Bank ID	*Accession*	*Family*	*Name*
822	93378A	Arecaceae	Phoenix dactylifera 'Zahidi'
1224	76843F	Arecaceae	Phoenix loureiroi
1221	92263B	Arecaceae	Phoenix loureiroi
990	92263C	Arecaceae	Phoenix loureiroi
1009	95255B	Arecaceae	Phoenix paludosa
1010	99875B	Arecaceae	Phoenix paludosa
1011	89800A	Arecaceae	Phoenix roebelenii
1230	P.129B	Arecaceae	Phoenix roebelenii
1005	88536C	Arecaceae	Phoenix rupicola
1222	88536C	Arecaceae	Phoenix rupicola
996	88536E	Arecaceae	Phoenix rupicola

Many of the world's most valuable food crops–including wheat, rice, maize and many pulses–produce seeds that are relatively easy to store. For conservation purposes, the seeds are cleaned, dried, and placed in a sealed jar or packet. For medium term storage (20 to 30 years), the seeds are maintained at a temperature of about 5°C. For long-term storage (up to 100 years or even more) they are kept in cold rooms, or even domestic deep freezers, at –18 to –20°C. Like plant tissue, isolated plant DNA can be stored, enabling the conservation of the genetic make up of plants like seedless banana varieties that would otherwise be difficult or impossible to maintain (de Vicente and Andersson, 2006).

As methods for extraction, purification and amplification of DNA improve, we actually need very little plant material in order to adequately supply molecular labs. In general, it has been recommended to collect about 3-5 square inches of fresh, young plant material, normally 3-5 leaves or cut parts of larger leaves. It is preferable to remove the midrib from leaf material before putting it into silica gel. Scissors or secateurs can be used to cut plant material. Tearing the material will bruise the leaves, releasing enzymes that will breakdown DNA. The most important protocol to follow when collecting material for DNA is getting the plant material to dry quickly and this is accomplished by increasing the surface area (cutting the leaves into smaller pieces) and using an excess of silica gel. The material is then placed in a reclosable plastic bag with enough silica gel to ensure drying of the material in a couple of hours. Once the material is dry, most of the silica gel can be removed from the sample and dehydrated for future use. Indicator silica gel should be used to help to determine the overall 'dryness' of the sample. Once has to check the sample for signs of mold and add fresh silica gel if necessary. Once dry, the sample can be stored at room temperature without degradation of DNA, although it is preferable to store samples at –20°C (de Vicente and Andersson, 2006).

In general, DNA is extracted using a modification of the CTAB protocol and the DNA banded in cesium chloride. Although more time consuming then the rapid methods available today, the quality of the DNA is very high and the quantity is sufficient that one can share it with many people (Liston *et al.*, 1990; Do and Adams, 1991; Adams, 1997). Leaf material is collected in the field and stored in self-sealing plastic bags at 4°C until processed. Voucher specimens collected are deposited at the local herbarium within the country. All extracted DNA is purified, quantified and maintained at –20°C to –80°C (Nickrent, 1994; Gachet *et al.*, 1998; Porteous *et al.*, 2003).

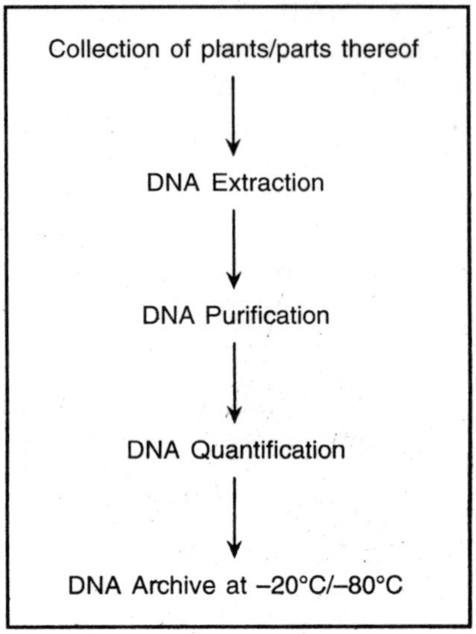

Figure 27.1: An Overview of Steps Involved in the Process of Plant DNA Banking at –20°C/–80°C

Plant DNA Banks Around the World

Several international and national efforts have been documented to establish DNA banks worldwide. As plant genes are now being sequenced and barcoded, access to continuous supply of DNA samples from individuals is of great deal in biodiversity and molecular research. Although, DNA banks exist in many countries and in different regions of the world, there is no effective coordination among DNA banks. Also the value of DNA banking to develop conservation and population management strategies is immense. DNA bank also provides the opportunity of analyzing data from the wild and helps in the development of conservation strategies (de Vicente and Andersson, 2006).

In the 1980s, attempts were made to create the plant 'DNA Bank-Net' (Adams, 1988; Adams and Adams, 1992). Unfortunately, the DNA Bank-Net initiative has remained inactive until recently (Ryder *et al.*, 2000; Savolainen and Reeves, 2004). Although there have been more than 1500 genebanks, less than twenty plant DNA banks are operational in various geographical regions of the world, with specific theme of collections of their own country's plant genetic resources and/or exotic germplasm if any (Table 1). Among them, most well known DNA banks are at the Royal Botanic Gardens, UK and the National Herbarium, Netherlands (Adams, 1992). One can deposit and request DNA of a kind of genetic material there are interested to study (Adams *et al.*, 1992). For students and scientists, this will be given at free of cost for research purpose and for others, a nominal charge of around 25USD will be charged for each DNA sample OF 25μL (Adams, 1993; Adams, 1994). These DNA banks will also do DNA extraction for outside people on specific requests (Adams *et al.*, 1999).

Plant DNA Bank: An Indian Scenario

Many of India's natural habitats are under threat and the species within them face potential extinction. Increased pressures from agricultural practices, industry, recreational pursuits and an expanding population has seen India's native flora decline. In order to better manage and preserve our botanical heritage, we need to understand the diversity within our native flora (Harlan, 1975). In this connection, it is necessary to build the research and conservation potential of plant scientists and

Table 27.2: List of Major Plant DNA Banks Around the World

Name of the DNA Bank	*Place and Country*	*Website*
Missouri Botanical Garden	Missouri, USA	http://www.mobot.org/MOBOT/research/diversity dna_banking.htm
Fairchild Tropical Botanic Garden	Florida, USA	www.fairchildgarden.org/
South African National Biodiversity Institute	Capetown, South Africa	http://www.nbi.ac.za/
Kew Royal Botanic Garden	Surrey, Great Britain	http://www.kew.org/data/dnaBank/homepage.html
Australian Plant DNA bank Ltd	Centre for Plant Conservation Genetics, Southern Cross University, Lismore, Australia	http://www.biobank.com
Irish DNA Bank	Trinity College, Dublin, Ireland	www.tcd.ie/
Brazil Plant DNA Bank	Instituo de Pesquisas, Jardim Botanico de Rio de Janeiro, Rio, Brazil	http://www.jbrj.gov.br/pesquisa/div_molecular/ bancodna/index_ing.htm
University of Hawaii	Hawaii	www.hawaii.edu
National Herbarium	Netherlands	www.nationaalherbarium.nl
The Berlin-Dahlem DNA bank	Germany	www.bgbm.org/bgbm/research/dna
NIAS DNA bank	National Institute of Agrobiological Sciences, Ibaraki, Japan	www.dna.affrc.go.jp/
The Plant DNA bank of Korea	South Korea	www.pdbk.korea.ac.kr
DNA Bank	International Rice Research Institute, Philippines	www.irri.org/grc
Italian Plant DNA Bank	Institute of Plant Genetics (CNR), Bari, Italy	http://www.inaav.ba.cnr.it
CATIE DNA Bank	Turrialba, Costa Rica	http://www.catie.ac.cr
RIPF DNA Bank	Pomologiczna, Poland	http://www.insad.pl
CIRAD DNA Bank	Montpellier, France	http://www.cirad.fr

Table 27.3: Potential Centers of Plant DNA Banking Possibilities within India

Institute and Place	*State*	*Special Interest*
National Botanical Research Institute, Lucknow	Uttar Pradesh	Wild plants
NRC for Banana, Trichy	Tamil Nadu	Bananas and Plantains
Central Institute of Medicinal and Aromatic Plants, Lucknow	Uttar Pradesh	Medicinal and Aromatic Plants
Central Drug Research Institute, Lucknow	Uttar Pradesh	Medicinal and Aromatic Plants
Tropical Botanic Garden and Research Institute, Palode	Kerala	Trees, woody lianas, orchids, bamboos, medicinal, aromatic, rare and threatened plants
Indian Institute of Horticultural Research, Bangalore	Karnataka	Wild and cultivated forms of fruit, vegetable, floricultural and horticultural crops
Institute of Forest Genetics and Tree Breeding, Coimbatore	Tamil Nadu	Tree, timber and forest species
Botanical Survey of India	Throughout India	Angiosperms, gymnosperms and the entire Indian flora
Central Potato Research Institute, Shimla	Himachal Pradesh	Potato
GB Pant Institute of Himalayan Environment & Development, Almora	Uttarakhand	Himalayan Species
National Research Centre for Soybean, Indore	Madhya Pradesh	Soybean
National Research Centre for Sorghum, Hyderabad	Andhra Pradesh	Sorghum
National Research Centre for Orchids, Pakyong	Sikkim	Orchids
National Research Centre for Onion and Garlic, Pune	Maharashtra	Onion and garlic
National Research Centre for Medicinal & Aromatic Plants, Boriavi	Gujarat	Medicinal and Aromatic Plants
National Research Centre for Grapes, Pune	Maharashtra	Grapes
National Research Centre for Cachew, Puttur	Karnataka	Cachew
National Bureau of Plant Genetic Resources, Pusa	New Delhi	Agri-horticultural crops
Sugarcane Breeding Institute, Coimbatore	Tamil Nadu	Sugarcane
Indian Institute of Vegetable Research, Varanasi	Uttar Pradesh	Vegetable Crops
Indian Institute of Sugarcane Research, Lucknow	Uttar Pradesh	Sugarcane
Indian Institute of Spices Research, Calicut	Kerala	Spices
Indian Institute of Pulses Research, Kanpur	Uttar Pradesh	Pulses
Indian Grassland and Fodder Research Institute, Jhansi	Uttar Pradesh	Grass and fodder species
Central Tuber Crops Research Institute, Thiruvananthapuram	Kerala	Tuber crops
Central Rice Research Institute, Cuttack	Orissa	Rice
Central Plantation Crops Research Institute, Kasaragod	Kerala	Plantation Crops
Central Institute for Cotton Research, Nagpur	Maharashtra	Cotton
Central Agricultural Research Institute, Port Blair	Andaman & Nicobar Islands	Agri-horticultural crops
National Research Centre for DNA Fingerprinting, Pusa	New Delhi	Agri-horticultural crops and their wild relatives

conservationists in India by setting up a unique archive of plant genetic resources in the form of a DNA bank, and to formulate a policy framework and supporting documents to facilitate its scientific use under strict control and in line with developing CBD policy on access to genetic resources. The establishment of the Indian Plant DNA Bank will provide a repository for DNA samples as a resource to be utilized in plant research by current and future generations. It serves as a historical record of plant diversity and a foundation for conservation genetics and biotechnology. India is one of the 12 megadiverse countries in the world, also provided with two hotspots (Reid, 1998). Hence, our ultimate aim should focus on to preserve representative genetic information from the entire Indian flora. In the field of plant diversity conservation, DNA storage is not very common and little information is available about its practice in India. Moreover, the technical capacity and financial resources for DNA banking are currently very limited in most of the developing countries including India.

Collective efforts have to be made to catalogue, characterize and preserve representative genetic information in the form of genomic DNA from the entire Indian flora, with the aim of facilitating taxonomic and evolutionary studies world-wide by way of inclusion of genetic diversity of plant genetic resources available in the country. By way of honoring the spirit of the Convention on Biological Diversity, emphasis has to be placed over the next few years to extract DNA (and voucher) of all Indian plant species at national institutes and research centers of excellence, universities and of course in botanic gardens. Efforts have also to be made in extracting DNA from the existing herbarium sheets of such extinct/extinct in the wild/endangered species. The focus of an exclusive Indian Plant DNA Bank should be on the collection of India native flora and aims not only to include a representative of each known species but also to capture the within and between species genetic variations.

Of the approximately 17,000 vascular plant species reported to exist in India, nearly 15 per cent are under threat (Ahmedullah, 1999). In Sikkim state of India alone, 60 per cent of the plant species are endemic (Myers, 1988), while 33 per cent of the total Indian flora is endemic to the country (TERI, 2004). With very high and varied diversity of flora, India possesses 11.9 per cent of the recorded world's plants including 49,219 higher plant species and is homeland of 167 cultivated species and 329 wild relatives of crop plants. It has about 30,000 to 50,000 indigenous land races of cultivated plants. In the past three decades, there has been an increasing concern towards the loss of biodiversity due to human and natural factors. Consequently, countrywide efforts have to be made towards conservation of wild and cultivated genetic resources. Since DNA extraction is the routine activity of numerous laboratories working in diverse areas of genetics, biochemistry, molecular biology and biotechnology across the country, DNA of the same can be deposited in any of the authorized DNA repositories in place of voucher specimens of conventional herbariums.

The very purpose of the DNA bank should be to collect, extract and store genomic DNA from all the available plant genetic resources in the country with their wide range of genetic and population diversity. This is in fact can be of special interest for the presence of useful their genes, which can be easily traced from the DNA available in the bank.

The Applications

The plant DNA banks have been solely for the purpose of supporting molecular phylogenetics and can be used only for the study of relationships of plants or for studies aimed at improving our understanding of evolutionary mechanisms. They may not be used for bioprospecting endeavors, screening for genes of interest in agricultural research, or any other commercial application (Henry, 2006). Although, the plant DNA banks can also be used by the scientific community for purposes such as conservation genetics, biodiscovery, plant breeding support, and as quality assurance standards in certain cases. A facility like this will also be advantageous to the research community by reducing the costs often associated with field trips to collect plant specimens. It will also make material from highly protected species available to a wider range of researchers. The DNA bank is one way of saving as many rare plants as possible from extinction (de Vicente and Andersson, 2006).

In course of time, it will become very easier to exchange genetic resources as DNA samples, rather than seed or vegetative propagules, which normally requires time consuming phytosanitary certification procedures and protocols. As molecular analyses are particularly useful in determining bottlenecks and historical events, population structure and phylogenetic and phylogeographic relationships in plants, extraction and storage of DNA is of immense practical value (Brown *et al.*, 1997; Frankham *et al.*, 2002). Herbaria normally maintain only specimens for the purpose of identification and classification based on morphology, whereas a DNA bank keeps tissues for the purpose of DNA extraction and use in molecular based research (de Vicente and Andersson, 2006).

Now a days, plant DNA banks have an enormous potential to collect, conserve and utilize a great deal of information about plant genetic diversity (Mattick *et al.*, 1992). Even more, the assessment of genetic variability within plant populations has made valuable contributions to conservation genetics of rare and threatened species (Byrne, 2005). There are a number of ways by which DNA banks could make an impact in the foreseeable future (de Vicente and Andersson, 2006). Most promising possibilities are to monitor changes in population genetic structure, to assess genetic diversity, to disclose taxonomic relationships, to search for new gene variants, to develop new markers, to identify and to trace typical plant products (Henry, 2006). Moreover, the applications of plant DNA banks on genetic diversity studies, germplasm management, genebank management and germplasm use are of immense practical utility (de Vicente and Andersson, 2006).

Conclusion and Future Prospects

This article focused on the utilitarian value of genetic diversity in terms of plant DNA bank. This is something novel and applicable to our economic and political world in the near future. Certainly the value of genetic diversity is immense, and DNA banking is one example. But some hold that a merely utilitarian view of biodiversity can never succeed in conserving it in that our orientation towards the use of nature is the root cause of the crisis. The refusal by the USA to sign ICBD indicates how a focus on the utility of conservation can hamper cooperation. This point of view has been articulated eloquently by many American thinkers (Thoreau, Leopold, Carson, Abbey and others). Recently, Ehrenfeld asks us to consider conservation not for its utility, but because it is simply the right thing to do.

Interestingly, roughly as many developing as developed countries conserve DNA, although developing countries report greater difficulties in procuring the necessary equipment. Responses from developing countries, in particular, displayed considerable lack of knowledge and insecurity concerning the applications of DNA banking and associated logistical issues. The crucial needs are more and better information about the possible advantages of DNA banking and broader discussion among the scientific community of the opportunities and limitations it provides. Also, issues related to DNA exchange, intellectual property rights and patent laws must be resolved. Policy research is urgently needed in these areas to devise solutions for closing legal gaps and to make DNA storage a more broadly viable crop diversity conservation practice (Correa, 1995; Leskien and Flitner, 1997; Lesser, 1998; Hoisington *et al.*, 1999; Serageldin, 1999; Swanson and Göschla, 2000).

Also, DNA banking is of crucial importance and contributes a pivotal role in the case of critically endangered/threatened species where very few populations or even individuals exist, the entire genepool can be collected and maintained as DNA extracts as one of the measure of *ex situ* collections, in order to represent the maximum available genetic diversity in these collections (Pflieger *et al.*, 2001; Dolezel and Bartos, 2005). However, the management and mobilization of such DNA materials should be well regulated to comply with the provisions of Convention on Biological Diversity and the International Treaty on Plant Genetic Resources for Food and Agriculture (Kate and Laird, 2000; FAO, 2004). Nevertheless, most felt that DNA storage would be very useful for future research, gene/genome conservation and safety duplication. Thus, plant DNA banks are the technology for the near future.

References

Adam, R.P. (1993). The conservation and utilization of genes from endangered and extinct plants: DNA Bank-Net. In: *Gene Conservation and Exploitation* (Ed. Gustafson, J.), pp. 35-52, Plenum Press, New York.

Adams, R.P. (1988). *The preservation of genomic DNA: DNA Bank-Net*. University of California, Davis, USA.

Adams, R.P. (1992). DNA Bank-Net to use DNA technology to save endangered germplasm. *Diversity* 8: 23.

Adams, R.P. (1993). DNA Bank-Net : An international network of DNA banks and *in vitro* storage institutions. *AgBiotech News and Information* 5: 317N-319N.

Adams, R.P. (1994). DNA Bank-Net: An overview. In: *Conservation of Plant Genes II. Utilization of Ancient and Modern DNA* (Eds. Adams, R.P., Miller, J.S, Goldenberg, E.M., Adams, J.E.), pp. 1-13, Missouri Botanical Garden Press, St. Louis.

Adams, R.P. (1997). Conservation of DNA: DNA banking. In: *Biotechnology and Plant Genetic Resources: Conservation and Use* (Eds. Callow, J.A., Ford-Lloyd, B.V., Newbury, H.J.), pp. 163-174, CAB International, Wallingford, UK.

Adams, R.P. (1997). Conservation of DNA: DNA banking. In: *Biotechnology and Plant Genetic Resources* (Eds. Callow, J.A., Ford-Lloyd, B.V., Newbury, H.J.), pp. 163-174, CAB International, Wallingford, UK.

Adams, R.P., Adams, J.E. (1992). Conservation of Plant Genes. *DNA banking and in vitro Biotechnology*. Academic Press, San Diego, USA.

Adams, R.P., Do, N., Gelin, C. (1992). Preservation of DNA in plant specimens from tropical species by desiccation. In: *Conservation of Plant Genes and In vitro Biotechnology* (Eds. Adams, R.P., Adams, J.E.), pp. 135-152, Academic Press, New York.

Adams, R.P., Zhong, M., Fei, Y. (1999). Preservation of DNA in plant specimens: Inactivation and re-activation of DNases in field specimens. *Molecular Biology* 8: 681-684.

Ahmedullah, A. (1999). Country report on red listing of plants in India. In: Using the IUCN red list criteria at the national level: A regional consultative workshop for South and South Asia. http://www.biodiversityasia.org/ebook/html

Altieri, M.A., Merrick, L. (1987). *In situ* conservation of crop genetic resources through maintenance of traditional farming systems. *Journal Economic Botany* 41: 86-96.

Armstrong, K.F., Ball, S.L. (2005). DNA barcodes for biosecurity: Invasive species identification. *Philosophical Transactions of the Royal Society B. Biological Sciences* 360: 1813-1823.

Avise, J.C. (1991). Ten unorthodox perspectives on evolution prompted by comparative population genetic findings on mitochondrial DNA. *Annual Review of Genetics* 25: 45-69.

Avise, J.C., Arnold, J., Ball, R.M., Bermingham, E., Lamb, T., Neigel, J.E., Reeb, CA., Saunders, N.C. (1987). Intraspecific Phylogeography: The Mitochondrial DNA Bridge Between Population Genetics and Systematics. *Annual Review of Ecology and Systematics* 18: 489-522.

Bragdon, S., Fowler, C., França, Z., Goldberg, E. (2005). Law and Policy of Relevance to the Management of Plant Genetic Resources. Learning Module with Review of Regional Policy Instruments, Developments and Trends. 2nd edn. The CGIAR System-wide Genetic Resources Programme (SGRP), the International Plant Genetic Resources Institute (IPGRI) and the International Food Policy Research Institute (IFPRI), IPGRI, Rome, Italy.

Brower, A.V.Z. (2006). Problems with DNA barcodes for species delimitation: 'ten species' of *Astraptes fulgerator* reassessed (Lepidoptera: Hesperiidae). *Systematics and Biodiversity* 4: 127-132.

Brown, A.H.D., Briggs, J.D. (1991). Sampling strategies for genetic variation in *ex situ* collections of endangered plant species. In: Genetics and conservation of rare plants (Eds. Falk, D.A., Holsinger, K.E.), pp. 99-119, Oxford University Press, New York.

Brown, A.H.D., Brubaker, C.L., Grace, J.P. (1997). Regeneration of germplasm samples: Wild versus cultivated plant species. *Crop Sci.* 37: 7-13.

Byrne, C.A. (2005). Genetic variation in plant populations: Assessing cause and pattern. In: *Plant Diversity and Evolution* (Ed. Henry, R.J.), pp. 139-165, CABI Publishing, Wallingford, UK.

Cohen, J.I., Williams, J.T., Plucknett, D.L., Shands, H. (1991). *Ex situ* conservation of plant genetic resources: Global development and environmental concerns. *Science* 253: 866–872.

Correa, C.M. (1995). Sovereign and property rights over plant genetic resources. *Journal Agriculture and Human Values* 12: 58-79.

Dale, P.J. (1992). Spread of Engineered Genes to Wild Relatives. *Plant Physiology* 100: 13-15.

de Carolis, D.M., Deeds, D.L. (1999). The impact of stocks and flows of organizational knowledge on firm performance: an empirical investigation of the biotechnology industry. *Strategic Management Journal* 20: 953–968.

de Vicente, M.C. (2004). The Evolving Role of Genebanks in the Fast-developing Field of Molecular Genetics. Issues in Genetic Resources No. 11, International Plant Genetic Resources Institute (IPGRI), Rome, Italy.

de Vicente, M.C., Andersson, M.S. (2006). DNA banks–providing novel options for genebanks? Topical Reviews in Agricultural Biodiversity. International Plant Genetic Resources Institute, Rome, Italy.

de Vicente, M.C., Carmen de Vicente, M., Andersson, M.S. (2006). DNA banks–providing novel options for genebanks? Bioversity International, Italy.

de Vicente, M.Cl., Andersson, M.S. (2006). DNA banks–providing novel options for genebanks? International Plant Genetic Resources Institute, Rome, Italy.

Deeds, D.L., Decarolis, D., Coombs, J.E. (1997). The impact of firmspecific capabilities on the amount of capital raised in an initial public offering: Evidence from the biotechnology industry. *Journal of Business Venturing* 1: 31-46.

Deutsche Forschungsgemeinschaft. (2008). DFG Biodiversity Research. Wiley-VCH Verlag, Weinheim, pp. 47-52.

Do, N., Adams, R.P. (1991). A simple technique for removing plant polysaccharide contaminants from DNA. *BioTechniques* 10: 162-166.

Dolezel, J., Bartos, J. (2005). Plant DNA flow cytometry and estimation of nuclear genome size. *Annals of Botany* 95: 99-110.

Dunwell, J.M. (2000). Transgenic approaches to crop improvement. *Journal of Experimental Botany* 51: 487-496.

Ellstrand, N.C., Elam, D.R. (1993). Population genetic consequences of small population size: Implications for plant conservation. *Annual Review of Ecology and Systematics* 24: 217-242.

ETC. (2002). GM pollution in the gene bank? *Biotechnology and Development Monitor* 49: 24.

FAO. (2004). The International Treaty on Plant Genetic Resources for Food and Agriculture. Food and Agriculture Organization of the United Nations (FAO), Rome, Italy.

Fowler, C., Hodgkin, T. (2004). Plant genetic resources for food and agriculture: Assessing global availability. *Annual Review of Environment and Resources* 29: 143-179.

Francisco-Ortega, J., Santos-Guerra, A., Kim, S.G., Crawford, D.J. (2000). Plant genetic diversity in the Canary Islands: a conservation perspective. *American Journal of Botany* 87: 909-919.

Frankel, O.H., Brown, A.H.D., Burdon, J.J. James. (1995). The Conservation of Plant Biodiversity. Cambridge University Press, Cambridge.

Frankham, R., Ballou, J.D., Briscoe, D.A. (2002). Introduction to Conservation Genetics. Cambridge University Press, Cambridge, UK.

Gachet, E., Martin, G.G., Vigneau, F., Meyer, G. (1998). Detection of genetically modified organisms (GMOs) by PCR: a brief review of methodologies. *Trends in Food Science and Technology* 9: 380-388.

Giesecke, S. (2000). The contrasting roles of government in the development of biotechnology industry in the US and Germany. *Research Policy* 29: 205-223.

Giri, C.C., Laxmi, G.V. (2000). Production of transgenic rice with agronomically useful genes: an assessment. *Biotechnology Advances* 18: 653-683.

Griliches, Z. (1991). The Search for R and D Spillovers. NBER Working Paper No. W3768. http://ssrn.com/abstract=248593

Griliches, Z. (1991). The search for R and D spillovers. *NBER Working Paper No. W3768.* http://ssrn.com/abstract=248593

Harlan, J.R. (1975). Our vanishing genetic resources. *Science* 188: 617–621.

Henry, R.J. (2006). Plant Conservation Genetics. Food Products Press, New York.

Hilder, V.A., Boulter, D. (1999). Genetic engineering of crop plants for insect resistance–a critical review. *Crop Protection* 18: 177-191.

Hodkinson, T.E., Waldren, S., Parnell, J.A.N., Kelleher, C.T., Salamin, K., Salamin, N. (2007). DNA banking for plant breeding, biotechnology and biodiversity evaluation. *Journal of Plant Research* 120: 17-29.

Hoisington, D., Khairallah, M., Reeves, T., Ribaut, J.M., Skovmand, B., Taba, S., Warburton, M. (1999). Plant genetic resources: What can they contribute toward increased crop productivity? *PNAS* 96: 5937-5943.

Jan Engels, L., Visser, A. (2003). Guide to effective management of germplasm collections. *Bioversity International*, Italy.

Jarvis, D.I., Hodgkin, T. (2002). Wild relatives and crop cultivars: detecting natural introgression and farmer selection of new genetic combinations in agroecosystems. *Molecular Ecology* 8: S159-S173.

Jauhar, P.P. (2006). Modern biotechnology as an integral supplement to conventional plant breeding: The prospects and challenges. *Crop Sci.* 46: 1841-1859.

Johnson, N.L., Lilja, N., Ashby, J.A. (2003). Measuring the impact of user participation in agricultural and natural resource management research. *Agricultural Systems* 78: 287-306.

Johnson, R.C. (2008). Gene banks pay big dividends to agriculture, the environment, and human welfare. *PLoS Biol.* 6: e148.

Kate, K.T., Laird, S. (2000). *The Commercial Use of Biodiversity- Access to Genetic Resources and Benefit-Sharing*. Earthscan Publications, London, UK.

Kennedy, K.L. (2004). The role of Federal Guidance and State and Federal partnership in *ex situ* plant conservation in the United States. In: *Ex situ plant conservation supporting species survival in the wild* (Eds. Guerrant, E.O., Havens, K., Maunder, M.), pp. 67-83, Island Press, Washington DC, USA.

Klenk, H.-P. (2008). DNA-Bank-Netzwerk etabliert. *BIOspektrum* 1: 86.

Komen, J., Persley, G.J. (1993). Agricultural Biotechnology in Developing Countries: A Cross-Country Review. ISNAR Research Report No. 2 (Intermediary Biotechnology Service). International Service for National Agricultural Research, The Hague.

Leskien, D., Flitner, M. (1997). Intellectual Property Rights and Plant Genetic Resources: Options for a Sui Generis System. Issues in Genetic Resources No. 6. International Plant Genetic Resources Institute, Rome, Italy.

Lesser, W. (1998). Intellectual property rights and concentration in agricultural biotechnology. *Ag. Bio. Forum* 1: 56-61. http://www.agbioforum.org

Li, Y., Wu, Y.H., McAvoy, R., Duan, H. (2001). Transgenics in crops. *Biotechnol Annu Rev.* 7: 239-260.

Liston, A., Rieseberg, L.H., Adam, R.P., Do, N. (1990). A method for collecting dried plant specimens for DNA and isozyme analyses, and the results of a field test in xinjiang, China. *Ann. Missouri Bot. Gard.* 77: 859-863.

Mattick, J.S., Ablett, E.M., Edmonson, D.L. (1992). The Gene Library: Preservation and analysis of genetic diversity in Australasia. In: Conservation of Plant Genes: DNA Banking and *in vitro* Biotechnology. (Eds. Adams, R.P., Adams, J.E.), pp. 15–35, Academic Press, San Diego, USA.

Maunder, M., Havens, K., Guerrant Jr. E.O., Falk, D.A. (2004). *Ex situ* methods a vital but underused set of conservation resources. In: *ex situ Plant conservation supporting species survival in the wild* (Eds. Guerrant Jr. E.O., Havens, K., Maunder, M.), pp. 3-20, Island Press, Washington DC, USA.

Maxted, N., Ford-Lloyd, B.V., Hawkes, J.G. (1997). *Plant genetic conservation: The in situ approach.* Chapman and Hall, London, UK.

McCullum, C., Benbrook, C., Knowles, L., Roberts, S., Schryver, T. (2003). Application of modern biotechnology to food and agriculture: Food systems perspective. *Journal of Nutrition Education and Behavior* 35: 319–332.

Moore, P.H., Ming, R. (2008). *Transgenics for new plant products, applications to tropical crops, Volume 1, Genomics of tropical crop plants*, Springer, New York.

Morrisa, M.L., Singh, R.P., Pal, S. (1998). India's maize seed industry in transition: changing roles for the public and private sectors. *Food Policy* 23: 55-71.

Neel, M.C., Cummings, M.P. (2003). Effectiveness of conservation targets in capturing genetic diversity. *Conserv. Biol.* 17: 219-229.

Nickrent, D.L. (1994). From field to film: Rapid sequencing methods for field-collected plant species. *Biotechniques* 16: 470-475.

Pasternak, D. (1987). Salt Tolerance and Crop Production-A Comprehensive Approach. *Annual Review of Phytopathology* 25: 271-291.

Peferoen, M. (1997). Progress and prospects for field use of *bt* genes in crops. *Trends in Biotechnology* 15: 173-177.

Pflieger, S., Lefebvre, V., Causse, M. (2001). The candidate gene approach in plant genetics: a review. *Journal Molecular Breeding* 7: 275-291.

Plucknett, D.L. (1987). *Gene Banks and the World's Food*. International Rice Research Institute, Manila, The Philippines.

Porteous, L.A., Seidler, R.J., Watrud, L.S. (2003). An improved method for purifying DNA from soil for polymerase chain reaction amplification and molecular ecology applications. *Molecular Ecology* 6: 787–791.

Pray, C.E. (2003). Public–private sector linkages in research and development: Biotechnology and the seed industry in Brazil, China and India. *American Journal of Agricultural Economics* 83: 742–747.

Purty, R.M., Kumar, G., Singla-Pareek, S.L., Pareek, A. (2008). Towards salinity tolerance in Brassica: An overview. *Physiology and Molecular Biology of Plants* 14: 39-49.

Raybould, A.F., Gray, A.J. (1993). Genetically modified crops and hybridization with wild relatives: A UK perspective. *Journal of Applied Ecology* 30: 199-219.

Reid, W.V. (1998). Biodiversity hotspots. *Trends in Ecology and Evolution* 13: 275-280.

Ribaut, J.M., Hoisington, D. (1998). Marker-assisted selection: new tools and strategies. *Trends in Plant Science* 3: 236-239.

Rubinoff, D. (2006). Utility of Mitochondrial DNA Barcodes in Species Conservation. *Conservation Biology* 20: 1026–1033.

Ryder, O.A., McLaren, A., Brenner, S., Zhang, Y.-P., Benirschke, K. (2000). DNA banks for endangered animal species. *Science* 288: 275-277.

Savolainen, V., Reeves, G. (2004). A plea for DNA banking. *Science* 304: 1445.

Serageldin, I. (1999). Biotechnology and Food Security in the 21st Century. *Science* 285: 387–389.

Sharma, H.C., Sharma, K.K., Seetharama, N., Ortiz, R. (2000). Prospects for using transgenic resistance to insects in crop improvement. *Electronic Journal of Biotechnology*. http://ejbiotechnology.ucv.cl/content/vol3/issue2/full/3/index.html

Smale, M. (2000). *Farmers, Gene Banks and Crop Breeding: Economic Analyses of Diversity in Wheat, Maize, and Rice*. Kluwer Academic, The Netherlands.

Smith, C.W., Dilday, R.H. (2002). *Rice: Origin, History, Technology, and Production*. John Wiley and Sons, Oxford.

Stewart, Jr, N.C., Halfhill, M.D., Warwick, S.I. (2003). Genetic modification: Transgene introgression from genetically modified crops to their wild relatives. *Nature Reviews Genetics* 4: 806-817.

Swanson, T., Göschla, T. (2000). Property rights issues involving plant genetic resources: implications of ownership for economic efficiency. *Ecological Economics* 32: 75-92.

TERI. (2004). Hot spots in India. http://www.teriin.org/biodiv/hotspot.htm

Uzogara, S.G. (2000). The impact of genetic modification of human foods in the 21st century A review. *Biotechnology Advances* 18: 179-206.

Virchow, D. (2003). *Efficient conservation of crop genetic diversity: Theoretical approaches and empirical studies*. Springer, New York.

Willis, C.K., van Wyk, E. (2006). Integrating *in situ* and *ex situ* conservation. *SANBI Biodiversity Series* 1: 35-39.

Wilson, E.O. (2003). The encyclopaedia of life. *Trends Ecol. Evol.* 18: 77-80.

Xiao, J., Lia, J., Grandillo, S., Ahn, S.N., Yuan, L., Tanksley, S.D., McCouch, S.R. (1998). Identification of trait-improving quantitative trait loci alleles from a wild rice relative, *Oryza rufipogon*. *Genetics* 150: 899-909.

Zetzsche, H., Gemeinholzer, B. (2007). Aufbau eines DNA-Bank-Netzwerkes. *GfBS Newsletter* 18: 32-36.

Microbial Biotechnology and Ecology (2011) *Pages* **501–519**
Editors: **Deepak Vyas, G.S. Paliwal, P.K.Khare and R.K Gupta**
Published by: **DAYA PUBLISHING HOUSE, NEW DELHI**

Chapter 28

Progress in Biotechnology of Guava (*Psidium guajava* L.)

☆ ***Nasim Akhtar***

Post Graduate and Research Department of Microbiology and Biotechnology, Al-Ameen Arts Science and Commerce College, Bangalore – 560 027

Introduction

Common guava (*Psidium guajava* L.; Family, Myrtaceae) is a diploid species (2n=22), but known to exist in some natural and artificial forms of triploidy (2n=3x=33), thereby giving seedless fruits. It occupies prime position among the furits crops of winter and stands 4th among the most important fruit crops grown in India after citrus, mango and banana (Annon, 1985). Guava is a potent alternative of apple in the tropical world, as it is a rich source of proteins, carbohydrates, fats, vitamins and minerals, and within the reach of a common man and other poorer sections of society, probably that's why it occupies the prime position among the fruit crops and proudly called as the 'poor men's apple'. It is indemic to tropical America, but has been thorougly naturalized throughout the tropics and subtropics from Mexico to Peru to India. It has widespread commercial exploitations in Florida and Hawain islands, Egypt, South Africa, Brazil, Columbia and West Indies (Purseglov, 1968).

The guava plant is a shallow rooted evergreen or subdeciduous large shrub or sometimes a small tree of 3-10 metre height and about 0.9 metre girth. The stem is irregular and fluted when old, branching close to the ground and often produces suckers from root near the base of trunk. Leaves are opposite with oblong to elliptical lamina with variable size (5-15 cm × 3-7 cm). Flowers are epigynous and bisexual (perfect), axillary, solitary or born in 2-3 flowered cyme. Calyx is adnate to bud and persistent. Stamens are numerous and inserted in rows on disc. The anthers are pale yellow in colour dehiscence

* E-mail: nasimakhtar01@yahoo.co.in

longitudinally. Ovary, which is 4-5 locular bears numerous bitegmic ovules in axile placentation. The fruit is a berry, surmounted by calyx lobes and is variable in shape (globose, ovoid or pyriform), skin colour, flesh, taste and flavour. Usually numerous bony, reniform seeds are present in pulp having curved embryos (Purseglove, 1968; Mitra and Bose, 1985; Singh, 1985).

Guava prefers to grow on deep, friable, well drained soil and known to tolerate a soil pH in the range of 4.5 to 8.2, as well as mild conditions of salinity, water-logging and high temepratures, but highly susceptible to low temperature and frost. On comparative ground, it has been found more resistant to drought than other perennial fruit crops. It has been successfully grown in the habitats ranging from tropics to subtropics and from sea level to 5,000 ft; and therefore, can be made adaptable for a wide ranging of agro-climatic zones.

Both self and cross pollination occur in *Psidium guajava* L. The flowers are visited by bees and other insects. Natural cross pollination has been shown to be about 30-35 per cent (Singh, 1985). Fruit setting takes place even in unopen or partially opened flower buds without exposing stigma and anthers conferring self pollination or closed pollination. The flower buds to be cross pollinated are fully opened exposing all floral organs and usually with styles longer than the stamens.

It is cultivated mainly for its delicious sweet, juicy fruits, which are rich in proteins, carbohydrates, minerals, sugars, oils and vitamins (mainly Vitamin-C) and therefore, canned for jams, jellies marmalades, pies, ketchups, chutney and nectar (Annon, 1985).

Guava produces as many as three crops in a year. In a mild tropical climate it can flower and fruit throughout the year continuously. But in regions like northern India where seasons are quite distinct due to variation in temperature or precipitation or both, three flowering seasons *viz.* summer, rainy and autumn with the corresponding harvesting periods as rainy, winter and spring, have been distinctly observed.

Guava cultivars display a great diversity in tree size, bearing habits and yield as well as in fruit shape, size, quality and ripening season. Some of the well grown Indian cultivars are Allahabad safeda, Lucknow-49, Banarsi, Harijha, Chittidar, red/pink fleshed and seedless (Purseglove, 1968; Mitra and Bose, 1985).

Guava cultivation is frequently threatened by wilting syndrome due to *Fusarium, Rhizoctonia,* anthracnos and canker. No definite control measures against the wilt by *F. oxysporum* f. Psidii is reported so far. The wilt (seedling blight) due to *Rhizoctonia bataticola* is equally dreaded. Another disease to which guava succumbs, is its canker caused by *Glomerella psidi, Colletotrichum psidii and Pestalotiopsis psidii.* The bark canker of guava is caused by *Physalospora psidii,* which is also prone to be eaten by a caterpillar *Indarbela queadrinotata;* other serious pests of guava in rainy seasons include mealy scale, *Pulvinaria psidii,* maggots of fruits fly (*Dacus ferrugineus* and *D. zonatus*). The old and neglected orchards of guava are damaged seriously by some partial parasites also (*e.g. Dendrophthaoe falcata*). The plant is also known to attacked by guava mosaic virus.

Traditionally, it has been a cross pollinated perennial fruit crop propagated through seeds (Purseglov, 1968), due to which considerable variabilities exist among its natural populations. It's seedless, triploid and other cultivars have been conventionaly propagated through budding onto the root stocks, inarching or approach grafting to maintain its clonal populations (Menzel, 1985; Singh, 1986).

Despite all these, there has been severe shortage of its suitable planting material that can be vegetatively propagated to get a good harvest of seedless guava having enhanced medicinal and nutritional qualities.

Fruit consumption in India is anticipated to increase by 4 per cent per year according to projected growth rates for income, population, and trends in food preferences. This would result in demand for all fruits reaching 66 million tonnes by 2010, with the bulk of this demand growth for tropical fruit. Mango is by far the most significant tropical fruit produced in India, followed by guava. Mango output was 9.8 million tonnes in 1999, with guava at 1.8 million tonnes. India is the one of the largest and most varied fruit producing nations in the world, accounting for 10 per cent of all fruits and nearly 40 per cent of tropical fruits produced globally. Fruit production in India increased at an average annual growth rate of 5.5 per cent between 1990 and 1999. Fruit tree production contributes 10 per cent on average to the gross value of total agricultural output in India, and 13 per cent of the export earnings attributable to major agricultural products, FAO (2001).

So the demand for good quality guava plants appears to be promising and bears enough potential for domestic and international trade with regards to its wide ranging edible commercial preparations and environment-friendly safe and effective medicinal products. This significance of fruit tree is now being realized globally and recently a few international symposiums have been organized exclusively on guava (Singh *et al.*, 2007); Annonymous (2008).

Breeding and Crop Improvement

Yet, the conventional breeding techniques have been responsible for the development of most delicious cultivar Lucknow-49. But presence of large number of seeds per fruit is a major disadvantage hindering its acceptability (Mitra and Bose, 1985). Only little success was achieved regarding the development of more productive seedless cultivar by corssing Allahabad safeda (a seeded but good quality heavy bearer) with a natural triploid (seedless but very shy bearer). De Cruz and Babu Rao (1962) were the first to isolate two aneuploids with 21 and 31 chromosomes form the progency of a triploid guava strain. Similarly, Majumdar and Mukherjee (1972a, b) have reported 26 diploids, 9 trisomics and 5 tetrasomics (2n+1+1) in the 73 plants raised by crossing the diploids and triploids through conventional breeding techniques.

Among the other conventional breeding work for the improvement of guava, the Fruit Research Stations, Basti, India, have developed 55 hybrid plants of guava by crossing Seedless × Allahabad safeda; Seedless × Lucknow-49; Allahabad safeda × Patillo; Apple coloured × Kothrud and Apple coloured × Red fleshed. The performances of these hybrids have been found promising for the semi-arid tropical areas of Telangana, Rayalseema etc. (Mitra and Bose, 1985). However, retention of their hybrid characteristics during subsequent generations still remained questionable.

The conventional technique of crop improvement for optimum land use, have succumbed to keep-up pace with rapid changes in the natural habitats of horticultural fruit crops, and appears inadequate to comply the demand of fast changing scenario. Biotechnology has opened-up unprecedented opportunities to give authoritative replies to such calamities.

In vitro Progress in Guava Improvements

The improvement of clonally propagated crops through somatic cells have been promising (Litz *et al.*, 1985; Carlson, 1987), which can also be used for guava improvement and its hybrid cultivars development. It bears potentials to incorporate the economically important traits through somatic cell manipulation into existing cultivars to produce improved varieties. Therefore, developing a suitable protocol for the vegetative clonal propagation of biotechnologically raised variants of guava, having better nutritional and medicinal properties in its seedless fruits, and its subsequent release as artificial seeds are the major highlights for the improvement of guava cultivars. In the following section a review of the recent progress in guava biotechnology is presented. The last part of the chapter is discussing the various parameters used in the studies of somatic embryogenesis of guava, followed by discussion on the importance of these parameters for the process in this and other species.

Organogenesis

The culture of excised tissuses of fruit mesocarp *in vitro* for manipulating the somatic cells/ tissues was the first report by Schroeder (1961). Babbar and Gupta (1986a, b) were the first to report the induction of callus and androgenesis in the *in vitro* cultured anthers. In the same year, the *in vitro* proliferation of shoots and plantlet formation from somatic tissues was reported by Jaiswal and Amin, (1986). Soon after, its *in vitro* propagation from shoot tip culture of mature tree was also reported (Jaiswal and Amin, 1987). Guava has been successfully micropropagated through its nodal explants, and it is also possible to obtain the adventitious shoots regenerated from the leaves of already micropropagated *in vitro* grown plantlets (Jaiswal and Amin, 1986, 1987; Amin and Jaiswal, 1987, 1988, 1989a, b). Subsequently, clonal propagation of guava from seedling and grafted plants was reported by Loh and Rao (1989). The very low frequency of regeneration in all these reports was because of high phenolic exudation and media browning. However, Fitchet (1989) has claimed the drying of explants prior to inoculation on the media during the establishment of nodal and shoot tips without oxidative browning. Subsequently, Papadatou *et al.* (1990) reported the rapid multiplication of guava seedlings through *in vitro* shoot tip culture with a success rate of more than 95 per cent, without any oxidative problem of phenolics or necrosis of the explants. Histological analysis of organogenesis in guava cultures was reported for the first time by Amin (1991). Recently, Yasseen *et al.* (1995) of USA have reported the *in vitro* shoot proliferation and propagation of guava from germinated seedlings. The first overview on somatic cell genetic and organogenesis of guava was published by Jaiswal and Amin (1992). Recently direct organogenesis on explants from germinated seddlings of guava has been reported by Singh *et al.* (2002).

The foregoing account on the organogenesis of guava appears to be promising but have almost failed to give any coherent picture of the work done so far on somatic embryogenesis, probably due to the fact that it is basically a highly recalcitrant fruit crop of tropics, and subtropical region of the World.

Somatic Embryogenesis

The first report on induction of somatic embryogenesis as unpublished data of the author and colleague was appreciated (see Jaiswal and Amin (1992). Further published work came out in 1993, when the authors became first in reporting the studies on induction and factors controlling somatic embryogenesis in guava using the zygotic embryo as explants (Jaiswal and Akhtar, 1993). Somatic embryogenesis from zygotic embryo culture of guava had subsequently been reported by the authors (Akhtar, 1996; Akhtar and Jaiswal, 1994, 1995; Jaiswal and Akhtar, 1994). The complete protocol for induction of somatic embryogenesis, dvelopment, maturation and germination of somatic embryos, production of artificial seeds and improvement of guava species had been achieved with commendable success by the author (Akhtar, 1997). Following similar protocols, Ramirez and Salazar (1998) have also reported the induction and development of somatic embryos from the zygotic embryo explant. An overview of the somatic embryogenesis and its applications in guava and other fruit tree species improvement has been overviewed in the recent past (Akhtar *et al.*, 2000; Akhtar and Jain, 2000). Subsequently Jaiswal and Jaiswal (2005) have overviewed the further progress in guava somatic cell genetics. The somatic embryogenesis work on guava is further being perfected at various locations and some new concepts are being published in the recent years by the author and other working on the problem (Chandra *et al.*, 2004; Rai *et al.*, 2007, 2008a, b, c).

The focus of the present report is mainly on recent progress for the improvement of guava species through somatic embryogenesis and development of protocol for micropropogation, artificial seed production.

Salinity Tolerance

As described in the preceding section that induction of somatic embryogenesis had partly been regulated through the stress management. Somatic embryos produced in presence of different concentrations of sodium chloride (NaCl) from both 3 per cent and 5 per cent sucrose containing medium were subjected for germination under the salt stress conditions. Somatic embryos from 75 mM NaCl containing induction and development medium showed better performance in 50 mM NaCl added half strength germination medium. Such plantlets were further grown to 75 and then 100 mM salt containing growth medium. The tolerant plant were subjected to *ex vitro* transfer in artificially made saline condition. As a result guava somatic plant line were selected which were tolerant to 100 mM NaCl containing medium (Akhtar, 1997; Akhtar *et al.*, 2000).

Guava had been found little tolerant to water logging and mild drought conditions. Since induction of water stress condition in the medium by addition of polyethylene glycol in the medium did not support the somatic embryogenesis, the process could not be applied for the selection of drought tolerant lines.

Artificial Seed Production

Various stages of somatic embryos from different week old cultures encapsulated in 2 per cent sodium alginate gel prepared in double distilled water proved to be the best matrix for artificial seeds production. The pH of the gelling medium either 5.8 or 7.0 had no distinct effect of conversion of encapsulated somatic embryos. Sodium alginate gel prepared in quarter or half strength MS medium alone or along with the sucrose had significantly reduced the germination response of artificial seeds. A 3 per cent sodium alginate gel prepared similarly in different medium showed slightly reduced conversion as compared to that encapsulated in 2 per cent sodium alginate gel (Akhtar, 1997; Akhtar *et al.*, 2000).

Artificial seeds encapsulating elongated torpedo stage somatic embryos from 8 weeks old culture showed maximum conversion frequency (Figure 28.1D). Short torpedo stage somatic embryos encapsulated similarly showed only 60-70 per cent conversion frequency. Synthetic seeds with lower (CHG) stages somatic embryos showed much reduced percentage of germination and never converted into plantlets. There was no significant variation in germination frequency of encapsulated and non-encapsulated somatic embryos of different stages (Akhtar, 1997; Akhtar *et al.*, 2000). Recently guava has been regenerated from encapsulated shoot tips of guava (Rai *et al.*, 2008b), alginate encapsulation of segments of guava (Rai *et al.*, 2008c), and encapsulated somatic embryos (Rai *et al.*, 2008a),

Cryopreservation

Cryopreservation as it also referred to freeze preservation or cryogenic storage involves transfer of biological materials for long term storage in liquid nitrogen (LN) (Bajaj, 1995). Cryopresevation of plant tissue culture can eliminate several problems of *in vitro* maintenance of culture in an active state.

There are several reports indicating that cryopreseved cells maintained their biosynthetic and morphogenetic potentials and no phenotypic, genotypic or epigenetic changes have been found as a result of cryopreservation process (Kartha and Engelemann, 1994; Wither and Engelemann, 1998). The cryopreservation studies in citrus and other species was reviewed earlier (Prez, 2000; Akhtar and Jain, 2000).

In case of guava, embryogenesis had been shown as the direct event from the zygotic embryo explants without any callus phase. Substantial efforts have been made in cryogenic preservation of somatic embryos and artificial seeds of guava. Somatic embryogenesis was induced with the average production of 100-150 embryos per zygotic embryo explants in 83.33 per cent culture frequency. Normally about 88–98 per cent elongated and short torpedo stages somatic embryos were converted into plantlets. Zygotic embryo explants (10–weeks post anthesis) treated with 1.0 mg/l of 2,4-D for 8–days in 5 per cent sucrose containing full strength MS medium were used for cryopreservation and regeneration after storage. A slow freezing followed by immersion into liquid nitrogen and subsequent storage at –70 °C of such embryogenic cultures as well as zygotic embryo explants (non induced) was adopted after the cryoprotectant treatment (Grout, 1995; Gupta *et al.*, 1987). Among cryoprotectants highest embryogenesis was observed with 7.5 per cent glycerol followed by in decreasing order by 10 per cent glycerol, 7.5 per cent polyethylene glycol, 10 per cent polyethylene glycol, 7.5 per cent dimethyle sulfoxide, 10 per cent dimethyle sulfoxide along with 5 per cent sucrose dissolved in half strength MS medium. Zygotic embryos pre-exposed to embryogenic treatments showed somatic embryogenesis in 41.67 per cent of cultures with only 25 per cent platlets recovery. Cryopreserved zygotic embryos without embryogenic induction were unsuccessful for somatic embryogenesis irrespective of thawing and induction treatment.

A mixture of cryoprotectant involving dimethylsulfoxide (DMSO) (5-15 per cent) and sucrose at different concentrations (0.1- 1.0 M) have also been used for the cryoprotection of both nacked and encapsulated somatic embryos from 8 week old cultures of guava. The cryoprotectants were added by incubating somatc embryos and synthetic seeds in a series of cryoprotective solution of increasing concentration as described by Gupta *et al.* (1987). Freezing of the material was carried out by dehydrative slow cooling methods of Sakai *et al.* (1991). The material was transferred to cryotubes and incubated first at 4 °C for 30-120 minutes, and then in deep freezer at –35°C for 20–40 minutes prior to direct immersion in liquid nitrogen. The cooling rate was roughly about 2 °C/min. The cryopreserved somatic embryos and artificial seeds were taken out after different periods of storage and thawing was performed by fast warming and immersing the cryotubes in a water bath at 37–40°C (Perez *et al.*, 1997).

The recovery of plantlets from somatic embryos and artificial seeds reduced drastically following the cryopreservation treatment and storage. The best plantlet recovery from somatic embryos and artificial seeds was noted in 28 per cent and 26 per cent frequency, respectively, up till one week of storage in 10 per cent DMSO along with 0.5M sucrose with dehydrative slow cooling and fast thawing procedure. In longer period of storage as well as with slow thawing at room temperature recovery of the plantlets was almost unsuccessful from both encapsulated and non- encapsulated somatic embryos. This very low success in cryopreservation of pretreated and post induced zygotic embryos as well as lower recovery of plantlets from cryopreserved somatic embryos and artificial seeds might be because the guava species is highly susceptible to frost and freezing. The other cryopreservation methods like vitrification could be tried as better alternative for the preservation of guava somatic embryogenesis as it was successful in novel orange (Sakai *et al.*, 1991), wasabi (Matsumoto *et al.*, 1994), strawberry (Hirai *et al.*, 1998).

Protocol of Somatic Embryogenesis in Guava

Material and Methods

Selection of Plant and Preparation of Explant

A 15-20 year old guava trees growing in the botanical garden of Banaras Hindu University were selected for zygotic embryo explants preparation and standardization of various factors for the local cultivar. Development of flower buds were observed regularly in each growing season. Flower buds were tagged and bagged (to ensure self/closed pollination) whenever necessary. Zygotic embryos for other cultivars like pink fleshed, Allahabad safeda, Lucknow-49 were collected from the guava trees growing at various locality of Banaras Hindu University, and subjected similarly to see effect of genotype on induction of somatic embryogenesis. Fruits of a particular age (7-14 weeks after anthesis) were picked up and brought to the laboratory and washed thoroughly under running tap water. The hard green exo- and greenish white mesocarp were removed with the help of a sharp knife or scalpel and the central pulp with seeds was cut into 4-6 vertical lobes. The pulp washed under running tap water for 30 mins and subsequently surface sterilized by gentle shaking in 2 per cent Cetavlon (v/v) with 2-4 drops of Tween-X for about 15-mins. The material then washed under running tap water followed by sterilization under the aseptic condition over a Laminar Flow Hood. Following a Short rinse in 70 per cent ethanol the material was sterilized with 0.05 per cent $HgCl_2$ solution (w/v) for about 20 mins then rinsed 4-5 times with sterile double distilled water.

Dissection of Zygotic Embryos

One of the pulp lobe along with seeds was taken into petridishes in sterile water. Zygotic embryo was dissected by picking up seeds from the pulp one by one. Seeds of guava are more or less J or U shaped with one arm slightly longer than the other. The seed was held with a forceps in left hand over a microscopic slide and a small piece of longer arm was removed with the help of a fine and pointed scalpel. The seed then turned opposite and a small piece of shorter arm was removed similarly. A blunt needle was inserted either through the longer or shorter arm side so that embryo came out of the other end. The embryo was inoculated immediately on the medium in a test-tube with the help of inoculation needle.

Culture Medium

The culture medium consisted of the normal strength of MS basal salts. The initial pH (4.2 ± 0.05) of the medium was raised to 6.2 ± 0.05 with 0.1 N NaOH. The medium was kept on continuous stirring on a magnetic stirrer. After about 1-2 hours the pH of the medium automatically stabilized to 5.8 ± 0.05 or if necessary adjusted with 0.01 N NaOH or HCl. Depending upon the requirement different amount of sucrose dissolved to one litre of medium. The medium was gelled with agar (0.8 per cent, w/v) by heating under pressure. The medium was modified with 0.001, 0.005, 0.01, 0.05, 0.1, 0.5, 1.0, 2.0, 3.0, 5.0, 10.0, 15.0 and 20.0 mg/l of 2,4-D prior to autoclaving. About 8-12 ml of medium was poured to each culture tubes. The medium was autoclave sterilized at 121 °C for 15 mins at 1.1 kg/cm^2 pressure. The medium modified with 2,4- D or any growth regulator(s) were reffered to as induction medium and the other without any growth regulator(s) and used for the development of somatic embryos were reffered to as development medium irrespective of sucrose conditions. All cultures were incubated in an air conditioned culture room maintained at 25 ± 2 °C temperature, 60-65 per cent relative humidity and 16 hrs photoperiod. The cultures were illuminated with 40W white fluorescent tube from a distance of 30-35 cm receiving 30-50 mEm^{-2}s^{-1} light intensity.

Induction of Somatic Embryogenesis

Some of the initial experiments had resulted the conclusions that treatments of zygotic embryos (10 weeks post anthesis) for a minimum of 8 days with 0.5 mg/l of 2,4-D in full strength MS basal medium supplemented with 3 per cent sucrose were the optimum condition for induction of somatic embryogenesis. Later a systematic studies were conducted for the development of efficient regeneration protocols for micropropagation of this species.

Development of Somatic Embryos

Somatic embryos were visible as small watery white transparent globules under stereozoom microscope usually at the end of the third week of culture initiation. In the fourth week of culture several new globular somatic embryos appeared while the older ones increased in size and proceeded in the next stage of development with simultaneous change in the colour from watery white transparent to translucent milksh (Figure 28.1B). In the fifth week of culture some new globular somatic embryos appeared while the older ones continued to grow towards advanced stages of development. After eight weeks of culture initiation most of the somatic embryos were hard solid and milky white irrespective of the developmental morphology.

The number of somatic embryos per explant varied considerably ranging from a single to usually several hundreds, sometimes more than a thousand depending upon the nature of treatment. Usually somatic embryos followed a normal development but sometimes anomalies were also observed as a function of treatment. However, the frequency of anomalous development and the secondary embryogenesis were very low.

Development of somatic embryos was found asynchronous showing 4-5 discernible stages from the same zygotic embryo explants in five weeks and older cultures (Figures 28.1B, C). Depending upon their sizes and post developmental responses *i.e.* response of somatic embryos during maturation, germination and encapsulation, they were grouped into three categories. The most preferred somatic embryos longer than 1.5 mm were called elongated torpedo stage (ET) somatic embryos; slightly less preferred ones ranging from 1.0-1.5 mm in size were termed as short torpedo stage (ST) somatic embryos; the least preferred ones at cotyledonary, heart and globular stages of development and smaller than 1.0 mm were grouped into third category and designated as lower or the CHG stage somatic embryos (Akhtar, 1997; Akhtar *et al.*, 2000).

The fitness of a treatment was calculated based on the six different embryogenic parameters *viz.* i. frequency of embryogenesis, ii. Intensity of embryogenesis *i.e.* average number of somatic embryos produced per explant per treatment, iii. frequency of ET stage somatic embryos, iv. frequency of ST stage somatic embryos, v. frequency of lower or CHG stage somatic embryos, and vi. efficiency of embryogenesis.

Evaluation of Somatic Embryogenesis

The cultures were observed daily and change in zygotic embryo explant were noted after an interval of 1-, 2-, 4-days or weekly. Embryognenic responses were recorded usually 8-10 weeks after culture initiation. Depending upon asynchrony in development and post developmental responses, during maturation germination and in recurrent embryogenesis various stages of somatic embryos were grouped in to three categories (Akhtar, 1997; Akhtar *et al.*, 2000). The results of various treatments has been presented and described under the following headings (Table 28.1).

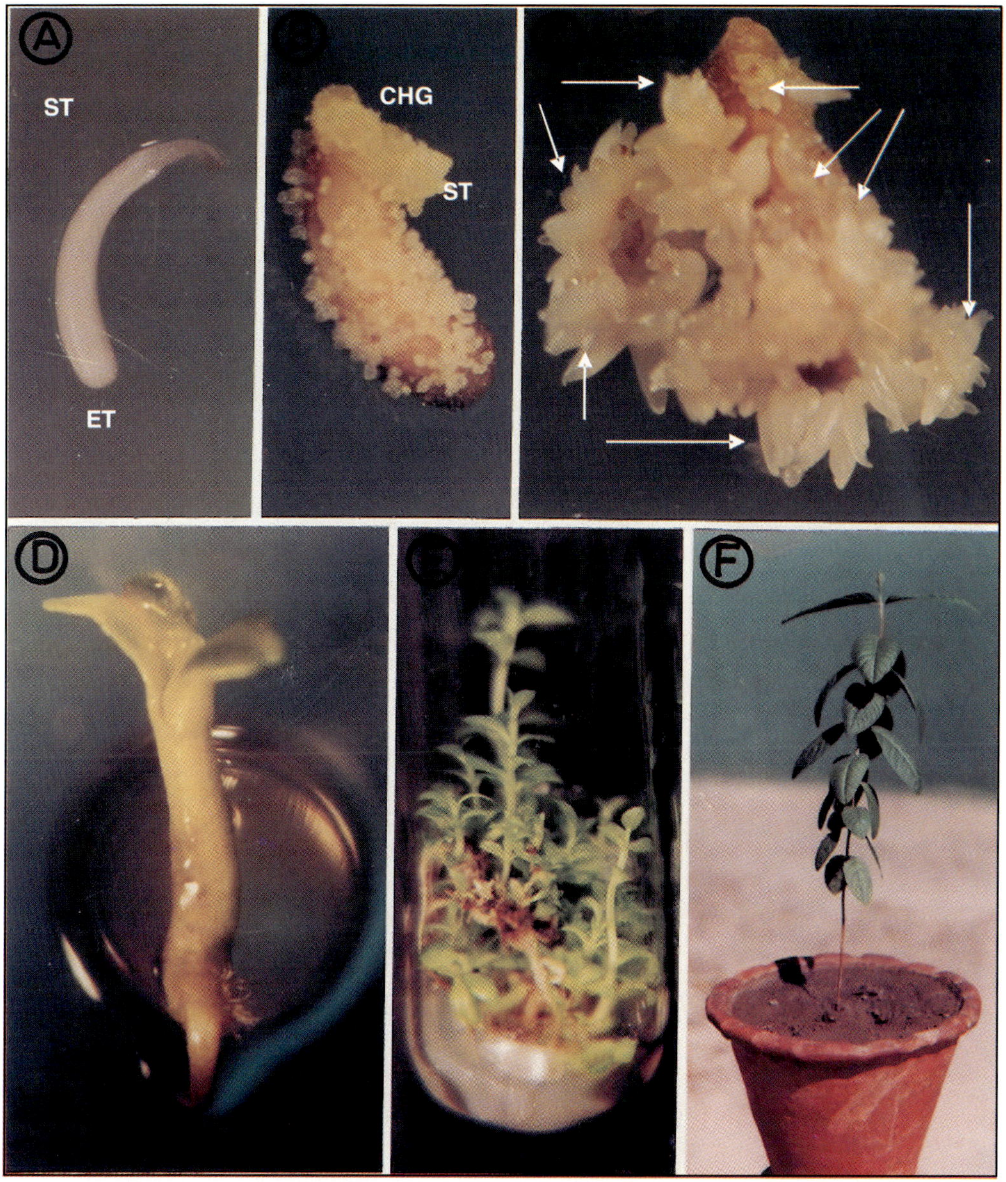

Figure 28.1: Induction of Somatic Embryos from Zygotic Embryos and Regeneration of Plantlet from Somatic Embryos and Artificial Seed of Guava (*Psidium guajava* L.)

(A) Ten weeks post anthesis zygotic embryo used as explants source. (B) Early stages of somatic embryos developing from zygotic embryo explant after four weeks of culture initiation. (C) Advanced stages of somatic embryos developing from zygotic embryo explants after six weeks of culture initiation. (D) A germinating artificial seed (eight weeks old elongated torpedo stage somatic embryos encapsulated in sodium alginate gel matrix) showing the emergence of shoot and root at opposite poles. (E) Germination of different stages of somatic embryo in half strength MS basal medium with 3 per cent sucrose. (F) A six weeks old guava plant (after soil transfer and hardening) regenerated through somatic embryogenesis.

Table 28.1: Induction of Somatic Embryogenesis from Zygotic Embryos of Guava (*Psidium guajava* L.) (Ten week old zygotic embryos of guava were treated continuously with different concentrations of 2,4-D in full strength MS medium added with 3 per cent sucrose. Embryogenic responses were noted after 10 weeks of culture initiation).

Sl. No.	*Conc. of 2,4-D (mg/l)*	*Frequency of Embryogenesis*	*Intensity of Embryogenesis*	*Frequency of Various Stages of Somatic Embryos*			*Efficiency of Embryogenesis*
		FE (per cent)	*IE (ANEPC)*	*ET (per cent)*	*ST (per cent)*	*CHG (per cent)*	*EE (Relative)*
1.	0.000	0.000	0.000	0.000	0.000	0.000	0.000
2.	0.001	9.99 ± 4.75	8.27 ± 5.24	30.21± 6.29	56.42 ± 8.54	13.37 ±11.67	0.65 ± 0.42
3.	0.005	28.97 ± 4.80	13.23 ±1.83	21.26 ± 8.99	47.54 ±12.55	31.30 ±18.50	1.58 ± 0.32
4.	0.01	63.66 ± 2.80	23.80 ± 6.19	10.52 ± 2.07	28.45 ± 6.85	46.50 ± 3.80	13.75 ± 1.02
5.	0.05	30.17 ±7.45	10.39 ± 3.98	6.69 ± 3.20	22.73 ± 10.06	63.03 ± 8.30	2.19 ± 0.52
6.	0.1	17.30 ± 5.09	10.10 ± 5.20	2.89 ± 3.17	8.82 ± 14.80	74.94 ± 13.21	0.39 ± 0.04
7.	0.5	8.33 ± 2.98	8.52 ± 3.65	2.27 ± 5.08	0.000	88.91 ± 19.70	0.18 ± 0.36
8.	1.0	5.55 ± 4.30	6.75 ± 4.16	0.000	0.000	100.00	0.000
9.	1.5	0.000	0.000	0.000	0.000	0.000	0.000
10.	2.0	0.000	0.000	0.000	0.000	0.000	0.000
11.	3.0	0.000	0.000	0.000	0.000	0.000	0.000
12.	4.0	0.000	0.000	0.000	0.000	0.000	0.000
13.	5.0	0.000	0.000	0.000	0.000	0.000	0.000
F (12, 65), *P 0.01*		122.29	19.799	30.215	41.76	21.78	438.702

CHG: Cotyledonary, Heart, Globular Stage Somatic Embryos; 2,4-D: 2,4–Dichlorophenoxy acetic acid; ET: Elongated Torpedo Stage Somatic Embryos; ANEPC: Average number of embryos per culture; ST: Short Torpedo Stage Somatic Embryos; MS: Murashige and Skoog's (1962).

Frequency of Embryogenesis (FE)

It is the proportion of explant showing at least one somatic embryo at any stage of development out of total number of explant inoculated in a particular treatment. It is calculated and presented as percentage of response by the following formula:

$$\text{FE (per cent)} = \frac{\text{Sum of the number of explants showing at least one somatic embryo per treatment}}{\text{Total number of explant inoculated per treatment}} \times 100$$

Intensity of Embryogenesis (IE)

It is the mean number of somatic embryos (irrespective of stage of development) produced per responsive explant in a particular treatment. It is calculated by the following formula and presented as the average number of somatic embryos produced per treatment (ANEPC).

$$\text{IE (ANEPC)} = \frac{\text{Sum of the total number of somatic embryos produced in each of the explant in a particular treatment}}{\text{Total number of explant showing at least one somatic embryo per treatment}} \times 100$$

Since proportion of various stages of somatic embryos varied considerably among different treatments hence, percentage of three different categories of somatic embryos were calculated as per following formulae for every treatment.

Proportion of various stages (categories) of somatic embryos. It is the proportion of a particular stage (category) of somatic embryo developed out of total number of somatic embryos produced in each of the responsive culture in a treatment. It is presented as the percentage of the total number of somatic embryo and calculated separately for each catagory of somatic embryos in the following way:

Proportion of Elongated Torpedo Stage Somatic Embryos (ET)

$$\text{ET (per cent)} = \frac{\text{Sum of the total number of elongated torpedo stage somatic embryos produced in each explant per treatment}}{\text{Sum of the total number of somatic embryos produced per treatmernt}} \times 100$$

Proportion of Short Torpedo Stage Somatic Embryos (ST)

$$\text{ST (per cent)} = \frac{\text{Sum of the total number of short torpedo stage somatic embryos produced in each explant per treatment}}{\text{Sum of the total number of somatic embryos produced per treamernt}} \times 100$$

Proportion of Cotyledonary, Heart and Globular Stage Somatic Embryos (CHG)

$$\text{CHG (per cent)} = 100 - [\,\text{ET (per cent)} + \text{ST (per cent)}\,]$$

Efficiency of Embryogenesis (EE)

This is the most important parameter for induction of somatic embryogenesis in guava which is defined as the potential of a treament to produce somatic embryos at elongated or short torpedo stage or both the stages of development, that are convertible into plantlets. It is entirely relative parameter without any unit and calculated as under:

$$\text{EE (Relative)} = \frac{\text{FE (per cent)} \times \text{IE (ANEPC)} \times [\text{ ET (per cent)} + \text{ST (per cent) }]}{10{,}000}$$

Results

Effect of 2,4-Dichlorophenoxy Acetic Acid on Induction of Somatic Embryogenesis

The result of continuous treatment of zygotic embryos with different concentration of 2,4-D in 3 per cent sucrose containing Murashige and Skoog's medium is presented in the Table 28.1. It is clear from the results that 2,4-D had a significant effect on induction of somatic embryogenesis between 0.001–1.0 mg/l concentration as compared to the control culture without any growth regulator. Concentration of 2,4-D higher than 1.0 mg/l had been inhibitory for induction of somatic embryogenesis in continuous presence in 3 per cent sucrose containing full strength MS medium.

The process of somatic embryogenesis in various studies have reported mostly either as frequency or intensity as average number of somatic embryos produced per explant or the both. These parameters however were insufficient to indicate the proportion of convertible somatic embryos. If we consider only frequency and intensity of embryogenesis 0.01 mg/l od 2,4-D showed the maximum response but it did reflect the pproportion of convertible somatic embryos. In addition the convertible somatic embryo (ET + ST), however, showed a decreasing trend in the presence of 2,4-D with maximum number produced in 0.001 mg/l concentration. The proportion of CHG stages somatic embryos showed a positive correlation with 2,4-D concentration. Still there was no clear image to indicate the best inductive treatment of 2,4-D. Hence, it had been imperative to calculate the efficiency of somatic embryogenesis in order to show the potential of each and every treatment of growth regulator (see above). With this single parameter it was easy to find out the potential of any treatment irrespective of types of growth regulators (data not shown). Therefore it is clear from the table that 0.01 mg/l of 2,4-D showed the highest efficiency of somatic embryogenesis with the optimum proportion of convertible embryo which is the utmost requirements for micropropagation of any species.

Growth of Plantlets

Somatic plantlets in liquid as well as agar in solidified germination medium ceased to grow after two weeks of subculture and showed very little percentage of survival upon transfer and hardening to *ex vitro* condition. When subcultured to full strength MS liquid medium with 3 per cent sucrose for an extended growth of two weeks, there was significant improvement of survival frequency. All the plant grew well in this growth medium and attained full height of the culture vessels within 2 weeks of subculture (Akhtar, 1997; Akhtar *et al.*, 2000).

Ex vitro Performance

Plantlets regenerated through somatic embryogenesis in guava could not survive well on non-sterile soil mixture because of fungal infection. Amin (1987) had also encountered such attack of damping off diseases on guava plantlets regenerated through organogenesis. Plantlet from liquid or agar solidified germination medium did not survive even on sterile soil mixture. However, plantlets from liquid growth medium initially required a more porous mixture of sand, soil and compost in a ratio of 5:1:2. Once the root system adjusted to this primary mixture, transfer to a mixture with elevated proportion in a ratio of 1:2:3 of the above component was found beneficial. A regular watering of plantlets at an interval of 1-2 days had resulted in 50-90 cm plant height within 8-10 weeks of soil transplantation (Figure 28.1F). These plants had also performed well in the field and fruited within a period of less than one and a half year.

Discussion

The zygotic embryos had been used as explants material in the present study had also served as the starting explant material for induction of somatic embryogenesis in pea (Stejskal and Griga, 1992), soybean (Barwale *et al.*, 1986, Lazzeri *et al.*, 1987), feijoa (Cruz *et al.*, 1990). There was considerable changes in the explants morphology *in vitro*, reflects the reorganization of structural components of somatic cells in the presence of 2,4-D and other growth regulator (data no shown). The importance of phytohormones in influencing the gene(s) to trigger differentiation of cells in culture has already been reviewed extensively (Vasil 1980, Evan *et al.*, 1983, Bhojwani, 1990, Hobbie *et al.*, 1994, Duddit *et al.*, 1991). The cell size, cytoplasmic content and the type of first division were distinctly different between cell committed to embryogenesis and the cell that form callus tissue (Thorpe, 1988). Asymmetric partition at first division was frequently observed in sub epidermal cell layers of guava zygotic embryo explants among histological preparations (not shown). While, the cells of epidermal layers showed vertical and equal divisions in similar sections resulting into large vacuolated cells with comparatively lesser amount of cytoplasm and developing into callus from the zygotic embryo explants of guava.

Development of somatic embryos among the yellowish or yellowish brown calli at surfaces distal to the medium reflect the formation gradation of growth regulator concentration among different cell layers of the zygotic embryo explants but requires an extensive biochemical investigations. This is further evident why only fraction of cells appears to responds to the embryogenic treatment. Indeed, the detailed comparison between embryogenic and non-embryogenic clones from some genotype of alfalfa has revealed considerably increased sensitivity to 2,4-D in protoplast derived cells or root explants of embryogenic genotype. Embryogenesis in this clone occurs in tissues or colonies grown in the presence of 2,4-D at concentrations that already inhibited the growth of callus tissue (Bogre *et al.*, 1990). Similarly yellowish to yellowish brown calli contributed to embryo formation, in the feijoa (Cruz *et al.*, 1990), potato (Pretoa and Dedicova, 1992), whereas, in soybean (Sellar *et al.*, 1990) almost 95 per cent of zygotic embryos produced calli but somatic embryos produced directly from explants surface than the calli. Although there is no uniformity in the callus, morphology contributing to the embryo formation among various species but their characterization would certainly form the basis for the recognition of cell types for the development of an automated system for the process of somatic embryogenesis.

High frequency of globular stage and embryos showing anomalies in mango (DeWald *et al.*, 1989), feijoa (Cruz *et al.*, 1990), cherry (March *et al.*, 19993) and other systems are linked to long and continuous exposure of explants with auxins. However, frequency of anamolous somatic embryos was invariably low in the present study. The low frequency of anomalies and development of embryos beyond globular stage might be due to the fact that in the present study the culture conditions might maintain the high intracellular concentrations and did not block its polar transport.

Induction of somatic embryogenesis has also been achieved in the presence of other auxins like indole acetic acid, indole butyric acid and naphthalene acetic acid. A minor induction of somatic embryogenesis has been achieved in the presence of cytokinins but presentation of result is beyond the subject of the present study.

In the present study we report the induction of somatic embryogenesis in the presence of broad range of 2,4-D. Evaluation of the process of induction of somatic embryogenesis under several parameters like frequency, intensity and efficiency as well proportions of different stages of somatic embryos have proved beneficial for the analysis of potential of various concentrations of 2,4-D and other growth regulators. This would help to understand the process for high efficiency micropropagation in other species.

Future Prospects

The present work described sufficiently the importance of various parameters on induction of somatic embryogenesis, development, maturation, encapsulation and germination of somatic embryos as well as *ex vitro* performance of regenerated plants for micropropagation of *Psidium guajava* L. species. However, to maintain the clonal fidelity of the cultivar, embryogenesis protocol from somatic tissue like leaves would be beneficial and have yet to develop. Further, a convincing protocol leading to large scale automated commercial production of propagules and their field trial is still lacking and will be the topic for the future research. Indirect production and synchronous development of somatic embryos from callus and cell suspension cultures have to come up to achieve this goals.

The *in vitro* manipulation of somatic cells or tissues for the regeneration of guava plants or for improving its fruit quality, has not been explored sufficiently, and will be the problem for the future research. The problem of large number of seeds per fruits can be minimized in general, by raising the triploid plants through endosperm culture (Johri and Bhojwani, 1977; Wang and Chang, 1978; Bhojwani, 1985; Garg *et al.*, 1996). There is no report available so far on regeneration of guava from endosperm culture in a bid to either induce aneuploidy to get the genetically engineered seedless ones, and hence, will be the focus of future research to get such results.

Somaclonal variations are of common occurrence in various tissue cultures methodologies, which can also be employed in selecting guava variants through different treatment modifications of embryogenic explants. Since the plant regeneration from callus cultures have been suggested to be helpful in selecting the tolerant strains (Jaiswal and Amin, 1992), and the same with embryogenic culture of guava could lead to the development of variants plants resistant to various environmental stress(es) and disease(s). Such developed guava variants can be grown on relatively poor/otherwise problematic soils and may become popular among the poor as a source of monetary earnings from their non-productive wastelands. In any case, it would provide the basis for further investigations on the little explored biotechnological aspects of guava to enrich the intellectual contribution of the Nation.

Isolation and culture of protoplasts offer *direct transformation of plants* through protoplast fusion and somatic hybridization for the generation of somaclonal (protoclonal) variants in fruit crops (Hammerschlag and Litz, 1992). This has led to the development of protoplasts-to-plant protocols in a number of herbaceous and a few woody perennial species (Ochatt *et al.*, 1992). Yet, somatic hybridization occupies the prime position among the aims of protoplast manipulations (Ochatt *et al.*, 1989; 1992), but such reports are not available for the production of somatic hybrids of guava. The isolation of protoplasts from the tissues of embryogenic cultures of existing guava cultivars will be the problem for the future research to produce its somatic hybrids having desired characters.

The improvement of guava for good quality fruits with delayed ripening through the genetic manipulation of somatic cells or tissues is yet to come up. Thus, the genetic transformations using the protoplasts of the embryogenic cultures/calli having high regeneration potential along with the gene(s) of interest are the key components of the future research.

Conclusion

Somatic cell genetic of fruit crops is contingent upon the efficient and reliable regeneration of plants from cell and tissue cultures for clonally selected plants. In our study a number of parameters have been found influencing the induction of somatic embryogenesis, development, maturation, encapsulation and germination of somatic embryos as well as the *ex vitro* performance of regenerated plants for micropropagation of *Psidium guajava* L. species. However, the controlling mechanism underlying the regulation of the process is still missing in this and other fruit species. A better understanding of the process and the factors that influence embryogenesis at molecular level, their interactions couples with the physiological and biochemical studies in guava would probably lead to success in other fruit crops.

The progress in guava somatic embryogenesis described here shows promise as a model system for biotechnological improvement of other fruit tree. Although somatic embryogenesis is a common feature of *in vitro* morphogenesis in many fruit species, this appears to be the first review on somatic embryogenesis for *Psidium guajava* L. Consequently, the establishment of sound protocol for the production of somatic embryos of *Psidium guajava* L. is the first step towards a methodology for production of clonal stocks. Further research should be focused on the development of somatic embryogenesis protocols from tissues of somatic origin like leaves etc. to maintain the clonal propagule in this and other fruit trees.

Acknowledgements

The financial support to the Laboratory of Morphogenesis, Centre of Advanced Study in Botany, Banaras Hindu University, Varanasi, for various research positions under different projects by the University Grant Commission, Council of Scientific and Industrial Research, Central Silk Board, Department of Biotechnology, Department of Science and Technology, Government of India is gratefully acknowledged. The laboratory facilities extended by Jain Institute of Vocational and Advanced Studies, Bangalore and PG Department of Microbiology and Biotechnology, Al-Ameen Arts Science and Commerce College, Bangalore for continuing tissue culture research to the author is gratefully acknowledged.

References

Akhtar, N. (1996). Recurrent embryogenesis as a mean of conservation of guava germplasm. In: *Proc. National Research Seminar in Botany: Environment, Biodiversity and Conservation*, Balaghat (MP), India, p. 17 (Abst.).

Akhtar, N. (1997). Studies on induction of somatic embryogenesis and production of artificial seeds for micropropagation of a tropical fruit tree guava (*Psidium guajava* L.). Unpublished *Ph.D. Thesis*, Banaras Hindu University, Varanasi; 203 pp.

Akhtar, N. and Jain, S. M. (2000). Application of somatic embryogenesis for the improvement of tropical fruit trees. In: *Somatic Embryogenesis in Woody Plants* Vol. 6., (eds). S. M. Jain, P. K. Gupta and R. J. Newton. pp. 215-247. Kluwer Academic Publisher, The Netherland.

Akhtar, N. and Jaiswal, V. S. (1994). Role of somatic embryogenesis and artificial seed production production for micropropagation and improvement of guava (*Psidium guajava* L.). In: *Proc. Symp. on Role of Plant Tissue Culture in Biotechnology in India*, Varanasi, India. p. 2 (Abst.).

Akhtar, N. and Jaiswal, V. S. (1995). Rapid multiplication of *Psidium guajava* L. through recurrent embryogenesis. In: *Proc. All India Symposium on Recent Advances in Biotechnological Application of Plant Tissue Culture*. Mysore, India. p. 49 (Abst.).

Akhtar, N., Kumari, N., Pandey, S., Ara, H., Singh, M., Jaiswal, U., Jaiswal, V. S. and Jain, S. M. (2000). Somatic embryogenesis in tropical fruit trees. In: *Somatic Embryogenesis in Woody Plants* Vol. 6., (eds). S. M. Jain, P. K. Gupta and R. J. Newton. pp. 93-140. Kluwer Academic Publisher, The Netherland.

Amin, M. N. (1987). *In vitro* clonal propagation of guava (*Psidium guajava* L.) and Jackfruit (*Artocarpus heterophyllus* Lamk.). *Ph.D. Dissertation*, Banaras Hindu University, Varanasi.

Amin, M. N. (1991). *Plant Tissue Culture*, 1: 13-18.

Amin, M. N. and Jaiswal, V. S. (1987). Rapid clonal propagation of guava through *in vitro* shoot proliferation on nodal explants of mature trees. *Plant Cell Tiss. Org. Cult.*, 9: 235-243.

Amin, M. N. and Jaiswal, V. S. (1988). Micropropagation as an aid to rapid cloning of a guava cultivar. *Sci. Hort.*, 36 : 89-95.

Amin, M. N. and Jaiswal, V. S. (1989a). *In vitro* propagation of guava (*Psidium guajava* L.): Effects of sucrose, agar and pH on growth and proliferation of shoots. *Bang. J. Bot.*, 18 : 1-8.

Amin, M. N. and Jaiswal, V. S. (1989b). Effects of phloroglucinol, sucrose, pH and temperature on *in vitro* rooting of guava (*Psidium guajava* L.) microcuttings. *Bang. J. Bot.*, 18: 129-139.

Annonymous (1985) : *Publication Information Directorate*, C. S. I. R., New Delhi.

Annonymous (2008). *II International symposium on guava and other Myrtaceae*, Meride Mexico.

Babbar, S. B. and Gupta, S. C. (1986a). Induction of androgenesis and callus formation in *in vitro* cultured anthers of a myrtaceous fruit tree (*Psidium guajava* L.). *The Botanical Magazine*, Tokyo, 99: 75-85.

Babbar, S. B. and Gupta, S. C. (1986b). Embryogenesis in stamen and hypocotyl explants of a myrtaceous fruit tree *Psidium guajva* L. In: *VI. International Congress on Plant Tissue and Cell Culture*. Minneapolis, p. 133. (Abstract).

Bajaj, Y. P. S. (1995). Cryopreservation of somatic embryos. In: *Biotechnology in Agriculture and Forestry. Vol. 30 : Somatic Embryogenesis and Synthetic Seed-I*. Y.P.S. Bajaj (Ed.). Springer Verlag, Berlin, Heidelberg, pp. 221-229.

Benson, E. E., Lynch, P. T. and Stacey, G. N. (1998). Advances in plant cryopreservation technology: current application in crop plant biotechnology. *AgBiotech News and Information*. 10 (5): 133-141.

Bhojwani, S. S. (1985) : In: (Ed.) I.K. Vasil; pp. 258-268, Academic Press London.

Carlson, P. S. (1987) : In : S. Natesh, V. L. Chopra and S. Remohandran (Eds); Oxford and IBH Publishers, New Delhi, Bombay, Calcutta, pp. 85-92.

Chandra, R., Bajpai, A., Gupta, S., and Tiwari, R. K. (2004). Embryogenesis and plant regeneration from mesocarp of *Psidium guajava* L. (guava). *Ind. J. Biotechnol.*; 3: 246- 248.

De'Cruz, R. and Babu Rao, G. (1962). *J. Indian Bot. Soc.*; 41: 316-321.

Doijode, S. D. (2001). Guava: Psidium guajava L. In: *Seed Storage of Horticultural Crops*. (Ed.). Doijode, S.D. pp. 65 -67. Haworth Press, New York.

FAO (2001). Market for Tropical Fruits in India. FAO report on the Second Session on Committee on Commodity Problem: Inter Governmental Groups on Banana and Tropical Fruits, San Jose, Costa Rica: CCP:BA/TF 01/11.

Fitchet, M. (1989). Citrus and Subtropical Fruit Research Institute: *Information Bulletin*; 201: 4-5.

Garg, L., Bhandari, N. N. Rani, V. and Bhojwani, S. S. (1996). *Plant Cell Rep.*, 15 : 855-858.

Grout, B. (Ed.) (1995). *Genetic Preservation of Plant Cells in vitro*: Springer Lab. Manual. Springer- Verlag, Berlin, 169 pp.

Gupta, P. K., Durzan, D. J., and Finkle, B. J. (1987). Somatic polyembryogenic cell masses of *Picea abies* (Norway spruce) and *Pinus taeda* (loblolly pine) after thawing from liquid nitrogen. *Can. J. For. Sci.*, 17: 1130- 1134.

Hammerschlag, F. A. and Litz, R. E. (1992) (Eds). CAB International, Wallingford. UK. 550 pp.

Hirai, D. Shirai, K. Shirai, S. and Sakai, A. (1998). Cryopreservation of *in vitro* grown meristem of strawberry (*Fragaria x ananassa* Duch.) by encapsulation- Vitrification. *Euphytica*; 101:109-115.

Jaiswal, U. and Jaiswal, V. S. (2005). *Psidium guajava. In: Biotechnology of Fruits and Nut Crops* , (Ed.) Litz, RE. pp. 394-401. CAB International, Wallingford, UK.

Jaiswal, V. S. and Akhtar, N. (1993). Studies on induction and factors controlling somatic embryogenesis in guava. In: *Proc. Symp. on Plant Tissue Culture and its Biotechnological Application*, Jaipur, India, p. 48 (Abst.).

Jaiswal, V. S. and Akhtar, N. (1994). Induction of somatic embryogenesis and artificial seed prioduction for micropropagation of guava (Psidium guajava L.). In: *Proc. VIII Int. Cong. Plant Tiss. Cell Cult.* June 12-17, Italy, p. 55.

Jaiswal, V. S. and Amin, M. N. (1986). *In vitro* shoot proliferation and plantlet formation from somatic tissues of guava (*Psidium guajava* L.). *VI International Congress on Plant Tissue and Cell Culture*, Minneapolis, p. 279. (Abstract).

Jaiswal, V. S. and Amin, M. N. (1987). *In vitro* propagation of guava from shoot cultures of mature trees. *J. Plant Physiol*, 130 : 7-12.

Jaiswal, V. S. and Amin, M.N. (1992). Guava and Jackfruit. In: *Biotechnology of Perennial Fruit Crops. Biotechnology in Agriculture*. (Eds.) Hammerschlag, F. A. and Litz, R.E. No. 8. pp. 421-431. CAB International, Wallingford, UK.

Johri, B. M. and Bhojwani, S. S. (1977) : In : J. Reinert and Y. P. S. Bajaj (Eds); Springer-Verlag, Berlin, pp. 398-411.

Kartha, K. K. and Engelmann, F. (1994). Cryopreservation and germplasm storage. NRCS publication no. 32487. In: *Plant Cell and Tissue Culture*. (Eds) I. K. Vasil and T. A. Thorpe. pp. 195-230.

Litz, R. E. (1985). Somatic embryogenesis in tropical fruit trees. In: *Tissue Culture in Forestry and Agriculture*. (Eds.) Henke, R. R., Hughes, K. W., Constantin, M. J. and Hollaender, A. pp. 179-193. Plenum Press, New York,

Loh, C. S. and Rao, A. N. (1989). Clonal propagation of guava (*Psidium guajava* L.) from seedling and grafted plants and adventitious shoot formation *in vitro*. *Sci. Hort.*, 39 : 31-39.

Majumdar, P. K. and Mukherjee, S. K. (1971). *Nucleus*; 14 : 119-131.

Majumdar, P. K. and Mukherjee, S. K. (1972a). *Cytologia*; 34 : 541-548.

Majumdar, P. K. and Mukherjee, S. K. (1972b). *Nucleus*; 15: 42-47.

Matsumoto, T., Sakai, A. and Nako, Y. (1994). Cryopreservation of *in vitro* grown apical meristem of wasabi (*Wasabia japonica*) by vitrification and subsequent high plant regeneration. *Plant Cell Rep.*, 13: 442- 446.

Menzel, C. M. (1985). *Queensland Agri. J.*, 111: 93-98.

Mitra, S. K. and Bose, T. K. (1985). Guava. In : Bose, T. K. (Ed.). *Fruits of India: Tropical and Subtropical*. Naya Prokash, Calcutta, pp. 277-279.

Murashige, T., and Skoog, F. (1962). A revised medium for rapid growth and bioassay with tobacco tissue culture. *Physiol. Plant.*, 15 : 473-497.

Nijjar, G. S. (Ed.) (1977). *Fruit Breeding in India*. Oxford and IBH Publishers, New Delhi, Bombay, Calcutta, 219 pp.

Ochatt, S. J., Patat-Ochatt, E. M. and Power, J. B. (1992). In : F. A. Hammerschlag and R. E. Litz (Eds); CAB International Wallingford, U.K. pp. 77-103.

Ochatt, S. J., Patat-Ochatt, E. M., Rech, E. L., Davey, M. R. and Power, J. B. (1989). *Theo. Appl. Genet.*, 78: 35-41.

Papadatou, P., Pantikis, C. A., Ephtimiadou, E. and Lydaki, M. (1990). Rapid multiplication of guava seedlings by *in vitro* shoot tip culture. *Sci. Hort.*, 45 : 99-103.

Perez, R. M. (2000). Cryostorage of *Citrus* embryogenic cultures. In: *Somatic Embryogenesis in Woody Plants*, Vol. 6. (Eds.) S. M. Jain, P. K. Gupta and R. J. Newton. pp. 687- 705. Kluwer Academic Publisher, The Netherland.

Perez, R. M., Novarro, L. and Duran-Vila, N. (1997). Cryopreservation and storage of embryogenic cullus cultures of several citrus species and cultivars. *Plant Cell Rep*. 17: 44- 49.

Purseglove, J. W. (1968). *Tropical Crops Dicotyledons*. Longman Group Ltd. London, pp. 414-419.

Rai, M. K., Akhtar, N. and Jaiswal, V. S. (2007). Somatic embryogenesis and plantlet regeneration in *Psidium guajava* L. cv. Banarsi Local. *Scientia Horticulturae*; 113: 129 -133.

Rai, M. K., Jaiswal, V. S. and Jaiswal, U. (2008a). Effect of ABA and sucrose on germination of encapsulated somatic embryos of guava (*Psidium guajava* L.). *Scientia Horticulturae*; 117: 302-305.

Rai, M. K., Jaiswal, V. S. and Jaiswal, U. (2008b). Encapsulation of shoot tips of guava (*Psidium guajava* L.) for short term storage and germplasm exchange. *Scientia Horticulturae*; 118: 33- 38.

Rai, M. K., Jaiswal, V. S. and Jaiswal, U. (2008c). Alginate encapsulation of segments of guava (*Psidium guajava* L.) for germplasm exchange and distribution. J. *Horticultural Sci. and Biotechnol.*; 83 (5): 569- 573.

Sakai, A., Kobayashi, S. and Oiyama, I. (1991). Survival by vitrification of nucellar cells of Novel orange (*Citrus sinensis* var. *brasileinsis* Tanaka) colled to–190 °C. *J. Plant Physiol.*; 137: 465-470.

Saxena, Sangeeta, Rajan, S., Chandra, R., Srivastava, N. and Bajpai, A. (2007). Molecular characterization of closely related open pollinated seedling selections in Guava. Proceedings of the First International Guava Symposium. *Acta Horticulturae* No. 735 March 49-55.

Schroeder, C. A. (1961). *Bot. Gaz.* 122: 198-201.

Singh, A. (1986). Kalyani Publishers, New Delhi, pp. 328-331.

Singh, G., Krishnu, R., and Chandra, R. (2007). First International guava symposium, Organized by Society for Development of Subtropical Horticulture and Central Institute for Subtropical Horticulture Lucknow in Collaboration with International Society for Horticultural Science, Nandi Prisad and Agricultural Agricultural Proceesed Food Product Export Authority, at Lucknow. ISHS, ACTA Horticulturae 735 (CDROM).

Singh, R. (1985). *Fruits*. 4th Edition. National Book Trust, India, 213 pp.

Singh, S. K., Meghwal, P. R., Sharma, H. C. and Singh, S. P. (2002). Direct shoot organogenesis on explant from germinated seedlings of *Psidium guajava* L. cv. Allahabad Safeda. *Scientia Horticulturae*; 95: 213–221.

Wang, T. and Chang, C. (1978). In : *Proc. Symp. Pl. Tiss. Cult*, Science Press, Peking, pp. 463-467.

Wilson, C. W. (1980). In : S. Nagy and P. E. Shaw, A. V. I. Publishing, West Port.

Wither, L. A. and Engelmann, F. (1998). *In vitro* conservation of plant genetic resources. *Agricultural Biotechnology*. Marcel Dekker, New York. pp. 57- 88.

Yadav, U. L. (1996). Guava production ij Georgia under cold protection structure. In: *Progress in New Crops*. (Ed.) Janick, J. pp. 451–457. ASHS Arlington, VA.

Yasseen, Y. M., Barringer, S. A., Schnell, R. J. and Splittstoesser, W. E. (1995). *In vitro* shoot proliferation and propagation of guava (*Psidium guajava* L.) from germinated seedlings. *Plant Cell Rep.*, 14: 525-528.

Microbial Biotechnology and Ecology (2011) *Pages* **520–524**
Editors: **Deepak Vyas, G.S. Paliwal, P.K.Khare and R.K Gupta**
Published by: **DAYA PUBLISHING HOUSE, NEW DELHI**

Chapter 29

Hepatoprotective Medicinal Plants: A Review

☆ *Siddhartha Singh*, A. Mehta and P. Mehta*

Department of Botany, Dr. H.S. Gour University, Sagar, M.P.

ABSTRACT

Use of herbal drugs in the treatment of liver diseases has a long tradition, especially in Eastern medicine. The chemical compounds present in the plant are responsible for their medicinal properties. The quality and quantity of phytochemicals present in the extract depends upon the solvent used for extraction and the method of extraction. The methanolic extract of *Silybum marianum, Apium graveolens* and *Hygrophila auriculata* seeds, aqueous extract of *Andrographis paniculata* leaves, aqueous and ethanolic extract of bark stem of *Pterocarpus* santalinus, aqueous seed extract of *Cleome viscosa* are some of the examples of hepatoprotective plant extracts. The potential of the active components present in these plant extracts are proved by CCl_4 induced liver damage models in mice.

Keywords: Herbal drugs, Phytochemicals, Extract and Carbon tetrachloride (CCl_4).

Introduction

Medicinal plants play a key role in the human health care. About 80 per cent of the world populations rely on the use of traditional medicine which is predominantly based on plant materials. Liver, the key organ of metabolism and excretion, is constantly endowed with the task of detoxification of xenobiotics, environmental pollutants and chemotherapeutic agents. Thus, disorders associated with this organ are numerous and varied. While a curative agent has not yet been found in modern

* E-mail: siddhartha.research@rediffmail.com

medicine, the current usage of corticosteroids and immunosuppressive agents only brought about symptomatic relief. Furthermore, their usage is associated with risk of relapses and danger of side effects. On the other hand, Ayurveda, an indigenous system of medicine in India, has a long tradition of treating liver disorders with plant drugs. Enhanced lipid peroxidation produced during the liver microsomal metabolism of ethanol may result in hepatitis and cirrhosis. To focus on a more practical and systematic approach towards the development of standardized medicine, many plant derived drugs used in modern medicine are developed by ethnomedical leads and subsequent ethno pharmacological studies. There are more than 100 drugs of known structure that are extracted from higher plants and used in allopathic medicine.

To assess the liver functions, biologists have chosen three basic biochemical parameters which are levels of SGOT, SGPT and serum bilirubin with CCl_4 induced liver damage models of albino rats.

Hepatoprotective Drug Development

Pharmacological validation of each hepatoprotective plant should include efficacy evaluation against liver diseases induced by various agents. Efficacy of the drug against viral hepatitis as well as liver damage induced by hepatotoxic chemicals by oxidative mechanisms or other mechanisms should be determined. Further, the plant drugs have to be evaluated for their effects on liver regeneration and bile secretion, the same drug is unlikely to be effective against hepatic damage caused by various agents.

Medicinal Plant and Hepatoprotection

There are numerous medicinal plants which have shown high efficiency for the hepatoprotective activity and have shown great potential as a natural remedy for the treatment of acute liver diseases and damages. Summary of some of the plants with proven hepatoprotective activity are discussed below.

Silybum marianum

Commonly known as 'Milk thistle' and belongs to family 'Compositae'. It occurs widely in Europe, Canada and South America. In India it is indigenous to Kashmir. Silymarin a flavonoid is an active compound obtained from *S. marianum* with hepatoprotective activity. Silymarin is able to neutralize the hepatotoxicity of several agents, including *Amanita phalloides*, ethanol, paracetamol (acetaminophen) and carbon tetrachloride in animal models. The protection against *A. phalloides* is inversely proportional to the time that has elapsed since administration of the toxin. Silymarin protects against its toxic principle α-amanitin by preventing its uptake through hepatocyte membranes and inhibiting the effects of tumor necrosis factor, which exacerbates lipid peroxidation. The active substances from *S. marianum* also stabilize the cell membrane, exert antioxidant properties and exhibit intrinsic toxicity when tested on primary culture of human hepatocytes.

Andrographis paniculata

It is an annual herb which is commonly called as 'Kalmegh' and belongs to family 'Acanthaceae'. It is widely distributed in Sri Lanka and throughout India especially in Maharashtra, Uttar Pradesh, Madhya Pradesh, Tamil Nadu and Andhra Pradesh. It is another proven hepatoprotective agent. The aqueous extract of which, when given orally to the animals with liver damage induced by hexa-chloro-cyclohexane (BHC) inhibited BHC induced liver toxicity in Swiss male mice as assessed by the biochemical values. The aqueous extract of *A. paniculata* has also role in delaying the hepatic turnerogenic condition by lowering the SGPT, SGOT and alkaline phosphatase level in BHC treated mice.

Apium graveolens and *Hygrophila auriculata*

A.graveolens is a biennial plant found in the regions of Himalaya, Punjab and North Africa. Its common name is 'Celery seeds' and belongs to family 'Umbelliferae'. Pretreatment of rats with 200 mg/kg p.o. of the methanolic extracts of the seeds of *A. graveolens* or *H. auriculata* 24, 48 and 2 h before paracetamol administration exhibited a significant reduction in the paracetamol induced increase in the levels of different SGOT, SGPT and serum bilirubin.

Ginkgo biloba

It is one of the oldest known trees and is widely distributed all over the world, also known as 'Maiden Hair Tree' and belongs to family 'Gingkoaceae'. It is reported that there was a significant increase in the serum hepatic enzyme levels and significant decrease in the TP and Alb levels after CCl_4, which was reversed with *G.biloba.*

Pterocarpus santalinus

P. santalinus belongs to 'Leguminoseae' family and also known by names like 'Bijasal' and 'Indian Kino Tree'. It is mainly found in hilly regions of Gujrat, Madhya Pradesh, Uttar Pradesh and Bihar. It is reported that there was a significant increase in serum levels of bilirubin, alanine transaminase, aspartate transaminase and alkaline phosphatase with a decrease in total protein level, in the CCl_4 treated animals, reflecting liver injury. In the aqueous and ethanol extracts treated animals there was a decrease in serum levels of the markers and significant increase in total protein, indicating the recovery of hepatic cells. Histological study of aqueous extract treated group exhibited moderate accumulation of fatty lobules and cellular necrosis where as ethanol extract treated animals revealed normal hepatic cords without any cellular necrosis and fatty infiltration.

Pergularia daemia

It is a perennial herb grows along the road sides in India and the studies on the hepatoprotective effect of alcoholic extract was carried out using carbon tetrachloride induced liver damage in wistar albino rats. Acetone sub fraction showed significant protective effect by lowering serum levels of various biochemical parameters. These biochemical observations were supplemented by histopathological examination of liver sections. Here, Silymarin was used as positive control. The results justify the use of *P. daemia* as a hepatoprotective agent.

Cleome viscosa

Research suggest that the aqueous seed extract of *C. viscosa* has hepatoprotective activity against carbon tetrachloride (CCl_4) induced liver damage in wistar rats. The aqueous seed extract of *C. viscosa* (200 mg/kg) was administered orally to the animals with hepatotoxicity induced by CCl_4. Silymarin (200mg/kg) was given as reference standard. There was significant reduction in serum enzyme aspartate aminotransferase (AST), alanine aminotransferase (ALT), alkaline phosphatase (ALP), Y-glutamyl transpeptidase and lipid peroxidase and increase in reduced glutathione (GSH). Various pathological changes like centrilobular necrosis and vacuolization observed in CCl_4 treated rats, which were prevented to a moderate extent in groups, treated with *C. viscosa* and Silymarin. It was concluded from the study that aqueous seed extract of *C. viscosa* possesses hepatoprotective activity against CCl_4 induced hepatotoxicity in rats.

Picrorhiza kurrooa

Commonly called as 'Indian Gentian' and 'Kutki'. This perennial herb of family 'Scrophulariaceae' is distributed in upper Himalayas and China and grows naturally in Kashmir and Himachal Pradesh. The hepatoprotective study of *P. kurrooa* was evaluated against alcohol-CCl_4, induced liver damage in rat at the doses of 3, 6 and 12 mg/kg/day, p. o. for 7 days. It showed a significant hepatoprotective activity by lowering of the elevated levels of glutamate oxaloacetate transaminase, glutamate pyruvate transaminase, acid phosphatase, alkaline phosphatase, glutamate dehydrogenase and bilirubin in the serum of alcohol-CCl_4 induced liver damage models.

Discussion

Medicinal plants are of great importance as they have been proven for their effectiveness against various diseases with less or no side effects. Disorders associated with liver are one of the most common problems faced by the human and medicinal plants hold the key for it. The active compounds present in the plant extract are responsible for their hepatoprotective potential. The type and concentration of the phytochemicals present in the extract depends upon the solvent used for extraction and the method of extraction. Flavonoids, diterpinoids, tannins etc. are the most common chemical compounds present in the plant which are responsible for their hepatoprotective potential.

References

Dvorjak, Zdenejk, Kosina, Pavel, Walterova, Daniela, Sjimanek, Vilym, Bachleda, Petr and Ulrichova, Jitka (2003). Primary cultures of human hepatocytes as a tool in cytotoxicity studies: cell protection against model toxins by flavonolignans obtained from *Silybum marianum*. *Toxicology Letters;* 137: 201-212.

Fransworth, N.R. (1990). The role of ethnopharmacology in drug development. In: *Bioactive compounds from plants*. Ciba Foundation Symposium, Chichester, John Wiley and Sons; 154: 2-21.

Fraschini, F., Demartini, G. and Esposti, D. (2002). Pharmacology of Silymarin. *Clin Drug Invest;* 22(1): 51-65.

Handa, S.S., Sharma, A. and Chakraborthi, K.K. (1986). Natural products and plants as liver protecting agents. *Fitoterapia,* LVII (5): 307–351.

Manjunatha, B.K. (2006). Hepatoprotective activity of *Pterocarpus santalinus* L.f., an endangered medicinal plant. *Indian J Pharmacol;* 38: 25-8.

Mitra, S.K., Venkataranganna, M.V., Sundaram, R. and Gopumadhavan, S. (1998). Protective effect of HD-03, a herbal formulation, against various hepatotoxic agents in rats. *Journal of Ethnopharmacology;* 63: 181–186.

Sengottuvelu, S., Duraisamy, R., Nandhakumar, J. and Sivakumar, T. (1998). Hepatoprotective activity of *Cleome viscosa* against Carbon tetrachloride induced hepatotoxity in rats. *Pharmacognosy Magazine,* ISSN: 0973-1296.

Shenoy, K.A. (2001). Hepatoprotective Effects of *Ginkgo biloba* against Carbon Tetrachloride Induced Hepatic injury in Rats. *Indian Journal of Pharmacology;* 33: 260-266.

Singh, A. and Handa, S S. (1995). Hepatoprotective activity of *Apium graveolens* and *Hygrophila auriculata* against paracetamol and thioacetamide intoxication in Rats. *Journal of Ethnopharmacology;* 49: 119-126.

Smuckler, E.A. (1998). Alcoholic drink: Its production and effects. *Fed Proe*; 34: 2038-2044.

Subramoniam, A. and Pushpangadan, P. (1999). Development of Phytomedicines for Liver Diseases. *Indian Journal of Pharmacology*; 31: 166-175.

Sureshkumar, S. V. and Mishra, S. H. (1996). Hepatoprotective activity of extracts from *Pergularia daemia* Forsk. against carbon tetrachloride-induced toxicity in rats. *Pharmacognosy Magazine*, ISSN: 0973-1296.

Tripathi, S.C., Patnaik, G.K. and Dhawan, B.N. (1991). Hepatoprotective activity of Picroliv against Alcohol-Carbon tetrachloride induced damage in Rats. *Indian J Pharmac;* 23: 143-148.

Trivedi, N. and Rawal, U.M. (1998). Effect of Aqueous Extract of *Andrographis Paniculata* on Liver Tumor. *Indian Journal of Pharmacology*; 30: 318-322.

Trivedi, N. and Rawal, U.M. (2000). Effect of *Andrographis paniculata* on Severe Liver Damage. *Indian Journal of Pharmacology*; 32: 288-293.

Microbial Biotechnology and Ecology (2011) *Pages* **525–543**
Editors: **Deepak Vyas, G.S. Paliwal, P.K.Khare and R.K Gupta**
Published by: **DAYA PUBLISHING HOUSE, NEW DELHI**

Chapter 30

Cyanobacterial Exploitation in Biotechnology

☆ ***B.D. Kaushik***

Division of Biotechnology
Netaji Subash Institute of Technology, Sector 3, Dwarka, New Delhi – 110078

While writing this chapter I am lost in the sweet memory of late Prof. K.M. Vyas, who dedicated his entire life to up bring the Department of Botany at Sagar University. This lecture is wholly devoted to Prof Vyas.

Introduction

Cyanobacteria are both unicellular and multicellular filamentous. Some of the forms have terminally differentiated specialized structures called heterocyst. The cyanobacteria are self-supporting system and some of the species are capable of fixing atmospheric nitrogen. These are ubiquitous in nature and are next to bacteria in distribution. They grow wherever water, sunlight and limited number of inorganic elements available and rice field ecosystem is the ideal niche for the growth and proliferation of cyanobacteria. Due to their wider occurrence in rice field ecosystem, these organisms have been exploited as a means to provide biologically fixed nitrogen. Some of the cyanobacteria have potential to colonize inhospitable ecosystem *e.g.* salty soils hence are exploited to reclaim such soils. Still some have great economic importance as a source of food, feed and substances of pharmaceutical importance.

Cyanobacteria in Agriculture

Photosynthesis and nitrogen fixation are the two most important biological processes on the earth. Though the atmosphere contains about 79 per cent nitrogen, this form of nitrogen cannot be utilized by most plants, but utilizes nitrogen only in the combined mineral form, such as ammonium

(NH_4^+) or nitrate (NO_3^-). Nutrient nitrogen as such is the most limiting nutrient in soil and use of inorganic nitrogen fertilizers to meet the crop requirements is unavoidable for intensive farming practices. However, the crop utilizes only 30 to 35 per cent of the applied nitrogen and the remaining 65-70 per cent is lost either by volatilization or leaching. These losses are further enhanced due to reduced or almost no use of organic amendments. Under such circumstance microorganisms play significant role in maintaining soil fertility. The dinitrogen reduction so far is confined to prokaryotic cellular forms in nature. Among these prokaryotes, the photosynthetic diazotrophs–cyanobacteria, more commonly known as blue-green algae, are free living as well as symbionts with lichens, ferns and cycads and their contribution in total nitrogen fixation by biological means probably comes next to the legume-*Rhizobium* symbiosis.

Rice is an important food crop as more than 40 per cent of the world's population depends on rice as a major source of calories. In India, rice crop is cultivated in about 44 million hectares with an approximate total production of 81.16 million tonnes. Per hectare rice yield in India is about 1928 kg/ha, which is quite low compare to China (5807 kg/ha) and Japan (6273 kg/ha). Rice crop accounts for the largest share of chemical fertilizers. In tropical rice fields' biological nitrogen fixation is a cyanobacterial process. N-fixing blue-green algae (BGA) have been shown to be the most important in maintaining and improving the productivity of rice fields (Venkataraman, 1979; Kaushik, 1994a). It has been demonstrated that the N fertility of soil is sustained better under flooded conditions than under dry land conditions (Watanbe and Roger, 1984). Favourable conditions for biological N_2 fixation by such BGA is considered to be one of the reasons for the relatively stable yield of rice under flooded condition.

Diazotrophic Cyanobacteria

Cyanobacteria are one of the earliest forms of life on Earth and known to be the primary colonizers in most inhospitable habitats. As early as 1936 (De, 1936) suggested that the fertility of Indian rice field is maintained due to the presence of cyanobacteria, since then large number of forms has been reported from diverse habitats. In Indian rice field soils on an average cyanobacteria accounts to about 33 per cent (Venkataraman, 1975) whereas reports show that up to 50 per cent of the total algae was BGA in some of the southern and eastern states. The aerobic filamentous heterocystous as well as colonial forms represent the common flora of arable lands and are considered to be the important in nitrogen economy of rice cultivation (Table 30.1). The most common filamentous heterocystous species include *Anabaena, Aulosira, Calothrix, Cylindrospermum, Fischerella, Hapalosiphon, Nostoc, Scytonema, Stigonema, Tolypothrix* and *Westiellopsis* (Stewart *et al.*, 1979; Prasanna and Kaushik, 1995) and the unicellular or colonial are the species of *Aphanothece, Gloeothece, Synechoococcus, Synechocystis* and *Xenococcus* (Ripkka *et al.*, 1979; Prasanna and Kaushik, 1994). The non-heterocystous filamentous forms capable of nitrogen fixation are species of *Oscillatoria, Trichodesmium, Microcoleus, Pseudanabaena, Plectonema* and *Lyngbya* (Wyatt and Silvey, 1969).

Nitrogen Contribution

Prasad (1949) estimated *in situ* addition of nitrogen in soils of South Bihar by scrapping encrustations of cyanobacterial mats on soil surface collected from different places and the estimated nitrogen input varied from 12 to 16 kg ha^{-1} after the harvest of the crop. Similarly De and Mandel (1956) estimated nitrogen fixation by cyanobacteria in six rice growing soils of West Bengal and reported values ranging from 13.8 to 44.4 kg ha^{-1} of nitrogen in cropped area. In Japan Watanabe *et al.* (1951) observed addition of 20 kg ha^{-1} nitrogen by *Tolypothrix tenuis*. The water layer with extensive floating colonies of *Gloeotrichia* sp. in Moroccan rice fields has been found to show high rates of nitrogen fixing activity (44 n moles C_2H_4 ml^{-1} hr^{-1}) (Reynault *et al.*, 1975). Using ^{15}N technique, Mac Rae and Castro (1969) demonstrated an addition of 10-15 kg N ha^{-1} in rice fields due to cyanobacteria. Using same technique Yoshida and Ancazas (1993) found an addition of 40-80 kg N ha^{-1}. Even in the temperate soils, nitrogen fixation by algae is substantial. Henriksson (1971) showed annual fixation rate of 15-51 Kg N ha^{-1} yr^{-1} in an agricultural field where *Nostoc* was abundant, and 4.44 Kg N ha^{-1} yr^{-1} in a lakeside meadow containing *Nostoc, Anabaena, Cylindrospermum* and *Calothrix*. The relative contribution of cyanobacteria as a percentage of the total nitrogen fixed in paddy fields varies widely and is mainly dependent on physico-chemical properties of soil, climatic and biotic factors. Quantitative estimations from different countries are: Philippines 18-33 kg/ha (Watanabe and Lee, 1975; Alimango and Yoshida, 1977), India 15-53 kg/ha (De and Biswas, 1952; Singh, 1961; Venkataraman, 1979), Senega 10-30 kg/ha (Renaud and Roger, 1978), Mali 50-80 kg/ha (Traore *et al.*, 1978) and Japan 11-23 kg/ha (Okuda and Yamaguchi, 1955).

Table 30.1: Nitrogen Fixing Genera of Blue-green Algae (Cyanobacteria)

Genera	*Species in India*	*No. of N_2 Fixer*	*Genera*	*Species in India*	*No. of N_2 Fixer*
	Non-heterocystous		*Fischerella*	3	2
	Colonial		*Gloeotrichia*	10	1
Chlorogloea	2	1	*Hapalosiphon*	10	2
Gloeocapsa	23	2	*Mastigocladus*	1	1
Synechocystis	-	1	*Mastigocladopsis*	1	1
	Filamentous		*Nodularia*	1	1
Lyngbya	65	2	*Nostoc*	25	25
Oscillatoria	85	4	*Nostochopsis*	3	1
Phormidium	48	1	*Rivularia*	7	1
Plectonema	11	2	Scytonema	49	2
	Heterocystous		*Scytonematopsis*	2	2
Anabaena	25	25	*Stigonema*	12	1
Anabaenopsis	3	2	*Tolypothrix*	19	1
Aulosira	8	1	*Westiella*	2	1
Calothrix	29	All	*Westiellopsis*	1	1
Cylindrospermum	14	14	*Wollea*	1	1

Higher concentrations of combined nitrogen have been shown to inhibit both the growth as well as the nitrogen fixation ability of the cyanobacteria, and ammonium being toxic at 75 ppm (Stewart, 1964; Mekonnen *et al.*, 2002). Nitrate nitrogen is preferred by the cyanobacteria and is not detrimental even at 100 ppm concentration. There are, however, numerous reports on the beneficial effects of algalization in the presence of nitrogenous fertilizers (Satapathy, 1999). Singh *et al.* (1972) reported that grain yield, when the field was inoculated with algae and 60 kg ha^{-1} urea, was comparable to that obtained with 120 kg nitrogen alone. Similar observations were also made with ammonium sulphate, at nitrogen level as high as 120 kg ha^{-1} (Venkataraman and Goyal, 1963). In an experiment conducted in Haryana, Kaushik (1998) reported about 10-15 per cent increase in rice yield in the presence of fertilizer nitrogen up to 150kg ha^{-1} (Table 30.2). Thus, the response to algalization was generally positive at any level of nitrogen applied in the field, though the response being lower at higher levels.

Table 30.2: Response of Rice to Cyanobacterial Biofertilizer Application–Demonstration Trial 1996-1997 at Farmer's Field in Haryana

Treatments		*No. of Trials*	*Soil Type*	*Variety*	*Yield Q ha^{-1}*		*N-Saving kg ha^{-1}*	*% Increase*
N: P	*N: P + BGA*				*N:P*	*N: P + BGA*		
58:0	29:0	4	Loam	Basmati	45	53	29	17.8
71:35	48:35	6	Loam	Basmati	42	49	23	16.6
80:57	57:57	5	Clay	PR-6	50	60	23	20.0
104:46	76:46	5	Clay	PR-6	45	52	28	15.5
100:35	71:35	5	Loam	Basmati	40	44	29	10.0
108:57	80:57	5	Loam	Basmati	40	45.5	28	13.7
108:57	80:57	5	Loam	PR-6	40	46	29	15.0
115:0	86:0	5	Loam	Terrycot	42	45	29	7.1
124:23	95:23	5	Loam	Basmati	56	62	29	12.7
137:57	109:57	10	Clay	HR-120	48	55	29	14.6
137:57	103:57	5	Clay	HR-120	46.5	52.7	29	13.2
137:57	108:57	5	Clay	Basmati	35	45	29	22.2
Total		65			44.13	50.77	27.83	14.87

Nutrients fixed by algae are released either through exudation or through microbial decomposition after the cells die (Roger and Reynaud, 1979). This fixed nitrogen is partly released extracellularly during growth and remaining through microbial decomposition after death. In paddy fields the death of algae biomass is associated with alternate desiccation and wetting during cultivation cycle or finally after the harvest of the crop. It results in gradual build-up of soil fertility with a residual effect on succeeding crop also. N-uptake studies indicated that 39 per cent of the nitrogen from ^{15}N-labelled *Aulosira* sp. spread on soil and 51 per cent from the algae incorporated into the soil was recovered in the rice crop. It shows that nitrogen fixed through cyanobacteria is readily available to rice (Wilson, 1980).

Growth-Promoting Effect

A large number of researchers have found better growth and germination of seeds of many crop plants after treating them with algal cultures or their extracts. The majority of them observed an enhancement in rice-seed germination, root and shoot growth, weight of rice grains and their protein content (Shukla and Gupta, 1967; Venkataraman and Neelakantan, 1967; Singh and Trehan, 1973; Jacq and Roger, 1977; Misra and Kaushik, 1989a, b); while others found similar stimulatory effects on wheat (Gupta *et al.*, 1967), tomato (Kaushik and Venkataraman, 1979), tomato and radish (Rodgers *et al.*, 1979), peas (Gupta and Gupta, 1972), banana (Ganapathi *et al.*, 1994) etc. Different opinions exist regarding the nature of these substances. Some have described them as hormones, *i.e.* gibberelin-like (Singh and Trehan, 1973), cytokinin-like (Rodgers *et al.*, 1979), auxin-like (Ahmad and Winter, 1968) or abscisic acids (Marsalek *et al.*, 1992); while others have described them either as vitamins particularly vitamin B (Grieco and Desrochers, 1978) or as amino acids (Watanable, 1951), antibiotics and toxins (Metting and Pyne, 1986). However, detailed quantitative and qualitative analyses of these substances produced by different cyanobacterial species were given by Kaushik and coworkers. They reported the different vitamins in the range of ng mg^{-1} biomass: niacin 80-320, pantothenic acid 5-10, folic acid 0.79-1.08 and vitamin B_{12} 3.6 × 10^{-3} in *Nostoc muscorum* and *Hapalosiphon fontinalis*, and stimulation in the root and shoot growth was observed with folic acid, vitamin B_{12} and pantothenic acid but not with nicotinic acid (Misra and Kaushik, 1989a). Similarly the growth hormones like indole-3-acetic acid and indole-3-propionic acid or 3-Methyl indole were identified and quantified in addition to extracellular sugars and amino acids (Misra and Kaushik, 1989b).

Soil Organic Matter

The most important constituent to determine the fertility of a soil is organic carbon. Soils rich in organic carbon are productive as it helps in better nutrient utilization efficiency and checks the nutrient loss through leaching and volatilization. As early as 1950 (De and Sulaiman, 1950) reported a build up of organic matter due to blue green algal inoculation. Later, a number of workers supported the work. However, a wide variation in the reported values of organic carbon or biomass addition by BGA exists (Table 30.3). A well developed continuous layer of colonies of different species of BGA in rice fields yield a significant amount of biomass. Using ^{15}N, Nekrasova and Aleksandrova (1982) confirmed that algal biomass contributed significantly to humus formation in soils despite the absence of typical lignin in them. Roger *et al.* (1987) indicated that under favourable conditions a good algal bloom in rice fields yields on an average about 6-8 t of fresh biomass. A 0.03 per cent (672 kg ha^{-1}) increased soil organic carbon content was reported due to encouragement of native algal flora alone under laboratory conditions in a six months period (Kaushik, 1985), whereas inoculation of halotolerant cyanobacterial strains to sodic soils led to an addition of 5.3–7.6 t carbon ha^{-1} in a cropping season (Subhashini and Kaushik, 1984).

Table 30.3: Addition of Organic Carbon to Soil Through Blue Green Algae (BGA)

Quantity t/ha	*References*
0.75	Prasad, 1949
6.70	Fuller and Roger, 1952
0.5-1.4	Gollerbach *et al.*, 1956
0.45	Rao and Burns, 1990
41.6	Reynaud and Roger, 1981
3.2-6.8	Das *et al.*, 1991

Improvement in Soil Physical Properties

At the time of puddling before transplanting rice seedlings, the soil structure is completely destroyed. The inoculation of rice fields with BGA may help to quickly regenerate and improve the soil structure (Roychoudhury *et al.*, 1979). BGA are known to excrete extracellularly a number of compounds like polysaccharides, peptides, lipids etc. during their growth in soil (Bertocchi *et al.*, 1990). These compounds possibly diffuse around soil particles, glue and hold them together in the form of microaggregates. Besides these compounds, particularly polysaccharides are made of fibers, which can also entangle clay particles and form clusters. These clusters or microaggregates, in turn, grow and take the shape of macroaggregates and subsequently of larger soil aggregates. The interwoven nature of growing algal filaments may also help in binding the soil particles along with the organic C added through algal biomass. The importance of these compounds in soil-aggregate formation or soil stabilization has been indicated by many workers (Roychoudhury *et al.*, 1979; Schulten, 1985; Rogers and Burns 1994). Some researchers, however, have considered polysaccharides as transient adhesives (Tisdall and Oades, 1982) and or not directly involved in the formation of aggregates, although the products of their microbial degradation like aliphatic and polyphenolic compounds are considered to be responsible for this (Haynes *et al.*, 1991). The quantity and quality of the excreted compounds also vary depending on the species of BGA, their physiological growth stages and also the associated environmental conditions (Saxena and Kaushik, 1992; Roychoudhury *et al.*, 1985). Some species give low yields of excreted compounds (*e.g. Anabaena*), while other give high yields (*e.g. Gloeotrichia, Westiellopsis*). Polysaccharides from different algal species also differ with respect to their protein, uronic acid and sugar composition (Bertocchi *et al.*, 1990) and thus in their stability in soils with respect to microbial and thermal degradation. Kaushik and his co-workers (Kaushik and Krishna Murty, 1981; Subhashini and Kaushik, 1984; Kaushik, 1985) observed significant increases in soil aggregate stability due to an increase in the polysaccharide content of soils as a result of algal inoculation (Table 30.4). Similarly, water-stable aggregates, which are an integral part of good (soil) aggregate formation, have also been shown to increase significantly due to algal inoculation, resulting in an improvement in the water-holding capacity and aeration status of soils (Singh, 1961; Roychoudhury *et al.*, 1979). Such improvement in soil aggregation due to algal inoculation ultimately favoured better seedling emergence of upland crops sown after the paddy harvest (Rogers and Burns, 1994). All these results apparently suggest that the algal species, which liberate higher amounts of complex and thermostable polysaccharides, will possibly be a better choice for inoculation in rice fields for the regeneration of the soil structure. However, such algal species are known to be less efficient N_2 fixers.

Table 30.4: Comparative Effect of Blue Green Algae and *Azolla* on Soil Aggregation and Particle Size Distribution (Roychoudhury *et al.*, 1979)

Treatments	*Particle Size Distribution*		*Per cent Aggregates*		*Per cent Increase > 50 μ Aggregates*
	> 250 μ	*50-250 μ*	*> 250 μ*	*50-250 μ*	
Sand (>50m)	0.5	44.5	-	-	-
Control	1.2	46.8	0.7	2.3	-
BGA	1.5	48.0	1.0	3.5	50
Azolla	1.0	47.0	0.5	2.5	-
60 kg N ha^{-1}	0.7	46.6	0.2	2.1	-
60 kg N ha^{-1}+BGA	1.6	49.0	1.1	4.5	70
60 kg N ha^{-1}+ *Azolla*	0.8	47.2	0.3	2.7	-

Transformation of Soil P

Phosphorus is estimated to be in insufficient amounts in most of the Indian soils as available phosphorus. According to one of the recent compilation, based on about 9.6 million soil tests for available phosphorus in Indian soils, 49.3 per cent area spread in various states is in the low category, 48.8 per cent is of medium and only 1.9 per cent has high phosphorus status (Hasan, 1994). The problem of P management in soil is highly intricate as the applied phosphate through fertilizers is often fixed and becomes unavailable to the crops.

Cyanobacteria, like P-solubilizing bacteria, are known to have the ability to mobilise bound phosphates. They have been shown to solubilise insoluble $(Ca)_3(PO_4)_2$ (Bose *et al.*, 1971), $FePO_4$ (Wolf *et al.*, 1985), $AlPO_4$ (Dorich *et al.*, 1985) and hydroxyapatite $(Ca_5(PO_4)_3OH$ (Cameron and Julian, 1988) in soils, sediments or in pure cultures. There are mainly two hypotheses to explain how cyanobacteria solubilise such bound phosphates. One of the mechanisms may be that cyanobacteria might synthesise a chelator/chelators? Ca^{2+} and drive the following dissolution reaction without changing the pH of the growth medium (Cameron and Julian, 1988; Roychoudhury and Kaushik, 1989):

$$Ca_{10}(OH)_2(PO_4)_6 = 10Ca^{2+} + 2OH^- + 6PO_4^{3+}$$

Others (Bose *et al.*, 1971) were however of the opinion that H_2CO_3 and other organic acids released by cyanobacteria during their growth could solubilise P from Ca sources following the reaction:

$$Ca_3(PO_4)_2 + 2H_2CO_3 = 2CaHPO_4 + Ca(HCO_3)_2$$

A third group (Arora, 1969; Saha and Mandal, 1979; Mandal *et al.*, 1992) believed that the above mechanisms operate simultaneously. Once solubilised, the PO_4^{3-} is taken up by the growing algal cells for their nutrition. After completing their growth cycle, when the cells undergo lysis the cell-bound PO_4^{3-} is released in the growth medium and becomes available to plants on mineralisation. Observance of an initial decrease in the available–P content in soils due to algal growth and an appreciable increase later during biomass decomposition (Saha and Mandal, 1979; Mandal *et al.*, 1992) supported this possible pathway (Table 30.5).

Table 30.5: Changes of Available P mg g^{-1} Soil Due to Incorporation of Blue Green Algae, *Azolla* and Urea

Days of Flooding	*Control*	*Blue Green Algae*	*Azolla*	*Urea*
7	7.50	7.92	9.37	9.08
14	3.58	3.94	4.89	4.48
21	5.32	13.55	13.97	6.29
28	6.97	14.06	18.45	7.90
35	7.45	10.89	11.62	7.85
LSD at 5 per cent	Treatment mean 1.19	Days mean 1.33	Interaction 2.66	

The solubilising effect of cyanobacteria on bound PO_4^{3-} may also be used for the efficient utilisation of low cost, low grade (in terms of P content) rock-phosphate fertilizers where PO_4^{3-} remains bound as various forms of apatite, like hydroxyapatite, carbonapatite, choloroapatite etc. These rock phosphates are generally used in acid soils as P fertilizer. In fact, Roychoudhury and Kaushik (1989) reported increased solubility of P in rock phosphate due to its inoculation with cyanobacteria.

Fuller and Roger (1952) observed a greater uptake of P by plants from algal materials than from inorganic phosphates when applied in equal amounts. They concluded that: (1) P in algal material is more available to plants than inorganic phosphates over longer periods, (2) chemical fixation is not so important a factor with respect to algal material as with respect to inorganic phosphates, and (3) temporary conversion of available soil or fertilizer P into cell materials by soil algae may be a desirable process from the stand-point of long-term availability. Their proposed hypothesis was that soil algae removed available P from the sphere of chemical fixation by converting it into cell materials for by absorbing it in "luxury" amounts that might be released gradually for crop use through the process of exudation, autolysis or microbial decomposition of old algal cells. Algae also have long been known to take up P in excess of their immediate needs, which may subsequently be released in the form of dissolved organic P (Lean and Nalewajko, 1976).

Technological Developments

The idea of utilizing algal biofertilizer to supplement the nitrogen requirement of rice crop was mooted as early as 1939. However the systematic approach was under taken at IARI and an algal biofertilizer technology was developed (Venkataraman, 1972). The technology was named as rural oriented mass multiplication technology and the application in field was often referred as algalization. In this process cyanobacteria were allowed to grow with soil as carrier and then utilized as inoculum in rice fields. Farmers adopted this all over India because of its simplicity and economic benefits. Depending upon the requirement of algal biofertilizer, the production technology could be scale up to–troughs, pits, and field plots/cemented tanks/polythene layered tanks. It involved -

However the technology could not become popular due to (1) non-availability of inoculum at the time of application, and (2) large variation in inoculum quality. Hence the cyanobacterial biofertilizer of assured quality in terms of known population density; nature of each strain with better competitiveness was necessary to develop. In view of this, a new technology was developed in which cyanobacteria were grown in a formulated medium in glass/polyethylene house round the year of assured quality (Kaushik, 1998; Prasanna *et al.*, 1998).

Amelioration of Saline and Sodic Soils

Saline and sodic soils constitute a large area of agricultural land in the world. Sodic soils have a high pH, high exchangeable Na, measurable amounts of carbonates and undergo extensive clay dispersion (deflocculation, due to the high zeta potential of active Na^+), leading to poor hydraulic conductivity and reduced soil aeration. While saline soils contain excess amounts of soluble salts imparting high osmotic tension to plant roots for absorption of water and nutrients. Primarily, the reclamation of sodic soils requires the replacement of exchangeable Na with Ca. This is normally done by the application of suitable soil amendments, like gypsum, followed by leaching. The improvement of saline soils, on the other hand, requires leaching of excess soluble salts from the rhizosphere by good quality water. Reports of biological reclamation of these soils through the use of cyanobacteria are available (Singh, 1950; Kaushik, 1989). These methods of reclamation were basically based on the observations of extensive growth of cyanobacteria on alkali soils of India (Singh, 1950) and the USSR (Gollerbach *et al.*, 1956). These observations indicated a considerable tolerance of cyanobacteria to salinity and/or alkalinity stress. Several physiological mechanisms underlying such tolerance have now been identified. Curtailment of Na^+ influx (Subhashini and Kaushik, 1986; Apte *et al.*, 1987) and accumulation of K^+ ion (Jha *et al.*, 1987) or organic (sugar, polyols and quaternary amines etc.) osmoregulators (Blumwald *et al.*, 1983; Reed *et al.*, 1984; Kumar and Kaushik, 1994) are the important mechanisms that provide adequate protection to cyanobacteria against such salt/sodicity stress. Synthesis of salt stress specific fatty acids (Senthil *et al.*, 1993; Venkataraman and Kaushik, 1994) and proteins (Goel *et al.*, 1997) has been detected in several species of algae and cyanobacteria. Cyanobacteria exposed to such salinity/sodicity stress may lose or have diminished nitrogenase activity possibly due to the diversion of their cellular energy towards the biosynthesis of osmoregulators (Apte *et al.*, 1987), while Jha *et al.* (1987) reported enhanced heterocyst frequency in *Westiellopsis prolifica* under salinity suggesting the increased nitrogen demand of organism to overcome the stress. In such cases, the addition of a few kilograms of N, particularly as NO_3^- per hectare together with algal inoculation has been shown to be very effective in protecting them from these stresses and allowing them to establish in the stressed environment, and subsequently enhances their potential as N biofertilizer (Reddy *et al.*, 1989; Fernandes *et al.*, 1993). Once established and subsequently acclimatized to the stress, they may act in ameliorating their surrounding environment. A reduction in electrical conductivity, exchangeable Na and soil pH (Kaushik *et al.*, 1981; Kaushik and Subhashini, 1985) and an increase in the soluble Ca and Mg content of soils (Jain and Kaushik, 1989) have also been reported due to such algalization. The increase in soluble Ca and Mg and decrease in exchangeable Na may ultimately lower the sodium adsorption ratio–an index of alkalinity in soil–of the soils and thus alleviate the negative effects of high sodicity. In fact, Kaushik and his co-workers (Kaushik and Krishna Murthi, 1981; Subhashini and Kaushik, 1981) observed a considerable increase in the hydraulic conductivity of sodic soils due to cyanobacterial inoculation and found that it compared favourably to the use of gypsum in such soils.

A significant reduction in soil salinity (12-35 per cent) due to repeated cultivation of cyanobacteria in soils rendered saline owing to bad farm management has also been reported. The rice yield in saline soils is also less affected when N is given in the form of algal inocula vis-à-vis urea (Antarikanonda and Amarit, 1991). Further, such algal inocula have proved to be almost equally effective in increasing the rice yield in saline soils when compared with gypsum and/or pyrite treatments (Sharma *et al.*, 1989; Kaushik 1994b). Even when irrigated with saline water, cyanobacterial inoculation was found to be useful in minimizing the deleterious effect of the increased salinity on the barley crop (Jain and Kaushik, 1989).

In an alkali soil almost all the problems with respect to rice cultivation are associated with the high concentration and/or activities of Na^+. If somehow its activity is checked, the whole problem is solved. Cyanobacteria secrete organic acids particularly under salt stress (Singh, 1961; Sprent and Sprent, 1990), which can act on $CaCO_3$ to dissolve Ca. They also excrete a number of biologically active compounds, *i.e.* bioflocculants (Jha *et al.*, 1987; Levy *et al.*, 1992), which can flocculate the dispersed clay particles in alkali soils by inactivating/scavenging Na^+. All these effects of cyanobacteria may at least temporarily cause inactivation of Na^+ in alkali soils and make the soil environment favourable to the growth of plants. Successive cultivation of cyanobacteria makes the environment more favourable and after a few years it may help to produce a reasonably good yield of crops as observed in sugarcane and paddy after 3 years of reclamation with cyanobacteria. Although such biological soil amelioration is a time-consuming process, it is considered to be a sustainable approach as compared to reclamation by chemical amendments (Oikarinen, 1996).

Cyanobacteria as a Source of Value Added Products

Spirulina, a filamentous cyanobacterium is one of the extensively exploited micro algae. Its high nutritional value in terms of high protein content *i.e.* about 60-70 per cent of dry weight (Richmond, 1992), high vitamin content (Switzer, 1980) (Table 30.6), presence of unsaturated fatty acids like γ-linoleic acid which do not form cholesterol (Mahajan and Kamat, 1995), low nucleic acid content, easy digestibility due to absence of cellulosic cell wall, confer upon it the status of ideal single cell protein. In addition *Spirulina* is a source of compounds of therapeutic value to treat wounds (Clement *et al.*, 1967), β-carotene known for its anticancer property (Peto *et al.*, 1981). The pigment phycocyanin is used as a natural colour in health food and cosmetic products (Switzer, 1980). Hence, *Spirulina* is commercially cultivated as a source of human food, animal feed, and natural products and for wastewater treatment.

Spirulina Cultivation

In summer possibly high light intensity is found to be the growth regulatory whereas, in winter the low daytime temperature imposes the major limitations (Vonshak *et al.*, 1982). The optimum temperature for photosynthesis for *S. platensis* has been shown to be 35°C and for dark respiration 45°C

(Torzillo and Vonshak, 1994). In India maximum temperature during the summer is about 40°±5°C and in Northern India during winter months the temperature goes down to as low as 2-5°C. While studying the effect of temperature and light intensity on seven strains of *Spirulina* it was observed that all the strains showed maximum growth at 25 Klux light intensity at 30°C temperature, whereas 62-69 per cent growth was reported at 45 Klux light and 20°C temperature and none of the strains survived at 40°C (Basirath and Kaushik, 2002).

Table 30.6: Vitamin Content of *Spirulina* vis-à-vis Other Sources mg kg^{-1} Dry Matter and Meat

Vitamins	*Spirulina*	*Spinach*	*Beef Liver (Fresh)*
Provitamin A	840		
Vitamin E	120	122	10
Thiamine	44	4.9	3
Riboflavin	37	9.7	290
Pyridoxine	3	9.7	7
Cobalamine	7		0.65
Vitamin C	80	2480	310
Biotin	0.3	0.34	1
Inositol	380		
Folic acid	0.4	3.65	2.9
Ca-Pantothenate			
	13	14.8	73
Nicotinate		29.2	136

The growth of *Spirulina* is also affected by the concentration of sodium bicarbonate and the lowest concentration in which *S. platensis* could be maintained has been reported to be 0.2 M (Vonshak *et al.*, 1983). We observed that a concentration of 8gL^{-1} $NaHCO_3$ (commercial grade) provides biomass statistically equivalent to 16 g L^{-1} as reported earlier.

Natural Colours

A large array of natural products of economic potential may be produced from cyanobacteria. These also represent an attractive source of natural pigments such as phycobiliproteins, amounting to about 20 per cent of total dry weight vis-à-vis phycocyanin and phycoerythrin and *allo*-phycocyanin (Borowitzka, 1988); and carotenoids (Walsh *et al.*, 1997). Phycoerythrins in particular have applications as natural dyes as well as fluorescence microscopy. Some strains of *Anabaena* and *Nostoc* are however, particularly rich in phycoerythrin (Guerrero *et al.*, 1990). These may be used as pigments or colourants in the food industry as well as for cosmetics. The Dai Nippon Inks and Commercial Co. of Japan markets phycocyanin powder under the trade name of "Lina Blue". The product is an odourless, nontoxic blue powder with slight sweetness and has a brilliant blue with faint reddish fluorescence in water. The blue pigment from *Spirulina* is being used for eye shadow, eyeliner and lipsticks. A preparation of *Spirulina* chlorophyll in a mixture containing iron oxide and a higher alcohol was patented as a strong deodorant (Yamaguchi, 1981). Carotenoids have the widest commercial application *e.g.* food colouring, feed additives to enhance flesh colour of salmonoid fish, as well as the colour of the egg yolk. Also these improve the health and fertility of cattle (Borowitzka, 1988).

The blue pigment phycocyanin given orally to mice injected with liver tumor cells showed that the survival rate of treatment group was significantly higher than that of controls (Iijima *et al.*, 1982). They suggested that phycocyanin might generally stimulate the immune system providing protection against a variety of diseases. It was possible to prevent experimental oral cancer in rats by extracts of *Spirulina* (Schwartz *et al.*, 1986). Relatively high β-carotene of *Spirulina* was suggested to reduce certain cancer risks when ingested in suitable amounts (Schwartz *et al.*, 1986).

Recently (Kaushik, 2000) forty one species of cyanobacteria including unicellular, colonial, non-heterocystous and heterocystous filamentous were screened for the phycobilin pigments. *Lyngbya versicolor* ARM 348 which produced 10.2 per cent total phycobilins including c-phycocyanin 74.37 mg g^{-1} of dry biomass was superior to other unicellular *Anacystis nidulans* ARM 336, *Synechocystis* sp. ARM 340, or colonial *Gloeocapsa rupestris* ARM 338, *Aphanocapsa pulchra* ARM 341 and *Chlorogloea fritschii* ARM 342. The twenty one species of *Calothrix* showed greater diversity and produced as low as < 1 per cent to as high as 18.21 per cent by *Calothrix brevissima* ARM 114. The other heterocystous forms *Anabaena variabilis* ARM 310, *Aulosira fertilissima* ARM 68, *Hapalosiphon* sp. ARM 419 and *Tolypothrix tenuis* ARM 76 are the probable cyanobacterial species these can be exploited as the source of natural phycobilin content which varied from 14.72 to 17.52 per cent. From a total of six heterotrichous forms examined barring *Hapalosiphon* sp. ARM 419 (14.72 per cent phycobilin) all produced very little quantity of pigments.

Transgenic Cyanobacteria

Growing of genetically modified algae (or other organisms) is still not popular, as it is feared that the artificially engineered organisms could be unsafe and harmful for the environment due to spread of the foreign gene to unmodified organisms. But, genetically modified organisms are specially created for introducing useful traits not existing in their natural counterparts and many genetically modified organisms have a revolutionary impact on the agricultural sector.

A large number of genes have been cloned in cyanobacteria, *e.g. cat* (Friedberg, 1988), *lac* Z (Subramanian and Kaushik, 2001) and *lux* A and B (Schmetterer *et al.*, 1986) that catalyze the production of chloramphenicol, acetyl transferase, β-galactosidase and luciferase, respectively. Cyanobacterial colonies expressing *lac* Z when treated with the indicator 5-bromo-chloro-3-indolyl-β-D-galactoside (X-Gal) turn from green to turquoise surrounded by the blue halos. The *lux* operon has been originally derived from the bacterium *Vibrio fischeri* and *V. harveyi*, and the *luc* operon from the American firefly *Photinus pyralis*. Cells containing luciferase produce light on addition of the aldehyde and the resultant light emission is quantified. Subramanian and Kaushik (2001) have used the technique of cloning of the *lac* Z and *luc* genes in cyanobacteria *Anacystis cylindrica* ATCC 29414 and *Synechocystis* sp. ARM 340 to assess the growth and establishment of inoculated species under field conditions as well as the stability of these reporter genes in soil microcosm.

Most insecticides currently used are chemicals, but there are successful biological insecticides also. Resistance to chemical insecticides has occurred in several insect species. Management of such resistance remains a challenge, and new methods of controlling such species are needed. Various proteins have been identified as possessing insecticidal activities, *e.g.* α-amylase, lectins and insecticidal proteins from *Bacillus* species. The proteins produced by the *Bacillus* species possess the most potent insecticidal activities. Use of transgenic cyanobacteria to control insects is a new technology. *Bacillus thuringiensis* subsp. *israelensis* has been widely used as a biological pesticide to control mosquitoes and black flies, which are vectors for infectious human diseases, without adverse effects on other organisms. The larvicidal activity of this organism is due to the production of d-endotoxin. The applications of *B. thuringiensis israelensis* for mosquito control are limited by the short half-life of existing preparations under field conditions. This problem can be overcome by the cloning of d-endotoxin genes into cyanobacteria, which inhabit the breeding zones of mosquitoes and are used by them as a food source. Several attempts have been made to produce transgenic mosquitocidal cyanobacteria (Soltes-Rak *et al.*, 1995) and some success has been achieved in expressing single *cry* genes in unicellular strains, but the transformed cells displayed lower levels of mosquito-larvicidal activity than the level required for effective and practical use as bioinsecticides in natural aquatic environment. Various combinations of the genes *cry* IVA, *cry* IVD and p20 from *Bacillus thuringiensis israelensis* were introduced into *Anabaena* sp. strain PCC 7120, which displayed toxicity against third-instar larvae of *Aedes aegypti* at levels greater than any level previously reported for transgenic cyanobacteria (Xiaoqiang *et al.*, 1997).

Conclusion

A yield improvement of rice of between 5 per cent and 25 per cent was found when fields were inoculated with cyanobacteria even in the presence of 100-150 kg N ha^{-1} as fertilizer.

Cyanobacteria bring about, directly or indirectly, certain changes in the physical, chemical and biological properties of the soil and soil-water interface in rice fields, which are of agronomic importance. The extracellular organic compounds liberated by the algae and the O_2 released due to photosynthesis during their active growth period, and the subsequent addition of biomass after their death, are likely to cause some important changes in the physico-biochemical properties of soil.

The work done at Indian Agriculture Research Institute, New Delhi can be summarised as: (*a*) in areas where chemical nitrogen fertilisers are not used for various reasons, algal application as biological input can give to the rice farmers the benefit of applying 20-30 kg N/ha, (*b*) where fertilisers are used, the dose can be reduced by 30kg N/ha by algal supplementation, and (*c*) even with high levels of nitrogen fertilisers, the yield per unit input can be increased by 10-15 per cent through algal complementation.

Saline and alkali soils can be effectively reclaimed by the use of specific salt tolerant cyanobacteria along with gypsum is amply demonstrated in the work done at author's laboratory.

These mosquitocidal cyanobacterial strains could be efficiently grown in the production tanks as well as in the rice fields without any application of the insecticides thus reducing the pollution effect in the environment.

Blue green algal biomass has been considered since long as an alternative source of protein that could supplement conventional food and feed production. Other practical uses include the reclamation of wastewater, and the production of variety of commercial chemicals on a renewable basis as lipids, polysaccharides, pigments, vitamins and other compounds of industrial and pharmaceutical interest. Besides this, cyanobacteria are also useful in the production of different bioactive compounds (antibiotics, enzymes, cell growth promoters, toxins etc.). The interest in these organisms as generators of pharmacologically active compounds has been stimulated by recent results from the group of workers related to anti-cancer and immuno-stimulation properties of phycobiliproteins, as well as to the inhibition by the blue green algal sulfolipids of cytopathic effects of the human immuno-deficiency viruses (HIV).

References

Ahmad, M.R. and Winter, A. (1968). Studies on the hormonal relationship of algae in pure culture. 1. The effect of indole-3-acetic acid on the growth of blue-green and green algae. *Planta*, 78: 277-286.

Alimango, B.V. and Yoshida, T. (1977). *In situ* acetylene-ethylene assay of biological nitrogen in lowland rice soils. *Plant and Soil*, 47: 239-244.

Antarikanonda, P. and Amarit, P. (1991). Influence of blue-green algae and nitrogen fertilizer on rice yield in saline soils. *Kasetsart J. Nat. Sci.*, 25: 18-25.

Apte, S.K., Reddy, B.R. and Thomas, J. (1987). Relationship between sodium influx and salt tolerance of nitrogen-fixing cyanobacteria. *Appl. Environ. Microbiol.*, 53: 1934-1939.

Arora, S.K. (1969). The role of algae on the availability of phosphorus in paddy fields. *Riso* 18: 135-138.

Basirath, R. and Kaushik, B.D. (2002) Influence of light and temperature on mass multiplication of *Spirulina* strains. *Indian J Plant Physiology*, 7: 52-55.

Bertocchi, C., Navarini, L., Cesarp, A. and Anastasio, M. (1990). Polysaccharides from cyanobacteria. *Carbohydr. Polym.*, 12: 127-153.

Blumwald, E., Mehlhorn, R.J. and Packer, L. (1983). Studies of osmoregulation in salt-adaptation of cyanobacteria with ESR spin-probe techniques. *Proc. Natl. Acad. Sci.* USA, 80: 2599-2602.

Borowitzka, M.A. (1988). Vitamins and fine chemicals from micro algae. In: *Micro-algal biotechnology* (M.A. Borowitzka and L.J. Borowitzka eds) pp.153-196, Cambridge University Press, Cambridge.

Bose, P., Nagpal, U.S, Venkataraman, G.S. and Goyal, S.K. (1971). Solubilisation of tricalcium phosphate by blue-green algae. *Curr. Sci.*, 40: 165-166.

Cameron, H.J, and Julian, G.R. (1988). Utilisation of hydroxyapatite by cyanobacteria as their sole source of phosphate and calcium. *Plant Soil*, 109: 123-124.

Clement, G., Robellar, M. and Zarrouk, C. (1967). Wound treating medicaments containing algae. *Fr. Med.* pp. 5279.

Das, S.C., Mandal, B. and Mandal, L.N. (1991). Effect of growth and subsequent decomposition of blue green algae on the transformation of iron and manganese in submerged soils. *Plant Soil*, 138: 75-84.

De, P.K. (1936). The problem of nitrogen supply of rice: Part I. fixation of nitrogen in the rice soils under waterlogged conditions. *Indian J Agr Sci*, 6: 1237-1242.

De, P.K. and Biswas, N.R.D. (1952). Fixation of nitrogen in rice soils in the dry period. *Indian J agric Sci.*, 22: 375-388.

De, P.K. and Mandal, L.N. (1956). Fixation of nitrogen by algae in rice. *Soil Sci* 81: 453-458.

De, P.K. and Sulaiman, M. (1950). The influence of algal growth in the rice fields on the yield of crops. *Indian J Agric Sci*, 20: 327-342.

Dorich, R.A., Nelson, D.W. and Sommers, L.E. (1985). Estimating algal-available phosphorus in suspended sediments by chemical extraction. *J. Environ Qual*, 14: 400-405.

Fernandes, T.A., Iyer, V. and Apte, S.K. (1993). Differential responses of nitrogen-fixing cyanobacteria to salinity and osmotic stresses. *Appl. Environ. Microbiol*, 59: 899-904.

Friedberg, D. (1988). Use of reporter genes in cyanobacteria. *Methods Enzymol.* 167: 736-747.

Fuller, W.H. and Roger, R.N. (1952). Utilisation of the phosphorus of algal cells as measured by the Neubauer technique. *Soil Sci*, 74: 417-429.

Ganpati, T.R., Suprasanna, P., Bapat, V.A. and Rao, P.S. (1994). Stimulatory effect of cyanobacterial extract on banana shoot tip cultures. *Trop Agric* (Trinidad), 71: 299-302.

Goel, S., Mamta, M. and Kaushik, B.D. (1997). Nitrogen fixation and protein profile of halotolerant *Nostoc muscorum*-R strain isolated from rice fields and ARM 221 strain. *Indian J exptl Biol.*, 35: 746-750.

Gollerbach, M.M., Novichkova, L.A. and Sdubrikova, N.Y. (1956). The algae of takyrs. In: *Takyrs of western Turkmenia and routes of their agricultural conquest. Nauk. Moscow*, pp. 22-29.

Grieco, E. and Desrochers, R. (1978). Production de vitamine B_{12} parune algae belue. *Can J. Microbiol*, 24: 1562-1566.

Guerrero, M.G., Vargas, A.M., Rivas, J., Rodriguez, H., Moreno, J. and Losada, M. (1990). Cultivation of nitrogen fixing micro-algae for the generation of products with commercial interest. In: *From genes to bioproducts*. (Ed. Organizing Committee of Biotech.) 90. 91-97, PPU DM, MURACIA.

Gupta, A.B. and Gupta, K.K. (1972). Effect of *Phormidium* extract on growth and yield of *Vigna catjang* (cowpea) T.5369. *Hydrobiologia*, 40: 127: 132.

Gupta, A.B., Agarwal, V. and Kushwaha, A.S. (1967). The effect of algal growth-promoting substances on wheat. *Proc. Natl. Acad. Sci.* India, 37B: 349-355.

Hasan, R. (1994). In: *Phosphorus research in India*. (Edt. G. Dev) *Proc Group Discussion* at IARI-PPICIP Gurgaon, India. Pub 7.

Haynes, R.J, Swift, R.S. and Stephen, R.C. (1991). Influence of mixed cropping rotations (pasturearable) on organic matter content, water-stable aggregation and cold porosity in a group of soils. *Soil Till Res*, 19: 77-87.

Henriksson, E. (1971). In: *Biological nitrogen fixation in natural and agricultural habitats*. (Edd. T.A. Lie and *E.G.* Mulder) pp. 415. *Plant and Soil*, Special Volume.

Iijima, N., Fugii, I., Shimamatsu, H. and Katoh, S. (1982) Anti-tumor agent and method of treatment therewith. U S Patent pending, ref. No. P1150-726-A 82679.

Jacq, V. and Roger, P.A. (1977). Decrease of losses due to sulphate reducing processes in the spermosphere of rice by pre-soaking seeds in a culture of blue-green algae. *Cah ORSTOM Ser Biol*, 12: 101-108.

Jain, B.L. and Kaushik, B.D. 1989. Effect of algalization on crop response under saline irrigation. *J. Indian Soc Soil Sci.*, 37: 382-384.

Jha, M.N., Venkataraman, G.S. and Kaushik, B.D. (1987). Response of *Westiellopis prolofica* and *Anabaena* sp. to salt stress. *Mircen J. Appl. Microbiol Biotechnol*, 3: 307-317.

Kaushik, B.D. (1985). Effect of native algal flora on nutritional and physico-chemical properties of sodic soils. *Acta Bot Indica*, 13: 143-147.

Kaushik, B.D. (1989). Reclamative potential of cyanobacteria in salt-affected soils. *Phykos*, 28: 101-109.

Kaushik, B.D. (1998). Use of cyanobacterial biofertilizer in rice cultivation: A Technology Improvement. *In*: Cyanobacterial Biotechnology (edt) G Subramanian, BD Kaushik, GS Venkataraman, pp. 211-222, Science Publs. Inc. USA.

Kaushik, B.D. (1994a). Blue green algae and sustainable agriculture. In: *Natural Resource Management for sustainable Agriculture and Environment*, (edt. DL Deb) Angkor Publishers, New Delhi, pp.404-416.

Kaushik, B.D. (1994b). Algalization of rice in salt-affected soils. *Ann.Agril. Res.*, 15: 105-106.

Kaushik, B.D. (2000). Screening of cyanobacterial strains for natural colours. *Indian Hydrobiology*, 3: 81-84.

Kaushik, B.D. and Krishna Murti, G.S.R. (1981). Effect of blue green algae and gypsum application on physico-chemical properties of alkali soils. *Phykos*, 20: 91-94.

Kaushik, B.D., Krishna Murti, G.S.R. and Venkataraman, G.S. (1981). Influence of blue-green algae on saline alkali soils. *Sci. Cult.*, 47: 169-170.

Kaushik, B.D. and Subhashini, D. (1985). Amelioration of salt affected soils with blue green algae. II. Improvement in soil properties. *Proc Indian natn Sci Acad*. B 51: 386-389.

Kaushik, B.D. and Venkataraman, G.S. (1979). Effect of algal inoculation on the yield and vitamin C content of two varieties of tomato. *Plant and Soil*, 52: 135-137.

Kumar, H. and Kaushik, B.D. (1994). Response of *Calothrix brevissima* to salt. *Indian J Microbiol.*, 34: 38-41.

Lean, D.R.S. and Nalewajko, C. (1976). Phosphate exchange and organic phosphorus excretion by freshwater algae. *J. Fish Res. Bd Can.*, 33: 1312-1323.

Levy, N., Magdassi, S. and Baror, Y. (1992). Physico-chemical aspects in flocculation of bentonite suspensions by a cyanobacterial bioflocculant. *Water Res.*, 26: 249-254.

Mac Rae, I.C. and Castro, T.F. (1969). Nitrogen fixation in some tropical rice soils. *Soil Sci* 103: 277.

Mahajan, G. and Kamat, M. (1995). Gamma-linolenic acid production from *Spirulina platensis*. *Appl. Microbiol. Biotechnol.* 43: 466-469.

Mandal, B., Das, S.C. and Mandal, L.N. (1992). Effect of growth and subsequent decomposition of blue-green algae in the transformation of phosphorus in submerged soils. *Plant Soil*, 143: 289-297.

Marsalek, B., Zahradnickova, H. and Hronkova, M. (1992). Extracellular abscisic acid produced by cyanobacteria under salt stress. *J. Plant Physiol.*, 139: 506-508.

Mekonnen, A.E., Prasanna, R and Kaushik, B.D. (2002). Cyanobacterial nitrogen fixation in presence of nitrogen fertilizers. *Indian J Experimental Biol.*, 40: 854-857.

Metting, B. and Pyne, J.W. (1986). Biologically-active compounds from microalgae. *Enzyme Microbial Technol.*, 8: 386-394.

Misra, S. and Kaushik BD. (1989a). Growth promoting substances of cyanobacteria.I. Vitamins and their influence on rice plant. *Proc Indian natn Sci Acad*, B55: 295-300.

Misra, S. and Kaushik, B.D. (1989b). Growth promoting substances of cyanobacteria. II. Detection of amino acids, sugars and auxins. *Proc Indian natn Sci Acad*, B55: 499-504.

Nekrasova, K.A. and Aleksandrove, I.V. (1982). Participation of *Collembola* and earthworm in the transformation of algal organic matter. *Sov Soil Sci.*, 14: 31-39.

Okuda, A. and Yamaguchi, M. (1955). Nitrogen fixing microorganisms in paddy soils. I. Characteristics of the nitrogen fixation in paddy soils. *Soil and Plant Food*, 1: 102-104.

Oikarinen, M. (1996). Biological soil amelioration as the basis of sustainable agriculture and forestry. *Biol Fertil Soils*, 22: 342-344.

Peto, R., Doll, R., Bukley, J.D. and Sporn, M.B. (1981). Can dietary Beta-carotene materially reduce human cancer rates? *Nature* 290: 201-208.

Prasad, S. (1949). Nitrogen recuperation by blue green algae in soils of Bihar and their growth on different types. *J. Proc. Inst. Chem India*, 21: 135-140.

Prasanna, R. and Kaushik, B.D. (1994). Physiological and molecular genetic aspects of nitrogen fixation in non-heterocystous cyanobacteria. *Ind J Exptl Biol*, 32: 248-251.

Prasanna, R. and Kaushik, B.D. (1995). Nitrogen fixation and *nif* gene organization in branched heterocystous cyanobacteria; variation in the presence of *xis*A. *Folia Microbiol*, 40: 176-180.

Prasanna, R., Sharma, B.K., Sharma, R.K. and Kaushik, B.D. (1998). Standardization of growth parameters and formulation of medium for cyanobacterial biofertilizer strains. *Indian J Microbiol.*, 38: 211-215.

Rao, D.L.N. and Burns RG. (1990). Use of blue Green algae and bryophyte biomass as a source of nitrogen for oilseed rape. *Biol Fertil Soils*, 10: 61-64.

Renaud, P.A. and Roger, P.A. (1978). Nitrogen fixing algal biomass in Senegal rice fields. *Ecol Bull.* Stockholm, 26: 148-157.

Reynaud, P.A. and Roger, P.A. (1981). Seasonal variations of algal flora and of nitrogen fixing activity in a waterlogged sandy soils. *Rev Ecol Biol Sol*, 18: 9-27.

Reynault, J., Sasson, A., Pearson, H.W. and Stewart, W.D.P. (1975). Nitrogen fixing algae in Morocco. In: *Nitrogen fixation by free living microorganisms* (WDP Stewart ed.) p 229-248, London, Cambridge Univ. Press.

Reddy, B.R., Apte, S.K. and Thomas, J. (1989). Enhancement of cyanobacterial salt tolerance by combined nitrogen. *Plant Physiol.*, 89: 204-210.

Reed, R.H., Richardson, D.L., Warr, S.R.C. and Stewart, W.D.P. (1984). Carbohydrate accumulation and osmotic stress in cyanobacteria. *J. Gen Microbiol.*, 130: 1-4.

Richmond, A. (1992). *Spirulina.* In: Microalgal Biotechnology. Borowitzka MA and Borowitzka LJ (eds.) pp. 85-121. Cambridge University Press, Cambridge.

Rippka, R., Deruelles, J., Waterbury, J.B, Herdmann, M. and Stanier, R.Y. (1979). Generic assignment, strain histories and properties of pure culture of cyanobacteria. *J Gen Microbiol*, 111: 1-61.

Rodgers, G.A., Bergman, B., Henriksson, E. and Udris, M. (1979). Utilisation of blue-green algae in biofertilizers. *Plant Soil*, 52: 99-107.

Roger, P.A. and Reynaud, P.A. 1979. Ecology of blue green algae in paddy fields. p.289-309 *In*: International Rice Res. Institute. *Nitrogen and Rice*. Los Banos, Philippines.

Roger, P.A., Grant, I.F., Reddy, P.M. and Watanabe, I. 1987. The photosynthetic aquatic biomass in wetland rice fields and its effect on nitrogen dynamics. In: *Efficiency of nitrogen fertilizers for rice.* IRRI, Manila, pp. 43-68.

Rogers, S.L. and Burns, R.G. 1994. Changes in aggregate stability nutrient status indigenous microbial populations and seedling emergence following inoculation of soil with *Nostoc muscorum. Biol Fertil. Soils*, 18: 209-215.

Roychoudhury, P. and Kaushik, B.D. (1989). Solubilization of Mussorie rock phosphate by cyanobacteria. *Curr Sci*, 58: 569-570.

Roychoudhury, P., Kaushik, B.D., Krishna Murty, G.S.R. and Venkataraman, G.S. (1979). Effect of blue green algae and *Azolla* application on the aggregation status of soil. *Curr Sci.*, 48: 454-455.

Saha, K.C. and Mandal, L.N. (1979). Effect of algal growth on the availability of P, Fe and Mn in rice soils. *Plant Soil*, 52: 139-149.

Satapathy, K.B. (1999). Comparative efficiency of blue-green algae, *Azolla* and other biofertilizers on growth of rice. *Indian J Plant Physiol* 4: 100-104.

Saxena, S. and Kaushik, B.D. (1992). Polysaccharides (biopolymers) from halotolerant cyanobacteria. *Indian J exptl Biol.*, 30: 433-434.

Schmetterer, G., Wolk, P. and Elhai, J. (1986). Expression of luciferases from *Vibrio harveyi* and *Vibrio fischeri* in filamentous cyanobacteria. *J Bacteriol* 167: 411-414.

Schulten, J.A. (1985). Soil aggregation by cryptogams of a sand prairie. *Am J. Bot.*, 72: 1657-1661.

Schwartz, J., Scklar, G. and Suda, D. (1986). Inhibition of experimental oral carcinogenesis by β-carotene. Harvard School of Dental Medicines. *Carcinogenesis*, 7: 711-715.

Senthil, C., Roychoudhury, P. and Kaushik, B.D. (1993). Lipid profile of halosensitive *Calothrix marchica bharadwajae*. *Indian J Microbiol.*, 33: 281-285.

Sharma, M.L., Bhardwaj, G.S. and Chauhan, Y.S. (1989). Study on the effect of biofertilizer, pyrite and gypsum on paddy in the salt-affected soils. *Indian J Agron.*, 34: 129: 130.

Shukla, A.C. and Gupta, A.B. (1967). Influence of algal growth-promoting substances on growth, yield and protein contents of rice plants. *Nature*, 213: 744.

Singh, D.N., Venkataraman, G.S., Rao, S.B.P., Bhattacharya, A. and Goyal, S.K. (1972). Effect of algal inoculation on jaya rice variety under field condition. *Labdev J Sci Technol*, 10B: 107-108.

Singh, R.N. (1950). Reclamation of "usar" lands in India through blue-green algae. *Nature*, 165: 325-326.

Singh, R.N. (1961). Role of Blue green algae in nitrogen economy of Indian agriculture. Indian Council of Agricultural Research, New Delhi. p.175.

Singh, V.P. and Trehan, T. (1973). Effect of extracellular products of *Aulosira fertilissima* on the growth of rice seedlings. *Plant Soil*, 38: 457-464.

Soltes-Rak, E., Kushner, D.J., Williams, D.D. and Coleman. J.R. (1995). Factors regulating *cry* IVB expression in the cyanobacterium *Synechococcus* PCC 7942. *Mol Gen Genet* 246: 301-308.

Sprent, J.I. and Sprent, P. (1990). Nitrogen fixing organism–pure and applied aspects. *Chapman and Hall*. London. pp. 256.

Stewart, W.D.P. (1964). Nitrogen fixation by Myxophyceae from marine environment. *J Gen Microbiol* 36: 415-422.

Stewart, W.D.P., Rowell, P., Ladha, J.K. and Sampio, M.J.A. (1979). Blue-green algae (cyanobacteria)–some aspects related to their role as source of fixed nitrogen in paddy soils. In: *Nitrogen and Rice*, International Rice Research Institute, Los Banos, Philippines, p. 263.

Subhashini, D. and Kaushik, B.D. (1981). Amelioration of sodic soils with blue-green algae. *Aust. J. Soils Res.*, 19: 361-367.

Subhashini, D. and Kaushik, B.D. (1984). Amelioration of salt affected soils with blue green algae. I. Influence of algalization on the properties of saline alkali soils. *Phykos*, 23: 273-277.

Subhashini, D. and Kaushik, B.D. (1986). Uptake of sodium and potassium by blue green algae (cyanobacteria). *Zbl fur Mikrobiol.*, 141: 177-180.

Subramanian, S. and Kaushik, B.D. (2001). Development of genetic markers in cyanobacteria and stability of genetically marked strains in soil. *World J. Microbial Biotech*, 17: 535-544.

Switzer, L. (1980). *Spirulina*–The whole food Revolution. Los Angeles, CA: Bantan Books.

Thomas, J. (1977). Biological nitrogen fixation. *Nucl India*, 15 : 2-8.

Tisdall, J.M. and Oades, J.M. (1982). Organic matter and water stable aggregates in soils. *J. Soil Sci.*, 33: 141-163.

Torzillo, G. and Vonshak, A. (1994). Effect of light and temperature on the photosynthetic activity of the cyanobacterium *Spirulina platensis*. *Biomass Bioenergy*. 6: 399-403.

Traore, T.M., Roger, P.A., Renaud, P.A. and Sasson A. (1978). Nitrogen fixation by blue green algae in paddy field in Mali. *Cab ORSTOM Ser Biol*, 13: 181-185.

Venkataraman, G.S. (1972). Algal biofertilizer and rice cultivation. Today and Tomorrow's publ., New Delhi.

Venkataraman, G.S. (1975). The role of blue green algae in tropical rice cultivation. In: *Nitrogen fixation by free living microorganisms*. (Edt. W.D.P. Stewart) Cambridge Uni Press, p.207-268.

Venkataraman, G.S. (1979). Algal inoculation of rice fields. *In*: International Rice Res. Institute. *Nitrogen and Rice*. Los Banos, Philippines, 311-321.

Venkataraman, G.S. and Goyal, S.K. (1963). Role of blue-green algae in nitrogen economy of Indian soils. *Agric Res* New Delhi. 3: 9.

Venkataraman, G.S. and Neelakantan, S. (1967). Effect of cellular constituents of nitrogen fixing blue-green alga *Cylindrospermum* on root growth of rice plants. *J. Gen. Appl. Microbiol.*, 13: 53-62.

Venkataraman, S. and Kaushik, B.D. (1994). Fatty acid profile of a marine *Nostoc calcicola* under saline and non-saline conditions. *Curr Sci.*, 67: 120-122.

Vonshak, A., Abelovich, A., Boussiba, S., Arad, S.M. and Richmond, A. (1982). Production of *Spirulina* biomass: Effects of environmental factors and population density. *Biomass* 2: 175-185.

Vonshak, A., Boussiba, S., Abelovich, A. and Richmond, A. (1983). Production of Spirulina biomass: Maintenance of monoalgal cultures outdoors. *Biotechnology and Bioengineering* 25: 341-349.

Walsh, K., Jones, G.J. and Dunstan, R.H. (1997). Effect of irradiance on fatty acid, carotenoid, total protein composition and growth of *Microcystis aeruginosa*. *Phytochemistry* (UK) 44: 817-824.

Watanabe, A. (1951). Production in cultural solution of some amino acids by the atmospheric nitrogen fixing blue-green algae. *Arch Biochem. Biophys.*, 34: 50.

Watanabe, A. (1973). On the inoculation of paddy fields in the pacific area with nitrogen fixing blue-green algae. *Soil Biol Biochem* 5: 161-162.

Watanabe, A., Nishigzaki, S. and Konishi, C. (1951). Effect of nitrogen fixing blue green algae on the growth of rice plants. *Nature* 108: 748.

Watnable, I. and Roger, P.A. (1984). Nitrogen fixation in wet land rice fields. In: *Current development in biological nitrogen fixation*. (Edt. N.S. Subba Rao) Oxford IBH New Delhi, pp. 237-276.

Wilson, J.T. (1980). In Blue green algae and Rice. PA Roger and SA Kulasooriya. International Rice Research Institute, Los Banos, Philippines.

Wyatt, J.T, and Silvey, J.K.G. (1969). Nitrogen fixation by *Gloeocapsa. science*, 165: 908-909.

Yoshida, T. and Ancazans, P.R. 1993. *Soil Sci Soc Amer Proc* 37: 42.

Wolf, A.M., Baker, D.E., Pionke, H.B. and Kunichi, H.M. 1985. Soil test for estimating labile, soluble and algal available phosphorus in agricultural soils. *J. Environ. Qual.*, 14: 341-348.

Xiaoqiang W, Vennison SJ, Huirong L, Ben-Dov E, Zaritsky A and Boussiba S 1997 Mosquito larvicidal activity of transgenic *Anabaena* strain PCC 7120 expressing combinations of genes from *Bacillus thuringiensis* sub. sp. *israelensis*. *Appl Enviro. Microbiol* 63: 4971-4975.

Yamaguchi, T. (1981). Strong Deodorant. *Japanese Patent*, 79 JP-116131.

Microbial Biotechnology and Ecology (2011) *Pages* 544–552
Editors: Deepak Vyas, G.S. Paliwal, P.K.Khare and R.K Gupta
Published by: DAYA PUBLISHING HOUSE, NEW DELHI

Chapter 31

Effect of Herbicides on Mitosis, Nucleic Acids and Protein Contents in *Corchorus* and Soybean Plants

☆ *A.K.S. Chouhan and S.K. Yadav*

Department of Botany, Dr. H.S. Gour Vishwavidyalaya, Sagar – 470 003, M.P.

ABSTRACT

Effects of three herbicides, *viz.*, Fluchloralin, Butachlor and Pendimethalin in varying concentrations (200, 400, 600, 800 and 1000 ppm) was studied on mitotic cell division, nucleic acids, (DNA and RNA) and protein contents of leaves of soybean crop and *Corchorus olitorius* Lin. a dominant weed of soybean field. All the three herbicides reduced cell division (Miotic Index) significantly in *Corchorus* root meristem at all the test concentrations. Further, the reduction of miotic index increased with the increase in the concentration of herbicides. In soybean lower concentrations of these herbicides showed no effect, where as higher concentrations reduced mitotic index significantly. No significant chromosomal abnormalities were observed in our experiment, however few abnormal cell showing sticky and eroded chromosoms were seen in both the plant samples treated with 1000 ppm concentration of fluchloralin.

All the three herbicides affect the DNA, RNA and protein contents of leaves in Soybean and *Corchorus* plants, but the magnitude of effect vary for the plants as well as for herbicides. Fluchloralin was found to be the most effective herbicide. It decreased DNA, RNA and protein contents considerably in both the plants, and the amount of these biomolecules decreased with the increase in treatment concentration. Butachlor and pendimethalin also decreased nucleic acids and protein contents considerably in *Corchorus* but with less efficiency as compared to fluchloralin. Lower concentrations of these herbicides *i.e.* 200 and 400 ppm showed a vary little or no effect on nucleic acids and protein in Soybean where as, higher concentrations reduced the amount of these molecules considerably. It is suggested that for the control of weeds in the crop field, lower concentrations of herbicides should be used to avoid the damage of crop plants.

Introduction

The continuous, consistent and indiscriminate use of agrochemicals for maximizing the crop yield is leading to senseless chemical saturation of environment and creating a vary serious human health as well as environmental problem (Odum 1971).

Herbicides are important agrochemicals used for controlling weeds in the crop. The use of these chemicals are increasing tremendously all over the world and a large number of them are being added annually to the market under different brand names. A survey of available literature on chemical weed control reveals that much of the work done so far pertains mostly on the morphological and anatomical aspects of the weeds as well as the crop plants. Workers like Sawamura (1964), Strove (1973), Amer and Ali (1974), Skorupaska (1975), Tomkins and Grant (1976), Sikka and Sharma (1976), Mousa (1982), Reddy and Rao (1982), Hess (1983), Soheir *et al.* (1989), Badr (1983), Soupuri and Mishra (1993), studied the effects of different herbicides on cell division in different plants and have reported various types of chromosomal aberrations and reduction in the number of dividing cells. A very few records are available on the biochemical effects of herbicides in weed and crop plants. Workers like West *et al.* (1960), Key and Hanson (1961), Key *et al.* (1966), Moreland *et al.* (1969), Chen *et al.* (1972), Bell *et al.* (1976), Chouhan and Yadav (2007). Kolhe (1979) have investigated the effect of different herbicides on nucleic acids and protein contents in different plants. Some of them observed increase and some of them observed decrease in nucleic acid and protein contents.

The forgoing reports indicate that several herbicides used in the control of unwanted weeds have the ability to inhibit cell division and induced chromosomal abnormalities. Chromosomal aberrations have been considerd as reliable indicators of mutagenic activity (Mohandas and Grant 1972) and there have been evidences for a correlation between chromosomal damage and toxic effects of a number of herbicides (Gichner *et al.*, 1968; Tomkins and Grant, 1976; Delcourt and Deysson 1976). Neilson *et al.* (1968) stated rightly that any new drugs or chemicals must be tested for their toxic effects on cell at cellular level before they are released for general use. For better and safe use of herbicides for increasing the agricultural productivity study of mode of action of these chemicals at cellular level is pre-requisite. Fluchloralin, Butachlor and Pendimethalin are the common herbicides, widely used for weed control in Soybean fields in this region. So for as the authers are aware these herbicides were not studied for their cytological and biochemical effects. It is with this view the present problem was undertaken to investigate the effects and the mode of action of these herbocides on cell division and nucleic acids and protien contents in *Corchorus* weed and Soybean crop.

Materials and Methods

Effect of these herbicides, *viz.*, Fluchoralin, Butachlor and Pendimethalin in five different concentrations (200, 400, 600, 800 and 1000ppm) on cell division, nucleic acids and protien contents in *Corchorus* and Soybean plants was studied. The details of methods are as follows:

Effect on Cell Division

Seeds of *Corchorus* and Soybean were treated with different concentrations of the test herbicides for 24 hours. Then they were washed thoroughly with water and kept for germination on moist filter paper in petridishes in a germination incubator at 27°±2°C. For control the seeds were soaked in water for 24 hours. Root tips of about 48 hours old germinating seeds were used for this study.

Roots in each case were examined to note the extent of cell division process and chromosomal abnormalities. Root squashes were made in 2 per cent acetoorcein following the method described by Sharma and Sharma (1965). The data obtained were analysed statisticaly and expressed as mean ±

standard error to gather with 't' significance test for mitotic index and as percentage of chromosomal abnormalities.

Effect on Nucleic Acids and Protein Contents

The plants of *Corchorus* and Soybean were sprayed with different concentrations of herbicides. The leaves were collected when morphological symptoms of the herbicides were seen on leaves.

Preliminary Extraction of Leaves

Leaves weighing one gram fresh weight were taken for estimation of nucleic acids and proteins. For extraction, the procedure of Nieman and Poulsan (1963) was followed with slight modification.

The Frozen samples were homogenized in chilled 10 ml of 95 per cent of ethanol. The homogenate was rinsed into a centrifuge tube with another 5 ml of cold ethanol and centrifuged. All centrifuging were done at room temperature at 3500 rpm. The supernatant was discarded each time. The sediment was extracted successively by suspending and centrifuging with 12 ml of each of the following reagents.

1. 95 per cent of ethanol at room temperature: Once
2. 50 per cent ethanol adjusted to pH 4.5 with glacial acitic acid at room temperature: Twice.
3. Chilled (4°C 0.2 N perchloric acid (PCA): Twice : The suspension was held at 4°C for 20 minutes before centrifuging.
4. 95 per cent ethanol at room temperature: Once.
5. Boiling absolute ethanol at room temperature: Once.
6. Ether at room temperature: Once.

Separation of Nucleic Acids and Proteins

The leaf powder obtained in the above procedure was suspended with 10 ml of 0.3N NaOH and held at 30°C for 18 hours. The sediment was removed by centrifuging and washed once with 5 ml of 0.3N NaOH. The extract and the wash were combined and made to 15 ml with 0.3N NaOH. Eight ml (8.0ml) aliquot, reserving 7 ml for protein estimation, was acidified to pH 1 with 15 per cent PCA, held at 4°C for 40 minutes and then centrifuged. The supernatant equals RNA fraction. The DNA-protein sediment was resuspended with 2 ml of water followed by 2 ml of 1 N PCA. The suspension was held at 4°C for 20 minutes, then centrifuged and the supernatant was added to the RNA fraction. The final RNA extract was made to 15 ml with water.

The DNA-protein was suspended with 3 ml of 0.5 N PCA, heated at 70°C for 15 minutes and centrifuged. The supernatant equals the DNA fraction. The protein sediment was washed with 2 ml of 0.5 N PCA and centrifuged and the wash added to the DNA fraction. The final DNA extract was made to 5 ml with 0.5 N PCA.

Estimation of Nucleic Acids

DNA

The samples of separated DNA from the extract ranging from 0.6 to 1.5 ml were taken in different test tubes and the volume was made to 2 ml with distilled water. Blank was prepared using 2 ml distilled water. Each tube was added with 4 ml of freshly prepared diphenylamine reagent and the mixture was heated for 5 minutes in boiling water bath. They were cooled to room temperature and read at 600 nm on spectrophotometer (UV 108 spectrophotometer). The DNA content of the samples was calculated from standard graph of calf-thymus DNA. The total DNA content was calculated on per gram-dry weight basis.

RNA

The samples of separated RNA from the extract ranging from 0.1 to 0.5 ml were taken in clean dry test tubes and the volume was made 2 ml with distilled water. Blank was prepared taking 2 ml distilled water only. In each test tube 5 ml of ferric chloride-hydrochloric acid was added, followed by 0.3 ml orcinol reagent. The tubes were placed in boiling waterbath for 20 minutes. The tubes were allowed to cool to room temperature and read at 660 nm on spectrophotometer. The RNA content in sample was calculated using standard graph of yeast RNA. The total RNA content was calculated on per gram dry weight basis.

Estimation of Proteins

The samples of proteins ranging between 0.1 to 0.4 ml were taken in clean dry test tubes and the volume was made to 1.2 ml with distilled water. Each tube was added with 6 ml freshly prepared alkaline copper solution. After 10 minute 0.3 ml Follin Ciacaltue's phenol reagent was added mixed thoroughly. Percent transmittance was read at 500 nm on spectrophotometer after 30 minutes. The protein content in the sample was calculated using a standard graph of egg albumin protein and one per gram dry weight basis.

Results and Discussion

The experimental data obtained after the treatment of plants were analysed in terms of following parameters:

1. Mitotic Index (MI) comparison between treated and control and 't' significance test.
2. Induced mitotic abnormalities–any abnormality in the dividing cells at chromosomal level.
3. Percentage increase or decrease in the amount of nucleic acids and protein contents after herbicidal treatment.

Effect on Cell Division

The results indicate that all the test herbicides inhibited cell division process significantly in *Corchorus* weed as well as in Soybean crop. Moreover, in *Corchorus* the inhibitory effect increased gradually with the increase in the concentration of herbicides, but in Soybean lower concentrations showed no effect where as higher concentrations of these herbicides reduced cell division (mitotic index) significantly. Fluchloralin was found to be the most effetive in both the plants followed by Butachlor and Pentimethalin. No significant chromosomal abnormalities were observed in our investigation, however, few abnormal cells showing sticky and eroded chromosomes were seen in both the plant samples treated with 1000 ppm of fluchloralin (Table 31.1).

Reduction in number of dividing cells due to herbicidal treatment has been reported by number of workers. Ryland (1948) observed a general inhibition of cell division in young root tips of *Allium* and *Narcissum* with the treatment of 2, 4-D. Kolhe (1979) studied the cytological effets of 2,4-D, Hyvar X and Gramoxone on farm weeds. He observed different chromosomal abnormalities and reduction in number of dividing cells in root tip by these herbicides. Many other workers also noticed similar observations in different weed species treated with different herbicides (Sawamura, 1964; Khosla, 1967; Bakale, 1972; Amar and Ali, 1974; Tomkins and Grant, 1976; Sikka and Sharma, 1976; Njagi and Gopalan, 1981; Mousa, 1982; Reddy and Rao, 1982). All these investigators noted changes in the mitotic stages with herbicides.

Table 31.1: Miotic Indices in Root Meristem of *Corchorus olitorius* and Soybean Treated with Different Concentrations of Herbicides

Species	*Herbicides*	*Concentration of Herbicides (ppm)*					
		Control	*200*	*400*	*600*	*800*	*1000*
Corchorus oliterius	F	1.83±0.36	1.12±0.18 (4.84)**	0.61±0.12 (8.03)**	0.48±0.09 (9.42)**	0.40±0.13 (8.48)**	0.33±0.15 8.39)**
	B	1.83±0.36	1.49±0.19 (2.01)	0.59±0.13 (7.94)**	0.42±0.12 (9.67)**	0.52±0.15 (8.08)**	0.58±0.20 (7.31)**
	P	1.83±0.36	1.18±0.11 (4.53)**	0.85±0.08 (7.13)**	0.59±0.06 (9.38)**	0.49±0.22	0.74±0.35 (4.27)**
Glycin max	F	2.34±0.41	2.02±0.22 (1.65)	1.91±0.12 (2.13)	1.69±0.15 (3.35)**	1.48±0.19 (5.84)**	0.62±0.18 (9.51)**
	B	2.34±0.41	2.18±0.08 (1.98)	2.04±0.30 (2.02)	1.94±0.24 (2.29)*	1.87±0.26 (2.88)*	0.93±0.13 (6.45)**
	P	2.34±0.41	2.08±0.31 (1.92)	1.95±0.17 (3.63)**	1.88±0.17 (3.63)**	1.64±0.31 (4.02)**	1.08±0.22 (5.98)**

*: Significant at P > 0.05 : 18 df values given in poven these one 't' value.

**: Significant at P > 0.01 18 df value.

An agent can disrupt the mitotic cycle in three ways (*i*) by inhibiting the process of cell division (*ii*) by disturbing the normal functioning of mitotic spindle and (*iii*) by producing chromosomal abnormalities. In this study reduction in percentage of dividing cells (MI) and increase in pre-metaphase/Postmetaphase ratio (Table 31.2) due to test herbicides indicate that the herbicides have some antimitotic activity and they are acting at both interphase and pre-metaphase levels affecting some vital functional units of the cell and the normal functioning of spendle mechanism respectively.

Effect on Nucleic Acids and Protein Contents

All the three herbicides affect the DNA, RNA and protein contents of leaves in Soybean and *Corchorus*, however the magnitude of effect vary for the plants as well as for herbicide (Table 31.3). Fluchloralin was found to be the most effetive herbicide. It decreased the DNA, RNA and protein contents (biomolecule) considerably in both plants. Further, the decrease in biomolecule contents was increased with the increase in concentration of this herbicide. Butachlor and pendimenthalin also decreased the amount of DNA, RNA and protein in *Corchorus* but with less efficiency as compared to fluchloralin. Lower concentrations of these herbicides showed a very little or no effect on biomolecule contents in Soybean, where as higher concentrations reduced the amount of these biomolecules considerably in Soybean also.

It is evident from the results that fluchloralin was the most effective herbicide as it caused maximum redution of biomoleule contents in both the plants followed by butachlor and pendimethalin. Some workers have studied the effect of different herbicides on DNA, RNA and protein contents in different weed and crop species. Key *et al.* (1966) reported inrease in DNA and protein contents in soybean seedling after spray with 2,4-D. Chen *et al.* (1972) also reported incrase in DNA and protein in cucumber and wheat seedling by 2,4-D. Inhibition of DNA synthesis by paraquat in *Vicia faba* seedling was reported by Bell *et al.* (1976). Decrease in DNA content in *Taphrosia hamiltonii* after treatment with 2,4-D and gramoxone was reported by Kolhe (1979). West *et al.* (1960) reported in corn and cucumber tissue that with one exception, low concentration of 2,4-D decreased and high concentrations increased the level of RNA. Moreland *et al.* (1969) studied the effects of 22 different herbicides on RNA and protein synthesis in maize mesocotyle and soybean hypocotyle and found that 14 of the herbicides used, inhibited RNA and protein synthesis *in vivo*. Similar results were also reported by other workers (Chrispeels and Hanson, 1962; Shannon *et al.*, 1964; Gulifoyle and Hanson, 1974; Bakale, 1972; Chouhan and Yadav, 2007).

Table 31.2: Premetaphase (A) and Post-metaphase (B) Ratio in Root Meristem of *Corchorus olitorius* and Soybean Treated with Various Concentration of Herbicides

Herbicides	*Concentration (ppm)*	*Corchorus olitorius*			*Glycine max*		
		A	*B*	*A/B Ratio*	*A*	*B*	*A/B Ratio*
Control	–	38	13	2.92	68	32	2.13
	200	18	10	1.80	42	24	1.75
	400	08	02	4.00	15	04	3.75
	600	05	01	5.00	13	03	4.33
	800	03	01	3.00	12	02	6.00
	1000	03	00		09	01	9.00
	200	27	09	3.00	51	15	3.40
	400	10	03	3.33	23	06	3.83
	600	07	01	7.00	18	03	6.00
	800	07	01	7.00	15	03	5.00
	1000	06	01	6.00	16	04	4.00
	200	18	12	1.50	38	17	2.24
	400	12	05	2.40	28	09	3.11
	600	07	01	7.50	18	05	3.60
	800	06	00	–	13	03	4.33
	1000	03	00	–	15	03	5.00

Table 31.3. Amount of nucleic acids and protein (mg/g) in *Corehorus* and soybean after spraying with different concentrations of Fluchloralin (F), Butachlor (B) and Pendimethalin (P)

Corchorus

Concentration (ppm)	*DNA*			*RNA*			*Protein*		
	F	*B*	*P*	*F*	*B*	*P*	*F*	*B*	*P*
Control	160.00	160.00	160.00	168.00	168.00	168.00	240.00	240.00	24.00
200	144.00 (–100.00)	148.00 (–07.00)	160.00 (00.00)	148.00 –11.91	156.00 –07.15	160.00 –04.77	167.00 (–30.21)	200.00 (–16.67)	202.50 (–15.53)
400	124.00 (–22.50)	140.00 (–12.50)	140.00 (–12.50)	132.00 (–21.43)	140.00 (–16.67)	140.00 (–16.67)	130.00 (–45.84)	175.00 (–27.09)	182.00 (–24.19)
600	112.00 (–30.00)	142.00 (–23.75)	125.00 (–21.88)	112.00 (–33.34)	120.00 (–28.58)	120.00 (–56.25)	105.00 (37.50)	150.00 (37.50)	157.00 (–34.59)
800	110.00 (–31.25)	110.00 (–31.25)	120.00 (–25.00)	92.00 (–45.24)	112.00 (–33.34)	110.00 (–34.53)	67.00 (–72.09)	132.00 (–45.00)	132.50 (–44.80)
1000	80.00 (–50.00)	108.00 (–32.50)	100.00 (–37.50)	64.00 (–61.95)	92.00 (–45.24)	92.00 (–45.24)	57.00 (–76.25)	120.00 (–50.00)	114.50 (–52.30)

Contd...

Table 31.3–Contd...

Soybean

Concentration (ppm)	DNA			RNA			Protein		
	F	B	P	F	B	P	F	B	P
Control	200.00	200.00	200.00	192.00	192.00	192.00	412.50	412.50	412.00
200	188.00 (–06.00)	200.00 (00.00)	200.00 (00.00)	160.00 (–16.67)	168.00 (–12.50)	168.00 (–12.50)	377.50 (–08.49)	382.50 (–07.28)	407.50 (–01.22)
400	152.00 (–24.00)	200.000 (00.00)	200.00 (00.00)	116.00 (–39.59)	132.00 (–31.25)	152.00 (–20.84)	340.00 (–17.58)	362.50 (–12.13)	382.50 (–07.28)
600	128.00 (–36.00)	160.00 (–20.00)	180.00 (–10.00)	100.00 (–47.92)	108.00 (–43.75)	120.00 (–37.50)	282.00 (–31.52)	330.00 (–20.00)	325.00 (–21.22)
800	100.00 (–50.00)	140.00 (–30.00)	160.00 (–20.00)	80.00 (–58.34)	88.00 (–54.17)	112.00 (–41.67)	200.00 (–51.52)	250.00 (–39.40)	300.00 (–27.48)
1000	84.00 (–58.00)	120.00 (–40.00)	140.00 (–30.00)	52.00 (–72.92)	68.00 (–64.59)	100.00 (–47.92)	119.00 (–70.16)	160.00 (–63.22)	272.00 (–33.94)

Values given in Parenthesis are increase (+) or decrease (–) above and below control level.

In many cases the herbicidal effects noted on protein and nucleic acids synthesis may be indirectly due to limitation of photosynthetic energy supply by reducing ATP synthesis (Moreland, *et al.*, 1969). It is known that approximately 90 per cent of all the ATP generated in the cell are used to produce macromolecules such as DNA, RNA and proteins. The treatment of cell with toxic substances at sublethal dose affects the enzyme system of cell (Dustin, 1947) and presence of organic compound in a cell may exert some fundamental effects on enzyme system (Wuu and Grant 1966a) might be resulting in disturbance of homeostasis of the cell by genetic variability induced by herbicides.

The forgoing results and discussions indicate that all the three test herbicides possessed the potential to inhibit the process of cell division and the synthesis of nucleic acids and protein in *Corchorus* and Soybean. The possible explanation and interpretation for the mechanism of action of these herbicides may be derived from the above observations. In all possibilities as the enzyme system of biochemical pathways leading to the generation of ATP and synthesis of macromolecules, are impaired, these herbicides may act as metabolic inhibitors. Metabolic disturbance due to these chemicals may generally effect the synthesis of nucleic acids and proteins, consequently resulting in decrease in rate of cell division and so also in the miotic index.

Acknowledgement

The authors are grateful to Head, Department of Botany, Dr. H.S. Gour Vishwavidyalaya, Sagar for providing necessary lab facilities.

References

Amar, S.M. and Ali, E.M. (1974). Cytological effects of pesticides V. effects of some herbicides on *Vicia faba. cytologia* 39: 633-643.

Badr, A. (1983). Mitodepressive and Chromotoxic activities of two herbicides in *Allium cepa. Cytologia* 48 : 451-457.

Bakale, V.L. (1972). Effect of Weedicides on cytomorphology of weeds. *Althernanthera polygonides var. erecta, Cressa cretica* and *Xanthium strumarium. Ph.D. Thesis*, Bombay University.

Bell, S.L., Schwarz, O.J. and Hughes, K.W. (1976). Studies on the herbicide paraquat, I. Effects on the cell cycle and DNA synthesis in *Vicia faba. Can. J. Genet. Cytol.* 18: 93-99.

Chen, L.G., Switzer, C.M. and Fletcher, R. 1972). Nucleic acid and protein changes induced by auxin like herbicides. *Weed Sci.*, 20: 53-55.

Chouhan, A.K.S. and Yadav, S.K. (2007). Effect of three herbicides on the nucleic acids and protein content in wheat crop. *J. Bot. Univ. Sagar*, 42: 12-19.

Chrispeels, M.S. and Hanson, J.B. (1962). The increase in RNA content of cytoplasmic particulates of soybean hypocotyl indued by 2,4-D. *Weeds* 10: 123-125.

Delcourt, A. and Daysson, G. (1976). Effet of trifluraline on the root meristem of *Allium sativum. Cytologia* 41(1): 75-89.

Dustin, P. (1947). Some new aspects of mitotic poisoning, *Nature*. 159: 794-796.

Gichner, T., Goul, H. and Omura, T. (1968). The influence of post treatment washing and redrying of barley seeds on the mutagenic activity of N-methyl N nitroso urea and N-ethyl N-nitroso urea. *Radiation Botany* 8: 499-537.

Gulifoyle, T.J. and Hanson, J.B. (1974). Greater length of ribonucleic acid synthesized by chromatin-bound polymerase from auxin-treated soybean hypocotyls. *Plant Physiol.* 110-113.

Hess, F.D. (1983). Mode of action of herbicides that affect the cell division. In :Pesticide chemistry, Human health welfare and environment Eds. J. Miyamoto and P.C. Kervey, Pergaman Press, Ltd., Headington, Hill Hall Oxford : 420-425.

Key, J.L. and Hanson, J.B. (1961). Some effects of 2,4-D on soluble nucleotides and nucleic acids of soybean seedlings. *Plant Physiol.* 36: 145-152.

Key, J.L., Lin, C.Y., Gifford, E.M. (Jr) and Dengler, R. (1966). Relation of 2,4-D induced growth aberrations to changes in nucleic acid metabolism in soybean seedlings. *Bot. Gaz.* 127: 87-94.

Khosla, S.N. (1967). Effect of herbicides on the cytomorphology of weeds. Ph.D. thesis, Bombay University.

Kolhe, R.L. (1979). Effect of herbicides on the cytomorphology of farm weeds. Ph.D. thesis, Nagpur University, Nagpur.

Mohandas, T. and Grant, W.F. (1972). Cytogenetic effects of 2,4-D and amitrole in relation to nuclear volume and DNA content in some higher plants. *Can. J. Genet. Cytol.*, 14: 773-783.

Moreland, D.E., Malhotra, S.S., Gruenhagen, R.D. and Shokraii, E.H. (1969). Effects of herbicides on RNA and protein synthesis. *Weed Sci.*, 17: 556-563.

Mousa, M. (1982). Effects of some herbicides on mitotic activity and mitotic aberrations in root tips of onion. *Egyption J. Gemet. Cytol.*, 11: 193-2007.

Neilson, J.W., Friedrich and Tsuboi, T. (1968). Chromosome abnormalities and Psychotropic drugs. *Nature*. 218: 488-489.

Nieman, R.H. and Poulsen, L.L. (1962). Spectrophotometric estimation of nucleic acid of plant leaves. *Plant. Physiol*, 38: 31-35.

Njagı, C.D.E. and Gopalan, H.N.B. (1981). Mutagenicity testing of herbicides fungicides and insecticides I. Chromosome aberration in *Vicia Faba. Cytologia*, 46: 169-172.

Odum, E.P. (1971). *Fundamentals of Ecology*. W.B. Saunders Co. Phildelphia, London.

Reddy, S.S. and Rao, G.H. (1982). Cytogenetic effects of agricultural chemicals II. Effects of herbicides 'Lasso' and 'Bassagram' on chromosomal mechanism in relation to yield in chilli : (Capsicum annum L.). *Cytologia*, 47: 257-267.

Ryland, A.G. (1948). A cytological study of the effect of colchicine, indoleacetic acid, potassium cynide and 2,4-Dichlorophenoxy acetic acid on plant cells. *J. Elisha Mitchell Sci. Soc.*, 64: 117-125.

Sawamura, S. (1964). Cytological studies on the effects of herbicides on plant cells *in vivo*. I. Hormonic herbicides. *Cytologia*, 29: 86-102.

Shannon, J.C., Hanson, J.B. and Wilson, C.M. (1964). Ribonuclease levels in the mesocotyl tissue of *Zea maize* as a function of 2,4-D application. *Plant Physiol.*, 39: 804-809.

Sharma, A.K. and Sharma, A. (1965). Chromosome Techniques: Theory and Practice, Butterworth and Co. Ltd. London.

Sikka, K. and Sharma, A.K. (1976). The effects of some herbicides on plant chromosome. *Proc. Indian Nat. Sci. Academy*, (B) 42: 299-307.

Skorupska, H. (1975). Radioimmetic effect of herbicides on the progress of mitosis in pea in comparison with those of chromomutagens and gamma rays. *Genetica Polonica*. 16: 301-311.

Sohier El-Khodary. Antoinette Habıt and Atif Haliem (1989). Cytological effects of the herbicide Garlon-4 on root mitosis of *Allium cepa*. *Cytologia*, 54: 465-472.

Sonpuri, A.K. and Mishra, S.K. (1993). Cytological effects of starthene (Acephate) in root meristems of *Allium cepa*. *Biome*: 6(2): 102-104.

Strove, V.S. (1973). The cytogenetic activity of the herbicides atrazine-chloro-LPC and Paraquat. *Sov. Genet.* 6: 293-297.

Tomkins, D.J. and Grant, W.F. (1976). Monitorning natural vegetation for herbicides induced chromosomal aberrations *Mut. Res.* 36: 73-83.

West, S.H., Hanson, J.B. and Key, J.L. (1960). Effect of 2,4-D on nucleic acid and protein content of seedling tissue. *Weeds*. 8: 333-340.

Wuu, K.D., Grant, W.F. (1966a). Morphological and somatic chromosomal aberrations induced by pesticides in barley (*Hordeum vulgare*) *Can. J. Genet. Cytol.* 8: 481-501.

Microbial Biotechnology and Ecology (2011) *Pages* **553–570**
Editors: **Deepak Vyas, G.S. Paliwal, P.K.Khare and R.K Gupta**
Published by: **DAYA PUBLISHING HOUSE, NEW DELHI**

Chapter 32

Biotechnological Potential of Cyanobacteria

☆ *Surendra Singh**

Centre of Advanced Study in Botany,
Banaras Hindu University, Varanasi – 221 005, U.P.

ABSTRACT

Cyanobacteria are ubiquitous in nature and considered to be next to bacteria as far as their distribution is concerned. This large and diverse group of Gram negative prokaryote having traditional applications, is now being utilized and further exploited for biotechnological development. These organisms also possess immense potentials for application towards human welfare. However, except for the use of *Spirulina* as a health food, there are no other major products from cyanobacteria in India and elsewhere. This is due to the limitations in the economic production of cyanobacterial products in low quantity. Furthermore, mass cultivation technology has not yet been evolved for such potential organisms and in many cases the methods of industrial extraction are not even optimized. Intensive research is thus warranted to understand many of the basic aspects pertaining to the production of a metabolite with the concurrent evolution of applied research towards the large scale production of the product.

In recent years people throughout the world are focusing their attention towards cyanobacteria for their possible use for the photobiological production of biofuel, ammonia, amino acids, various metabolites, vitamins, toxins, therapeutic substances, aqua or animal feed. Various chemicals including restriction enzymes, pharmacological probes and labeled compounds for research, as well as fluorescent probes for clinical diagnostics are now commercially available. Potentially useful compounds, including pharmaceuticals and industrial chemicals, have been identified and are

* E-mail: surendrasingh.bhu@gmail.com

currently being developed. This review summarizes the biotechnological potentials of cyanobacteria, and highlights various areas of applications mainly nutraceutical, pharmaceutical and bioactive potential.

Introduction

Cyanobacteria are oxygenic photosynthetic prokaryotes, which have long been recognized as having potential in biotechnology. These organisms have been evolved nearly about three billion years ago and have been responsible for the initial oxygenation of the Earth's atmosphere. They are the large and diverse group of prokaryotic organism occupying unique taxonomic position. They combine an autotrophic mode of nutrition that is common to eukaryotic algae and plant cells with a metabolic system that is generally regarded as bacterial rather than the plant like. These prokaryotic organisms thus offer the growth potential of microbial cells growth with the light harvesting capabilities of plant cells, making them ideal candidates for biosolar energy conversion programs. Another advantage of cyanobacteria over eukaryotic organisms is the capacity of several strains to fix atmospheric nitrogen (N_2). Cyanobacteria perform various metabolic functions like other prokaryotic bacteria. Similarly they are also exposed to diverse conditions in their natural habitats and hence have evolved alternate metabolic pathways and elegant regulatory mechanisms for adaptation.

Cyanobacteria are now increasingly attracting the attention of the biologists in view of their importance in agriculture, industry, environmental pollution abatement, conservation, biodiversity and pharmaceutical markets. Cyanobacteria in general are one of the richest sources of known and novel bioactive compounds including toxins with wide pharmaceutical applications (Becker 1994; Patterson 1996). This has led to the systematic screening of cyanobacterial strains both to test their potential pharmaceutical applications and also to know their general effects on animal systems. In recent years people throughout the world are focusing their attention towards cyanobacteria for their possible use for the photobiological production of biofuels, ammonia, amino acids, various metabolites, vitamins, toxins, therapeutic substances and aqua or animal feed (Singh 1992, 1993; Mohanthy 1996). Although, the screening of cyanobacterial strains having the ability to produce novel compounds with antimicrobial and antineoplastic activities were started as early as 1970 (Moore 1981; Rinehart *et al.*, 1981), the systematic screening for therapeutic substances from micro-algae particularly cyanobacteria has, however, received greater attention recently.

Pharmaceuticals

Microorganisms play a significant role in medical sciences as they not only cause infections, but also produce organic substances that can cure infections. The discovery of penicillin in 1929 heralded the era of antibiotics and led to the realization that microorganisms can be considered as a rich source of clinically useful natural products. Since then several natural products with bioactive potentials have been discovered from microorganisms. The tremendous rate of discovery is a testament to the inherent ability of microorganisms to produce bioactive metabolites.

However, despite many discoveries, the pharmaceutical industries have focused their attention on the same group of microorganisms, mainly the actinomycetes as their primary source for new drugs. Although actinomycetes continue to be studied extensively, it is clear that the rate of discovery of novel metabolites from this group is decreasing and hence new sources of bioactive natural products must be explored. One such resource is cyanobacteria, which are recently being exploited for their potential as producers of biomedically efficient metabolites.

Cyanobacterial bioactive compounds will contribute significantly to drug development either directly or as models for developing analogs for increased activity and as research tools for studying physiological processes and diseases. Pharmacological evalutions of cyanobacterial products have been revolutionized over the past two decades, beginning with the early investigation of toxins, followed by studies of cytotoxic and antitumor activity, to the present where a myriad of activities based on whole animal models and receptor binding assays are being pursued.

Antimicrobials

Any compound, natural, synthetic or semi synthetic (*i.e.*, chemically derived from natural substances) that is clinically used for the treatment of microbial infections is generally known as antimicrobials. These antimicrobials include antibacterials, antifungals and antialgals and antivirals.

Antibacterials

The antibacterial activity of cyanobacteria in general has been poorly studied (Reichelt, and Borowitzka 1984; Moore *et al.*, 1988).The cell-extract fractions from *Spirulina subsalsa* resulted in the complete inhibition of the growth of a clinical isolate of *Pseudomonas aeruginosa*, which is considered to be one of the difficult pathogens to treat (Sundararaman 1992). *Nostoc muscorum* also produces an antibacterial agent having broad-spectrum activity (Bloor and England 1991). The chloroform-methanol extracts of several *Oscillatoria* species inhibit the Gram-positive bacterial pathogen, *Staphylococcus aureus*. Whereas, the compounds obtained from *Phormidium angustissimum* and *Lyngbya* sp. were found to be effective against Gram negative and Gram positive pathogenic bacteria, the compounds obtained from *S. subsalsa* were, however, found to possess antibacterial activity against Gram negative organisms only (Sundararaman 1992). Similarly, the crude extracts of *P. angustissimum* and *Lyngbya* sp. were found to be potential sources of new broad-spectrum antibacterial compounds (Sundararaman 1992).

Antifungals

A large number of cyanobacteria are reported to possess the ability to produce the compounds which exhibit the antifungal activity. The cyanobacterial extracts that inhibit the proliferation of tissue cultured mammalian cells also inhibit the fungal growth, indicating that the mechanism of inhibition involves a function similar to eukaryotic cells. Scytophycin, a potent bioactive metabolite from cyanobacteria also exhibited antifungal activity (Ishibashi *et al.*, 1986; Patterson and Carmeli 1992). Tjipanazoles, N-glycosides of indole-carbazoles isolated from the cyanobacterium *Tolypothrix tjipanasensis* exhibited significant fungicidal activity against phytopathogenic fungi (Bonjouklian *et al.*, 1991). The laxaphycins are a large family of cyclic undeca and dodecapeptides responsible for antifungal activity caused by crude extracts of *Anabaena laxa*. The laxaphycins structurally and biologically resemble to a group of cyclic peptides known as hormothamnins, which have recently been isolated from the *Hormothamnion enteromorphoides* (Gerwick *et al.*, 1989). Extracts of *Synechococcus elongatous*, *P. corium* and *Dichothrix baueriana* were found to be active against *Candida* species (Sundararaman 1992). A compound, presumably a methoxylated disaccharide, purified from *Oscillatoria latevirens* exhibited anti-*Candida*, anti-*Mucor* and antibacterial activity (Deth 1999).

Antialgals

One hundred and ninety-eight cyanobacterial strains, isolated mainly from diverse habitats in Australia, Indonesia, Nepal, Thialand and Vietnam, were screened for their antibiotic activity against green algal species of the genera *Coelastrum*, *Scenedesmus* and *Monoraphidium* (Schlegel *et al.*, 1999). Out of these, ten strains of *Fischerella*, seven strains of *Nostoc* and three strains of *Calothrix* produce antialgal compounds with a broad spectrum activity. Some of these bioactive cyanobacteria are reported to inhibit the growth of *Coelastrum*, *Scenedesmus* and *Monoraphidium*, whereas others inhibited the growth of either two or one algae (Schlegel *et al.*, 1999). *P. valderianum* completely eliminated a dense population of marine diatoms inducing the extracellular production of antidiatom compound.

Antivirals

To discover a new antiviral agent from cyanobacteria, extracts from cyanobacteria were screened for anti-HIV properties (Boyd 1988). Cellular extracts of *Lyngbya langerhimii* and *Phormidium tenue* protected human lymphoblastoid T cells from the cytopathic effect of HIV- 1 infection. With the use of an *in vitro* anti-HIV-1 assay, a new class of HIV-1 inhibitory compound, the sulfonic acid containing glycolipids was isolated (Gustafson *et al.*, 1989). Recently a tetrazolium based microculture assay was developed to screen the extracts of cultured cyanobacteria for inhibition of the cytopathic effects of HIV (Gustafson *et al.*, 1989). At non-cytotoxic concentration, pure cyanobacterial compounds were found to be active against HIV-1 in cultured human lymphoblastoid T cell-lines (Gustafson *et al.*, 1989).

An aqueous extract of *Spirulina platensis* inhibits HIV-1 replication in human derived T-cell lines and in human peripheral blood mononuclear cells. The extracts also blocked Raucher Murine Leukemia Virus (RLV) induced plaque (Ayehunie *et al.*, 1998). The aqueous extract of *Spirulina* also exhibited a dose- dependent inhibition of the replication of *Herpes simplex* virus type 1 (HSV-1) in Hela cells within the concentration range of 0.08-50 mg ml^{-1}. However, this extract does not show any virucidal activity and could not even interfere the adsorption of the virus to the host cells (Hayashi *et al.*1993). Bioactivity-directed fractionation of a hot water extract from *S. platensis* led to the isolation of a novel sulfated polysaccharide, calcium spirulan (Ca-SP), as an antiviral principle. This polysaccharide was composed of rhamnose, ribose, mannose, fructose, galactose, xylose, glucose, glucuronic acid, galacturonic acid, sulfate and calcium. Ca-SP was found to inhibit the replication of several enveloped viruses, including *Herpes simplex* virus type 1 (HSV 1), Human Cytomegalo Virus (CMV), Measles Virus, Mumps Virus, Influenza A Virus and HIV 1. Ca-SP also selectively inhibits the penetration of viruses into the host cells. Retention of the molecular conformation by chelation of calcium ion with sulfate groups was suggested to be indispensable to its antiviral effect (Hayashi *et al.*1996).

A large number of cyanobacterial extracts also exhibited antiviral activity against *Herpes simplex* virus II (HSV-2) (Rinehart *et al.*, 1981). Whereas 10 per cent of the cyanobacterial extracts showed anti HSV-2 activity, 2.5 per cent of them, however, exhibited activity against respiratory syncytial virus (Patterson *et al.*, 1993). When antiviral activity of lipophilic and hydrophilic extracts of cyanobacteria against three pathogenic viruses were examined, a significant reduction in cytopathic effect normally associated with viral infection was recorded (Patterson *et al.*, 1993). Sulfoglycolipids isolated from cyanobacteria also exhibit strong antiviral properties. In *in vitro* studies, Helper T cells exposed to cyanobacterial sulfoglycolipids were protected from HIV-1 infection (Gustafson *et al.*, 1989). *P. valderianum*, *Plectonema terebrans* and *Pseudanabaena schimidelei* also exhibited HbsAg-clearing activity (Sundararaman 1992), indicating their potential use against Hepatitis B virus.

Cyanovirin derived from the cultures of the freshwater cyanobacterium *Nostoc ellipsosporum* binds to GP-120 protien of HIV and thus inhibits its replication. *N. ellipsosporum* producing cyanovirin-N and cyanovirin-V is also capable of inhibiting the growth of HIV-I and HIV-II (Chauhan *et al.*, 2004). Recently, allophycocyanin, a red fluorescent protein isolated from *S. platensis* is found to inhibit enterovirus 71-induced apoptosis, which may have a significant impact on protecting hosts from severe consequences associated with infection with enterovirus 71 (Shih *et al.*, 2003).

Other Potential Drugs

The hypoglycemic effect of *Spirulina* in non-insulin dependent diabetes mellitus (NIDDM) patients has been reported (Venkataraman, 1983). A significant reduction in the fasting and postprandial blood and urine sugar level has been reported following the uptake of about 2g of *Spirulina* (Venkatraman 1983). This has also been confirmed in experiments using albino rats with allaxan-induced diabetes. The reduction in the sugar level following *Spirulina* uptake has been attributed to the possible stimulation of prostaglandins. However, more studies on the similar lines are necessarily required not merely to establish the levels of *Spirulina* tablets, but also to understand the mechanisms of its action (Venkataraman, 1983). β–carotene as a dietary supplement has been reported to inhibit the development of 7, 12 dimethyl benzanthrazene (DMBA)-induced salivary gland carcinomas in rats.

Nutraceuticals

Cyanobacteria have found many tradition applications. However, its biotechnological applications have recently been recognized. *Spirulina*, a filamentous and non-heterocystous cyanobacterium has been used as food and food supplement and considered to be the major product from cyanobacteria (Puniya *et al.*, 1995). The high contents of protein, vitamin B_{12}, iron, β-carotene, iodine and γ-linolenic acid make the *Spirulina* protein as " naturally made" promising dietary substitute that could provide significant nutritional input to the Third World (Puniya *et al.*, 1995). A number of different cyanobacteria, mostly *Nostoc* strains have traditionally been used as food or food supplements, however, these strains are not extensively exploited (Jassby *et al.*, 1988).

Protien Efficiency Ratio (PER) of *Spirulina* protein has been reported to be higher than vegetables, cereals and soya proteins. The digestibility coefficient and biological values of *Spirulina* protein are only marginally lower than the milk protein, casein. Supplementation of *Spirulina* with cereals like rice and wheat on isoproteinic levels improved the protein quality (Venkataraman 1983). The bioavailability of total carotenes and in particular β-carotene from *Spirulina* was found to be comparable to the values reported for carotenes from other plant sources like leafy vegetables and carrots (Annapurna *et al.*, 1991). *Spirulina* is also a good source of vitamin A and hence it can be used as a source of vitamin A in the diet (Annapurna *et al.*, 1991). *Spirulina* is the most extensively researched food microalga documented in numerous books, journals, articles and periodicals.

Phycobiliproteins

Cyanobacteria in general possess all the known phycobiliproteins: phycocyanin, phycoerythrin, phycoerythrocyanin and allo-phycocyanin. Among them, phycocyanin and phycoerythrin are commercially valuable. 'Linablue' a phycocyanin product from Dainippan ink and Chemicals Inc, Japan is an odourless, non-toxic blue powder and used for colouring candy, ice cream, dairy products and soft drinks (Cohen 1986). Phycocyanin is also obtained in a water insoluble form from *Spirulina* and used in eye shadow, eyeliner and lipstick preparations (Dainippon Patent 1980). The blue or red chromophores are isolated by enzymatic or acid hydrolysis of the protein to yield more concentrated pigment which is used in cosmetics (Dainippon Patent 1981). Fhycoerythrin from *Spirulina* and other cyanobacteria is used as a food colour for products like icecream, yoghurt and it could also be used in cosmetics.

C-phycocyanin is a major biliprotein of *S. platensis*, the cyanobacterium having antioxidant (Upasani, *et al.*, 2001), anti-arthritic, and anti-inflammatory properties (Ramirez *et al.*, 2002). Phycocyanin acts as a hepatoprotective agent and as a hydroxyl radical scavenger (Madhava *et al.*,

2000). This pigment also inhibits oxidative damage in DNA and hence may be used as a therapeutic agent (Bhat and Madyastha 2001). However, the mechanism of its action is not yet clearly understood.

C-phycocyanin, which has been shown to be a selective COX-2 inhibitor, induces dose-dependent apoptosis in BC- 8 cells with a concomitant decrease in cell viability. However, BC-8 cells transfected with Bcl-2 gene were found to be resistant to phycocyanin mediated apoptosis. These observations were further confirmed by staining of these cells with Annexin-V. Annexin-V staining is observed in BC-8 cells after treatment with phycocyanin, which was, however, not observed in Bcl-2-transfected cells. These observations clearly demonstrate the phycocyanin-induced apoptotic death in BC-8 cells and its inhibition by Bcl–2 (Bobbili *et al.*, 2003).

β-Carotene

The β-carotene level in *Spirulina* at over 300 mg per cent is perhaps the highest content available from any other natural sources (Venkataraman, 1983). The β-carotene content of *Spirulina* is more than 18 times greater than carrot, which is the normally known source of β-carotene. Furthermore, the natural β-carotene of *Spirulina* is different from the synthetic β-carotene, since it contains a higher percentage of 9-cis-isomer as compared to over 47 per cent of all trans in the synthetic form. β-carotene as a dietary supplement has been reported to inhibit the development of DMBA-induced salivary gland carcinomas in rats and UV-induced skin cancer in hairless mice. Based on epidemiological studies, it has been suggested that β-carotene can also reduce the chances of occurrence of cancer in human (Normal and Temple 1988).

Amino Acids

Overproduction of amino acids is common feature of microbial resistance to analogs of amino acids. In general, the overproducing organisms are heterotrophic and production of metabolites is supported by the organic sources supplied with the medium. To our knowledge, such a phenomenon has been reported rarely for obligate photoautotrophs, a category which includes most cyanobacteria. Cyanobacteria liberate small quantities of amino acids into their medium (Fogg 1971; Fogg *et al.*, 1973). The most abundant amino acids liberated by cyanobacteria were phenylalanine, threonine, glutamate and glycine, irrespective of the nitrogen source available on which the cells were grown. Amino acid production was obtained with the mutant strains of *Anabaena variabilis,* which liberated alanine as the major abundant amino acid along with phenylalanine and tyrosine (Kerby *et al.*, 1987). Anolog-resistant mutants of *S. platensis* also produce the corresponding amino acids (Riccardi *et al.*, 1981). Continuous photobiological production of amino acids was achieved following immobilization of the mutant cells in calcium alginate gel (Kerby *et al.*, 1988). However, the effective isolation of strains for commercial production of amino acids is dependent on our knowledge of transport and metabolism of amino acids. Characterization of such strains can increase our basic knowledge thus permitting further strain improvement. Advances are being made in both fundamental and applied aspects by use of mutant strains of cyanobacteria. Recent advances in cyanobacterial molecular genetics can be expected to have a major impact in this field.

Ammonia

One of the biotechnological aspects receiving considerable interest, especially by using cyanobacteria, is the photobiological production of ammonia. Light-driven synthesis of ammonia at the expense of N_2 and water via photosynthesis is a photobiological process for the conversion of solar energy into chemical energy, namely ammonia which can be used as fertilizer. Since, cyanobacteria in natural conditions release limited or no fixed-N (ammonia), attempts are being made to generate strains which can fix N_2 at higher rate and release most of the fixed-N in the form of ammonia. In symbiosis about 50-90 per cent of the N_2-derived ammonia is liberated by the cyanobiont (Rai 1990). Attempts have also been made for the photobiological production of ammonia at the expense of N_2 and nitrate by using L-Methionine -DL-sulphoximine (MSX), a glutamate analogue and an irreversible inhibitor of GS (Vincenzini *et al.*, 1986; Singh and Rai 1989; Musgrave *et al.*, 1982; Guerrero *et al.*, 1982; Ramos *et al.*, 1987; Singh 1992, 1993). However, this approach seems to be impractical because MSX rapidly leads to deficiency of glutamine and other vital nitrogenous compounds with cessation of ammonia production and finally lysis of the cells. An alternative approach is to select mutant strains particularly deficient in GS activity which would liberate ammonia resulting from N_2 fixation into the medium. Mutant strains of *A. variabilis* and *Nostoc* ANTH fixed N_2 and liberated most of the fixed N_2 in the form of ammonia into the external medium (Singh 1992, 1993). Following immobilization in calcium alginate gel, sustained photoproduction of ammonia was obtained in reactors. The rate of ammonia production was comparable to the rate of its MSX-treated freely suspended wild type counterpart.

Fatty Acids

Fatty acids, in general, are of commercial value and many are pharmaceutical agents. Fatty acid of *Spirulina* possesses palmitic acid, γ-linolenic acid (GLA), linolenic acid and oleic acid predominantly. The concentration of GLA ranged between 8 to 31.7 per cent of the total fatty acids. Since, *Spirulina* is one of the best sources of GLA, a rare essential fatty acid and the key to good health, efforts are being made to produce GLA from other cyanobacterial strains.

Furthermore, the presence of essential fatty acids in *Spirulina* is very important. Essential fatty acids protect the cell membranes from destruction by short wave length electromagnetic radiations, carcinogens, bacteria and viruses (Horrobin 1990). Essential fatty acids decrease the aggregation of blood platelets and coronary thrombosis and also inhibit the salt-induced hypertension (Horrobin 1990). Essential fatty acids are also recommended for the prevention and treatment of multiple sclerosis (Horrobin 1990).

In diabetes mellitus, essential fatty acids reduce the diabetic retinopathy. Diabetes induces the capillary permeability and fragility whereas essential fatty acids increase the membrane intregrity. Because of their fortifying action in cellular membranes, essential fatty acids are important in the control of gastrointestinal, connective tissue and skin disorders (Horrobin 1990).

Vitamins

Although, a number of cyanobacteria are known to be rich source of vitamins and many are known even to excrete them into their milieu (Borowitzka and Borowitzka 1988), no report is, however, available on the commercial exploitation of these organisms for vitamin production. Some of the marine cyanobacteria appear to be the potential source of vitamin B complex group and vitamin E (Sundararaman 1992). *Spirulina* is considered to be the potential source of vitamin B_{12}, which is essential for healthy nerves and tissues especially for vegetarians (Henrikson 1997). Studies among preschool children in India have demonstrated *Spirulina fusiformis* to be an effective source of dietary vitamin A (Annapurna *et al.*, 1991).

Cyanobacteria as a Source of Polysaccharides

The exploitation of cyanobacteria as polysaccharides is emerging as new area of biotechnological application for the human welfare. The polysaccharides of microbial origin show advantages over the polysaccharides extracted from plants or macroalgae. The cyanobacteria have been considered for polysaccharide production due to their capability to excrete mucilaginous material with potential useful properties. In fact, most of the cyanobacterial strains possess additional surface structures, mainly of a polysaccharide nature, that comprise of sheaths, capsules and slimes. During cell growth in batch cultures aliquots of the polysaccharides material of both capsules and slimes may be released as water soluble material into the surrounding medium.

Bioactive Compounds

The discovery of new sources of bioactive natural products are of paramount importance to assure the supply of new drugs needed to treat those infectious diseases that have become resistant to currently available drugs and also to combat newly emergent diseases. Considerable efforts have been made in the past to identify interesting bioactive compounds from different sources including cyanobacteria (Table 32.1). The potential impact of this field on cyanobacterial biotechnology is immense (Patterson 1996). The wide array of cyanobacterial strains and the range of habitats that they occupy guarantee statistically to predict success in finding a valuable compound.

Table 32.1: Bioactive Compounds from Cyanobacteria (Burja *et al.,* 2001)

Organisms	*Compounds*	*Activity*
Microcystis aeruginosa	Aeruginosin, Kawaguchipeptin, microcystilide, microcystin micropeptin, microviridin	Enzyme inhibitor, cytotoxic, cell-differentiation, hepatotoxin, promoter, endotoxic, tumor antibiotic, anticancer
Microcystis viridis	Cyanoviridin RR	Toxic
Synechocystis trididemni	Didemnin	Anticancer, antiviral, Immunosuppressive
Lyngbya langerheimii	Sulfolipid	Anti–HIV activity
Lyngbya majuscula	Antillatoxin, aplysiatoxin, Aparamide, barbamide, Sulfolipid, lyngbyatoxin	Anticancer, anti-inflammatory, antiviral, anti-HIV, antimicrobial, cytotoxic, toxin
Oscillatoria acutissima	Acutiphycin	Anticancer activity
Oscillatoria agardhii	Agardhipeptin, microcystin	Enzyme inhibitor, hepatotoxin
Oscillatoria nigroviridis	Oscillatoxin	Anticancer
Phormidium tenue	Sulfolipid	Anti-HIV activity
Plectonema radiosum	Radiosumin, tubercidin	Enzyme inhibitor
Spirulina platensis	Calcium spirulan, poly-β-hydroxybutyrate, phycocyanin	Anticancer, anti-HIV activity, free-radical scavenger
Anabaena basta	Bastadin, bastadin *O*-sulfate esters	Antibiotic, anti-inflammatory, cytotoxicity
Anabaena circinalis	Circinamide, microcystin	Enzyme inhibitor, toxic
Aphanizomenon flos-aquae	Aphanorphine, siatoxin	Antibiotic, anticancer toxic
Nostoc commune	Nostodione, microsporine, diterpenoid	Antifungal, antibiotic, antimitotic, cytotoxic, suncreen pigment
Nostoc ellipsosporum	Cyanovirin	Anti-HIV, antiviral
Nostoc muscorum	Muscoride	Antibiotic
Nostoc sphaericum	Staurosporine, indolecarbazole	Antiviral, cytotoxic
Tolypothrix nodosa	Tolyporphin	Antibiotic
Tolypothrix tenuis	Toyocamycin, tubercidin	Antifungal, cytotoxic

Screening for a wide variety of potentially useful bioactivities, including cytotoxic, multi-drug resistance reversal, anti-fungal and antiviral effects, has led to the discovery and identification of numerous novel bioactive compounds or secondary metabolites possessing unprecedented structures. Structural novelty can be combined with exceptionally high incidence of biological activity.

Enzymes

Microbial enzymes have been used in various industries for many centuries. Recently, with the advent of biotechnology, there has been a growing interest and demand for enzymes with novel properties. Cell bound and extracellular phosphatase activities of cyanobacterial isolates have also been reported (Whitton *et al.*, 1991). A number of studies have been conducted on the detection of cyanobacterial enzymes including phosphatase, arylsulfatase, chitinase, L-asparginase, L-glutaminase, amylase, protease, lipase, cellulase, urease and lactamase (Chandrasekaran 1997). Freely-suspended and immobilized cells of *Anacystis nidulans* contain amino acid oxidase activity, which is further increased by illumination with red light (Wikstrom *et al.*, 1982). Superoxide dismutase exists in all aerobic organisms and has been purified from *Spirulina*.

Restriction Enzymes

Although more than 500 restriction enzymes with at least 100 different site specificities have been reported (Roberts 1985), enzymes with new specificities are continuously sought to increase the possibilities of DNA manipulation. Many species of cyanobacteria, both filamentous and unicellular, appear to be a promising source of such enzymes (Table 32.2).

Table 32.2: Important Restriction Enzymes Identified in Cyanobacteria (Ciferri *et al.*, 1989)

Organisms	*Enzymes*	*Isoshizomer of*	*Recognition Sequence*
Anacystis nidulans	Ani I	–	CCNNGG
Anabaena variabilis	Ava I	–	C ↓ PyCGPuG
	Ava II	—	G ↓ G(A/T)CC
	AvaIII	–	ATGCAT
Anabaena cylindrica	Acy I	–	Gpu ↓ CGPyC
Nostoc sp. 6705	Nsp BI	AsuII	TTCGAA
	Nsp BII	–	C(G/C)G ↓ C(T/G)G
Nostoc sp. 8009	NspMACI	BglII	A ↓ GATCT
Nostoc sp. 7413	NspHI	Nsp(7524)I	PuCATG ↓ Py
	NspHII	AvaII	GG(A/T)CC
Spirulina platensis	SplI	–	C ↓ GTACG
	SplII	TthIIII	GACNNNGTC
	SplIII	HaeIII	GGCC
Tolypothrix tenuis	TtnI	HaeIII	GGCC

Cyanobacteria contain different endonucleases in different combinations, some of which are isoschizomers even in unrelated strains. For example, the isoschizomers *Aqu* I and *Ava* I, that recognize and cleave the sequence CPyCGPuG are produced by a unicellular cyanobacterium *Synechococcus* and a filamentous cyanobacterium *Anabaena*. Similarly, besides other restriction endonucleases, *Fremyella* and *Nostoc* contain enzymes with the same specificity as *Ava* II (Whitehead and Brown 1985), the enzyme which was first identified in *Anabaena*. This is not unusual as the same restriction enzyme has been found in different bacterial genera (Roberts 1985). Such findings may indicate that some of the activities are plasmid or phage-borne, although, at the moment, there is no evidence correlating restriction activity and presence of plasmids or phages.

Immunomodulators

Spirulina is commercially produced for human consumption as well as for feed ingredients. Recently, whole *Spirulina* and cell extracts were found to enhance immunity in animals by increasing phagocytic activity. The whole cells of cyanobacteria and lipopolysaccharides (LPS) isolated from cyanobacterial cells were shown to stimulate the production of macro and micro globulin antibodies (Rathore *et al.*, 2004).

C-phycocyanin and polysaccharides isolated from *Spirulina* was found to possess high erythropoeitin (EPO) activity. Effect of these compounds on peripheral blood and bone marrow hematopoietic stem and progenitor cell in normal, irradiated and anemic mice showed the unique capacity of C-phycocyanin and polysaccharide to influence the differentiation and proliferation of committed hematopoietic progenitor cell. In addition to this, they also lower the anemic degree of mice combined with gamma-ray irradiation and Benz hydrazine hydrochloric acid and peritoneri injection (Zhang-Chang-Wu 1994).

Mice fed with *Spirulina* diet showed increased numbers of spleenic antibody producing cells in the primary immune response to sheep red blood cells (SRBC). *Spirulina* enhances the immune response, particularly the primary response by stimulating the macrophages function, phagocytosis and IL-1 production (Hayashi *et al.*, 1994).

Bronchoalveolar lavage macrophages isolated from cats, when cultured on a glass cover slip with water-soluble extracts of *S. platensis* in a concentration range of 0-60 mg ml^{-1} for 2 h could not cause significant macrophage cytotoxicity over untreated controls. The *S. platensis* extract enhances macrophage phagocytic fuction which implies that *Spirulina* supplementation might improve the disease resistance potential in cats (Qureshi *et al.*, 1996).

The exposure of chicken macrophage to *S. platensis* extract enhances selected effector functions of cells of the chicken immune system. However, the ability of spleenic natural killer cells to kill tumour cell targets was not affected by *Spirulina* treatment. Macrophage cultures exposed to *Spirulina* produced a factor in their culture supernatant with tumoricidal potential, which was similar in reactivity to the factor produced by macrophage following exposure to lipopolysaccharide (Qureshi *et al.*, 1995).

Chicken macrophage exposed to water soluble extracts of *Spirulina* showed enhanced phagocytic activity *in vitro*, suggesting activation of mononuclear phagocytic system function. Furthermore, dietary supplemention of *Spirulina* (1,000 to 10,000 ppm), improved thymic weights, enhanced CBH response, increased tumor cell killing by NK cells and doubled the macrophage phagocytic potential over chickens fed a basal diet. *Spirulina* enhances several immunological end points in chickens both during *in vitro* and in *in vivo* exposures (Qureshi *et al.*, 1995). *Spirulina* consumption lowers the amount of igE in the blood, which in turn normalizes and reduces allergies in the body without any side effect (Evets *et al.*, 1994). *Spirulina* also shows the immunomodulatory and biomodulatory effects (Rathore *et al.*, 2004).

Anticancer/Antitumer/Anti-Proliferative and Cytotoxic Compounds

Micro algae in general and cyanobacteria in particular have been identified as the most significant group of organisms from which new anticancer type natural products have been isolated (Moore *et al.*,

1988). Many of these bioactive compounds possess unprecedented structures. Structural novelty is combined with exceptionally high incidences of biological activity.

The extracts of some cultured cyanobacteria exhibited good anticancer activity (Barchi *et al.*, 1983). Ethanolic extract of cultured *Tolypothrix byssoideae* was found to show significant inhibitory activity *in vitro* against KB and NIH/3T3 cells and *in vivo* against P-388 lymphocytic leukemia in mice (Barchi *et al.*, 1983). Bioassay directed isolation and identification of the active constituent (tubericidin) have also been reported from *T. byssoideae* (Barchi *et al.*, 1983).

Bioassay directed fractionation of extract of *Phormidium tennue* led to the isolation of three classes of glycolipids *viz.* monogalactosyl diacylglycerols (MGDGs) as antitumor factors. Among 17 tested MGDGs and digalactosyl diacylglycerols (DGDGs), three of the DGDGs showed more potent inhibitory activity towards tumor proliferation than the others (Shirahashi *et al.*, 1993).

The extract of *Lyngbya majuscula* was found to have strong cytotoxic effect against a Vero cell line ATCC CCI 81 (Gerwick *et al.*, 1994). The extract was also found to be toxic to brine shrimp (LC 50=25ng ml^{-1}). Using the brine shrimp assay to guide fractionation, a unique metabolite was isolated (8-10 per cent of the crude extract) which was found to be responsible for the potent brine shrimp toxicity (LC 50=3ng ml^{-1}) and mammalian cell antiproliferative activity (IC 50=6.8ng ml^{-1}) in the Chinese hamster Aux B1 cell line.

Bioassay guided fractionation of the organic extract of *L. majuscula* led to the isolation of a new lipid, curacin A, with exceptional brine shrimp toxic and tumor antiproliferative activities. Pure curacin A is an antimitotic agent (IC 50 values in three cell lines ranging from 7-200 nm) that inhibits microtubule assembly and the binding of colchicines to tubulin (Gerwick *et al.*, 1994).

In discovering new efficient drugs for the treatment of solid tumors in clinical conditions, Corbett *et al.* (1992) developed a rapid inexpensive disk diffusion assay to find agents that exhibit greater cytotoxicity against tumour cells than leukemia cells. Aulosirazole, the major cytotoxin of *Aulosira fertillissima* shows solid tumor selective activity in the Corbett assay. Its structure has been determined to be 5-hydroxy-3 methoxinapthol (2,3-d) 1,2-Thiazole-4,9-dione (Startman *et al.*, 1994).

The recently discovered antiproliferative agents cryptophycin-A, curacin A (Gerwick *et al.*, 1994) and notodion A (Kobayashi *et al.*, 1990) produced by *Nostoc* sp., *L. majuscula,* and *Nostoc commune*, respectively, inhibit microtuble assembly function and show promise as potentially useful anticancer agents.

The extracts of *Spirulina-Dunaliela* in which β- carotene is a major component inhibited the growth and destroyed oral cancer cells *in vitro* (Schwartz *et al.*, 1988). β- carotene, a nontoxic carotenoid had a cytostatic and cytotoxic dose-dependent effect on hamster and human oral squamous cell carcinoma cell lines (Schwartz and Shklar 1987). The chemo preventative activity of *Spirulina* (1g day^{-1} for 12 months) in reversing oral leukoplakia in pan tobacco chewers in Kerala, India showed complete regression of lesions in evaluable subjects supplemented with *Spirulina* (Mathew *et al.* 1995).

Polysaccharide of *Spirulina* can inhibit the proliferation of ascetic hepatoma cells of mice at a concentration of 200 mg Kg^{-1}. It can inhibit the incorporation of H-thymidine, H-uridine and H-leucine into DNA, RNA and protein synthesis of sarcoma 180 and ascetic hepatoma cells during the period of 24 h after exposure *in vitro*. The degree of inhibition increases with the increase in incubation time. Polysaccharide of *Spirulina* can inhibit DNA synthesis of Sarcoma 180 and ascetic hepatoma cells. The inhibition is due to the alteration in DNA metabolism (Lisheng *et al.*, 1991).

An extract of *Spirulina-Dunaliella* was shown to prevent tumor development in hamster buccal pouch when a solution was applied topically three times weekly for 28 weeks (Schwartz *et al.*, 1988). Anti-cancer properties and antineoplastic activity have also been demonstrated using compounds from marine cyanobacteria (Murray and Mitsui 1982).

Antitoxin

Mercury induced kidney toxicity in rats was suppressed by feeding *Spirulina*. Para aminophenol and cisplatin induced renal toxicity in rats was significantly reduced by phycocyanin extract of *Spirulina*. (Fukino *et al.*, 1990). Thus, phycocyanin plays a major role in the protective effect of *Spirulina* against mercury and pharmaceutical drugs induced renal failure and that *Spirulina* might be applicable to the reduction of general renal dysfunction (Fukino *et al.*1990).

Radioprotection

The radioprotective effect of *Spirulina* extract was studied using the micronucleus test in polychromatic erythrocytes of bone marrow in mice. The extract caused a significant reduction of the micronucleus frequencies induced by gamma radiation (Qishen *et al.*, 1989).

Spirulina reduces urine radioactivity levels by 50 per cent within 20 days. Use of *Spirulina* also decreases radiation load received from food contaminated with radio nucleotides, Cesium-137 and Strontium-90. *Spirulina* is favourable for normalizing the adaptive potential of children's bodies in conditions of long lived low dose radiation (Loseva and Dardynskaya 1993). The intake of easily digestible micro and macro elements of *Spirulina* also has a positive influence on many functions of the immune system (Loseva 1999).

Reproductive Modulators

Carotenoids are used to improve the health and enhance the fertility of cattle (Jackson 1981). Crude ethanolic extract of *Oscillatoria willei* causes abnormal changes in the sperm of mouse (Thirunalasundari and Subramanian 2000). An immune potentiating property coupled with murine antifertility activity, without being toxic to other systems in mice models was exhibited by the extracts of *O. willei* (Krishnaprema 1996; Jagdeshwari 1998; Ansarudin 1999) and *O. late-virens* (Deth 1999).

Conclusion

Cyanobacteria are oxygenic, photosynthetic prokaryotes which have long been recognized as having potential in biotechnology as biofertilizers, amino acid production, fuel production, antibiotic production, single cell protein production and for the production of other secondary metabolites and pigments used in food industries. These interesting features have fascinated the Scientists all over the world to understand and explore in detail the basic mechanisms involved for the biotechnological exploitation of cyanobacteria for the betterment of mankind.

The traditional use of cyanobacteria including food, energy source and biofertilizers flourish today and are beginning to reap the benefits of modern molecular biology to enhance their performance. The first generation of high value added byproducts produced by cyanobacteria, including phycoflours are currently available in the market and is quite popular. However, only a few cyanobacterial strains producing intresting compounds have so far been exploited commercially. Screening programs must therefore identify useful products and the cyanobacterial strains having the ability to produce the products in significant amounts should be cultured in mass. Projection for future research on pharmaceutical potential of cyanobacteria includes discovery of novel bioactive metabolites, investigation of the pathways by which selected metabolites are synthesized and regulated and a precise elucidation of the mechanism by which these novel compounds exert their biological effects.

Acknowledgements

I am thankful to Dr. Narendra Kumar and Mr. Pawan Kumar for their help in preparation of the manuscript.

References

Annapurna, V., Shah, N., Bhaskaram, P., Bamji, M.S. and Reddy, V. (1991). Bioavailability of *Spirulina* carotene. *Journal of Clinical Biochemistry and Nutrition* (10). : 145–151.

Ansarudin, P.A. (1999). Bioactive compound from a marine cyanobacterium *Oscillatoria willei* BDU 130511. *M.Sc. Thesis*, Bharthidasan University, Tiruchirappalli, India.

Ayehunie, S., Belay, A., Baba, T.W. and Ruprecht, R.M. (1998). Inhibition of HIV-1 replication by an aqueous extract of *Spirulina platensis* (*Arthrospira platensis*). *Journal of Acquired Immune Deficiency Syndromes and Human Retrovirology* (18): 7-12.

Barchi, J.J., Norton, T.R., Furusawat, E., Patterson, G.M. and Moore RE (1983). Identification of a cyatotoxin from *Tolypothrix byssoideae* as tubericidin. *Phytochemistry* (22). : 2851-2855.

Becker, E.W. (1994). Microalgae. In: *Biotechnology and Microbiology*. Cambridge University Press, Cambridge, U.K.

Bhat, V.B. and Madyastha, K.M. (2001). Scavening of peroxynitrite by phycocyanin and phycocyanobilin from *Spirulina platensis*: protection against oxidative damage to DNA.. *Biochemical and Biophysical Research Communications* (285): 262-266.

Bloor, S. and England, R. R. (1991). Elucidation and optimization of the medium constituents controlling antibiotic production by the cyanobacterium *Nostoc muscorum*. *Enzyme Microbial Technology* (13): 76-81.

Bobbili, V. V., Pardhasaradhi, A., Mubarak Ali, A., Kumari, L., Reddanna, P. and Khar, A. (2003). Phycocyanin-mediated apoptosis in AK-5 tumor cells involves down-regulation of Bcl-2 and generation of ROS. *Molecular Cancer Therapy* (2): 1165–1170.

Bonjouklian, R., Smitra, T.A., Doolin, L.E., Molly, R.M., Debono, M., Shaffer, S.A., Moore, R.E., Stewart, J.B. and Patterson, G.M.L. (1991). Tjipanazoles, new antifungal agents from the blue-green alga *Tolypothrix tjipanasensis*. *Tetrahedron*. (47): 7739-7740.

Borowitzka, M.A. and Borowitzka, L.J. (1988). *Microalgal Biotechnology*. Cambridge University Press, Cambridge, U.K.

Boyd, M.R. (1988). Strategies for the identification of new agents for the treatments of AIDS: A national program to facilitate development of new drug candidates for clinical evaluation. In : *AIDS Etiology, Diagnosis, Treatment and Prevention*. (Eds. V.J. De Vita Jr., S. Hillman and S.A. Rosenberg). pp. 305-315, Philadelphia, Lippincott.

Burja, A. M., Banaigs, B., Mansour, E.A., Burgess, J.G. and Wright, P.C. (2001). Marine cyanobacteria–a prolific source of natural products. *Tetrahedron* (57). : 9347-9377.

Chandrasekaran, M. (1997). Industrial enzymes from marine microorganisms: The Indian Scenario. *Journal of Marine Biotechnology* (5): 86-89.

Chauhan, V. S., Singh. B., Singh, S. and Bisen, P.S. (2004). Screening of cyanobacterial strains producing cyanovirin–N and cyanovirin–V, protein capable of inhibiting growth of HIV-I and HIV-II. In: *Proceedings of the International Conference on Recent Advances in Biomedical and Therapeutic Sciences*, pp. 247, Bundelkhand University, Jhansi, India.

Ciferri, O., Tiboni, O. and Sanangelantoni, A.M. (1989). The genetic manipulation of cyanobacteria and its potential uses. In: *Algal and Cyanobacterial Biotechnology*. (Eds. R.C. Cresswell, T.A.V., Rees and N. Shah). pp. 239-262, Longman Group, UK.

Cohen, Z. (1986). Products from microalgae In: *CRC Handbook of Microalgal Mass Culture*. (Eds. A. Richmond). pp. 421-454, CRC Press, Boca Raton, Florida, USA.

Corbett, T.H., Valeriote, F.A., Polin, L.C., Pugh, S. and White, K. (1992). Models and concepts for discovery and development. In : *Cytotoxic Anticancer Drugs*. (Eds. F.A. Valeriote, T.H. Corbett and L.H. Baker). pp. 35-41, Kluwer Academic Publishers. Norwell.

Dainippon Ink and Chemicals Inc. (1981). Cosmetics containing water soluble phycocyanin. *Japanese Patent* (79). : 138755.

Dainippon Ink and Chemicals Inc. (1980). Production of highly purified alcoholophilic phycocyanin. *Japanese Patent* (80). : 77890.

Deth, S.K. (1999). Antimicrobial compound from marine cyanobacteria with special reference to the bioactivity of a purified compound from *Oscillatoria latevirens*. *Ph.D. Thesis*, Bharthidasan University, Tiruchirappalli, India.

Evets, L.B., Belookaya, T., Lyalikov, S., Orehov, S.D. and Shipulin, E. (1994). Means to normalize the levels of immnoglobulin E. Russian Federation Committee of Patents and Trade. Patent Number (19). RU (11)20005486 C1 (51). 5 A 61K35/80.

Fogg, G.E. (1971). Extracellular products of algae in freshwater. *Archives of Hydrobiology* (5): 1-25.

Fogg, G.E., Stewart, W.D.P., Fay, P. and Walsby A.E. (1973). *The Blue Green Algae*. Academic Press, London.

Fukino, H., Takagi, Y. and Yamane, Y. (1990). Effect of *Spirulina* (*S. platensis*). on the renal toxicity induced by inorganic mercury and cisplatin. *Eisei Kagaku* (36): 5.

Gerwick, W.H., Mrozek, C., Moghaddam, M.F. and Agrawal, S.K. (1989). Novel cytotoxic peptides from the tropical marine cyanobacterium *Hormothamnion enteromorpoides*. 1. Discovery, isolation and initial chemical and biological characterization of the hormothamnins from wild and cultural material. *Experientia* (45). : 115-117.

Gerwick, W.H., Proteau, P.J., Nagle, D.G., Hamel, E., Blokhin, A. and Slate, D.I. (1994). Structure of curacin A, a novel antimitotic, antiproferative and brine shrimp toxic natural product from the marine cyanobacterium *Lyngbya majuscula*. *Journal of Organic Chemistry* (59): 1243-1247.

Guerrero, M.G., Ramos, J.L. and Losada, M. (1982). Photosynthetic production of ammonia. *Experientia* (38): 53-55.

Gustafson, K.R., Cardellina, J.H. II, Fuller, R.W., Weslow, O.S., Kiser, R.F., Snader, K.M., Paterson, G.M. and Boyd, M.R. (1989). AIDS-antiviral sulfolipids from cyanobacteria (blue-green algae). *Journal of National Cancer Institute* (81): 1254-1258.

Hayashi, K., Hayashi, T. and Morita, N. (1993). An extract from *Spirulina platensis* is a selective inhibitor of *Herpes simplex* virus type 1 penetration into HeLa cells. *Phytotherapy Research* (7): 76-80.

Hayashi, K., Hayashi, T. and Kojima, I. (1996). A natural sulfated polysaccharide, calcium spirulan, isolated from *Spirulina platensis*: *in vitro* and *ex vivo* evaluation of anti-*Herpes simplex* virus and anti-human immunodeficiency virus activities. *AIDS Research and Human Retroviruses* (12): 1463-1471.

Hayashi, O., Katoh, T. and Okuwaki, Y. (1994). Enhancement of antibody production in mice by dietary *Spirulina platensis. Journal of Nutritional Science and Vitaminology* (40): 431-441.

Horrobin, D.F. (1990). *Omega-6 Essential Fatty acids : Pathophysiology and Roles in Clinical Medicine.* Wiley-Liss, New York.

Ishibashi, M., Moore, R., Patterson, G.M.L., Xu, C. and Clardy, J. (1986). Scytophycins: cytotoxic and antimycotic agents from the cyanophyte *Scytonema pseudohojmanni. Journal of Organic Chemistry* (51): 5300-5305.

Jackson, P.S.A. (1981). Note on a possible association between plasma β-carotene levels and conception rate in a group of winter housed dairy cattle. *Animal Production* (32): 109-112.

Jassby, A., Lembi, C.A. and Waaland, J.R. (1988). *Algae and human affairs.* Cambridge University Press, Cambridge, UK.

Jagdeshwari (1998). Effect of purified compound from *Oscillatoria willei* BDU 130511 on the testis and accessory reproductive organs of swiss albino mice. *M.Sc. Thesis*, Bharthidasan University, Tiruchirappalli, India.

Kerby, N.W., Niven, G.W., Rowell, P. and Steward W.D.P. (1987). Photoproduction of amino acids by mutant strains of N_2- fixing cyanobacteria. *Applied Microbiology and Biotechnology* (25): 547-552.

Kerby, N., Niven, G., Rowell, P., Steward, W.D.P. (1988). Ammonia and amino acid production by cyanobacteria. In: *Algal Biotechnology.* (Eds T. Stadler, J. Mollion, M.C.Verdus, Y. Karamanos, H. Morvan and D. Christiasen). pp. 277-283, Elsevier Applied Science, London, New York.

Kobayashi, J., Ishibashi, M., Murayama, J., Takamatsu, M., Iwamura, M., Ohizumi, Y. and Sasaki, T. (1990). Amphidinolide-E, a novel antileukemic 19-membered microlide from the cultured symbiotic dinoflagellate *Amphidinium* sp. *Journal of Organic Chemistry* (55): 3421-3425.

Krishnaprema, K. (1996). Effect of the crude ethanolic extract of *Oscillatoria willei* BDU 130511 on male BALB C mice. *M.Sc. Thesis*, Bharthidasan University, Tiruchirappalli, India.

Lisheng, L., Baojiang. G., Jihong. R., Guangquan, Q. and Botang, Wu (1991). Inhibitive effect and mechanism of polysaccharide of *Spirulina platensis* on transplanted tumor in mice. *Marine Science* (5): 33-38.

Loseva, L.P. and Dardynskaya, I.V. (1993). *Spirulina*- natural sorbent of radionucleides. In : *Proceedings of 6th International Congress of Applied Algology*, Czech Republic, Belrus.

Loseva, L.P. (1999). *Spirulina patensis* and specialties to support detoxifying pollutants and to strengthen the immune system. In : *Proceedings of 8th International Congress of Applied Algology*, Italy, Republic, Belrus.

Madhava, C., Reddy, J., Bhat, V. B., Kiranmai, G. M., Reddy, S., Reddanna, P. and Madyastha, K. M. (2000). Selective Inhibition of cyclooxygenase-2 by C-phycocyanin, a biliprotein from *Spirulina platensis. Biochemical and Biophysical Research Communications* (277): 599-603.

Mathew, B., Sankaranarayanan, R., Nair, P.P., Varghese, C., Somanathan, T., Amma, B.P., Amma, N.S. and Nair, M.K. (1995). Evaluation of chemoprevention of oral cancer with *Spirulina. Nutrition and Cancer* (24). : 197-202.

Mohanthy, P. (1996). Cyanobacterial photosynthesis: concepts and applications. *Journal of Scientific and Industrial Research* (55). : 553-570.

Moore, R.E. (1981). Constituents of blue- green algae. In: *Marine Natural Products : Chemical and Biological Perspectives* (Eds. P.J. Scheuer). Academic Press, New York.

Moore, R.E., Patterson, G.M.L. and Carmichael, W.W. (1988). New pharmaceuticals from cultured blue green algae. In : *Biomedical Importance of Marine Organisms*. (Eds. D.G. Fautin). pp. 143-146, California Academy of Sciences.

Murray, R.L. Jr and Mitsui, A. (1982). Growth of hybrid Tilapia fry fed nitrogen fixing marine blue green algae in sea water. *Journal of World Mariculture Society* (13): 198-203.

Musgrave, S.C., Kerby, N.W., Codd, G.A. and Stewart, W.D.P. (1982). Sustained ammonia production by immobilized filaments of the nitrogen-fixing cyanobacterium *Anabaena* 27893. *Biotechnology Letters* (4): 647-52.

Normal, J. and Temple, I. (1988). Does β-carotene prevents cancer : a critical appraisal. *Nutrition Research* (8): 685-690.

Patterson, G.M.L., Baker, K.K., Baldwin, C.L., Bolis, C.M., Caplan, F.R., Larson, L.K., Levine, I.A., Moore, R.E., Nelson, C.S., Tschappat, K.D. and Tuang, G. (1993). Antiviral activity of cultured blue-green algae (cyanophyta). *Journal of Phycology* (29): 125-130.

Patterson, G. M. L. and Carmeli, S. (1992). Biological effects of toly-toxin (6-hydroxy-7-O-methyl-scytophycin B), a potent bioactive metabolite from cyanobacteria. *Archives of Microbiology* (157): 406-411.

Patterson, G. M. L. (1996). Biotechnological application of cyanobacteria. *Journal of Scientific and Industrial Research* (55). : 669-679.

Puniya, A.K., Singh, S., Ganeshkumar, C. and Singh, K. (1995). Single cell protein : a promising dietary substitute. *Indian Journal of Experimental Biology* (33). : 545-551.

Qishen, P., Baojiang, G. and Kolman, A. (1989). Radioprotective effect of extract from *Spirulina platensis* in mouse bone marrow cells studied by using the micronucleus test. *Toxicology Letters* (48): 165-169.

Qureshi, M. and Ali, R. (1996). *Spirulina platensis* exposure enhances macrophage phagocytic function in cats. *Immunopharmacology and Immunotoxicology*. (18). : 457-463.

Qureshi, M.A., Kidd, M.T. and Ali, R.A. (1995). *Spirulina platensis* extract enhances chicken macrophage functions after *in vitro* exposure. *Journal of Nutritional Immunology* (3): 35-44.

Qureshi, M.A., Garlich, J. and Kidd, M. (1996). Dietary *Spirulina platensis* enhances humoral and cell-mediated immune function in chickens. *Immunopharmacology and Immunotoxicology* (18): 465-476.

Rai, A. N. (1990). *Handbook of Cyanobacteria*. CRC Press Boca Raton, Florida, USA.

Ramirez, D., Gonzalez, R., Merino, N., Rodriguez, S. and Ancheta, O. (2002). Inhibitory effects of *Spirulina* in zymozan-induced arthritis in mice. *Mediators of Inflammation*. (11): 75-79.

Ramos, J. L., Guerrero, M.G. and Losada, M. (1987). Factors affecting the photoproduction of ammonia from dinitrogen and water by the cyanobacterium *Anabaena* sp. Strain ATCC 33047. *Biotechnology and Bioengineering* (29): 566-571.

Rathore, N.K., Singh, S. and Singh, V. (2004). *Spirulina* as immuno- enhancer and biomodulator. *Journal of Immunology and Immunopathology* 6 (2): 1-10.

Riccardi, G., Sora, S. and Cifferi, O. (1981). Production of amino acids by analog-resistant mutants of the cyanobacterium *Spirulina platensis*. *Journal of Bacteriology* (147): 1002-1007.

Rinehart, K. L. Jr, Shaw, P. D., Shield, L. G., Gloer, J. B., Harbour, G. C. and Koker, M. E. S. (1981). Marine natural products as sources of antiviral, antimicrobial and antineoplastic agents. *Pure and Applied Chemistry* (53). : 795-799.

Reichelt, J. L. and Borowitzka, M.A. (1984). Antibiotics from algae: results of a large scale screening programme. *Hydrobiologia* (116/117). : 143-147.

Roberts, R. J. (1985). Restriction and modification enzymes and their recognition sequences. *Nucleic Acids Research* (13): 2165-2200.

Schwartz, J. and Shklar, G. (1987). Regression of experimental hamster cancer by beta-carotene and algae extracts. *Journal of Oral Maxillofacter Surgery* (45). : 510-515.

Schwartz, J., Shklar, G., Reid, S. and Trickler, D. (1988). Prevention of experimental oral cancer by extracts of *Spirulina-Dunaliela* algae *Nutrition and Cancer* (11): 127-134.

Schlegel, I., Thanh Doan, N., Nola de, C. and Smith, G. D. (1999). Antibiotic activity of new cyanobacterial isolates from Australia and Asia against green algae and cyanobacteria. *Journal of Applied Phycology* (10). : 471-476.

Shih, S. R., Tsai, K. N., Li, Y. S., Chueh, C. C. and Chan, E. C. (2003). Inhibition of enterovirus 71-induced apoptosis by allophycocyanin isolated from a blue-green alga *Spirulina platensis*. *Journal of Medical Virology* (70): 119–125.

Shirahashi, H., Murakami, N., Watanabe, M., Nagatsu, A., Sakakibara, J., Tokuda, H., Nishino, H. and Iwashima, A. (1993). Isolation and identification of antitumour promoting principle from freshwater cyanobacterium *Phormidium tunne*. *Chemical and Pharmaceutical Bulletin* (41): 1664-1667.

Singh, S. (1992). Photobiological production of ammonia by an L-methionine-DL-Sulfoximine -resistant mutant of the cyanobiont *Nostoc* ANTH. *Indian Journal of Experimental Biology* (30): 284-287.

Singh, S. (1993). Immobilization-caused physiological and biochemical changes in the cyanobiont *Nostoc* ANTH. *Indian Journal of Experimental Biology* (31): 446-449.

Singh, S. and Rai, A.K. (1989). Methionine sulphoximine induced changes in the cyanobacteria *Anabaena doliolum* and *Anacystis nidulans*. *Indian Journal of Experimental Biology* (27): 227-229.

Startman, K., Belli, J., Jensen, C.M., Moore, R.E. and Patterson, G.M.L. (1994). Aulosirasole, a novel solid tumor selective cytotoxin from the blue green alga *Aulosira fertilissima* *Journal of Organic Chemistry* (59): 6279-6285.

Sundararaman, M. (1992). Intra and extra cellular substances in marine cyanobacteria. *Ph.D. Thesis*, Bharathidasan University, Tiruchirappalli, India.

Thirunalasundari, T. and Subramanian, G. (2000). Antifertility compound from marine cyanobacterium. In: *Proceedings of International Conference of Reproductive Biology*. 9th French- Czech- Slovak Colloquium. Slovak Academy of Sciences, Stara Lesna.

Upasani, C.D., Khera, A. and Balaraman, R. (2001). Effect of lead with vitamin E, C or *Spirulina* on malondialdehyde, conjugated dienes and hydroperoxides in rats. *Indian Journal of Experimental Biology* (39): 70-74.

Venkataraman, L.V. (1983). *Spirulina* in India. In : *Proceedings of National Seminar on Cyanobacterial Research-Indian Scene* (Eds. G. Subramanian). pp. 92, National Facility of Marine Cyanobacteria, Tiruchirappalli.

Vincenzini, M., De Philippis, R., Ena, A. and Florenzano, G. (1986). Ammonia photoproduction by *Cyanospira rippkae* cells "entrapped" in dialysis tube. *Experientia* (42): 1040-43.

Whitehead, P.R. and Brown N.L. (1985). Three restriction endonucleases from *Anabaena flos-aquae*. *Journal of General Microbiology* (131): 951-958.

Whitton, B. A., Grainger, S. L. J., Hawley, G. R. W. and Simon, J. W. (1991). Cell bound and extracellular phosphatase activities of cyanobacterial isolates. *Microbial Ecology* (21): 85-98.

Wikstrom, P., Szwajcer, E., Brodelius, P., Nilsson, K. N. and Mosbach, K. (1982). Formation of alpha keto acids from amino acids using immobilized bacteria and algae. *Biotechnology Letters* (4): 153-156.

Zhang-Chang-Wu (1994). Effects of polysaccharides and phycocyanins from *Spirulina* on peripheral blood and haemopoietic system of bone marrow in mice. In: *Proceedings of Second Asia Pacific Conference on Algal Biotechnology*. pp. 58, University of Malasiya.

Microbial Biotechnology and Ecology (2011) *Pages* **571–576**
Editors: **Deepak Vyas, G.S. Paliwal, P.K.Khare and R.K Gupta**
Published by: **DAYA PUBLISHING HOUSE, NEW DELHI**

Chapter 33

Plants and Rhizobium Techniques

☆ *Pooja Agrawal*

Department of Applied Microbiology and Biotechnology,
Dr. H.S. Gour University, Sagar – 470 003, M.P.

Soils are the excellent culture media for the growth of different groups of microorganisms *e.g.* bacteria, virus, fungi, algae and actinomycetes. The number and kinds of soil microorganism found in a habitat, depends upon its nature, chemical composition, temperature, pH etc.

Maximum microbial growth and activity in the soil is found around the roots of plants (Hiltner, 1904). This region is called *Rhizosphere.* Practically all the interactions such as symbiosis, syntrophism, antagonism, synergism and commensalisms between plants and microorganisms and among different microorganisms are operative in this region. Their interactions may be favourable, indispensable, unfavourable or can be lethal. The most important functions of soil microflora is to decompose various kinds of organic matter and to mineralize them into inorganic/mineral constituents.

Some soil microorganisms are supporting the plant growth and are called "plant growth promoting rhizobacteria" or PGPR. These bacteria are generally found in symbiotic association with the leguminous plants (Pea, Gram, Mung, Soybean, Arhar and many other) and possess ability to fix atmospheric nitrogen.

The root nodule symbiosis with leguminous plant is generally developed by *Rhizobium, Bradyrhizobium* and *Azorhizobium* belonging to *Rhizobiaceae* family (Elkan, 1984). Bousingault (1834) proved that nodules on legume roots are responsible for fixing atmospheric nitrogen. Based on our knowledge of known leguminous plants it has been proved that not all the legumes bear nodules on their roots, only 10-12 per cent of legumes has been so far examined for nodulation. Beside this, some of the non-leguminous plants also bear nodule and has the ability to fix atmospheric nitrogen *e.g. Parasponia* (Trinick, 1973). *Rhizobium* is the well known genus known for fixing atmospheric nitrogen in symbiotic association with leguminous plants by way of forming nodules on the plant roots. Apart from legumes nodulated by genus *Rhizobium*, roots of some plants belonging to different families also nodulated by member's of actinomycetes classified in the genus, *Frankia*.

The *Rhizobium* is the most exploited genus of bacteria known for the development of bioinoculant. The genus has 7 known species *i.e. R. leguminosarum, R. phaseoli, R. trifoli, R. japonicum, Rhizobium* spp., *R. meliloti, R. lupini* forming cross inoculation groups with Pea group, Bean group, Clover group, Soyabean group, Cowpea group, Alfalfa group, Lupini group. (The cross-inoculation groups are based on the classical studies of Fred *et al.*, 1932). The principal of cross inoculation grouping is based on the ability of an isolate of *Rhizobium* to form nodules in a limited genera of species of legumes *i.e. Pisum, Phaseolus, Trifolium, Glycine, Vigna, Arachis, Melilotus, Lupinus* related to one another.

Nodulation and Mechanism of Nitrogen Fixation

The nodules formed on plants are agronomically significant as they provide an alternative to the use of expensive ammonium fertilizer's. Symbiotic nitrogen fixation takes place in a specialized structure called nodule that originate in a root as a consequence of microsymbiont infection. During nodulation the bacteria are transformed into bacteriods which are able to carry out transformation of nitrogen to ammonia ($N_2 \rightarrow NH_3$) later the ammonia is incorporated to the plant metabolism, is called nitrogen metabolism (Dilworth, 1991).

Symbiotic nitrogen fixation is that which occurs by the symbiosis of both plant and bacteria *e.g. Rhizobium* that can fix atmospheric nitrogen only when it gets associated with roots of leguminous plants. Some microorganisms are also capable of fixing nitrogen non-symbiotically *i.e. Azotobacter, Cynobacteria, Azospirullum and Clostridium.*

The reduction of nitrogen to ammonia requires a valency change *i.e.* from initial oxidation state 0 to –3 oxidation state. Six electrons are required to reduce one mole of nitrogen to Ammonia.

$$N_2 + 6H^+ + 6e^- \longrightarrow 2NH_3$$

This requires the break up of nitrogen bonds. It is agreed that the atoms of oxygen are separated in the biological oxidation state through the change in the valency of metal ion (Fe^{++}) in the enzyme, Cytochrome oxidase and that the atoms of nitrogen are separated through the change in the valency of metal ion Molybdenum (Mo^{+2}), bound to the enzyme involved in the reduction of nitrogen. The main enzyme involved in the nitrogen fixation is nitrogenase, it binds to nitrogen and converts it into free ammonium ions. It consist of two component, component I consist of Mo-Fe protein while component II consist of Fe protein, combination of two component reconstitute nitrogenase activity. If no suitable electron acceptor is available nitrogenase forms molecular hydrogen which generates ATP and consequently improves the efficiency of nitrogen fixation (Pawar and Dagniwala, 2002). While moving on to the genetic level and searching for the genes required for nodulation, *Rhizobium* species that nodulate the legumes were studied and it was proposed that genes for nodulation (nod) genes and genes for nitrogen fixation (nif) are clustered on large symbiotic plasmid. Nif genes are found in clustered form in a particular region of DNA. They are not found in scattered form (Roberts, 1978). The genes represent host range specificity.

Isolation and Characterization of *Rhizobium*

Beijernick (1888) was the first to isolate and cultivate a microorganism from the nodules of leguminous plants. He named it as *Bacillus radicicola*. Later on Frank in 1889 renamed the nodulated bacteria as *Rhizobium*. It is now placed in Bergey's manual of determinative bacteriology under the genus *Rhizobium*. *Rhizobium* falls under Rhizobiaceae family which consist of four genera, two of which include the symbiotic nitrogen fixing bacteria *Rhizobium* and *Bradyrhizobium*. *Rhizobia* is a non spore forming free living bacteria found in soil. Bergey's manual described rhizobia as cells without endospore, normally rod shaped, motile, aerobic and gram negative.

Rhizobia contains one polar, subpolar or peritrichous flagella. It produces extracellular slime on carbohydrate containing media. It can survive at low temperature and can tolerate temperatures upto 50°C. Rhizobia cells are generally bacilli, some times coccobacilli and formed poly-β-hydroxy granules which are also visible under the light microscope.

Yeast Extract Mannitol Agar (YEMA) is the specific medium for the culture of *Rhizobium* (Date, 1974). Vitamins such as thiamine and biotin are required for *Rhizobium* growth. *Rhizobium* produces white, translucent, glistening, elevated and comparatively smaller colonies with entire margin.

Method of Isolation

Collection of Roots of the leguminous plants will be done carefully. The process of isolation of *Rhizobium* from root nodules includes many steps which are described in the following text.

Surface Sterilization of Nodules

Healthy, unbroken, large and pinkish nodules should be preferred from the plant roots after thorough washing with the tap water. Nodules will be washed with sterile water. The nodules will then immersed in 3 per cent H_2O_2 for about 5 minutes to surface sterilize them. There after the nodules will washed in sterilized water to get rid of sterilizing agent. The nodules will then placed in 70 per cent ethyl alcohol for 2-3 minutes and then again washed with distilled water, these nodules will crushed with the help of glass rod in 1 ml of sterile distilled water. This suspension will then used for the isolation of bacteria using YEMA medium having congo red.

Preparation of Medium

Yeast extract mannitol agar (YEMA) with congo red (as differential reagent) should prepared using following ingradients as suggested by Subbarao (1995).

Mannitol	10.0 gm
K_2HPO_4	0.5 gm
$MgSO_4 . 7H_2O$	0.2 gm
NaCl	0.1 gm
Agar	20.0 gm
Yeast extract	1.0 gm
Distilled water	1000 ml
Congo red 1 per cent solution	2.5 ml

The media without congo red will be autoclaved at 121°C for 20 minutes. 1 per cent cango red solution will be prepared and sterilized separately. 2.5 ml of this will be added to 1000 ml sterile medium before pouring into the petriplates.

Isolation of Bacteria

The isolation of *Rhizobium* will be carried out by serial dilution of the nodule suspension followed by a streak plate method (Waksman, 1927; Vincent, 1970). Serial dilutions of the nodule suspension have to prepare in sterile distilled water. The suspension will dilute upto 10^{-5} and this dilution was used for streaking. The plates having YEMA medium with congo red will streaked with a loopful suspension using disposable plastic loops. The plates were then incubated at 37°C and were observed after 24 hours for bacterial growth. The colonies with typical *Rhizobium* growth (small, white, translucent, glistening) were picked up on a separate medium for further examination.

Purification

All the colonies picked up for the isolation of *Rhizobium* were purified. Purification was done by adopting a streak plate technique in which a particular colony will be picked up from the streaked plates showing bacterial colonies, using a loop and then streake onto a plate containing YEMA with congo red (Waksman, 1927). These plates are then incubated at 37°C to obtain the pure culture or specific colonies of *Rhizobium*.

Preservation

The purified cultures obtained during the isolation will be preserved and maintained on agar slants having YEMA medium. These should keep in a refrigerator at 4°C for future use.

Morphological Characterization

To study the morphological characteristics, each isolate was studied for gram staining and carbol fuchsin staining. The procedure adopted for gram staining and carbol fuchsin staining are as follows:

Gram's Staining

The gram stain procedure is the most widely used differential staining procedure in bacteriology. The gram positive bacteria appears violet since they retain the primary stain and gram negative bacteria appears red or pink.

The stained slide was then examined under the oil immersion objective to determine which organism is gram positive and which is gram negative.

Carbol Fuchsin Staining

A loop full of the suspension of selected bacterium is spread on a microscopic slide in a drop of water and allowed to dry in air. The slide is then passed through the top position of the flame of a Bunsen burner to heat fix the bacteria. The slide is then allowed to cool, and then flooded with dilute carbol fuchsin stain for 10-20 seconds. The slide is then washed in running water and allowed to dry. It is then examined under microscope for the presence or absence of poly-β-hydroxy butyrate granules.

Lactose-Agar Test

This test is used to differentiate *Agrobacterium* from *Rhizobium*. *Agrobacterium* utilize lactose to form a reduced product ketolactose, when Benedict's reagent is poured over Agar medium containing lactose on which nodule bacteria are growing, the development of yellow colour due to Cu_2O around the colony indicates the presence of *Agrobacterium*. The medium used was having following composition:

Lactose	10.0 gm
K_2HPO_4	0.5 gm
$MgSO_4.7H_2O$	0.2 gm
NaCl	0.1 gm
Yeast	1.0 gm
Distilled water	1000 ml
Agar	20.0 gm

Citrate Utilization Test

This Test is used to differentiate bacteria on the basis of ability to utilize citrate or ferment citrate as a sole carbon source. The utilization of citrate depends on the presence of an enzyme citrate permease produced by the organism that helps its transport into the cell. The composition of the medium used for testing the citrate utilization was as follows:

K_2HPO_4	1 gm
$NH_4H_2PO_4$	1 gm
NaCl	5 gm
Sodium citrate	2 gm
Magnesium sulphate	0.2 gm
1.5 per cent Bromothymol blue	10.0 ml
Distilled water	1000 ml
Agar	20 gm
pH	6.8

Root nodulating bacteria (*Rhizobium*) has been classified based on their growth on definite substrate as (i) Fast grower's and (ii) Slow grower's (Elkan, 1984). Fast grower include rhizobia associated with clover, alfalfa, pea and bean while slow grower include soybean and cowpea rhizobia. Fast grower are acid producing (Vincent, 1970) contains peritrichous flagella and generally called as *Rhizobium* while slow grower's are alkali producing, contain sub-polar flagella and are generally called as *Bradyrhizobium*. *Rhizobium* consist of five species *R. leguminosarum, R. meliloti, R. loti, R. galegae* and *R. fredii.* (Elkan, 1984).

A new genus called *Azorhizobium* has also been officially recognized (Dreyfus *et al.*, 1988). This bacterium was first isolated from the stem nodules of *Sesbania rostrata* and was found to be phenotypically and genotypically different from the other two genera *i.e., Rhizobium* and *Bradyrhizobium. Azorhizobium* consist of the species *A. caulinodans* which falls into taxa different from slow and fast growing *Rhizobium.* A unique photosynthetic *Rhizobium* is also known and named as *Photorhizobium* (Hardy, 1989).

Mass Production of *Rhizobium* Bioinoculant

Rhizobial broth is grown on YEM medium in large flasks on a shaker for small requirements or in fermenters for large requirements. The powdered carrier (passing through 100 mesh sieve) is neutralized with $CaCO_3$ and autoclaved at 15 lbs pressure for 4 hours. After cooling, a high count broth is mixed so as to attain a 40 per cent moisture holding capacity of the carrier. The carrier and the broth are mixed either manually or by means of a mechanical mixer, cured in trays for 2-5 days and packeted in polythene bags.

References

Beijernick, M.W. (1888). Die bacterien der popilionaceen knollchen. *Bot. Ztg*, 46 : pp. 726-804.

Bergey's Manual of discriminative bacteriology Vol. 1 (1984). pp. 261-267.

Boussingault, J. B. (1834). The nitrogen cycle. *Proceedings of American Philosophical Society*, Vol. 114, No. 6, pp. 435-479.

Date, R.A. (1974). Principles in Rhizobium strain selection. In: *Symbiotic Nitrogen Fixation in Plants*. Ed. P.S. Nutman Cambridge University, pp. 137-150.

Dilworth, H.J. (1991). Biology and biochemistry of nitrogen fixation. *Biochem. Biophys*, pp. 127, 285-294.

Dreyfus, B.L. (1981). Nitrogen fixing nodules induced by Rhizobium on stem of tropical legumes *S. rostrata. FEMS Microbial, Lett.*, pp. 313-317.

Elkan, G.H. (1984). Taxonomy and metabolism of Rhizobium and its genetic relationship in biological. *Nitrogen*, pp. 1-38.

Fred, E.B. Baldwin, I.L. and McCoy, E. (1932). *Root Nodule Bacteria and Leguminous Plants*. Univ. Wisconsin, Madison.

Hardy, R.W.F., Burns R.C and Parshall, G.W. (1971) . The biochemistry of nitrogen fixation. In: *Advances in Chemistry*, pp. 218-247.

Hellriegel, H. and Wilfarth, H. (1896). Die Lendwirtchaftlichen versuchs stationen Bd.46, S- 1-8. (Mil Bild Und verzeichris der wichtingsten vero foentlichugen).

Hiltner, L. (1904). Iifer neuere Erfahrungen iind probleme aufodem gebiet der Bodenbakteriologie iind unter besonderer Berucksichtingung der Griindiinging iind Brache, Art.Dtsch., Landw. Ges; 98, pp. 58-78.

Pawar, C.B. and Daginawala, H.F. (2000). *General Microbiology*, Ed. 2, pp. 520-540.

Roberts, D. G. (1978). RRS discovery cruise 90-20, institute of oceanographic science, Cruise report, Vvormeley, U.K., 68.

Subba Rao, N.S. (1993). *Biofertilizer's in Agriculture and Forestry*, Ed. 3, pp. 1-40.

Subba Rao, N.S. (1995). *Soil Microorganism and Plant Growth*, Ed. 3, pp. 170-190.

Trinick, J.J. (1973). Symbiosis between Rhizobium and non Legume, Trema aspera. *Nature*, London, pp. 224, 459-460.

Vincent, J.M. (1970). *A Manual for the Practical Study of the Root Nodule Bacteria*. IBP, Handbook No. 15, Backwell scientific publication, Oxford.

Waksman, S.A. (1927). *Principles of Soil Microbiology*. Williams and Wilkins Co., Baltimore.

Microbial Biotechnology and Ecology (2011) *Pages* **577–590**
Editors: **Deepak Vyas, G.S. Paliwal, P.K.Khare and R.K Gupta**
Published by: **DAYA PUBLISHING HOUSE, NEW DELHI**

Chapter 34

Quest for a Novel Tuberculosis Vaccine: A Global Endeavor

☆ *Umesh Dutta Gupta*[1]*, *Vijender Singh*[2] *and Pushpa Gupta*[1]

[1]*National JALMA Institute for Leprosy and Other Mycobacterial Diseases, Agra – 282 001*
[2]*Institute of Microbial Technology, Chandigarh – 160 036*

Tuberculosis (TB): A Global Health Problem

Tuberculosis (TB) is an archaic scourge that has distressed mankind for millennia and *Mycobacterium tuberculosis*, the causative agent, is one of the oldest and most successful of human pathogens (Collins, 1997). The etiological agent for TB was discovered by Robert Koch in the last of 19th century, is a gram-positive bacterium known as tubercle bacilli or *Mycobacterium tuberculosis*. In 1993, the World Health Organization (WHO) has declared TB as 'a global emergency'. Despite a more than a half century of extensive vaccination, TB claims a high burden of morbidity and mortality throughout the world (Korenromp *et al.*, 2009) (WHO 2009). Estimates of PPD skin-test indicate that tubercle bacilli infect a one-third of the world's population, leading to approximately 8.9 new cases of the disease and 1.7 million casualities each year (Korenromp *et al.*, 2009). Despite the worldwide efforts to control TB, the worldwide problem of TB is deteriorating, particularly in developing countries. Further the situation exacerbated by HIV-TB co-infection and the global emergence of multidrug-resistant strains of *M. tuberculosis* (WHO 2009). Although drug-susceptible TB can be effectively treated by current chemotherapy with a 6-month regimen which involves three or four drugs (WHO, [Directly Observed Treatment and Short-course drug therapy] DOTS strategy, with >95 per cent cure rate) (Acheson *et al.*, 2000) but for a definite control and elimination of this disease, there is a serious need and challenge for scientific community to develop a novel and effective vaccine against TB.

* E-mail: gupta.umesh95@gmail.com

Epidemiology of Tuberculosis

On the basis of surveillance and survey, WHO has reported approximately 9.2 million new cases of TB occurred in 2007. Out of these 9.2 million new cases, about 44 per cent (~ 4.1 million) account for new smear positive cases. India, China, Indonesia, Nigeria, and South Africa order first to fifth in terms of the total number of incidences. Among all global cases Asia responsible for 55 per cent and the African region for 31 per cent. The American, European and Eastern Mediterranean regions are accountable only for small fractions of global cases. The African region is at the rank first with 13 countries out of total 15 countries with the highest estimated TB incidence rates. This phenomenon is directly linked to high rates of HIV co-infection. In 2007 amongst total 9.2 million incident cases of TB, 1.37 million (14.8 per cent) were estimated as HIV-positive. The African region accounted for most (79 per cent) HIV-positive TB cases, where South Africa alone responsible for 31 per cent of cases in the African region. The occurrence of HIV-positive TB cases in the South-East Asia Region (mainly India) is about 11 per cent of total cases. Globally, around 0.5 million cases of MDR-TB were found, of which 0.3 million were among people not previously treated for TB and 0.2 million were among previously treated TB patients. Indian and China are at top among all countries with the largest number of cases of MDR-TB (WHO 2009).

The total numbers of global cases are still escalating in an absolute terms due to population growth. There is anticipated occurrence of new 9.8 million cases of TB in 2010, more than in any previous year in history. About 80 per cent of the cases will be found in the highest-burden countries like India and China which will be account for one third of 80 per cent cases (Dye *et al.*, 2009). Nonetheless, due to various strategies of WHO like based on Directly Observed Treatment and Short-course drug therapy (DOTS) which was reinforced by Stop TB strategy in 2006, the number of incident cases per capita is falling globally at an estimated 1 per cent per year (WHO 2009).

Pathogenesis of the Tuberculosis

Various observations from human TB infection and experimental animal models give raise the pattern of various events that start with the infection with mycobacterium (Flynn and Chan, 2005; Russell, 2007). The droplet nuclei containing virulent *M. tuberculosis* enters the lungs via the aerosol route on inhalation. The bacteria in inhaled droplets are engulfed primarily by resident alveolar macrophages which trigger innate and adaptive immune response in the host. This leads to the induction of a local inflammatory response that resulted in to recruitment of mononuclear cells from neighboring blood vessels. These cells are the basic ingredient of the granuloma, which is the characteristic pathological feature of this disease. In the initial stage, the granuloma is an unstructured mass of macrophages, monocytes, and neutrophils. In granuloma the macrophages also differentiate into several specialized cell types, like multinucleated giant cells, epithelioid macrophages and foamy macrophages. With the arrival of lymphocytes during acquired immune response, the granuloma acquires a more structured and stratified structure. In a classical granuloma, the macrophage rich center is surrounded by a layer of lymphocytes which may be restricted within a fibrous sheath that marks the periphery of the structure. The arrival of mycobacterium-specific lymphocytes which takes about 2 to 3 weeks post-infection, check the rapid bacterial multiplication and set a containment state which is in mice, characterized, by relatively stable bacterial numbers. At this particular stage, the extensively vascularization of granuloma results in to active recruitment of cells at the site of infection. The pathology results as disease progress; the fibrous cuff at the periphery of granulomas becomes more firms and blood vessels that penetrate granuloma diminish and vascularity of the granuloma reduces to minimum. Next, number of foamy macrophages also increase, which may be accountable for the caseous debris in the center of the granuloma (Russell *et al.*, 2009). In the next stage of progression, the caseous center of the granuloma becomes hypoxic (Via *et al.*, 2008). This condition can induce a state of persistence with non-replicative state of *M. tuberculosis*. In the last stage, active granulomas exhibit extensive pathology which ultimately lead to the rapture of the granuloma and liberate thousands of viable, infectious bacilli into the airways (Kaplan *et al.*, 2003) and a productive cough develops, in human that facilitates aerosol spread of infectious bacilli ready to infect healthy hosts.

Host Response to Tuberculosis

M. tuberculosis primarily infects macrophages of the lungs. This bacillus is an enormously successful pathogen that has evolved numerous mechanisms for evading elimination by the host immune response. After infection, mycobacteria persist within macrophages in early phagosomes and escape from the immune system notably by preventing phagosome acidification, arresting phagosome maturation and fusion with lysosomes (Russell, 2003).The immunological control of *M. tuberculosis* infection is a complex and multi-faceted. However, given the intracellular nature of the pathogen, the role of cellular immunity in the control of *M. tuberculosis* infection has been established beyond doubt. T-lymphocytes play a central role in the protection against TB. The protection against TB critically depends on T-helper 1 (Th1) type CD4 T cells which secrete the key cytokines IFN-g and TNF-α (Quesniaux, Jacobs *et al.*, 2010) Moreover, IFN-g and TNF-α are the important cytokines which activate macrophages to control bacterial multiplication by increasing fusion of the phagosome with the lysosome (Schaible *et al.*, 1998; MacMicking *et al.*, 2003). In addition, CD8 T cells also contribute to by secreting similar cytokines and by producing effector molecules like perforin and granulysin, which not only kill *M. tuberculosis* infected macrophages but also directly attack on the mycobacteria (Woodworth *et al.*, 2008).

During tuberculosis infection, in addition to the conventional CD4 T cells and CD8 T cells, CD1-restricted unconventional T cells are also give specific response against glycolipids of *M. tuberculosis* (Ulrichs *et al.*, 2003). Recent evidence also indicated the contribution of newer T cell subset, T-helper 17 (Th17) which secrete IL-17. These T cells mainly attract monocytes to the site of *M. tuberculosis* replication where they activate them. It has been also shown that Th17 cells are recruited to the lungs after immunization and rapidly respond to TB infection by increasing the enrolling Th1 cells into the lung (Khader *et al.*, 2007). Recent studies also showed that presence of regulatory T cells (Treg) can diminish an effective T-cell response required for protection against TB (Kursar *et al.*, 2007). Hence for novel vaccination strategy, restriction of Tregs cells and enrichment of IL-17 producing Th17 cells may help in generation of effective Th1 response for the control of TB infection.

Bacillus Calmette-Guerin (BCG): The Current Vaccine

Albert Calmette and Camille Guerin, in 1908 isolated *Mycobacterium bovis* from cattle suffering from tuberculous mastitis. They made an attenuated strain from this isolated mycobacterium (*M. bovis*) which a close relative of *M. tuberculosis*, by growing it for 13 years and 230 subcultures until it lost the ability to cause disease in inoculated individuals (Huebner 1996).Today more than 3 billion people have vaccinated with BCG, which makes this vaccine the most widely used throughout world. Further, in 1992, approximately 100 million newborns and children vaccinated with BCG through the World Health Organization (WHO)/United Nations Children's Emergency Fund (UNICEF) expanded program for immunization (Andersen *et al.*, 1995) (WHO 1992). At present, BCG is the only vaccine available for prevention of TB and estimated that it prevent only 5 per cent of the all vaccine preventable fatalities due to TB (Bishai and Mercer, 2001). In spite of the fact that BCG is the most widely administered of all vaccines, TB still continues to scourge the human life. It is however, been shown to be protective against the meningeal TB in young children (Udani, 1994). Nevertheless, its efficacy in preventing adult pulmonary TB, that is responsible for the major burden of morbidity and mortality, varies dramatically as revealed by carefully conducted studies throughout the world from 0 per cent in Chinglepet, India to approximately 80 per cent in the UK (Fine, 1995).

Although, first BCG vaccine has been administered successfully in human in 1921(Sakula 1983), yet TB continues to kill approximately 2 million people annually (Korenromp *et al.*, 2009).

Although BCG is enough good vaccine but due to number of unidentified host as well as environmental factors badly affect its protective efficiency (Smith *et al.*, 2000). There are many hypotheses that can be explained the insufficient protective efficiency of BCG against pulmonary tuberculosis and probable new strategies for its improvement. Some of the probable reasons for inadequate efficacy of the BCG are directly related to the vaccine itself like a number of genes related to protective immunity had been lost during attenuation procedure (Behr, 2002); inappropriate treatment, storage and also inadequate strains of BCG are few more direct correlates (Fine, 1995). In addition, other hypotheses suggest defects in basic immunological character. As discussed various T cells play an important role in immunity to TB; failure of BCG to stimulate all required T cells types effectively also had been suggested (Behr *et al.*, 1999).

Besides probable reasons mentioned above, many others also have been attributed like methodology of clinical trials, genetic difference within and among population and pre-sensitization of host immune response by environmental mycobacteria. Further, the comparative genomic analysis of different strains of BCG has shown a lot of variability among all six BCG vaccine sub-strains (Copenhagen-1331, Pasteur-1173P, Glaxo-1077, Tokyo-172, Russian, Moreau) and this genetic variability may contribute to the disparity in protective efficacy in different geographic regions of the world (Behr, 2001).

Nevertheless, as several reports indicated that neonatal vaccination with BCG is effective in reducing the incidence of childhood TB in endemic regions and reducing the rate of complications from disseminated TB (Fine, 2001). The protection imparted by BCG in children indicates that it expresses antigenic repertoire necessary to defend against TB. In contrast, its failure to protect against pulmonary TB in adults signifies that BCG cannot induce the generation of long-lasting memory response (Brewer and Colditz, 1995; Orme *et al.*, 2001; Triccas and Nambiar, 2009). Thus, the development of more effective novel vaccines and immunization strategies should be designed to protect against primary infection and also to boost waning BCG-induced protective immune responses are of urgent need to facilitate world-wide control of tuberculosis.

Novel Vaccines and Vaccination Strategies

To reduce the global burden of TB, new vaccination strategies against *M. tuberculosis* need to induce a more efficient immunity than that attained with BCG vaccine, not only in infants but also in adolescents and adults. With the emergence of understanding about BCG vaccine ineffectiveness in protecting adult form of TB has impelled a major global effort to develop new candidate vaccines (Izzo *et al.*, 2005; Orme, 2005; Orme, 2006).

Present Vaccine Candidates

The crucial fundamental mechanism of action of present vaccine candidates is stimulation of Th1 cells, which activate macrophages capacities for antimycobacterial activity (Kaufmann, 2006; Kaufmann *et al.*, 2006). The multifunctional memory T cells that produce several cytokines concomitantly especially, IFN-γ, TNF-α, and IL-2 are of central importance (Foulds *et al.*, 2006). In addtion, CD8 T cells that can directly attack *M tuberculosis* by perforin and granulysin (Stenger *et al.*, 1998) and Th17 cell support Th1 responses (Khader *et al.*, 2007) are also play a very considerable role.

Live Vaccines

Live vaccines are based either on improvement of BCG vaccine through addition of protective genes or on deletion of virulence genes of *M. tuberculosis.*

Recombinant BCG Vaccines

The protective efficacy of BCG vaccine against TB was always a matter of dispute. But some basic features of BCG for instance, enough immunogenic nature, can be given at the time of birth and single dose is enough. These factors have encouraged many investigators to focus their efforts on approaches based on recombinant DNA technology to modify BCG into a recombinant BCG vaccine against TB.

Two superior recombinant BCG (rBCG) vaccines have already started initial clinical evaluation. In the preclinical studies in different animal models, the potency and safety of both rBCG vaccines was found to be superior to the currently used BCG vaccine (Grode *et al.*, 2005; Tullius *et al.*, 2008). The rBCG30 that overexpresses Ag85B, a secretory and immunogenic protein of *M. tuberculosis* and BCG was the first vaccine candidate that showed substantially greater protection than parental BCG in different animal models (Horwitz *et al.*, 2000; Horwitz and Harth, 2003). In Phase I clinical trial which was conducted over 30 healthy adult volunteers, no significant safety issues was observed and also compared to parent BCG strain an enhanced immunogenicity was reported (Hoft, 2005).

The second vaccine, rBCGΔureC:Hly, is a recombinant BCG strain that secretes listeriolysin (of *Listeria monocytogenes* origin which is a cytolysin pores froming protein) which punches holes in the endosome and allows some leakage of BCG into the cytoplasm of infected host cells (Hess *et al.*, 1998; Grode *et al.*, 2005). Because optimal activity of Hly requires an acidic environment, urease gene (*ureC*) of the BCG, which facilitates mycobacteria to block the acidification of the early phagosome was also deleted in the construct (Grode *et al.*, 2005). In addition to CD4 T cells, protective immunity against *M. tuberculosis* infection seems to also involve both CD8 T cells, so antigens must be presented by MHC class I. Perforation of the phagosomal membrane probably promotes antigen translocation into the cytoplasm and facilitates cross-priming and stimulation of Th17 cells (Winau *et al.*, 2006; Torchinsky *et al.*, 2009). In preclinical studies in mice, a better efficacy of rBCGΔUreC:Hly than the BCG strain was observed (Grode *et al.*, 2005). The Aeras Global TB Vaccine Foundation (a non-profit Product Development Partnership) constructed different rBCG strains by utilizing pH-independent perfringolysin (rBCG-pfo), instead of Hly (Portnoy *et al.*, 1992), with overexpression of several immunodominant antigens including Ag85A, Ag85B and TB10.4 (Sadoff, 2005).

Attenuated *M. tuberculosis* Vaccines

Alternatively, *M. tuberculosis* that contains above 120 (some are directly related to virulence) genes that were deleted in BCG during development, can be attenuated by the genetic manipulation (Behr *et al.*, 1999; Gordon *et al.*, 1999; Salamon *et al.*, 2000). MTBVAC01 vaccine was constructed by disruption of the transcriptional regulator *phoP* gene associated with *M. tuberculosis* virulence) and the *fadD26* genes (Martin *et al.*, 2006; Verreck *et al.*, 2009). Other attenuated *M. tuberculosis* vaccine candidates includes *M. tuberculosis* deleted mutants ΔRD1ΔpanCD (Sambandamurthy *et al.*, 2006). Some of these live vaccines probably enter in clinical trials in the next few years (StopTB Partnership Working Group on New TB Vaccines).

Subunit and Live Vector-Based Vaccines for Prime-Boost Strategy

Numerous vaccines for prime-boost strategy have been used to enhance the immune response evoked by primary BCG vaccine. The main desirability from the heterologous prime–boost approach is that it should preferentially expand TB-reactive pre-existing memory T cells against common antigens of the prime and booster vaccines. Subunit vaccine candidates are based on antigens of *M. tuberculosis* that trigger T cells in infected humans (cured or latently infected) and experimental animals are believed better vaccine candidates.

Subunit vaccine candidates were recognized by various biochemical techniques. Initially, major secretary proteins of *M. tuberculosis* like Ag85 complex, MPT32, PhoS, DnaK, GroES, MPT46, MPT53, MPT63 and the 19-kDa lipoprotein were identified (Andersen and Hansen, 1989; Nagai *et al.*, 1991; Young and Garbe, 1991). Next, low molecular mass antigens in the range of 6–10 kDa (ESAT-6 family) and 26–34 kDa (Ag85B and MPT59) were defined (Andersen *et al.*, 1995; Skjot *et al.*, 2000). Other important immune-dominant antigens like Mtb32 and Mtb39 (Skeiky *et al.*, 1999) were recognized by T-cell and serological expression cloning (Alderson *et al.*, 2000; Dillon *et al.*, 2000; Skeiky *et al.*, 2000). Potential antigen like Rv3407 (Mollenkopf *et al.*, 2004) was identified by proteomic approach (Covert, *et al.*, 2001; Mattow *et al.*, 2003).

Two kinds of candidates have been developed for boosting primary immune response generated by BCG vaccine: (*i*) recombinant fusion proteins consisting of two or three immuno-dominant *M. tuberculosis* or BCG protective antigens; (*ii*) live viral vectors that express one or several mycobacterial proteins. The fusion proteins needs an adjuvant that enhances Th1 immune responses and can be given repeatedly while live viral vector vaccines do not require an adjuvant but pre-sensitization of the vector can clear the construct. These vaccine candidates induced a dominated T helper-1 (Th1) immune response to the expressed heterologous antigen and also generate Cytotoxic T cells (CD8 T cells) responses. In preclinical studies in mice and guinea pigs model of tuberculosis, almost all prime-boost candidate vaccines exhibited a similar level of protection from TB, obtained with BCG vaccine. In addition, several vaccine candidates found to be protective in non-human primates (Verreck *et al.*, 2009).

Recombinant fusion proteins based candidate vaccines were selected by their strong recognition by immune cells of experimentally vaccinated or infected animals and infected individuals. Following fusion proteins either with appropriate adjuvant or live virus vector were employed as vaccine candidates for testing against TB.

Ag85B-ESAT-6 (Hybrid 1)

Early secreted antigen target (ESAT-6) which is absent from all BCG, but present in *M. tuberculosis* strains. Mycobacterium reactive T cells from infected individuals or in animal models of TB recognized this protein very strongly. The second major protein secreted by *M. tuberculosis* is Ag85 complex, which is a mycolyltransferase consist of three closely related proteins, Ag85A, Ag85B and Ag85C. Among, Ag85A and Ag85B are the most potent antigens and main targets of *M. tuberculosis* specific T cell in human (Huygen *et al.*, 1996; Baldwin *et al.*, 1998). A fusion protein of two antigens was given with IC31 (by Intercell, Vienna, Austria) which is a strong Th1 adjuvant and its potency was assessed in two phase 1 clinical trials (Dietrich *et al.*, 2006; Lingnau *et al.*, 2007). This vaccine candidate was well tolerated and found to be very immunogenic in human volunteers without any adverse effects (Aagaard *et al.*, 2009).

Ag85B-TB-10.4 (H4, AERAS-404)

Alternatively, TB10.4 which also a strong immuno-dominant protein in BCG vaccinated donors was fused with Ag85B to minimize the interference with ESAT-6-based diagnostic assays (Skjot *et al.*, 2002; Dietrich *et al.*, 2005). This fusion protein elicited comparable protection against TB to Ag85B-ESAT-6 and BCG vaccine.

Mtb72

A hybrid protein of two TB antigens, Mtb32 and Mtb39 which is formulated in an adjuvant system consist of monophosphoryl lipid A and QS21 to strengthen the immune response required for protective immunity again TB (Skeiky *et al.*, 2004). This vaccine candidate also had reasonable acceptability and induced required T-cell responses in healthy adults (van Dissel *et al.*, 2010).

MVA85A

It is a recombinant, modified Vaccinia Ankara virus (replication-deficient vaccinia), that expresses Ag85A. Priming with BCG and then boosting with MVA85A regimen, induced a enhance level of both CD4 and CD8 T cells. Protection imparted this regimen was found better that either BCG or MVA85A alone (Goonetilleke *et al.*, 2003). In phase I clinical trials in human beings (immunized with BCG from 6 months to 36 years earlier), MVA85A boosting showed persisting CD4 T cells which secrete high level of IFN-γ. In contrast, only transient response was observed in case of BCG or MVA85A alone (McShane *et al.*, 2004; Sander *et al.*, 2009). Prior exposure with vaccinia virus might diminish the effectiveness of vaccines based on this viral vector (Rooney *et al.*, 1988).

Ad35 or AERAS-402 and AdAg85A

These vaccine constructs are based on replication deficient (E1-deleted) recombinant adenoviruses (Santosuosso *et al.*, 2006; Radosevic *et al.*, 2007).

Adenoviral vectors are potent delivery vehicles for recombinant vaccines but prior natural exposure with homologous or cross-reacting strain might hamper the required responses (Bangari and Mittal, 2006). The elicitation of superior CD8 T-cell responses is the main feature of vaccine constructs. Mice immunized by intranasal route with AdAg85A (recombinant Adenovirus 5 expressing Ag85A) exhibited substantial protection (mediated by bothCD4 and CD8T cells) compared to BCG (Wang *et al.*, 2004). Highly prevalence of Adenovirus 5 in human populations can affect the potency of vaccine due to preexisting antibodies. Adenovirus 35 based vaccines might be attractive choice because of it low sero-prevalence (5 per cent–20 per cent in different geographical regions). Areas Global TB Vaccine Foundation and Crucell (a global biopharmaceutical company) together constructed various vaccine candidates with non-replicating Ad35 vectors expressing Ag85A, Ag85B and TB10.4. Single immunizations of with rAd35-TB construct conferred protection against TB (Radosevic, 2006). In a prime-boost study this construct responded with enhanced level of mycobacterium specific CD4 and CD8 T-cell responses compared with rAd35-B or BCG alone immunized mice.

Novel Tuberculosis Vaccines in Clinical Trials

New vaccines are urgently needed to reach the ultimate goal of reducing global incidence of tuberculosis considerably by 2050 (Young and Dye, 2006). Promising finding of various vaccine candidates in pre-clinical testing in various animal models of tuberculosis, stipulate for their extension for approval in initial phase I (safety) by regulatory authority for testing in human subjects. The first objective of initial clinical trials for new tuberculosis vaccines is to assess their safety in a small study group including healthy, PPD-negative individuals (usually adults). Populations at risk of TB-HIV co-infection should be considered particularly during trials with live vaccines. Additional Phase I trials may be include PPD+ individuals, children, infants, or other groups for which the vaccine may be given ultimately. Usually these trials are combined with a phase IIA trial in which generation of immunological responses to the vaccine was studied. Biomarkers to assess the immunity generated by the vaccine like assays measuring T-cell responses to peptide epitopes contained within the vaccine should be checked at this stage. Unfortunately, present biomarker assays are not directly correlates with protective efficiency of vaccines. The demonstration of safety in phase I and immunogenicity in phase II are prerequisites for any new TB vaccine candidate to go forward into phase III (efficacy) trials. Table 34.1 lists the novel TB vaccines, which are currently at various levels of clinical trials in human being.

Table 34.1: Novel Tuberculosis Vaccines at Different Steps in Clinical Trials

Candidate	*Description*	*Clinical Trail Stage*	*Sponsor or Funding Agency*
H1-CAF01	Recombinant fusion protein of Ag85B-ESAT-6 in CAF01 adjuvant	Phase 1 continuing	Statens Serum Institute, Tuberculosis Vaccine Initiative
AdAg85A	Ag85A expressing Ad5 (Non-replicating)	Phase 1	McMaster University
AERAS-402	Ag85A, Ag85B, and TB10.4 expressing Ad35 (Non-replicating)	Phases 1 and 2 continuing	Aeras
H4-IC31 (AERAS-404)	Recombinant fusion protein of Ag85B-TB10.4 in IC31 adjuvant	Phase 1 completed	Statens Serum Institute, Aeras
rBCGΔUreC:Hly (VPM1002)	BCG with an endosome escape	Phase 1 completed	Tuberculosis Vaccine Initiative, Max Planck Institute
H1-IC31	Recombinant fusion protein of Ag85B-ESAT-6 in IC31 adjuvant	Phase 1 completed	Statens Serum Institute, Tuberculosis Vaccine Initiative
M72	Recombinant fusion protein of Mtb39 and Mtb32 in AS02 and AS01 adjuvants	Phases 1 and 2 completed	GlaxoSmithKline, Aeras, Tuberculosis Vaccine Initiative
rBCG30	Ag85B over-expressing BCG	Phase 1 completed	University of California, Los Angeles; Aeras
MVA85A	Attenuated vaccinia expressing Ag85A	Phase 1 completed and phase 2 continuing; phase 2b in infants started	Wellcome Trust, Aeras, Emergent BioSolutions
RUTI	Detoxified *M. tuberculosis* in liposomes	Phase 1 completed	Archivel Farma
M. vaccae	Inactivated *M. vaccae*	Phase 3 completed	National Institutes of Health

Ag: Antigen; ESAT-6: Early secretory antigenic target-6; CAF: Cationic adjuvant formulation; Ad: Adenovirus; Hly: Haemolysin; AS: Adjuvant system; MVA: Modified vaccinia Ankara.

Challenges Continues

TB vaccine research has gained impetus in recent years; still there are some major barriers. The efficacy of novel TB vaccines must be greater than current BCG and importantly should be safe enough so can be used in HIV-endemic countries. Although, several subunit vaccines are in various phages of clinical trials but they could not show better efficacy than BCG in pre-clinical studies in various animal models, which is the reason to considered them as a booster but rather than a replacer for BCG. In addition, the safety issue of various recombinants BCG and attenuated *M. tuberculosis* strains remains in question. As WHO has recommend that BCG should not be given to HIV-positive children, even if they are asymptomatic, because of high risk of BCG-induced disseminated disease in immune-compromised individuals with HIV infection.

Regulatory authorities of developing nations should actively participate in both product and clinical assessment to tested TB vaccines within their country. But many developing countries do not have their regulatory authorities. So some global systems are considered necessary for sharing information among regulatory authorities of all countries engaged in the production and clinical evaluation of candidate vaccines. To the surprise, till the date there is no comprehensible global regulatory pathway for new TB vaccine candidates; hence a "regulatory plan" should be developed before the start of phase III of clinical trials. So there is a need for global competent and efficient local and national ethics and regulatory authorities for all countries engaged in TB vaccine development.

On the basis of outcome of Phase II of clinical trials the promising TB vaccine candidates may enter in phase III of clinical trials which require a large number (10 000, or more) of individuals to check the efficacy to vaccines candidates. These clinical trials should be done in various geographical regions of the world so the generated data can be translated to other countries with genetically, culturally, and environmentally diverse populations. This is the main obstacle for the development of a site for the assessment of new tuberculosis vaccine candidates.

In addition, there is paucity of distinct surrogate bio-markers for the assessment of protective efficacy of vaccine candidates. So, clinical studies are crucial to ascertain the efficacy of vaccine candidates. This needs continuous follow-up, high cohort maintenance, and especially good surveillance systems. Scientific research is an indispensable but inadequate facet in the reduction or elimination of TB. Any successful vaccine needs to be accompanied by strong political determination, clear policies and objectives, a facilitative regulatory authority, and sufficient funding, to considerably reduce the disease.

References

Aagaard, C., J. Dietrich, *et al.* (2009). "TB vaccines: current status and future perspectives." *Immunol Cell Biol* 87(4): 279-86.

Acheson, D., G. A. Alleyne, *et al.* (2000). "Round table discussion. Health inequalities and the health of the poor." *Bull World Health Organ* 78(1): 75-85.

Alderson, M. R., T. Bement, *et al.* (2000). "Expression cloning of an immunodominant family of Mycobacterium tuberculosis antigens using human CD4(+) T cells." *J Exp Med* 191(3): 551-60.

Andersen, A. B. and E. B. Hansen (1989). "Structure and mapping of antigenic domains of protein antigen b, a 38,000-molecular-weight protein of Mycobacterium tuberculosis." *Infect Immun* 57(8): 2481-8.

Andersen, P., A. B. Andersen, *et al.* (1995). "Recall of long-lived immunity to Mycobacterium tuberculosis infection in mice." *J Immunol* 154(7): 3359-72.

Baldwin, S. L., C. D'Souza, *et al.* (1998). "Evaluation of new vaccines in the mouse and guinea pig model of tuberculosis." *Infect Immun* 66(6): 2951-9.

Bangari, D. S. and S. K. Mittal (2006). "Development of nonhuman adenoviruses as vaccine vectors." *Vaccine* 24(7): 849-62.

Behr, M. A. (2001). "Correlation between BCG genomics and protective efficacy." *Scand J Infect Dis* 33(4): 249-52.

Behr, M. A. (2002). "BCG–different strains, different vaccines?" *Lancet Infect Dis* 2(2): 86-92.

Behr, M. A., M. A. Wilson, *et al.* (1999). "Comparative genomics of BCG vaccines by whole-genome DNA microarray." *Science* 284(5419): 1520-3.

Bishai, D. M. and D. Mercer (2001). "Modeling the economic benefits of better TB vaccines." *Int J Tuberc Lung Dis* 5(11): 984-93.

Brewer, T. F. and G. A. Colditz (1995). "Relationship between bacille Calmette-Guerin (BCG) strains and the efficacy of BCG vaccine in the prevention of tuberculosis." *Clin Infect Dis* 20(1): 126-35.

Collins, F. M. (1997). "Tuberculosis research in a cold climate." *Tuber Lung Dis* 78(2): 99-107.

Covert, B. A., J. S. Spencer, *et al.* (2001). "The application of proteomics in defining the T cell antigens of Mycobacterium tuberculosis." *Proteomics* 1(4): 574-86.

Dietrich, J., C. Aagaard, *et al.* (2005). "Exchanging ESAT6 with TB10.4 in an Ag85B fusion molecule-based tuberculosis subunit vaccine: efficient protection and ESAT6-based sensitive monitoring of vaccine efficacy." *J Immunol* 174(10): 6332-9.

Dietrich, J., C. Andersen, *et al.* (2006). "Mucosal administration of Ag85B-ESAT-6 protects against infection with Mycobacterium tuberculosis and boosts prior bacillus Calmette-Guerin immunity." *J Immunol* 177(9): 6353-60.

Dillon, D. C., M. R. Alderson, *et al.* (2000). "Molecular and immunological characterization of Mycobacterium tuberculosis CFP-10, an immunodiagnostic antigen missing in Mycobacterium bovis BCG." *J Clin Microbiol* 38(9): 3285-90.

Dye, C., K. Lonnroth, *et al.* (2009). "Trends in tuberculosis incidence and their determinants in 134 countries." *Bull World Health Organ* 87(9): 683-91.

Fine, P. E. (1995). "Variation in protection by BCG: implications of and for heterologous immunity." *Lancet* 346(8986): 1339-45.

Fine, P. E. (2001). "BCG: the challenge continues." *Scand J Infect Dis* 33(4): 243-5.

Flynn, J. L. and J. Chan (2005). "What's good for the host is good for the bug." *Trends Microbiol* 13(3): 98-102.

Foulds, K. E., C. Y. Wu, *et al.* (2006). "Th1 memory: implications for vaccine development." *Immunol Rev* 211: 58-66.

Goonetilleke, N. P., H. McShane, *et al.* (2003). "Enhanced immunogenicity and protective efficacy against Mycobacterium tuberculosis of bacille Calmette-Guerin vaccine using mucosal administration and boosting with a recombinant modified vaccinia virus Ankara." *J Immunol* 171(3): 1602-9.

Gordon, S. V., R. Brosch, *et al.* (1999). "Identification of variable regions in the genomes of tubercle bacilli using bacterial artificial chromosome arrays." *Mol Microbiol* 32(3): 643-55.

Grode, L., P. Seiler, *et al.* (2005). "Increased vaccine efficacy against tuberculosis of recombinant Mycobacterium bovis bacille Calmette-Guerin mutants that secrete listeriolysin." *J Clin Invest* 115(9): 2472-9.

Hess, J., D. Miko, *et al.* (1998). "Mycobacterium bovis Bacille Calmette-Guerin strains secreting listeriolysin of Listeria monocytogenes." *Proc Natl Acad Sci U S A* 95(9): 5299-304.

Hoft, D. F. (2005). "Results of the 1st Phase I trial of a recombinant BCG TB vaccine (rBCG30." Presented at US–Japan Cooperative Medical Science Program; *40th Tuberculosis and Leprosy Research Conference.* (Seattle, Washington).

Horwitz, M. A. and G. Harth (2003). "A new vaccine against tuberculosis affords greater survival after challenge than the current vaccine in the guinea pig model of pulmonary tuberculosis." *Infect Immun* 71(4): 1672-9.

Horwitz, M. A., G. Harth, *et al.* (2000). "Recombinant bacillus calmette-guerin (BCG) vaccines expressing the Mycobacterium tuberculosis 30-kDa major secretory protein induce greater protective immunity against tuberculosis than conventional BCG vaccines in a highly susceptible animal model." *Proc Natl Acad Sci U S A* 97(25): 13853-8.

Huebner, R. E. (1996). "Bacillus of Calmette and Guerin (BCG) vaccine." In *Tuberculosis* (W. N. Rom, S. Garay, eds.), New York, Little, Brown and Company: 893–904.

Huygen, K., J. Content, *et al.* (1996). "Immunogenicity and protective efficacy of a tuberculosis DNA vaccine." *Nat Med* 2(8): 893-8.

Izzo, A., L. Brandt, *et al.* (2005). "NIH pre-clinical screening program: overview and current status." *Tuberculosis* (Edinb) 85(1-2): 25-8.

Kaplan, G., F. A. Post, *et al.* (2003). "Mycobacterium tuberculosis growth at the cavity surface: a microenvironment with failed immunity." *Infect Immun* 71(12): 7099-108.

Kaufmann, S. H. (2006). "Envisioning future strategies for vaccination against tuberculosis." *Nat Rev Immunol* 6(9): 699-704.

Kaufmann, S. H. (2006). "Tuberculosis: back on the immunologists' agenda." *Immunity* 24(4): 351-7.

Kaufmann, S. H., S. Baumann, *et al.* (2006). "Exploiting immunology and molecular genetics for rational vaccine design against tuberculosis." *Int J Tuberc Lung Dis* 10(10): 1068-79.

Khader, S. A., G. K. Bell, *et al.* (2007). "IL-23 and IL-17 in the establishment of protective pulmonary CD4+ T cell responses after vaccination and during Mycobacterium tuberculosis challenge." *Nat Immunol* 8(4): 369-77.

Korenromp, E. L., A. L. Bierrenbach, *et al.* (2009). "The measurement and estimation of tuberculosis mortality." *Int J Tuberc Lung Dis* 13(3): 283-303.

Kursar, M., M. Koch, *et al.* (2007). "Cutting Edge: Regulatory T cells prevent efficient clearance of Mycobacterium tuberculosis." *J Immunol* 178(5): 2661-5.

Lingnau, K., K. Riedl, *et al.* (2007). "IC31 and IC30, novel types of vaccine adjuvant based on peptide delivery systems." *Expert Rev Vaccines* 6(5): 741-6.

MacMicking, J. D., G. A. Taylor, *et al.* (2003). "Immune control of tuberculosis by IFN-gamma-inducible LRG-47." *Science* 302(5645): 654-9.

Martin, C., A. Williams, *et al.* (2006). "The live Mycobacterium tuberculosis phoP mutant strain is more attenuated than BCG and confers protective immunity against tuberculosis in mice and guinea pigs." *Vaccine* 24(17): 3408-19.

Mattow, J., U. E. Schaible, *et al.* (2003). "Comparative proteome analysis of culture supernatant proteins from virulent Mycobacterium tuberculosis H37Rv and attenuated M. bovis BCG Copenhagen." *Electrophoresis* 24(19-20): 3405-20.

McShane, H., A. A. Pathan, *et al.* (2004). "Recombinant modified vaccinia virus Ankara expressing antigen 85A boosts BCG-primed and naturally acquired antimycobacterial immunity in humans." *Nat Med* 10(11): 1240-4.

Mollenkopf, H. J., L. Grode, *et al.* (2004). "Application of mycobacterial proteomics to vaccine design: improved protection by Mycobacterium bovis BCG prime-Rv3407 DNA boost vaccination against tuberculosis." *Infect Immun* 72(11): 6471-9.

Nagai, S., H. G. Wiker, *et al.* (1991). "Isolation and partial characterization of major protein antigens in the culture fluid of Mycobacterium tuberculosis." *Infect Immun* 59(1): 372-82.

Orme, I. M. (2005). "Current progress in tuberculosis vaccine development." *Vaccine* 23(17-18): 2105-8.

Orme, I. M. (2005). "Tuberculosis vaccines: current progress." *Drugs* 65(17): 2437-44.

Orme, I. M. (2006). "Preclinical testing of new vaccines for tuberculosis: a comprehensive review." *Vaccine* 24(1): 2-19.

Orme, I. M., D. N. McMurray, *et al.* (2001). "Tuberculosis vaccine development: recent progress." *Trends Microbiol* 9(3): 115-8.

Portnoy, D. A., R. K. Tweten, *et al.* (1992). "Capacity of listeriolysin O, streptolysin O, and perfringolysin O to mediate growth of *Bacillus subtilis* within mammalian cells." *Infect Immun* 60(7): 2710-7.

Quesniaux, V. F., M. Jacobs, *et al.* (2010). "TNF in host resistance to tuberculosis infection." *Curr Dir Autoimmun* 11: 157-79.

Radosevic, K., C. W. Wieland, *et al.* (2007). "Protective immune responses to a recombinant adenovirus type 35 tuberculosis vaccine in two mouse strains: CD4 and CD8 T-cell epitope mapping and role of gamma interferon." *Infect Immun* 75(8): 4105-15.

Radosevic, K. (2006). "Presented at TB vaccines for the world (April 19–21, Vienna, Austria)."

Rooney, J. F., C. Wohlenberg, *et al.* (1988). "Immunization with a vaccinia virus recombinant expressing herpes simplex virus type 1 glycoprotein D: long-term protection and effect of revaccination." *J Virol* 62(5): 1530-4.

Russell, D. G. (2003). "Phagosomes, fatty acids and tuberculosis." *Nat Cell Biol* 5(9): 776-8.

Russell, D. G. (2007). "Who puts the tubercle in tuberculosis?" *Nat Rev Microbiol* 5(1): 39-47.

Russell, D. G., P. J. Cardona, *et al.* (2009). "Foamy macrophages and the progression of the human tuberculosis granuloma." *Nat Immunol* 10(9): 943-8.

Sadoff, J. (2005). "Public private partnership approach to vaccine development." *Presented at New Approaches to Vaccine Development*. (Sept. 8–10, Berlin, Germany).

Sakula, A. (1983). "BCG: who were Calmette and Guerin?" *Thorax* 38(11): 806-12.

Salamon, H., M. Kato-Maeda, *et al.* (2000). "Detection of deleted genomic DNA using a semiautomated computational analysis of GeneChip data." *Genome Res* 10(12): 2044-54.

Sambandamurthy, V. K., S. C. Derrick, *et al.* (2006). "Mycobacterium tuberculosis DeltaRD1 DeltapanCD: a safe and limited replicating mutant strain that protects immunocompetent and immunocompromised mice against experimental tuberculosis." *Vaccine* 24(37-39): 6309-20.

Sander, C. R., A. A. Pathan, *et al.* (2009). "Safety and immunogenicity of a new tuberculosis vaccine, MVA85A, in Mycobacterium tuberculosis-infected individuals." *Am J Respir Crit Care Med* 179(8): 724-33.

Santosuosso, M., S. McCormick, *et al.* (2006). "Intranasal boosting with an adenovirus-vectored vaccine markedly enhances protection by parenteral Mycobacterium bovis BCG immunization against pulmonary tuberculosis." *Infect Immun* 74(8): 4634-43.

Schaible, U. E., S. Sturgill-Koszycki, *et al.* (1998). "Cytokine activation leads to acidification and increases maturation of Mycobacterium avium-containing phagosomes in murine macrophages." *J Immunol* 160(3): 1290-6.

Skeiky, Y. A., M. R. Alderson, *et al.* (2004). "Differential immune responses and protective efficacy induced by components of a tuberculosis polyprotein vaccine, Mtb72F, delivered as naked DNA or recombinant protein." *J Immunol* 172(12): 7618-28.

Skeiky, Y. A., M. J. Lodes, *et al.* (1999). "Cloning, expression, and immunological evaluation of two putative secreted serine protease antigens of Mycobacterium tuberculosis." *Infect Immun* 67(8): 3998-4007.

Skeiky, Y. A., P. J. Ovendale, *et al.* (2000). "T cell expression cloning of a Mycobacterium tuberculosis gene encoding a protective antigen associated with the early control of infection." *J Immunol* 165(12): 7140-9.

Skjot, R. L., I. Brock, *et al.* (2002). "Epitope mapping of the immunodominant antigen TB10.4 and the two homologous proteins TB10.3 and TB12.9, which constitute a subfamily of the esat-6 gene family." *Infect Immun* 70(10): 5446-53.

Skjot, R. L., T. Oettinger, *et al.* (2000). "Comparative evaluation of low-molecular-mass proteins from Mycobacterium tuberculosis identifies members of the ESAT-6 family as immunodominant T-cell antigens." *Infect Immun* 68(1): 214-20.

Smith, D., E. Wiegeshaus, *et al.* (2000). "An analysis of some hypotheses related to the Chingelput bacille Calmette-Guerin trial." *Clin Infect Dis* 31 Suppl 3: S77-80.

Stenger, S., D. A. Hanson, *et al.* (1998). "An antimicrobial activity of cytolytic T cells mediated by granulysin." *Science* 282(5386): 121-5.

Torchinsky, M. B., J. Garaude, *et al.* (2009). "Innate immune recognition of infected apoptotic cells directs T(H)17 cell differentiation." *Nature* 458(7234): 78-82.

Triccas, J. A. and J. K. Nambiar (2009). "Challenge of developing new tuberculosis vaccines to generate life-long protective immunity." *Expert Rev Vaccines* 8(7): 823-5.

Tullius, M. V., G. Harth, *et al.* (2008). "A Replication-Limited Recombinant Mycobacterium bovis BCG vaccine against tuberculosis designed for human immunodeficiency virus-positive persons is safer and more efficacious than BCG." *Infect Immun* 76(11): 5200-14.

Udani, P. M. (1994). "BCG vaccination in India and tuberculosis in children: newer facets." Indian J Pediatr 61(5): 451-62.

Ulrichs, T., D. B. Moody, *et al.* (2003). "T-cell responses to CD1-presented lipid antigens in humans with Mycobacterium tuberculosis infection." *Infect Immun* 71(6): 3076-87.

van Dissel, J. T., S. M. Arend, *et al.* (2010). "Ag85B-ESAT-6 adjuvanted with IC31 promotes strong and long-lived Mycobacterium tuberculosis specific T cell responses in naive human volunteers." *Vaccine* 28(20): 3571-81.

Verreck, F. A., R. A. Vervenne, *et al.* (2009). "MVA.85A boosting of BCG and an attenuated, phoP deficient M. tuberculosis vaccine both show protective efficacy against tuberculosis in rhesus macaques." *PLoS One* 4(4): e5264.

Via, L. E., P. L. Lin, *et al.* (2008). "Tuberculous granulomas are hypoxic in guinea pigs, rabbits, and nonhuman primates." *Infect Immun* 76(6): 2333-40.

Wang, J., L. Thorson, *et al.* (2004). "Single mucosal, but not parenteral, immunization with recombinant adenoviral-based vaccine provides potent protection from pulmonary tuberculosis." *J Immunol* 173(10): 6357-65.

WHO (1992). "*World Health Organization. Expanded program for immunization: program report.*" Geneva: World Health Organization."

WHO (2009). "World Health Organization, Tuberculosis fact sheet number 104, March 2010. Available at http://www.who.int/mediacentre/factsheets/fs104/en/index.html. Accessed on June 14, 2010".

Winau, F., S. Weber, *et al.* (2006). "Apoptotic vesicles crossprime CD8 T cells and protect against tuberculosis." *Immunity* 24(1): 105-17.

Woodworth, J. S., Y. Wu, *et al.* (2008). "Mycobacterium tuberculosis-specific CD8+ T cells require perforin to kill target cells and provide protection *in vivo*." *J Immunol* 181(12): 8595-603.

Young, D. and C. Dye (2006). "The development and impact of tuberculosis vaccines." *Cell* 124(4): 683-7.

Young, D. B. and T. R. Garbe (1991). "Lipoprotein antigens of Mycobacterium tuberculosis." *Res Microbiol* 142(1): 55-65.

Microbial Biotechnology and Ecology (2011) *Pages* **591–605**
Editors: **Deepak Vyas, G.S. Paliwal, P.K.Khare and R.K Gupta**
Published by: **DAYA PUBLISHING HOUSE, NEW DELHI**

Chapter 35

Silver Nanoparticles: A New Generation of Antimicrobial Agents

☆ ***Mukesh K. Upadhyay*, Devendra Jain, Abhijeet Singh, Sushma Rishi and Narayan Rishi***

School of Life Sciences, Jaipur National University, Jaipur – 302 025, Rajasthan

ABSTRACT

Nanotechnology is gaining tremendous impetus in the present century due to its capability of modulating metals into their nanosize, which drastically changes the chemical, physical and optical properties of metals. Silver has been in use since ancient time in the form of metallic silver, silver nitrate, silver sulfadiazine for the treatment of burns, wounds and several infections. Nanosilver has developed as a potent antibacterial, antifungal, anti-viral and anti-inflammatory agent. Due to the outbreak of the infectious diseases caused by different pathogenic microorganisms developing antibiotic resistance, the pharmaceutical companies and the researchers are searching for new antibacterial agents. In the present scenario, nanoscale silver has emerged as novel antimicrobial agents. Hence, silver nanoparticles have emerged up with diverse medical applications ranging from silver based dressings, silver coated medicinal devices, such as nanogels, nanolotions etc.

Introduction

Silver is a naturally occurring pre-historic element. It has been positioned as the 47th element in the periodic table having an atomic weight of 107.8. For centuries silver has been in use for the treatment of burns and chronic wounds. It's use for medicine or local antibacterial agents was not

* E-mail: mukeshfungi@gmail.com

recognized until the 19th century (Oloffs *et al.*, 1994). Since then the antimicrobial property of silver has been investigated and employed more extensively than any other inorganic antibacterial agent (Russell and Hugo, 1994). Ag is known to attack a broad range of biological processes in microbes including the alteration of cell membrane structure and functions (Pal *et al.*, 2007). Ag inhibit expression of proteins associated with ATP production (Yamanaka *et al.*, 2005) although it's specific antimicrobial mechanism are still not completely understood. Micromolar doses of Ag are sufficient to kill bacteria in water (Liu *et al.*, 1994) while Ag can be toxic at high doses to mammals (Conrad *et al.*, 1999), freshwater and marine organisms (Morgan *et al.*, 2005). Such micromolar concentrations of Ag have no harmful effects on humans (Berger *et al.*, 1976) hence Ag is widely used for development of many biological and pharmaceutical products and coating materials for medical devices (Raad and Hanna, 2002). With the emergence and increase of microbial organisms resistant to multiple antibiotics, and the continuing emphasis on health care costs, many researchers have tried to develop new, effective anti-microbial reagents free of resistance and with reduced cost. Ag has long been known to exhibit a strong toxicity to a wide range of microbes (Liau *et al.*, 1997), for this reason silver based compounds have been used extensively in many bactericidal applications (Nomiya *et al.*, 2004). Silver nanoparticles show efficient antimicrobial property compared to other salts due to their extremely large surface area (Song and Kim, 2008), which provides better contact with microorganisms. The nanoparticles get attached to the cell membrane and also penetrate inside the bacteria. The bacterial membrane contains sulfur-containing proteins and the silver nanoparticles interact with these proteins in the cell as well as with the phosphorus containing compounds like DNA. When silver nanoparticles enter the bacterial cell it forms a low molecular weight region in the centre of the bacteria to which the bacteria conglomerates thus, protecting the DNA from the silver ions (Song *et al.*, 2006). The nanoparticles preferably attack the respiratory chain, cell division finally leading to cell death. The nanoparticles release silver ions in the bacterial cells, which enhance their bactericidal activity. Therefore, this review is emphasized for the development of silver nanoparticles as new generation medicine.

Nanotechnology

The term nanotechnology was coined by Professor Norio Taniguchi of Tokyo Science University in 1974 (Taniguchi, 1974). Nanotechnology is not a single thing rather a collective term that implies capacity to work with materials at nanoscale level. Nanotechnology enables us to expand the surface area of silver particles markedly. It is emerging as a rapidly growing field with its application in science and technology (Albrecht *et al.*, 2006). Nanotechnology involves the creation and manipulation of materials at nanoscale levels (1–100 nm) to create products that exhibit novel properties. Till recent times nanotechnology has been a domain of physics, chemistry, electrical engineering and material sciences, there has been a considerable progress in the visualization, construction and manipulation of material at nanometer scale. It is creating a growing sense of excitement in life sciences especially bio-medical devices and biotechnology.

Nanoparticles

Nanoparticles are clusters of atoms with a size up to 100 nm. They have high surface area to volume ratio (Song *et al.*, 2008). Specific surface area is relevant for catalytic activity and other related properties such as antimicrobial activity in silver nanoparticles. Their unique size dependent properties make these materials superior and indispensable as they show unique physical, chemical and biological properties. Use of nanoparticles is gaining impetus in the present century as they posses defined chemical, optical and mechanical properties. The metallic nanoparticles are the most promising as they show good antibacterial properties.

Biomimetics

A number of approaches are available for synthesis of silver nanoparticles for example reduction in solution, thermal decomposition of silver compounds, chemical and photochemical reactions in reverse micelles, ultrasonication irradiation etc. but these methods are proving harmful for the environment hence, it is our need to learn about the secrets inherently present in the nature and it's products, led to the development of biomimetic approaches for the growth of advancements in the synthetic process of nanoparticle synthesis (Song and Kim, 2008). Biomimetics is a technology in which we use biological systems such as yeast, fungi, bacteria and plants for the synthesis of nanostructures. This technology is an advancement over chemical methods as it is cost effective and eco-friendly (Singh *et al.*, 2010).

Metallic Silver

Silver is inert in its metallic state but reacts with the moisture in the skin and the fluid of the wound and gets ionized. The ionized silver is highly reactive, as it binds to tissue proteins and brings structural changes in the bacterial cell wall and nuclear membrane leading to cell distortion and death. Silver also binds to bacterial DNA and RNA by denaturing and inhibits bacterial replication (Castellano *et al.*, 2007).

Silver as an Antimicrobial Agent

The use of silver for medicine or local antibacterial agent was not recognized until the nineteenth century. Since then the antimicrobial property of silver has been investigated and employed more extensively than any other inorganic antibacterial agent. As early as 1000 B.C. silver was used to make water potable. Silver nitrate was used in its solid form for the treatment of venereal diseases, fistulae from salivary glands, and bone and perianal abscesses (Klasen, 2000; Landsdown, 2002). In the19th century granulation tissues were removed using silver nitrate to allow epithelization and promote crust formation on the surface of wounds. Varying concentrations of silver nitrate was used to treatment of fresh burns (Castellano *et al.*, 2007; Klasen, 2000). In 1881, Carl S. F. Crede cured opthalmia neonatorum using silver nitrate eye drops. Crede's son, B. Crede designed silver impregnated dressings for skin grafting (Klasen, 2000; Landsdown, 2002). In the 1940s, after penicillin was introduced the use of silver for the treatment of bacterial infections minimized (Hugo and Russell, 1982; Demling and DeSanti, 2001; Chopra, 2007). Silver again came in picture in the 1960s when Moyer introduced the use of 0.5 per cent silver nitrate for the treatment of burns. He proposed the antibacterial property of silver against *Staphylococcus aureus*, *Pseudomonas aeruginosa* and *Escherichia coli* (Moyer *et al.*, 1965; Bellinger and Conway, 1970). In 1968, silver nitrate was combined with sulfonamide to form silver sulfadazine cream, which served as a broad-spectrum antibacterial agent and was used for the treatment of burns. Silver sulfadazine is effective against bacteria like *E. coli*, *S. aureus*, *Klebsiella* sp., *Pseudomonas* sp. Due to the increasing bacterial resistance to classic antibiotics, the investigations on the antibacterial activity of silver nanoparticles have increased.

Nanosilver Synthesis

The biosynthesis of nanoparticles emerging as an highlight of the combination of nanotechnology and biotechnology has received increasing attention. Increasing attention due to a growing need to develop environmentally benign technologies in material synthesis (Huang *et al.*, 2007). Silver nanoparticles can be synthesized through an array of methods like chemical, biological etc. like spark discharging, electrochemical reduction, solution irradiation and cryochemical synthesis.

Chemical Synthesis

Chemical reduction is the most frequently applied method for the preparation of silver nanoparticles (AgNPs) as stable, colloidal dispersions in water or organic solvents as stable, colloidal dispersions in water or organic solvents. Kazachenko *et al.* (2000) investigated the synthesis and antimicrobial activity of silver complexes with histidine and tryptophan. The toxicity of silver complexes of tryptophan and histidine was tested on a group of white mongrel mice. The histidine complex with silver compound showed good antimicrobial activity against Gram-negative bacteria while, the tryptophan complex with silver compound showed higher antimicrobial activity and broad spectrum of action. Panigrahi *et al.* (2004) synthesized different metal nanoparticles (Au, Ag, Pt, Pd) using sugars like glucose, fructose and sucrose as reducing agents. Song *et al.* (2006) reported the synthesis of nano-sized Ag colloid using silver nitrate dissolved in ammonia, water, formaldehyde as a reducing agent and polyvinyl pyrrolidone (PVP) as the stabilizing agent. Nanoparticles around 5 nm in size were obtained. They showed excellent antibacterial activity against *S. typhi, E.coil, P. aeruginosa, S. aureus* and *M. tuberculosis*. Andreescu *et al.* (2007) prepared silver nanoparticles by reducing silver diamine ions $[Ag(NH_3)^2]^+$ with D-glucose in the presence of a stabilizing agent. Uniform Ag nanoparticles ranging in size from 30 to 120 were obtained. Kim *et al.* (2007) reported the use of silver nanoparticles for the control of infections occurring due to microorganisms. The silver nanoparticles were synthesized using silver nitrate and sodium borohydride (reducing agent). The antimicrobial efficiency of silver nanoparticles was tested against yeast and bacteria isolated from *Bovine mastitis*. The silver nanoparticles showed absorption band at ~391 nm which is typical for spherical nanoparticles. Shape and size distribution of the silver nanoparticles was studied using TEM, which depicted the presence of monodisperse particles in the size range of 13.5 nm. The synthesis of silver nanoparticles with different sizes (7, 29, and 89 nm mean values) using gallic acid in an aqueous chemical reduction method was reported by various researchers and the nanoparticles were characterized using TEM, dynamic light scattering (DLS), XRD, and UV–Vis absorption spectroscopy. The antibacterial activity of these nano silver particles was assessed using the standard microdilution method (Kim *et al.*, 2007; Kim *et al.*, 2008). From the microscopies studies (TEM) it was observed that silver nanoparticles have spherical (7 and 29 nm) and pseudospherical shape (89 nm) with a narrow size distribution. Das *et al.* (2009) reported the preparation of stable, uniform silver nanoparticles by reduction of silver ions by ethanol. Samples were characterized by (XRD) and (TEM), which revealed that the particles are mostly spherical in shape and have an average size of 16 nm. The UV-Vis. spectra showed an absorption peak at 410 nm.

Drawback of Chemical Methods

These are energy intensive processes and a large amount of capital is involved. The use of hazardous chemicals such as hydrazine eliminates the method from being an eco-friendly one. Moreover, nanoparticles produced by these methods exhibit aggregation with time, there compromising with the size factor upon storage (Mukherjee, 2008). The growing need to develop clean, non-toxic and eco-friendly methods for the synthesis of nanoparticles has resulted in the use of biological systems such as bacteria, fungi, plants etc.

By Bacteria

Biosynthetic methods employing either biological microorganisms or plant extracts have emerged as a simple and viable alternative to chemical synthetic procedures and physical methods (Saifuddin *et al.*, 2009). Although it is known that microorganisms such as bacteria play an important role in remediation of toxic metals through reduction of the metal ions, this was considered interesting as nanofactories very recently. Some bacteria reduce Fe^{3+} oxides by producing and secreting small, diffusible redox compounds that can serve as electron shuttle between the microbe and the insoluble iron substrate. This extracellular enzyme shows an excellent redox properties and it can act as an electron shuttle in the metal reduction. It was evident that electron shuttles or other reducing agents (e.g. hydroquinones) released by bacteria is capable of reducing ions to nanoparticles. However, biological nanoparticles synthesis generates nanoparticles at a much slower rate. In the studies synthesis of silver nanoparticles using the cell mass of bacteria or their leached cell components, the time required completing the reaction ranged from 24 to 120 h; this lengthy reaction is the one major drawback of the biological synthesis. Biological synthesis would have greater commercial viability if the nanoparticles could be synthesis more rapidly and in the large production scale. Shahverdi *et al.* (2007) investigated the combination effects of silver nanoparticles with antibiotics. The silver nanoparticles were synthesized using *Klebsiella pneumoniae* and evaluated its antimicrobial activity against *S. aureus* and *E. coli*. From the experimental work it was observed that the antibacterial activity of antibiotics like penicillin G, amoxicillin, erythromycin, clindamycin and vancomycin increased in the presence of silver nanoparticles against *E. coli* and *S. aureus*. The highest activity was observed with erythromycin against *S. aureus*. Minaeian *et al.* (2007) produced nanoparticles by *Klebsiella pneumoniae, Escherichia coli, Enterobacter cloacae, Staphylococcus aureus, Bacillus subtilis, Lactobacillus acidophilus* and *Candida albicans.* The silver nanoparticles were characterized by scanning electron microscopy and UV-visible spectroscopy. Saifuddin *et al.* (2009) described a novel combinatorial synthesis approach for the synthesis of metallic nanostructures of noble metals such as silver (Ag), by using a combination of culture supernatanant of *Bacillus subtilis* and microwave (MW) irradiation in water in absence of a surfactant or soft template. It was found that exposure of culture supernatanant of *Bacillus subtilis* and microwave irradiation to silver ion lead to the formation of silver nanoparticles. The silver nanoparticles were in the range of 5-60 nm in dimension. The nanoparticles were examined using UV-Visible Spectroscopy, and Transmission Electron Microscopy (TEM). The formation of nanoparticles by this method was extremely rapid, requires no toxic chemicals and the nanoparticles are stable for several months. Pugazhenthiran *et al.* (2009) synthesized AgNPs by *Bacillus* sp. TEM and energy dispersive X-ray (EDX) analyses confirmed that silver nanoparticles of 5–15 nm in size were deposited in the periplasmic space of the bacterial cells. Natarajan *et al.* (2010) demonstrated a simple biotechnological process for the intracellular and extracellular synthesis of silver nanoparticles using *E.coli*. The formation of nanoparticles using different basal medium was also investigated; the LB medium has the ability to maximize the synthesis of silver nanoparticles than NB medium. The characterization of silver ion exposed to microbial strain and the reduction of silver nitrate to silver nanoparticles was confirmed by UV-visible spectroscopy. Natarajan *et al.* (2010) also investigated that maximum synthesis of silver nanoparticles is proportional to bacterial biomass which occurred in the stationary phase of incubation period. It revealed that the particles formed from the distinct basal medium were associated with growth. The extracellular formation of silver nanoparticles was more expressed than intracellular formation.

By Fungi

Mukherjee *et al.* (2001) synthesized silver nanoparticles using the fungus *Verticillium*. Exposure of the fungal biomass to aqueous Ag^{+} ions resulted in the intracellular reduction of the metal ions and formation of silver nanoparticles of dimensions 2.5 -12 nm. Fungal cells exposed to Ag^{+} ions show broad absorption band at 450 nm. Electron microscopy analysis of thin sections of the fungal cells indicated that the silver particles were formed below the cell wall surface. The metal ions were not toxic to the fungal cells and the cells continued to multiply after biosynthesis of the silver nanoparticles.

Kowshik *et al.* (2002) synthesized silver nanoparticles in the size range of 2–5 nm extracellularly by a silver-tolerant yeast strain MKY3. Optical absorption, transmission electron microscopy, X-ray diffraction and X-ray photoelectron spectroscopy investigations confirmed that metallic (elemental) silver nanoparticles were formed. Duran *et al.* (2005) carried out the extracellular production of metal nanoparticles by several strains of the fungus *Fusarium oxysporum*. It was found that aqueous silver ions when exposed to several *F. oxysporum* strains are reduced in solution, thereby leading to the formation of silver hydrosol. The silver nanoparticles were in the range of 20–50 nm in dimensions. The reduction of the metal ions occurs by a nitrate-dependent reductase and a shuttle quinone extracellular process. The silver nitrate solution incubated with SMS changed to a yellow color from 24 hr onward, indicating the formation of silver nanoparticles. The purified solution yielded the maximum absorbance at 436 nm. X-ray analysis of the freeze-dried powder of silver nanoparticles confirmed the formation of metallic silver. Transmission electron microscopic analysis of the samples showed a uniform distribution of nanoparticles, having an average size of 30nm. These silver nanoparticles showed excellent antibacterial activity against two representative bacteria, *Staphylococcus aureus* and *Klebsiella pneumoniae*. Durán, *et al.* (2005) performed the extracellular production of silver nanoparticles by *F. oxysporum* strain and studied its bactericide effect in cotton and silk cloth against *S. aureus*. To incorporate silver nanoparticles in the cloth, these were immersed in the filtrate fungal, centrifuged and dried. The reduction of metal ions occurred by a nitrate dependent reductase and quinone extracellular process. Antibacterial activity was observed when silver nanoparticles were incorporated in cotton cloth. However, in silk cloth this activity was not observed demonstrating a low incorporation of the silver nanoparticles possibly due to the size of the silk pores. Their work demonstrated the possible use of biologically synthesized silver nanoparticles and its incorporation in cloths leading them to sterilization. Mukherjee *et al.* (2008) for the first time reported a biosynthetic route for the formation of nanocrystalline silver particles by using cell-free aqueous filtrate of a non-pathogenic and commercially viable biocontrol agent *Trichoderma asperellum*. The kinetics of the reaction was studied using UV–Vis spectroscopy. An intense surface plasmon resonance band at 410 nm in the UV–Vis. spectrum clearly revealed the formation of silver nanoparticles. The size of the silver particles using TEM and XRD studies was found to be in the range 13–18 nm. These nanoparticles were highly stable and even after prolonged storage for over 6 months they do not showed significant aggregation. Ingle *et al.* (2008) investigated the use of *Fusarium accuminatum*, isolated from infected ginger, for the synthesis of silver nanoparticles. The antibacterial activity of silver nanoparticles was checked against *E. coli*, *S. aureus*, *Salmonella typhi*, *S. epidermis*. The UV–Vis. spectra showed absorption at 420 nm, which is characteristic for spherical nanoparticles. The spherical shape of the nanoparticles was confirmed by the TEM study and the size of the nanoparticles ranged from 5–40 nm. Gade *et al.* (2008) exploited *Aspergillus niger* for the synthesis of silver nanoparticles. The silver nanoparticles were detected with the help of UV–Vis spectrophotometer while, the characterization was done using TEM. The antibacterial activity of silver nanoparticles was assessed against *E. coli* and *S. aureus*. The UV–Vis spectra denoted a peak at 420 nm. The TEM study of silver nanoparticles confirmed the presence of spherical shaped nanoparticles in the size range of 20 nm. The silver nanoparticles showed efficient antibacterial activity against *E. coli* and *S. aureus*, which, was also analyzed using TEM. The electron microscopic study demonstrated presence of silver nanoparticles in the cell membranes of test bacteria. The synthesis of nanoparticles from some fungi is summarized in Table 35.1.

By Plants

Synthesis of nanoparticles by plant extracts is at present under exploitation as some researchers worked on it (Prabhu *et al.*, 2009; Jain *et al.*, 2009; Singh *et al.*, 2010). The employ of *Azadirachta indica*

(Neem), *Medicago sativa* (Alfalfa), *Aloe vera, Jatropha curcas* (Ratanjoot) and *Emblica officinalis* (Amla, Indian Gooseberry) has previously been reported (Chandran *et al.*, 2006; Bar *et al.*, 2009). Consistent with preceding reports, the polyol components and the water-soluble heterocyclic components are largely accountable for the reduction of Ag ions and the stabilization of the nanoparticles, respectively. There are also reports on reductases and polysaccharides as factors involved in biosynthesis and stabilization of the nanoparticles, respectively (Bar *et al.*, 2009). Singh *et al.* (2010) states that some factors jointly conclude the nanoparticle synthesis, as well as the plant source, the organic compounds in the crude leaf extract, the concentration of Ag nitrate, the temperature and other than these, even the pigments in the leaf extract. The Ag is only characterized by UV-Vis spectroscopy, TEM, and atom force microscopy (AFM). TEM and AFM of the sample showed uniform and monodispersive particle distribution in the colloids. The synthesis of metal nanoparticles synthesized from some plants are summarized in Table 35.2.

Table 35.1: List of Metal Nanoparticles Synthesized from Different Fungi (Gade *et al.*, 2009)

Sl.No.	*Name of Organism*	*Nanoparticle Synthesized*	*Intracellular/Extracelluar*	*Size (nm)*
1.	*F. oxysporum*	Silver	Extracellular	5–15
2.	*F. oxysporum*	Silver	Extracellular	20–50
3.	*Trichothecium* sp.	Gold	Intracellular and Extracellular	–
4.	*F. oxysporum*	Silica and Titania	Extracellular	5–15
5.	*Colletotrichum* sp.	Gold	Extracellular	20–40
6.	*F. oxysporum f.* sp. *lycopersici*	Platinum	Intracellular and Extracellular	10–100
7.	*F. oxysporum*	Magnetite	Extracellular	20–50
8.	*Schizosaccharomyces* pombe	CdS	Intracellular	2–2.5
9.	*Torulopsis* sp.	PbS	Intracellular	2–5
10.	*Verticillium* sp.	Silver and Gold	Intracellular	2–20

Table 35.2: List of Metal Nanoparticles Synthesized from Different Plants (Kumar and Yadav, 2008)

Sl.No.	*Plants*	*Silver/Gold/Copper*	*Shape*	*Size (nm)*
1.	*Medicago sativa*	Silver	Spherical	6–20
2.	*Medicago sativa*	Gold	fcc tetrahedral, hexagonal platelets, icosahedral multiple twinned, decahedral multiple twined and irregular shaped.	15–200
3.	*Brassica juncea*	Silver, gold, copper	Triangular, spherical	2–20
4.	*Azadirachta indica*	Silver, gold and Ag core–Au shell	Polydisperse, flat, plate-like spherical, peculiar core–shell structure	5–35
5.	*Cinnamommum camphora*	Silver and gold	Triangular, spherical	55–80
6.	Rice paper plant stem	Silver	–	Below 100
7.	*Triticum aestivum*	Gold	fcc tetrahedral, hexagonal platelets, irregular shaped, rod shaped	10–30
8.	*Pelargonium graveolens*	Gold	Spherical rods, flat, sheets and triangular	21–70
9.	*Aloe vera*	Silver and gold	Triangular, spherical	6–20
10.	*Argemone mexicana*	Silver	Polygonal	20-30

Prabhu *et al.* (2009) made the first attempt of introducing silver herbal nanoparticle isolation and antimicrobial assessment. The herbal leaves like *Ocimum sanctum* and *Vitex negundo* were included to analyze the productivity of nanoparticles. The silver phyto nanoparticles were collected from each herbal plant and tested their antibacterial activity. The test cultures included in this study were *Staphylococcus aureus, Vibrio cholerae, Proteus vulgaris* and *Pseudomonas aeruginosa.* The antibacterial activities of all the herbal nanoparticles obtained from *Ocimum sanctum* showed maximum inhibitory rate using 150μg of these plants extract compared with *Vitex negundo.* The silver nanoparticles from herbal leaves showed a good antibacterial activity than the plants used. Parashar *et al.* (2009a) described the synthesis of silver nanoparticles using *Parthenium* leaf extract. UV-Vis spectroscopy was used to investigate shape and size of nanoparticles Transmission electron microscopy analysis of these particles showed that they are ~50 nm in size and very irregular in shape. SAED pattern indicate that the silver nanoparticles were single crystalline in nature. Chandran *et al.* (2006) synthesized gold nanotriangles and spherical silver nanoparticles by a simple procedure using *Aloe vera* leaf extract as the reducing agent. The kinetics of gold nanotriangle formation was followed by UV-Vis-NIR absorption spectroscopy and transmission electron microscopy (TEM). The effect of reducing agent concentration in the reaction mixture on the yield and size of the gold nanotriangles was studied using transmission electron microscopy. It was observed that the slow rate of the reaction along with the shape directing effect of the constituents of the extract are responsible for the formation of single crystalline gold nanotriangles. Reduction of silver ions by *Aloe vera* extract however, led to the formation of spherical silver nanoparticles of 15.2 nm. X-ray diffraction analysis showed that the nanoparticles were of face centered cubic structure. A comparison of radius of nanoparticles obtained from HRTEM image with the optimized cavity radius of the cyclic peptides present within the latex revealed that the particles having radius 10–20 nm are mostly stabilized by the cyclic peptides. Jain *et al.* (2009) for the first time synthesized nanoparticles using plant fruit extract. Extract of papaya fruit was used as reducing and capping agent. Nanoparticles were characterized using UV–Vis absorption spectroscopy, FTIR, XRD and SEM. Further these biologically synthesized nanoparticles were found to be highly toxic against different multi drug resistant human pathogens.

Parashar *et al.* (2009 b) and Farooqui *et al.* (2010) synthesized silver nanoparticles from leaf extracts of *Mentha piperita* and *Clerodendrum inerme.* They observed a rapid reduction of silver ions leading to the formation of stable crystalline silver nanoparticles in the solution. Nanoparticles were analysed by TEM and UV-Vis Spectroscopy. Atomic force microscopy (AFM) analysis of the nanoparticles revealed differences in sizes for the nanoparticles synthesized from different leaf conditions.

Mechanism of Action

The exact mechanism of action of silver on the microbes is still not known but the possible mechanism of action of metallic silver, silver ions and silver nanoparticles have been suggested according to the morphological and structural changes found in the microorganisms.

Mechanism of Action of Silver/Silver Ions

The mechanism of action of silver is linked with its interaction with thiol group compounds found in the respiratory enzymes of bacterial cells. Silver binds to the bacterial cell wall and cell membrane and inhibits the respiration process (Klasen, 2008). In case of *E. coli*, silver acts by inhibiting the uptake of phosphate and releasing phosphate, mannitol, succinate, proline and glutamine from *E. coli* cells (Yamanaka *et al.*, 2005). The mechanism for the antimicrobial action of silver ions is not properly understood however, the effect of silver ions on bacteria can be observed by the structural and morphological changes. It is suggested that when DNA molecules are in relaxed state the replication of DNA can be effectively conducted. But when the DNA is in condensed form it loses its replication ability hence, when the silver ions penetrate inside the bacterial cell the DNA molecule turns into condensed form and loses its replication ability leading to cell death. Also, it has been reported that heavy metals react with proteins by getting attached with the thiol group and the proteins get inactivated (Feng *et al.*, 2000).

Mechanism of Action of Silver Nanoparticles

The silver nanoparticles show efficient antimicrobial property compared to other salts due to their extremely large surface area, which provides better contact with microorganisms. The nanoparticles get attached to the cell membrane and also penetrate inside the bacteria. The bacterial membrane contains sulfur-containing proteins and the silver nanoparticles interact with these proteins in the cell as well as with the phosphorus containing compounds like DNA. When silver nanoparticles enter the bacterial cell it forms a low molecular weight region in the center of the bacteria to which the bacteria conglomerates thus, protecting the DNA from the silver ions. The nanoparticles preferably attack the respiratory chain, cell division finally leading to cell death. The nanoparticles release silver ions in the bacterial cells, which enhance their bactericidal activity (Feng *et al.*, 2000; Sondi and Salopek-Sondi, 2004; Morones *et al.*, 2005; Song *et al.*, 2006).

Effect of Size and Shape on the Antimicrobial Activity of Nanoparticles

The size of the nanoparticle implies that it has a large surface area to come in contact with the bacterial cells and hence, it will have a higher percentage of interaction than bigger particles (Kreibig and Vollmer, 1995; Mulvaney, 1996; Morones *et al.*, 2005; Pal *et al.*, 2007). The nanoparticles smaller than 10 nm interact with bacteria and produce electronic effects, which enhance the reactivity of nanoparticles. Thus, it is corroborated that the bactericidal effect of silver nanoparticles is size dependent (Raimondi *et al.*, 2005; Morones *et al.*, 2005). The antimicrobial efficacy of the nanoparticle depend on the shapes of the nanoparticles also, this can be confirmed by studying the inhibition of bacterial growth by differentially shaped nanoparticles (Morones *et al.*, 2005) (Figure 35.1). According to Pal *et al.* (2007) truncated triangular nanoparticles show bacterial inhibition with silver content of 1 μg. While, in case of spherical nanoparticles total silver content of 12.5 μg is needed. The rod shaped particles need a total of 50 to 100 μg of silver content. Thus, the silver nanoparticles with different shapes have different effects on bacterial cell.

Nanosilver as a Potent Bactericidal Agent

Microorganisms (bacteria, yeast and fungi) play an important role in toxic metals remediation through reduction of metal ions. Silver nanoparticles have been known to have inhibitory and bactericidal effect (Chu *et al.*, 1988 Deitch *et al.*, 1987; Margraff and Covey, 1977; Silver, 2003; Atiyeh *et al.*, 2007; Law *et al.*, 2008). The combined effect of silver nanoparticles with antibiotics has proven to be productive. Such effects were first observed in *Staphylococcus aureus* and *E. coli* using disc diffusion method. The antibacterial activities of penicillin G, amoxicillin, erythromycin, clindamycin, and vancomycin increased in the presence of AgNPs against both test strains. The same effect was not observed in the case of antibiotics tested. The effects of AgNPs on the antibacterial activity of the above mentioned antibiotics for *E. coli* were lower than *S. aureus*. In contrast, the most synergistic activity was observed with erythromycin against *S. aureus* (Hope *et al.*, 1989).

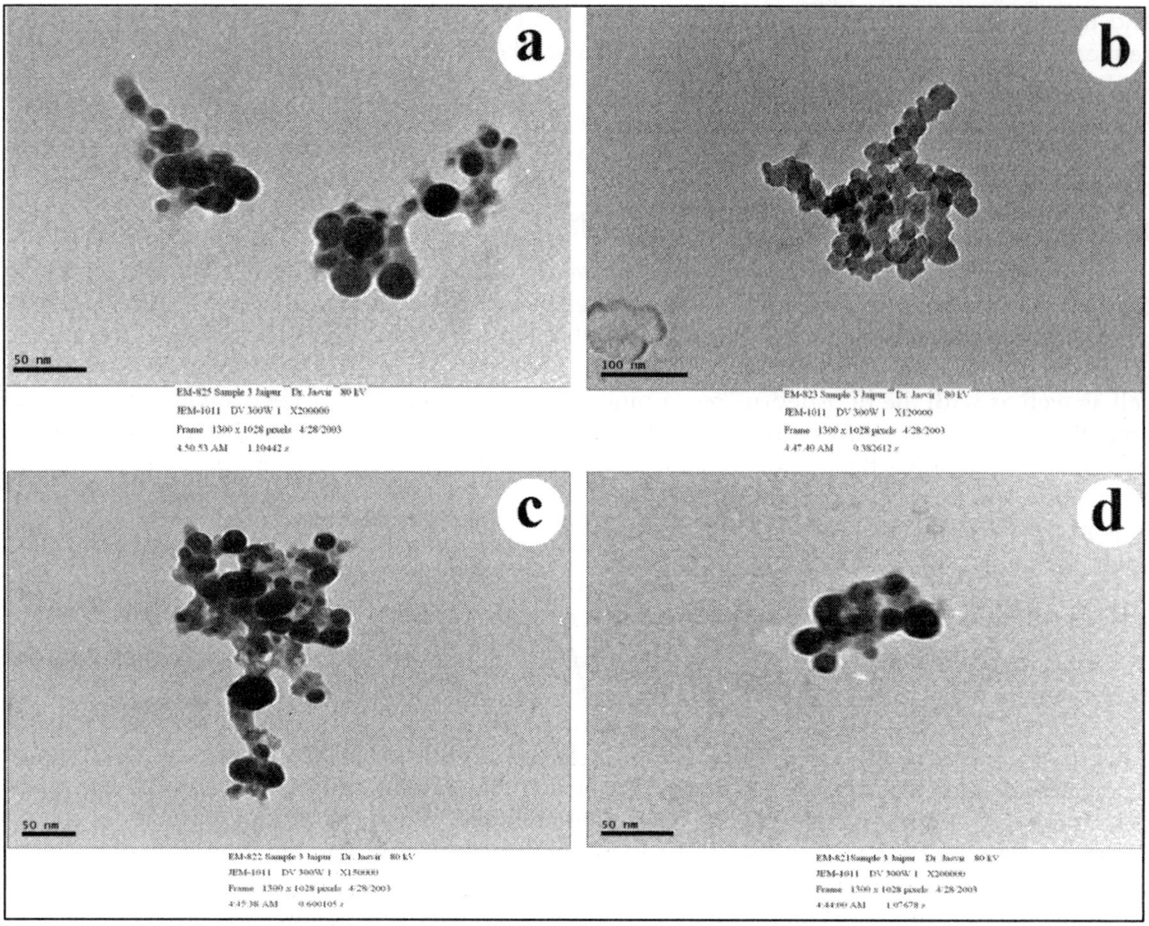

Figure 35.1: a,b,c and d are Different Shape of Nanoparticles

Nanosilver as an Antifungal Agent

Fungal infections have become more common in the recent years. In particular, fungal infections are more frequent in patients who are immunocompromised because of cancer chemotherapy, or organ or human immunodeficiency virus infections (Groll *et al.*, 1996). The availability of the antifungal drugs is limited because prophylaxis with antifungals may lead to the emergence of resistant strains. By 1980, members of the four major classes of antifungal agents–polyenes, azoles, morpholines, and allylamines–had been identified, yet the only new drug introduced for the treatment of systemic fungal infections was oral ketoconazole (Kauffman and Carver, 1997). The development of new antifungal drugs is creating new potential combination therapies. Amphotericin B, an antifungal agent used to treat serious systemic infections was used as a positive control to compare with AgNPs (Hartsel and Bolard, 1996). AgNPs showed significant antifungal activity against *T. mentagrophytes* and *Candida* species. Towards all fungal strains, AgNPs exhibited similar activity with amphotericin B, but more potent activity than fluconazole. However, this compound exhibited less potent activity than amphotericin B for *C. parapsilosis* and *C. krusei*. The ability of AgNPs to disrupt the fungal envelope structure was documented using Transmission Electron Microscope. The treated fungal cells showed significant damage, which was characterized by the formation of a "pit" in their cell walls and pores in their plasma membrane.

Nanosilver as an Antiviral Agent

Silver nanoparticles have been used to exhibit the antimicrobial efficacy against viral particles. They have been shown to exhibit promising cytoprotective activities towards HIV-infected T-cells; however, the effects of these nanoparticles towards other kinds of viruses remain largely unexplored. The research on the nanoparticles HIV-1 interaction demonstrated that silver nanoparticles inhibited HIV-1 infectivity *in vitro* by binding to the disulfide bond regions of the CD4 binding domain within the gp120 glycoprotein subunit. Binding of these nanoparticles to the gp120 subunit appeared to be size dependent as particles greater than 10 nm were not observed attached to the viral envelope (Elechiguerra *et al.*, 2005).

Conclusion

Numerous synthetic methods for Ag NPs using inexpensive biological materials were summarized and experimental approaches under different conditions which are responsible for morphology and size of the Ag particles. Rapid and green synthetic methods using extracts of bio-organisms have shown a great potential in Ag NP synthesis. However, understanding the mechanism by which the biomolecules of these systems are involved in nano material synthesis is still lacking.

References

Albrecht, M.A., Evan, C.W. and Raston, C.L. (2006). Green chemistry and the health implications of nanoparticles. *Green Chem*. 8: 417–432.

Andreescu, D., Eastman. C., Balantrapu, K. and Goia, D.V. (2007). A simple route for manufacturing highly dispersed silver nanoparticles. *J Mater Res*. 22(9): 2488-2496.

Atiyeh, B.S., Costagliola, M., Hayek, S.N. and Dibo, S.A. (2007). Effect of silver on burn wound infection control and healing: review of the literature. *Burn*. 33: 139–148.

Bar, H., Bhui, D.K., Sahoo, G.P., Sarkar, P. and De, S.P. and Misra, A. (2009). *Colloids and Surfaces A: Physicochemical and Engineering Aspects*. 339: 134–139.

Bellinger, C.G. and Conway, H. (1970). Effects of silver nitrate and sulfamylon on epithelial regeneration. *Plast Reconstr Surg*. 45: 582–505.

Berger, T.J., Spadaro, J.A., Chapin, S.E. and Becker, R.O. (1976). Electrically generated silver ions: Quantitative effects on bacterial and mammalian cells. *Antimicrob. Agents Chemother*. 9: 357-358.

Castellano, J.J., Shafii, S.M., Ko, F., Donate, G., Wright, T.E. and Mannari, R.J. (2007). Comparative evaluation of silver-containing antimicrobial dressings and drugs. *Int Wound J*. 4(2): 114-122.

Chandran, S.P., Chaudhary, M., Pasricha, R., Ahmad, A. and Sastry. M. (2006). Synthesis of gold nanotriangles and silver nanoparticles using *Aloe vera* extract. *Biotechnology Progress*. 22(2): 577-583.

Chopra I. (2007). The increasing use of silver-based products as antimicrobial agents: a useful development or a cause for concern? *J Antimicrob Chemother*. 59: 587–590.

Chu, C.S., McManus, A.T., Pruitt, B.A. and Mason, A.D. (1988). Therapeutic effects of silver nylon dressing with weak direct current on *Pseudomans aeruginosa* infected burn wounds. *J Trauma*. 28: 1488–92.

Das, R., Nath, S.S., Chakdar, D., Gope, G. and Bhattacharjee, R. (2009). Prepration of silver nanoparticles and their characterization. *Journal of Nanotechnology Online*. 10: 1343-1348.

Deitch, E.A., Marin, A., Malakanov, V. and Albright, J.A. (1987). Silver nylon cloth: *in vivo* and *in vitro* evaluation of antimicrobial activity. *J Trauma*. 27: 301–4.

Demling, R.H. and De Santi, L. (2001). Effects of silver on wound management. *Wounds*. 13: 4.

Duran, N., Marcato, P.D., Alves, O.L., D'Souza, G. and Esposito, E. (2005). Mechanistic aspects of biosynthesis of silver nanoparticles by several *Fusarium oxysporum* strains. *J Nanobiotechnol*. 3: 8–14.

Elechiguerra, J.L., Burt, J.L., Morones, J.R., Bragado, A.C., Gao, X. and Lara, H.H. (2005). Interaction of silver nanoparticles with HIV-1. *J Nanobiotechnol*. 3: 1-10.

Farooqui, M.D., Chauhan, P.S., Krishnamoorthy, P. and Shaik. J. (2010). Extraction of Silver nanoparticles from the leaf extract of Clerodendrum inerme. *Digest Journal of Nanomaterials and Biostructures*. 5(1): 577-583.

Feng, Q.L., Wu, J., Chen, G.Q., Cui, F.Z., Kim, T.N. and Kim, J.O. (2000). A mechanistic study of the antibacterial effect of silver ions on *Escherichia coli* and *Staphylococcus aureus*. *J Biomed Mater*. 52(4): 662–668.

Groll, A.H., Shah, P.M., Mentzel, C., Schneider, M., Just-Nuebling, G. and Huebner, K. (1996). Trends in the postmortem epidemiology of invasive fungal infections at a University Hospital. *J Infect*. 33: 23–32.

Hartsel, S. and Bolard, J. (1996). Amphotericin B: new life for an old drug. *Trends Pharmacol Sci*. 17: 445–9.

Hope, P.G., Kristinsson, K., Norman, P. and Elson, R.A. (1989). Deep infection of cemented total hip arthroplasty caused by coagulase-negative staphylococci. *J Bone Jt Surg Br*. 71: 851–5.

Huang, J., Li, Q., Sun, D., Lu, Y., Su, Y. and Yang, X. (2007). Biosynthesis of silver and gold nanoparticles by novel sundried *Cinnamomum camphora* leaf. *Nanotechnology*. 18:104. 1045.

Hugo, W.B. and Russell, A.D. (1982). Types of antimicrobial agents. In: *Principles and practice of disinfection, preservation and sterilization*. 106-108.

Ingle, A., Gade, A., Pierrat, S., Sonnichsen, C. and Rai, M. (2008). Mycosynthesis of silver nanoparticles using the fungus *Fusarium accuminatum* and its activity against some human pathogenic bacteria. *Curr Nano*. 4: 141–144.

Jain, D., Daima, H.K., Kachhwaha, S. and Kothari, S.L. (2009). Synthesis of plant mediated silver nanoparticles using papaya fruit extract and evaluation of their antimicrobial activities. *Digest Journal of Nanomaterials and Biostructures*. 4(3): 557-563.

Kauffman, C.A. and Carver, P.L. (1997). Antifungal agents in the 1990s: current status and future developments. *Drugs*. 53: 539–49.

Kazachenko, A.S., Legler, T.A.V., Peryanova, T.O.V. and Vstavskaya, Y.A. (2000). Synthesis and antimicrobial activity of silver complexes with histidine and tryptophan. *Pharm Chem J*. 34(5): 34–35.

Kim, J.S., Kuk, E., Yu, K.N., Kim, J.H., Park, S.J. and Lee, H.J. (2007). Antimicrobial effects of silver nanoparticles. *Nanomed Nanotechnol Biol Med.* 3: 95-101.

Kim, K.J., Sung, W.S., Suh, B.K., Moon, S.K., Choi, J.S. and Kim, J.G. (2008). Antifungal activity and mode of action of silver nano-particles on *Candida albicans*. *Biometals*. 22(2): 235-242.

Klasen, H.J. (2008). A historical review of the use of silver in the treatment of burns. Part I early uses. *Burns*. 30: 1–9.

Kreibig, U. and Vollmer, M. (1995). *Optical properties of metal clusters*. Berlin, Germany: Springer.

Landsdown, A.B.G. (2002). Silver I: its antibacterial properties and mechanism of action. *J Wound Care*. 11: 125–138.

Law, N., Ansari, S., Livens, F.R., Renshaw, J.C. and Lloyd, J.R. (2008). The formation of nano-scale elemental silver particles via enzymatic reduction by *Geobacter sulfur reducens*. *Appl Environ Microbiol*. 74: 7090–3.

Liau, S.Y., Read, D.C., Pugh, W.J., Furr, J.R. and Russell, A.D. (1997). Interaction of silver nitrate with readily identifiable groups: relationship to the antibacterial action of silver ions. *Lett. Appl. Microbiol.* 25: 279-283.

Liu, Z., Stout, J.E., Tedesco, L., Boldin, M., Hwang, C., Diven, W.F. and Yu, V.L. (1994). Controlled evaluation of copper-silver ionization in eradicating *Legionella pneumophila* from a hospital water distribution system. *J. Infect. Dis.* 169: 919-922.

Margraff, H.W. and Covey, T.H. (1977). A trial of silver–zinc-allantoine in the treatment of leg ulcers. *Arch Surg*. 112: 699–704.

Minaeian, S., Shahverdi, A.R., Nohi, A.S. and Shahverdi, H.R. (2008). Extracellular biosynthesis of silver nanoparticles by some bacteria. *JSIAU*. 17(66) :1-4

Morgan, T.P., Guadagnolo, C.M., Grosell, M. and Wood, C.M. (2005). Effects of water hardness on toxicological responses to chronic waterborne silver exposure in early life stages of rainbow trout (Oncorhynchus mykiss). *Environ. Toxicol. Chem.* 24: 1642-1647.

Morones, J.R., Elechiguerra, J.L., Camacho, A. and Ramirez, J.T. (2005). The bactericidal effect of silver nanoparticles. *Nanotechnology*. 16: 2346--53.

Moyer, C.A., Brentano, L., Gravens, D.L.,Margraf, H.W. and Monafo, W.W. (1965). Treatment of large human burns with 0.5 per cent silver nitrate solution. *Arch Surg*. 90: 812–867.

Mukherjee, P., Ahmad, A., Mandal,D., Senapati, S., Sainkar, S.R. and Khan, M.I. (2001). Fungus-mediated synthesis of silver nanoparticles and their immobilization in the mycelial matrix: A novel biological approach to nanoparticle synthesis. *Nano Lett. 1*: 515–609.

Mukherjee, P., Roy, M., Mandal, B.P., Dey, G.K., Mukherjee, P.K. and Ghatak, J. (2008). Green synthesis of highly stabilized nanocrystalline silver particles by a non-pathogenic and agriculturally important fungus *T. asperellum*. *Nanotechnology*. 19: 7.

Mulvaney, P. (1996). Surface plasmon spectroscopy of nanosized metal particles. *Langmuir*. 12: 788–800.

Natarajan, K., Selvaraj, S. and Murty, V.R. (2010). Microbial production of silver nanoparticles. *Digest Journal of Nanomaterials and Biostructures*. 5(1): 135-40.

Nomiya, K., Yoshizawa, A., Tsukagoshi, K., Kasuga, N.C., Hirakawa, S. and Watanabe, J. (2004). Synthesis and structural characterization of silver(I), aluminium(III) and cobalt(II) complexes with 4-isopropyltropolone (hinokitiol) showing noteworthy biological activities. Action of silver(I)-oxygen bonding complexes on the antimicrobial activities. *J. Inorg. Biochem.* 98: 46-60.

Oloffs, A., Grosse-Siestrup, C., Bisson, S., Rinck, M., Rudolph, R. and Gross, U. (1994). Biocompatibility of silver-coated polyurethane catheters and silver-coated Dacron® material. *Biomaterials.* 15: 753-758.

Pal, S., Tak, Y.K. and Song, J.M. (2007). Does the antibacterial activity of silver nanoparticles depend on the shape of the nanoparticle? A study of the Gram-negative bacterium *Escherichia coli. Appl. Environ. Microbiol.* 73: 1712-1720.

Panigrahi, S., Kundu, S., Ghosh, S.K., Nath, S. and Pal, T. (2004). General method of synthesis for metal nanoparticles. *Journal of Nanoparticle Research.* 6: 411-14.

Parashar, V., Parashar, R., Sharma, B. and Pandey, A.C. (2009 a). Parthenium mediated synthesis of silver nanoparticles: A novel approach towards weed utilization. *Digest Journal of Nanomaterials and Biostructures.* 4(1): 45-50.

Parashar, U.K., Saxena, P.S. and Srivastava, A. (2009 b). Bioinspired Synthesis of Silver nanoparticles. *Digest Journal of Nanomaterials and Biostructures.* 4(1): 159–166

Prabhu, N., Raj, D.T., Yamuna, G. K., Ayisha, S.S., Joseph, P. and Innocent, D. (2010). Synthesis of Silver phytonanoparticles and their antibacterial efficacy. *Digest Journal of Nanomaterials and Biostructures.* 5(1): 185-189.

Pugazhenthiran, N., Anandan, S., Kathiravan, G., Prakash, N.K.U., Crawford, S. and Kumar, M.A. (2009). Microbial synthesis of silver nanoparticles by *Bacillus* sp. *J Nanoparticle Research.* 11: 1811-15.

Raad, I.I. and Hanna, H.A. (2002). Intravascular catheter-related infections: New horizons and recent advances. *Arch. Intern. Med.* 162: 871-878.

Raimondi, F., Scherer, G.G., Kotz, R. and Wokaun, A. (2005). Nanoparticles in energy technology: examples from electochemistry and catalysis. *Angew. Chem., Int. Ed.* 44: 2190–209.

Russell, A. D. and Hugo, W.B, (1994). Antimicrobial activity and action of silver. *Prog. Med. Chem.* 31: 351-370.

Saifuddin, N., Wong, C.W. and Nuryasumira, A.A. (2009). Rapid biosynthesis of silver nanoparticles using culture supernatant of bacteria with microwave irradiation. *E-J Chem* 1149. 6: 61–70.

Silver. S. (2003). Bacterial silver resistance: molecular biology and uses and misuses of silver compounds. *FEMS Microbiol Rev.* 27: 341–53.

Sondi, I. and Salopek-Sondi, B. (2004). Silver nanoparticles as antimicrobial agent: a case study on *E. coli* as a model for Gram-negative bacteria. *J Colloid Interface Sci.* 275: 177- 182.

Song, H.Y., Ko, K.K., Oh, L.H. and Lee, B.T. (2006). Fabrication of silver nanoparticles and their antimicrobial mechanisms. *Eur Cells Mater.* 11: 58.

Song, J.Y. and Kim, B.S. (2008). Rapid biological synthesis of silver nanoparticles using plant leaf extracts. *Bioprocess Biosystem Engineering.* 32: 79-84.

Taniguchi, N. (1974). On the Basic Concept of Nano-Technology. *Proc. Intl. Conf. Prod. Eng. Tokyo, Part II*. Japan Society of Precision Engineering.

Vigneshwaran, N., Kathe, A.A., Varadarajan, P.V., Nachane, R.P. and Balasubramanya, R.H. (2007). Silver Protein (Core-Shell) Nanoparticle Production Using Spent Mushroom Substrate. *Langmuir*. 23: 7113-117.

Yamanaka, M., Hara, K. And Kudo, J. (2005). Bactericidal actions of a silver ion solution on *Escherichia coli*, studied by energy-filtering transmission electron microscopy and proteomic analysis. *Appl. Environ. Microbiol.* 71: 7589-93.

Microbial Biotechnology and Ecology (2011) *Pages* **606–615**
Editors: **Deepak Vyas, G.S. Paliwal, P.K.Khare and R.K Gupta**
Published by: **DAYA PUBLISHING HOUSE, NEW DELHI**

Chapter 36

Phenolic Quantification and Anti-Aging Activity of *Morchella esculenta*

☆ ***Javed Ahmad Wagay*, Ashok Shukla, Anjuli Choubey, Poonam Dehairya and Deepak Vyas***

Lab of Microbial Technology and Plant Pathology, Department of Botany, Dr. H.S. Gour Central University, Sagar – 470 030, M.P.

ABSTRACT

The present investigation was designed for the analysis to determine the *in vitro* antioxidant potential of a pharmaceutically important mushroom (*Morchella esculenta*). Four basic test systems *viz.*, DPPH (1,1-diphenyl-2-picrylhydroxyl radical) free radical scavenging, nitric-oxide scavenging activity, super-oxide scavenging activity and total phenolic compounds were used. Among all the scavenging tests a significant correlation existed between concentrations of the extract and per cent inhibition of free radicals. Measurement of total phenolic content of the extract was achieved using Folin-Ciocalteau reagent containing 23.42±0.89mg g^{-1} of phenolic content, which was found significantly higher when compared to reference standard gallic acid. DPPH scavenging activity of methanolic mushroom extract showed highest per cent inhibition value, which was found to be 70.8±0.89, followed by super oxide scavenging activity (52.9±0.76). The least value was found to be that of nitric oxide *i.e.* 49.04±0.23.

Keywords: DPPH, Phenols, Gallic acid, Antioxidant, Morchella esculenta.

Introduction

Morchella esculenta (Guchi or sponge mushroom) belongs to family Helvellaceae. The most therapeutic and highly priced mushroom found in Himalayan ranges of extreme north western

* E-mail: ufaq07@rediffmail.com/javediqbal49@gmail.com

extremity of Indian subcontinent–Kashmir (J and K).The state is located between 32°-17 and 37°-5 North latitude and 73°-26 and 80°-30 East longitude and its average height from the mean sea level varies from 300 to 7200 meters with average rain fall 150 cm. The temperature range is between –5°C to 28°C as minimum and maximum in January and June. Due to suitable environmental conditions, the region is rich in mushroom diversity particularly different species of *Morchella*. Its growing season is usually from 15-March to July and is mostly used as food and ethenomedicine after sun dried by the tribals of Kashmir.

Majority of the diseases / disorders are mainly linked to oxidative stress due to free radical formation (Gutteridgde, 1995). Free radicals are fundamental to any biochemical process and represent an essential part of aerobic life and metabolism (Tiwari, 2001). The most common reactive oxygen species (ROS) include superoxide, anion, hydrogen peroxide (H_2O_2), peroxyl (ROO-) radicals, and reactive hydroxyl (OH-) radicals. The nitrogen derived free radicals are nitric oxide (NO_2) and peroxynitrite anion (ONOO). ROS have been implicated in over a hundred of diseases which range from arthritis and connective tissue disorders to carcinogenesis, aging, physical injury, infection and acquired immunodeficiency syndrome (Joyce, 1987). In treatment of these diseases, antioxidant therapy has gained an immense importance. Current research is now directed towards finding naturally occurring antioxidants of plant origin. Antioxidants have been reported to prevent oxidative damages by quenching free radical formation, and may prevent the occurrence of diseases such as, cancer and aging which are directly related to reactive free radical production. It can interfere with the oxidation process by reacting with free radicals, chelating catalytic metals, and also by acting as oxygen scavengers (Buyukokuroglu, 2001). Mushrooms are being used as a source of medicine since long. The medicinal properties of mushrooms have been investigated in the recent scientific developments throughout the world, due to their potent antioxidant activities, no side effects and have economic viability (Auudy, 2003). Phenolic compounds widely distributed in mushrooms which have been reported to exert multiple biological effect, including antioxidant, free radical scavenging abilities, anti-inflammatory, anti-carcinogenic etc. (Miller, 1996). They were also suggested to be a potential iron chelators (Boyer, 1998; Harsteen, 1983). The antioxidative and free radical scavenging properties of the phenolic content of mushroom methanolic extracts have been reported, suggesting possible protective roles of these compounds, due to their ability to capture metals, inhibit lipoxygenase and scavenge free radicals (Mau, Chang, Huang, and Chen, 2004). Recently,Valentao *et al.* (2005) identified the presence of six phenolic compounds (3-, 4- and 5-O-caffeoylquinic acid, caffeic acid, p-coumaric acid and rutin) and five organic acids (citric, ascorbic, malic, shikimic and fumaric acids) in wild edible mushroom *Cantharellus cibarius* Fr., kept under four different conditions (dried, frozen, preserved in olive oil and in vinegar). The organic acids citric, ketoglutaric, malic, succinic, oxalic, ascorbic, quinic, shikimic and fumaric were also found in the edible mushrooms *viz., Amanita caesarea* (Scop.) Pers., *Boletus edulis* Bull., *Gyroporus castaneus* (Bull.) Que´l., *Lactarius deliciosus* (L.) Gray, *Suillus collinitus* (Fr.) Kuntze and *Xerocomus chrysenteron* (Bull.) Que´l. (Valentao *et al.*, 2005b).

Kashmir being the temperate region with suitable climatic conditions which favours luxuriant growth of different mushroom species including different species of *Morchella*. The aim of present work is to analyse the anti-oxidant capacity of one of the dominantly found species *Morchella esculenta* var. as little or no work has been done so far, to check its antioxidant property to the best of our knowledge is mostly used by the tribals of the region to treat different diseases particularly for chronic arthritis, general weakness and feed *Morchella esculenta* to women along with milk during and after pregnancy and claims that it provides quick recovery and resistence to the women after birth.

Material and Methods

Preparation of Mushroom Extract

Morchella esculenta fruiting bodies were collected in March–May 2009 from coniferous forests of Kashmir (J and K). The fruiting bodies were dried in an oven at 40 °C for 4 hours. The specimens were taken to Mushroom Culture Collection Centre–Sherie Kashmir Agriculture University of Science and Technology (SKAUST) Srinagar (Kashmir) for identification. The dried fruiting bodies were crushed to powder by using electronic blender. About 50g of powder were taken in 500ml of methanol in soxhlet extraction unit for extraction at 30°C for 16-18 hours and filtered through Whatman No. 4 filter paper. The residue was then, extracted with two additional 500 mL of methanol as described above. The combined methanolic extract was then rotary evaporated at 40°C to dryness, redissolved in methanol to a concentration of 10 mg/ml and stored at 4°C for further use.

Antioxidant Activity

Chemicals

1,1 Diphenyl -2- picrylhydroxyl (DPPH), butylated hydroxytoluene (BHT), L-ascorbic acid, Trolox (6-hydroxy-2-5-8-tetramethyl chroman-2-carboxylic acid), Gallic acid, Tween-20, Folin ciocalteu's phenol reagent (FCR), Sodium carbonate, Sodium nitroprusside, methanol, chloroform and the other chemicals and reagents purchased from Biochem Pharmaceutical Industries (Mumbay). All other unlabeled chemicals and reagents were of analytical grade.

Scavenging Effect on 1,1-Diphenyl-2-Picrylhydroxyl (DPPH)

The hydrogen atom or electron donating ability of the methanolic mushroom extract was measured from the bleaching of the purple-coloured methanol solution of 1,1-Diphenyl-2-picrylhydroxyl (DPPH). This spectrophotometric assay uses the stable radical DPPH as a reagent (Burits M and Bucar F., 2000). 1000µl of various concentrations (20-200µg/ml) of the extract were added to 4 ml of 0.004 per cent methanol solution of DPPH. After 20-30 min incubation period at room temperature, the absorbance was read against a blank at $\lambda = 517$ nm. Inhibition of free radical by DPPH in per cent (per cent I) was calculated by the following formula.

$$\% \text{ inhibition} = (A_{blank} - A_{sample} / A_{blank}) \times 100$$

where, A_{blank} is the absorbance of the control (Solvent), and A_{sample} is the absorbance of the test compound. Extract concentration providing 50 per cent inhibition (IC_{50}) was calculated from the graph plotted inhibition percentage against extract concentration. Tests were carried out in triplicate.

Total Phenolic Compounds Determination

Total soluble phenolics in the mushroom methanolic extract was determined with Folin-Ciocalteu reagent according to the method of Slinkard and Singleton (1977) by using gallic acid as a standard. 1.0 ml mushroom extract solution was taken in a volumetric flask and was diluted with 45ml of distilled water. 1ml Folin-Ciocalteu reagent was added and mixed thoroughly. After 5 min, 2ml of Na_2CO_3 (2 per cent) was added and the mixture was allowed to stand for 2½ hours with intermittent shaking. The absorbance of developed blue colour was measured at 760 nm. The concentration of total phenols was expressed as mg/g of the dry extract (Kim *et al.*, 2003). The concentration of total phenolic compounds in the mushroom methanolic extract was determined as µg of gallic acid equivalent by using an equation that was obtained from standard gallic acid graph.

$$\text{Absorbance} = 0.0009 \times \text{gallic acid } (\mu g)$$

Nitric Oxide Scavenging Activity Assay

Sodium nitroprusside in aqueous solution at physiological pH spontaneously generates nitric oxide, which interacts with oxygen to produce nitrite ions.This can be determined by the use of the Griess Illosvoy reaction. 2 ml of 10 mM sodium nitroprusside in 0.5 ml phosphate buffer saline (pH 7.4) was mixed with 0.5 ml of extract at various concentrations and the mixture was incubated at 25°C for 2½ hours. From the incubated mixture 0.5 ml was taken out and added into 1.0 ml sulfanilic acid reagent (33 per cent in 20 per cent glacial acetic acid) and incubated at room temperature for 5 min. Finally, 1.0 ml naphthyl ethylenediamine dihydrochloride (0.1 per cent w/v) was mixed and incubated at room temperature for 30 min. The absorbance at 546 nm was measured by UV-Vis spectrophotometer (Figure 36.11). The nitric oxide radicals scavenging activity was calculated according to the following equation:

$$\text{Per cent Inhibition} = (A_0 - A_1)/A_0 \times 100$$

where,

A_0 is the absorbance of the control and A_1 is the absorbance of test sample.

Superoxide Radical Scavenging Activity Assay

This activity was measured using NBT (nitroblue tetrazolium reagent) method as described by Sabu and Ramadasan (2002). The method is based on generation of superoxide radical (O^-_2) by auto-oxidation of hydroxylamine hydrochloride in presence of NBT, which gets reduced to nitrite. Nitrite in presence of EDTA gives a color that was measured at 560 nm. Test solutions of extract (20–200 µg/ml) were taken in a test tube. To this, reaction mixture consisting of 1 ml of (50 mM) sodium carbonate, 0.4 ml of (24 mM) NBT and 0.2 ml of 0.1 mM EDTA solutions were added to the test tube and immediate reading was taken at 560 nm. About 0.4 ml of (1 mM) of hydroxylamine hydrochloride was added to initiate the reaction then reaction mixture was incubated at 25°C for 15 min and reduction of NBT was measured at 560 nm. Ascorbic acid was used as the reference compound. Decreased absorbance of the reaction mixture indicates increased superoxide anion scavenging activity. All the extracts of mushroom extract were treated in the similar manner, absorbance was recorded and the percentage of inhibition was calculated according to the following equation:

$$\% \text{ Inhibition} = (A_0 - A_1)/A_0 \times 100$$

where,

A_0 is the absorbance of the control (blank) and A_1 is the absorbance of test samples. All the tests were performed in triplicate and the graph was plotted with the mean values.

Statistical Analysis

Data are expressed as mean ± SD. Means of triplicate analyses were calculated. A difference was considered to be statistically significant when $p < 0.05$.

Results and Discussion

Antioxidant Activity of Methanolic Mushroom Extract

Inhibition of DPPH Radical

DPPH is a stable free radical at room temperature and accepts an electron or hydrogen radical to become a stable diamagnetic molecule (Soares *et al.*, 1997). The reduction capability of DPPH was determined by the decrease in its absorbance at λ=517 nm, which is induced by antioxidants. Positive

DPPH test suggests that the methanolic extract of the mushroom was free radical scavenger. The scavenging effect of mushroom extract and ascorbic acid on DPPH radical was compared. On the DPPH radical, mushroom extract had significant scavenging effects with increasing concentrations in the range of 20–100 µg/ml when compared with that of ascorbic acid (Figure 36.1). 100 µg/ml of mushroom extract and ascorbic acid exhibited 79.8 per cent and 99.8 per cent inhibition respectively (Table 36.1). The IC50 values were found to be 67.12 and 28.42 µg/ml for mushroom and ascorbic acid, respectively (Table 36.2). The different concentrations of mushroom extract (20, 40, 60, 80 and 100 µg/ml) showed antioxidant activities in a dose dependent manner on DPPH radical.

Table 36.1: Antioxidant Activity of *Morchella esculenta* of Different Concentrations of the Extract by Per cent Inhibition on Different Free Radical Formation and Ascorbic acid (Standard)

Group	*Concentration (µg/ml)*	*Per cent Inhibition*		
		Nitric -oxide	*DPPH*	*Super-oxide*
Methanolic extract	20	37.2±0.45	39.7±0.87	36.12±0.77
of mushroom	40	40.01±0.36	49.8±0.81	40.10±0.87
	60	42.04±0.53	58.9±0.96	44.9±0.34
	80	46.02±0.65	67.9±0.94	48.10±0.65
	100	49.04±0.23	70.8±0.89	52.9±0.76
Ascorbic acid	20	48.3±0.56	64.2±0.65	46.10±0.98
(Standard)	40	50.02±0.76	74.2±0.77	50.9±0.88
	60	52.04±0.87	84.2±0.98	54.10±0.79
	80	57.03±0.74	94.2±0.86	60.8±0.87
	100	63.02±0.84	104.3±0.87	64.9±0.86

Values are: Mean ± Standard Deviation (n = 3). $p < 0.05$.

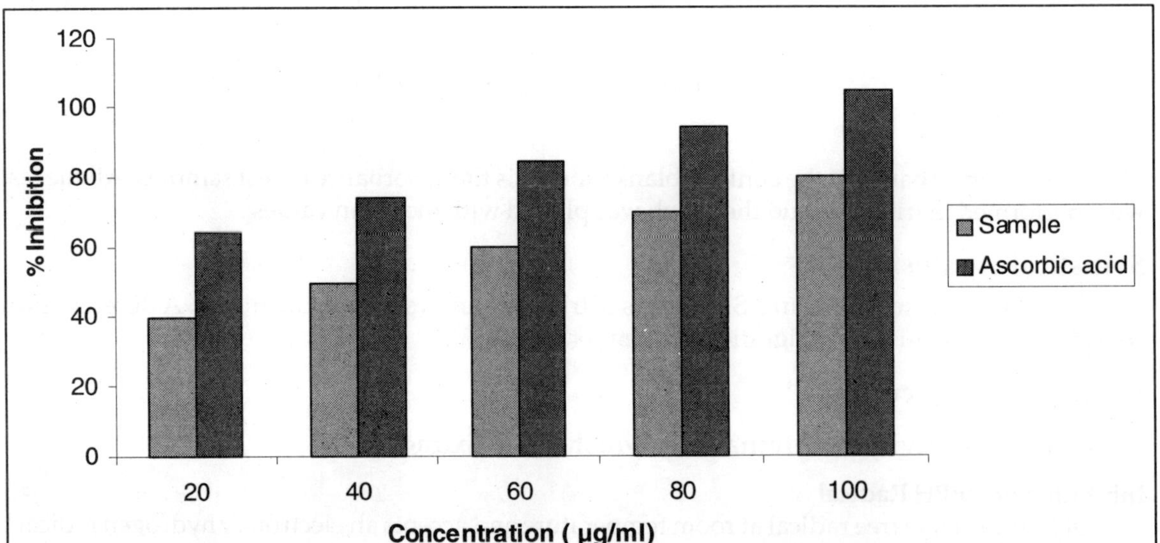

Figure 36.1: DPPH Radical Scavenging Activity of the Methanolic Extract of *Morchella esculenta* at Different Concentrations

Table 36.2: Total Phenolic Content (µg GAEs/mg extract) and IC50 (µg/ml) Values of DPPH, Super-oxide and Nitric-oxide of Methanolic Extract of *Morchella esculenta* and Ascorbic Acid (Standard)

Morchella esculenta	*Total Phenolic Compounds (µg GAEs/mg extract)*	*IC50 Values (µg/ml)*	*Mushroom methanolic Extract*	*Ascorbic Acid*
Methanolic extract	23.42±0.89	DPPH	67.12	28.42
		Super-oxide	102.14	36.23
		Nitric-oxide	114.34	67.13

Values: Mean ± Standard Deviaton (n = 3). $p < 0.05$.

Determination of Total Phenolic Content

The total amount of phenolic content present in the mushroom extract is shown in Figure 36.1, which was detected to be 47.01 ± 0.29 µg/mg gallic acid equivalent of phenols. These results suggest that the higher levels of antioxidant activity is due to the presence of phenolic components. Phenols are very important plant constituents because of their scavenging ability due to their hydroxyl groups (Hatano *et al.*, 1989). The phenolic compounds may contribute directly to antioxidative action (Duh *et al.*, 1999). It is known that polyphenolic compounds have inhibitory effects on mutagenesis and carcinogenesis in humans when ingested up to 1 g daily from a diet rich in fruits and vegitables (Tanaka *et al.*, 1998). Phenolic compounds from mushrooms are known to be good natural antioxidants. However, the activity of synthetic antioxidants was often observed to be higher than that of natural antioxidants (Ningappa *et al.*, 2007). Phenolic compounds, at certain concentrations, markedly slowed down the rate of conjugated diene formation. The interests of phenolics are increasing in the food industry because they retard oxidative degradation of lipids and thereby improve the quality and nutritional value of foods (Aneta *et al.*, 2007).

Nitric-oxide Radical Scavenging Assay

Nitric oxide plays an important role in various types of inflammatory processes in the animal body. In this study, the methanolic mushroom extract of *M. esculenta* was checked for its inhibitory effect on nitric oxide production. Nitric oxide radical generated from sodium nitroprusside at physiological pH was found to be inhibited by the mushroom extract. The various concentrations of the extract (20–100 µg/ml) showed 37.2 per cent, 40.01 per cent, 42.04 per cent, 46.02 per cent and 49.04 per cent inhibition, respectively. Results showed the percentage of inhibition in a dose dependent manner (Figure 36.2). A 100 µg/ml of the extract and ascorbic acid exhibited 49.04 per cent and 63.02 per cent inhibition, respectively. The concentration of the extract needed for 50 per cent inhibition (IC50) was found to be 114.34 µg/ml, whereas 67.13 µg/ml was needed for ascorbic acid (Table 36.2). These results were found to be statistically significant ($P < 0.05$).

Superoxide Radical Scavenging Assay

Superoxide is a reactive oxygen species, which can cause damage to the cells and DNA leading to various diseases. It was therefore proposed to measure the comparative interceptive ability of the antioxidant extracts to scavenge the superoxide radical. Several in vitro methods are available for generation of superoxide radicals (Vani *et al.*, 1997). In our study, superoxide radicals were generated by auto-oxidation of hydroxylamine in presence of NBT (nitroblue tetrazolium). The reduction of NBT in presence of antioxidants was measured. The decrease of absorbance at 560 nm with antioxidants

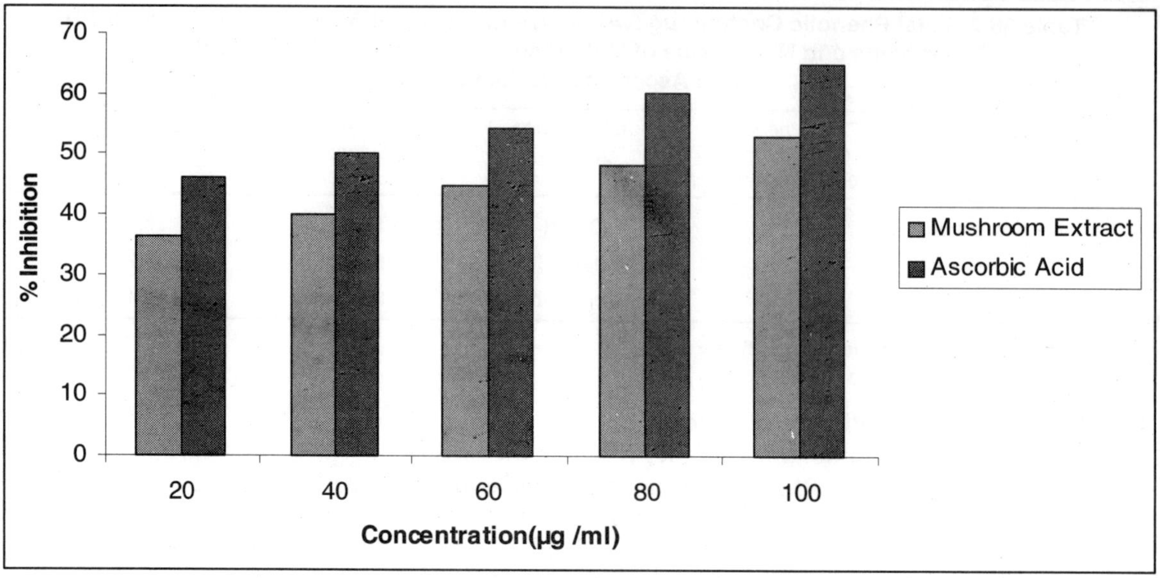

Figure 36.2: Nitric-oxide Radical Scavenging Activity of the Methanolic Extract of *Morchella esculenta*

thus indicates the consumption of superoxide anion in the reaction mixture. Different concentrations of methanolic mushroom extract (20–100 µg/ml) had strong superoxide scavenging activity (36.12 per cent,40.10 per cent, 44.9 per cent, 48.10 per cent and 52.9 per cent inhibition, respectively). A 100 µg/ml of methanolic mushroom extract and ascorbic acid exhibited 52.9 per cent and 64.9 per cent inhibition, respectively (Figure 36.3). IC50 value of methanolic mushroom extract on superoxide radical scavenging activity was found to be 102.14 µg/ml, whereas the IC50 value of ascorbic acid was found

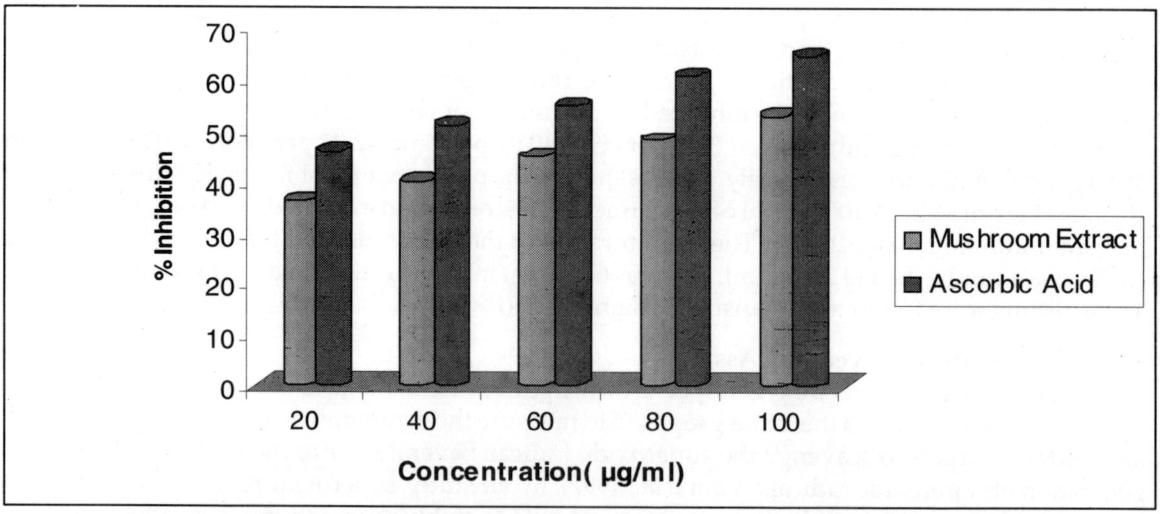

Figure 36.3: Super-oxide Radical Scavenging Activity of the Methanolic Extract of *Morchella esculenta* at Different Concentrations

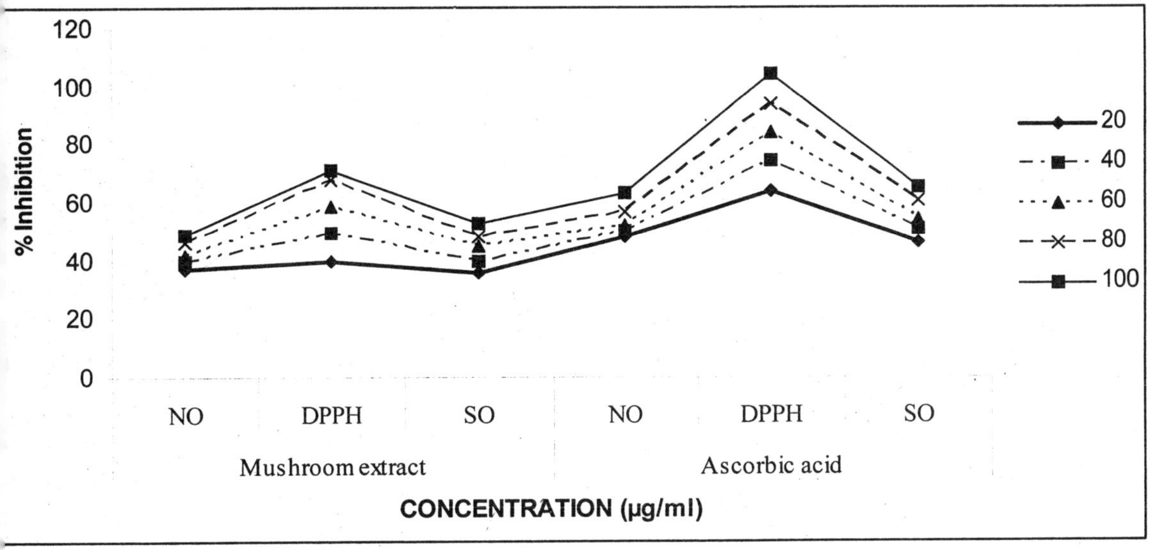

igure 36.4: The Graph Represents Comparative Study of Per cent Inhibition of Different Free Radical Formation by Different Mushroom Concentrations along with Standard (Ascorbic acid)

Values are: NO = Nitric oxide, DPPH = 1,1-Diphenyl-2-picrylhydroxyl, SO = Superoxide

ɔ be 36.69 µg/ml (Table 36.2). All of the extracts had a scavenging activity on the superoxide radicals ı a dose dependent manner (20–100 µg/ml in the reaction mixture). Nonetheless, when compared to scorbic acid, the superoxide scavenging activity of the extract was found to be low. This could be due ɔ the presence of reactive concentration of bioactive constituents and mixture of other nutrients in the xtract. Results were found statistically significant ($P < 0.05$).

Oxidative stress has been implicated in the pathology of many diseases such as 'arkinson disease, Alzheimer, diabetes, cardiovascular disoders, aging and inflammatory conditions tc. The results obtained in the present studies may be attributed to several reasons *i.e.,* scavenging of uperoxide radicals, nitric oxide radicals and DPPH radicals generated in the mammalian cells, ıvolved in the regulation of various physiological processes. However, excess production of these adicals are associated with several diseases (Ialenti, 1993). In the present study the nitrite produced y the incubation of solutions of sodium nitroprusside in standard phosphate buffer at 25 °C was educed by the methanolic extract of the investigated mushroom. This may be due to the antioxidant rinciples in the extract which compete with oxygen to react with nitric oxide thereby inhibiting the eneration of these free radicals. Superoxide dismutase catalyses the dismutation of the highly reactive uperoxide anion to oxygen and hydrogen peroxide (Kamalakkannan, 2003). Superoxide anion is the irst reduction product of oxygen (Ray, 2002), which is measured in terms of inhibition of generation f O_2.

According to the several references which reported antiinflammatory,antibacterial and antitumoral ctivities of this mushroom have previously been investigated (Mau *et al.*, 2004; Elmasta *et al.*, 2006; Jitha *et al.*, 2007; Ramirez-Anguiano *et al.*, 2007). Based on a detailed report on the antioxidant ctivity of *M. esculenta*, it has exhibited excellent activity patterns in beta-carotene/linoleic acid, reducing ower, and metal chelating effect systems (Mau *et al.*, 2004). In the light of these reports, our results are parallel and more excellent as investigated the antioxidant activity of the test mushrooms in mo advanced scavenging test systems *viz.*, nitric oxide, super oxide and DPPH systems and find best ar positive results.

References

Aneta, W., Jan, O., Renata, C. (2007). Antioxidant activity and phenolic compounds in 32 selecte herbs. *Food Chem.* 105, 940–949.

Auudy B, Ferreira F, Blasina L, Lafon F, Arredondo F, Dajas R and Tripathi PC. (2003). Screening antioxidant activity of three Indian medicinal plants traditionally used for the management neurodegenerative diseases. *J. Ethanopharmacol.*, 84: 131-138.

Buyukokuroglu ME, Oktay M and Kufrevioglu (2001). *In vitro* antioxidant properties of dantroler sodium. *Pharmacol. Res.*, 44: 491-95.

Boyer RF, Clark HM and Leroche AP (1998). Reduction and release of ferritin iron by plant phenols, *Inorq. Biochem.* 32: 171-181.

Burits M and Bucar F. (2000). Antioxidant activity of Nigella sativa essential oil. Phytother Re 14:323–328.

Cuendet M, Hostettmann K and Potterat O. (1997). Iridoid glucosides with free radical scavengir properties from Fagraea blumei. *Helv Chim Ac.* 80: 1144-1152.

Duh, P.D., Tu, Y.Y., Yen, G.C. (1999). Antioxidative activity of water extracts of Hamgjy (Chrysanthemum morifolium). Lebensm.-Wiss. *Technol.* 32: 269–277.

Dapkevicius, A., Venskutonis, R., Van Beek, T.A., Linssen, P.H. (1998). Antioxidant activity of extrac obtained by different isolation procedures from some aromatic herbs grown in Lithuania. *Journ of the Science of Food Agriculture.* 77: 140–146.

Hatano, T., Edamatsu, R., Mori, A. (1989). Effects of interaction of tannins with coexisting substance *Chem. Pharm. Bull.* 37, 2016–2021.

Joyce DA. (1987). Oxygen radicals in disease. *Adv. Drug Reac. Bull.*, 127: 476-479.

Kim, D., Jeong, S., Lee, C.H. (2003). Antioxidant capacity of phenolic phytochemicals from vario cultivars of plums. *Food Chem.* 81, 321–326.

Mau, J. L., Chang, C.-N., Huang, S.-J., and Chen, C.-C. (2004). Antioxidant properties of methanol extracts from *Grifola frondosa, Morchella esculenta* and *Termitomyces albuminosus mycelia. Fo Chemistry*, 87, 111–118.

Miller AL (1996). Antioxidant flavonoids: structure, function and clinical. *Alt. Med. Rev* 1: 103-111.

Ningappa, M.B., Dinesha, R., Srinivas, L. (2007). Antioxidant and free radical scavenging activities polyphenol-enriched curry leaf extract (*Murraya koenigii* L.). *Food Chem.*. doi:10.1016 j.foodchem.2007.06.057.

Oyaizu, M. (1986). Studies on products of browning reactions: Antioxidative activities of products browning reaction prepared from glucosamine. *Jpn. J. Nut.*, 44: 307-315.

Slinkard K and Singleton V L (1977) Total phenol analyses: automation and comparison with manu methods. *American J. Enolo. Viticul.* 28: 49–55.

Tiwari A. (2001). Imbalance in antioxidant defense and human diseases: Multiple approach of natur antioxidants therapy. *Curr. Sci.*, 81: 1179-1187.

ınaka, M., Kuei, C.W., Nagashima, Y. (1998). Application of antioxidative maillrad reaction products from histidine and glucose to sardine products. *Nippon Suisan Gakkaishi* 47: 1409–1414.

ıwari A. (2001). Imbalance in antioxidant defense and human diseases: Multiple approach of natural antioxidants therapy. *Curr. Sci.*, 81: 1179-1187.

ıbu, M.C., Ramadasan, K. (2002). Anti-diabetic activity of medicinal plants and its relationship with their antioxidant property. *J. Ethanopharmacol.* 81: 155–160.

alentao, P., Lopes, G., Valente, M., Barbosa, P., Andrade, P. B., Silva, B. M., Baptista, P., and Seabra, R.M. (2005b). Quantitation of nine organic acids in wild mushrooms. *Journal of Agricultural and Food Chemistry*, 53: 3626–3630.

Microbial Biotechnology and Ecology (2011) Pages 616–62
Editors: Deepak Vyas, G.S. Paliwal, P.K.Khare and R.K Gupta
Published by: DAYA PUBLISHING HOUSE, NEW DELHI

Chapter 37

Antibacterial Potential of *Morchella esculenta* Against Some Human Pathogenic Bacteria

Javed Ahmad Wagay*, Poonam Dehariya, Anjali C. and Deepak Vya

Lab of Microbial Technology and Plant Pathology, Department of Botany, Dr. H.S. Gour Central University, Sagar – 470 03, M.P.

ABSTRACT

Methanolic extract of *Morchella esculenta* was investigated for its *in vitro* antibacterial activity.The antibacterial activity of this mushroom was evaluated against eight human pathogenic bacteria on the basis of inhibition zones by using disc diffusion and well diffusion methods.The minimum inhibitory concentration (MIC) and minimum bacterial concentration (MBC) of the extract (µg/ml)were also determined. The lowest value of MIC and MBC were obtained for *Vibrio cholariae* (MIC-78.52µg/ml) and *E. coli* (MBC-88.76 µg/ml) and maximum value of MIC and MBC were found against *Streptococcus* sp. (MIC-97.66 µg/ml) and *Vibrio* sp. (MBC-151.05 µg/ml). The mushroom was found to have no antagonistic effect against *Enterococcus* sp.

Introduction

Morchella esculenta (Guchi or sponge mushroom) belongs to family Helvellaceae is the mos therapeutic and highly priced mushroom found in Himalayan ranges of extreme north wester extremity of Indian subcontinent–Kashmir (J and K). The state is located between 32°-17 and 37°- north latitude and 73°-26 and 80°-30 east longitude and its average height from the mean sea leve

* E-mail: ufaq07@rediffmail.com/javediqbal49@gmail.com

varies from 30 to 7200 meters. The average rain fall is 150cm and the temperature remains moderate from 20 to 30°C through out the year. Due to suitable environmental conditions the region is rich in mushroom diversity especially different species of *Morchella*. The mature ascocarp of *Morchella* consists of a stalk like portion–the stipe surrounded by hollow, conical cap. The cap is 3-6 cm in diameter, 4-8 cm long, pale brownish cream, globose to ovoid. Surface covered with irregularly interwoven pits of various shapes, framed by irregular ridges following the pits, edges of the ridges usually not darker than the pits. The stem (stipe) is 2-8 cm long, 2-4 cm in diameter, glabrous, round, much larger at the base, wrinkled, grooved longitudinally, covered with small scurfy tufts, fragile and hollow. The Asci long, cylindrical, hyaline, 280-320 × 18-22 µm in diameter. Flesh is waxy, thin, whitish and produce spores. Each spore is 18-25 × 115 µm in diameter, and ellipsoid in shape with smooth surface. The honey combed surface of pileus gives it the appearance of a sponge. It is pitted and ridged by and is about the same length as the stipe. It mostly grows in soil rich in organic matter usually in dense deciduous forest litler and fruits in the spring. Its growing season is usually from 15 April to July and is mostly used as food and ethenomedicine after sun dried by the tribals of Kashmir.

The usage of this mushroom as medicine has a long history dates back 3000 BC in the Roman and Greek civilisation (Singh, 2000). They produce a series of biologically active compounds to prevent themselves from being victimised to different microorganisms such as bacteria and pathogenic fungi. They provide a rich and varied source of terpenoids that have antibiotic and antifungal activities (Ayer and Browne, 1981). Benedict and Brady (1972) tested the potential of some mushrooms on some bacteria and reported that the best inhibitory responses were seen against some gram +ve organisms including acid fast bacterium (*M. smegnatis*) and pathogenic strains of yeast (*C. albicans*). *Morchella esculenta* is being used in traditional Chinese medicine to treat indigestion, phlegm, croup, and shortness of breath (Ying *et al.*, 1987). The anti-tumour activity of ethanolic extract of *Morchella esculenta* mycelium grown in submerged culture was determined by the mouse solid tumour model induced by Daltons Lymphoma ascite cells. Oral administration of the morel extract resulted in a 74.1 per cent inhibition in tumor volume and 79.1 per cent decrease in tumor weight 30 days after tumor cell implantation (Nitha and Janardhanan, 2005). Later research further confirmed the antitumor activity of the extract against both ascites and solid tumours (Nitha *et al.*, 2007). Methanolic extracts prepared from the mycelia of *M. esculenta* showed high antioxidant activity as compared to some antioxidants such as ascorbic acid, and α-tocopherol respectively (Mau *et al.*, 2004).

Although bioactive polysaccharides are wide spread among mushrooms, different species can produce different types of polysaccharides having different antimicrobial properties. Antimicrobial drugs have long been used for prophylactic and therapeutic purposes, but the recent increase in the drug resistant bacterial strians is creating serious problems. Consequently the antimicrobial activities of various antibacterial polysaccharides from different mushrooms are being evaluated in terms of their clinical efficacy. A lot of review is available which proved that research has been going the different aspects of *Morchella esculenta* especially its anticancerous and antidiabetic properties but little work has been reported on its antibacterial activity in Indian sub continent. So the present research was therefore designed to investigate antibacterial potiential of methanolic extract from whole ascocarp including stipe of *Morchella esculenta* against different human pathagenic bacteria.

Materials and Methods

Extraction

Morchella esculenta fruiting bodies were collected in March–April, 2009 from coniferous forests of Kashmir. The fruiting bodies were dried in an oven at 40°C for 4 hour. The specimens were taken to

Mushroom Culture Collection Centre–Sher-e-Kashmir Agriculture University of Science and Technology (SKAUST), Srinagar (Kashmir) for identification.The dried fruiting bodies were crushed to powder by using electronic blender. About 20g of powder were taken in 200ml of 95 per cent methanol in soxhlet extraction unit for extraction at 30°C for 12-18 hours. The extraction was filtered through Whatman no.4 filter paper. The methanolic extract was then vacuum rotary evaporated at 30°C to dryness. The extract was redissolved in 95 per cent methanol to a concentration of 10mg/ml and stored at 4°C for further use.

Test Microorganisms

Th antibacterial investigation of *M. esculenta* was measured against eight bacterial cultures namely *Klebsiella pneumoniae,Vibrio cholariae* (MTCC.1168), *E. coli* (MTCC.1581), *Pseudomonas auroginosa* (MTCC.1568), *Vibrio* sp., *Streptococcus* sp., *Bacilus* subtilis (MTCC.1789) and *Enterococcus* sp. Bacterial cultures were obtianed from Microbial type culture collection (MTCC) Chandigarh (Haryana) and Chandraka Pathological Laboratory, Jabalpur (M.P).The disc diffusion method outlined by the Natural committee for clinical laboratory standards (NCCLS) was used with 40µg/ml chloramphenical disc as reference antibiotic. The size of inhibitory zones were measured in millimeters after incubation period of 24 hours at 37±1°C in thermostate incubator.

Screening of Antibacterial Activity of Mushroom Crude Extract

Antibacterial property of *Morchella esculenta* methanol extract was determined by disc diffusion method (Brown *et al.*, 2005). All the bacterial strains mentioned were incubated at 37±0.1°C for 24-48 hours in Nutrient broth (NB). The culture suspension were prepared and adjusted by comparing against 0.4-0.5 MC Farland standard turbidity tubes. Nutrient agar was poured in to each sterilised Petri dish (10×100 mm) and allow it to solidify for overnight in refrigerator. The bacterial cultures (100µl of each) were inoculated on sterilized agar surface of each plate. The bacterial culture was being spread evenly on the surface by using a glass swab to and fro by rotating the plates at 60°C to ensure homogeneously distribution of inoculam. After allowing the inoculam to dry at room temperature, the disc loaded with the extract of desired concentrations and placed on the agar plates. The positive control was maintained using chloramphenicol (Antibiotic) solution of 0.40µg/ml concentration. The plates were then inoculated at 37±0.1°C for about 16-24 hours. Subsequently the plates were examined for bacterial growth inhibition and the radius of zones of inhibition around each disc was measured in millimeters respectively. The plates were made in trilicates to get the accurate antibacterial results.

Minimum Inhibitory Concentration (MIC) and Minimum Bacterial Concentration (MBC)

MIC and MBC were determined by varying the concentration of mushroom extract followed by Agar well diffusion assay (Prescottet, 2000). For MBC, the concentration ranging from 1-1000µg/ml were prepared that showed antibacterial activity during the preliminary screening. Four different concentrations were 100 µg/ml, 250 µg/ml, 500 µg/ml and 750 µg/ml respectively. A drop of each concentration of each extract was dispensed in to each well using calibrated dropper.The plates were incubated at 37±0.1°C for 18-24 hours. The minimum concentration that showed inhibition was taken as MIC (Table 37.2). Similarly the solution for MBC determination was prepared by 2 times serial decrease dilution from 290 to 100µg/ml. Wells of 6mm diameter were bored in each agar plate and each well was loaded with different concentrations to determine its MBC after incubation period of 24

48 hours at 37±0.1ºC The lowest dilution yielding no growth at all was taken to be its MBC.The MIC and MBC assay were performed thrice to make accurate results.The different values of MIC and MBC are given in Table 37.2.

Results and Discussion

The *M. esculenta* used in the present investigation has been found to have broad antibacterial potential against different tested human pathogenic bacteria.This has been proved by clear zones of inhibition. Produced by methanol extract impregnated discs of different concentrations (Figure 37.1). The results proved that the mushroom extract was highly effective against all investigated test bacteria. The best *in vitro* antibacterial activity of extract was found against *Streptococcus* sp. in which zone of inhibition was measured to be 28.33 mm at an concentration of 100 µg/ml (Figure 37.1 and Table 37.1). The second bacteria which found to be highly susceptible against extract was *Klebsiella pneumoniae* in which zone of inhibition was measured to be 25.99 mm which is followed by *Bacllus subtilis, E. coli, Vibrio* sp., *Pseudomonas aeruginosa* and *V. cholariae* with zone diameters 25.83mm, 22.3mm, 22.30mm, 18.65mm and 16.66mm respectively (Table 37.1). The results clearly shows that extract has highest antigonastic activity against *Streptococcus* sp. and lowest against *Vibrio cholariae* respectively shown by graphical representation in Figure 37.2.

Table 37.1: Antibacterial Activity of Methanolic Extract (100µg/ml) of *Morchella esculenta* Against Different Human Pathogenic Bacteria

Name of Bacteria	*Zones of Inhibition (mm)*		
	Mushroom Extract	*Positive Control*	*Negative Control*
Streptococcus sp.	28.33±0.22	36.66±0.71	12.33±0.80
K. pneumoniae	25.99±0.5	33.83±1.82	9.16±0.88
Bacillus subtilis	25.83±0.33	32.0±0.16	9.83±0.5
E. coli	22.33±0.33	30.33±0.25	12.99±0.32
Vibrio sp.	22.30±0.43	31.83±0.5	0
P. auroginosa	18.65±0.67	28.8±1.67	0
V. cholariae	16.66±0.65	24.3±1.25	6.33±0.93
Enterococcus sp.	0	28.2±0	0

Values: Mean ±Standard deviation (SD).

Positive control = Chloramphenicol (Antibiotic).

Negative control = 95 per cent methanol.

The antibacterial activity of the extract was quantitively assessed by determining the minimum inhibitory concentration (MIC) and minimum bacterial concentration (MBC) shown in Table 37.2. It had been considered that if the MIC of the extract was less than 100µg/ml, the antibacterial property was best, if MIC value is ranging from 100-500 µg/ml was value found to better and if the MIC value was found to be ranging from 500-750µg/ml was considered weak and similarly if MIC was found to be ranging between 750-1000µg/ml was considered inactive respectively (Khan *et al.*, 2009).The present study on antibacterial activity of *M. esculenta* has been found to have broad antibacterial activities as its MIC against each test bacteria was below 100µg/ml respectively. However, the highest MIC value was found against *Streptococcus* sp. (97.66µg/ml) and least against *Vibrio cholariae* (78.52µg/ml).

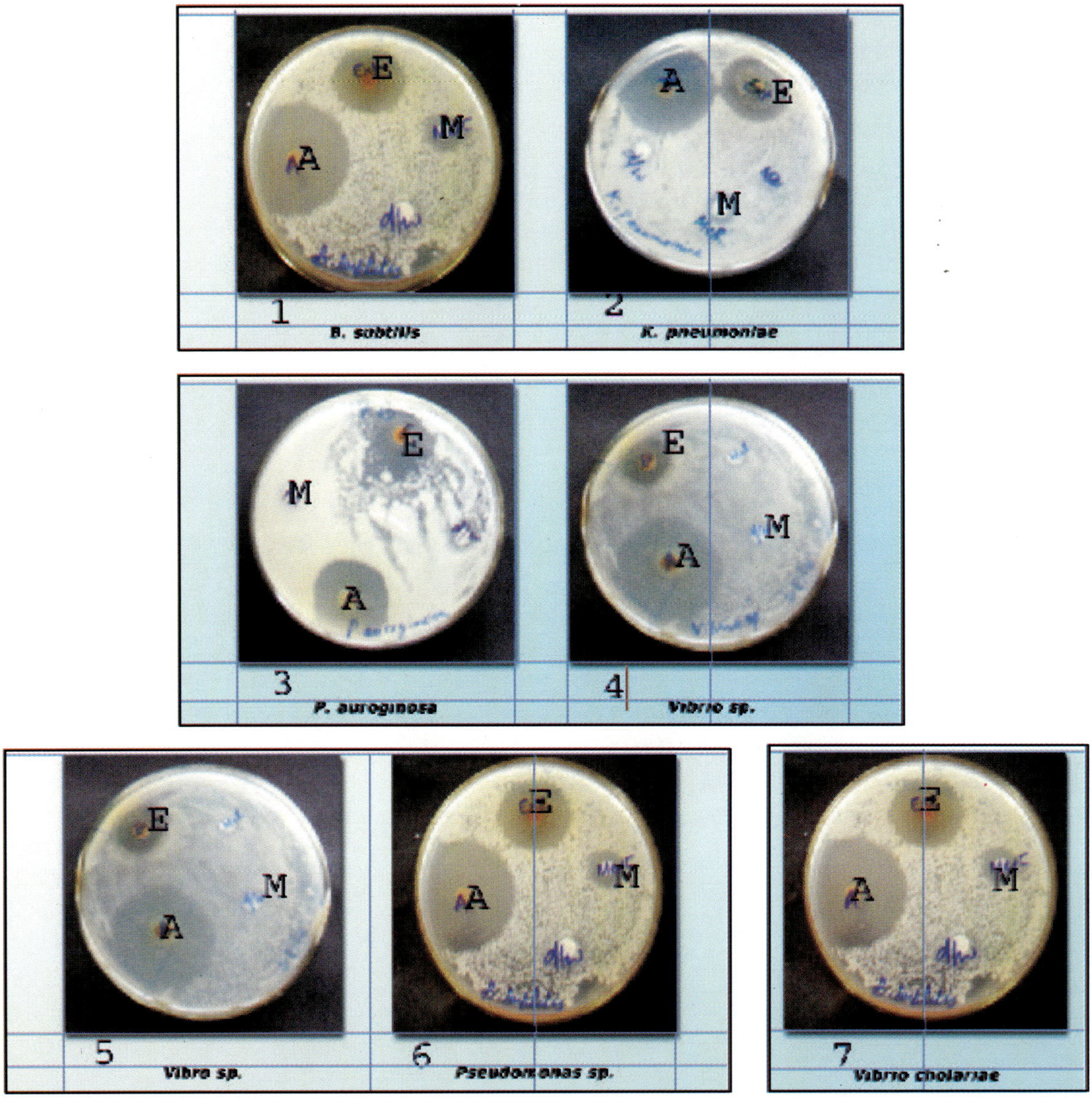

Figure 37.1: Plates Showing Zones of Inhibition by Mushroom Extract (mm), Positive and Negative Control Against Different Bacteria

Values: A = Positive control, E= Extract, M = Negative control

1: *Streptococcus* sp., 2: *K. pneumoniae*, 3: *B. subtilis*, 4: *E. coli*, 5: *Vibrio* sp, 6: *P. auroginosal* 7: *V. cholariae*

Similarly the minimum concentration at which no bacterial growth was observed was considered to be its MBC, the minimum value of MBC was found against *E. coli* (88.76µg/ml) and maximum value of MBC was found against *V. cholariae* (133.09 µg/ml). Table 37.2 also shows MIC and MBC values of mushroom extract against other bacteria as well. This clearly proved that the extract highly suppressed

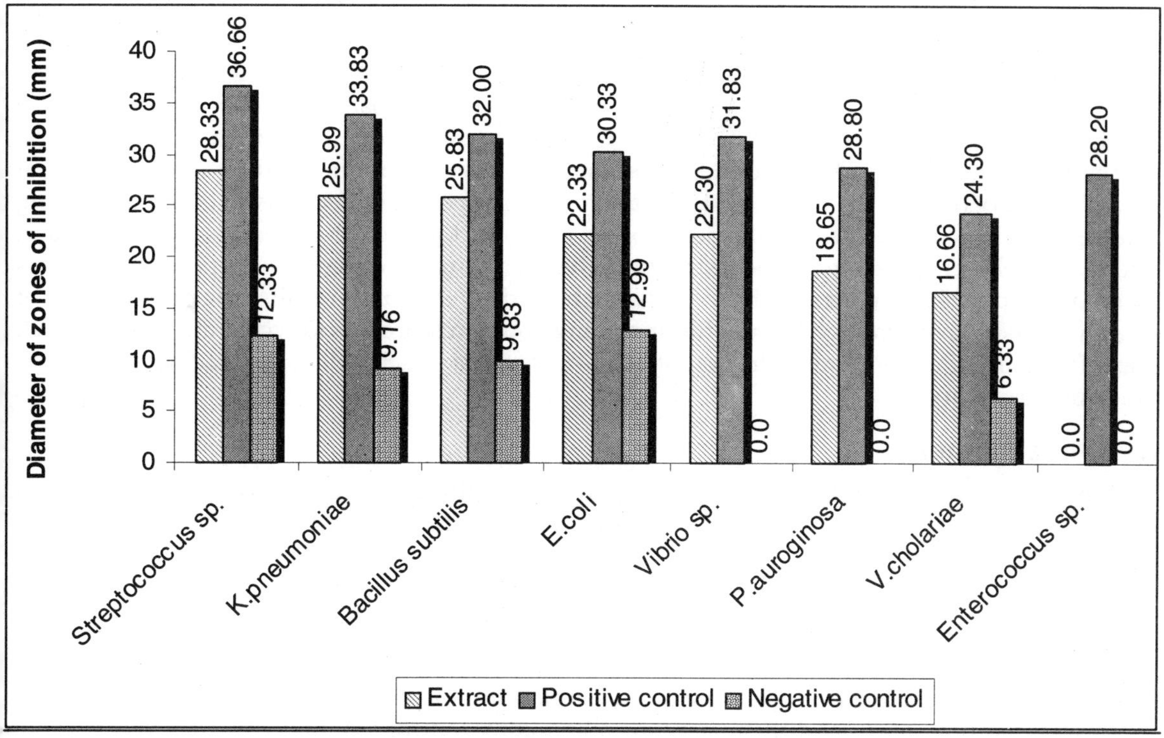

Figure 37.2: Graphical Representation of Extract Showing Different Values Against Each Bacteria

the growth and may not allow bacterial genome either to undergo normal processes of transcription and translation or may not allow for the synthesis of its cell wall which resulted the death of its strains simultaneously at very low concentrations which clearly proved beyond any doubt that *M. esculenta* is a valuable mushroom having sufficient therapeutic potential especially antibacterial activity. It also proved that most of the bio active compounds of this mushroom are highly soluble in methanol, hence methanol is a best solvent for extraction. The result also depicts that disc diffusion method is a satisfactory and result oriented method for the detection of antibacterial properties of mushrooms.

Table 37.2: Mushroom Extract of *M. esculenta* Shows Minimum Inhibitory Concentration (MIC in µg/ml) and Minimum Bacterial Concentration (MBC in µg/ml)

Bacteria	*MIC µg/ml*	*MBC µg/ml*
Streptococcus sp.	97.66	99.43
K. pneumoniae	95.87	101.12
B. subtilis	89.92	112.07
E. coli	89.68	88.76
Vibrio sp.	87.86	151.05
P. auroginosa	78.66	106.85
V. cholariae	78.52	133.09

Conclusion

The result of present investigation on the antibacterial activity of *M. esculenta* against some human pathogenic bacteria has proved that the mushroom has efficient antibacterial potential against some investigated tested human pathogenic bacteria which cause different diseases in human beings and other animals. The biological compounds possessing by this mushroom are thus acting as natural antibiotics compensating the drug resistance and can cure bacterial infections with out disturbing normal physiological processes of the body, but it needs further investigation at biochemical and molecular level for analyzing chemical structural details of bioactive compounds possessed by this mushroom responsible for degredation of different bacteria in the above results. The present piece of research will be proved enough helpful for upcoming young researchers to accelerate their research temperament in different therapeutic potential of *M. esculenta* which will be a big achievement and miracle of the century.

Acknowledgement

The authors are highly thankful to Microbial Type Culture Collection (MTCC), Chandigarh (Haryana) and Chandraka Pathological Laboratory, Jabalpur (M.P) for providing bacterial cultures.

References

Ayer, W.A and Browne, L.M. (1981). Terpenoid metabolites of mushrooms and related Basidomycetes.*Tetrahedron Report* 110(37): 2199-2248.

Brown, S.D., Traczewski M.M. (2005). Comparative *in vitro* antimicrobial activity of a new carbepenem, doripenem: Tentative disc diffusion criteria and quality control.*J. Antimicobial Chemotherapy*. 55: 944-949.

Benedict, R.C. and Brady, L.R. (1972). Antimicrobial activity of mushroom metabolites. *J. Pharm. Sci.* 61: 180-182.

Bon Chang, S.T. and Mshigeni, K.E. (2001). *Mushroom and human health: their growing significance as potent dietary supplements*. University of Namibia, Windhoek, Namibia, 79.

Bose, S.R. (1946). Antibiotics in a Polypours (*Polystictus sanguineus*). *Nature* 158: 292-296.

Ikekawa, T. (2001). Beneficial effects of edible and medicinal mushrooms on health care. *Int. J. Med. Mush.* 3: 291-298.

Kan, N.A, Qureshi, S, Pandey, A.K and Ashutosh, S. (2009). Antibacterial activity of some medicinal mushrooms of central India. *Tur. J. Biol.* 33: 1-5.

National Committee for Clinical Laboratory Standards (NCCLS). Performance Standards for antimicrobial susceptibility tests, Approved standard NCCLS Publication M2-A5, Villanova, PA, USA, 1993.

Prescott, L.M, Harley, J.P. and Klein, D.A. (2002). *Microbiology* 5th edn; MC Graw-Hill. New York, N.Y. 10020: pp. 809-811.

Singsh, S.K., Upadhyay, R.C. and Verona, R.N. (2000). Physico-Chemical preferences for efficient mycelial colonization in edible mushrooms. *Mush. Res.* 9(2): 85-89.

Nitha, B., Janardhanan, K.K. (2005). Antioxidant, anti-inflammatory and antitumor activities of cultured mycelia of morel mushroom, *Morchella esculenta*. *Ind J. Med. Res.* 121: 133.

Nitha, B., Meera, C.R, Janardhanan, K.K. (2007). Anti-inflammatory and antitumor activities of cultured mycelium of morel mushroom, *Morchella esculenta. Curr. Sci.* 92(2): 235-239.

Mau, J.L, Chang, C.N, Huang, S.J, Chen, C.C. (2004). Antioxidant properties of methanolic extracts from *Grifola frondosa, Morchella esculenta* and *Termitomyces albuminossus* mycelia. *Food Chem.* 87(1): 11-18.

Ying J, Mao X, Ma Q, Zong Y, Wen H. (1987). *Icones of Medicinal Fungi from China.* Science Press, Beijing. 14: pp. 38-45.

Microbial Biotechnology and Ecology (2011) *Pages* **624–665**
Editors: **Deepak Vyas, G.S. Paliwal, P.K.Khare and R.K Gupta**
Published by: **DAYA PUBLISHING HOUSE, NEW DELHI**

Chapter 38

Production and Partial Characterization of Fungal and Bacterial Cellulases on CMC and Cellulose Powder

☆ *Deepak Vyas*[1] *and Ashish Vyas*[2]

[1]*Lab of Microbial Technology, Department of Botany, Dr. H.S.G. University, Sagar*
[2]*Department of Biotechnology, Lovely University, Jalandhar, Punjab*

ABSTRACT

The endoglucanases, exoglucanases and β-glucosidases (cellulases complex) of fungi and bacteria were determined and presented in chapter 5. It was observed that fungal spp. have much more potential to produce cellulase enzymes than bacterial spp. Among fungi, *Aspergillus terreus, Aspergillus nidulans, Aspergillus flavus, Fusarium oxysporum, Penicillium rubrum, Phoma* sp., *Trichoderma viride* were found to be effective producer of all the three enzymes tested. Among bacterial genera, *Bacillus marinus, Bacillus* sp. (V2) and *Bacillus subtilis* were promising candidate for cellulolytic activity. *Aspergillus terreus, Aspergillus nidulans, Fusarium oxysporum, Bacillus marinus, Bacillus* sp. (V2) and *Bacillus subtilis* were selected for detailed study for their enzymatic potential. All the three enzymes (endoglucanases, exoglucanases and β-glucosidases) were found optimally active at acidic pH range in selected three fungi and bacterial spp. However, *Fusarium oxysporum* was found to show stable endoglucanase activity at alkaline pH range too. 60°C was found optimum for the maximum activity of endoglucanase and β-glucosidase and 50°C temperature for exoglucanases by *Aspergillus nidulans. Aspergillus terreus* was optimally active at pH 4 and 40°C for all the three enzymes tested. The increasing importance of endoglucanase enzyme in different industrial applications has placed this enzyme to be more important among other cellulases enzymes. Therefore to know the functioning of endoglucanase enzyme an attempt has been made to know the effect of some chemical reagent (EDTA, 2, 4 DNP, $HgCl_2$, $SnCl_2$). EDTA was found to be moderately affecting the endoglucanase activity in all the test organisms. 2, 4 DNP and $HgCl_2$ were found to inhibit almost complete endoglucanase activity (more than 90 per cent) at 5mM in all the test organisms. $SnCl_2$ showed moderate inhibitory effect on endoglucanase activity at 5mM concentration in the test organism.

Introduction

Cellulases refer to a group of enzymes which acting together, hydrolyze cellulose. Cellulose, is a homopolymer of about 8,000-12,000 glucose unit linked linearly by β-1,4-gycosidic linkages forming a crystalline unit held together by hydrogen bonding (Brown, 1982). Based on this structure, Reese *et al.* (1950) proposed a hypothesis that enzymatic hydrolysis of cellulose requires at least two enzymes C_1 and Cx. According to this hypothesis, C_1 enzymes render the crystalline cellulose accessible for hydrolytic attack and a Cx enzyme then hydrolyze the modified cellulose interaction of the original model of the C_1–Cx cellulolytic systems, as proposed by Reese *et al.*, 1950.

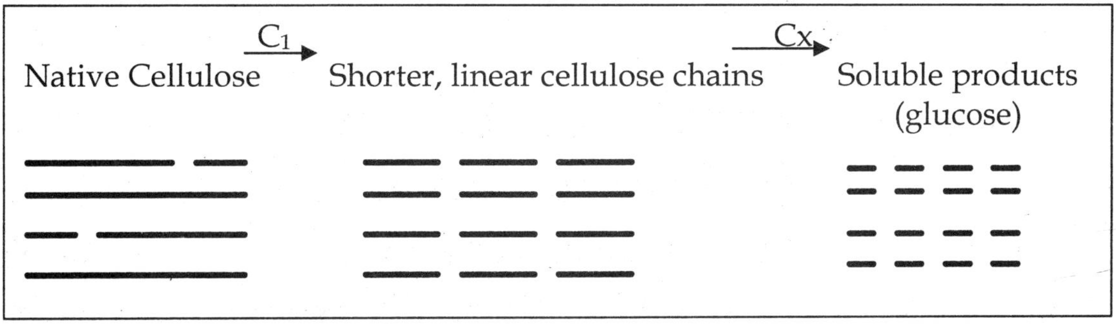

However, the original model of the C_1–Cx cellulolytic system as proposed by Reese *et al.* (1950). However, the precise action of C_1 was not made clear. Subsequently, Reese's suggestion was misinterpreted Cx was an endoglucanase that acted before C_1, a cellobiohydrolase (Wood *et al.*, 1989). The C_1 factor in 60s and 70s year of decade was not isolated and purified isolated and this model of cellulose degradation is now regarded as antiquated (Din *et al.*, 1994) and existence of C_1 enzyme is still obscure. However, furthermore considerable progress has been made in the understanding of the nature and function of cellulolytic enzymes therefore, the definition was modified. There are atleast three major types of hydrolytic enzymes involved in the degradation of native cellulose to glucose (Bhat and Bhat, 1997).

1. The endo-(1,4)-β-D-glucanase (synonyms: endoglucanase, endocellulase, carboxymethyl cellulase [EC 3.2.1.4]), which cleaves β-linkages at random, commonly in the amorphous parts of cellulose.
2. The exo-(1,4)-β-D glucanase (synonyms: cellobiohydrolases, exocellulase, microcrystalline cellulase, avicelase [EC 3.2.1.91]), which releases cellobiose from either the nonreducing or the reducing end, generally from crystalline part of cellulose.
3. The β-glucosidase (synonym: cellobiase [EC 3.2.1.21]), which releases glucose from cellobiose and short chain oligosaccharides.

Cellulases are distributed throughout the biosphere; they are most manifest in fungi and bacterial sources (Enari, 1983). In microbial sources these cellulolytic enzymes are either cell bound or extracellular. The ability to produce extracellular cellulolytic enzymes is widespread in fungi and bacteria and these systems have been extensively studied (Enari, 1983; Wood, 1985), however fungi are the most studied organisms with respect to cellulose degradation. Generally fungi hydrolyse cellulose to simple soluble sugar and utilize them as a carbon source in their metabolism. Cellulase biosynthesis is generally assumed to be subject to genetic and biochemical regulation. Cellulases are extracellular, inducible enzymes, the best inducer being cellulose, cellobiose, lactose and sophorose

(2-O-β-D-glycopyranosyl-D-glucose). Synthetic compounds as palmitate and acetate esters o disaccharides and thiocellobiose have also been shown to function as inducers of cellulases (Reese e *al.*, 1969). In virtually all microorganisms examined the synthesis of cellulases is induced by presenc of cellulose and repressed in the presence of glucose. The most generally accepted view of the inductio process is that the organism produced a basic level or a constitutive amount of cellulase thus givin rise to soluble hydrolysis product of cellulose, which enter the microbial cells and functions as induce (Eriksson *et al.*, 1990). Cellulases catalyze the breakdown of the β-1,4 glucosidic linkages. It has bee proposed that cellulase hydrolysis occurs by an acid catalysis mechanism at the active site (Rees *et al.*, 1950; Knowles *et al.*, 1987) proposed a two step mechanism that for a cellulase to attack crystallin cellulose two factors were required. Cx factor or carboxymethylcellulase is able to attack solubl derivatives and amorphous cellulose, but without activity against crystalline cellulose (*e.g.* cotton). C was a factor which enabled Cx to act on cotton or native cellulose mechanism being by disrupting th structure of cellulose by breaking H-bonds. It is now well established fact that the C_1 fraction is an exo acting β-1,4-glucanase (often referred as exocellulase or cellobiohydrolase) removing cellobiose fron non reducing end of the polymer. It is now widely accepted that C_1 and cellobiohydrolase are identica (Gong *et al.*, 1979). The third member of the complex is β-1,4 glucosidase (cellobiase) which hydrolyz cellobiose to glucose. According to mechanism proposed by Eriksson (1969), amorphous region in th cellulose fibre are first attacked by endoglucanases, thereby exposing free chain ends. Exo-glucanase then come into action by hydrolytically removing cellobiose from the chain ends. The free cellobiose i hydrolyzed further to glucose through action of β-glucosidases. In general, cellulases are bimodula proteins with large catalytic and small cellulose binding module linked by short highly glycosylate protein sequence rich in Serine and Proline (Gilkes *et al.*, 1991). Most of the fungal cellulases ar glycoproteins and exist in multiple forms (Coughlan and Ljunghdahl, 1988). These multicomponen enzymes are required to effectively participate in enzymatic reactions and physical homogeneity o the substrate. Proteolytic cleavage of cellulase separates the cellulose binding and catalytic fragment demonstrated the presence of a two-domain organisation in these enzymes (Gilkes *et al.*, 1988; Tomm *et al.*, 1988). The genes encoding cellulases are chromosomal in both fungi and bacteria. In fung cellulase genes are randomly distributed over the genome, with each gene having its ow transcriptionally regulatory elements (Tomme *et al.*, 1995). In bacteria, the cellulase genes are eithe randomly distributed in *Clostridium thermocellum* (Guglielmi and Beguin, 1998) or clustered on th genome in *C. cellulolyticum, C. cellulovorans, C. acetobutylicum* (Belaich *et al.*, 1997; 1999). Cellulase enzymes are extracellular enzyme systems, strictly defined as those enzymes that have passed throug the cytoplasmic membranes of the cell originally responsible for synthesis (Priest, 1984). Curren theory indicates that these enzymes pass through the cell membranes in linear form and assume thei tertiary structure outside the cell known as cotranslational secretion (Davis and Tai, 1980).

The best known fungal forms producing complete cellulase system is *Trichoderma reesei* (Mande and Reese, 1964; Fagerstam and Pettersson, 1980) *Trichoderma koningi* (Wood and McCrae, 1972, 1975 1986a; Fan *et al.*, 1987), *Penicillium funiculosum* (Wood *et al.*, 1980; Wood and McCrae, 1982; Wood an McCrae, 1986b), *Fusarium solani* (Wood, 1969, 1975; Wood and McCrae, 1977), *Fusarium oxysporun* (Christakopoulas *et al.*, 1989; Christakopoulas *et al.*, 1995a,b) and *Aspergillus niger* (Singh *et al.*, 199C 1991) are other well known cellulolytic fungi. Anaerobic rumen inhabiting fungi also have the capabilit to degrade cellulose. The cellulolytic enzyme system is one of the rumen anaerobic fungi, *Neocalimasti frontalis*, has been studied by Wood *et al.* (1986).

Cellulose degradation in nature by bacterial system occur aerobically and anerobically. Bacteria degradation of cellulose seems to take by cluster of enzymes rather than by individual endo an exoglucanases as in the case of fungi (Ljungdahl and Eriksson, 1985). Cellulases of aerobic bacteria such as the *Cellulomonas* sp., appears to be tightly bound to their cellulose substrate and appear in the culture medium only in stationary phase (Langsford *et al.*, 1984). Various cellulolytic spp. of *Bacillus* such as *B. subtilis, B. polymyxa, B. brevis, B. licheniformis* and *B. cereus* have been described by number of workers (Knosel, 1971; Priest, 1977; Thayer, 1978; Robson and Chambliss, 1984, 1987).

Endoglucanase (CMCase) hydrolyze Carboxy Methyl Cellulose (CMC) or swollen cellulose in random fashion resulting in rapid decrease in chain length (Wood, 1985). Endoglucanases are thought to be primarily responsible for decreasing degree of polymerization by internally cleaving cellulose chain at relatively amorphous regions, thereby generating new cellulose chains ends susceptible to the action of cello biohydrolases (Teeri *et al.*, 1998). *Trichoderma reesei* produced three different endoglucanases *i.e.* EGI, EGII and EGIII, which also occur in multiple forms (Labudova and Farkas, 1983).

Exoglucanases act in a processive manner on the reducing or non-reducing ends of cellulose polysaccharides chains, liberating either glucose (glucanohydrolases) or cellobiose (cellobiohydrolases) as major products. Exoglucanases acts on microcrystalline cellulose, presumably peeling cellulose chain from the microcrystalline structure (Teeri, 1997). Cellobiohydrolases or exo-glucanases degrade cellulose by hydrolyzing cellobiose units from non reducing end of the chain and have very limited action on substituted celluloses such as CMC and hydroxy ethyl cellulose (Bisaria and Mishra, 1989). Exoglucanases exhibit highly cooperative synergistic action in the presence of endoglucanases (Enari and Niku-Paavola, 1987; Bisharia and Mishra, 1989). Heterogeneity of cellobiohydrolases components from *T.reesei* and *Penicillium pinophillum* has been studied in detail (Wood and McCrae, 1986a; Wood *et al.*, 1988). A general feature of most cellulases is a modular structure of including both catalytic and carbohydrate modules (CBMs). The presence of CBMs is particularly important for the initiation and processivity of exoglucanases (Teeri *et al.*, 1998). Cellobiohydrolases activity is essential for the hydrolysis of microcrystalline cellulose. CBH I and CBH II are the principle components of the *T. reesei* cellulase systems representing 60 per cent and 20 per cent respectively, of the total cellulase protein produced by the fungus on the mass basis (Wood, 1992).

β-glucosidases or cellobiase (EC 3.2.1.21), one of the component of cellulase complex represent very small portion of the total extracellular protein secreted by *T. reesei*. β-glucosidase hydrolyzes degradation products of cellulose to glucose. β-glucosidases are characterized by their transferase activity, retention of the β-configration of glucose on hydrolysis, inhibition by low concentration of glucanolactone, and high activity toward cellotriose and less activity toward larger cellodextrins (Bisaria and Mishra, 1989). The purification of β-glucosidases from extracellular culture filtrates, intracellular extracts, and plasma membranes of organisms has reported by several groups (Bergem and Pettersson, 1973; Singh *et al.*, 1990). However, the exact genetic and biochemical relationship among these different β-glucosidases is not yet clear. Glycosides such as salicin (Mandel and Reese, 1964) are commercially available and have been used as substrate for determining β-glucosidase activity.

Cellulase activity in fungi and bacteria were greatly dependent upon factors like pH and temperature, and incubation period. In order to obtain economic hydrolysis of cellulose with maximum saccharification in short span of time it is essential to determine the optimum conditions. Enzymes being proteins contain ionizable groups and consequently the pH of the culture medium affects their structures and function.

In view of the above, an attempt has been made to investigate the cellulolytic capabilities of selected fungal and bacterial cultures. The present chapter is planned as follows:

- Determination of endoglucanase (CMCase), exoglucanase (FPase) and β-glucosidase enzyme activity in some fungal and bacterial cultures.
- Partial characterization of selected fungal and bacterial cultures *i.e.* determining its optimum pH and temperature range.
- Effect of some chemical reagents on endoglucanase (CMCase) activity of fungal and bacterial cultures.

Materials and Methods

Collection of Samples

The samples were collected from previously decomposing organic cellulosic and lignocellulosic waste subjected for degradation. For this different cellulosic substrates *viz.* Sugarcane bagasse waste, Sagaun saw dust, Mushroom wheat compost, Mushroom soyabean compost, decomposing *Polyalthia pendula* leaves manure, decomposing organic cellulosic waste, decomposing cellulosic laboratories waste, like cotton pieces, potato pith, paper waste etc were taken. For the above mentioned substrates (5 kg. each) were degraded in the pits of size 2ft. × 2ft. × 2ft. in the Botanic Garden of Department of Botany, Dr. Hari Singh Gour University, Sagar. These substrates were environmentally degraded for the period of 5 months from 28.09.01 to 28.02.02. After the appropriate time interval, 100 gm of decomposing residue is collected in the sterilized polythene bags and brought in the laboratory. The fungi were also isolated from Farmyard dung manure, nadib manure, chicken manure, decomposing newspaper, decomposing logs of wood, bark of *Magnifera indica* and garden soil and deseased portion of Lemon fruits (*Citrus* sp.), *Cicer arietinum, Saccharum officinarum.*

Isolation of Fungal and Bacterial Cultures

The decomposing residue brought in the laboratory was kept at 4°C±1. Isolation of fungi from the samples of different substrates was carried out using Czapek Dox Agar Medium (Dox, 1910).

Czapek Dox Agar Medium (gm/lt.)

Sodium nitrate $NaNO_3$	3.00
Dipotassium hydrogen phosphate K_2HPO_4	1.00
Magnesium sulphate $MgSO_4.7H_2O$	0.50
Potassium chloride KCl	0.50
Ferrous sulphate $FeSO_4.7H_2O$	0.01
Sucrose	30.00
Agar	15.00
Distilled Water	1000.00 ml
pH	5.5

Fungi are isolated by direct agar plate method as per the method of Warcup (1955). A small amount of the sample (0.005-0.015 g) was taken from the sample with the help of sterile flattered tip needle and was dispersed in a drop of sterile water in the bottom of petridish. The sterilized Czapek

Dox Agar medium was then poured into the petri dish. The plated sample was dispersed in the plate by rotating the medium before its solidification. After this plates were kept in incubator at 28±1°C for their growth.

Isolation of cellulolytic bacteria was performed by enrichment techniques suggested by Hans and Srinivasan (1968). About 1 gm mixture of rotting cellulosic substrates was inoculated into the isolation media. Isolation media consisted of mineral salts solution supplemented with 0.1 per cent yeast extract and strip of filter paper. The mineral salt solution contain : NaCl, 6.0 gm; $(NH_4)_2SO_4$, 1.0 gm; KH_2PO_4, 0.5 gm; K_2HPO_4, 0.5 gm; $MgSO_4$, 0.1 gm; $CaCl_2$, 0.1 gm; and 1 lt. of water.

After 3 to 7 days of incubation at 30°C on reciprocal shaker, a patch of a yellow pigmented material appeared at the liquid air interface on the filter paper. As soon as the pigmented material appeared, a portion of filter paper was transferred with sterile wire and inoculated into fresh medium. The process was repeated several times to enrich aerobic and mesophilic cellulose utilizing organisms. The filter paper from the enriched culture was removed, macerated in a small amount of sterile water, and streaked on the plates containing carboxy methyl cellulose agar plates.

Representatives of the various colonies which developed on each of this media were further purified by pure culture techniques.

Purification of Fungal and Bacterial Isolates

The contaminated cultures of fungal isolates were purified and derived through single spores, using dilution plating. To ensure purity, cultures were derived from single spore in each case. The bacterial contamination was prevented by using antibiotic streptomycin in the medium during isolation and purification of fungi. After ensuring purity, the cultures were maintained in PDA slants at 4°C.

Purification of bacterial cultures was done by pour plate method. In these method four test tubes, each containing 10 ml of distilled water was taken. Test tubes were labelled from 1-4. Inoculate 1 ml. of mixed bacterial suspension into test tube no. 1 with the help of sterile pipette. Test tubes were shaked slowly to obtain uniform suspension from test tube no. 1 to test tube no. 2 and further 1ml. of bacterial suspension each from test tube no. 2 to test tube no. 4 each time by repeating the step. Prepare sterilized petri plates from no. 1-4. Transfer 1 ml of bacterial suspension each from test tube no. 1-4 to petri plates no. 1-4. Prepare nutrient agar medium and pour into petri plates and rotate the plates gently to ensure uniform distribution in plates. Invert the Petri plates after solidification and incubate at 37°C±1 for 24 hrs.

Congo Red (CR) Test

Cellulase (endoglucanase) activity is determined by method of Teather and Wood (1982) with slight modification. In this method, Mandels and Reese medium is prepared with following constituents:-

Mandels and Reese medium (gm/lt.)

☆ Proteose peptone	1.0
☆ Ammonium sulphate $(NH_4)_2SO_4$	1.4
☆ Potassium dihydrogen phosphate KH_2PO_4	2.0
☆ Urea NH_2-CO-NH_2	0.3
☆ Magnesium sulphate $MgSO_4.7H_2O$	0.3
☆ Calcium chloride $CaCl_2$	0.002

☆ Ferrous sulphate $FeSO_4.7H_2O$	0.005
☆ Mangnese sulphate $MnSO_4.H_2O$	0.001
☆ Zinc chloride $ZnCl_2$	0.017
☆ Carboxy Methyl Cellulose (CMC)	10
☆ Distilled water	1000 ml
☆ pH	5.3

Mix all the constituent in flask and autoclave the medium at 15 psi for 15 min., 25 ml of Mandel and Reese medium was poured in 150 ml of sterilized flask. The flask was than inoculated with a spore/mycelial suspension by adding 2 ml. of sterile distilled water to freshly grown fungal agar slants (Reddy *et al.*, 1998) and incubated at 28±°C for 6 days. After appropriate incubation period, sample was withdrawn and centrifuged at 9500 rpm for 20 min in Centrifuge. The cell free supernatant is used as crude enzyme source. Cellulase activity is then revealed by pouring Mandel and Reese agar plates medium supplemented with CMC (1.2w/v). After solidification of medium, cylindrical probes were made in the centre of Petri plates of 6 mm diameter with the help of cork borer. These wells are then filled with crude enzyme of test organisms. Petri plates were then incubated at 28°C±1 for 2 days. After incubation the Petri plates are flooded with 10 ml of 0.1 per cent aqueous solution of Congo Red. The colouration is terminated after 20 min by pouring off the dye and flooding the plates with 10 ml of a 5 M NaCl solution. After an additional 20 min, the salt solution is discarded and endoglucanase activity is revealed by pale reddish zone around the well. In case of bacterial cultures, petri plates of Mandels and Reese agar medium supplemented with CMC (1.2 w/v) was streaked by bacterial test organism and incubated at 37±2°C for 48 hrs. After incubation period, the same process as followed in fungi were followed but instead of 0.1 per cent, 0.5 per cent Congo red is taken to reveal pale reddish zone around the bacterial colony. The zone around the well (in fungi) and bacterial colony were measured in mm. A control consisting of Mandels and Reese agar medium without CMC was also run simultaneously with the test.

Viscometric Method

CMCase (β-1, 4 glucanase) activity of the test organism was evaluated using Fenske Ostwald Viscometer (efflux time for water = 15 sec). The preparation of crude enzyme for fungi is mentioned under congo red test. In bacterial cultures, Mandel and Reese broth medium supplemented with CMC (1.2 w/v) is prepared and 25ml. of medium is dispensed in 150 ml. of flask and autoclaved at 15 psi for 15 min. The flask was then inoculated with 2 ml. of test bacterial suspension in nutrient broth which gave the absorbance of 1.2 at 660 nm in spectrophotometer. The flasks were then incubated in BOD incubator at 37±2°C for 2 days. After incubation, aliquot (5 ml.) is taken and centrifuged at 9500 rpm for 20 min. The supernatant obtained after centrifugation is used for viscometric analysis.

Reaction mixture consists of CMC solution (1.2 per cent w/v), citrate buffer (pH 4.8) and crude enzyme of test organism in the ratio of 4:1:2. The mixture is poured into the small arm and efflux time (ET_O) is immediately determined. Viscometer is then kept at water bath at 50°C. The efflux time is then determined after 60 min of incubation. A control consisting of enzyme boiled at 100°C is also run simultaneously with the test. Efflux time of reaction mixture is taken after 60 min and percent loss in viscosity of CMC is calculated using the formula of Capelline (1966):

$$V = \frac{ETo-ETt}{ETo-ETw} \times 100$$

where,

V: Percent loss in viscosity

ETo: Flow time of reaction mixture at zero time (*i.e.* before start of reaction)

ETt: Flow time of reaction mixture at time T intervals

ETw: Flow time of distilled water

Growth on Mandel and Reese agar Medium Containing Carboxy Methyl Cellulose (CMC) and Cellulose Powder

The growth of fungal and bacterial cultures is observed on Mandel and Reese medium agar medium containing CMC (1.2w/v) and cellulose powder (1 per cent w/v) as the sole source of carbon. The growth of fungal colonies (in mm) were measured on 3rd, 6th and 9th day at 28±1°C and growth of bacterial colony were measured on 2nd, 4th, and 6th day at 37±2°C. The test organisms showing growth on these polysaccharide was assumed to be positive for cellulolytic activity.

Thermophilic Property of Fungi and Bacteria

Cellulase positive fungi and bacteria were also screened for there thermophilic property. In each case fungal and bacterial cultures were grown on PDA medium and Nutrient Agar Medium plate at 45±1°C respectively.

Identification of Fungal and Bacterial Cultures

The fungal forms found to be positive for cellulolytic activity identified with the help of available literature (Thom and Raper, 1945; Subramanian, 1971; Barnett and Hunter, 1987; Gilman, 2001) and also with the help of online photographs and identification service from internet websites:

(www.mycologyonline.adelaide.ac.edu), (www.aspergillusman.ac.uk).

The bacteria positive for cellulolytic activity were identified on the basis of morphological and epigenetic characters *i.e.* biochemical test with the help of following keys:- Bergey's Manual of Systematic Bacteriology (Kreig and Holt, 1984; Sneath *et al.*, 1986), Bergey's Manual of Determinative Bacteriology (Holt *et al.*, 1994) and one species was further identified and confirmed from Microbial Type and Culture Collection and Gene Bank (MTCC) at Institute of Microbial Technology (IMTECH), Chandigarh, India.

Cellulase Enzyme Assay

Extracellular cellulases complex *viz.* endoglucanase, exoglucanase and β-glucosidase activity of test organisms *i.e.* fungal and bacterial cultures were determined by following methods:

Production Medium

Each test organism for endoglucanase activity was grown on Carboxy Methyl Cellulose (CMC), exoglucanase and β-glucosidase activity on cellulose powder as sole source of carbon in Mandel and Reese medium (Mandels and Reese, 1957).

Inoculum Preparation

Fungal cultures were grown on Potato Dextrose Agar (PDA) medium and the spores were harvested aseptically from 5 day old PDA slants. 2 ml of sterile distilled water is added to the fungal agar slants and shaked vigorously for preparing uniform suspension was used as inoculum (Reddy *et al.*, 1998). The bacterial inoculum consists of 2 ml of 24 hour old bacterial culture grown on nutrient broth giving absorbance of 1.2 at 660 nm.

Preparation of Buffer Solution

Citrate Buffer

Stock Solution

0.1 M solution of Citric Acid = 21.01 gm in 1 lt. of distilled water

0.1 M solution of Sodium Citrate = 29.4 gm in 1 lt. of distilled water

Table 38.1

pH	*0.1 M Citric Acid (in ml.)*	*0.1 M Sodium Citrate (in ml.)*
3.0	16.4	3.6
3.4	14.6	5.4
3.8	12.7	7.3
4.2	10.8	9.2
4.6	8.9	11.1
5.0	7.0	13.0
5.4	5.1	14.9

Potassium Phosphate Buffer

Stock Solution

0.2 M solution of KOH = 11.22 gm. in 1 lt. of water

0.2 M solution of KH_2PO_4 = 27.22 gm. in 1 lt. of water

Table 38.2

pH	*0.2 M KOH (in ml.)*	*0.2 M KH_2PO_4 (in ml.)*
5.8	3.5	50
6.0	5.8	"
6.2	9.1	"
6.4	13.0	"
6.6	18.0	"
6.8	24.0	"
7.0	30.0	"
7.2	35.0	"
7.4	40.0	"
7.6	43.0	"
7.8	45.0	"
8.0	47.0	"

Borax–NaOH Buffer (0.05 M Borate)

Stock Solution

0.05 M solution of Borax

0.2 N solution of NaOH

Mix stock solution and dilute to 200 ml. with water.

Table 38.3

pH	*0.05 M Borax (in ml.)*	*0.2 N NaOH (in ml.)*
8.3	50	0
9.3	"	11
9.6	"	23
9.8	"	34
10.0	"	43
10.1	"	46

Preparation of Enzyme Sample

Twenty five ml of Mandels and Reese medium was dispensed in each 150 ml of flask. All the flasks were then sterilized at 15 psi for 15 minutes. Fungal and bacterial inoculum is inoculated in Mandels and Reese medium containing CMC and cellulose powder separately. The experiment was done in duplicates and inoculated flasks were incubated at 28±1°C for fungi and 37±1°C for bacteria in an incubator under stationary condition. Aliquots (5ml.) were removed periodically from flasks under aseptic condition after 4th, 8th, 12th day in fungi and 2nd, 4th, 6th day of incubation in bacteria. The fungal mycelium was removed by filtering through whatmann filter paper no. 1 and supernatant obtained was centrifuged at 9500 rpm for 20 minutes. The supernatant obtained was used for enzymatic activity.

Preparation of Di-Nitro Salicyclic Acid (DNS) Reagent

3, 5 di-nitro salicyclic acid reagent was prepared for determining the reducing sugar as per the method of Miller (1959).

- ☆ Distilled water — 1416 ml
- ☆ 3, 5 di-nitro salicyclic reagent — 10.6 gm
- ☆ Sodium hydroxide — 19.8 gm

Dissolve the above chemicals, and then add

- ☆ Sodium Potassium Tartarate (Rochelle salt) — 306 gm
- ☆ Phenol (melted at 50°C) — 7.6 ml
- ☆ Sodium metabisulfite — 8.3 gm

This reagent should be stored at 4°C in stoppered bottle untill used for enzymatic activity.

Assay of Endoglucanase Activity (CMCase)

Endo glucanase activity was measured as per the method described by Mandels (1974). Reaction mixture consist of 0.5 ml. of 1 per cent carboxy methyl cellulose (CMC) in citrate buffer (pH 4.8), 0.5 ml.

of enzyme sample, diluted with citrate buffer (pH 4.8). Several dilutions is made of enzyme such that one dilution will release from CMC substrate, slightly more than 0.5 mg. of glucose in the reducing sugar assay and other slightly less. The reaction mixture is incubated at 50°C for 30 minutes. After appropriate period of incubation, the reaction is stopped by adding 3 ml. of DNS reagent and the reaction mixture is boiled for exactly 5 minutes in vigorously boiling water bath. The tubes are cooled at room temperature and the absorbance is then measured at 540nm. A control consisting of substrate blank is also run simultaneously with the test so as to correct any amount of sugar present in the crude enzyme. The amount of reducing sugar or glucose formed after substraction from control is translated using linear glucose standard prepared by plotting glucose used against absorbance at 540nm, into microgram (μg) of glucose produced during enzymatic reaction. Enzyme activity is presented in term of International Units (IU), which is micromole (μm) of glucose release/min./ml.

$$\text{IU/ml.} = \text{mg. glucose} \times 0.37 \times \text{dilution factor}$$

or

$$\text{IU/ml.} = \frac{0.185}{\text{Enzyme dilution to give 0.5 mg. glucose}}$$

Assay of Exo Glucanase (FPU) Activity (FPase)

Exoglucanase activity determined as per the method of Mandels (1974). In this method, 1 ml. of citrate buffer (pH 4.8) is added to a test tube, followed by 0.5 ml of enzyme suitably diluted in the same buffer and 1 × 6 cm. strips of whatmann no. 1 filter paper (50 mg). At least two dilution of enzyme must be used. One dilution should be capable of releasing more than 2 mg. of glucose; the other should release slightly less. The reaction mixture is incubated at 50° C for 60 min., after the incubation period, reaction is stopped by adding 3 ml. of DNS reagent. The suspension is mixed well and the tubes are transferred to a boiling water bath for exactly 5 min. Tubes are immediately cooled in the cold water and absorbance is then measured at 540nm. A control consisting of substrate blank is also run simultaneously with test so as to correct any amount of sugar present in the crude enzyme. The amount of reducing sugar or glucose formed after subtraction from control is translated using linear glucose standard prepared by plotting glucose used against absorbance at 540 nm, into microgram (μg) of glucose produced during enzyme reaction. Enzyme activity is presented in term of International Units known as Filter Paper Units (FPU) which is micromoles (μm) of glucose released/min./ml.

$$\text{IU/ml.} = \text{mg. glucose} \times 0.185 \times \text{dilution factor}$$

or

$$\text{FPU/ml.} = \frac{0.37}{\text{Enzyme dilution to give 0.5 mg. glucose}}$$

Assay of β-glucosidase Activity

β-glucosidase activity is measured as per the method described by Mandels (1974). In this method, 0.5 ml of 1 per cent salicin prepared in citrate buffer (pH 4.8), 0.5 ml of enzyme sample, diluted with citrate buffer (pH 4.8) several dilutions are made of enzyme such that one dilution will release from salicin substrate, slightly more than 0.5 mg of glucose in the reducing sugar assay and other slightly ess. The reaction mixture is incubated at 50°C for 30 min. After appropriate period of incubation, the eaction is stopped by adding 3 ml. of DNS reagent and the reaction mixture is boiled for exactly 5 min. n vigorously boiling water bath. The tubes are cooled immediately in cold water and the absorbance s then measured at 540 nm. A control consisting of substrate blank is also run simultaneously with est so as to correct any amount of sugar present in the crude enzyme. The amount of reducing sugar glucose formed after subtraction from control is translated using linear glucose standard prepared by plotting glucose used against absorbance at 540nm into microgram (μg) of glucose produced during enzyme reaction. Enzyme activity is presented in term of International Units (IU) which is micromole μm) of glucose released/min/ml.

$$\text{IU/ml.} = \text{mg. glucose} \times 0.37 \times \text{dilution factor}$$

or

$$\text{IU/ml.} = \frac{0.185}{\text{Enzyme dilution to give 0.5 mg. glucose}}$$

Effect of pH on Activity of Cellulase Enzymes

The effect of pH was determined by measuring activity at standard assay over a pH range 3-10. Substrates CMC (1 per cent) for endoglucanase and salicin (1 per cent) for β-glucosidase activity was prepared in three buffer *viz.* citrate buffer (pH 3-5), Potassium Phosphate buffer (6-8), and Borax-NaOH buffer (9-10). Exoglucanase activity is assayed in buffers of different pH 3-10. Enzyme of test organism was mixed with substrate and cellulase assay was performed. Cellulase activity was determined in IU/ml.

Effect of Temperature on Activity of Cellulase Enzymes

The activity was determined by carrying out the standard assay at various temperatures between 30-90°C at optimum pH obtained for test organism. Enzyme mixed with substrate CMC for endoglucanase activity and salicin for β-glucosidase activity prepared at buffer of optimum pH activity obtained, exoglucanase activity was also performed in buffer of optimum pH. Endoglucanase, exoglucanase and β-glucosidase activity was carried out at different temperature range of 30-90°C. Cellulase activity was determined in IU/ml.

Effect of Chemical Reagents on Endoglucanase Activity

The effect of chemical reagents was evaluated on the endoglucanase activity of fungal and bacterial cultures as per the method of Ozaki and Ito (1991) with slight modification. EDTA, 2,4 DNP, $HgCl_2$ and $SnCl_2$ were prepared of three concentration *viz.* 0.1mM, 1mM and 5mM. The crude enzyme (1ml.) of test organism and each chemical reagents (1 ml.) were mixed and incubated at 30°C for 30 min. and an aliquot (0.5 ml.) was used for the assay. The endoglucanase activity was measured in IU/ml. A control was run consist of enzyme without reagent.

Chemicals

All chemicals which are used in the present investigation were of analytical grade. The majority of chemicals were of Ranbaxy, India followed by Himedia (CDH, Chemical India).

(Merck company, Darmstcidt, Germany), (Fluka Chemical Ltd. Gillingham, Dorset, UK)

Results

Cellulase Enzymes of Fungal Cultures

The endoglucanase (CMCase), exoglucanase (FPase), and β-glucosidase activity was determined on 4th, 8th and 12th day of incubation.

The data depicted in the Table 38.4 shows the activities of endoglucanase, exoglucanase and β-glucosidase activity in forty fungal cultures. In Zygomycotina class, two spp. of *Mucor* (AV69, AV62) shows best endoglucanase, exoglucanase and β-glucosidase activity on 8th day of incubation. *Rhizopus* sp. (AV 15) shows highest endoglucanase, exoglucanase and β-glucosidase activity on 12th day of incubation. In Ascomycotina class, among the spp. of *Chaetomium*, the endoglucanase, exoglucanase and β-glucosidase activity differs. In case of *Chaetomium cochliodes*, endoglucanase, activity was higher on 4th day but decreases gradually as the incubation period increased upto 12th day, whereas, exoglucanase activity was not observed on 4th day, it appeared on 8th day and show little increase on 12th day. No β-glucosidase activity was found. In case of *Chaetomium globosum*, the endoglucanase activity increased gradually upto 12th day of incubation. This result was total opposite of *Chaetomium cochloides*. Exoglucanase activity was found to increase gradually as the incubation period increases upto 12th day, for β-glucosidase, the activity increases upto 8th day and then reduces. In case of *Chaetomium* sp. (AV56), endoglucanase activity increases upto 8th day and become almost more than two fold of 4th day activity, but on 12th day significant reduction *i.e.* almost half of the 8th day of activity was observed. Exoglucanase activity was found increases upto 8th day and then became stable and no further increase in activity was observed. In β-glucosidase activity initially there was no change in the activity upto the 8th day a little higher activity was obtained in 12th day of incubation.

Table 38.4: Cellulases Enzymes of Fungal Cultures

Fungal Cultures	*Isolate Number*	*Incubation Time (in days)*	*Endoglucanase Activity (CMCase) (IU/ml)*	*Exoglucanase Activity (FPase) (FPU/ml)*	*β-glucosidase Activity (IU/ml)*
		ZYGOMYCOTINA			
Mucor sp.	AV62	4	0.515	0.010	0.084
		8	0.982	0.014	0.150
		12	0.881	0.013	0.148
Mucor sp.	AV69	4	0.440	0.030	0.199
		8	1.169	0.049	0.401
		12	0.924	0.040	0.180
Rhizophus sp.	AV15	4	0.201	0.011	0.029
		8	0.347	0.038	0.033
		12	0.593	0.066	0.090
		ASCOMYCOTINA			
Chaetomium cochliodes	AV6	4	0.210	ANO	ANO
		8	0.192	0.012	ANO
		12	0.186	0.018	ANO
Chaetomium globosum	AV17	4	0.919	0.013	0.053
		8	1.396	0.020	0.064
		12	1.515	0.021	0.054

Contd..

Table 38.4–Contd...

Fungal Cultures	Isolate Number	Incubation Time (in days)	Endoglucanase Activity (CMCase) (IU/ml)	Exoglucanase Activity (FPase) (FPU/ml)	β-glucosidase Activity (IU/ml)
Cheatomium sp.	AV56	4	0.510	0.015	0.088
		8	1.212	0.019	0.088
		12	0.688	0.019	0.093
		DEUTEROMYCOTINA			
Aspergillus flavus	AV23	4	0.380	0.010	0.202
		8	0.528	0.010	0.279
		12	1.065	0.010	0.217
Aspergillus nidulans	AV14	4	0.353	0.058	0.083
		8	0.636	0.102	0.189
		12	0.516	0.074	0.055
Aspergillus fumigatus	AV4	4	0.370	0.025	0.090
		8	0.485	0.029	0.103
		12	0.502	0.037	0.115
Aspergillus niger	AV8	4	0.102	0.024	0.112
		8	0.220	0.026	0.186
		12	0.362	0.030	0.203
Aspergillus sp.	AV36	4	0.056	0.019	0.059
		8	0.081	0.022	0.070
		12	0.082	0.025	0.070
Aspergillus sp.	AV51	4	0.179	0.010	0.134
		8	0.331	0.028	0.282
		12	0.567	0.029	0.483
Aspergillus terreus	AV49	4	0.444	0.069	0.062
		8	1.167	0.140	0.080
		12	0.833	0.097	0.067
Aspergillus ochraceous	AV65	4	0.251	ANO	0.009
		8	0.503	ANO	0.014
		12	0.455	ANO	0.015
Aspergillus sp.	AV70	4	0.515	0.092	0.045
		8	1.207	0.109	0.062
		12	1.118	0.078	0.060
Alternaria alternata	AV32	4	0.116	0.001	0.006
		8	0.214	0.024	0.005
		12	0.352	0.002	0.006
Curvularia sp.	AV39	4	0.156	0.007	0.037
		8	0.148	0.007	0.042
		12	0.147	0.007	0.046

Contd...

Table 38.4–Contd...

Fungal Cultures	Isolate Number	Incubation Time (in days)	Endoglucanase Activity (CMCase) (IU/ml)	Exoglucanase Activity (FPase) (FPU/ml)	β-glucosidase Activity (IU/ml)
Colletotrichum sp.	AV46	4	0.226	0.061	0.055
		8	1.063	0.089	0.065
		12	0.817	0.088	0.123
Drechslera sp.	AV31	4	0.556	0.070	ANO
		8	1.011	0.096	ANO
		12	0.897	0.091	ANO
Fusarium sp.	AV40	4	0.134	0.010	0.018
		8	0.703	0.013	0.087
		12	0.737	0.022	0.106
Fusarium oxysporum	AV43	4	0.425	0.010	0.022
		8	0.742	0.034	0.038
		12	0.541	0.022	0.034
Fusarium sp.	AV45	4	0.420	0.013	0.070
		8	0.823	0.016	0.103
		12	0.866	0.016	0.097
Geotrichum candidum	AV28	4	0.073	0.011	0.052
		8	0.148	0.013	0.058
		12	0.366	0.014	0.059
Helminthosporium sp.	AV59	4	0.083	0.042	ANO
		8	0.331	0.065	ANO
		12	0.320	0.062	ANO
Penicillium rubrum	AV18	4	0.493	0.018	0.062
		8	1.030	0.026	0.127
		12	0.752	0.045	0.193
Penicillium nigricans	AV1	4	0.181	0.009	0.031
		8	0.415	0.011	0.034
		12	0.714	0.029	0.055
Penicillium chrysogenum	AV10	4	0.182	0.011	0.026
		8	0.375	0.028	0.064
		12	0.614	0.031	0.092
Penicillium sp.	AV57	4	0.243	0.034	0.058
		8	0.553	0.068	0.120
		12	0.565	0.081	0.123
Penicillium sp.	AV41	4	0.390	0.007	0.120
		8	0.756	0.010	0.127
		12	0.679	0.008	0.119

Contd.

Table 38.4–Contd...

Fungal Cultures	*Isolate Number*	*Incubation Time (in days)*	*Endoglucanase Activity (CMCase) (IU/ml)*	*Exoglucanase Activity (FPase) (FPU/ml)*	*β-glucosidase Activity (IU/ml)*
Phoma sp.	AV50	4	0.821	0.013	0.023
		8	1.187	0.015	0.032
		12	1.110	0.025	0.045
Rhizoctonia sp.	AV67	4	0.275	ANO	0.011
		8	0.344	ANO	0.014
		12	0.361	ANO	0.011
Trichoderma viride	AV13	4	0.657	0.089	0.031
		8	1.267	0.152	0.810
		12	1.058	0.155	0.729
Trichoderma sp.	AV21	4	0.826	0.022	0.187
		8	2.168	0.084	0.509
		12	1.947	0.117	0.399
Trichoderma koningi	AV34	4	0.490	0.018	0.125
		8	0.589	0.032	0.159
		12	0.596	0.034	0.151
Trichothecium roseum	AV54	4	0.301	0.006	ANO
		8	0.704	0.009	ANO
		12	1.810	0.013	ANO
Mycelia sterlia (Black)	AV5	4	0.599	0.011	0.123
		8	1.079	0.011	0.157
		12	0.479	0.012	0.157
Mycelia sterlia (Yellow)	AV29	4	0.173	0.012	0.070
		8	0.617	0.021	0.081
		12	0.316	0.021	0.078
Unidentified 1	AV55	4	0.144	0.003	0.009
		8	0.189	0.004	0.023
		12	0.217	0.004	0.023
Unidentified 2	AV48	4	1.695	0.015	0.076
		8	1.295	0.016	0.106
		12	0.906	0.022	0.116
Unidentified 3	AV53	4	0.271	0.055	ANO
		8	0.542	0.089	ANO
		12	0.452	0.069	ANO

ANO: Activity Not Observed

All data given are the mean value of the two independent experiments.

In Deuteromycotina, twenty nine cultures were evaluated for cellulase complex. Among nine spp. of *Aspergillus*, three spp. of *Aspergillus viz. A.terreus, Aspergillus* sp. (AV70) and *A.nidulans* shows good activity of endoglucanase, exoglucanase and β-glucosidase on 8th day of incubation, whereas in *Aspergillus* sp. (AV51), *A. flavus, A. fumigatus, A. niger* and *Aspergillus* sp. (AV36) shows good endoglucanase, exoglucanase and β-glucosidase activity on 12th day of incubation. Interestingly, in *A.ochraceous* maximum endoglucanase and β-glucosidase activities was obtained on 8th day, however no exoglucanase activity was found in it. Out of five spp. of *Penicillium*, except *Penicillium rubrum* and isolate no. AV41, all the three *Penicillium* spp. *i.e. P. nigricans, P. chrysogenum* and *Penicillium* sp. (AV57) shows higher activity of endoglucanase on 12th day whereas, in case of *P. rubrum* and *Penicillium* sp. (AV41) higher activity of endoglucanase was observed on 8th day. Higher activity of exoglucanase and β-glucosidase was observed on 12th day of incubation in *P.rubrum, P.nigricans, P.chrysogenum* and *Penicillium* sp. (AV57), whereas in *Penicillium* sp. (AV41) best activity of exoglucanase and β-glucosidase was observed on 8th day. In three spp. of *Trichoderma, Trichoderma* sp. (AV21) shows maximum endoglucanase and β-glucosidase activity on 8th day of incubation, whereas, activity of exoglucanase was found maximum on 12th day of incubation. Almost similar trend in activity of endoglucanase, exoglucanase and β-glucosidase was observed in *Trichoderma viride* and *Trichoderma koningi*.

During the present investigation, it was found that three spp. of *Fusarium* shows good amount of endoglucanase, exoglucanase and β-glucosidase activity. Among the three species of *Fusarium, Fusarium oxysporum* shows best endoglucanase, exoglucanase and β-glucosidase activity on 8th day of incubation, whereas, best endo- and exoglucanase activity in *Fusarium* spp. (AV45, AV40) was observed on 12th day of incubation. However, β-glucosidase activity in *Fusarium* sp. (AV 45 and AV 40) was observed maximum on 8th and 12th day of incubation respectively. *Alterneria alternata* was found to show maximum activity of endoglucanase and β-glucosidase on 12th day and exoglucanase activity on 8th day of incubation. *Curvularia* sp. showed maximum endoglucanase and exoglucanase activity on 4th day, however endoglucanase activity decreased upon with the increase in incubation period whereas exoglucanase activity was found to be static. β-glucosidase activity was found to be maximum on 12th day of incubation. In *Colletotrichum* sp. maximum endoglucanase and exoglucanase activity was found on 8th day and β-glucosidase activity on 12th day of incubation. In *Drechslera* sp., maximum endo- and exoglucanase activity was observed on 8th day of incubation. β-glucosidase activity was found to be absent. In *Geotrichum candidum* maximum endoglucanase activity was observed on 12th day, whereas exoglucanase and β-glucosidase activity was almost similar on 8th and 12th day of incubation. In *Helminthosporium* sp. maximum endoglucanase and exoglucanase activity was observed on 8th day of incubation and β-glucosidase activity was found absent. *Phoma* sp. shows maximum endoglucanase activity on 8th day of incubation. Exoglucanase and β-glucosidase activity was found to be maximum on 12th day of incubation. *Rhizoctonia* sp. shows maximum endoglucanase activity on 12th day of incubation. β-glucosidase activity was found to be best on 8th day of incubation. Exoglucanase activity was found to be absent. *Trichothecium roseum* shows maximum endoglucanase and exoglucanase activity on 12th day of incubation, while β-glucosidase activity was found to absent. Mycelia sterlia (AV5) and (AV29) showed maximum endoglucanase, exoglucanase and β-glucosidase activity on 8th day of incubation respectively.

In case of Unidentified 1 (AV55) endoglucanase activity was highest was 12th day. Interestingly no change in exoglucanase and β-glucosidase activity was recorded after 8th day of incubation. In case of Unidentified 2 (AV48) endoglucanase activity was highest on 4th day and then decreased after further period of incubation, whereas exoglucanase and β-glucosidase activity were observed maximum in 12th day of incubation. Unidentified 3 (AV53) showed optimum endoglucanase and exoglucanase activity on 8th day of incubation. β-glucosidase activity was found to be absent.

Effect of pH on Activity of Cellulase Enzymes in Three Fungal Cultures

Two *Aspergillus* spp. *viz. A.nidulans* and *A.terreus* and one sp. of *Fusarium oxysporum* was selected for their endoglucanase, exoglucanase and β-glucosidase activity for their optimum pH range. In *A.nidulans* it was found in range of pH 5-9, 4-6, 4-6 for endoglucanase, exoglucanase and β-glucosidase activities respectively. For endoglucanase optimum pH was 6, and for exoglucanase and β-glucosidase optimum pH was 5. Above and below this optimum range, the endoglucanase, exoglucanase and β-glucosidase activity declines rapidly.

Aspergillus terreus showed very broad pH range from 3-6. For endoglucanase and exoglucanase activities, optimum pH was 4. β-glucosidase activity was found optimum at pH 5. *Fusarium oxysporum* also showed a very broad range of tolerance for pH from 4-9 and pH 5 was optimum for endo- and exoglucanase activities and for β-glucosidase activity optimum pH was 4 (Table 38.2 and Figure 38.1).

Effect of Temperature on Activity of Cellulase Enzymes in Three Fungal Cultures

In the present investigation, fungal cultures were taken to evaluate the optimum and active temperature ranges. The data were depicted in Table 38.3 and Figure 38.2. It is clearly evident from results that *A.nidulans* shows a very broad stable temperature range for endoglucanase, exoglucanase and β-glucosidase activity ranging from 30-80°C, 30-60°C and 40-80°C respectively with maximum 60°C for endoglucanase and β-glucosidase activities, 50°C for exoglucanase activity. In *A.terreus,* endoglucanase activity was found to be active from 30-70°C with optimum at 40°C. The exoglucanase and β-glucosidase activities range 30-50°C and 40-50°C respectively. *Fusarium oxysporum* shows temperature range 30-60°C for endoglucanase, 30-50°C for exoglucanase and β-glucosidase activities. However, optimum temperature 60°C, 50°C and 50°C respectively for endo-, exoglucanase and β-glucosidase activity was observed.

Effect of Chemical Reagents on Endoglucanase Activity of Three Fungal Cultures

Figures 38.3 and 38.4 shows the effect of chemical reagent on endoglucanase activity of *Aspergillus nidulans*. It is clearly evident from the result that EDTA and $SnCl_2$ did not affect much, whereas 2, 4 DNP and $HgCl_2$ were found to reduce endoglucanase activity more severely. The lower concentration (0.1 mM) of 2, 4 DNP reduces the endoglucanase activity more than 50 per cent and higher concentration (5 mM) was found to inhibit activity more then 90 per cent. Lower concentration (0.1 mM) of $HgCl_2$ was more inhibitory and found to inhibit endoglucanase activity more than 80 per cent and its higher concentration causes 93 per cent inhibition in the activity.

Figures 38.5 and 38.6 depicts the effect of chemical reagents on endoglucanase activity of *Aspergillus terreus*. The result shows that EDTA and $SnCl_2$ did not affect much as 5 mM concentration of EDTA causes 37 per cent and same concentration of $SnCl_2$ inhibits only 22 per cent activity. 2, 4 DNP and $HgCl_2$ were found to reduce endoglucanase activity more severely. The lower concentration (0.1 mM) of 2, 4 DNP reduce the endoglucanase activity upto 70 per cent and higher concentration (5 mM) was found to inhibit 94 per cent activity. Concentration of 0.1 mM of $HgCl_2$ inhibited 87 per cent of endoglucanase activity and higher concentration (5 mM) of $HgCl_2$ causes 97 per cent inhibition in activity.

The results of effect of chemical reagents on endoglucanase activity of *Fusarium oxysporum* are shown in Figures 38.7 and 38.8. It is evident from the result that unlike *Aspergillus nidulans* and *Aspergillus terreus*, EDTA inhibit (48 per cent) of endoglucanase activity at (0.1 mM) and 63 per cent inhibition at 5 mM. $SnCl_2$ also shows effect on endoglucanase activity as 5 mM concentration causes

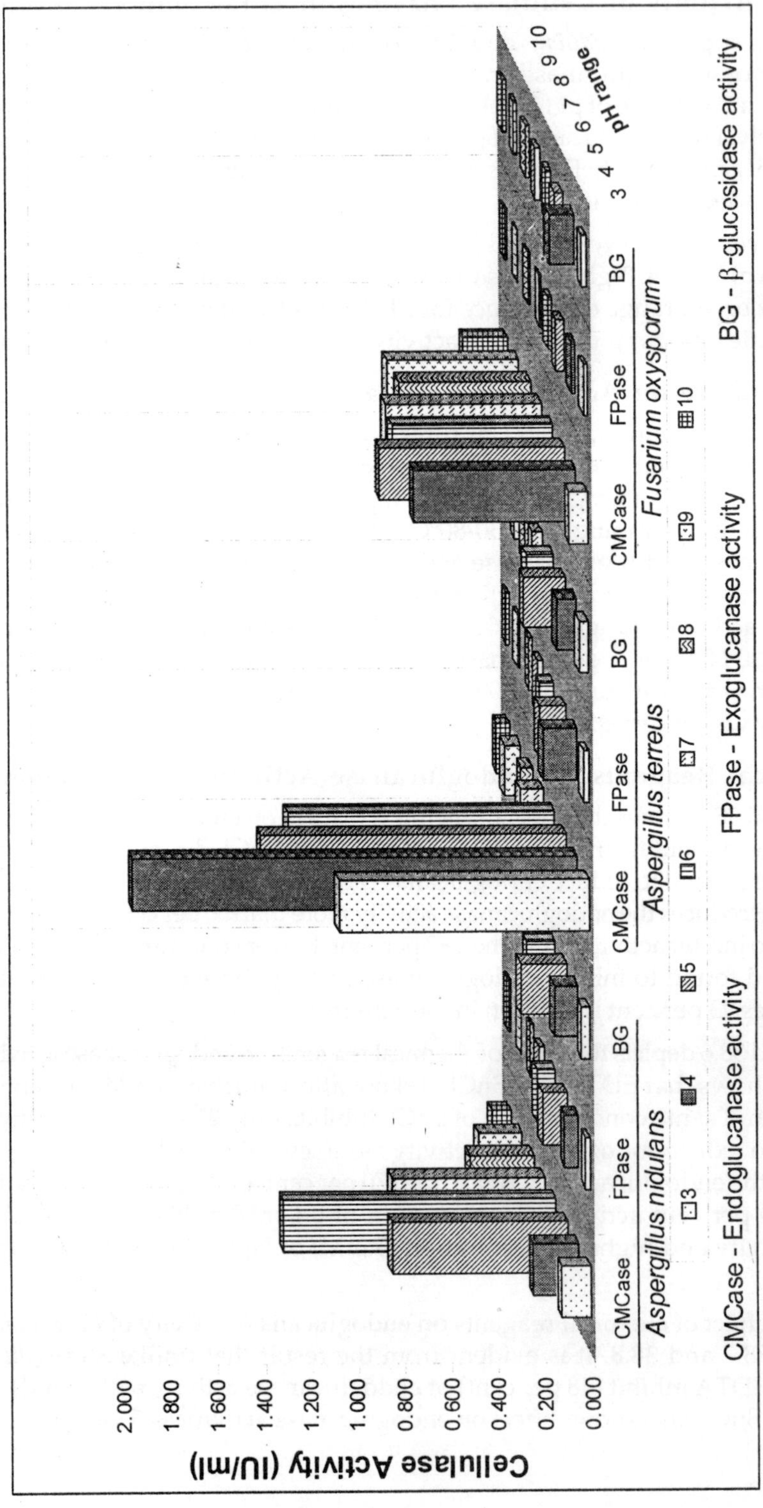

Figure 38.1: Effect of pH Range on the Activity of Cellulase Enzymes in Three Fungal Cultures

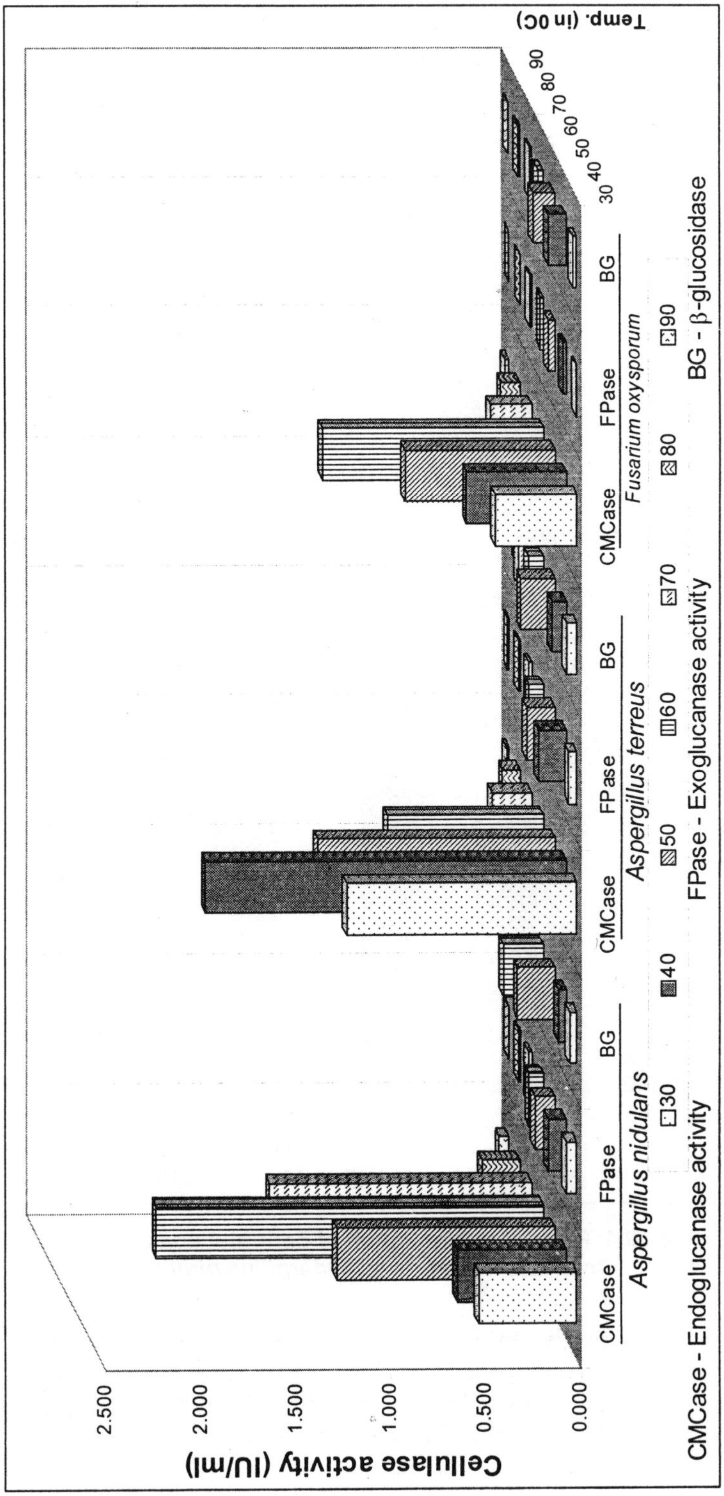

Figure 38.2: Effect of Temperature Range on the Activity of Cellulase Enzymes in Three Fungal Cultures

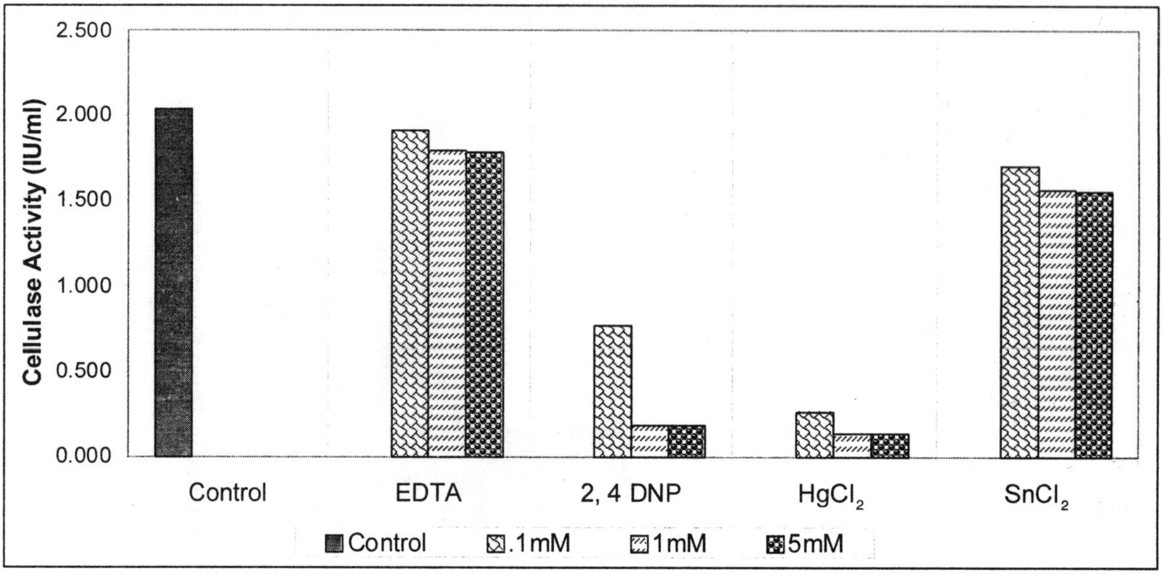

Figure 38.3: Effect of Chemical Reagents on Endoglucanase Activity of *Aspergillus nidulans*

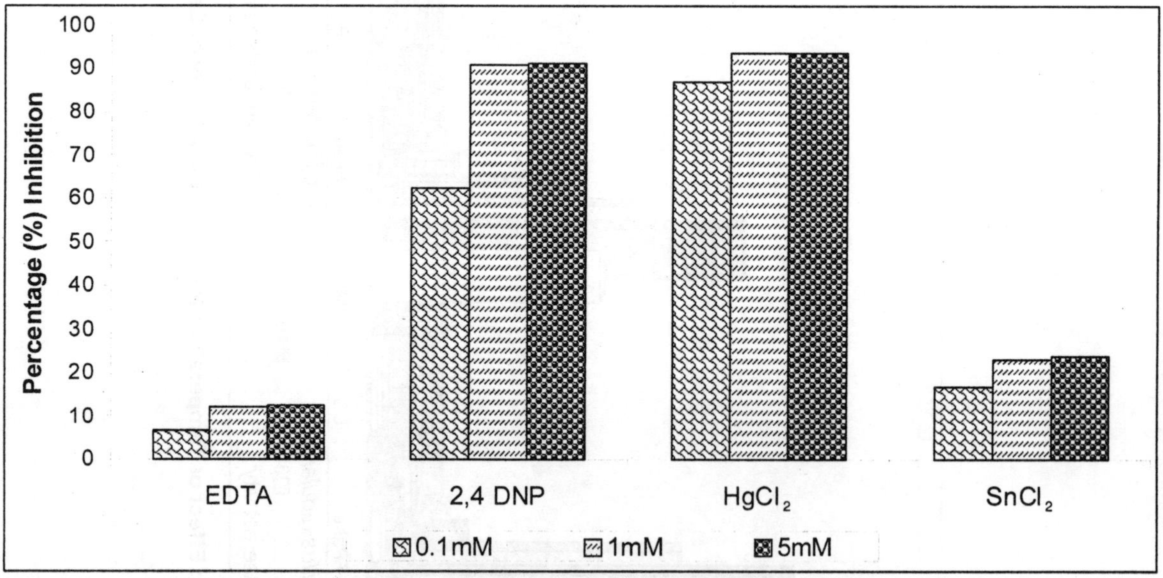

Figure 38.4: Percentage Inhibition of Endoglucanase Activity by Chemical Reagents in *Aspergillus nidulans*

35 per cent inhibition. Interestingly, only 38 per cent inhibition in endoglucanase activity was found at (0.1 mM) concentration and higher concentration (5 mM) causes 83 per cent inhibition by 2, 4 DNP. The effect of $HgCl_2$ was also evaluated and found 80 per cent inhibition at (0.1 mM) concentration and 5 mM concentration causes 100 per cent inhibition in the activity.

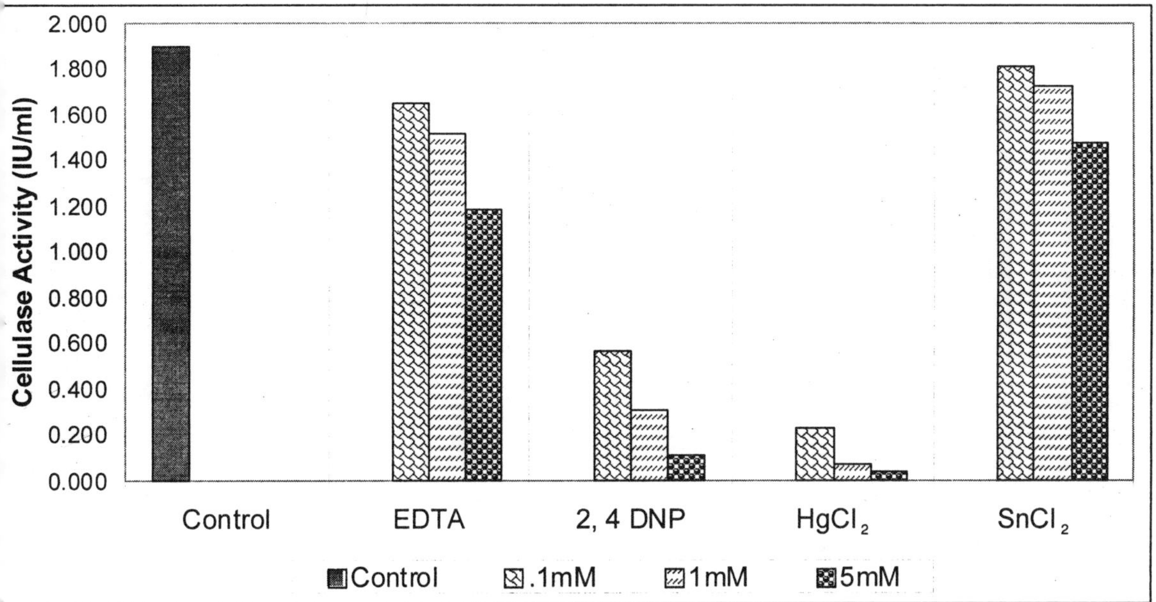

Figure 38.5: Effect of Chemical Reagents on Endoglucanase Activity of *Aspergillus terreus*

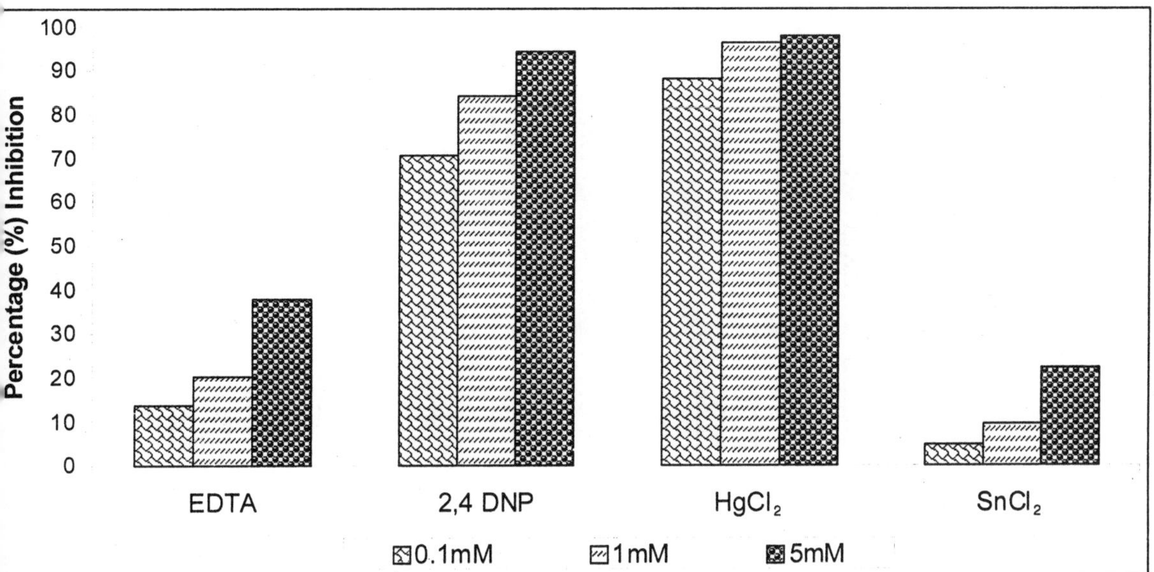

Figure 38.6: Percentage Inhibition of Endoglucanase Activity by Chemical Reagents in *Aspergillus terreus*

;ellulase Enzymes of Bacterial Cultures

The results of cellulase activity of bacterial cultures were presented in Table 38.4. The cellulolytic .ctivity was measured as endoglucanase, exoglucanase and β-glucosidase. Out of the five bacterial

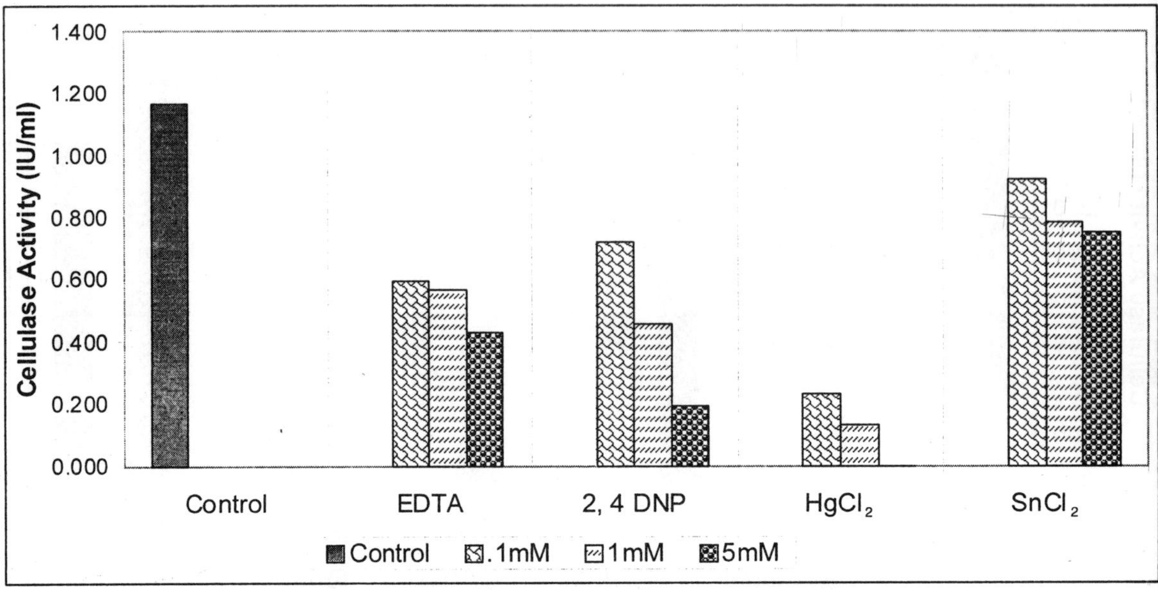

Figure 38.7: Effect of Chemical Reagents on Endoglucanase Activity of *Fusarium oxysporum*

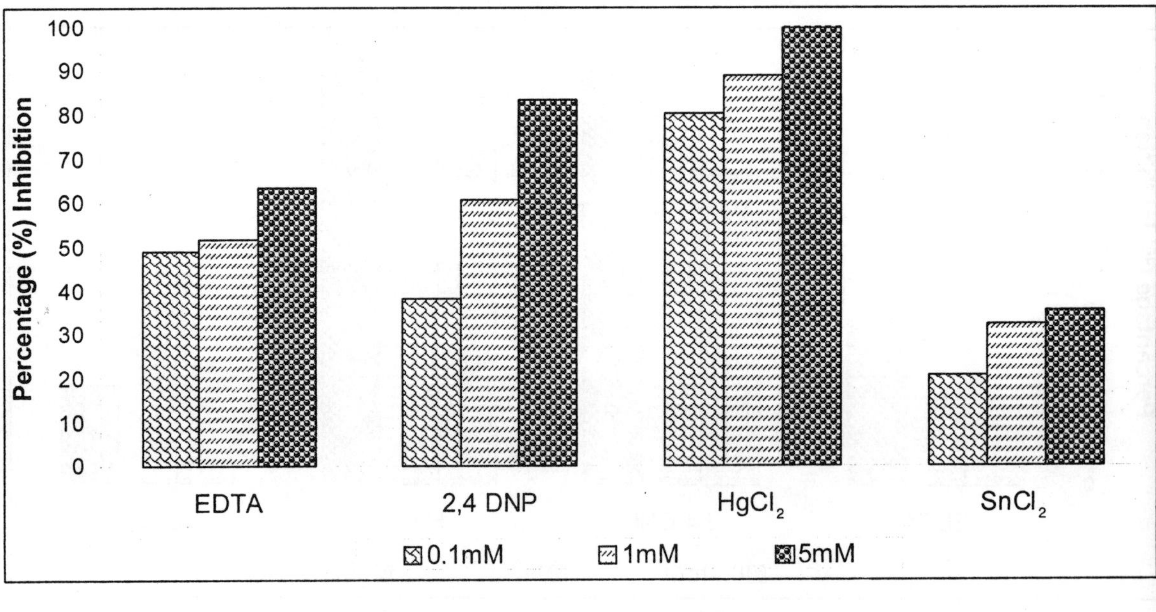

Figure 38.8: Percentage Inhibition of Endoglucanase Activity by Chemical Reagents in *Fusarium oxysporum*

cultures, four belongs to genus *Bacillus*. All the test *Bacillus* species shows endoglucanase, exoglucanase and β-glucosidase activities. In *Bacillus marinus*, *Bacillus* sp. (V2), *Bacillus subtilis* and *Bacillus* sp. (J2) best endo-, exoglucanase and β-glucasidase activity was observed on 6th day of incubation. *Bacillus marinus* was found best because it shows maximum endo- and exoglucanase activity on 6th day of incubation. Interestingly, β-glucosidase activity was recorded higher in *Bacillus* sp. (J3). *Aeromonas hydrophila* shows best endoglucanase, exoglucanase and β-glucosidase activity on 4th day of incubation.

Effect of pH on Activity of Cellulase Enzymes in Three Bacterial Cultures

The data presented in Table 38.5 and Figure 385.9 shows the active pH range for cellulase complex of three selected bacterial cultures. In *Bacillus marinus*, endoglucanase activity was observed at broader pH range *i.e.* 4-9 with optimum at pH 6, whereas, exoglucanase and β-glucosidase activity was found to show pH range 4-8 and 3-8 respectively, with optimum at pH 5 and 6 respectively. *Bacillus* sp. (V2) shows pH range from pH 3-8, for endoglucanase activity with optimum at pH 6. For exoglucanase activity pH range 4-7 with optimum was observed at pH 6 whereas, for β-glucosidase activity pH range was 3-7 with optimum at pH 5. In *B.subtilis*, for endoglucanase activity range of pH lies between 4-8 and optimum was observed at pH 5. For exoglucanase activity pH range was 3-8 and optimum was observed at pH 5.

Table 38.5: Effect of pH Range on the Activity of Cellulase Enzymes in Three Fungal Cultures

Sl.No.	Fungal Culture	pH Range	Endoglucanase Activity (CMCase) (IU/ml)	Exoglucanase Activity (FPase) (FPU/ml)	β-glucosidase Activity (BG) (IU/ml)
1.	*Aspergillus nidulans*	3	0.122	0.018	0.032
		4	0.191	0.060	0.101
		5	0.742	0.105	0.192
		6	1.165	0.078	0.118
		7	0.649	0.022	0.071
		8	0.270	0.009	0.020
		9	0.186	0.004	0.018
		10	0.082	0.005	0.017
2.	*Aspergillus terreus*	3	1.060	0.022	0.034
		4	1.892	0.145	0.083
		5	1.300	0.111	0.175
		6	1.138	0.059	0.115
		7	0.099	0.023	0.041
		8	0.046	0.008	0.029
		9	0.070	0.009	0.020
		10	0.055	0.009	0.022
3.	*Fusarium oxysporum*	3	0.076	0.013	0.019
		4	0.683	0.016	0.108
		5	0.786	0.035	0.036
		6	0.683	0.018	0.022
		7	0.665	0.005	0.020
		8	0.559	0.004	0.011
		9	0.563	0.004	0.013
		10	0.186	0.004	0.014

All data given are the mean value of the two independent experiments.

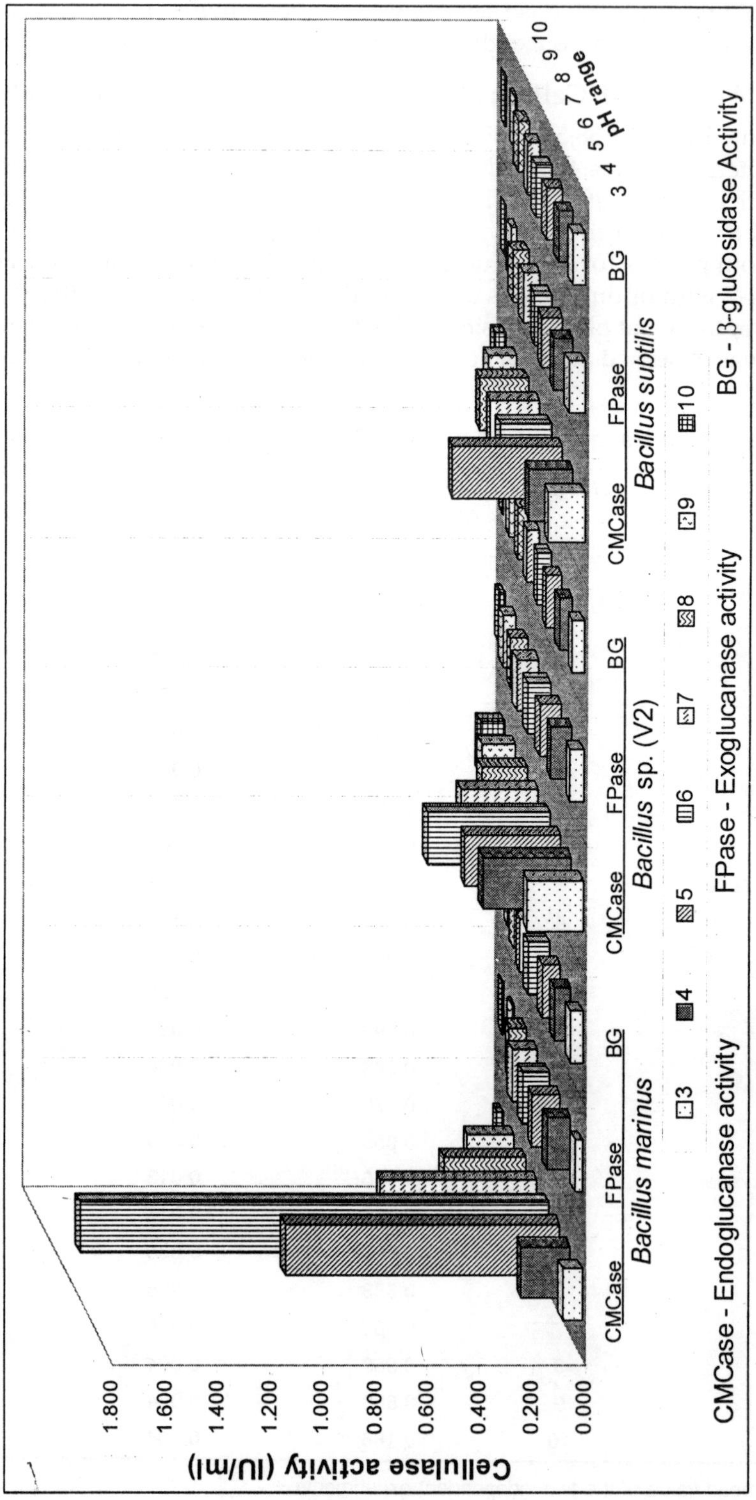

Figure 38.9: Effect of pH Range on the Activity of Cellulase Enzymes in Three Bacterial Cultures

Effect of Temperature on Activity of Cellulase Enzymes in Three Bacterial Cultures

The cellulase enzymes of selected bacterial cultures were subjected to different temperature range to evaluate optimum temperature for cellulolytic activities. The data obtained from the experiment are depicted in Table 38.6 and Figure 38.10. It is evident from the results that *Bacillus marinus* has very narrow range of temperature tolerance for endoglucanase, exoglucanase and β-glucosidase activity *i.e.* from 30-50°C. The maximum activity of all the three enzymes complex was best at 50°C. Any further increase in the temperature was found detrimental for the cellulases activity. The *Bacillus* sp. (V2) shows much broader range of temperature tolerance ranging from 30-80°C for endoglucanase activity whereas 30-70°C was found good for exoglucanase and β-glucosidase activities, 60°C was found best and optimum for all the three enzymes. *B.subtilis* shows broader range of temperature tolerance *i.e.* 30–70°C temperature for endoglucanase, exoglucanase and β-glucosidase activities. 50°C temperature was found best for all the three enzymes.

Table 31.6: Effect of Temperature Range on Activity of Cellulase Enzymes in Three Fungal Cultures

Sl.No.	Fungal Culture	Temperature Range (°C)	Endoglucanase Activity (CMCase) (IU/ml)	Exoglucanase Activity (FPase) (FPU/ml)	β-glucosidase Activity (BG) (IU/ml)
1.	*Aspergillus nidulans*	30	0.518	0.057	0.040
		40	0.567	0.090	0.039
		50	1.144	0.105	0.193
		60	2.040	0.074	0.210
		70	1.386	0.018	0.100
		80	0.202	0.009	0.087
		90	0.050	0.007	0.019
2.	*Aspergillus terreus*	30	1.210	0.045	0.055
		40	1.900	0.145	0.073
		50	1.250	0.146	0.177
		60	0.818	0.074	0.078
		70	0.211	0.018	0.071
		80	0.093	0.009	0.021
		90	0.013	0.009	0.015
3.	*Fusarium oxysporum*	30	0.430	0.007	0.025
		40	0.520	0.015	0.092
		50	0.788	0.035	0.110
		60	1.165	0.014	0.024
		70	0.218	0.009	0.012
		80	0.101	0.002	0.010
		90	0.022	0.001	0.003

All data given are the mean value of the two independent experiments.

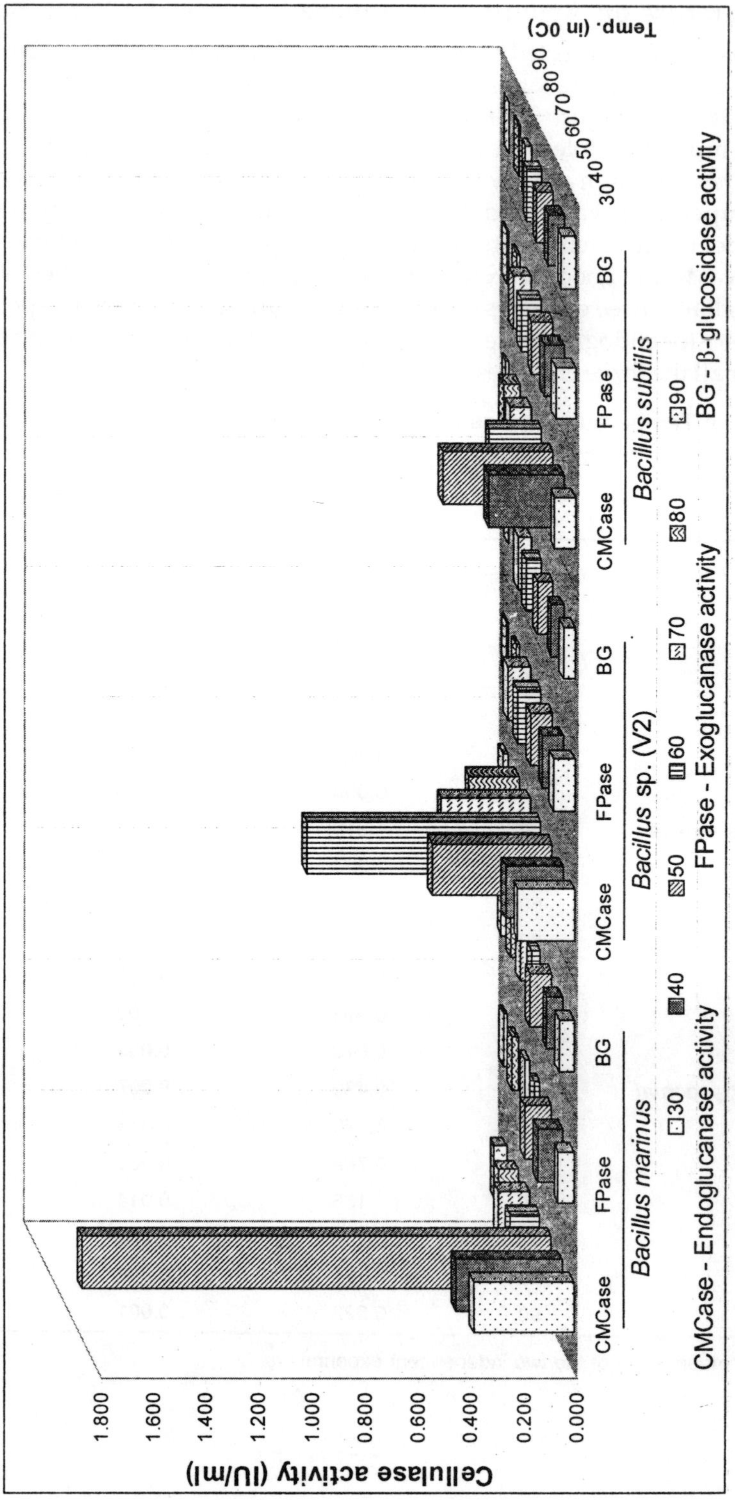

Figure 38.10: Effect of Temperature Range on the Activity of Cellulase Enzymes in Three Bacterial Cultures

Table 38.6: Cellulase Enzymes of Bacterial Cultures

Bacterial Cultures	Isolate Number	Incubation Time (in days)	Endoglucanase Activity (CMCase) (IU/ml)	Exoglucanase Activity (FPase) (FPU/ml)	β-glucosidase Activity (IU/ml)
Aeromonas hydrophila	J2	2	0.116	0.025	0.046
		4	0.238	0.069	0.057
		6	0.192	0.058	0.051
Bacillus marinus	O1	2	0.190	0.044	0.048
		4	0.404	0.086	0.065
		6	0.606	0.101	0.067
Bacillus subtilis	M1	2	0.204	0.030	0.044
		4	0.307	0.073	0.042
		6	0.393	0.074	0.056
Bacillus sp.	V2	2	0.189	0.042	0.036
		4	0.191	0.068	0.049
		6	0.425	0.081	0.057
Bacillus sp.	J3	2	0.102	0.024	0.112
		4	0.220	0.026	0.186
		6	0.362	0.030	0.203

All data given are the mean value of the two independent experiments.

Table 38.7: Effect of pH Range on the Activity of Cellulase Enzymes in Three Bacterial Cultures

Sl.No.	Bacterial Culture	pH Range	Endoglucanase Activity (CMCase) (IU/ml)	Exoglucanase Activity (FPase) (FPU/ml)	β-glucosidase Activity (BG) (IU/ml)
1.	*Bacillus marinus*	3	0.073	0.028	0.049
		4	0.187	0.09	0.065
		5	1.046	0.103	0.069
		6	1.783	0.100	0.077
		7	0.594	0.095	0.066
		8	0.311	0.063	0.044
		9	0.178	0.009	0.028
		10	0.022	0.004	0.025
2.	*Bacillus* sp. (V2)	3	0.208	0.051	0.045
		4	0.334	0.081	0.050
		5	0.362	0.081	0.057
		6	0.463	0.085	0.049
		7	0.294	0.081	0.044
		8	0.173	0.062	0.035
		9	0.128	0.048	0.028
		10	0.094	0.020	0.009

Contd...

Table 38.7–Contd...

Sl.No.	Bacterial Culture	pH Range	Endoglucanase Activity (CMCase) (IU/ml)	Exoglucanase Activity (FPase) (FPU/ml)	β-glucosidase Activity (BG) (IU/ml)
3.	*Bacillus subtilis*	3	0.136	0.061	0.049
		4	0.168	0.074	0.056
		5	0.415	0.077	0.061
		6	0.190	0.063	0.059
		7	0.185	0.061	0.044
		8	0.182	0.060	0.037
		9	0.109	0.022	0.009
		10	0.040	0.002	0.004

All data given are the mean value of the two independent experiments.

Table 38.8: Effect of Temperature range on Activity of Cellulase Enzymes in Three Bacterial Cultures

Sl.No.	Bacterial Culture	Temperature Range (°C)	Endoglucanase Activity (CMCase) (IU/ml)	Exoglucanase Activity (FPase) (FPU/ml)	β-glucosidase Activity (BG) (IU/ml)
1.	*Bacillus marinus*	30	0.379	0.059	0.058
		40	0.407	0.096	0.060
		50	1.776	0.100	0.083
		60	0.116	0.020	0.033
		70	0.114	0.019	0.033
		80	0.074	0.021	0.031
		90	0.046	0.012	0.021
2.	*Bacillus* sp.(V2)	30	0.215	0.080	0.046
		40	0.220	0.080	0.048
		50	0.456	0.085	0.057
		60	0.885	0.090	0.058
		70	0.336	0.084	0.046
		80	0.188	0.010	0.003
		90	0.020	0.004	0.001
3.	*Bacillus subtilis*	30	0.078	0.077	0.054
		40	0.288	0.077	0.061
		50	0.418	0.080	0.062
		60	0.199	0.078	0.056
		70	0.080	0.068	0.020
		80	0.061	0.009	0.005
		90	0.009	0.003	0.002

All data given are the mean value of the two independent experiments.

Effect of Chemicals Reagents on Endoglucanase Activity of Three Bacterial Cultures

Figures 38.11 and 38.12 shows the effect of chemical reagents *viz.* EDTA, 2, 4 DNP, $HgCl_2$ and $SnCl_2$ on *Bacillus marinus*. It is clearly evident from the results that EDTA and $SnCl_2$ did not much affected the endoglucanase activity. On the other hand, 2, 4 DNP and $HgCl_2$ were found to inhibit severely. The lower concentration (0.1 mM) of 2, 4 DNP reduces the endoglucanase activity upto 67 per

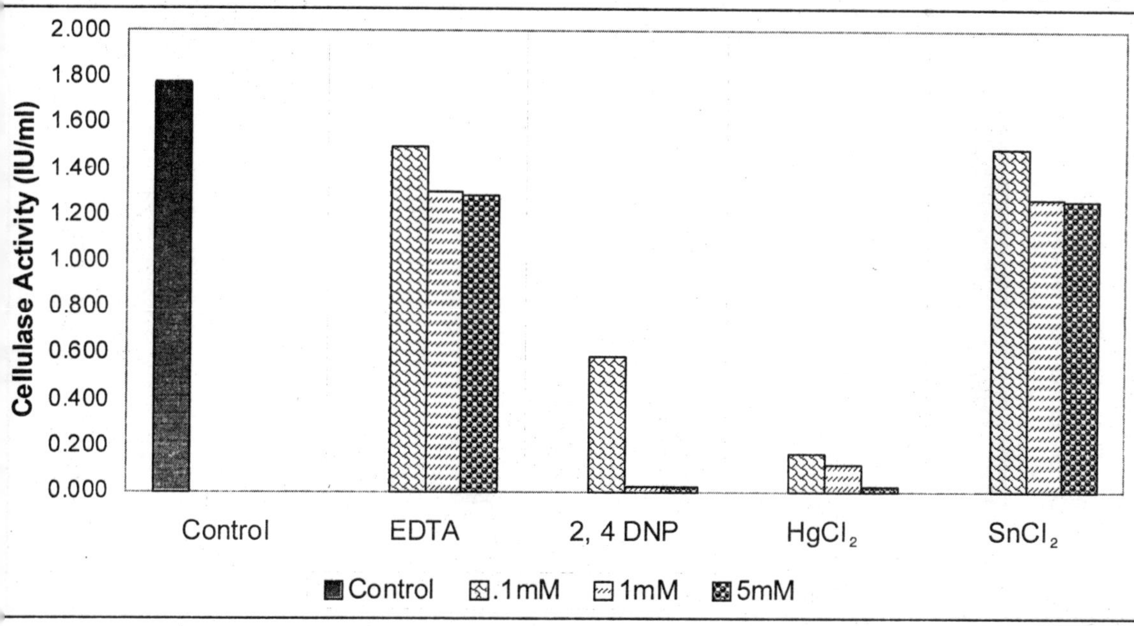

Figure 38.11: Effect of Chemical Reagents on Endoglucanase Activity of *Bacillus marinus*

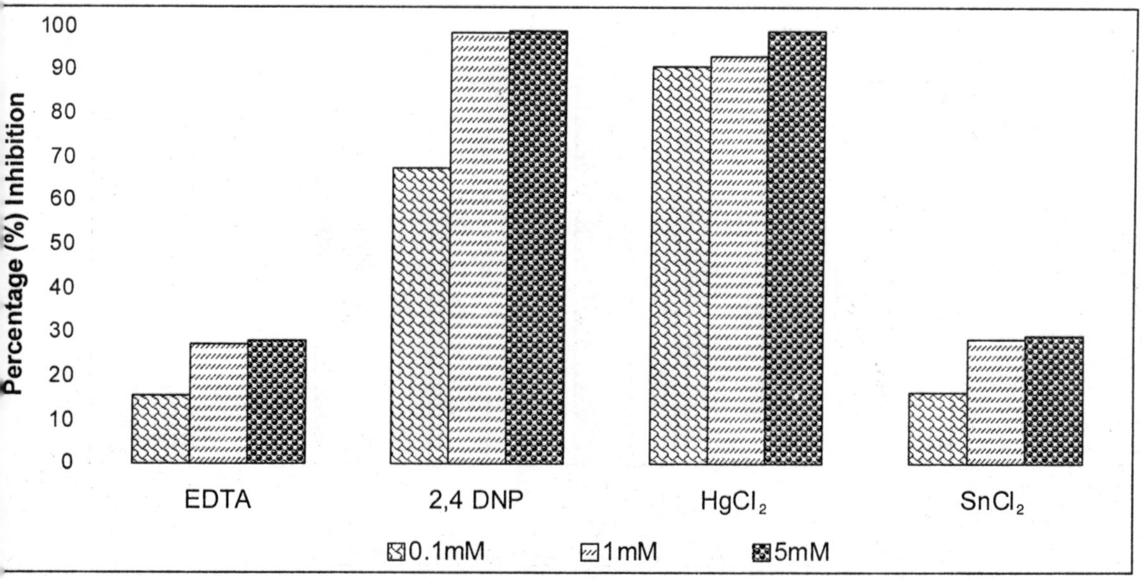

Figure 38.12: Percentage Inhibition of Endoglucanase Activity by Chemical Reagents in *Bacillus marinus*

cent and higher concentration (5 mM) were found to inhibit activity more than 95 per cent, and 0.1 mM concentration of $HgCl_2$ causes 95 per cent inhibition and higher concentration (5 mM) inhibited 100 per cent activity.

The results of effect of chemical reagent on endoglucanase activity of *Bacillus* sp. (V2) were presented in Figures 38.13 and 38.14. The data of EDTA and $SnCl_2$ suggest that they have lesser negative influence on the endoglucanase activity. 2, 4 DNP at low concentration (0.1 mM) reduces (60 per cent) of

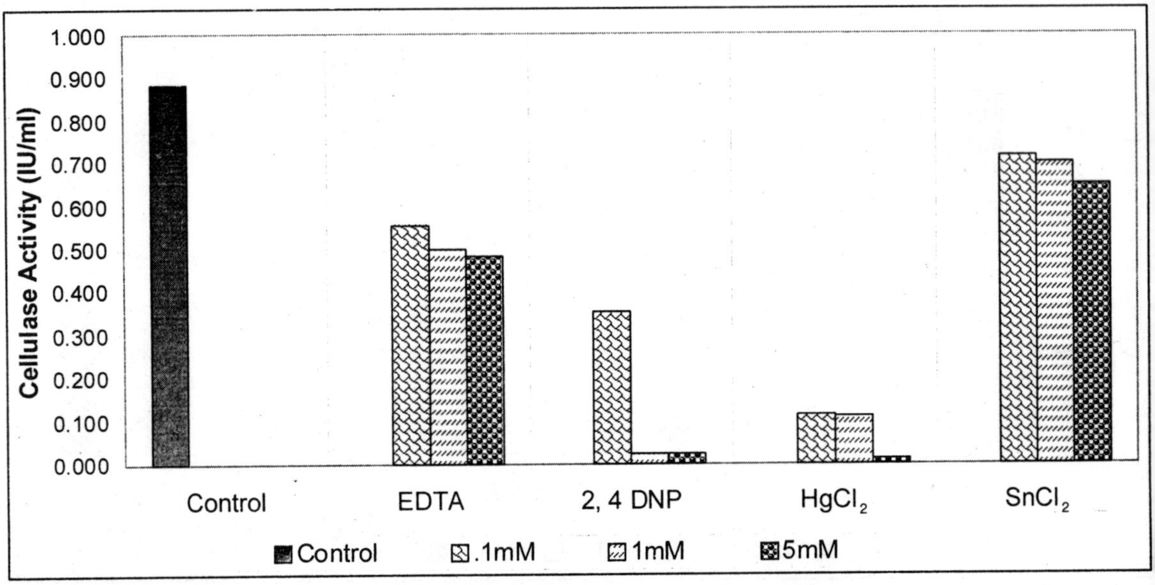

Figure 38.13: Effect of Chemical Reagents on Endoglucanase Activity of *Bacillus* sp. (V2)

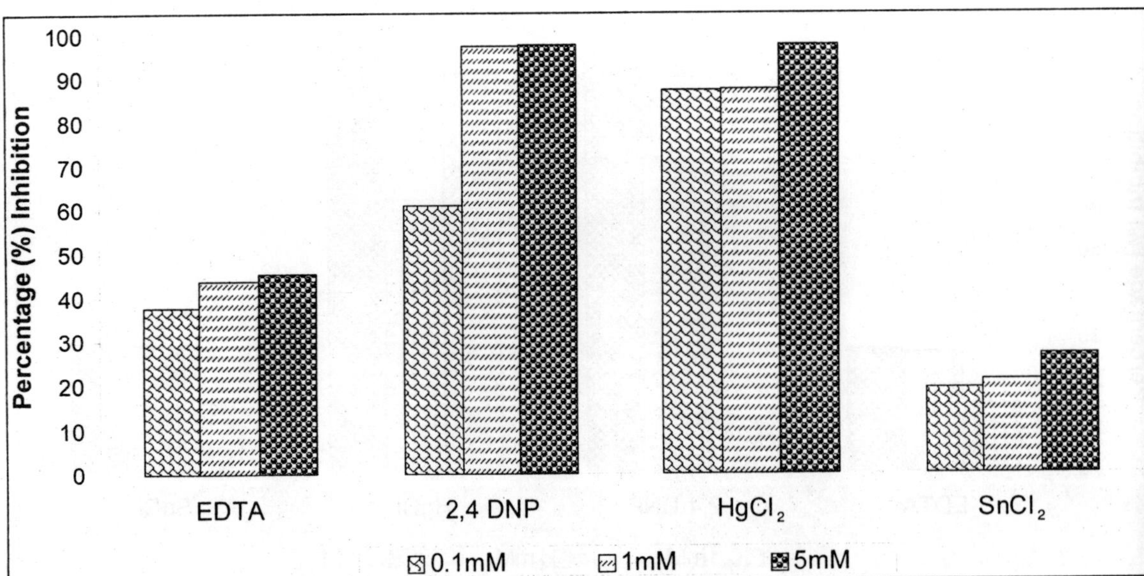

Figure 38.14: Percentage Inhibition of Endoglucanase Activity by Chemical Reagents in *Bacillus* sp. (V2)

endoglucanase activity and at higher concentration (5 mM) causes 97 per cent inhibition. Lower concentration (0.1 mM) of $HgCl_2$ causes 87 per cent inhibition and inhibited 98 per cent at 5 mM concentration.

Figures 38.15 and 38.16 shows the effect of chemical reagent on endoglucanase activity of *Bacillus subtilis*. The results clearly reveal that EDTA moderately inhibited the activity (40 per cent) at 0.1 mM and at 5 mM concentration causes 58 per cent inhibition. $SnCl_2$ made no significant adverse affect on

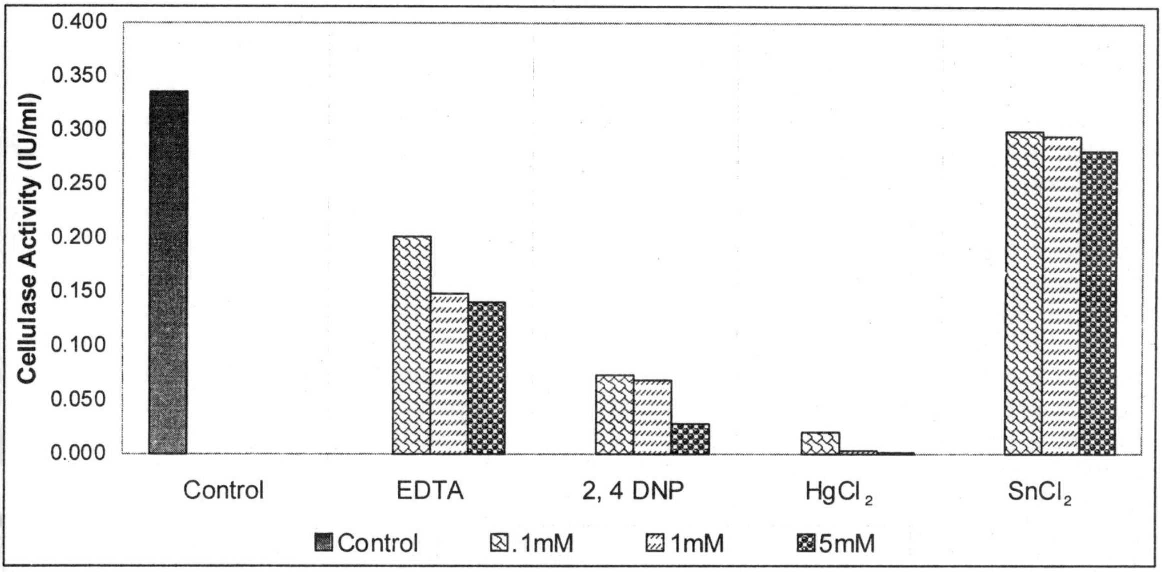

Figure 38.15: Effect of Chemical Reagents on Endoglucanase Activity of *Bacillus subtilis*

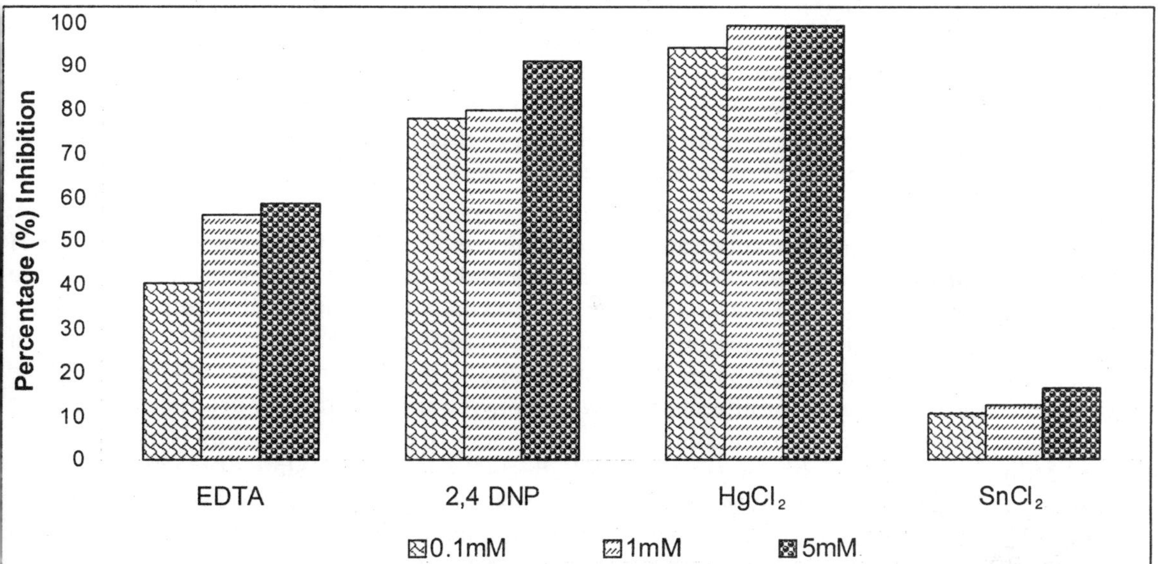

Figure 38.16: Percentage Inhibition of Endoglucanase Activity by Chemical Reagents in *Bacillus subtilis*

the activity. 2,4 DNP inhibited 77 per cent endoglucanase activity at 0.1 mM concentration and 5 mM concentration causes 90 per cent inhibition. $HgCl_2$ was found to the potent inhibitor as 94 per cent inhibition of endoglucanase activity at 0.1 mM concentration and 5 mM concentration causes 100 per cent inhibition of the activity.

Discussion

The tremendous range of catabolic diversity among microorganisms is one of the distinguishing features of the microbial world. The range of this diversity also differs widely among species utilizing one or few substrates to number of compounds as sole carbon and energy sources; however cellulolytic microbes lie near the specialist end of this continuum. The cellulose catabolism involves both enzymatic depolymerization of insoluble cellulose and cellular utilization of the hydrolytic products. Many cellulolytic microbes especially fungi when cultured on cellulosic carbon source produce extracellular cellulase systems consisting of endoglucanases, exoglucanases and β-glucosidase in free and multiple form (Wood, 1992). In comparison with fungi, most bacteria produce mainly endoglucanase enzymes capable of solubilizing amorphous and swollen cellulose (Wood, 1985). Although both endo and exoglucanases show very high activity toward amorphous cellulose, only endoglucanases can hydrolyze CMC. In the present investigation all the screened microbial forms secreted appreciable amount of cellulases in culture filterates. Species of *Aspergillus, Trichoderma, Chaetomium, Fusarium, Penicillum, Bacillus* produced high amount of endoglucanase, exoglucanase and β-glucosidase activity. Endoglucanase was found to be higher in activity when compared with that of exoglucanase and β-glucosidase. In most of the species, optimum activity was observed on 8th day (fungi) and 6th day (bacteria) of incubation respectively. It is generally accepted view that both bacteria and fungi are subjected to an inducer–repressor system with catabolite repression involved in the regulation mechanism of cellulase systems (Lao and Wilson, 1994). In catabolite repression the end products of cellulose hydrolysis interact with the cellular proteins to form a complex which interact with particular gene at the transcription level and represses cellulase synthesis (Lewin, 1987). The proof for catabolite repression is based on the fact that no cellulase in formed during the growth of a microorganisms on glucose, glycerol and other carbon sources related to glycolytic metabolism (Bhat and Bhat, 1997). However, general carbon repressor protein CRE 1 represses the transcription of cellulase genes in *Trichoderma reesei* have been isolated (Strauss *et al.*, 1995) probably support our finding indirectly that with the increasing incubation period, the accumulated product of cellulose hydrolysis catabolically repress the cellulase synthesis leading to the decrease in the enzyme activity.

The cellulase activity in fungi and bacteria is greatly dependent on factors like pH, temperature, incubation period, nutritional salt, substrates concentration etc. In present investigation effect of pH on cellulase activity was evaluated. *Aspergillus nidulans* and *Aspergillus terreus* showed a cellulase activity in acidic range whereas *Fusarium oxysporum* showed somewhat alkaline cellulase activity upto pH 9 in case of endoglucanase enzyme. According to Garg and Neelkanthan (1981), the optimum pH in *Aspergillus terreus* ranges from 3 to 6. It was suggested by many workers that pH may affect the velocity of reaction by changing the attraction between enzyme and substrate or products (Gascoigne and Gascoigne, 1960). The effect of different pH on the velocity of enzymatic reaction is mainly due to changed affinity of enzyme with its substrate, pH effects the ionization state of reaction mixture. Accordingly to Bagga *et al.* (1990) two endoglucanases exist endo I and II. Endo I shows maximum activity at pH 4 and whereas, endo II shows maximum activity at pH 5. Exoglucanases also occurs in three forms I, II, III but detail work was done on exo I and it shows maximum activity at pH 5.5. Similarly, β-glucosidase also exist in two forms I and II. β-glucosidase I shows maximum activity at pH 6 and pH 5 was enough for maximum activity of β-glucasidase II in *Aspergillus nidulans*. Tavolibov *et al.* (1988), Buachidze *et al.* (1987), Rodinova *et al.* (1987) have reported optimum pH of 4.5 for β-glucosidase in *Aspergillus terreus*. Cellulase activity has previously been reported in number of *Aspergillus* spp. (Rodionova *et al.*, 1987; Murao *et al.*, 1988; Tabolibov *et al.*, 1988; Bagga *et al.*, 1990; Himmel *et al.*, 1993; Galas and Romanowska, 1997; Chikamatsu *et al.*, 1999), *Fusarium oxysporum* (Christakopoulos, 1995 a,b), *Penicillium* sp. (Wood and McCrae, 1982; Bhat *et al.*, 1989; Reddy *et al.*, 1998).

Temperature is another important parameter for enzyme activity as lower temperature retard and higher temperature often inactivates the enzyme proteins by denaturing. The optimum temperature of an enzyme varies with factors such as enzyme purity, substrate, presence of activator or inhibitors, its usefulness for the purpose of characterizing enzyme is limited. *Aspergillus nidulans* showing temperature stability from 30-70°C indicate this fungi to exhibit thermophilic properties. *Aspergillus terreus* and *Fusarium oxysporum* was found to show mesophilic properties. Bagga *et al.* (1990) has reported maximum endoglucanases activity at 65°C, exoglucanase activity at 50°C and β-glucosidase activity at 65°C. However, in our finding maximum endoglucanase activity is observed at 60°C, exoglucanase and β-glucosidase activities are found at 50°C. Buachidze *et al.* (1987), Rodionova *et al.* (1987), Tavbolibov *et al.* (1988) reported the temperature optima of 55-60°C for β-glucosidase activity in *Aspergillus terreus*. *Fusarium oxysporum* showed optimum temperature 60°C (endoglucanase), 50°C (exoglucanase and β-glucasidase activities). Christakopoulos *et al.* (1995a) reported 55°C as optimum for endoglucanase activity in *Fusarium oxysporum*.

The number of bacterial genera was reported to produce cellulase enzymes. Among these *Bacillus* spp. were found to be dominant spp. in decomposing cellulosic waste. In nature *Bacillus* spp. are ubiquitous. There is a great diversity in physiology among member of the genus *Bacillus*, whose collective features include degradation most of all substrates derived from plant and animal sources including cellulose, starch, pectin, protein, agar, hydrocarbons and other antibiotic production, nitrification, denitrification, nitrogen fixation facultative lithotrophy, autotrophy, acidophily, alkaliophily, psychrophily, thermophily. Member of *Aeromonas* sp. were reported to produce cellulases (Ohya *et al.*, 1977; Nihlani and Satanarayana, 1992; Joshi and Bal, 1993). In the present finding, *Bacillus marinus* has broad pH and narrow temperature range for endoglucanase, exoglucanase and β-glucosidase activities, maximum enzymatic activities, at pH 6 and 50°C temperature were observed and therefore safely can be designated as mesophilic. This is the first ever report about the cellulolytic property of *Bacillus marinus*. *Bacillus subtilis* showed pH 5 and temperature optima 50°C for the three enzymes tested. *Bacillus* sp. (V2) showed optimum pH 6 for endo and exoglucanase activities and 5 pH for β-glucosidase activity. Temperature 60°C for endoglucanase, exoglucanase and β-glucosidase activities was observed optimum in *Bacillus* sp. (V2). However, Ozaki and Ito (1991) have reported the optimum pH 5.2 and 45°C temperature for cellulase activity in *Bacillus* sp. KSM-330. Robson and Chambliss (1984) observed optimum pH 4.8 and temperature 58°C for cellulase activity in *Bacillus* isolate. The capacity of selected *Bacillus* strains to produce and secrete large quantities of extracellular enzymes has placed them among the most important industrial enzyme producers. The ability of different *Bacillus* spp. to ferment in the acid, neutral and alkaline pH ranges, combined with the presence of thermophily in the genus, has lead to the development of variety of new commercial product with the desired temperature, pH activity and stability properties to address a variety of specific applications like deinking of pulp and paper industry and biostoning in textile industry. The endoglucanases enzyme system of bacteria seem to be quite complex. Yamne *et al.* (1970) identified two extra cellular (A and B) and one cell bound (c) endoglucanase components of bacteria *Pseudomonas fluorescens* var. *cellulosa*, which were purified and characterized. There are relatively few reports

(Sashihara *et al.*, 1984) of *Bacillus* spp. producing endoglucanases. Endoglucanases act synergistically with cellobiohydrolases and β-glucosidase during solubilization of crystalline cellulose (Nidetzky *et al.*, 1994; Bhat and Bhat, 1997). The best characterized example is *Bacillus subtilis* DLG (Robson and Chambliss, 1987). A number of *Bacillus* spp. were studied for their cellulase activity (Tewari and Chahal, 1977; Dhillon *et al.*, 1985; Robson and Chambliss, 1987; Nihlani and Satanarayana, 1992; Joshi and Bal, 1993; Gorska *et al.*, 2001).

Previously, there has been attempt to identify the active sites of endoglucanases, as these enzymes of cellulases complex have been widely used in the industrial processes. Active sites of cellulases have been carried out by chemical modification and inhibition studies (Dinur *et al.*, 1986). Chelator EDTA was found to effect endoglucanase activity in all fungi and bacterial cultures moderately at 5 mM concentration (not more than 45 per cent inhibition) accept *Fusarium oxysporum* and *Bacillus subtilis* in which 63 per cent and 53.86 per cent inhibition in endoglucanase activity was observed at 5 mM. The results probably suggest that EDTA is a metal chelator and probably might not chelate a possible divallent(s) required for endoglucanase activity. However, Robson and Chambliss (1984) reported that EDTA at concentration 50 mM inhibited only 10 to 20 per cent of TNP-CMC degrading activity in *Bacillus* sp. DLG and concluded that these compounds were not chelating cations required TNP-CMC degrading activity. Kotchoni *et al.* (2003) also reported 20 per cent inhibition in endoglucanase activity at 1 mM of EDTA in wild strain of *Bacillus pumillus* BPCRI. However in our finding 27.08 per cent of endoglucanase activity was inhibited in *Bacillus marinus*, 43 per cent and 55.65 per cent in *Bacillus* sp (V2) and *Bacillus subtilis* respectively. The most notable inhibition was caused by 2,4 DNP and $HgCl_2$ in all the microbial form tested and inhibited almost complete endoglucanase activity at 5 mM concentration. 2,4 DNP, however has been reported as an uncoupler for oxidative phosphorylation (Ali and Sayed, 1992) indicating that a series of phosphorylation steps may play a crucial role in signaling transduction leading to cellulase synthesis. 2,4 DNP is a phenolic compound and phenol compound are well known repressor of cellulases enzymes (Varadi, 1972). Kotchoni *et al.* (2003) reported 100 per cent inhibition of endoglucanase activity in *Bacillus pumillus*. Metal shows both stimulatory as well as inhibitory effect on microorganisms while, iron ions are essential for growth of microorganisms. Although desired in minute concentration, some ions like mercury and silver are toxic even at small concentration. In our results $HgCl_2$ was found to inhibit almost complete endoglucanase activity in all the test organisms. The sensitivity toward $HgCl_2$ may be results of the interaction with a trypthophan residues(s) at the active site of endoglucanase as earlier suggested by Eriksson and Pettersson (1968) and Ozaki and Ito (1991). $SnCl_2$ also moderately inhibit the endoglucanase activity at high concentration. $SnCl_2$ is the member of p block, IV group metals (De 1999). Lead, is also the member of same block and was found to be inhibiting cellulase activity in *Trichoderma reesei* (Kim *et al.*, 2001). $SnCl_2$ is also heavy metal like lead and probably might be inhibiting cellulase activity due to non-specific salt formation. Generally speaking cellulases enzymes are inhibited by Hg^{+}, Ag^{++}, Co^{++}, Cr^{++}, Pb^{++}, Zn^{++} salts at 10^{-3} M (Mandels and Reese, 1963). Similarly Okada (1977) also reported the complete inhibition of endoglucanase (3.2.1.4) activity at 1 mM by Hg^{+} in *Trichoderma viride*.

Acknowledgement

Authors are thankful to Head, Department of Botany, Dr. H.S.G. Uinversity, Sagar, M.P. for providing laboratory facilities. AV greatfully acknowleged Head, Department of Biotechnology, Lovely University, Jalandhar, Punjab.

References

Ali, S., and A. Sayed. (1992). Regulation of cellulase biosynthesis in *Aspergillus terreus. World J. Microbiol. Biotechnol.* 8: 73-75.

Bagga, P.S., D.K. Sandhu, and S. Sharma. (1990). Purification and characterization of cellulolytic enzyme produced by *Aspergillus nidulans. J. Appl. Bacteriol.* 68: 61-68.

Belaich, J.P., A. Belaich, H.P. Fierobe, L. Gal, C. Gaudin, S. Pages, C. Reverbel-Leroy, and C. Tordif. (1999). The cellulolytic system of *Clostridium cellulolyticum*. p. 479-487. In: K. Ohmiya, K. Hayashi, K. Sakka, Y. Kobayashi, S. Karita, and T. Kimura ed., *Genetics, biochemistry and ecology of cellulose degradation*. Uni. Publishers, Tokyo, Japan.

Belaich, J.P., C. Tordif, A. Belaich, and C. Gaudin. (1997). The cellulolytic system of *Clostridium cellulolyticum. J. Biotechnol.* 57: 3-14.

Bergem, L.E.R., and L.G. Pettersson. (1973). The mechanism of enzymatic cellulose degradation, purification of a cellulolytic enzyme from *Trichoderma viride* active on highly ordered cellulose. *Eur. J. Biochem.* 37: 21-30.

Bhat, M.K. (2000). Cellulases and related enzymes in biotechnology. *Biotechnol. Adv.* 18: 355-383.

Bhat, M.K., and Bhat, S. (1997). Cellulose degrading enzymes and their potential industrial application *Biotechnol. Adv.* 15: 583-620.

Bhat, M.K., Gaikwad, J.S. and Maheswari, R. (1993). Purification and characterization of an extracellular β-glucosidase from thermophilic fungus *Sporotrichum thermophile* and its influence on cellulase activity. *J. Gen. Microbiol.* 139: 2825-2832.

Bhat, M.K., S.I. McCrae, and T.M. Wood. (1989). The endo-(1-4)-α-D-glucanase system of *Penicillium pinophillum* cellulase: isolation, purification and characterization of five major endoglucanase components. *Carbohyd. Res.* (190: 279-297.

Bisaria, V.S., and Mishra, S. (1989). Regulatory aspects of cellulase biosynthesis and secretion. *CRC Crit. Rev. Biotechnol.* 9: 61-103.

Brown, R.M. (1982). *Cellulose and Natural Polymer Systems: Biogenesis, Structure and Degradation*. Plennum. New York.

Buachidze, T.S., I.M. Tavabilov, N.A. Rodinova, and G.I. Kvesitadze. (1987). Isolation and purification of β-D-glucosidase from the thermophilic mutant *Aspergillus terreus. Appl. Biochem. Microbiol.* 23: 187-196.

Chikamatsu, G., K. Shirai, M. Kato, T. Kobayashi, and N. Tsukagoshi. (1999). Structure and expression properties of the endo-β-1, 4-glucanase A gene from the filamentous fungus *Aspergillus nidulans. FEMS Microbiol. Lett.*, 175: 239-245.

Christakopoulos, P., B.J. Macris, and D. Kekos. (1989). Direct fermentation of cellulose to ethanol by *Fusarium oxysporum. Enzyme Microb. Technol.*, 11: 236-239.

Christakopoulos, P., D. Kekos, B.J. Macris, M. Claeyssens, and M.K. Bhat. (1995a). Purification and characterization of a less randomly acting endo-1, 4-β-D-glucanase from culture filterates of *Fusarium culmorum. Arch. Biochem. Biophys.* 316 (1): 428-433.

Christakopoulos, P., D. Kekos, B.J. Macris, M. Claeyssens, and M.K. Bhat. (1995b). Purification and mode of action of low molecular mass endo-1, 4-α-D-glucanase from *Fusarium oxysporum. J. Biotechnol.* 39: 85-93.

Coughlan, M.P., and L.G. Ljungdahl. (1988). Comparative biochemistry of fungal and bacterial cellulolytic enzyme systems. p. 11-30. *In* J.P. Aubert, P. Beguin, and J. Millet ed. *Biochemistry and genetics of cellulose degradation*. FEMS Sym. 43. Academic Press, London.

Davis, B.D., and P.C. Tai. (1980). The mechanism of protein secretion across membranes. *Nature* (London). 283: 433-438.

De, A.K. (1999). *A text book of inorganic chemistry* Eighth edition. New Age Intel.

Dhillon, N., S. Chhibber, M. Saxena, S. Pajni, and D.V. Vadehera. (1985). A constitutive endoglucanase (CMCase) from *Bacillus licheniformis*-1. *Biotechnol. Lett.* 7: 695-697.

Din, N., H.G. Damude, N.R. Gilkes, R.C. Miller, R.A.J. Warren, and D.G. Kilburn. (1994). C_1-Cx revisited: intramolecular synergism in a cellulase. *Proc. Natl. Acad. Sci. USA*. 91: 11383-11387.

Dinur, T., K.M. Osiecki, G. Legler, S. Gatt, R.J. Desnick, and G.A. Grabourki. (1986). *Proc. Natl. Acad. Sci. U.S.A.* 83: 1660-1664.

Eggins, H.O.W., and G.J.F. Pugh. (1962). Isolation of cellulose decomposing fungi from the soil. *Nature* London 193: 94-95.

Ehrlich, P.R., and E.O. Wilson. (1991). Biodiversity studies. Science and Policy. *Science*. 253: 758-762.

Eidsa, G. (1972). Cellulase from *Coniophora cerebella* (Pers.) Duby. *Thesis*. University of Bergen.

Elberson, M.A., F. Malekzadeh, M.T. Yazdi, N. Kameranpour, M.R. Noori-Dioii, M.H. Matt, M. Shahamat, R.R. Colwell, and K.R. Sowers. (2000). *Cellulomonas persica* sp. nov. and *Cellulomonas iranensis* sp. nov. mesophilic cellulose degrading bacteria isolated from forest soil. *Int. J. Syst. Evol. Microbiol.* 50: 993-996.

Enari, T.M. (1983). Microbial cellulases, p. 83-223. *In* W.F. Forgaty (ed.) *Microbial enzymes and biotechnology*. Applied Sciences Publishers, London.

Enari, T.M., and M.L. Niku-Paavola. (1987). Enzymatic hydrolysis of cellulose: is the current theory of the mechanisms of hydrolysis valid? *CRC Crit. Rev. Biotechnol.* 53: 67-87.

Eriksson, K.E. (1969). New methods for the investigation of cellulases. *Adv. Chem. Ser.* 95: 88-104.

Eriksson, K.E. (1975). Enzymatic mechanism involves in the degradation of wood component. 12-14 *March. Symposium on enzymatic hydrolysis of cellulose*. Aulan Ko. Finland.

Eriksson, K.E. (1978). Enzyme mechanism involved in cellulose hydrolysis by rot fungus *Sporotrichum pulverulentum*. *Biotechnol. Bioeng.* 70: 317-332.

Eriksson, K.E., and G. Pettersson. (1968. Studies on the cellulolytic enzymes V. some structural properties of the cellulase from *Penicillium notatum*. *Arch. Biochem. Biophys.* 124: 160-166.

Eriksson, K.E., R.A. Blanchette, and P. Ander. (1990). "*Microbial and Enzymatic Degradation of Wood and Wood Components*". Springer Verlag, Berlin. Heidelberg. Germany.

Fagerstam, L.G., and L.G. Pettersson. (1980). The 1, 4-β-Glucan cellobiohydrolases of *Trichoderma reesei* QM 9414. A new type of cellulolytic synergism. *FEBS Lett.* 119: 97-100.

Fan, L.T., M.M. Gharpuray, and Y.H. Lee. (1987). *Cellulose Hydrolysis*. Berlin, Germany: Springer–Verlag 3: 1-68.

Fan, L.T., Y.H. Lee, and D.H. Beardmore. (1981). The influence of major structural features of cellulose on rate of enzymatic hydrolysis. *Biotechnol. Bioeng.* 23: 419-424.

Galas, E., and I. Romanowska. (1997). Purification and some properties of β-glucosidase from *Aspergillus niger* IBT-90. *Acta Microbiol. Pol.* 46: 241-252.

Garcia-Martinetz, D.V., A. Shinmyo, A. Madia, and A.L. Demain. (1980). Studies on cellulase production by *Clostridium thermocellum. Eur. J. Appl. Microbiol. Biotechnol.* 9: 189-197.

Gardner, R.M., K.C. Doerner, and B.A. White. (1987). Purification and characterization of exo-β-1, 4 glucanase from *Ruminococcus flavifaciens* FD-1. *J. Bacteriol.* 169: 4581-4588.

Garg, S.K., and S. Neelkanthan. (1981. Effect of cultural factors on cellulase activity and production by *Aspergillus terreus. Biotechnol. Bioeng.* 23: 1653-1659.

Gascoigne, J.A., and M.M. Gascoigne. (1960). *Biological degradation of cellulose.* Butterworths and Co. (Publ.) Limited. London. 264.

Ghose, T.K. (1969). Continuous enzymatic saccharification of cellulose with culture filterate of *Trichoderma viride* QM6a. *Biotechnol. Bioeng.* 11: 239.

Ghosh, M., and B. Nandi. (1995). p. 53 A.K. Roy, and K.K. Sinha ed. In: *Recently Advances in Phytopathology.* MD Publication, New Delhi.

Ghosh, S., B.P. Kapadnis, and N.B. Singh. (2000). Composting of cellulosic hospital waste: A potentially novel approach. *Int. Biodeter. Biodegrad.* 45: 89-92.

Ghosh, T.K., T. Panda, and V.S. Visaria. (1984). In: *VIIth Int. Biotechnol. Sym.* IIT, New Delhi. 620.

Gilkes, N.R., B. Henrissat, D.G. Kilburn, R.C. Miller Jr., and R.A.J. Warren. (1991). Domains in microbial β-1, 4-glycanases: sequence conservation function and enzyme families. *Microbiol. Rev.* 55: 303-315.

Gilkes, N.R., R.A.J. Warren, R.C. Miller Jr., and D.G. Kilburn. (1988). Precise excision of the cellulose binding domains from two *Cellulomonas fimi* cellulases by a homologous protease and effect on catalysis. *J. Biol. Chem.* 263: 10401-10407.

Gilman, J.C. (2001). *A manual of soil fungi.* Biotech Book, New Delhi.

Gong, C.S., M.R. Ladisch, and T.G. Tsao. (1979). Biosynthesis, purification and mode of action cellulases of *Trichoderma reesei. Adv. Chem. Ser.* 181: 261-287.

Gorska, E., B. Tudek, and S. Russel. (2001). Degradation of cellulose by nitrogen-fixing strain of *Bacillus polymyxa. Acta Microbiol. Pol.* 50: 129-137.

Guglielmi, G., and P. Beguin. (1998). Cellulase and hemicellulase genes of *Clostridium thermocellum* from five independent collections contain few overlap and are widely scattered across the chromosome. *FEMS Microbiol. Lett.* 161: 209-215.

Hans, Y.W., and V.R. Srinivasan. (1968). Isolation and characterization of cellulose utilizing bacterium. *Appl. Microbiol.* 16: 1140-1145.

Hawksworth, D.L., and R.R. Colwell. (1992). Biodiversity amongst microorganisms at its relevance. *Biodiv. Conserv.* 1: 221-345.

Hawksworth, D.L., P.M. Kirk, B.C. Sutton, and D.N. Peglar. (1995). *Dictionary of Fungi.* CAB Intl.

Himmel, M.E., W.S. Adney, J.W. Fox, D.J. Mitchell, and J.O. Baker. (1993). Isolation and characterization of two forms of β-D-glucosidase from *Aspergillus niger. Appl. Biochem. Biotechnol.* 39/40: 213-225.

Joshi, A., and B. Ball. (1993). Extracellular alkaline enzymes of facultative bacteria of $CaCO_3$ kilns near Jabalpur. *Ind. J. Microbiol.* 33: 179-183.

Kim, D.W., Y.H. Jang, C.S. Kim, and N.S. Lee. (2001). Effect of metals ions on the degradation and adsorption of two cellobiohydrolases on microcrystalline cellulose. *Bull.*

Knosel, D. (1971). Continued investigation for pectolytic and cellulolytic activity of different *Bacillus* species. Zentralblatt. fur. *Bacteriologie* 126: 604-609.

Knowles, J.K.C., P. Lehtovaara, T.T. Teeri, M. Penttila, I. Salovuori, and L. Andre. (1987). *Philos. Trans. R. Soc.* London. 321: 449-454.

Kotchoni, O.S., O.O. Shonukan, and W.E. Gachomo. (2003). *Bacillus pumilus* BpCRI 6, a promising candidate for cellulase production under condition of catabolite repression. *Afr. J. Biotechnol.* 2: 140-146.

Labudova, I., and V. Farkas. (1983). Multiple enzyme forms in the cellulase system of *Trichoderma reesei* during its growth on cellulose. *Biochem. Biophys. Acta.* 744: 135-140.

Langford, M.L., N.R. Gialkes, W.W. Wakarchuk, D.G. Kilburn, R.C. Miller, and R.A.J. Warren. (1984). The cellulase system of *Cellulomonas fimi. J. Gen. Microbiol.* 130: 1367-1376.

Lewin, B. (1987). *Genes III, An up-to-date treatment of the molecular genetics of both eukaryotes and prokaryotes, along with related cell and molecular biology.* John Wiley, New York.

Ljungdahl, L.G., and K.E. Eriksson. (1985). Ecology of microbial cellulose degradation. *Adv. Microb. Ecol.* 8: 237-299.

Mandels, M., and E.T. Reese. (1963). Inhibition of cellulase and β-glucosidase. p. 115-158. *In* E.T. Reese, ed. *Advances in Enzymatic Hydrolysis of Cellulose and Related Materials.* Pergamon Press, London.

Mandels, M., and E.T. Reese. (1964). *Fungal cellulases and the microbial decomposition of cellulosic fabric. Development in Industrial* (Amer. Inst. Biol. Sci.) Washington D.C. Microbiol. 5: 5-20.

Murao, S., S. Sakamoto, and M. Arai. (1988). Cellulases of *Aspergillus aculeatus. Methods Enzymol.* 160: 275-299.

Nidetzky, B., W. Steiner, M. Hayn, and M. Claeyssens. (1994). Cellulose hydrolysis by the cellulases from *Trichoderma reesei::* a new model for synergistic interactions. *Biochem. J.* 298: 705-710.

Nihlani, D., and T. Satanarayana. (1992). Isolation and characterization of extracellular enzyme producing bacteria. *Ind. J. Microbiol.* 32 : 434-449.

Ohya, T., N. Yokoi, and T. Mase. (1977). "Process for preparation of enzymes". *UK Patent.* 1465307.

Okada, G. (1977). Enzymatic studies on a cellulase systems of *Trichoderma viride* II Purification and properties of two cellulases. *J. Biochem.* 77: 33-42.

Ozaki, K., and S. Ito. (1991). Purification and properties of an acid endo-1, 4-β-glucanase from *Bacillus* sp. KSM–330. *J. Gen. Microbiol.* 137: 41-48.

Priest, F.G. (1984). *Extracellular enzymes.* American Soc. Microbiol, Washinghton, D.C.

Reddy, B.R., G. Narsimha, and G.V.A.K. Babu. (1998). Cellulolytic activity of fungal culture. *J. Scient. Ind. Res.* 57: 617-620.

Reese, E.T., J.E. Lola, and F.W. Parrish. (1969). Modified substrates and modofied products as inducers of carbohydrolases. *J. Bacteriol.* 100: 1151-1154.

Reese, E.T., R.G.H. Siu, and H.S. Levinson. (1950). The biological degradation of soluble cellulose derivatives and relationship to the mechanism of cellulose hydrolysis. *J. Bacteriol.* 59: 485-497.

Robson, L.M., and G.H. Chambliss. (1984). Characterization of the cellulolytic activity of *Bacillus* isolate. *Appl. Environ. Microbiol.* 47: 1039-1046.

Robson, L.M., and G.H. Chambliss. (1987). Endo-β-1, 4 glucanase gene of *Bacillus subtilis* DLG. *J. Bacteriol.* 169: 2017-2025.

Rodionova, N.A., I.M. Tavobilov, L.I. Martinovich, T.S. Buachidze, G.I. Kvesitadze, and A.M. Bezborodov. (1987). β-Glucosidases from cellulolytic fungi *Aspergillus terreus, Geotrichum candidum* and *Trichoderma longibrachiatum* as typical glycosidases. *Biotechnol. Appl. Biochem.* 9: 239-250.

Sashihara, N., T. Kudo, and K. Horikoshi. (1984). Molecular cloning and expression of cellulase genes of alkaliphilic *Bacillus* sp. strain N-4 in *Escherchia coli. J. Bacteriol.* 158: 503-506.

Singh, A., A.B. Abidi, N.S. Darmwal, and A.K. Agrawal. (1991). Single cell protein production by *Aspergillus niger* and its evaluation. *Zentbl. Mikrobiol.* 146: 181-184.

Singh, A., A.K. Agrawal, A.B. Abidi, and N.S. Darmwal. (1990). Properties of cellobiose from *Aspergillus niger. Appl. Microbiol. Biotechnol.* 34: 356-358.

Singh, J.S., and E. Khurana. (2002). Paradigms of Biodiversity: An overview. *Proc. Ind. Natn. Sci. Acad. (PINSA).* B68 (3): 273-296.

Strauss, J., R.L. Mach, S. Zeilinger, G. Hartler, G. Stoffler, M. Wolschek, and C.P. Kubichek. (1995). Cre1, the carbon catabolite repressor protein from *Trichoderma reesei. FEBS. Lett.* 376: 103-107.

Subramanian, C.V. (1971). *Hyphomycetes: An account of Indian species except Cercosporae.* Indian Council of Agricultural Research. New Delhi. India.

Tavolibov, I.M., T.S. Buachidze, N.A. Rodionova, G.I. Kvesitadze, and A.M. Bezborodov. (1988). Substrate specificity and certain catalytic characteristic of β-glucosidase from *Aspergillus terreus. Biochemistry* 49: 847-853.

Teeri, T.T. (1997). Crystalline cellulose degradation: new insight into function of cellobiohydrolases. *Trends Biotechnol.* 15: 160-167.

Teeri, T.T., A. Koivula, M. Linder, G. Wohlfahrt, C. Devine, and T.A. Jones. (1998). *Trichoderma reesei* cellobiohydrolases: why so efficient on crystalline cellulose? *Biochem. Soc. Trans.* 26: 173-178.

Tewari, H.K., and D.S. Chahal. (1977). Growth and cellulase formation by *Bacillus* sps. *Ind. J. Microbiol.* 17: 23.

Tewari, H.K., S.S. Marwaha, and L. Singh. (1988). Studies on cellulase production from groundnut shell. *J. Res. (PAU)* 25: 424-432.

Thayer, D.W. (1978). Carboxymethylcellulase produced by facultative bacteria from the hind gut of the termite *Reticulitermes hesperus. J. Gen. Microbiol.* 106: 13-18.

Tomme, P., R.A.J. Warren, and N.R. Gilkes. (1995). Cellulose hydrolysis by bacteria and fungi. *Adv. Microbiol. Physiol.* 37: 1-81.

Varadi, J. (1972). In *"Biodetorioration of materials"*. pp. 129-135 *In.* A.H. Walters and E.H. Hueck van der plas (eds). Applied Science, London.

Vyas, A., and D. Vyas. (2002). Cellulolytic activity of fungal cultures isolated from cellulosic laboratory waste. *J. Basic Appl. Micol.* 1: 183-185.

Vyas, A., D. Vyas, and K.M. Vyas. (2003a). Screening of Extracellular cellulases producing fungi from different lignocellulosic waste. *J. Basic Appl. Mycol.* 2: 41-43.

Vyas, A., D. Vyas, and K.M. Vyas. (2003b). Cellulolytic strains of *Aspergillus* and *Penicillium*. *J. Bot. Soc. University of Saugor. (Formerly Bull. Bot. Soc.)* 39: 61-77.

Warcup, J.H. (1955). The soil plate method for isolation of fungi from soil. *Nature* (London): 116-117.

Wood, P.J. (1980). Specificity in the interaction of direct of direct dyes with polysaccharides. *Carbohydr. Res.* 85: 271-287.

Wood, T.M. (1969). The cellulase of *Fusarium solani*. Resolution of the enzyme complex. *Biochem. J.* 115: 457-464.

Wood, T.M. (1975). Properties and mode of action of cellulase. pp. 111-137 *In* G.R. Wilki ed. *Cellulose as chemical, energy source*, N. York. John Wiley. New York.

Wood, T.M. (1985). Properties of cellulolytic enzyme system. *Biochem. Soc. Trans.* 13: 407-410.

Wood, T.M. (1992). Microbial enzymes involved in the degradation of cellulose components of plant cell walls. pp. 10-24. *In*: Rowett Research Institute Annual Report.

Wood, T.M., and S.I. McCrae. (1972). The purification and properties of the C_1 component of *Trichoderma koningii* cellulases. *Biochem. J.* 128: 1183-1192.

Wood, T.M., and S.I. McCrae. (1975). pp. 231-254. *In* M.J. Bailey, T.M. Enari, and M. Linko (ed.). *"Enzymatic hydrolysis of cellulose"* SITRA Helsinki.

Wood, T.M., and S.I. McCrae. (1977). Cellulase from *Fusarium solani*: Purification and properties of the C_1 components. *Carbohydr. Res.* 53: 117-133.

Wood, T.M., and S.I. McCrae. (1982). Purification and some properties of (1, 4)-β-D-glucan glucanohydrolase associated with the cellulase from the fungus *Penicillium funiculosum*. *Carbohydr. Res.* 110: 291-303.

Wood, T.M., and S.I. McCrae. (1986a). The cellulase of *Penicillium pinophilum*: synergism between enzyme components in solubilizing cellulosic with special reference to the involvement of two immunologically distinct cellobiohydrolases. *Biochem. J.* 234: 93-94.

Wood, T.M., and S.I. McCrae. (1986b). Purification and properties of a cellobiohydrolases from *Penicillium pinophillum*. *Carbohydr. Res.* 148: 331-344.

Wood, T.M., C.A. Wilson, S.I. McCrae, and A.L. Joblin. (1986). A highly active extracellular cellulase from anaerobic rumen fungus *Neocallimastix frontalis FEMS Microbiol. Lett.* 34: 37-40.

Wood, T.M., S.I. McCrae, and C.C. MacFarlane. (1980). The isolation purification and properties of the cellobiohydrolase components of *Penicillium funiculosum* cellulase. *Biochem. J.* 169: 51-65.

Wood, T.M., S.I. McCrae, and M.K. Bhat. (1989). The mechanism of fungal cellulase action. Synergism between enzyme components of *Penicillium pinophillum* cellulase in solubilizing hydrogen bond ordered cellulose. *Biochem. J.* 260: 37-43.

Wood, T.M., S.I. McCrae, C. Wilson, M.K. Bhat, and L. Gow. (1988). Aerobic and anaerobic fungal cellulases with special reference to their mode of attack on crystalline cellulose. pp. 31-53. *In* J.P. Aubert, P. Beguin, and J. Millet (ed.). "*Biochemistry and genetics of cellulose degradation*. FEMS Symp. 43 Academic Press London.

www.aspergillus.man.ac.uk

www.mycologyonline.adelaide.ac.edu

www.textbookofbacteriology.net/Bacillus.html

Yamane, K., H. Suzuki, and K. Nisizawa. (1970). Purification and properties of extracellular and cell bound cellulase components of *Pseudomonas fluorescens* var. *cellulosa*. *J. Biochem*. 67: 19-35.

Ecology

Microbial Biotechnology and Ecology (2011) *Pages* **669–682**
Editors: **Deepak Vyas, G.S. Paliwal, P.K.Khare and R.K Gupta**
Published by: **DAYA PUBLISHING HOUSE, NEW DELHI**

Chapter 39

Lichen Bioindicator Communities in Achanakmar: Amarkantak Biosphere Reserve, Madhya Pradesh and Chhattisgarh

☆ ***Satya and D.K. Upreti****

Lichenology Laboratory, National Botanical Research Institute (CSIR)
Rana Pratap Marg, Lucknow – 226 001, U.P.

ABSTRACT

Achanakmar-Amarkantak area, an interstate Biosphere Reserve falling in Madhya Pradesh and Chhattisgarh state, has been designated as 14th Biosphere Reserve by Government of India in the year 2005. Based on the habitat together with some morphological and anatomical characters the lichen taxa of the Achanakmar–Amarkantak Biosphere Reserve, categorized into 11 lichen communities. The biosphere reserve exhibit maximum cover of Physcioid lichen communities represented by 18 per cent followed by 16 per cent Pertusarioid, 13 per cent of Lecanoroid and 12 per cent of Lecideoid communities. On the basis of different lichen communities the study area can be clearly demarcated into undisturbed, moderately disturbed and disturbed areas. The Attaria, Kurkhuredader and near to Kabirchabutra area falls under the undisturbed area. The Kapildhara, Mai ki bagia, 7 km before Karanjia from Gorakhpur, Chauradader, Khurkhuridader VALCO mining site vergin forest area are moderately disturbed while 5 km before Chaparwa from Amarkantak, Pataleshwar, Durgadhara, Dindori forest rest house, 13 km before Kabirchabutra from Chauradader, are the disturbed area of the biosphere.

Keywords: *Bioindicator, Lichen Communities, Achanakmar-Amarkantak Biosphere Reserve.*

* E-mail: upretidk@rediffmail.com

Introduction

Conservation agencies are interested in promoting and maintaining biodiversity, based on rapidly measurable forest characteristics for effective implementation of protective measures (Neitlich and McCune, 1997). The lichens are one of the important components of biodiversity as they are well known as bioindicators. Lichens are vital components of the ecosystem serving as substrates, shelter, food, nutrient cycles and succession (Gradstein, 1992). Lichens are sensitive and widely accepted as indicators of air quality, biodiversity and climate change. Large scale data on lichens and environmental variables that determine their distribution are an integral component of many global conservation programmes.

Plants and allies can serve as bioindicators of environmental conditions and ecosystem health. Lichens communities are currently used as indicators of forest ecosystem function in several contexts. Region wide studies have goals of monitoring the relation of major factors such as climatic, air quality and forest tree species composition to variation in lichen communities, and how these relations interact and change over time. Studies of lichens of particular forest type often have goals of monitoring effects of forest management practices and landscape context, including a variety of indirect human impacts on forest environment. The presence of some particular communities of lichens in a forest indicates the status of the habitat. In the recent years lichen community are included in the forest health monitoring programs because they help to answer several key assessment questions. Repeated sampling of lichen communities in permanent plots can be utilized to document the changes in the condition of lichen communities overtime which also reflects the environmental condition of the area.

Owing to the vast natural resources, diverse and rich flora and fauna, unique topography and geographical situation, the Achanakmar–Amarkantak area an interstate Biosphere Reserve has been designated as 14th Biosphere Reserve by Government of India in the year 2005. Though, a number of information regarding the higher group of plants are available from the Biosphere Reserve, however, the studies on most of cryptogamic plants are quite neglected. Recently, Satya (2006), enumerated 130 species belonging to 44 genera and 25 families of lichens from the Biosphere Reserve. In the present paper an attempt has been made to provide the detailed account of Bioindicator lichen communities in the Biosphere Reserve. The available account of bioindicator communities and their distribution pattern can be used as an useful tool for defining and presenting the disturbed and undisturbed areas of the Biosphere Reserve. The bioindicators communities also used in indicating areas of conservation interest.

Materials and Methods

Study Area

The Achanakmar-Amarkantak Biosphere Reserve (AABR) lies between 22º15' to 22º58' N and 81º25' to 82º5' E, having an area of 3835.51 sq km (Figure 39.1 and Table 39.1). It comprises part of Bilaspur district of Chhattisgarh, Dindori and Anuppur districts of Madhya Pradesh falling in agro-climatic zone. Out of 3835.51 sq km area, 1224.98 sq km falls in Madhya Pradesh and remaining area of 2610.53 sq km in Chhattisgarh state. The AABR area includes one core zone *i.e.* Achanakmar Wildlife Sanctuary lying in Bilaspur district with a total geographical area of 551.15 sq km.

The AABR has a varied topography, the districts of Bilaspur and Anuppur has plane rice fields. The Dindori has undulating hillocks, while Maikal range of Satpuras culminating in the Amarkantak plateau of bauxite rocks. Numerous streams and "nallas", many of which are perennial, cut up the whole area. Thus variety of landforms is one of the major attributes of AABR. The Biosphere Reserve is one of the major watershed of the peninsula and it is the source of three major river systems namely

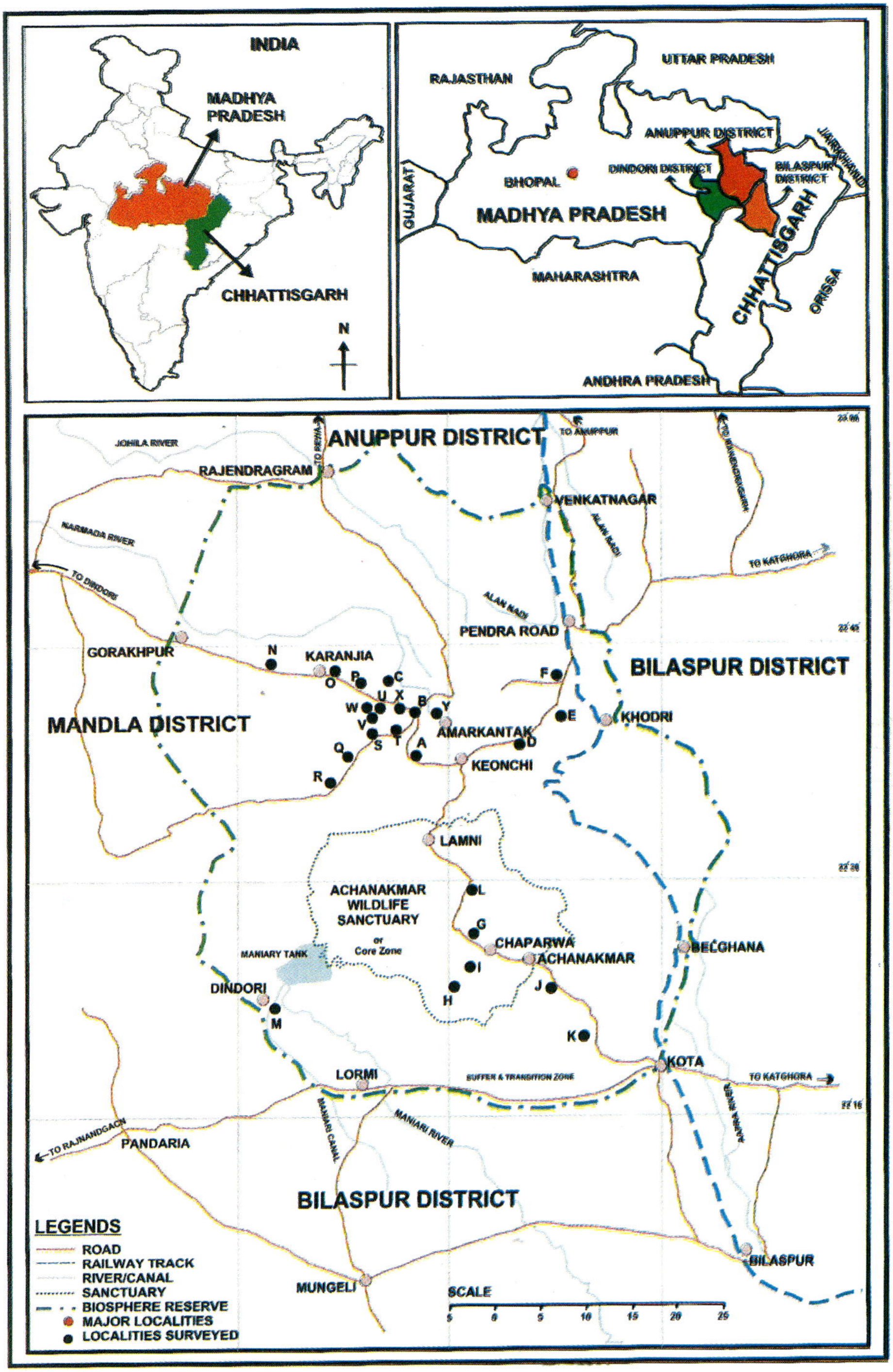

Figure 39.1: Map Showing the Localities Surveyed for Lichen Study, for Legends Please see Table 39.1

Narmada, Johilla and Sone (Ganga basin) and Ama nallah stream that join Arpa river (Mahanadi basin).

Table 39.1: The Sites Selected for Collection of Lichens, their Altitude and Major Phorophytes

Locality Codes	*Localities*	*Altitudes (m)*	*Phorophytes*
A	Attaria	680	*Shorea robusta*
B	Kabirchabutra	808	*Shorea robusta, Mangifera indica, Syzygium cuminii, Ficus* sp.
C	Kapildhara	808	*Shorea robusta, Syzygium cuminii, Mallotus philippensis*
D	Mai ki Bagia	804	*Shorea robusta*
E	Jwaleshwar	804	*Shorea robusta, Emblica officinalis, Ficus* sp, *Terminalia arjuna, Mallotus philippensis*
F	Durgadhara	784	*Mangifera indica,* Unidentified tree, *Syzygium cuminii*
G	5 km before Chaparwa from Amarkantak	476	*Mallotus philippensis, Shorea robusta,* Unidentified tree
H	Gabhighat	488	*Mallotus philippensis, Ficus racemosa, Terminalia arjuna, Mangifera indica, Syzygium cuminii,* Unidentified tree
I	Chaparwa naala	457	*Mallotus philippensis, Shorea robusta, Ficus racemosa*
J	5 km away from Chaparwa towards Kota	439	*Mallotus philippensis, Syzygium cuminii, Shorea robusta*
K	Border of Achanakmar Wildlife Sanctuary	448	*Mallotus philippensis, Shorea robusta, Bauhinia* sp., *Syzygium cuminii,* Unidentified tree
L	22 km before Keonchi from Chaparwa	488	*Mallotus philippensis, Syzygium cuminii, Shorea robusta, Mangifera indica, Emblica officinalis,* Unidentified
M	Dindori forest rest house	303	*Acacia nilotica*
N	7 km. before Karanjia from Gorakhpur	400	*Mangifera indica, Pongamia Pinnata, Grewia* sp., *Azadirachta indica, Acacia nilotica*
O	Karanjia forest rest house	400	*Shorea robusta, Mangifera indica*
P	Jagatpur forest rest house	393	*Shorea robusta*
Q	Chauradader	454	*Shorea robusta*
R	Tarwartola	448	*Shorea robusta*
S	13 km. before Kabirchabutra from Chauradader	454	*Grewia* sp., *Pongamia pinnata* sp.
T	8 km. before Kabirchabutra from Chauradader	448	*Shorea robusta, Emblica officinalis*
U	Khurkhuridader VALCO mining site plantation area	448	*Shorea robusta*
V	Khurkhuridader VALCO mining site vergin forest	448	*Shorea robusta, Mangifera indica, Eucalyptus* sp., *Grewia* sp.,
W	Khurkhuridader	454	*Shorea robusta.*
X	Near to Kabirchabutra	424	*Shorea robusta.*
Y	Patalashwar temple	808	Lime plaster of monuments.

Sampling Method

The lichen samples were collected from rocks and trees from 25 localities of AABR Madhya Pradesh and Chhattisghar. The lichen specimens were identified by studying their morphology, apothecial anatomy, chemistry and following the literature of Awasthi (1988, 1991) and Upreti (1991). The chemical investigations of the specimens were performed according to Walker and James (1980).

Results and Discussion

Bioindicator Lichen Communities

The presence of some particular communities of lichens in a forest indicate the status of the habitat. In the recent years lichen communities are included in the forest health monitoring programs because they help to answer several key assessment questions. Repeated sampling of lichen communities in permanent plots can be utilized to document the changes in the condition of lichen communities overtime which also reflects the environmental condition of the area.

Lichens respond to the environmental changes in the forest, including changes in forest structure, air quality and climate. The disappearance of lichens in a forest may indicate environmental stress, such as high level of sulphur dioxide based pollutants and nitrogen.

The lichen taxa so far known from the biosphere reserve represents eleven lichen community types, based on the habitat together with some morphological and anatomical characters of the lichens (Figure 39.2).

Pyrenocarpous Lichen Community

The lichen community is represented by the perithecia bearing lichen taxa and growing mostly on the smooth barked trees in moist evergreen phorophytes. *Dermatocarpon miniatum* a single saxicolous

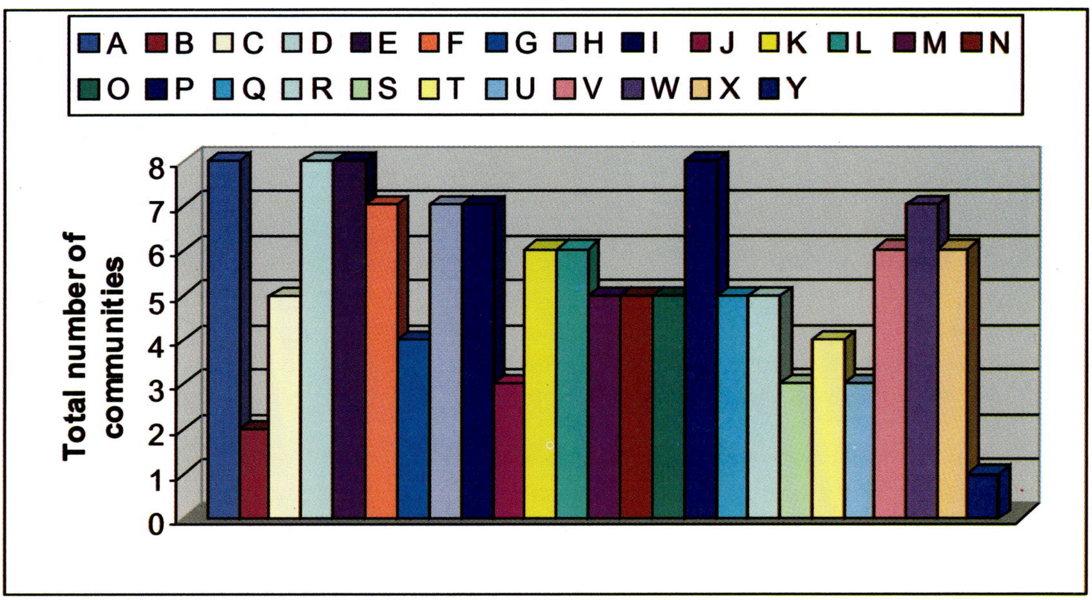

Figure 39.2: Bioindicator Lichen Communities in Different Localities of AABR

species of pyrenocarpous lichens is known from the Kapildhara falls area found growing on moist boulders under shade. The Chaparwa area, 5 km before Chaparwa from Amarkantak, 22 km before Keonchi from Chaparwa are the sites exhibit occurrence of *Pyrenula fuscoolivacea* and *P. subglobriscula*, dence forest patches with evergreen phorophytes (Plate 39.1).

Graphidaceous Lichen Community

The fruiting bodies (apothecia) in Graphidaceous lichen taxa are modified into elongate, simple to branched structures called lirellae, represented by the lichen genera *Graphis*, *Phaeographis* and *Arthonia*. The graphidaceous lichens grow both on dry and moist places. In dry well lit areas of the forest they grow in association with Lecanoroid communities while in moist habitats, pyrenocarpous lichens are their companions.

The Jagatpur forest rest house, Khurkhuridader, Near to Kabirchabutra, Attaria, Kapildhara, Mai ki Bagia, Jwaleshwar, Durgadhara and Gabhighat area clearly indicates the moist evergreen forest patches with in the dry deciduous forest, and hence represent excellent growth of graphidaceous lichen communities (Plate 39.1B).

Cynophycean Lichen Community

Collema, *Leptogium* and *Peltula* are the major blue-green alga containing lichen taxa of the biosphere reserve. Both *Collema* and *Leptogium* have ability to grow on soil, rocks and barks of the trees while *Peltula* species only grows on rocks. The cynophycean lichen indicates the nutrient poor, moist and shady locations of the biosphere reserve (Plates 39.1C). The Kapildhara, Gabhighat, Chaparwa nala, 5 km away from Chaparwa towards kota, Border of AWLS, Tarwartola, Khurkhuridader VALCO mining site plantation area, Khurkhuridader VALCO mining site vergin forest, near to Kabirchabutra area are with perennial streams and have shallow vallies which provides a moist and shady habitat for cynophycean lichen communities to colonize. The rocks both exposed and under forest, bear luxuriant growth of *Peltula*.

Lecanoroid Lichen Community

The Lecanoroid community is characterized by the apothecia bearing crustose lichens. Presence of Lecanoroid communities indicates the well lit habitats of the Biosphere Reserve. Twelve species of the Lecanoroid lichen communities comprising *Lecanora, Lecidella* and *Vainoria* sp., are found growing on both thick, thin, dry and moist bark of trees. The Chaperwa nala area exhibit the maximum diversity of the Lecanoroid communities (Plate 39.1D).

Parmelioid Lichen Community

The Parmelioid lichen communities are characterized by the leafy growth formed lichens, best developed in well lit, moderately acidic substrates (Plate 39.1E). The rough-barked trees of *Shorea robusta* both in more open and closed canopy forest area provide a suitable habitat for Parmelioid lichens. *Shorea robusta* tree is a suitable phorophyte for the luxuriant growth and development of crustose lichen. Out of 64 lichen, species recorded from *Shorea robusta* trees in various forests sites of India, 42 species fall under the crustose lichens (Satya *et al.*, 2005). The Tarwartola and Jagatpur Forest Rest House area exhibit the best habitat for colonization of this community.

Some Parmelioid member also grows on rocks both in exposed and on rocks covered under thick forest canopy. Such rocks in Kapildhara, Jwaleshwar forest and Kurkhuridader area exhibit the growth of saxicolous Parmelioid lichens in association with *Caloplaca, Peltula, Buellia, Staurothele* and *Diploschistes* species.

Plate 39.1 A–E: Bioindicator Lichen Communities

A: Pyrenocarpous lichen *Pyrenula* sp.; B: Graphidaceous lichen, *Graphis* sp.; C: Cynophycean lichen, *Leptogium* sp.; D: Lecanoroid lichen, *Lecanora* sp.; E: Parmelioid lichen, *Parmotrema* sp.

Plate 39.2

A: Water Fall in Kapildhra Area; B: *Peltula* sp.; C: *Collema* sp.; D: *Dermatocarpon miniatum*

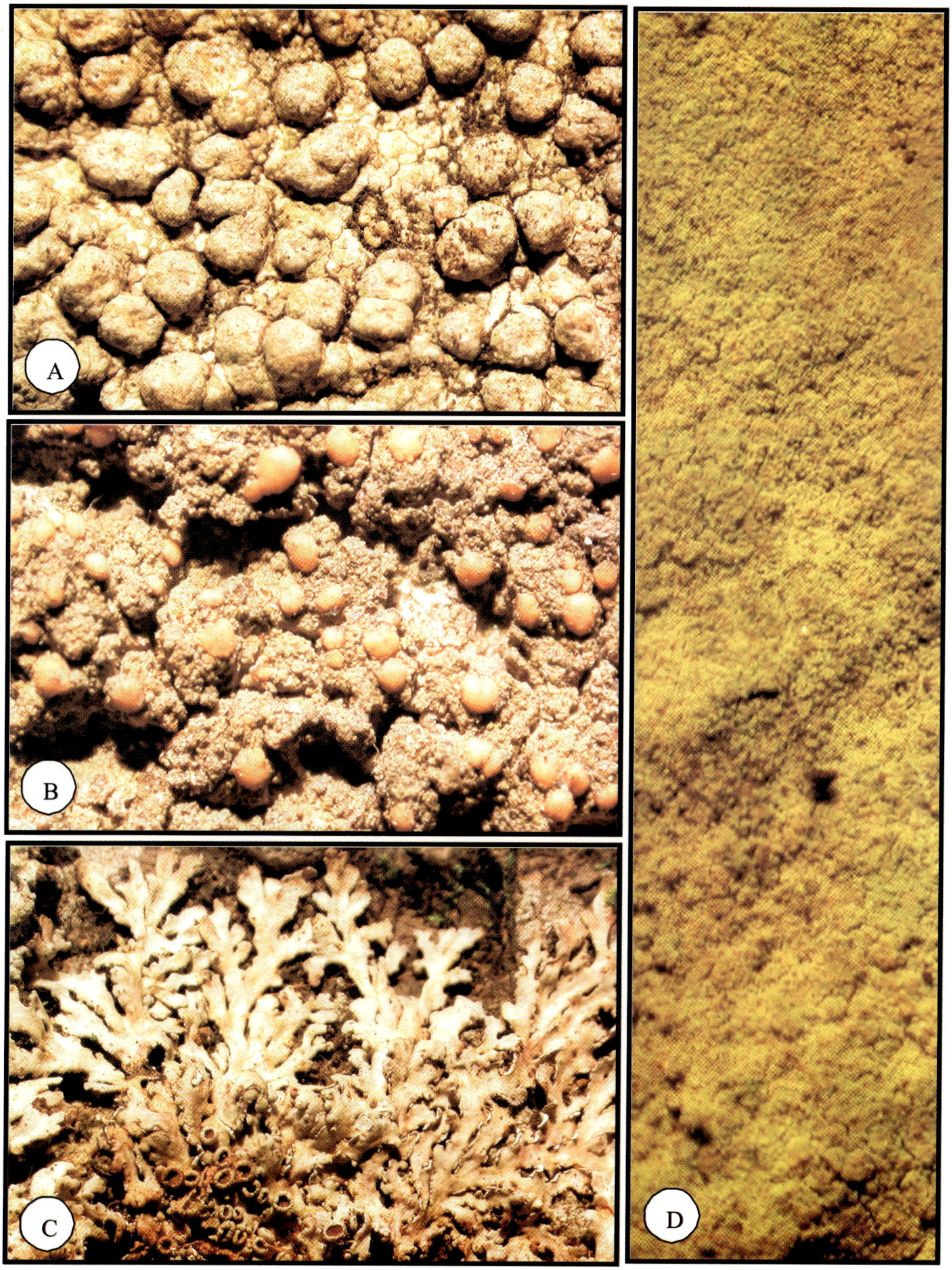

Plate 39.3 A–D: Bioindicator Lichen Communities

A: Pertusarioid lichen, *Pertusaria* sp.; B: Lecideoid lichen, *Bacidia* sp.; C: Physcioid lichen, *Hetetodermia* sp.; D: Leprarioid lichen, *Chrysothrix* sp.

Plate 39.4 A–B: Bioindicator Lichen Communities

A: Teloschistaceous lichen, *Letrouitia* sp.; B: Lichinioid lichen, *Phylliscum* sp.

Pertusarioid Lichen Community

The community is characterized by crustose, verrucose-fertile and verrucose-sorediate thallus. The community indicates old tree forest. The thick, rough-barked trees of *Shorea robusta* in the dry deciduous forest appears exellent host for the Pertusarioid communities with verrucose-fertile taxa as 19 such species are reported on the *Shorea robusta* trees, while the *Pertusaria amara* (a verrucose-sorediate) taxon often colonize on rather more smooth-bark trees of *Grewia*. The community member grows on exposed roots of *Shorea* on trunk at base, on branches and twigs in association with the members of both foliose and crustose lichens. The common foliose taxa are the species of *Dirinaria* while *Buellia*, *Lecanora* and lirelliforem species of Graphidaceous lichen are the common crustose lichens (Plate 39.2A).

Lecideoid Lichen Community

The community is characterized by crustose to squamulose thalloid lichens bearing apothecia without thalline cover. The members of the community colonized on the bark of deciduous trees in sheltered and well lit sites. The genera *Bacidia, Phyllopsora* and *Schadonia* have similar morphology of thallus which grow on thick, rough-barked trees at their basal parts in moist, shady places while *Haematomma* and *Protoblastenia* prefers smooth barked trees in open area. The rock inhabiting lecideoid community belongs to a single genus *Trapeliopsis* sp., which grows on moist rocks in Kabirchabutra area. The members of Graphidaceous lichens, *Buellia*, *Lecanora* are the common associates of this community (Plate 39.2B).

Leprarioid Lichen Community

The community has the members of lichens having powdery thallus of varnish like crust over substrates. *Chrysothrix, Cryptothecia* and *Lepraria* are the common genera of the Leprose lichen in the Biosphere Reserve. The member of this community grows both on rocks and trees. The rocks in vertical moist places exhibit luxuriant growth of Lepraria while *Chrysothrix* and *Cryptothecia* mostly colonized on dry, rough-barked trees in moist and dry places in association with Lecanoroid and Teloschistaceous lichens. After forest fire, the *Chrysothrix* grows mostly on mature *Shorea robusta* trees at basal parts and trunks as the pioneer colonizer (Plate 39.2D).

Physcioid Lichen Community

The members of the family Physciaceae belongs to this community. Comprising of both to the crustose and foliose form of lichens, growing on a varied range of substrates. The foliose genera *Dirinaria, Pyxine, Physcia, Heterodermia* and *Phaeophyscia* under this community exhibit the maximum host diversity. The saxicolous taxa of the community mostly grows on rocks in moist places. The crustose taxa of the community such as *Buellia* and *Rinodina* are mostly grows in moist places. Mature *Shorea robusta* trees in moist places exhibit luxuriant growth of *Dirinaria* in association with *Heterodermia* in the Khurkhuridader area while Kabirchabutra and Kapildhara area has luxuriant growth of *Heterodermia diademata* on soil, rocks and barks of different trees (Plate 39.2C).

Teloschistaceous Lichen Community

The community is characterized by anthraquinone containing yellow crustose thallus and lichen genera *Caloplaca* and *Letrouitia* belongs to this community. *Letrouitia* is a corticolous genus while *Caloplaca* grows both on rocks and bark of trees. The members of *Caloplaca* indicate the moist as well as well lit area of the biosphere reserve. The members grow both on rough and smooth surfaced rocks in exposed and mostly on the semi shaded rocks. The Kapildhara area with semi shaded moist rocks are the excellent habitat for *Caloplaca* to colonize. *Letrouitia* species exhibit their luxuriant growth on smooth, hard barked trunks of *Mallotus* trees (Plate 39.3A).

Lichinioid Lichen Community

The lichen genus *Phylliscum* is the only member of this community known from the biosphere reserve and found growing on man-made substrate in Amarkantak area. This genus has dark brown black crustose to squamulose thallus, which forms large black patches on the lime plaster of buildings and monuments. The member of the community also indicates the presence of calcareous rocks in the Biosphere Reserve. The member of Pyrenocarpous lichens such as *Endocarpon* and *Staurothele* also grow in association of this community.

The biosphere reserve exhibit the maximum 18 per cent cover of Physcioid lichen communities followed by 16 per cent Pertusarioid, 13 per cent Lecanoroid and 12 per cent of Lecideoid communities. The members of Physciaceae family are more tolerant to extreme environment (Backor *et al.*, 2003).

The Cynophycean, Graphidaceous and Pyrenocarpous lichen communities are represented each by 7 per cent, while the Lichinioid lichen community has the minimum cover of 1 per cent (Figure 39.3).

Based on the occurrence of the different lichen communities the study area can be clearly demarcated into undisturbed, moderately disturbed and disturbed areas (Figure 39.3). The Attaria, Kurkhuredader and near to Kabirchabutra area falls under the undisturbed area. The Kapildhara, Mai ki bagia, Border of AWLS, Jwaleshar, 7 km before Karanjia from Gorakhpur, Karanjia forest rest house, Jagatpur forest rest house, Chauradader, Khurkhuridader VALCO mining site plantation area, Khurkhuridader VALCO mining site vergin forest area are moderately disturbed while 5 km before Chaparwa from Amarkantak, Pataleshwar, Durgadhara, Dindori forest rest house, 13 km before Kabirchabutra from Chauradader, 8 km before Kabirchabutra from Chauradader falls under the disturbed category.

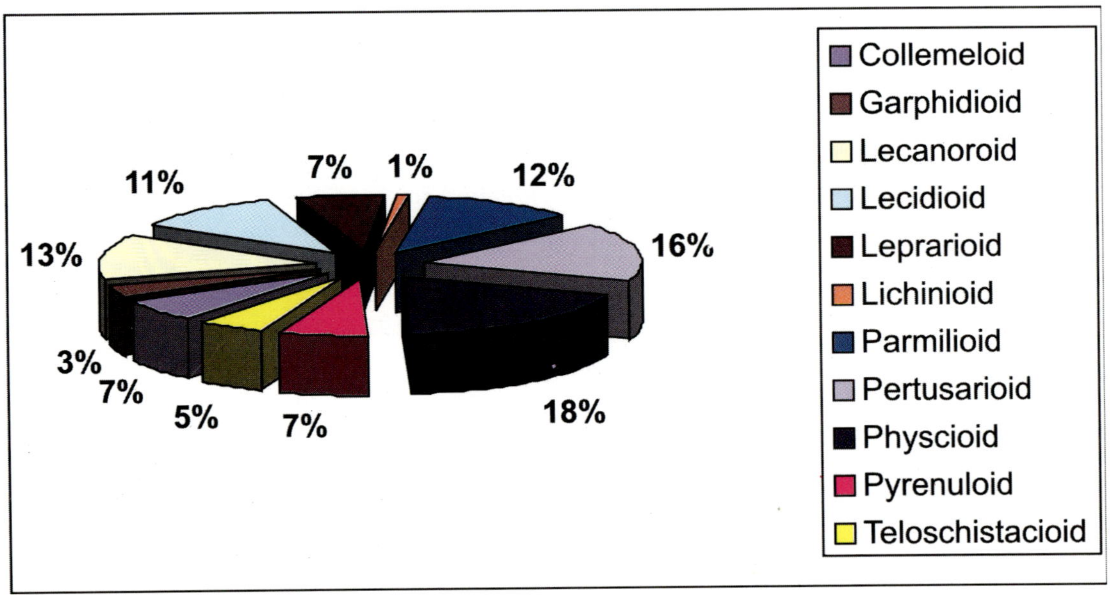

Figure 39.3: Frequency of Bioindicator Lichen Communities in AABR

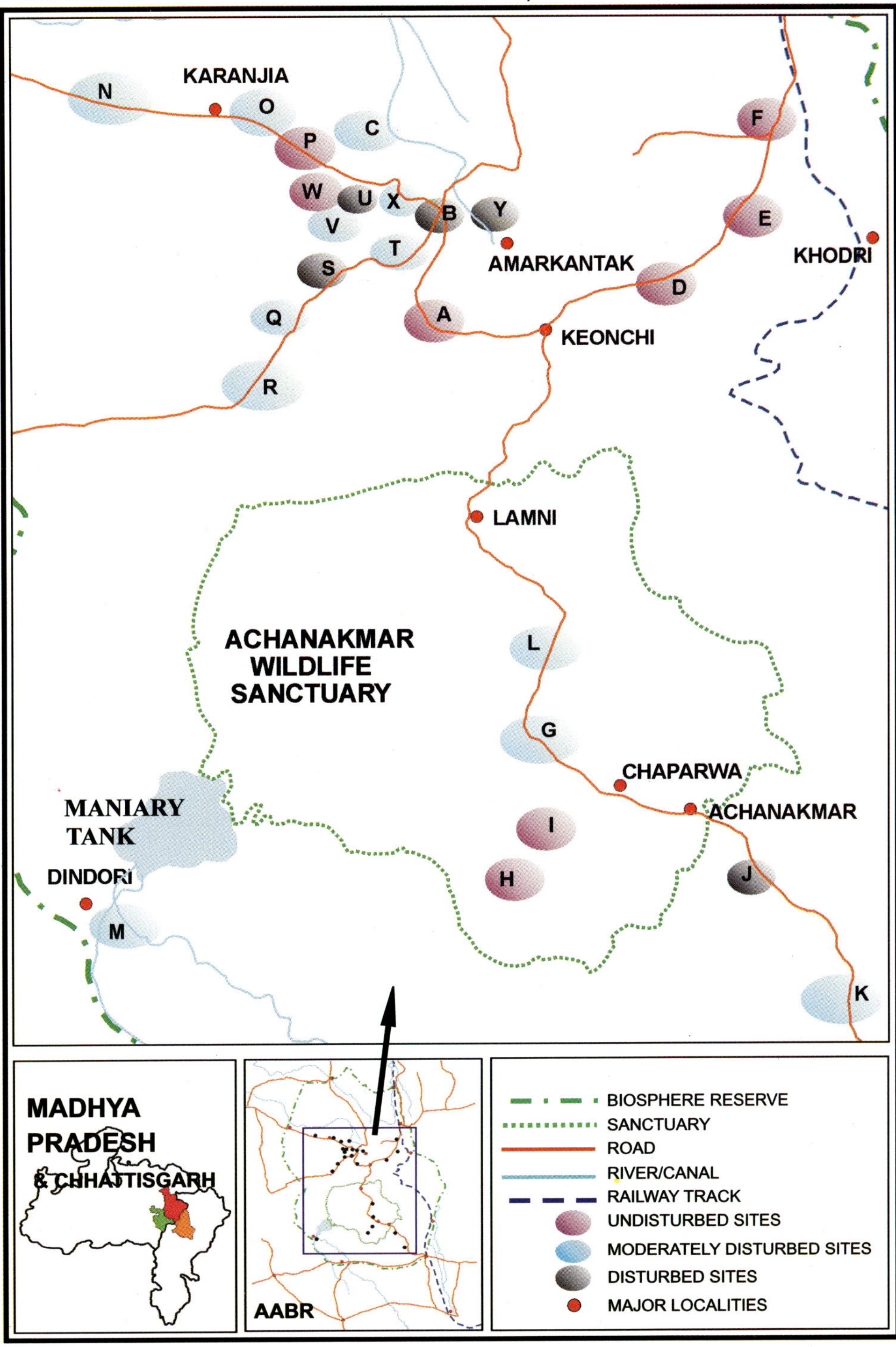

Figure 39.4: Lichen Exploration Sites Showing Different Level of Disturbance in AABR

Conclusion

In future to conduct studies, for the forest health monitoring progrmmes the available account of the lichen communities will help to answer several key assessment questions. The periodical monitoring of lichen communities in the are can be utilized to document the changes in the condition of the lichen communities overtime which also reflect the change in environmental condition of the area.

Acknowledgements

We are thankful to Director, National Botanical Research Institute for providing necessary laboratory facilities to work, to Ministry of Environment and Forests for financial support, to forest officers of Achanakmar-Amarkantak Biosphere Reserve for their cooperation.

References

Awasthi, D. D. (1988). A key to the macro lichens of India and Nepal. *Journal of Hattori Botanical Laboratory* 65: 207-302.

Awasthi, D.D. (1991). A key to the microlichens of India, Nepal and Sri Lanka. *Bibliotheca Lichenologica*. 40: 1-337.

Backor, M., Fahselt, D., Davidson, R.D. and Wu, C.T. (2003). Effect of copper on wild and tolerant strains of the lichen photobiont *Trebouxia erici* (Chlorophyta). and possible tolerant mechanisms. *Archives of Environmental Contamination and Toxicology* (45): 159-167.

Gradstein, S. R. (1992). The vanishing tropical rain as an environment for bryophytes and lichens. In *Bryphytes and Lichen in Changing Environment*, (Eds. J. W. Bates and A. M. Farmer). pp. 234-253. Clarendon, Oxford.

Neitlich, P. N. and B. McCune (1997). Hot spot of Epiphytic Lichen diversity in Two Young Managed Forests. *Conservation Biology* (11): 172-182.

Satya, Upreti, D.K. and Nayaka, S. (2005). *Shorea robusta*, an excellent host tree for lichen growth. *Current Science* 89 (4): 594-595.

Satya (2006). Ecological Studies of lichens in Amarkantak Biosphere Reserve. *Ph.D. Thesis*. University of Lucknow.

Upreti, D.K. (1991). Lichen genus *Pyrenula* from India: The species with spores of *Pyrenula brunnea* type. Bull. Soc. Bot. Fr. 138. *Letters Bot.* 3: 241-247.

Walker, F.J. and James, P.W. (1980). A revised guide to microchemical technique for the identification of lichen products. *Bulletin of British Lichen Society*. 46: 13-29.

Microbial Biotechnology and Ecology (2011) Pages 683–707
Editors: Deepak Vyas, G.S. Paliwal, P.K.Khare and R.K Gupta
Published by: DAYA PUBLISHING HOUSE, NEW DELHI

Chapter 40

Silvipastoral Systems: The Solution for the Sustainable Production in Semi Arid Tropics

☆ R.K. Bhatt, A.K. Mishra and H.S. Tiwari

Indian Grassland and Fodder Research Institute, Jhansi – 284 003, U.P.

Introduction

The traditional land use system in India have become fragile and unproductive due to over exploitation of natural resources, resulting in large scale degradation and consequent occurrence of wastelands. In India land degradation is a continuous process primarily due to increase in human and animal population pressure on a limited land resource. According to the estimates given by NBSSLUP (Sehgal and Abrol, 1994), 57 per cent of land (187.7 million ha) in the country is potentially exposed to various degradation forces; of which a dominant fraction is under water erosion 45 per cent (149.9 million ha), wind erosion 4.1 per cent (13.5 million ha), chemical deterioration 4.2 per cent (13.8 million ha) and physical deterioration 3.6 per cent (11 million ha). As per record of Ministry of Agriculture (1994), 37 per cent of total geographical area of the country (107.43 million ha) is degraded due to various causes of land degradation.

The problem of overgrazing is faced and presently the grazing pressure is 3.34 ACU/ha against the normal carrying capacity of 1 ACU/ha for a productive grassland. The over grazing resulted replacement of suitable grass cover by poor quality grasses and weeds (UNESCO, 1979). The grazing pressure on the rangelands under some semi arid region particularly in Bundelkhand region is still higher nearly 5 ACU/ha. It is higher than the national is (2.5 ACU/ha) average (Singh and Singh, 1994). The carrying capacity of a well managed rangeland in the region does not exceed 1 ACU/ha (Shanker *et al.*, 1988). Increasing demand of fuel wood, fodder, timber and lack of cash and infrastructure in many developing countries are some most relevant arguments for tree integration into farming anc pastoral areas. It is aims at sustained productivity and conservation, rather than destructive exploitatio of natural resources. The consequences of deforestation and desertification call for immediate action Agro forestry systems have a great potential to solve a number of such problems. The semi-arid anc arid lands with mostly marginal soil cover the largest area of land in the tropics and subtropics. Th use of trees and shrubs to enhance organic matter production to maintain soil fertility, and to reduc erosion and to create more even microclimate has been found very effective. Thus, for sustainabl production, the best alternative is to promote appropriate integrated agroforestry systems in the ari and semi-arid areas.

Silvipastoral land use has been identified as a holistic approach to rehabilitate the degrade lands and rangelands for meeting the needs of fodder and fire wood by rural masses and conservatio of natural resources (Paroda 1990). Silvipastoral system is a suitable technology for the managemen of degraded and rangelands as this have been described as a low input technology (Deb Roy an Pathak, 1974, 1984). The establishment of silvipastoral system on wastelands/degraded an rangelands is not only to improve the soil health and microclimate but also facilitate the needs o human and livestock. The basic concepts of silvopastoral system based on sound ecological principle (productivity + conservation of resources) and are very much similar to natural approach.

Silvipastoral System

The term silvipasture essentially refers to an integrated system of growing and managing th tress/shrubs in association with grass or grasses and/or grass–legume mixture simultaneously o the same piece of land, where animals may or may not become a part of the system. Thus the syster will include tree + pasture and/or animals. Tejwani (1987) used two different terms "Silvopasture and "Pastoral Silviculture" for this type of integrated system of land use. It has been suggested that combination of tree and grasses is referred to as "Silvipasture". If the grazing land is in a forest; wher as the system is referred to as "pastoral silviculture" if tree are scattered in a pasture or wasteland.

A multi tired silvopasture systems will be near natural comparatively stable and what is mor important is that it will also ensure better solar energy flow through the food chain. A Silvopastur structured on the functional pattern of natural ecosystem will, therefore, be a viable propositio (Paroda and Jatasra, 1985). Trees and shrubs are used as forage either through browsing or loppin and other forage like grasses and legumes as under story seasonal crops. The tree species also provid timber and other minor products to the inhabitants as well as shade and shelter to the grazing animal The tree and shrubs grow luxuriantly even during the drought years as their deep tap root syster enables them to grow independently of moisture. Trees improve soil through leaves and other litt and in addition to decomposition of old roots besides its capacity to fix its atmosphere nitrogen. The act as a nutrient and moisture pumping machine by drawing nutrients and moisture from deep layers of soil. The presence of trees acts as a protective barrier against soil erosion or as wind break Tree reduces the soil moisture evaporation by lowering the soil surface temperature. Grasses whi conserving soil and water also utilize off season precipitation. The legumes enrich soil by nitroge fixation and in the mixture they help growth of the grasses and trees besides improving the forag quality. Compared with other land uses, in marginal and sub marginal lands the silvopastoral syste provides forage/grazing availability and quality for a longer period of the year.

Silvopastoral system is a mixed community composed of trees and crops, which form entire different pattern for the utilization of light, water and nutrients. The growth and productivity c component crops in such system depends upon the use of available resources. The available so moisture, mineral nutrients and light are the main resources governing plant productivity. The infiltered light affect the growth and other biological phenomenon of the crops and as well as the physical parameters under the canopy and to the surroundings. The solar radiation transmission a tree crop system is the major basis for understanding the principle of yield patterns and construction of a model for optimum yield and sustainability.

A suitable tree species to be planted for silvopastoral system should be of extreme economic, social and ecological significance because these systems maintain the socio-economic and environmental balances of ecosystem. For these programmes there is a need to focus research attention on identifying the potential tree species which are required to be grown along with the crops. It has been recommended that the tree species should be multipurpose, easy availability and wider distribution, fast growing, nitrogen fixing capacity, deep root system, good coppicing ability, higher rate of leaf fall during deciduous period with rapid decomposition of leaf, water stress tolerant, pest resistant, capacity to withstand in intensive frequent lopping and promising better under canopy growth.

Silvipastoral Development

Realising the importance of silvipasture systems for meeting the biomass requirements, the research work was initiated during the sixties at several institute in the country. The result obtained during past decades indicate that fodder and fuel wood production could be increased several folds from the existing levels through proper management of wasteland. In India, silvipasture research was initiated during the early sixties at the Central Arid Zone Research Institute (CAZRI), Jodhpur. This institute worked out the technologies for establishment of shelterbelts, stabilization of sand dunes and development of agri-silviculture systems. Later on comprehensive studies on such system were taken up around 1970s at Indian Grassland and Fodder Research Institute (IGFRI), Jhansi. The studies included on designing and development of silvopastoral system on degraded lands with special emphasis on fodder production one hand and the conservation on the other hand. Gradually the other institutes *viz.*, Central Soil and Water Conservation Research and Training Institute (CSWCRTI), Dehra Dun and Central Institute for Dry Land Agriculture (CRIDA), Hyderabad initiated work on such systems to develop appropriate land use system in their respective region. Besides these, an All India Coordinated Research Project on Agroforestry (AICRPAF) with 20 centers was initiated by ICAR during the year 1983. Agroforestry research in India was further strengthened during the seventh plan (1985-90) by adding 11 more centers to the AICRPAF and by creating a National Research Center for Agroforestry (NRCAF) at Jhansi during the year 1988. Now studies on various aspects of Agroforestry are also under taken by most of the state Agricultural Universities (SAUs) and some traditional universities as well.

Land Development

In the wasteland, fellow lands, sloppy and uneven topography the rain water harvesting techniques with a watershed management approach is the pre-requisite for providing the conservation of water, soil and soil moisture and checking erosion and loss of nutrients. The small staggered contour bunds, trenches and intermittent pitting is essential to provide this effect and channelize the water on one side for slow disposal and storage. The trenches of 3 × 0.5 × 0.3 m size (length adjusted depending upon the slope) are the requirement for soil moisture conservation. Their number per hectare was determined based upon the rainfall intensity and storms and plan to check maximum of the peak storm intensity erosion.

The cattle proof trench (CPT) of 1.2 × 1 × 0.7 m size depositing the earth on one side and sowing seeds of non palatable woody species *viz.*, *Prosopis juliflora*, *Euphorbia* sp. etc. mixed with *Acacias*, *Dichrostachys cineria* and *Stylos* make a very effective green impenetrable wall to keep off the animals. Besides it produces desirable biomass in time.

Planting Technique

Multiple purpose trees were planted in the existing pastures/grazing lands or on wastelands or denuded lands followed by sowing/planting grasses, legumes or cereals in between the lines of trees. The 45 cm^3 pits were dug at a spacing 3 × 3 or 3 × 4 or 4 × 5 m. With the onset of monsoon, the tree seedlings were planted in these pits after applying 100 g mixture of urea, single super phosphate and potash in a ratio of 1:4:1 with 10 g BHC around the seedling to protect it from termites. The intermittent space between these seedlings was sown with the seeds of adaptable grasses and legumes in alternate lines. The small seeds were either pelleted with soil and fertilizers mixture or with a small dose of super phosphate and urea were sown. The grass and legume lines were at a distance of 0.5 m. The grass seedlings/root slips were intermittently planted in line providing good establishment.

Where the grass seeds are not available, Pathak (1983) advanced the simple technique for larger area of wastelands. The technique involves surface seeding of legumes *viz.*, *Stylosanthes hamata* and *Macroptilium atropurpureum* at a line spacing of 1 m. On the trench mounds, contours, around the pits and earth deposits of cattle proof trenches its early colonization, growth and formation of a compact mat saves the soil from erosion and provides a good association with native grasses suppressing most of the seasonal weeds. While sowing, the seeds are mixed with single super phosphate @ 40 kg phosphorus/ha. This method gave better success on very poor undulating sites and even in ravenous areas.

In the first year the area was given a top dress of 25-30 kg/ha of nitrogen during early August. The grasses and legumes were harvested during November in the first year when the legumes have shed their seeds. In the subsequent years the harvesting could be done twice depending upon the area. The trees when form good canopy after 5-7 years they were lopped seasonally to provide additional forage and also light canopy cover to provide optimum light at the grass growth level.

Afterwards the area was maintained either as seasonal hay plot followed by aftermath grazing by animals or grazing the animals as per the carrying capacity of the land in a rotational or deferred rotational grazing system.

Choice of Woody Species

The species are selected based on tolerance to high incidence of pruning, multiple purpose *viz.*, forage, firewood, nitrogen fixing, low crown diameter to bole diameter ratio, tolerant to side shade, phyllotaxis to permit penetration of light to the ground, the phenology specially leaf flushing and fall to be advantageous to the growth of annual crops, litter fall and decomposition to be advantageous to ground flora and soil. They should be compatible to, other species with ideal root growth habit and should be efficient nutrient pumps most of these have been desired for agroforestry species also (King 1979). Most of the native grasses have been found to associate well with trees (Deb Roy and Pathak 1974; Wilson 1951 and Krishnaswamy 1942).

Micro-environmental Changes in Silvipastoral System

Although much work has been done on the production aspects of grasses in the experimental field trials, in the grassland system and also in silvopastoral system but little work is available on tree-crop interaction with respect to micro-environmental changes and physiological processes of the grasses under tree canopies in the silvopasture systems in the semi arid tropics. Therefore, the work on tree–crop interaction with respect to the micro-environmental changes in the system, growth behavior, productivity, assimilatory processes and quality of the under storey grasses and also to find out relationships between these factors with the productivity was started by Bhatt (1988) at IGFRI, Jhansi.

The extent and amount of effect of trees on productivity and adaptation of under growing grasses in silvipastoral system can be explained by the physical and physiological influences in the system and/or on the surrounding environment. In such systems the components and environmental parameters influences one another by which the environment causes a response in plant functions and growth, and the plant then show the effect on environment. Researches on tree-crop system will, therefore, primarily concern on the interaction of components and influences on the surroundings. The law of temporal and spatial distribution of the tree shading and variation in light intensity may help in improving the farming technique in order to bring in to full play the productive potential of tree-crop inter cropping system (Liu, 1991). The radiation distribution pattern is variable in different plantation with spacing. The solar radiation received by crop varied with the shading effect of trees. To see the effect of the light it is best account the daily sum of shady area and shade light intensity. The vertical and horizontal distribution of radiation is also important for crop productivity. Radiation intensity of a point varies greatly from place to place, depending on the sun elevation angle, which leads to corresponding changes in the horizontal distribution in the daily course. The physical parameters such as temperature of the canopy and to its surroundings, relative humidity, wind velocity, light transmission, soil temperature, soil moisture and energy balance such as net radiation, latent heat flux, soil heat flux etc. are the important parameters, which interact in the system. It is very important to study the seasonal and diurnal light transmission and distribution in silvipastoral system throughout the growth of under canopy crop. The spatial and temporal distributions of available light under tree canopies may help in improving the productivity in the tree-crop inter cropping system.

Light Transmission Under Different Tree Canopies

Mishra and Bhatt (1994) studied light transmission under different silvopastoral systems in semi arid tropics. *Acacia tortilis* and *Hardwickia binata* allowed 60 per cent and 55 per cent Photosynthetically Active Radiation (PAR) infiltration of the total PAR whereas *Leucaena leucocephala* infiltered least amount of PAR (33 per cent). The diurnal photosynthetically active radiation (PAR) distribution patterns varied under tree canopies. Figure 40.1 showed diurnal variation in the PAR transmission under full grown tree canopies of *L. Leucocephala* and in open and revealed that PAR transmission increased from 6 AM and reached at peak during 12 to 14 hours and then decreased gradually (Bhatt *et al.*, 2003). On an average, *Leucaena* canopies allowed only 28-32 per cent PAR at the tree spacing 4 × 4 m at full canopy growth. The amount and intensity of PAR infiltered by the *Leucaena* tree canopy reduced to a great extent. The seasonal variation in light transmission under tree canopies during different months depends on the canopy growth and different phenological phases (Figure 40.2). The PAR transmission increased gradually from January to July because of leaf senescence and reached maximum is May–June when the canopy was leaf less.

Studies on microclimate variations around the tree cover by Ramakrishna (1984) revealed under 5 year old *Acacia tortilis* based Silvopastoral system, the total incident radiation just beneath the tree canopy was only 14 to 30 per cent of that received in open area. It was insufficient for good growth and productivity of pasture grasses like *Cenchrus ciliaris*. Similarly micro metrological parameters reported

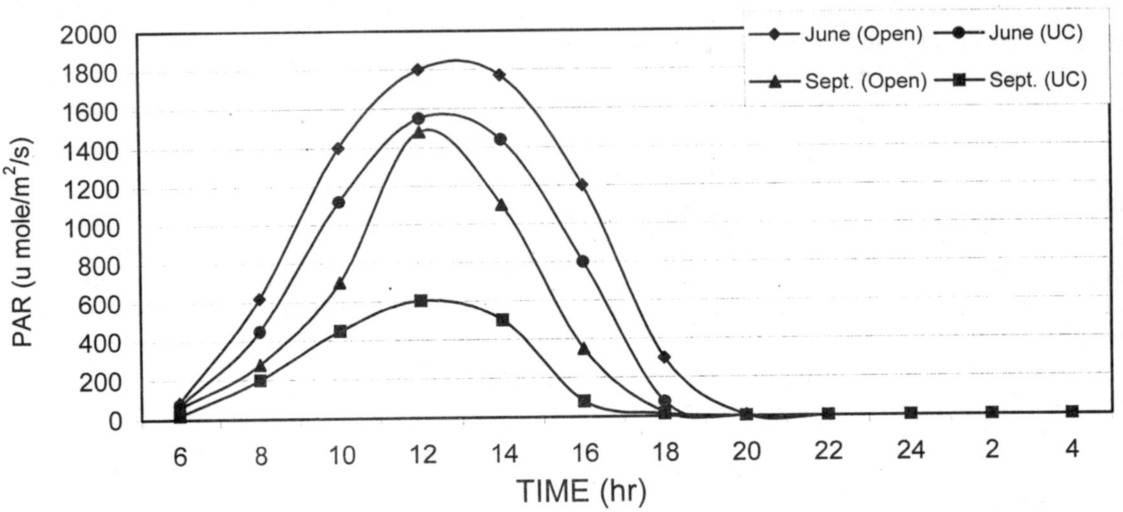

Figure 40.1: Diurnal Variation in Light Transmission Under *Leucaena leucocephala* Tree Canopies

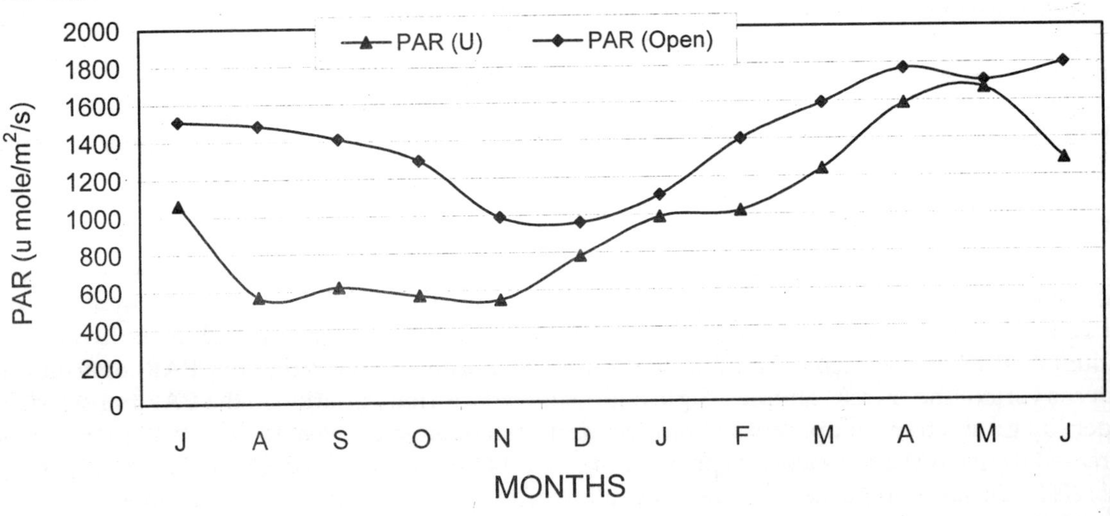

Figure 40.2: Monthly Variation in Light Transmission Under *Leucaena leucocephala* Tree Canopies

by Hazra (1994) in different silvopastoral systems revealed that under the *A. nilotica* canopies, th light infiltration was minimum of 59.2 per cent of PAR while it was maximum in association wit *H. binata* (83.1 per cent). Studies on micro metrological parameters by Hazra and Patil (1986) in 4 tre species (*A. lebbek, A. procera, L. leucocephala* and *A. tortilis*) at the age of 4-5 years under silvopastora condition revealed that the light filtration under different tree canopies varied from 74 to 93 per cent c photosynthetic active radiation (PAR). The tree cover maintained higher relative humidity (62-70 pe cent) as compared to open filed (56 per cent).

Temperature Variations Under Different Tree Canopies

Leucaena trees have major influence on temperature pattern (Figure 40.3). The shading of the canopy and reduction in radiation led to decreased under canopy air temperature. The under canopy temperature was lower to air temperature of open field during June–July whereas in winter months it was higher to air temperature (Figure 40.4). Higher canopy–air temperature difference was recorded under *Leucaena leucocephala*. In Open field at the grass canopy the temperature was significantly higher whereas in the silvipasture system the canopy temperature was lower to air temperature. The under canopy temperature was lower to air temperature in *Leucaena leucocephala–Panicum maximum* silvipasture system in all the months except in December. *Acacia tortilis–Cenchrus* based silvipasture system maintains relatively higher canopy temperature as compared to *L. leucocephala–Panicum* based silvipasture system. The canopy temperature during the full canopy growth was 32 °C and 33.5 °C as recorded in the month of Sept.–Oct. in *L. leucocephala* and *Acacia tortilis* based silvipasture systems respectively. On average of two years data the canopy–air temperature difference was highest (–4.5°C) under *L. leucocephala* canopies where as under *A. tortilis* canopies the canopy temperature difference was about –1.7 °C (Table 40.1). In open the temperature of grass canopy was 35.8 °C with canopy air temperature difference of –1.2 °C.

Table 40.1: Average Value of Canopy Temperature (CT), CT–AT, and PAR Transmission in the Silvopastoral Systems

Parameters	*Canopy*	*1999*	*2000*	*Average*
CT (°C)	*A. tortilis*	30.2	29.8	30.0
	L. leucocephala	28.8	28.4	28.6
CT–AT	*A. tortilis*	-1.6	-1.8	-1.7
	L. leucocephala	-4.6	-4.4	-4.5
PAR μmole/m²/s	*A. tortilis*	845	825	835
	L. leucocephala	520	475	498

The change is soil temperature under tree canopies at different soil depth was gentle and following the same trends during different months and day time. The soil temperature decreases with decreasing in soil depth. The maximum soil temperature (41°C) was recorded in June and minimum (18 °C) in January. The night soil temperature was higher in *Leucaena* field than in the control and under canopy air temperature. During the day time variation in soil temperature was due to shading under *Leucaena* trees canopies. The light transmission under tree canopies was low because of shading of tree canopies and grass canopies which caused decrease in soil temperature during full canopy growth of trees and grasses, whereas it increases in summer months.

The mean daily maximum soil temperature beneath the tree canopy of *Acacia tortilis* and grass was lower by as much as 10 °C to 16 °C in the top soil zone (0.5 cm) and 4 °C to 5 °C at 30 cm depth than recorded in the open which indicates a better soil thermal region under the tree plantation (Ramakrishna and Sastri 1977 b). Studies carried in a 7 year old *Acacia tortilis* plantation during the monsoon season at Jodhpur showed that during the morning hours air temperature beneath the tree canopy was lower by 0.1 °C to 0.7 °C than these recorded in the open, while during the afternoon period the temperatures underneath the tree canopy remained lower by as much as 0.6 °C to 2.0 °C than those recorded in the open (Rama Krishna and Sastri, 1977 a).

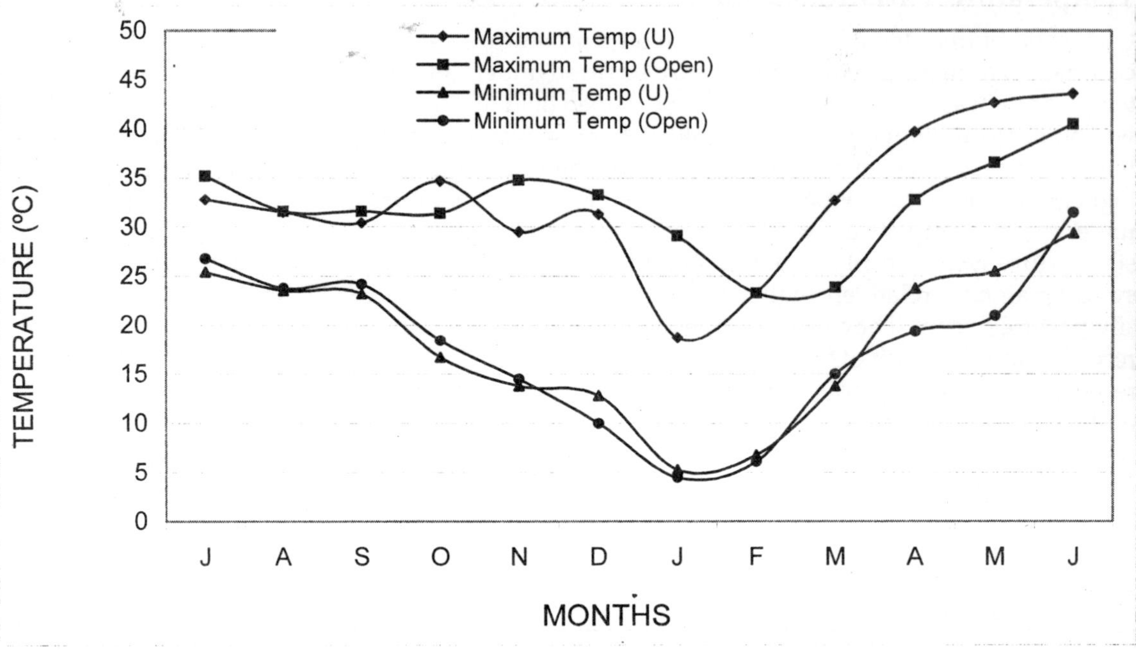

Figure 40.3: Seasonal Variation in Canopy Temperature Under the Trees of *L. leucocephala*

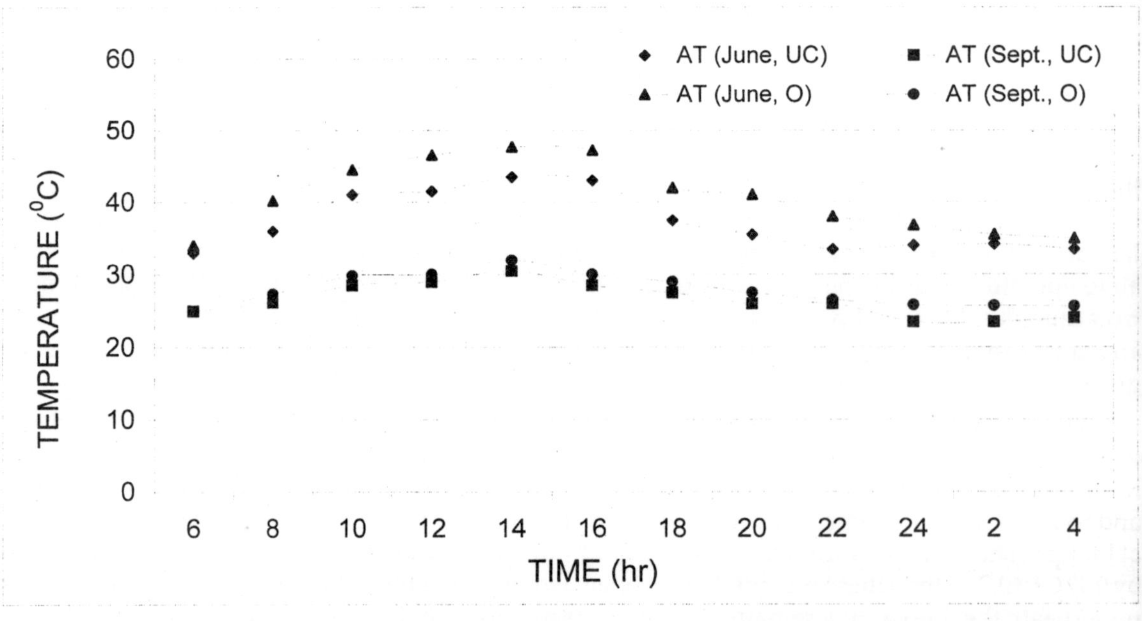

Figure 40.4: Diurnal Variation in Canopy Temperature Under the Trees of *L. leucocephala*

Photosynthetic Pigments

Bhatt *et al.* (1994) studied the accumulation of photosynthetic pigments in various grasses under the canopies of ten year old tree of *Hardweckia binata*, *Acacia tortilis*, *Albizia lebbeck*, *A. amara* and in open field condition and they found that (Chl-a and Chl-b) in *Sehima nervosum* and *Heteropogon contortus* was higher under the tree canopies than in open field. Higher synthesis of Chl b in these grasses under canopies indicates their shade adaptation ability. *S. nervosum* accumulated higher amount of Chl-b as compared to *H. contortus*. The high chlorophyll content (Chl a and Chl b) of the *Cenchrus ciliaris* and *Panicum maximum* under tree canopies of *L. leucocephala* also estimated by Bhatt *et al.* (1994) and reported that maximum chlorophyll a:b was observed in *C. ciliaris* indicating its fast adaptation to light intensity. The accumulations of photosynthetic pigments in grasses were higher under shading of tree canopies. The more accumulation of chlorophyll b under tree canopies indicates shade adaptation efficiency of these grasses (Table 40.2).

Table 40.2: Photosynthetic Pigments in Grasses Under Different Tree Canopies

Canopy	*Grasses*	*Photosynthetic Pigments (mg/g f.wt.)*			
		Chl-a	*Chl-b*	*Chl a+b*	*Chl a:b*
Open	*S. nervosum*	0.34	0.29	0.63	1.17
	H. contortus	0.33	0.12	0.45	2.75
A. tortilis	*S. nervosum*	0.49	0.36	0.85	1.36
	H. contortus	0.43	0.37	0.8	1.16
H. binata	*S. nervosum*	0.69	0.55	1.24	1.25
	H. contortus	0.3	0.29	1.54	1.03
L. Leucocephala	*S. nervosum*	0.38	0.39	0.77	0.97
	H. contortus	0.25	0.27	0.52	0.93
A. amara	*S. nervosum*	0.35	0.29	0.64	1.21
	H. contortus	0.32	0.38	0.7	0.84
A. lebbeck	*S. nervosum*	0.31	0.32	0.63	0.97
	H. contortus	0.18	0.18	0.55	0.49

Total chlorophyll and carotenoid content were more in the shade grown plant on the other hand the chl a/b ratio was lower in shade grown plants as compared to sum plants (Naidu and Swamy, 1993). Shade plants have a higher relative content of chl b than chl a (Egle 1960, Goodchild *et al.*, 1972). The decrease in chl a/b ratio was usually accompanied by an increase in chl b and specific leaf weight, which shows the shade adaptive feature in the grass species as observed by Bhatt and Shinha (1990). On the basis of high chl a:b *Chrysopogon fulvus*, *Paspalum notatum*, *Cenchrus ciliaris* and *Setaria sphacelate* were identified for higher productivity. Some shade tolerance grasses (*Chloris gyana* and *Brachiaria*) were selected on the basis of their chlorophyll accumulation (Bhatt and Sinha, 1990). Chlorophyll content (a + b) was maximum under 50-75 per cent shading in most of the grass species (Bhatt, *et al.*, 2006).

Physiological Processes of Grasses Under Tree Canopies

The production potentiality of the system components can be determined by physiological phenomenon, *viz.* Pigment adaptation, photosynthesis, transpiration, carboxylation efficiency, water use efficiency, and biomass accumulation, partitioning of biomass and mineral accumulation. These physiological processes in the silvipastoral systems are controlled by under-canopy micro-environmental parameters. Therefore, this research was concentrated on the interface between tree and crops, the micro-environmental changes and resultant physiological processes to understand the adaptation behavior, sustainability, productivity and feasibility of a tree-crop system.

Physiological Processes of *Cenchrus ciliaris* and *Panicum maximum* Under Leucaena Tree Canopies

The rate of photosynthesis, transpiration, stomatal conductance and accumulation of photosynthetic pigments was evaluated in grasses (*Panicum maximum* and *Cenchrus ciliaris*) in open and under *Leucaena leucocephala* tree canopies. Rate of photosynthesis, transpiration, stomatal conductance and leaf temperature of grass species decreased significantly under *Leucaena* tree canopies due to high degree of shading. In *C. ciliaris* maximum rate of photosynthesis was 39.28 µmole/m²/s at 1806 µmole/m²/s PAR and 10.27 µmole/m²/s at 408 µmole/m²/s PAR in open field and under tree canopies respectively. Whereas in *P. maximum* higher rate of assimilation (37.44 µmole/m²/s) was recorded at 1800 µmole/m²/s PAR and 16.51 µmole/m²/s at 350 µmole/m²/s PAR (Tables 40.3 and 40.4). Variation in the rate of photosynthesis in grass species was observed at different places under the tree canopies which were due to changing in PAR intensity. The rate of photosynthesis is linearly related to the available PAR (r=0.91) (Figures 40.5 and 40.6).

Table 40.3: Physiological Parameters in *Cenchrus ciliaris* in Open and Under *Leucaena* Canopies

Months	PN (µ mole/m²/s)		LT (°C)		CINT (ppm)		CS (cm/s)		PN/CINT		PN/TR (µ mole CO_2/ µmole H_2O)		TR (µmole H_2O)	
	Open	Shade	Open	Shade	Open	Shade	Open	Shade	Open	Shade	Open	Shade	Open	Shade
July	38.25	10.09	39.66	34.69	153	206	1.02	0.56	0.25	0.049	3.64	1.72	10.50	5.86
Aug	39.28	10.27	39.11	34.71	135	239	1.05	0.55	0.291	0.043	3.63	1.71	10.82	6.00
Sept	39.28	10.80	39.13	35.25	135	225	1.21	0.57	0.291	0.048	3.62	1.79	10.85	6.03
Oct	29.24	5.12	37.31	34.62	170	233	0.976	0.235	0.172	0.022	2.14	1.29	13.66	3.96
Nov	15.74	7.67	28.82	25.84	254	294	0.579	0.474	0.062	0.023	2.69	1.77	5.85	4.33
Dec	5.59	5.08	28.53	24.63	245	254	0.364	0.278	0.031	0.02	1.86	1.99	4.08	2.55

Table 40.4: Physiological Parameters in *Panicum maximum* in Open and Under *Leucaena* Canopies

Months	PN (µ mole/m²/s)		LT (°C)		CINT (ppm)		CS (cm/s)		PN/CINT		PN/TR (µ mole CO_2/ µ mole H_2O)		TR (µ mole H_2O)	
	Open	Shade	Open	Shade	Open	Shade	Open	Shade	Open	Shade	Open	Shade	Open	Shade
July	34.34	14.36	37.55	35.61	162	266	1.194	0.768	0.212	0.054	2.89	1.83	11.88	7.84
Aug	36.19	16.63	37.41	35.69	154	264	1.26	0.768	0.235	0.063	3.02	1.97	11.98	8.44
Sept	37.44	16.51	37.41	35.66	144	254	1.28	0.796	0.26	0.065	3.07	1.95	12.19	8.46
Oct	31.05	9.93	35.74	32.55	115	191	0.829	0.344	0.27	0.052	4.12	2.15	7.53	4.61
Nov	13.55	11.04	27.49	26.4	154	181	0.294	0.271	0.088	0.061	3.15	4.14	4.30	2.66
Dec	10.16	9.28	28.69	25.20	242	211	0.177	0.319	0.042	0.044	3.70	3.33	2.74	2.78

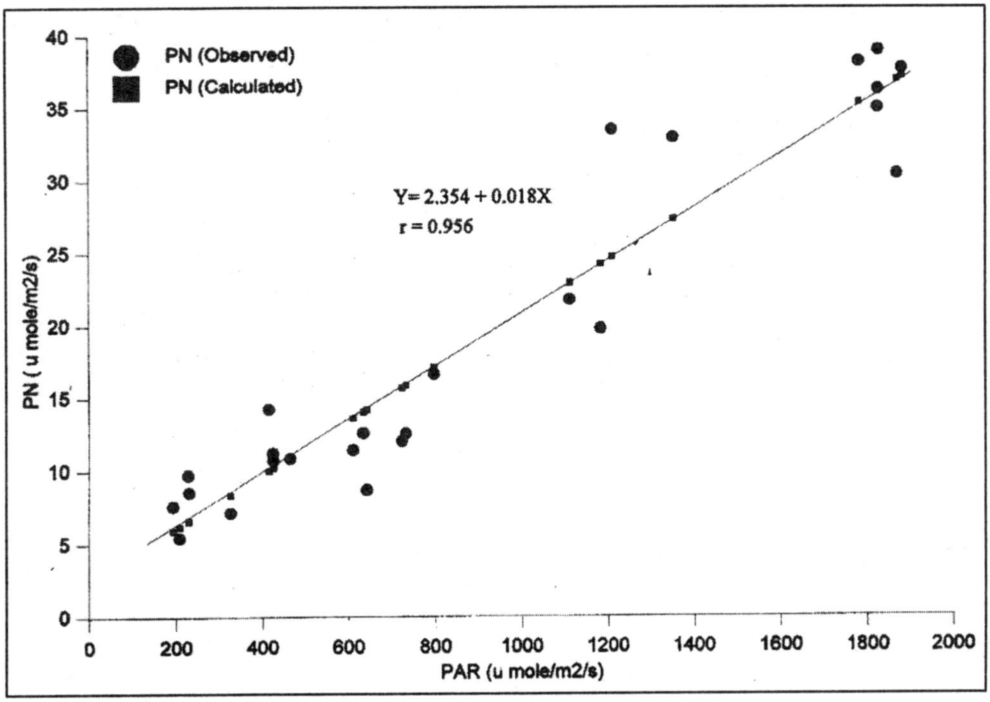

Figure 40.5: Rate of Photosynthesis (PN) in *P. maximum* at Various Levels of PAR in Open and Under *Leucaena* Species

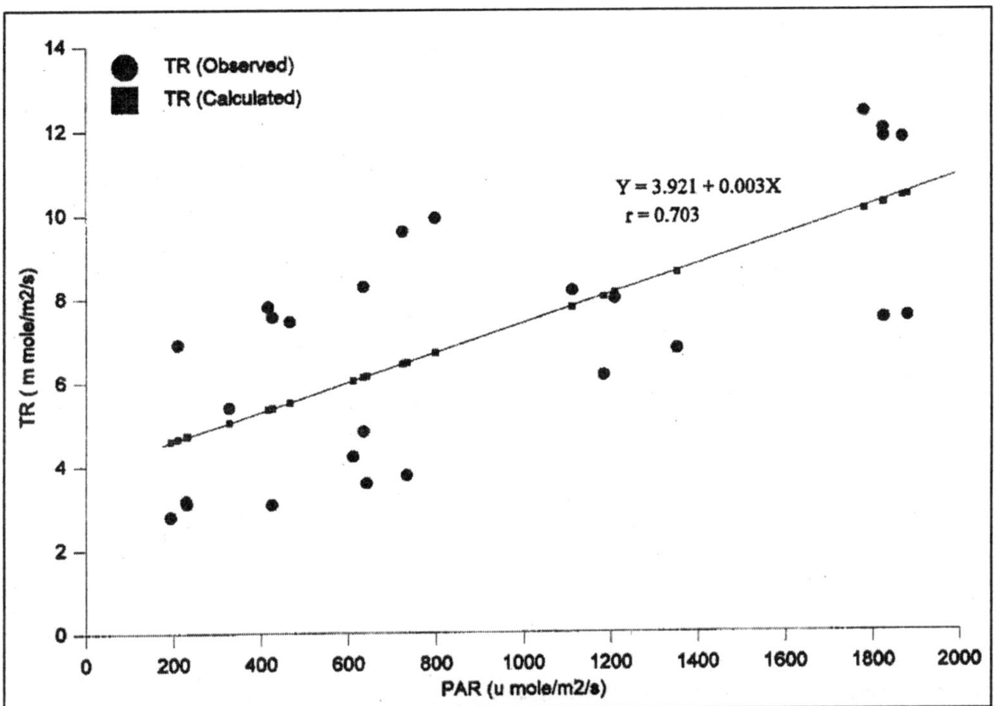

Figure 40.6: Transpiration Rate (TR) in *P. maximum* at Various Levels of PAR in Open and Under *Leucaena* Species

Intercellular CO_2 concentration enhanced two fold in both the grass species under tree canopie exhibiting lower assimilation of CO_2. The rate of CO_2 assimilation was intimately associated wit stomatal conductance as it is evidenced by highly significant correlation coefficient (r = 0.8796) betwee these two characters. The ratio of PN/CINT which represents the carboxylation efficiency was high in open grown grasses and decreased significantly under high shading of tree canopies. Transpiratic rate also decreased under shading of tree canopies. Photosynthesis and transpiration ratio (PN/TI which indicated the water use efficiency was decreased in grasses under tree canopies as compared open grown grasses (Tables 40.3 and 40.4).

The reduction in these physiological activities under tree canopies of *L. leucocephala* might be du to decrease in PAR and through which partial closure of stomatal pore occurred (Bhatt *et al.*, 1994 The highly significant correlation of these parameters with PAR is suggestive of the fact that the physiological processes are mainly influenced by these micro-environmental parameters. Fo maintaining the optimum functioning of all physiological parameters and sustaining the productivi under *L. leucocephala* based silvopastoral system it is essential to manage the canopy size either b pollarding, lopping and maintaining the ideal spacing between the trees or to workout threshold tre density per hectare for optimum PAR distribution in the system.

Physiological Processes of *Cenchrus ciliaris* Under *Acacia* Based Silvipastoral

Cenchrus ciliaris were evaluated for productivity, adaptational behaviour and physiologic activities under canopies of *Acacia tortilis* and in open field condition. The full grown canopy c *A. tortilis* allowed 50–55 per cent PAR. The rate of transpiration, photosynthesis, leaf temperature an stomatal conductance of *C. ciliaris* decreased under tree canopies as compared to open field. Similar t these attributes PN/TR, PN/CI and PN/COND also decreased under the shade environment of tre canopies. The reduction in these characters under tree canopies occurred due to low availability c PAR for assimilation of CO_2. The highly significant correlation of physiological characters and DM with PAR is suggestive the fact that this micro-environmental parameter is the only major facto which governed the yield and adaptation of grasses under silvipastoral system. The lowest reductio in stomatal conductance and higher specific leaf mass, chl b and chl a:b in *C. ciliaris* under *A. tortil* exhibiting its associability and sustainability in this system under semi-arid region.

Biomass Production Under Silvipastoral Systems

The Silvipastoral system has been studied for their advantage over existing agricultural practice for their growth and productivity (Deb Roy and Pathak 1983, Pathak and Roy 1994, Pathak *et al.*, 199 Grewal *et al.*, 1996). Also there have been a few attempts to monitor productivity (In ecosystem contex of Silvopasture system based on *Albizia lebbeck. Albizia procera* and *Hardweckia binata* in Bundelkhan region (Deb Roy 1986, Roy 1995). The information's available on forage production under differer silvopastoral systems in different agroclimatic conditions were generally on the basis of 4 to 6 year studies (Singh and Puri, 1975; Deb Roy *et al.*, 1980 Singh and Roy, 1991; Hazra *et al.*, 1994; Dogra *et al* 1997, Arya *et al.*, 1998, etc.). A long term studies conducted on silvipastoral systems at Jhansi an Datia in red gravelly soils with five tree species at each location (*Albizia amara, Albizia lebbek, Acac tortilis, Hardwickia binata* and *Leucaena leucocephala*) at Jhansi and (*A. amara, A. lebbek, A. tortilis, Dalberg sissoo* and *L. leucocephala*) at Datia under 3 spacing (4 × 3, 4 × 4 and 4 × 6 m) revealed that trees did nc have any adverse effect on forage production up to 11-12 years at any spacing (IDRC Reports, 1982-8 to 1993-94). Thus, the results showed that trees may be grown at the rate of 825 plants/ha up to 11-1 years without reduction in understorey biomass.

In arid condition, it has been observed that in addition to indigenous species like *Prosopis cineraria* and *Zizyphus numularia*, the introduced species such as *A. tortilis, D. cinerea, Colophospermum mopane* showed very good performance in silvopastoral systems (Ahuja, 1980). Forage production from lower storey crops under 14-18 years old plantation of 4 desert trees (*P. cineraria, Tecomella undulata, A. lebbek* and *Acacia senegal*) ranged from 0.6 to 1.5 t/ha/year (Ahuja, 1980).

The relative forage yield under *A. procera* was minimum (70.7 per cent) while relative dry forage production of 94.6 per cent was recorded higher in association with *A. lebbeck*. It was also observed that reduction in forage yield of understory crop, light transmission through canopy is not only responsible factor but moisture contents seems to be another factor as relative humidity under *A. nilotica* was maximum (70 per cent) although lowest relative yield of 70.7 per cent was recorded in association with *A. procera*. (Hazra 1994)

On the basis of more than 20 years studies in shallow red gravelly soils under semi arid condition at Jhansi, Pathak *et al.* (1996) reported that the degraded lands producing hardly up to 1 t/ha/year biomass which can be improved up to 10 t/ha/year at a 10 years rotation through silvopastroal systems. Besides yield improvement by 8 to 10 times, the quality of mixed forage has also improved by 6 to 7 times.

A fifteen years studies in Yamuna ravine near Agra, revealed that a silvipastoral system consisting of *A. nilotica* + *C. ciliaris* and *A. tortilis* + *C. ciliaris* planted at 3 × 3 m spacing produced on an average the total biomass yield of 5.0 t/ha/year (2.5 t fuel + 2.5 t forage) and 6.9 t/ha/year (4.2 t fuel + 2.7 t forage), respectively (Prajapati *et al.*, 1993). Studies on the production potential of pasture alone (*C. ciliaris*), fodder trees alone (*H. binata* + *C. mopane*) and silvopastoral systems (*C. ciliaris* + *H. binata* + *C. mopane*) were done at CAZRI, Jodhpur for nine years. On the basis of nine years results, it was observed that silvopastoral systems was better for higher average forage production (1.73 t/ha/yr) and livestock maintenance (4.1 Acu/ha) followed by pure pasture and pure trees block (Harsh *et al.*, 1992).

Grass production in 14–18 year old plantation of four desert trees was monitored by Ahuja *et al.*, 1978 in arid situation. The peak forage yield was reported under *Prosopis cineraria* (1.54 t DM/ha) and the least production was under *Acacia Senegal* (0.69 t DM/ha). The increase in grass yield was attributed to increased availability of nutrients under *P. cineraria*. Deb Roy *et al.* (1978) studied the forage production of *Cenchrus ciliaris* and a mixture of *Chrysopogon fulvus*, + *Sehima nervosum*, under different tree spacing of *Hardweckia binata* and *Albizia amara* from a typical degraded site in Bundelkhand. Forge production from the mixture of *C. fulvus* + *S. nervosum* (4.23 t DM/ha) was higher than *C. ciliaris* (3.74 t DM/ha). Deb Roy (1989) studied average forage production (5 year) of *Cenchrus ciliaris* and *Cenchrus setigerus* in association with *Acacia tortilis* gave an average forage production of 3.5 t DM/ha while in association with *Leucaena leucocephala* it gave higher production (4.1 t DM/ha). In contrast, *Cenchrus setigerus* gave higher production (3.1 t DM/ha) in association with *Acacia tortilis* when compared to *Leucaena Leucocephala* (3.0 t DM/ha).

In the silvopastoral systems under semi arid tropics crop growth rate (CGR) and dry matter yield (DMY) of grasses (*Sehima nervosum* and *Heteropogon contortus*) were higher under the canopies of *A. tortilis* and *H. binata* while the maximum reduction in yield of these grasses were recorded under *L. leucocephala* trees. The DMY was closely associated with available PAR (r= 0.71) under the tree canopies indicating that for the optimum production of these grasses about 55-60 per cent PAR is required (Misra and Bhatt, 1994). Thus on the basis of physiological adaptation and biomass production it is indicative that *A. tortilis* and *H. binata* are suitable tree species with *H. contortus* and *S. nervosum* under silvipastoral systems in the semi-arid region. About 98 per cent dry matter yield of grasses was obtained under *Hardwickia binata* and *Acacia tortilis* canopy when compared with the dry matter yield in open field Bhatt *et al.* (1994).

In one of the study *C. ciliaris* and *P. maximum* have produced 15.78 t. and 27.83 t/ha forage production under the tree canopies of *A. tortilis* and *L. leucocephala* respectively in the 5th year of growth (Bhatt *et al.*, 2006). The dry matter yield was 4.72 t/ha and 8.37 t/ha in *C. ciliaris* and *P. maximum* under the tree canopies (Figure 40.7). The biomass yield and leaf area index was increased continuously up to 5th year and then sustained up to 8th year in both the grass species in open and under the tree canopies (Table 40.5). Plant height and tiller production per plant varied from year to year depending on the environmental conditions but the new tiller production continued every year. On average of all the years, the biomass production was 5.29 t/ha and 4.20 t/ha in open and under tree canopies of *A. tortilis* in *C. ciliaris* whereas it was 7.56 t/ha and 6.84 t/ha in *P. maximum* in open and under *L. leucocephala* trees. Overall the biomass production was reduced by 21 per cent in *C. ciliaris* under the tree canopies of *A. tortilis* and 10 per cent in *P. maximum* under *L. leucocephala* canopies indicating that the productivity of these grasses sustained for longer time (Figure 40.7). The dry matter yield was significantly correlated (r = 0.931) with the available PAR under tree canopies exhibiting that for the optimum production of these grasses under the shading environment of trees, about 40–60 per cent PAR is required.

Table 40.5: Growth, Physiological and Quality Characters of Grasses Grown in Open and Under Tree Canopies (Figure in parenthesis indicates percent reduction and gain)

Characters	*C. ciliaris*		*P. maximum*	
	Open	*UC*	*Open*	*UC*
Photosynthetic rate (μmole/m²/s)	32.36	16.52(–49.0)	30.51	18.69 (–48.2)
Transpiration (μmole/m²/s)	10.33	4.8(–53.5)	9.57	5.82 (–39.2)
Stomatal conductance (cm/s)	0.96	0.48(–50.0)	0.97	0.58 (–40.2)
Canopy Photosynthesis (PN X LAI)	153.06	69.71(–54.4)	215.09	127.65(–40.6)
Chlorophyll a (mg/g f.wt.)	2.28	2.32(+1.7)	2.48	2.66 (+7.2)
Chlorophyll b (mg/g f.wt.)	1.02	1.22(+19.6)	1.10	1.32 (+20.0)
Sugar (per cent)	2.13	2.53(+18.8)	2.12	2.62(+23.8)
Starch (per cent)	6.22	6.38(+2.6)	6.12	6.56(+7.2)
Crude Protein (per cent)	6.39	7.12(+11.4)	10.12	10.48(+14.9)
Plant height (cm)	129	136(+5.4)	153	152(+6.5)
Tillers/plant	34	31(–8.8)	28	26(–7.1)
Leaf Area Index (LAI)	4.73	4.22(–10.8)	7.05	6.83(–3.1)
Forage yield (t/ha)	16.78	13.27(–20.9)	22.83	21.74(–4.8)
Crude Protein Yield (kg/ha)	333.65	268.6(–19.47)	683.1	716.83(+4.9)

UC: Under tree canopies.

Production of fuel wood at 8 years of growth under silvipasture produced 28t/ha in *A. tortilis* and 38t/ha/yr in *A. amara* (3.5 and 4.75 t/ha/year on an average). For most of the short rotation species on such lands on a cycle of 10 years one could expect from 4-6 t/ha/year fuel wood production at a tree density of 500/ha. In many trees annual lopping at 6-7 years has been found to give 2.3-3.5 t/ha

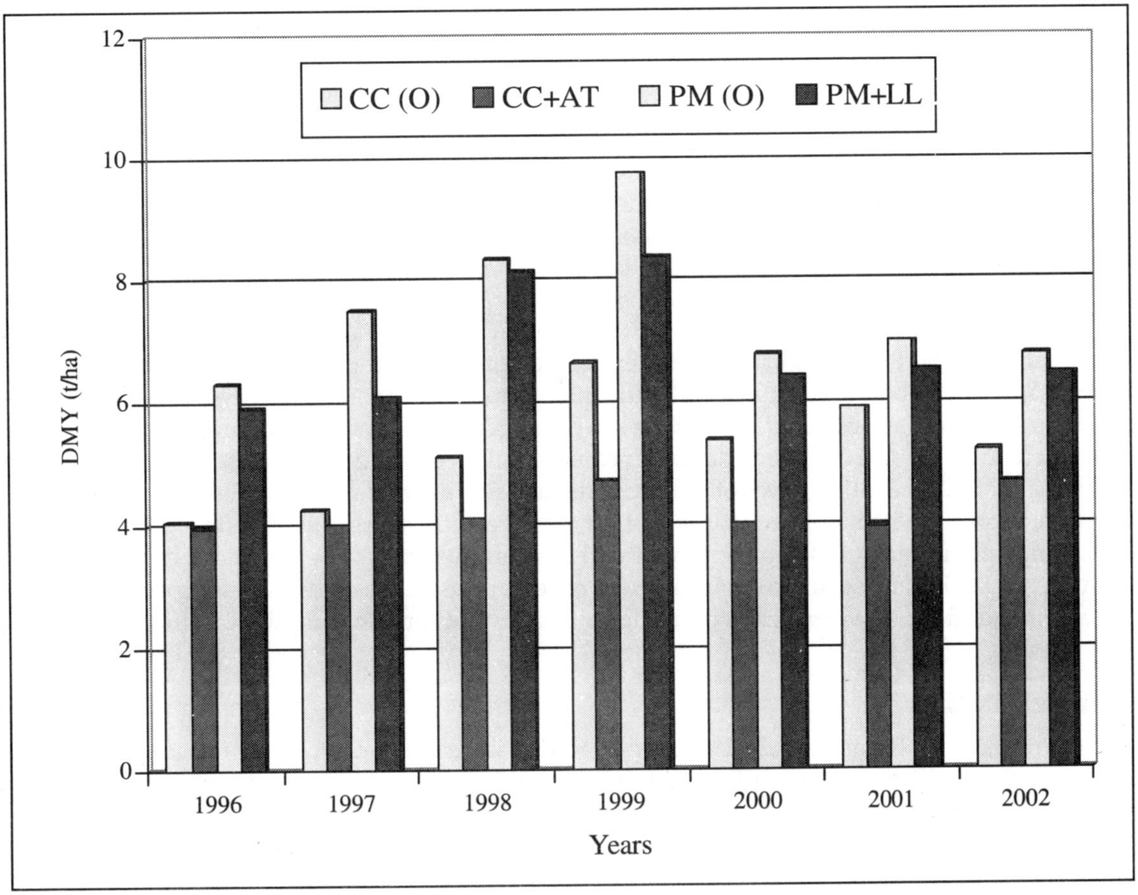

Figure 40.7: Dry Matter Yield in *C. ciliaris* (CC) and *P. maximum* (PM) in Open (O) and Under Tree Canopies of *A. tortilis* (AT) and *L. leucocephala* (LL)

fodder and 4.5-6.5 t/ha firewood. Thus, forage and top feed combined yield produced 5.3-9.5 t/ha forage and 5 t/ha firewood or 10-14.5 t/ha total harvestable biomass.

Biophysical Modelling of Tree-Crop Intercropping System

The adaptation and ability to survive of plants with in a given environment depends upon the extent of their interaction with the surroundings through phenomena such as transpiration, light absorption and energy exchange. PAR adversely affects the relative humidity, stomatal diffusion resistance and therefore, transfer of water vapour increased (Nobel, 1977). Transpiration is influenced by environmental factors such as temperature, light intensity, wind velocity, relative humidity, soil moisture and resistance to water movement. Transpiration increase with increase in temperature and decrease in relative humidity during the day time (Kaul, 1974). The rate of transpiration and the water content of leaves have an inverse relationship with each other and mainly governed by the relative humidity and temperature variation under tree canopies. The crop components interact with each other through the coupling with the environment and causing response on plant functioning and then these has the effect on microenvironment by changing its factor to a required level. Therefore, research on tree–crop system will primarily concern the ways in which a plant can influence its neighbours by changing their environment. Interaction between species are mediated by environment through the "response and effect principle"(Goldberg and Werner, 1983).

On the basis of all micrometeorological parameters, physiological characters and biophysical principles a hypothetical flow model have been developed for tree–crop intercropping system/ agroforestry system (Figure 40.8). From schematic diagram and flow chart and preceding discussion, it is evident that all parameters are intimately associated with the physiological economy and adaptation of system components which interact each other for sustainability of the system. By the process of radiation, convection and transpiration, the absorbed energy exchanges between canopy surface and environment. The main assumption here is that the net radiation and sensible heat flux will reduced due to reduction in radiant energy load and thermal load under tree canopies whereas latent heat flux will increase. On the basis of biophysical principles if most of the absorbed energy is lost by re-radiation and transpiration then there will be lower/negative flow of convectional energy through which heat load of the surrounding and under canopy will decrease and ultimately the atmosphere will become cool. After cooling down of atmosphere there will be condensation of surface water due to low evaporation by which soil moisture will increase resulting less water deficit. Therefore, the surrounding and under canopy environment will be more suitable for better under canopy crops's growth. On the basis of these micro environmental observations in *Leucaena* based silvipastoral system it is recommended that *Leucaena* trees planting can achieve favorable climatic conditions for under canopy growth.

Soil Amelioration

In semi-arid conditions Pathak (1992) reported that in *A. lebbek–Cenchrus–Stylo* and *A. procera–Cenchrus–Stylo* based systems, significant improvement in soil organic matter build up was 38 per cent in former compared with 25 per cent in latter system. Available phosphorus and potash were 22 and 16, and 9.5 and 11.0 per cent, respectively in the above systems. In another study, it has been found that the initial organic matter, available nitrogen and phosphorus and field capacity of 0.32 per cent, 131 kg/ha, 6.2 kg/ha and 11 per cent were enhanced to 0.78 per cent, 267 kg/ha, 15.5 kg/ha and 15.5 per cent, respectively over a period of 10 years of establishment of silvopastoral system (Pathak, 1994).

Yadava and Varsheny (1997) reported on the basis of five years that silvopastoral system reduced the run off and soil loss about 6 and 11 times, respectively against a bare land. Similarly loss of total soluble salts, dissolved nitrogen and potassium was reduced at the tune of 69, 67 and 43 per cent, respectively as compared to barren land. While improvement in organic carbon and available nitrogen and phosphorus in silvipastoral system was 53, 23 and 8 per cent, respectively against the initial status of the nutrients of this area. In case of potassium status in the system was almost maintained. However, in case of bare land all the nutrients were decreased after 4 years as compared to initial status.

In arid condition, Singh and Lal (1969) investigated profile characteristics and level of fertility under *P. cineraria* and *A. nilotica* canopies and concluded that silvopasture systems enhanced organic matter, total nitrogen, available phosphorus, soluble calcium and down the pH. Better mechanical composition of soil up to 120 cm depth due to silvopasture systems promoted the growth and yield of crops grown in the vicinity of *P. cineraria*. Aggarwal *et al.* (1975) concluded that status of available micronutrients generally improved under plantation especially under *P. cineraria*. Comparative study

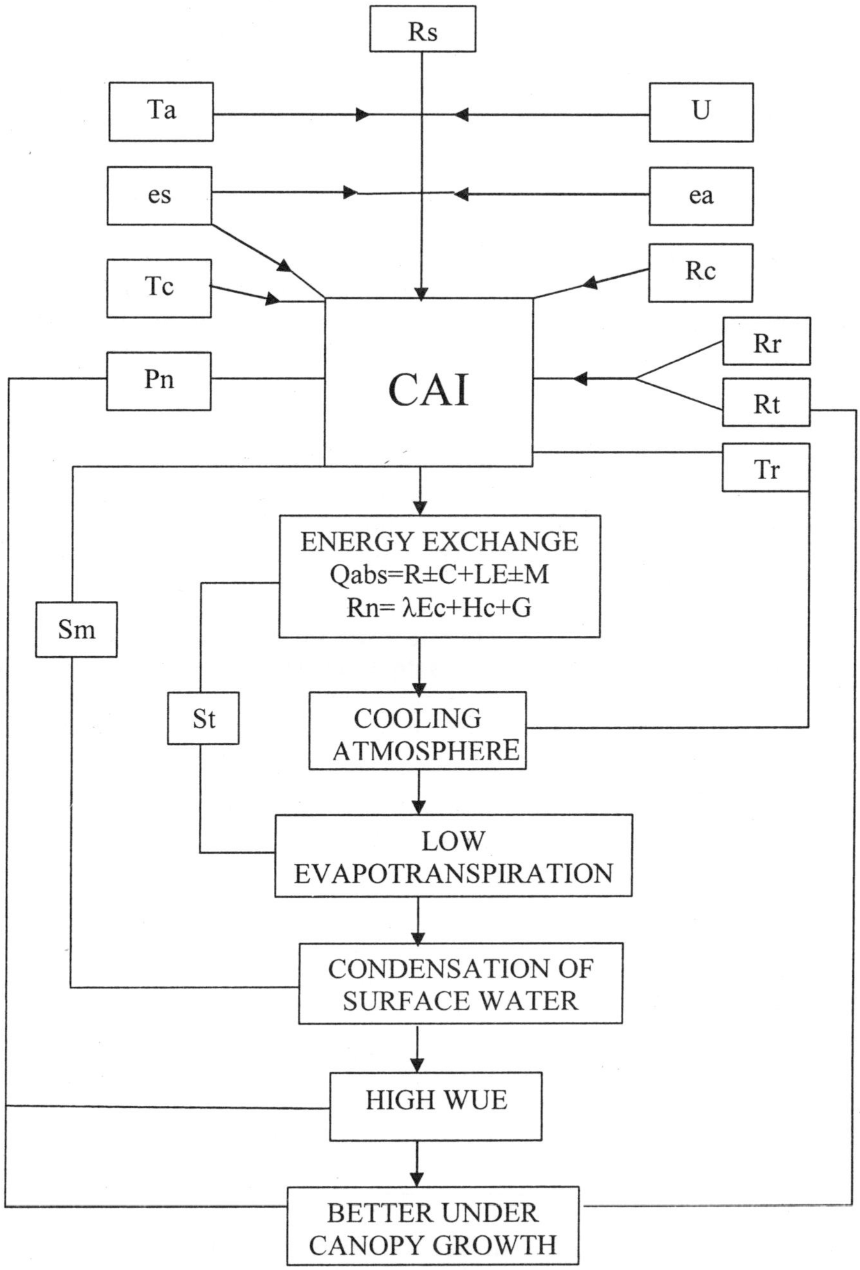

Figure 40.8: Energy Flow Chart for Agroforestry

Rs: Solar radiation; Ta: Air temperature; U: Wind velocity; es: Air humidity; ea: Boundary layer resistance; Tc: Canopy temperature; Rc: Canopy resistance; CAI: Canopy Area Index; Rr: Reflected light; Rt: Transmitted light; Pn: Photosynthesis; Tr: Transpiration; Sm: Soil moisture; St: Soil temperature; WUE: Water Use Efficiency; Qabs: Energy absorbed; R: Radiation energy; C: convection energy; LE: latent energy; M: Metabolic energy; Rn: Net radiation; λEc: Long wave radiation; Hc: Heat convection; G: Ground energy.

(Gupta and Saxena, 1978) on the soil moisture status up to 120 cm soil profile under five tree plantation suggested that the soil moisture regime remained generally higher below P. cineraria and T. undulat than that in P. juliflora, A. lebbek and A. senegal. The soil moisture under P. cineraria vicinity ranged fror 27 to 50 mm during the dry period from November to June and 115 to 140 mm from July to Septembe

The soil properties, in general improve with tree cropping as compared to non-tree situatio Hazra (1990) found that field capacity; wilting point, organic carbon, cation exchange capacity an available N and P contents of soils were greatly improved, whereas bulk density, pH and EC value were appreciably decreased under *Albizia lebbek* plantation as compared to normal cropping. Th canopy structure and also the density of tree species influenced the grass production underneath. Th leguminous shrubs and trees had great influence in building up soil organic matter (0.32 to 0.91 pe cent) available soil nitrogen (131 to 293 kg/ha), available phosphorus (6.2 to 18.5 kg/ha) and fiel capacity (11 to 15.8 per cent).

Sharma *et al.* (1996) analyzed the soil profile under tree canopy and open field conditions. Bette nutrient status in 0-20 cm and 20-60 cm soil depth was reported under tree canopy. The organi carbon, nitrogen and potassium in these profile depth was 0.15 per cent, 512 Kg/ha, 309 Kg/ha an 0.13 per cent, 522 Kg/ha and 242 Kg/ha, respectively under *Prosopis cineraria* canopy respectively. I open situation, the value of organic carbon, nitrogen and potassium were 0.13, 449 Kg/ha and 29 Kg/ha and 0.13 per cent, 506 Kg/ha and 295 Kg/ha, respectively.

Table 40.6: Organic Carbon at Different Soil Depth in Pure Grasses and Under Silvopastoral Systems

Grass Species	*Systems*	*Soil depth*	*Initial Year*	*After Two Years*
Cenchrus ciliaris	Open	0-15	0.426	0.505
		15-30	0.406	0.44
		30-45	0.421	0.45
	C. ciliaris + A. tortilis	0-15	0.69	0.88
		15-30	0.499	0.64
		30-45	0.606	0.60
Panicum maximum	Open	0-15	0.600	0.63
		15-30	0.541	0.60
		30-45	0.556	0.59
	P. maximum + L. leucocephala	0-15	0.714	0.80
		15-30	0.556	0.65
		30-45	0.531	0.67

The pattern of soil moisture varied in different months under the A. *tortilis* + *C. ciliaris* an *L. leucocephala* + *P. maximum* based silvipastoral systems (Mishra, 2002). The moisture availabilit under the tree canopies was higher for longer time for sustainability of the system productivity. O average the soil moisture at 0–30 cm soil depth was 7.5 per cent and 8.57 per cent in open field grasses and 8.87 and 9.73 per cent under the tree canopies of *A. tortilis* and *L. leucocephala* respectivel during the active growth stage of the crop (July to November). The organic carbon content in the sc was 0.45 per cent in open field of *C. ciliaris* and 0.67 per cent in *A. tortilis* + *C. ciliaris* while it was 0.5 per cent under sole *P. maximum* in open field and 0.79 per cent in *L. leucocephala* + *P. maximum* base silvipastoral system. Higher relative humidity, soil moisture and organic carbon content under the tree canopies as compared to the open grown grasses (without trees) indicating that the system has improved the microclimate and soil health for the sustainable production of the system for longer period.

Mishra (2002) reported that organic carbon was maximum at the top layer (0-15 cm) followed by (30–45 cm) soil depths in pure grasses in open field as well as in the *C. ciliaris* + *A. tortilis* and *P. maximum* + *L. leucocephala* based silvipastoral systems. The increase in organic carbon by 10 to 40 per cent under the silvopasture system as compared to sole pastures may be because of more litter decomposition and available root biomass. The increase in organic carbon under the silvopasture systems has indicated the improvement in soil health, which intern higher system productivity and sustainability. *Leucaena leucocephala* + *Panicum maximum* based silvipasture system maintained higher organic carbon and therefore, the sustainability of system production were maintained over the years. Maximum increase was estimated in case of available nitrogen, phosphorus and potassium content under the tree pasture canopies than the sole pasture.

Maximum nutrients were estimated at the top layer of the soil than at different depths. Generally, higher level of nutrients under tree + pasture based silvopasture systems during two years may be due to decay of litter, root biomass and other organic carbon matter (Mishra, 2002). The availability of phosphorus and potassium significantly improved under the tree canopies exhibiting the more available nutrients for maintaining the sustainable production of the systems in the degraded land. Many workers have also reported the more availability of nutrients under tree cover (Sanchez *et al.*, 1985, 1987; Nair, 1989). Also symbiotic nitrogen fixations by trees often result in increase of soil nitrogen availability (Alpizar *et al.*, 1986; Dommer gues 1987). Root decay may be another source of adding organic carbon and nutrients in the soil (Young, 1991; George and Kumar, 1998).

The analysis of soils after 6 years of silvipasture establishment indicated increase in the nutrients like nitrogen, phosphorus, organic carbon and potash in time even when grasses were harvested and removed every year (Singh *et al.*, 1977). Misra *et al.* (1982) also observed improvement in soil physical properties, pH, nitrogen and organic carbon with 4 years silvipasture plantation on a calcareous soil. In another study it has been found that the initial organic matter, available nitrogen, available phosphorus and field capacity of 0.32 per cent, 131 kg/ha, 6.2 kg/ha and 11 per cent were changed to 0.78 per cent, 267 kg/ha, 15.5 kg/ha and 15.5 per cent respectively over a period of 10 years. From these studies it was amply indicated that poor soils on the wastelands could restore their fertility under such systems.

Carbon Sequestration

Effective carbon sequestration in wasteland can be achieved by means of silvopastoral, pastoral and watershed approach. R and D in the wasteland development programme may increase the carbon sequestration rate by considering all the ecological, social, and economic implications. There are two fundamental approaches to sequestering carbon in such ecosystems: (1) protection of ecosystems that store carbon so that sequestration can be maintained or increased; and (2) enhancement of the ability of ecosystems to increase carbon sequestration beyond current conditions. Restoration of degraded lands offers significant benefits and carbon sequestration potential in both below and above ground systems.

Early estimates of huge losses of carbon from soils due to degradation and erosion of soils gave the idea to develop wastelands as carbon storage. The major strategy for increasing carbon sequestration is to increase afforestation. Such increases in productivity will likely lead to greater amounts of carbon sequestration in both above- and below-ground plant parts, as well as in soils. Reclamation of wastelands increased the vegetation cover and biomass production along with the soil carbon. However, sequestration of carbon implies not only increasing the amount of carbon entering the soils but also a decrease with amount leaving either through decomposition or erosion (Johnson and Kern, 1991). The silvopastoral systems of *Leptochloa* along with *Acacia* and *Prosopis* in degraded lands in 4 to 7 years rotations have produced 20 Mg/ha/yr and 4 Mg/ha/yr biomass of tree and grass along with increase in soil carbon (Singh *et al.*, 1988). Growth of high nitrogen fixing tree species like *Sesbania* rapidly increased soil carbon even in moderate sites (Rao *et al.*, 1990).

The organic carbon content of rakar soils in degraded lands of Bundelkhand region was estimated only 0.155 per cent with bulk density of 1.5 Mg/m^3 at the soil depth of 0–15 cm. The initial organic carbon stock of such soils was calculated as 3.49 Mg C/ha. The initial productivity of such land was only 1 Mg/ha/yr with carbon content of 0.5 Mg/ha/yr (assuming that vegetation contain 50 per cent C out of the total biomass). After the establishment of silvopasture systems the organic carbon stock of the soils was estimated to 10.125 Mg C/ha after ten years of rotation. This land has produced 11mg/ha biomass at the 10th years of growth of the system. The carbon stock in vegetation was calculated as 5.5 Mg/ha/yr (Bhatt, 2008). Therefore, the total carbon sequestration in such lands was 11.60 Mg/ha which have improved the productivity of the such lands. This has suggested that integration of trees with naturally growing grasses could be a viable land use option for improving biological productivity and carbon sequestration of highly degraded soils.

Agroforestry has tremendous potential in carbon sequestration because of carbon storage potential in its multiple plant species and soil as well as its applicability in agricultural lands and in reforestation (Montagnini and Nair, 2004). Proper development of agroforestry systems can make them effective carbon sinks. In total the extent of C sequestered depends on the amounts of C in standing biomass, in roots, and C sequestered in wood products. Average carbon storage by agroforestry practices has been estimated as 9, 21, 50, and 63 Mg C ha^{-1} in semiarid, subhumid, humid, and temperate region. For smallholder agroforestry systems in the tropics, potential C sequestration rates range from 1.5 to 3.5 Mg C ha^{-1} yr^{-1}. Agroforestry can also have an indirect effect on C sequestration when it helps decrease pressure on natural forests, which are the largest sink of terrestrial C. Another indirect avenue of C sequestration is through the use of agroforestry technologies for soil conservation, which could enhance C storage in trees and soils. Agroforestry systems with perennial crops may be important carbon sinks. In order to exploit this vastly unrealized potential of C sequestration through agroforestry in both subsistence and commercial enterprises in the tropics and the temperate region, innovative policies, based on rigorous research results, have to be put in place.

References

Aggarwal, R.K., Gupta, J.P., Şaxena, S.K. and Muthana, K.P. (1975). Studies on soil Physico–chemical and ecological changes under twelve year old five desert species of west Rajastha, *Indian For.*, 102; 863-872.

Ahuja, L.D. (1980). Grass production under Khejri. *In:* Khejri (*Prosopis cineraria*) in Indian Deserts its role in agroforestry (*Eds.* H.S. Mann and S.K. Saxena) *Monograph* No. 11, CAZRI, Jodhpur, pp. 28-30.

Ahuja, L.D., Verma, C.M, Sharma, S.I. and Lamba, T.R. (1978). Range management studies on the contribution of ground storey (grass) in afforested areas in arid region. *Ann. Arid Zone*, 17: 304-331.

ılpizer, L.H.W., Fassbender, Heuveldop, H.P. and Enriquez, G. (1986). Modelling agroforestry system of Coco (*Theobroma cocoa*) with Laurel (*Cordia alliodora*) and Poro (*Erythrina poeppigiana*) in Costa Rica. I. Inventory of organic matter and nutrients. *Agroforestry system*, 4: 175-189.

ırya, R.L., Singh, A., Yadav, R.B., Singh, S.P., Gangwar, M.K. and Lal, Banwari. (1998). Early performance of trees and grasses under silvipastoral systems in salt affected soil of Indo Gangetic plains of Uttar Pradesh. *Range Mgmt. and Agroforestry* 19: 53-57.

hatt, R. K. (2008). Climate change and mitigation options through silvipastoral systems. Paper presented in the National Seminar of National Institute of Ecology, New Delhi, 26-27 Aug.

hatt, R. K., H. S. Tiwari, Vandana and L.P. Misra. (2006). Photosynthesis and shade tolerance in tropical range grasses and legumes. *Indian J. Plant Physiol*, Vol. 11, No. 2, (N.S.) pp. 172-177.

hatt, R. K., Khan, T.A., Pathak, P.S. and Misra, A.K. (2006). Sustainable forage production under silvipastoral systems in semiarid tropics. *Indian J. of Agroforestry* Vol. 8 No.1:1, 22-25.

hatt, R. K., Suresh, G., and Ram Newaj (2003). Biophysical aspects of tree crop interactions in silvipastoral system. Pub. In Agroforestry potentials and Opportunities. (ed.) P. S. Pathak and Ram Newaj, Pub. By Agrobios, Jodhpur pp. 249-265.

hatt, R. K. (1988). IGFRI, Annual report.

hatt, R. K., Misra, L. P. and Pathak, P. S. (1994). Forage yield and related plant processes in *Cenchrus* and *Panicum* under silvipastoral system. In: *Agroforestry system for Degraded Lands.* (ed. Panjab Singh, P.S. Pathak and M.M. Roy), Oxford and IBH Publishing Co. Vol.2: 732-738.

hatt, R.K. and Sinha, N.C. (1990). Photosynthetic pigment–a pre-requisite of grass productivity. *Photosynthetica* 24(1): 147-150.

eb Roy, R. (1986). Studies on some aspects of renewable energy (fuel) and pasture production in an ecosystem of *Albizia lebbek* (L.) Benth. With grass and legumes. *Ph.D.Thesis*, Jiwaji University Gwalior (India).

eb Roy, R. (1989). Biomass production potential of fodder cum fuel tree under silvipastoral system. In: Promotion of Fodder and Fuelwood Tree (eds N.G. Hedge, L.L. Relwani and V.D. Kelkar). BAIF, Pune: 70-76.

eb Roy, R. and Pathak, P. S. (1974). Silvipastoral management, *Indian Farming* 24:41-42.

eb Roy, R. and Pathak, P. S. (1983). Silvipastoral research and development in India. *Indian Rev. of Life Science* 3:247-264.

eb Roy, R. and Pathak, P. S. (1984). Silvipastoral research and development in India. *Indian Rev. of Life Science* 3: 247-64.

eb Roy, R., Patil, B.D. and Pathak, P.S. (1978). Silvipastoral farming for amelioration and increased productivity for aird and semi arid region. In: Proceeding of the International symposium on Arid Zone Research and Development.CAZRI, Jodhpur, India: 531p.

eb Roy, R., Patil, B.D. Pathak, P.S. and Gupta, S.K. (1980). Forage production of *Cenchrus ciliaris* and *Cenchrus setigerus* under silvipasture system. *Indian J. Range Mgmt., 1*: 113-120.

ɔgra, K., Katoch, B.S., Sood, B.R. and Singh, Gurudev (1997). Production potential and quality of silvi–herbage system vis-à-vis natural grasslands.

Dommergues, Y.R. (1987). The role of biological fixation in agroforestry. *In: Agroforestry: A Decade* *Development* (eds. H.A. Skppler and P.K.R. Nair). ICRAF. Nairobi (Kenya): 245-271.

Egle, K. (1960). Menge and verhaltnis der pigment–In: Ruhland, W(*ed.*): *Handbuch der flanzen Physiologi* Vol. VII pp. 444-496. Springer-Verlag, Berlin–Gottingen–Heidelberg.

George, S.J. and Kumar, B.M. (1998). Liter dynamics and soil fertility improvement in silvopastor systems of the humid tropics of southern India. Intt. *Tree crops J.*, 9: 267-282.

Goldberg, D. E. and Werner, P. A. (1983). Equivalence of competitors in plant communities: a nu hypothesis and a field experimental approach. *Am. J. Bot.* 70: 1098-1104.

Good Child, D. J., Bojorkman, D. and Pyliotis, N. A. (1972). Chloroplast ultrastructure, leaf anatom and content of chlorophyll and soluble protein in rain forest species. In: *Carnegic Institute Ye* *Book* 71:102-107.

Grewal, S.S., Singh, K., Juneja, M. L. And and Singh, Shivcharan (1996). Relative growth, fuel woc yield, litter accumulation and conservation potential of seven Acacia species and an under store grass on a sloping bouldery soil. *Indian J. Forestry*. 19: 174-182.

Gupta, J.P. and Saxena, S.K. (1978). Studies on monitoring of the dynamics of moisture in the soil ar the performance of ground flora under desertic communities of trees. *Indian J. Ecol.*, 5: 30-36.

Harsh, L.N., Tiwari, J.C., Berman, U. and Sharma, S.K. (1992). Agroforestry for arid region. *Indian Fm* 42: 32-37.

Hazra, C.R. (1990). Revitalization of degraded land resources through silvipastoral system for herbag production. In: *Silvipasture system in India* (eds. P.S. Pathak and Punjab Singh). RMSI, IGFI Jhansi: 81-86.

Hazra, C.R. (1994). Role of agroforestry for amelioration as edaphclimatic conditions for increased f productivity. *In:* Advances in Agroforestry and its Role for Sustainable Agriculture ar Environment. (*Eds.* R. Deb Roy, A.K. Bisaria and P.Rai). NRCAF, Jhansi. pp. 105-118.

Hazra, C.R. and Patil, B.D. (1986). Forage production under silvipastural system–light and temperatu interaction. *Indian J. Range Mgmt.*, 7: 33-36.

Hazra, C.R., Singh, D.P. and Kaul, R.N. (1994). Greening of Common Ilands in Jhansi through Villa Resource Development–a case study. National Aforestation and Eco-development Board, Ne Delhi.

IDRC (1982-1994). Annual Report of Silvipasture Operational Research Project for the Bundelkhar Region, IGFRI, Jhansi.

Johnson, M. G. and Kern, J. S. (1991). Sequestering Carbon in Soil. In: *Proc. EPA Workshop*. p. USPEAERL. Corvallis, USA.

Kaul, R. (1974). Study of the seasonal variation of the water content of the leaves and the rate transpiration in *Hemigraphis dura. J. Indian Bot. Soc.* 53: 97-99.

King, K.F.S. (1979). Key note address: Some principles of agroforestry, Proc. *Agroforestry Semin* ICAR, 17–26.

Krishnaswamy, M.H. (1942). Revised working plan of Kurnool East forest division for the peri 1940–51, Govt Press Madras, 181 P.

.iu Naizhaung. (1991). Light distribution in tree intercropping area and its agricultural value. In agroforestry system in China (ed. Zhu Zhaohua *et al.,*). *China Academy of Forestry and IDRC.* Publ. Pp. 14-20.

/lishra, A. K. (2005). Growth, productivity and physiological process of grasses under silvopastoral system in semi-arid region. *Ph. D. Thesis,* Bundelkhand University, Jhansi.

/lisra, L.P. and Bhatt, R.K. (1994). Productivity and adaptation of grasses under silvipastoral system in semiarid region. In *Proc. Agroforestry systems for Degradedlands* (ed. Panjab Singh, P.S. Pathak and M.M. Roy), Oxford and IBH Publishing Col. Vol.2, PP.727-731.

/lisra, L.P. and Bhatt, R.K. (1994). Productivity and adaptation of grasses under silvipastoral system in semi arid region. *In:* Agroforestry systems for Degraded Lands (*Eds.* P.Singh, P.S. Pathak and M.M. Roy). Oxford and IBM Publishing Co. Pvt. Ltd., New Delhi. pp. 727-731.

/lisra, S.M., Patil, B.D. and Pathak, P.S. (1982). Comparative differences in soil properties as a result of growing different grass, legume and tree combinations. *Indian J. Range Mgmt.* 3(1 and 2) : 71–74.

/lontagnini, F. and Nair, P. K. (2004). Carbon Sequestration: An under-exploited environmental benefit of agroforestry systems. *Agroforestry Systems* 61: 281-295.

Naidu, C.V. and Swamy, P.M. (1993). Effect of shade on growth, biomass production and associated physiological parameters in *Pongamia pinnata* (Linn.) Pierre. *Indian J. Plant Physiol.* XXXVI (4) : 212-214.

Nair, P.K.R. (1989). The role of trees in soil productivity and protection. *In: Agroforestry systems in the Tropics* (ed. P.K.R. Nair). ICRAF, Nairobi and Kluwer Academic Publishers, Dordoecht, (the Netherlads): 567-589.

Nobel, P.S. (1977). Water relations and photosynthesis of a barrel cactus, *Ferocactus acanthodes* in colorado desert, *Oecologia* 27: 117.

'aroda, R. S. (1990). Silvipastoral approach for wasteland development: present options and future strategy. In: *Silvipastoral systems in India* (eds. PS Pathak and Panjab Singh). Range Management Society of India, Indian grassland and Fodder research Institute, Jhansi, pp. 1-3.

'aroda, R.S. and Jatasra, D.S. (1985). Forages for arid lands. In: *Current Practices in Dryland Resources and Technology.* (eds. Alani Singh and G.R. Chowdhary). Scientific and Geo–environ Academia, Jodhpur, 256p.

'athak, P. S. and Roy, M. M. (1994). *Silvipastoral System of Production.* IGFRI, Jhansi: 55 p.

'athak, P.S. (1983). Silvipasture through natural vegetation manipulation- an ecological assessment, Nat. Seminar on Silvipastoral System, 8–10 Oct., IGFRI, Jhansi.

'athak, P.S. (1992). Silvopastoral technology for wasteland development. NARP training program on agroforestry, forage production and animal nutrition. pp. 14-20.

'athak, P.S. (1994). Silvipastoral system of production for degraded lands in India. *In:* Advances in Agroforestry and its Role for Sustainable Agriculture and Environment (*Eds.* R.Deb Roy, A.K. Bisaria and P. Rai). NRCAF, Jhansi. pp. 186-201.

'athak, P.S., Gupta, S.K. and Singh Punjab (1995). Rehabilitation of degraded lands. IGFRI Jhansi : 23.

'athak, P.S., Gupta, S.K. and Singh, P. (1996). IGFRI Approaches Rehabilitation of Degraded lands. IGFRI, Jhansi, pp. 1-23.

Prajapati, M.C. K., Nambiar, T.N., Puri, D.N., Singh, J.P. and Malhotra, B.M. (1993). *Acacia nilotica* an *Cenchrus ciliaris* silvipasture for sustainable productivity of fuel and fodder in ravines.I Agroforestry for 2000 AD for the semi arid and arid tropics. Extended abstract (Late arrivals) (*Ed:* A.S. Gill, R. Deb Roy and A.K. Bisaria). NRCAF, Jhansi. pp. 33-34.

Ramakrishana, Y.S. and Sastri, A.S.R.A.S. (1977b). Microclimate under *Acacia tortilis* plantation *Annual Report*. CAZRI, Jodhpur: 69-70.

Ramakrishna, Y.S. (1984). Forest Microclimate and its importance in agroforestry. *In: Agroforestry i Arid and Semi-Arid Zones* (ed K.A. Shankarnarayan). CAZRI, Jodhpur : 9-14.

Ramakrishna, Y.S. and Sastri, A.S.R.A.S. (1977a). Climatological studies on rainfall interception, diurna variation in the total and net radiant energy available under *Acaci tortilis* trees. *Annual Repor* CAZRI, Jodhpur: 69 p.

Rao, D.L.N., Gill, H.S. and Abrol, I.P. (1990). Regional experience with perennial Sesbania in India. In *Perennial Sesbania Species in Agroforestry Systems* (Eds. Macklin, Bill and Evans, D.O.), NFTA Hawail, USA pp. 189-98.

Roy, M.M. (1995). Productivity of *Hardwickia binata* based silvopastures in Bundelkhand region. *Ph.D Thesis*, Bundelkhand University, Jhansi: 182 p.

Sanchez, P.A. (1987). Soil productivity and sustainability in agroforestry. In: *Agroforestry : A Decade o Development* (eds. H.A. Steppler and P.K.R. Nair) ICRAF, Nairobi (Kenya): 205-223.

Sanchez, P.A., Palm, C.A., Davey, C.B., Szott, L.T. and Russell, C.E. (1985). Tree crops as soil improver in humid tropics. In : *Attributes of Tree as Crop Plants* (eds. M.G.R. Cannell and J.E. Jackson) Institute of Terrestrial Ecology, National Environmental Research Council, Abbets Ripton Huntingdon, England: 327-350.

Sehgal, J.L. and Abrol, I.P. (1994). *Soil Degradation in India status and impact*. 80 Oxford and IBH Publishing Co., Pvt. Ltd. New Delhi, India.

Shankar, V., Kumar, Suresth and Tyagi, R.K. (1988). Grazing resources of semi arid and arid regions In: *Pasture and Forage Crop Research–A state of knowledge* Report (ed. Punjab Singh). RMSI, Jhansi 63-65.

Sharma, S.K. Datta, B.K. and Tiwari, J.C. (1996). *Prosopis cineraria* (L) Druce in silvipastoral system in arid regions of western Rajasthan. *Range Mgmt. and Agroforestry*, 17: 81-85.

Singh, D., Mannikar, N.D., Srivas, N.C. and Saxena, D.C. (1977). Changes in soil fertility under long term cultivation of fodder crops and grasses. *J.Indian Soc. Soil* 25: 28–32.

Singh, J.P. and Puri, D.N. (1975). Economic utilization of ravines for fuel cum fodder. Annual Report CSWCR and TI, Dehra Dun. pp. 39-40.

Singh, K.S. and Lal, P. (1969). Effect of khejri (*Prosopis cineraria*) and Babul (*Acacia arabica*) trees and soi fertility and profile characteristics. *Ann. Arid Zone*, 8: 33-36.

Singh, P. and Roy, M.M. (1991). Forage production through agroforestry constraints and priorities *Range Mgmt. And Agroforestry* 12: 169-178.

Singh, V.P., Dey, S.K. and Murthy, K.S. (1988). Effect of low light stress on growth and yield of rice *Indian J. Plant Physiol.* 31: 84-91.

Tajwani, K.G. (1987). Agroforestry practices and research in India. In: *Agroforestry: Realities, Possibilities and Potentials* (ed. H.L. Gholz), Martinus Nijhoff Publisher, Dordrecgt, Netherlands, pp. 109-136.

UNESCO (1979). *Tropical Grassland Ecosystem- A state of Art Report.* United National Educational Scientific and Cultural Organization, Paris: 65p.

Wilson, J.A. (1951). Grazing management and improvement in Madras Province. Proc. 7th Silvicultural Conf. FRI, Dehradun.

Yadava, R.B. and Varshney, R.B. (1997). Multistorey silvipastoral approach for sustaining soil fertility in degraded rangelands of Bundelkhand. *Range Mgmt. and Agroforestry* 18: 195-200.

Young, A. (1991). Soil Fertility. In: *Biophysical Research for Asian Agroforestry* (eds M.E. Avery, M.G.R. Cannell and C. Ong). Winrock International, USA: 187-207.

Microbial Biotechnology and Ecology (2011) *Pages* **708–723**
Editors: **Deepak Vyas, G.S. Paliwal, P.K.Khare and R.K Gupta**
Published by: **DAYA PUBLISHING HOUSE, NEW DELHI**

Chapter 41

Agroforestry: Conservation of Natural Resources, Biomass Production and Climate Moderations

☆ *Ashok Shukla* and V.D. Tripathi*

National Research Centre for Agroforestry, Jhansi – 284 003, U.P.

ABSTRACT

Agroforestry, a land use system/technology in which trees are deliberately planted on the same units of land with agricultural crops and/or animals, has been recognized as one of the most promising strategy for rehabilitating the already degraded areas. It has been a way of life in India ever since the evolution of human beings through the forests. Uses of trees for different purposes include, wood, shelter and, medicines have been the common feature. The benefits of agroforestry includes, it optimizes production, provide supplement food and fodder, increase farm income, amelioration of soil chemical and physical properties, reduction of soil erosion, utilization of waste and degraded lands, improve environment, minimize adverse effect of climatic factors etc. The present chapter enlighten the brief account on the importance of agroforestry systems *viz.*, system productivity and enhancement, amelioration of problematic soil, conservation of natural resources and climate moderation.

Introduction

Agroforestry is an intensive land use management system that integrates trees, shrubs, crops and livestock on a landscape level to achieve optimum benefits (Garret *et al.*, 1994). It has been a way of life in India ever since the evolution of human beings through the forests. Uses of trees for different

* E-mail: ashokshukla06@gmail.com

purposes include, wood, shelter and medicines have been the common feature. Agroforestry systems have several advantages and some of these advantages are optimizes production, gives supplement food and fodder, improve soil, utilize wasteland and degraded lands, provide employment opportunities, increase farm income, minimize adverse effect of climatic factors and aid industrial growth.

Agroforestry research is more than hundred years old in India beginning with introduction of shade trees in tea gardens, but organized research was initiated during fifties in few ICAR institutes. India has been at the forefront since organized research in agroforestry started worldwide. In 1979 first seminar on agroforestry was organized by the ICAR at Imphal to accumulate and compile the data on the R and D of agroforestry in India. The All India Coordinated Research Project (AICRP) on Agroforestry with 20 centers all over the country was launched in 1983. At present 36 centers are under AICRP on agroforestry. These centers represent almost all the agro-climatic zones of the country. In addition to ICAR, ICFRE, private organizations (WIMCO, ITC, BAIF, IFFDC, West Coast Paper Mills Ltd, Hindustan Paper Mills Ltd and National Tree Growers Cooperatives) and NGOs are also engage in research and promotion of the agroforestry in the country. Agroforestry research and education has been started in more than 15 State Agricultural Universities. The major thrust of agroforestry research in the beginning was as follows:

1. Diagnostic survey and appraisal of existing agroforestry practices.
2. Collection and evaluation of promising tree species/cultivars of fuel, fodder and small timber.
3. Studies on management practices of agroforestry systems.
4. Environment, wasteland and community land development through agroforestry interventions, and
5. Integration of livestock, fishery, apiculture, lac etc. as a component of agroforestry.

Table 41.1: Area Under Various Types of Tree Plantation in India

Plantation Type	*Area (M ha)*
Agroforestry and social forestry	23.00
Externally aided social forestry	
(a) Farm forestry/agroforestry	1.24
(b) Village woodlots	0.50
(c) Strip plantation	0.07
(d) Rehabilitation of degraded forests	0.24
Sub total	2.06
By NGOs/Vas	0.04
Tree growers co-operatives	0.04
Conservative forestry	0.19
Afforestation (windbreaks/shelterbelt), Under irrigation region development program of the Indira Gandhi Canal Project	0.40
Total	25.72

As per one of the estimate about 25.72 M ha area is under various types of tree plantations (Tabl 41.1), which include agroforestry, social forestry, farm forestry and others. According to FSI (2005) th total forest cover in India is 67.8 M ha. Of this 5.1 M ha is very dense forest, 33.9 M ha is moderatel dense forest and 28.8 M ha is open forest. Recently Dhyani *et al.*, 2006 estimated that 7.45 M ha area i under agroforestry. There is scope for using modern tool and products like RS and GIS to ascertain th information. National Research Centre for Agroforestry, Jhansi has taken a lead in this regard recently

Dhyani *et al.* (2007) has illustrated the potential of agroforestry contributions like fuel wood timber, fodder, fruits, livestock and bio-energy. The scenario of timber for domestic and industria purposes is given in Table 41.2. Currently the gap between supply and demand is met througl imports. The deficit of 16 M cum as on date (2008) could be met by promoting fast growing exotics lik *Eucalyptus*, Subabul, *Poplar*, *Acacia mangium* etc. The major contribution of agroforestry has been in th supply of small timber, where more than half of the demand is being met from agroforestry.

Table 41.2: Demand and Supply of Timber for Furniture, Agriculture and Industry (M cum)

Particulars	*2001*	*2006*	*2010*[a]	*2020*[a]	*2025*[a]
Demand	73	82	89	107	116
Supply from forest	12	12	12	12	12
Supply from plantation and agroforestry	47	53	61	90	100
Deficit	14	17	16	5	4

[a]Estimates; (Note: 2.2 cum = 1 tone); Ganpathy (1997) and Saxena (1990)

Table 41.3: Prominent Agroforestry Systems in Different Regions of Country

Agroclimatic Zone/Region	*Agroforestry Systems*					
	AS[a]	*AHS*	*AH*	*HP*	*SP*	*Specialized*
Western Himalayan	4	4	5	4	4	-
Eastern Himalayan	5	3	5	4	3	Jhum (shifting cultivation)
Lower gangetic plain	5	-	3	2	2	Energy plantation
Middel gangetic plain	4	-	2	-	-	Homestead
Upper gangetic plain	5	-	5	-	-	Agri-horticulture
Upper gangetic plain	5	-	5	-	-	Agrihorti-silviculture
Trans-gangetic plain	5	-	4	3	4	Block plantation
Eastern plateau and hills	4	-	5	4	4	Block plantation
Western plateau and hills	4	-	5	2	4	-
Southern plateau and hills	5	-	3	2	4	-
East coast plateau and hills	4	5	4	4	2	Home garden/shelterbelt
West coast plains and ghats	4	5	4	2	2	Home garden
Gujarat plains and hills	4	2	4	3	5	Block plantation
Western plains dry area	5	3	4	-	5	Shelterbelt
The islands	4	5	4	1	2	Home garden

AS: Agrisilviculture; AHS: Agrihortisilviculture; AH: Agrihorticulture; HP: Hortipastoral; SP: Silvipastoral (Solanki, 2006).

The research and development efforts undertaken during the last two decades have clearly demonstrated the potential of agroforestry for resource conservation, improvement of environmental quality, rehabilitation of degraded lands and providing multiple outputs to meet the day to day demand of the rural population. The diagnostic survey and appraisal of existing agroforestry practices in the country revealed that there are enumerable agroforestry practices prevalent in agroclimatic zones (Pathak *et al.*, 2000, Sharda *et al.*, 2001, Solanki, 2006). On the basis of the design (D and D) exercise, the prominent agroforestry systems from different regions of the country are given in Tables 41.3 and 41.4, respectively.

Table 41.4: Prominent Agri-horticulture Systems in India

State	*Based Agri-horti Systems*
J and K, Himachal Pradesh, Uttranchal	Apple
Punjab, Haryana	Citrus
Uttar Pradesh	Mango, Aonla
Bihar	Litchi
NEH region	Citrus
Maharashtra, Goa	Pomegranate, Mango, Coconut
Karnataka, Andhra Pradesh	Mango, Coconut
Kerala, Tamil Nadu	Coconut, Arecanut, Palmira
Rajasthan	Ber
Gujarat	Mango
Orissa, West Bengal	Coconut, Palmira

Present chapter enlighten the brief account on the importance of agroforestry systems *viz.*, system productivity and enhancement, amelioration of problematic soil, conservation of natural resources and climate moderation.

System Productivity and Enhancement

Solanki *et al.* (2000) assessed the biomass production of different trees after 7 years of plantation (Table 41.5). Highest biomass was recorded in *Casuarina equisetifolia* (171 t ha^{-1}), followed by *Leuceana leucocephala* (159 t ha^{-1}), *Eucalyptus* hybrid (152 t ha^{-1}) and *Dalbergia sissoo* (65 t ha^{-1}).

Table 41.5: Biomass Production of Different Tree Species

Species	*Productivity*
Casuarina equisetifolia	171
Leuceana leucocephala	159
Eucalyptus hybrid	152
Dalbergia sissoo	65

Comparative Biomass of Poplars in Different Localities

The performance of Poplar plantation with same and different ages in various locations is being presented in Table 41.6. Comparing the growth performance of Poplars in terms of productivity, it was higher in tropical countries compared to temperate countries. The growth greatly depends on edapho-climatic conditions of the particular region. For instance in India, the productivity varied from 3.2 Mg ha^{-1} yr^{-1} (North central) to 17.9 and 22.3 Mg ha^{-1} yr^{-1} (Alluvial plain and Tarai region, respectively).

Table 41.6: Comparative Account of Distribution of Above Ground Biomass of Poplar

Species	*Location*	*Age (yrs)*	*Density (ha^{-1})*	*Biomass ($Mg\ ha^{-1}$)*	*Reference*
Populus deltoides	India (Tarai)	1-4	666	7.4-89.3	Lodhiyal and Lodhiyal, 1997
P. deltoides	India (Tarai)	5-8	400	67.4-134.3	Lodhiyal *et al.*, 1995
P. tremuloides	North Central	8	-	25.4	Ruark and Bokheim, 1988
P. tremuloides	Wiscosin Canada	5	-	8.3-34.8	Wang *et al.*, 1995
P. deltoides	Albana	8	-	89.7	Carker and Tandon, 1997
P. deltoides	India Dehra Dun	5	-	22.5	Negi and Tandon, 1997
P. deltoides (Clone IC)	India	10	500	105.4	Singh, 1989
P. deltoides	India	14	-	44.5	Raizada and Srivastava, 1989
P. deltoides	USA	16	-	133.7	Shelton *et al.*, 1982
P. deltoides	India (Hisar)	9	208-2250	715-251.5	Puri *et al.*, 1994
P. deltoides (Agrisilvi)	India (Raipur)	4	666	20.9-35.8	Puri and Swamy, 2005

The large variability in aboveground biomass among different clones is probably due to genetic make up of individual clones. Among different components, the proportion of biomass was highest in stem followed by branches, bark and leaves. The proportion of biomass stored in roots were found to be 15-20 per cent of the total biomass, which are higher than the value 10-12 per cent reported for eucalyptus (Feller 1980) and equal to 5 years old poplar plantations (Lodhiyal *et al.*, 1992).

Eucalyptus

Chaturvedi (1981) assessed the yield potential of *Eucalyptus* hybrids in different site qualities (I-V) on a 6-9 years rotation (Table 41.7). The air dry tonnage yield varied from 6.9 $ha^{-1}yr^{-1}$ (site quality I) to 0.7 $ha^{-1}yr^{-1}$ (site quality V).

Table 41.7: Yield Potential of *Eucalyptus hybrid* in Different Site Quality

Site Quality	*Rotation (yrs)*	*No. of Stems ha^{-1}*	*ADT ($ha^{-1}yr^{-1}$)*
I	6	964	7.9
II	6	1038	5.1
III	7	1105	3.0
IV	8	1202	1.6
V	9	1345	0.7

The productivity potential of seven years old *E. tereticornis* managed under short rotation high density system were studied in the mid hill zone of H.P. (Saralch, 2005). The productivity per plant in all plant components declined significantly with increase in plant densities where as results on one hectare basis for the same parameters were exactly opposite to it (Table 41.8).

Mixed plantations of *Acacia mearnsii* (black wattle) and *Eucalyptus globulus* (blue gum) made in alternative rows at different spacing resulted in increased yield of wood and leaves over pure plantation in a 10 years rotation (Table 41.9).

Table 41.8: Above Ground Biomass and Productivity of *E. tereticornis*

Parameters	Density		
	27,778 trees ha^{-1} (60 x 60 cm)	12,346 trees ha^{-1} (90 x 90 cm)	6,944 trees ha^{-1} (120 x 120 cm)
Biomass (kg $tree^{-1}$)	4.57	5.74	7.13
Biomass (t ha^{-1})	127.07	70.85	49.49
Productivity (kg $tree^{-1}$ yr^{-1})	0.65	0.82	1.02
Productivity (t ha^{-1} yr^{-1})	18.15	10.12	7.07

Table 41.9: Biomass of Black Wattle and Blue Gum in Mixed Plantation (Cu mha^{-1})

Tree Species	3.5 x 3.5 m	3.65 x 3.65 m	3.65 x 4.57 m	Mean
Eucalyptus globules	274	226	318 460	330
Acacia mearnsii	180	228	108 90	152

Source: Forest Department, Tamil Nadu.

Leucaena leucocephala

The yield of three different varieties of *L. leucocephala* at Jhansi was studied by Pathak and Patil (1980) over a period of 10 years. The total productivity of Salvador registered 82 per cent and 64 per cent over K-8 and K-28 respectively (Table 41.10).

Table 41.10: Biomass Production of *L. leucocephala* Varieties

Variety	Main Stem ADT$ha^{-1}yr^{-1}$	Branchwood	Total
Salvador	31.6	11.8	43.4
K-8	21.8	3.0	23.8
K-28	22.7	3.7	26.4

In 1994, Pathak and Gupta assessed the growth and biomass production of *L. leucocephala* provenances and promising selection at IGFRI, Jhansi. They found that selection S24 was quite promising and better than K8 and it can be used in northern India (Table 41.11).

Table 41.11. Growth and Production of *L. leucocephala* Provenances and Selections

Provenance	DBH (cm)	Height (m)	Biomass (kg $tree^{-1}$)
K8	9.10	12.80	41.34
K 217	8.90	12.70	37.07
K 601	9.20	12.20	41.27
K 636	8.90	12.90	31.26
K 397	8.30	12.50	31.30
K 340	8.70	11.60	34.93

Contd...

Table 41.11–Contd...

Provenance	DBH (cm)	Height (m)	Biomass (kg $tree^{-1}$)
K 29	8.90	11.70	41.98
IGFRI 123-1	9.30	11.40	40.19
SP 1	8.50	11.60	39.94
S 10	10.10	12.50	50.16
S 14	9.90	12.50	43.18
S 15	9.00	11.70	36.58
S 22	8.90	11.70	44.73
S 23	7.70	10.70	23.91
S 24	9.50	12.80	56.36

Dalbergia sissoo

Gupta (1993) evaluated the performance of *D. sissoo* in the site quality class of VII at two different spacing *viz.*, 3 × 3 m and 2 × 2m. He found that at 6th year the plantation with 2 × 2 m spacing yielded 60.22 m^3ha^{-1} but it was achieved only in 10th when spaced 3 × 3m (Table 41.12).

Table 41.12: Standing Volume of *D. sissoo* on Class VII Land

Age (yrs)	DBH (cm)	Height (m)	Volume ha^{-1} (m^3)
Spacing 3 x 3m			
4	4.80	5.36	4.29
5	6.88	6.02	9.88
6	8.58	6.82	17.40
7	9.81	7.22	24.05
8	11.47	7.77	35.69
9	12.20	8.40	44.29
10	13.90	9.00	60.16
11	15.14	9.82	77.32
Spacing 2 x 2m			
4	7.90	7.21	35.21
5	8.70	8.20	48.51
6	9.20	9.08	60.22
7	9.40	10.03	69.20

Amelioration of Problematic Soil

In addition to biomass production, mitigating climate change, the trees plays a vital role in ameliorating the problematic soil by improving physical, chemical and biological properties of the soil. Nitrogen fixing trees ameliorate alkali soils at much faster than non nitrogen fixing trees.

Soil salinity and alkalinity have degraded about 7.41 M ha of land in India (Sharma, 1998). A rge acreage in the canal irrigated tracts of arid and semi arid regions of the country suffer from water ogging and salinity and the resultant loss to crop production is increasing at an alarming rate. stablishment of tree plantations on salt affected soils sets in amelioration process by favorably ffecting physical, chemical and biological properties of soil (Sing *et al.*, 1993). As per an estimate 45.3 er cent land degradation of the total degraded land of India is caused by water erosion (Table 41.13).

Table 41.13: Land Degradation in India

Type of Degradation	*Area (M ha)*	*Per cent of Total Area*
Water erosion	148.9	45.3
Wind erosion	13.5	4.1
Chemical deterioration (salinity etc.)	13.8	4.2
Water logging	11.6	3.5
Total	187.8	57.1

Some woody trees differ in their tolerance to alkali soil conditions at the genetic as well as species vel. Under saline conditions, only those which can withstand high salt content and thrive under igh water table conditions should be selected for planting. The promising tree species for plantation nder salt affected soils are listed here according to their degree of tolerance (Table 41.14).

Table 41.14: Promising MPTs for Reclamation

Alkali Soils	*Saline Soils*	*Bio-drainage*
Prosopis juliflora	*Casuarina equisetifolia*	*Populus deltaidus*
Acacia nilotica	*Leucaena leucocephala*	*Terminalia arjuna*
Tamarix articulata	*Prosopis juliflora*	*Eucalyptus tereticornis*
Eucalyptus tereticornis	*Acacia nilotica*	*Acacia auriculiformis*
Pongamia pinnata	*Tamarix troubii*	*Syzigium cuminii*
Terminalia arjuna	*Tamarix articulata*	*Albizia lebbeck*
Dalbergia sissoo	*Acacia farnesiana*	*Pongamia pinnata*
	E. Camaldulensis	*Dalbergia sissoo*

Singh *et al.* (1997) evaluated agriculture, forestry and agroforestry practices in a moderately alkali ils in northwestern India over a period of six years at CSSRI, Karnal. The soil conditions were much nproved in terms of the build up of soil organic matter, N, P and K when trees were associated with ops (Table 41.15).

The ameliorating effects of 20 years old plantations of *Prosopis juliflora, Acacia nilotica, Albizia beck* and *Terminalia arjuna* were reported by Singh (2000). The ameliorating effect on the soil appeared be in the order: *Prosopis*>*Acacia*>*Terminalia*>*Albizia* (Table 41.16).

An experiment conducted by Kumar and Warsi (2000) in salt affected soils of Daleep Nagar due the presence of hand par (Kanpur pan) between 70 to 125 cm depth. The planted trees *viz., D. sissoo, leucocephala, A. indica* and *A. nilotica* were substantially improved the physico-chemical property er a period of 4 years due to extraction of salt by plants, addition of organic matter (Table 41.17).

Table 41.15: Role of Agroforestry for Changes in some Soil Properties

Land Use System	*pH*	*EC (dSm⁻¹)*	*OC (%)*	*Available Nutrients (Kg ha⁻¹)*		
				N	*P*	*K*
Initial soil properties						
Sole crops	9.26	0.56	0.22	65	14.3	238
Eucalyptus based system	9.32	0.61	0.25	65	13.8	224
Acacia based system	9.34	0.67	0.28	72	13.6	212
Poplar based system	9.27	0.60	0.26	68	14.4	237
After five years						
Sole crops	**8.81**	**0.45**	**0.29**	**75**	**13.6**	**20.8**
Eucalyptus based system	8.65	0.32	0.37	86	18.5	259
Acacia based system	8.71	0.41	0.48	103	31.9	286
Poplar based system	8.47	0.35	0.43	93	21.7	321

Table 41.16: Ameliorating Effect of Tree Plantation on an Alkali Soil

Species	*Original*		*After 20 years*	
	pH	*Organic C (per cent)*	*pH*	*Organic C (per cent)*
A. nilotica	10.3	0.12	9.03	0.55
A. lebbek	10.3	0.12	8.67	0.47
T. arjuna	10.3	0.12	8.15	0.58
P. juliflora	10.3	0.12	8.03	0.58

Table 41.17: Changes in Physico-chemical Properties of Sodic Soils

Parameters	*Initial*	*After 4 yrs*
pH	10.5	8.6
ESP	65.2	36.8
EC (dSm⁻¹)	2.93	1.92
Org carbon (per cent)	0.48	0.52

Singh and Lal (1969) reported that there is an increase in nutrient content of soil under *Prosop* species in arid environments (Table 41.18).

Conservation of Natural Resources

Agroforestry systems have the potential to contribute significantly to maintaining or improvi soil and water quality. The degree to which these and other ecological functions can be provided w depend on plant species composition and their physical structure both above and below groun Agroforestry systems are generally perceived to be sustainable and to enhance soil properties. Growi trees in conjunction with annual crops or pastures is believed to provide a more thorough plant cov to protect the soil from erosion and a deeper or more prolific root system to enhance nutrient cycling. Appropriate agroforestry systems improve soil physical properties, maintain soil organic matter and promote nutrient cycling. Narain *et al.* (1994) studied the soil and water conservation efficiency of the different agroforestry systems (Tables 41.19 and 41.20).

Table 41.18: Nutrient Content of Soil Under Two *Prosopis* Species in Arid Region

Tree	*Site*	*Org C (per cent)*	*Total N (per cent)*	*Total P*	*Total K*
P. juliflora	Under tree	0.73	0.075	37	296
	Open area	0.25	0.027	31	294
P. cineraria	0-30 cm				
	Under tree	0.37	0.045	3.82	12.2
	Open area	0.25	0.038	1.52	7.52
P. cineraria	31-60 cm				
	Under tree	0.11	0.02	1.95	9.31
	Open tree	0.04	0.01	1.23	6.36

Table 41.19: Soil and Water Conservation by Different Agroforestry Systems

Land use	*Runoff %*	*Soil Loss (t)*	*% Reduction in*	
			Runoff	*Soil Loss*
Maize alone	18.3	17.7	-	-
Maize + Subabul	8.9	5	51.9	71.8
Maize + Eucalyptus	3.6	0.91	80.3	94.9
Chrysopogon fulvus	1.6	0.33	91.3	98.1
Grass + Subabul	0.6	0.13	96.7	99.3
Subabul	0.4	0.04	97.8	99.8
Grass + Eucalyptus	0.1	0.02	99.5	99.9
Eucalyptus	0.1	0.01	99.5	99.9

Table 41.20: Runoff, Soil and Nutrient Loss from Different Land Use Systems in Shivalik Region

Land Use Systems	*No. of Years of Observations*	*Runoff*	*Soil Loss (t ha^{-1})*	*Nutrient loss (Kg ha^{-1})*	
				N	*K*
Eucalyptus-Bhabar grass	6	0.05	0.07	0.46	0.90
Acacia catechu-forage grass	3	2.0	0.24	6.97	0.52
Leucaena-Napier grass	3	4.4	0.28	6.60	1.20
Teak-Leucaena-Bhabar grass	3	3.3	0.43	2.08	0.55
Eucalyptus-*Leucaena*-Turmeric	5	2.6	0.59	2.47	0.73
Poplar-*Leucaena*-Bhabrar grass	5	4.8	1.54	5.90	1.10
Sesamum-rape seed	3	20.5	2.69	42.5	3.00
Cultivation fallow	3	23.0	5.65	51.3	5.00

Climate Moderation

Carbon Sequestration by Trees

Intergovernmental Panel on Climate Change (2000) defines carbon sequestration is an increase in carbon stocks in any non-atmospheric reservoir. Carbon sequestration by trees from the atmosphere is a function of the land use, time for growing and technological progress. By two ways, trees can be used to reduce carbon omissions. One is direct carbon sequestration by reforestation and afforestation that yields a stock of carbon in shading trees. The other is the use of forest products as substitutes for fossil fuel or fossil fuel intensive goods such as steel and concrete.

Direct carbon sequestration by trees has been proposed as an emergency stop-gap measure to stop or reverse the increase in atmospheric CO_2, while allowing time to develop and implement clean energy technologies. The use of tree biomass for fossil fuel substation can be a long-term measure because harvesting and replanting in a given piece of land can be carried out in perpetuity. Additionally, the use of biomass for energy production helps rural economics to grow and enhances energy security.

Mechanism

Trees take atmospheric CO_2 during photosynthesis and fix it in woody (branches, stems and woody roots) and non-woody (foliage and fine roots) parts. Woody and non-woody parts have different growth dynamics and decay characteristics (Dewar and Cannel, 1992). At the end of the each rotation, a portion of biomass transferred to wood product, or energy feedstocks. The energy feedstocks release carbon back to the atmosphere upon combustion. The wood products eventually decay to release the carbon either by aerobic or anaerobic decomposition (Figure 41.1).

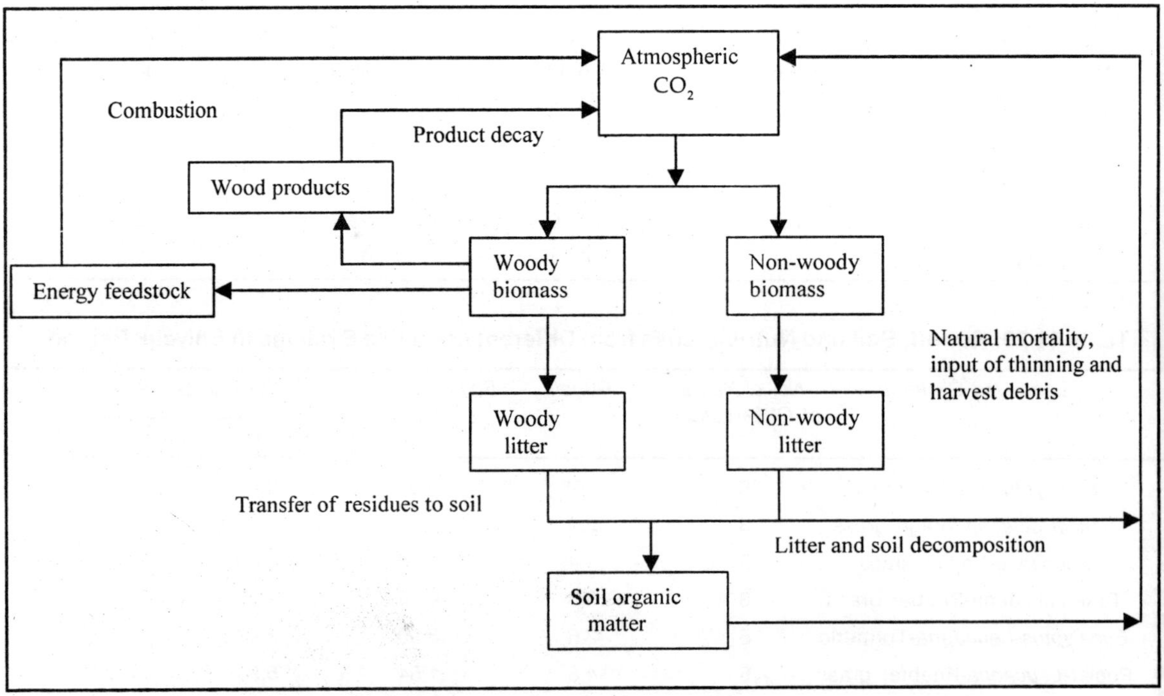

Figure 41.1: Carbon Mitigation by Trees, which are Regularly Harvested and Planted (*Source*: Dewar and Cannel, 1992)

The forest floor receives dead woods and non-woody litter continuously as tree components die because of natural mortality throughout each rotation and also at each thinning and harvest. A portion of litter is decomposed by microorganisms in to CO_2, which eventually returns to the atmosphere.

The amount of carbon that can be sequestered by trees depends on the biomass accumulation rate and rotation length. For long rotation forests, the present forest management can provide harvestable yields of about 4-8 dry metric tons/ha/yr in temperate and 10-12 metric tons/ha/yr in tropical areas (Hall *et al.*, 1991). For short rotation woody crops (SRWC), yields as harvestable woody biomass in range of 9-16 metric tons $ha^{-1} yr^{-1}$ (Wright *et al.*, 1992). The yields of SRWC largely depend on genotype, cultural practices and site quality (Graham *et al.*, 1992).

Carbon Storage and Flows in India

The carbon emission and sequestration estimates for India are given in Table 41.21. The standing biomass (above and below ground) is estimated to be 8357 Mt for the year 1986. The corresponding carbon storage is 4179 Mt (assuming 50 per cent of woody biomass as carbon). The total carbon stored in forests, including soil, is estimate to be 9578 Mt. When the net annual carbon emission in forest ecosystem is considered (uptake-release) for India, the carbon emission is nearly offset by sequestration in forests under succession and in tree plantations.

Table 41.21: Carbon Storage and Flows in India (Mt of C) Estimated for the Year 1986

Carbon Flux	*1986*
Carbon storage#	
Forest vegetation	4179.0
Soil	5399.0
Total	9578.0
Carbon emissions$	
Shifting cultivation (Combustion and decomposition of woody litter)	1.5
Combustion of wood (from clear felling, forest degradation) during the year	31.9
From past long term use (as structural timber 30 years earlier)	1.4
From past short term use (as pulp, packaging etc. with a life of 3 years)	3.1
Current short term use (as pulp)	2.8
Release from soil in clear felled forests	3.9
Biomass decomposition of woody litter in forests	18.8
Total carbon emission	63.4
Carbon uptake in 64 million ha forests and plantations*	68.8
Total C uptake	68.8

#: Carbon storage is estimated for different forest types by taking area under forest types and standing biomass from literature. The tree crown cover and standing biomass relations are considered. C storage in soil is estimated using area under forest types and organic carbon content of different forest types; $: C emissions are estimated by considering various land conversions, end use of felled forest biomass and timing of C emissions. Combustion; same year, long term; structural timber with 30 year storage, short term; pulp, rayon, packing with C release over 3 years. C emissions from combustion of wood from non-forest sources (farm trees, homestead gardens, shrubs in commons) are not included, only the forest source is included; *C uptake: C sequestration in tree trunk and large branches in forest under succession and tree plantations in the range of 1.28 to 2.84 t of wood biomass/ha/year based on productivity values.

Source: Ravindranath *et al.* (1992).

Brown *et al.* (1996) estimated that by 2050, plantation in tropical countries have the potential to capture as much as 16.4 Gt C whereas agroforestry has the potential to capture 6.3 Gt C. The estimates of Trexler and Haugen (1994) are much lower than this, especially for agroforestry (Table 41.22), indicating that there is large degree of uncertainty regarding estimates of potential C sequestration at a global scale.

Table 41.22: Potential for C Sequestration in Tropical Countries in Gt C by 2050

Production System	*Texler and Haugen (1994)*	*Browen et al. (1996)*
Plantations	2.0–5.0	16.4
Agroforestry	0.7–1.6	6.3
Forest re-growth	9.0–23.0	11.5–28.7

Carbon Stocking in Different Agroforestry Systems

The amount of C sequestration largely depends on the agroforestry system being practiced. Factors influencing C storage in agroforestry systems include tree species, system management, environment and socio-economic aspects. The C storage potential of agroforestry system in different regions is presented in Table 41.23. Its varies from 12 to 228 Mg C ha^{-1}.

Table 41.23: Potential C Storage for Agroforestry Systems in Different Eco-regions of the World

Country	*Eco-region*	*System*	*Mg C ha⁻¹*	*References*
Africa	Humid tropic high	Agro-silvicultural	29-53	Dixon *et al.*, 1993
South America	Humid tropic low Dry low land	Agro-silvicultural	39-102[a] 39-195	
Southeast Asia	Humid tropical Dry low land	Agro-silvicultural	12-228 68-81	Krankina and Dixon 1994
Australia	Humid tropic low	Silvopastoral	28-51	
North America	Humid tropic high Humid tropic low Dry low land	Silvopastoral	133-154 104-198 90-175	Schroeder 1994
Northern Asia	Humid tropical low	Silvopastoral	13-18	Winjum *et al.*, 1992

a: Carbon storage value were standardize to 50 year rotation.

Many reports are available that the tree component in agroforestry has a higher capacity of biomass production than natural vegetation (Table 41.24). The table summarizing the prominent agroforestry practice for higher organic C content for the different regions. Das and Itnal (1994) reported that organic C content was about double in agri-horticultural and agroforestry systems as compared to sole cropping.

Total C Storage Under Agroforestry Systems in Different Regions of Country

C storage capacities of different agroforestry system under different agro-ecological regions have been given in Table 41.25. In semi arid region of U.P, C dynamics was studied by Ram Newaj *et al.* (2006) under agrisilviculture involving three pruning regimes and two crop sequences. C content in tree and crop was slightly higher in greengram-wheat (35.13) then blackgram-mustard (34.77) crop sequence.

Table 41.24: SOC Content in Some Agroforestry Systems

Region	Agroforestry Practices	SOC (g Kg⁻¹ soil)	References
Semi-arid	Alley cropping		Kang *et al.* (1990)
Tropics	*Leucaena leucocephala* (pruning retained)	10.7	
	L. leucocephala (pruning removed)	6.5	
	L. leucocephala (pruning + 80 Kg N ha^{-1})	11.9	
Dry-land region	Sole cropping	3.9	Das and Itnal (1994)
	Agroforestry	7.2	
	Agri-horticulture	7.3	
	Agri-silviculture	4.7	
Bay Islands and	Monoculture systems (Coconut)	9.6	Liyange and Dessanayake
Kerala	Mixed farming with coconut and cacao/coffee/ pepper/clove/cinnamon	16.3 12-14.6	(1993)

Table 41.25: Total C Storage Under Agroforestry Systems in Different Regions of Country

Region	Agroforestry System and Components	Total Storage (t C ha^{-1})	References
Semi-arid	Silvipastoral (age 5 yrs)		Rai *et al.*, 2001
	A. nilotica + natural pasture	9.5-17.0	
	A. nilotica + established pasture	19.7	
	D. sissoo + natural pasture	12.4	
	D. sissoo + established pasture	17.2	
	H. binata + natural pasture	16.2	
	H. binata + established pasture	17.0	
North-western India	Silvipastoral (age 6 yrs)		Kaur *et al.*, 2002
	Acacia/Dalbergia/Prosopis + *Desmostacya*	6.8-18.55	
	Acacia/Dalbergia/Prosopis + *Sporobolus*	1.5-12.32	
Central India	Block plantations (age 6 yrs)		Swamy *et al.*, 2003
	Gmelina arborea	24.12-31.12	
Arid region (Rajasthan)	Agri-silvicultural (age 8 yrs)		
	E. officinalis + *V. radiata*		
Semi-arid region	Agri-silvicultural (age 11 yrs)		NRCAF 2005
	D. sissoo + crop	26.02	
Semi-arid region	Agri-silvicultural (age 5 yrs)		Ram Newaj *et al.*, 2006
	A. procera + black-gram-mustard	34.77	
	A. procera + green-gram-wheat	35.13	
North-western Himalaya	Silvipastoral system	2.17	AICRPAF 2006
	Agri-hortipastoral	1.15	
	Hortipastoral	1.08	

References

Brown, S., Sathaye, J, Cannell, M. and Kauppi, P. (1996). Management of forest for mitigating o greenhouse gas emissions. In: *Climate Change 1995* (Eds. Watson, R.T., Zinyovera, M.C. anc Moss, R.H.), Impact Adaptation and Mitigation of Climate Change: Scientific-technical Analysis Cambridge University Press, NY.

Chaturvedi, A.N. (1981). Yield tables for *E. hybrid* Forest Research Development Circle, U.P. Publ. No 44.

Das, S.K. and Itnal, C.J. (1994). Capability based land use systems: role in diversifying drylanc agriculture. In: *Soil Management for Sustainable Agriculture in dryland Area Bulletin*. Indian Society of Soil Science, 16:107-128.

Dewar R.C., Cannel, M.G.R. (1992). Carbon sequestration in the trees, products and soils of fores plantations: an analysis using UK examples. *Tree Physiology*, 11: 49-71.

Dhyani, S. K, Kareemulla, K. and Handa, A.K. (2006). *Agroforestry potential and returns across agroclimati zones*. NRCAF, Jhansi.

Dhyani, S. K., Samra, J.S, Ajit, Handa, A.K. and Uma. (2007). Forestry to Support to Increase Agricultura Potential: Focus on Employment Generation and Rural Development. *Agricultural Economic Research Review*, 20: 179-202.

FSI 2005. State of Forest Report (2005). *Forest Survey of India*. Ministry of Environment and Forest Dehradun, India.

Feller, M.C. (1980). Biomass and nutrient distribution in two eucalypt forest ecosystems. *Aust. J. Ecc* 51: 309-333.

Garret, H.E., Buck, L.E., Gold, M.A., Hardest, L.H., Kurt, W.B., Lassoie, J.P., Pearson, H.A. and Slusher J.P. (1994). *Agroforestry: An integrated land-use management systems for production and farmlan conservation*. USDA SCS 68-3A75-3-134. Washington, DC: US Department of Agriculture.

Graham, R.L., Wright, L.L. and Turhollow, A.F. (1992). The potential for short-rotation woody crops to reduce US CO_2 emissions. *Climatic Change*, 22: 223-38.

Gupta, R.K. (1993). *Multipurpose trees for agroforestry and wasteland utilization*. Oxford and IBH Publishing Co. Pvt. Ltd., New Delhi, pp. 562.

Hall, D.O., Mynic, H.E. and Williams, R.H. (1991). Alternative roles for biomass in coping with greenhouse warming. *Science and Global Security*, 2: 113-51.

Intergovernmental Panel on Climate Change (IPCC) (2000). *Land use, land-use change and forestry*. IPCC Special Report: Summery for Policymakers.

Kumar, M. and Warsi, A.S. (2000). Role of MPTs in management of wastelands with special referenc to sodic soils. In: *MPTs: Research-Retrospect and Prospect*, (Eds.) K.R. Solanki, A.K. Bisaria and A.K Handa) Agrobios (India), Jodhpur. (198-202 pp.

Lodhiyal, L.S., Singh, R.P. and Rana, B.S. (1992). Biomass and productivity in an age series of shor rotation Populus deltoids plantation. *Tropical Ecol.* 33(2): 214-222.

Pathak, P.S. and Patil, B.D. (1980). Fuel wood and forage production from *L. leucocephala*. *Leucaen Newsletter*, 1: 11.

Pathak, P.S. and Gupta, V.K. (1997). In: *Agroforestry System Degraded Lands* (Eds. P. Singh, P.S. Pathak and M.M. Roy), Oxford and IBH Publishing Co. Pvt. Ltd., New Delhi, India

Pathak, P.S., Pateria, N.M. and Solanki, K.R. (2000). *Agroforestry Systems in India: A Diagnosis and Design Approach*. National Research Centre for Agroforestry. ICAR, Pp. 166.

Ram Newaj, Shanker, A.K. and Yadav, R.S. (2006). *Carbon and nitrogen dynamics in Albizia procera based agroforestry systems*. Annual report of ICAR ad-hoc scheme, NRCAF, Jhansi (Unpublished data).

Sharda, V.N., Dhyani, S. K. and Sharma, A.R. (2001). *Role of agroforestry in disaster management*. HPC (DM Plan) DOA and C, Govt of India, New Delhi, pp.37.

Sharma, R.C. (1998). Nature, extent and classification. In: *Agricultural Salinity Management in India*. CSSRI, Karnal.

Singh, G. (2000). Management of MPTs for salt affected soils in agroforestry system. In: *MPTs: Research-Retrospect and Prospect* (Eds.: K.R. Solanki, A.K. Bisaria and A.K. Handa) Agrobios (India), Jodhpur. 190-197 pp.

Singh, G. and Lal, P. (1969). *Ann. Arid Zone*, 8: 33-36.

Singh, G., Singh, N.T. and Tomar O.S. (1993). *Agroforestry in salt affected soils*. Bulletin No. 17, CSSRI, Karnal, 65.

Singh, G. Singh, N.T., Dagr, J.C., Singh, H. and Sharma, V.P. (1997). An evaluation of agriculture, forestry and agroforestry practices in a moderately alkali soil in northwestern India. *Agroforestry Systems*, 37: 219-295.

Solanki, K.R., Bisaria, A.K. and Handa, A.K. (2000). *Multipurpose tree species: Research Retrospect and Prospect*. Agrobios (India), Jodhpur, pp. 337.

Solanki, K.R. (2006). Agroforestry: Opportunity and challenges. In: *Production Technology and Management of Agroforestry Models* (Eds; S.K. Chauhan, S.S. Gill, H.N. Khajuria and R. Chauhan). Agrotech Publishing Agency, Udaipur, pp.14-15.

Trexler, M.C. and Haughen, C. (1994). Keeping it Green: Tropical Forestry: A Partial Literature Review. Report to GTZ for Enquette Commission, German.

Wright, L.L., Graham, R.L., Turhollow, A.F. and English, B.C. (1992). Opportunities to mitigate carbon dioxide build-up using short rotation woody crops. In: Sampson RN, Hair D, editors. *Forest and Global Change*, Washington, DC: American Forestry Association; 1: 123-156.

Microbial Biotechnology and Ecology (2011) *Pages* **724–728**
Editors: **Deepak Vyas, G.S. Paliwal, P.K.Khare and R.K Gupta**
Published by: **DAYA PUBLISHING HOUSE, NEW DELHI**

Chapter 42

Soil Mycoflora of Forest Field at Jagdalpur Region

☆ *K.L. Tiwari, S.K. Jadhav and S. Kunjam*

School of Studies in Biotechnology
Pt. Ravishankar Shukla University, Raipur – 492 010, C.G.

ABSTRACT

Soil is a rich habitat for the growth of microorganisms, among these microorganisms, fungi are one of the dominant group present in soil. Soil samples are collected from different places of forest fields at Koprab and Kolawal forest area of Jagdalpur. From soil samples fungi are isolated with the help of serial dilution methods on Potato Dextrose Agar (PDA) medium. During present study 29 fungal species (251 fungal colonies) are recorded. Among them, the percentage contributions of different fungal groups are as follows, Zygomycotina (13.79 per cent), Anamorphic Fungi (68.96 per cent), Unknown groups (13.79 per cent) and sterile mycelia (3.44 per cent) are observed. Anamorphic fungi are dominated to the total fungal groups. During present study, the maximum percentage contribution of *Aspergillus awamori* (11.15 per cent) followed by *Penicillium* sp. IV (10.75 per cent), *Aspergillus niger* (9.96 per cent), *Aspergillus nidulans* (9.56 per cent) and *Penicillium* sp. V while minimum percentage contribution of *Drechslera* sp., *Aspergillus flavus, Cladosporium* sp., and *Penicillium purpurogenum* (0.79 per cent) similarly *Aspergillus versicolor, Cunninghamella* sp. and *Penicillium rugulosum* (0.39 per cent) are recorded.

Introduction

Soil forms a rich and dynamic medium for all microorganisms. Fungi occur in soil either in mycelial state or reproductive stage. Soil is a rich habitat for the growth of microorganisms. Among these microorganisms, fungi are one of the dominant groups present in soil.

An exhaustive work has been done on the taxonomy of forest soil fungi in India by the different scientist *i.e.* Madhusudan Rao and Manoharachary (1981), Manoharachary *et al.* (1990), Venugopal Rao (1984), Mohanty and Panda (1994) etc. have studied the soil fungi of different forests. Very little information is available regarding diversity of forest soil fungi of Jagdalpur region.

Materials and Methods

Soil samples are collected from different places of forest fields at Koprab and Kolawal forest area of Jagdalpur with the help of sterile plastic tubes. Fungi are isolated using serial dilution methods on Potato Dextrose Agar (PDA) medium and plates are incubated at 28±1º for incubation periods. After the incubation period fungal colonies are counted, isolated and maintained pure culture. The percentage contribution and percentage frequency are assessed by given formula. Fungi are identified with the help of available literatures (Nagmani *et al.*, 2006; Barnett, 1969; Ellis, 1971; Gilmans, 1959) and finally identified by authentic authority.

$$\text{Percentage Frequency} = \frac{\text{No. of observations in which a sp. observed}}{\text{Total no. of observations}} \times 100$$

$$\text{Percentage Contribution} = \frac{\text{No. of fungal colonies of individual sp. in all the observations}}{\text{Total no. of colonies of all sp.}} \times 100$$

Results and Discussion

29 fungal species of 251 fungal colonies are recorded during present investigation (Table 42.1). Among them, the percentage contributions of different fungal groups are as follows, Zygomycotina (13.79 per cent), Anamorphic Fungi (68.96 per cent), Unknown groups (13.79 per cent) and sterile mycelia (3.44 per cent) are observed (Figure 42.1). Anamorphic fungi are dominated to the total fungal groups.

During present study, the maximum percentage contribution of *Aspergillus awamori* (11.15 per cent) followed by *Penicillium* sp. IV (10.75 per cent), *Aspergillus niger* (9.96 per cent), *A. nidulans* (9.56

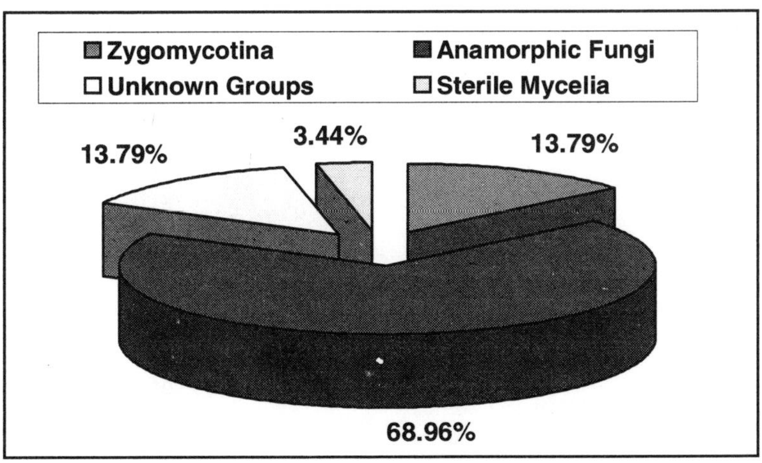

Figure 42.1: Percentage Contributions of Fungal Groups of Forest Soil Mycoflora

per cent) and *Penicillium* sp. V (7.56 per cent) (Figure 42.2) while *Mucor* sp. I (6.77 per cent), *Aspergillu* sp. (6.37 per cent) and *Pycnidial* sp. I (5.97 per cent) etc. are moderate contributions and *Cunninghamell* sp., *Aspergillus versicolor*, *Fusarium* sp. II, *Penicillium rugulosum* and *Penicillium* sp. VI (0.39 per cent) et are least contributions (Table 42.1).

Table 42.1: Fungal Distribution and Percentage Contributions of Soil Mycoflora of Forest Fields

S.N.	Name of Fungi	Soil Sample								Total No. of Fungal Colonies	Percentage contributions
		I	II	III	IV	V	VI	VII	VIII		
	Zygomycotina										
1.	*Absidia* sp.	–	–	–	–	03	–	–	–	03	1.19
2.	*Cunninghamella* sp.	–	–	–	–	–	–	01	–	01	0.39
3.	*Mucor* sp. I	–	–	01	01	14	–	–	01	17	6.77
4.	*Mucor* sp. II	–	–	01	01	–	–	–	–	02	0.79
	Anamorphic Fungi										
5.	*Aspergillus awamori*	–	10	02	08	01	05	02	–	28	11.15
6.	*A. flavus*	–	–	01	01	–	–	–	–	02	0.79
7.	*A. fumigatus*	–	–	01	–	–	01	–	–	02	0.79
8.	*A. nidulans*	–	02	–	17	05	–	–	–	24	9.56
9.	*A. niger*	–	03	10	10	01	–	01	–	25	9.96
10.	*A. versicolor*	–	01	–	–	–	–	–	–	01	0.39
11.	*Aspergillus* sp.	–	–	07	–	–	07	02	–	16	6.37
12.	*Cladosporium* sp.	–	–	–	–	–	01	01	–	02	0.79
13.	*Drechslera* sp.	–	02	–	–	–	–	–	–	02	0.79
14.	*Fusarium* sp. I	–	01	–	01	01	–	01	02	06	2.39
15.	*Fusarium* sp. II	–	–	01	–	–	–	–	–	01	0.39
16.	*Penicillium lilacinum*	–	–	–	–	–	02	–	04	06	2.39
17.	*P. purpurogenum*	–	–	–	–	–	–	02	–	02	0.79
18.	*P. rugulosum*	–	–	–	–	–	–	–	01	01	0.39
19.	*Penicillium* sp. IV	–	–	08	04	12	–	03	–	27	10.75
20.	*Penicillium* sp. V	–	–	02	04	12	–	–	01	19	7.56
21.	*Penicillium* sp. VI	–	–	–	01	–	–	–	–	01	0.39
22.	*Pycnidial* sp. I	–	–	03	06	–	06	–	–	15	5.97
23.	*Pycnidial* sp. II	–	–	–	04	–	04	–	06	14	5.57
24.	*Trichoderma viridae*	–	–	04	–	–	03	03	01	11	4.38
	Unidentified group										
25.	Unknown sp. I	03	–	–	–	–	–	–	–	03	1.19
26.	Unknown sp. II	01	–	01	02	–	–	–	11	14	5.57
27.	Unknown sp. III	–	–	01	–	–	–	–	–	01	0.39
28.	Unknown sp. IV	–	–	–	–	–	–	–	01	01	0.39
	Sterile mycelia										
29.	Sterile mycelia	–	03	01	–	–	–	–	–	04	1.59

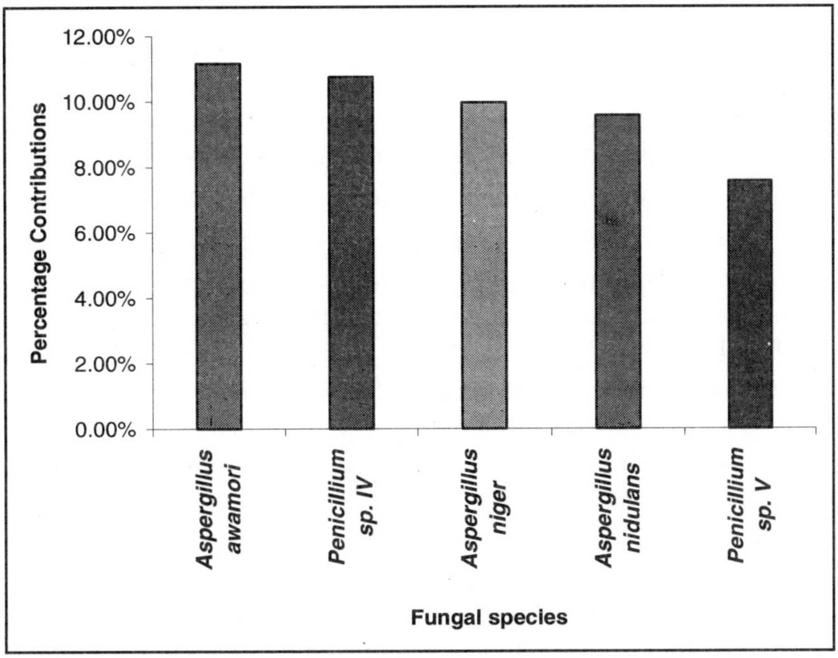

Figure 42.2: Maximum Percentage Contribution Fungal Species from Forest Soil Mycoflora

It is also observed that *Aspergillus awamori, A. niger, Fusarium* sp. I are most frequent fungi while *Cunninghamella* sp., *Aspergillus versicolor*, *Drechslera* sp., *Fusarium* sp. II etc are showing least frequent fungi.

Chaudhary and Sachar (1934), Saksena (1955), Miller *et al.* (1957) studied seasonal variation in forest soil fungi and pointed out seasonal changes in soil mycoflora and fungal population, which drastically differ from season to season. Tresner *et al.* (1954), Persiani *et al.* (1998) also observed seasonal variations in forest soil mycoflora. Dwivedi (1966) discussed seasonal variations of fungal population in some soil types.

Among all isolated fungal species, the Anamorphic fungi are dominated in forest soil. *Aspergillus* and *Penicillium* sp. are the most dominated genus to the total fungal flora. These observations in the present investigation are similar to Manoharachary *et al.* (1990), Kamal and Bhargava (1973), Rane and Gandhe (2006), who have reported that the *Aspergillus* occur more frequent in soil mycoflora, similar results are also reported by Amin and Abdalla (2004) and observed that *Aspergillus niger, A. fumigatus* and *A. nidulans* are most dominant fungi. Most frequent genera are *Aspergillus* and *Penicillium* is also recorded by Abdel Hafez (2004) at Saudi Arabia.

References

Abdel Hafez, S.I.I (2004). Survey of the mycoflora of desert soils in Saudia Arabia. *Mycopathologia*. 80 (1): 3–6.

Amin, E.N.E and Abdalla, M.H. (2004). Survey of soil fungi from the Sudan Gezira. *Mycopathologia*. 71 (3): 131–136.

Barnett, H.L. (1969). *Illustrated genera of Imperfect fungi.* (Mins Burgess Pub. Co.).

Chaudhary, H. and Sachar, G.S. (1934). A study of fungus flora of Punjab soils. *Annales of Mycology*. 32: 90--110.

Dwivedi, R.S. (1966). Ecology of soil fungi of some grasslands of Varanasi–II, distribution of soil mycoflora. *Bulletin of International Society of Tropical Ecology*. 7: 84–99.

Ellis, M.B. (1971). Diatomaceous Hyphomycetes. Kew, England Common Wealth Mycological Institute.

Gilmans, J.C. (1959). *A manual of soil fungi*. Oxford and IBH Pub. New Delhi, India.

Kamal and Bhargava, K.S (1973). Studies on soil fungi from tak forest of Gorakhpur. Edaphic factors and distribution of soil micro fungi in teak stands of different ages. *Proceedings of National Academy of Science*. 43(B): 9–16.

Madhusudan Rao, M. and Manoharachary, C. (1981). Studies on the myco-ecology of two pond muds and forest soils of Mannaur forest, Andhra Pradesh. *Indian Journal of Botany*. 4(2): 137-148.

Manoharachary, C., Reddy, V.R.T. and Prasad, U. (1990). Distribution and phenology of soil fungi supporting forest vegetation. *Geophytology*. 20 (2): 58–70.

Miller, J.H, Giddins, J.E. and Foster, A.A. (1957). A survey of fungi of forest and cultivated soils of Georgia. *Mycologia*. 49: 779–808.

Mohanty, R.B. and Panda, T. (1994). Ecological studies of the soil micro fungi in a tropical forest of South Orissa in relation to deforestation and cultivation. *Journal of Indian Botanical Society*. 73: 213–215.

Nagmani, A, Kunwar, I.K. and Manoharachary, C. (2006). *Handbook of Soil fungi*. I. K. International Pvt. Ltd. New Delhi.

Persiani, A.G., Maggi, M.A. and Pineda, F.D.C. (1988). Diversity and variability in soil fungi from a distributed tropical rain forest. *Mycologia*. 90 (2): 206–214.

Rane, G. and Gandhe, R.V. (2006). Seasonal distribution of soil fungi from forest soils of Jalgaon district, Maharashtra. *Zoos Print Journal*. 21 (9): 2407–2409.

Saksena, S.B. (1955). Ecological factors governing the distribution of soil micro fungi in some forest soils of Sagar. *Journal of Indian Botanical Society*. 34: 262–298.

Tresner, H.D., Bakus, M.P. and Curtis, J.T. (1954). Soil micro fungi in relation to hardwood forest continuum in Southern Wiskinson. *Mycologia*. 46: 314–333.

Venugopal Rao, K., Rao, N.K. and Manoharachary, C. (1984). Thermophilic fungi from forest soils of Andhra Pradesh. *Indian Botanical Report*. 3 (2): 158-161.

Microbial Biotechnology and Ecology (2011) *Pages* 729–740
Editors: Deepak Vyas, G.S. Paliwal, P.K.Khare and R.K Gupta
Published by: DAYA PUBLISHING HOUSE, NEW DELHI

Chapter 43

Study of Life-Forms and Biological Spectrum of Forest of Sagar District, Madhya Pradesh

☆ *A.S. Thakur[1] and P.K. Khare[2]**

[1]*Department of Botany, Govt College, Khurai, Sagar – 470 117, M.P.*
[2]*Department of Botany, Dr. H.S. Gour Vishwavidyalaya, Sagar – 470 003, M.P.*

ABSTRACT

Present study was carried out in tropical dry deciduous vegetation of Sagar district selecting ten study sites having uniform composition. Vegetation was studied taking samples of three different sizes for trees, shrubs and herbs. Life forms occurring in the vegetation were recorded and biological spectrum of the entire district was constructed. Life forms of the angiosperms belonging to 54 families, 130 genera and 160 species have been studied. Life forms in order of importance were phanerophytes (55 per cent), therophytes (31.87 per cent), geophytes (6.25 per cent), chamaephytes (5 per cent), epiphytes (1.25 per cent) and hemicrytophyte (0.62 per cent). The dominance of phanerophytes and therophytes reveals that phytoclimate of the area as Phanero-therophytic.

Keywords: Biological spectrum, Phanero-therophytic phytoclimate, Dry deciduous forest.

Introduction

Life form is an important character in describing vegetation which represents a preliminary picture of the ecological character of the vegetation (Cain, 1950). Life forms also help in ecological

---

E-mail: pramodpkk@rediffmail.com; p.k.khare@gmail.com

interpretation of the vegetation of a region (Sharma and Singh, 1995). Raunkiaer (1934) defined the life forms as the sum of adaptation of plants to the climate. Adaptation of plants to the survival of unfavorable seasons, especially with regard to the position of perennating buds on plants and degree of their protection during adverse condition. The life form classes reflects the stratification of different layers above and further indicate how a plant passes the unfavorable seasons for growth (Rao, 1968).

The concept of life forms in the study of vegetation proposed for the first time by Humboldt (1806) who suggested groupings of vegetational types on the physiognomic basis. Later, several workers like Kerner (1890), Grisebach (1873) and Drude (1890) have emphasized the dependence of life forms on climate and assessed the role of species in vegetation, with special reference to duration of protection to the perennating organs and mode of propagation. Raunkiaer's system (1934) of classification of life forms is the most widely accepted one and has been universally followed. Raunkiaer's life form spectrum was modified by Braun-Blanquet (1951) and Ellenberg and Muller-Dombois (1967).

The ratio of life forms of different species in terms of number or percentage in any floristic community is called "Biological spectrum" or "the spectrum of life forms" (Milne and Milne, 1971) which can be used to calculate numerical results quickly (Lemon, 1962) to indicate the stratification and layering pattern of community (Rao, 1968; Krebs, 1994), to indicate the prevailing environment (Kotiwar *et al.*, 1996), its aridity or humidity (Meher-Homji, 1964), to monitor the impact of ambient stress factors on climate (Palit *et al.*, 2002) and to determine the nature of bioclimate or phytoclimate (Malik *et al.*, 2006).

Geographically, widely separated plant communities can be usefully compared with one another on the basis of their biological spectrum. Since life forms are related to the environment, biological spectrum is also an indicator of the prevailing environment (Meher-Homji, 1981). The occurrence of similar biological spectra in different regions indicate similar climatic conditions. The bioclimate of the regions according to Raunkiaer is characterized by life form(s) whose percentage value is much higher than the normal spectrum. This indicates the predominance of a particular type of climate and other condition that favour the development of the life form(s) in higher proportions; however due to biotic disturbance, life form may change.

The present chapter describes the biological spectrum of tropical dry deciduous forests in Sagar district of Madhya Pradesh which can be regarded as well as for the Central Indian region.

Study Area

Sagar district is located in the north of Narmada valley. Total geographical area of the district is 10,252 sq. km. The general elevation of the district varies from *Ca.* 400m to 580m above mean sea level (MSL). It is situated between 23°10' and 24°27' latitude North and from 78°04' to 79°22' longitude East. The climate of Sagar district is monsoonal. The whole year can be divisible into three well marked seasons; *i.e.* rainy (mid June to September), winter (October to February) and summer (March to mid June). Average annual rainfall is *Ca* 1200 mm, almost 90 per cent of the rainfall is received in rainy season and winter rains are of common occurrence. The maximum temperature is recorded during summer (45°C) and mean minimum temperature during winter (12.5°C).

Natural vegetation of the district comprises of tropical dry deciduous type after Champion (1936) and Champion and Seth (1968). Forests envelope 26.95 per cent of geographical area. In general, vegetation of the district is teak dominating. Mainly two major rock formation *viz.* Basalt or deccan trap and Vindhyan sandstone are found in the district. Black soil derived from weathering of basalt rock covers a wide area in district. This soil supports good vegetation with dominance of teak (*Tectona* randis) (Thakur and Khare, 2006). Red soil occur in the area covered by Vindhyan sandstone, it enerally support poor vegetation.

laterial and Methods

Floristic analysis of all the study sites was carried out after sampling of vegetation by taking 100 andom quadrats of 10 × 10 m, 5 × 5 m and 1 × 1 m for trees shrubs and herbs species, respectively. A st of all angiospermic species including climbers and epiphytes encountered in sampling area was repared. For the construction the biological spectrum life forms of all plant species were determined n the basis of the position of perennating buds following Raunkiaer (1934) and Muller-Dombois and llenberg (1974). The number of species in each life form class were calculated. The percentage of pecies belonging to each life forms was determined by the following formula.

$$\text{Percentage of species in life form} = \frac{\text{Number of species in the life form}}{\text{Total number of species of all life forms}} \times 100$$

Biological spectrum for the district was prepared on the basis of percentage species composition ı each life form and taking all the study sites.

esults and Discussion

Floristic composition of vegetation showed that a total number of 160 species of angiospermic lants were encountered during the sampling of vegetation. Out of these, 73 species were of trees, 14 f shrubs, 65 of herbs, 11 of climbers and 2 of epiphytes. They belong to 54 families out of which 47 ere dicotyledons and 07 were monocotyledons. Depending upon the number of inclusion of species, mily wise top position upheld by Fabaceae (22 species). Other dominating families are Euphorbiaceae, steraceae and Rubiaceae (Table 43.1).

The vegetation of Sagar district showed highest percentage of phanerophytes (55 per cent), other roups of life forms are in the order of importance were therophytes (31.87 per cent), geophytes (6.25 er cent), chamaephytes (5 per cent), epiphytes (1.25 per cent) and hemi-cryptophytes (0.62 per cent) able 43.2 and Figure 43.1). Similar proportions were found at different study sites of the district. esults showed dominace of phanerophytes and therophytes having contributions of 47.05-59.59 ıd 30.30-42.18 per cent respectively (Table 43.2).

In the present investigation, phanerophytes showed maximum divergence from the normal ectrum as given by Raunkiear, accordingly the phytoclimate of the area may be termed as phanero-erophytic. Similar phytoclimatic association has also been reported by Rajendraprasad *et al.* (1998) om sacred groves of Kerala, Lakshmanan (1962) for Vihar lake Bombay, Misra *et al.* (1979) for ahmyoni hill, Gaya, Saxena (1980) for North west catchment of Gola river, Nainital, Saxena *et al.* 982) for oak forest of Kumaon and Khatri (2000) for Satpura National Park M.P.

According to Meher-Homji (1964) the life forms are reflected by bioclimate of the area. Thus in ımid regions the bioclimate should be phanerophytic, in arid regions therophytic and in semiarid gions thero-chamaephytic. In Sagar district, both phanerophytes and therophytes share considerably ıportance in depicting the phytoclimate of dual character extremes *i.e.* warm-moist and warm-dry imate.

In the present study therophytes stand next to phanerophytes with approximately more than half e phanerophtes at most of sites. The predominance of therophytes due to grazing is a common ıenomenon (Yadav and Singh, 1977). It is also an indicator of biotic pressures (Barucha and Dave,

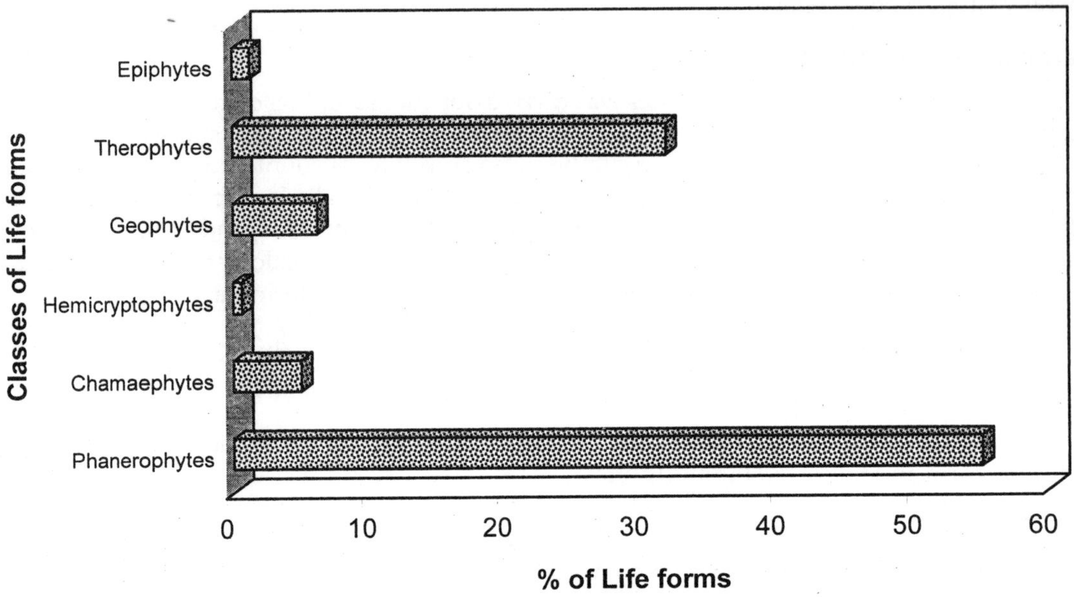

Figure 43.1: Biological Spectrum of Sagar District

1944). Pandeya (1964) observed that the life forms of the flora of different grassland association c Sagar were maintained by intensity of grazing. The high percentage of therophytes in grassland is du to overgrazing resulting in the introduction and spread of weedy grasses (Cain, 1950). Down (197 also recorded higher number of therophyte in bare and early successional stands. According t Dansereau (1957) the predominance of therophytes indicates warm climate.

The growth of therophyte was much favoured in disturbed areas (Keeley and Albert, 1977; Vor and George, 1987). The dominance of therophytes has also been reported by several workers i grassland and disturbed forests (Srivastava, 1944; Sarup, 1952; Pandeya, 1964; Trivedi and Sharm 1965; Singh, 1967; Singh and Yadava, 1974; Siddiqui 1972; Singh and Ambasht, 1975; Chara *et al.*1978; Ansari and Singh, 1979; Misra and Misra, 1979; Verma and Das 1980; Ali and Dixit, 198 Rajwar and Gupta, 1981; Srivastava *et al.*, 1981; Sharma and Dhakre, 1993; Singh, 1995; Kunhikanna 1999; Baruah, 2003; Kakati and Yadava, 2005 and Subramani *et al.*, 2007).

Percentage composition of chamaephytes varied from 1.66 to 7.84 at different study sites (Tab 43.2). They retain their regeneration capacity in favorable season from the above ground and we protected perennating buds and showed hardy nature of the plants for which they can withstan long period of adverse climatic conditions. A number of species such as *Tectona grandis, Diospyr melanoxylon, Terminalia tomentosa, Butea monosperma* and other showed similar behaviour as the sprout after cutting in rainy season. Coppicing is a common character in woody species found i tropical dry deciduous forests.

Table 43.1: Life Forms and Habit Classification of Plants Found in Different Study Sites

Sl.No.	Family	Sl.No.	Name of Plant Species	Life Form	Habit
			Dicotyledons		
1.	Annonaceae	1.	*Annona squamosa* Linn.	Ph	Tree
		2.	*Miliusa tomentosa* (Roxb.) J.Sinclair	Ph	Tree
2.	Menispermaceae	3.	*Cissampelos pareira* Linn.	Ph	Climber
		4.	*Cocculus hirsutus* (Linn.) Diels	Ph	Climber
3.	Bixaceae	5.	*Casearia tomentosa* Roxb.	Ph	Tree
		6.	*Flacourtia indica* (Burm.F.) Merr.	Ph	Tree/Shrub
4.	Malvaceae	7.	*Hibiscus solandra* L' Herist.	Th	Herb
		8.	*Kydia calycina* Roxb.	Ph	Tree
		9.	*Malvastrum tricuspidatum* A. Grey.	Th	Herb
		10.	*Salmalia malabarica* (DC) Schott and Endl.	Ph	Tree
		11.	*Sida rhombifolia* Linn.	Ch	Herb
		12.	*Sida spinosa* Linn.	Th	Herb
		13.	*Sida veronicaefolia* Lamk.	Ch	Herb
5.	Sterculiaceae	14.	*Helicteres isora* Linn.	Ph	Shurb
		15.	*Sterculia urens* Roxb.	Ph	Tree
6.	Tiliaceae	16.	*Corchorus actangulus* Lam.	Th	Herb
		17.	*Triumfetta neglecta* W. and A.	Th	Herb
		18.	*Triumfetta rhomboidea* Jacq.	Th	Herb
7.	Euphorbiaceae	19.	*Acalypha indica* Linn.	Th	Herb
		20.	*Bridelia retusa* Spreng.	Ph	Tree
		21.	*Emblica officinalis* Gaerth.	Ph	Tree
		22.	*Euphorbia geniculata* Orteg.	Th	Herb
		23.	*E. hypericifolia* Linn.	Th	Herb
		24.	*E. hirta* Linn.	Th	Herb
		25.	*Flueggea microcarpa* Baill.	Ph	Tree
		26.	*Phyllanthus deblis* Ham.	Th	Herb
		27.	*P. simplex* Retz.	Th	Herb
		28.	*P. urinaria* Linn.	Th	Herb
8.	Geraniaceae	29.	*Biophytum sensitivum* DC.	Th	Herb
9.	Rutaceae	30	*Aegle marmelos* Correa.	Ph	Tree
		31.	*Feronia elephantum* Correa.	Ph	Tree
10.	Simarubaceae	32.	*Ailanthus excelsa* Roxb.	Ph	Tree
		33.	*Balanites roxburghi* Planch.	Ph	Shrub
11.	Burseraceae	34.	*Boswellia serrata* Roxb.	Ph	Tree
		35.	*Garuga pinnata* Roxb	Ph	Tree
12.	Meliaceae	36.	*Azadirachta indica* Juss.	Ph	Tree

Contd...

Table 43.1–Contd...

Sl.No.	Family	Sl.No.	Name of Plant Species	Life Form	Habit
13.	Celastraceae	37.	*Gymnosporia spinosa* Forsk.	Ph	Shrub
		38.	*Elaeodendron glancum* Pers.	Ph	Tree
14.	Rhamnaceae	39.	*Ventilago maderaspatana* Gaerth.	Ph	Climber
		40.	*Zizyphus oenoplia* Mill.	Ph	Shrub
		41.	*Zizyphus jujuba* Lamk.	Ph	Tree
		42.	*Zizyphus rotudifolia* Lamk.	Ph	Shrub
		43.	*Zizyphus xylopyrus* Willd.	Ph	Tree/Shrub
15.	Vitaceae	44.	*Vitis trifolia* Linn.	G	Climber
	(Amplidaceae)	45.	*Leea aspera* Edgw.	G	Climber
16.	Sapindaceae	46.	*Sapindus laurifolius* Vahl.	Ph	Tree
		47.	*Schleichera oleosa* (Lour.)	Ph	Tree
17.	Anacardiaceae	48.	*Buchanania lanzan* Spreng.	Ph	Tree
		49.	*Lannea coromandelica* (Houth.)	Ph	Tree
		50.	*Semecarpus anacardium* Linn.	Ph	Tree
18.	Papilionaceae	51.	*Abrus precatorius* Linn.	Ph	Climber
		52.	*Alysicarpus monilifer* DC.	Th	Herb(legume)
		53.	*Alysicarpus tetragonolobus* Edgew.	Th	Herb(legume)
		54.	*Butea monosperma* (Lamk) Taub.	Ph	Tree
		55.	*Crotalaria prostrata* Roxb.	Th	Herb
		56.	*Dalbergia lanceolaria* L.F.	Ph	Tree
		57.	*Dalbergia paniculata* Roxb.	Ph	Tree
		58.	*Desmodium diffusum* D.C.	Th	Herb(legume)
		59.	*Desmodium gangeticum* D.C.	Ch	Herb(legume)
		60.	*Desmodium triflorum* D.C.	Th	Herb(legume)
		61.	*Erythrina indica* Lamk.	Ph	Tree
		62.	*Indigofera cordifolia* Heyne.	Th	Herb(legume)
		63.	*Indigofera glandulosa* Willd.	Th	Herb(legume)
		64.	*Indigofera trifoliata* Linn.	Th	Herb(legume)
		65.	*Ougeinia dalbergioides* Benth.	Ph	Tree
		66.	*Pongamia pinnata* (Linn.) Pierre.	Ph	Tree
		67.	*Pterocarpus marsupium* Roxb.	Ph	Tree
		68.	*Tephrosia purpurea* Pers.	Ch	Herb(legume)
		69.	*Tephrosia villosa* Pers.	Ch	Herb(legume)
		70.	*Teramnus labialis* (Linn.F.) Spreng.	Th	
		71.	*Xylia xylocarpa* (Roxb.) Tanb.	Ph	Tree
		72.	*Zornia diphylla* (Linn.) Pers.	Th	Herb (legume)

Contd...

able 43.1–Contd...

Sl.No.	Family	Sl.No.	Name of Plant Species	Life Form	Habit
19.	Caesalpiniaceae	73.	*Bauhinia retusa*	Ph	Tree
		74.	*Cassia fistula* Linn.	Ph	Tree
		75.	*Cassia pumila* Lamk.	Th	Herb(legume)
		76.	*Cassia tora* Linn.	Th	Herb(legume)
		77.	*Delonix regia* Raf.	Ph	Tree
		78.	*Tamarindus indica* Linn.	Ph	Tree
20.	Mimosaceae	79.	*Acacia catechu* Willd.	Ph	Tree
		80.	*Acacia leucophloea* Willd.	Ph	Tree
		81.	*Acacia nilotica* (Linn.) Del.	Ph	Tree
		82.	*Albizzia lebbek* Benth.	Ph	Tree
		83.	*Albizzia odoratissima* Benth.	Ph	Tree
21.	Combretaceae	84.	*Anogeissus latifolia* Wall.	Ph	Tree
		85.	*Anogeissus pendula* Edgew.	Ph	Tree/Shrub
		86.	*Terminalia arjuna* W. and A.	Ph	Tree
		87.	*Terminalia belerica* (Gaearth) Roxb.	Ph	Tree
		88.	*Terminalia tomentosa* W. and A.	Ph	Tree
22.	Myrtaceae	89.	*Eugenia heyneana* Wall.	Ph	Tree
23.	Lecythidaceae	90.	*Careya arborea* Roxb.	Ph	Tree
24.	Lythraceae	91.	*Lagerstroemia parviflora* Roxb.	Ph	Tree
		92.	*Woodfordia fruticosa* Kurg.	Ph	Shrub
25.	Onagraceae	93.	*Ludwigia parviflora* Roxb.	Th	Herb
26.	Cucurbitaceae	94.	*Bryonopsis laciniosa.*	Th	Herb
27.	Rubiaceae	95.	*Adina cordifolia* Hook.F.	Ph	Tree
		96.	*Borreria stricta* Linn. F.	Th	Herb
		97.	*Gardenia latifolia* Aiton.	Ph	Tree
		98.	*Ixora parviflora* Vahl.	Ph	Tree
		99.	*Mitragyna parvifolia* Korth.	Ph	Tree
		100.	*Morinda tinctoria* Roxb.	Ph	Tree
		101.	*Randia spinosa* (Thumb.) Keay.	Ph	Tree/Shrub
		102.	*Wendlandia puberula* D.C.	Ph	Tree
28.	Asteraceae	103.	*Ageratum conyzoides* Linn.	Th	Herb
		104.	*Bidens biternata* (Lour.) Merr. and Sherff	Ch	Herb
		105.	*Elephantous scaber* Linn.	G	Herb
		106.	*Lagascea mollis* Cav.	Th	Herb
		107.	*Pulicaria angustifolia* DC.	Th	Herb
		108.	*Sphaeranthus indicus* Linn.	Th	Herb
		109.	*Tridax procumbens* Linn.	Th	Herb

Contd...

Table 43.1–Contd...

Sl.No.	Family	Sl.No.	Name of Plant Species	Life Form	Habit
		110.	*Xanthium strumarium* Linn.	Th	Herb
		111.	*Vernonia cinerea* Linn.	Th	Herb
29.	Sapotaceae	112.	*Madhuca indica* Gmel.	Ph	Tree
30.	Ebenaceae	113.	*Diospyros melanoxylon* Roxb.	Ph	Tree
31.	Oleaceae	114.	*Nyctanthes arbortristis* Linn.	Ph	Tree
		115.	*Schrebera swietenoides* Roxb.	Ph	Tree
32.	Apocynaceae	116.	*Carissa spinarum* Linn.	Ph	Shrub
		117.	*Holarrhena antidysentrica* Wall.	Ph	Tree/Shrub
		118.	*Wrightia tinctoria* Linn.	Ph	Tree
33	Asclepiadaceae	119.	*Gymnema sylvestre* (Retz) R.Br.	Ph	Climber
		120.	*Hemidesmus indicus* (Linn.) Schultz	Ph	Climber
34.	Gentianaceae	121.	*Canscora diffusa*	Th	Herb
35.	Boraginaceae	122.	*Cordia vestita*	Ph	Tree
		123.	*Ehretia laevis* Roxb.	Ph	Tree
36.	Convolvulaceae	124.	*Convolvulus pluricaulis* Chois.	G	Herb
		125.	*Evolvulus alsinoides* Linn.	G	Herb
		126.	*Ipomaea coccinea* Linn.	G	Climber
		127.	*Merremia emarginata* (Burm.F.) Hau.F.	Th	Herb
37.	Bignoniaceae	128.	*Stereospermum suaveolens* DC.	Ph	Tree
38.	Acanthaceae	129.	*Barleria prionitis* Linn.	Ch	Herb
		130.	*Blepharis boerhaavifolia* Pers.	Ch	Herb
		131.	*Justicia diffusa* Willd.	Th	Herb
		132.	*Justicia simplex* Don.	Th	Herb.
39.	Verbenaceae	133.	*Gmelina arborea* Roxb.	Ph	Tree
		134.	*Lantana camara* Linn.	Ph.	Shrub
		135.	*Tectona grandis* Linn.	Ph.	Tree
		136.	*Vitex negundo* Linn.	Ph	Shrub
40.	Labiatae	137.	*Leucas aspera* Spreng.	Th	Herb
41.	Nyctaginaceae	138.	*Boerhaavia diffusa* Linn.	G	Herb
42.	Amarantaceae	139.	*Achyranthes aspera* Linn.	Th.	Herb
		140.	*Celosia argentea* Linn.	Th	Herb
43.	Proteaceae	141.	*Loranthus longiflorus* Desr.	E	Epiphyte
44.	Ulmaceae	142.	*Holoptelea integrifolia* Planch.	Ph	Tree
45.	Santalaceae	143.	*Santalum album* Linn.	Ph	Tree
46.	Moraceae	144.	*Ficus bengalensis* Linn.	Ph	Tree
		145.	*Ficus glomerata* Roxb.	Ph	Tree
		146.	*Ficus infectoria* Roxb.	Ph	Tree
47.	Loganiaceae	147.	*Mitreola oldenlandioidea* Wall.	Th	Herb

Con

Table 43.1–Contd...

Sl.No.	Family	Sl.No.	Name of Plant Species	Life Form	Habit
			Monocotyledons		
48.	Orchidaceae	148.	*Vanda roxburghii* R.Br.	E	Epiphyte (orchid)
49.	Dioscoreaceae	149.	*Dioscorea bulbifera* Linn.	G	Climber
50.	Commelinaceae	150.	*Murdannia nudiflorum* (Linn.) Brenon.	G	Herb
51.	Palmaceae	151.	*Phoenix sylvestris* Roxb.	Ph	Tree
52.	Cyperaceae	152.	*Cyperus aristatus* Rottb.	Th	Sedge
53.	Zinziberaceae	153.	*Curcuma amada* Roxb.	G	Herb
54.	Gramineae	154.	*Eragrostis interrupta*	Th	Grass
		155.	*Eragrostis pilosa* (Linn.) P. Beauv.	Th	Grass
		156.	*Eragrostis tenella* Linn.	Th	Grass
		157.	*Iseilema anthephoroides* Hack.	H	Grass
		158.	*Ophlismemus burmannii* (Retz. P. Beauv.)	Th	Grass
		159.	*Setaria glauca* (Linn.) P. Beauv.	Th	Grsss
		160.	*Sporobolus diander* (Retz.) P. Beauv.	Th	Grass

Table 43.2: Biological Spectra of All the Study Sites

Sl.No.	Name of the Sites	Life Forms					
		Ph (%)	Ch (%)	H (%)	G (%)	Th (%)	E (%)
1.	Patharia	59.59	4.04	1.01	2.02	33.33	-
2.	Mainpani	59.09	4.54	-	6.06	30.30	-
3.	Jaruwakheda	54.54	6.81	-	2.27	36.36	-
4.	Rahatgarh	47.05	7.84	1.96	5.88	37.25	-
5.	Bahadurpur	52.45	6.55	-	6.55	34.42	-
6.	Ramna	58.33	1.66	-	5.00	35.00	-
7.	Ranipura	52.23	7.46	1.49	4.47	34.32	-
8.	Gopalpura	48.43	4.68	1.56	3.12	42.18	-
9.	Baraytha	56.09	6.09	1.21	3.65	31.70	1.21
10.	Nauradehi	57.14	5.71	-	2.85	31.42	2.85
	Total Vegetation of Sagar District	55.00	5.00	0.62	6.25	31.87	1.25

Ph: Phanerophytes; Ch: Chamaephytes; H: Hemicryptophytes; G: Geophytes; Th: Therophytes; E: Epiphytes.

Percentage of geophytes in the present study ranged from 2.02 to 6.55 showing a relatively long period of dryness. More humid areas with fertile upper soil horizon favour the development of perennating organ of geophytes. Agarwal and Paul (1996) confirmed this observation while studying the biological spectrum of flora of Chhotanagpur. Geophytes are good indicators of soil moisture of the regime. This group is first emergents in any vegetation after first rain and adapted to withstand long period of adverse conditions. The storage of food in perennating organs is also important for it survival (Sharma and Singh, 1995).

Hemi-cryptophytes recorded at some of the sites with low percentage value from 1.01 to 1.96 an in total vegetations it was 0.62 per cent (Table 43.2).

Due to alternation of long dry period with moderate rainfall, warm dry and warm moist climat are characteristics of this area. These climatic conditions are responsible for phanero-therophyti phytoclimate. It is experienced that vegetation in a stress of biotic pressure gradually increase th percentage of therophytes. It is pertinent to state that the composition of phanerophytes and therophyte is close in this area, an increase in biotic pressure would change the biological spectrum to therophytic phanerophytic.

References

Agarwal, V. and S.R. Paul, (1996). Biological spectrum of Chotanagpur flora, *Indian J. Forestry*, 19(1) 35-39.

Ali, S.J. and Dixit, S.N. (1980). Biological spectrum of Lachhmipur forests, *Geobios*, 7(4), 187-188.

Ansari, A.A. and Singh S.K. (1979). Biological spectrum of the Madhaulia forest of Gorakhpur, *Indian J. Forestry*, 2(2), 153-157.

Baruah, P.P. (2003). Biological spectrum of the flora of sand bars (Chapories) of Brahmaputra river *Indian J. Forestry*, 26(2), 172-176.

Bharucha, F.R. and Dave, R.N. (1944). The biological spectrum of grassland association, *J. Univ Bombay*, 13, 15-16.

Braun-Blanquet, J. (1951). *Planzensoziologie*. Springer-Verlag, Vienna, 879 pp.

Cain, S.A. (1950). Life-forms and Phytoclimates, *Bot. Rev.*, 16, 1-32.

Champion, H.G. (1936) A preliminary survey of the forest types of India and Burma, *Indian For. Rec* (N.S.), Vol. 1.

Champion, H.G. and S.K. Seth (1968). A revised survey of forest types of India, Govt. of India Publication, Delhi.

Charan, A.K., Sen, D.N. and Roypurohit, K.S. (1978). Biological spectrum of the vegetation of western Rajasthan Desert India. *Indian J. Forestry*, 1, 313-315.

Dansereau, P. (1957). *Biogeography: An Ecological Perspective*, Ronand Press, New York.

Down, C.G. (1973). Life form succession in plant communities on colliery waste, *Environmental Pollution* 5, 19-22.

Drude, O. (1890). *Hondbuch der Pflanzengeographie*, Stuttgart J. Engelhorn, 582 pp.

Ellenberg, H. and D. Muller-Dombois (1967). A key to the Raunkiaer plant life-forms with revised subdivision, *Ber. Geobat. Inst. Eidg. Tech. Hochsch. Stift. Rubel.*, 37: 56-73.

Grisebach, A. (1873). Die Vegetation der Erde nach ihrer, klimatischen in del Schweiz, Beiter Geol *Schweiz. Geotch Ser.*, 17.

Humboldt. A.V. (1806). *De distributione geographica plantanum secundum coeli temperiem et altitudinem montiuna prolegomena* Paris, 28 pp.

Kakati, L.N. and Yadava, P.S. (2005). Floristic composition, Life form and Biological Spectrum of a Grassland Ecosystem at Imphal, Manipur, *Indian Journal of Ecology*, 32(2), 176-179.

Keeley, S.C. and Albert, W.J. (1977). A comparison of pattern of herb, shrub growth in comparable sites of Chili and California, *Am. Hidl. Nat.*, 97, 120-132.

Kerner, M.A. (1890). *Das Pflanzenleben der Donaulander,* 2nd ed. by F. Vierhapper, 1929. University Verlag Wagner, Innsbruck 452 pp.

Khatri, P.K. (2000). Study on biodiversity in tropical forest ecosystem of Satpura National Park, Madhya Pradesh, Ph.D. Thesis Forest Research Institute, Dehra Dun.

Kotiwar, O.S., Pandit, B.R. and Pahurkar, A.J. (1996). Life forms and biological spectrum of the flora of Gir forest, Gujarat, *Geobios New Reports*, 15, 17-20.

Krebs, C.J. (1994). *Ecology: The Experimental Analysis of Distribution and Abundance,* Harper Collins, California, 467 pp.

Kunhikannan, C. (1999). Studies on vegetation ecology of Tadoba National Park, Chandrapur, Maharastra, Ph.D. Thesis Forest Research Institute, Dehra Dun.

Lakshmanan, N.K. (1962). The application of Raunkiaer's life forms, *Journal of Indian Botanical Society,* 41, 585-589.

Lemon, P.C. (1962). *Field and Laboratory Guide for Ecology,* Burgess, Minn, 154 pp.

Malik, S., Bhattacharya, A. and Mukherjee, A. (2006). Biological spectrum of Chandur Forest in Hooghly District, West Bengal, *Indian Journal of Forestry*, 29(2), 145-148.

Meher-Homji, V.M. (1964). Life-forms and biological spectra as epharmonic criteria of aridity and humidity in tropics, *J. Indian Bot. Soc.*, 43(3), 424-430.

Meher-Homji, V.M. (1981). Environmental implications of life form spectra from India, *Journal of Economic Taxonomic Botany*, 2, 23-30.

Milne, L. and Milne, M. (1971). *The Arena of Life. The Dynamics of Ecology.* Double day\Natural History Press, Garden City, New York, 240 pp.

Mishra, M.K. and Mishra, B.N. (1979). Biological spectrum of tropical grassland community of Berhampur, *Indian J. Forestry,* 2, 313-315.

Misra. M.P., Mishra P.N. and Pandey B.N. (1979). An ecological account of the vegetation of Brahmyoni Hills (Gaya), *Proceedings of Indian Science Congress,* 3, 108-109.

Mueller-Dombois, D. and Ellenberg, H. (1974). *Aims and methods of vegetation ecology,* John Wiley and Sons, New York, 525 pp.

Palit, D., Ganguly, G. and Mukherjee, A. (2002). Ecological studies in Ramnabagan Wildlife Sanctuary Burdwan I. Biological spectrum, *Sci. and Cult.*, 68(5-6), 147-149.

Pandeya, S.C. (1964). Ecology of grassland of Sagar M.P. Composition of the fenced grassland association, *J. Indian Bot. Soc.*, 43, 577-605

Rajendraprasad, M., Krishnan, P.N. and Pushpangandan, P. (1998). The life from spectrum of sacred groves-a functional tool to analyse the vegetation, *Tropical Ecology*, 39(2), 211-217.

Rajwar, G.S. and Subodh Gupta, (1981). Flora and biological spectrum of Khoh valley district Pauri (Garhwal), *Indian J. Forestry*, 4(1), 49-55.

Rao, C.C. (1968). Biological spectrum of Karamnasa watershed flora (Varanasi, India). *Proc. Symposium on recent advances in Tropical Ecology*, Part-II (R. Misra and B. Gopal, eds). Banaras Hindu University, Varanasi.

Raunkiaer, C. (1934). *The life form of plants and statistical plant geography*, Being the collected papers of C. Raunkiaer, Oxford, Clarendon Press, 632 pp.

Sarup, S. (1952). The biological spectrum of the flora of Mt. Abu. *Univ. Rajasthan Stub. Biol. Sci.*, 1.

Saxena, A.K., Pandey, P. and Singh, J.S. (1982). Biological spectrum and other structural functional attributes of vegetation of Kumaun Himalaya *Vegetatio.*, 49, 111-119.

Saxena. A.K. (1980). Ecology of vegetation complex of north western catchment of river Gola, Progress report Ecology Research Circle, Dept. of Botany, Kuamaon University, Nainital, pp. 1-4.

Sharma, M. and Singh, H. (1995). *Life forms and biological spectrum of the flora of Champa district (Himachal Pradesh)*, In: Panday, A.K (ed). *Taxonomy and Biodiversity*. CBS Publishers and Distributors, Delhi, pp. 81-86.

Sharma, S.C. and Dhakre, J.S. (1993). Life form classification and biological spectrum of the flora of Shahjahanpur district, Uttar Pradesh (India), *Ind. J. Forestry*, 15(4), 366-371.

Siddiqui, M.O. (1972). Biological spectrum of the forests of Gorakhpur, *Trop. Ecol.*, 3(2), 234-235.

Singh, J.S. (1967). *Seasonal variation in composition, plant biomass and community productivity in the grassland of Varanasi*, Ph.D. thesis, B.H.U., Varanasi.

Singh, J.S. and Yadava, P.S. (1974). Seasonal variation in composition, plant biomass and net primary productivity of a tropical grassland at Kurukhetra, India, *Ecol. Monogr.*, 44, 351-376.

Singh, L.B. (1995). Biological spectrum of the flora of Munger district Bihar, *Taxonomy and Biodiversity*, Ed. A.K. Pandey CBS Publishers and Distributors, Delhi, pp. 105-107.

Singh, V.N. and Ambasht, R.S. (1975). Biotic stress and variability in structure and organic (net primary) production of grass land communities at Varanasi, India, *Trop. Ecol.*, 16, 85-95.

Srivastava, G.D. (1944). The biological spectrum of the Allahabad flora, *J. Ind. Bot. Soc.*, 23, 1-7.

Srivastava, M.M., Srivastava, P.K. and Kukreti, U.D. (1981). Biological spectrum of a marshy grassland at Rajpur, Dehradun (U.P.) India. *Indian J. Forestry*, 4(3), 220-223.

Subramani, S.P., Jishtu V., Verma, R.K. and Kapoor, K.S.(2007). Floristic composition, life-forms and biological spectrum of renuka wildlife sanctuary, Himachal Pradesh, *The Indian Forester*, 133(1), 79-92.

Thakur, A.S. and Khare, P.K. (2006). Species diversity and dominance in tropical dry deciduous forest ecosystem, *J. Environ. Res. Development*, 1, 26-31.

Trivedi, B.S. and Sharma, P.C. (1965). Biological spectrum of Lucknow flora, *Proc. Nat. Acad. Sci. India*, 35B(1), 15-20

Verma, R. and Das R.R. (1980). A new approach towards the use of Raunkiaer's biological spectrum, *Tropical Ecology*, 21(1), 9-15.

Vora, A.B. and George, V.C. (1987). The distribution of various life forms in the ground flora under different canopies of Panchamahal forests: Gujarat, India, *Indian J. Forestry*, 10, 223-225.

Yadav, J.S.P. and Singh, J.S. (1977). *Grassland Vegetation, its structure, function utilization and management*, Today and Tomorrows Printers and Publishers, New Delhi, 132 pp.

Microbial Biotechnology and Ecology (2011) Pages 741–765
Editors: Deepak Vyas, G.S. Paliwal, P.K.Khare and R.K Gupta
Published by: DAYA PUBLISHING HOUSE, NEW DELHI

Chapter 44

Environmental Impact of Fly Ash on Soil Health, Yield and Nutrient Uptake by Rice

☆ *T. Thomas*[1], *Ram Bharose*[1], *Arun A. David*[1], *H. Thomas*[2]* *and R.K. Tamphasana*[1]

[1]*School of Forestry and Environment, SHIATS-Deemed University, Allahabad*
[2]*Department of Applied Geology, Dr. H.S. Gour University, Sagar – 470 003, M.P.*

ABSTRACT

The research was carried out in the farm of Allahabad Agricultural Institute Deemed University-Allahabad UP. The result generated from the present study show the beneficial effect of fly ash due to essential plant nutrients and fly ash applied @ 40 t ha^{-1} along with $N_{120}P_{60}K_{60}$ was found the most effective. Hence, it is reported that the application of fly ash could be an alternative source of plant nutrients with fertilizers and biofertilizers for sustaining soil fertility status vis-à-vis crop productivity. However, on the basis of one season experimental results, recommendations can not be made; hence further research on this aspect needs to be done for recommendation to the farmers.

Keywords: *Fly ash, Nutrients, Soil health.*

Introduction

Agriculture is the key to the global agenda on population, environment, poverty reduction, and food security. The world's affluent obtain vital protein through foods such as meat, eggs and dairy

* E-mail: harelthomas@yahoo.com

products, but millions of world poor get most of their protein through rice. It is the foremost cereal crop of the world providing 22 per cent of the world's supply and 17 per cent of protein. About 60 per cent people rely on rice as their staple food.

Rice has 13.7 per cent moisture, 6.8 per cent protein, 0.6 per cent minerals, 0.5 per cent fat, 0.2 per cent fibre, 78.2 per cent carbohydrates and 10mg calcium, 160 mg phosphate, 3.1 mg iron and small amount of vitamin B complex (Bhakra 1996).

Rice is grown in an area of more than 135 million hectares in the world. India has the largest area of about 42 million hectares under rice, with irrigation facilities to cover 43 per cent of the total area. The rice production during 1999-2000 was 88.0, million tones against 84.74 million tones in 1998-99 (71st Annual General Meeting of I.C.A.R. Society, New Delhi, 15 September 2000).

The crop removes the major portion of plant nutrients from soil by crop harvesting. Unless the soils are supplied with nutrients removed by the crop, it will be great threat to sustain crop production. The use of chemical fertilizers is the quickest and surest way of boosting crop production. However, their cost and other constraints frequently deter farmers from using them in recommended quantities and in balanced proportions. As consequence of this and other constraints, there seems to be no option but fully exploit potential alternative sources of plant nutrients. Complementary use of plant nutrients from waste material and biofertilizers along with mineral fertilizers is of great importance for soil productivity, *i.e.* soil structure, soil bio-activity, cation exchange capacity and water holding capacity (Roy 1994).

Fly ash, grey in colour, abrasive, acidic/alkaline and refractory in nature, is a by-product of thermal power stations where electrical energy is obtained by firing finely powdered coal. Huge amount of fly ash is produced everyday at every thermal power station in the country. As it stands today, about 100 million tones of fly ash are generated per year form the 82 thermal power stations in the country. It has established beyond doubt that fly ash can be advantageously used in agriculture as soil conditioner-improving upon some important physico-chemical properties of soil, such as hydraulic conductivity, bulk density, porosity, water holding capacity, etc. fly ash has also been successfully used as a source of essential plant nutrients like phosphorous, potassium, calcium, magnesium, sulphur, zinc, copper, chlorine, boron, iron, manganese, molybdenum and thus, boosting crop growth and yields (Aitken *et al.*, 1984; Kene *et al.*, 1991).

Fly ash has been successfully applied in different agro-climatic conditions and soil types in different parts of the country using different doses (25 to 500 t/ha). It has been used for the cultivation of different types of crops like paddy, wheat, maize, and mustard, til, sugarcane, cotton, vegetables and fruits. Repeated field trails and residual increase in the crop yield (20 to 50 per cent) with high nutritional value, *i.e.* protein and oil, etc. without any significant carry out of trace/heavy metals and radioactivity in the soil, water bodies and crop produce. Scientists of central fuel research institute have established and successfully demonstrated gainful utilization of fly ash as modifier of agricultural soil and as a source of some plant nutrients. It has the potential to alter the soil texture by increasing the percentage of silt sized particle and thus permanently increasing the soil moisture holding capacity. Fly ash inhibits the infiltration rate of water in sandy type soils and enhances the infiltration rate of clay rich soil due to higher proportion of silt content in fly ash.

Fly ash for its abundant availability and in addition, due to its ameliorating property and a good source of nutrients imparts the demand of an inexpensive product management technology for its fruitful utilization. Its integrated use with chemical fertilizers and biofertilizers would open new avenues to improve soil potential and would boost up the agricultural productivity, simultaneously saving the environment from ill hazards.

Hence, the present investigation entitles "ENVIRONMENTAL IMPACT OF FLY ASH ON SOIL HEALTH, YIELD AND NUTRIENT UPTAKE BY RICE" was undertaken to explore the feasibility of using fly ash as component of integrated plant nutrients system (IPNS) for sustaining soil productivity and crop yield, with the following objectives:

1. To study the effect of various levels of fly ash on yield and nutrient uptake by rice.
2. To study the changes in chemical properties of soil due to fly ash.

Materials and Methods

The details of the materials used and technologies adopted during the course of this investigation with brief description of soil, cropping history of field, planting materials used and climatic conditions prevalent in the locality while this experiment was being conducted is being dealt in this chapter.

Experimental Site

This experiment was conducted during kharif/monsoon season of the year 2001 in the crop research farm of department of agronomy, Allahabad Agricultural Institute- Deemed University, Allahabad and was carried out in plot number 100. The experimental area is very near to the river Yamuna and approximately 5 Km from Allahabad city.

Soil

Soil samples (0 to 15 cm depth) were collected randomly prior to tillage operations, air dried and passed through 2 mm sieve and a composite soil sample was taken for mechanical and chemical analysis. Bouyoucos Hydrometer method were used for the mechanical analysis of soil, to determine sand, silt and clay fraction in the sample. The result of mechanical and chemical analysis is summarized in Table 44.1.

Table 44.1

Analysis	*Particulars*	*Result*	*Method Mechanical and Chemical Analysis of Soil (Pre-sowing)*
Mechanical Analysis	Sand	60.80 per cent	Hydrometer method (Bouyoucos 1952)
	Silt	24.10 per cent	
	Texture of soil	Sandy Loam	
Chemical Analysis	Soil pH (1 : 2)	8.00	Digital pH meter. MK VI
	EC	0.29 dSm^{-1}	Electrical conductivity meter.
	Organic carbon	5.70 g kg^{-1}	Walkley and Black method (1934).
	Available N	270 kg ha^{-1}	Kjeldahl method (Subbiah and Asija 1956).
	Available P	20.40 kg ha^{-1}	OLSEN spectrophotometric method (OLSEN *et al.*, 1954)
	Available K	225.50 kg ha^{-1}	Flame photometric method (Jackson 1973).

Climate

Allahabad has a subtropical and semi-arid climate with temperature reaching the extremes during summer and winter. Rainfall is mostly experienced from July to September. Geographically Allahabad is situated at an altitude of 98 meter sea level at 25.57° North latitude and 81.5° East longitude.

Cropping History

The sequential cropping history of experimental plot is recorded for last years to get an idea about the different species grown. The cropping history summarized in Table 44.2.

Table 44.2: Cropping History

Year	*Cropping History*		
	Kharif	*Rabi*	*Zaid*
1996-97	Maize	Wheat	Fallow
1997-98	Rice	Mustard	Urd
1998-99	Moong	Winter Maize	Fallow
1999-2000	Tile Niger	Wheat	Moong
2000-2001	Maize	Mustard and Maize	Moong

Experimental Details

Plan of Layout

The experiment was laid out in a 3 × 3 factorial design with three levels of fly ash and three levels of fertilizers and their combination. The treatments were replicated three times and were allocated randomly in each replication.

Details of Treatment

The details of the treatments with the notation used under each of them are given in Table 44.3.

Table 44.3

Factors	*Treatments*	*Notation*
Levels of Fly ash	0 t ha^{-1}	F_0
	20 t ha^{-1}	F_1
	40 t ha^{-1}	F_2
Levels of Fertilizers	0 kg ha^{-1}	M_0
	N_{120} P_{60} K_{60} kg ha^{-1}	M_1

Details of Treatment Combinations

Treatments and combination are given in Table 44.4.

Table 44.4

Treatment	*Combination*	*Description*
T_0	F_0M_0	Control
T_1	F_0M_1	$N_{120}\,P_{60}\,K_{60}$ kg ha^{-1}
T_2	F_1M_0	20 t ha^{-1}
T_3	F_1M_1	20 t ha^{-1} + $N_{120}\,P_{60}\,K_{60}$ kg ha^{-1}
T_4	F_2M_0	40 t ha^{-1}
T_5	F_2M_1	40 t ha^{-1} + $N_{120}\,P_{60}\,K_{60}$ kg ha^{-1}

Details of Field Operation

Raising of Nursery

The site of nursery was chosen at Crop Research Farm, Allahabad Agricultural Institute-Deemed University, Allahabad. The nursery was prepared by wet bed method. The field was irrigated, ploughed and puddle. Later the field was demarcated and raised into narrow strips. Needed amount of Urea, DAP and MOP was applied to the strips.

Planting Materials

The rice variety "Pusa Basmati-I" was used in the experiment. It is achieved from a cross between IR- 8 × TKM- 6 at IARI, New Delhi. It is an early maturing rice variety with good plant type responsive to fertilizers management. It possesses good tillering ability with highly synchronous habit. It is resistant to blast and moderately resistant to stem borer but highly susceptible to bacterial leaf blight and tungro virus. Yield potential is about 45 to 50 quintals per hectare of grain yield under optimum conditions of crop management.

Broadcasting of Incubated Seed

After the preparation of nursery the soaked and incubated seeds of the rice were uniformly broadcasted on the standing water in the strips. The nursery was irrigated whenever needed to keep the soil moist to facilitate the germination of rice seeds.

Before sowing of nursery by wet method, seeds were treated with streptocycline + carbendazim @ 0.1 per cent.

Pre-Transplanting Operations of Land

The experimental field was ploughed once with a tractor drawn disc plough followed by harrowing. Stubbles and weeds picked up from the field and puddling was done with the help of tractor drawn puddle. Then the plots were demarcated according to the layout and planking was done. The calendar of operations was given in Table 44.5.

Fly Ash Application

The fly ash used in this experiment was brought from IFFCO, Phulpur. The physical and chemical characteristics of the fly ash are given in Table 44.6.

Table 44.5: Calendar of Pre-transplanting Operations

Operations Done	*Date*	*Remarks*
Ploughing	25.07.2001	By tractor
Harrowing	27.07.2001	By tractor
Puddling	28.07.2001	By tractor
Demarcation and layout	29.07.2001	By manual labour
Fly ash application	31.07.2001	By manual labour
Fertilizer application (Basal dressing)	01.08.2001	By manual labour
Transplanting	02.08.2001	By manual labour

Table 44.6: Physical and Chemical Analysis of Fly Ash to be Observed

Mechanical Analysis	*Unit*	*Results*
Physical characteristics		
Sand	per cent	77.80
Silt	per cent	19.99
Clay	per cent	2.21
Soil texture		Sandy
Porosity	per cent	35.61
Bulk density	g cm^{-3}	0.84
Hydraulic conductivity	m/day	11.04
Chemical characteristics		
pH (1:2)		7.70
EC	dSm^{-1}	0.41
Organic Carbon	per cent	0.64
Total Nitrogen	per cent	0.07
Total Phosphorus	per cent	0.35
Total Potassium	per cent	2.50
Total Sulphur	per cent	2.00
Total Calcium	per cent	0.05
Copper	ppm	33.40
Zinc	ppm	167.00
Manganese	ppm	33.40
Iron	ppm	5.40
Boron	ppm	22.50
Cobalt	ppm	33.40
Molybdenum	ppm	10.24

Fly ash was applied at the rate of 0, 20 and 40 t ha^{-1} to the fly ash treated plots respectively. The fly sh applied was added in surface (0 to 15 cm) soil and mixed thoroughly with the soil within the plots fter the demarcation and layout.

'ertilizer Application

Nitrogen, phosphorus and potassium were applied at the rate of 100, 60 and 60 kg per hectare, espectively. Urea and DAP was used as a source of nitrogen and phosphorus. Potassium chloride MOP) was used as a source of potassium. Half of the nitrogen and total quantity of phosphorus and otassium were applied before the transplanting. The remaining dose of nitrogen was top-dressed in wo equal splits before panicle initiating and flowering stage.

'ransplanting

Seedlings were uprooted from the nursery and transplanted on the standing water using 2 to 3 eedling per hill at a depth of 2 cm with a spacing of 20 cm between row-to-row and 15 cm between lant-to-plant.

Table 44.7: Post-transplanting operations

Sl.No.	Operation	Date
1.	1st Irrigation	07.08.2001
2.	Gap filling	09.08.2001
3.	2nd Irrigation	13.08.2001
4.	3rd Irrigation	16.08.2001
5.	4th Irrigation	18.08.2001
6.	5th Irrigation	26.08.2001
7.	Hand weeding	27.08.2001
8.	6th Irrigation	28.08.2001
9.	1st top dressing	30.08.2001
10.	$ZnSO_4$ spraying	07.09.2001
11.	7th Irrigation	17.09.2001
12.	2nd top dressing	24.09.2001
13.	8th Irrigation	29.09.2001
14.	9th Irrigation	11.10.2001
15.	10th Irrigation	23.10.2001
16.	11th Irrigation	01.11.2001
17.	Harvesting	24.11.2001
18.	Threshing and winnowing	27.11.2001

bservation Taken in Crop

Random sampling technique was adopted for recording the observation on various morphological aracters of the plant during the course of study. Five plants from each parameter at each stage was mputed and recorded. The frequency of plants on which the observations were taken on the plant owth, yield and nutrient uptake are given in Table 44.8.

Table 44.8: Observations Taken in Crop

Sl.No.	Observation	Frequency
1.	Plant height	5
2.	Tiller count	5
3.	Dry weight	4
4.	Length of panicle (cm)	1
5.	Number of spikelets per panicle	1
6.	Grain yield (q ha^{-1})	1
7.	Straw yield (q ha^{-1})	1
8.	Uptake of Nitrogen by rice grain (kg ha^{-1})	1
9.	Uptake of Phosphorus by rice grain (kg ha^{-1})	1
10.	Uptake of Potassium by rice grain (kg ha^{-1})	1

Pre-Harvest Observation in Crop

Height of Plant

Height was taken from the five tagged plants in each plot. The observations were taken at 20, 4 60, 80 and 100 DAT (at an interval of 20 days). The height of the main stem or culm (according to IAR was taken as the height of plants.

Number of Tiller Per Hill

The numbers of tillers per hill were taken from the same tagged plants whenever the plant heig was recorded.

Dry Weight

Two hills from each plant were uprooted carefully. These hills were air-dried first and kept in t oven at 70°C for complete drying, to get a constant weight. Mean values of these plants were taken dry weight per plant at 25, 50, 75 and 100 DAT.

Length of Panicle

Average length of panicle per hill of five samples was recorded from the panicle base to the tip the same.

Number of Spikelets Per Panicle

From the same plants the number of spikelets per panicle is taken at 100 DAT.

Postharvest Observation

Grain Yield

In order to know the response of different treatments on the grain yield, grain from harvested ar (1 square metre from each plot) was dried in the sun, threshed, winnowed and weighed separatel

Straw Yield

After threshing of grain, the straw was dried and weighed individually for individual plots.

Nitrogen, Phosphorous, Potassium Content and its Uptake by Grain

Grain samples were dried in oven and ground in a Wiley Mill, and sieved. These samples were analysed for nitrogen by modified Kjeldahl's method, total phosphorus and potassium was estimated spectrophotometrically and flame photometrically method, respectively after diacid (HNO_3 and $HClO_4$) digestion as described by Jackson (1973).

The Kjeldahl method for total nitrogen content (per cent) essentially involves: digestion of the sample to convert N compounds in the sample to the ammonium (NH_4^+) form. The grain sample is digested with Sulphuric acid and hydrogen peroxide. Hydrogen peroxide destroys organic material material to minimize foaming. Salt/catalyst mixture (K_2SO_4 + $CuSO_4$ + selenium power) is added to each digestion tube to raise the boiling temperature of the digestion mixture and to shorten the digestion time. After the digestion, digest is cooled to room temperature before dilution. Determination of ammonium in digest: (Kjeldahl method). Few ml of digest (10 ml) is distilled with 40 per cent NaOH of the Kjeldahl flask, which gives ammonium. This liberated ammonia is absorbed by the boric acid solution with mixed indicator. After completion of distillation, the boric acid is back titrated against H_2SO_4.

Total phosphorus content (per cent) of the grain sample was carried out using a di-acid 9:4 mixture of HNO_3 : $HClO_4$. Plant phosphorus is converted to orthophosphates during digestion. These orthophosphates react with molybdate and vanadate and give yellow coloured unreduced vanado-molybdo-phosphoric heteropoly complex in acid medium. The intensity of this colour is directly proportional to the concentration of phosphates present in the sample which can be read on colorimeter or spectrophotometer at 430 nm using blue filter. Lastly phosphorus concentration was found out with the help of standard curve.

For total potassium content (per cent) the procedure of plant sample digestion is same as that of phosphorus using diacid mixture. Potassium in the digested is determined using flame photometer.

The nitrogen, phosphorus and potassium uptake in the rice grain was determined by multiplying the N, P, K content (per cent) with the grain yield.

Results and Discussion

The results of the present investigation entitled "ENVIRONMENTAL IMPACT OF FLY ASH ON SOIL HEALTH, YIELD AND NUTRIENT UPTAKE BY RICE" have been presented and discussed accordingly in this chapter along with its confirmation with previous researchers.

Results for all parameters have been taken, summarized and discussed under the following sub-headings:

1 Pre-harvest studies

2. Postharvest studies

The tables of analysis of variance for different parameter have been given in the appendix.

Pre-Harvest Studies

Plant Height

Plant height was recorded at 20, 40, 60, 80 and 100 days after transplanting and the data recorded were analyzed and results were given below:

Effect of Fly Ash Levels on Plant Height

The statistical analysis of data indicate that there was a significant difference in plant height at 20, 40, 60, 80 and 100 DAT, when the plant were supplied with 20 and 40 tonnes fly ash per hectare against the control.

Table 44.9: Effect of Fly Ash Levels on Plant Height

Treatment	*Plant Height (cm)*				
	20 DAT	*40 DAT*	*60 DAT*	*80 DAT*	*100 DAT*
F_0	18.72	24.75	39.73	77.19	77.25
F_1	20.97	27.31	41.29	79.42	79.18
F_2	24.45	30.18	44.02	81.51	82.22
S.E.	0.35	0.43	0.38	0.55	0.45
C.D at 5 per cent	0.78	0.93	0.80	1.16	0.95

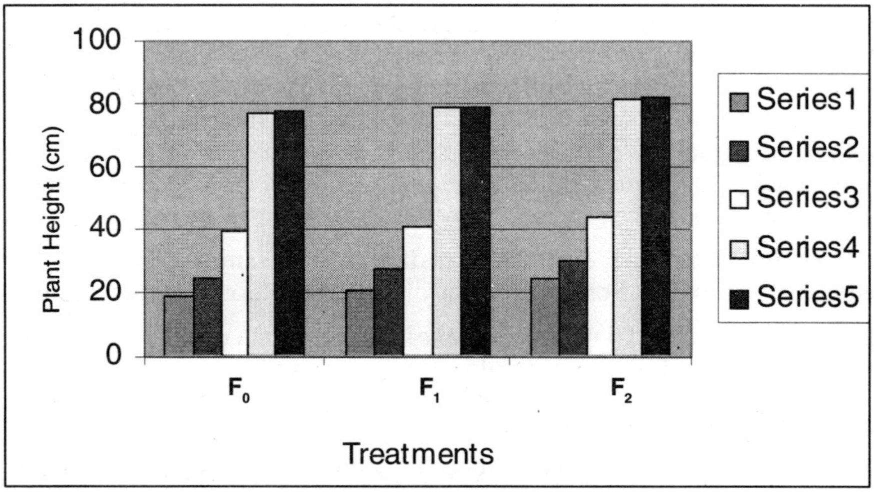

Figure 44.1: Effect of Fly Ash Levels on Plant Height

The effect of fly ash levels on plant height is presented in Table 44.1 and Figure 44.1. The result revealed that the plant height growth was maximum between 60-80 DAT. At 100 DAT, F_2 (Fly ash @ 40 t ha^{-1}) had maximum plant height of 82.22 cm, which was greater than the control (77.25 cm). F_1 (Fly ash @ 20 t ha^{-1}) had also a significant difference over the control. Thus, the plant height showed an increasing trend by increasing levels of fly ash. Increase in plant height with fly ash has also been reported by Dubey *et al.* (1982). Tripathi and Sahu (1997) also reported that 50 per cent fly ash applied to soil increases the plant height.

Number of Tillers Per Hill

The numbers of tillers per hill were recorded at 20, 40, 60, 80 and 100 DAT and results were summarized below:

fect of Fly Ash Levels on Number of Tillers per Hill

The statistical analysis of number of tillers per hill indicate that there was a significant difference . number of tillers per hill at 20, 40, 60, 80 and 100 DAT when the plot were supplied with 20 and 40 nnes of fly ash per hectare against the control.

Table 44.10: Effect of Fly Ash Levels on Number of Tillers per Hill

Treatment	*Number of Tillers per Hill*				
	20 DAT	*40 DAT*	*60 DAT*	*80 DAT*	*100 DAT*
F_0	7.23	12.53	12.66	12.05	11.25
F_1	7.65	13.36	13.38	13.13	12.43
F_2	8.25	14.13	14.03	13.74	13.07
S.E.	0.25	0.40	-	0.31	0.43
C.D at 5 per cent	0.54	0.84	NS	0.65	0.91

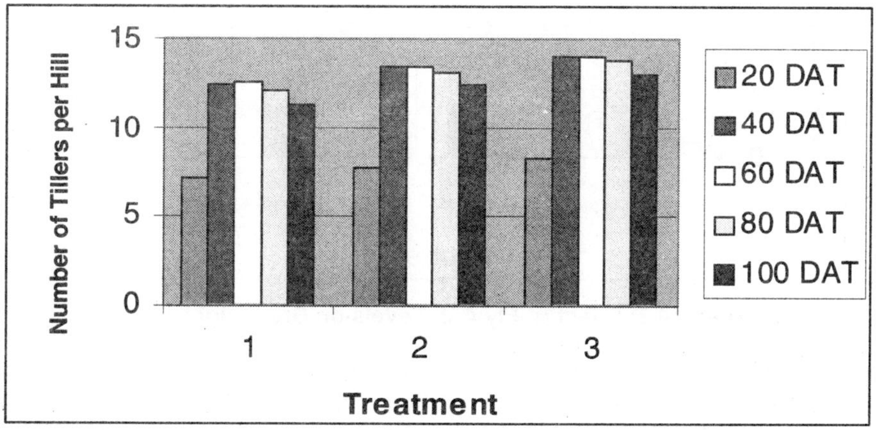

Figure 44.2: Effect of Fly Ash Levels on Number of Tillers per Hill

From the data in Table 44.10 and Figure 44.2. It is revealed that maximum number of tiller (13.07) as recorded with the higher fly ash level (F_2) of 40 t ha^{-1} and minimum (11.25) with the control. The (Fly ash @ 20 t ha^{1}) had also a significant difference from the control, the number of tiller showed a ›sitive response to the increased fly ash level. This might be due to the fact that fly ash increased the ntents of N, P, K in plants at all growth stages. Uptake of these nutrients by the crop also increased ith the application of fly ash. These findings are in conformity with the findings of previous workers ingh *et al.*, 1986 and Kene *et al.*, 1991).

ry Weight

The dry weight was recorded at 25, 50, 75 and 100 DAT and results are present below:

fect of Fly Ash Levels on Dry Weight of Plants

The data presented in Table 44.11 and Figure 44.3 shows that all the successive stages of growth rice shows significant difference in dry weight of plants per hill.

Table 44.11: Effect of Fly Ash Levels on Dry Weight of Plants

Treatment	*Dry Weight (g)*			
	25 DAT	*50 DAT*	*75 DAT*	*100 DAT*
F_0	3.78	11.26	28.91	51.90
F_1	4.30	13.45	32.43	57.56
F_2	7.08	17.03	35.00	60.44
S.E.	0.30	0.17	0.78	0.81
C.D at 5 per cent	0.63	0.36	1.65	1.71

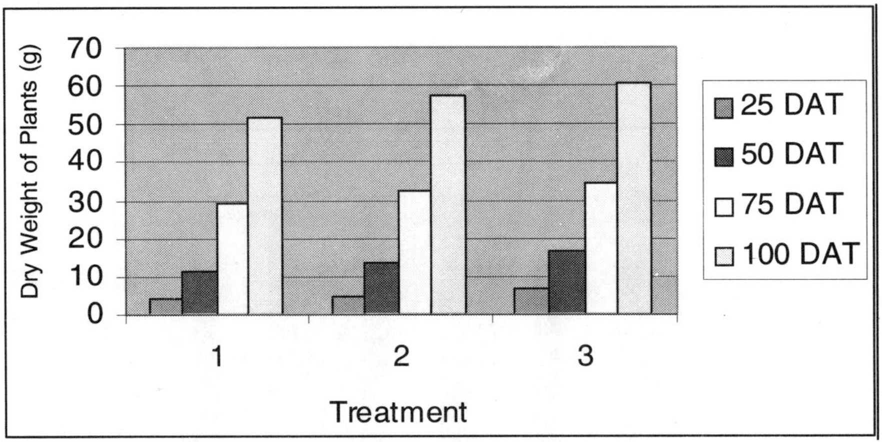

Figure 44.3: Effect of Fly Ash Levels on Dry Weight of Plants

Among the levels tried in this experiment, fly ash @ 40 t ha^{-1} increased the dry weight of plant which was significant over control. The plot supplied with fly ash @ 20 t ha^{-1} was also significant ove control. The results revealed that dry weight growth was maximum between 75-100 DAT. In genera F_2 (fly ash @ 40 t ha^{-1}) had maximum plant dry weight than F_1 (fly ash @ 20 t ha^{-1}). The results shov that at 100 DAT, the maximum dry weight (60.44 g) was recorded in F_2 (fly ash @ 40 t ha^{-1}) plots an the minimum of 50.90 g with F_0 (control) plots. Rajeev *et al.* (1992) reported that mean dry matter yiel increased significantly due to fly ash treatment due to increased inputs of N, S and Fe from fly ash Sikka and Kansal (1995) also reported that fly ash application had a beneficial effect on the dry matte yield of rice.

Length of Panicle

Effect of Fly Ash Levels on Panicle Length

The data presented in Table 44.12 and Figure 44.4 shows that the effect of different levels of fly as was statistically significant.

From Table 44.12 is revealed that there was a significant difference in the panicle length betwee the plots supplied with fly ash @ 20 t ha^{-1} (F_1) and 40 t ha^{1} (F_2) over control. The maximum was foun in F_2 (fly ash 40 @ t ha^{-1}) plots with 28.07 cm followed by 27.40 cm in F_1 plots (fly ash 20 @ t ha^{-1}) an minimum in control 26.73 cm panicle length of rice crop was significantly influenced by differen

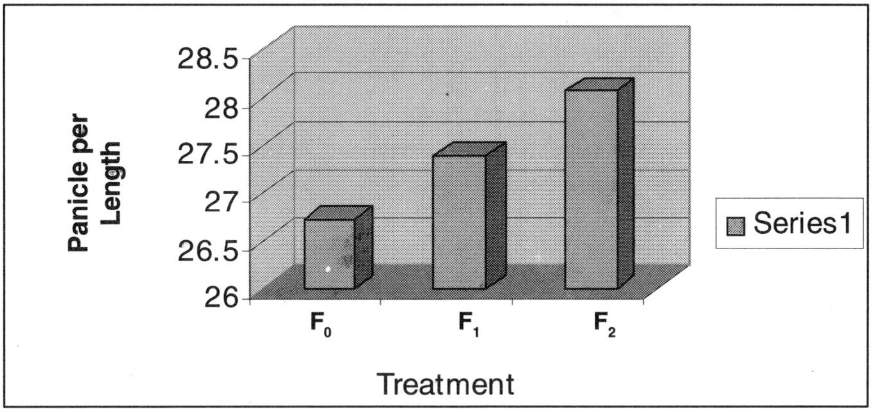

Figure 44.4: Effect of Fly Ash Levels on Panicle Length

levels of fly ash. The maximum length was found with 40 t ha^{-1} fly ash application and minimum in control.

Table 44.12: Effect of Fly Ash Levels on Panicle Length

Treatment	*Panicle Length (cm)*
F_0	26.73
F_1	27.40
F_2	28.07
S.E	0.32
C.D at 5 per cent	0.67

Table 44.13: Effect of Fly Ash Levels on No. of Spikelets per Panicle

Treatment	*No. of Spikelets/Panicle*
F_0	10.27
F_1	10.65
F_2	10.94
S.E	–
C.D at 5 per cent	NS

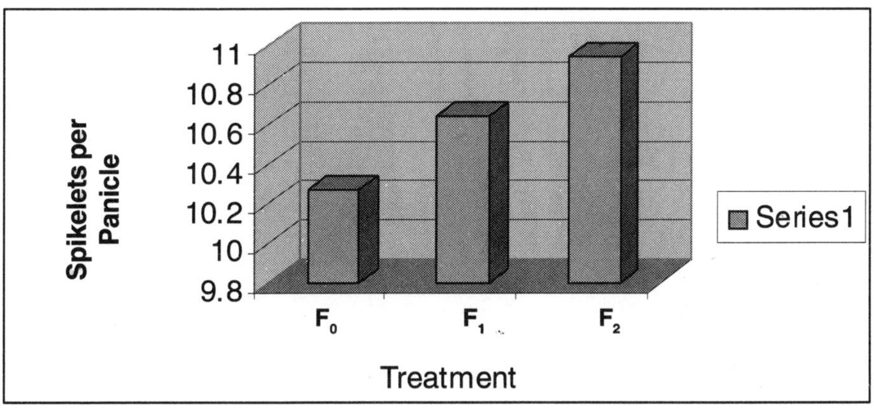

Figure 44.5: Effect of Fly Ash Levels on No. of Spikelets per Panicle

Number of Spikelets Per Panicle

Effect of Fly Ash Levels on Number of Spikelets per Panicle

The number of spikelets per panicle was recorded at final stage of crop growth and it wa statistically analysed. It was revealed that the effect of different levels of fly ash was non-significan From the data in Table 44.13, it is revealed that increasing levels of fly ash from 0-20 t ha^{-1} and from 2 to 40 t ha^{-1} had no effect on the number of spikelets per panicle. These findings are in conformity wit the results reported by Tripathy and Sahu (1997).

Postharvest Studies

Grain Yield

Effect of Fly Ash Levels on Grain Yield

Grain yield was statistically analysed and the results are given in Table 44.14.

The grain yield as influenced are presented in Table 44.14 and Figure 44.6. It is clear from the Table that different levels of fly ash influenced the grain yield of rice significantly over the control. The crop showed increasing trend in yield with increasing level of fly ash. The maximum yield (44.53 q ha^{-1}) was recorded when the crop was treated with fly ash @ 40 t ha^{-1} whereas fly ash @ 20 t ha^{-1} produced 41.84 q ha^{-1}. It observed that the treatment received fly ash @ 20 t ha^{-1} and 40 t ha^{-1} recorded a yield increase of 3.64 and 10.30 percent, respectively over the control treatment. Similar results were reported by Matte and Kene (1995). They observed that application of 20 t fly as per ha gave the best result which increased grain yield by 7 to 30 per cent. Anjali *et al.* (2000) reporte that available N P K and micronutrients like Cu, Fe, Zn, Mn and exchangeable Ca and Mg increase

Table 44.14: Effect of Fly Ash Levels on Grain Yield

Treatment	*Grain Yield (q ha^{-1})*
F_0	40.37
F_1	41.84
F_2	44.53
S.E	0.34
C.D at 5 per cent	0.75

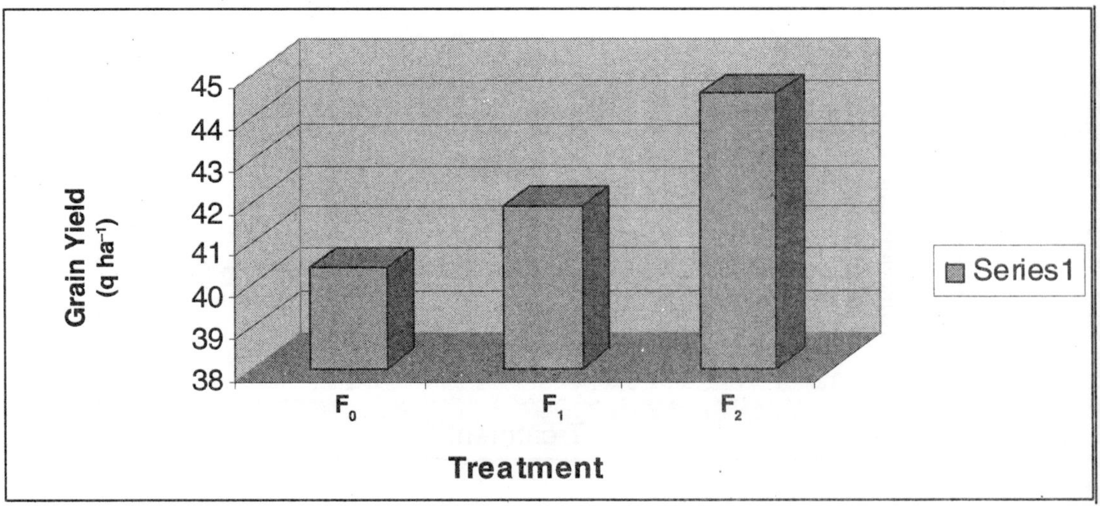

Figure 44.6: Effect of Fly Ash Levels on Grain Yield

with fly ash application which increased the grain yield. Selvakumari *et al.* (2000) also found that treatments which received fly ash at 20 and 40 t ha^{-1} increased the grain yield by 10.3 and 16.3 per cent, respectively over no fly ash application. Also Tiwari *et al.* (1992), Patil *et al.* (1996), Arvind *et al.* (1998), Bharti *et al.* (2001) reported that application of fly ash increased grain yield.

Straw Yield

The effect of different levels of fly ash on straw yield was statistically analysed and the result are given in Table 44.15.

The straw yields are presented in Table 44.15 and Figure 44.7 different levels of fly ash influenced the straw yield of rice significantly over the control. The crop showed increasing trend in yield with each level of fly ash. The maximum straw yield (76.80t ha^{-1}) was recorded with fly ash @ 40 t ha^{-1} whereas fly ash @ 20 t ha^{-1} produced 73.50 q ha^{-1}. There is significant increase in straw yield with both levels of fly ash over control.

Table 44.15: Effect of Fly Ash Levels on Straw Yield

Treatment	*Straw Yield (q ha^{-1})*
F_0	66.35
F_1	73.50
F_2	76.80
S.E	0.45
C.D at 5 per cent	0.95

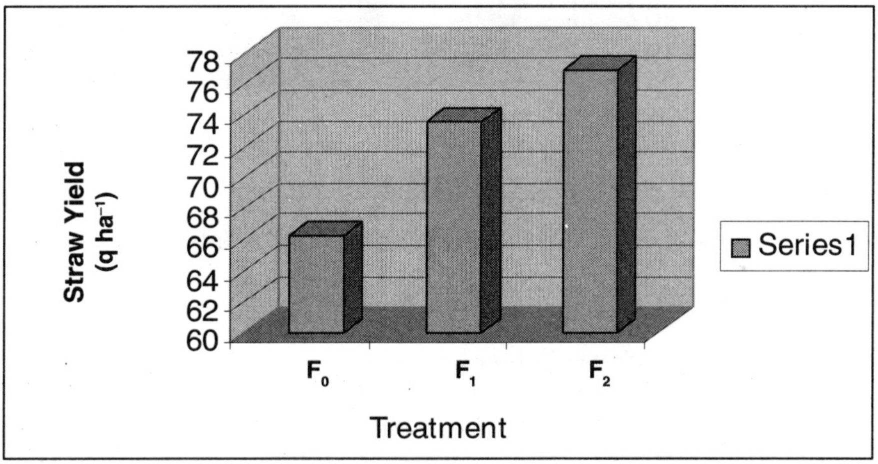

Figure 44.7: Effect of Fly Ash Levels on Straw Yield

Nitrogen, Phosphorus and Potassium Content (per cent) and Uptake by Rice Grain

Nitrogen content in rice grain ranged between 0.98 to 1.59 per cent. Phosphorus content in rice grain ranged between 0.38 to 0.54 per cent. Potassium content in rice grain ranged between 0.22 to 0.41 per cent.

Nitrogen Uptake by Grain

From Table 44.16 and Figure 44.8, the results revealed that different levels of fly ash influenced the N-uptake by rice grain significantly. There was a significant difference in N-uptake between the plot treated with fly ash @ 20 t ha^{-1} and 40 t ha^{-1} over control. The maximum N-uptake (64.33 kg ha^{-1}) was found in plots treated with fly ash @ 40 t ha^{-1} followed by plots treated with fly ash @ 20 t ha^{-1}

(56.82 kg ha^{-1}) and minimum (49.64 kg ha^{-1}) in control. The analysis showed that application of fly ash @ 20 and 40 t ha^{-1} recorded an increase of 7.18 and 14.69 kg ha^{-1} of N-uptake over control. Singh *et al*. (1986) reported that application of fly ash significantly increased the content of N P K at all growth stages. Thus, uptake of nutrients increases both in grains and straw, due to fly ash addition. Sikka and Kansal (1995) also reported that addition of fly ash resulted in a significant increase in the content of N.

Table 44.16: Effect of Fly Ash Levels on Nitrogen Uptake by Grain

Treatment	*N-uptake (kg ha^{-1})*
F_0	49.64
F_1	56.82
F_2	64.33
S.E	0.53
C.D at 5 per cent	1.06

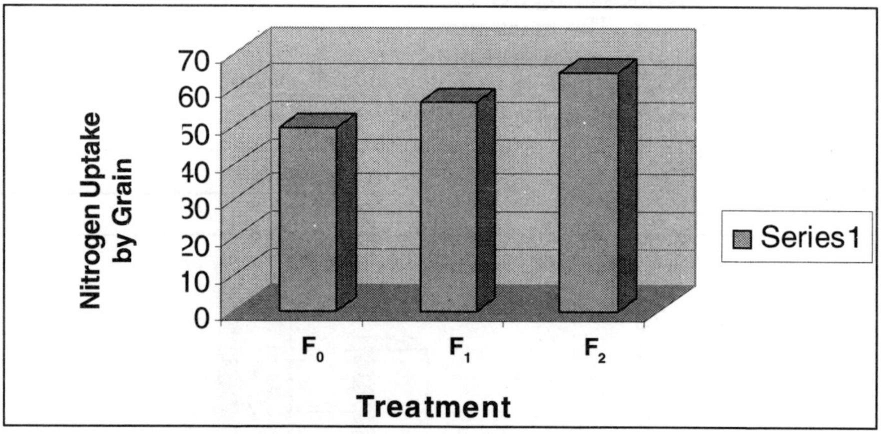

Figure 44.8: Effect of Fly Ash Levels on Nitrogen Uptake by Grain

Phosphorus Uptake by Grain

Effect of Fly Ash Levels on Phosphorus Uptake by Grain

From Table 44.17 and Figure 44.9 the results revealed that different levels of fly ash influence the P-uptake by rice grain significantly. There was a significant difference in P-uptake between the plots treated with fly ash @ 20 and 40 t ha^{-1} over control. The maximum P-take (22.34 kg ha^{-1}) was found in plots treated with fly ash @ 40 t ha^{-1} followed by (20.41 kg ha^{-1}) in plots treated with fly ash @ 20 t ha^{-1} and minimum (18.94 kg ha^{-1}) in control. The analysis showed that the application of fly ash @ 20 and 40 t ha^{-1} recorded an increase of 1.47 and 3.62 kg ha^{-1} of P-uptake over control. Lanjewar (1989), Warambhe *et al*. (1992) and Raghupathy (1993) also reported that increased uptake of P by plant with the application of fly ash.

Table 44.17: Effect of Fly Ash Levels on Phosphorus Uptake by Grain

Treatment	*P-uptake (kg ha^{-1})*
F_0	18.94
F_1	20.41
F_2	22.34
S.E	0.38
C.D at 5 per cent	0.80

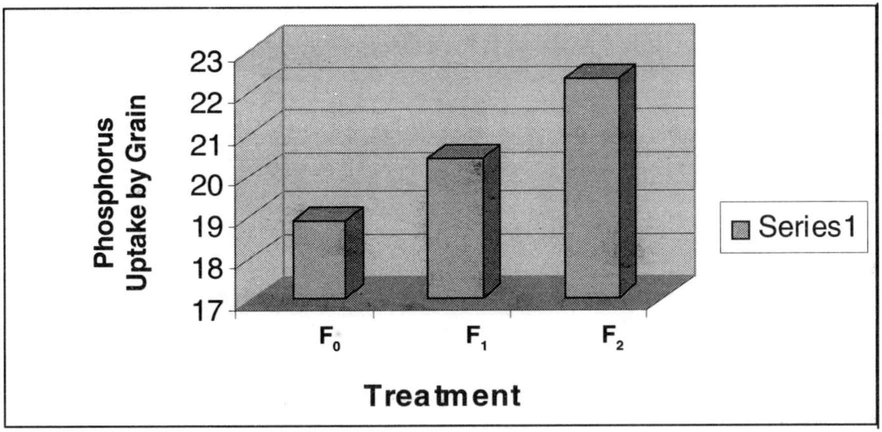

Figure 44.9: Effect of Fly Ash Levels on Phosphorus Uptake by Grain

Potassium Uptake by Grain

Effect of Fly Ash Levels on Potassium Uptake by Grain

From the Table 44.18 and Figure 44.10, the result revealed that different levels of fly ash influenced the K-uptake by rice grain significantly. The plots supplied with fly ash @ 40t ha^{-1} showed a significant increase in K-uptake over control. The maximum K-uptake (16.31 kg ha^{-1}) was found in plots treated with fly ash @ 40 t ha^{-1} and minimum (12.19 kg ha^{-1}) in control. The difference in K-uptake with fly ash @ 20 t ha^{-1} over control was non-significant, but the difference between fly ash @ 20 t ha^{-1} and 40 t ha^{-1} remained significant. The application of fly ash @ 40 t ha^{-1} recorded an increase of 4.12 kg ha^{-1} of K-uptake over control. Grewal *et al.* (2001) reported that application of fly ash increased uptake of N P K both in grain and straw of pearl millet and wheat crops. Increase in K-uptake of corn and bean due to application of fly ash was also reported by Adriano *et al.* (1978).

Table 44.18: Effect of Fly Ash Levels on Potassium Uptake by Grain

Treatment	*K-uptake (kg ha^{-1})*
F_0	12.19
F_1	12.78
F_2	16.31
S.E	0.45
C.D at 5 per cent	0.95

Soil pH

Effect of Fly Ash Levels on pH of the Postharvest Soil

The initial pH of the soil was 8.00. there was decrease in pH of the postharvest soils due to fly ash (pH 7.70) addition as compared to control. The pH decreased from 7.87 to 7.71 due to addition of fly ash @ 20 t ha^{-1}. However, the decrease in pH was not significant. The above findings are in accordance with that reported by Tiwari *et al.* (1992) and Khandar *et al.* (1996). Also Naveen *et al.* (2000) reported that pH of soil decreased with ash content for sandy-loam soils.

Table 44.19: Effect of Fly Ash Levels on pH of the Postharvest Soil

Treatment	*Soil pH*
F_0	7.87
F_1	7.71
F_2	7.81
S.E	-
C.D. at 5 per cent	NS

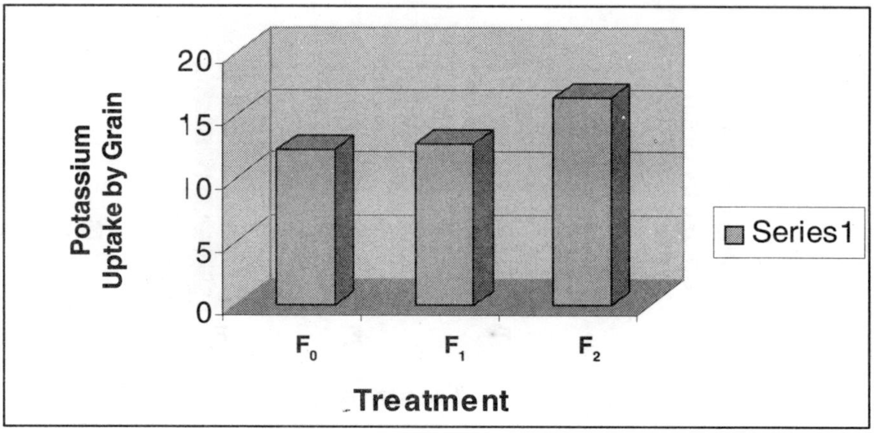

Figure 44.10: Effect of Fly Ash Levels on Potassium Uptake by Grain

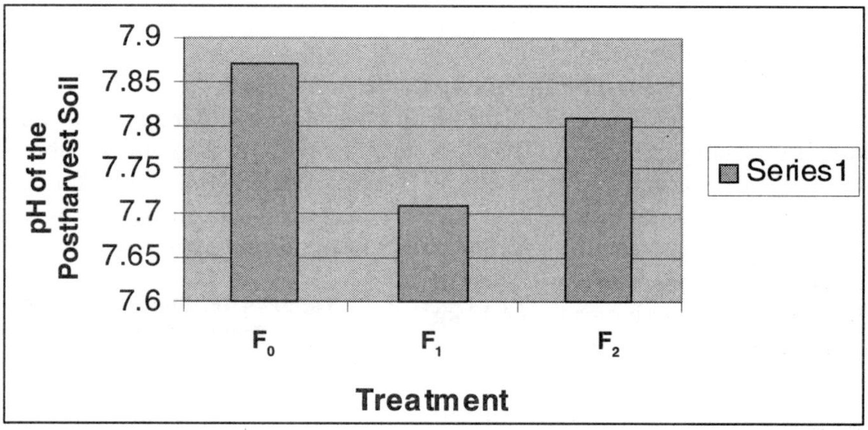

Figure 44.11: Effect of Fly Ash Levels on pH of the Postharvest Soil

Electrical Conductivity (EC)

Effect of Fly Ash Levels on EC of the Postharvest Soil

The initial EC of soil was 0.29 dSm^{-1}. The EC of fly ash was 0.41 dSm^{-1}. The analysis of data indicated non-significant difference in EC for fly ash. The addition of graded levels of fly ash showed no marked influence on EC of the postharvest soil. The EC of plots supplied with fly ash @ 20 t ha^{-1} was 0.20 dSm^{-1} and in control 0.21 dSm^{-1}. This application of fly ash has no marked effect on EC in postharvest soils. These observation corroborate with the earlier work reported by Anjali *et al*. (2000).

Table 44.20: Effect of Fly Ash Levels on EC of the Postharvest Soil

Treatment	*EC (dSm^{-1})*
F_0	0.21
F_1	0.20
F_2	0.21
S.E	-
C.D. at 5 per cent	NS

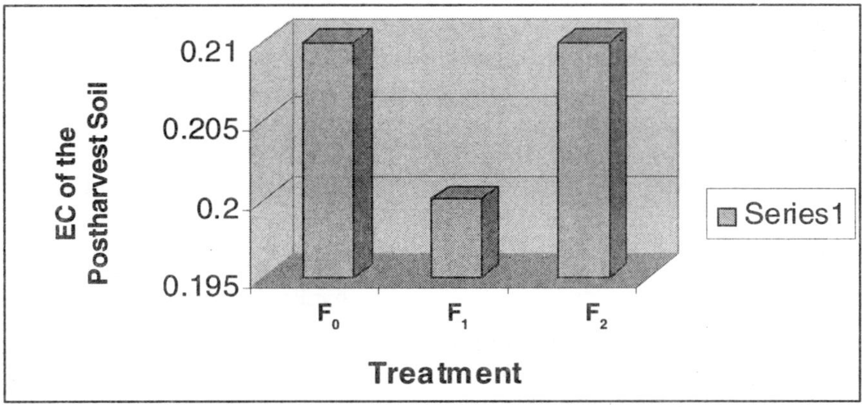

Figure 44.12: Effect of Fly Ash Levels on EC of the Postharvest Soil

Organic Carbon

Effect of Fly Ash Levels on Organic Carbon Content of Postharvest Soil

The analysis of data indicated significant increase in organic carbon content of the postharvest soils due to fly ash addition as compared to control. The organic carbon content in postharvest soil ranged from 5.34 to 5.86 g kg^{-1} with lowest in control and highest in plots treated with fly ash @ 40 t ha^{-1}, as against initial value of 5.70 g kg^{-1} (Pre-sowing). Increasing in organic carbon content in fly ash treated plots over control might be due to higher content of burnt carbon in fly ash. Naveen *et al.* (2000) also reported that organic carbon content of sandy-loam soil increased with the application of fly ash.

Table 44.21: Effect of Fly Ash Levels on Organic Carbon Content of Postharvest Soil

Treatment	*Organic Carbon (g kg^{-1})*
F_0	5.34
F_1	5.74
F_2	5.86
S.E	0.09
C.D. at 5 per cent	0.19

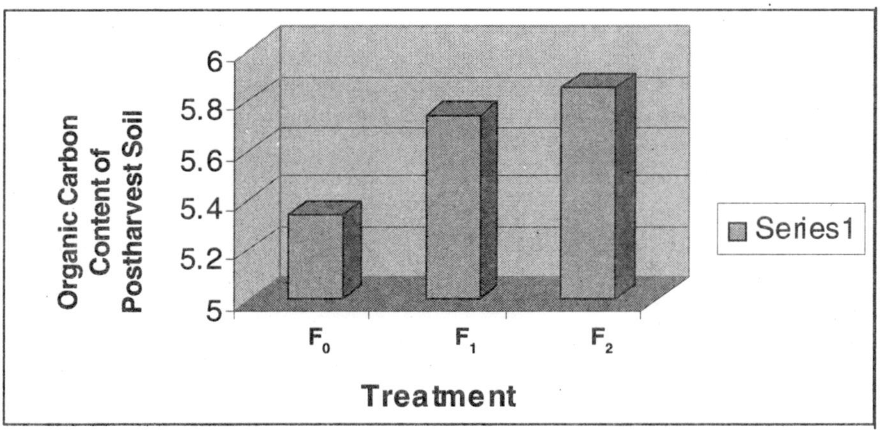

Figure 44.13: Effect of Fly Ash Levels on Organic Carbon Content of Postharvest Soil

Available Nitrogen

Effect of Fly Ash Levels on Available Nitrogen of the Postharvest Soil

From the results revealed that different levels of fly ash did not produce marked variation in available nitrogen of the postharvest soils. The very little total nitrogen content of fly ash (0.07 per cent total N) might not have been sufficient enough to cause any increase in the level of available nitrogen in the postharvest soils besides crop removed more N from the treated plots. Thus, it was revealed that fly ash did not result in the build-up of available N in the postharvest soils.

Table 44.22: Effect of Fly Ash Levels on Available Nitrogen of the Postharvest Soil

Treatment	*Available Nitrogen (kg ha^{-1})*
F_0	263.20
F_1	265.27
F_2	265.71
S.E	-
C.D. at 5 per cent	NS

Available Phosphorus

Effect of Fly Ash Levels on Available Phosphorus of the Postharvest Soil

From the results revealed that available P was significantly increased by the graded doses of fly ash. The available P content in postharvest range from 20.95 to 22.90 kg ha^{-1} with lowest in control and highest in plots treated with fly ash @ 40 t ha^{-1}(F_2). The application of fly ash @ 20 and 40 t ha^{-1} recorded an increase of 1.77 and 2.50 kg ha^{-1} in postharvest soil, as against initial value of 20.40 kg ha^{-1} (pre-sowing). The favourable effect of fly ash on P availability was ascribed to its effect on biotic activity and the P release via biotic activity. The silicon which is present in fly ash would also have played a major role in releasing insoluble sources in fly ash as well in soil. Anjali *et al.* (2000) reported similar increase in available P content of soil. Koter *et al.* (1984) also observed increase in available P status and they attributed it to the P content of fly ash.

Table 44.23: Effect of Fly Ash Levels on Available Phosphorus of the Postharvest Soil

Treatment	*Available Phosphorus (kg ha^{-1})*
F_0	20.95
F_1	22.17
F_2	22.90
S.E	0.33
C.D. at 5 per cent	0.69

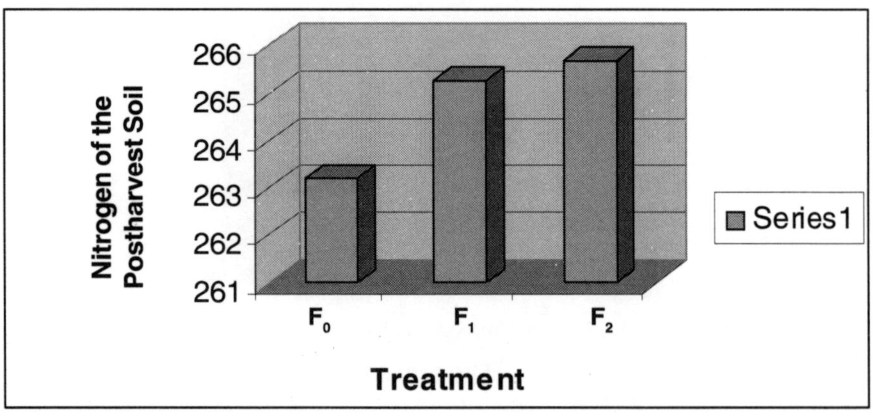

Figure 44.14: Effect of Fly Ash Levels on Available Nitrogen of the Postharvest Soil

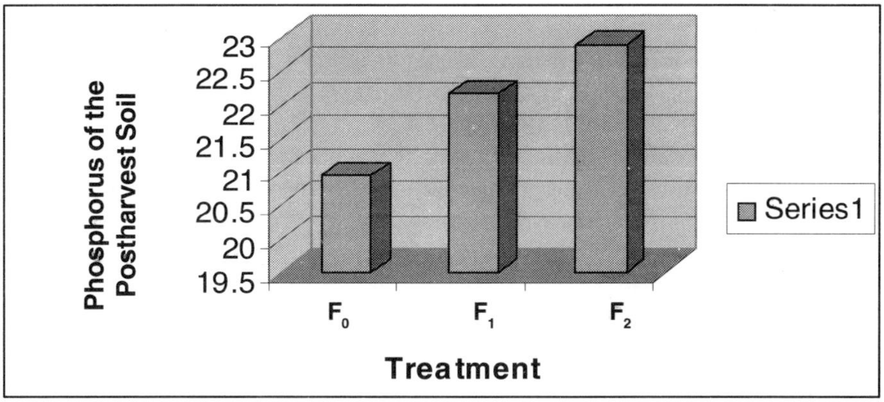

Figure 44.15: Effect of Fly Ash Levels on Available Phosphorus of the Postharvest Soil

Available Potassium

Effect of Fly Ash Levels on Available Potassium of the Postharvest Soil

From the results revealed that available K was significantly increased by the graded doses of fly ash. The increase was 19.16 and 25.50 kg ha^{-1} of the postharvest soil due to addition of fly ash @ 20 and 40 t ha^{-1} as against initial value of 225.50 kg ha^{-1} (pre-sowing). Positive effect of fly ash addition on available K content of soil was also reported by Warambhe *et al.* (1992). Also Anjali *et al.* (2000) and Selvakumari *et al.* (2000) reported that application of fly ash increased K content of soil.

Table 44.24: Effect of Fly Ash Levels on Available Potassium of the Postharvest Soil

Treatment	*Available Potassium (kg ha^{-1})*
F_0	219.33
F_1	244.66
F_2	251.00
S.E	0.76
C.D. at 5 per cent	1.61

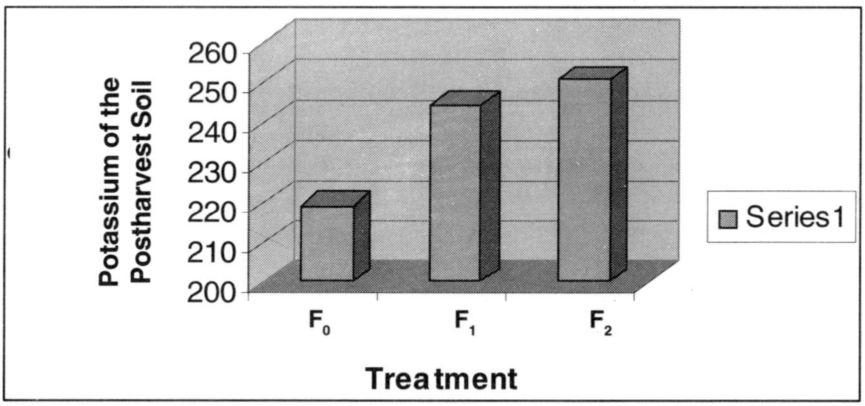

Figure 44.16: Effect of Fly Ash Levels on Available Potassium of the Postharvest Soil

Summary and Conclusion

Field experiment was conducted during kharif season of 2001 at the Crop Research Farm of Department of Agronomy, Allahabad Agricultural Institute- Deemed University, Allahabad, on sandy loam soil, to study the "ENVIRONMENTAL IMPACT OF FLY ASH ON SOIL HEALTH, YIELD AND NUTRIENT UPTAKE BY RICE". The experiment was laid out in 3 × 3 factorial design and the treatments were replicated thrice. The result obtained during the field during the field study are summarized and discussed in previously. The general findings are summarized below:

Pre-Harvest Findings

- ☆ The levels of fly ash significantly increased the plant height.
- ☆ The number of tillers per hill are significantly increased by increasing the levels of fly ash.
- ☆ The dry weight of the plants at successive stages of growth were significantly affected by different levels of fly ash.
- ☆ Panicle length of rice crop was significantly influenced by different levels of fly ash.
- ☆ The number of spikelets per panicle was significantly not influenced by the levels of fly ash.

Postharvest Findings

- ☆ Addition of fly ash at graded doses increased grain yield of rice crop significantly over control. Fly ash applied @ 20 and 40 t/ha increased the grain yield by 3.64 and 10.30 per cent, respectively over the treatment without any addition of fly ash.
- ☆ The straw yield was also significantly increased by the graded doses of fly ash. Apparently, the positive effect of fly ash was found. The maximum grain and straw yield was recorded in the treatment F_2 fly ash @ 40 t/ha.
- ☆ The N P K uptake by rice grain was significantly higher in treatments that received fly ash @ 20 and 40 t/ha than in without fly ash treatment. The maximum N P K uptake by rice grain was observed with the application of fly ash @ 40 t/ha (F_2).
- ☆ There was a decrease in soil pH in the postharvest soil due to fly ash addition as compared to control. The pH decreased from 7.87 to 7.71 due to application of fly ash @ 20 t/ha, however, the decrease in soil pH was statistically non-significant.
- ☆ The addition of fly ash at graded doses did not produce marked influence on EC in the postharvest soil. The organic carbon content range form 5.34 to 5.86 g/kg with highest at (F_2) fly ash @ 40 t/ha level of application.
- ☆ Addition of fly ash at graded levels did not result in the build-up of available N in the postharvest soils.
- ☆ There was an increase in available phosphorous and potassium status of the postharvest soil with fly ash.

Observation Taken in the Soil

Chemical Analysis of Postharvest Soil

Postharvest soil samples were taken to the plough level, *i.e.* upto 0 to 15 cm depth from each plot for determination of important chemical properties of soil, *viz.*, pH, EC, organic carbon, available Nitrogen, Phosphorus and Potassium.

)etermination of Soil pH

Digital Electric pH meter MK VI (Systronics) was used for the determination of pH of the 1:2 soil vater suspensions. The instrument being a potentiometer requires to be calibrated before use with uffer solution of known pH values of 4.0, 7.0 and 9.2 at 25°C.

)etermination of Electrical Conductivity in Soil

Amount of soluble salts in saturated extract or 1:2 soil:water suspension is measured by Direct eading Conductivity meter-304 (Systronics). The instrument is to be calibrated or checked with .01 M standard potassium chloride solution at 25°C.

)etermination of Organic Carbon in Soil

For easily oxidizable carbon determination, most widely acceptable method in India is Walkley nd Black method (1934). This method is mainly based on the principle of wet oxidation of organic arbon in an acid dichromate solution followed by back titration of the remaining dichromate with errous ammonium sulphate.

)etermination of Available Nitrogen in Soil (Kjeldahl Method)

Available nitrogen is mainly determined by alkaline permanganate method (Subbiah and Asija 956). In this method soil is distilled with alkaline potassium permanganate solution (0.32 per cent) nd 40 per cent NaOH which gives ammonia. This liberated ammonia is absorbed by boric acid olution with mixed indicator, which is determined volumetrically and serves as an index of available itrogen status.

etermination of Available Phosphorus by Olsen's Method or Sodium Bicarbonate ($NaHCO_3$) xtractable Phosphorus (Olsen *et al.*, 1954)

Phosphorus determination in soil is a two steps process, *i.e.* extraction of phosphorus from the oil by sodium bicarbonate. After extraction from the soil, phosphate in the extract is measured by the eaction of phosphate with ammonium molybdate in an acid medium to form molybdophosphoric cid. The molybdophosphoric acid is then reduced to a blue coloured complex (reduced hosphomolybdenum blue) through reaction with ascorbic acid. The ascorbic acid method has proved be reliable and less subjected to interference in colour development than the $SnCl_2$ method. The olour is stable for 24 hours. Absorbance readings are taken at a 730 nm wavelength using a pectrophotometer. A standard curve constructed from absorbance readings standards is used to educe phosphate concentration of sample.

etermination of Available Potassium (Jackson, 1973)

Potassium is extracted from the soil with the help of suitable extractant neutral normal ammonium cetate (1N CH_3COONH_4) by shaking followed by filtration and is determined in the extract using ame photometer (Mediflame 127-Systronics). The analysis of flame photometer is based on the easurement of the intensity of characteristic wavelength given by the element to be determined. 'hen a solution of a salt is sprayed into a flame, the salt gets separated into its component atoms ecause of high temperature. The energy provided by flame excites the atom to higher energy levels. 'hen the electrons return back to the ground or unexcited state, they emit radiation of characteristic ave length (Line emission spectrum). The intensity or these radiations is proportional to the ncentration of particular element in solution which is measured through a photocell in the flame hotometer.

Statistical Analysis

The data recorded during the course of investigation will subjected to statistical analysis as per the method of "analysis of variance technique". The significant and non-significant treatment effects were judge with the help of 'F' table. The significant differences between the means were tested against the critical difference at 5 per cent probability level. For testing the hypothesis the following ANOVA table will used.

The 'F' test will use for judging the significant of treatment means. The significant difference between different means were judge by using least significant difference (C.D) at 5 per cent level of significance, which will calculated by the formula.

Table 44.25: Skeleton of ANOVA Table

Source of Variation	*d.f.*	*S.S*	*M.S.S*	*F(cal)*	*F (tab) at 5 per cent*
Replication	(r-1)	SSR	$\frac{SSR}{d.f}$	$\frac{MSSR}{MESS}$	
Flyash (F)	(f-1)	SSF	$\frac{SSF}{d.f}$	$\frac{MSSF}{MESS}$	
Fertilizer (M)	(m-1)	SSM	$\frac{SSM}{d.f}$	$\frac{MSSM}{MESS}$	
Interaction (F × M)	(f-1)(m-1)	SSFM	$\frac{SSFM}{d.f}$	$\frac{MSSI}{MESS}$	
Error	(r-1)(t-1)	SSE	$\frac{SSE}{d.f}$		
Total	(rfm-1)	TSS			

Sem ± for Flyash levels (F) = $\sqrt{\frac{2 \times MESS}{r \times f}}$

C.D. for Flyash levels = SEm × t_{16} value for error d.f at 5 per cent

SEm ± for Fertilizer levels (M) = $\sqrt{\frac{2 \times MESS}{r \times M}}$

C.D. for Fertilizer levels = SEm × t_{16} value for error d.f at 5 per cent

SEm ± for interaction (F × M) = $\sqrt{\frac{2 \times MESS}{r \times M}}$

C.D. for interaction (F × M) = SEm × t_{16} value for error d.f at 5 per cent

where,

r: Number of replications; t: Number of treatments; d.f: Degree of freedom; S.S: Sum of squares; SSR: Sum of squares due to replication; MSSR: Mean sum of squares due to replication; SSE: Error sum of squares; TSS: Total sum of squares; S (m): Standard error of means; F (cal): Calculated value of 'F'; F (tab): Value of 'F' from variance table; C.D: Critical difference; SSM: Sum of squares due to fertilizer; SSF: Sum of squares due to fly ash; SSFM: Sum of squares due to interaction (F × M).

References

Anjali, Deshmukh, Matte, D.B., Bhaisare, Bharti, Deshmukh, A. and Bhoisare, B. (2000). Soil properties as influenced by flyash application. *Jour. Soil and Crops* 10 (1) 69-71.

Bharti, B., Matte, D.B., Batole, W.P., Anjali, D., Bhaisare, B. and Deshmukh, A. (2000). Effect of flyash on yield uptake of nutrients and quality of green gram growth on vertisol. *J. Soil and Crops,* 10 (1) 122–124.

Bharti, B., Matte, D.B., Batole, W.P., Anjali, D., Bhaisare, B. and Deshmukh, A. (1999). Effect of fly ash on physico-chemical properties of vertisols and yield of green gram. *J. Soil and Crops,* 9(2): 255-257.

Buck, J.K., Honston, R.J. and Beimborn,W.A. (1990). Direct seedling of anthrotic refuge using coal fly ash a major soil amendment in a process of mining and reclamation conference and exhibition West Virgina University public service.

Deshpande, V.P., Kole, C.K., Godkari, S.K. and Badrinath (1993). Environmental impacts due to fly ash from coal fired thermal power plants. *Paper presented at Nat. sem. Env. Asp. Thermal Power Plants Org.* by R and D center Noida,U.P. from 9-10 Dec.

Dubey, P.S., Pawar, K., Shrigi, S.K. and Trivedi, L. (1982). The effect of flyash deposition on photosynthetic pigments and dry matter production of wheat and grains. *Agro, Ecosystem,* 8(2): 137-140.

Garg, R.N., Singh, G., Kalra, N., Das, D.K. and Singh, A. (1996). Utilization of applied phosphorous by wheat as influenced by fly ash application. *Annals of Biology,* Ludhiana, 12(1): 19-24.

Kuchanwar, T., Matte, D.B. and Kane, D.R. (1997). Effect of fly ash application on physico chemical properties of soil. Nat. Env. Res. Ins. (NEERI) Nehru Marg Nagpur, India.

Kumar, W. and Annanaric, A. (2000). Effect of green manure on growth and yield attributes in pea (*Pisum sativum*). *J. Plant Nutri.,* 24(6): 240-248.

Kene, D.R., Lanjewar, S.A., Ingole, M. and Chaphale, S.P. (1991). Effect of application of fly ash on physico-chemical properties of soil. *J. Soil and Crops,* 1(1): 118.

Microbial Biotechnology and Ecology (2011) *Pages* **766–772**
Editors: **Deepak Vyas, G.S. Paliwal, P.K.Khare and R.K Gupta**
Published by: **DAYA PUBLISHING HOUSE, NEW DELHI**

Chapter 45

Eutrophication: Causes, Consequences and Control Measures

☆ ***Ashu and Mukesh Kumar***

Department of Botany, Sahu Jain P.G. College, Najibabad – 246 763, U.P.

ABSTRACT

Eutrophication is an increase in the concentration of chemical nutrients in an ecosystem to an extent that increases in the primary productivity of the ecosystem. Depending on the degree of eutrophication, subsequent negative environmental effects such as anoxia and severe reductions in water quality, fish, and other animal populations may occur. Eutrophication poses a problem not only to ecosystems, but to humans as well. Reducing eutrophication should be a key concern when considering future policy, and a sustainable solution for everyone, including farmers and ranchers, seems feasible. While eutrophication does pose problems, humans should be aware that natural runoff (which causes algal bloom in the wild) is common in ecosystems and should thus not reverse nutrient concentrations beyond normal levels.

Introduction

Eutrophication is a process whereby water bodies, such as lakes, estuaries, or slow moving streams receive excess nutrients that stimulate excessive plant growth (algae, periphyton attached algae, and nuisance plant weeds). This enhanced plant growth, often called an algal bloom, reduces dissolved oxygen in the water when dead plant material decomposes and can cause other organisms to die. Nutrients can come from many sources, such as fertilizers applied to agricultural fields, golf courses, and suburban lawns; deposition of nitrogen from the atmosphere; erosion of soil containing nutrients; and sewage treatment plant discharges. Water with a low concentration of dissolved oxygen is called hypotoxic.

In Lakes, Eutrophication is frequently a result of nutrient pollution, such as the release of sewage ffluent, urban storm water run-off, and run-off carrying excess fertilizers into natural waters. :utrophication generally promotes excessive plant growth and decay, favors certain weedy species ver others, and may cause a severe reduction in water quality. In aquatic environments, Eutrophication nhances growth of problems such as lack of oxygen in the water, needed for fish and shellfish to urvive. The water then becomes cloudy, coloured a shade of green, yellow, brown, or red. Human ociety is impacted as well: Eutrophication decreases the resource value of rivers, lakes, and estuaries uch that recreation, fishing, hunting, and aesthetic enjoyment are hindered. Health-related problems an occur where eutrophic conditions interfere with drinking water treatment (Bartram *et al.*, 1999).

Eutrophication was recognized as a pollution problem in European and North America lakes nd reservoirs in the mid-20th century (Rodhe 1969), since then, it has become more widespread. urveys showed that 54 per cent of lakes in Asia are eutrophic; in Europe 53 per cent; in North-merica 48 per cent; in South-America 41 per cent; and in Africa 28 per cent. (ILEC/Lake Biwa esearch Institute (Eds.) 1988-1993).

Although Eutrophication is commonly caused by human activities, Eutrophication can also be a atural condition for many lakes. Paleolimnologists now recognize that climate change, geology, and ther external influences are critical in regulating the natural productivity of lakes. Some lakes also emonstrate the reverse process (meiotrophication), becoming less nutrient rich with time (Walker, 006; Whiteside, 1983).

Phosphorus is often regarded as the main culprit in case of Eutrophication in lakes. Studies onducted in the Experimental Lakes Area in Ontario have shown a relationship between the addition f phosphorus and the rate of Eutrophication. Humankind has increased the rate of phosphorus ycling on Earth by four times, mainly due to agricultural fertilizer production and application. etween 1950 and 1995; 600,000,000 tonnes of phosphorus were applied to Earth's surface, primarily n croplands (Carpenter and Smith, 1998).

The World Resources Institute has identified 375 hypoxic coastal zones in the world, concentrated coastal areas in Western Europe, the Eastern and Southern coasts of the US, and East Asia, particularly Japan (Selman and Mendy, 2007). In addition to runoff from land, atmospheric anthropogenic fixed itrogen can enter the open ocean. A study in 2008 found that this could account for around one third f the ocean's external nitrogen supply and up to 3 per cent of the annual new marine biological roduction (Duce, 2008).

Although traditionally thought of as enrichment of aquatic systems by addition of fertilizers into kes, or other semi-enclosed waters (even slow moving rivers), terrestrial ecosystems are subject to milarly adverse impacts. Increased content of nitrates in soil frequently leads to undesirable changes vegetation composition and many plant species are endangered as a result of Eutrophication in rrestic ecosystems, *e.g.* majority of orchid species in Europe (Pullin, 2002). Ecosystems (like some eadows, forests and bogs that are characterized by low nutrient content and species-rich, slowly owing vegetation adapted to lower nutrient levels) are overgrown by faster growing and advantage unnaturally elevated nitrogen level and the area may be changed beyond recognition and vulnerable ecies may be lost. Example, Species-rich fens are overtaken by reed or reedgrass species, spectacular rest undergrowth affected by run-off from nearby fertilized field is turned into a thick nettle and amble shrub.

Chemical forms of nitrogen are most often of concern with regard to Eutrophication becaus plants have nitrogen requirements so that additions of nitrogen compounds stimulate plant growt (primary production). Nitrogen is not readily available in soil because N_2, a gaseous form of nitrogen is very stable and unavailable directly to higher plants. Terrestrial ecosystems rely on microbia nitrogen fixation to convert N_2 into other physical forms (such as nitrates). However, there is a limit t how much nitrogen can be utilized. Ecosystems receiving more nitrogen than the plants require ar called nitrogen-saturated. Saturated terrestrial ecosystems contribute both inorganic and organi nitrogen to freshwater, coastal, and marine Eutrophication, where nitrogen is also typically a limitin nutrient (Hornung *et al.*, 1995). However, in marine environments, phosphorus may be limiting becaus it is leached from the soil at a much slower rate than nitrogen, which are highly insoluble (Smith *et al.* 1999).

Causes of Eutrophication

The two major causes of Eutrophication are excess nitrates and excess phosphates in water. It i important to study these nutrients because often times, human activity is responsible for their negativ effect on the environment.

Nitrates

Nitrates (NO_3^-) are water-soluble (they dissolve easily in water), and are commonly applied t agricultural fields as fertilizer. Once applied, nitrates may leach into groundwater or erode and en up in surface runoff. Eventually, they may enter in lake, river, or stream and contribute to Eutrophicatio Nitrates can also vaporize into the atmosphere, where they become a major source of acid rain. Whe ingested by organisms in drinking water, nitrates bind to haemoglobin and reduce the oxygen-carryin capacity of the bloodstream. This form of nitrate poisoning seems particularly prevalent amon amphibians, and may be contributing to a worldwide decline in the biodiversity of these species.

28 per cent of the annual Nitrogen fixation of the global biosphere is caused by nitrogen fertiliz production, which is energetically expensive and largely based on fossil fuel consumption. A thre fold increase in nitrogen utilization by agriculture in Western Europe and USA has been recorded o agricultural land that actually declines because increased efficiency or over-exploitation. Anim waste from intensive husbandry is of particular significance for nutrient point sources: considerab amounts are directly supplied to freshwater and fields. There are also significant losses by NH emission and denitrification on fields, rivers or shallow estuaries, connecting the agricultural lan directly to the atmosphere. Agricultural run-off has given rise to significant Eutrophication in mo estuaries, but also in entire coastal seas such as the southern North Sea, Baltic Sea, Kattegat, northe Adriatic Sea, Chesapeake Bay and Seto Inland Sea in Japan.

Annual global nitrogen fertilizer consumption for 1960-1995 (1 Tg = 10^{12} g; data from FAO 199 The rate of increase was relatively steady until the late 1980s, when the collapse of the Soviet Uni reduced fertilizer use in Russia and former Soviet republics. Fertilizer use is growing again, driven large part by use in China

Phosphates

Unlike nitrates, phosphates (PO_4^{3-}) are not water-soluble; they do not usually dissolve in wat However, they do adhere to soil particles, and as such often accumulate in soil and erode along wi soil into aquatic environments. Phosphates also form a major component of most fertilizers, and conjunction with nitrates, they have made agriculture the largest source of nonpoint water polluti in the United States.

Effects of Eutrophication

Ecological Effects

Following are the adverse effects of Eutrophication on lakes, reservoirs, rivers and coastal marine waters

1. Increased biomass of phytoplankton
2. Toxic or inedible phytoplankton species
3. Increases in blooms of gelatinous zooplankton
4. Changes in macrophyte species composition and biomass
5. Decreases in water transparency
6. Taste, odour, and water treatment problems
7. Dissolved oxygen depletion
8. Increased incidences of fish kills
9. Loss of desirable fish species
10. Reductions in harvestable fish and shellfish
11. Decreases in perceived aesthetic value of the water body

Many ecological effects can arise from stimulating primary production, but there are three particularly troubling ecological impacts: decreased biodiversity, changes in species composition and dominance, and toxicity effects.

Decreased Biodiversity

When an ecosystem experiences an increase in nutrients, primary producers reap the benefits first. In aquatic ecosystems, species such as algae experience a population increase (called an algal bloom). Algal blooms limit the sunlight available to bottom-dwelling organisms and cause wide swings in the amount of dissolved oxygen in the water.

Oxygen is required by all respiring plants and animals and it is replenished in daylight by photosynthesizing plants and algae. Under eutrophic conditions, dissolved oxygen greatly increases during the day, but is greatly reduced after by the respiring algae and by microorganisms that feed on the increasing mass of dead algae. When dissolved oxygen levels decline to hypoxic levels, fish and other marine animals suffocate. As a result creatures such as fish, shrimp, and especially immobile bottom dwellers die off (Horrigan *et al.*, 2002). In extreme cases, anaerobic conditions ensue, promoting growth of bacteria such as *Clostridium botulinum* that produces toxins deadly to birds and mammals. Zones where this occurs are known as dead zones.

Effect of Fertilizers

Fertilizers containing plant nutrients are sprayed onto fields to make plants grow faster and boost crop yields. When it rains the nutrients may get washed down from the fields and into rivers and lakes (this is called run-off). The result is Eutrophication–which can kill almost everything living in the aquatic environment.

Effect of Pesticides

Pesticides are chemicals used to kill insects, weeds and microorganisms that might damage crops. However, pesticides damage other organisms apart from those they are intended to kill–for example, depriving insect-eating birds of food. Pesticides can also enter local food chains. Organism that ingest them cannot break them down, so they persist in their bodies. The pesticides may the build up at ever-higher levels until they become toxic to much larger organisms.

New Species Invasion

Eutrophication may cause competitive release by making abundant a normally limiting nutrient This process causes shifts in the species composition of ecosystems. For instance, an increase in nitrogen might allow new, competitive species to invade and out-compete original inhabitant species This has been shown to occur in New England salt marshes.

Toxicity

Many algae produce potent toxins. Toxic compounds they produce can make their way up the food chain, resulting in animal mortality (Anderson, 1994). When the algae die or are eaten, neuro and hepatotoxins are released which can kill animals and may pose a threat to humans (Lawton and Codd, 1994; Shumway, 1990). Toxigenics blue green algae like Anabaena, Microcystis and Nosto have been noted whose toxins can kill fish. The toxin may also contaminate the drinking wate supplies.

Sources of High Nutrient Runoff

Point Sources

Point sources are directly attributable to one influence. In point sources the nutrient waste travel directly from source to water. Point sources are relatively easy to regulate.

Nonpoint Sources

Nonpoint source pollution (also known as diffuse or runoff pollution) is that which comes from ill-defined and diffuse sources. Nonpoint sources are difficult to regulate and usually vary spatially and temporally (with season, precipitation, and other irregular events).

It has been shown that nitrogen transport is correlated with various indices of human activity in watersheds, (Cole *et al.*, 1993; Howarth *et al.*, 1996) including the amount of development. There are three reasons that nonpoint sources are especially troublesome:

Soil Retention

Nutrients from human activities tend to accumulate in soils and remain there for years. It has been shown that the amount of phosphorus lost to surface waters increases linearly with the amoun of phosphorus in the soil (Raimammake *et al.*, 2003). Thus much of the nutrient loading in soi eventually makes its way to water. Nitrogen, similarly, has a turnover time of decades or more.

Run Off to Surface Water and Leaching to Groundwater

Nutrients from human activities tend to travel from land to either surface or groundwater. Nitrogen in particular is removed through storm drains, sewage pipes, and other forms of surface runoff Nutrient losses in runoff and leachate are often associated with agriculture. Modern agriculture often involves the application of nutrients onto fields in order to maximise production. However, farmers frequently apply more nutrients than are taken up by crops or pastures (Buol, 1995).

Atmospheric Deposition

Nitrogen is released into the air because of ammonia volatilization and nitrous oxide production. The combustion of fossil fuels is a large human- initiated contributor to atmospheric nitrogen pollution. Atmospheric deposition (*e.g.*, in the form of acid rain) can also affect nutrient concentration in water, (Paerl, 1997) especially in highly industrialized regions.

Controlling Eutrophication

Eutrophication can be controlled by adopting the following methods:

1. Planting vegetation along streambeds to slow erosion and absorb nutrients
2. Controlling application amount and timing of fertilizer
3. Controlling runoff from feedlots
4. Researching use of biological controls; for example, the process of denitrification uses specialized bacteria that convert nitrates to harmless molecular nitrogen
5. Laws regulating the discharge and treatment of sewage have led to dramatic nutrient reductions to surrounding ecosystems, but it is generally agreed that a policy regulating agricultural use of fertilizer and animal waste must be imposed. In Japan the amount of nitrogen produced by livestock is adequate to serve the fertilizer needs for the agriculture industry (Kumuzawa, 2002).
6. Soil Nitrogen Testing (N-Testing) is a technique that helps farmers optimize the amount of fertilizer applied to crops. By testing fields with this method, farmers saw a decrease in fertilizer application costs, a decrease in nitrogen lost to surrounding sources, or both (Haung, Lu and Uri, 2001). By testing the soil and modelling the bare minimum amount of fertilizer needed, farmers reap economic benefits while the environment remains clean and pure.
7. Reseachers at the National Academy of Sciences found that organically fertilizing fields "significantly reduce harmful nitrate leaching" over conventionally fertilized fields.

Acknowledgements

The authors are thankful to the Principal, Sahu Jain P. G. College, Najibabad for providing necessary facilities.

References

Anderson, D.M. (1994). Red tides. *Scientific American*, 27: 62–68.

Bartram, J., Wayne, Carmichael, Ingrid Chorus, Gary Jones, and Olav Skulberg, (1999). Chapter 1 Introduction. In: *Toxic Cyanobacteria in Water: A guide to their public health consequences, monitoring and management*. World Health Organization. URL: WHO document.

Buol, S.W. (1995). Sustainability of Soil Use. *Annual Review of Ecology and Systematics*, 26: 25–44.

Carpenter, S.R., Caraco, N.F. and Smith, V.H. (1998). Nonpoint pollution of surface waters with phosphorus and nitrogen. *Ecological Applications*, 8: 559-568.

Cole, J.J., Peierls, B.L., Caraco, N.F. and Pace, M.L. (1993). Nitrogen loading of rivers as a human-driven process. In: McDonnell, M.J. and Pickett, S.T.A. (Eds) *Humans as Components of Ecosystems*. Springer Verlag, New York, USA, pp. 141-157.

Duce, R.A. *et al.* (2008). Impacts of atmospheric anthropogenic nitrogen on the open ocean. *Science*, 320: 893–897.

Hornung, M., Sutton, M.A. and Wilson, R.B. (Eds.) (1995). Mapping and modelling of critical loads for nitrogen a workshop report. Grange over Sands, Cumbria, UK. UNECE.

Horrigan, L., Lawrence, R.S. and Walker, P. (2002). How sustainable agriculture can address the environmental and human health harms of industrial agriculture. *Environmental Health Perspectives,* 110: 445–456.

Howarth, R.W., Billen, G., Swaney, D., Townsend, A., Jaworski, N., Lajtha, K., Downing, J.A., Elmgren, R., Caraco, N., Jordan, T., Berendse, F., Freney, J., Kudeyarov, V., Murdoch, P. and Zhu Zhaoliang (1996). Regional nitrogen budgets and riverine inputs of N and P for the drainages to the North Atlantic Ocean: natural and human influences. *Biogeochemistry* 35: 75–139.

LEC/Lake Biwa Research Institute (Eds) (1988-1993). Survey of the State of the World's Lakes. Volumes IV. International Lake Environment Committee, Otsu and United Nations Environment Programme, Nairobi.

Lawton, L.A. and Codd, G.A. (1991). Cyanobacterial (bluegreenalgae) toxins and their significance in UK and European waters. *Journal of Soil and Water Conservation*, 40: 87–97.

Paerl, H.W. (1997). Coastal eutrophication and harmful algal blooms: Importance of atmospheric deposition and groundwater as "New" nitrogen and other nutrient sources. *Limnology and Oceanography*, 42:1154-1165.

Pullin, Andrew S. (2002). *Conservation Biology*. Cambridge University Press.

Raimammake, A., Pietilainen, O.P., Rekolainen, S., Kauppila, P., Pitkanen, H., Niemi, J., Raateland, A. and Vuorenmaa, J. (2003). Trends of phosphorus, nitrogen, and chlorophyll *a* concentrations in Finnish rivers and lakes in 1975-2000. *The Science of the Total Environment*, 310: 47–59.

Rodhe, W. (1969). Crystallization of eutrophication concepts in North Europe. In: *Eutrophication: Causes, Consequences, Correctives*. National Academy of Sciences, Washington D.C.

Selman Mindy, (2007). *Eutrophication: An Overview of Status, Trends, Policies, and Strategies.* World Resources Institute.

Shumway, S.E. (1990). A review of the effects of algal blooms on shellfish and aquaculture. *Journal of the World Aquaculture Society*, 21: 65–104.

Smith, V.H., Tilman, G.D. and Nekola, J.C. (1999). Eutrophication: impacts of excess nutrient inputs on freshwater, marine, and terrestrial ecosystems. *Environmental Pollution*, 100: 179–196.

Walker, I.R. (2006). Chironomid overview. In: S.A. Elias (ed.) *Encyclopedia of Quaternary Science.* Elsevier, Amsterdam, 1: 360-366.

Whiteside, M.C. (1983). The mythical concept of Eutrophication. *Hydrobiologia*, 103: 107-111.

Microbial Biotechnology and Ecology (2011) *Pages* 773–779
Editors: Deepak Vyas, G.S. Paliwal, P.K.Khare and R.K Gupta
Published by: DAYA PUBLISHING HOUSE, NEW DELHI

Chapter 46

Aquatic Biodiversity Assessment should Trigger Freshwater Body Monitoring and Conservation

☆ *S.D. Guru*[1] *and H.K. Goswami*[2]*

[1]*Department of Botany, Ranchi Women's College, Ranchi, Jharkhand*
[2]*24, Kaushalnagar, P.O. Misrod, Bhopal – 462 047, M.P.*

ABSTRACT

Extensive studies are being carried out under the umbrella of Biodiversity which principally constitute enumeration and listing of species from a particular area and or water body but negligible attempts point out to real interplay of biodiversity which is inherent among related and unrelated species within the ecosystems. Mollusks, fish, sponges and many species of insects invite birds to rest and even nest in the nearby bushes which in turn release their faeces to change pH of the water. Growth of algae is directly related to available chemical and pH factors with in the soil and water. A freshwater lake is bound to be protected because aquatic biodiversity- ecosystems can not be dug and enriched soon; we have enormous wealth in a lake more so when that lake supplies drinking water. Briefly, this is suggested that botanists, zoologists, geochemists and physiologists will be needed to recover a sick lake and only comprehensive and multidisciplinary study, as documented in the past two decades, at Bhojwetland Bhopal, will be needed. The most important follow up step often ignored, is repetition of deweeding programmes. Lake or a waterbody need to be taken care of on regular basis by multidisciplinary teams.

Keywords: *Wetland conservation, Biodiversity assessments and drinking water, Biological monitoring of freshwater lakes, Conservation and management of freshwater.*

---

* E-mail: hkgbhopal@hotmail.com

Introduction

Water has become synonym to life. Life arose in water and more than 90 per cent contents of eac functional and dividing cell is water. Since the time immemorial man has been settling around river and lakes and in later period, migrating human populations have had dug water bodies to nurture a life forms. Needless to mention, forests, water resources and ever increasing human populations ar inseparably interdependent on each other. A greater part of the world has one major biological threa we will need more water than food to survive for a few more hundred years because rise in huma populations is currently inversely proportional to depleting water resources.

Therefore conservation of water bodies with basic biological out put is the major challenge befor biologists, environmentalists and planners world over, more so in most Asian countries. In mos townships in many Asian countries big lakes are one of the major sources of drinking water supply t millions of inhabitants of that area. Such a situation is most common in Indian subcontinent. Based o International parameters we designate such huge lakes and water bodies as wetlands which hav water flooded land mass as well. One of the major wetlands in India was identified in Bhopal (Madhy Pradesh) as bhojwetland. This short communication intends only to emphasize that what we need i to conserve each and every lake and we can not continue to describe floristic composition in the forr of biodiversity and end our endeavour before a noble beginning. There have been hundreds of pape and dozens of conferences focusing on need of this exhaustive work but not a single multidisciplinar and real collaborative effort to save and enrich a lake with quality and quantity of water has come t beginning in Central India. Our published monographic treatise on Bhojwetland (Agarker *et al.*, 199 has been based on comprehensive studies for which seeds were sown in 1984 and major results coul be seen after a decade. Further studies have been slowly in progress on this wetland, and als simultaneously, one of us (HK) has been conducting field work in various states to impress upo teachers in particular, who have a good background of limnological and floristic studies, to shar their expertise with chemists, geologists and administrative officials. Academic excellence in at a places has to be brought to safety and security of water availability for our generations.

Background

The bhoj wetland has an area of more than 40 sq kms and supplies drinking water to more tha a million people in Bhopal township. The multidisciplinary efforts on conservation of this wetland fc more than a decade have had made this as a model for biological conservation and such an exhaustiv treatise by any research group have been never documented as was presented by Agarker *et al.* (1994 To begin with, this was estimated that the original upperlake has lost about its 30 per cent area withi past 40 years on account of silting, increasing human encroachments, water loss, prolific growth weeds and erstwhile dragging water for small ponds. The density gradient of weeds (excessive growt of weeds) was found to be biologically supported by in flow of pouring sewage fed water pipes fro different nearby localities including wastewater discharge from nearby hospitals and local colonie Our primary studies had revealed that the lake was biologically very sick and exhibited totally disrupte complex ecosystems. Subsequent follow up studies were carried out to expose the major problems an find out possible remedial efforts. We worked on all possible parameters like enumeration phytoplankton, zooplankton, a large number of algal forms and aquatic insects, angiospermou weeds including aquatic ferns, sponges, mollusks, fish and birds (Total about 900 species: very ric biodiversity excluding birds) and distribution with their possible correlation, of metals in water an soil samples from around the lake.

Well planned remedial, both physical (deweeding programmes; removal of top soils and weed assemblages on dry land areas formed due to silting etc in side the lake) as well as biological (strategic conservation of many mollusks, reptiles, amphibians and even bird fauna in adjoining grassland) was undertaken and plantations to check silting was carried out at a large scale. The results were alarmingly fruitful leaving a caution that lakes will continuously need care; leaving at people's mercy would again bring discredit to the water. Condemning deweeding operations will be harmful.

The above background has been emphatically mentioned for a purpose to emphasize the importance of such a massive attempt to be made on every water body which feeds drinking water to a fraction of a population. This will be a matter of only going to the water body for assessments; remedial efforts have to be then urgently planned and executed with the automatic consent of enlightened masses.

Present Reference

Recently, we have initiated collaborative study on conservation and monitoring Kanke reservoir at Ranchi (Jharkhand) which supplies drinking water on multidisciplinary approach involving physico-chemical parameters and their comparative assessments on the same and different water bodies as well as on exhaustive coverage of phytoplankton, zooplankton, higher algal forms. Enumeration of macrophytic vegetation, is being carried out with a reference to explaining utility of each species in the water body. In other words, does these plants offer specific shelter or support in completing life cycles of mollusks/insects/or act as hangers for algal mass? Why are some plants more common in some and not in other parts? Additionally, can we account by metallic separation from different ash samples of these plants and also from soil samples as to why and what severely dangerous microelements are slowly poisoning drinking water resource? We can detect what are the undeclared sewage or other drains pouring effluent are discharged without notice? We know very well from meticulous studies of a large number workers that pollution status and metallic status is reliably demonstrated by selective presence of plant forms.

Our Approach

The exploratory module of research on conservation and wetland management as has been documented by Agarker *et al.* (1994) is by far the best approach applicable in principle to all water bodies in tropics seeking conservation after repair. Hereunder, we mention those major steps on which we are proceeding with a battery of information of publications on limnological and algal biodiversity by various authors in that area.

These steps are:

Zonal Cassification of the Water Body

The lake should be divided in to minimum of two or may be in four zones and identified/marked for further demarcations of sampling spots. For example, North zone, Nzone in to N1/N2– etc depending upon the area of the lake.

Decision of Parameters

There are certain essential parameters which must be investigated on every water body. Physicochemical characteristics of water and soil from different sampling zones are absolutely essential with a reference to seasonal variation.

Depending upon the manpower and finances available for field and equipments' expenses, we plan and also place demand for future repetition. Sampling tools can be designed as per the needs along with standard ones. In order to take out smaller and larger weeds we had designed (Agarker *et al.*, 1994) various shapes with dentate margins from thin iron rods which could catch even floating threads and filaments. Small boats carrying 2 or 3 persons are the best moving agents on the water. Incumbents must know swimming.

Biodiversity Assessments

Sampling and sorting of collection requires a real test of patience. Preparing slides and examining phytoplankton-zooplankton and other forms on the same day before fixation gives more precise information than examining on fixation; whenever this was not actually possible, we preferred to repeat. In order to assess total biodiversity (which is actually needed now) we must share taxonomic expertise of many experts; it is bound to be a collaborative study demanding unshakable faith among workers. Every water body, in and around, has wealth of medicinal plants (Agarker, 1994; Bajpai *et al.*, 1996; Bajpai, 2000) which can be taken up for a proper plantations.

Experimental Trials

In order to prove or rather suggest with confidence that particular macrophyte is positively correlated with the excessive presence of a heavy metal or in other words a species is better adapted to water containing specific chemical has to be ascertained by experimental studies. Bajpai and Goswami (2000) have conducted on metallic tolerance of strontium and chromium (trivalent) on *Polygonum glabrum* under tissue culture. Results indicate the reason why *Polygonum glabrum* is more found at those sites where soil does contain strontium-chromium complexes. Mere presence is not important but it is the absorption and utilization of metallic ions, a problem which would further need well planned physiological study.

Algal samples or any samples collected are being examined and attempts are designed to grow them in sterilized Knop's solution under identical temperature and light source. Hasty description of new species or strains due to abnormal features is not called for; we should see if these were ecophenes? We guess that even ecotypes of certain alga (example, *Scenendesmus, Chlorella, Cosmarium* etc) may be validated by experimental trials. Attributing abnormal features to pollution should be needed to be taken up by a different experimental approach.

Now a days algal population is rightly being used as pollution indicator during bio-assessment of the water resources (Phillips and Rainbow, 1993; Cairns *et al.*, 1994; De Lang, 1994). Structural abnormalities (Misra *et al.*, 2003) in genera *Oocystis, Oedogonium, Euastrum* and *Cosmarium* and many other algae as well as diversity in distribution in the same and different areas need a careful follow up (Guru and Sahu, 2006; 2007 a,b, c,d,e,f; Guru, 2008 a,b). Among these, the genus *Scenendesmus* alone (An *et al.*, 1999) is taking up a position of water purifier seeking a support for its inoculation studies in different (Guru, 2008) polluted waters.

Shore Line Conservation by Medicinal Plants

Many herbaceous medicinal plants are also hyper accumulators of lead, cadmium and mercury (pteridophytes: *Ceratopteris, Marsilea, Pteris* spp etc.). The Classical medicinal plants are also found in some lake side areas; angiosperms:- *Bacopa, Centella* spp; there is a good list, which can be cultivated around the catchment area or in the marginal fringes of the water body or may even be a small multiplication garden for local medicinal plants can be erected. Initial attempts have already been made by us.

A Practical Suggestion

There are many very large lakes and water bodies in many remote areas, which need biological assessments. Teachers in the Universities should pose very sound and meaningful proposals and convince administrators and related local politicians to realize basic needs of water bodies. Fortunate are those who have lakes and unfortunate are those who can not maintain them?

Lakes are open treasures for survival of every species; plants, mollusks, fish, rodents, arboreal animals and birds all feed on lake water. Must not we save them?

Acknowledgements

We are grateful to Dr Jawed Ahmed, the Principal and other faculty members at Marwari College, Ranchi for accepting our joint endeavour in presenting a two day workshop organized at their institution. In this semiscientific presentation emphasis is laid on seriously taking up such studies on those water bodies which supply drinking water. We acknowledge that our present write up has been mainly guided by following publications mentioned hereunder.

References

Agarker, M.S., Goswami H.K., Kaushik, S., Mishra, S. M., Bajpai, A.K. and Sharma, U.S. (1994). Biology, Conservation and Management of Bhojwetland. I. Upper Lake Ecosystems in Bhopal. *(Bionature) Monograph*, pp. 1-119.

An S. S., Friedel T. and Hegewald E. (1999). Phylogenetic relationships of *Scenedesmus* and *Scenedesmus*–like coccoid green algae as inferred from ITC-2 rDNA sequence comparisons. *Plant Biology* 1: 418-428.

Bajpai, A.K (2000). *In vitro* multiplication of *Bacopa monnieri* and *Centella asiatica*. *Bionature*. 20: 75-79.

Bajpai, A.K. and Agarker, M.S. and Goswami, H.K. (1996). Biology, Conservation and Management of Bhojwetland III. Floristic composition in and around upperlake ecosystem in Bhopal. *Bionature*, 16: 53-64.

Bajpai, A.K. and Goswami, H.K. (2000). Assessment of metallic tolerance in *Polygonum glabrum*. *Physiol. Mol. Biol. Plants*, 6: 149-152.

Becker E. W. (1994). *Microalgae, Biotechnology and Microbiology*. Cambridge Univ. press. Cambridge, U.K.

Bohra C. and Kumar A.(2004). Plankton diversity in the wetlands of Jharkhand. In: *Biodiversity and Environment* (Ed. A. kumar) ABD Pub. New Delhi. pp. 91.

Cairns J. Jr. Mc Cormic P. V. and Nitherlehner B. R. (1994). Bioassay and field assessment of pollution effects. In: *Algae and water pollution*. (*Eds*. Rai L. C, Gaur J. P. and Soider C. J.) Schweizerbort'sch Verlagsbuchhanlung. Stuggert. pp. 267.

Chapman V. J. and Chapman D. J.(1973). *The Algae*. ELBS Publication, Macmillan Press. London. pp. 54.

Das R. N. and Sahu R. (1989). Chloroccocales from freshwater habitats of Ranchi. *Vegetos* (2) 2: 223-229.

De Lang E.(1994). *Mannual of simple water quality analysis.* International Water Tribunal (IWT) Foundation, Amesterdam.

Guru, S.D. and Sahu Radha (2006). Morphological variations in Stigeoclonium (Chlorophyceae) in polluted water in water. *Bionature*, 26 (2): 55-59.

Guru, S. D. (2007a). Effect of Xenobiotic pollution on the algal (Chlorophyceae) population of some freshwater lakes of Jharkhand. *Biospectra*, 2 (1): 143-147.

Guru,S. D.(2007b). A comparative study of the species richness and diversity of chlorophycean assemblage of two city based polluted tanks of Ranchi. *Biospectra*, 2 (2): 343-345.

Guru, S. D. (2007c). Survival Strategies for Medicinal Plants. *Recent Advances in Biological science* (in press).

Guru, S.D. (2007d). Limnological studies of freshwater bodies of Ranchi. *Trans. Zool. Soc. East India*, (in press).

Guru,S. D. (2007e). Prospect of Biomonitoring of Pollution in aquatic system by Algal (Chlorophyceae) Assessment. *Indian J. Environ. and Ecoplan*. 14 (3): 651-654.

Guru, S.D. (2007f). Distribution and biodiversity of desmids in a freshwater lake of Ranchi. *Vegetos*, 20 (2): 81-85.

Guru, S. D. (2008a). Distribution and Biodiversity of Chlorococclean Flora in the Phytoseston of a Freshwater lake. *Journal of Phytological Research*, 21(1).

Guru, S. D. Sushma (2008). Species Richness of *Scenedesmus* (Chlorophyceae) in some sewage fed water bodies of Ranchi. *Bionature*, 28 (1): 1-4.

Hegewald E. and Silva P. (1998). Annotated catalogue of *Scenedesmus* and nomenclaturally related genera including original descriptions and figures. *Bibliotheca phycologica* 80. J. Cramer, Berlin-Stuttgart. Pp. 587.

Johnson J. L., Fawley M. W. and Fawley K. P.(2007). The diversity of *Scenedesmus* and *desmodesmus* (Chlorophyceae) in Itasca-State Park, Minnesota, USA. *Phycologia* 46: 214-229.

Kumar, N., Singh, N., and Agarker, M.S. (1986) Some rare and interesting chlorococcales from Jabalpur, M.P. Bionature, 6: 35-36.

Lewis L. A. and Flechtner V. R. (2004). Criptic species of *Scenedesmus* (Chlorophyta) from desert soil communities of western North America. *Journal of Phycology* 40: 1127-1137.

Misra P. K., Mehrotra R. K., Jai Prakash, Srivastava A. K. and Kishore S. (2002). Freshwater Planktonic algae from Basti. Uttar Pradesh. *Geophytology*. 30(1 and 2): 61-73.

Misra P.K., Srivastava,A.K. and Prakash, J (2003-2004). Structural abnormalities in some green algae from north-eastern part of Uttar Pradesh. *Geophytology* 33 (1 and 2):17-20.

Phillips D. J. H. and Rainbow P. S. (1993). *Biomonitoring of Trace Aquatic contaminants*. Elsevier Applied Science. New York.

Prescott G. W. (1962). *Algae of the Western Grate Lakes Area.* Koenigstein, West Germany. Pp. 977.

Shon Yun-Hee, Nam Kyung-Soo, Kim mi-kyung (2004). Cancer chemopreventive potential of *Scenedesmus* spp. Cultured in medium containing Gioreacted swine urine. *J. microbiol. Giotechnol.* 14: 158-161.

useela M. R. and Dwivedi S. (2005). Species diversity of Tola Lake, Uttar Pradesh. In "*Glimpses of Indian phycology*. (Eds. Keshri J. P and Kargupta A. N.) B.S.M.P.S. Publication. Dehra Dun. P. 333-340.

'rainor F. R. (1998). Biological aspects of *Scenedesmus* (Chlorophyceae)–phenotypic plasticity. *Nova Hedwigia* Beiheft 117, J. Cramer, Berlin. Pp. 367.

'oltolina D. Gomez-villa H, Correa G. (2004). Biomass production and nutrient removal in semicontinous culture of *Scenedesmus* sp. (chlorophyceae) in artificial wastewater under stimulated day night cycle. *VIE MILIEU*. 54: 21-25.

Microbial Biotechnology and Ecology (2011) *Pages* **780–82**
Editors: **Deepak Vyas, G.S. Paliwal, P.K.Khare and R.K Gupta**
Published by: **DAYA PUBLISHING HOUSE, NEW DELHI**

Chapter 47

Eco-Physiological Effects of Red Mud Waste of an Aluminium Industry on the Seed Biology of a Crop Plant

☆ *Alka Sahu, Anita R. Patnaik, R.K. Behera and A.K. Panigrah*

Department of Botany, Berhampur University, Berhampur – 760 007, Orissa

Introduction

Pollution sources can be classified in various ways depending on the origin of pollutants, the main component, property and the effects etc. Very often the classification based on the origin of th pollutants is generally used, which categorizes the pollution sources into domestic, agricultura industrial and mining. Domestic sewages mostly contain metabolic wastes, various deterger formulations, consumer products and significant amounts of trace elements, making them rich i organic matters and nutrients. The organic matters increase the BOD load of the receiving wate bodies and render them unsuitable for the survival of the aquatic animals. The organic matters in th domestic sewages are mostly biodegradable and cause problem only when the rate of input into th environment exceeds the rate of degradation or dispersion. The addition of nutrients by the way c domestic sewage into the water bodies lead to eutrophication, which brings change in the communi and the biological composition of inhabiting communities and populations.

Modern agriculture with its rapid mechanization and spreading of fertilizers, besides the use protective treatments such as herbicides, insecticides, and fungicides exerts an increasing pressure c the natural environment. Some of these chemicals are non-biodegradable, and hence, persist in th ecosystem. These chemicals can be absorbed and concentrated by the living organisms via ar mechanism or pathway. This phenomenon is known as bioaccumulation and creates various problem especially when the pollutants are of high toxic nature with high biological half-life time. This because the level of pollutants in an organism increases with the increase in its distance from the primary producer in the food chain, *i.e.* with increase in the trophic level. Thus, as a result of biomagnifications, the organisms placed at higher trophic level usually accumulate a persistent pollutant in its tissues to a concentration much greater than those present in the surrounding habitat. Industry is the third and the most important source of pollution. Over 75,000 chemicals are in common use today and several thousands of new compounds are being added to this figure each year (Cairns, 1980 and Miller, 1984). These chemicals are now, an inevitable part of the process of industrialization and mostly they are produced and used by the industry to deliver finished goods. Some of these chemicals, the by-products and the waste discharges of the industry are released inadvertently or accidentally into the environment to such an extent that they really threaten the ecosystem in global scale. The mining operations also contribute significantly to the cause of pollution. The exploitation of ore deposits invariably exposes fresh rock surfaces and large quantities of waste rock or soil to accelerate weathering. In addition, ore processing, smelting and refining operations caused dispersion and deposition of large quantity of chemicals into the water, air and soil of the surrounding environment.

Ramade (1977) has pointed out three principal causes of contamination of the environment. These are: (*i*) the production of energy from combustible fossils or from atom, currently this forms the basis of almost all human activity in industrialized countries and leads to many types of pollution, (*ii*) the chemical industry and its extraordinary expansion over the last few decades, more than half of million chemicals are currently in use, all over the world. (*iii*) Modern agriculture releases substantial quantities of chemicals into the environment by the use of fertilizers, herbicides, fungicides and mechanization. Entries of chemicals into the environment can be classified into three categories (Harris, 1976): point spills, chronic local releases and wide spread releases. By point spill it is meant that a significant amount of chemical has entered an ecosystem at a point in both space and time and the effects of contamination are expected in a well defined more or less local area. Generally this happens when a substance stored in large quantity is released suddenly into the environment. The Bhopal gas tragedy can be an example of this category, wherein 2500 people died and thousands were seriously affected by the accidental release of methyl iso-cyanate, a deadly poisonous gas, from a pesticide industry. Chronic local releases are cases in which discharges have taken place over such periods of time, and in such quantities, that a large region (for example, a river system, a catchments basin or the landscape downwind of a source) has been contaminated. Such a situation is in association with a large industrial or municipal source (Holdgate, 1979). There can be no better example than the Minamata Bay and the Niigata tragedy of mercury poisoning, for this kind of release. Wide spread releases are the release of substances in sufficient quantity and over a wide area. Such releases could result in a noticeable pollution of a significant part of the entire earth's surface. Though this type of release is yet to happen, there are examples of manufacture of a thousand substances in such quantities as to be capable of polluting the entire globe (Butler, 1978). One such situation of this type has arisen due to the increased release of Chlorofluorocarbons (CFCs) and Fluro-bromo-carbons into the atmosphere. This has resulted in progressive destruction of the ozone layer in the last decade, posting a serious threat to result global warming and greenhouse effect. This threat has been responded well in time by the world in the Montreal Protocol, an international treaty which will eliminates CFC use by the year 2000 AD (Hoffman, 1990). Another situation, very close to this, has occurred in the recent Gulf war, where several oil wells in Kuwait caught fire emitting huge quantities of smoke and shoot. These were carried by the wind to the neighboring states of Iraq and Iran covering most of their sky. Even shoot and carbon particles were found in the snowfall on the Himalayas of the Indian sub-continent.

The emission of pollutants into the environment is a complex phenomenon and cannot be limited to the fixed image of waste pipe spilling out its effluents into a lake or river. In almost all cases, substances discharged into the environment are going to be carried a long way from their source. Atmospheric and hydrologic circulation systems then disperse them progressively throughout the biosphere. The pollutants act in many ways on the living systems. Acute and/or long-term actions are the most obvious eco-toxicological impacts of a given pollutant. It results demo ecological effects at population levels, which are displayed through immediate or premature death, reduced reproductive success, reduced growth, and/or increased loss at the juvenile stages. These are ultimately reflected in the lower abundance and perturbed distribution of the exposed populations of sensitive species. The effect of the sub-lethal concentrations of some pollutants is by far the most frequent eco-toxicological problem and it can be far more noxious in the long run for the exposed species resulting in chronic deterioration of the ecosystem. The persistent non-biodegradable pollutants are absorbed by living organisms and are concentrated many times more than the concentration of the surrounding medium. Further, these bio-accumulated pollutants are passed from one trophic level to another and exhibit increasing concentration in organisms related to their trophic status resulting in bio-magnification.

Industrial Pollution and India

Industry is responsible for creating a fantastic array of new chemicals every year all of which eventually find their way into the environment. For most of these chemicals, not even the chemical formula is known and much less is known about their acute, chronic or genetic effect on the living organisms. At present, the industry is the focus of attention, the world-over, as the most important source of pollution of the environment. Chemical industry in India has grown up phenomenally since independence. There are today about 4000 chemical factories in India. They release large quantities of chemicals in the form of gas, liquid and solid wastes, into the environment. Many of these chemicals are toxic and create pollution problem. The problem of toxic hazard has already reached alarming proportions in this country and is bound to grow with increasing industrialization. India is the largest manufacturer of pesticides in the whole of South Asia and Africa. As many as 139 organic chemicals, heavy metals like zinc, lead, chromium, copper, mercury and various other compounds are used in the manufacture of dyes only. Sundaresan *et al.* (1983) have given the growth of industries dealing with toxic chemicals and generating toxic and hazardous wastes during 1950s, 1970s and 1980s, in India.

Industries are known to produce potentially toxic and hazardous wastes are the pesticides, dyes and pigments, organo-chemicals, fertilizers, non-ferrous metals, steel and chloro-alkali plants. The major foci of such industries in India are Bombay, Calcutta, Kanpur, Delhi, Chandigarh, Jamshedpur, Bokaro, Hyderabad, Vishakhapatnam, Madras, Baroda, Cochin etc. The wastes from the industry are generally disposed by land filling or released into water bodies. Generally the toxic effluents from industry are neutralized before their discharge, but still they contain substantial amount of toxic substances that can cause pollution. At present it is believed that rivers are most severely polluted by industries followed by estuaries, lakes and ocean in declining order. Extensive literature on the pollution of surrounding biota through the discharge of effluents and sludge from chlor-alkali industries is available (Weiss *et al.*, 1971; Suckcharoen, 1979 and 1980; Lodenius, 1980; Suckcharoen and Nourteva, 1982; and Lodenius and Tulisalo, 1984). Studies on loss of mercury into the environment from the chlor-alkali industry were carried-out by, Bouveng (1968 and 1972), and Ross (1974). Turney (1970) reported contamination of river Detrait by mercury discharges from a chlor-alkali plant, similar studies were also conducted by Matida and Kumada (1969) on the river Agano of Japan, by Suobodova and Hejtmanek (1976) on the Ohre river and Carlos (1979) on the Jab river of Papua, Newguinea. Hattula t al. (1978) indicated a severe pollution of the lake Paijane, which had received in past, the waste lischarge from a caustic soda factory.

The purpose of acute toxicity test is to assess various abnormalities caused due to administration f a chemical to live systems on one occasion or other and to determine the order of lethality of the hemical. Many of the schemes are available at present, to test for adverse effects that may result when hemicals enter into the environment. In aquatic toxicology, acute lethal toxicity tests with the fish, eed germination of crop plants and invertebrates are usually intended to assess the numerical value f toxicity, to compare potentials of toxicants, to assess the effects of environmental variables on oxicity. Besides, statistical analysis of toxicity is essentially a tool in all these studies. The rapid ncrease in contamination of aquatic environments with industrial pollutants in recent years has esulted in an escalation of scientific interests in the biological effects of pollution. Because of the mportance of crop in the aquatic environment, any deleterious effect of the toxicant on this is likely to e reflected on the entire ecosystem. Little information is available pertaining to the toxicity studies on rop plants due to the waste of industries. Metallic cations are highly essential for many enzymes to chieve their catalytic activity. These cations maintain the acid base balance; and are also highly ssential for the ionic composition of the body fluids. Disturbance in electrolyte concentration are uite common in toxicity. Different studies on electrolyte change resulting from toxic status have roduced a variety of diverse effects, because of differential size of toxic actions in the body. The eaction of mercury compounds with sulfahydryl groups has been well established. Bruin (1976) eported that cation movements within tissue cells originate from toxic effects on cellular membranes. ungicides of diverse chemical structure are capable of changing mitochondria function, thereby naking the membrane permeable to cation. Such cations are the mobile co-factors which either firmly ound to metal are called metaloenzymes or metal activated enzymes. Mercurials have long been ecognised as agents interact with the poison proteins in general and enzymes in particular (Fox *et al.*, 975). The DNA molecule is a target site of most, if not all, carcinogenic and mutagenic agents. dditionally, a number of agents are acutely cytotoxic to cells because, they damage the DNA. Numerous hemical agents damage DNA in vivo (Ahmad *et al.*, 1977; Ahuja and Subrahamanyam, 1978). Geener nd Jeener (1952) reported the reduced RNA synthesis due to methyl mercury intoxication. Reports re available pertaining to the effects of pesticides on rat; pesticides induced DNA damage in cultured uman cells (Ahmad *et al.*, 1977), reappraisal of lead on cell proliferation and influence of lead on ucleic acid metabolism in rat, effect of inorganic mercury on different biochemical parameters of eshwater fish (Panigrahi, 1980). Different aspects concerning the biochemical assessment of functional npairment in vital organs were explored in toxicological studies. Untoward effects can usually be aced back in the functional incompetence of organs. Whenever dysfunction occurs in a tissue it has s origin in biochemical abnormality may become apparent before the onset of morphological changes nd precedes the development of chronic degenerative disease. Explanation of biochemical tests acilitates diagnosis of pre-toxic condition in man and it may be valuable in the routine screening of emicals in animal experimentation. Biochemical studies are the useful tools in aiding diagnostic athology, which covers a wide range embracing various organs and systems. Pesticides act either as selective toxicant or may display rather a broad spectrum of adverse biological activities. Toxicant fects on growth may be studied by examining macromolecules involved in growth, such as DNA, NA, amino acid and protein. Life depends on a complex network of chemical reactions brought bout by a specific enzyme. The enzymes are the primary instruments for the expression of gene ction. Enzymes are of biological importance in metabolic functions, which have the most highly pecialized class of proteins. The change in metabolism can well be correlated with the disturbance in ehavior and action of the enzymes (Panigrahi, 1984). The reduction in active metabolism has been correlated with the toxic effective different types of toxicants. The disturbance in enzymatic action can be related to the specific action of certain toxicants by inhibiting the enzymes to act, bind the active sites of the enzyme and enzymes become non functional due to disintegration. Inhibition of enzymes by chemical agents may be reversible or irreversible, competitive or noncompetitive. There are many different mechanisms through which thiol groups of enzymes can be acted upon by heavy metals. The importance of enzymatic studies in the system lies in the functional interpretation of the pollutant causing different types of disorders. Membranes play a role in the structure and function of the cell and its organelles. One of the way by which membranes can affect cellular metabolism is by interacting with enzymes. Some membrane function as the organizing matrix for enzyme systems, the micro environment in all these cases at or near the membrane surface may modulate the activities of the enzymes located in the region.

No information is available on the effects of Red Mud wastes of Aluminium industry on biota. Few preliminary reports and news publications are only available on pollution problems of NALCO plant. Unfortunately, NALCO plant has received the best award for pollution control and management.

The Industry Under Study

The products of his own activities have always burdened man. Hence, the present situation that exists is different from the historic one, in several important ways. Amongst which, the most important remarkable difference is the way at which man has contributed to the environment, both good and bad. This difference is mostly due to inter related phenomena *i.e.* growth of population, increasing urbanization and rapid industrialization. Every underdeveloped or developing country aiming to be highly developed, explore nature and tried to find out methods and adopted industrial revolution leaving behind the consequence of such trend. When the problems of environmental status and quality was taken into consideration it became apparent that, it was not enough to alleviate undesirable consequences of man's activities by technological means or to prevent them forbidding certain practices of the are of certain chemicals.

Rapid industrialization and exploitation of natural resources on massive scale, accumulate undesirable substances in huge quantity thus, pollute the environment. Pollution is caused due to deliberate or unavoidable discharge of industrially produced wastes. These pollutants causing health hazards and definite and undesirable problems to flora and fauna, apart from influencing the ecological balance and brings significant climatic changes. The field under study refers to the Allumina industry specifically to Mining and Refinery complex, NALCO, Damanjodi in the district of Koraput, Orissa, India. For analysis of various factors of pollution in Allumina industry it becomes imperative for us to discuss the manufacturing process from ore mining to refining in brief with reference to NALCO. The allumina is extracted in two main stages *i.e.* Mining operation and Ore refinery. The ore of Alluminium *i.e.* Bauxite mine is situated at Panchaputamali in the district of Koraput, Orissa. It is at the height of 400ft from sea level. NALCO has marked 17sq. Kilometers of its area for getting deposited Bauxite. 37 million tones deposit in estimated. The mining capacity of NALCO is 24,00,000 tpa and being expanded to 48,00,000 tpa. The method used to mine Bauxite includes exploration and mine planning, removal of topsoil and over burdened and excavation of ore blasting operation. The hard cap rock at the surface is blasted to allow it to be mined together with the ore beneath. Heavy earth moving equipment's like dozer, back-end hoe, front-end loader and dampers etc. are used for mining and transporting operations. Then the ore is crushed by single state double roll tooth crusher of 900 tpa for each transportation and dispatched by various means to the refinery complex, which is done by a long cable conveyer belt systems of 14.6 Km long at single figat multy curve with 900tph capacity at M ar

R complex. Damanjodi. It is the longest conveyer belt of the world. The entire distance from the mining site to refinery complex, the air of the area looks cloudy due to heavy dust loading from operation and open transportation by conveyer belt. In the course of mining exploration and transportation, minute dust particles looked like the labourers working in that area as well as the inhabitants residing nearby inhaled few particles. This air heavily loaded with dust or suspended particles cause lung cancer and tuberculosis. A health report indicated that high percentage of the inhabitants suffer from these two diseases and also respiratory distress and bronchitis: Not only the human beings are affected but also this pollution spoils very valuable flora of the environment due to heavy deposition of red coloured dust. The dust are deposited on the green leaves of the plants, reducing the incidence and trapping of sunlight, coming from sun, the only and ultimate sources of energy. Thus reducing the photosynthetic efficiency of plants. It is also noted that during the transportation by conveyer belt, huge amount of crushed Bauxite powders are dispersed, mixed with air, thus, the inhabitants are suffering a lot. Many news items appeared on this issue in different Newspaper and public reacted very badly opposing the way the area is being polluted.

Due to the set up of the industry, not only 17 Km^2 of PANCHAPUTAMALI hills will be deforested but also 1742acres of acquired land were subjected to deforestation, disappearance of valuable flora and disappearance of wild life. The NALCO project-plantation program in barren lands thus reclaimed by back filling and on the two sides of the newly constructed roads. But yet this did not fulfill the purpose of reappearance of the forest species and lead to degradation of biodiversity. The noise and vibration due to continuous blasting and heavy machinery equipment caused noise pollution in the mining area. 'Environment Management Plan' At Allumina Refinery Complex. NALCO has taken all possible measures during construction and also taking all possible measures for monitoring the pollution control parameters so that no pollutant is allowed to be discharged from plant as process effluent or township effluent to the water body without treatment, conforming the IS norms.

Tree Plantation for Ecological Balance

Extensive tree plantation program has been started around for reforestation of the same area. It is very important as trees and forest moderate the climate. Micro- climate in the forested area is distinctly different from the adjoining non-forested area. This occurs mainly through their influence on wind, temperature and humidity. Evapo-transpiration process increased humidity and also lowered atmospheric temperature in the area. Trees not only assimilate CO_2 and release O_2 but also play an important role in trapping some toxic gases and particulate matter in the air. Hence forests function both as filters and sinks for air contaminants. It also reduced noise pollution and enrich aesthetic beauty of the area. For soil conservation and soil stabilisation work, small diversion dams and cross bounds are planned in down flow hills. As well as Lemon grass turfing, Mulberry plantation and cashew-nut plantation are planned. Yet after 14 years, no good forest has been developed. The Alumina plant is surrounded by hills in three sides. Under its "Green belt program" plantation was planned in three concentric rings. First ring of plantation was done in the surrounding hills. Second ring of plantation was carry out in the valley lands and third ring of plantation was done beyond the township area. In this program, first of all soil stabilization work by the sides of newly constructed road, crusher house pump and near conveyer were under taken. Over these dumps and hills sloped so far 1.5 lakhs of plants of different species were planted. Till so far the above green belt plantation is a dream and remained in a planning stage. Land scaping was take-up by planting flowering species like *Jacaranda* and *Cassia siama* of 50mts. width near bauxite storage area and in the township.

Proper program was made for heavy plantation of the hills surrounding the red mud pond and ash pond to keep the ecological balance of the area. The following steps were taken for afforestation of these hills after expansion of Red Mud Pond. (*i*) The D.F.O. Rayagada division under took the afforestation program in central hills. The species selected were mostly of every green in nature and a mixture of monocultures of cashew-nut, mango and jack fruit etc. were planned. (*ii*) With regard to the surrounding hill near Sundhipara and Kantagada afforestation will be taken up by planting:

1. *Accacia auriculiformis*
2. *Hybrid Eucalipatus* sp.
3. *Pongamia pinata* and
4. *Anacardium occidentale*

(*iii*) For stabilisation of the slopes by the sides of the guard land drains in RMP, Lemon grass and Palm rose oil grass will be planted Hills. But in these areas, the dry mud of the industry is being dumped in huge quantities. In future small hillocks of dry red mud will be formed. In rainy season, the leachates of those wastes might enter into the RMP pond. Hills surrounding RM Pond, Dumurigude and Dengajanigada villages are also planned for afforestation program. The workers NALCO Damanjodi as well as the tribal inhabitants are affected by the poisonous gases and ash emitted from the chimney. This also affects the natural life in that area. It is also seen that micro particles of allumina and ash emitted during the process are deposited on the leaves of the herb and trees and cause chlorosis. The physicians suspect the increasing number of patients suffering from asthma and lung cancer may also be due to inhalation of powdered allumina and ash from the plant along with the ore dust spread over the area.

Ore Refinery

The capacity of Allumina plant at NALCO, Damonjodi is 8,00,000 tpa and being expanded to capacity of 15,75,000 tpa. The ore refining is done by a classical process called BAYER's Process, in which the aluminum in the Bauxite is digested in caustic at 105°C and ultimately recovered by precipitation with the removal of impurities. After successive operation of digestion, desilication setting and washing the precipitated hydrate *i.e.* $NaAl(OH)_4$ is filtered and calcined at 1000°C. There are two production stream of 4,00,000tpa capacity each atmospheric pressure digestion process i carried out at 107°C with energy efficient fluidised bed calciners of 1400tph (3 × 18.5MW) power is co generated by use of back pressure turbine in steam generation plant. The residual mud is separated and pumped to the red mud pond.

Red Mud Pond

In the process of bauxite refining the red mud, which is generated after digestion is the major waste product. For 8 lakhs MT allumina production 10 lakhs MT of red mud is generated and is being discharged in red mud pond by 2nd stage pump in slurry form at the rate of $370m^3/hr$ with 150gp solid consistency and 2.3 gpl Na_2O the dry red mud flow in the RMP is 110MT/hr. To store such a huge quantity of red mud slurry, the Red Mud Pond has been constructed towards the west of refinery complex. Three earthen dams called Orissa Dam-1, 2 and 3, flank the Red mud pond. Fortunately or unfortunately natural hill ridges with high elevation surround the red mud pond. These three earthen dams connect the natural ridges. The Orissa Dam-1 is towards east *i.e.* towards the refinery complex the Orissa Dam-2 towards north, the Orissa Dam-3 is towards west *i.e.* opposite to dam-1 to dam1 thi dam separated RMP from ash pond area. Hill ridges with high elevation covered South of the pond

Orissa dam-2 has a natural hill ridge with 930 meters elevation. No spill way is provided in the dams to avoid spillage contamination to natural nalaha. Elevation of dams during the first phase is 930 meters the anticipated life is for 3.5 years. It is planned to expand the elevation up to 945 mts and 957 meters during second and 3rd phase operation. The anticipated life of the pond is 7 and 11 years respectively.

The initial elevation of RMP was 905 mts at different increased elevations the catchment area and the capacity of RMP are given below. At 920 mt elevation catchment area is 16.544 hectares with total capacity of 301×10^4 m^3 at present elevation *i.e.* system which needs very less area for the disposal and stacking of 930 meters the area is raised up to 108.6 Hectares and 1016.17×10^4 m^3 capacity. After second phase expansion up to 945 meters; the area will be raised to 187 Hectares with capacity of 3305.87×10^4 m^3. At final phase of expansion up to 957 meters elevation the total catchments area of the pond will be 212.40 hectares with total capacity of 5728.82×10^4 m^3 with 2mts free board. After the life of the red mud pond is over, one area has been already ear marked during the feasibility study for the construction of second pond. Horizontal expansion of red mud pond is feasible to certain extent. During last 3 years it is noted that the pond level is rising abnormally due to more precipitation than evaporation at Damanjodi. In order to check the pond water level the alkaline water is returned back from the pond and is recycled for use in plant various process mainly to carry fresh generation of red mud slurry as a result alkalinity of the pond water is increasing day by day. The shore of this pond is extremely affected. The water of the pond is toxic to the plants. Although the life of the dam in phase-I is over, yet the second phase expansion work has not started, rather in order to explore the long-term solution to the wet tailing system, Dry mud disposal system, which needs very less area for the disposal and stacking of dry mud. It does not call for construction of large earthen dams. The system multiplies alkalinity of the area of dispersal. The run-off leachates of red mud stacked near RMP are entering into the pond with rainwater. This is also an addition to the alkali strength to the RMP water instead of diluting it.

Table 47.1: Analysis of Mud of RMP

Parameter (per cent)	*Sample–I*	*Sample–II*	*Sample–III*
Al_2O_3	19.11	16.60	18.56
Fe_2O_3	56.88	57.93	58.77
TiO_2	4.56	4.03	4.15
SiO_2	5.97	6.67	6.94
Na_2O	3.86	4.40	4.10
pH	9.90	10.02	10.06
Return water pH	14	–	–
NaOH	30-35 per cent	–	–
Na_2CO_3	65-70 per cent	–	–

NALCO the Industry

National Aluminium Co. Limited (NALCO) is considered to be a turning point in the 50 years old history of Indian Aluminium Industry. In a major leap forward NALCO has addressed itself to the country's need for self sufficiency in aluminium as well given the country a technology edge in making this strategic metal on the best of world standard. NALCO was set up in 1981, in the public

Table 47.2: Physical Properties of Alumina

	Typical	*Guaranteed Range*
LOI (300-1000° C) per cent	0.6–0.8	0.5–1.0
BET surface area (M^2/gm)	60–70	60–80
Alpha Aluminium content per cent	–	10 Max.
Granulometry		
+125 Micron (per cent)	–	15 Max.
–45 Micron (per cent)	10	12 Max.

Chemical Properties

	Typical	*Limit*
Al_2O_3(per cent)	98.7	98.5 Min.
Na_2O (per cent)	0.38	0.5 Min.
Fe_2O_3 (per cent)	0.01	0.015 Max.
SiO_2 (per cent)	0.012	0.02 Max.
CaO (per cent)	0.042	0.050 Max.

Alumina Hydrate: Physical Properties

	Typical	*Maximum*
LOI (110–1000°C) per cent	34–36	36
Moisture	3–6	8

Granulometry

	Typical	*Maximum*
–45 Micron (per cent)	3–6	7

Chemical Properties

	Typical	*Max/Min*
Al_2O_3 (per cent)	65+0.5	64 Min.
Na_2O (per cent) Total	0.23–0.30	0.33 Max.
Na_2O (per cent) Soluble	0.015–0.025	0.30 Max.
SiO_2 (per cent)	0.007–0.010	0.016 Max.
Fe_2O_3 (per cent)	0.006–0.008	0.016 Min.
Hydrate Content (per cent)	99.0	98.5 Min.

sector, following a major investment decision of government of India in 1980,to exploit a part of large deposit of Bauxite, discovered in the east coast in 1975. Aluminium Pechiny of France provided the technology and basic engineering for Bauxite refinery, smelter and mine. The project cost of Rs.2408 crore was partly financed by Rs.1119 crore equivalent of Euro- dollar lone extended by a consortium of international banks. The project was most expeditiously implemented, different, segments of NALCO want into production in a phased manner from November 1985 sale of Aluminium was commercial on May 1987 and export was started on September 1988. Initially 1987-88 the Aluminum production was 160,000 tones the production rate remarkably increased to 807000 tones in 1995-96, in 1996-97, 840,000 MT and in 1997-98 it was 883,000 MT and has chartered a course of international confidence in India's industrial capability. To day NALCO has emerged as the largest integrated Bauxite Alumina -Aluminium complex in Asia, enabling India to witness a quantum jump in Alumina and Aluminium production. NALCO produces calcined allumina at refinery complex, Damanjodi with following specifications.

Topology and Location

The mines and refinery complex of NALCO, Damonjodi is situated Similiguda block, under Potangi tahasil in the district of Koraput, Orissa state, India. From the district head quarter Koraput, it is 38 kilometers towards southeast on road, *i.e.* 27 kms towards south in NH-43 up to Similigude junction and further 11kms towards east on project road. It is 60 kms from Jeypore, the oldest city of Koraput district. It is 560 Kms away from Bhubaneswar the state capital city of Orissa and 172 kms from Visakhapatnam (A.P), nearest air port. Damonjodi is at a highest of about 1300 mts from sea level, located at latitude 18°-6′–18°-58′ towards North and longitude 82°.57′-83°.04′ East. The area enjoys an annual rainfall of 1723 mm in 1998. The relative humidity of area ranges from a minimum of 47 per cent in April and maximum of 85 per cent in July 1998. The temperature of the area varied from a minimum of 15°C to maximum 35°C in 1997-98 and from a minimum of 9°C to maximum of 42°C in 1998-99. Southeast side is covered by Deomali hill range. The slope of the area is towards the northwest. A small river Karandia is flowing near by the Refinery complex and the Damanjodi township. It carries almost all the emissions of the plant and township. At its down stream mixed with the Kolab River at its catchment area of Upper Kolab project near Sunabeda.

Affected Area

As we know the Red Mud pond is situated in such a place that the pond water is restricted by high hill ridges at south, ash pond at west separated Orissa dam-3. But at its north seepage is marked as a natural nalaha at its down stream about 100mts away from Orissa Dam-2 of RMP. This seepage might be due to the storage of huge quantity of effluent at a high level from the original point of leaching. The leachates gradually flow downwards mixed with natural nalaha and ultimately enter into the crop field. These leachates from RMP gone down in its slope towards the river Karandi in the north west direction and finally mixes with the river Kolab near Sunabeda. As a result, cultivators of villages, Sukuriguda, Janigada, Dengajaniguda, Kharaguda, Khajuriguda, Kalapadar and other village in the vallery, depending on the water source for their agricultural purpose are being harassed. The contaminated water in the catchment area of upper kolab project is being lifted for Sunabeda as well as Damanjodi town-ship for all domestic use and human consumption. Ultimately this polluted water flows down along the river and being used by the inhabitants of Jeypore town as well as the village at the riverbank.

Toxic Effects

The leachates at its initial point is highly alkaline in nature with Fe, Al, Ti, Na ions the pH of i ranging from 9 to 10 even more. Gradually when the leachates entered into the rice crop field, i influenced the rice plants very badly. The cultivators generally use the medium high yield and resistan variety (Lalat). Due to the effects of toxic chemicals rice plants do not grow well. It is found that the soi colour of the crop field changed to violet red with high caustic smell. On the stem, and leaf lamin radish colouration is developed ultimately resulting chlorosis. Straw-yellow long slender grains o lalat variety of rice do not reproduce; the size became smaller with development of reddish brow colour on the seed grains as well as on the panicle. The peduncle was weak to bear the load of th inflorescence. At the alternate year the seedlings could not be established due to less rainfall durin month of the crop (May-June), the concentration of the leached chemicals became very high with pH more than 8.

Year Wise Study

In 1995, it is studied that the rice plant become very short, reddish colour spots developed on lea lamina and on the stem. Roots became thickened and curled and number of lateral roots is relativel less. The grain size is remarkably reduced and the colour of the grain was brick red instead of straw yellow. The husk production is relatively more. In 1996, from the study in the field it is observed that maximum seedlings after germination attained an height of 6″ to maximum 1ft (at low density o pollution area), the rottening of roots started the seedling could not be established within a period o 25 to 30 days. These effects ultimately resulted nil production that year. This happened due to increase alkalinity on the crop field (pH –<10) for less rail fall during first month (April-May). The colour of th soil in the crop fill was violet red with high caustic smell.

In the year 1997, the rainfall during the period of seed germination was better then the last year Hence alkalinity of the leachate contaminated crop field was also relatively low (pH–7.68). It als found that the typical effects due to the toxic leachate were lesser as we go far from the RMP. In tha year the typical symptoms reappeared as on the 1995. Through personal interview with the cultivators of the nearby village Sukriguda, it was reported that-

1. In general the production rate is decreasing every year.
2. Before the construction of RMP, the production of Lony baele was good.
3. Rice grain size became smaller.
4. Seed colour turned reddish.
5. Red colouration developed on the leaf lamina and the stem.
6. Root was not well developed, no lateral roots are less and those are thickened and curled.
7. Most of the rice plants are affected by chlorosis.
8. Peduncles were so week that, they do not bear the load of the inflorescence.

Other Effects

The general public opinion flashed several times in local dailies and weekly news papers that the undesirable waste emissions of NALCO mixing with water bodies, pollute the river and the reservoir, which man used. It was reported in one of the newspaper that, there is every possibility of loosing the memory power as well as reproduction ability by animals including man, if the polluted water is used for consumption. It is published in a newspaper that by using the contaminated water for drinking purpose, many people suspected their deafness and some have lost their reproductive ability. Now users report different types of skin diseases caused by polluted water is to be because of the contamination. On the present issue/problem many physicians have expressed their fear and also apprehension that in future, this industry causing pollution may create wide spread health hazards. Not only the human community is facing hazards but also animals, aquatic flora and fauna are also affected. It is expected that the toxic pollutants might be accumulated in the body of the contaminated fishes, which can be observed from the symptoms like the scale souring. Some time heavy emission lead to fatal condition of fish and aquatic organism. The abnormal deposit of red mud in the red mud pond and increasing in alkalinity of red mud pond water may become a future threat for the people near by locality in Koraput district, because of heavy leaching from the red mud pond.

Keeping in view; the leaching of chemicals from the Red mud pond of the industry (Damonjodi) and entry of these chemicals into paddy fields of the locality and runoff water of the dumps during rainy seasons and its entry into paddy fields and their possible effect on the crop plant in the crop fields, this project was undertaken to evaluate the eco-toxicological effects of the leached wastes on the germination, seedling growth and establishment of the rice plant, *Oryza sativa*, L.

Materials and Methods

Study Material

The seeds of *Oryza sativa* L. (Cult: LALAT, OR 26.2014-14) [Parentage-OBS 677/IR 2071/VIKRAM/W1268, Released by Orissa] was selected as test material for this study. The pure line seeds were collected from O.U.A.T. Rice Research Center, Jeypore Branch, Koraput. Before conducting any investigation, the seeds were further sampled by hand sieving and sorting, to obtain healthy and uniform seeds.

Choice of the Test Material

1. A popular early and high yielding variety of rice, found to be suitable for cultivation in medium and high lands.
2. Grains of this variety show high percentage of germination both in the field and in laboratory conditions.
3. It is cultivated in all seasons and in various agro-climatic conditions because of its special type of physiological adaptations such as: Weak photosensitivity and minimal variation in physiological processes in response to climatic changes.
4. Seed disinfections is recommended for this variety as seeds and seedlings/plants are very much susceptible to fungal, bacterial and viral pathogens rather than insects and pests. This variety is resistant to Gall fly, BPH, sheath rot and medium resistant to stem borer.
5. Adopted to direct seed sowing and also by transplantation methods.

Some important information about Cult: Lalat of *Oryza sativa* L. as relevant to this piece of work is detailed as follows:

Important Technical Information About the Seed Material (Rice)

A) Cultivation Manual

1. Name of the cultivar: Lalat
2. Type: Medium high yielding resistance variety (Irrigated type). This can also be used in non-irrigated land, widely cultivated by the cultivators of Koraput.
3. Suitability: For both high and medium land

4. Duration: 125-130 days
5. Sensitivity: Weakly photoperiod sensitive can be grown in various agro-climatic conditions.
6. Tillering ability: 299/25 plants or 10-12/plant
7. Average plant height: 85-90 cm (80.7cm (av.)
8. Plantation practice: 15 × 15 cm. Distance to be maintained between plants during plantation.
9. Seed quantity/acre: 30 Kg.
10. Leaf sheath colour: Yellowish green; Spikelet colour: Straw yellow; Kernel colour: White
11. Panicle length: 22.2cm
12. Yield: 45 quintal/hectre (Average)

B) About Grain Character

1 Average no. of grains/panicle: 124 ± 08
2. Grain type: Long and slender
3. Length/breadth of the grain: 8.5/2.5
4. Number of grains/Kg.: 45000

Test Chemical

The leached chemicals of the red mud pond of the NALCO plant prepared in the laboratory have been used as the test chemical for the experimental study. The red mud collected from RM pond is air dried, powdered and sieved. A known (2kg.) quantity of dried powder is mixed with known amount (2lit.) of distilled water and stirred for 15 days. The mixture was allowed to stand for one day and the supernatant was decanted and again passed through a multisieve soaking and filtering system and the obtained extract is the leached chemical of the waste, which is stored for future experiments. This is the red mud extract and known as the test toxicant.

Study of Seed Viability

The seeds were soaked with different concentrations of test solution from 12h to 96h. After scheduled period of exposure, the grains were washed with tap water, followed by distilled water and their viability were tested according to the procedure of Moore (1969). The result was expressed as percentage of mortality.

Germination Study

Before the utilization of seeds in the field, the germination studies must be conducted under controlled and standardized laboratory conditions, because:

1. Results of germination test should be known before sowing.
2. Care can be taken to reproduce the results of laboratory conditions in the field condition.
3. The maximum germination potential of the seed lot can be known easily.
4. Undesirable factors can be eliminated, so that influence/effect of test extract can be sorted and without any interference.

But all these above conditions are neither possible nor practicable in field conditions (Annonymous, 1989). Hence for germination, the sterilized petridishes with sterilized absorbent cotton and blotting paper were utilized. Always twenty-five seeds were kept in each petridish and each treatment including control. For germination and growth studies mainly two types of treatments were carried out:

1. Short-term treatment in recommended concentrations of toxicant in petriplates both in racks and seed germinator.
2. Long term treatment in experimental concentrations of the toxicant in pot culture.

Short-term experiments were conducted in petriplates. A selected concentration of the extract was applied and the constancy of the medium was maintained both in open rack culture condition and in Seed Germinators. Long-term experiments were conducted in pot cultures. Where exact amount of dried red mud powder was mixed with soil: biological fertilizer mixture and water was added daily to maintain constancy.

In short term experiments readings were noted after 144hrs of exposure. In long term experiments, the parameters were studied after 15, 30, 45 and 60 days of exposure. In the third type of investigation, the seeds were treated with different concentrations of RM leached chemical for 24 hours and then kept for germination in those concentration media, as the seeds were soaked previously. For this experiment, a graded series of concentrations of the toxicant was prepared for this pilot test considering the actual amount of the test chemical present and the amount of this test chemical available in the affected areas at Damonjodi, Koraput. On the next day, 2nd dose was given *i.e.* petridishes were supplied with same quantity of freshly prepared solutions of different concentrations. On the next day, 3rd dose was given to compensate the possibility of biological detoxification, oxidation or loss of the toxicant in the process of evaporation. Then after the set was regularly supplied with distilled water at an interval of 12 hour. In all sets, the seeds were allowed to germinate in normal photo inductive cycles, normal humid atmosphere and at a temperature of $30 \pm 2^{\circ}$ C for 48 hours. After 48 hours of normal soaking and incubation, all seedlings were transferred to growth chamber (Culture rack and Seed Germinator [Remi].

To calculate the percentage of germinations, periodical observations were made at an interval of 24 hours from 48 hours to 144 hours. First emergence of coleorhiza (Referred as Root) about 2 mm. in length was considered as normal germination, so also appearance of coleoptile (Referred as Shoot) before emergence of coleorhiza was considered as abnormal germination. From this pilot test, a range was selected by visual screening and percentage of germination and percentage of seedling establishment was noted. The seeds were allowed to grow in petridishes in growth chamber. Visual screening was carried out basing on the growth and the percentage of seed germination and percentage of seedling establishment was also calculated. A test was run within the expected range to select the concentrations for the study.

Study of Seedling Growth (Petriplate Culture)

In petriplate culture, first the petriplates are cleaned with tap water then with chromic acid and then tap water and finally with double distilled water, and dried in an oven. Twenty-five soaked seeds were sown in petriplates at uniform distance in all the sets. All petriplates were kept at room temperature and normal photo-inductive cycle was maintained by providing light from fluorescent tubes in daytime. For all other studies, seeds were grown up to seedling establishment stage (up to 7th day of exposure to the toxicant), inside petriplates in different concentrations of the toxicant including control set, as it was done in germination studies. On each 7th day of exposure, seedlings were harvested for respective experimental analysis. From the above-recorded parameters, three concentrations were selected. Future experiments were conducted at LC_0 (MAC value–Maximum Allowable Concentration) LC_{10}, LC_{50} and

LC_{90} values, to find out the effects of the toxicant on the growth performance of crop plant, *Oryza sativa* L. Same concentrations were selected for pot culture experiments and three doses of the toxicant was applied.

Table 47.3

Lethal Concentration	*Concentration of the RM Leachate*
LC_0 (A)	0.21 per cent
LC_{50} (B)	1.25 per cent
LC_{90} (C)	2.02 per cent

Values were selected from the toxicity test.

In future experiments always ten seeds were kept in each petridish and each treatment including control instead of twenty five seeds as was done in pilot tests.

Study of Morphological Parameters

Shoot and Root Length

From each replicate shoot and root lengths were measured, with the help of a scale.

Shoot and Root Weight

Seedlings from each replicate were taken. The shoots were separated from the roots. These were washed thoroughly, surface dried with the help of blotting paper. Then fresh weight of roots and shoots was taken separately by a single pan electric balance.

Pigment Studies

The amount of total chlorophyll and total phaeophytin was calculated by using the formula given by Vernon (1960). The amount of carotene was calculated by using the formula given by Davies (1976).

$$\text{Total chlorophyll (mg g}^{-1}) = (6.45 \times \text{OD at 665 nm}) + (17.22 \times \text{OD at 649 nm})$$

$$\text{Total phaeophytin (mg g}^{-1}) = (6.75 \times \text{OD at 666 nm}) + (26.03 \times \text{OD at 655 nm})$$

$$\text{Carotene content (mg)} = \frac{\text{D V K}}{2500 \times 100}$$

where,

D = Optical density at 475 nm, V = Volume of the extract, K = Dilution factor, 2500 = Specific extinction coefficient at 475 nm

Oxygen Evolution and Consumption Measurements

The evolution of oxygen due to photosynthesis and consumption of O_2 in respiration were measured manometrically with the help of a Photo-warburg's apparatus (New Paul, India) following the procedure of Hannan and Patouillet (1972) and Oser (1965). Rate of photosynthesis was expressed in ml of O_2 evolved hr/1 g of the plant and respiration rate was expressed in ml of oxygen consumed hr^{-1} g^{-1} of leaf of the plant. The GPP (Gross Primary Production) and NPP (Net Primary Production) values were calculated from the values obtained from both the oxygen evolution and oxygen consumption studies.

ATPase activity studies: The enzyme activity was studied following the method described by ahu (1987). Estimation of inorganic phosphate by molybdenum blue method (Martineck, 1970) was ɔllowed.

Estimation of Nitrogen: Nitrogen was estimated by Nesslerization method (Herbert *et al.*, 1971).

Estimation of Glucose: The glucose was estimated following Herbert's method (Herbert, 1971).

Results

Healthy selected seeds were exposed to graded series of concentrations of leached chemicals of ne red mud waste in sterilized petriplates, to find out the lethal concentration and percent survival alues for the present set of experiments. Series of pilot tests were conducted and the experiments were epeated at least thrice to determine the lethal concentration values. The Table 47.4 indicate the percent urvival (PS) and lethal concentration (LC) values for the leached chemicals of the red mud waste.

Table 47.4: Per cent Survival and Lethal Concentration values, when Rice Seeds were Exposed to Leached Chemicals of the Red Mud Waste

Percent (PS) Survival Value	*Concentration of the Toxicant (per cent, v/v)*	*Lethal Concentration (LC) Value*
PS_{100}	0.075	LC_0
PS_{90}	0.2	LC_{10}
PS_{50}	1.25	LC_{50}
PS_{10}	2.0	LC_{90}
PS_{00}	2.2	LC_{100}

During lethal concentration studies, seeds and seedlings were carefully watched to find out the bnormalities during seed germination and seed establishment. No significant morphometric changes ere observed. Blackening of the shoot at the base of the exposed seedlings was marked. The entire ɔot became black after few days of exposure. Some of the plants became pure white, the green colour f the plant totally disappeared. Control seedlings remained clinically healthy. After prolonged xposure, the exposed seedlings became brown and dry. At very high concentrations, roots became nall and less in number.

Cent percent germination after 144hrs of exposure was recorded in the control and at 0.1 and 0.2 er cent RM extract. The percentage of seed germination decreased with the increase in RM extract ɔncentration. No germination was recorded beyond 2.6 per cent RM extract. Cent percent seedling stablishment was recorded in control and 0.1 and 0.2 per cent RM extract. Seedling establishment ecreased with the increase in the RM extract. No seedling establishment was recorded beyond 2.2 per ent RM extract (Table 47.5). Basing on the data, 0.2 per cent RM extract concentration was considered s 'A', where cent percent seed germination and seedling establishment was recorded, as maximum lowable concentration for the toxicant under study. Concentration 'B' and concentration 'C' were elected basing on the same data, where approximately 50 per cent and 10 per cent germination and eedling establishment were recorded. The correlation coefficient analysis indicated the existence of ighly significant ($P \leq 0.001$) correlation between percentage of seed germination and toxicant ɔncentration in petriplate culture. But in case of pot culture, non-significant negative correlation xisted at 24 and 48 hrs of germination. At 72 hrs of exposure, significant ($P \leq 0.05$) correlation existed nd beyond this period higher significant level ($P \leq 0.01$) was observed.

Table 47.5: Changes in Seed Germination, Seedling Establishment, Shoot and Root Length, Ratio Value and Per cent Change Value, when Compared to Control, of 144 hrs Old Rice Seedlings, Grown in Different Concentrations of the Toxicant in Petriplate Culture. Data calculated from the mean of the samples ± standard deviation.

Concentration of the Toxicant	*Percentage of Seed Germination*	*Percentage of Seedling Establishment*	*Shoot Length, in cms*	*Root Length, in cms*	*Root/ Shoot Length Ratio*	*Percent Change in Shoot Length*	*Percent Change in Root Length*
Control	100	100	8.12±0.31	8.96±0.22	0.91	--	--
0.2 per cent (A)	100	100	8.16±0.42	7.11±0.08	1.15	+0.9	-20.64
1.25 per cent (B)	48	32	3.14±0.14	2.42±0.04	1.29	-61.33	-72.99
2.0 per cent (C)	20	12	1.14±0.32	0.84±0.06	1.36	-85.96	-90.62

In the control set, 100 per cent seed germination and 100 per cent seedling establishment wa observed. In concentration A (0.2 per cent), all the seeds germinated and 100 per cent seedling established. To conduct experiments in the seed germinator the selected concentrations were: A–0. per cent, B–1.25 per cent, C–2.0 per cent. In concentration B, 48 per cent seeds germinated, out of whic only 32 per cent germinated seedlings established. The rest died because of the toxicant. In concentratio C, seeds swelled and the seeds started germinating but 72 per cent seeds could not reach to 2mm siz to be counted as germinated. At this stage, no further germination or prolongation of plumule wa marked. In case of 'C' only 14 per cent seedlings established with stunted growth. No germinatio was marked beyond 2.2 per cent of the toxicant. The rest 14 per cent seeds though germinated but thes germinated seeds could not establish altogether (Tables 47.5 and 47.6). The percentage of germinatio decreased showing negative significant correlation at all exposure periods except 24h and 144 where non-significant correlation was marked. The two-way analysis of variance ratio test indicate that there was no significant difference between rows and columns.

Table 47.6: Changes in Seed Germination, Seedling Establishment, Shoot and Root Length, Ratio Value and Per cent Change Value when Compared to Control, of 144 hrs Old Rice Seedlings, Grown in Different Concentrations of the Toxicant in Pot Culture. Data calculated from the mean of the samples ± standard deviation.

Concentration of the Toxicant	*Percentage of Seed Germination*	*Percentage of Seedling Establishment*	*Shoot Length, in cms*	*Root Length, in cms*	*Root/ Shoot Length Ratio*	*Percent Change in Shoot Length*	*Percent Change in Root Length*
Control	100	100	9.6±0.4	9.1±0.3	1.05	--	--
0.2 per cent (A)	100	100	10.2±0.8	8.2±0.7	1.24	+6.25	-9.89
1.25 per cent (B)	48	32	5.4±0.2	3.2±0.4	1.69	-43.75	-64.83
2.0 per cent (C)	28	14	2.1±0.3	1.1±0.4	1.91	-48.12	-87.91

The shoot length declined from 8.12 ± 0.31 cm to 1.14 ± 0.32 cm at 0.2 per cent of the toxicant. N shoot was present in concentration beyond 2.2 per cent of the toxicant. A maximum of 85.96 per ce decrease in shoot length was marked at concentration C, when compared to the control value. The ro length of the control seedling was 8.96 ± 0.22 cm and the root length declined 7.11 ± 0.88 cm at cor A and 2.42 ± 0.04 cm at conc. B and the length further decline to 0.84 ± 0.06 cm conc. C. In concentrati A, 20.64 per cent, in conc. B., 72.99 per cent and in conc. C. 90.62 per cent decrease in root length w noted, when compared to the control value (Tables 47.5 and 47.6). The root/shoot length ratio increased from 0.91 to 1.15 in conc. A and to 1.29 in conc. B and 1.356 in conc. C. A maximum of 49.45 per cent increase in R/S ratio value was observed at conc. C of leached chemicals of the red mud waste. The shoot fresh weight increased from 32.91 ± 3.14 mg to 34.22 ± 5.82 mg at concentration A, where a maximum increase by 3.98 per cent was recorded. Where as, at higher concentration of the toxicant, the shoot weight decreased showing a maximum decrease by 74.63 per cent at concentration C. The changes in physiological variables in control and toxicant exposed rice seedlings in petriplate cultures. All the experiments were conducted in a seed germinator under controlled conditions, to evaluate the toxicity of the toxicant, excluding the effect of environmental variables. The shoot fresh weight declined from 32.91± 3.14 cm declined to 12.81±4.02 mg in conc. B and further depleted to 9.35 ± 2.14 mg in conc. C, where a maximum of 74.63 per cent decrease was recorded in conc. C. The root fresh weight decreased significantly from 18.5 ± 3.1 mg to 1.2 ± 0.4 mg at concentration C, where a maximum of 93.5 per cent decrease in root fresh weight was marked (Table 47.7). With the increase in toxicant concentration, the root weight significantly declined. Beyond 2.2 per cent toxicant concentration of the red mud waste, no root was seen. The root/shoot fresh weight ratio showed significant increase, when compared to the control value. The ratio value increased at all the three selected concentrations and the ratio value increased from 1.78 to 6.96 at concentration C. The shoot dry weight increased from 6.9 ± 0.2 mg to 7.10.4 mg at conc. A. With the increase in toxicant concentration, the shoot dry weight gradually decreased, when compared to control. A maximum of 88.4 per cent decrease at conc. C was recorded, when compared to control (Table 47.8).

Table 47.7: Changes in Shoot and Root Fresh Wt., Ratio Value, and Shoot and Root Dry Weight and Per cent Change when Compared to Control, of 144 hrs Old Rice Seedlings, Grown in Different Concentrations of the Toxicant in Petriplate Culture. Data calculated from the mean of the samples ± standard deviation.

Concentration of the Toxicant	*Shoot Fresh Wt., in gms*	*Root Fresh Wt., in gms*	*Root/Shoot Weight Ratio*	*Shoot Dry Wt., in gms*	*Root Dry Wt., in gms*	*Per cent Change in Shoot Weight*	*Per cent Change in Root Weight*
Control	32.9±3.14	18.5±3.1	1.78	6.9±0.2	3.6±0.4	---	---
0.2 per cent (A)	34.2±5.82	13.6±2.8	2.52	7.1±0.4	3.1±0.3	+3.98	-26.49
1.25 per cent (B)	12.8±4.02	4.9±1.4	2.61	2.1±0.3	1.2±0.1	-61.07	-73.51
2.0 per cent (C)	8.35±2.14	1.2±0.4	6.96	0.8±0.1	0.3±0.08	-74.63	-93.51

Table 47.8: Changes in Shoot and Root Fresh Wt., Ratio Value, and Shoot and Root Dry Weight and Per cent Change when Compared to Control, of 144 hrs Old Rice Seedlings, Grown in Different Concentrations of the Toxicant in Pot Culture. Data calculated from the mean of the samples ± standard deviation.

Concentration of the Toxicant	*Shoot Fresh Wt., in gms*	*Root Fresh Wt., in gms*	*Root/Shoot Weight Ratio*	*Shoot Dry Wt., in gms*	*Root Dry Wt., in gms*	*Per cent Change in Shoot Weight*	*Per cent Change in Root Weight*
Control	36.8±1.8	26.4±2.8	1.39	8.6±0.8	6.2±0.6	---	---
0.2 per cent (A)	41.6±2.8	14.9±4.2	2.79	9.2±0.4	4.1±0.3	+13.4	-43.56
1.25 per cent (B)	16.4±1.9	5.2±1.8	2.15	4.4±0.7	1.9±0.4	-55.43	-80.03
2.0 per cent (C)	9.2±0.8	1.8±0.6	5.11	1.6±0.3	0.7±0.2	-75.0	-93.18

The root dry weight also decreased with the increase in toxicant concentration. The root dry weight decreased from 3.6 ± 0.4 to 0.3 ± 0.08 mg at concentration C, when compared to control. The root/shoot dry weight ratio increased from 1.92 to 2.67 at conc. C. The increase in ratio value was not linear with the increase in toxicant concentration. A maximum 39.06 per cent increase in the dry weight of the root was recorded at concentration C (Tables 47.7 and 47.8). Only 8.85 per cent decrease in the root/shoot dry weight was recorded at conc. B, when compared to the control value. A careful examination of the percent values revealed that the variables studied showed lesser values in the exposed seedlings, when compared to the control value. The correlation co-efficient analysis revealed the existence of negative correlation between variables with the toxicant concentration and the percent decrease showed the existence of a positive correlation with the toxicant concentration. All the values were statistically significant. Another set of experiment was conducted to study the effect of the toxicant in pot culture. The obtained data can be compared with the data obtained in the seed germinator, when the environmental parameters were controlled. The shoot length declined from 9.6 ± 0.4 to 2.1 ± 0.3 cm at conc. C, where, a maximum of 78.12 per cent decrease was recorded. In petriplate culture, the decrease was 85.96 per cent and in pot culture the decrease was 78.12 per cent in shoot length. So here, 7.84 per cent benefit in shoot length was available in pot culture (Table 47.7). The benefit might be due to the interaction of environmental parameters and most important was the availability of nutrients in the soil. And in addition, the advantage of toxicant natural detoxification or binding of the toxicant with the soil might have played a crucial role in giving such advantage in the pot culture. The root length in pot culture declined from 9.1 ± 0.3 cm to 1.1 ± 0.4 cm at concentration C. The decrease was linear and steady with the increase in toxicant concentration, where a maximum of 87.91 per cent decrease was recorded. In petriplate culture, a maximum 90.62 per cent decrease was recorded. A difference of 2.71 per cent decrease, when compared to petriplate culture was marked at conc. C. In pot culture, this is not significant. The non-significant difference observed in the pot culture indicates significantly the adverse effect of the toxicant. The root/shoot length ratio increased showing a maximum increase by 81.90 per cent at concentration C. The shoot fresh weight increased from 36.8 ± 1.8 to 41.6 ± 2.8 mg at concentration A and than the fresh weight value decreased to 16.4 ± 1.9 mg at conc. B. The shoot fresh weight declined at concentration C, showing a maximum decline by 75 per cent, when compared to control. The decrease in shoot fresh weight was again linear, with the increase in toxicant concentration except at conc. A (Table 47.8). The shoot weight in exposed seedling declined indicating decrease in growth induced by the toxicant. The root fresh weight declined at concentration C, where a maximum of 98.18 per cent decrease was recorded, when compared to the control value, in pot culture. The root/shoot weight ratio increased from 1.39 to 5.11 at concentration C, where a maximum of 267.62 per cent increased was recorded. The shoot dry weight decreased from 8.6 ± 0.8 mg to 1.6 ± 0.3 mg at concentration C. A maximum of 81.39 per cent decrease when compared to control was recorded. The root dry weight declined from 6.2 ± 0.6 to 0.7 ± 0.2 mg at conc. C. Here, a maximum of 88.71 per cent decrease in root dry weight was recorded at concentration C, when compared to control value. A maximum of 66.19 per cent and 64.03 per cent increase in the ratio value was recorded at conc. B and conc. C, respectively (Table 47.8). All the exposed parameters were found to be less than the control values except at few cases, where an increase was recorded. Higher toxicity values were recorded in petriplate cultures and less toxicity values were recorded in pot culture. The difference in both was the environmental influence and the common factor in both the experiment was the application of the toxicant at the same concentration.

The changes in pigments content of control and leached chemicals of the red mud waste exposed seedlings of a rice plant at different concentrations of the toxicant, after 144 hour of exposure in petriplate culture. The chlorophyll-a content increased from 0.566 ± 0.08 to 0.58 ± 0.11 $mg.g^{-1}$ fresh weight at concentration A, decreased from 0.56 ± 0.08 to 0.22 ± 0.04 mg.g^{-1} fresh weight at concentration B, and from 0.56 ± 0.08 to 0.06 ± 0.02 mg.g^{-1} fresh weight at conc. C (Table 47.9). The decrease was steady and linear. A maximum of 89.28 per cent decrease (Table 47.10) was recorded at con. C. The chlorophyll-b increased from 0.54 ± 0.11 to 0.59 ± 0.06 at conc. A decrease from 0.54 ± 0.11 to 0.19 ± 0.22 at conc. B and from 0.54 ± 0.11 to 0.22 ± 0.01mg.g^{-1} fresh weight at concentration C. Where a maximum of 96.29 per cent decrease was recorded at conc. C of the toxicant. The total chlorophyll content increased from 1.19 ± 0.22 to 1.24 ± 0.31 mg.g^{-1} fresh weight at conc. A; decrease from 1.14 ± 0.3 to 0.48 ± 0.2 mg.g^{-1} fresh weight at conc. B and from 1.19 ± 0.22 to 0.09 ± 0.2 mg.g^{-1} fresh weight at concentration C after 144 hours of exposure (Tables 47.9 and 47.10). A maximum of 92.44 per cent decrease in total chlorophyll content was recorded at concentration C, when compared to the control value. The total phaeophytin content decreased from 0.96 ± 0.06 to 0.91 ± 0.04 mg.g^{-1} fresh weight at concentration A; from 0.96 ± 0.06 to 0.21 ± 0.05 mg.g^{-1} fresh weight at concentration B and from 0.96 ± 0.66 to 0.08 ± 0.01 mg.g^{-1} fresh weight at concentration C. A maximum of 91.67 per cent decrease was recorded at concentration C when compared to the control value. The carotene decreased from 0.0146 ± 0.0014 to 0.0141 ± 0.0008 mg.g^{-1} fresh weights at concentration B and from 0.0146 ± 0.0014 to 0.0008 ± 0.0001 mg.g^{-1} fresh weights at concentration C. A maximum of 94.52 per cent decrease over the control value was recorded at concentration C (Table 47.9). The chlorophyll a/b ratio increased with the increase in toxicant concentration from 1.04 to 3.0. The total chlorophyll/phaeophytin ratio decreased from 1.24 to 1.12 at conc. C and the total chlorophyll/total carotene ratio increased from

Table 47.9: Changes in Chlorophyll-a and b; Total Chlorophyll, Phaeophytin and Carotene Content of Control and Exposed 144 hrs Old Rice Seedlings, Grown in Different Concentrations of the Toxicant in Petriplate Culture. Data calculated from the mean of the samples ± standard deviation.

Concentration of the Toxicant	*Chlorophyll-a mg.g^{-1} Fresh Weight*	*Chlorophyll-b mg.g^{-1} Fresh Weight*	*Total Chlorophyll mg.g^{-1} Fresh Weight*	*Phaeophytin mg.g^{-1} Fresh Weight*	*Carotene mg.g^{-1} Fresh Weight*
Control	0.56±0.08	0.54±0.11	1.19±0.22	0.96±0.06	0.0146±0.0014
0.2 per cent (A)	0.58±0.11	0.59±0.06	1.24±0.31	0.91±0.04	0.0141±0.0008
1.25 per cent (B)	0.22±0.04	0.19±0.02	0.42±0.03	0.21±0.05	0.0032±0.0004
2.0 per cent (C)	0.06±0.02	0.02±0.01	0.09±0.02	0.08±0.01	0.0008±0.0001

Table 47.10: Percent Changes in Chlorophyll-a and b; Total Chlorophyll, Phaeophytin and Carotene Content of Exposed Rice Seedlings, when Compared to the Control Value, Grown in Different Concentrations of the Toxicant in Petriplate Culture. Data calculated from the mean of the samples.+- increase;– decrease.

Concentration of the Toxicant	*Per cent Change in Chlorophyll-a Content*	*Per cent Change in Chlorophyll-b Content*	*Per cent Change in Total Chlorophyll Content*	*Per cent Change in Phaeophytin Content*	*Per cent Change in Carotene Content*
Control	---	---	---	---	---
(A)	+3.57	+9.26	+4.20	-5.21	-3.42
(B)	-60.71	-64.81	-64.70	-78.12	-78.08
(C)	-89.28	-96.29	-92.44	-91.67	-94.52

81.51 to 112.5 at conc. C, where a maximum increase of 38.02 was recorded at conc. C. The correlation coefficient values of pigment content of the 144-hour-old seedlings in petriplate culture, where significant negative correlations were marked between toxicant concentrations and pigments like chl-a, chl-b, total chlorophyll, phaeophytin and carotene content. The analysis of variance ratio test conducted basing on the data indicated the existence of significant difference between columns and non-significant difference between rows. The changes in total chlorophyll in control and leached chemicals of the red mud waste exposed rice plant leaves at different period of exposure in pot culture (Table 47.11). The total chlorophyll content was 2.28 ± 0.16 $mg.g^{-1}$ fresh weight on 15th day of exposure in the control rice plant. The total chlorophyll content increased from 2.28 ± 0.16 to 2.44 ± 0.24 $mg.g^{-1}$ fresh weight at concentration A of the toxicant and on 15th day of exposure in pot culture, where a maximum 7.02 per cent increase was recorded. With the increase in toxicant concentration, the percent decrease of total chlorophyll increased showing a significant correlation on 30th day of exposure, the total chlorophyll content decreased steadily from 3.14 ± 0.62 to 0.41 ± 0.02 $mg.g^{-1}$ fresh weight at concentration C and where, a maximum of 86.94 per cent decrease was recorded. On 45th day of exposure, the total chlorophyll content declined from 3.08 ± 0.44 to 0.08 ± 0.02 $mg.g^{-1}$ fresh weight at concentration C. On 45th day of exposure, a maximum of 97.4 per cent decrease, when compared to control value was recorded at concentration C. The decrease in values with the increase in toxicant concentration and the increase in percent decrease with the increase in toxicant concentration indicated stress. On 60th day of exposure, the total chlorophyll content decreased from 1.01 ± 0.11 to 0.01 ± 0.005 $mg.g^{-1}$ fresh weight at concentration C, where, a maximum 99.02 per cent decrease was recorded (Table 47.11).

Table 47.11: Changes in Total Chlorophyll Content (mg. g^{-1} fresh weight) of Leaves of Rice Plant at Different Concentrations of the Toxicant at Different Days of Exposure in Pot Culture. Values in parentheses indicate percent change values when compared to control values. Data calculated from the mean of samples ± standard deviation.

Concentration of Toxicant	*Exposure in Days*			
	15	*30*	*45*	*60*
Control	2.28±0.16	3.14±0.62	3.08±0.44	1.02±0.11
0.2 per cent (A)	2.44±0.24 (+7.02)	3.26±0.18 (+3.82)	2.84±0.22 (–7.79)	0.88±0.24 (–13.72)
1.25 per cent (B)	1.26±0.08 (–44.74)	1.36±0.29 (–56.69)	1.65±0.14 (–46.43)	0.28±0.08 (–72.54)
2.0 per cent (C)	0.38±0.04 (–83.33)	0.41±0.02 (–86.94)	0.08±0.02 (–97.40)	0.01±0.005 (–99.02)

The changes in total phaeophytin content in control and leached chemicals of the red mud waste exposed rice plant leaves at different days of exposure and at different concentrations of the toxicant. The same table and figure indicate the percent change in phaeophytin content in the exposed seedlings in pot culture. The total phaeophytin content declined from 1.82 ± 0.27 to 1.74 ± 0.38 $mg.g^{-1}$ at concentration A, from 1.82 ± 0.27 to 0.91 ± 0.29 $mg.g^{-1}$ at concentration B, and from 1.82 ± 0.27 to 0.52 ±0.18 $mg.g^{-1}$ at concentration C on 15th day of exposure (Table 47.12). With the increase in toxicant concentration, the phaeophytin content decreased and the percent decrease increased with the increase in toxicant concentration and 95.95 per cent maximum decrease was recorded at concentration C. After 30 days of exposure the total phaeophytin content decreased from 1.86 ± 0.28 to 1.31 ± 0.47 mg. g^{-1} at concentration A, from 1.86 ± 0.28 to 0.76 ± 0.38 $mg.g^{-1}$ at concentration B, and from 1.86 ± 0.28 to

0.43 ± 0.28 $mg.g^{-1}$ at concentration C. A maximum of 76.88 per cent decrease in phaeophytin content was recorded at concentration C, on 30th day of exposure (Table 47.12). The total phaeophytin content decreased from 2.12 ± 0.54 to 1.42 ± 0.51 $mg.g^{-1}$ at concentration A, from 2.12 ± 0.54 to 0.54 ± 0.28 $mg.g^{-1}$ at concentration B and from 2.12 ± 0.54 to 0.28 ± 0.08 $mg.g^{-1}$ at concentration C on 45th day of exposure. A maximum of 86.79 per cent decrease was recorded at conc. C and on 45th day of exposure. The total phaeophytin content decreased from 2.72 ± 0.82 to 1.14 ± 0.22 $mg.g^{-1}$ at concentration A, from 2.72 ± 0.82 to 0.31 ± 0.07 $mg.g^{-1}$ at conc. B (Table 47.12) and from 2.72 ± 0.82 to 0.11 ± 0.02 $mg.g^{-1}$ at conc. C and was recorded on 60th day of exposure in pot culture. A maximum of 95.95 per cent decrease was recorded at concentration C and on 60th day of exposure. With the increase in toxicant concentration and day of exposure, the phaeophytin content decreased showing a negative and significant correlation. With the increase in exposure period and toxicant concentration, the percent decrease in phaeophytin content showing the existence of positive correlation. Both the assessments indicated the toxicant stress.

Table 47.12: Changes in Total Phaeophytin Content (mg. g^{-1} fresh weight) of Leaves of Rice Plant at Different Concentrations of the Toxicant at Different Days of Exposure in Pot Culture. Values in parentheses indicate percent change values when compared to control values. Data calculated from the mean of samples ± standard deviation.

Concentration of Toxicant	*Exposure in Days*			
	15	*30*	*45*	*60*
Control	1.82±0.27	1.86±0.28	2.12±0.54	2.72±0.82
0.2 per cent (A)	1.74±0.38 (−4.39)	1.31±0.47 (−29.57)	1.42±0.51 (−33.02)	1.14±0.22 (−58.09)
1.25 per cent (B)	0.91±0.29 (−50.0)	0.76±0.38 (−59.14)	0.54±0.28 (−74.53)	0.31±0.07 (−88.6)
2.0 per cent (C)	0.52±0.18 (−71.43)	0.43±0.28 (−76.88)	0.28±0.08 (−86.79)	0.11±0.02 (−95.95)

The changes in carotene content in control and leached chemicals of the red mud waste exposed rice plant leaves at different exposure periods and at different toxicant concentrations. The carotene content declined from 0.0148 ± 0.0016 to 0.0088 ± 0.0016 $mg.g^{-1}$ fresh weights at 15th day of exposure, where 40.54 per cent decrease, when compared to control was recorded. On 30th day of exposure, 1.78 per cent increase in carotene content over the control value was recorded at concentration A and 45.83 per cent decrease was recorded at concentration C (Table 47.13). On 45th day of exposure, 5.75 per cent increase in carotene content over control value was recorded in concentration A. However, 59.16 per cent decrease was recorded in concentration C, in contrast to concentration a. On 60th day of exposure in concentration A, 6.08 per cent increase in carotene content was recorded over the control value and in contrast, 58.01 per cent and 82.32 per cent decrease in carotene content over the control value was recorded on concentration B and concentration C, respectively. At lower concentration of the toxicant, increase in carotene level and at higher concentrations decrease in carotene level was marked at all exposure periods. Significant negative correlation was marked at all exposure periods for total chlorophyll. For total phaeophytin non-significant correlation at 15th day exposure and significant negative correlation at higher exposure period was marked. Non-significant negative correlation was marked at all exposure periods for carotene content change. The chlorophyll/carotene ratio value declined at all exposure periods and highest depletion was recorded at concentration-C. Lowest value was recorded at concentration C and on 60th day of exposure in pot culture (Table 47.13). The carotene/phaeophytin ratio indicated that the ratio values increased with the increase in exposure periods and with the increase in toxicant concentration in all exposed plants. No trend was recorded in the control set of the pot culture experiments. The chlorophyll/phaeophytin ratio decreased with the increase in toxicant concentration at all exposure periods. No definite trend was recorded with the exposure period. The phaeophytin/carotene ratio indicated significant depletion of the ratio value at higher exposure period and at higher concentration of the toxicant (Table 47.13).

Table 47.13: Changes in Carotene Content (mg. g^{-1} fresh weight) of Leaves of Rice Plant at Different Concentrations of the Toxicant at Different Days of Exposure in Pot Culture. Data in parentheses indicate percent change values, when compared to control values. Data calculated from the mean of samples ± standard deviation.

Concentration of Toxicant	*Exposure in Days*			
	15	*30*	*45*	*60*
Control	0.0148±0.0016	0.0168±0.0024	0.0174±0.0031	0.0181±0.0014
0.2 per cent (A)	0.0151±0.0008 (+2.03)	0.0171±0.0020 (+1.78)	0.0184±0.0051 (+5.75)	0.0192±0.0022 (+6.08)
1.25 per cent (B)	0.0092±0.0012 (–37.84)	0.0099±0.0012 (–41.07)	0.0096±0.0026 (–44.83)	0.0076±0.0012 (–58.01)
2.0 per cent (C)	0.0088±0.0016 (–40.54)	0.0091±0.0008 (–45.83)	0.0071±0.0004 (–59.19)	0.0032±0.0002 (–82.32)

Table 47.14: Changes in the Respiration Rate (ml of CO_2 evolved g^{-1} hr^{-1}), Rate of Photosynthesis (ml of O_2 evolved g^{-1} hr^{-1}) and Gross Production (ml of O_2 evolved g^{-1} hr^{-1}) of Control and Exposed Rice Seedlings After 144hrs of Germination in Petriplate Culture. Data calculated from the mean of samples ± standard deviation

	After 144 hrs of Germination (ml of CO_2/O_2 evolved g^{-1} hr^{-1})		
Concentration of Toxicant	*Rate of Respiration*	*Rate of Photosynthesis (Net Primary Production, NPP)*	*Gross Primary Productivity (GPP)*
Control	236.5±18.4	224.8±16.4	461.3
0.2 per cent (A)	208.4±19.6	246.2±28.5	454.6
1.25 per cent (B)	102.6±28.8	126.4±18.5	229.0
2.0 per cent (C)	64.5±12.6	74.6±8.5	131.1

Interesting changes in rate of respiration, rate of photosynthesis (NPP) and Gross Primary Productivity (GPP) in control and toxicant exposed 144-hour-old seedlings in petriplate culture, grown in a seed germinator, under controlled conditions, were observed (Table 47.14). The rate of respiration decreased from 236.5 ± 18.4 to 208.4 ± 19.6ml of CO_2 evolved hr^{-1} g^{-1} dry weight of the seedling at concentration A, from 236.5 ± 18.4 to 102.6 ±28.8ml of CO_2 evolved hr^{-1} g^{-1} at concentration B and from 236.5 ± 18.4 to 64.5 ± 12.6ml of CO_2 evolved hr^{-1} g^{-1} at conc. C in 144 hour old rice seedlings. The rate of photosynthesis increased from 224.8 ± 16.4 to 246.2 ± 28.5 ml of O_2 evolved hr^{-1} g^{-1} dry weight of the seedlings in conc. A; the rate of photosynthesis decreased from 224.8 ±16.4 to 126.4 ± 18.8 ml of O_2 evolved hr^{-1} g^{-1} dry weight of the seedlings in conc. B and from 224.8 ± 16.4 to 74.6 ± 8.5 ml of O_2

evolved hr^{-1} g^{-1} dry weight of the 144 hour old seedlings in petriplate cultures (Table 47.14). The gross primary productivity decreased in exposed seedlings, when compared to control seedlings. The GPP value decreased from 461.3 to 454.6 at concentration A, from 461.3 to 229.0 and from 461.3 to 139.1ml of O_2 evolved hr^{-1} g^{-1} dry weight of the 144-hour-old seedlings in petriplate culture (Table 47.14).

Change in photosynthetic rate, respiration rate and GPP values at different concentrations of the toxicant and at different days of exposure in pot culture indicated interesting features. The photosynthetic rate (NPP) decreased from 184.6 to 91.2ml of O_2 evolved hr^{-1} g^{-1} dry weight on 15th day of exposure, where a maximum of 50.59 per cent decrease in NPP value was recorded. On 30th day of exposure, the NPP value depleted from 191.2 to 82.4ml of O_2 evolved hr^{-1} g^{-1} dry weight at concentration C, where a maximum of 56.9 per cent decrease was recorded (Table 47.15). On 45th day of exposure, the NPP value depleted from 202.4 to 71.6 ml of O_2 evolved hr^{-1} g^{-1} dry weight of the seedlings at concentration C, where a maximum of 64.62 per cent decrease was recorded. On 60th day of exposure, the NPP value depleted from 212.6 to 48.5 ml of O_2 evolved hr^{-1} g^{-1} dry weight of the seedling at concentration C, where a maximum of 77.19 per cent decrease in NPP value was recorded (Table 47.15). With the increase in exposure period and toxicant concentration, the NPP value declined indicating a negative correlation. With the increase in exposure period and toxicant concentration,

Table 47.15: Changes in the respiration rate (ml of CO_2 evolved g^{-1} hr^{-1}), rate of photosynthesis (ml of O_2 evolved g^{-1} hr^{-1})and gross production (ml of O_2 evolved g^{-1} hr^{-1}) of control and exposed rice seedlings at different days of exposure in pot culture. Data calculated from the mean of samples ± standard deviation. Data in parentheses indicate percent increase (+)/percent decrease (–) in the values, when compared to control.

Concentration of Toxicant	*Exposure Period in Days (Respiration rate–ml of CO_2 evolved g^{-1} hr^{-1})*			
	15	*30*	*45*	*60*
Control	142.3	151.4	158.6	161.2
0.2 per cent (A)	144.6(+1.62)	136.2(–10.04)	101.4(–36.06)	88.9(–44.85)
1.25 per cent (B)	114.2(–19.75)	91.6(–39.63)	74.8(–52.84)	36.6(–77.29)
2.0 per cent (C)	82.4(–42.09)	42.7(–71.79)	21.6(–86.38)	8.4(–94.79)
Concentration of Toxicant	*Exposure Period in Days (Photosynthetic rate–ml of O_2 evolved g^{-1} hr^{-1})*			
	15	*30*	*45*	*60*
Control	184.6	191.2	202.4	212.6
0.2 per cent (A)	188.4(+2.06)	198.7(+3.92)	218.2(+7.81)	260.4(+22.48)
1.25 per cent (B)	128.6 (–30.33)	171.4(–10.35)	145.8(–27.96)	126.2(–40.6)
2.0 per cent (C)	91.2(–50.59)	82.4(–56.90)	71.6(–64.62)	48.5(–77.19)
Concentration of Toxicant	*Exposure Period in Days (GPP–ml of O_2 evolved g^{-1} hr^{-1})*			
	15	*30*	*45*	*60*
Control	326.9	342.6	361.0	373.8
0.2 per cent (A)	333.0(+1.87)	334.9(–5.02)	319.6(–11.47)	349.3(–6.39)
1.25 per cent (B)	242.8(–25.73)	262.8(–25.47)	220.6(–38.89)	162.8(–56.37)
2.0 per cent (C)	173.6(–46.89)	125.1(–64.52)	93.2(–74.18)	56.9(–84.75)

the percent decrease in NPP value increased showing a positive and significant correlation in pot culture. The respiration rate increased from 142.3 to 144.6ml of CO_2 evolved $hr^{-1}g^{-1}$ dry weight in conc A; from 142.3 to 114.2ml of CO_2 evolved $hr^{-1}g^{-1}$ dry weight conc. B and from 142.3 to 82.4 ml of CO_2 evolved $hr^{-1}g^{-1}$ dry weight at conc. C, on 15th day of exposure, where a maximum of 42.09 per cent decrease was recorded in conc. C. The respiration rate on 30th day of exposure declined from 151.4 to 42.7 ml of CO_2 evolved $hr^{-1}g^{-1}$ dry weight at conc. C, where a maximum of 71.79 per cent decrease was recorded at conc. C and on 30th day of exposure. On 45th day of exposure, the respiration rate of the exposed seedlings depleted from 158.6 to 21.6ml of CO_2 evolved $hr^{-1}g^{-1}$ dry weight at conc. C, showing a maximum depletion by 86.38 per cent in pot culture experiment. On 60th day of exposure, the respiration rate significantly declined from 161.2 to 8.4 ml of CO_2 evolved $hr^{-1}g^{-1}$ dry weight at conc. C in pot culture, where a maximum of 94.79 per cent decrease was recorded (Table 47.15). The gross primary productivity of the exposed seedlings drastically reduced, when compared to the control value. The GPP value decreased from 326.9 to 173.6 ml of O_2 evolved $hr^{-1}g^{-1}$ dry weight at conc. C and on 15th day of exposure; the GPP value depleted from 342.6 to 125.1ml of O_2 evolved $hr^{-1}g^{-1}$ dry weight at conc. C and on 30th day of exposure; the GPP value depleted from 361.0 to 93.2 ml of O_2 evolved $hr^{-1}g^{-1}$ dry weight at conc. C and on 45th day of exposure and the GPP value depleted from 373.8 to 56.9 ml of O_2 evolved $hr^{-1}g^{-1}$ dry weight at conc. C and on 60th day of exposure in pot cultures (Table 47.15). With the increase in toxicant concentration and exposure period, the GPP value significantly declined at all exposure periods. Highest depletion in GPP value was recorded in concentration C. With the increase in exposure period, the GPP value declined significantly on 15th day 46.89 per cent; on 30th day 64.52 per cent; on 45th day 74.18 per cent and on 60th day 84.75 per cent decrease was recorded, when compared to the control value. On 15th day the percent decrease was 25.73 per cent and 46.89 per cent in conc. B and C, respectively. On 30th day, the percent decrease was 5.02 per cent, 25.47 per cent and 64.52 per cent in conc. A, B and C, respectively. On 45th day, the percent decrease was 11.47 per cent, 38.89 per cent and 74.18 per cent in conc. A, B and C, respectively. On 60th day, the percent decrease was 6.39 per cent, 56.37 per cent and 84.75 per cent in conc. A, B and C, respectively. Maximum depletion in GPP value was recorded on 60th day of exposure and in concentration C (Table 47.15). The data indicated the existence of significant ($p< 0.05$) negative correlation between toxicant concentration and NPP, GPP, rate of respiration values in 144-hour-old seedlings in petriplate culture. The analysis of variance ratio test related to rate or respiration, NPP value and GPP value in 144 hour old seedlings and rice plants in pot culture at different exposure periods indicated the existence of non significant difference between rows and columns was quite clear from the analysis of variance ratio test. The NPP value showed non-significant correlation at 15 and 30 days of exposure and significant correlation at 45 and 60 days exposure. The respiration rate showed all significant negative correlation values at all exposure periods. The GPP values showed significant ($p<0.05$) negative correlation values at all exposure periods. Changes in glucose content of both control and exposed leaves of rice plants at different exposure periods and at different selected concentrations of the toxicant, in pot culture showed the effects caused by the toxicant.

The glucose content declined from 17.4 ± 2.8 to 6.2 ± 0.8 mg/g fresh weight after 15th day of exposure; from 18.9 ± 3.5 to 5.1 ±1.2 mg/g fresh weight after 30th days of exposure; from 20.4 ± 5.5 to 3.6 ± 0.9 mg/g fresh weight after 45 days of exposure and from 22.2 ± 1.4 to 1.8 ±0.4 mg/g fresh weight after 60 days exposure. Highest depletion was recorded at conc. C (2.0 per cent of the toxicant) at all exposure periods. Drastic depletion in glucose level was recorded at all concentrations and at all exposure periods, except 45th and 60th day of exposure in conc. A, when compared to control values (Table 47.16). Significant ($p<0.001$) negative correlation existed between the parameter with exposure period and between the parameter and toxicant concentration. Steady and linear increase in percent decrease in glucose content with the increase in toxicant concentration and exposure period was marked. Maximum of 54.37 per cent, 73.01 per cent, 82.35 per cent and 91.89 per cent decrease was observed at 15, 30, 45 and 60 days of exposure at concentration C. All the values were statistically significant. Positive significant ($p<0.001$) correlation exists between the percent decrease and toxicant concentration and exposure period. Changes in nitrogen content in shoot and root of both control and toxicant exposed rice plants grown in pots, at different exposure period and at different toxicant concentrations revealed interesting sequences (Table 47.16). Highest depletion in nitrogen content of the shoot of the exposed plant was recorded at conc. C and at 60th day of exposure, when compared to the rest of the values. The nitrogen contents depleted from 3.8 ± 0.8 to 1.4 ± 0.7 from 4.1 ± 1.1 to 1.4 ± 0.6; from 4.4 ± 1.4 to 1.2 ±0.6 and from 5.1 ± 0.9 to 1.1 ± 0.5 mg/g fresh weight was recorded at 15, 35, 45 and 60 days of exposure in shoot of the exposed seedlings. Drastic depletion in nitrogen content was marked in the exposed roots. The nitrogen content decreased from 4.1 ± 0.4 to 1.3 ± 0.9 mg/g dry weight at 15 days of exposure; from 4.5 ± 0.6 to 1.4 ± 0.3 mg/g dry weight after 30 days of exposure; from 5.1 ±0.6 to 0.8 ± 0.2 mg/g dry weight after 45 days of exposure and from 5.6 + 0.8 to 0.2 + 01 mg/g dry weight after 60 days of exposure (Table 47.16). Highest depletion in nitrogen content was recorded at concentration C in root and shoot of the exposed seedlings, when compared to control, conc. A and conc. B rice plants. At all exposure periods and toxicant concentration, the nitrogen content significantly declined. Highest depletion was marked at conc. C both in shoot and root. Maximum depletion by 63.16 per cent, 65.85 per cent, 72.72 per cent and 78.43 per cent was recorded at 15, 30, 45 and 60 days exposure period at conc. C. Drastic depletion in nitrogen content in the root of the exposed plants when compared to control plants was marked. Maximum decrease by 68.29 per cent, 68.89 per cent, 84.31 per cent and 96.43 per cent was noted at noted at 15, 30, 45 and 60 day exposure period at conc. C (Table 47.16). The values of shoot did show significant trend and significant correlation was marked. In case of root, significant positive ($p<0.01$) correlation was marked between the percent decrease in nitrogen content with the exposure period and with the toxicant concentration.

Table 47.16: Changes in Glucose Content and Nitrogen Content in Control and Exposed Rice Leaves at Different Concentrations of the Toxicant and at Different Days of Exposure. Data calculated from the mean of the samples ± standard deviation. Data in parentheses indicate percent decrease(–) and percent increase(+) in the parameters, when compared to control values.

Concentration of Toxicant	*Exposure Period in Days (Glucose content, mg. g^{-1})*			
	15	*30*	*45*	*60*
Control	17.4±2.8	18.9±3.5	20.4±5.5	22.2±1.4
0.2 per cent (A)	16.2±4.9(–6.89)	18.1±4.2(–4.23)	21.6±3.2(+5.88)	23.6±3.6(+6.31)
1.25 per cent (B)	11.5±2.2(–33.91)	12.4±3.9(–34.39)	11.2±1.8(–45.09)	9.1±1.2(–59.01)
2.0 per cent (C)	6.2±0.8(–64.37)	5.1±1.2(–73.01)	3.6±0.9(–82.35)	1.8±0.4(–91.89)
Concentration of Toxicant	*Exposure Period in Days (Nitrogen content, mg. g^{-1})*			
	15	*30*	*45*	*60*
Control	3.8±0.8	4.1±1.1	4.4±1.4	5.1±0.9
0.2 per cent (A)	3.6±0.3(–5.26)	4.0±0.8(–2.44)	3.9±2.6(–11.36)	3.7±1.1(–27.45)
1.25 per cent (B)	2.6±1.1 (–31.58)	2.8±1.4(–31.71)	2.4±1.1(–45.45)	1.8±0.2(–64.70)
2.0 per cent (C)	1.4±0.7(–63.16)	1.4±0.6(–65.85)	1.2±0.6(–72.72)	1.1±0.5(–78.43)

The root was more affected than the shoot of the exposed rice plants. Control plants remained throughou the experimental period and served as standard for comparison. The data of petriplate culture was more prominent than the pot culture. As the pot culture experiments were subjected to environmenta influence. The changes in ATPase activity of 144-hour-old rice seedlings in different concentrations o the toxicant in petriplate culture, grown in a seed germinator under controlled conditions showed features related to toxicity. The ATPase activity of the 144-hour-old seedlings, decreased with the increase in toxicant concentration, when compared to the control value. The ATPase activity decreased from 3.4 + 0.3 to 2.6 + 0.6 at conc. A; from 3.4 + 0.3 to 1.4 + 0.5 at conc. B and from 3.4 + 0.3 to 0.3 + 0.2 mg i.p liberated/hr/100 mg dry weight in concentration C. A significant ($r = -0.993, p<0.01$) negative correlation between the parameter with the toxicant concentration was marked. The ATPase activity decreased by 23.53 per cent in conc. A, 58.82 per cent in conc. B and 91.18 per cent decrease in conc. C when compared to the control value. All the obtained values were statistically significant. ATPase activity of the 144-hour-old seedlings showed the existence of significant ($p<0.01$) negative correlation with the toxicant concentration.

When all the data were considered along with the statistical data it can be inferred that the toxicant was more effective in petriplate culture, when compared to pot culture experiment. Higher percent of inhibition was recorded in petriplate culture experiments. Petriplate experiments were conducted in side a seed germinator, where all environmental factors were controlled. But in pot culture experiments, the environmental factors played a crucial role and reduced the effect of the toxicant on the crop plant. In both the cases the root was more affected than the shoot, in all the variable parameters. The residual accumulation of the toxicants in the exposed rice plants could have provided much better information. But due to lack of facilities and equipments, residual analysis part was not carried out. In presence of residual accumulation of toxic ant data, the obtained data can be analyzed, correlated and interpreted properly. However, in absence of residual accumulation data concrete conclusion may not be drawn but to a greater degree of accuracy, the data was presented and interpreted. All the obtained data were statistically significant. Further, higher-level work is necessary to assess the toxic nature and site of the action studies. Further work is needed to find out possible ways of detoxification of the leached chemicals of the red mud. No doubt, leaching will be there, in addition groundwater contamination cannot be avoided, as there is no bottom lining. At least, care could be taken to stop surface leaching leading to run off water, entering into water bodies.

Discussion

The eco-toxicological assessments of toxic chemicals depend on two basic items of information First, the environmental available concentrations of the chemical which occur as a result of discharge and distribution in the natural environment and secondly, the toxicological properties of the chemical at that concentration and its possible impact on the biota including humans. Environmental concentrations and toxicological properties also include the concentrations of by-products or transformation products as well. Different methods and procedures are used to draw all the available eco-toxicological information's together to arrive at an assessment. The basic requirements for such procedures are the pattern of use of chemicals; their disposal and release; physico-chemical characteristics; degradation and bioaccumulation; and biological factors such as growth inhibition toxicity testing etc. Thus, the eco-toxicology of chemicals is an important aspect of environmental management and the management of pollution.

High percentage of seed germination and seedling establishment, disease free seedling and healthy growth of the plant are considered to be some of the reliable indications about high yield potential of any crop plant. However, in nature, crop plants experience extensive damage due to insect pests, fungi, and mainly because of different type of industrial effluents and wastes etc. The infected seeds spread the disease when grown in fields. Thus, to avoid decline in food grain productions, killer chemicals are applied to crop plants (Butler, 1978). It is essential to understand the amount of toxicant present, its movement and biological magnification along the critical pathways finally reaching to man. Compared to other chemicals, specially, the unrestricted use of pesticides in agriculture towards the last past of the 18th century has increased substantial amount of pesticides in the aquatic and terrestrial environment, and has created a serious threat to biotic components of the earth (Fiskesjo, 1979). Now it stands as a global problem. It is highly essential to understand the amount of toxic pesticides present, its movement and biological magnification along the critical path to man. It was realized that from pesticide-contaminated environment, the entry and accumulation into the biological system, occurs at a very slower rate and also it takes a longer time period for its dreadful expressions. But, the pesticide, which enters into, the agricultural soil in the form of seed dresser, or by mass spray finds its way into the living system through direct consumption of food grain.

The industrial wastes once thrown into the environment, these wastes spread all over and affect the flora and fauna. For example, the red mud waste discharged from the NALCO plant was stored in a natural pond surrounded by hills from all sides except one side, where the industry people have constructed three dams. Unfortunately, they have never visualized probably that the leached chemicals from the red mud pond, will create future problems. The data obtained in this piece of investigation revealed that the extract of the red mud is deadly toxic and can seriously affect the crop plants. We have selected a resistant variety, still then, the extract effected the crop plant seriously, particularly to the production/yield of the crop plant. The morphological changes were clearly visible and effects were more pronounced. Hence, care should be taken to check leaching of the chemicals from the red mud pond or they must find out ways and means for safe disposal of the waste. We are not conservative enough to request/direct the closure of the industry but some safe disposal methods are ought to be adopted to safe guard the interest of the public in general and environment in particular. According to Baicu and Diaconu (1972), smaller seeds abosorb/retain more amount of the treated chemicals due to exposure of more surface area, resulting in high transfer/entry of mercury/pesticide/toxicant to the agricultural soil through smaller seeds. Hence, the agricultural soil increases substantial amount of toxicant within a notable period of time making it dangerous for human health, when red mud wastes enter nutritional cycle through dressed seeds or other treated parts of plants intentionally or due to carelessness (Koronowski,1973). However, literature explaining the following aspects in details is lacking.

1. Toxicant dynamics in different seeds and seed types disinfected with different types of insecticides formulations and leached chemicals of red mud wastes.
2. Red mud waste dynamics in seeds in relation to seed biology.
3. Amount of red mud leached chemical transfer to agricultural soil in relation to seed type, seed size, seed quantity etc.
4. Leached chemicals of the red mud transfer from seed to soil and again from soil to seed in crop rotation is a usual practice in modern agriculture.
5. Interferences of the toxicant in seed physiology during germination and establishment and its biochemistry.

In the present investigation, the red mud waste dynamics in seed and seed biology during germination and seedling establishment was studied in RM waste treated seedlings. The decrease in root and shoot length of the treated seedlings at higher concentration was in agreement with Lipsey (1978) and Pradhan (1998) who observed inhibition in growth of maize seedlings grown in methy mercury hydroxide (MMH) and rice seedlings in pesticide exposure. Retardation in growth of crop plants treated with organo-mercurial formulation was reported by Sorteberg (1974), Weaver *et al* (1984) and Semu *et al.* (1985). At low concentrations of PMA and ethyl-mercuric chloride the winte wheat showed changes in the seedling vigour (Bateman, 1975). The seedlings of tomato, corn and bean grown in a complete nutrient solution containing MMH (Haney and Richard, 1973), showed th tomato seedlings were killed within 4-8 days at concentration of 0.05 mg/l. The growth was inhibite at concentration of 0.01 mg/l for corn and beans. In this study, growth of the seedlings was inhibite beyond 2.2 per cent of red mud leached chemicals in only petriplate culture. At 1.2 per cent, roo growth significantly declilned when compared to shoot growth. Although the seeds germination wa inhibited beyond 2.6 per cent of red mud waste but seedling establishment was inhibited at lowe doses (beyond 2.0 per cent) of RM leached chemicals. It seems that these rice seeds were more effective when compared to mustard grains (Padhi, 1990) in respect of seed germination. Padhi (1990) reporte significant variation or no germination at all, when rice grains were exposed to MEMC (an organo mercurial compound), such an observation was marked here. When sugar beet seeds were treate with ethyl mercuric phosphate, phytotoxic effects were observed (Byford, 1971). The same autho reported that methyl mercury- dicyano-diamide was found to be more phyto-toxic than the less volatil ethyl and phenyl mercury compounds. Toubey (1972) reported that the emergence of summit whea seeds was reduced significantly when the seed was stored for a long period after being treated wit normal doses of PMA, a fungicide. Here also, with higher concentrations of red mud leached chemical and longer period of exposure reduced the percent germination and seedling establishment. Hasse *et al.* (1976) stated that inhibition of root elongation is one of the sensitive parameters to assess th toxicity of toxicants. Further they stated that compared to the growth rate, pesticide was found to b more toxic than mercury, cadmium and zinc. Godbold and Huttermann (1985) observed that the roo elongation was severely inhibited by 0.1 to 0.5 m mole of mercury after 24 hours. At 5 and 15 m mol the root elongation was ceased completely and radial shrinkage was observed in 1.5 mm from behin the root tip. Application of $HgCl_2$ to primary root tip of *Zea mays* in the form of droplets, showe reduction of root growth and also inhibited their gravy response (Pilet and Jean, 1981). Treatment c various concentrations of metallic compounds (both inorganic and organic); Mercury (Mukherji an Ganguly, 1974) on different varieties of rice have shown that, towards lower concentration the inhibitio of germination was not at all marked rather stimulation was noticed (Padhi, 1990).

Siegel *et al.* (1984) studied the seed germination and early growth of the seedling of 15 specie treated with mercury vapour and observed that, more than 50 per cent plants did not show inhibitio of root and shoot length, but occasional stimulation was noticed. Sai (1998) reported an initial decreas by 11.3 per cent in shoot length and 22.3 per cent decrease in root length at 100 ppm of pesticide i petriplate culture. It was the indication of no stimulation. No stimulation was observed in the preser investigation with red mud extract. The length of ratio shoot and root calculated for all types c treatments, showed positive trend in relation to concentration, which indicated that root length wa much affected than the shoot system (Sai,1998 and Pradhan,1998). This might be due to the fact th the root being the primary absorbing organ, was in direct contact with a toxic medium and henc resulted higher reduction in different parameters of root. The distribution of red mud leached chemic in plant was asymmetrical with greater amount of elements in the roots than the shoots. Although th level of red mud leached chemical gradually increases in plant tissues, with increasing amount i external levels, the proportion retained in the roots to that of shoots was almost constant (*i.* approximately 95 per cent). The initial blackening followed by browning and the observed chloros was only due to the red mud extract chemical waste. Similar observations were not found in pesticide poisoning or mercury poisoning. The type of deposition should be identified and characterised to reach to a logistic conclusion. As the apical meristem of *Elodia densa* showed deposition of more methyl mercury, so also the deposition of more mercury in the region of the root might have caused inhibition. When mercuric chloride was applied in droplets to the cap of primary roots of the maize, it reduced their elongation (Pilet and Jean, 1981). Sharma (1982) reported that seeds of different varieties *Phaseolus aureus* treatment with mercuric acetate solution showed inhibition of seedling growth. Sharma (1983 and 1985) also reported that the seeds of *Pisum sativum* C. V. Bonneville, when treated with mercuric acetate solution shown promotion of radicle and epicotyle length at lower concentration (10 mg/l) but in higher concentration at 50 mg/l, a significant reduction in seedling growth was observed. The decrease in shoot, root length and seedling establishment with the increase in the concentration of the stress probably suggests and supports the idea that (1) the toxicant increases the osmotic pressure of the medium by reducing or preventing the availability of water required for germination, (2) the toxicity of the chemical and (3) presence of higher concentrations of ions and active ingredients, help jointly for the reduction of the seedling growth and establishment (Donovan and Dey, 1969; Panigrahi and Misra, 1978 and Pradhan, 1998). Experiments were conducted to study the toxic effect of various mercury compounds such as inorganic monovalent mercurous chloride, divalent mercuric chloride, mercuric acetate and organic compound of mercury as phenyl mercuric acetate on seed germination, pigment content, root and shoot lengths, fresh and dry weights of barley. It was observed that phenyl mercuric acetate was more toxic, followed by mercuric acetate and mercuric chloride (Mukhiya *et al.*, 1983). In case of red mud leached chemicals poisoning, drastic depletion in the parameters were marked but the extent of pollution was far less when compared to above mentioned compounds. Interestingly in case of insecticide poisoning no stimulation was marked at low concentrations and also at low exposure periods but Patra (1993) reported that at low exposure period, inhibition was not marked rather stimulation was observed. The stimulation might be due to change in certain biochemical parameters caused due to mercury toxicity. Inhibition of root growth was considered to be important and an indication of metal toxicity (Godbold *et al.*, 1984; Godbold and Hutterman, 1985), the promotion of the shoot might be the secondary effect reflected due to root inhibition. The differential effect of PMA on root and shoot, might be due to the asymmetrical distribution of mercury (Baeuford *et al.*, 1977), which might be imparting a differential toxicity to the seedling parts. Our interpretation agrees fully with Padhi (1990) and partly with Baeuford *et al.* (1977). It is very difficult at this stage to infer the toxic effects of red mud leached chemical on the seedling establishment of the rice seed. The pigment analysis, physiological change studies of the seedling, might throw light on detailed aspects of RM waste toxicity. Photosynthetic pigments of the plant system play a vital role in trapping solar energy and CO_2 fixation (Kashyap and Gupta, 1981). The chlorophylls have long been recognized as the primary light acceptors in plants and they are invariably present in every organism, which carries out photosynthesis with absorption of CO_2 and evolution of molecular oxygen. Zingmark and Miller (1975), De Filippis and Pallaghy (1976 a), Rai and Khatonair (1980), and Rai *et al.* (1981 b) reported that heavy metals reduce chlorophyll content and inhibit photosynthesis. De *et al.* (1985) suggested that at 20 ppm the Hg (II) lowered the chlorophyll by decreasing the synthesis of chlorophyll as well as possibly by increasing chlorophyllase activity in *Pistia*. De Filippis and Pallaghy (1976 b) reported that among the heavy metal solution PMA and $ZnCl_2$ caused the greatest and least inhibition respectively. These levels were reflected in the levels of phaeophytin (De Filippis and Pallaghy, 1976 c), a degradation product of chlorophyll (Wolken *et al.*, 1955). The above authors added that the chlorophyll and phaeophytin a/b ratios were similarly affected. Cultures treated with heavy metal solutions increased their carotenoid/chlorophyll ratios. Rai *et al.* (1981 a) in their review article mentioned the acute toxicity of cadmium to green algae, which inhibited photosynthesis [$^{14}CO_2$ uptake] (Stratton and Carke, 1979). Conway (1978) found a significant lowering of pigment content in *Asterionella formosa* after addition of cadmium. Excessive concentrations of copper cause loss of photosynthetic pigments in *Chlorella* and in a natural phytoplankton community dominated by *Aphonizmenon flos-aque* (Whittakar *et al.*, 1978).

Sai (1998) reported significant variations in chlorophyll content in the exposed plants indicated the presence of stress. Chlorosis was marked in the exposed plants. Similar type of chlorosis was observed by Padhi (1990) and Sai (1998) in exposed rice seedlings. Hence, it can be indicated that the observed chlorosis in exposed seedlings might be due to breakdown of chlorophyll pigments by the toxicant. The growth rate depleted significantly, even at very low concentration of the red mud leached chemicals. It indicated that red mud leached chemicals at lower concentrations acts as a growth inhibitor by way of affecting the biochemical mechanism during seed germination and at higher concentrations acts as a growth inhibitor due to the red mud leached chemicals toxicant. As red mud leached chemicals at lower concentrations, it reduced transpiration, thus decreasing growth of the plant. However, with increase in concentration of red mud leached chemical, the amount of pigment content decreased significantly, with slight increase in the initial concentration applied over the control values (Sai, 1998). An initial increase in pigment content might be due to the stimulatory effect of the pesticide (Sai, 1998). Pesticide, Monocrotophos showed dichotomous behaviour in the investigation carried out by Sai (1998), we could not notice such a change in behaviour in the present investigation. The pigment change in exposed seedlings was due to only red mud leached chemical. Harris *et al.* (1970) and Overneil (1975) reported the reduction of chlorophyll pigment as a result of reduction in growth and suggested that the chlorophyll estimation can be considered as a parameter of measuring growth of the system. The phaeophytin content in the leaf of plant has an important role for the fact that any unfavourable change in the environment inhabited by the plant is reflected through a change in phaeophytin level. Chlorophylls are known to be converted to phaeophytins as a consequence of exposure to weak acids, by replacement of Mg^{2+} with two atoms of hydrogen and thereby changing the spectral properties (Rao and Le Blanc, 1966 and Singh and Singh, 1984 a). Degradation to phaeophytins might be the first step towards the breakdown of chlorophyll, which was evident from the increased level of phaeophytin in the initial concentration in the treated culture. In high concentration of red mud leached chemical, the amount of carotenoid content also decreased significantly as the case of chlorophyll content. It clearly indicates that red mud leached chemical in lower concentration degrades chlorophyll pigments to phaeophytin, but in higher concentration it also destroys phaeophytin pigments alongwith chlorophyll pigments. That means red mud leached chemical is highly toxic in higher concentrations and also at lower concentrations. Carotenoids play a vital role as a protector of photosynthetic tissues against photosensitized oxidation. The decrease in carotenoid content in the leaves exposed to red mud waste stress lead to a decrease in protection from the stress to the photosynthetic tissue. Here, in our experiment also there was decrease in chlorophyll and phaeophytin content with the decrease in carotenoid content. So clearly the decrease in carotenoid content in shoot exposed to red mud leached chemicals stress lead to a decrease in protection from the stress to the photosynthetic tissue. The ratio of chlorophyll to carotenoid has long been cosidered as a valuable parameter for defining environmental conditions unfavourable for growth of the plant. The variation in the ratio value with the increase in the concentration of the toxicant indicated stress. When the nutrients in the medium are exhausted or a toxicant was introduced into the medium the ratio increased due to decrease in the chlorophyll content (Rai *et al.*, 1981 b). The same result was found in higher concentration (2.0 per cent) when compared to that of control. Increased ratio indicated inhibition of chlorophyll biosynthesis, inactivation of enzyme system and disruption of many physiological and biochemical processes (Leland *et al.*, 1979; Sorentino, 1979). The insecticide at relatively low concentrations also effect the energy transfer by selectively affecting the phycocyanin and phycobilisomes of intact cells of *Spirulina*, which was reported by Murthy and Mohanty (1991). It is a common observation that toxicity of metals showed great variation under laboratory and field conditions (Whitton, 1970). The phaeophytin/carotenoid ratio increased at lower concentrations and also at higher concentrations of red mud leached chemical. A significant increase at higher concentration might be either due to the interaction of environmental parameters or due to the increase in red mud leached chemical toxicity. According to Pradhan (1998) the seedlings grown under field condition showed significant growth than those grown under laboratory-controlled conditions. Hence it becomes explicit that laboratory based information cannot solely be used to stimulate field conditions, because many environmental and nutritional factors operate to bring about metal/pesticide/industrial waste toxicity in field conditions (Harding and Whitton, 1977; Gadd and Griffiths, 1978). Pure petriplate culture in open racks, provide some sort of environmental influences. In the present investigation, petriplates culture experiments were conducted in a seed germinator, under total controlled conditions. The obtained data will vary strongly to open rack culture experiments. Field experiments were conducted in pots and the culture will have restricted environmental influence. Pot was selected to restrict the leaching and drainage of the red mud waste after application, which generally happens in field conditions.

Almost all the cellular chlorophyll and carotenoid pigments are localized in the photosynthetic lamellae of the cells, the thylakoids (Wolk, 1973). The lamellae are proteinaceous (Schmitz, 1967) and increase in thylakoid surface area with metal exposure may indicate an increase in photosynthetic potential by active synthesis of thylakoidal protein and associated pigments or it may reflect an uncoupling of the bonds in protein causing the lamellae to stretch (Rachlin *et al.*, 1982). The metals cause a reduction in the volume of the intra-thylakoidal spaces. The significance of such reduction is not understood but the result is to bring the thylakoids into closer approximation. This may have some value in facilitating photosynthetic activity in the metal stresses cell so that the necessary ATP is manufactured to drive the metabolic machinery required to effect detoxification (Rachlin *et al.*, 1982). Rothstein (1959) has demonstrated that metals interact with sulphyhydryl groups in cell membranes to produce–S–metal–S–bridges, and Simkis (1979) has discussed the importance of proteins and lipoproteins in binding with metal as means of inactivating them within the cell. Lipsey (1978) also tested under *in vitro* conditions, MMH inhibited photosynthesis of chloroplast and phosphate uptake by mitochondria. The same author also opined that MMH showed stunted growth and respiration of isolated mitochondria were inhibited. Haumann (1987) showed that mercury affects the chloroplast by causing considerable increase of the grana stacks in *Chara vulgaris*. In addition, mercury caused dilation of the endoplasmic reticulum and of the mitochondrial tubuli. Many of the above results and interpretations tally with our findings and ideas. Hence, it might be a common man's observation that red mud leached chemicals significantly affects the pigments and cause great physiological changes evinced from the observation of NPP, GPP and respiration studies.

The observed toxic changes in both the sets of experiments might be due to red mud leached chemicals only. Considering the reports that PMA inhibits electron transport and photophosphorylations, its effectiveness as an anti-transparirant may depend primarily on its restriction to the epidermal layer of the leaf. The disturbances in growth and growth substances are always related with metabolic changes inside the plant body. Biochemical changes proceed any change in growth because growth is the outcome of many biochemical processes; and physiological and biochemical effects, are the underlying cause of measurable whole organism effects. Complexes between Hg^{2+} and certain purine and pyrimidines especially thymine, are quite stable (Katz, 1962) and complexes with phosphates are probably formed readily; thus one would expect nulceotides and nucleic acids also to bind Hg^{2+} rather well. Inogaki (1940) found various nucleotides, such as AMP, GMP and IMP to be precipitated by mercuric compounds. Hg^{2+} has been found to produce complex with nucleic acids from thymus, plants (Trim, 1959), Pneumococci (Dove and Yamane, 1960), and TMV (Katz and Santilli, 1962 b). Interference with cellular functions via interaction with various nucleic acid components is undoubtedly in large part responsible (Ramel, 1969; Mulvihilli, 1972) for the mutagenecity of mercurials. The excessive discharge of red mud leached chemicals in recent years tends to contaminate the environment. Controversy has centered around both the mode of action of red mud leached chemical in effecting plant metabolism and the extent of its toxicity resulting from penetration into the mesophyll. The literature on residual accumulation of red mud leached chemical and its effects on crop like rice is scanty. Vascular plants accumulate red mud leached chemicals by three routes of uptake: through the root from the soil (ionic), through the stomata from the leaf surface deposition and by the retentions of particulate dust, atmospheric uptake predominating in the above ground (mostly leaves) parts of herbaceous plants. As the Hg concentrations prevailing in soils, plants retain Hg almost exclusively within the root, where it seem to be relatively tightly bound to acidic groups of cell walls only at exceptionally high soil mercury contents, there occurred significant translocation to shoot (Lorenz, 1979). Higher concentrations of red mud leached chemical in leaves than in other tissues of the plants as well as in grasses, draw the attention of ecologists because, grazers mostly depend on grasses and leaves of the plants. Thus the grazers of the area (cows, sheep, goat) must be accumulating red mud leached chemicals in their tissues via food chain since they are totally dependent on the live flora of the area. Hamilton and Ruthvein (1967) analyzed potato seeds, which had been treated with organo-mercurial disinfectants, for export purposes. The potatoes grown from the seed contained very small amount of mercury and were indistinguishable from those grown from untreated seeds. Plants treated with red mud leached chemical compounds may have high red mud leached chemical concentrations in certain parts but it is not known whether the chemical was absorbed and translocated by plant tissues to these parts, or only occurred as a superficial deposit. Plants are capable of absorbing red mud leached chemical and translocating it through their tissues. Thus, red mud leached chemicals deposited/transported to leaves might be translocated to the fruit and other parts. Red mud leached chemical concentrations in water, due to the ultimate discharge of Red Mud from the NALCO plant cause a considerable reduction in the photosynthetic capability of phytoplankton. High percentage of seed germination and seedling establishment, disease free seedling and healthy growth of the plant are considered to be some of the trustworthy indications about high yield potential of any crop plant. In nature, as the plants are very much exposed to different environmental changes, wastes, insect and pests, they suffer from variety of diseases at different stages of their life cycle. So, the crop plant cannot maintain a perfect toxicant/disease free life, unless proper care is being taken. The seed produced from the affected plants cannot be utilized in agriculture or for food purposes, because these lack germination vigor and nutritional value. Rather these seeds contain a lot of toxicants that also gets distributed throughout the plant and finally the seeds of this plant again accumulate a lot of toxicants. Really, the cultivation of this crop is a challenging task for a cultivator, because, the cultivation needs a lot of experience and much care. All phases of its life cycle (*i*) Germination phase, (*ii*) Seedling phase, (*iii*) Growth and tillering phase, (*iv*) Flowering phase, (*v*) Fruiting phase and (*vi*) Harvesting phase are very much sensitive to toxicant stress, industrial wastes, pest infection and also minor changes in the climate and environment. Out of all the phases, the germinating phase needs more attention, because it is the basic and most sensitive phase from which the life cycle of the plant starts. If seed and soil borne pathogens, will be eradicated before or at this stage, the future seedling become healthy and can resist/tolerate up and downs of natural changes by application of specific insecticides/ pesticides. On the other hand, compared to other chemicals, specially, the unrestricted use of organo-mercurial compounds/formulations in industry and agriculture towards the last part of the nineteenth century has increased substantial amount of mercury in the aquatic and terrestrial environment, and has created a serious threat to biotic component of the earth (Fiskesjo, 1979). Now it stands as a global problem. According to Sahu (1987) the toxic hazards of red mud leached chemicals and mercury has now gained sociological and political importance. Hence, it is highly essential to understand the amount of toxic red mud leached chemical present, it's movement and biological magnification along the critical path to the man. Though the process of bio-accumulation, bio-magnification of the heavy metal, mercury, and red mud leached chemical was recently understood, but it was realized that, from the red mud leached chemical contaminated the entry and accumulation into the biological system, occurs at a very slower rate and also it takes a longer time period for its dreadful expressions. But, the pesticide, which enters into the agricultural soil by leaching and through irrigation, finds its way into the living system through direct consumption of food grain. Hence the pesticide burden of the biological system had increased in an astounding rate in a short time. Recently it has been reported that, besides the seeds, the largest in-put of pesticide into U.K. agricultural soil arise from atmospheric deposition after mass fumigation. The cultivators of the area reported several times, the reporters, the NGO's reported and protested several times regarding the effects of the leached chemicals of the red mud pond and also demonstrated several times regarding the destruction of crop plant and crop production many a times, but neither the government nor the industry people show any interest on the issue. Rather NALCO plant gets the award as the best non-polluting industry obeying all the norms for the protection of the environment and the surrounding.

The prediction of Koronowski (1973) become a reality and opined that the agricultural mercury becomes dangerous for human health, when mercury containing fungicides enter into nutritional cycle through dressed seeds or other treated parts by mistake, ignorance and carelessness. Similar predictions can also be attributable for the red mud leached chemical and mercury present in the irrigated water or the leached chemicals of the red mud waste. Siegel *et al.* (1977) have shown that he use of mercurial seed dresser, year after year has not only increased the mercury burden or the agricultural soil but also contaminated the abiotic and biotic components of the ecosystem. Mercury based pesticides treated seed grains, which is flooded after planting (*e.g.* rice) can directly contribute this toxic metal to water and soil. Including above statements, further seed dressing can introduce approximately 10g/ha of mercury into the soil. As all most all the components of the environment are polluted with pesticide and mercury, the entry of it into the living system through food chain is a must and cannot be avoided (Jernelov and Lann, 1971). Hence, the idea of mistake, ignorance and carelessness as quoted by Koronowski (1973) at this stage are found to have no meaning. In Poland, it has been first realized that the mercury content of water (River and Stream) draining the agricultural soils was found to be increasing during sowing time of cereal crops. In future, Damonjodi area and its surrounding will face the consequences of the leached chemicals of the red mud pond. A big poison lake is under preparation for the destruction of the future of Damonjodi, Koraput area. Towards last part of the present century the agricultural contamination of red mud leached chemical and its entry through food chain, have not only drawn the attention of the scientists of all over the world, but also a lot of research work in this line have already been undertaken. Many countries have enacted their legislation to restrict the use of pesticide and mercury both in industry and in agricultural sectors. Some developed countries like Japan, Sweden, USA, Canada, and France etc. have totally banned the use of mercurial compounds in agriculture. Here it is interesting to note that in Sweden, the mercury content of freshwater (arising from leaching and non-off of fungicide etc.) is still increasing even though the use of mercury compound was banned some ten years ago. The restriction and also the ban on the use of mercurial compounds and pesticides by advanced/developed countries have indirectly compelled the industrial entrepreneurs of those countries to shift their market and dump those in developing and underdeveloped countries of the world. Many compounds banned by those countries due to their hazardous effect are dumped in the third world countries in the form of charity or in liberal soft business terms/loans. So the people of developing countries usually suffer from various types of dreadful diseases. To have better, non-toxic, potent and bio-degradable pesticide chemical, the desire and unending search of man through decades no doubt have produced/introduced various chemical formulations from time to time, but no chemical was found to be satisfactory to man till to-day. The banning of one, and introduction of new chemical has become the common practice. Hence, it is a general feeling in developing countries that all chemicals introduced till to-today are/were in the experimental phase only. The same condition has also happened for mercurial compounds as these were proved to be hazardous both to the nature and also to the biotic component of the ecosphere. Therefore, at present, it is logical to predict that the seed dressing chemicals which are introduced in recent past (non-mercurials) after some more years again these will be obsolete and may be banned by assigning some other negative reasons. In future, a similar step might be taken by the government to ban the industry but by the time sufficient harm might have occurred in the affected area. Here it can be stressed that, before introduction of any chemical proper scientific scrutiny, from all possible aspects is essential, so the reckless handling of the chemical, cannot cause any residual toxicity problem in future. Secondly, sound knowledge on chemical behavior and toxicity will be very much essential prior to introduction of the chemicals either through irrigation or natural entry through leaching. Literature explaining the following aspects in details is lacking:

1. Leached chemicals of the red mud waste dynamics in seeds in relation to seed biology. 2. Amount of leached chemical transfer to agricultural soil and its transfer to seedlings. 3. Leached chemical of the red mud pond transfer from soil to seed from seed to seedling and from seedling to plant and from the plant again to the seeds. 4. Interferences of leached chemical of red mud in seed physiology during germination and establishment and its biochemistry. 5. Comparative study of RM leached chemical toxicity in petriplate culture, pot culture and field culture. The agricultural contaminated waste poses to be a threat to the present generation. The study of the above aspects certainly add additional information pertaining to the contribution of man for enriching the agricultural soil. Further, it can be said that, the degree and the process of bioaccumulation of such added chemicals through various crop plant could be identified and investigated. Hence, the present piece of work was designed to study the effects of Red mud leached chemical on a rice seed, seedling and the plant. Usually the protrusion of the radical through seed coat though considered as germination, but this simple process comprises of four phases: (*i*) Imbibition phase, (*ii*) Activation phase, (*iii*) Phase of mitosis and (*iv*)The radical protrusion phase. Each distinct phase can be influenced by an external or internal factor. During germination, the seeds not only imbibe water, but also absorb inorganic and organic substances present in the imbibing medium other physiological processes of the seedling. But actually the nature of the chemical substance entered and also nature of the seed, both combined determine the degree of response and direction of the effect. During normal course of germination according to Young (1938), the emergence of primary root to the external environment precedes to that of coleoptile. But in the investigation conducted by Padhi (1990), a reverse process was observed in the long term treated seeds or seeds treated with high concentration of MEMC. The coleoptile was observed not only to emerge prior to radical but also there was a considerable delay in radical emergence. This type of germination reported by Padhi (1990) was designated as abnormal germination. This finding was similar to the findings of Taylor (1942 a and b). Delayed germination as observed in the present

investigation was also reported in case of beans, cabbage, carrot, millet, onions, tomato (Bache *et al.*, 1973), Oats (Sorteberg, 1974), Maize (Bache *et al.*, 1973 and Lipsey, 1975) and Bermuda grass (Bache *et al.*, 1973 and Weaver *et al.*, 1984). High MEMC content in the soil (50 mg Kg^{-1}, Aretan) also delayed the germination of wheat for 5 days, but at this concentration the seedlings of $HgCl_2$ treatment, did not show delay in germination (Semu *et al.*, 1985). The delay and reduction in germination might be due to accumulation of more amounts of toxic ions in the seed or toxic effect of the chemical. It was claimed that the reduced percentage of germination was directly proportional to the osmotic pressure it produces and thus reducing/preventing the availability of water needed for imbibition or toxic effect of specific ions in the medium where the seeds were exposed. We agree with the findings of the above authors. At still higher concentrations, the seed germination was totally stopped. According to Semu *et al.* (1985), the germination in bean failed completely, when the soil contained 50 mg of Hg Kg^{-1}. Even after repeated sowing no germination was recorded in that contaminated soil. Similar cases were reported in pesticide and waste contaminated agricultural fields.

The characteristic malformations of seedlings of cereals were produced from organic mercury exposure (ethyl mercury phosphate) when applied at high doses. The maize seedling grown in methyl mercury hydroxide showed severe decrease in growth (Lipsey, 1975). At 50 mg of Hg Kg^{-1} Aretan (MEMC) in the soil, the wheat plant not only showed stunted growth of the seedling but also remained shorter throughout the life period. The wheat plants grown in soil containing similar concentrations of Hg treated with $HgCl_2$, showed stunted growth up to flowering but the plants could recover to its normal size at the time of maturity (Semu *et al.*, 1985). Growth retardation of crop plants were also reported by Sorteberg (1974), Lipsey (1975) and Weaver *et al.* (1984) as the seeds were treated with different types of organo-mercurial formulations. At low concentrations of PMA and ethyl mercuric chloride the winter wheat showed changes in the seedling vigor (Bateman, 1975). The seedlings of tomato, corn and bean grown in complete nutrient solution containing MMH (Haney and Richard, 1973) showed that tomato seedlings were killed within 4-8 days at concentrations of 0.05 ppm. The growth was inhibited at concentration of 0.01 ppm for tomatoes and 0.05 ppm for corn and beans. Hassett *et al.* (1976) and Godbold *et al.* (1984) stated that inhibition of root elongation is one of the sensitive parameters to assess the toxicity of the heavy metals and pesticides. Godbold and Huttermann (1985) observed that the root elongation was severely inhibited by 0.1 to 0.5 mM of mercury after 24 hours. At 5 and 15 mm the root elongation was ceased completely and radial shrinkage was observed in 1.5 mm from behind the root tip. The treatment of PMA to the root tip Secale cereal (Sokolowska *et al.*, 1975) and treatment of Methoxy-beta ethyl mercuric chloride to *Allium cepa* (Venot and Glachhero, 1971) showed irregularity in cell division in root and also showed growth inhibition. The cytological study revealed that the treated seeds of Apel-*pine* (*Pinus halepensis*) showed precocious mitosis, accelerated lipolysis, proteolysis and inhibition of synthesis of organelles, especially in radical portion of the seedling. According to Ganesan and Monoharan (1983) the effect of cadmium and mercury on root length, stem length, leaf area and dry matter production *of A. esculantus* is inversely proportional to their concentration. When mercuric chloride applied in droplets to the cap of primary roots of the maize it reduced their elongation (Pillet and Jean, 1981). Sharma (1982) also reported that seeds of different varieties *Phaseolus aureus* treated with mercuric acetate solution showed inhibition of seedling growth. The same author, Sharma (1983 and 1985), reported that the seeds of *Pisum sativum* C.V. Boneville when treated with mercuric acetate solution shown promotion of radical and epicotyle length at lower concentration (10 mg l^{-1}), but in higher concentration at 50 mg l^{-1}, a significant reduction in seedling growth was observed. Inhibition of seedling growth was also observed by Mukherji and Ganguly (1974) in rice. The root inhibition was much more pronounced than the shoot inhibition, which was recorded in lettuce seedling (Mukherji and Das Gupta, 1972) in rice (Das Gupta and

Mukherji, (1977) treated with copper and rice (Banerjee and Mukherji, (1980) treated with Nicke Increase in duration of pretreatment, increased the inhibition of rice seedlings in case of Bariur exposure. Further, depending on the root elongation of ryegrass, the toxicity of various heavy metal was compared (Wong and Bradshaw, 1982). The length ratio of shoot and root calculated for all type of treatments, showed the negative trend in relation to concentration and period of treatment. Thi indicated that root length was much affected than the shoot system. This might be due to the fact tha the root being the primary absorbing organ, was in direct contact with a toxic medium and henc resulted higher reduction in different parameters of root. No doubt it is a confirmatory test whic depicts the nature of growth of the root and shoot. According to Baeuford *et al.* (1977), the uptake c mercury in plants growth in hydroponic solutions was a function of external concentration. Over th concentration range considered, the accumulation of mercury in root was linear on a log-log basi although the uptake of elements in shoots appeared to be biphasic. The distribution of mercury i plant was asymmetrical with greater amounts of element in the roots than the shoots. As the apica meristem of *Elodi densa* showed deposition of more methyl mercury (Czuba and Mortimer, 1982), s also the deposition of more mercury in the apical region/meristematic region of the root might hav caused inhibition. The observation made in the present investigation agrees with the findings of a the workers reported here except Siegel *et al.* (1984) and Padhy (1990). But in the present observatio water content of the seedling decreased in shoot and root portion of the seedling, the toxicity of th chemical might be the real reason of growth inhibition. Hence at this stage it will be premature to inf but it can be predicted that if growth studies of seedlings are undertaken at very low doses of leache chemicals of the red mud waste, at low exposure periods and also in case of very young seedling significant and notable inhibition of shoot system could be marked. The stimulation observed migl be due to change in certain bio-chemical parameters caused due to mercury toxicity. So far, most of th research efforts were concentrated after 144 h of seed germination in normal and toxic medium References pertaining to the physiology and biochemical changes during germination and seedlir establishment before 144 h of germination is scanty and the only report of Padhy (1990) and Pradha (1998) is interesting. In absence of sufficient literature pertaining to the early phases of germinatio satisfactory explanations and logical conclusions might face strong criticism. However, Padhy (199 made an effort to illustrate the probable mechanism in a simpler way and the work indicated separate line of thinking and suggested the real mechanism during early phases of germination ar seedling growth.

In this study, inhibition of root was more pronounced than the shoot, which is probably th characteristic of toxicity as root is the first site to absorb the toxicants. Hassete *et al.* (1976); ar Godbold *et al.* (1984) indicated with positive confirmation of the idea and support of findings present in this piece of investigation. On seedling basis, it become clear that the decrease in fresh weight w only due to decrease in water content, as the dry weight of the seedling decreased significantl Increase in shoot/root dry weight ratio confirmed the above explanation. Hence, in the treated seedlir water stress condition was created. Inhibition of root growth was considered to be an indication pesticide toxicity (Godbold *et al.*, 1984), depletion of the shoot might be the secondary effect reflect due to root inhibition. Hence, at this stage it was difficult to infer, whether stimulation of shoot w due to toxicity of the leached chemical or this chemical had no effect on the shoot system in low concentrations and also in very young seedlings (Patra, 1993). According to present investigatio acquisition of more amounts of toxic ions by the root from the leached chemical contaminated s might be imparting high toxic effect on root cells. On the other hand the low amount of leach chemical probably deposited in shoot might have shown inhibitory action. Very low concentrations pesticide induce stimulation in growth and here in case of red mud waste such a trend was n marked. The above finding was confirmed by Sahu (1987), Shaw (1987) and Padhy (1990). As in lower concentration, the stimulation in seed germination and shoot growth was not observed, so also, the chlorophyll content of the seedling decreased with the dose. It can be said that inhibition of shoot growth might have inhibited the chlorophyll biosynthesis which in turn showed decrease in total chlorophyll content of the seedling. No doubt the inhibition of root growth and the stimulation of shoot growth are the consequences of mercury toxicity at lower concentrations, so also the increase in chlorophyll was the consequence of toxic behavior of mercury in sub-acute concentrations (Padhi, 1990). The effect of $HgCl_2$, MMC, MEMC, PMA on this species also have shown striking decrease in chlorophyll, and dry matter content. Lee and Young (1972) also reported that when compared to cadmium and zinc, mercury depressed the growth rate and chlorophyll content of *Chlorella ellipsoidea*. Rai *et al.* (1981) had suggested that specific growth rate and final yield of *Chlorella* treated with zinc, mercury and methyl mercury, showed decrease in total chlorophyll content and increase in carotenoid/chlorophyll ratio with increase in heavy metal concentration. Lipsey (1975) also stated that *in vitro* conditions, MMH inhibited photosynthesis of chloroplast and phosphate uptake by mitochondria. The same author reported that when maize seedlings were treated with MMH, showed stunted growth and respiration of isolated mitochondria were inhibited. In addition, the mercury caused dilation of the endoplasmic reticulum and of the mitochondrial tubuli. Lee *et al.* (1973) demonstrated that binding of Hg to chloroplast depend on concentrations and nature of the compound. The binding of Hg governs the blockage of electron flow by altering the Hills reaction. Mercury abolishes the communication between photosystems and associated redox pool. But according to Kimimura and Sakae (1972)–the $HgCl_2$ completely inhibits methyl viologen Hill reaction as well as methyl viologen photo-reduction with reduced DCPIP as electron donor. The extent of inhibition of Hill activity was much higher under high intensity of light. Here it can be added that in activation of photosynthesis and loss of leaf protein are both related to total mass of leaf chloroplast during senescence of rice seedling. Both actinomycin–A and $HgCl_2$ inhibited the development of ATP producing system, linked with most part with the development of non-cyclic photo-phosphorylation, which may be due to participation of ATP, synthesized in cyclic electron transport in the formation of chloroplast during greening. According to Huising and Kline (1972) it has been reported in Barley that the absorption and translocation of mercury from treated seeds to seedling leaves and roots showed a definite dosage–concentration dependence. Further, the same authors have stated that, the anionic moiety to which mercury is attached, influence absorption and translocation. Mercury levels in plants decreased with time. The level in leaves decreased more rapidly than that in roots. Actually the entry of mercury in relation to seed behaviour during germination could not be explained in a better way (Radha, 1993). The residual mercury was found to be located in the rice bran (specially in protein components of the bran) portion of the grain as the plants were sprayed with inorganic mercury during vegetative and flowering stages of growth. From the observations and available literature, it clearly indicated that the organo-mercurial fungicide–MEMC induced significant alternation in germination and in normal morphology of the seedling. Simultaneously the entry of mercury into the endosperm was also realized. Hence the findings of the present investigation have not only induced to study the bio-chemical changes of the germinating seedling but also their behaviour in the growth and productivity of the crop plant. Significant changes in respiration rate of the different aged seedlings, when exposed to different concentrations of the red mud waste leachates indicated clearly the effect of the test chemical. At lower exposure periods, the low aged seedlings, the effect seems to be more drastic than the higher aged seedlings. The effect of the toxicant at higher doses (2 per cent) was much more drastic when compared to low doses of the test chemical. Debnath and Mukherji (1982), reported identical observations under $BaCl_2$ stress in different types of aged seedlings. The decrease in respiration rate in young germinating seedlings might be due to the stress, the red mud waste. Because, at this stage, the body burden of Red mud leached chemical might be the maximum, which gets diluted with the aging of the seedling. But the significant variation observed at 144 h old seedling was interesting. At lower concentrations of the toxicant, decrease in rate was observed. Sahu (1987), Shaw (1987), Padhy (1990), and Pradhan (1998) reported an initial increase followed by a decrease in respiration rate both in inorganic and organic mercurial compound exposure. This result did not support the biphasic or dual behavior of the toxicant in different biosystems. After a decrease, again an increase in respiration rate can be correlated with the dilution of the toxicant in the seedlings and also at higher exposure period. This supports the idea that low stress causes depression and high stress induces inhibition.The germination of seeds was subjected to activity changes, which are regulated by exogenous and endogenous factors. In the moment of soaking, the complex reaction began in dormant embryo and also in its nutritive tissue. RNA, Protein and DNA conversions are set in a predetermined way of germination and seedling development. All these processes are bio-chemically reflected in alteration of enzyme status. Investigation on changes in enzyme activities and pattern, in germinating seeds, suggest not only the information's about the course of germination process but also they may contribute to a deeper understanding of gene activators. Since all cellular processes are essentially enzyme catalyzed, the intersection of a toxicant agent with living matter generally causes disturbance in the enzymatic reactions. Further such interference at the enzymatic level provides clue to the mode of action of the toxic substance (Bruin, 1976). Hence in view of the above fact the effect of an exogenous toxic chemical the Red mud leached chemical on bio-chemical and enzymatic changes of young rice seedling in different cultures needs proper investigation.

At very high concentrations of the toxicant significant bimolecular changes can be marked. But at very low concentrations of the toxicant, the observed insignificant inhibition might be a transitional short stayed phase. Pradhan (1998) and Sahu (1987) stressed the dichotomous behaviour of the mercurial toxicant. Both the authors emphasized stimulation at very low concentrations of the toxicant and shorter periods of exposure and inhibition both at higher concentrations of the toxicant and longer period of exposure. When we overview the works of Rath (1984), Sahu (1987), Shaw (1987) and Panigrahi (1990) which gives clear indication of dichotomous behaviour of the mercury compounds on different plant systems. It becomes rather too difficult here to explain the dual behavior in germinating seedlings, where as, we have only observed one type of effect *i.e.* inhibition. The only probable note, we can leave here just to escape from the present situation might be due to the following reasons: (*i*) the real detailed bio-chemical and physiological mechanisms in germinating seedlings are yet to be ascertained, (*ii*) non-existence of a dynamic sequence of events occurring in germinating seedlings, (*iii*) insufficient amount of bio-macromolecule production, (*iv*) phasic changes in the rates of metabolism in germinating seedlings, (*v*) role of water and its highest absorption disturbing the osmotic resistance and mechanical fragility of the cells of the germinating seedlings, (*vi*) fast and unstable macromolecular conversions in a very limited time. Here it can be stressed that, the inhibited amount of RNA produced certainly takes part in the protein synthesis, so that low amount of protein formation occurs in the root and shoot portion. But in the shoot, only a significant decrease in protein content was noticed. Hence it is suggested that the low amount of protein synthesized might have been derived towards the formation of more hydrolysing enzymes, such as proteases abundantly seen in rice seeds. It is known that protease is related to dynamic state or << turnover >> of protein in the cell than to the absolute protein level at any one time (Das Gupta and Mukherji, 1977). In copper treated lettuce seedlings, a considerable rise in the level of soluble protein was marked (Mukherji and Das Gupta, 1972). The same authors opined that, this type of behavior of soluble protein formation probably synchronizes with the similar increase in enzyme activity. In addition, they suggested that the increase in enzyme activity. In addition, they suggested that the increase in the activity of the enzymes studied (Mukherji and Das Gupta, 1972) was due to enhanced protein synthesis and the entire happenings were probably due to copper (a heavy metal) toxicity. However, the present investigation did not agree 100 per cent with the above findings. But at this stage, it will be premature to reach to a firm conclusion. Mukherji and Das Gupta (1972) and Das Gupta and Mukherji (1977) conducted the experiments in old seedlings of rice. Secondly, the type of toxicant was different, which shows differential activity based on their site of action and site of stimulation or inhibition. Further detailed work on enzymes explained late, might suggest again a total different picture and even the story may take a different turn. On this basis, protein levels may rise even though the protease shows increasing activity. To put it in the other way, enzymatic degradation by protease may be rapid but the coincident rate of protein synthesis may also be rapid (Das Gupta and Mukherji, 1977). Our findings agree with the above-mentioned workers. Hence accumulation of free amino acids in the system is facilitated and DNA/protein ratio is almost maintained at a steady state. However, this phase is a purely temporary one and in later phase (increase in exposure period to toxic chemical) all above parameters will show differential trends, because stimulation of RNA and synthesis of protein will be ceased and degradation of the bio-macromolecules will set in. Salt sensitive cultivars (TKM 9) of rice showed accumulation of excess soluble protein in shoots and roots. Lead toxicity on growth and metabolism of germinating rice (Mukherji and Maitra, 1977), mercury toxicity in rice seedling (Mukherji and Nag, 1977 a) are available. They have reported that the oxidising enzymes *viz.*, catalase, peroxidase, IAA oxidase and ascorbic acid oxidase, all showed constant enhancement and have been envisaged as general biochemical symptoms of metal toxicity. Consequently, a close correlation between reduced growths and increased activity of these enzymes has been established (Debnath and Mukherji, 1982). Mukherji and Das Gupta, (1972) showed in lettuce seedlings, that, respective of age of the seedling, the catalase activity showed about 2 fold increase at the highest copper dose. Enzyme level decreased both in control and copper with aging, which was presumed to be result of an impaired metabolic status of the older seedlings. Steeman and Wium-Anderson (1969), Steeman and Kamp-Nelson (1970) recorded that excess copper in the growing medium caused drastic reduction in the growth of an alga and attributed this phenomenon to a permanent injury to the photosynthetic mechanism. The same authors opined that product change might be due to the imbalance of the enzyme systems responsible for growth. They have suggested that Ni, Hg, Co, Cd application brought changes in quantity of citric, malic and aconitic acids. The amphiphilic nature of Me-Hg favoured the interactions of lipophilic cell structure, above all with the membranes, which may represent the primary target of organo-metallic attack. Further, both $HgCl_2$ and CH_3HgCl_2 (MMC) crossed the bio-molecular lipid membranes and entered into the cell and consequently act on different substrates and enzymes of the cell. Most of the results obtained in the present investigation do not tally with the results as reported by earlier workers as pointed out during discussion. Most of the references cited here were related to either heavy metal toxicity or pesticide toxicity. We do not find sufficient literature on this type of poisoning related to leached chemicals of red mud waste coming from the Alluminium industry, which might contain several other metals and might be heavy metals. The peculiar behavior shown by this toxicant (Red mud waste) might be due to synergistic effect of all types of chemicals available in the red mud waste.

Most of the results of different parameter were found to be statistically significant; it does mean that, the trend or changes observed in the present piece of work are to be taken into confidence. In this connection, here it can be stressed that as the study was undertaken during transition phase of growth or at active growth phase or the phase, in between emergence (embryo) and establishment (seedling), hence most of the parameters did confirm to the statistical 'significance' test, as observed by other workers. From the above observations and interpretations, erecting a model becomes a difficult task in absence of residual level data. However, many interferences and predictions raised during discussion, provides gaps or fields unexplored yet. In this piece of investigation, though we felt to a link a gap but after completion of this piece, we find many questions remain unanswered. In this piece of investigation, we tried to find out the effects of Red mud leached chemical, on seed germination and seeding establishment of rice crop, *Oryza sativa*, L. From the results obtained in present study, it is clearly evident that, Red mud leached chemical, which is now well known for its interference in biochemical metabolism, becomes phyto-toxic when used in lower and moderate concentrations. Thus, we can conclude that even though many advantages have been derived from the industry but we must not neglect the environment by way of constructing poison ponds like RED MUD PONDS near NALCO. Many similarities exist in biological process among all forms of life, and investigative effects of this nature are not only important to plants, fish and wildlife, but to all life forms, including man. However, as long as these chemicals are thrown into the environment without proper check, this type of leaching will always be there. These leached chemicals will always enter into crop fields; contaminate the groundwater and water bodies. There is always the possibility in future that unexpected and undesirable complication may arise even though minimum immediate hazards are occurring at present. Safe disposal of these wastes and proper care to avoid leaching of these chemicals can protect the non-target organisms, the blue-green algae. These tiny particles perform a fantastic job of atmospheric nitrogen fixation, increasing the fertility of paddy fields, maintaining the balance, used as biofertilizer. Due to careless handling of chemicals, unplanned disposal of wastes, future and fate of the waste chemicals in the environment, non adoption of safety measures, non-maintenance of ecological balance, improper adoption of plantation, non-selection of pollution reducing plantation, will significantly harm the man made agricultural ecosystem in particular and the ECOSYSTEM in general. Hence, protection of non-target systems is equally essential like protection of the crop plants. Due to heavy and constant discharge of these waste chemicals without following the directions given by Center and State Pollution Control Boards, Department of Environment, many a times, failure of the crop was marked. This failure was always attributed and pointed towards the seed banks. They are always accused of supplying bad seeds or non-viable seeds. But the fact lies at the site, where the seeds are sown to grow a crop. Hence, proper care and proper and safe disposal of these killer chemicals is essential to protect the agricultural ecosystem, hence, protecting the main users of these agricultural produce, *i.e.* M A N.

Summary

In petriplate culture, with the increase in toxicant concentration in the soil, the percentage of germination significantly declined and above 2.0 per cent leached Red mud waste chemical, germination of the seeds was not marked. The percentage of germination increased with the increase in exposure period and the same parameter decreased with the increase in toxicant concentration. The control set showed normal germination and a maximum of 100 per cent germination of seeds was marked after 144 hrs of exposure. Out of which, 100 per cent seedlings established. With the increase in toxicant concentration, the percent of seed germination reduced significantly. The difference between control and exposed sets was significant at higher exposure periods. Higher percentage of seed germination was marked in the pot, when compared to petriplate culture experiments. At higher conc. of the toxicant, the percentage of seedling establishment showed significant decline in pot culture experiments and the decrease showed the existence of negative correlation between the percentage of seedling establishment and toxicant concentration. A maximum of 85.96 per cent decease in shoot length was marked at 2.0 per cent of the toxicant exposed seedlings when compared to control seedlings, with the increase in the toxicant the shoot length decreased significantly. The length of the roots was more than the length of the shoots in the control set. The length of the root decreased from 8.96 ± 0.22cms in the control set to 0.84±0.06 cm in conc. C (2.0 per cent). In petriplate culture experiments, the changes observed in shoot, root length and fresh weight, root and shoot dry weight were highly significant, where the correlation coefficient values were all negative and significant. The gradual decrease in shoot fresh wt with the increase toxicant concentration dose was probably due to the toxicant in laboratory controlled petriplate culture. A maximum of 88.4 per cent and 82.35 per cent decrease was marked in shoot and root dry weight respectively in exposed seedlings exposed to toxicant. In pot culture, with the increase in toxicant concentrations, the shoot length decreased significantly, showing a maximum of 47.22 per cent decrease, when compared to the control value. The root and shoot fresh weight ratio did show significant variation and the exposed values were less than the control value. Significant variation was observed in pot culture, when compared to the petriplate culture. Significant differences were marked between petriplate culture and pot culture data. A maximum of 89.28 per cent decrease in chlorophyll 'a' content was marked in seedlings exposed to conc. C, when compared to the control value in petriplate culture. At 2.0 per cent of red mud leached waste, 92.44 per cent depletion in total chlorophyll content was marked. Interestingly the total phaeophytin content declined significantly with the increase in toxicant concentration in exposed rice seedlings and 91.67 per cent decrease in phaeophytin content when compared to control value was marked in 144 hr old seedlings in petriplate culture, at conc. C. The total carotene content decreased with the increase in toxicant concentration significantly, when compared to the control value. In pot culture, in each exposure period, the total chlorophyll content of the leaves, decreased from the control value and maximum decrease was marked at conc. C. In conc. A on 15th and 30th day of exposure, increase in chlorophyll content by 7.02 per cent and 3.82 per cent was recorded. In pot culture, in each exposure period and at all the three concentrations of the toxicant, the phaeophytin content depleted significantly and a maximum of 58.09 per cent, 88.6 per cent and 95.95 per cent decrease was marked at concentration A, B and C, respectively indicated highly toxic nature of the toxicant, leached chemicals of the red mud. Changes in carotenoid content followed, almost the same trend indicating the toxic nature of the toxicant, red mud waste.

The respiration rate decreased with the increase in the conc. of the toxicant. Significantly, the net primary productivity decreased in conc. C of the toxicant. Similarly the gross primary productivity decreased with the increase in concentration of the toxicant in 144 hr old rice seedlings. Maximum depletion in net primary productivity was marked at conc. C of the toxicant on 60th day of exposure. The changes in gross primary productivity in control and exposed seedlings in pot culture followed an identical trend. At lower exposure periods and lower doses of the toxicant, stimulation in the net primary productivity values were marked when compared to the control values. Significant negative correlation coefficient values were marked between the respiration rates of the rice plant leaves at different concentrations of the toxicant. Under field conditions in pot culture, the respiration rate of exposed plants declined with the increase in time period and concentration of the toxicant. Highest decrease in NPP value was recorded at conc. C and 60th day of exposure. The NPP value decreased significantly. The effects were more in pot culture in prolonged period of experiment. In petriplate culture, probably less amount of the toxicant was absorbed. But in prolonged period of experiment, higher amount of toxicant was absorbed and accumulated in the exposed plant. The body burden might have increased significantly showing the effects. The gross primary productivity was the calculated value of NPP and RR. Hence GPP values showed an identical trend like that of NPP and RR. Highest depletion was recorded at conc. C and on 60th day of exposure. Glucose content decreased significantly by 88.36 per cent at conc. C, when compared to the control value. Approximately 90.32 per cent decrease in nitrogen content was marked in seedlings exposed to the toxicant, when compared to control seedlings. With increase in the toxicant, the glucose content of the exposed leaves, declined significantly except in conc. A. The decrease in glucose content in the leaves of control plant at 60th day of exposure might be due to the collaborative effort of the toxicant and aging and senescence of the exposed rice plants. A maximum of 78.43 per cent and 96.43 per cent decrease in nitrogen content was marked at conc. C after 60th day of exposure in shoot and root, respectively. The ATPase activity of the exposed seedlings showed significant depletion and 91.18 per cent decrease was recorded in conc. C in petriplate culture.

From the above observations, it can be concluded that the toxicant, leached chemicals of the red mud pond is deadly toxic. The variation in toxicity was only due to the concentration of the toxicant Every year during the rainy season, these toxicants leach heavily and leaching was continuous in all seasons and in all days, enters into the paddy fields, thereby increasing the toxicant concentration. It was observed that toxicants enter into the plant and there might be a gradual build up of toxicant level in the rice plant. From the plants, toxicant might be getting deposited in the rice grains. Toxicant build up in the plants and than in the grains, and consequent build up in grazers may lead to availability of toxicant in milk and meat of herbivores or grazers. From milk, meat and rice grain, toxicant(s) may enter into human body. This build up of toxicant in the human system may lead to another significant incident, like many other incidents related to pollution.

References

Ahmad, F. E., R. W. Hart and N. J. Lewis (1977). *Mut. Res.*, 42: 161.

Ahmad, M. R., P. N. Saxena, D. V. Amla and R. Shyam (1978). *Beitr. Biol. Pflanzen*, 54: 33.

Ahuja, A. K. and D. Subrahmaniyam (1978). *Proc. Ind. Acad. Sci.*, 87: 189.

Annonymous (1989). Special short Course Training on Seed Testing. Seed testing Laboratory, Directorate of Agrl. and Food Production, Bhubaneswar.

Bache, C.A., W.H. Gutenmann and D.J.Hisk (1971). *Science*, 172: 951.

Baicu, T. and T. Diaconu (1972). *An Inst. Cercet. Prot. Plant.*, 10: 381.

Banerjee, A. and A.S. Mukherji (1980). *Indian Journal of Exp.Biol.*18: 438.

Barron, M.G and Adelman, I.R. (1984). Nucleic acid. Protein content, and growth of larval fish sub-lethally exposed to various toxicants. *Can. J. Fish. Aquat. Sci.*, 41: 141.

Barrineau, P., P. Gillbert, W.J.Jackson, C.S.Jones, A.O. Summers and S. Wisdom (1984). *Mol. Appl Genet.*, 2: 601.

Beauford, W. J. Barber and A.R.Barringer (1977). *Physiol. Plant.*, 39: 261.

Bouveng, H.O.(1968). The chlorine industry and mercury problem. *Modern. Kemi.*, 3: 45.

Bruin, A. De. (1976). Nucleic acid metabolism. In: *Biochemical Toxicity of Environmental Agents*. Ed Bruin. A. De. pp. 605-655. Elsevier/North Holland Biomedical Press, The Netherlands.

Butler, G.C. (1978). *Principles of Ecotoxicology*. John Wiley and Sons, Chisester, New York, Brisbane, pp 350.

Byford, W.J. (1971). *Ann. Appl.* 79(3). 221.

Cairns, J. (1980). *Bioscience*, 30: 101.

Cairns, J. Jr., D. Gruber, K. L. Dickson, A. C. Hendrick and S. W. H. Van Deer (1977). Developing an on site continuous biological monitoring system for chemical industry. In : *Proc. 5th Annual Industrial Pollution Conf. 1*. Delpino and Krigman (eds.), 285.

Cairns, J. Jr., D. K. L. Dickson and G. F. Wastelake (1976). Biological monitoring of water and effluent quality. Philadelphia, American Society for Testing of Materials, STP, 607.

Cairns, J. Jr., G. R. Lanza and B. C. Parker (1972). *Proc. Acad. Nat. Sc.*, Philadelphia, 124: 79.

Cairns, J. Jr., K. L. Dickson and A. W. Maki (eds.) (1978). Estimating the hazard of chemical substances to aquatic life. *ASTM*, Philadelphia.

Czuba, M. And D.C.Mortimer (1982). *Can J. Bot.*, 60(5). 657.

Das Gupta, B. And S. Mukherji (1982). *Z. Pflanzen Physiol.*, 82: 95.

De Filippis, L. F. (1979). *Zietschrift fur Pflanzen Physiologie*, 92: 39.

De Filippis, L. F. and C. K. Pallaghy (1976 a). *Z. Pflanzen-physiol. Bd.*, 78: 197.

De Filippis, L. F. and C. K. Pallaghy (1976 b). *Z. Pflanzen-physiol. Bd.*, 78: 323.

De Filippis, L. F., R. Hamp and H. Zeigler (1981). *Z. Pflanzen Physiol. Bd.*, 101: 37.

De, A.K., A.K. Sen, D.P. Modak and S. Jana (1985). Studies of toxic effects of Hg(II) on *Pistia stratioles*. *Water, Air and Soil Pollution*, 24: 351.

Debnath, R. and S. Mukherji (1982). *Biologia Plantarum (PRAHA)*, 24(6). 423.

Desiah, D., L.K.Cutkomp, R.B.Koch and G. A. Garvinean (1975). *Arch. of Environ. Contam. Toxicol.*, 3(2). 132.

Dove, W.F. and T. Yamane (1960) : *Biochem. Biophys. Res. Commun.*, 3: 608.

Donovan, T.J. and A.D.Day (1969). Some effects of high salinity on germination and emergence of barley (Hordeum vulgare, L. and Lati) *Agron. J.*, 61: 236.

Dove, W.F. and T.Yamane (1960). *Biochem. Biochem. Biophys. Res. Commun.*, 3: 608.

Filip, D. S. and R. T. Lynn (1972). *Chemosphere*, 6: 251.

Fox, J.H.K. Patel-Mandlik and M.M. Cohen (1975). Comparative effects of organic and inorganic mercury on brain slice respiration and metabolism. *J. Neuro-chemistry*, 24: 757.

Frenkel, G.D. and K. Randles (1982). Specific stimulation of a-amanitin sensitive RNA synthesis in isolated HeLa Nuclie by methyl mercury. *J. Biol. Chem.* 257-6275.

Frenkel, G.D., R.Cain and E.S-E. Chao (1985). Exposure of DNA to methyl mercury results in an increase in the rate of its transpiration by RNA polymerase-II. *Biochemical and Biophysical Research Communications*, 127: 849.

Furukawa, K., T. Suzuki and K. Tonomura (1969). *Agric. Biol. Chem.*, 33: 128.

Gadd, G.M. and A.J. Griffith (1978). *Microbial Ecol.*, 4: 303.

Geener, H. and J.R.Jeener (1952). *Exptl. Cell Res.*, 3: 675.

Gill, T.S., J.C. Pant and H.Tewari (1988). *Bull. Environ. Contam. Toxicol.*, 41: 241.

Godbold, D.L., and A.Hutterman (1985). Environmental Pollution (Ser.A.). *Ecological and Biological*, 38: 375.

Gross, R. E., P. Pungo and W. M. Dugger (1970). *Plant Physiology*, 46: 183.

Hamilton, G. and A. Ruthvein (1967). *Journal of Science of Food and Agriculture*, 18 (12) : 558.

Hannan, P.J. and C. Patouillet (1972). Effects of Pollutants on growth of algae, *Reports of Naval Res. Lab. Progress*, p.1-8.

Hannan, P. J., P. E. Wilkniss, C. Patouillet and R. A. Carr (1973). Measurement of mercury sorption by algae, *NRL Report*, 7623.

Hanumann, H.G. (1987). *Potoplasma*, 136 (1). 37.

Harris, R.C., D.B. White and R.B. Mac Farlane (1970). Mercury compounds reduce photosynthesis of plankton. *Science*, 170: 736.

Hattula, M.L., J. Sarkka, J.Janatuinen, J. Passivirta and A.Roos (1978). *Env. Pollut.*, 17: 19.

Holdgate, M.W. (1979). *A perspective of environmental pollution.* Cambridge University Press, Cambridge, London, New York, Melbourne, p.278.

Huckabee, J. W., J. W. Elwood and S. G. Hildebrand (1979). In: *Biogeochemistry of Mercury in the Environment*, Nriagu, J. O. (Ed.) Elsevier, Holland, 277-302.

Huckabee, J.W., S.D.Francisco, S.A. Janzen and J.Solmon (1983). *Environ. Pollut. Ser.* A. 30: 211.

Jernelov, A. and H. Lann (1971). *Oikos*, 22: 403.

Joy, C. M. (1990). *Bull. Environ. Contam. Toxicol.*, 45: 915.

Kar, S. and P. K. Singh (1978). *Bull. Environ. Contam. Toxicol.*, 20: 707.

Kashyap, A.K. and S.L. Gupta (1981). *Acta Botanica Indica*, 9: 265.

Katz, S. (1962). *Nature*, 194: 569.

Katz, S and V. Santilli (1962). *Biochem. Biophys. Acta.*, 55: 621.

Kimamura. M. and S. Kotoh (1972). *Biochem. Biophys. Acta.*, 283: 178.

Koronowski, K. (1973). *Mitt. Biol. Bundesanst.* Land Forstwirtsch Berling-Dahlem. 153: 1-21.

Kumar, K. B. (1984). Studies on physiology of senescence in crop plants. II. Effect of growth regulators and photo period on leaf senescence in *Eleusine coracana*, Gaertn. *Ph. D. Thesis*, Berhampur University, India.

Leland, H. V., S. N. Luoma and J. M. Fielden (1979). *J. Wat. Pollut. Contr.Fed.*, 51: 1592.

Levitt, J.(1972). *Response of plants to environmental stresses.* Academic Press, Inc. N. Y.

Lodenius, M. (1980). *Ann. Bot. Fenn.*, 17: 336.

Lodenius, M. and E. Tulisalo (1984). Environmental mercury contamination around a Chlor-alkali plant. *Bull. Environ. Contam. Toxicol.*, 32: 439.

Misra, S.R. (1984). Analysis of the effects of solid waste extracts of a caustic-chlorine factory on growth and physiology of rice seedlings. *Ph.D. thesis.* Berhampur University, Orissa, India.

Mukherji, S. and G. Ganguly (1974). *Indian J. Exp. Biol.*, 12: 432.

Mukherji, S. and P. Nag (1977). *Biochem. Physiol. Pflanzen.*, 171: 227.

Mukherji, S. and P. Maitra (1977). *Z. Pflanzen. Physiol.*, 81: 26.

Murthy, S. D. S. and P. Mohanty (1991). *Plant Cell Physiol.*, 32(2). 231.

Odum, E.P. (1971). *Fundamentals of Ecology*, 3rd Edition. W.B.Saunders Co., Philadelphia, London.

Oser, B. L. (1965). *Hawks Physiological Chemistry*, 14th edn., Edited by B. L.Oser, TMH Publishers, New Delhi, reprinted 1979, p. 179, 224 and 444.

Padhi, N. P. (1990). Eco-physiological response of germinating rice seedlings to an organo-mercurial fungicide, MEMC., *Ph.D. Thesis*, Berhampur University, India.

Panigrahi, A.K. and B.N .Misra (1978). *Comp. Physiol. Ecol.*, 8: 253.

Patra, R.R. (1993). Eco-physiological effects of an organomercury based pesticide on a crop plant. *Ph.D. Thesis*. Berhampur University. Orissa. India.

Phillip, J.R. (1958). *Plant Physiol.*, 33, 23.

Poljakoff-Mayber, A. and J. Gale (1975). *Plants in Saline Environments (Ecological studies–15)*. Springer-Verlag. Berlin.

Pradhan, P. (1998). Eco-toxicological effects of a pesticide on a crop plant. *Ph.D. Thesis*. Berhampur University. Orissa. India.

Rai, L. C. and N. Khatoniar (1980). *Indian J. Environ. Hlth.*, 22: 113.

Rai, L. C., J. P. Gaur and H. D. Kumar (1981, a). *Biol. Rev.* 56: 99.

Rai, L. C., J. P. Gaur and H. D. Kumar (1981, b). *Environmental Research*, 25: 250.

Sai Babu, T. (1998). Eco-physiological effects of a pesticide on a crop plant. *M.Phil. Thesis*, Berhampur University. Orissa, India.

Sax, I.N. (1974). *Industrial pollution*. Van Nostrand Reinhold Company, New York.

Semu, E., B.R. Singh, A.R. Selmen-Olsen and K. Steenberg (1985). *Plant and Soil.*, 87: 347.

Sharma, B.K. and H. Kaur (1993). *Environmental Chemistry*. Goel Publishing House, Krishna Prakashan Media (P) Ltd., Meerut.

Shaw, B.P. (1987). Eco-physiological studies of industrial effluent of a cholr-alkali factory on biosystems. *Ph.D. Thesis*. Berhampur University, Orissa. India.

Siegel, B.Z., M. Lasconta, E.Yaeger and S.M. Siegel (1984). *Water Air Soil Pollut.*, 23(1). 15.

Siegenthaler, P.A. and L. Packer (1965). Light dependent volume changes and reactions in chloroplasts.I. action of alkenylsuccinic acids and phenyl mercuric acetate and possible relation to mechanism of stomatal control. *Pl. Physiol. Lancaster*, 40: 785.

Smith, L. and M.Baltscheffsky (1959). Respiration and light induced phosphorylation in extracts of *Rhodospirillum rubrum*. *J. Biol. Chem.*, 234: 1575.

Srivastava, A.K., V.K.Sharma and Ahuja, K.L. (1972). Effect of salt stress on seed germination and survival of soyabean seedlings. *Trop. Ecol.*, 18: 27.

Suckcharoen, S. (1979). *Ann. Bot. Fenici.*, 16: 173.

Sunderesan, B.B., P.V.R. Subrahmanyam and A.D. Bhinde (1983). An overview of toxic and hazardous waste in Indian Industry and Environment. Special issue (1983). In: The State of India's Environment 1984-85. *The Second Citizen's Report*, p. 196.

Toubey, C.L. (1972). *Aust. J. Exp. Agric. Anim. Husk.*, 12(55). 176.

Weaver, R.W., J.R. Melton, D.Wang and R.L.Dable (1984). *Environmetal Pollution*. Series-A. *Ecol. Biol.*, 33(2). 133.

Whitton, B.A. (1970). Toxicity of heavy metal to freshwater algae. A review. Phykos, 9: 116.

Wolk, C. P. (1973). *Bacteriol. Rev.*, 37: 32.

Wong, M.H. and A.D.Bradshaw (1982). *New Phytol.*, 91(2). 255.

Yung, Chi-Tung (1938). Development anatomy of the seedling of rice plant. *Bot. Gaz.*, 99: 786.

Zingmark, R.C. and T.G. Miller (1975). *Belle W. Baruch Libr. Marine Science*, 3: 45.

Microbial Biotechnology and Ecology (2011) Pages 827–836
Editors: Deepak Vyas, G.S. Paliwal, P.K.Khare and R.K Gupta
Published by: DAYA PUBLISHING HOUSE, NEW DELHI

Chapter 48

Effect of Pollution in Groundwater Quality and Soil Health of Naini Area at Allahabad City

☆ *Saurabh Upadhyay*[1], *Arun A. David*[1], *T. Thomas*[1], *H. Thomas*[2]*, *R.K. Tamphasana*[2] *and Ram Bharose*[1]

[1]*School of Forestry and Environment, SHIATS-Deemed University, Allahabad*
[2]*Department of Applied Geology, Dr. H.S. Gour University, Sagar – 470 003, M.P.*

ABSTRACT

The experiment was conducted to assess the quality of soil and groundwater in Naini area of Allahabad, U.P. India. The 4 sampling sites were selected S_1–AAI-DU research farm of Soil Science and Agricultural Chemistry Farm as center, S_2–Arail, S_3–Cotton mill, and S_4–Kharkauni (village) and observation were taken in pre–monsoon and post–monsoon seasons for groundwater and soil properties respectively. pH 7.1-7.82, Turbidity 0.02-0.20 NTU, EC 0.47-0.88 dS m^{-1}, Hardness 430.6-490.4 mg l^{-1}, DO 2.8-4.5 mg l^{-1}, BOD 0.31-0.67 mg l^{-1}, Alkalinity 37-84 mg l^{-1}, Chloride 25.48-69.9 mg l^{-1} and for soil pH 7.65-7.26, EC 0.46-0.43 dS m^{-1} Bulk density 1.62-1.83 g cm^{-3}, Particle density 2.6-3.6 g cm^{-3}, per cent OC 0.27-0.8, Available Nitrogen 239-271 $kgha^{-1}$, Available Phosphorus 27.4-31.6 $kgha^{-1}$, Potasium 256-306 $kgha^{-1}$, Cu 0.32-0.95 $mgkg^{-1}$ were recorded respectively. These values were analyzed by appropriate statistical desined *viz.*, two way analysis of variance and determination and analysis were done with the standard methods. According to the results obtained by analyzing various physico-chemical parameters of groundwater and soil quality the sites S_1–AAI- DU and S_2–Arail was safe as compared to other locations *viz.* S_3–Cotton mill and S_4–Kharkauni (village).

* E-mail: harelthomas@yahoo.com

Introduction

Soil quality is the key to agricultural productivity, soil organic matter and nutrient levels, prim indicators of soil quality, readily decline when soils are taken under cultivation (Lobe *et al.*, 2001 Fertilizer use will improve the nutrient status of soils, but rarely over comes the soil organic matter an few ideas are around to remedy this crucial step in soil degradation (Marticus *et al.*, 2001). It is widel recognized that water and soil degradation problem is man made problem. It has been taking plac since the down of agriculture (Barrow 1991; Hillel 1991). What is now the intensity of degradation i recent times where is the soil of the world, on average, lost 25.3 million tonnes (Mt) of Humus per yea since agriculture began some 10,000 years ago, these loses were, on average, 300 Mt yr^{-1} in the last 30 years and 760 Mt in the past 50years (Rozauov *et al., 1990*). Thus the growth of population, expansio of crop lands, and destruction of vegetation, global warming emergence of yield-enhancing technologie all seem to have played a role. In essence, the last 50 years have been a saga of economic growth an ecological losses.

There is an urgent need to develop indicators, threshold limits or symptoms, potential indicator should be quantitative, change, widely applicable, able to asses the present status and trend, able t distinguish change due to natural cycles as apposed to anthropogenic interventions, and relevant t ecologically significant phenomena (Rubio and Bichet 1998). The assessment of degradation on th basis of yield trend is however difficult due to the use of improved varieties and fertilizer maskin ongoing degradation (Oldewau 1994; Swaling and Braun 1998). May elaborate methods such a classical soil testing or modern technologies such as soil genomics may eventually after new means c assessing the health of our soil (Torsvik and Oureas 2002).

Keeping in view of the problem stated above mentioned about may be a suitable solution for th Allahabad region. The problem was taken during the period 2007-2008 on topic entitled "Effect c pollution in groundwater quality and soil health of Naini area at Allahabad city", were studied wit the following objectives:

To check the quality of groundwater and soil in Naini area of Allahabad City. To compare groundwater quality with Indian standards. To compare soil quality with standards of Soil testin Laboratory, State Agricultural Department (U.P.).

Materials and Methods

Study Area

The area selected for the present study was Naini town located in Allahabad city. Which located at 25.57° N latitude, 81.56°E longitude and 98 m elevation above the mean sea level. Naini an industrial town of Allahabad lodging various types of industries manufacturing products. comes under sub tropical climate receiving the mean annual rainfall of about 807.4 mm, major rair are received from July to end of September. However, occasional precipitation is also not uncommo during winter. The winter months are cold while summer months are very hot and dry. The minimu temperature during the crop season was 7.3°C and maximum was 31.4°C. The maximum humidit was 93 per cent and minimum was 21 per cent during study period.

***Source*: Metrological data of College of forestry AAI-DU (Allahabad)**

Water and Soil sampling were done in pre–monsoon (March, April, and May) and post-monsoc (October, November and December) period 2007–2008, with the standard methods, groundwat samples were collected from the different sites/points of Naini area Soil sampling were done with help of soil auger from the upper soil layer having depth of (0–15cm) from different Sites *viz.* S_1-AAI–DU Allahabad, S_2- Arail, S_3–Cotton mill and S_4–Kharkauni village, Naini Allahabad.

Table 48.1: General Information about Allahabad District- 2008

Area	2,261 sq.km
Population	49,21,313 (census 2001)
STD. Code	0532
Best Season	November–March
State	Uttar Pradesh
Language	Hindi, Urdu, English
Altitude	98 meters above sea level
Temperature	Minimum 4 degree Celsius to max 48 degree Celsius
Local Transport	Taxi, Bus, Tempo, Auto-rickshaw, Cycle rickshaw
Accessibility	Air, Rail, Road

Source: ADA Allahabad (2008).

Table 48.2: Experimental Sites/Locations and their Distance from AAI-DU

Sl.No.	*Site Symbol*	*Site: Name*	*Distance from AAI–DU (Km)*
1	S_1	AAI–DU, Alld.	0
2	S_2	Arail	2
3	S_3	Cotton mill	4
4	S_4	Kharkauni village	3

Table 48.3: Different Biochemical Parameters Taken for Groundwater Study their Respective Methods Used for Analysis

Parameters	*Methods*
pH	Jackson (1958)
Turbidity (NTU)	APHA (1985)
EC (dS m^{-1})	APHA (1995)
Hardness (mg l^{-1})	APHA (1985)
Alkalinity (mg l^{-1})	APHA (1985)
Chloride (mg l^{-1})	APHA (1985)
Dissolved oxygen (mg l^{-1})	APHA (1985)
Biochemical oxygen demand (mg l^{-1})	APHA (1985)

Table 48.4: Different Soil Parameters and their Respective Methods Used for Analysis

pH 1:2 (w/v)	Jackson (1958)
EC (dS m^{-1})	Digital Electrical Conductivity Meter
Bulk density (g cm^{-3})	Black (1965)
Particle density (g cm^{-3})	Black (1965)
Percentage Organic Carbon (per cent)	Walkley (1947)
Available Nitrogen (kg ha^{-1})	Walkley (1947)
Available Phosphate (kg ha^{-1})	Olsen *et al.* (1954)
Available Potash (kg ha^{-1})	Toth and Prince (1949)
Copper (mg kg^{-1})	Cheng and Bray (1953)

Results and Discussion

Groundwater

In the pre-monsoon pH of investigated sites varied from 7.1-7.82 given in Table 48.5, pH value c groundwater to be was recorded at four different sites was found to be maximum in Cotton mill 7.8 followed by Kharkauni 7.30, Arail 7.21 and the lowest in AAI-DU, Alld. 7.10. As represented in th Table 48.5 the Turbidity value of groundwater to be was recorded at four different locations was foun to be maximum in Cotton mill 0.18 followed by Arail 0.13, Kharkauni 0.07 and the lowest in AAI.–DU 0.01 NTU respectively. Electrical Conductivity concentration of groundwater was recorded at fou different sites were found to be maximum in Kharkauni 0.87 followed by Arail 0.84, Cotton mill 0.8 and the lowest in AAI–DU 0.46 dS m^{-1}. This may be found due to leachate infiltration from the refus dumps. Hardness value of groundwater to be were recorded at four different sites was found to b maximum in Kharkauni 483.20, followed by Arail 451.00, Cotton mill 438.10 and the lowest in AAI. DU 423.00 mg l^{-1} respectively. Alkalinity concentration of groundwater were recorded at four differer sites was found to be maximum in Kharkauni 82.00 followed by Cotton mill 81.30, Arail 64.00 and th lowest in AAI–DU 35.60 mg l^{-1} respectively, this alkalinity values having lower concentration a compared to BIS standards recommendations. Chloride value of groundwater were recorded at fou different sites were found to be maximum in Cotton mill 68.70 followed by Kharkauni 56.50, Ara 43.90, and the lowest in AAI.–DU 22.20 mg l^{-1} respectively. DO concentration of groundwater we recorded at four different sites was found to be maximum in AAI.–DU 5.20 followed by Cotton mi 4.40, Kharkauni 3.60 and the lowest in Arail 2.70 mg l^{-1} respectively. BOD concentration of groundwat were recorded at four different sites was found to be maximum in Kharkauni 0.63 followed by Cotto mill 0.51, Arail 0.38 and the lowest in AAI–DU 0.28 mg l^{-1} respectively.

In the post-monsoon pH of investigated sites varied from 7.31-7.95 given in Table 48.6, p concentration of groundwater to be was recorded at four different sites were found to be maximum i Cotton mill 7.95 followed by Kharkauni 7.65, Arail 7.31 and the lowest in AAI-DU, Alld. 7.40. Th neutral to alkaline pH may be attributed to the reaction of applied fertilizer material with soil colloi which resulted in the reaction of basic cations on the exchangeable complex of the water. The p concentration values were found under the permissible limits of the standards. As represented in th Table 48.6 the Turbidity value of groundwater to be was recorded at four different sites were found be maximum in Kharkauni 0.98 Cotton mill 0.50 followed by Arail 0.18, and the lowest in AAI.–D 0.02 NTU respectively. This turbidity indicating presence of suspended and colloidal matterial su

Table 48.5: Effect of Groundwater on Physico-chemical Properties of Water in Pre-monsoon Period at Different Sites

Sl.No.	*Parameters*	*Sites*				*Permissible Limit (WHO)*
		AAI-DU	*Arail*	*Cotton mill*	*Kharkauni*	
1.	pH	7.1	7.21	7.82	7.3	6.5-8.5
2.	Turbidity (NTU)	0.01	0.13	0.18	0.07	5NTU
3.	E.C. (dS m^{-1})	0.46	0.84	0.82	0.87	
4.	Hardness (mg l^{-1})	423	451	438.1	483.2	500
5.	Alkalinity (mg l^{-1})	35.6	64	81.3	82	
6.	Chloride (mg l^{-1})	22.2	43.9	68.7	56.5	250
7.	Dissolved oxygen (mg l^{-1})	5.2	2.7	4.4	3.6	
8.	Biochemical oxygen demand (mg l^{-1})	0.28	0.38	0.51	0.63	

Table 48.6: Effect of Groundwater on Physico-chemical Properties of Water in Post-monsoon Period at Different Sites

Sl.No.	*Parameters*	*Sites*				*Permissible Limit (WHO)*
		AAI-DU	*Arail*	*Cotton mill*	*Kharkauni*	
1.	pH	7.4	7.31	7.95	7.65	6.5-8.5
2.	Turbidity (NTU)	0.02	0.18	0.50	0.98	5NTU
3.	EC (dS m^{-1})	0.52	0.92	0.95	0.98	
4.	Hardness (mg l^{-1})	426.00	456.00	445.40	490.60	500
5.	Alkalinity (mg l^{-1})	36.40	66.50	85.40	87.80	
6.	Chloride (mg l^{-1})	28.70	55.20	75.60	65.80	250
7.	Dissolved oxygen (mg l^{-1})	4.90	2.30	3.90	3.20	
8.	Biochemical oxygen demand (mg l^{-1})	0.31	0.42	0.58	0.82	

as clay, silt and fibrous particle such as asbestos minerals. Electrical Conductivity, concentration of groundwater was recorded at four different sites were found to be maximum in Kharkauni 0.98, followed by Cotton mill 0.95, Arail 0.92, and the lowest in AAI.–DU 0.52 dS m^{-1} respectively. This may be found due to leachate infiltration from the refuse dumps. Hardness value of groundwater to be were recorded at four different sites was found to be maximum in Kharkauni 490.60, followed by Arail 456.00, Cotton mill 445.40 and the lowest in AAI.–DU 426.00 mg l^{-1} respectively. Alkalinity concentration of groundwater were recorded at four different sites were found to be maximum in Kharkauni 87.80 followed by Cotton mill 85.40, Arail 66.50 and the lowest in AAI.–DU 36.40 mg l^{-1} respectively, this alkalinity values having lower concentration as compared to BIS standards recommendations. Chloride value of groundwater were recorded at four different sites were found to be maximum in Cotton mill 75.60, followed by Kharkauni 65.80, Arail 55.20, and the lowest in AAI.–DU 28.70 mg l^{-1} respectively. A direct correlation between the chloride content and EC was recorded in the present study high level of chloride can also be contributed by mineral deposits, and industrial wastes. DO concentration of groundwater were recorded at four different sites was found to be maximum in AAI.–DU 4.90 followed by Cotton mill 3.90, Kharkauni 3.20, and the lowest in Arail 2.30 mg l^{-1} respectively. BOD concentration of groundwater were recorded at four different sites was found to be maximum in Kharkauni 0.82, followed by Cotton mill 0.58, Arail 0.42 and the lowest in AAI.–DU 0.31 mg l^{-1} respectively. This BOD values having lower concentration as compared to WHO/BIS standards recommendations.

Soil

As represented in the Table 48.7, the pH concentration of soil was recorded in pre-monsoon period, at four different sites was found to be maximum in Cotton mill 7.62 followed by Kharkauni 7.58, Arail 7.57 and the lowest in AAI–DU 7.23. Alkaline range of pH in most of the soil sample may be due to the general alkaline nature of the effluents being released into sampling sites/locations. EC concentration of soil to be were recorded at four different sites was found to be maximum in Cotton mill 0.44 followed by AAI.–DU 0.42 and the lowest were in Kharkauni 0.41, and Arail 0.41dS m^{-1} respectively. Particle density values of soil were recorded at four different sites was found to be maximum in Kharkauni 2.80, followed by Cotton mill 2.70, Arail 2.70 and the lowest in AAI.–DU 2.40g cm^{-1} respectively. Bulk density values of soil were recorded at four different sites was found to be maximum in Cotton mill 1.80, followed by Kharkauni 1.73, Arail 1.72 and the lowest in AAI.–DU 1.57g cm^{-1} respectively. Per cent Organic Carbon value of soil were recorded at four different sites was found to be maximum in Kharkauni 0.81 followed by Cotton mill 0.76, Arail 0.27 and the lowest in AAI.–DU 0.25 per cent respectively. Available Nitrogen concentration of soil to be were recorded at four different sites was found to be maximum in Kharkauni 271.00 followed by Cotton mill 263.00,

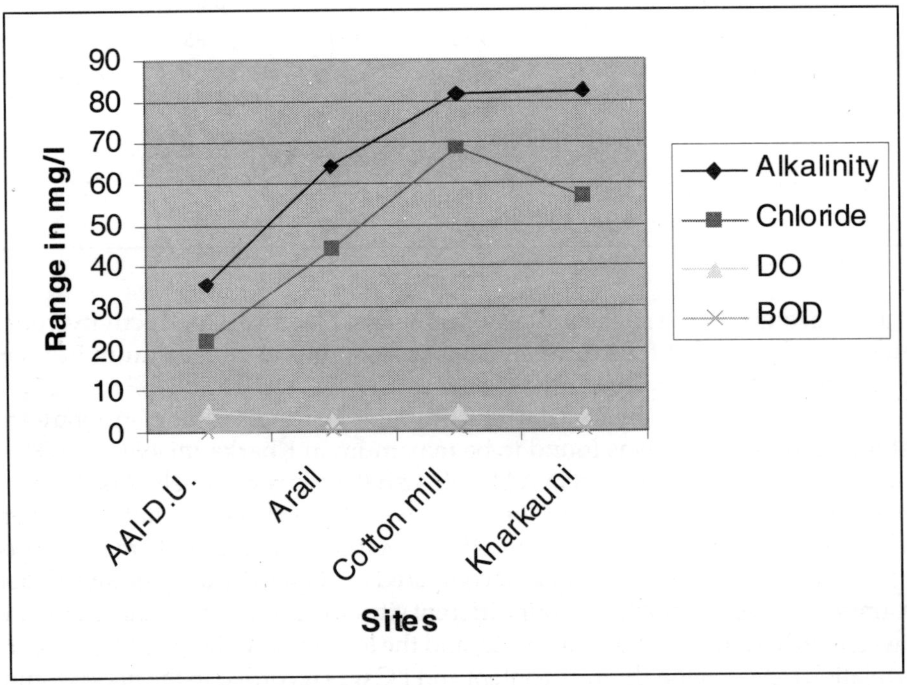

Figure 48.1: Response of Alkalinity, Chloride, DO and BOD on Water at Different Sites of Naini Area in Pre-monsoon Period

Arail 252.00 and the lowest in AAI–DU 241.00kg ha^{-1} respectively. Available Phosphorus concentration of soil were recorded at four different sites was found to be maximum in Kharkauni 31.60 followed by Cotton mill 35.00, Arail 29.40 and the lowest in AAI–DU 28.20kg ha^{-1}kg ha^{-1} respectively. Available Potassium concentration of soil were recorded at four different sites was found to be maximum in Khaekauni 280.30 followed by Cotton mill 303.30, Arail 313.00 and the lowest in AAI.–DU 256.00kg ha^{-1} respectively Cu concentration of soil were recorded at four different sites was found to be maximum in Kharkauni 0.93 followed by Cotton mill 0.83, Arail 0.72 and the lowest in AAI–DU 0.30mg kg^{-1} respectively.

Table 48.7: Effect on Soil on Physico-chemical Properties of Soil in Pre-monsoon Period at Different Sites, Naini Allahabad

Sl.No.	Parameters	Sites				Permissible Limit (Agricultural Deptt. U.P. 2008)
		AAI-DU	Arail	Cotton mill	Kharkauni	
1	pH 1:2 (w/v)	7.23	7.57	7.62	7.58	7.0-8.5
2	EC (dS m^{-1})	0.42	0.41	0.44	0.41	1.40-1.50
3	Particle density (g cm^{-3})	2.40	2.70	2.70	2.80	2.60-2.70
4	Bulk density (g cm^{-3})	1.57	1.72	1.80	1.73	1.40-1.50
5	Percentage Organic Carbon (per cent)	0.25	0.27	0.76	0.81	0.6-0.80
6	Available Nitrogen (kg ha^{-1})	241.00	252.00	263.00	271.00	240-320
7	Available Phosphorus (kg ha^{-1})	28.20	29.40	30.50	31.60	28-34
8	Available Potassium (kg ha^{-1})	256.00	313.00	303.30	280.30	240-320
9	Copper (mg kg^{-1})	0.30	0.72	0.83	0.93	0.4-1.0

As represented in the Table 48.8, the pH concentration of soil were recorded in post-monsoon period, at four different sites was found to be maximum in Kharkauni 7.89 followed by Cotton mill 7.73, Arail 7.62 and the lowest in AAI–DU 7.28. Alkaline range of pH in most of the soil sample may be due to the general alkaline nature of the effluents being released into sampling sites/locations. EC concentration of soil to be were recorded at four different sites was found to be maximum in Kharkauni 0.52 followed by Cotton mill 0.49, Arail 0.48 and the lowest in AAI.–DU 0.43dS m^{-1} respectively. This E.C may be due to mineralization by soil organism and conductivity increase is due to leachate infiltration from the refuse dump. Particle density values of soil were recorded at four different sites was found to be maximum in Kharkauni 2.90, followed by Arail 2.80, Cotton mill 2.70 and the lowest in AAI.–DU 2.50g cm^{-1} respectively. Bulk density value of soil to be were recorded at four different sites was found to be maximum in Cotton mill 2.10 followed by Arail 1.78, Kharkauni 1.73 and the lowest in AAI.–DU 1.59g cm^{-1} respectively. Per cent OC concentrations of soil were recorded at four different sites was found to be maximum in Kharkauni 0.83 followed by Cotton mill 0.78, Arail 0.29 and the lowest in AAI.–DU 0.27 respectively. Available Nitrogen concentation of soil to be were recorded at four different sites was found maximum in Kharkauni 273.00 followed by Cotton mill 264.00, Arail 253.00 and the lowest in AAI–DU 242.00kg ha^{-1} respectively. Available Phosphorus concentration of soil were recorded at four different locations was found to be maximum in Kharkauni 32.10 followed by Cotton mill 31.20, Arail 29.80 and the lowest in AAI–DU 28.40kg ha^{-1} respectively. Available Potassium concentration of soil were recorded at four different sites was found to be maximum in

Arail 314.00 followed by Cotton mill 304.00, Kharkauni 281.00 and the lowest in AAI.–DU 257.00kg ha^{-1} respectively. Cu concentration of soil to be were recorded at four different sites was found to be maximum in Kharkauni 0.97 followed by Cotton mill 0.85, Arail 0.74 and the lowest in AAI–DU 0.32mg kg^{-1} respectively.

According to the result obtained by analyzing various physico-chemical parameters groundwater and soil quality in AAI.–DU. and Arail were safe as compared to other locations *viz.* Cotton mill and Kharkauni.

Table 48.8: Effect of Soil on Physico-chemical Properties of Soil in Post-monsoon Period at Different Sites

Sl.No.	*Parameters*	*Sites*				*Permissible Limit (Agricultural Deptt. U.P. 2008)*
		AAI-DU	*Arail*	*Cotton mill*	*Kharkauni*	
1	pH1:2 (w/v)	7.28	7.62	7.73	7.89	8.0–8.5
2	EC (dS m^{-1})	0.46	0.48	0.49	0.52	1.40–1.50
3	Particle density (g cm^{-3})	2.50	2.80	2.70	2.90	2.60–2.70
4	Bulk density (g cm^{-3})	1.59	1.78	2.10	1.73	1.40–1.50
5	Percentage Organic Carbon (per cent)	0.27	0.29	0.78	0.83	0.6–0.80
6	Available Nitrogen (kg ha^{-1})	242.00	253.00	264.00	273.00	240–320
7	Available Phosphorus (kg ha^{-1})	28.40	29.80	31.20	32.10	28–34
8	Available Potassium (kg ha^{-1})	257.00	314.00	304.00	281	240–320
9	Copper (mg kg^{-1})	0.36	0.74	0.85	0.97	0.4 -1.0

Summary and Conclusion

Monitoring the health of our soil is urgently needed and will require a long term commitment, a dense network of monitoring sites and well functioning laboratory facilities. It will require the development of indicators, threshold limits or symptoms that can predict the on set of land degradation. ombined with the use of satellite imagery as discussed in this paper; such a system may be made cost effective.

Human needs of change the natural system of new one that is stable and virtually closed to one that is degrading and more or less open, dependent on the severity of the human interferences. The rapid decline in water and productivity of land following conservation to agriculture has been demonstrated again and again. The present study represents an in-depth investigation of the current status of Groundwater and Soil quality of Naini area of Allahabad city (U.P.). The findings of the research have resulted in environmental degradation with regarding to water and soil. As such the present environmental status of Naini area is not so alarming. The Industrial Complex and its surroundings have been growing fast due to rapid urbanization and industrialization in the last four decades. However, if proper environmental management is not followed, then the situation may degrade further. The groundwater in some localities are found to be polluted due to various industrial activities as well as domestic activities. Some water quality parameters have already exceeded the permissible limits and some parameters are approaching towards the permissible limits because of leaching, percolating and other such phenomenon for groundwater. There are various measures which can be made to control the water and soil pollution; however the following measures are important depending on the topographic and metrological consideration of the area under study.

Municipal and sewage water should be treated before it is released to the water bodies and irrigation use in the farmer's field. For industrial houses, provision of zero according to the guideline of the pollution control board/bodies/departments/Govt. (Central and State) emission or minimum emission should be enforced. Proper planning should be implemented before allowing any body to draw the water from under ground. Groundwater used for infant feed must be pretreated and reanalyzed so as to ensure the safety for live-stock and human mankind.

Acknowledgements

Authors sincerely acknowledge and express thanks to Prof. (Dr.) S.B.Lal, Director Research/ Dean College of Forestry, SHIATS-DU Allahabad for providing guidance and facilities for conducting research work and preparing manuscript.

References

Adekunlel, M., *et al.* (2007). "Assessment of Groundwater Quality in a Typical Rural Settlement in Southwest Nigeria". *Int. J. Environ. Res. Public Health* 4(4), 307-318.

Bharambe, P.R. *et al.* (2005). Water table fluctuations, quality of groundwater and soil health in Jayakwadi command". *Environment and Ecology*,. 34(4): 431 521.

Black, C.A (ed). (1965). *Methods of Soil Analysis Vol. 1.* Am. Soc Agron.Madison, Wisconsin, U.S.A.

Brady, N. C. (1980). *The Nature and Properties of Soils.* Eurasia Publishing House Private Limited, Ram Nagar, New Delhi-110 055.

Cheng, K.L., and R.H. Bray (1953). Two specific methods of determining Cu in soil and plant material. *Anal. Chem.* 25: 655-659.

Chaudhary, S. *et al.* (2004). Studied that "Soil and groundwater pollution in Faridabad." Environment and Ecology, 2004. 22(3): 636-641.

Hillel, D.J. (1991). *Out of the earth: civilization and the life of the soil.* The free press, New York, NY, USA.

Hutson, J.L. (1997). The Soil Quality Indices Literature Compiler's Report; Appendix D-"W and Listings *Journal of Environmental Quality* 26: 41–48.

Jackson, M. L. (1958). *Soil Chemical Analysis.* Prentice Hall of India. Private Ltd. New Delhi.

Lobe, I., Amelung, W. and Dee Preez, C.C. (2001). Losses of carbon and nitrogen with prolonged arable cropping from sandy soil of the South African Highveld. *E uropean Journal of Soil Science* 52 pp. 93-101.

Mayan Kutty, P.C. *et al.* (1995). examined that "Monitoring of trace metals in desalinated Drinking water and their permissible levels". Paper presented at the IDA conference, Abu Dhabi, held during 18–24 Nov., 1995.

Marticus, C., Tiessen, H. (2001). The management of organic matter in tropical soils; What are the priorities ? *Nutrient Cycling in Agroecosystems* 61, 1-6.

Moti R. Sharma (2004). Assessment of Groundwater quality of Hamirpur area in Himachal Pradesh. 23(1): 131-134, *Enviromedia.*

Muhr, G.R. N.P. Datta, H. Sankarasubramoney, V.K. Leley, and R.P. Donahue (1965). *Soil Testing in India.* U.S.Agency for Intern. Development Mission to India, New Delhi.

Oldman L. R. (1994). The global extent of soil degradation. In: *Soil Resistance and Sustainable Land Use.* (D. J. Greenland and I, Szaboles, Eds.) pp. 99-118 CAB International, Wallingford, U. K.

Olsen, S.R., Cole, Cv., Watnahe,F.S. and Dean, L.A. (1954). Estimation of available phosphorus in soils by extraction with sodium bicarbonate *U.S. Deptt.Agr. Circ.* 939.

Palanisamy, P. N., Geetha, A., Sujatha, M., Sivakumar, P. and Karunakaran, K. studied that Groundwater samples collected from different localities in and around Gobichettipalayam town, Erode District, Tamil Nadu. Journal of environmental biology USEPA.

Rozanov, B. G., Targulian V., Orlov D. S. (1990). Soils in the Earth as transformed by human action (B. L. Turner, W. C. Clark, J.f. Richards, J. T. Mathews and W. B. Meyer, Eds.). Cambridge University Press, Cambridge, U. K.

Rubio J. L. and Bichet, E. (1998). Desertification indicators as diagnostic criteria for desertification in Europe. *Journal of Arid Environments.* 39: 113-120.

Torsvik, V. and Ovreas, L. (2002). Microbial diversity and function in soil: from genes to ecosystems. *Current Opinion in Microbiology,* 5: 240-245.

Toth, S.J. and A.L. Prince (1949). Estimination of cation exchange capacity and exchangeable Ca, K and Na content of soil by flame photometer technique. *Soil Sci.* 67: 439-445.

Walkley, A. (1947). Critical examination of rapid method for determining organic carbon in soils. Effect of variation in digestion conditions and of inorganic soil constituents. *Soil Sci.* 632: 251.

Wick, B., R. F. Kuhne, K. Vielhauer, and P. L. G. Vlek. (2002). Temporal variability of selected soil microbiological and biochemical indicators under different soil quality conditions in south-western Nigeria. *Biology and Fertility of Soils* 35, No. 3: 155-167.

Wilcox L.V (1950). Electrical conductivity. *Amer. Water Works Assoc.* J. 42: 775-776.

Microbial Biotechnology and Ecology (2011) *Pages* 837–845
Editors: Deepak Vyas, G.S. Paliwal, P.K.Khare and R.K Gupta
Published by: DAYA PUBLISHING HOUSE, NEW DELHI

Chapter 49

Modifying Influence of Woody Perennials on Atmospheric Deposition of Nutrient Ions in a Dry Tropical Region of India

☆ *J. Pandey*

Department of Botany, Banaras Hindu University, Varanasi – 221 005, U.P.

ABSTRACT

Atmospheric depositions of SO_4^{2-}, NO_3^-, PO_4^{3-}, NH_4^+, Na^+, K^+, Ca^{2+} and Mg^{2+}, both as clearfall and throughfall were measured in a seasonally dry tropical region of India. Significant spatial variations in the deposition of these cations and anions were observed. In general, depositions of these ions were high at source oriented sites (industrial and traffic oriented locations), and lower at comparatively remote forested sites. The pH of bulk deposition was alkaline at sites influenced by local sources of basic cations. Among the anions, maximum deposition was observed for SO_4^{2-} followed by NO_3^- and minimum of PO_4^{3-}. For cations, the deposition of Ca^{2+} appeared maximum followed by NH_4^+, Na^+, K^+ and Mg^{2+}. The pH as well as the deposition of cations and anions were higher in throughfall received under woody perennials than in clearfall samples. This modifying influence of woody perennials on atmospheric deposition may leads to long-run consequences on nutrient cycling and on soil characteristics even for areas void of direct human interference and has relevance in adopting practices towards forest management.

Introduction

Increased atmospheric deposition of nutrient elements as a result of anthropogenic activities such as fossil fuel combustion in urban–industrial areas and fertilizer application in agriculture has become a common phenomenon in many parts of the world. Atmospheric emission coupled with wind action is adding pollutants and nutrients even in areas far away from source oriented sites (Pandey and Pandey, 2005). Although developed countries, such as United States, have made considerable progress during past few decades reducing the amount of pollutants discharged from point sources, the control of pollutants generated by dispersed or non–point sources is still a big challenge. The problem is more acute in developing countries where, due to new establishing industries and extensive urban growth, atmospheric loading and deposition of pollutant aerosols is continued to rise.

India is a rapidly developing country where air pollution levels are steadily rising in urban industrial areas (Pandey and Agrawal, 1994a; 1994b; Pandey, 2005). Our country has been rapidly increasing its food and energy needs to sustain exponentially growing population and industrial development. For instance, due to continued population pressure, increase in fertilizer use has become unavoidable to achieve maximum agricultural harvest. This has led to rapid growth of fertilizer factories in India. Furthermore, at present, coal based electric generation accounts for about 64 per cent of the total installed capacity. These activities coupled with massive number of automobiles continue to increase the atmospheric emission in metropolitan cities and in the industrial areas.

Chemicals from atmosphere can stimulate, inhibit or have variable effects on growth and productivity of agricultural crops and on forest vegetation. Studies have shown stronger dependence of forest maintenance of nutrients derived from atmosphere and even in some ecosystems, atmospheric input of nutrients provide an important mechanism for replacing hydrologic losses as well as losses due to timber extraction (Friedland and Miller, 1999). However, the relative importance of such deposition often depends upon specific nutrient and site. For instance, N in bulk deposition is equivalent to about 70 per cent of nitrogen incorporated annually in above ground woody tissues in some temperate hardwood forests (Swank, 1984). The contribution of atmospheric deposition in many part of our country has been shown to be significantly high to modify the accumulation pattern in vegetation and soil (Singh and Agrawal, 2005; Pandey and Pandey, 2009) and productivity in aquatic ecosystems (Pandey and Pandey, 2005). This has particular relevance for dry tropical regions where removal of nutrients by leaching remains very slow. Under such condition, atmospheric deposition could shorten the time of nutrient saturation. In agroecosystem, excessive atmospheric nutrient input may lead nutrient cycle to remain open and unsustainable. Relatively less information however, are available on atmospheric inputs of nutrient elements with modifying influences of woody perennials in dry regions of India. The main objectives of this study were to characterize source and non–source oriented variations in deposition of nutrient elements and the role of vegetation in modifying the deposition in a dry tropical region of India.

Materials and Methods

Study area: This study site lies between 24°35' N lat and 73°42' E long and 510 to 580 m above msl in Udaipur district of Rajasthan. The climate of the region is tropical with three distinct seasons, a hot and dry summer (March–June), a short rainy season (July–September) and a cold and dry winter season (November–February). Mean monthly maximum temperature ranges between 23°C in January and 42°C in May and mean monthly minimum temperature ranges between 5°C in January and 25°C in May. Mean annual rainfall remains below 650 mm, about 90 per cent of which occurs during rainy season. Wind direction shifts from predominantly south–westerly during November through March and south–westerly to southerly for most of the remaining months.

Ten sites were selected in and around urban settlement of Udaipur considering the source of emission, relief features and vegetation. Three of the sampling sites were located adjacent to three different lakes, two under partial influence of road emission (Fateh Sagar and Udai Sagar) and third situated in the midst of a deciduous forest with least human influence. Two of the sites, Sukher and Madri, receive emission from industries. Rest of the sites represents traffic oriented urban and sub-urban locations. The details of study sites, emission characteristics, vegetation types etc are presented in Table 49.1. The sampling period consisted of November 15, 2003 to October 14, 2005.

Table 49.1: Characterization of Study Sites

Site No.	*Site*	*Location/source*	*Vegetation*
1.	Delhi Gate	City center, high frequency of petrol driven vehicles	Few avenue trees
2.	Sukher	Sub-urb, marble industry and heavy duty vehicles	Avenue trees and shrubs
3.	Fateh Sagar	Lake site, NW margin of city, petrol driven vehicles	Deciduous trees and shrubs on uplands
4.	Shilpgram	3 km NW from the city, Low frequency of petrol driven vehicles	deciduous forest with ground vegetation
5.	New Campus	3 km NE from the city, residential, low frequency of petrol driven vehicles	Plantation of avenue trees, agricultural field
6.	Sector 5	5 km from the city, sub-urb, low frequency of petrol driven and heavy duty vehicles	Avenue trees and shrubs
7.	Savina	3 km S from the city, sub-urb, heavy duty vehicles	Avenue trees and shrubs
8.	Madri	8 km SE from the city, industrial area	Plantation of avenue trees
9.	Udai Sagar	Lake site, 14 km E from the city, highway, agricultural field	Avenue trees, agricultural land and grasses
10.	Baghdara	Lake site, 20 km SE from the city, remote but partial industrial influence	Deciduous forest, shrubs and grasses

Dust Deposition

The settleable particulates were collected using dry polyethylene collectors having 30 cm height and 15 cm diameter. Collectors were placed at heights of 5 m for known period of time at all sites in open locations for collection of clearfall and under the canopy of *Cassia fistula* for throughfall measurements. *Cassia fistula*, selected as a representative of tropical woody perennials for this study, is an ornamental tree found uniformly at all sites including traffic–orientated urban locations.

$$\text{Dust deposition rate } (g\ \text{m}^{-2}\ \text{d}^{-1}) = \frac{g \text{ particulates}}{a \times d}$$

where,

a is the cross–sectional area (πr^2) of the mouth of the particulate collector and *d* is the number of days for which the collector was exposed to atmospheric deposition.

Deposition of Nutrient Ions

Bulk atmospheric depositions were collected by using properly washed and dried five liter polyethylene bulk collectors mounted with stainless steel funnel of 30 cm diameter. These collectors were exposed to atmosphere for a period of 30 days to collect bulk deposition as clearfall in open locations and throughfall under the canopy of *Cassia fistula*. A protective guard ring was fitted around each funnel to avoid materials from birds, insects etc. A nylon membrane (240 µm) was placed between the collector and funnel. At each site, triplicate sampling collectors were placed. At the end of each sampling period (30 d), atmospheric deposits were quantitatively transferred from the collectors to properly washed polythene containers. The wall of the funnel and nylon membrane was carefully rinsed with deionized water to incorporate the deposits adsorbed on these surface. The final volume was made 100 ml and after proper stirring, these were left for overnight to extract the maximum water soluble ions. During the rain events, the wet samples were preserved with chloroform. These were filtered through Whattman filter paper and kept under refrigerated condition until the completion of chemical analysis.

Chemical Analysis

The pH was measured with the help of a digital photovolt pH meter (EI, model 101E) using a glass electrode standardized with pH 4.7 and 9.2 reference buffers. Conductivity was measured using a conductivity meter (Electronic India). All the atmospheric depositions were analyzed for SO_4^{2-} NO_3^-, NH_4^+, PO_4^{3-}, Na^+, K^+, Ca^{2+} a nd Mg^{2+}. Sulphate–S content was determined by turbiditimetric method (Rossum and Villaruz, 1961). $NO_3^-$and NH_4^+ contents were determined by using nitrate and ammonium electrodes attached to an ion analyzer. Concentrations of Na^+, K^+ and Ca^{2+} were measured using a flame photometer (ELICO, India). Magnesium concentration was determined by Atomic Absorption Spectrophotometer (Perkin Elmer, USA).

Quality Assurance and Statistical Analysis

For quality assurance, data comparisons were made between calculated vs measured conductivity and cation *vs* anion balance of individual samples. The data were treated as reliable when the differences were below 10 per cent. Blank analyses were performed for about 10 per cent of all samples according to standard operating procedures. The analytical precision expressed as relative standard deviation was below 10 per cent in all the cases.

Effects of sites were tested using analysis of variance (ANOVA). Means were separated by using Duncan's multiple range tests. Correlation coefficients were calculated between the atmospheric deposition parameters and pollutant concentrations.

Results and Discussion

The dust deposition rates, both as clearfall as well as throughfall, were maximum at site 8 and minimum at site 10 (Figure 49.1). Site–wise variation in dust deposition rates was significant ($P < 0.01$, ANOVA). Throughfall dust deposition rates appeared uniformly higher than the clearfall. Sitewise variations in atmospheric dust depositions clearly indicated the influence of local sources including urban and industrial activities as well as the partial influence of local soil sources.

The pH of the clearfall varied between 6.00 at site 10 to 7.80 at site 2 (Figure 49.2). Site–wise variation in pH of bulk deposition was significant ($p < 0.01$). The pH of the clearfall remained lower than that of the throughfall deposition. The pH of the deposition was comparatively lower in forested areas (site 4 and 10). The low pH at site 1 may be ascribed to be due to relatively higher SO_4^{2-}- S and comparatively lower proportion of base cations. At site 2 on the other hand, high proportion of base cations could neutralize the acidity of depositions even at higher SO_4^{2-}- S. Jain *et al.* (2000) have reported atmospheric bulk deposition pH between 6.3 to 6.5 at urban sites, 6.6 to 6.7 at urban industrial sites and 7.1 at certain rural areas in India.

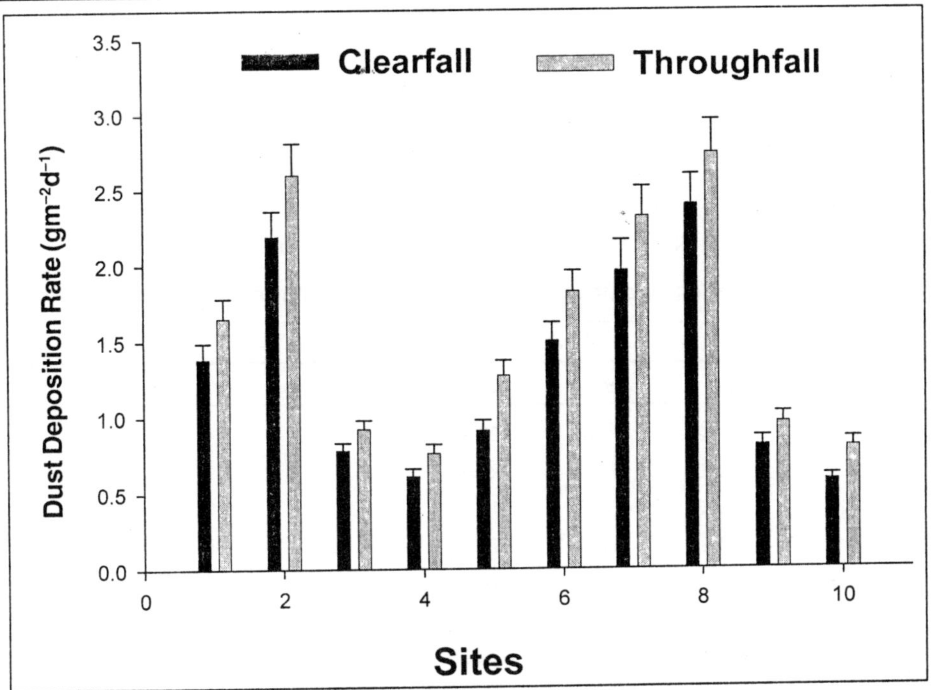

Figure 49.1: Dust Deposition Rates Measured as Clearfall and Throughfall at Different Study Sites. Bars represent mean of three replicates ± 1SE.

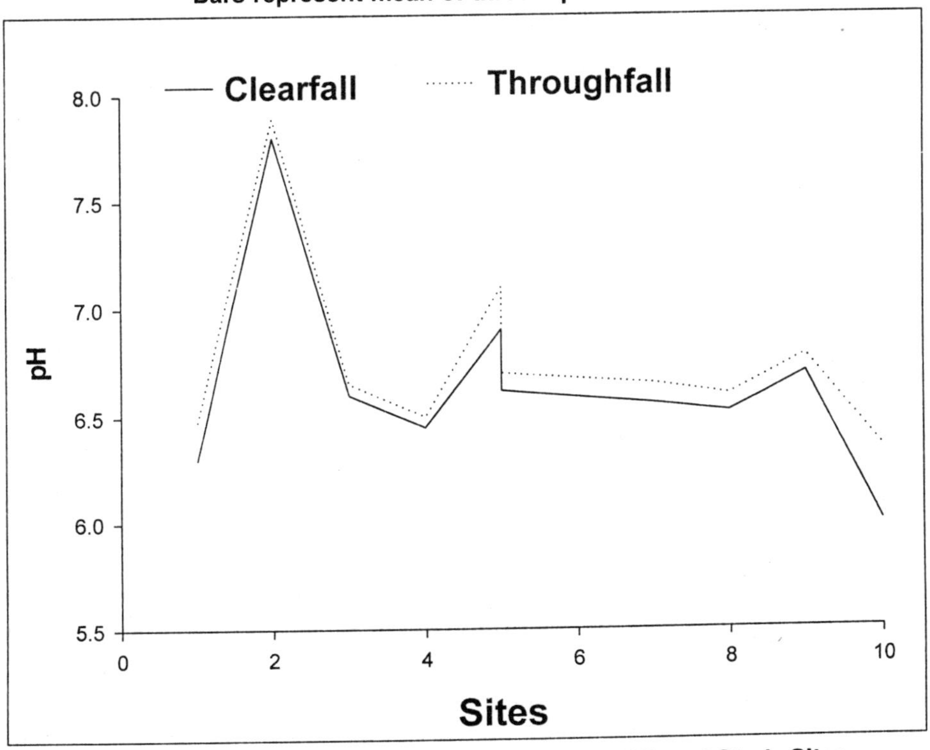

Figure 49.2: pH of Clearfall and Throughfall at Different Study Sites

The nature of atmospheric particulates often contributes in maintaining the alkaline pH of atmospheric deposition. In remote areas, atmospheric particulates are often enriched with soil oriented basic cations (Jain *et al.*, 2000). In forested areas, as the case of sites 4 and 10, with no local dust sources long range transport of alkaline dust remain smaller and hence the pH of the atmospheric deposition remain low. On the contrary, at site 2 characterized by marble based industries, potential influence of base cations specially Ca^{2+} originating from local sources can be seen on the pH of atmospheric deposition (Table 49.2). Substantially high Ca^{2+} in atmospheric deposits of this site could make the pH alkaline. The concentration of NH_4^+ in bulk deposition also contributes to alkalinity.

Table 49.2: Atmospheric Deposition of Nutrient Ions as Clearfall at Different Study Sites

Site	Atmospheric Deposition (kg ha⁻¹ yr⁻¹)*							
	PO_4^{3-}	NO_3^-	NH_4^+	SO_4^{2-}	Na^+	K^+	Ca^{2+}	Mg^{2+}
1.	0.48	18.92	7.32	23.90	3.90	2.24	18.61	4.20
2.	0.59	19.80	7.80	25.12	11.25	5.19	48.90	10.66
3.	0.75	12.85	5.06	16.02	2.87	2.08	14.26	3.72
4.	0.46	7.92	2.90	11.80	2.00	1.05	10.15	2.66
5.	0.78	13.74	5.76	18.05	4.98	4.00	18.20	3.96
6.	0.70	14.28	5.10	17.15	4.95	3.05	14.51	4.85
7.	0.78	16.63	5.90	20.38	6.01	3.14	18.95	6.80
8.	1.08	22.45	8.78	26.95	9.65	5.36	26.86	10.25
9.	1.12	14.50	4.91	17.64	4.18	5.80	20.98	5.18
10.	0.98	5.83	2.01	7.25	3.02	1.12	7.35	2.15

*Between site differences were significant at $p < 0.01$ (ANOVA).

Bulk deposition of SO_4^{2-} as clearfall was maximum at site 8 and minimum at site 10 (Table 49.2). Variation in this anion was significant due to site and showed significantly positive correlations with dust deposition rate ($r = 0.71$; $p < 0.01$) and SO_2 concentration ($r = 0.79$; $p < 0.01$). Sulphur dioxide (36 to 167 μg m^{-3}) was found to be one of the major air pollutants in the area (Pandey and Chasta, 2006). As observed for site 2, alkaline particulates could favour the adsorption of SO_2. Atmospheric bulk deposition of SO_4^{2-} observed at source oriented sites in this study was higher than the values measured in urban–industrial areas of Delhi (Jain *et al.*, 2000).

Atmospheric bulk deposition of NO_3^- was typically high at industrial sites (2 and 8) and traffic oriented urban site (Table 49.2). Bulk deposition of NO_3^- showed significant positive correlations with dust fall rate ($r = 0.82$; $p < 0.01$) and NO_2 concentration in the ambient air ($r = 0.72$; $p < 0.01$). This indicates that the industrial activities are mainly responsible for high atmospheric NO_3^- deposition in the area (Pandey and Chasta, 2006). Further, high NO_3^- deposition at site 1 may be due to high frequency of vehicles. Miller *et al.* (1993) have reported atmospheric N input of 16.7 kg ha^{-1} y^{-1} at a high elevation forest of Whiteface mourtain, New York receiving regional transport of pollutant aerosols. This value is higher than those observed for remote forested site (5.83 kg ha^{-1} y^{-1}) but lower than those recorded for industrial site (22.45 kg ha^{-1} y^{-1}) of this study. Dry deposition of HNO_3 could be a factor contributing to high bulk deposition of NO_3^-. The climate of this region is characterized by extended period of dryness terminated by a short wet period. Formation of HNO_3 is favoured by high day temperature and light intensity which raise the concentration of OH^- radicals. The NO_3^- particles formed by the reaction of atmospheric HNO_3 with particulates could lead to high deposition of NO_3^- at site 1, 2 and 8 (Table 49.2).

Atmospheric deposition of PO_4^{3-} varied from 0.46 kg ha^{-1} y^{-1} at site 4 to 1.12 kg ha^{-1} y^{-1} at site 9. The bulk deposition of orthophosphate observed in this study remained lower in comparison to all other nutrient ions (Table 49.2). Pandey and Pandey (2005) reported relatively lower deposition of P in some remote areas of dry tropics. These observations indicate that the forested remote sites of this study are also under strong industrial influence. In particular, sites 9 and 10, although away from direct anthropogenic activity, receive atmospheric loading of P from phosphate factories by temporal flushing of north–westerly winds. Furthermore, site 9 is also under the influence of vast stretch of agricultural land as far as nutrient loading is concern.

Atmospheric deposition of Ca^{2+} was observed maximum at site 2 (48.90 kg ha^{-1} y^{-1}) and minimum at site 10 (7.35 kg ha^{-1} y^{-1}). A similar trend appeared for bulk atmospheric deposition of Na^+ and Mg^{2+}. The deposition of K^+ was maximum at site 8 (Table 49.2). Atmospheric deposition of Ca^{2+} showed significant positive correlation with Na^+ ($r = 0.86; p < 0.01$), Mg^{2+} ($r = 0.89; p < 0.01$) and K^+ ($r = 0.79; p < 0.01$). Correlation coefficients between cations indicate their common source of origin. Values of correlation coefficient between Ca^{2+} and SO_4^{2-} ($r = 0.75; p < 0.01$) and between Ca^{2+} and NO_3^- ($r = 0.80; p < 0.01$) were also significant.

Table 49.3: Atmospheric Deposition of Nutrient Ions as Throughfall at Different Study Sites

Site	*Atmospheric Deposition (kg ha^{-1} yr^{-1})**							
	PO_4^{3-}	NO_3^-	NH_4^+	SO_4^{2-}	Na^+	K^+	Ca^{2+}	Mg^{2+}
1.	0.54	21.26	8.20	26.18	4.16	2.51	20.69	4.80
2.	0.67	23.05	8.71	28.01	13.00	5.71	56.18	11.19
3.	0.84	14.01	6.80	18.10	3.70	2.39	15.90	4.09
4.	0.52	8.50	3.25	12.51	2.65	1.26	11.05	3.10
5.	0.88	14.80	5.98	19.19	5.60	4.88	20.10	4.18
6.	0.85	15.90	6.11	20.00	5.19	3.62	16.02	5.60
7.	0.90	18.05	6.78	22.01	6.70	3.76	20.18	7.56
8.	1.17	25.01	9.60	30.15	10.80	5.94	29.50	11.80
9.	1.30	16.15	5.75	20.18	4.78	6.61	23.15	6.20
10.	1.09	6.74	2.69	8.00	3.48	1.33	8.76	2.40

*Between site differences were significant at $p < 0.01$ (ANOVA).

Bulk atmospheric deposition as measured in terms of throughfall (under the influence of woody perennial) also showed significant between site variations in pH as well as in anions and cations (Figure 49.2 and Table 49.3). Higher pH of throughfall than that of the clearfall may be ascribed to continued deposition and neutralizing effects of basic cations on leaf surfaces. Furthermore, according to Liu *et al.* (2002), the hydrogen ion exchange with other cations at canopy exchange site increases the cations on leaf surfaces. Higher pH of throughfall has also been reported in earlier studies (Singh and Agrawal, 2005). Throughfall deposition of anions and cations were also higher than the clearfall at all sites. Such enrichment effects of plant leaves have also been reported in other studies including the Silent Valley forest in south India (Rao *et al.*, 1995) and subtropical montane moist forest on Ailao

Mountains, China (Liu *et al.*, 2002). Chemical enrichment observed in throughfall can be attributed to impaction, leaching of substances from plant tissues and the influence of foliar microflora. Throughfall enrichment can be major pathway in nutrient cycling in natural ecosystems especially for those areas where direct anthropogenic flushing of nutrient is almost negligible.

In the present study, N deposition at industrial site 8 (31.23 kg ha^{-1} y^{-1}) was about 4 times higher than the distantly situated forested site 10 (7.84 kg ha^{-1} y^{-1}). The former value is more than 10 times higher than that reported for non-source oriented northern forested ecosystems (Aber *et al.*, 1989). This suggests that, site 10 in this study, although situated away from direct human influence, is under the strong influence of atmospheric loading through airshed. Miller *et al.* (1993) reported annual N and S input at the rate of 16.7 kg ha^{-1} y^{-1} and 17.5 kg ha^{-1} y^{-1} respectively, at high elevation forest site at Whiteface mountain of New York receiving regional transport of pollutants.

The SO_4^{2-}- S deposition as clearfall was 26.95 kg ha^{-1} y^{-1} at heavily polluted site and 7.25 kg ha^{-1} y^{-1} at distantly situated forested site (Table 49.2). Singer *et al.* (1996) have reported atmospheric bulk deposition of SO_4^{2-}- S to be 14.3 kg ha^{-1} y^{-1} at a forest site on Mt. Carmel. These values appear close to SO_4^{2-}- S deposition measured at suburban sites of this study. For cations with few exceptions, bulk deposition was maximum for Ca^{2+} followed by Mg^{2+}, Na^+ and K^+ (Table 49.2). The depositions of all the nutrient elements were uniformly higher in throughfall than in clearfall depositions (Tables 49.2 and 49.3). In a study, conducted in a deciduous forest on the Walker branch watershed, Lindberg and Harriss (1981) recorded 75 per cent of dryfall input when hard woods were in leaf, indicating the importance of aerosol impaction on foliage. Liu *et al.* (2002) have also reported enrichment of throughfall bulk deposition of Ca^{2+}, Na^+, Mg^{2+}, K^+ and SO_4^{2-}- S.

Conclusions

The present study indicates that the bulk atmospheric deposition, both the clearfall and throughfall as observed at different sites are greatly influenced by anthropogenic sources including urban-industrial activities and agricultural practices and by regional wind patterns. Among anions, the deposition was recorded maximum for SO_4^{2-}- S followed by NO_3^- and PO_4^{3-}, whereas among cations, the deposition was maximum for Ca^{2+} followed by NH_4^+ and then Na^+, Mg^{2+} and K^+. The pH of deposition was greatly influenced by the atmospheric loading of anions and base cations. Source oriented loading of base cations showed strong neutralizing effect on the pH of the deposition. Both, pH and nutrient elements remained significantly higher in throughfall than in clearfall. This strong modifying influence of vegetation on atmospheric deposition may leads to long–run consequences on biogeochemical cycling of nutrients with consequent ecosystem level implications even for those areas situated far away from direct human interference. Continuous atmospheric deposition of nutrient aerosols will alter tree leaf microflora, composition of herbaceous vegetation, microbial biomass and activity and mineralization and release of nutrients with long–term effects on structure and composition of terrestrial ecosystems. In addition, direct atmospheric input of nutrient aerosols will change the trophic status of aquatic ecosystems with subsequent change in phytoplankton composition, species diversity and water use. Thus, quantification of air–driven nutrient input need serious attention for forest management as well as for restoration of aquatic ecosystems.

References

Aber, J. D., Nadelhoffer, K. J., Steudler, P. and Melillo, J. M. (1989). Nitrogen saturation in northern forest ecosystems. *Bioscience* (39): 378–386.

Friedland, A. J. and Miller, E. K. (1999). Major elemental cycling in a high elevation Adirondack forest: patterns and changes. 1986–1996. *Ecological Applications* (9): 958–967.

ain, M., Kulshrestha, U. C., Sarkar, A. K. and Parashar, D. C. (2000). Influence of crustal aerosols on wet deposition at urban and rural sites in India. *Atmospheric Environment* (34): 5129–5137.

Lindberg, S. E. and Harriss, R. C. (1981). The role of atmospheric deposition in an eastern United States deciduous forest. *Water, Air and Soil Pollution* (16): 13–31.

Liu, W., Fox, J. E. D. and Xu, Z. (2002). Nutrient fluxes in bulk precipitation, through fall and stream flow in montane subtropical moist forest on Ailao Mountains in Yunnan, south–west China. *Journal of Tropical Ecology* (18): 527–548.

Miller, E. K., Panek, J. A., Friedlnd, A. J., Kadlecek, J. and Mohnen, V. A. (1993). *Atmospheric deposition to a high elevation forest at Whiteface mountain*, New York, USA. Tellus (45): 209–227.

Pandey, J. and M. Agrawal (1994a). Evaluation of air pollution phytotoxicity in a seasonally dry tropical urban environment using three woody perennials. *The New Phyologist* (126): 53–61.

Pandey, J. and M. Agrawal (1994b). Diurnal and seasonal variations in air pollutant concentrations in a seasonally dry tropical urban environment. *Current Science* (66): 293–303.

Pandey, J. (2005). Evaluation of air pollution phytotoxicity downwind of a phosphate fertilizer factory in India. *Environmental Monitoring and Assessment* (100): 249-266.

Pandey, U. and J. Pandey (2005). The influence of catchment modifications on two freshwater lakes of Udaipur. In: *Urban Lakes in India: Conservation, Management and Rejuvenation* (Part-I). K.K.S.Bhatia and S.D. Khobragade (Eds.) National Institute of Hydrology, Roorkee. pp. 256-262.

Pandey, J. and A. Chasta (2006) Evaluation of air pollution phytotoxicity in urban environment of Udaipur, Rajasthan. *Plant Archives* 6 (1): 93-96.

Pandey, J. and U. Pandey (2009). Accumulation of heavy metals in dietary vegetables and cultivated soil horizon in organic farming system in relation to atmospheric deposition in a seasonally dry tropical region of India. *Environmental Monitoring and Assessment* 148 (1): 61–74.

Rao, P. P. S., Momin, G. A., Safai, P. D., Pillai, A. G. and Khemani, L. T. (1995). Rain water and throughfall chemistry in the Silent Valley forest in South India. *Atmospheric Environment* (29): 2025–2027.

Rossum, J. R. and Villarruz, P. (1961). Suggested methods for turbidimetric determination of sulphate in water. *Journal of American Water Works Association* (53): 873.

Singer, A., Ganor, E., Fried, M. and Shamay, Y. (1996). Throughfall deposition of sulfur to a mixed Oak and Pine forest in Israel. *Atmospheric Environment* (30): 3881–3889.

Singh, R. K. and Agrawal, M. (2005). Atmospheric depositions around a heavily industrialized area in a seasonally dry tropical environment of India. *Environmental Pollution* (138): 142–152.

Swank, W. T. (1984). Atmospheric contributions to forest nutrient cycling. *Water Resources Bulletin* (20): 313–321.

Microbial Biotechnology and Ecology (2011) *Pages* **846–858**
Editors: **Deepak Vyas, G.S. Paliwal, P.K.Khare and R.K Gupta**
Published by: **DAYA PUBLISHING HOUSE, NEW DELHI**

Chapter 50

Trends and Determinants of Health Practices Profile Among the Baigas

☆ ***Arun K. Jain[1] and A.N. Sharma[2]***

[1]*Department of Anthropology and Tribal Development, G.G. University, Bilaspur, C.G.*
[2]*Department of Anthropology, Dr. H.S. Gour University, Sagar – 470 003*

ABSTRACT

The Baigas of Mandla district has a good account of indigenous medicines and they know the medicine for the cure of most each and every type of disease. There are folk healers called guniya who cure disease of village people by giving medicines or performing magico-religious acts. The objective of the present chapter is to study health practices profile among the Baigas of Mandla district (M.P.), India. The present investigation is based on interviews of 400 Baiga households from five different blocks of Mandla district. A sizable medicinal plant wealth exists in Mandla district, which still awaits scientific exploration. The indiscriminate deforestation should be checked, so that rare plants could not become extinct. Promoting the indigenous medicines system of Baigas, can fulfill the World Health Organization's important issue, *i.e.*, people must be involved as partners in the decision making and management process for health at all levels.

Keywords: *Health practices, Indigenous medicines, Baiga, Mandla district.*

Introduction

The history of evolution of medicine is a long story stretching from prehistoric to modern times. The 'art of healing' began thousands of years ago with the primitive man trying to provide relief with his limited knowledge, to the kind in sickness and suffering. In the absence of explanation, disease

* E-mail: arun_raj_jain@yahoo.com

and calamites were attributed to wrath of gods, the invasion of the body of evil spirits or malevolent influence of stars and planets (Park and Park, 1997).

Anthropologist's interest in the study of disease and cure is as old as their interest in the study of tribal culture. Rivers (1924) stated that art of medicine of the people was in some respect more rational than our own, in that its mode of diagnosis and treatment follow more directly from the ideas concerning the causation of disease, even though there are wrong ideas from the civilized point of view (Dwivedi, 2003).

Since long tribal and forests are intertwined related to each other. Forests are not only the source of major and minor forest produces, but they depend much on forest for their day-to-day needs. From the very beginning of human civilization, men depend on nature for his food, shelter and medicine.

It is true that use of medicinal plants for curing disease in human society is as old as existence itself. The earliest mention of use of medicinal plants is found in 'Rigveda', which happens to be the oldest scripture. In 'Rigveda' and 'Charak Samhita', 2000 names of medicinal plants have been listed. Out of which, 1600 species of medicinal plants have so for been established.

Indigenous medicine is to be studied, as medicine in a primarily spoken not written tradition. The importance of tribal medicine and like could draw the attention of Mittal (1979), Mathur (1982), Tiwari (1984), Banerjee and Jalota (1988), Rizvi (1991), Sharma and Sharma (2000), Jain and Sharma (2006), Jain and Sharma (2006), Shukla, Chakravarty and Gautam (2008).

The Baigas of Mandla district has a good account of indigenous medicines and they know the medicine for the cure of most each and every type of disease. There are folk healers called *guniya* who cure disease of village people by giving medicines or performing magico-religious acts. The local people depend for medical cure on indigenous methods. They initially do not prefer to go to allopathic doctors. In case of such diseases, which are beyond the cure of their indigenous remedies, they consult allopathic practitioners, occasionally.

Materials and Methods

The present investigation is based on interviews of 400 Baiga households from five different blocks of Mandla district. Mandla district is located in the east central part of the Madhya Pradesh. It lies between the latitude 220.2′ and 230.22′ north and longitude 800.18′ and 810.50′ east. The total area of the district is 13,269 Sq. Km. The block and villages were selected randomly as per the concentration of blocks. In this regard 80 samples were selected randomly from the five villages of each block and approximately 16 samples were selected randomly from each village. The data were collected from the households of the Baigas of Mandla district, Madhya Pradesh. One sample was collected randomly from each household of randomly selected villages. The present study has been conducted through interview schedule. Simultaneously, group discussions and informal interview methods has been used. Observations have been conducted through semi-participants methods.

Results and Discussions

Keeping in mind these views, an attempt has been made to investigate the indigenous medicines for the cure of different diseases among the Baigas. The observations and results of the study are presented in the following tables:

Table 50.1 shows information regarding type of treatment received by the family members, who feel sick during last one year, it could be seen from the table that maximum individuals received self treatment (36.00), rather than allopathic treatment (33.00) and other peoples were also noticed who do their indigenous or ethno medicines during sickness (31.00). It could be concluded that, 67.00 per cents Baigas used the indigenous or ethno-medicine and self treatment and 33.00 percents Baigas are received allopathic treatment, but allopathic treatment received secondary. The Baigas are firstly preferred indigenous or ethno-medicines and self treatment and secondary preferred allopathic treatment.

Table 50.1: Information Regarding Type of Treatment Received by the Family Members, during Last One Year

Sl.No.	*Type of Treatment*	*AB.*	*PC.*
1.	Allopathic	132	33.00
2.	Indigenous or ethno-medicines	124	31.00
3.	Self treatment	144	36.00
	Total	400	100.00

Table 50.2 shows indigenous medicine used for the cure of fever. It could be seen from the table that Baigas use the root of Jhilmili to get relief from fever.

Table 50.2: Information Regarding Medicine Used for the Cure of Fever

Sl.No.	*Name of Indigenous Medicine or Herb (Local name)*	*Botanical Name*	*Part of the Herb Used*	*Method of Applying*	*Probable Dosages*	*Duration of Application*
1.	Jhilmili	–	Root	Crush it and boil in water	Two times Two spoon	2 Days

Table 50.3 reveals information regarding indigenous medicine used for the cure of Malaria. It could be seen from the table that the different medicines, which are used for the cure of Malaria, are the leaves of Neem, bark of Jamun, seeds and leaves of Banjeet and an equally made mixture of grinded roots of Mainer and Kadi Kudaiya. The method of application can be seen from the table.

Table 50.3: Information Regarding Medicine Used for the Cure of Malaria

Sl.No.	*Name of Indigenous Medicine or Herb (Local name)*	*Botanical Name*	*Part of the Herb Used*	*Method of Applying*	*Probable Dosages*	*Duration of Application*
1.	Neem	*Azadirachta indica* (Meliaceae)	Leaves	Boil the leaves in water and take it	2-3 Times One spoon	2 Days
2.	Jamun	*Eugenia jambolana* (Myrtaccae)	Bark	Grind well to make powder and take it with Gur (Jaggery)	Once in the morning Two spoon	2 Days
3.	Banjeet	–	Seed or leaves	Grind well or boil in water and take the water	Two Times Two Spoon	2 Days
4.	Mainar+ Kadi Kudaiya	–	Root+ Root	Grind well equally both the roots and dissolve in water take the solution orally	Two Times Two spoon	3 Days

Table 50.4 shows information regarding indigenous medicines used for the cure of Typhoid. It could be observed from the table that root of Khirrari and leaves of Amla are used for the cure of Typhoid. Both these are boiled in water, in a clay pot (Khapra) and the resultant solution is taken as such by the patient.

Table 50.4: Information Regarding Medicine Used for the Cure of Typhoid

Sl.No.	Name of Indigenous Medicine or Herb (Local name)	Botanical Name	Part of the Herb Used	Method of Applying	Probable Dosages	Duration of Application
1.	Khirrari+ Amla	– *Phyllanthus emblica* (Euphorbiaceae)	Root+ Leaves	Take same water in a clay pot (Khapra) and boil the root and leaves in it. Take the solution as Medicine.	Once time Two spoon	2-3 Days

Table 50.5 exhibits information regarding indigenous medicines used for the cure of Cough and Cold. It can be observed from the table that root of Akuvan can be used for the cure of Cough and Cold. The leaves of Harrah are also an effective medicine to cure cough and cold.

Table 50.5: Information Regarding Medicine Used for the Cure of Cough and Cold

Sl.No.	Name of Indigenous Medicine or Herb (Local name)	Botanical Name	Part of the Herb Used	Method of Applying	Probable Dosages	Duration of Application
1.	Akuvan (Akaua)	*Calotropis gigantea* (Asclepiadeceae)	Root	Crush the root and boil in one liter water Add sugar and take it	Once time Two spoon	1 Day
2.	Harrah	*Terminalia-chebula* (Combretaceae)	Leaves	Boil the leaves in water and take the water	Two times Two spoon	2 Days

Table 50.6 shows information regarding indigenous medicines used for the cure of wounds, cuts and burns. It can be observed from the table that the bark of Khamher, leaves of Kubbi and leaves of Chirchitta are all effective medicines for the cure of wounds, cuts and burns. All these are applied externally on the wounds etc. and the use of these medicines varies from individual to individual or folk healer to folk healer.

Table 50.6: Information Regarding Medicine Used for the Cure of Wounds, Cuts and Burns

Sl.No.	Name of Indigenous Medicine or Herb (Local name)	Botanical Name	Part of the Herb Used	Method of Applying	Probable Dosages	Duration of Application
1.	Khamher (Gurja)	*Cimelina arborea*	Bark	Grind it to make its powder and apply it on the wound externally (only on the skin).	Two Times	Applied up to the relief
2.	Kubbi	–	Leaves	Crush it to make paste like, and apply it on the wound	Two times	Applied up to the relief
3.	Chirchitta	*Lyciuin barbatum*	Leaves	Crush and apply it on the wound	Three Times	Applied up to the relief

Table 50.7 exhibits information regarding indigenous medicines used for the cure of toothache or caries. It could be seen from the table that the lactin of papaya is an effective medicine for the cure of toothache or caries.

Table 50.7: Information Regarding Medicine Used for the Cure of Toothache

Sl.No.	Name of Indigenous Medicine or Herb (Local name)	Botanical Name	Part of the Herb Used	Method of Applying	Probable Dosages	Duration of Application
1.	Papita	*Carica papaya*	Lactin of the Papaya	Apply the lactin on teeth and fill it in the Tooth depressions/ carries	Two or Three times	2 Days

Table 50.8 reveals information regarding indigenous medicine used for the cure of stomachache or diarrhea. It can be observed from the table that there are mainly five type of medicines used for the same. These are–root of Tinkapari, root of Padhin, root of using these medicines varies from individual to individual or folk healer. The use of specific type of medicine also depends on the availability of herbs etc.

Table 50.8: Information Regarding Indigenous Medicine Used for the Cure of Stomachache or Diarrhoea

Sl.No.	Name of Indigenous Medicine or Herb (Local name)	Botanical Name	Part of the Herb Used	Method of Applying	Probable Dosages	Duration of Application
1.	Tinkhapari	*Oxalis* sp.	Root	Grind it to make the powder and take with water	One or Two times One spoon	Applied up to relief
2.	Padhin	*Mentha arvensis*	Root	Give the patient to chew it and suck the juice	Two times Two spoon	2 Days
3.	Phulchuhi	–	Root	Crush the root and boil in water. Take the water as medicine	Two times Two spoon	1 Day
4.	Tinpariya	*Oxalis corniculata Holarina*	Root	Chew the root to take the juice Grind it to make	Two times One cm. above Two times	2 Days
5.	Kadi Kudaiya	*Intidisen-erica*	Root	The powder and take with water	One spoon	1 Day

Table 50.9 shows information regarding indigenous medicines used for the cure of head-ach. It could be seen from the table that Baigas use the leaves of Bantulsi for the cure of head-ache. The medicine was reported to be very effective for the cure of head-ache.

Table 50.9: Information Regarding Indigenous Medicines Used for the Cure of Head-ache

Sl.No.	*Name of Indigenous Medicine or Herb (Local name)*	*Botanical Name*	*Part of the Herb Used*	*Method of Applying*	*Probable Dosages*	*Duration of Application*
1.	Bantulsi/ Mahatulsi	*Hypits suaveolens*	Leaves	Crush the leaves to get the juice and apply it on the forehead	2-3 times	Apply up the relief

Table 50.10 shows the information regarding about indigenous medicines used for the cure of earache. It can be seen from the table that powder of dried burnt shrub of Sanpavur is an effective medicine, used by the Baigas for the cure of earache.

Table 50.10: Information Regarding Indigenous Medicines Used for the Cure of Earache

Sl.No.	*Name of Indigenous Medicine or Herb (Local name)*	*Botanical Name*	*Part of the Herb Used*	*Method of Applying*	*Probable Dosages*	*Duration of Application*
1.	Sanpavur	–	Dried Bush	Burn the plant to get the dried bush of ash and mix it in mustard oil Pour it in ear of the patient	Two times and Two drops	3 Days

Table 50.11 exhibits information regarding indigenous medicines used for the cure of Oral ulcer. It can be seen from the table that the three important medicines used by the Baigas for the cure of ulcers are bark or leaf of Khirrari, leaves of Amla and honey. The medicines were reported to be very effective for the cure of mouth ulcers.

Table 50.11: Information Regarding Indigenous Medicines Used for the Cure of Dial Ulcer

Sl.No.	*Name of Indigenous Medicine or Herb (Local name)*	*Botanical Name*	*Part of the Herb Used*	*Method of Applying*	*Probable Dosages*	*Duration of Application*
1.	Khirsari	*Euphorbia titirucalli*	Bark or Leaves	Grind and apply on ulcers	Two times	2-3 Days
2.	Amla	*Phyllanthus emblica* (Euphorbiaceae)	Leaves	Eat the leaves orally to take the juice	2-3 times	2 Days
3.	Honey	–	As such	Eat a spoonful honey and take the juice slowly	2-3 times	One dose

Table 50.12 reveals information regarding indigenous medicines used for the cure of blood dysentery. It could be seen from the table that the lactin of Badh and sugar when mixed in equal amount, proves to be a good medicine for the cure of blood dysentery.

Table 50.12: Information Regarding Indigenous Medicines Used for the Cure of Blood Dysentery

Sl.No.	Name of Indigenous Medicine or Herb (Local name)	Botanical Name	Part of the Herb Used	Method of Applying	Probable Dosages	Duration of Application
1.	Badh+ Sugar	*Bengalensis* (Urticaceae)	Lactin of Badh Sugar grains	Mix equal amount of both and take it with warm water	Once	2 Days

Table 50.13 shows information regarding indigenous medicines used for the cure of Jaundice. It could be seen from the table that root of Jhilmili is an effective medicine for the cure of Jaundice, which is used in different ways.

Table 50.13: Information Regarding Indigenous Medicines Used for the Cure of Jaundice

Sl.No.	Name of Indigenous Medicine or Herb (Local name)	Botanical Name	Part of the Herb Used	Method of Applying	Probable Dosages	Duration of Application
1.	Jhilmili	–	Root	Grind it to make its powder and take with water	Two times Two spoons	2 Days

Table 50.14 shows information regarding indigenous medicines used for the cure of vomiting. The Baigas use the medicine for the cure of vomiting, *i.e.*, root of Dooba.

Table 50.14: Information Regarding Indigenous Medicines Used for the Cure of Vomiting

Sl.No.	Name of Indigenous Medicine or Herb (Local name)	Botanical Name	Part of the Herb Used	Method of Applying	Probable Dosages	Duration of Application
1.	Dooba	*Cynodon dactylon*	Root	Grind it and add in water. Take it as medicine	Once time One spoon	Once day

Table 50.15 reveals information regarding indigenous medicines used for the cure of worms of stomach. It could be seen from the table that the seeds of Chewla are used by the Baigas for the cure of stomach worms. The medicine is reported to be very effective.

Table 50.15: Information Regarding Indigenous Medicines Used for the Cure of Worms of Stomach

Sl.No.	Name of Indigenous Medicine or Herb (Local name)	Botanical Name	Part of the Herb Used	Method of Applying	Probable Dosages	Duration of Application
1.	Chewla	*Butea monosperma*	Seeds	Grind seeds to make the power and take with water	Once time One spoon	2 Days

Table 50.16 exhibits information regarding indigenous medicines used for the cure of weakness or body-ache. It could be seen from the table that there are two medicines, which are used by the Baigas to remove weakness. These are very effective medicines.

Table 50.16: Information Regarding Indigenous Medicines Used for the Cure of Weakness

Sl.No.	Name of Indigenous Medicine or Herb (Local name)	Botanical Name	Part of the Herb Used	Method of Applying	Probable Dosages	Duration of Application
1.	Okhad (Badi Davai)	–	Root or Bark	Boil the root or bark in water and add some gur take the solution as medicine	Two times One spoon	2 days
2.	Bijnauri	–	Root	Chew the root to suck the juice	Two times One cm. above	1 Day

Table 50.17 shows information regarding indigenous medicines used for the cure of skin diseases. It could be seen from the table that fat of Goh animal is an effective medicine for the cure of skin diseases. This medicine is very effective.

Table 50.17: Information Regarding Indigenous Medicines Used for the Cure of Skin Diseases

Sl.No.	Name of Indigenous Medicine or Herb (Local name)	Botanical Name	Part of the Herb Used	Method of Applying	Probable Dosages	Duration of Application
1.	Goh (Varanus)	–	Fat of Goh	Apply the fat on affected part externally	Applied up to the relief	Apply up to the relief

Table 50.18 Shows information regarding indigenous medicines used for the cure of wounds, which due to itching, forms locally known as Biskuti, or Baltodh. It could be seen from the table that leaves and seed of Kharenta are applied by the Baigas as an effective medicine.

Table 50.18: Information Regarding Indigenous Medicines Used for the Cure of Wounds Due to Itching

Sl.No.	Name of Indigenous Medicine or Herb (Local name)	Botanical Name	Part of the Herb Used	Method of Applying	Probable Dosages	Duration of Application
1.	Kharenta	*Dodonoea visocosa* (Sapindaceae)	Seed and leaves	Crush the seeds and leaves and apply juice on the part	Two times	2 Days

Table 50.19 shows information regarding indigenous medicines used for the cure of scorpion sting. It could be seen from the table that leaf of grass and leaves or Ber are used as medicine for the cure of scorpion sting among the Baigas.

Table 50.20 shows information regarding indigenous medicines used for the cure of snake bite. It could be seen from the table that the root of Bariyari, which is uprooted in one attempt, and a combination of three medicines–root of Mua, root of Kansa and root of Tumba, used for the cure of snake bite. These medicines are very effective.

Table 50.19: Information Regarding Indigenous Medicines Used for the Cure of Scorpion Sting

Sl.No.	*Name of Indigenous Medicine or Herb (Local name)*	*Botanical Name*	*Part of the Herb Used*	*Method of Applying*	*Probable Dosages*	*Duration of Application*
1.	Grass	–	Leaves	Rub and crush the grass and apply on the bite	Once time	Once
2.	Ber	*Zizyphus-jujube* (Rhamnaceae)	Leaves	Take 1 grm. leaves of Ber	Once time	Once
3.	Guard (bitter)	*Benincasa-hispida*	Root	Crush the root and apply on the bite	Once time	Once

Table 50.20: Information Regarding Indigenous Medicines Used for the Cure of Snake Bite

Sl.No.	*Name of Indigenous Medicine or Herb (Local name)*	*Botanical Name*	*Part of the Herb Used*	*Method of Applying*	*Probable Dosages*	*Duration of Application*
1.	Bariyari	*Sida spinosa*	Root	Take the root and eat it to get the juice	Once a time Two spoons	One day
2.	Mua+ Kansa+ Tumba	–	Root + Root + Root	Grind all the three roots properly. Mix them well and take it.	Once time Two spoon	One day

Table 50.21 reveals information regarding indigenous medicines used for the cure of dog bite. It could be seen from the table that the seeds of Khutlaiya are used by the Baigas as the medicine of dog bite, which is reported to very effective.

Table 50.21: Information Regarding Indigenous Medicines Used for the Cure of Dog Bite

Sl.No.	*Name of Indigenous Medicine or Herb (Local name)*	*Botanical Name*	*Part of the Herb Used*	*Method of Applying*	*Probable Dosages*	*Duration of Application*
1.	Khutlaiya	–	Seeds	Eat the seeds as the medicine	Once a time One spoon	One day

Table 50.22 shows information regarding indigenous medicines used for the cure of bleeding. It can be observed from the table that bark of Gurja is used as an effective medicine to cure the bleeding, where as Chuna and Kattha is also applied by many people for the same purpose. These medicines are very effective.

Table 50.23 shows information regarding use the medicines for anti fertility purpose. It could be seen from the table that bark of white flowers of Tilya are the effective medicine. This medicine is very effective.

Table 50.22: Information Regarding Indigenous Medicines Used for the Cure of Bleeding

Sl.No.	*Name of Indigenous Medicine or Herb (Local name)*	*Botanical Name*	*Part of the Herb Used*	*Method of Applying*	*Probable Dosages*	*Duration of Application*
1.	Gurja	*Lania coromandilica*	Bark	Rub or Grind the bark and apply on the wound	Once a time	One day
2.	Kattha + Chuna	*Acacia catechu*	Powdered bark As it is	Tie the mixture on the bleeding wound	Once a time	One day

Table 50.23: Information Regarding Use the Medicines for Anti Fertility Purpose

Sl.No.	*Name of Indigenous Medicine or Herb (Local name)*	*Botanical Name*	*Part of the Herb Used*	*Method of Applying*	*Probable Dosages*	*Duration of Application*
1.	Tiliya	–	Bark or the white flower	Bark or crush the flower and make its tablets with gur. Take with water or milk	Two tablets in the evening	2 days

Table 50.24 shows information regarding indigenous medicines used for re-fertility purpose. It could be seen from the table that *talka* fruit are the effective medicine of re-fertility purpose. This medicine is very effective.

Table 50.24: Information Regarding Use the Medicines for Re-fertility Purpose

Sl.No.	*Name of Indigenous Medicine or Herb (Local name)*	*Botanical Name*	*Part of the Herb Used*	*Method of Applying*	*Probable Dosages*	*Duration of Application*
1.	Talka	–	Fruit	Eat the Vega	Two times above 50 gm	1 day

Table 50.25 shows information regarding indigenous medicines used for the cure of swelling of body and stomach, which is locally known as *'Fulni'*. It could be seen from the table that this is the common indigenous medicine, which are used for the cure of 'Fulinf is root of Chittawar'. The medicines are very effective.

Table 50.25: Information Regarding Indigenous Medicines Used for the Cure and Stomach Swelling

Sl.No.	*Name of Indigenous Medicine or Herb (Local name)*	*Botanical Name*	*Part of the Herb Used*	*Method of Applying*	*Probable Dosages*	*Duration of Application*
1.	Chittawar	*Plumbago zelaniya*	Root	Grind the root and make its tablets in gur. Take with water	Time 10 mg.	2 Days

Table 50.26 shows information regarding indigenous medicines used for the cure of fever with unconsciousness and tight teeth, which are locally known as Maura. It could be seen from the table that to cure such fever, Kalihari is tied in the neck of the patient. The medicines are very effective.

Table 50.26: Information Regarding Indigenous Medicines Used for the Cure of Fever with Unconsciousness

Sl.No.	*Name of Indigenous Medicine or Herb (Local name)*	*Botanical Name*	*Part of the Herb Used*	*Method of Applying*	*Probable Dosages*	*Duration of Application*
1.	Kalihari	*Gloriosa suberva*	Dried bark	The kalihari in the neck	Once	One day

Table 50.27 shows information regarding use of indigenous medicine used for the cure of 'heat of winter'. In this disease itching and burning sensation is felt in the urine. It could be observed from the table that the root of Tejraj proves to be a good medicine for the cure of this disease, which is reported to be very effective.

Table 50.27: Information Regarding Use of Indigenous Medicine Used for the Cure of Heat of Winter

Sl.No.	*Name of Indigenous Medicine or Herb (Local name)*	*Botanical Name*	*Part of the Herb Used*	*Method of Applying*	*Probable Dosages*	*Duration of Application*
1.	Tejraj	*Peucedanum dhana*	Root	Eat the root as the medicine	Once a time, one cm-above	2 Days

Table 50.28 shows information regarding use the medicines for the cure of cracks. It could be seen from the table that oil of the *wilma* are the effective medicine of cure of the cracks of the heals. This medicine is very effective.

Table 50.28: Information Regarding Use the Medicines for Cracks of the Heals

Sl.No.	*Name of Indigenous Medicine or Herb (Local name)*	*Botanical Name*	*Part of the Herb Used*	*Method of Applying*	*Probable Dosages*	*Duration of Application*
1.	Wilma	–	Fruit	Cut the Wilma fruit, burn the mud pot and collect in the lower side. Lape the hill	Two time	2 days

It could be concluded from the foregoing brief reporting of indigenous practices that:

1. The 46.07 per cent individuals received indigenous or ethno medicines.
2. There are a large number of herbal medicines and plant wealth which should be explored through modern techniques and pharmaceutical investigations.
3. These herbal resources could generate a good source of income for the Baigas of Mandla district.
4. A socio-psychological and therapeutic investigation should be made to find out the reliability of the usefulness of magi-co-religious acts.

5. A proper and reliable list of indigenous medicines with diseases should be proposed for the awareness of the tribals, so that they can uplift their health status without modern medical facilities.
6. The proper and reliable list of indigenous medicines with diseases should be evolved within same ecological conditions.
7. The reported medicinal plants should be further investigated concerning the availability of various species and botanical identification of the therapeutic values through scientific analysis.
8. The Baigas have a good knowledge about the indigenous medicines for anti-fertility.
9. The medicines are used for fever, malaria, cough and cold, wounds, cuts, burns, toothache, stomachache, diarrhoea, headache, earache, ulcers, blood dysentery, jaundice, vomiting, worms of stomach, weakness, body-ache, skin diseases, wounds due to itching, scorpion bite, snake bite, dog bite, bleeding, anti fertility, reflectivity, body or stomach swelling, fever with unconsciousness and tight teeth, typhoid, crack etc. and need extensive scientific investigations for wider spectrum of human welfare.

In short, a sizable medicinal plant wealth exists in Mandla district, which still awaits scientific exploration. The indiscriminate deforestation should be checked, so that rare plants could not become extinct. Promoting the indigenous medicines system of Baigas, can fulfill the World Health Organization's important issue, *i.e.*, people must be involved as partners in the decision making and management process for health at all levels. In the words of Nakajima (1990), "The time has now come to galvanize our efforts. The knowledge and technology already exist, what we need now is determination, courage and foresight. Health is a product of special action. Active community participation and supportive social policies are necessary for progress in health and here in lies our challenge". So, it may be stated from the foregoing brief reporting of indigenous health practices that there are number of indigenous medicines and plant wealth, which should be explored through modern techniques and pharmaceutical investigation. Such attempts will give new material and direction in the field of pharmaceutical researchers.

References

Banerjee, B.G. and Jalota, R. (1977). *Folk Illness and Ethno-medicine*. Northern Book Centre, New Delhi.

Dwivedi, P. (2003). Demographic Profile and Health Care Practices among the Baigas of Samanpur Block of Dindori District, Madhya Pradesh, *Unpublished Dissertation*, Department of Anthropology, Dr. H.S. Gour University, Sagar (M.P.).

Jain, A.K. and Sharma, A.N. (2006). *The Berias (Rai Dancers), Socio-Demographic and Reproductive Health Care Practices*. Sarup and Sons, New Delhi.

Jain, A.K. and Sharma, A.N. (2006). Indigenous Health Care Practices among the Beria Women of Sagar District, M.P. India. In: *Indigenous Health care and Ethno-Medicine*, Edited By, A.N. Sharma, R.K. Goutam and A.K. Gharami, Sarup and Sons, New Delhi, p. 50-53.

Mathur, P.R.G. (1982). Anthropology of Tribal Medicine: Disease and Curing Techniques among the Tribal of North Wynad (Kerla), *Man in India*, 62: 295-312.

Mittal, K. (1979). *Primitive Medicine Versus Modern Medicine among the Santhal*. Dr. of Social Research, 22: 49-54.

Park, J.E. and Park, K. (1977). *Textbook of Preventive and Social Medicine*, Banarsidas Bahanot, Jabalpur.

Rizvi, S.N.H. (1991). Health practices of the Jounsaris: A socio-cultural analysis. In: *Tribal Health: socio-Cultural Dimensions*, Edited by Budhadev Chouhari. Inter India Publications, New Delhi.

Sharma, A.N. and Sharma, N.M. (2000). Indigenous health practices related to fever among Bharias of Patalkot, M.P. *Tribal Health Bulletin*, 6(1): 6-8.

Shukla, R., Chakravarty, M. and Gautam, M.P. (2008). Indigenous medicine used for treatment of gynecological disorders by tribal of Chhattisgarh, India. *Journal of Medicinal Plants Research*, 2(12): 356–360.

Tiwari, S.C. (1984). Socio-cultural and genetico-environmental determinants of tribal health: Measures for health development. In: *Tribal Health in India*, Edited by Salil Basu. Inter India Publications, New Delhi.

Author Index

Subject Index

BR 2912
RS. 4500
(2 Vol set)